Himalayan Tectonics

This volume is dedicated to the memory of three young Himalayan geologists who died in the mountains.

R. Paul Metcalfe

Ingrid Reuber

Ugo Pognante

GEOLOGICAL SOCIETY SPECIAL PUBLICATION NO. 74

Himalayan Tectonics

EDITED BY

P.J. TRELOAR
School of Geological Sciences,
Kingston University, UK

M.P. SEARLE
Department of Earth Sciences,
Oxford University, UK

1993

Published by

The Geological Society

London

Geological Society Special Publications
Series Editor J. BROOKS

THE GEOLOGICAL SOCIETY

The Society was founded in 1807 as the Geological Society of London and is the oldest geological society in the world. It received its Royal Charter in 1825 for the purpose of 'investigating the mineral structure of the Earth'. The Society is Britain's national learned society for geology with a Fellowship of 7500 (1992). It has countrywide coverage and approximately 1000 members reside overseas. The Society is responsible for all aspects of the geological sciences including professional matters. The Society has its own publishing house which produces the Society's international journals, books and maps, and which acts as the European distributor for publications of the American Association of Petroleum Geologists and the Geological Society of America.

Fellowship is open to those holding a recognized honours degree in geology or cognate subject and who have at least two years relevant postgraduate experience, or who have not less than six years relevant experience in geology of a cognate subject. A Fellow who has not less than five years relevant postgraduate experience in the practice of geology may apply for validation and subject to approval, may be able to use the designatory letters C Geol (Chartered Geologist).

Further information about the Society is available from the Membership Manager, The Geological Society, Burlington House, Piccadilly, London W1V 0JU, UK.

Published by The Geological Society from:
The Geological Society Publishing House
Unit 7
Brassmill Enterprise Centre
Brassmill Lane
Bath BA1 3JN
UK
(*Orders*: Tel. 0225 445046
Fax 0225 442836)

First published 1993

British Library Cataloguing in Publication Data

A catalogue record for this book is available from the British Library

ISBN 0–903317–92–3

Typeset by Type Study, Scarborough
Printed by Alden Press, Oxford, UK

Distributors

USA
AAPG Bookstore
PO Box 979
Tulsa
Oklahoma 74101–0979
USA
(*Orders*: Tel. (918)584–2555
Fax (918)584–0469)

Australia
Australian Mineral Foundation
63 Conyngham Street
Glenside
South Australia 5065
Australia
(*Orders*: Tel. (08)379–0444
Fax (08)379–4634)

India
Affiliated East-West Press PVT Ltd
G-1/16 Ansari Road
New Delhi 110 002
India
(*Orders*: Tel. (11)327–9113
Fax (11)331-2830)

Japan
Kanda Book Trading Company
Tanikawa Bldg
3–2 Kanda Surugadai
Chiyoda-ku
Tokyo 101
Japan
(*Orders*: Tel. (03) 3255–3497
Fax (03) 3255–34957

Contents

Preface ix
Acknowledgments x

SEARLE, M.P. & TRELOAR, P.J. Introduction 1

Karakoram and Afghanistan

CAPORALI, A. Recent gravity measurements in the Karakoram 9

ZANCHI, A. Structural evolution of the North Karakoram cover, North Pakistan 21

GAETANI, M., JADOUL, F., ERBA, E. & GARZANTI, E. Jurassic and Cretaceous orogenic events in the North Karakoram: age constraints from sedimentary rocks 39

CRAWFORD, M.B. & SEARLE, M.P. Collision-related granitoid magmatism and crustal structure of the Hunza Karakoram, north Pakistan 53

TRELOAR, P.J. & IZATT, C.N. Tectonics of the Himalayan collision between the Indian Plate and the Afghan Block: a synthesis 69

North and West Pakistan

AHMED, Z. Leucocratic rocks from the Bela ophiolite, Khuzdar district, Pakistan 89

ARIF, M. & JAN, M.Q. Chemistry of chromite and associated phases from the Shangla ultramafic body in the Indus suture zone of Pakistan 101

JAN, M.Q., KHAN, M.A. & QAZI, M. S. The Sapat mafic–ultramafic complex, Kohistan arc, North Pakistan 113

KHAN, M.A., JAN, M.Q. & WEAVER, B.L. Evolution of the lower arc crust in Kohistan, N Pakistan: temporal arc magmatism through early, mature and intra-arc rift stages 123

SULLIVAN, M.A., WINDLEY, B.F., SAUNDERS, A.D., HAYNES, J.R. & REX, D.C. A palaeogeographic reconstruction of the Dir Group: evidence for magmatic arc migration within Kohistan, N. Pakistan 139

POGNANTE, U., BENNA, P. & LE FORT, P. High pressure metamorphism in the High Himalayan Crystallines of the Stak Valley, northeastern Nanga Parbat–Haramosh syntaxis, Pakistan Himalaya 161

GEORGE, M.T., HARRIS, N.B.W. & BUTLER, R.W.H. The tectonic implications of contrasting granite magmatism between the Kohistan island arc and the Nanga Parbat–Haramosh massif, Pakistan Himalaya 173

CRONIN, V.S., SCHURTER, G.J. & SVERDRUP, K.A. Preliminary Landsat lineament analysis of the northern Nanga Parbat-Haramosh massif, northwest Himalaya 193

DIPIETRO, J.A., POGUE, K.R., LAWRENCE, R.D., BAIG, M.S., HUSSAIN, A. & AHMAD, I. Stratigraphy south of the Main Mantle Thrust, Lower Swat, Pakistan 207

GRECO, A. & SPENCER, D.A. A section through the Indian plate, Kaghan Valley, NW Himalaya, Pakistan 221

Tethyan Himalaya

VANNAY, J.C. & SPRING, L. Geochemistry of the continental basalts within the Tethyan Himalaya of Lahul–Spiti and SE Zanskar, northwest India 237

SPRING, L., BUSSY, F., VANNAY, J.-C., HUON, S. & COSCA, M.A. Early Permian granitic dykes of alkaline affinity in the Indian High Himalaya of Upper Lahul and SE Zanskar: geochemical characterization and geotectonic implications 251

STECK, A., SPRING, L., VANNAY, J.-C., MASSON, H., BUCHER, H., STUTZ, E., MARCHANT, R. & TIECHE, J.C. The tectonic evolution of the Northwestern Himalaya in eastern Ladakh and Lahul, India 265

GARZANTI, E. Sedimentary evolution and drowning of a passive margin shelf (Giumal Group; Zanskar Tethys Himalaya, India): palaeoenvironmental changes during final break-up of Gondwanaland 277

ROBERTSON, A.H.F. & DEGNAN, P.J. Sedimentology and tectonic implications of the Lamayuru Complex: deep-water facies of the Indian passive margin, Indus Suture Zone, Ladakh Himalaya 299

High Himalaya

POGNANTE, U. & BENNA, P. Metamorphic zonation, migmatization and leucogranites along the Everest transect of Eastern Nepal and Tibet: record of an exhumation history 323

LOMBARDO, B., PERTUSATI, P. & BORGHI, S. Geology and tectonomagmatic evolution of the eastern Himalaya along the Chomolungma–Makalu transect 341

SCHNEIDER, C. & MASCH, L. The metamorphism of the Tibetan Series from the Manang area, Marsyandi Valley, Central Nepal 357

REDDY, S.M., SEARLE, M.P. & MASSEY, J.A. Structural evolution of the High Himalayan Gneiss sequence, Langtang Valley, Nepal 375

HARRIS, N.B.W., INGER, S. & MASSEY, J.A. The role of fluids in the formation of High Himalayan leucogranites 391

ENGLAND, P.C. & MOLNAR, P. Cause and effect among thrust and normal faulting, anatectic melting and exhumation in the Himalaya 401

GUILLOT, S., PÊCHER, A., ROCHETTE, P. & LE FORT, P. The emplacement of the Manaslu granite of Central Nepal: field and magnetic susceptibility constraints 413

SEARLE, M.P., METCALFE, R.P., REX, A.J. & NORRY, M.J. Field relations, petrogenesis and emplacement of the Bhagirathi leucogranite, Garhwal Himalaya 429

PATEL, R.C., SINGH, S., ASOKAN, A., MANICKAVASAGAM, R.M. & JAIN, A.K. Extensional tectonics in the Himalayan orogen, Zanskar, NW India 445

BROWN, R.L. & NAZARCHUK, J.H. Annapurna detachment fault in the Greater Himalaya of central Nepal 461

Main Central Thrust Zone

GRASEMANN, B. Numerical modelling of the thermal history of the NW Himalayas, Kullu Valley, India 475

METCALFE, R.P. Pressure, temperature and time constraints on metamorphism across the Main Central Thrust zone and High Himalayan Slab in the Garhwal Himalaya 485

MEIER, K. & HILTNER, E. Deformation and metamorphism within the Main Central Thrust zone, Arun Tectonic Window, eastern Nepal 511

MORRISON, C.W.K. & OLIVER, G.J.H. A study of illite crystallinity and fluid inclusions in the Kathmandu Klippe and the Main Central Thrust zone, Nepal 525

Main Boundary Thrust, Lesser Himalaya and beyond

NAJMAN, Y., CLIFT, P., JOHNSON, M.R.W. & ROBERTSON, A.H.F. Early stages of foreland basin evolution in the Lesser Himalaya, N India 541

PIVNIK, D.A. & SERCOMBE, W.J. Compression- and transpression-related deformation in the Kohat Plateau, NW Pakistan 559

MCDOUGALL, J.W., HUSSAIN, A. & YEATS, R.S. The Main Boundary Thrust and propagation of deformation into the foreland fold-and-thrust belt in northern Pakistan near the Indus River 581

JADOON, I.A.K., LAWRENCE, R.D. & LILLIE, R.I. Evolution of foreland structures: an example from the Sulaiman thrust lobe of Pakistan, southwest of the Himalayas 589

FRANCE-LANORD, C., DERRY, L. & MICHARD, A. Evolution of the Himalaya since Miocene time: isotopic and sedimentological evidence from the Bengal fan 605

Index 623

Preface

The Himalaya, resulting from continental thickening following the Tertiary collision of India and Asia, is an ideal field laboratory in which to study the, still continuing, orogenic related processes of deformation, crustal thickening, metamorphism and melting and foreland basin evolution. These processes have been the theme of a number of Himalayan Workshops held within Europe since 1985. The first Himalayan workshop was held in Leicester University in 1985, and has been followed by subsequent, generally informal discussion meetings, in Vandoeuvre-les-Nancy (Spring 1987), Imperial College, London (Autumn 1987), Lausanne (1988), Milan (1990), and Grenoble (1991). The 7th of these meetings was held in Oxford in 1992.

Since 1985, there has been a tremendous upsurge of interest and research within the Himalayan region and the workshops have expanded to include studies of the Tibetan Plateau north of the Himalaya as well as all of the mountain ranges of Central Asia. Although most of the fieldwork is still concentrated in the Himalayan countries of Pakistan, India and Nepal where access and logistics are relatively easy, field work is increasingly being undertaken in the remoter ranges of the Karakoram, Pamirs, Hindu Kush, Kun Lun and the Tien Shan. It is a, not unreasonable, contention that the increasing amount of data being derived from multidisciplinary studies of the India-Asia collision zone, and which span all the fields of geological research, are significantly adding to our understanding of the mechanisms and processes of orogeny.

Although papers presented at previous workshops have been included in special issues of regularly published journals (*Journal of Metamorphic Geology*, 1989, Vol. 7, no. 1 after the Nancy meeting; and *Eclogae Geologicae Helvetiae* 1989, vol. 82, part 2, after the Lausanne meeting) this is the first time that a full conference volume has arisen out of a Himalayan workshop. This volume covers all aspects of the geological sciences and has been divided into six parts encompassing the geology of the Karakoram and Afghanistan; Kohistan and the internal zones of the Indian Plate in Northern Pakistan; the Tethyan Himalaya; the High Himalayan crystallines; the Main Central Thrust zone; and the foreland basins and beyond.

Acknowledgments

This volume arises out of the seventh Himalaya–Karakoram–Tibet workshop held at Oxford University from 6–8 April 1992, under the auspices of the Geological Society of London. We are grateful to the Geological Society and British Gas for financial support. The conveners gratefully acknowledge the support and assistance of the Department of Earth Sciences, Oxford University especially Professors J.D.C. McConnell and J.F. Dewey, John O'Sullivan and Sally Thompson; of Alison Currington at Kingston University; and the bursar and staff of Keble College. We also thank all the participants of the Oxford Conference for a fascinating series of talks and posters, and for lively discussion. We would also like to thank all contributors for dealing with editorial requests and decisions courteously and promptly, and the Geological Society Publishing House for their assistance throughout. A special vote of thanks goes to Pam Nieto-Treloar who has tolerated and survived having her home snowed under by abstracts, manuscripts and referees comments for over a year.

We gratefully acknowledge the efforts of the following colleagues and friends who kindly helped with the reviewing of manuscripts submitted for this volume. Without their prompt response publication would have been far less rapid than has proved the case.

C.J. Banks
A.C. Barnicoat
A. Baud
J-M. Bertrand
M.J. Bickle
K.H. Brodie
M. Brunel
D.W. Burbank
J-P. Burg
J.M.L. Cater
A.F. Chambers
J.C. Clemens
P. Copeland
M.P. Coward
K.G. Cox
M.A. Crawford
F. Debon
L. Derry
J.F. Dewey
E.H. Derbyshire
P.C. England
R.W. England
C. France-Lanord
M. Gaetani
E. Garzanti
R.H. Graham
P. Henderson
N.B.W. Harris
I.F. Hazell
K.V. Hodges
A.J. Hurford
M.S. Hubbard
D.H.W. Hutton
C.N. Izatt
M.Q. Jan
H.C. Jenkins
G.D. Johnson
M.R.W. Johnson
M.A. Khan
N.J. Kusznir
X. Le Pichon
B.E. Leake
B. Lombardo
A.M. Macfarlane
J.D. McDougall
E.L. McClelland
P. Matte
P. Molnar
J. McC. Moore
B.J. Murton
Y. Najman
G.J.H. Oliver
L.A. Owen
I. Parsons
M.G. Petterson
J. Platt
U. Pognante
D.J. Prior
H.M. Pritchard
C.J. Pudsey
S.M. Reddy
A.H.F. Robertson
D. Robinson
J. Rose
D.A. Rothery
L.H. Royden
A.D. Saunders
A.M.C. Sengor
H. Smith
D.A. Spencer
J.G. Spray
L. Spring
M.A. Sullivan
P. Tapponnier
U. von Rad
C. Wake
D.J. Waters
A.B. Watts
J. Wheeler
G.D. Williams
M.P. Williams
P.K. Zeitler

Himalayan Tectonics – an introduction

M. P. SEARLE[1] & P. J. TRELOAR[2]

[1] *Department of Earth Sciences, Oxford University, Parks Road, Oxford OX1 3PR, UK*

[2] *School of Geological Sciences, Kingston University, Kingston-upon-Thames, Surrey KT1 2EE, UK*

The Himalaya are renowned as an excellent example of the effects of continent–continent collision, in this case, the results of the closure of Tethys and the subsequent collision and indentation of the Indian plate into the collage of plates or terranes that made up Central Asia some 50 Ma ago. Although the Himalayan chain has long been recognized as a spectacular example of the result of this collision, its effects are recognized over a much larger region than that of the Himalaya alone.

Since initial collision with Asia, India has moved northwards for a further 2000 km, resulting not only in compressional tectonics within the Himalayan chain to the south of the Indus suture, but in compression in the Karakoram, Hindu Kush, Pamir, Kun Lun and Tien Shan mountain ranges, as well as uplift of the 5 km high Tibetan plateau. Earlier Palaeozoic or Mesozoic suture zones across the Tien Shan, Kun Lun and Tibet have been reactivated by the collision of India with Asia as crustal thickening, with associated regional metamorphism and anatexis, spread north of the collision zone into the Karakoram and Pamirs, as well as south across the Himalaya. The totality of crustal shortening required by continued indentation of India into Asia is accommodated by thrusting and folding in the Himalaya as well as in the mountain ranges to the north of the suture, internal homogeneous thickening of Tibetan lithosphere, partial underthrusting of the Indian Plate beneath Tibet, and lateral eastward extrusion of Tibet and southeast Asia.

The Indus suture zone extends for at least 2500 km along the northern margin of the Himalaya in northern Pakistan, northwest India and south Tibet, and marks the actual zone of collision. Deep water Tethyan sediments, melanges with both serpentinite and shaley matrix, remnant ophiolites, blueschists and thick continental molasse deposits are all exposed along the suture zone. In the Pakistan and western Ladakh sector a large island-arc terrane, the Dras–Kohistan–Nuristan arc, is sandwiched within the Indus suture zone. West of Nanga Parbat, Kohistan is bounded along the south by the Main Mantle Thrust, deformation along which began at about 55 Ma ago, and along the north by the Shyok suture zone which closed at about 100 Ma ago.

The northern continental margin of the Indian plate is exposed along the Zanskar Range, south of the Indus suture zone and north of the High Himalaya. It continues eastward into the Spiti region and along the south Tibet border region with both India and Nepal. These strongly folded and thrust rocks are composed of a thick, Mesozoic, dominantly shelf carbonate sequence which overlie Palaeozoic sediments and Carboniferous to Permian Panjal Trap volcanic rocks. Early Tertiary marine limestones of Palaeocene to Lower Eocone age are the youngest marine sediments exposed, after which continental red-bed deposition characterized both the Indian shelf and the Indus suture zones. The Tethyan Himalayan zone is bounded along the south or southwest by the major South Tibetan detachment system, a north-dipping extensional fault that separates the sedimentary succession above from the High Himalayan Central Crystalline complex below.

The High Himalaya, or Tibetan Slab, is the zone of regional metamorphic rocks, migmatites and crustal melt derived leucogranites that corresponds to the belt of topographically highest mountains. This northward-dipping slab is bounded along its top, in the north, by the normal fault, and along its base in the south by the Main Central Thrust. Although the High Himalaya is extremely difficult terrain in which to work, there is surprisingly good accessibility along the many valley systems which cut southwards through the High Himalaya. Many such antecendent rivers rise on the Tibetan Plateau, and drain south through the Himalaya into either the Indus drainage system in the west or the Ganges in the east. Metamorphic isograds are the right way-up along the top of the slab where they are exposed in Zanskar, and are everywhere inverted along the Main Central Thrust zone at the base of the slab. The larger leucogranite bodies tend to be concentrated along the top of the High Himalayan slab, frequently forming some of the highest mountains along the Himalaya, although in the east,

From TRELOAR, P.J. & SEARLE, M.P. (eds), *Himalayan Tectonics*
Geological Society Special Publication, No. 74, pp. 1–7.

particularly in Bhutan, they intrude up into the Tethyan sedimentary zone.

The Lesser Himalaya comprises the Precambrian and Palaeozoic sedimentary cover to the Indian basement, intruded by early Palaeozoic, mainly Cambrian granites, and exposed in a series of south-vergent thrust sheets. The Lesser Himalayan zone is bounded above by the Main Central Thrust zone of inverted metamorphism, and along its base by the Main Boundary Thrust that marks the southern topographic limit of the Himalaya. The Main Boundary Thrust places the Palaeozoic and Precambrian rocks of the Lesser Himalaya over the late Tertiary Siwalik molasse deposits of the Sub-Himalayan zone. Thrusting has continued to propagate southwards into the foreland, and is particularly well-exposed in the Pakistan sector in the west, notably in the Potwar Plateau and Salt Ranges. Within the foreland basins, late Tertiary, Quaternay and Recent folds and thrusts, often related to spectacular geomorphic features, extend around the Himalayan arc to the Sulaiman and Kirthar Ranges in western Pakistan, east of the Chaman fault system.

Karakoram and Afghanistan geology

Over the years numerous Italian expeditions have visited the Karakoram region. The first three papers include data derived from the more recent of these. **Caporali** describes recent gravity measurements from two new gravimetric profiles, one across the Aghil Range and Shaksgam region, north of K2, the other from Skardu to Gilgit along the Biafo and Hispar glaciers. Using GPS (Global Position System) satellites, several astrogeodetic stations were established for measuring the deflection of the vertical. The data indicate a northward increase in depth to the Moho across the Main Karakoram Thrust into the Karakoram mountains. **Zanchi** describes the structural evolution of the north Karakoram Permian to Cretaceous sediments from the Chapursan and Shimshal valleys. Post-Cretaceous north and northeast vergent folds and thrusts have been cut by later south vergent breakback thrusts and an important set of strike-slip faults. East–west aligned faults show left-lateral, sinistral slip, while NW–SE aligned faults are right-lateral, showing dextral displacements, similar to the Karakoram fault. **Gaetani *et al*.** used the ages of the sedimentary rocks to define the major deformation periods in the northern Karakoram. The earliest is Liassic to early-middle Jurassic and is probably related to the Cimmeride collision of Gondwana-derived micro-plates. A mid-Jurassic marine transgression then covered the Karakoram and marine conditions lasted up to the late Cretaceous. New discoveries of Campanian nannofossils date the youngest marine sediments known from the Karakoram.

Crawford & Searle describe the field relations, geochemistry and isotope chemistry of the post-collisional granites of the Hunza Karakoram. The Hunza dyke swarm consists of a cogenetic suite of granodiorites, monzogranites and leucogranites which are volatile-depleted, non minimum melts similar to the Baltoro pluton further east. The dykes intrude the early phase of the Karakoram batholith as well as the sillimanite-grade metamorphic rocks to the south. The Sumayar pluton is a homogeneous, water-saturated, minimum melt leucogranite with abundant tourmaline and evolved Sr isotope ratios, similar to the Manaslu leucogranite in the High Himalaya of Nepal. **Treloar & Izatt** provide a synthesis of the tectonics of Afghanistan and western Pakistan. They argue that two phases of Himalayan collision can be recognised in the western part of the arc. The first one, at about 55 Ma, involved collision of India with Asia. However, collision between India and Afghanistan did not occur until the Pliocene, a conclusion indicated by the post-Miocene age of deformation across much of southern and central Afghanistan as well as in the Sulaiman Ranges and Kohat Plateau of western Pakistan.

North and West Pakistan

Ahmed describes various leucocratic rocks, in particular trondhjemites, from the Bela ophiolite complex in western Pakistan that is one of the ophiolites emplaced onto the western and northern margins of the Indian Plate during the Palaeocene. He distinguishes between a trondhjemitic suite fractionated from a basic magma in a marginal basin, and a potassic suite generated by anatexis during ophiolite emplacement. As acidic plagiogranites from both suites have similar U-Pb zircon ages of 65 ± 1 Ma, he concludes that ophiolite emplacement must have closely followed volcanism in the marginal basin.

The geochemical evolution of rocks of the Kohistan island arc are described in a number of papers. **Arif & Jan** describe the chemistry of podiform chromite within the serpentinized ultramafic rocks of the Mingora ophiolitic melange located within the Main Mantle Thrust zone in the Shangla area, to the west of the well known Jijal Complex. Chromite chemistry is consistent with a transition from ocean floor to island arc, although these rocks have subsequently undergone amphibolite to granulite facies regional metamorphism. **Jan *et al.*** describe the Sapat layered mafic–ultramafic complex which outcrops on the hanging wall of the Main Mantle Thrust to the east of the Jijal Complex. These rocks have many petrographical and mineralogical similarities to those of the Jijal Complex, and like them are interpreted as forming the base of the Kohistan terrane being derived from an island-arc related magma which, in Jijal, has been typified as a high-Mg tholeiitic type that shows some form of interaction between oceanic lithosphere and island arc magmatism. **Khan *et al.*** summarize the trace element and rare earth element geochemistry of both the Kamila amphibolite belt, which marks the middle to lower island arc crust, and the Chilas Complex. Approximately 40 Ma of subduction-related intra-oceanic magmatism is preserved from the early MORB-related Kamila amphibolite belt to the transitional tholeiite to calc-alkaline plutons intruded during early stages of arc growth. The extensive Chilas complex gabbronorites were derived from partial melting of a mantle diapir emplaced, most probably, during early stages of sub-arc splitting. Later stages of back-arc splitting led to the eruption of picrites and high-Mg basalts. **Sullivan *et al.*** reconstruct the palaeogeography of the Dir Group, which marks the last stages of arc magmatism and which forms the upper part of the Kohistan arc. Following emplacement of granitoid intrusions at about 80 Ma, there was a hiatus in igneous activity until about 55 to 60 Ma, when sediments of the Baraul Banda Slates were deposited unconformably on the unroofed plutons. During this hiatus from the late Cretaceous to early Eocene, the locus of magmatism had migrated northwards within the Kohistan arc. The Baraul Banda Slates were deposited in an extensional basin, deposition ceasing at about the time that extrusion of the Utror volcanics commenced. Basin development and the abundance of rhyolites and pyroclastic flows suggests that extensional tectonics predominated within Kohistan during the early Eocene despite a regionally compressive environment.

Five papers describe aspects of the geology of the internal zones of the Indian Plate in Northwest Pakistan. **Pognante *et al.*** describe some high-pressure regional metamorphic rocks along the northeastern margin of the Nanga Parbat – Haramosh syntaxis. These are part of a regionally developed high to very high pressure metamorphism, locally eclogitic, developed on the footwall of the Main Mantle Thrust. This metamorphism followed shortly after collision and was followed by a rapid isothermal exhumation by processes as yet unidentified. This metamorphic history is very different from that of the High Himalaya to the east. **George *et al.*** describe the northwestern margin of the Nanga Parbat – Haramosh syntaxis. Post collisional magmatic histories and magma source regions in Kohistan are significantly different from those of the Indian Plate rocks within the syntaxis. Following the collision of Kohistan with the Indian Plate, biotite granite sheets followed by muscovite granite sheets were emplaced into the Kohistan batholith at around 30 Ma. Tourmaline lecuogranite dykes and plutons were emplaced into the Nanga Parbat gneisses belonging to the Indian plate at 2 to 12 Ma. Geochemistry and Sr-Nd isotope systematics suggest that these were the products of vapour-absent batch melting of pelites, probably Precambrian metasediments. The very recent, rapid exhumation of the Nanga Parbat region may be linked to decompression melting associated with high heat productivity. **Cronin *et al.*** present a lineament map of the northern Nanga Parbat – Haramosh massif based on Landsat imagery. Although some of the lineaments relate to known neotectonic faults, the majority appear not to and there is clear scope for future extensive fieldwork. **DiPeitro *et al.*** describe the metamorphosed sedimentary sequence of the Lower Swat region of north Pakistan, south of the Main Mantle Thrust, which they relate to a depositional history from the late Palaeozoic breakup of Gondwana to Mesozoic passive margin development and finally to drowning of the shelf at the start of the Himalayan orogeny. **Greco & Spencer** describe the stratigraphy, structure and metamorphism of the Kaghan Valley to the north of the Hazara Syntaxis and provide a coloured map of much of this region of the Northwest Himalaya.

Tethyan Himalaya

The Carboniferous to Permian rifting of Tethys and the breakup of Gondwana is recorded in the Panjal Traps of SE Zanskar and NW Spiti, which, as described by **Vannay & Spring**, are slightly evolved continental flood volcanics. In addition to these lavas, the region contains the Baralacha-la dyke swarm, a cogenetic suite of tholeiitic to alkalic basalts which evolved by fractional crystallisation and may have undergone some crustal contamination. They are chemically dissimilar to the Panjal Traps with which they cannot be comagmatic. The dykes may predate extrusion of the Panjal Traps basalts and could represent magmatism during early stages of rifting of Tethys. **Spring *et al*.** describe the Yunam granitic dykes that intrude the Ordovician Thaple Fm. in SE Zanskar and Lahoul. U-Pb zircon dating suggests an early Permian age of magmatism, which they relate to anorogenic magmatism during the Permo-Triassic rifting of Tethys. **Steck *et al*.** present a structural and stratigraphic synthesis of a section across the eastern Zanskar shelf and Indus suture zone in east Ladakh, based on detailing mapping. They also provide an excellent colour map of this transect and several detailed cross-sections. **Garzanti** describes the sedimentary evolution of the early Cretaceous Guimal Group which he relates to the drowning of a passive continental margin in the Zanskar Himalaya, as well as being related to major magmatic and tectono-eustatic events which affected the Tethys Himalaya at this time. **Robertson & Degnan** reconstructed the deep-water passive margin rocks of the Lamayuru complex, exposed along the Indus suture zone in Ladakh, and made comparisons with other such passive margin sequences, for example in the Oman mountains. The Lamayuru complex rocks, despite being structurally very complicated, can be restored to their sedimentary position outboard of the Zanskar shelf sediments.

High Himalaya

Papers by **Pognante & Benna** and **Lombardo *et al*.** both deal with regional metamorphism in the Everest–Makalu area of Nepal and south Tibet. Crustal thickening and subduction caused the main metamorphism in the High Himalaya zone with migmatisation and in situ melting of the leucogranites formed at medium pressure and aided by fluids released during prograde metamorphism. Previously proposed models, discussed here, for both the location of the leucogranites and the thermal history of the High Himalaya, include the condutivity contrast across the normal fault at the top of the slab, and the recumbent folding and thrust imbrication of isograds. The melting is interpreted by Lombardo *et al*. as being synchronous with extension along the South Tibetan Detachment Zone and as being dominated by the dehydration melting of muscovite during isothermal decompression. **Schneider & Masch** describe the metamorphism in the Manang area in the Marysandi valley of central Nepal, where the metamorphic zonation increases in grade down-structural section, at the top of the slab. The fluid composition at the base of the Tibetan Series was CO_2 rich and maximum temperatures based on carbonate solvus thermometry were around 510–530° C. Moving east to the Langtang area north of Katmandu, **Reddy *et al*.** describe the metamorphic units and structural evolution of the High Himalayan gneisses above the MCT zone. Kyanite is the stable alumino-silicate phase above the MCT, whereas sillimanite is present at higher structural levels. Microstructures show that foliation development occurred during peak metamorphism and was consistent with shearing towards the SW. The apparent metamorphic inversion resulted from tectonic juxtaposition along high-temperature shear zones and later brittle thrusts. Two sets of leucogranites are recognised. An early set, intruded into sillimanite-grade gneisses, are essentially layer parallel in situ melts emplaced during top to SW shearing. A later set, which intrude extensional features, are more probably decompression melts. Mapping of the upper Langtang valley also showed the normal fault to be located entirely within the Shisha Pangma region of Tibet to the north of Langtang.

The High Himalayan leucogranites are one of the most characteristic geological features of the Himalaya. Although there are some 15–20 large-scale individual plutons, such as Manaslu or Makalu in Nepal and the Chomolhari and Monlakarchung plutons in Bhutan, vast areas of the High Himalaya at the top of the Tibetan slab show migmatisation and intrusion of numerous dykes and sills of leucogranite. Many of the papers included in this volume (see Pognante & Benna, Lombardo *et al.*, Reddy *et al.*, Harris *et al.*, Guillot *et al.*, Patel *et al.*) argue that many of these are decompression melts, intrusion of which was synchronous with, or postdated, extension within the upper parts of the High Himalayan crystallines and along the normal fault zone above them. If melting was truly synchronous with extension, then thermodynamics of melting reactions would be consistent with melting being through vapour absent or dehydration melting rather than vapour present melting. The former reactions have positive slopes in P–T space, readily crossed during isothermal decompression, whereas the latter have negative slopes. **Harris *et al*.** show, by using isotope systematics and metamorphic phase equilibria, that micaceous metasediments are the most likely source rocks for the leucogranites, and that incongruent melting of muscovite is the most likely melt reaction. These authors argue that as Rb/Sr ratios are between 3 and 6 for most Himalayan leucogranites, these high ratios could not have formed under fluid-present conditions. They therefore suggest that fluid-absent melting occurred during fractional melting, and that the migration of the granite was aided by decompression melting. **England & Molnar** discuss the problem of extensional faulting in an overall compressive regime and provide a number of possible explanations for its occurrence. Decrease of shear stresses on the MCT caused by melting along the fault; reduction of strength within the High Himalayan wedge through devolatization reactions; elevation of the Himalaya due to lithospheric delamination, or by the transference of active motion onto another structurally lower, fault could each have resulted in normal faulting. They suggest that normal faulting was responsible for most of the exhumation and, consequently, also for anatexis through decompression melting. On balance, the most likely probable cause of normal faulting was either uplift of the Himalaya or cessation of slip along the MCT and transference of slip to a new underlying fault.

Guillot *et al*. investigated the emplacement mechanism of the Manaslu leucogranite using field and magnetic data. They suggest that granite emplacement post-dated the north-dipping collapse structures and were more directly related to E–W dextral shearing. **Searle *et al*.** working on the emplacement structures in and around the Bhagirathi leucogranites in the Garhwal Himalaya describe a series of low-angle, en echelon, lensoidal intrusions around the mountains of Shivling, Bhagirathi and Thalay Sagar. Early coaxial mineral growth was re-oriented by later non-coaxial deformation associated with a major zone of ductile NNE–SSW extension across the normal fault at the top of the slab. The final emplacement level was constrained by the density contrast between melt and country rock, the thermal blanketing of the Tethyan sediments above the fault and the extensional stress field at higher structural levels.

Along much of its strike length, the Higher Himalayan Crystallines are bounded above by a normal fault with downthrow to the north or NE. In the Zanskar region, this fault is represented by the Zanskar shear zone which is known to have acted as a major NE-dipping normal fault during culmination of the High Himalayan Central crystalline rocks along the footwall. **Patel *et al*.** reinterpret this structure as having had an earlier phase of top-to-SW thrusting superposed by layer-parallel NE–SW extension during the later stages of deformation associated with ramping onto the Lesser Himalayan foreland. Field photos in their paper imply that melting and emplacement of leucogranite sheets was synchronous with extension. **Brown & Nazarchuk** studied the normal fault in the Thakkola region of central Nepal. Here the Annapurna normal fault is marked by a 2–3 m thick ductile to brittle shear zone which separates kyanite-bearing gneisses and leucogranites on the footwall from greenschist facies Tethyan rocks on the hanging wall. Two stages of extensional shearing are recognized, the first a mid-crustal ductile one and the second an upper crustal brittle one which may have been accompanied by orogen parallel extension. Earlier SW-verging isoclinal folds in the Tethyan sediments above the detachment have been refolded by spectacular NE-vergent backfolds themselves cut by later SW-vergent folds and thrusts. The origin of the NE-vergent folds is unclear as they could be related to orogen thickening rather than gravitational collapse of the High Himalaya. If the latter were to be the case then a period of compression must have separated two distinct phases of extension.

Main Central Thrust

The problem of the inverted metamorphic gradient across the Main Central Thrust zone remains one of the most fascinating geological phenomena in the Himalaya. Ever since the recognition of an inverted metamorphic sequence along the MCT in Nepal, models for the thermal evolution of the thrust have been developed in terms of the thrusting of hot rocks over cold. What these models have tended to overlook are significant differences in the metamorphism and structural geometry of the thrust along strike. In places there is a sharp discontinuity from amphibolite facies rocks on the hanging wall to greenschist facies rocks on the footwall, whereas elsewhere it has been described as a ductile shear zone across which there is no clear metamorphic break. Recent studies have provided evidence for syn-metamorphic thrusting within the fault zone, post-metamorphic thrusting within the fault zone, and diachronous movement that spans the regional metamorphic event. These seeming disagreements about the relative age of thrusting and metamorphism may result from a number of possible explanations: it is possible that the MCT has become a generic term for any fault zone that separates the physiographic lower Himalaya from the higher; it may have become a generic term for any fault zone that separates higher grade rocks from lower grade ones regardless of the age of metamorphism; it may be the same structure all along the chain, but the age and extent of metamorphism may vary along strike; the true structural relationship may be difficult to establish due to poor exposure in forested areas. It has been shown that dissipative frictional heating at shear stresses of >100 MPa may contribute significantly to the development of inverted thermal sequences and consequent inverted metamorphic isograds. However, the logical consequence of this is that differences in temperature between footwall and hanging wall and of shear stress and slip rate may result in substantial differences in the scale of the thermal inversion. Part of the MCT problem may lie in the resolution of these differences.

A number of papers in this volume are relevant to this discussion. **Grasemann** shows, through numerical modelling of the thermal history of the Kulu Himalaya, that in a two-dimensional solution using realistic convergence rates of several centimetres a year, a temperature inversion does not occur. This contrasts with the temperature inversion assumed by one dimensional instantaneous thrust models. He assumes that the High Himalayan metamorphism occurred before movement on the MCT and that there must be a significant conductivity contrast between different rock types which then acted as a thermal barrier to the transport of heat. **Meier & Hiltner** describe the MCT zone in eastern Nepal as a zone, several hundred meters wide, of intensively mylonitised rock showing non-coaxial shearing. Although the main metamorphism in the high grade rocks on the MCT hanging wall predating shearing, metamorphism on the lower grade footwall both increases upwards and, on the basis of garnet rotations, was synchronous with the thrusting along the MCT. **Metcalfe** presents constraints on the pressure, temperature and timing of metamorphism and movement on the MCT zone, along the Bhagirathi River transect in the Garhwal Himalaya. Pressure and temperature both increase sharply across the Vaikrita Thrust (MCT) from south to north from c. 500° C to 770° C and from 6 to 12 kbars, but decrease slightly with increasing structural height to the north. The top 9 km (horizontal distance) of the slab remained at roughly isothermal and isobaric conditions. K-Ar muscovite cooling ages decrease to the south and $^{40}Ar/^{39}Ar$ hornblende ages show that the MCT cooled through 500° C during the early Miocene, but that structurally lower rocks have not been heated above 500° C since the Precambrian. **Morrison & Oliver** emphasize that metamorphism in the Lesser Himalaya was largely Palaeozoic in age, both in Kumaun and central Nepal. The Kathmandu klippe forms an outlier of the Tibetan crustal slab resting on the MCT and consists mainly of Precambrian and Palaeozoic metasediments and granites where the metamorphic zones are normal or right way-up. As pressure increases upwards across the MCT they argue that, in this most southerly exposed part of the MCT, metamorphic inversion is a structural feature and not a thermal one. Two faults mark the MCT zone in this vicinity. The structurally higher one that outcrops to the north, the MCT1 which is equivalent to the Vaikrita Thrust, shows the classic inverted Himalayan metamorphism. By contrast the structurally lower thrust, the MCT2 which outcrops to the south, shows no effects of inverted metamorphism.

Main Boundary Thrust, Lesser Himalaya and beyond

The Main Boundary Thrust marks the southern limit of the Lesser Himalaya and places Precambrian and Palaeozoic rocks southwards over the Tertiary molasse of the Siwalik fold belt. Thrusting has propagated even further south, into the Indian foreland and this is most spectacularly displayed in the Salt Ranges and Potwar Plateau region in the west. **Najman *et al.*** describe the early stages of foreland basin evolution in the north Indian segment of the Lesser Himalaya. During the late Cretaceous, a shallow shelf carbonate marine transgression onlapped the Precambrian to Palaeozoic basement of the Indian plate. During the Late Eocene, clastic sedimentation replaced marine carbonates, marking the onset of uplift of the Himalaya. Thrusting during the late Miocene to Quaternary has placed these rocks over the fluvial Siwalik Group sediments of the Sub-Himalaya along the Main Boundary Thrust.

Much of the work recently done on the development and deformation of foreland basins in the Himalayan external zones has concentrated on northern Pakistan where the foreland basin is at its widest, where there is extensive outcrop of Mesozoic and Cainozoic sequences, good logistical support and access to subsurface data acquired by petroleum exploration companies. The emergent Main Boundary Thrust system, exposed throughout India, Nepal and around the Hazara syntaxis, cannot be followed across northern Pakistan. Instead the boundary between the Potwar Plateau and the hill ranges to the north is now recognised as the erosional effect of a stratigraphic break. The emergent MBT trace presumably outcrops further south, but is as yet unrecognized. A new phase of deformation, that folds the MBT and thrust imbricates on its hanging wall, has been recognised although the cause of this folding remains uncertain. Mapping and interpretation of industry seismic data and well data across the Potwar Plateau has highlighted a number of palaeomountain fronts and deformation geometries perpendicular to the strike of the basin. Lateral changes in shortening geometry as well as unconformities in the Mio-Pliocene molasse have been recognised, although opinions conflict as to how these variations, breaks and oblique fronts are related.

McDougall *et al*. constructed cross-sections across the Main Boundary Thrust zone and estimated over 60 km of internal shortening, with 40 km of displacement along the thrust itself, with a further minor amount accommodated by blind thrusting south of the MBT. The MBT is interpreted as having a very low angle of dip. Although the MBT is currently inactive, shortening of the MBT allochthon from the rear, for instance in the Peshawar Basin, is currently building up the critical taper of the thrust wedge. **Pivnik & Sercombe** describe complex structural patterns in the Kohat Plateau region of northwestern Pakistan that arise from the overprinting of Miocene compression-related south-verging structures by Plio-Pleistocene transpressive structures that may date the initiation of left lateral displacements across the Chaman Fault between Afghanistan and the western edge of the Indian Plate and which die out eastward.

Further south in Baluchistan, the Triassic-Pliocene sections of the Kirthar and Sulaiman Ranges are interpreted as a shortened passive margin. There is growing disagreement, though, as to the geometry (thin skinned thrusting or thick skinned basin inversion), amount (300 or 23 km) and timing (Oligocene or late Pliocene) of this shortening. **Jadoon *et al*.** draw on surface geology, well and seismic data to construct cross-sections across foreland structures within the Sulaiman Lobe of northwestern Pakistan. A stratigraphic section about 10 km thick at the deformation front is detached from the crystalline basement of the Indian Shield at the base of the wedge. Structures within the Sulaiman Lobe suggest an evolution from incipient fault-propagation folds to broader concentric buckle folds to duplex structures.

Much of the erosional detritus removed from the southern flanks of the Himalaya are transported along river systems and deposited in the two major fans of the Indus River in the west and the Ganges river in the east. The Bengal Fan occupies much of the Bay of Bengal between the Coromandel coast of southeast India and the Andaman Islands. **France-Lanord *et al*.** report Sr, Nd, O and H isotope data from ODP Leg 116 cores, located at 1° S in the Bay of Bengal. They conclude that the source of the Bengal Fan sediments has not changed since the early Miocene and that the major source was the High Himalayan crystallines. The High Himalaya must therefore have been a significant topographic feature since at least the early Miocene. Minor variations in sedimentation and style were probably a result of tectonic activity, coupled with the effects of climate and sea-level changes.

Recent gravity measurements in the Karakoram

ALESSANDRO CAPORALI
Dipartimento di Geologia, Paleontologia e Geofisica Universitá di Padova. Via Giotto 1, 35137 Padova, Italy

Abstract: Two new gravimetric profiles have been obtained in the Karakoram as part of the activities of the Ev–K2–CNR Project. One profile, established in 1988, starts from Bazar Dara (China), north of the Aghil range, and covers the portion of the Shaksgam valley between the confluence of the Urdok-Gasherbrum glaciers to the SE and, along the Shaksgam river, Tek ri-Sughet Jangal to the NW. The second profile, carried out in 1990, starts from Skardu and ends in Gilgit via the Biafo Hispar glaciers. During both expeditions a number of astrogeodetic stations were established for the measurement of the deflection of the vertical, using the satellites of the Global Positioning System (GPS) to determine geodetic reference coordinates. Two astronomic stations were established in 1988 and fourteen in 1990. The new data, when added to those collected by Marussi and co-workers from 1954 to 1979, improve the coverage north of the Main Karakoram Thrust, where the Bouguer anomalies attain their largest negative values. For all available data Bouguer and Airy isostatic corrections have been computed using the ETOPO5 height grid. A mean crust density of 2.85 g cm^{-3}, a density contrast of 0.3 g cm^{-3} and a normal thickness $T = 30$ km have been assumed, so that the surface of isostatic compensation locally best fits the Moho. Two profiles have been selected, one continuing northward of the Nanga Parbat – Haramosh massif the deep seismic sounding profiles done in the 1970s. The other is such that it crosses the Indus Suture Zone, the Main Karakoram Thrust and the Karakoram Fault nearly at right angles. In both profiles, from Pakistan to China, the horizontal gradients of the Bouguer anomalies are in the interval -1.5 to -2 mgal km^{-1}, comparable to gradients found in the region between the Indo-Gangetic Plains and the Himalaya. A marked steepening of the Moho is implied here with increased detail. Isostatic anomalies underneath the Karakoram are found to be more negative than those computed in the earlier work of Marussi, who made different assumptions of density and density contrast. Clearly the new data do not uniquely constrain the local structure of the lithosphere but may be of help in better understanding non-negligible deviations from isostatic equilibrium.

Based on deep seismic sounding profiles done within international co-operative projects (Marussi 1983*a*) and on gravity data collected by Marussi and co-workers (Ebblin *et al.* 1983), McGinnis (1971) and Gulatee (1950), a number of authors have provided evidence that the crustal thickness varies, south to north, from 60–65 km in the Karakoram, to 75 km in northern Pamirs and 50–55 km in Ferghana, north of the loftiest mountain ranges (Beloussov *et al.* 1983; Krestnikov *et al.* 1984). Using values of seismic parameters along profiles and the density versus velocity relation, a density model of the crust and upper mantle has been constructed and the gravity field characterised (Beloussov *et al.* 1983). Large negative Bouguer gravity anomalies have been reported to be strongly correlated with the increased thickness of the crust, but the mechanism of isostatic compensation to support the weight of the topographic masses seems insufficient to explain the structure of the gravity anomalies. East of the Karakoram, in the western Himalayas and Ganga Basin, Lyon-Caen & Molnar (1985) found large deviations from local isostatic equilibrium which have led to speculation about a regional model of flexure (Karner & Watts 1983) of the Indian plate along the boundary with Eurasia. Departures from these broad Moho features have been suggested in the NW Himalayas and Kohistan arc by Verma & Prasad (1987), who analysed the gravity anomalies available from the Indian shield northwards, across the upper course of the Indus, along deep seismic sounding profiles between Chitral and Ladakh. The presence of high density rocks in the lower crust/upper mantle was suggested to explain additional features of the gravity anomalies not accountable for by lithospheric dipping. More recently Molnar (1988) has discussed a possible interpretation of the oscillatory character of the isostatic anomalies from Pir Panjal to the Tarim Basin pointed out by Marussi (1964). The depth of approximately -60 mgal corresponding to the axis of Tethys Karakoram would be the resultant of two opposite effects: a

From TRELOAR, P.J. & SEARLE, M.P. (eds), *Himalayan Tectonics*
Geological Society Special Publication, No. 74, pp. 9–20.

negative anomaly associated with overthickening of the crust (with respect to isostatic equilibrium) and a positive anomaly associated with cold material downwelling into the mantle.

The gravimetric data used in most of these investigations suffer from severe geographical limitations, especially within and beyond the Karakoram. In the literature there is a lack of new gravity data that extends the profiles done by Marussi and colleagues. The distribution of the gravity data is rather dense as far as Gilgit, in the north, and Skardu, to the east. However very few data are available to study those structures which cross the Karakoram range, such as the Shyok suture and the Karakoram Fault, and beyond the Pakistani border, in Chinese territory, where the lowest negative values of the Bouguer anomalies are expected. The purpose of this paper is to present new data recently obtained, in the form both of raw observations and of Bouguer anomalies, and to show the results of a preliminary analysis of two profiles particularly relevant to the local structure of the lithosphere. Only gravimetric (no deflections of the vertical) data will be used in this analysis.

The field work in 1988 and 1990

The lack of traverses across important tectonic subdivisions and a special permission granted to Professor Ardito Desio by the Chinese Authorities to access the Karakoram range (Fig. 1) from the north provided the guidelines for the selection of the itineraries to be followed in the framework of the Ev–K2–CNR project, which aims to promote scientific research at high altitudes, particularly in the Karakoram and Himalayas.

In 1988 we worked in the Aghil range and Shaksgam valley. Using signals broadcast by the satellites of the Global Positioning System (GPS) the gravimetric stations could be pre-

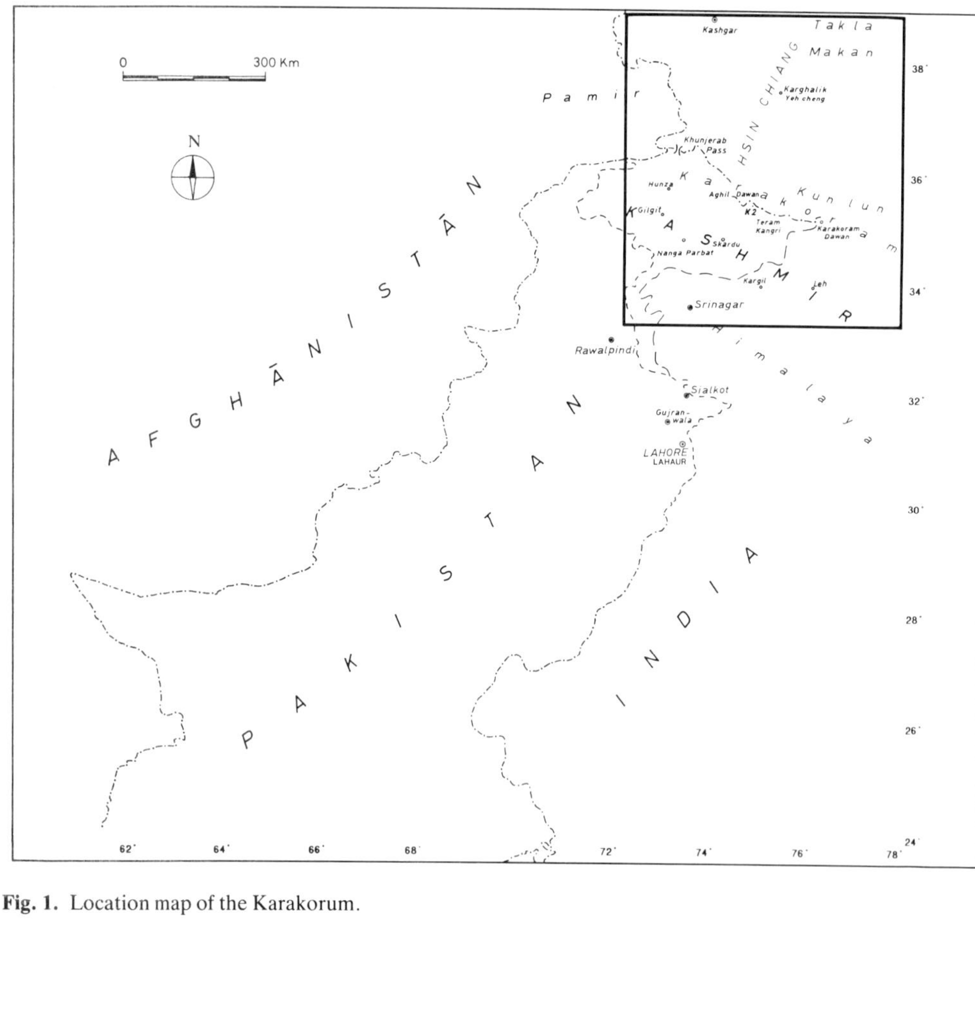

Fig. 1. Location map of the Karakorum.

cisely positioned. The same technique gave ellipsoidal coordinates in the WGS84 geocentric ellipsoid of two astronomic stations and, consequently, the deflection of the vertical at the two endpoints of the profile. At these two points, one located at the confluence of the Urdok and Gasherbrum glaciers, the other near the Tek-ri hill, on the Shaksgam river, we measured vertical angles to K2 which, being a trigonometric vertex, is directly connected to the Indian triangulation network. In this way the aneroid altimeters used to estimate height differences could be calibrated at two endpoints of the profile and estimated heights could be assigned to each gravimetric station. The gravimetric data set included, besides the profile along the upper Shaksgam river, an additional profile through the Aghil pass northwards, to Bazar Dara, and a number of control points in the Chinese cities of Yecheng, Yarkand, Kashgar, Tashkurgan Thajik up to the Khunjerab pass where we overlapped with Marussi's profiles. We could not recover the gravity stations measured by the De Filippi expedition of 1913–14, but Marussi (1964) fortunately does report the De Filippi pendulum observations, as well as those done by Niels Amboldt with the Sven Hedin expedition of 1931. Thus a number of gravity data points are available northeast of the Karakoram, in the Kun Lun and the Tarim basin, providing at least some kind of constraint.

In 1990 we made astrogeodetic, gravimetric and magnetometric observations from Skardu to Gilgit, ascending the Biafo glacier, crossing the main watershed on the Hispar pass and then descending the Hispar glacier down to Karimabad and Gilgit. Again we used GPS to provide coordinates to astronomic and gravimetric stations. To estimate the heights of the stations we used simultaneous thermobarometric observations locally and at the airports of Gilgit and Skardu and the familiar Laplace formula to convert to height differences.

It is very unfortunate that gravimetric measurements with an internal precision better than 0.1 mgal are complemented with heights which could be erroneous by as much as some tens of metres. Consequently the uncertainty in the gravity anomalies due to uncorrect height could be as large as 10 mgal. The assumption of the atmosphere being uniformly layered both at the remote and reference sites is unreliable in these mountain areas and with such large distances involved. The technique of differential GPS, which will be used in future work, gives however a reasonable hope to improve the accuracy of relative heights to better than one metre. Further improvement could be obtained by converting to gravity anomalies the observed astrogeodetic deflections of the vertical, which tend to be smoother and more immune from height errors than gravimetric data.

Data reduction

In the appendix a list is given of the gravity data measured in the Karakoram area in 1988 and 1990. For all gravimetric data points the following information is given:

(i) approximate geographic coordinates (latitude, longitude and height);

(ii) the normal gravity computed at the observation point; therefore the given value includes the free air correction;

(iii) the observed gravity, in the IGSN71 System;

(iv) the model gravity disturbance computed on the basis of the spherical harmonic expansion of the earth's gravity field OSU91A derived from a combination of satellite orbit perturbation, satellite altimetry and regional mean surface gravity anomalies (Rapp 1991);

(v) a Bouguer anomaly, inclusive of terrain effects, computed with the ETOPO5 digital terrain model assuming a mean density of 2.85 g cm^{-3};

(vi) an isostatic anomaly according to Airy, assuming a normal thickness $T = 30$ km and a density contrast of 0.3 g cm^{-3}:

The adopted values of the density and density contrast have been shown to be consistent with observations (Malinconico 1986; Baticci & Poretti 1990) and yield a ratio between the density of the crust and the density contrast between crust and mantle of 9.5, which best fits a linear model of the depth of the Moho (inferred from seismic studies) in terms of the average topographic height (Verma & Prasad 1987). Topographic and isostatic corrections have been computed up to Hayford zone 0 (166.7 km) for all the available data, including Marussi's, using the same program, the Digital Terrain Model and the assumptions on density and density contrast. The model free air anomalies are given to highlight the behaviour of the long wavelength components predicted by the global model OSU91A complete to degree and order 360. Thus its resolution is nominally 55 km, but because of the lack of local observations and the use of geophysically predicted anomalies it is safer to confine the attention to the features with the lowest frequencies. The model anomalies (Fig. 2) consist primarily of a low of over −200 mgal south of Nanga Parbat and of two regions of positive anomalies: one located over

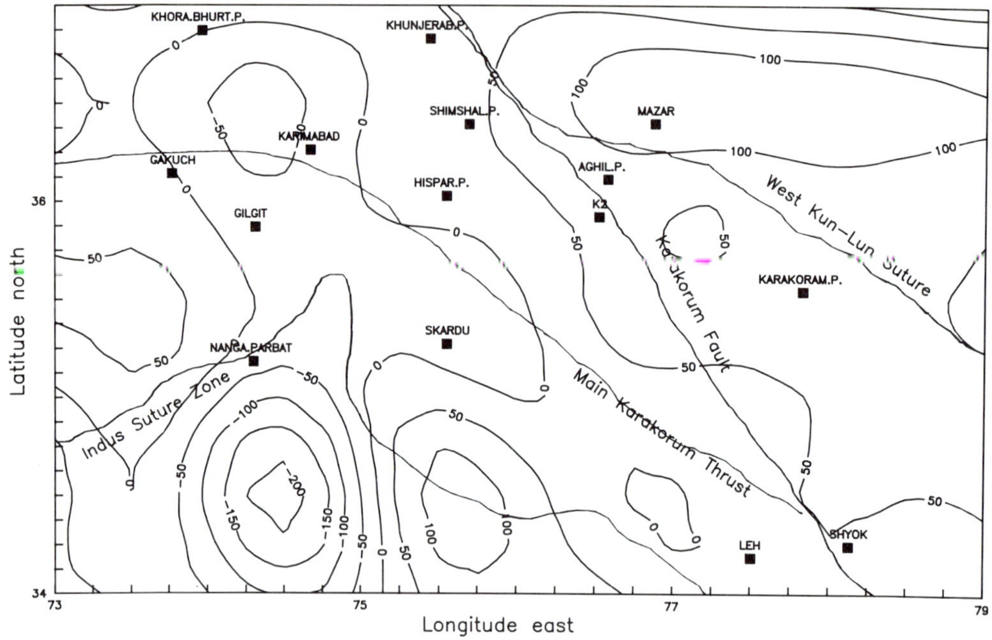

Fig. 2. Free air gravity anomalies computed using the OSU91A global model. The border of the Karakoram range is also shown. Contour interval is 50 mgal.

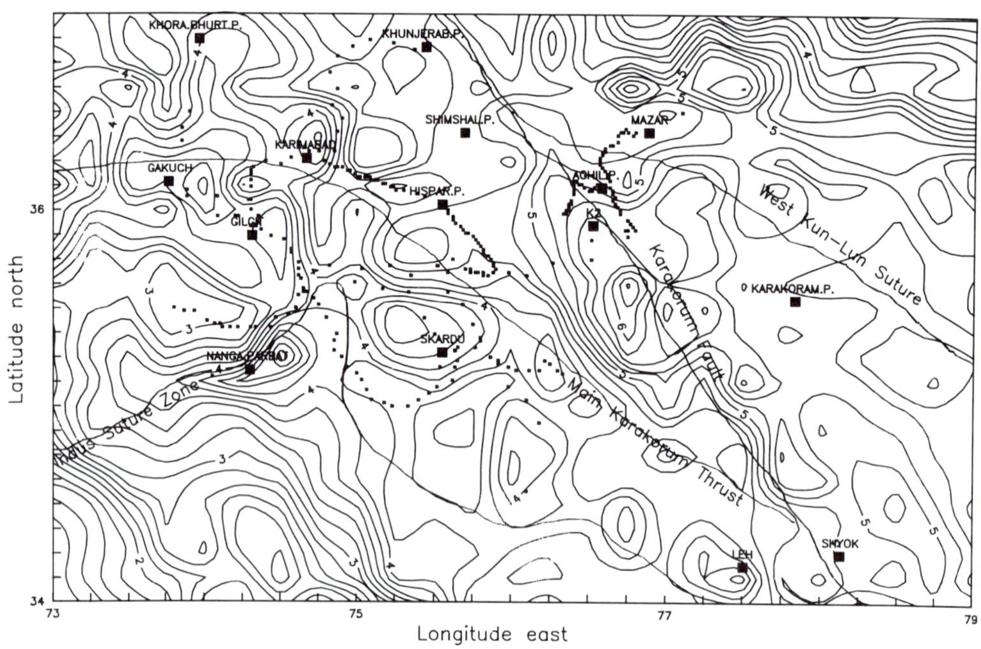

Fig. 3. Contour plot of the ETOPO5 Digital Terrain Model. Contour interval is 0.2 kn.

the Deosai Plains, south of Skardu, the other in the NE corner, located over the Tarim basin. On average the predicted anomalies appear weakly correlated with topography. The local minimum south of Nanga Parbat is remarkable but there is no match in the local anomaly maps. This feature could then be an example of a high frequency structure not of geophysical origin. The mean heights are shown in Fig. 3 by contouring the values of the ETOPO5 Digital Terrain Model of National Oceanic and Atmospheric Administration. This model is credited with a nominal resolution of 5 minutes of arc. Measured heights can significantly differ from model predictions, hence spline smoothing techniques have been used to make, in a neighbourhood of each gravity station, the model heights closest to the observed heights. In the absence of detailed and reliable cartography to a scale smaller than 1:250 000, ETOPO5 has been assumed as the basis for the computation of corrections to the gravity data. The main tectonic subdivisions are taken from Gaetani *et al.* (1991). The location of the data points is also plotted. The data points include those given in Ebblin *et al.* (1983) and the De Filippi and Hedin points given by Marussi (1964). During our 1988 and 1990 work we located a number of the Marussi stations and then took the opportunity to repeatedly cross-calibrate our measurements with the older ones, finding a systematic mean difference of 13.81 mgal (Marussi's values greater than ours) consistent with the difference between our IGSN71 and Marussi's Potsdam reference system. We used a La Coste Romberg mod. G gravimeter, whereas Marussi and co-workers had used a Worden gravimeter. Further details on station description and the measurements are given by Caporali *et al.* (1991).

Discussion of the gravity data

The Bouguer anomalies (Fig. 4) indicate that the profile from Skardu to Gilgit via the Hispar Pass carried out in 1990 covers an area immediately north of a rather steep negative slope of the anomaly, closely aligned with the Main Karakoram Thrust (MKT) line. The cross section labeled AA′ extends north of the 36th parallel in such a way that it extends the one studied by Verma & Prasad (1987). North of the MKT the Bouguer gravity anomaly stops decreasing and maintains a steady low value up to the Khunjerab Pass (Figs 5 & 6). This pattern is supported both by our 1990 data taken along the Biafo and Hispar glaciers, and by data taken by the Marussi group along the Karakoram Highway in the Hunza valley. The cross section BB′

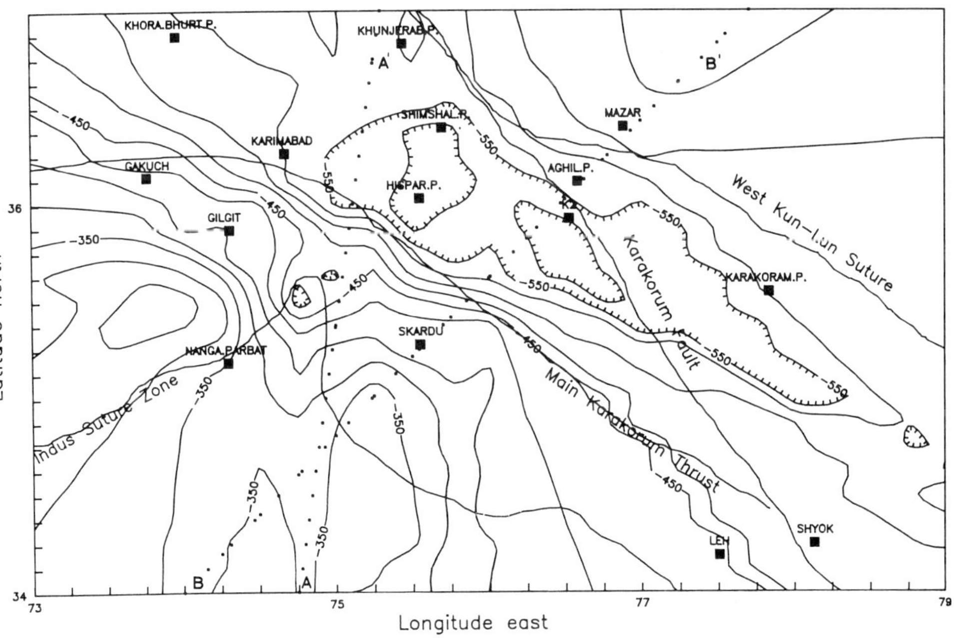

Fig. 4. Bouguer gravity anomalies. Also shown the location of the main tectonic features and two profiles (AA′ and BB′) studied in this analysis. Contour interval is 25 mgal.

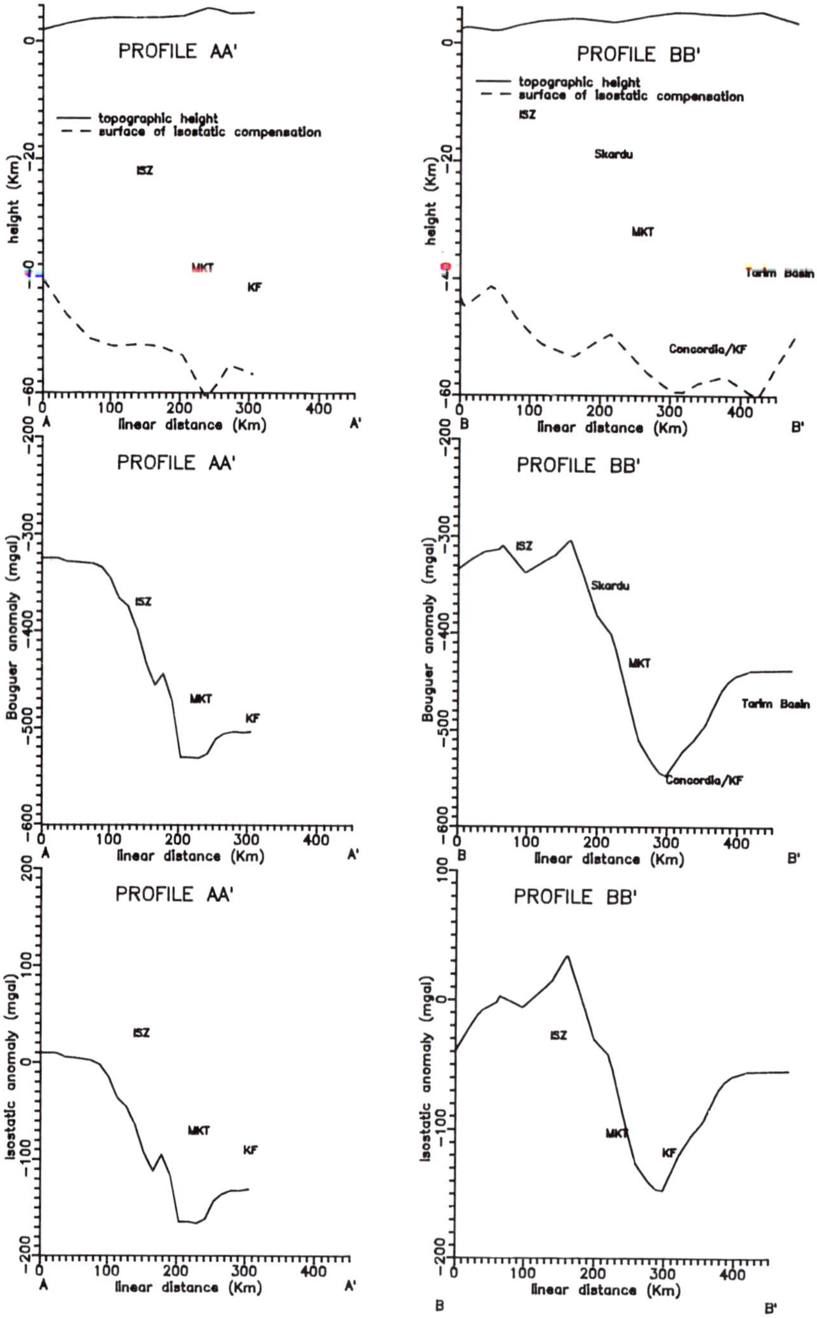

Fig. 5. The cross sections labeled AA′ and BB′ in Fig. 4. The section AA′ is aligned with the DSS profile through Nanga Parbat–Haramosh. The section BB′ is such that the main tectonic features are crossed orthogonally. ISZ, Indus Suture Zone; MKT, Main Karakoram Thrust; KF, Karakoram Fault. Each section is characterized by the mean topographic heights and corresponding surface of isostatic compensation (top); the Bouguer anomalies (centre) and the isostatic anomalies (bottom). The parameters used in the reduction are: crustal density 2.67 g cm^{-3}, density contrast 0.6 g cm^{-3}, normal thickness $T = 30$ km.

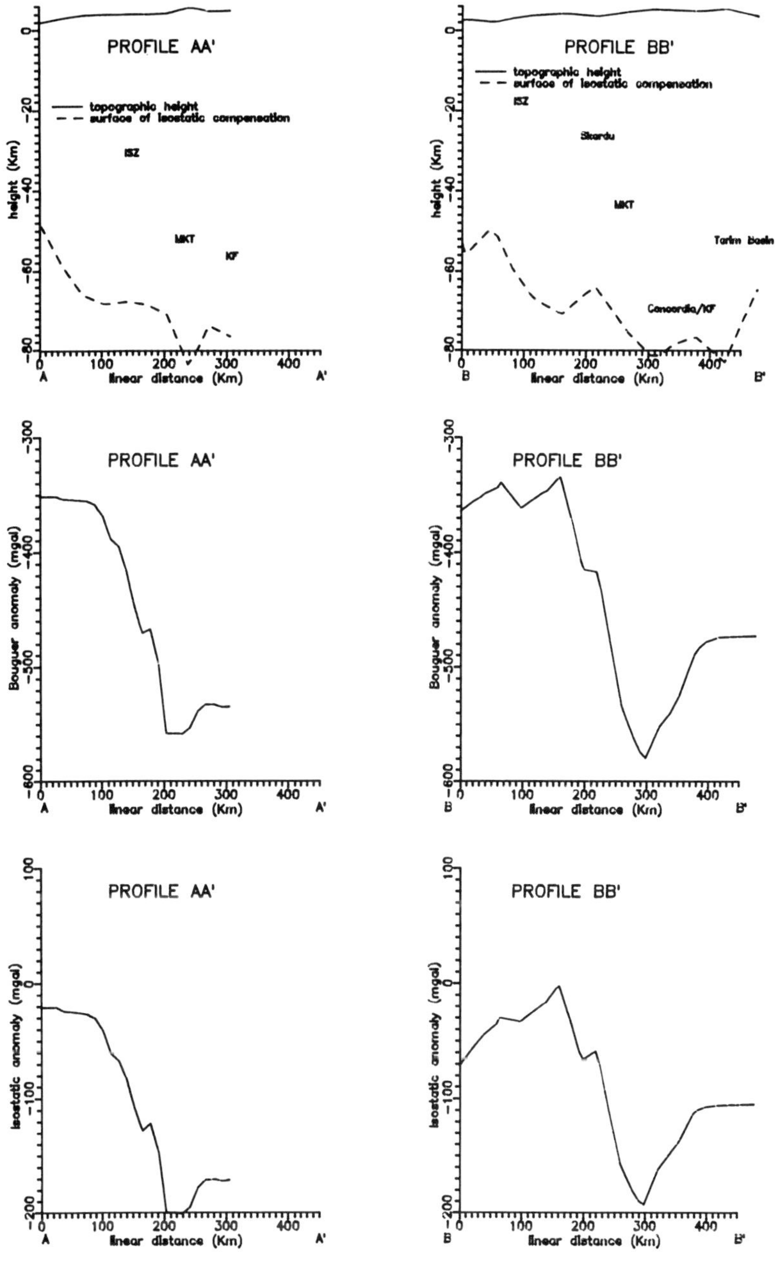

Fig. 6. as Fig. 5 but assuming a density and density contrast of 2.85 and 0.3 g cm^{-3} respectively.

cuts the Indus Suture Zone (ISZ), MKT, KF and West Kun Lun Suture. As for profile AA′ there is again a sharp decrease of the anomaly across the MKT, then a low corresponding to the axial Karakoram, Aghil and Kun Lun mountains, and eventually a very steep increase corresponding to the Eurasian plate, in particular the Takla Makan desert and Tarim basin (Figs 5 & 6). This pattern is supported by the data we collected in 1988 while travelling from Kashgar to the Kun Lun, and are consistent with values reported by De Filippi and Hedin. Both profiles AA′ and BB′ indicate an increase in dip of the Bouguer anomalies to values of approximately

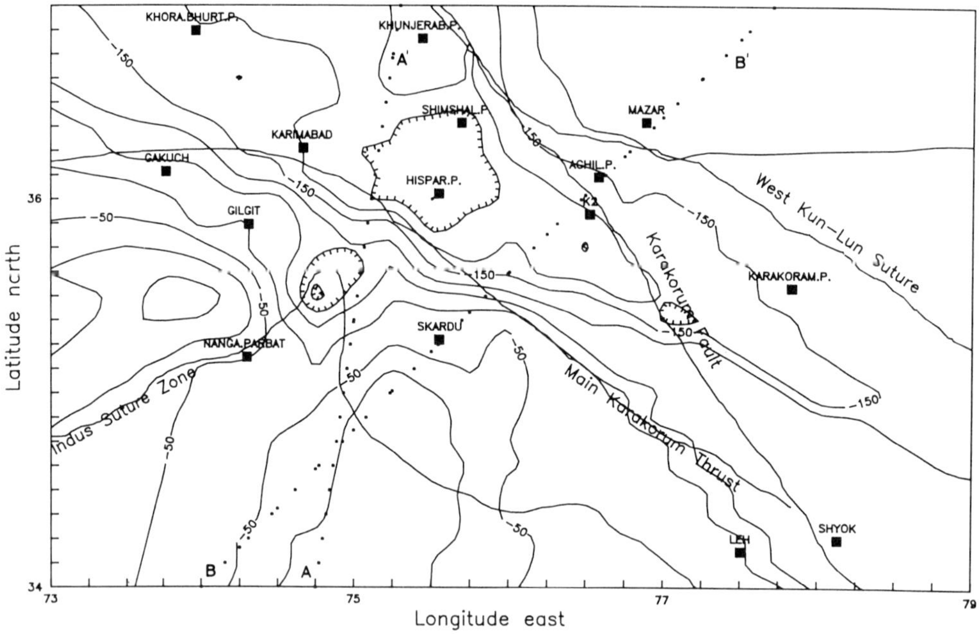

Fig. 7. Airy isostatic anomalies ($T = 30$ km). Contour interval is 25 mgal.

-1.8 mgal km^{-1}. This is close to the gradients found in the Himalayas (Molnar 1988) and interpreted there as due to a marked steepening of the Moho.

For local Airy isostatic equilibrium the thickness of the crustal root should be proportional to the topographic elevation. Airy isostatic anomalies are plotted in Fig. 7. The isostatic anomalies are negative and correlated with topography (Figs 5 & 6). This could indicate excess root formation underneath the Karakoram or, as suggested by Marussi (1964), the presence of low density rocks in the crust. Molnar (1988) has pointed out that the negative isostatic anomaly could result from a combination of effects of opposite sign: a negative anomaly associated with the overthicking of the crust could be partially compensated by the positive anomaly produced by cold material downwelling in the mantle. This model explains an observed negative anomaly corresponding to the Karakoram characterized by a width of 120 km and a maximum negative value of -60 mgal (Marussi 1964, plate IX). The profile BB′ in Fig. 5 confirms the tendency toward a positive isostatic anomaly over the Himalaya and negative values over the Karakoram. However, in contrast to the Bouguer anomalies which are very similar, the entire plot appears to be shifted down with respect to Marussi's by approximately 50 mgal. As may be expected the shift is larger in the corresponding plot of isostatic anomalies in Fig. 6 which is based on values of the crustal density and density contrast such that the surface of isostatic compensation is 75 to 80 km deep. Consequently, if the isostatic reduction is constrained by such a depth of compensation, a mechanism of positive anomaly production is needed to reproduce the observed gravity.

Conclusion

After having merged the gravimetric data collected in two surveys in 1988 and 1990 with those of Marussi, a Bouguer gravity anomaly map has been drawn for the Karakoram area (Fig. 4). Relative to previously published maps there is more detail north of the Main Karakoram Thrust, where the largest negative Bouguer anomalies are found. This behaviour is highlighted with two cross sections, one aligned with the deep seismic sounding profile through the Nanga Parbat–Haramosh massif, the other orthogonal to the main tectonic units from Baltistan (Skardu) to the Takla Makan Desert (Sinkiang). Bouguer and isostatic corrections to the raw data have been computed using the ETOPO5 Digital Terrain Model. The assumed density of 2.85 g cm^{-3} and density contrast of 0.3 g cm^{-3} are such that the surface of isostatic compensation has a depth consistent with values of the Moho inferred from seismic data. In both sections the Bouguer anomalies follow a steep descent across the Main Karakoram Thrust, to rise again to less negative values proceeding NE, past the Karakoram Fault, in the Sinkiang Province. The isostatic anomalies are found to be negative underneath the Karakoram, confirming original findings by Marussi. However the isostatic anomalies presented here are more negative than Marussi's, probably because of increased detail and because of different values of crustal density and density contrast used in the isostatic reductions, for a given normal thickness of $T = 30$ km. The added data bring to light more details on the width and depth of the isostatic anomaly over Tethys Karakoram, the interpretation of which still remains a challenge.

This research has been done as part of the Ev–K_2–CNR project. A. Desio, Project Leader, stimulated this work. I am grateful to P. Molnar and A.B. Watts for critically reviewing the manuscript, and to C. Tscherning for the program used to compute the topographic and isostatic corrections. F. Palmieri is thanked for his unvaluable field work which forms the basis of this paper.

References

BATICCI, G. & PORETTI, G. 1990. Gravity models of the deep structures of the Karakoram range. *Instituto Lombardo (Rendiconti Scientifici),* **B124**, 123–140.

BELOUSSOV, V.V., VOLVOSKY, B.S., VOLVOSKI, I.S., KAILA, K.L., MARUSSI, A., NARAIN, H., TALVIRSKY, B.B., FINETTI, I. & KHAMRABAEV, I.KH. 1983. General features of the lithospheric structure of southern Tien Shan, the Pamirs, Karakoram and the Himalayas. *Bollettino di Geofisica Teorica e Applicata*, **25**, 151–161.

CAPORALI, A., MARZARI, F. & PALMIERI, F. 1991. Geodetic and Geophysical Report. *In*: *Geodesy, Geophysics and Geology of the Upper Shaksgam Valley (North East Karakoram) and South Sinkiang*. Scientific Reports of the Italian Expedition to Karakoram 1988, Prof. A. Desio Leader. Ev K^2 CNR, Milano.

EBBLIN, C., MARUSSI, A., PORETTI, G., RAHIM, S. M. & RICHARDUS, P. 1983. Gravity measurements in the Karakoram. *Bollettino di Geofisica Teorica e Applicata*, **25**, 303–316.

GAETANI, M., GOSSO, G. & POGNANTE, U. 1991. Geological Transect from Kun Lun to Karakoram. *In*: *Geodesy, Geophysics and Geology of the Upper Shaksgam Valley (North East Karakoram) and South Sinkiang*. Scientific Reports of the Italian Expedition to Karakoram 1988, Prof. A. Desio Leader. Ev K^2 CNR, Milano.

GULATEE, B.L. 1950. *Gravity data in India*. Technical Paper, **10**. Survey of India, Dehra Dun.

KARNER, G.D. & WATTS, A.B. 1983. Gravity anomalies and the flexure of lithosphere at mountain ranges. *Journal of Geophysical Research*, **88**, 10449–10477.

KRESTINOKOV, V.N., NERSESOV, I.L. & STANGE, D.V. 1984. The relationship between the deep structure and Quaternary tectonics of the Pamir and Tien Shan. *Tectonophysics*, **104**, 67–83.

LYON-CAEN, H. & MOLNAR, P. 1985. Gravity anomalies, flexure of the Indian plate, and the structure, support and evolution of the Himalaya and Ganga basin. *Tectonics*, **4**, 513–538.

MALINCONICO, L.L. 1986. The structure of the Kohistan-arc terrane in northern Pakistan as inferred from gravity data. *Tectonophysics*, **124**, 297–307.

MARUSSI, A. 1964. Geophysics of Karakoram. *In*: Desio, A. (ed.) *Scientific Reports of the Italian Expeditions to Karakoram (K^2) and Hindu Kush. II Geophysics*. E.J. Brill-Leiden.

—— 1983*a*. The Pakistani Italian Karakoram Geophysical project in the frame of the Pamirs Himalaya International Project. *Bollettino di Geofisica Teorica e Applicata*, **25**, 137–141.

—— 1983*b*. Geophysical trends and evolution of the Pamirs syntaxis and Karakoram. *Bollettino di Geofisica Teorica e Applicata*, **25**, 443–461.

MCGINNIS, L.D. 1971. Gravity field and Tectonics in the Hindu Khush. *Journal of Geophysical Research*, **76**, 1894–1904.

MOLNAR, P. 1988. A review of geophysical constraints on the deep structure of the Tibetan Plateau, the Himalaya and the Karakoram, and their tectonic implications. *Philosophical Transactions of the Royal Society of London*, **A326**, 33–88.

RAPP, R.H. 1991. *The Ohio State 1991 Geopotential and Sea Surface Topography Harmonic Coefficient Models*. Dept. of Geodetic Sciences, The Ohio State University.

VERMA, R.K. & PRASAD, K.A.V.L. 1987. Analysis of gravity fields in the northwestern Himalayas and Kohistan region using deep seismic sounding data. *Geophysical Journal of the Royal Astronomical Society*, **91**, 869–889.

Appendix

Site	Lat	Lon	h	γ	g_{obs}	ΔT_{360}	Δg_B	Δg_I
Gravimetic stations of the 1988 Aghil–Shaksgam Profile								
PON.BAZ.DAR.	36 24	76 48	3545	978755.50	978684.50	118	−486	−106
MUR.SOST	36 23	76 47	3490	978771.00	978675.30	116	−504	−123
OMET. 01	36 24	76 45	3442	978787.20	978695.30	117	−493	−114
OMET. 02	36 25	76 44	3529	978761.80	978691.10	118	−483	−105
OMET. 03	36 24	76 43	3517	978764.10	978678.10	116	−497	−118
OMET. 04	36 24	76 42	3558	978751.40	978678.50	115	−490	111
ST 7	36 23	76 41	3576	978744.50	978671.80	113	−493	−113
ST 8 EX	36 23	76 41	3569	978746.60	978677.80	113	−488	−108
CASA ILICA	36 23	76 41	3519	978762.10	978688.80	113	−486	−106
C.B.2/ILICA	36 23	76 41	3497	978768.80	978691.40	113	−487	−108
ST 9	36 21	76 41	3548	978750.30	978669.80	109	−498	−116
ST 10	36 21	76 41	3579	978740.70	978657.80	109	−504	−123
ST 11	36 20	76 39	3603	978731.90	978650.20	105	−506	−125
ST 12	36 20	76 38	3629	978723.80	978644.50	105	−507	−126
ST 13 GPS	36 20	76 38	3650	978717.40	978641.50	105	−507	−125
ST 14	36 19	76 36	3715	978695.90	978630.80	101	−504	−121
C.B.3	36 18	76 36	3784	978673.20	978608.30	99	−512	−129
ST 15	36 18	76 35	3878	978644.30	978586.50	98	−517	−134
ST 16	36 17	76 35	3923	978628.90	978574.30	95	−519	−136
ST 17	36 16	76 34	4029	978594.90	978548.70	92	−524	−140
ST18	36 16	76 34	4116	978568.10	978533.00	92	−523	−140
ST 19	36 15	76 34	4179	978547.30	978522.10	90	−521	−136
ST 20 C.B.4	36 14	76 34	4264	978519.60	978501.80	88	−524	−139
ST 21	36 13	76 35	4418	978470.80	978464.80	86	−531	−145
ST 22	36 12	76 36	4614	978408.90	978422.30	84	−535	−149
ST 23 GR.SASSO	36 11	76 37	4762	978361.90	978388.90	82	−539	−152
ST 24 PS. AGHI	36 11	76 37	4780	978356.40	978387.70	82	−537	−150
VERT. CINESE	36 11	76 37	4815	978345.60	978375.80	82	−542	−155
ST 25	36 10	76 37	4430	978462.80	978429.30	80	−560	−172
ST 26 GR.MS.	36 9	76 38	4273	978509.70	978472.60	78	−544	−155
ST 27 C.B.5	36 6	76 39	4069	978568.20	978525.80	72	−523	−132
ST 41	36 5	76 39	4086	978561.50	978521.80	70	−522	−131
C.B.6 DRB.JN	36 4	76 40	4105	978554.30	978512.50	68	−526	−134
ST 40	36 3	76 39	4100	978554.40	978510.10	66	−528	−136
ST39	36 2	76 40	4095	978554.40	978506.90	64	−531	−138
ST 38	36 1	76 41	4114	978547.20	978495.10	62	−537	−143
ST 37	36 0	76 42	4129	978541.10	978488.30	61	−540	−145
ST 30 GPS 4	35 59	76 42	4134	978538.10	978479.50	59	−546	−151
ST 31	35 57	76 45	4140	978533.40	978473.40	56	−549	−153
ST 32	35 56	76 46	4129	978535.40	978468.90	55	−554	−157
ST 33	35 56	76 46	4151	978528.60	978463.50	55	−555	−159
ST 34	35 55	76 47	4187	978516.10	978456.40	54	−554	−157
ST 36 C.B.7 GP	35 54	76 48	4232	978500.80	978441.80	53	−559	−161
GPS 1	35 54	76 45	4382	978454.60	978419.90	54	−553	−156
GPS 2	35 52	76 45	4321	978470.60	978427.50	54	−553	−155
GPS 3	35 54	76 44	4348	978465.10	978421.70	55	−557	−161
ST 42	36 7	76 38	3992	978593.40	978529.90	74	−535	−145
ST 43	36 8	76 36	3983	978597.60	978534.10	75	−535	−145
ST 44	36 8	76 35	3974	978600.30	978532.10	75	−538	−149
ST 45	36 8	76 33	3931	978613.60	978541.60	75	−537	−148
ST 46	36 7	76 32	3926	978613.70	978544.10	72	−533	−144
ST 47	36 6	76 31	3884	978625.20	978536.60	70	−547	−157
C.B.8	36 6	76 30	3860	978632.60	978535.50	70	−552	−163
ST 48	36 6	76 30	3850	978635.70	978538.90	70	−550	−161
ST 49	36 7	76 28	3833	978642.30	978544.00	70	−550	−162
ST 50	36 8	76 27	3821	978647.40	978550.90	72	−547	−159
ST 51	36 8	76 26	3806	978652.10	978555.40	71	−545	−158
ST 52	36 9	76 25	3795	978656.90	978561.80	72	−545	−156

Appendix Continued.

Site	Lat	Lon	h	γ	g_{obs}	ΔT_{360}	Δg_B	Δg_I
ST 53	36 9	76 24	3796	978656.60	978569.60	71	−534	−148
C.B.9 GPS	38 8	76 24	3809	978651.20	978565.40	70	−535	−148
ST 54	36 7	76 24	3822	978645.80	978555.20	68	−541	−154
ST 55	36 6	76 24	3849	978636.00	978543.80	66	−546	−159
ST 56	36 5	76 24	3881	978624.70	978535.80	65	−546	−159
ST 57	36 3	76 23	3906	978614.10	978527.30	61	−547	−160
ST 58	36 3	76 23	3921	978609.50	978518.80	61	−553	−165
ST 59	36 2	76 23	3937	978603.10	978510.10	60	−557	−169
C.B.10 GPS	36 2	76 22	3969	978593.30	978507.90	59	−554	−166
ST 60	36 1	76 22	3979	978588.80	978483.80	58	−574	−186
ST 61	36 0	76 22	4012	978577.20	978473.80	57	−557	−189
ST 61B	36 0	76 22	4025	978573.20	978471.80	57	−577	−189
ST 62	36 0	76 21	4050	978565.40	978467.30	56	−577	−189
ST 63	35 59	76 21	4103	978547.70	978451.00	55	−582	−195
ST 64	35 59	76 20	4233	978507.70	978428.10	54	−581	−194
Gravimetic stations of the 1990 Hispar Biafo Profile								
K2 MOTEL	35 18	75 39	2288	979048.60	978895.50	−29	−418	−63
101A MAR	35 17	75 38	2310	979040.40	978890.10	−27	−418	−65
179A MAR	35 25	75 45	2295	979056.40	978878.80	−32	−441	−79
179B MAR	35 25	75 45	2319	979049.00	978875.00	−32	−440	−78
172A MAR	35 35	75 35	2345	979055.20	978826.30	−29	−497	−138
142A MAR	35 43	75 31	2447	979035.10	978771.00	−17	−542	−182
GOYUNGO	35 41	75 35	2500	979015.90	978740.80	−20	−562	−200
CHOKPIONG	35 45	75 37	2616	978985.90	978734.60	−11	−551	−185
HOTO BR.	35 42	75 42	2812	978921.20	978683.10	−14	−565	−199
PAKORA	35 42	75 44	2845	978911.10	978679.90	−12	−563	−195
210 TELEF.	35 41	75 44	2857	978905.90	978679.90	−14	−560	−193
310 SINO	35 41	75 46	2893	978894.80	978681.10	−12	−552	−183
410 N.B.	35 41	75 47	2939	978880.60	978680.80	−11	−544	−175
ASKOLE GPS	35 41	75 49	3070	978840.30	978658.30	−8	−542	−172
153B MAR	35 40	75 49	3063	978841.00	978659.90	−10	−541	−171
111 T.B.	35 40	75 51	3056	978843.10	978658.90	−7	−542	−171
211 STESTE	35 41	75 53	3071	978839.90	978664.10	−3	−535	−162
112 STOPKA	35 41	75 53	3075	978838.80	978662.00	−3	−536	−163
212 KOROPH	35 42	75 54	3211	978798.30	978634.20	1	−541	−167
312	35 43	75 53	3332	978762.40	978607.50	1	−547	−174
412	35 44	75 53	3414	978738.60	978588.50	2	−553	−180
512 NAMLA	35 44	75 53	3470	978721.30	978587.90	3	−543	−170
114	35 45	75 52	3523	978706.40	978556.60	3	−567	−194
214	35 46	75 51	3599	978684.40	978541.80	3	−569	−197
314	35 46	75 50	3657	978666.50	978530.80	2	−570	−198
414 MANGO	35 46	75 50	3707	978651.10	978541.90	2	−549	−177
115	35 47	75 49	3685	978659.30	978523.30	3	−573	−202
215	35 48	75 49	3773	978633.60	978504.80	5	−577	−205
315	35 49	75 48	3856	978609.50	978484.10	6	−584	−212
415	35 50	75 47	3947	978582.90	978468.60	7	−583	−212
515	35 51	75 46	3974	978576.00	978488.20	8	−560	−189
117	35 52	75 45	3988	978573.10	978482.20	9	−565	−195
217	35 53	75 43	4042	978557.90	978453.00	9	−585	−216
317	35 53	75 42	4131	978530.50	978425.30	9	−596	−227
417	35 54	75 41	4205	978509.10	978406.10	10	−603	−235
118	35 55	75 40	4265	978492.10	978396.30	11	−603	−235
218	35 56	75 38	4354	978466.10	978379.30	12	−605	−238
318	35 57	75 37	4391	978456.10	978384.50	14	−594	−228
418	35 58	75 37	4544	978410.40	978377.80	16	−573	−206
119	35 59	75 35	4545	978411.60	978339.90	17	−612	−245
219	35 60	75 35	4609	978393.30	978316.20	18	−624	−258
319	36 0	75 35	4634	978385.60	978315.10	18	−621	−254

Appendix Continued.

Site	Lat	Lon	h	γ	g_{obs}	ΔT_{360}	Δg_B	Δg_I
419	36 1	75 34	4674	978374.70	978318.40	20	−611	−245
121	36 1	75 33	4787	978339.90	978318.10	19	−589	−224
222	36 4	75 26	4427	978455.10	978409.60	22	−572	−207
324	36 5	75 23	4267	978505.80	978446.80	22	−566	−202
625	36 6	75 18	4185	978532.50	978463.90	21	−565	−203
226	36 6	75 15	4039	978577.40	978491.90	19	−563	−202
127	36 7	75 14	4027	978582.60	978492.60	19	−565	204
227	36 7	75 13	3820	978646.30	978529.80	18	−565	−205
327	38 8	75 11	3882	978628.70	978516.80	17	−565	−205
128	36 8	75 10	3808	978651.50	978529.30	16	−566	−207
228	36 9	75 8	3724	978678.80	978549.10	14	−560	−202
328	36 9	75 7	3670	978695.40	978565.40	13	−553	−196
428	36 9	75 6	3526	978739.80	978593.40	12	−552	−195
528	36 10	75 5	3459	978761.90	978611.80	11	−547	−191
628	36 10	75 4	3398	978780.70	978621.90	10	−549	−193
129	36 10	75 3	3319	978805.00	978634.50	8	−551	−195
229	36 10	75 1	3187	978845.70	978660.70	5	−551	−196
339	36 10	75 0	3129	978863.60	978673.80	3	−549	−194
130	36 11	74 59	3031	978895.20	978692.80	2	−548	−193
230	36 12	74 57	2863	978948.40	978715.80	−2	−558	−203
330	36 12	74 56	2730	978989.40	978734.90	−4	−564	−209
430	36 13	74 54	2627	979022.60	978753.80	−8	−566	−212
530	36 13	74 52	2566	979041.40	978770.60	−12	−562	−209
730 HURA	36 13	74 51	2720	978993.90	978762.00	−14	−544	−191
131	36 14	74 49	2763	978982.10	978759.80	−19	−540	−188
231	36 15	74 47	2476	979072.00	978816.60	−24	−536	−183
331	36 18	74 41	2125	979184.50	978874.90	−38	−548	−197
Additional gravimetic stations of the 1988 Aghil–Shaksgam Profile								
0988 PAK	36 11	74 18	1791	979277.40	978989.80	−53	−488	−155
Khunjerab Pass	36 51	75 27	4740	978426.30	978451.30	31	−538	−177
Mazar	36 30	77 6	4800	978377.50	978471.10	124	−475	−107

Structural evolution of the North Karakoram cover, North Pakistan

ANDREA ZANCHI

Dipartimento di Scienze della Terra, Project Ev-K2 CNR, Via Mangiagalli 34, 20133 Milano, Italy

Abstract: The oldest deformational structures recognized in the Karakoram sedimentary cover north of the Karakoram Axial Batholith (KAB) consist of an E–W deformed belt composed of Permian–early Cretaceous sediments. This belt is overlain with a stratigraphical unconformity by the Cretaceous Tupop conglomerates and subsequent Late Cretaceous marine sediments. In the Chapursan Valley a complex and polyphase antiformal stack consisting of a Permian to Cretaceous succession was recognized. Post-Cretaceous northward and northeastward directions of transport are shown by slip vectors along thrust planes and by fold axes in the Chapursan Valley, whereas NW–SE-trending folds due to northeastward transport were observed in the Shimshal Valley. The north-vergent structure is crossed by successive north-dipping thrusts along the northern side of the Chapursan Valley and along the contact between the sedimentary cover and the northern continuation of the KAB.

The last tectonic phase recognized in the Chapursan Valley caused the activation of an important set of E–W-orientated sinistral strike-slip faults. In the Shimshal Valley, large E–W sinistral and NW–SE dextral faults are present as well, and may be interpreted as a consequence of a NW–SE-directed simple-shear due to deformation along major NW–SE dextral strike-slip faults.

The sedimentary cover of the Karakoram was deformed under anchimetamorphic conditions in the Chapursan Valley, whereas sericite to muscovite low-grade metapelites with occurrence of chloritoid, epidote and biotite are present in the lower Shimshal Valley. Near the Shimshal Pass, andalusite, garnet, muscovite and biotite were found in phyllites at the contact aureole of an important belt of large previously unknown granitic bodies.

In order to understand the tectonic evolution of the North Karakoram region, structural analyses were carried out in the sedimentary cover cropping out north of the Karakoram Axial Batholith (KAB) in the Chapursan and Shimshal Valleys (Upper Hunza, Pakistan). The stratigraphic framework of the North Karakoram, firstly investigated by Desio & Martina (1972), was recently defined by Gaetani *et al.* (1990*a*). These authors reconstructed the stratigraphic evolution of the area, showing that the sedimentary cover consists of Permian to Jurassic mainly marine successions also including discordant conglomerates, possibly Cretaceous in age. The 1991 Ev-K2 expedition visited the Chapursan Valley for the second time. Reconnaissance work was also carried out in the Shimshal Valley up to the Shimshal Pass.

The aim of this paper is to provide new structural data on the main tectonic elements. A complete structural section across the sedimentary cover from the KAB to the Misgar Slates was reconstructed along the Yashkuk Glacier. General reconnaissance mapping and structural observations were carried out in the whole investigated area, leading to a general interpretation of the tectonic framework of the sedimentary cover. A relative chronology of the main tectonic events was established, although no precise dating is available at present. Structural data were obtained on the basis of previous results (Gaetani *et al.* 1990*a*) and of new stratigraphic data (Gaetani *et al.* this vol.).

Geological setting of the North Karakoram

The north Karakoram region (Fig. 1), also named Northern Karakoram Terrain, is located between two main suture zones: the Shyok (Northern) Suture Zone to the south, which separates the Kohistan Arc from the Karakoram (Coward *et al.* 1986), and the still problematic Rushan–Pshart Suture to the north, separating Southeast Pamir from Central Pamir (Shvolman 1981). Three main structural belts were distinguished by Gaetani *et al.* (1990*a*) from the KAB to the Wakhan–Sinkiang borders; south to north they consist of the Guhjal, Sost and Misgar Units (Fig. 2). The Guhjal Unit, directly north of the KAB, is composed of dolomitic to calcareous marbles included in dark slates and phyllites. A southwestward dipping reverse fault places the unit above the Sost Belt. The Sost Unit is composed of an antiformal stack

From TRELOAR, P.J. & SEARLE, M.P. (eds), *Himalayan Tectonics*
Geological Society Special Publication, No. 74, pp. 21–38.

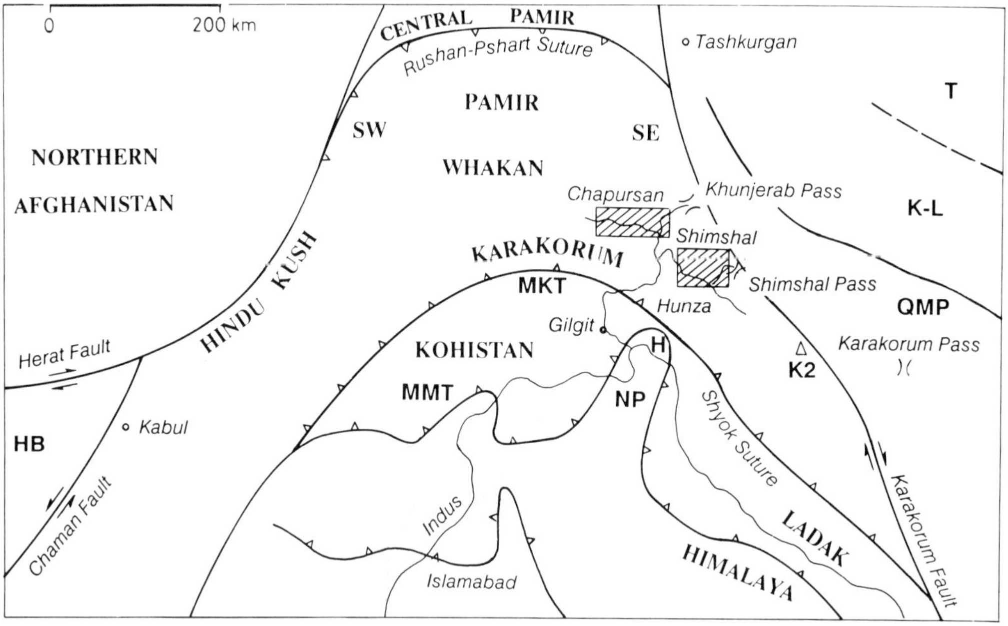

Fig. 1. Tectonic sketch map of Northern Pakistan and surrounding regions. Squared areas refer to Figs 2 and 12. T is the Tarim Basin, K-L is Kun-Lun, QMP is the Qiangtang microplate, HB is the Helmand Block. NP is Nanga Parbat, H is Haramosh Peak. MKT is the Main Karakoram Thrust, MMT is the Main Mantle Thrust. Modified from Gaetani *et al.* (1990*a*) and Searle (1991).

including Permian to Cretaceous sediments. Important E–W-trending sinistral strike-slip faults cut through the central part of the unit. The Misgar Unit, structurally overlying the Sost Unit, consists of a monotonous sequence of dark slates (Misgar Slates of Desio & Martina 1972), deformed at low metamorphic grade. Whereas in the Chapursan Valley the stratigraphic and structural setting were previously reconstructed, in the Shimshal Valley basic geological work was restricted to reconnaissance work carried out by Casnedi & Nicora (1985) up to the Shimshal village. The valley was recently crossed by Searle (1991), who includes part of the region in his geological map of the Karakoram. During our visit, we recognized a stratigraphic succession similar to that of the Chapursan Valley. Three main units were also crossed in this area, although the general tectonic setting of the Shimshal Valley shows some differences. The southern belt consists of low grade intensely folded metapelites. The central sector includes the non-metamorphic Permo-Mesozoic succession, which is cut by wrench faults. The northern belt includes black quartz phyllites intruded by large granodioritic bodies cropping out around the Shimshal Pass and the Shuijerab village.

Deformation in the Chapursan Valley

Pre-Upper Cretaceous deformation event

The oldest recognized tectonic phase is represented by an E–W-trending folded belt consisting of Triassic to earliest Cretaceous sedimentary rocks. These structures are unconformably overlain by the Tupop conglomerates of Cretaceous age. The unconformity was observed in the Borom Valley by Gaetani *et al.* (1990*a*), and is also present in the nearby Kundil Valley (Fig. 2). Here, steeply dipping Jurassic marls and limestones are overlain by the Tupop conglomerates (Fig. 3). The Tupop Formation is covered with a second unconformity (10–20°) by reddish marly limestones, which were recently dated as Campanian (Gaetani *et al.* this vol.). The Jurassic marls cropping out beneath the unconformity display closely-spaced slaty cleavage, probably associated to folding, indicating a very low grade metamorphic imprint.

Northward stacking of the sedimentary cover

A stack of imbricated thrust sheets bounded by south to southwest-dipping fault planes crops

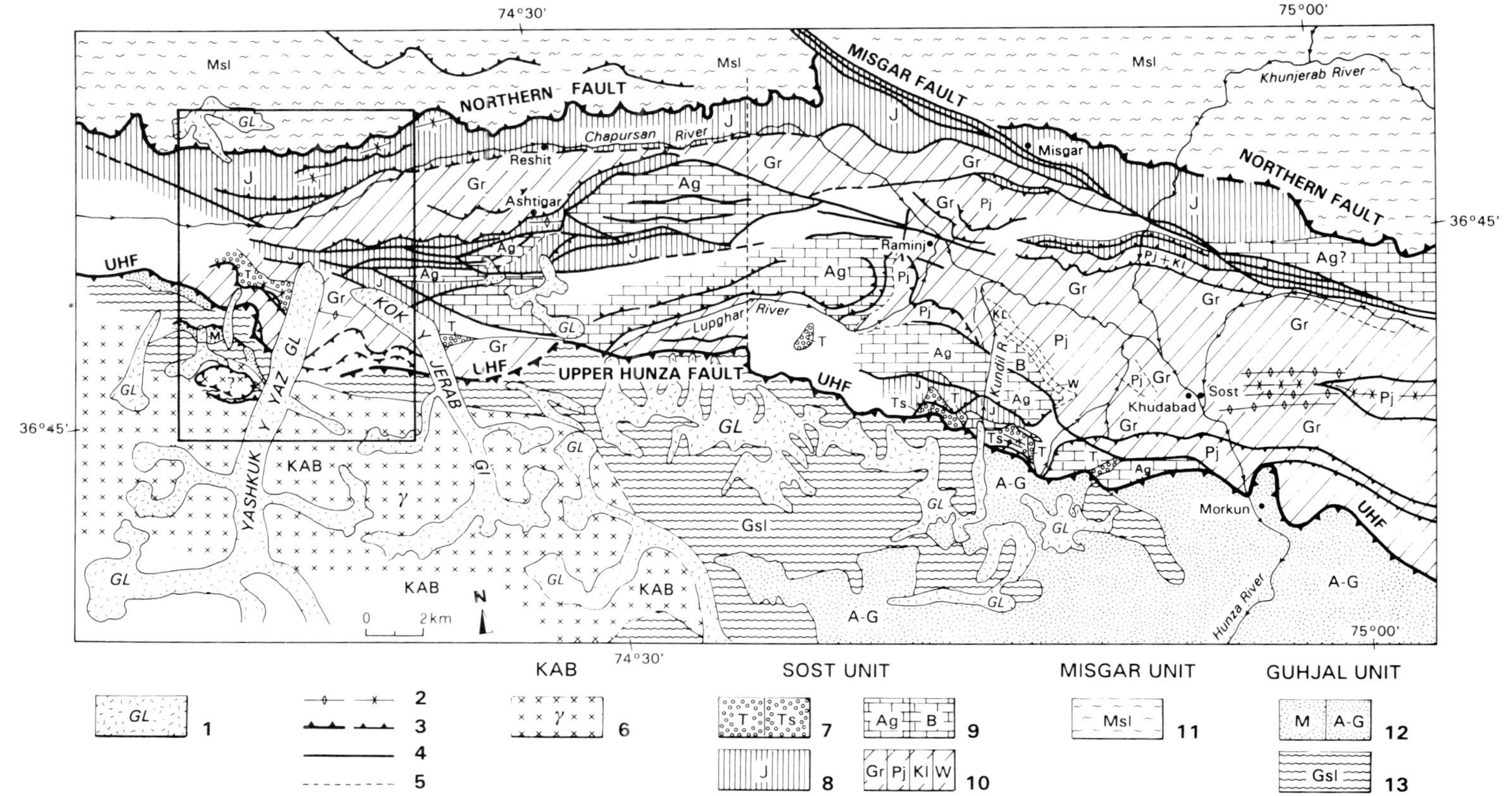

Fig. 2. Simplified geological map of the Chapursan Valley. (1) glaciers; (2) fold axes; (3) major and minor thrust planes; (4) strike-slip faults; (5) stratigraphic contacts; (6) Karakoram Axial Batholith; (7) Cretaceous: Tupop Fm. (T), Late Cretaceous marls (Ts); (8) Jurassic sediments undifferentiated (J); (9) Triassic: Aghil Fm. (Ag), Borom Fm. (B); (10) Permian: Gircha Fm. (Gr), Panjshah Fm. (Pj), Kundil Fm. (Kl), Wirokhun Fm. (W); (11) Misgar Slates (Msl); (12) marbles (M), Aghil-Guhjal dolomitic limestones (A–G); (13) black slates of the Guhjal Unit (Gsl). The rectangle corresponds to the area of Fig. 4.

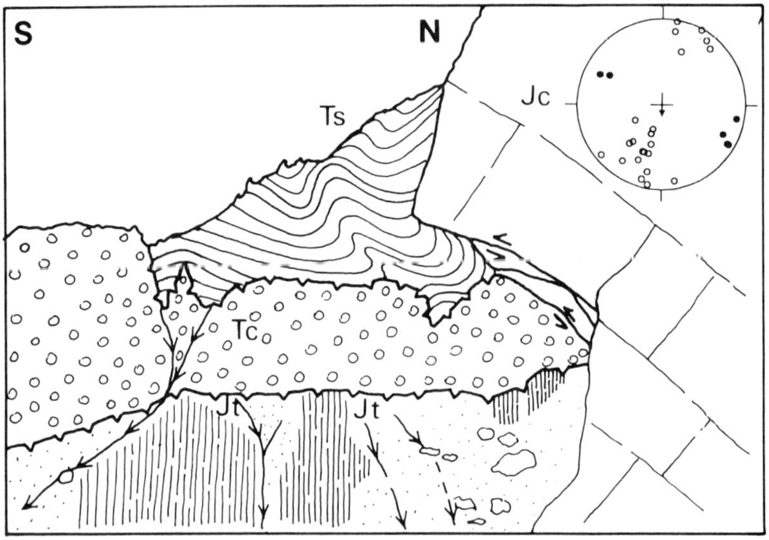

Fig. 3. Cretaceous unconformities in the Kundil Valley, (see also Gaetani *et al.* this vol., Fig. 5.) Sketch view of the two unconformities and subsequent high-angle, north-vergent thrusting. Jurassic limestones (Jc), Jurassic marls (Jt), Tupop Fm. (Tc), Late Cretaceous marls (Ts), lines with teeth are angular unconformities, vertical lines below the unconformity represent cleavage in Jt. The stereogram (Schmidt's lower hemisphere projection) refers to drag folds measured along the ramp fault in Jc; circles are poles to bedding, triangles are poles to cleavages, large black dots are fold axes.

out along the western side of the Yashkuk Glacier. The thrust fan includes intrusives probably associated with the KAB, the Guhjal and most of the Sost Unit. A general map of the area is shown in Fig. 4. In the cross section of Fig. 5, the north-verging stack is mainly preserved in the central part; successive important strike-slip faults cut the northern side of the section, whereas the southern boundary is crossed by a north-dipping late-stage thrust.

The uppermost unit seems to form a klippe consisting of dark dioritic to gabbroic intrusives, probably associated with Eocene Batura-type granitoids (Debon *et al.* 1987). Stacking of this unit on the top of the structure is mostly inferred from the geometry of contacts as deduced by long distance observations (Fig. 6) and satellite imagery interpretations. Below the diorites a large sheet consisting of the Guhjal Unit is defined by NNW–SSE to E–W thrust surfaces. This unit also includes pinky granodiorites very similar to those which are present southward along the Yashkuk Glacier and which represent the northern continuation of the KAB. Brucite marbles crop out along the contact with these intrusives, just north of the main north-dipping tectonic contact with the KAB. Marble layers are included within dominant slates and display mostly stratigraphic contacts. Attitude of bedding along the Yashkuk Glacier indicates the presence of complex folds in the marble layers.

The Sost Unit is composed of thin sheets of Permian to Cretaceous sediments. The Aghil Fm. and Jurassic marly limestones form duplex structures among dominant Lower Permian Gircha slates and sandstones. Tupop sediments and successive marine Campanian beds are also included in the fan (Fig. 5), indicating post-Cretaceous thrusting.

Mesoscopic structures such as slip vectors, cleavages and fold axes were measured along thrust planes and within thrust sheets (Fig. 7). Slip vectors were tested for stress tensor determinations using the methods described by Angelier (1990). Due to the high rates of deformation, they must be considered only as representative of incremental strain tensors (Table 1). One of the most characteristic fault assemblages was measured along the south-dipping thrust plane which bounds one of the small imbricate slices composed of Aghil limestones at site Y6B-7-8 (Fig. 7A). Faults, which are mostly north-vergent, are consistent with a N10° trending $\sigma1$ axis and a vertical $\sigma3$ axis. Most of the folds and cleavages measured in the Chapursan region have trends consistent with this phase of transport (Fig. 7). E–W-trending flexural slip folds were observed in the Jurassic limestones south

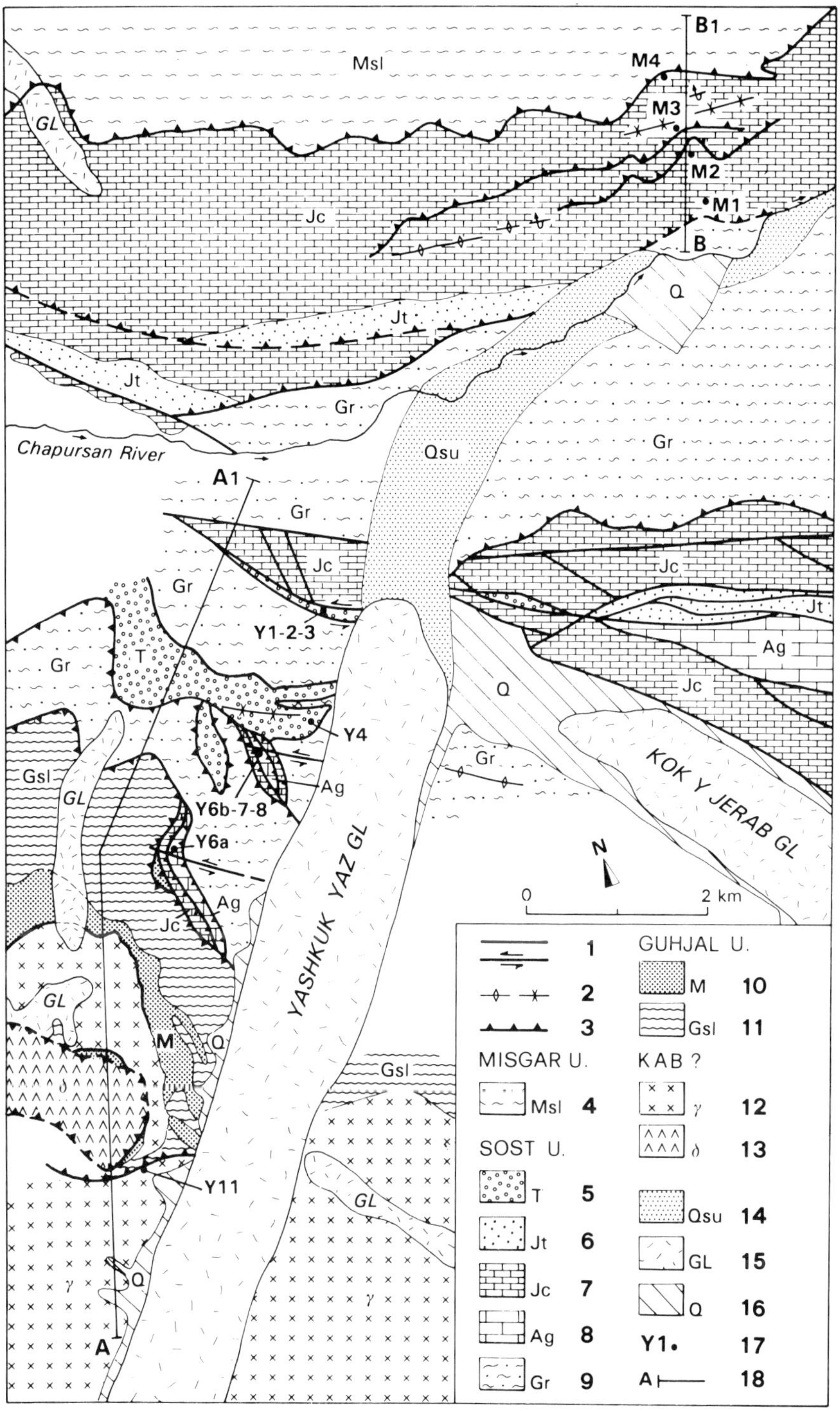

Fig. 4. Geological map of the western side of the Yaskuk Glacier. (1) strike-slip faults; (2) fold axes; (3) thrust planes; (4) Misgar Slates (Msl); (5) Tupop Fm. (T); (6) Jurassic terrigenous (Jt); (7) Jurassic to Early Cretaceous limestones (Jc); (8) Aghil Fm.; (9) Gircha Fm. (Gr); (10) marbles (M); (11) black slates of the Guhjal Unit (Gsl); (12) Batura-type granodiorites (γ); (13) dioritic intrusives (δ); (14) Recent glacial surge of the Yaskuk Glacier (Qsu); (15) glaciers (GL); (16) Quaternary (Q); (17) structural site; (18) Trace of cross-section of Fig. 5.

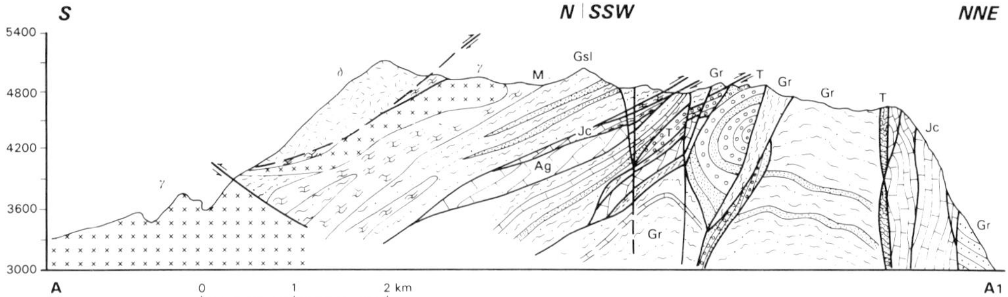

Fig. 5. Geological cross-section across the western side of the Yaskuk Glacier. Same symbols as in Fig. 4.

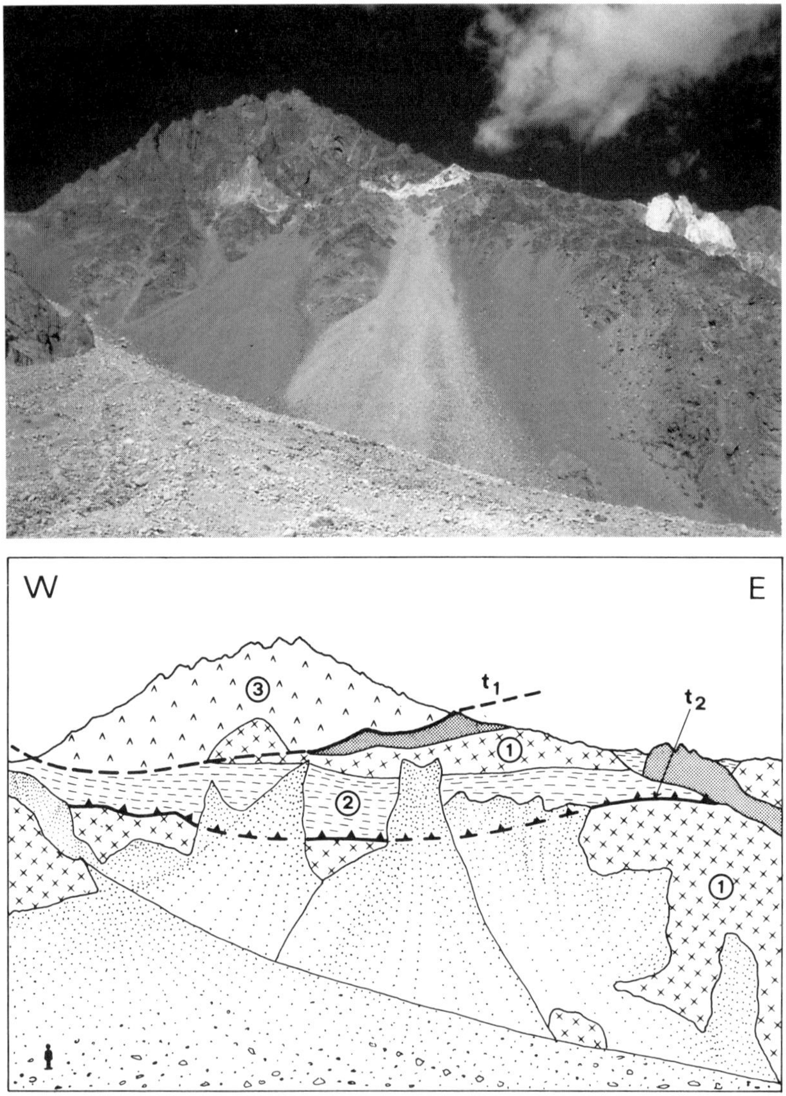

Fig. 6. Sketch view of the contact between the sedimentary cover and the KAB on the left (western) side of the Yaskuk Glacier. (1) Batura-Type Granodiorites; (2) Guhjal Unit with slates and marbles; (3) dioritic intrusives. t1 is an inferred south-dipping thrust plane probably stacking basic intrusives above the Guhjal Unit; t2 is a north-dipping thrust separating the KAB from the Guhjal Unit. Structural data in Fig. 10F refer to this plane.

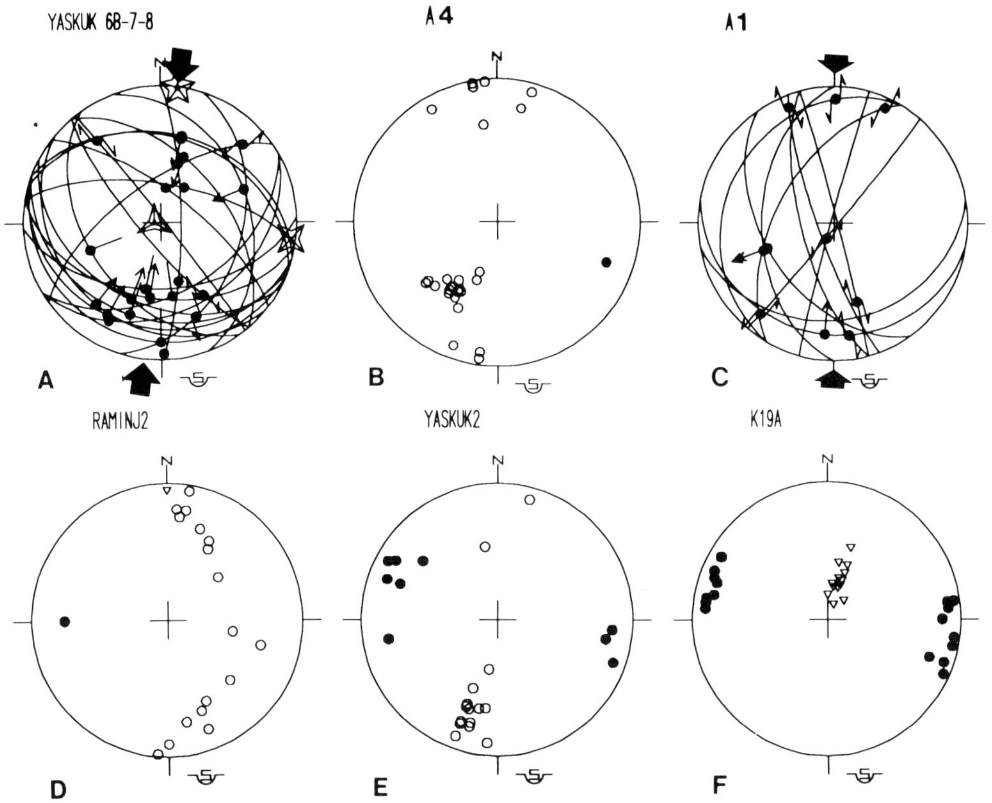

Fig. 7. Mesoscopic structures related to northward thrusting in the Sost stack. Great circles are faults, slickenside lineations are dots with arrows indicating the sense of motion. Convergent arrows correspond to the horizontal direction of the $\sigma 1$ axis; other symbols as in Fig. 3. (**A**) Fault populations and relative stress tensor determinations for site Y6B-7-8; location of site in Fig. 4. Stress axes are represented as stars with five ($\sigma 1$), four ($\sigma 2$) and three branches ($\sigma 3$). (**B**) Poles to bedding and fold axis in an E–W fold (A4), (**C**) small fault planes developed during folding (A1) in Jurassic Limestone in the valley south of Ashtigar (Fig. 2). (**D**) Poles to bedding and fold axis of a large fold in Gircha slates near Raminji (Raminji2; Fig. 2). (**E**) Folds in Tupop sandstones and marls along the Yashkuk Glacier (Yaskuk2; Fig. 4). (**F**) Fold axes and poles to axial plane cleavages (triangles) in the Late Cretaceous marls along the right side of the Kundil Valley (K19A; Fig. 2).

of Ashtigar (Fig. 7B). Here minor fault planes developed during folding also suggest a N–S compression (Fig. 7C). Slaty cleavages formed along axial planes of these folds in the dark slates belonging to Gircha Formation and in the Late Cretaceous marly sediments overlying the Tupop beds mainly trend E–W (Fig. 7D, E and F). Large concentric folds developed in the Late Cretaceous sediments of the Kundil Valley (Fig. 2) support the northward direction of transport related to this phase. Tight E–W folds with subvertical axial planes and cleavages also crop out south of Sost along the Karakoram highway.

Cataclasis and pressure solution were mainly active in the fault rocks developed along these thrust planes. Pressure solution cleavages and thin bands of calc-mylonites are present in limestones along the thrusts which bound the Aghil and Jurassic rocks forming thin sheets under the contact with the Guhjal Unit (Fig. 8). In some cases they indicate a northeastward transport, consistent with the regional geometry of the thrust plane which bound the Guhjal Unit. These different trends seem to suggest polyphase thrust motions.

Significant changes in metamorphic conditions were not noticed through the entire structure. Slaty cleavages, reorientated fine-grained sericite and quartz recrystallization were found in most of the slate samples collected in the three different units across the studied

Table 1. *Details of stress determinations measured on faults in the Chapursan and Shimshal valleys, North Karakoram, Pakistan*

Site	Method	Number of data	σ1 trend	σ1 plunge	σ2 trend	σ2 plunge	σ3 trend	σ3 plunge	ratio Φ	ANG (deg)	RUP (%)
Chapursan Valley (Yashkuk Glacier)											
Y6B-7-8	R4DT	17	007	01	097	04	264	86	0.55	13	31
Y1-2-3	R4DT	27	062	17	277	70	156	11	0.57	12	42
Y6B-7-8	R4DT	17	057	02	216	88	327	01	0.89	11	33
Y4	R4DT	18	192	12	283	06	37	76	0.02	13	41
Shimshal Valley (Zargdar Bin – Targil Dur)											
S9B	INVD	5	234	12	059	78	324	01	0.29	14	46
S10A	R4DT	14	239	17	123	55	339	30	0.26	16	42
S6-7	R4DT	20	007	09	221	79	098	06	0.22	10	38
S9A	R4DT	7	348	07	243	62	082	27	0.15	7	46
S10B	R4DT	6	008	06	173	84	277	01	0.21	8	38

INVD, R4DT, inversion methods described by Angelier (1990) σ1, σ2, σ3, maximum compressive stress, intermediate stress and minimum stress, with trend and plunge in degrees. Φ, ratio between principal stress differences (σ2–σ3)/(σ1–σ3). ANG, angle between computed shear stress and actual slip, in degrees (0°–180°). RUP, quality estimated ranging from 0 to 200% (Angelier 1990).

section. Nevertheless, the Misgar and Guhjal Units clearly show a major rate of grain recrystallization and re-orientation.

Southward thrusting

The north-vergent stack is crossed by north-dipping medium to low-angle thrust planes. Along the northern side of the Chapursan valley an E–W-thrust (Northern Fault after Gaetani *et al.* 1990*a*) extends for tens of kilometres, stacking the Misgar Slates over the Jurassic limestones and sandstones of the Sost Unit (Figs 2 and 4). A detailed N–S section from the Chapursan Valley up to the main fault plane is shown in Fig. 9. Large recumbent folds, minor thrust planes in association with strike-slip faults, fault propagation folds and solution cleavages indicate a southward motion of the Misgar Unit. Kinematic indicators (Fig. 10A, B and C) suggest a S to SSE direction of motion. Pressure solution and cataclastic flow probably due to high fluid pressures are the dominant deformation mechanisms. Mylonitic limestones occur along the main contact between Jurassic limestones and Misgar Slates. Closely-spaced twin lamellae developed in calcite crystals point to a deformation at low temperature conditions.

North-dipping thrust planes were also observed along the studied section in the Yashkuk area. At site Y4 minor faults cutting through vertical Tupop sandstones indicate a N–S direction of the σ1 axis and a vertical σ3 axis (Fig. 10D). E–W-trending low-angle cataclastic thrust planes define the contact between the south-dipping imbricated fan and the northernmost granitoids of the KAB (Fig. 10E).

In front of Raminji, on the left side of the Chapursan Valley, low-angle north-dipping thrusts stack several sheets of the Permian Gircha and Panjshah Formations with development of fine grained calc-mylonites. Faults in association with tension gashes related to sub-horizontal shear zones were measured near the thrust planes (Fig. 10F).

High-angle south-vergent thrusting is also evident in the Kundil valley, where Jurassic limestones are thrust above the Tupop Formation and younger Cretaceous sediments (Fig. 3). In many sites the previously folded and tectonized sedimentary successions are crossed by conjugate sets of small thrust planes due to a N–S-trending horizontal σ1 axis related to this phase.

Wrench faulting

One of the most important structural features of the study area is due to strike-slip motions, possibly active up to Neogene and Quaternary times. Desio & Martina (1972) recognized in the area a major WNW–ESE vertical fault system, named the Misgar Fault by Gaetani *et al.* (1990*a*). The fault was interpreted as a dextral strike-slip structure by Gaetani *et al.* (1990*a*) although no detailed structural analysis was carried out by these authors. In the central part of the Sost Unit along the hinge of the antiformal stack consisting of Permo-Mesozoic carbonatic massifs, tectonic contacts are often displaced by

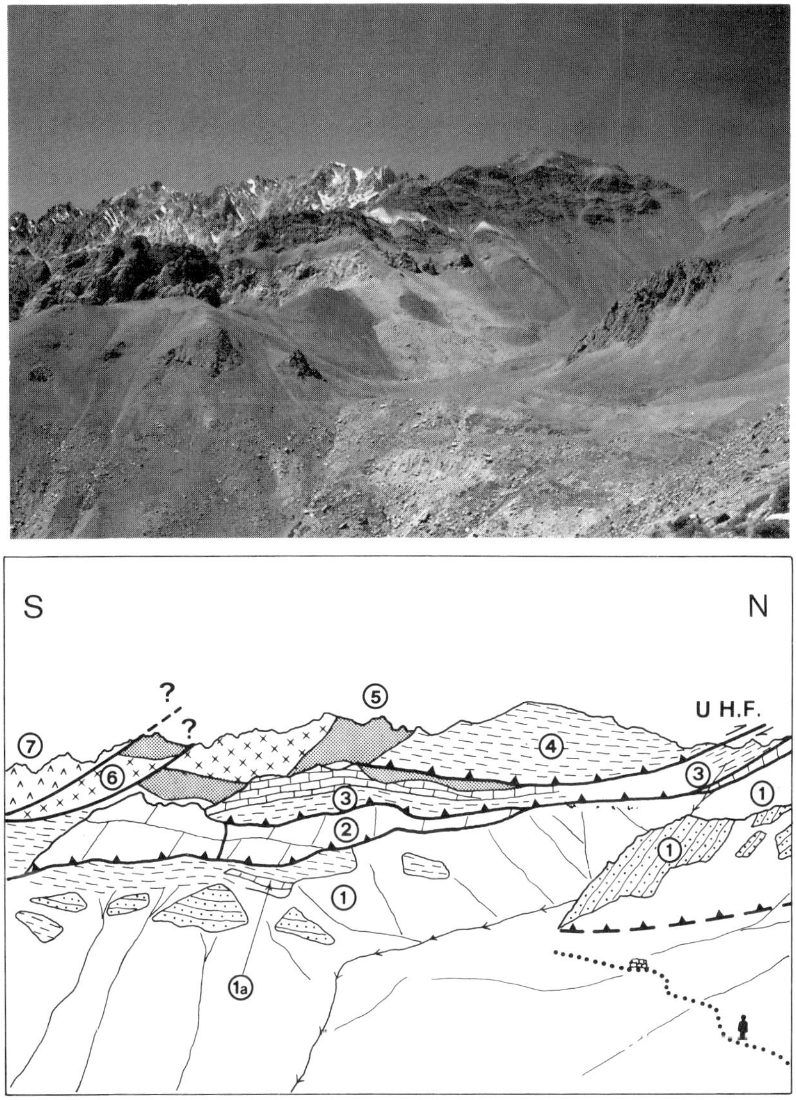

Fig. 8. Sketch view of thrust sheets along the left side of the Yaskuk Glasier; compare with Figs 4 and 5. Lines with teeth are thrust planes. (1) Gircha Fm. with fossiliferous limestones (1A); (2) Aghil Fm.; (3) Jurassic marls and limestones; (4) black slates of the Guhjal Unit; (5) marbles of the Guhjal Unit; (6) pinky granodiorites; (7) dioritic intrusives.

E–W-trending subvertical faults with horizontal striations indicating a left-lateral slip. In the Yashkuk zone, at least four major E–W-trending strike-slip faults cross the studied section (Figs 4 and 5). Strike-slip shear zones form thin and strongly brecciated slices often including different sedimentary successions. Slip vectors along fault planes, as indicated by mechanical striations due to friction and calcite growth fibres, suggest dominant E–W-trending sinistral strike-slip motion in association with NNE–SSE minor dextral shears. Stress tensor determinations obtained with faults measured along the Yashkuk glacier (sites Y1-2-3, Y6-7-8) indicate an ENE-WSW horizontal $\sigma 1$ axis, a vertical $\sigma 2$ and a horizontal NNW–SSE $\sigma 3$ axis

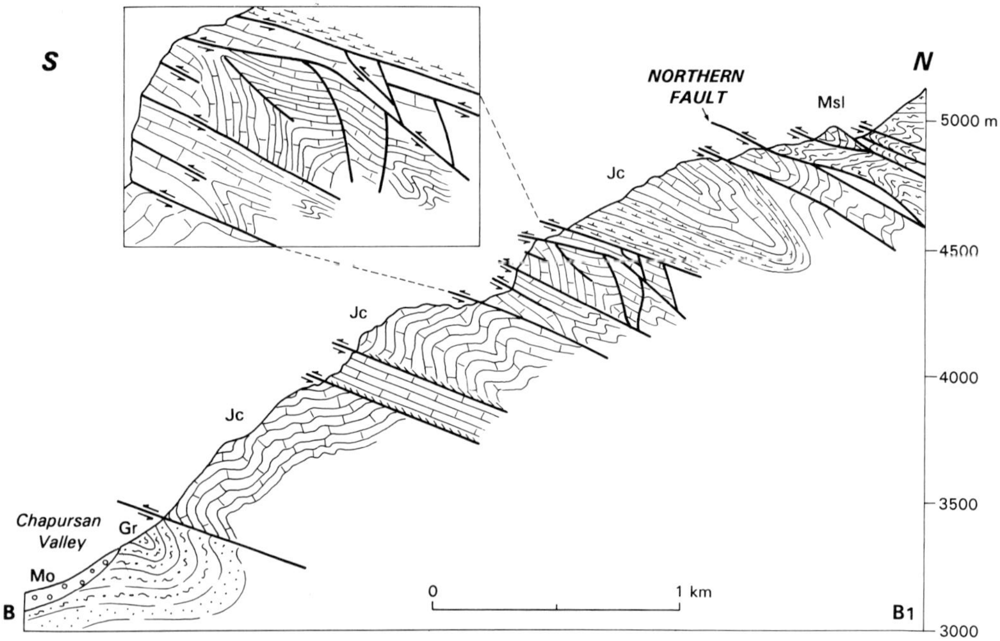

Fig. 9. Cross-section along the Northern Fault; see Fig. 4 for location. Msl are Misgar Slates, Jc are Jurassic (?) limestones, Gr is Gircha Fm, Mo are recent glacial deposits. Note south-vergent tectonic transport along thrusts.

(Fig. 11, Table 1). Strike-slip faults are clearly subsequent to thrust emplacement and folding, as shown by structural relationships. At site Y6-7-8, for example, the Aghil thrust sheet is clearly crossed by successive strike-slip faults, displacing the tectonic contact between Aghil limestones and Jurassic marls (Figs 4 and 5).

Fault breccias and cemented gouges often occur along the major fault planes. Remobilization of strike-slip fault with normal dip-slip motion was also noticed.

Deformation in the Shimshal Valley

During our expedition in the Shimshal Valley, reconnaissance work was carried out along the main track from the Hunza Valley (Jurjur camp) up to the Shimshal Pass (Pamir Pir), through the Shimshal village and the Zardgar Bin Pass (Win Sar Pir). A general geological map of the area (drawn at 1:250 000 scale) is presented in Fig. 12. Three main structural zones were distinguished which roughly correspond to the three belts developed in the Chapursan area.

Southern Zone

The Southern Zone mainly consists of slates and phyllites including in the western part the dolomitic marbles of the Tupopdan Peak (Karun Kuh II Peak) belonging to the Guhjal Unit. The unit is intruded southward by the KAB, here probably represented by post-collisional Eocene Batura-type granitoids (Searle 1991). Northward, the Southern Zone seems to overthrust the Permian to Cretaceous sedimentary cover. The slates also include fossiliferous sandstones ascribed to the Gircha Formation by Casnedi & Nicora (1985), rounded conglomerates and thick beds of quartzites. Casnedi & Nicora (1985) introduced the informal name 'Shimshal Slates' for these successions, which, in the whole, are very similar to the Gircha and Pasu slates. An increase in metamorphic grade in the lower part of the Shimshal Valley seems to be the only significant difference with the Gircha slates. Along the Shimshal gorge up to Dut and Ziarat, tight isoclinal folds with axial trends ranging from NW–SE to WNW–ESE are present (Fig. 13), as also reported by Searle (1991). Schistosity is defined by sericite, muscovite and fine-grained biotite; epidote occasionally occurs. Southwest-dipping cataclastic-mylonitic shear zones developed near Dut and east of Ziarat, showing muscovite and chloritoid assemblages along S–C foliations. Post-kinematic biotite, occurring in most of the samples, is probably related to a subsequent temperature

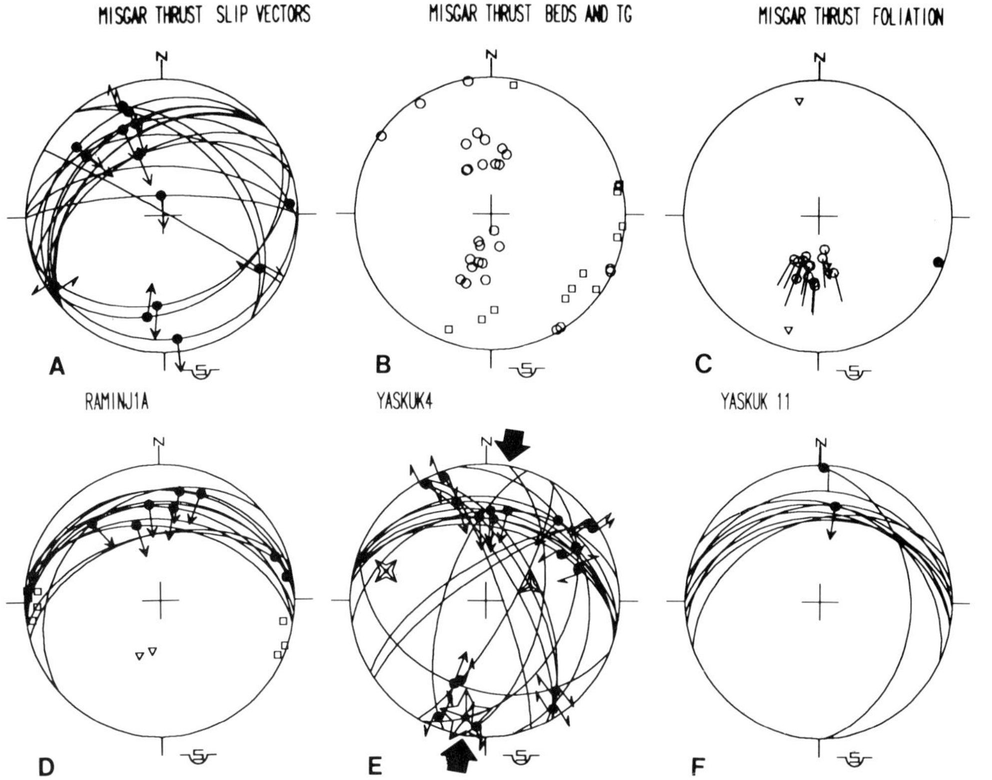

Fig. 10. Examples of fault populations and stress tensor determinations related to north-dipping thrust faults; symbols as in previous figures. (**A, B, C**) Sites M1-2-3-4 (Fig. 4); structures were measured along the section of Fig. 9, and refer to the Northern Fault. Small squares are poles to tension gashes in (B) and (D); open circles with lines in (C) are poles to mylonitic foliation along the Northern Fault. (**D**) North-dipping thrusts located in front of the Raminji village on the left side of the Chapursan Valley (Raminj1A; Fig. 2). Fold axes, poles to cleavages and tension gashes developed along the shear zone are also represented. (**E**) Fault planes crossing previously folded vertical beds of the Tupop Fm. (Yaskuk4) measured on the southern side of the Yaskuk Glacier (Fig. 4). (**F**) Fault planes measured along the main tectonic contact between the KAB and the Guhjal Unit along the southern part of the Yaskuk Glacier (Yaskuk11; t2 in Fig. 8).

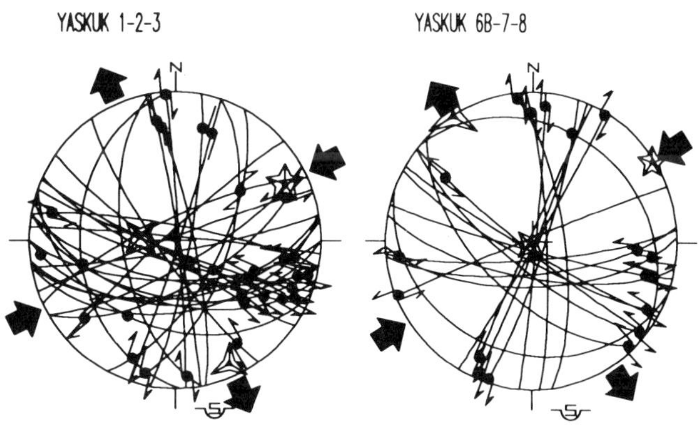

Fig. 11. Examples of fault populations and stress tensor determinations related to E–W strike-slip faults along the Yashkuk Glacier. Location of sites in Fig. 4; symbols as in Fig. 7A.

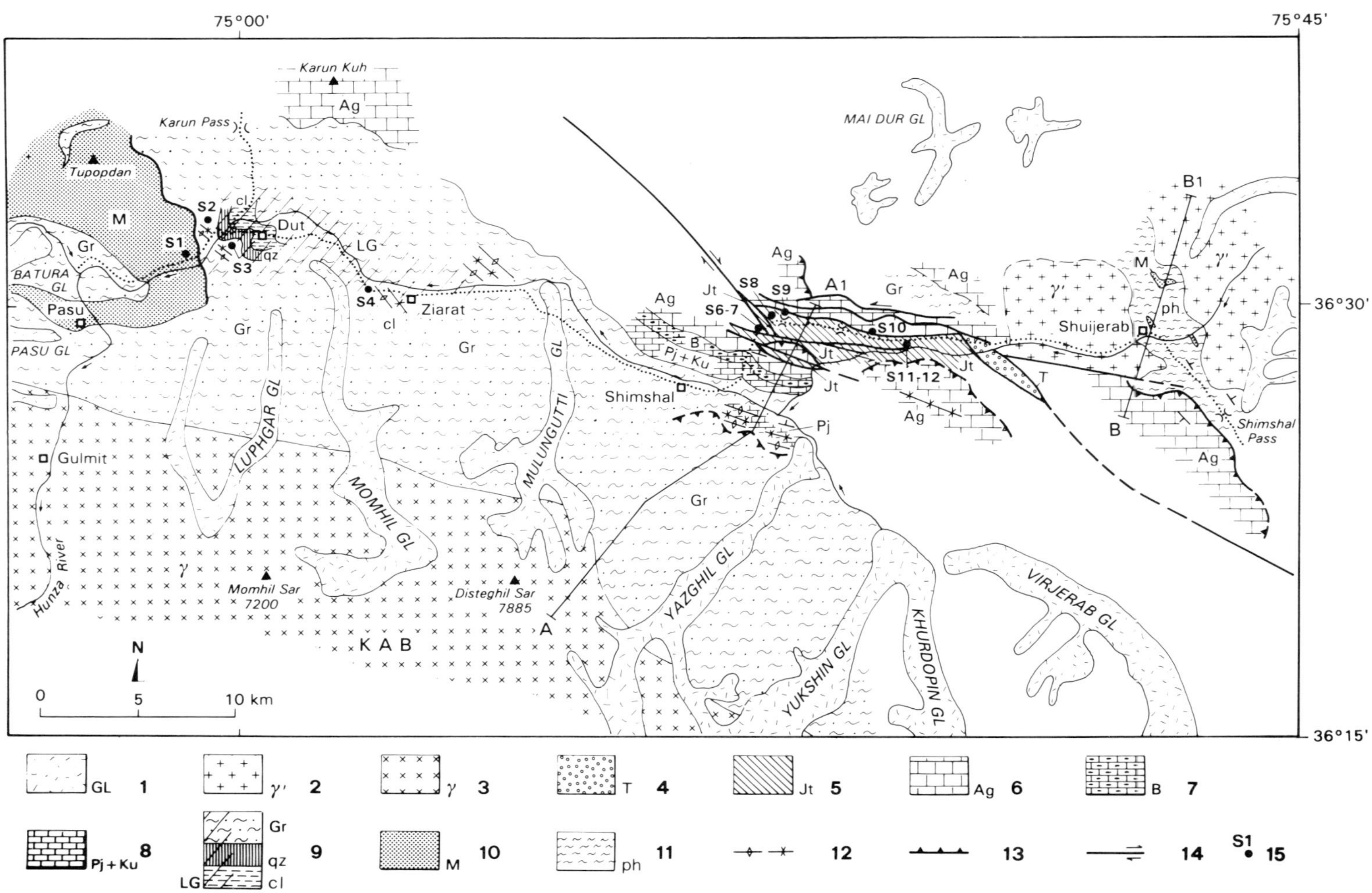

Fig. 12. Tentative geological map of the Shimshal Valley. (1) glacier; (2) granodiorites of Shuijerab and Shimshal Pass; (3) granodiorites of the KAB; (4) Tupop Fm.; (5) Jurassic sandstones and marls; (6) Aghil Fm.; (7) Borom Fm.; (8) Panjshah + Kundil Fm.; (9) Gircha Fm. (gr), quartzites (qz), chloritoid mylonitic schists (cl), Gircha Fm. (Gr), locally with low grade metamorphic imprint (LG); (10) dolomitic marbles of the Tupopdan Peak (Karun Kuh II Peak) belonging to the Guhjal Unit; (11) Shimshal Pass quartz phyllites; (12) fold axes; (13) thrust planes; (14) strike-slip faults, (15) structural sites.

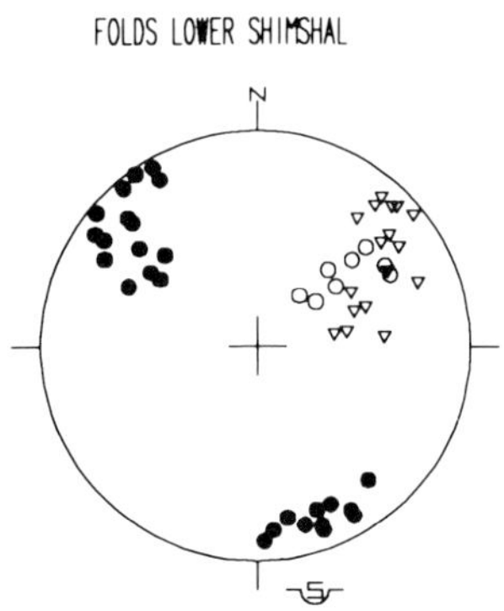

Fig. 13. Fold axes and related cleavages along the lower Shimshal Valley. See Fig. 12 for location. Large black dots are fold axes, triangles are poles to axial plane cleavages, small circles are poles to bedding.

increase due to the emplacement of the KAB forming the nearby Disteghil and Momhil Sar peaks. Crenulation cleavage, re-folding the previously described foliation, possibly indicate the presence of a subsequent deformational event.

Axial plane foliations, transposition, and boudinage of quartz layers, as well as the occurrence of mylonitic shear zones suggest an important increase in metamorphism with respect to the Chapursan sedimentary successions. The geometry of the axial planes of folds and extensional cleavages (S–C structures) formed in ductile shear zones indicate a northeastward direction of transport. Folding is here interpreted as due to overthrusting of the Southern Unit above the Intermediate belt, as schematically represented in the cross section of Fig. 14A. This shear zone may correspond to the

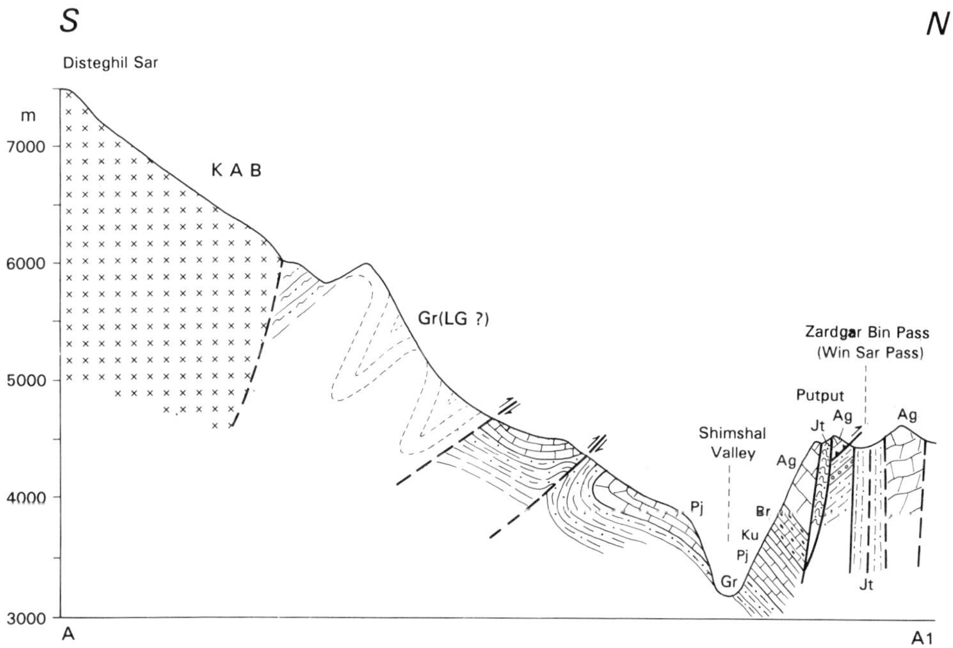

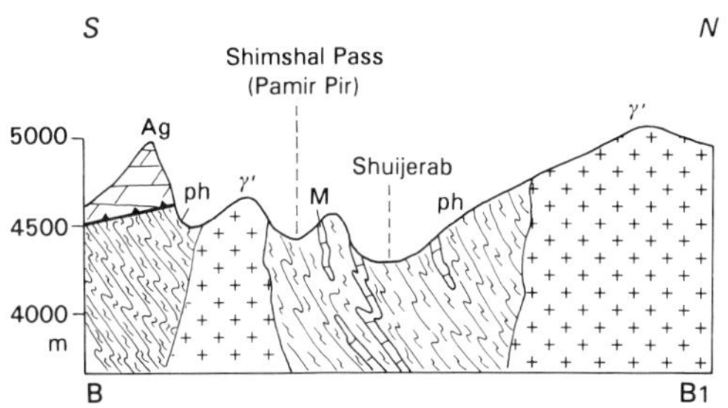

Fig. 14. Tentative geological cross-sections through the (**A**) Shimshal Valley (A-A1) and the (**B**) Shimshal Pass (B-B1). Location of sections and symbols as in Fig. 12. Vertical scale is exaggerated.

Upper Hunza Fault of the Hunza and Chapursan Valleys.

Intermediate Zone

The intermediate belt consists of the intensely faulted and folded but non-metamorphic Permian to Cretaceous sedimentary succession, which corresponds to the Sost Unit of the Chapursan area (Fig. 12). A continuous section, including all of the Permian–Triassic sequence from the Gircha to the Aghil Formations is well exposed along the southern flank of the Putput Peak in front of the Shimshal village (Fig. 14A).

Further north from Zardgar Bin up to the contact with the granitoids of the northern sector, the Intermediate Zone includes a belt of subvertical tectonic slices, consisting of Jurassic sandstones and cataclastic dolomitic limestones tentatively attributed to the Aghil Formation. Strike-slip faults generally define the contacts along this system of tectonic slices. Fault populations related to strike-slip motion (Fig. 15) were measured in Zardgar Bin (S6-7, S9A and S9B) and Targil Dur (S10A, S10B), the latter located a few kilometres east of the Zardgar Bin Pass. Two main fault assemblages were distinguished on the basis of mesoscopic analysis; they are respectively related to NW–SE dextral strike-slip faults and to E–W sinistral strike-slip faults. Stress tensor determinations obtained from these faults suggest two different stress directions, respectively related to a N–S trending horizontal σ1 axis and to an ENE WSW horizontal σ1 axis. The latter stress direction is very similar to the ones obtained in the Chapursan Valley relatively to major E–W-trending sinistral strike-slip faults. No clear chronology was distinguished between these two fault systems, due to the complexity and intensity of deformation. NW–SE faults seem to have an en echelon disposition, including domains intensely deformed by E–W sinistral faults. The observed strike-slip faults form wide belts of cataclastic rocks and gouge, indicating that fault motion occurred at very shallow crustal levels.

Earlier thrust structures are also present in this zone, stacking Aghil limestones and dolo-

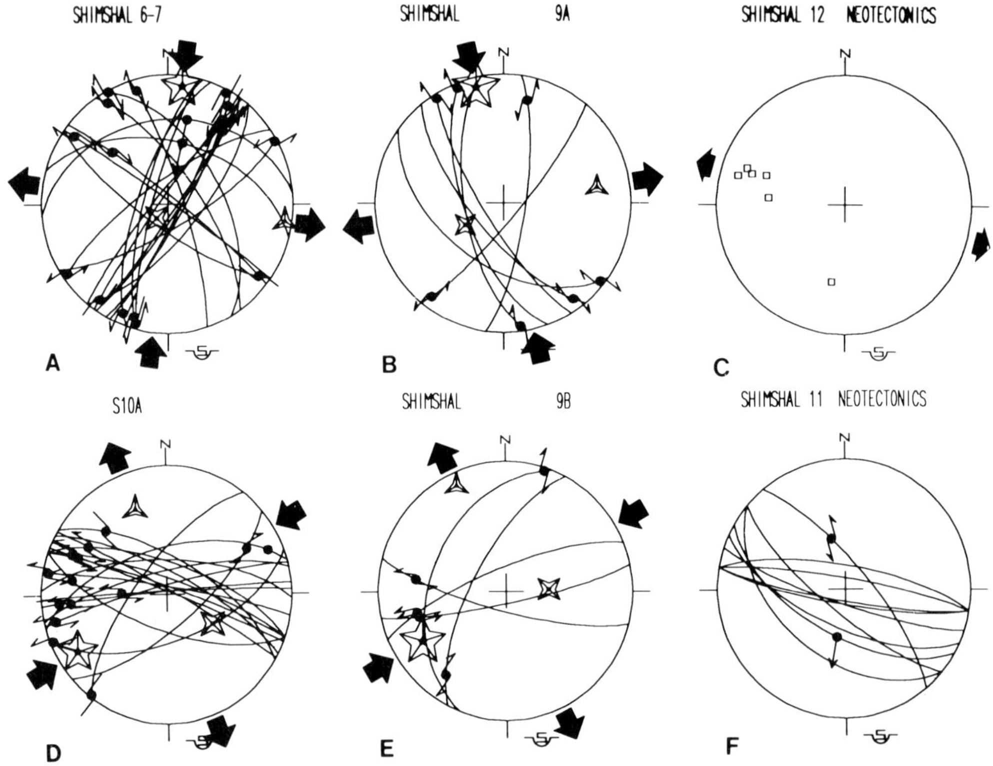

Fig. 15. Examples of fault populations and stress tensor determinations related to E–W strike-slip faults and NW–SE dextral faults near Zardgar Bin, upper Shimshal Valley. Location of sites in Fig. 12; symbols as in Fig. 7A.

mites over Jurassic sandstones. Folds measured in Jurassic sandstones mainly trend E–W, although subsequent strike-slip faulting caused important rotation and dispersion of axial trends.

Quaternary glacial and fluviatile deposits lying above the northern part of this belt show tectonic structures as faults, tension gashes and folds, also indicating different stress directions consistent with very recent reactivation of NW–SE dextral and E–W sinistral strike-slip faults (Fig. 15). N–S-trending vertical tension gashes measured at site S12 are consistent with a N–S-trending $\sigma 1$ axis, whereas faults of site S11 are due to a roughly N–S-trending $\sigma 3$ axis. Small mud diapirs observed near the Quaternary tectonized strata possibly suggest deformation due to seismic wave shaking with liquefaction of still soft sediments.

Northern Zone

The Northern Zone includes at least two distinct granodioritic bodies, intruded within low-grade quartz phyllites. Aghil rocks, overthrusting phyllites, are present east of the southern pluton. The two intrusives crop out south and north of Shuijerab village and are separated by a belt of quartz phyllites containing marble layers. The granodioritic bodies chiefly display intrusive contacts (Fig. 14B) with formation of metamorphic aureoles and pegmatitic dike swarms. The southern contact of the unit seems to be tectonic, and it is defined by an E–W-trending vertical, possibly faulted contact, juxtaposing Aghil dolomites with the intrusives. A contact belt with mylonitic muscovite-bearing marbles is preserved along the main track to Shuijerab just south of the intrusives. Granodiorites are generally massive and do not show superposed foliations. This important belt of granitoids can be correlated with the Sughet granodiorite, cropping out west of the Shaksgam sedimentary belt north of the K2 (Gaetani *et al.* 1990*b*). Another belt of similar plutons occurs around the Khunjerab Pass and near Sost in the Upper Hunza Valley and seems to be mid Cretaceous in age (Searle 1991; Crawford & Searle 1992).

The metapelites exposed between Shuijerab and the Shimshal Pass are intensively transposed, with axial plane foliations defined by biotite and muscovite, and with occasional occurrence of post-kinematic garnet. Folding of previous foliation and syn- to post-kinematic growth of andalusite is apparent in samples collected around the contact aureole of the Shimshal Pass granodiorites.

Discussion and conclusions

Preliminary data on the North Karakoram sedimentary cover indicate a very complex and polyphase structural evolution of the region. Several phases of folding, thrust imbrication, and late-stage extensive wrench tectonics are the most distinctive characteristics of the structural framework. The oldest tectonic event recognized on the basis of structural analysis is represented by a belt with a very low-grade metamorphic imprint, underlying with angular unconformity Cretaceous sediments. As discussed in Gaetani *et al.* (1990*a*) this phase is probably related to the early collision of the Kohistan island arc with the Karakoram microplate occurring in the late Early Cretaceous (Coward *et al.* 1986; Searle *et al.* 1987). Collisional phenomena are also indicated by roughly coeval intrusions of granitoids forming the Hunza Plutonic Complex (Debon *et al.* 1987), dated at about 100–95 Ma and by the Khunjerab Pass plutons, which generally bear similar K-Ar cooling ages (Searle 1991; Crawford & Searle 1992).

The antiformal stack recognized in the Chapursan Valley includes tectonic units with slightly different metamorphic imprint and older folds, and thus seems to be generally post-metamorphic. Late Cretaceous sedimentary rocks included in the fan indicate that thrust stacking was active also in post-Cretaceous times. Furthermore, KAB-related intrusives, probably associated with Eocene Batura-type granitoids (Debon *et al.* 1987) are possibly included in the southern part of the stack. Radiometric dating of these rocks could provide a further constraint on the time of emplacement of thrust sheets.

The formation of the thrust fan was probably polyphase as suggested by different trends of folds (cf. Fig. 16a and b), cleavages and slip-vectors. The different phases can be described as follows.

(1) Northeastward tectonic transport (Fig. 16B) is most evident in the lower part of the Shimshal Valley, where deformation is coeval to low-grade metamorphism, and in the corresponding Guhjal Unit forming the southern part of the Chapursan stack. The presence of post-kinematic biotite in the lower Shimshal Valley metamorphics, probably grown during the post-collisional instrusion of the Disteghil Sar granodiorites (Eocene Batura Unit of Debon *et al.* 1987, or Baltoro Unit of Searle 1991) give only a minimum age for deformation.

(2) The Sost Unit mostly indicates northward motion of thrust sheets (Fig. 16A) possibly

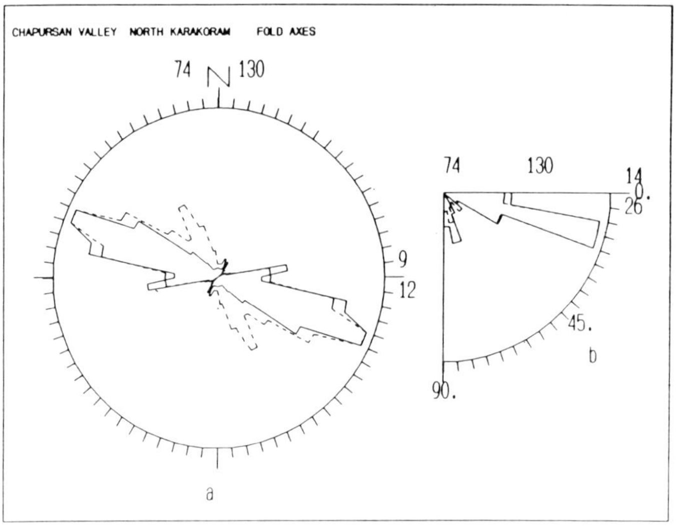

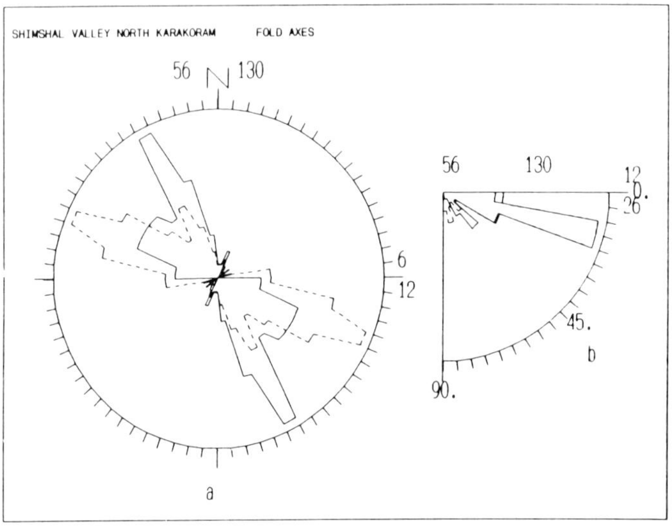

Fig. 16. Rose diagrams relative to fold axes in the antiformal stack of the Chapursan Valley (**A**) and to fold axes measured mainly in the lower part of the Shimshal Valley in the southern belt (**B**). Dashed roses correspond to all the measured folds (130). Numbers above the rose refer to subtotal and total of fold axes.

associated with a present-day Andean-type margin, as a consequence of north-directed subduction beneath the southern boundary of the Karakoram microplate (Debon *et al.* 1987). Northward thrusting of the cover might then represent the result of back-arc compression north of the magmatic arc, represented by the Cretaceous granitoids of the KAB.

(3) North-dipping thrust of the Misgar Slates onto the Sost Unit. Thrust faults with the same attitude also cut through the southern part of the Chapursan stack, which overthrusts the KAB in the Yashkuk area. This event may represent a distinct successive phase of south-vergent tectonic transport due to crustal thickening after the India–Eurasia continental collision, with late-stage out-of-sequence blind thrusting along the frontal part of the Chapursan antiformal stack. This south-vergent thrust event may be coeval with southward motion of the Main Karakoram Thrust (MKT) and related splays. The MKT seems to have been active since closure of the Indus Suture and at least until the Early Oligocene in the Baltoro region, where it

cuts pre-37 Ma M2 isograds (Rex *et al.* 1988 Searle & Tirrul 1991). High rates of deformation along these thrusts are indicated in the Chapursan Valley by the presence of calc-mylonites.

(4) During the last important tectonic phase large strike-slip faults cut through previous thrust fans. These fault systems were probably connected to the progressive tectonic indentation between India and Eurasia following continental collision (Molnar & Tapponier 1975). In the Chapursan Valley, large E–W sinistral strike-slip faults were observed in the central part of the Sost Unit. In the Shimshal Valley this fault system in association with large NW–SE dextral strike-slip faults was distinguished in the Zardgar Bin area, forming a very complex structural pattern. No clear indications about the chronology and age of motion relative to the two different fault systems was obtained during this preliminary phase of work. Nevertheless, the presence of small tectonic structures consistent with motion of both E–W and NW–SE faults in Quaternary sediments may suggest close relationships between these systems, as well as reactivation phenomena until recent times.

The NW–SE dextral faults of the Shimshal Valley can be interpreted as important pre-Quaternary dextral shears, probably related to the Karakoram Fault (Fig. 1). Moreover, dextral motion along these NW–SE strike-slip faults could produce large-scale right-lateral simple-shear. This may be, at least in part, absorbed by the formation of domains with E–W-trending sinistral strike-slip faults such as those recognized in the Chapursan and Shimshal Valleys (Fig. 17). These deformational domains are largely known along major wrench systems developed in continental crust as for the San Andreas Transform Fault in California or along the Dead Sea Transform (Garfunkel & Ron 1985; Nur *et al.* 1989). Such a model of crustal deformation was proposed for Eastern Tibet by England & Molnar (1990) as an alternative view to lateral extrusion of Tibet and southeast China along the Karakoram and Altyn Tagh faults (Molnar & Tapponnier 1975; Tapponnier *et al.* 1986). In this model, north-directed simple shear caused by the India–Eurasia collision is absorbed by formation of E–W-trending sinistral faults which cause clockwise rotation of blocks, with little material extruded eastward. The different stress directions obtained in the study area may thus be explained by the complexity of deformation processes occurring in the continental crust along collisional zones.

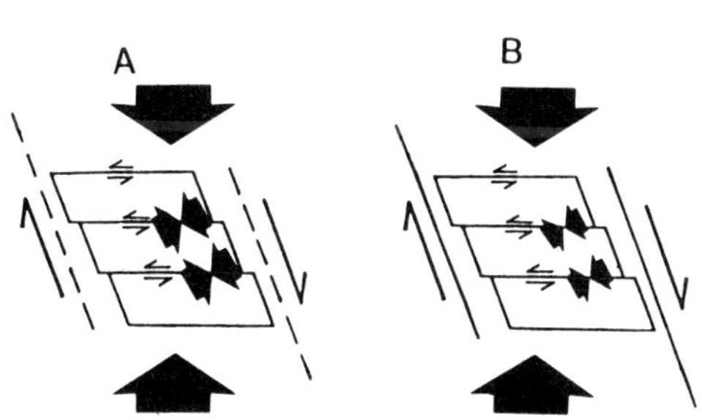

Fig. 17. Formation of E–W sinistral strike-slip faults as a consequence of NW–SE simple shear caused by major dextral strike-slip faults. Black arrows refer to relative importance of regional and local stress fields.

This work was supported by the project 'Geology of Karakoram', led by M. Gaetani (EEC Project CI*CT 90/0852.). Field work was carried out during a two years M.P.I. post-PhD scholarship. M. Gaetani strongly encouraged this work and read the preliminary manuscript and E. Garzanti improved English in the final manuscript. G.B. Siletto gave help at the microscope. The author is indebted to M. Gaetani, F. Jadoul and L. Angiolini, who provided excellent stratigraphy and wonderful, special days in the Karakorum. I am also grateful to the critical reviews of M.P. Searle and P. Matte for helpful suggestions. Drawings were prepared by M. Minoli. The Government of Pakistan is thanked for permitting field work in the northern restricted areas.

References

ANGELIER, J. 1990. Inversion of field data in fault tectonics to obtain the regional stress-III. A new rapid direct inversion method by analytical means. *Geophysical Journal International*, **103**, 363–376.

CASNEDI, R. & NICORA, A. 1985. Short notes on the Shimshal Valley geology (western Karakorum-Pakistan). *Rivista Italiana di Paleontologia e Stratigrafia*, **90**, 463–480.

COWARD, M.P., WINDLEY, B.F., BROUGHTON, R., LUFF, I.W., PETTERSON, M.G., PUDSEY, C., REX, D. & KHAN, M.A. 1986. Collision tectonics in the NW Himalaya. *In*: COWARD, M.P. & RIES, A.C. (eds) *Collision Tectonics*. Geological Society, London, Special Publications, **19**, 203–219.

CRAWFORD, M.B. & SEARLE, M.P. 1992. Field relationships and geochemistry of pre-collisional (India–Asia) granitoid magmatism in the central Karakorum, Northern Pakistan. *Tectonophysics*, **206**, 171–192.

DEBON, F., LE FORT, P., SONET, J. & ZIMMERMANN, J.L. 1987. Granites of western Karakoram and northern Kohistan (Pakistan): a composite Mid-Cretaceous to Upper Cenozoic magmatism. *Lithos*, **20**, 19–40.

DESIO, A. 1979. Geological Evolution of the Karakoram. *In:* FARAH, A. & DE JONG, K.A. (eds)

Geodynamics of Pakistan. Geological Survey of Pakistan, Quetta, 11–124.

—— & MARTINA, E. 1972. Geology of the Upper Hunza Valley, Karakorum, West Pakistan. *Bollettino della Società Geologica Italiana*, **91**, 283–314.

ENGLAND, P.C. & MOLNAR, P. 1990. Right-lateral shear and rotation as the explanation for strike-slip faulting in Eastern Tibet. *Nature*, **344**, 140–142.

GAETANI, M., GARZANTI, E., JADOUL, F., NICORA, A., TINTORI, A., PASINI, M. & SABIR ALI KAN, K. 1990*a*. The north Karakorum side of the Central Asia geopuzzle. *Geological Society of America Bulletin*, **102**, 54–62.

——, GOSSO, G. & POGNATE, U. 1990*b*. A geological transect from Kun Lun to Karakorum (Sinkiang, China): the western termination of the Tibetan Plateau. Preliminary note. *Terra Nova Research*, **2**, 23–30.

——, JADOUL, F., ERBA, E. & GARZANTI, E. 1993. Jurassic and Cretaceous orogenic events in the North Karakorum: age constraints from sedimentary rocks. *This volume*.

GARFUNKEL, Z. & RON, H. 1985. Block rotation and deformation by strike-slip faults 2. The properties of a type of macroscopic discontinuous deformation. *Journal of Geophysical Research*, **90**, 8589–8602.

MOLNAR, P. & TAPPONNIER, P. 1975. Cenozoic tectonics of Asia: effects of a continental collision. *Science*, **189**, 419–426.

NUR, A., RON, H. & SCOTTI, O. 1989. Mechanics of distributed fault and block rotation. *In*: KISSEL, C. & LAJ, C. (eds) *Paleomagnetic Rotations and Continental Deformation*. Kluwer Academic Publishers, 209–228.

REX, A.J., SEARLE, M.P., TIRRUL, R., CRAWFORD, M.B., PRIOR, D.J., REX, D.C. & BARNICOAT, A. 1988. The geochemical and tectonic evolution of the central Karakoram, north Pakistan. *Philosophical Transactions of the Royal Society of London*, **A326**, 229–255.

SEARLE, M.P. 1991. *Geology and Tectonics of the Karakoram Mountains*. John Wiley & Sons.

—— & TIRRUL, R. 1991. Structural and thermal evolution of the Karakoram crust. *Journal of the Geological Society, London*, **148**, 65–82.

——, WINDLEY, B.F., COWARD, M.P., COOPER, D.J.W., D.J., REX, A.J., REX, D.C., LI TINGDONG, XIAO XUCHANG JAN, M.Q., THAKUR, V.C. & KUMAR, S. 1987. The closing of Tethys and the tectonics of the Himalayas. *Geological Society of America Bulletin*, **98**, 678–701.

SHVOLMAN, V.A. 1981. Relicts of the Mesotethys in the Pamirs. *Himalayan Geology*, **8**, 369–378.

TAPPONNIER, P., PELTZER, G. & ARMIJO, R. 1986. On the mechanics of the collision between India and Asia. *In*: COWARD, M.P. & RIES, A.C. (eds) *Collision Tectonics*. Geological Society, London, Special Publications, **19**, 115–157.

Jurassic and Cretaceous orogenic events in the North Karakoram: age constraints from sedimentary rocks

MAURIZIO GAETANI, FLAVIO JADOUL, ELISABETTA ERBA & EDUARDO GARZANTI

Dipartimento di Scienze della Terra, Project Ev-K2 CEE, Via Mangiagalli 34, 20133 Milano, Italy

Abstract: At least two major episodes of tectonic deformation are recorded in the Mesozoic sedimentary succession of the North Karakoram. The older event took place between the Liassic and the early part of the Mid-Jurassic, as documented by tens to hundreds of metres thickness of dark grey litharenites (Ashtigar Fm.). The latter are sharply overlain by 35 m to 150 m thick red sandstones, interbedded with shales and siltstones (Yashkuk Fm.). Terrigenous bodies increase in thickness from west to east, and from north to south. Petrographic composition indicates the orogenic character of these quartzo-lithic sandstones, characterized by mafic volcanic and serpentinite detritus, being replaced upwards by sedimentary and subordinate metasedimentary clasts. The Ashtigar Fm. may have been deposited in a collisional basin close to a newly-formed suture belt. Contribution from allochthonous oceanic rocks was significant in the first stages of collision, whereas detritus was subsequently derived mainly from uplifted sedimentary successions. Terrigenous units were conformably transgressed by shallow-water limestones in the Aalenian, and later on marine conditions persisted until the earliest Cretaceous.

The second event of tectonic deformation occurred between the earliest and the Late Cretaceous, when the fan delta Tupop Conglomerate, mostly derived from uplifted carbonate and clastic rocks, sealed the strongly folded and faulted underlying succession. Mid-Cretaceous granitoid plutons are also widespread in the North Karakoram. This orogenic event is interpreted as related to the final welding of the Kohistan arc and Karakoram microplate to Asia.

The highly irregular topography of the Tupop Conglomerate is onlapped unconformably by reddish to grey nodular limestones, some hundred metres thick and containing phosphatized inoceramids and nannofossils of Campanian age (Darband Fm.). These are the youngest marine sediments detected in the Karakoram to date.

The aim of the present paper is to provide new evidence to constrain the age of the early deformations that affected the North Karakoram during the Mesozoic, before final welding of the Indian Plate against Eurasia, and thus before the Karakoram Range was built into its present shape by Himalayan collision tectonics. The sedimentary succession has been studied in a narrow belt extending from the head of the Chapursan valley in the west to the Shimshal pass in the east (Fig. 1). Since this belt (Sost Unit of Gaetani *et al.* 1990*a*) has been disrupted into several tectonic sheets, stratigraphic analysis is not always straightforward (Fig. 2). For details about this mainly Cenozoic complex structural evolution, the reader is referred to Zanchi (this volume).

Stratigraphic knowledge of the area was poor before our 1986 and 1991 expeditions. Only Desio & Martina (1972) provided some data for the area around Sost, and Casnedi & Nicora (1985) reported a few Palaeozoic fossils from the Shimshal valley. Gaetani *et al.* (1990*a*) presented the first stratigraphic scheme for the Chapursan area.

The Jurassic succession

Rocks of Jurassic age crop out both in middle Chapursan, from the Kundil valley to the Yashkuk area, and in the Shimshal area, from Zardinarbin to Parien. A thick and complete terrigenous succession crops out in most of the investigated region, whereas a reduced sequence is confined to the northwest (Yashkuk Fm. of Gaetani *et al.* 1990*a*).

Lithostratigraphy

From base to top, the following units were recognized.

(1) Up to 700 m thick, massive dark grey limestones and thick-bedded peritidal dolomites (Aghil Group, Desio 1980; Gaetani *et al.* 1991). Metre-scale regressive cyclothems contain large megalodontids of Late Triassic affinity. Blocks

From TRELOAR, P.J. & SEARLE, M.P. (eds), *Himalayan Tectonics*
Geological Society Special Publication, No. 74, pp. 39–52.

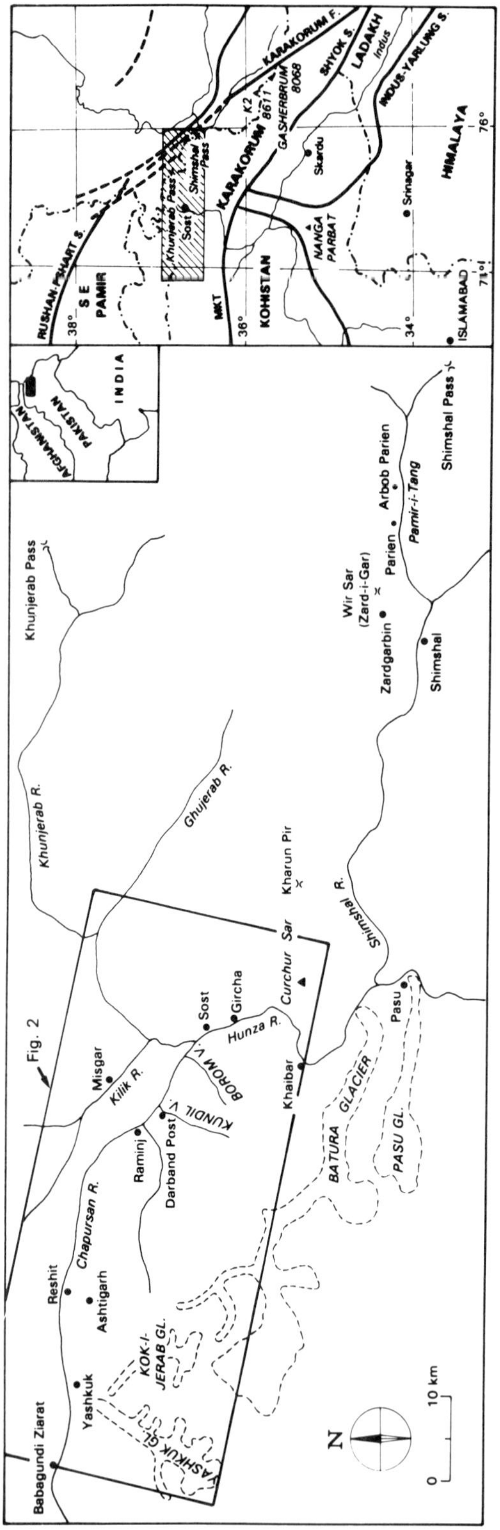

Fig. 1. Location map of the investigated area.

packed with *Lithiotis* occur in the scree, but could not be positioned within the succession. The Aghil Group interfingers towards the southeast with breccias and cherty limestones deposited in deeper waters.

(2) Up to 70 m thick, thin-bedded, dark-grey biocalcarenites and biocalcirudites very rich in crinoids, interbedded upward with thin grey shales and platy laminated mudstones.

(3) Dark-grey splintery shales and marls, tens to hundreds of m-thick, occurring in the South Chapursan and Kundil valleys and Shimshal district (Zardgarbin to Parien). Distal turbiditic beds, 15 to 80 cm thick and displaying Tb-Te Bouma intervals, pass upwards into a more continuous shaly succession. Overall thickness is between 100 and 200 m in Chapursan (Fig. 3), and perhaps more in Shimshal, where greater thickness may be ascribed to tectonic repetitions. We propose here the term Ashtigar Formation for this unit (Fig. 4).

The type-section was studied at 4350 to 4450 m altitude, on the ridge southeast of the Ashtigar meadows. The lower boundary is marked by a sharp transition from the calcarenites/calcirudites and dark laminated limestones of the topmost Aghil Fm. to black shales with intercalated lithoclastic calcarenites yielding volcanic debris. These 20 m thick basal Ashtigar shales are overlain by at least 70 m thick, dark grey marly shales with interbedded, fine-grained turbidites. The upper 20 m of the unit consist of alternating dark grey and greenish shales, with subordinate sandstones. The upper boundary corresponds to the erosional base of a 5 m thick interval of amalgamated conglomerates with abundant clasts of deformed Aghil carbonates and rare volcanic grains, followed by greenish pelites and greenish-reddish sandstones (Yashkuk Fm.).

In North Chapursan, the Ashtigar Fm. is missing. It is apparently replaced by very fine-grained hybrid arenites and dark grey, well-bedded, bioclastic packstones/wackestones rich in crinoids, brachiopods and foraminifera. This interval (hereafter called unit 2A), which is about 40–50 m thick and arranged in thickening-upward sequences, has no counterparts in the southern successions, and its base is covered.

(4) Fine-grained, red and green sandstones and siltstones in decimetre-thick beds, commonly intercalated within thin red to greenish-grey shales (Yashkuk Fm. of Gaetani *et al.* 1990*a*). Their base is locally marked by calcareous breccias or intraformational con-

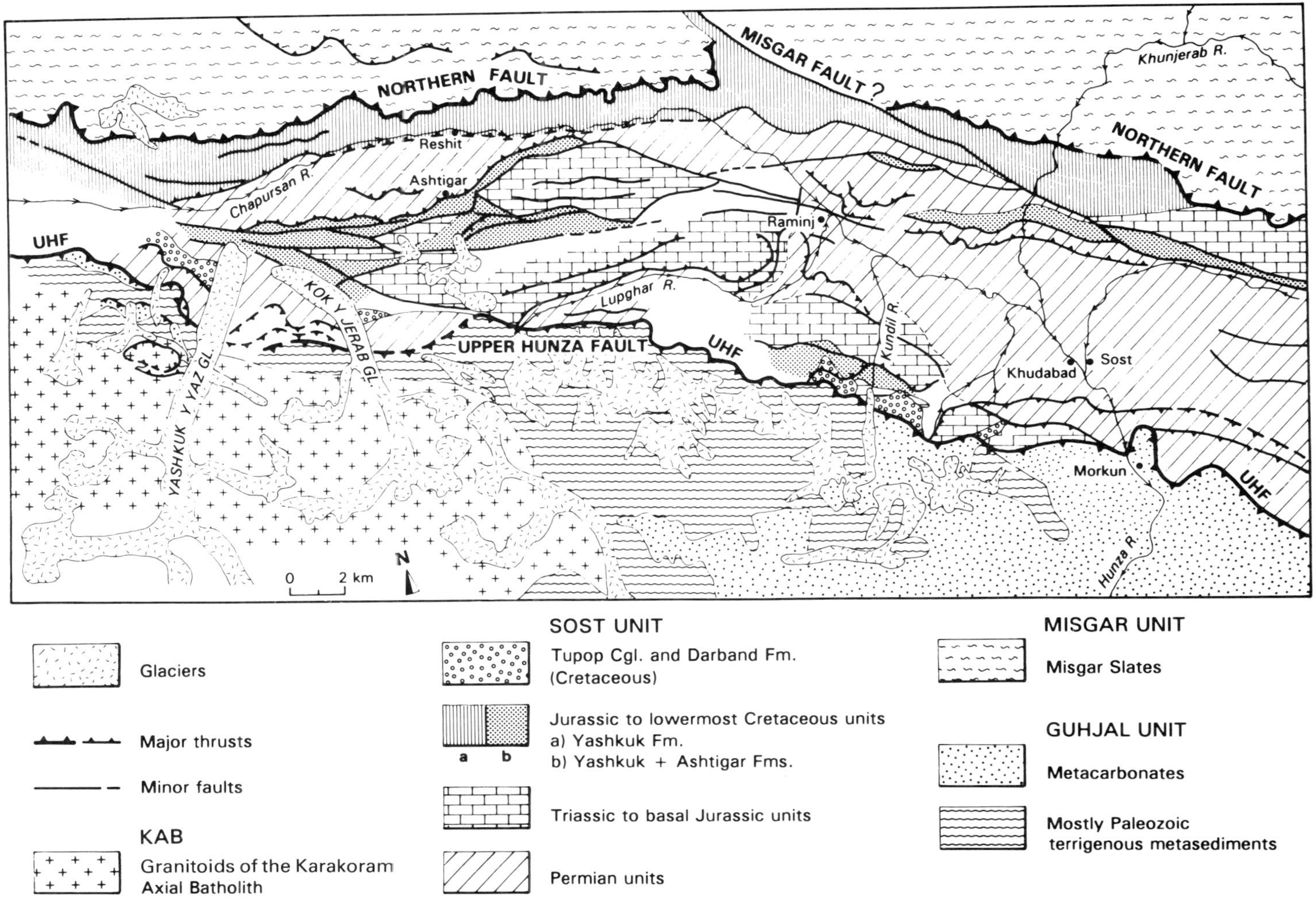

Fig. 2. Distribution of Jurassic and Cretaceous formations in the Chapursan valley, North Karakoram. (Simplified after Zanchi, this volume). Tectono-stratigraphic subdivisions after Gaetani *et al.* (1990*a*). KAB, Karakoram Axial Batholith. UHF, Upper Hunza Fault.

Fig. 3. The two Jurassic sandstone units in the Ashtigar section. Ash, Ashtigar Fm.; Yas, Yashkuk Fm.

glomerates, 3 to 6 m thick, which may erode the underlying Ashtigar Fm. Metre-scale lithic conglomerate lenses with red matrix also occur in the Pamir-i-Tang area below Wir Sar (Zard-i-Gar according to Shipton 1938). Red sandstones are rare or absent altogether in the east, where the unit may reach a thickness of 150 m (Fig. 3), and become more abundant in the west where, in the northern belt of Jurassic outcrops from north of the Yashkuk glacier snout to the Karakoram Highway north of the Kilik-Hunza river confluence, they predominate but are reduced to a few tens of metres (Fig. 4).

(5) The calcareous succession overlying the terrigenous units can be subdivided into two parts, from bottom to top: (5A) dark grey, oolitic or oncolitic, subtidal bioturbated packstones (about 160 m thick); (5B) dark grey, well-bedded mudstones and wackestones, locally with chert nodules (at least 150 m thick; Reshit Fm. of Gaetani *et al.* 1990*a*).

Units 5A and 5B may be locally replaced by dark grey, thick-bedded, amalgamated packstones/wackestones with rare chert nodules (thickness up to 200 m).

Superposition of sub-units 5A and 5B is locally doubtful, because of extensive disruption by closely-spaced strike-slip faults. Also the correct stratigraphic position in this part of the succession of anhydrite layers, stromatolitic mounds interbedded with grey-green shales, and *Cayeuxia*-bearing dark mudstones could not be ascertained due to faulting.

(6) The youngest carbonate unit, cropping out near the snout of the Yashkuk glacier, apparently consists of metre-scale megabreccias containing clasts of shallow-water carbonates, and calcarenites intercalated within thin beds of pinkish hemipelagic wackestones.

In summary, a complex carbonate platform (Aghil Group) was gradually drowned; the newly-formed basin received increasing terrigenous detritus, episodically transported by gravity flows (unit 3). At the flanks of the trough, shallow-water hybrid carbonates possibly continued to be deposited (unit 2A). Later on, the terrigenous Yashkuk Fm. (unit 4) spread over the area, which was characterized by deltaic to marginal marine conditions. Gradually, the sea returned over the area and shallow-water carbonates were deposited.

Biochronological data

Unit 1: Aghil Group. Biostratigraphic data are poor in the lower part of the unit. In the middle part, megalodontids and carbonate platform

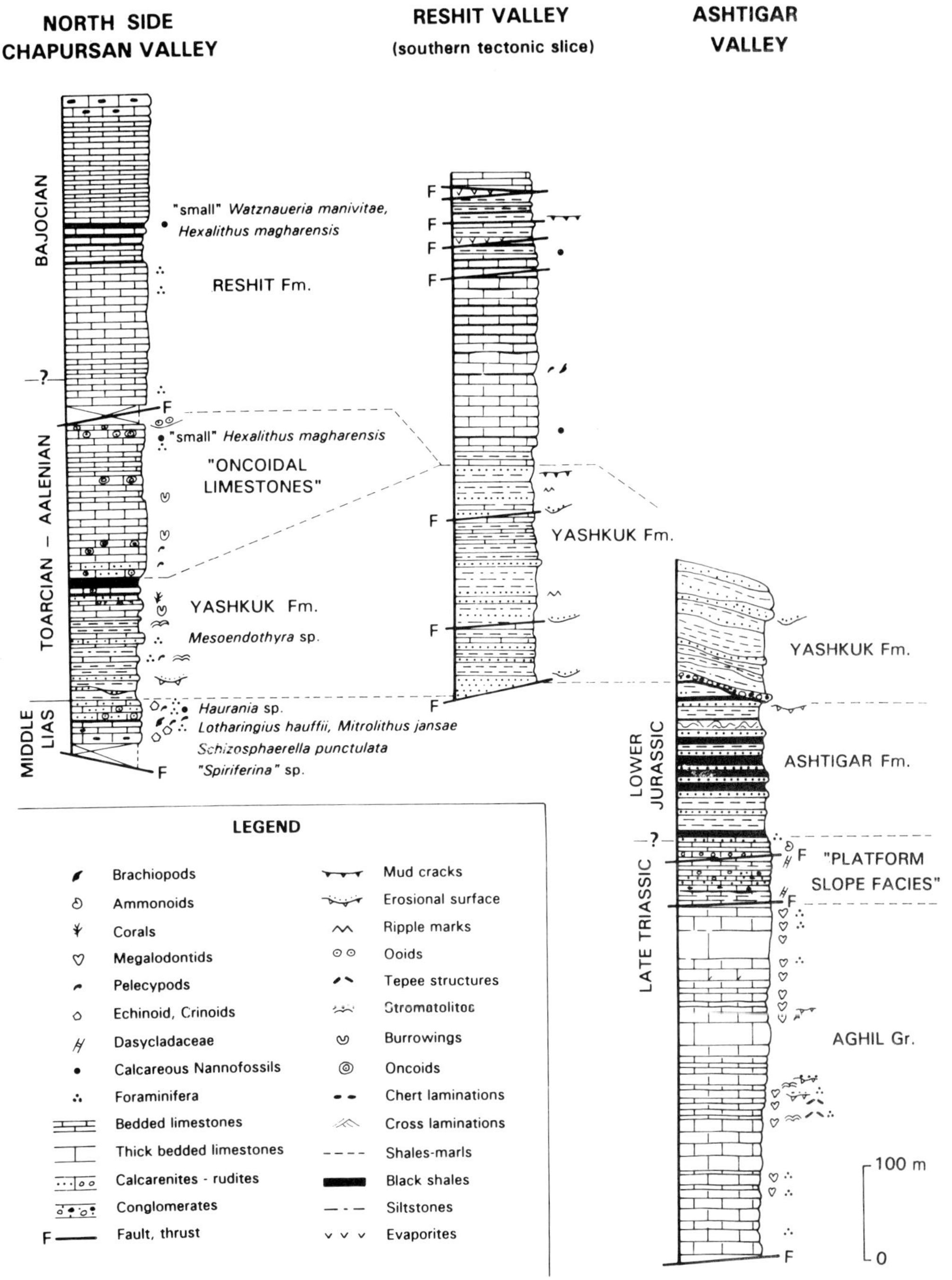

Fig. 4. Representative stratigraphic columns measured in the Jurassic units.

faunas of Norian/Rhaetian affinity have been recognized (*Aulotortus* sp., *Ophtalmidium* sp., *Miliolipora* sp., *Thaumatoporella* sp.). Since in the Reshit side valley *Lithiotis* has been found in loose blocks, a Late Triassic–earliest Jurassic age is inferred.

Unit 2. Bioclastic packstones/rudstones contain within the clasts a microfacies usually characterizing Upper Triassic carbonate platforms, including benthic foraminifera (*Aulotortus* sp., *Galeanella* sp., *Ophtalmidium* sp., *Glomospirella* sp., Nodosaridae and Miliolidae), Dasycladacean algae like *Thaumatoporella*, crinoids, bryozoans and a few Radiolaria.

Unit 2A. We found two intervals containing age-diagnostic fossils. The lower one consists of 6 to 7 m thick marly limestones, marls and bioclastic packstones. Its base yielded a brachiopod fauna with *Zeilleria* sp. and '*Spiriferina*' sp., whereas in the marls 1.7 m above a nannoflora with *Mitrolithus jansae* (Wiegard), *Lotharingius hauffii* Grün & Zweili and *Schizosphaerella punctulata* Deflandre was found. The brachiopod assemblage does not reach into the Toarcian, because of the presence of '*Spiriferina*'. Nannofossils point more precisely to a latest Pliensbachian–Early Toarcian age.

The upper interval, forming the topmost 15 to 20 m of the unit overlain by the Yashkuk Fm., is a regressive sequence characterized by bioclastic crinoidal packstones grading into oolitic bioclastic packstones enriched in quartz grains. It yielded foraminiferal assemblages including *Haurania* and *Involutina*. The latter apparently does not range above the Pliensbachian. A latest Pliensbachian age may thus be assigned to unit 2A.

Unit 3: Ashtigar Formation. Age data could not be obtained from this unit. Along the trail between Targir dur and Parien, some fragments of *Trigonia*-like bivalves have been extracted by L. Angiolini.

Unit 4: Yashkuk Formation. Age data could not be obtained where the unit is thicker. In the Yashkuk section a bioclastic packstone intercalated in the middle part contains abundant *Mesoendothyra* sp., pointing to a Toarcian or early Mid-Jurassic age.

Unit 5A. In the Yashkuk section, oncolitic limestones include the foraminiferal fauna listed in Gaetani *et al.* (1990*a*, p. 59). Two new levels with nannofossils have been found (Fig. 4). The lower one, 150 m above the top of the Yashkuk Fm., contains *Schizosphaerella punctulata* and small *Hexalithus magharensis* Moshkowitz & Ehrlich, and could thus be assigned to the Late Aalenian. In the pelagic limestones above, a 10 m marly intercalation yielded *Schizosphaerella punctulata, Hexalithus magharensis* and small *Watznaueria manivitae* Bukry. Age: Bajocian, because of the appearance of small *Watznaueria*.

Unit 5B: Reshit Fm. Some brachiopods (*Burmirhynchia* sp., a new genus of tererebratulid close to *Monsardithyris*) and a single specimen of the ammonoid family Perisphinctidae, pointing to a Mid-Jurassic age, have been found. Nannoflora, with small *Watznaueria manivitae, Cyclagelosphaera margarelii* Noël, *Discorhabdus criotus* Bown, *Hexalithus magharensis* confirm that suggestion.

Within the shales intercalated with the anhydrite lenses, a nannoflora with *Watznaueria* cf. *communis* Reinhart has been recognized. The genus *Watznaueria* spans the interval Bajocian–Cretaceous.

Unit 6. Small *Protopeneroplis* sp., indicating a latest Jurassic–earliest Cretaceous age, are fairly common in resedimented beds. The presence of a nannoflora comprising *Watznaueria barnesae* (Black), *Rucinolithus terebrodentarius* Applegate *et al.*, *Nannoconus* sp., and *Braarudosphaera* sp. in hemipelagic intervals points to the earliest Cretaceous. Orbitolinids have not been found in the Chapursan valley sedimentary rocks so far.

In summary, the Ashtigar Fm. can only be generally ascribed to the Early Jurassic, since it overlies the Aghil Fm. which may contain *Lithiotis* beds. The Yashkuk Fm. in contrast is fairly well bracketed between the latest Pliensbachian and the Late Aalenian. In Gaetani *et al.* (1990*a*, p. 58–59), the foraminiferal assemblages below and above the sandstones were considered as similar, and thus the unit was ascribed to the Pliensbachian; this conclusion is amended by new data.

Sandstone petrography

The Jurassic sandstones are litharenites (only a few samples with grain size around or below 150 μm are feldspathic litharenites) with common volcanic, sedimentary and metedimentary rock fragments. Main petrographic parameters (Dickinson 1970) of 15 fine to lower medium-grained analysed samples are comprised in the range Q 48 to 72; F 0 to 17; L 17 to 47. Secondary parameters are C/Q up to 21, P/F up to 89, V/L up to 89, S/L up to 62, M/L up to 27.

Ashtigar Formation. Very fine to lower medium-grained sandstones (detrital modes Q 55 ± 5, F 9 ± 4, L 35 ± 4; Fig. 5) contain common radiolarian chert, chlorite schist, serpentine schist and basaltic rock fragments, strongly testifying to erosion of an oceanic subduction complex (one eighth to one fourth of total detritus). Volcanic lithics are common (V/L 49 to 61), and comprise both 'basaltic' types, displaying plagioclase laths interspersed within a chloritized glassy groundmass, and vitric to felsitic grains. A few particles contain both quartz and plagioclase phenocrysts embedded in glassy groundmass, and were derived from the products of felsic explosive activity. Quartzose terrigenous (siltstone to sandstone) rock fragments also invariably occur along with extrabasinal carbonate grains (dolostone and limestone). The sedimentary to total lithic ratio (S/L 20 to 41, including chert grains) tends to increase upward; in some samples it may have been overestimated due to intrabasinal reworking, indicated by local finding of oversized serpentine schist-bearing quartzose silty clasts and by common occurrence of clayey intraclasts.

Detrital quartz is mainly monocrystalline (occasionally displaying straight extinction or rounded outlines), but polycrystalline grains, with mainly fine crystal size and irregular to stretched intercrystalline boundaries, are significant (C/Q 13 to 20, tending to increase with grain size). Feldspars are not abundant, with predominating plagioclase grains (P/F 67 to 89). Among the alkali-feldspars, chessboard-albite and more rarely orthoclase and microcline occur. Phyllite to paragneiss rock fragments (M/L 8 to 24, including serpentine schist grains) are invariably present, suggesting that low-grade metasedimentary rocks were exposed in the source areas and supplied a significant amount of quartz and subordinately albite grains. Blue-green to brown tourmaline, zircon, rutile and muscovite were sporadically recorded among the accessories, along with single radiolarians with opaque rims.

Yashkuk Formation. Detrital modes of fine-grained sandstones are Q 59 ± 8, F 6 ± 6, L 35 ± 10 (Fig. 5). Sandstones are moderately sorted and very fine to coarse grained. They contain insignificant amount of 'basaltic' and serpentine schist grains. Carbonate and subordinate terrigenous and radiolarite sedimentary lithics (S/L 43 to 70 in the lower part, passing upward to S/L 27 to 28; chert grains included) are instead more abundant, as well as phyllite to paragneiss rock fragments (M/L 6 to 27, with values over 50 reached in the Misgar area). The abundance of sedimentary lithic fragments is strongly controlled by grain size (r = 0.84; 1% sign. level), and the rare microconglomerate layers have calclithite compositions. Felsic rock fragments are as common as in the Ashtigar Fm.

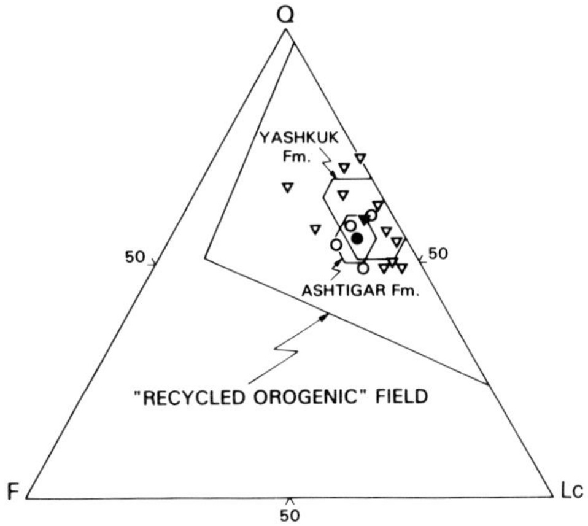

Fig. 5. Detrital modes of selected Jurassic sandstones (Yashkuk and Ashtigar Fms.), which plot in the 'recycled orogenic' field (Dickinson, 1985). Q quartz; F feldspars; Lc aphanite lithic fragments (carbonate grains and chert included). Polygons are one standard deviation each side of the mean. Abundance of chert, basalt, and serpentine schist fragments in the Ashtigar Fm. (circles) and predominant sedimentary–metasedimentary detritus in the Yashkuk Fm. (triangles) point to 'subduction complex' and 'collision orogen' provenance respectively (Dickinson & Suczek 1979).

but, due to absence of lathwork 'basaltic' grains, the volcanic lithic population is generally reduced (V/L 24 to 50). Only at Yashkuk, where sedimentary lithics are lacking, rock fragments are mostly volcanic felsites (V/L 80 to 84).

Polycrystalline quartz is subordinate (C/Q 14 to 21 in samples coarser than 150 μ and 3 to 9 in finer-grained samples). Detrital plagioclase is not very abundant (P/F < 60); alkali-feldspars include orthoclase and albite. Feldspar content tends to decrease from lower fine-grained samples (F 5 to 17) to sandstones coarser than 200 μm (F less than 5). Euhedral zircon, blue tourmaline and muscovite were recorded.

Intrabasinal mudclasts (illite–chlorite rip-up clasts and caliche fragments) may be locally abundant (up to 20% of the rock). Marine bioclasts (echinoderms, bivalves, benthic foraminifera, corals, bryozoans) sometimes occur in reddish layers. Single radiolarian tests were also recorded.

In the upper part of the Yashkuk Fm., where sandstones are invariably very fine-grained and interbedded with oolitic sublitharenites, hybrid oo-biocalcarenites, peloidal micrites, calcareous siltstones and subordinate stromatolitic bindstones, the mineralogy of the extrabasinal fraction is more difficult to assess quantitatively.

Source areas. The quartzo-lithic petrographic composition of the Jurassic terrigenous formations is typical of orogenic clastics. Detrital modes, plotted on a standard QFL triangle, fall in the middle of the 'recycled orogen' provenance field (Dickinson & Suczek 1979; Dickinson 1985). Mineralogical composition of the Ashtigar Fm., characterized by common serpentine schist, basalt and chert lithic fragments, as well as by detritus from sedimentary, metasedimentary and felsic volcanic rocks, documents erosion of a tectonically-uplifted thrust belt in the proximity of an arc-trench system. The source area included a subduction complex composed of deformed ophiolites and other oceanic rocks, emplaced adjacent to a foreland basin that developed during the first stages of arc–continent collision.

The Toarcian–Aalenian Yashkuk Formation documents the structural evolution of the foreland fold–thrust belt, with further uplift of the accretionary prism and extensive erosion of thrust sheets made of sedimentary and subordinately of low-grade metasedimentary rocks. Supply from ophiolitic melanges was strongly reduced. Neither uplifted basement block nor granitoid intrusive sources are ever indicated.

The Cretaceous succession

Information on the Cretaceous deformation event was obtained mainly in the Kundil and Borom side valley, in lower Chapursan (Fig. 2). In other areas the Tupop Conglomerate is present as well, but sections are disrupted by fold-thrust and strike-slip deformation (Zanchi, this volume) and reliable stratigraphic data could not be obtained.

Tupop Conglomerate

Gaetani *et al.* (1990*a*) introduced the term Tupop Conglomerate for the lithic conglomerate which seals the underlying sedimentary sequence, including basaltic rocks of unknown age, with spectacular angular unconformity (Fig. 6a).

The Tupop Conglomerate occurs from the Ziarat of Babagundi in the west, to the heads of the Kundil and Borom valleys, and up to the Shimshal area in the east (Pamir-i-Tang), where it may be observed in the gully where the hut of Arbob Parien is built.

The Tupop Conglomerate lacks age-diagnostic fossils. Petrographically, terrigenous layers are invariably calclithites, containing predominant sparitic carbonate rock fragments along with frequent sandstone, siltstone and shale clasts (Fig. 7). Also felsic volcanic lithics are common. Monocrystalline quartz occurs and commonly shows good roundness or embayments due to magmatic resorption, pointing respectively to polycyclic sedimentary and felsic volcanic provenance. Most calcareous lithic grains are extensively recrystallized, but oolitic or bio-intraclastic grainstones and biomicrites are sometimes recognizable. Calclithite layers yielding abundant reworked echinoderm fragments and Permian fusulinids are also found.

(a) (b)

Fig. 6. Stratigraphic relationships in the west side of the Upper Kundil valley, (**a**) The angular unconformity at the base of the Tupop Conglomerate, resting on shales of the Ashtigar Fm. Thickness of the Tupop bank is nearly 120 m. (**b**) Back-view of the same Tupop Conglomerate, onlapped by the Darband Fm.

Fig. 7. Close-up view of the Tupop Conglomerate, Kundil valley, West side. Scale bar is 15 cm long.

Among the cobbles, grey micritic limestones with Dasycladacean algae, bivalves, gastropods or foraminifera, reddish fine-grained quartzose sandstones or rarer sedimentary litharenites, grey or reddish oncoidal limestones, dark-cherty limestones and calcareous microbreccias, and a few peritidal carbonates with stromatolites, fenestrae or vadose pisoids, are recorded in this order of abundance. Composition of clasts compares well with the lithologies of the underlying Permian to Jurassic units. Thorough recrystallization of limestone clasts and well-developed cleavage in argillite rock fragments point to very low-grade metamorphic deformation in the source rocks.

Notable are: (a) lack of crystalline basement clasts, in contrast to the similar Urdok (Gaetani *et al.* 1990*b*, 1991) and Reshun Conglomerates (Pudsey *et al.* 1985); (b) lack of *Orbitolina*-bearing clasts, in contrast to the similar Reshun Conglomerate of the nearby Karambar valley (Casnedi 1980) and Chitral (Hayden 1915); (c) lack of pachyodont bivalve fragments, which are reported in Chitral (Pudsey *et al.* 1985).

Interpretation of depositional environments is problematic. Fluviatile, fining-upward festooned bars were observed near the Ziarat of Babagundi (Chapursan valley); in the Borom valley, instead, red nodular unfossiliferous limestones are intercalated also in the basal part of the unit, below the up to 60 m thick conglomerate bank, possibly suggesting deposition in deep-marine environments. Very steep cross-bedding in the basal conglomerates (Borom valley) suggests a fan delta environment for this part of the unit.

In summary, the Tupop Fm. is a lithic conglomerate derived from erosion of sedimentary and very low-grade metasedimentary source rocks ('recycled orogen' provenance of Dickinson & Suczek 1979). Since detritus contains all of the lithologies characteristic of the underlying Permo-Mesozoic stratigraphic units, local supply from the deformed and uplifted Karakorum sedimentary succession is indicated.

Darband Formation

The Tupop Conglomerate is overlain with a spectacular onlap by red and grey nodular wackestones in amalgamated beds (Figs 6 & 8). For this unit we propose here the term Darband Formation, from the Darband tower at the end of the Kundil valley, where we first observed loose blocks of these Scaglia-like mudstones.

Bedding forms an angle of 10° to 15° to the underlying conglomerates (Fig. 9). Total thickness is more than 300 m. At the top, these limestones are once more interbedded and overlain by concordant lithic conglomerates with abundant red matrix.

The unit crops out only at the head of the Borom and Kundil valleys, possibly extending up to the head of some side valleys south of the Lupghar valley, which were not reached during our expeditions. Younger units are not exposed in the studied area.

Microfacies have been largely destroyed by pervasive deformation and slaty cleavage. Only phosphatized fragments of inoceramids and ghosts of possible globotruncanids were observed. Search for coccoliths was more fruitful, and two specimens were found in sample KK191b: *Ceratolithoides verbekii* Perch-Nielsen (spread in the Early and Middle Campanian, CC18 to CC21) and *Calculites obscurus* (Deflandre) (Campanian to Early Maastrichtian, CC17 to CC24).

The Tupop Conglomerate is thus bracketed between two units of earliest Cretaceous and Campanian age, being separated from them by angular unconformities.

Fig. 8. Stratigraphic relationships in the east side of the Upper Kundil valley. Spectacular onlap of the Darband Fm. on vertical Jurassic limestones. The rock wall is about 400 m high. The normal fault seen in the centre of the photo is interpreted as syngenetic to deposition of the Darband Fm. (Zanchi, pers. comm.).

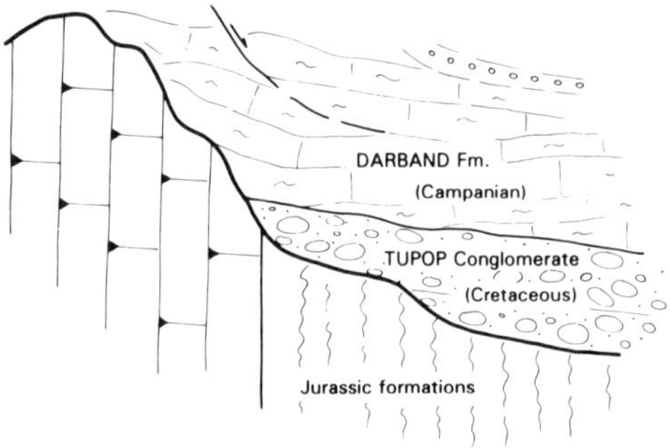

Fig. 9. Geometrical relationships between Jurassic and Cretaceous units at the head of the Kundil valley.

The Early Jurassic orogenic event

Early Jurassic turbiditic sandstones containing detritus from oceanic rocks are interpreted as derived from a subduction complex and deposited in a collisional basin close to a suture belt. A nearby carbonate ramp was apparently not affected by this episode. Turbiditic deposits occur in the south and east, and the carbonate ramp in the north and northwest. In the Toarcian–Aalenian, a deltaic to marginal marine 'recycled orogenic' red sandstone unit prograded onto the whole area and sealed the previous palaeogeography. The unit is thicker where turbidites were previously deposited, and thinner and finer-grained where platform carbonates persisted. Facies distribution and palaeocurrent directions suggest that detritus was derived from the southeast; later extensive fold-thrust and strike-slip deformation, however, may have severely changed the original palaeogeography.

Comparisons with proximal areas

Shaksgam valley, East Karakoram and Aghil Range. Red conglomerates and sandstones (Marpo Fm.) have been recently described in the Shaksgam valley (Gaetani *et al.* 1990*b*, 1991). Benthic foraminifera suggest a late Mid-Jurassic age, thus younger than that of the Yashkuk Fm. Similar quartzo-lithic sandstones/ microconglomerates associated with anhydrites and mudstones occur in the Surukwat area (Gaetani *et al.* 1990*b*, 1991).

West Tibet. In the southwestern Lingzi-Tang area, De Terra (1932) reported red sandstones and conglomerates overlying fossiliferous Norian 'limestones' and followed by grey-yellowish calcareous marls containing bivalves of Mid-Jurassic age. Brown sandstones of Mid-Jurassic age were observed by Norin (1946, 1976) on the Loqzung Mountains.

A limited strip of 'coarse littoral clastic rocks, intercalated with terrestrial sediments' occurs in West Qiangtang (Shi & Yang 1986).

Matte *et al.* (1990, 1992) reported basalts and thin-bedded marine Jurassic limestones onlapping the Permian sediments south of Lake Longmu. Marine conditions recovered during Aalenian–Early Bajocian times.

Southeast Pamir and Wakhan. Conglomerates and sandstones characterize the succession of the so called 'Axial Zone' of Southeast Pamir in the latest Triassic (Dronov *et al.* 1982). Coal and sandstones occur in the Toarcian–Aalenian, whilst the whole Early Jurassic is commonly missing (Dronov *et al.* 1959). Marine transgression took place in the Bajocian or later in the Callovian (Dronov *et al.* 1959).

In some areas external to the 'Axial Zone', like the Hyrymdinsk Zone, sandstones are confined to the Hettangian (Dronov *in* Dronov & Leven 1976). According to Leven (pers. comm., 1992) in the Southeast Pamir 'Intermediate Zone' of Dronov *et al.* (1982, p. 110, Fig. 5) Sinemurian sandstones overlie with angular unconformity a flyschoidal succession of latest Triassic age. Sandstones are in turn covered by Toarcian–Aalenian shallow-marine carbonates.

In the Alichur mountains, Middle and Upper Triassic basalts are covered by a melange with Triassic calcareous olistholiths, in turn overlain with angular unconformity by Early Jurassic marine sediments (Dronov & Leven 1990).

All of these data from Southeast Pamir point to deformation and uplift, shifting through time from the Late Triassic to the Early Jurassic.

In Wakhan, Kafarski & Abdullah (1976) describe a succession of thin terrigenous deposits, passing upward to oncolitic limestones of Early–Mid-Jurassic age, which rests unconformably on Lower Triassic pelites.

Rushan–Pshart area and Central Pamir. The Jurassic succession up to the Bajocian consists of sandstones and conglomerates; a shallow-marine ingression is recorded in the Bathonian (Dronov *in* Dronov & Leven 1976). In the Rushan–Pshart area, Lower Jurassic siltstones contain plant fragments (E. Leven, pers. comm. 1992).

Central Afghanistan. The complex evolution of Central Afghanistan has been recently reviewed by Boulin (1981, 1988, 1990). Three main tectono-stratigraphic elements may be distinguished south of the Herat Fault, which may be considered as the Palaeotethys suture (Sengör 1984): (1) the Turkman and Band-e-Bayan zones to the north, where the Permo-Triassic succession is folded and unconformably overlain by a terrigenous apron of Early to Mid-Jurassic age, fed by the so called 'Hercynian Belt' emerged in the north; (2) the Farah Rud zone, interpreted as a complex back-arc area floored by oceanic crust (Waser–Panjao Ocean); (3) the Helmand block, which during the Early Jurassic recorded rifting and platform drowning, but not compressional movements.

Discussion

Everywhere in neighbouring areas there is evidence of an erosional event linked to compression and uplift, which may be ascribed to the 'Eo-Cimmerian' orogeny (Sengör 1984). Open questions are of two types.

Timing. Terrigenous deposition began either in the latest Triassic or in the late Early Jurassic to early Mid-Jurassic. Although there is need for greater precision of age information, two distinct events are recognized. Sandstone petrographic data are unfortunately largely lacking from these adjoining areas.

The North Karakoram succession recorded both episodes. Provenance of detritus from an accretionary prism containing oceanic rocks suggests an arc–continent collision scenario for the older Early Jurassic episode. In Southeast Pamir, uplift and erosion started in the latest Triassic, i.e. somewhat earlier. The younger Pliensbachian–Bajocian episode in the North Karakoram is documented by 'molassic' deposition sealing the deformed continental margin between the latest Pliensbachian and the Late Aalenian.

Provenance. The overall picture we gain from Afghanistan and Pamir is that subduction and uplift occurred in the north (present position) with terrigenous detritus carried to the south, onto the southern passive margin represented by the Helmand Block. In Chapursan and partly in East Karakoram and Shaksgam, the available data suggest instead that deformation began and was more active in the south. Can this be due to late-stage strike-slip faults which inverted the original palaeogeography? Or do we have to start searching for another 'Eo-Cimmerian' orogenic belt in the south?

The Permian evolution of the Chapursan area strongly resembles that of the Helmand Block in Central Afghanistan, even though the latter received terrigenous detritus only at a late stage and from the north, by-passing the Farah Rud Zone (Boulin 1988). The problem of where the accretionary prism was located remains open, with the further possibility of long-distance longitudinal transport of detritus.

The mid-Cretaceous orogenic event

The Chapursan sedimentary succession was strongly faulted and folded at very low metamorphic grade prior to deposition of the fan delta Tupop Conglomerate, which sealed an highly irregular topography. Noteworthy is the lack of *Orbitolina* limestones and granitoid clasts.

Marine sedimentation continued until the Late Cretaceous in a very active geodynamic setting.

Comparisons with nearby areas

North Karakoram. Two belts of granitoid bodies with late Early Cretaceous radiometric ages occur south and north of the Chapursan sediments. The northern belt (Khunjerab, Giraf, Shuijerab, Sughet Jangle granitoids, plus the K2 Orthogneiss) has been the subject of only preliminary studies. Available ages (K/Ar, Ar/Ar on biotite or hornblende, U/Pb on zircon) cluster between 120 and 105 MA for crystallization, and up to around 70 MA for cooling (Treloar *et al.* 1989; Searle *et al.* 1990; Le Fort *in* Gaetani *et al.* 1991; Crawford & Searle 1992; Debon 1992). In the southern belt, representing the earliest known magmatic pulses of the Axial Batholith (e.g. Hunza pluton), U/Pb zircon ages of 95 ± 5 MA were obtained (Le Fort *et al.* 1983; Debon *et al.* 1987).

Hindu Raj and East Hindu Kush. The presence of *Orbitolina* limestones and derived pebbles in the Reshun Conglomerate has been known since Hayden (1915). Additional occurrence of Aptian–Albian rudist remains was reported by Talent *et al.* (1981) and Pudsey *et al.* (1985). In the Karambar valley, Casnedi (1980) reported *Orbitolina* pebbles in the Reshun Conglomerate, but no *Orbitolina* limestones.

Shaksgam valley and East Karakoram. The Urdok Conglomerate has been tentatively reconsidered as Cretaceous by Gaetani *et al.* (1990*b*, 1991). The earlier Desio (1980) assignment to the Middle Triassic seems untenable, since deep-water carbonates were deposited in the area at those times. We consider the Urdok as a possible equivalent of the Qizil Lunghar Conglomerate, cropping out at the southern border of the Depsang Plateau. Dainelli (1933, Chisil Lungur in his spelling) suggested a Norian age, since the conglomerate rests on fossiliferous Carnian deposits. Cretaceous microfossils, however, were found in the matrix (Gergan & Pant 1983; Gergan 1989 pers. comm.).

West Tibet. Strong deformation affected the sedimentary succession prior to marine transgression in the Cenomanian (Parona 1928; Dainelli 1933). Shallow-marine rudist-bearing limestones were deposited at least until the Turonian (Parona 1928). In other areas, continental red beds directly cover the folded succession.

Southeast Pamir and Wakhan. Lowermost Cretaceous marine sediments are unknown in this area (Shvolman 1973), where marine conditions probably ended in the Oxfordian (Dronov *in* Dronov & Leven 1976). The sequence mostly consists of tuffs, volcanoclastics, sandstones and conglomerates deposited in subaerial environments.

In the Tas Kupruk Zone of Wakhan, Kafarski & Abdullah (1976) describe red conglomerates (200 m) followed by some 500 m of purplish siltstones, tentatively attributed to the Early Cretaceous. They rest with marked angular unconformity onto Palaeozoic rocks.

Central Afghanistan. The Central Afghanistan blocks were uplifted in the Barremian–Aptian with emergent areas and *Orbitolina*-rich shallow seas occurring both in the Band-e-Bayan and Helmand blocks after closing of the Waser-Panjao Ocean in the course of the Neocomian (Boulin 1990). Tectonically-activated trenches started to be filled with terrigenous sediments in the Cenomanian. In the Senonian, shallow-water limestones prograded onto the peneplaned 'Neo-Cimmerian' range, both from north and south (Boulin 1990).

Discussion

In the Chapursan area there is undisputable evidence of the same mid-Cretaceous event which strongly affected all neighbouring blocks ('Neo-Cimmerian orogeny'; Sengör 1984). The lack of *Orbitolina* limestones might be ascribed to a still too deep environment during the Barremian. The finding of small *Protopeneroplis* in unit 6, however, may suggest that shallow seas existed in the earliest Cretaceous, but they did not last until the Barremian. The area thus possibly emerged at an early stage, suggesting an intermediate position between Southeast Pamir and East Hindu Kush.

Notable is the apparent lack of major related volcanism, which is instead widespread in Pamir and Helmand, even though the conglomerates directly overlie basalts of unknown age and do contain common felsitic volcanic rock fragments.

At the present state of knowledge we are still uncertain whether deformation was caused by northward subduction of Neotethys in the south, leading to collision with the Kohistan–Dras arc and formation of the Kohistan Northern Suture during the mid-Cretaceous (Petterson & Windley 1985; Coward *et al.* 1986; Searle 1991), or to subduction underneath a northern arc, feeding the Khunjerab Granitoid belt (as suggested by Debon *et al.* 1987).

Peculiar to the Chapursan area is the deposition of relatively deep marine sediments after the main deformation event. They document still active faulting, possibly related to the first stages of back-arc compression and strike-slip motions that characterized the structural evolution of the Karakoram Range during the Tertiary.

Paper supported by the project 'Geology of Karakorum', leader M. Gaetani. EEC Project CI*CT 90/0852. L. Angiolini and A. Zanchi assisted in the 1991 field work. M. Gallo organized logistics with great capability. I. Premoli Silva kindly identified some of the benthic foraminifera. Field-work in 1991 was carried out by M. Gaetani and F. Jadoul. E. Erba studied the nannofossils and E. Garzanti, who also did field-work in the 1986 expedition, the sandstone petrography. The manuscript has benefitted from careful comments by C. Pudsey and A. Zanchi. Special permission given by the Government of Pakistan to visit restricted areas is acknowledged.

References

BOULIN, J. 1981. Afghanistan structure, Greater India concept and eastern Tethys evolution. *Tectonophysics*, **72**, 261–287.

—— 1988. Hercynian and Eocimmerian events in Afghanistan and adjoining regions. *Tectonophysics*, **148**, 253–278.

—— 1990. Neocimmerian events in Central and Western Afghanistan. *Tectonophysics,* **175**, 285–315.

CASNEDI, R. 1980. Stratigraphical outline of the area between Karakorum and Hindu Kush with probable occurrence of a Hercynian geosynclinal stage. *Memorie della Societa' Geologica Italiana,* **20** (1979), 277–287.

—— & NICORA, A. 1985. Short notes on the Shimshal Valley geology (Western Karakorum, Pakistan). *Rivista Italiana di Paleontologia e Stratigrafia*, **90** (1984), 735–742.

COWARD, M.P., WINDLEY, B.F., BROUGHTON, R.D., LUFF, I.W., PETTERSON, M.G., PUDSEY, C.J., REX, D.C. & ASIF KHAN 1986. Collision tectonics in the NW Himalayas. *In*: COWARD, M.P. & RIES, A.C. (eds) *Collision Tectonics*. Special Publication Geological Society, London, **19**, 219–219.

CRAWFORD, M.B. & SEARLE, M.P. 1992. Field relationships and geochemistry of pre-collisional (India–Asia) granitoid magmatism in the Central Karakoram, northern Pakistan. *Tectonophysics,* **206**, 171–192.

DAINELLI, G. 1933. La Serie dei terreni. *In*: *Relazioni Scientifiche della Spedizione Italiana de Filippi nell'Himalaia, Caracorum e Turchestan Cinese (1913–1914)*, **II, 2**, Zanichelli, Bologna, 458–542.

DEBON, F. 1992. Geological setting of Eocene "Batura Granites" (Western Karakorum axial batholith, Northern Pakistan) (Abstract). *Himalayan Workshop*, Oxford.

——, LE FORT, P., DAUTEL, D., SONET, J., ZIMMERMANN, J.L. 1987. Granites of Western Karakorum and northern Kohistan (Pakistan): A composite Mid-Cretaceous to Upper Cenozoic magmatism. *Lithos,* **20**, 19–40.

DESIO, A. 1980. Geology of the Shaksgam Valley. *Scientific Reports of the Italian Geological Expeditions to Karakorum and Hindu Kush.* **III, 4**, 196 pp., Brill, Leiden.

DESIO, A. & MARTINA, E. 1972. Geology of the Upper Hunza Valley, Karakorum, West Pakistan. *Bollettino della Societa' Geologica Italiana,* **91**, 283–314.

DE TERRA, H. 1932. *Geologische Forschungen im Westlichen K'un-Lun und Karakorum-Himalaya.* Reimer & Vohsen, Berlin.

DICKINSON, W.R. 1970. Interpreting detrital modes of greywacke and arkose. *Journal of Sedimentary Petrology,* **40**, 973–985.

—— 1985. Interpreting provenance relations from detrital modes of sandstones. *In*: ZUFFA, G.G. (ed.) *Provenance of arenites* NATO ASI Series, **148**, Dordecht, Reidel, 333–361.

DICKINSON, W.R. & SUCZEK, C.A. 1979. Plate tectonics and sandstone composition. *American Association Petroleum Geologists, Bulletin,* **63**, 2164–2172.

DRONOV, V.I. & LEVEN, E.YA. 1976. *Stratigraphy and intrusive rocks of Tadjikistan.* Nedra, Dushanbe. (In Russian).

—— & —— 1990. Permian foraminiferal assemblages in the olistholithes of SE Pamir. *Doklady Akademii Nauk SSSR,* **311**, 694–697. (In Russian).

——, KARAPETOV, S.S. & LEVEN, E.YA. 1959. On the age of the coals in the Eastern Pamir. *Doklady Akademii Nauk SSSR,* **127**, 690–691 (English translation, 1960).

——, GADZDIZCKI, A. & MELNIKOVA, G.K. 1982. Die triadische Riffe in südöstlicher Pamir. *Facies,* **6**, 107–128.

GAETANI, M., GARZANTI, E., JADOUL, F., NICORA, A., TINTORI, A., PASINI, M. & KANWAR, S.A.K. 1990*a*. The North Karakorum side of the Central Asia geopuzzle. *Geological Society of America, Bulletin,* **102**, 54–62.

——, GOSSO, G. & POGNANTE, U. 1990*b*. A geological transect from Kun Lun to Karakorum (Sinkiang China): the western termination of the Tibetan Plateau. Preliminary note. *Terra Nova,* **2**, 23–30.

——, —— & —— 1991. Geological report. *In*: DESIO, A. *et al.* (eds) *Geodesy, Geophysics and Geology of the upper Shaksgam Valley (North-East Karakorum) and South Sinkiang*. Consiglio Nazionale delle Ricerche, Milano, 99–190.

GERGAN, J.T. & PANT, P.C. 1983. Geology and stratigraphy of Eastern Karakoram, Ladakh. *In*: THAKUR, V.C. & SHARMA, H. (eds) *Geology of Indus Suture Zone of Ladakh*. 99–106.

HAYDEN, H.H. 1915. Notes on the geology of Chitral, Gilgit and the Pamirs. *Records of the Geological Survey of India,* **45**, 271–320.

KAFARSKI, A.KH. & ABDULLAH, J. 1976. Tectonics of North-east Afghanistan (Badakshan, Wakhan, Nurestan) and relationship with the adjacent territories. *Atti Convegni Lincei,* **21**, 87–113.

LE FORT, P., MICHARD, A., SONET, J. & ZIMMERMANN, J.L. 1983. Petrography, geochemistry and geochronology of some samples from the Karakorum axial batholith (N. Pakistan). *In*: SHAMS, F.A. (ed.) *Granites of Himalayas, Karakorum and Hindu Kush*. Punjab University, Lahore, 377–387.

MATTE, P., TAPPONNIER, P., BOURJOT, L., PAN, Y. & WANG, Y. 1990. A section through the Paleozoic Kunlun Belt, north of the Altyn Tagh Fault (Western Tibet). *Colloque Kunlun-Karakorum 90 Paris,* 38.

——, ——, ——, —— & —— 1992. Tectonics of Western Tibet from the Kunlun to the Karakorum. Abstract. *Himalayan Workshop*, Oxford.

NORIN, E. 1946. *Geological exploration in Western Tibet*. Reports of the Scientific Expeditions Sven Hedin No. 29, Stockholm.

—— 1976. The "Black Slates" Formations in the Pamirs, Karakoram and western Tibet. *Atti Convegni Lincei,* **21**, 243–264.

PARONA, C.F. 1928. Faune Cretaciche del Caracorum e degli altipiani tibetani. *In*: *Relazioni scientifiche della Spedizione Italiana De Filippi, nell'Himalaya, Caracorum e Turchestan Cinese* (1913–1914). **II, VI**. Zanichelli, Bologna, 101–147.

PETTERSON, M.G. & WINDLEY, B.F. 1985. Rb-Sr dating of the Kohistan arc-batholith in the Trans-Himalaya of N. Pakistan, and tectonic implications. *Earth and Planetary Sciences Letters,* **74**, 45–57.

PUDSEY, C.J., COWARD, M.P., LUFF, I.W., SHACKLETON, R.M., WINDLEY, B.F. & JAN, M.Q. 1985. Collision zone between the Kohistan arc and the Asian plate in NW Pakistan. *Transactions of the Royal Society of Edinburgh: Earth Sciences,* **76**, 463–479.

SEARLE, M.P. 1991. *Geology and Tectonics of the Karakoram Mountains*. J. Wiley, Chichester.

——, PARRISH, R.R., TIRRUL, R. & REX, D.C. 1990. Age of crystallization and cooling of the K2 gneiss in the Baltoro Karakoram. *Journal of the Geological Society, London,* **147**, 603–606.

SENGÖR, C.A.M. 1984. *The Cimmeride orogenic system and the Tectonics of Eurasia.* Geological Society of America, Special Papers, **195**.

SHI, XIAOYING & YANG, ZUNYI 1986. *The Jurassic Brachiopod sequences of the Qinghai-Tibetan Plateau*.

SHIPTON, E. 1938. *Blank on the map*. Diamond books, 1985 reprint. Edinburgh.

SHVOLMAN, V.A. 1973. Cretaceous and Paleogene deposits in the SE Pamir and their position in the present structure. *Moscow Society of Naturalists, Bulletin,* **48**, 41–55 (In Russian).

TALENT, J.A., CONAGHAN, P.J., MAWSON, R., MOLLOY, P.D. & PICKETT, J.W. 1981. Intricacy of tectonics in Chitral (Hindu Kush): faunal evidence and some regional implications. *Himalayan Geology Seminar (1976), section IIA, Geological Survey of India, Miscellaneous Publication,* **41**, 77–101.

TRELOAR, P.J., REX, D.C., GUISE, P.G., COWARD, M.P., SEARLE, M.P., WINDLEY, B.F., PETTERSON, M.G., JAN, M.Q. & LUFF, I.W. 1989. K-Ar and Ar-Ar geochronology of the Himalayan collision in NW Pakistan: constraints on the timing of suturing, deformation, metamorphism and uplift. *Tectonics,* **8**, 881–909.

ZANCHI, A. 1993. Structural evolution of the North Karakorum cover, North Pakistan. This volume.

Collision-related granitoid magmatism and crustal structure of the Hunza Karakoram, North Pakistan

M.B. CRAWFORD[1] & M.P. SEARLE[2]

[1] *Fluid Processes Group, British Geological Survey, Keyworth, Nottingham NG12 5GG, UK*

[2] *Department of Earth Sciences, University of Oxford, Oxford OX1 3PR, UK*

Abstract: The Cretaceous to Miocene Karakoram Axial Batholith runs E–W across the middle Hunza Valley in north Pakistan and separates a dominantly sedimentary terrain to the north from a regional metamorphic terrain to the south. A strong phase of deformation and metamorphism associated with the *c*.65–50 Ma collision of India with Asia affected the area. The southern part of the batholith and the metamorphic rocks to the south are affected by strong south-vergent thrusting which has brought deeper level, hotter rocks over higher level, cooler rocks so that metamorphic grade increases with structural height. In contrast, the northern part of the batholith is relatively undeformed and shows an intrusive contact with an andalusite-bearing thermal aureole.

Two groups of post-collisional magmatic rocks can be recognised in the area. The first group, the Hunza dykes, occurs as a massive dyke complex intruding the early phase of the batholith as well as the sillimanite-bearing gneisses to the south. These dykes consist of a co-genetic suite of granodiorites, monzogranites and leucogranites which appear to be related by feldspar-biotite fractionation. They are volatile-depleted, non-minimum melts with unevolved crustal isotopic ratios and enriched Sr and Ba signatures similar to the Miocene Baltoro pluton further east. They are considered to represent lower-crustal magmas derived by dehydration melting of metasediment, probably promoted by intrusion of mantle-derived magmas.

The second group is represented by the Sumayar pluton which intrudes the staurolite-grade metamorphic rocks and is a homogeneous, water-saturated minimum-melt leucogranite, with a major element composition similar to that of the Manaslu leucogranite of central Nepal. Abundant tourmaline and evolved Sr isotope ratios indicate a metapelitic source. Analogy with Manaslu suggests that this source was either fluxed by fluids to promote melting or produced melt under vapour-absent conditions via muscovite breakdown. In the case of the latter, low melt fractions require that an extensive area was tapped to generate a pluton the size of Sumayar. Consideration of the petrogeneses of the two groups of magmas allow constraints to be placed on the deeper crustal structure of the Hunza Valley.

Intruded along the length of the High Himalaya is a belt of at least 12 Miocene leucogranitic plutons, together with numerous smaller plutons, sheets and injection dyke-sill complexes. The plutons outcrop towards the top of the Tibetan Slab above the Main Central Thrust (MCT) and some intrude up into the overlying sediments of the north Indian Plate margin (e.g. Gansser 1964; Dietrich & Gansser 1981; LeFort *et al.* 1987; Searle *et al.* 1987). To the north of these, a second intrusive belt crops out in the cores of a series of granite-gneiss domes (Kangmar – Lhagoi Kangri domes; Burg *et al.* 1984) piercing the sedimentary series of the Indian plate south of the Indus-Tsangpo Suture Zone (ITSZ). All these leucogranites are primary, crustal melts which have been related to either wet melting or vapour-absent dehydration melting of sillimanite-grade rocks during crustal thickening and possible fluid fluxing along large-scale intra-continental thrust zones caused by the collision and continued convergence of India with Asia since *c.* 50 Ma (e.g. Le Fort 1981; Vidal *et al.* 1982; Le Fort *et al.* 1987; Harris & Inger 1992). As such the magmas are intimately linked with the crustal structure of the area.

The ITSZ marks the junction between the former Indian and Asian plates. In north Pakistan the ITSZ bifurcates into the Shyok Suture Zone (SSZ) or Northern Suture and the Main Mantle Thrust (MMT) (Fig. 1 inset; e.g. Desio 1979; Searle *et al.* 1987). Sandwiched between the SSZ and MMT lies the Kohistan island arc terrane. The former Asian plate *sensu strictu* is represented by the Karakoram to the north of the SSZ. The Karakoram Axial Batholith (KAB) runs for some 600 km from east to west across this region, and provides a magmatic

From TRELOAR, P.J. & SEARLE, M.P. (eds), *Himalayan Tectonics*
Geological Society Special Publication, No. 74, pp. 53–68.

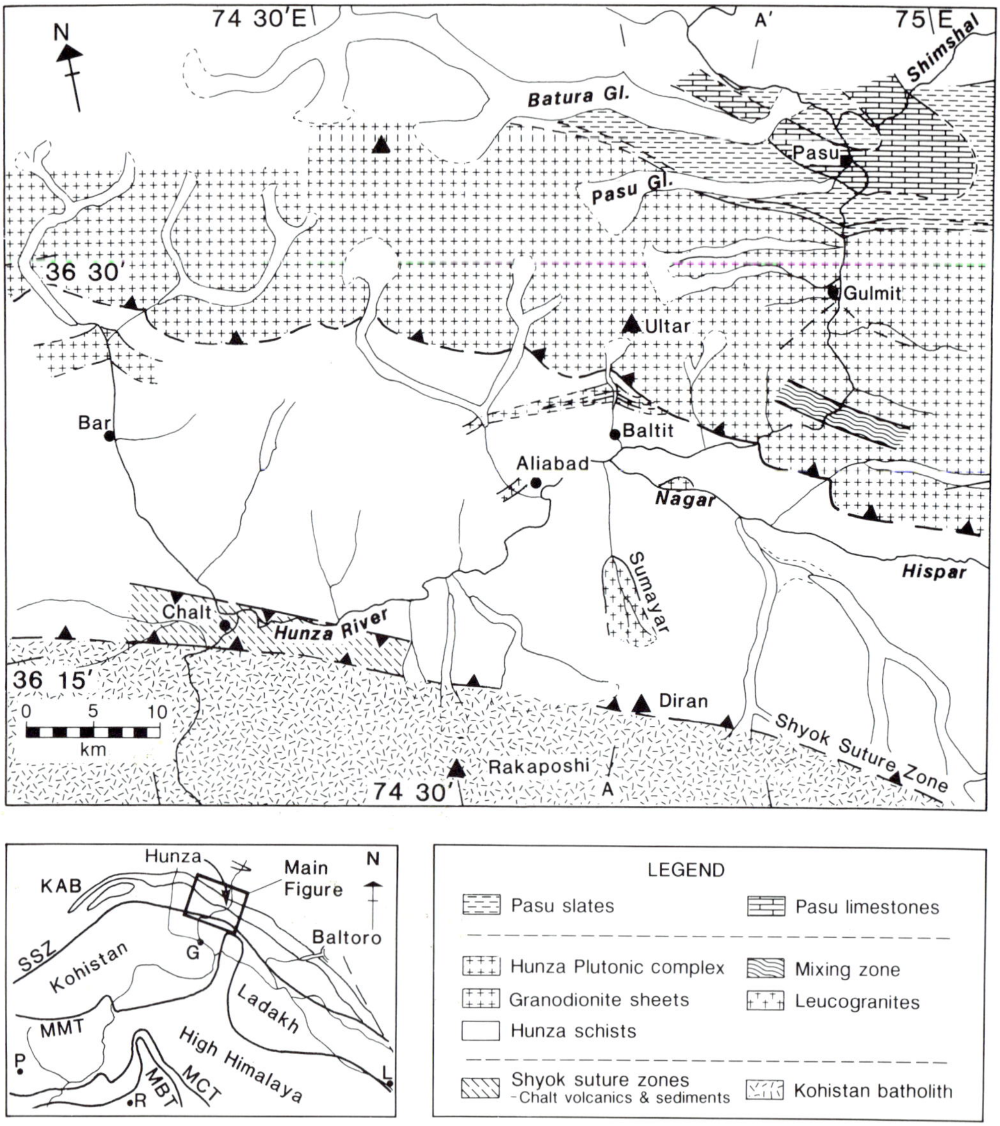

Fig. 1. Geological map of the middle Hunza Valley, north Pakistan. The unornamented area between the Hunza plutonic complex and the Shyok Suture Zone is the Karakoram metamorphic complex which increases in grade northwards. Line A–A′ shows the section taken in Fig. 2. Inset: KAB, Karakoram Axial Batholith; SSZ, Shyok Suture Zone; MMT, Main Mantle Thrust; MCT, Main Central Thrust; MBT, Main Boundary Thrust; P, Peshawar; R, Rawalpindi; G, Gilgit; L, Leh.

history of the area from the Cretaceous to the present (Debon *et al.* 1987; Crawford & Searle 1992). Intruded during the Miocene, at around the same time as the High Himalayan leucogranites, is the granitic Baltoro plutonic unit (BPU) (Fig. 1 inset; Desio & Zanettin 1970; Searle *et al.* 1986, 1992; Rex *et al.* 1988). Thermal re-equilibration and displacement of the former subduction-related mantle wedge downwards, through thickening of the crust above during convergence, led to contemporaneous upper mantle and lower crustal melting, and the Baltoro leucogranites have been linked to lower crustal melting promoted by intrusion of mantle material (Searle *et al.* 1992). Intrusion position and location of the crustal melting are again linked to the geometry of the collision zone.

The High Himalayan and BPU leucogranites represent the end products of two different petrogenetic mechanisms, both of which have been related to the collision of the Indian and Asian plates (Crawford & Windley 1990). The type localities of these two granite types lie on opposite sides of the Indus Tsangpo Suture Zone. However, this paper describes the occurrence of both types of granite in the Hunza Valley of north Pakistan which lies to the north of the ITSZ on the southern margin of the Asian plate (Fig. 1). The field relations, mineralogy and chemistry of the granites of Hunza can be used to relate these intrusions to the High Himalayan or BPU granites and, consequently, to infer their petrogenesis and relationship to the crustal structure of the region.

Field setting and age relationships

The middle Hunza Valley lies to the north of the SSZ and the rocks considered here all formed part of the southern margin of Asia prior to collision of the Kohistan island arc between 102 and 87 Ma (Treloar *et al.* 1989). The geology of the area has been detailed elsewhere (e.g. Desio & Martina 1972; Crawford 1988; Searle 1991) and only the salient points are described below. Low-grade metapelites and limestones in the north are separated from sillimanite-grade metamorphic rocks to the south by a large calc-alkaline quartz diorite–granodiorite pluton (first phase of the Hunza plutonic complex (HPC) within the KAB; Crawford & Searle 1992) which was intruded approximately 95 Ma ago (U-Pb zircon date; Le Fort *et al.* 1983). The southern part of the pluton and pelitic, calcareous and basic volcanic country rocks as far south as the SSZ were affected by southward thrusting related to the continental collision and continued convergence of India and Asia since *c.* 50 Ma (e.g. Searle *et al.* 1987). Deformation was accompanied by metamorphism, although not necessarily synchronously (Powell & Vernon 1979; Prior 1987), and thrusting brought deeper-level rocks southwards over cooler, higher-level rocks so that metamorphic grade increases with structural height. Despite some later refolding, the structure of the valley is still dominated by this event. The whole of this southern area from the pluton around Gulmit south to the SSZ (Figs 1 and 2) shows a foliation dipping north at various angles (generally >45° and often near vertical) and striking between E–W and 110°. Bedding is sub-parallel to the foliation and metamorphic grade increases northwards from greenschist-grade phyllites near the SSZ through staurolite-grade metapelites to sillimanite-grade metapelites and calc-silicates containing forsterite and diopside at the contact with the HPC. The main changes in metamorphic grade occur across thrust zones. The main phase of deformation ceased by 37 Ma (Searle *et al.* 1989).

A series of post-collisional magmatic phases intrude the 95 Ma granodiorite forming the Hunza plutonic complex. To the east, within the Karakoram Axial Batholith, the Baltoro plutonic unit (BPU) was also intruded after collision at 21 ± 0.5 Ma (U-Pb zircon; Parrish & Tirrul 1989), and the Baltoro and Hunza areas show similar field relationships (see Searle *et al.* 1989, 1992). The localization of magmatism to a narrow E–W band across the Karakoram since the Cretaceous is attributed to an inferred major crustal structure, the Karakoram Batholith Lineament (KBL; Rex *et al.* 1988). This is interpreted as a major, N-dipping, low-angle, normal fault caused by culmination collapse during crustal thickening of the rocks to the

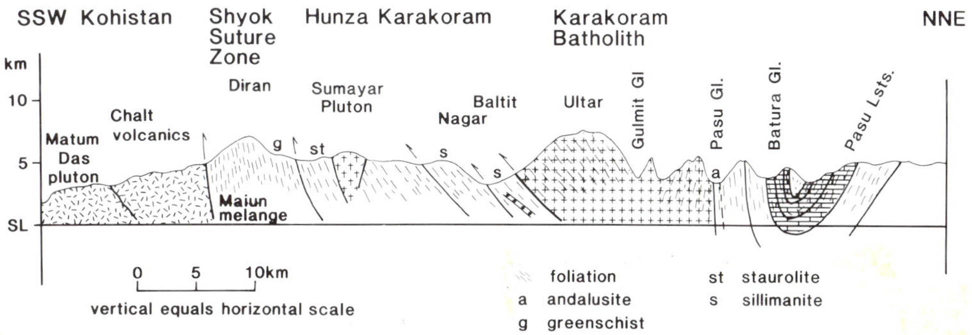

Fig. 2. Sketch section through the middle Hunza Valley along the line A–A′ shown in Fig. 1. Legend is as for Fig. 1.

south. As such it forms the junction between high-grade rocks to the south and lower-grade rocks to the north, and has a natural analogue located on the Indian plate to the south and east in Zanskar. There, the Zanskar Shear Zone bounds the northeastern margin of the High Himalaya and separates Tethyan sediments to the northeast from Tibetan Slab gneisses to the southwest (Searle 1986; Herren 1987).

On the basis of mineralogy, two post-collisional magmatic groups, the Hunza dykes and the Sumayar pluton, can be distinguished. The Hunza dykes usually form a dense, anastomosing network intruding the deformed part of the HPC and the sillimanite-grade metamorphic rocks to the south. The dykes are on a scale of tens of centimetres to a few metres wide, and constitute up to 40% of the rock by volume.

(b)

(a)

Fig. 3. (**a**) Cliff section in the Hunza plutonic complex 2 km south of Gulmit showing the Hunza dyke swarm intruding an earlier granodiorite phase of the complex. This section is close to the proposed position of the KBL, and the dykes are intruded as gently-dipping parallel sheets as opposed to the anastomosing networks seen further to the south. (**b**) Cliff section in the Sumayar valley just south of the Sumayar Glacier snout. The roof of the Sumayar pluton (light-coloured rock) can be seen below the ridge which is composed of staurolite-grade metasediments (dark rock). The contact dips into the cliff and the pluton has only been exposed due to the incision of the glacial valley. Vertical height is around 500 m.

Leucogranites are associated, and often layered, with aplitic granodiorites and monzogranites as is also seen in the Baltoro area. The dykes have sharp, intrusive contacts, and are often layered with pegmatitic margins and aplitic centres. Where the pegmatites are well developed they show a mineral zonation with biotite-rich margins and garnet-muscovite-rich centres. There is more than one intrusive phase as later dykes cross-cut earlier ones but only a short time span is envisaged between each phase. Two intrusions, the Nagar and Aliabad sheets (Fig. 1) are on a larger scale than the dykes, being over 100 m thick, and consist of homogeneous garnet–muscovite leucogranites. The southern limit of the dykes is approximately defined by the thrust marking the southern limit of sillimanite-grade rocks. Their northern limit coincides with the deformation front in the HPC, and near to this boundary the nature of the dykes changes from being anastamosing to form subparallel, gently dipping sheets (Fig. 3a). No accurate zircon age is available for these granites and K-Ar ages (WR, Mu and Bi) give variable results indicating a prolonged thermal history (2.7 to 16.9 Ma; Debon *et al.* 1987). However, they are interpreted to be mid-Miocene in age (Debon *et al.* 1987) and their range of ages correspond to those given by the Baltoro pluton to the east (Searle *et al.* 1989).

The Sumayar pluton consists of a homogeneous leucogranite, and differs from all the other granitoids in that it is intruded outside the sillimanite-grade rocks into the staurolite-grade zone (Figs 1 and 3b). It forms a large, homogeneous body about 4 km across, rather than a dyke network, and has sharp, intrusive contacts and a small contact aureole of baked slates about 50 m wide. Its structural position is very similar to other minor plutons intruding the Karakoram metamorphic complex south of the Baltoro Glacier, notably the Mango Gusar (37.0 ± 0.8 Ma, U-Pb zircon; R. Parrish *in* Searle *et al.* 1989) and Chingkiang plutons (36–34 Ma, K-Ar hornblende; Searle *et al.* 1989). The roof zone appears to have large rafts of country rock enclosed by granite. These were not observed at close hand due to inaccessibility, but this feature is similar to that seen in the roof zone of the Bhagirathi leucogranite in the Garhwal Himalaya (Searle *et al.* this volume). Otherwise there is no xenolithic material or evidence of assimilation. No age data are available for the Sumayar pluton and it can only be stated that it must be post *c.* 37 Ma because it cuts syn-metamorphic fabrics associated with the main M2 metamorphic event in the Baltoro area dated by Searle *et al.* (1989).

Mineralogy

All the granites are usually granular in habit and are dominated by quartz, alkali feldspar and plagioclase (oligoclase to andesine). The euhedral shape and non-oscillatory zonation of the crystals suggests that they crystallized from a melt and are not restitic in origin. The main differences between the two groups are in the type and amount of accessory phases present.

Within the Hunza dykes there is a range of compositions from biotite-bearing aplitic granodiorites and monzogranites through two-mica granites to pegmatitic, garnet–muscovite leucogranites. Biotite has an interstitial habit indicating that it was a late crystallizing phase which, in turn, indicates a low water content in the magma (Wyllie *et al.* 1976). Apatite is a common accessory in all rock types but sphene and opaque phases are confined to the monzogranites and granodiorites. A little primary tourmaline was observed in some of the larger pegmatites but, in general, it is absent and most observed tourmaline is a secondary deposit on fracture surfaces. This is in strong contrast to the leucogranite dykes in the High Himalaya (Le Fort *et al.* 1987; Searle & Fryer 1986), but is comparable to the leucogranites of the BPU (Searle *et al.* 1992).

The Sumayar pluton contains >2% tourmaline occuring as evenly distributed 'splotchy' growths with inclusions of quartz. Tourmaline appears to be a late, but primary phase and this distinguishes Sumayar from the Hunza dykes. Garnet is not common except near the margins of the pluton. A little biotite is included within large muscovite laths and appears to be an early phase suggesting a higher magmatic water content than in the Hunza dykes.

Garnet rim analyses from leucogranite compositions in the Hunza dykes (Table 1) are almandines with a high spessartine component. Garnets from the sillimanite-grade country rocks (Table 1) have much lower spessartine contents suggesting that the leucogranite garnets are magmatic and not xenocrystic in origin. Miller & Stoddard (1981) showed that garnets in silicic liquid precipitate in response to high Mn relative to Mg and Fe and therefore usually contain >10% spessartine as is observed here. The garnet is precipitated at the expense of biotite which explains the rare occurrence of biotite in the garnet–muscovite leucogranites.

The core composition of garnet from the Sumayar pluton is similar to the garnet rim analyses from the Hunza dykes (Table 1). However, towards the rim MnO shows a sharp increase whereas FeO shows an antithetic

Table 1. *Average garnet rim (except for Sumayar pluton core analysis) compositions from some Hunza leucogranites and sillimanite-grade country rocks*

Sample (no. of analyses)	Almandine	Pyrope	Grossular	Spessartine
Sumayar Pluton rim	31.2%	0.3%	3.2%	65.2%
Sumayar Pluton core	61.0%	0.6%	2.4%	36.0%
Hunza leucogranite dyke (n = 8)	64.6%	2.1%	4.8%	28.5%
Nagar leucogranite sheet (n = 4)	76.4%	2.6%	5.2%	15.8%
Hunza leucogranite dyke (n = 6)	59.1%	4.8%	7.2%	28.9%
Metapelite (n = 11)	68.8%	20.5%	4.9%	5.6%
Metapelite (n = 6)	78.9%	11.6%	8.7%	0.7%
Metabasite (n = 4)	51.8%	9.5%	33.9%	4.8%

Analyses were carried out at the University of Leicester on a JEOL JXA-8600S Superprobe using an operating voltage of 15 kV, probe current 3×10^{-8}A and a beam diameter of 20 μm. Count times of 10 seconds on peaks with a background count of 5 seconds were repeated twice for each element.

decrease. Green (1977) demonstrated that the amount of MnO (or spessartine) in garnet strongly affects the stability of the phase with regard to pressure. Garnets with <10 mol% spessartine indicate depths of >18 km whereas those with >25 mol% can be in equilibrium with a magma as shallow as 12 km or 3 kbar. The Sumayar garnet rim compositions are consistent with the observation of Pichavant & Manning (1984) that tourmaline will not crystallize above pressures of 3 kbar, and this suggests that the Sumayar pluton must have been emplaced at a shallower level than the Hunza dykes. As the pluton rose and crystallised, the garnets would have had to sequest more Mn from the magma to remain in equilibrium.

Chemistry: general trends and observations

(a) The Hunza dykes

Representative analyses of the Hunza dykes and the Sumayar average are given in Table 2. The Hunza dykes cluster in the monzogranite or adamellite field on a Debon & Le Fort (1982) classification diagram, but trend from granodiorites through to granites. On a quartz–albite–orthoclase phase diagram, the bulk of the data plot in a similar position to the Baltoro granites (Crawford 1988) close to the haplogranitic eutectic but trending away from it indicating either that they are non-minimum melts or that melting was not in equilibrium with the protolith.

The more basic monzogranite and granodiorite samples of the Hunza dykes have very similar compositions to the Baltoro monzogranites (Fig. 4 and Table 2; Crawford & Windley 1990), with high large ion lithophile (LIL) to high field strength (HFS) element ratios and negative Nb, P and Ti anomalies. Going from the monzogranites to the leucogranites (i.e. a general increase in silica) Fe_2O_3, Al_2O_3, CaO, MgO, TiO_2, Zr, Th, Ba, Sr and the light rare earth elements (LREE) all decrease, whereas Rb, K_2O and the heavy rare earth elements (HREE) generally increase. However, all the trends are scattered. Such scatter is inherent in granitic liquids since, during crystallization, the magma is viscous and small quantities of intercumulus liquid are trapped in each sample (e.g. Tindle & Pearce 1981). Similar chemical trends are seen in the BPU where the scatter is less, and they are explained by fractionation of biotite, plagioclase, K-feldspar and accessory phases including monazite (Searle *et al.* 1992). Modelling trace element evolution for the Hunza dykes using major phases gives a fractionating assemblage of 10–50% biotite, 26–60% K-feldspar and 24–30% plagioclase, with the exact assemblage being strongly dependent on which distribution coefficients are used (e.g. Hanson 1978; Henderson 1982). The trends cannot, however, be related to partial melting since the large increase in Ba and Sr from leucogranite to monzogranite would be precluded by residual feldspar in a crustal source.

The decrease of P, Ti and Zr suggest that the observed apatite, sphene and zircon are also important crystallising accessory phases. The precipitation of such phases is dependent on the bulk chemistry of the melt (e.g. Watson & Harrison 1984) and Watson & Harrison (1983) showed that the Zr saturation level is dependent on temperature and the cation ratio M (= (Na + K + 2Ca/(Si × Al)) of the melt. Taking the highest Zr abundance in the Hunza dykes (Zr = 273 ppm; M = 1.5), and assuming

Table 2. *Representative analyses of the Hunza dykes, Sumayar pluton and Nagar leucogranite sheet*

	Baltoro Mzg Av. (n = 20)	Hunza Mzg Av. (n = 18)	Hunza SLV14 Mzg Ap.	Hunza SLV20 Lcg Pg.	Hunza B62 Mzg Ap.	Hunza B67 Lcg Pg.	Nagar Lcg Av. (n = 5)	Sumayar Lcg Av. (n = 5)
SiO_2	71.43	71.28	70.7	74.6	70.9	74.6	73.5	74.09
TiO_2	0.26	0.27	0.34	0.02	0.31	0.04	0.01	0.05
Al_2O_3	14.61	14.72	15.1	14.9	14.7	13.8	14.3	14.83
Fe_2O_3t	1.59	1.68	1.8	0.5	1.8	0.5	0.86	0.72
MnO	0.03	0.03	0.02	0.06	0.02	0.04	0.11	0.03
MgO	0.34	0.37	0.5	0.0	0.4	0.0	0.0	0.01
CaO	1.74	1.99	2.0	0.9	2.2	1.4	1.1	0.79
Na_2O	3.96	3.93	3.9	4.6	4.5	4.2	3.9	4.37
K_2O	4.25	4.39	4.58	4.59	3.24	4.31	4.40	4.54
P_2O_5	0.11	0.12	0.18	0.04	0.09	0.0	0.08	0.08
LOI	0.38	0.37	0.68	0.43	0.52	0.24	0.39	0.51
Total	98.7	99.15	99.73	100.63	98.80	99.11	98.63	100.02
Trace elements in ppm								
Rb	157.0	149.3	172.3	168.4	83.5	144.7	139.3	429.9
Ba	1038.0	1018.5	1175.1	177.8	1444.2	471.4	128.3	68.0
Th	21.1	13.6	16.2	2.7	7.2	4.8	6.3	10.7
Pb	44.2	37.4	33.0	82.0	na	na	43.9	56.1
Nb	10.5	8.7	4.6	5.6	4.3	6.3	18.1	18.2
Sr	514.0	337.3	385.3	85.9	425.9	187.0	101.4	32.6
Zr	135.1	155.1	189.6	17.0	198.1	40.5	22.0	32.5
Y	9.0	12.2	7.9	16.8	6.3	9.5	19.8	22.5
V	18.6	21.4	31.7	0.0	22.9	2.0	0.0	0.0
Cr	12.3	7.0	14.0	9.3	7.1	0.4	7.2	3.5
Ni	3.4	1.7	0.3	0.0	1.3	0.0	0.8	1.1
Zn	39.7	38.4	47.5	5.4	49.0	11.7	22.8	37.5
Ga	18.2	19.9	19.3	15.8	20.2	19.5	18.9	25.1
Rare earth elements in ppm								
	(n = 8)	(n = 1)					(n = 1)	(n = 2)
La	51.74	55.90	45.1	2.8	56.73	12.3	12.4	12.02
Ce	102.20	100.40	76.5	5.63	91.53	20.1	17.45	19.9
Pr	na	9.83	7.58	0.87	9.24	2.14	2.02	na
Nd	39.26	32.50	24.8	2.35	21.08	6.6	7.75	8.49
Sm	5.78	4.81	2.96	0.62	3.67	1.04	2.06	2.43
Eu	1.07	0.81	0.74	0.39	0.9	0.4	0.19	0.17
Gd	na	3.38	2.09	1.14	na	1.5	2.79	1.95
Dy	na	1.96	1.05	1.84	0.77	1.14	2.43	1.7
Er	na	0.82	0.6	1.42	1.21	1.0	0.75	na
Yb	0.60	0.46	0.22	1.16	0.07	1.0	0.7	0.58
Lu	0.09	0.08	0.07	0.22	0.01	0.18	0.11	0.08

Compositions are an average where n (number of samples) is given. Baltoro monzogranite average from Crawford & Windley (1990). Analyses were performed at the University of Leicester on a Philips PW1400 X-ray fluorescence spectrometer and analytical details are similar to those described by Tarney *et al.* (1979) and Marsh *et al.* (1980). The rare earths were determined by ICP techniques (see Walsh *et al.* 1981) at Royal Holloway and Bedford New College. Fe_2O_3t is total iron expressed in terms of Fe^{3+}; na, not analysed; Mzg, monzogranite; Lcg, leucogranite; Ap, aplite; Pg, pegmatite; Av, average.

all Zr was dissolved in the melt and not contained in xenocrystic zircon, gives a temperature of about 825° C.

The REE patterns of the Hunza monzogranites are very similar to those of the Baltoro with steep LREE-enriched patterns and small, negative Eu anomalies (Fig. 5). The leucogranites have lower LREE and higher HREE abundances which can be attributed to a small amount of either monazite or allanite crystallization and the presence of cumulus garnet which sequesters HREE. The size of the Eu anomaly in the leucogranites can be attributed to the relative importance of allanite or monazite since

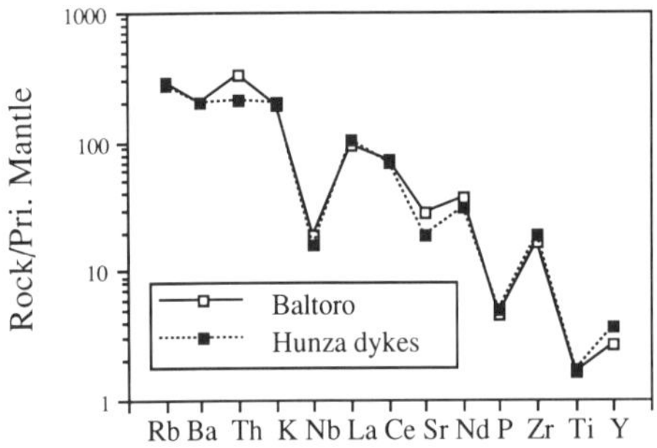

Fig. 4. Multi-element variation plots normalized to primitive mantle to compare the chemistry of the Hunza dyke monzogranite and Baltoro pluton monzogranite averages (Table 2). Primitive mantle from Taylor & McLennan (1985).

monazite will reduce the size of the anomaly as it does not remove Eu, whereas allanite will increase it (Henderson 1984).

The isotopic signature of the Hunza dykes is very similar to the Baltoro pluton (Fig. 6 and Table 3) with very low Sr isotope ratios compared to the High Himalayan leucogranites. This, combined with the other similarities in their chemistry, mineralogy, and temporal and spatial position suggests that the magmas in the two areas have a common origin.

(b) The Sumayar pluton

The difference in isotopic signature (Table 3 and Fig. 6), mineralogy and intrusion position all suggest that the Sumayar pluton has a different petrogenesis to the Hunza dykes. The high boron content of the magma, reflected by the abundant tourmaline, and the high-Sr–low-Nd isotopic ratios (Fig. 6) indicate a mature metapelitic source since pelites may contain up to 2000 ppm B_2O_3 due to the substitution of B for Al in clay minerals (Benard *et al.* 1985).

The trend of the Sumayar samples on a Debon

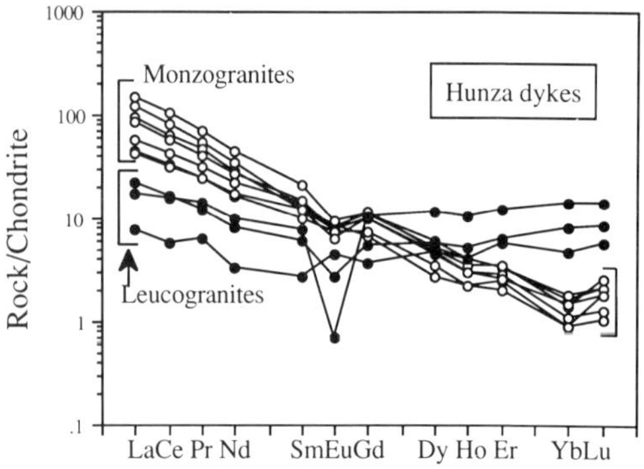

Fig. 5. Chondrite-normalized rare earth element variation diagram for the Hunza dykes. The monzogranites show consistent LREE-enriched patterns similar to the Baltoro monzogranites (see Searle *et al.* 1992) whereas the leucogranites show lower LREE and higher HREE abundances. Chondrite from Taylor & McLennan (1981).

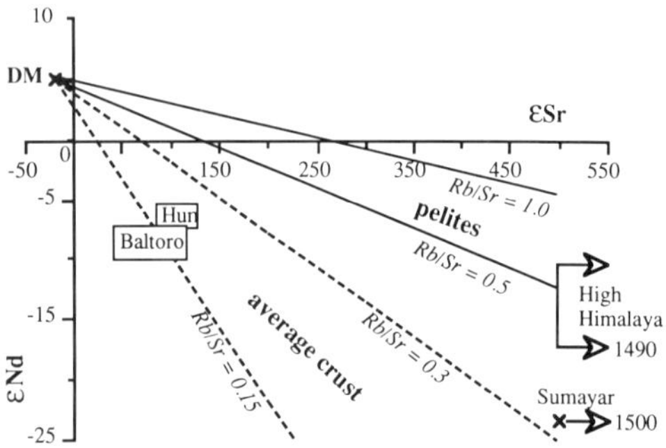

Fig. 6. Epsilon isotope diagram for the Hunza post-collisional granites. Data from the Baltoro (Searle *et al.* 1992) and High Himalaya (Allègre & Othman 1980) are shown for comparison. The evolution lines for different Rb/Sr ratios are taken from Miller (1985). ε values are calculated relative to bulk earth using the following: $(^{87}Sr/^{86}Sr)_{UR} = 0.7045$; $(^{87}Rb/^{86}Sr)_{UR} = 0.0839$ (DePaolo 1979); $(^{143}Nd/^{144}Nd)_{CHUR} = 0.51264$ (Hawkesworth & Norry 1983); $(^{147}Sm/^{144}Nd)_{CHUR} = 0.1967$ (Jacobssen & Wasserburg 1980). Hun, Hunza dykes.

Table 3. *$^{87}Sr/^{86}Sr$ and $^{143}Nd/^{144}Nd$ isotopic ratios for some Hunza Valley leucogranites*

Sample	Locality	Rb (ppm)	Sr (ppm)	$^{87}Rb/^{86}Sr$	$^{87}Sr/^{86}Sr$	Sm (ppm)	Nd (ppm)	$^{147}Sm/^{144}Nd$	$^{143}Nd/^{144}Nd$
B120	HPC dyke	221.3	296.3	2.1616	0.711242 ± 10	1.77	6.00	0.17833	0.512278 ± 30
B121	HPC dyke	225.7	43.5	15.02	0.713361 ± 10	–	–	–	–
SLV20	Sillimanite Zone	168.4	85.9	5.6801	0.722537 ± 10	–	–	–	–
SLV33	Sillimanite Zone	202.1	198.5	2.9474	0.713772 ± 10	3.40	19.40	0.10584	0.512258 ± 18
NSG1	Nagar Sheet	131.8	95.1	4.0123	0.714359 ± 10	1.79	6.45	0.16777	0.512378 ± 22
SG2	Sumayar Pluton	383.2	41.4	27.07	0.818818 ± 10	2.65	10.50	0.15254	0.511461 ± 24

Analyses were performed at the University of Leeds using an automated VG54E mass spectrometer for Sr and a manual Micromass 30 spectrometer for Nd. Sr was separated using standard cation exchange techniques and Nd was isolated using a cation column followed by a Teflon anion column. Sr isotope ratios are normalized to $^{86}Sr/^{88}Sr = 0.1194$ and Nd ratios to $^{144}Nd/^{146}Nd = 0.7219$. $^{87}Rb/^{86}Sr$ ratios are calculated from Rb and Sr analysed by XRF using $^{87}Rb/^{86}Sr = (2.6929 + (0.283*(^{87}Sr/^{86}Sr))) \times Rb/Sr$. $^{147}Sm/^{144}Nd$ ratios are calculated from Sm and Nd analysed by ICP (see Table 2) using $^{147}Sm/^{144}Nd = (0.531497 + (0.142521 \times (^{143}Nd/^{144}Nd))) \times Sm/Nd$.

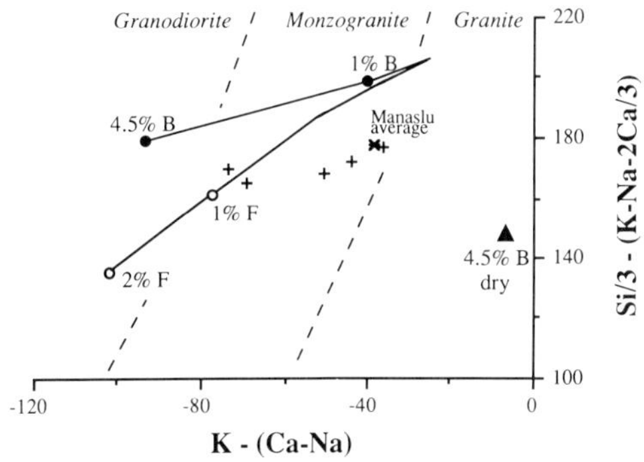

Fig. 7. Debon & Le Fort (1982) *Q–P* classification diagram for the Sumayar pluton (crosses) where Q = Si/3 − (K + Na + 2Ca/3) and P = K − (Na + Ca) calculated from molecular proportions. Closed circles show the position of the granitic eutectic at 1 kbar with 0, 1, and 4.5% boron present, and open circles show the position with 1 and 2% fluorine added. The triangle shows the eutectic under dry conditions with 4.5% boron, and the star represents the average composition of the Manaslu leucogranite (High Himalaya). After Le Fort *et al.* (1987).

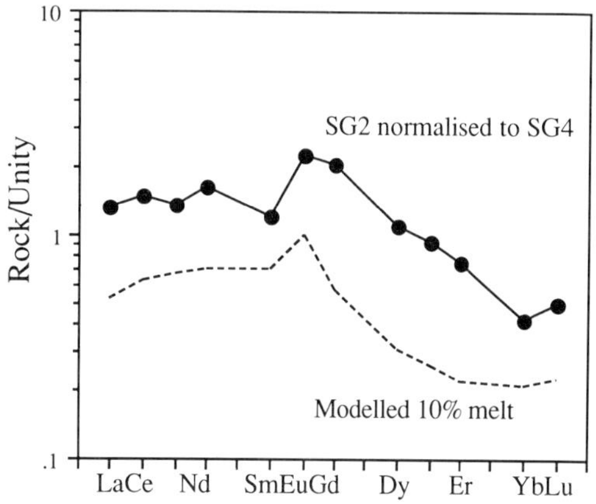

Fig. 8. Modelled REE pattern produced by partial melting of a metapelite, compared with the pattern of a sample from the Sumayar pluton normalised to a less-evolved sample from the same pluton. Metapelite source mineralogy was taken as 40% biotite, 30% plagioclase, 9.8% garnet, 10% quartz, 10% K-feldspar and 0.2% monazite. Melting was in the proportion 5% biotite, 30% plagioclase, 30% K-feldspar and 35% quartz. Distribution coefficients were taken from Henderson (1984) and estimated for monazite from data given in the same source.

& Le Fort (1982) granitoid classification diagram is shown on Fig. 7. Also shown is the position of the haplogranitic eutectic under various conditions and the minimum melt Manaslu leucogranite from the High Himalaya (after Le Fort *et al.* 1987). The Sumayar pluton defines a similar trend to the eutectic migration with increasing B content, and plots in the same position as the Manaslu pluton. Although the exact position of the eutectic would have been dependent on the pressure, water and boron contents of the system, small degrees of partial melting giving a eutectic composition would account for the homogeneity of the major element analyses from this pluton. Pichavant (1987) states that addition of boron will not shift the position of the eutectic significantly to alter the major element composition of a minimum melt drastically and the average analyses from Sumayar and Manaslu (Le Fort *et al.* 1987) show a remarkable similarity within error (Crawford & Windley 1990).

Trace element analyses from Sumayar show no consistent variation with silica, compatible with a minimum melt, but do show linear relationships between each other. The straight line relationship between Ba and Sr can be related directly to the amount of K-feldspar in the sample, and similar arguments relating to phase abundance can be made for other trace elements. The REE patterns are LREE-enriched with a large, negative Eu anomaly and are consistent with control during partial melting by an accessory phase in the protolith such as monazite (Fig. 8).

The lack of zircon, and in particular xenocrystic zircon, in the Sumayar pluton allow the equation of Watson & Harrison (1983) relating Zr abundance to temperature to be applied. The highest Zr abundance measured (40.4 ppm; M = 1.32) gives a magmatic temperature estimate of about 680° C. This is considerably lower than the estimate for the Hunza dykes and, considering the low pressures of emplacement predicted above, places the magma close to the granite wet melting curve.

Source and petrogenesis

(a) The Hunza dykes

We suggest that the Hunza dykes and BPU probably have a common origin. The isotopic signature and silicic nature of the magmas suggest a crustal origin. The high magmatic temperature estimates and volatile-depleted nature do not suggest wet melting and the size of the Baltoro pluton (Searle *et al.* 1989, 1992) indicates a reasonable degree of mobility for the magma. In order for a viscous, granitic magma to separate from its source a critical melt fraction of around 30%, depending on local conditions, must be exceeded (Wickham 1987). Temperatures in excess of 1000° C would be required to produce large quantities of granitic magma under dry conditions. However, granitic melt can be produced through the breakdown of hydrous phases during metamorphism of sedimentary rocks (e.g. Clemens & Wall 1981; Thompson 1982). The breakdown of muscovite occurs at around 665° C at 5 kbar, but this rarely produces large quantities of magma (Clemens & Vielzeuf 1987). At higher temperatures, dehydration and breakdown of biotite can produce up to 40% melt from a metapelite. There is some textural evidence for a localized biotite breakdown reaction

biotite + sillimanite + quartz = garnet
+ K-feldspar + liquid

in the sillimanite-grade zone (D. Prior pers. comm.). However, this continuous reaction which, dependent on pressure, occurs at around 700° C, takes place only in Fe-rich pelites and will cease as the evolving composition becomes more Mg-rich (Thompson 1982). In pelites, intermediate Fe–Mg compositions are most common and the main breakdown reaction which produces significant quantities of melt is discontinuous and occurs at temperatures between 850–875° C (Vielzeuf & Holloway 1988).

Model T_{DM} ages for the Hunza dykes are in the region of 1.5 Ga. If a biotite-rich sedimentary protolith was the source rock then it would have a high Rb/Sr ratio as a result of a high biotite content and would have evolved a much higher initial Sr isotope ratio over 1.5 Ga than that seen in Hunza. Therefore, the pelitic protolith itself must have been relatively immature and have been recently derived from an older Rb-depleted crustal source, such as a granulite, with the observed T_{DM} age. The pelite would inherit the isotopic signature of the granulite and there must have been insufficient time for this signature to alter before the pelite underwent amphibolite-facies metamorphism during crustal thickening related to the Himalayan collision. A major problem with this model is that calculations by England & Thompson (1986) for the thermal evolution of thickened crust show that temperatures some 20–30 Ma after collision are unlikely to exceed 800° C. Although such calculations are highly dependent on local conditions, temperatures are unlikely to be high enough for significant biotite dehydration. Additionally, melting of a sedimentary protolith would almost certainly involve residual feldspar which would buffer the Ba and Sr contents of the melt and preclude the high values seen in the Hunza dykes.

Similar problems encountered for the BPU were explained by the input of heat from the mantle via the intrusion of ultra-potassic magmas represented at the surface by biotite minette dykes (Searle *et al.* 1992). These minettes are enriched in large ion lithophile elements (LILE) such as Ba and Sr, and selective contamination of the granitic melts during ponding at the base of the crust (see experiments by Huppert & Sparks 1988) can also explain the LILE enriched signature. Similar ultra-potassic dykes are found to the north of the HPC in the Shimshal Valley (Fig. 1; Crawford 1988) and have been reported in Hunza by Schneider (1957) and in the Shaksgam Valley further north (Pognante 1990), so a similar petrogenesis could be proposed. However, irrespective of this, the need for either high temperatures or input of heat from the mantle requires that the source region for the Hunza dykes was close to the base of the crust. The evidence presented above suggests that the protolith was an immature metapelite derived from a much older domain passing through the amphibolite–granulite facies transition (marked by biotite dehydration) during crustal thickening.

(b) The Sumayar pluton

Minimum melt composition, low crystallization temperature and water-saturated nature suggest melting of a pelitic source, and the comparable chemistry and tectonic position suggest a similar petrogenesis to that of the High Himalayan leucogranites. The High Himalayan leucogranites are associated with sillimanite-grade metamorphism and subsequent migmatization in Formation I of the Tibetan Slab (e.g. Le Fort 1981; Vidal *et al.* 1982; Le Fort *et al.* 1987). Whereas little migmatisation is seen in Hunza due to the predominance of calcareous lithologies in the sillimanite-grade zone, it is likely that the Sumayar pluton is underlain by sillimanite-grade pelitic rocks, and so the potential of the metasedimentary rocks in this zone as a source for the granite needs to be considered.

The position in *P–T* space of two sillimanite-grade metapelites from just south of the HPC (Fig. 9) has been calculated using the garnet–biotite geothermometer of Hodges & Spear (1982) and the garnet–biotite–muscovite–plagioclase geobarometer of Hodges & Crowley (1985). Since chemical variation was small within the same phases, average rim analyses were used in an effort to reduce errors (Table 4). The temperature was checked by employing the garnet–hornblende geothermometer of Graham & Powell (1984) on another sample, and this

Table 4. *Average compositions for mineral phases from Hunza sillimanite-grade metapelites, SL20 and SL21, used for geothermometry and barometry calculations*

	GNT	BIOT	MUSC	PLAG
SL20	$N = 6$	$N = 6$	$N = 6$	$N = 8$
SiO_2	37.23 ± 1.36	35.54 ± 0.96	45.67 ± 1.64	63.33 ± 2.13
TiO_2	–	2.29 ± 0.63	0.71 ± 0.40	–
Al_2O_3	21.31 ± 0.40	18.51 ± 1.64	33.31 ± 1.74	22.59 ± 0.73
Cr_2O_3	–	0.10 ± 0.09	–	–
FeO	32.16 ± 0.72	16.50 ± 0.57	1.11 ± 0.24	–
MnO	4.19 ± 0.71	0.13 ± 0.06	–	–
MgO	4.20 ± 0.43	10.37 ± 0.61	0.81 ± 0.21	0.01 ± 0.01
CaO	0.96 ± 0.09	–	–	5.17 ± 0.59
Na_2O	–	0.36 ± 0.15	0.73 ± 0.10	9.40 ± 0.33
K_2O	–	9.40 ± 0.32	10.49 ± 0.31	0.12 ± 0.04
Total	100.05	93.20	92.83	100.62
O	24.000	22.000	22.000	8.000
Si	5.962	5.465	6.247	2.794
Ti	–	0.265	0.073	–
Al	4.022	3.355	5.371	1.175
Cr	–	0.012	–	–
Fe_2	4.307	2.122	0.127	–
Mn	0.568	0.017	–	–
Mg	1.003	2.377	0.165	0.001
Ca	0.165	–	–	0.244
Na	–	0.107	0.194	0.804
K	–	1.844	1.831	0.007
Sum	16.027	15.563	14.007	5.024
SL21	$N = 8$	$N = 5$	$N = 5$	$N = 6$
SiO_2	37.68 ± 1.27	35.84 ± 1.11	46.24 ± 2.42	59.78 ± 1.03
TiO_2	–	2.44 ± 0.54	–	–
Al_2O_3	21.59 ± 0.67	18.24 ± 1.45	33.28 ± 4.24	24.03 ± 0.71
Cr_2O_3	–	–	–	–
FeO	31.83 ± 0.78	15.63 ± 2.27	1.08 ± 0.72	–
MnO	3.24 ± 1.40	–	–	–
MgO	4.80 ± 1.73	11.28 ± 1.34	0.87 ± 0.14	–
CaO	1.44 ± 0.32	–	–	7.29 ± 0.85
Na_2O	–	0.23 ± 0.06	0.61 ± 0.26	8.26 ± 0.42
K_2O	–	9.61 ± 0.24	10.55 ± 0.72	0.11 ± 0.07
Total	100.58	93.27	92.63	99.47
O	24.000	22.000	22.000	8.000
Si	5.965	5.482	6.325	2.688
Ti	–	0.281	–	–
Al	4.029	3.288	5.365	1.273
Cr	–	–	–	–
Fe_2	4.214	1.999	0.124	–
Mn	0.434	–	–	–
Mg	1.133	2.572	0.177	–
Ca	0.244	–	–	0.351
Na	–	0.068	0.162	0.720
K	–	1.875	1.841	0.006
Sum	16.020	15.565	13.994	5.039

See Table 1 for analytical details.

gave good agreement (655° C). Some error may be inherent in the pressure calculation since, although the garnets are Fe-rich as required, the calibration of the barometer did not allow for non-ideality (Hodges & Crowley 1985). However, the results are in good agreement with those of Broughton (1981) for the same area (632° C, 5.5 kbar) which were obtained using a

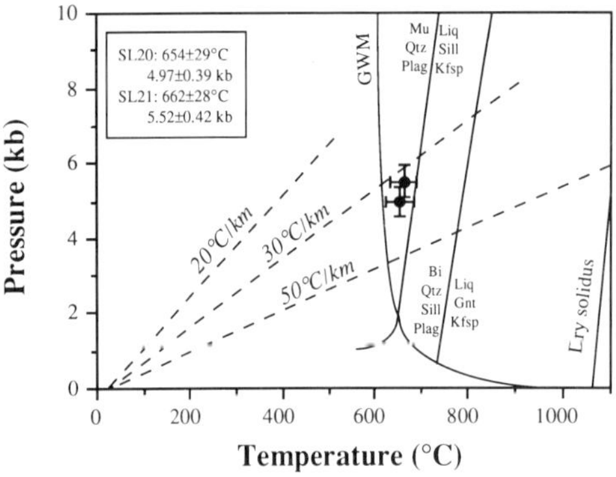

Fig. 9. Position in *P–T* space of two sillimanite-grade metasedimentary gneisses from just south of the HPC in Hunza. Also shown are the granitic wet melting curve, dehydration melting curves for muscovite and biotite (Wyllie 1983) and geothermal gradients. The stable mineral assemblages are shown for the breakdown reactions. See text for discussion.

different geobarometer based on the garnet–plagioclase–sillimanite–quartz equilibrium.

It can be seen from Fig. 9 that if the system were saturated with water then melting would occur along the wet, minimum-melting curve. The position of this curve is probably at even lower temperatures due to the presence of boron (Pichavant 1987), but extensive melts would still only be produced if the source composition was close to that of the eutectic since the temperature is not sufficiently high to allow melting to continue for a long period after one of the eutectic phases (usually K-feldspar) becomes exhausted. The melts would lie just above solidus temperatures and would not migrate large distances before crystallization.

It is unlikely that water is abundant in the middle-lower crust due to the *P–T* conditions and low porosities (e.g. Clemens & Vielzeuf 1987). For vapour-present melting to occur, an external source of fluid, as has been proposed for the source regions of the High Himalayan leucogranites (e.g. Le Fort 1981; Le Fort *et al.* 1987), has to be involved. One possible source of such a fluid may have been dehydration of underthrust Kohistan crust during convergence. Alternatively, vapour-absent melting through the breakdown of muscovite could be invoked (e.g. Harris & Inger 1992). It can be seen from Fig. 9 that the Hunza metasediments lie close to the muscovite breakdown curve, and the exact position of this curve may well be at lower temperatures depending on the composition of the Hunza pelites. The sillimanite-grade pelites exposed at the surface in Hunza do not contain muscovite and show localized in situ melting through the breakdown

$$\text{muscovite} + \text{plagioclase} + \text{quartz} = \text{K-feldspar} + \text{sillimanite} + \text{liquid}.$$

Similar reactions in rocks beneath Sumayar could have generated the pluton.

The high Rb/Sr ratios (>13) and low Sr/Ba ratios (≈0.5) indicate vapour-absent melting (Harris & Inger 1992). However, the volume of melt represented by Sumayar indicates a high melt fraction unless the magma was somehow aggregated from a very large area of low-fraction partial melting. Vapour-present melting can produce a melt fraction of around 40% whereas vapour-absent melting via the breakdown of muscovite gives much lower fractions (around 6%; Clemens & Vielzeuf 1987). The limitation of Sumayar-type magmatism in Hunza to just one pluton may be a result of this widespread aggregation. Alternatively, the spatial restriction of Sumayar may be explained by analogy to the High Himalaya where a similar observation is explained by the need for local concentrations of high-heat producing elements contained in minor phases in the source rocks (e.g. Pinet & Jaupart 1987; Cuney *et al.* 1984). This is supported by the observation that Sumayar has a much higher U content than the Hunza dykes (17 ppm compared to 7–9 ppm; Rex *et al.* 1988).

Crustal structure beneath the middle Hunza Valley

Consideration of the processes involved in the generation of the two types of collision-related granitoid magmatism can be used to make certain inferences about the crustal structure of the Hunza Valley. The following facts must be accounted for.

(1) The present surface structure observed in the field and shown in Fig. 2.
(2) If the Sumayar pluton has a similar origin to the High Himalayan leucogranites then it must be underlain by sillimanite-grade metapelites, probably similar to those seen at the surface to the north.
(3) Intrusion of the dykes at a deeper level than the Sumayar pluton.
(4) The localization of the Hunza dykes to the sillimanite-grade rocks and the deformed part of the HPC.

The juxtaposition of upper- and mid-crustal rocks on opposite sides of the BPU was suggested to be related to northward downthrow during culmination collapse along an inferred normal fault; the Karakoram Batholith Lineament (KBL; Searle *et al.* 1989). It was also suggested that this crustal structure spatially controlled magmatism along the Karakoram Axial Batholith (Rex *et al.* 1988). Across the KBL, if present, there would be a local high temperature anomaly due to the thermal conductivity contrast between the rock types on either side, with the sedimentary rocks to the north acting as a barrier to lateral heat transfer (see Jaupart & Provost 1985). This temperature anomaly would channel the Baltoro/Hunza dyke magma which would then be emplaced in the footwall of the KBL. If the magma rose by a series of feeder dykes which coalesced to form a pluton, then these dykes would only be found in the footwall. The transition from the mid-crustal metamorphosed rocks in the south to the upper-crustal low-grade rocks in the north in Hunza lies in the HPC along the deformation front where the foliation and recrystallization due to southward thrusting die out. This coincides with the position where the voluminous dyke swarms die out, with only a few dykes being found to the north of this. It is conceivable that the magmas in Hunza rose to a higher level than in the Baltoro area without coalescing to form a pluton. Alternatively, the smaller volume of magma in Hunza relative to the Baltoro may have only formed a small pluton which has since been eroded. As well as explaining the location of the Hunza dykes with respect to the deformation front, this model could also account for the change in nature from anastomosing to parallel intrusion patterns near the front. Close to the KBL, local isotherms due to the thermal contrast would be parallel to the fault and these would facilitate intrusion of the dykes by maintaining a constant temperature.

This tentative model leads to the crustal structure proposed in Fig. 10. The section is drawn at 37 Ma since this is when the deformation and metamorphism are assumed to have ceased based on the age of the cross-cutting Mango Gusar pluton (Searle *et al.* 1989). However, it is recognized that this is only a maximum age for the Hunza magmas. The depth of the

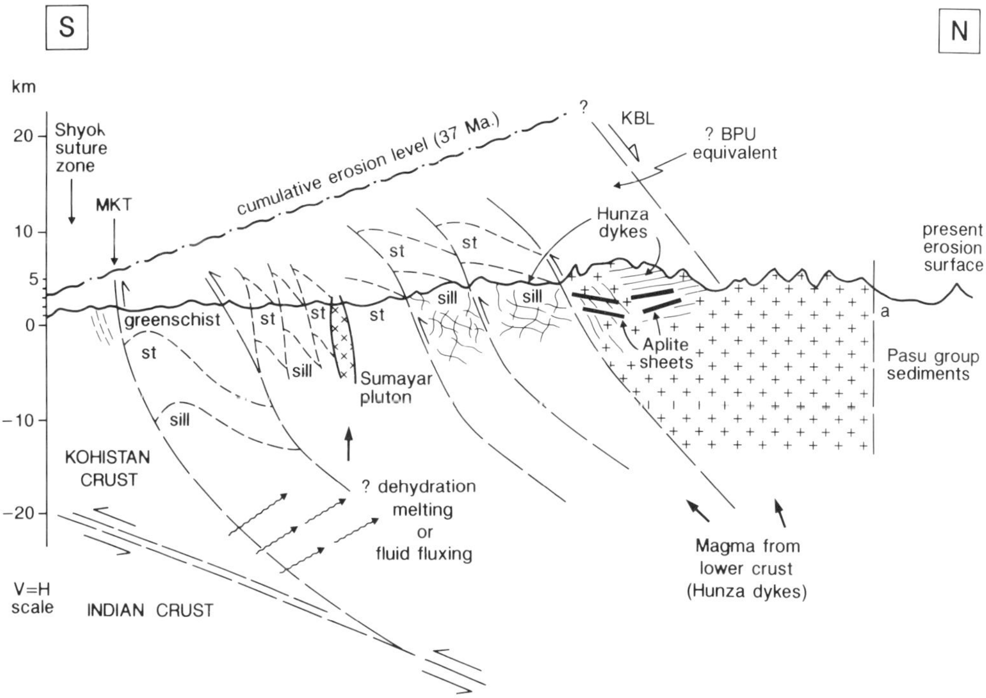

Fig. 10. Schematic crustal section through Hunza showing the relationship between the Hunza dykes and the Sumayar pluton. The Main Karakoram Thrust (MKT) represents the boundary between the Karakoram and underplated Kohistan crust. The thickness of the greenschist, staurolite (St), and sillimanite (Sill) metamorphic zones are based on a geothermal gradient of 30° C km^{-1} (Fig. 9). Dips are based on those seen at the surface (Fig. 2). Thick lines are major thrusts observed in the field, and thinner lines are minor thrusts necessary to explain the surface outcrop. KBL, Karakoram Batholith Lineament; horizontal scale is equal to the vertical scale.

sillimanite-grade rocks beneath the 37 Ma palaeosurface is based on a 30° C km^{-1} geothermal gradient (from Fig. 9 cf. England & Thompson 1986). The palaeosurface is sloping due to greater uplift-erosion rates (relative to mean sea-level) in the north, as shown by the thrusting of deeper-seated, hotter rocks over the top of shallower level, cooler rocks, and fission track dating (Zeitler 1985). The palaeosurface drops sharply across the proposed KBL to the sedimentary zone in the northern Karakoram, although this drop cannot be accurately quantified.

The section shows the position of Sumayar above sillimanite-grade rocks, presumably similar to those seen at the surface further north. Possible fluid sources for vapour-present melting are water originally present in the sediments or subducting Kohistan crust. There is also a need for large volumes of pelitic metasediment at the base of the crust to generate the Hunza dykes and BPU. These sediments must be sufficiently immature not to have developed significant Sr isotope signatures. Coward & Butler (1985) record a minimum shortening of 470 km of Indian plate upper crustal rocks based on balanced cross sections south of the MMT. Underthrusting of Indian plate lower crust to balance this section would result in the lower 20–30 km of the present day crust beneath Hunza being composed of Indian shield-type lower crustal granulites. These rocks would have already passed through the mica-dehydration reactions and, under dry conditions, would only melt at very high temperatures. Therefore they would not provide a suitable source for the Hunza dykes.

Another possible source is Kohistan crust. However, the supposed shallow depth of the MMT beneath Kohistan (*c.* 20 km; Malinconico 1986) makes this unlikely. Crustal thickening of the Karakoram by stacking, ductile shearing and anatexis has been invoked to balance shortening in the Kashmir–Ladakh area to the south and east of Hunza (Fig. 1 inset; Searle *et al.* 1987, 1988). Thickening of the crust would result in immature sediments passing through the amphibolite to granulite-facies transition for the first time and these could generate the Hunza dykes. However, such observations cannot be ratified on the basis of geochemistry alone, although they are supported by recent observations across the Tibetan plateau, along strike to the east, that there has not been significant underthrusting beneath the 65–70 km thick crust (Dewey *et al.* 1988).

M.B.C. acknowledges NERC grant GT4/85/GS/47 and a CNRS grant at the Université de Rennes. M.P.S. acknowledges NERC grants GT5/F89/GS/6 and GR3/4242 to B.F. Windley. We thank A. Saunders, T. Rex, Bor-ming Jahn and B. Windley for discussions, R. Wilson for help with the thermobarometry, and N. Marsh and G. Davies for help with analytical work. U. Pognante and D. Prior provided constructive criticisms of earlier drafts of the paper.

References

ALLÈGRE, C.J. & OTHMAN, B.D. 1980. Nd-Sr isotopic relationship in granitoid rocks and continental crust development: a chemical approach to orogenesis. *Nature,* **286**, 335–342.

BENARD, F., MOUTOU, P. & PICHAVANT, M. 1985. Phase relations of tourmaline leucogranites and the significance of tourmaline in silicic magmas. *Journal of Geology,* **93**, 271–291.

BROUGHTON, R.D. 1981. *Aspects of the geochemistry of some igneous and metamorphic rocks from the Hunza area, Karakorum, Pakistan.* MSc thesis, University of Birmingham.

BURG, J.P., GUIRAUD, M., CHEN, G.M. & LI, G.C. 1984. Himalayan metamorphism and deformations in the North Himalayan belt (southern Tibet, China). *Earth and Planetary Science Letters,* **69**, 391–400.

CLEMENS, J.D. & VIELZEUF, D. 1987. Constraints on melting and magma production in the crust. *Earth and Planetary Science Letters,* **86**, 287–306.

—— & WALL, V.J. 1981. Origin and crystallisation of some peraluminous (S-type) granitic magmas. *Canadian Mineralogist,* **19**, 111–131.

COWARD, M.P. & BUTLER, R.W.H. 1985. Thrust tectonics and the deep structure of the Pakistan Himalaya. *Geology,* **13**, 417–420.

CRAWFORD, M.B. 1988. *Leucogranites of the NW Himalaya: crust–mantle interaction beneath the Karakoram and the magmatic evolution of collisional belts.* PhD thesis, University of Leicester.

—— & SEARLE, M.P. 1992. Field relationships and geochemistry of pre-collisional (India–Asia) granitoid magmatism in the central Karakoram, northern Pakistan. *Tectonophysics,* **206**, 171–192.

—— & WINDLEY, B.F. 1990. Leucogranites of the Himalaya/Karakoram: implications for magmatic evolution within collisional belts and the study of collision-related leucogranite petrogenesis. *Journal of Volcanology and Geothermal Research,* **44**, 1–19.

CUNEY, M., LE FORT, P. & ZHIXIANG, W. 1984. Uranium and thorium geochemistry and mineralogy in the Manaslu leucogranite (Nepal, Himalaya). *In*: *Geology of Granites and their Metallogenic Relations.* Proceedings of an International Symposium at Nanjing University, China, 853–873.

DEBON, F. & LE FORT, P. 1982. A chemical mineralogical classification of common plutonic rocks and associations. *Transactions of the Royal Society of Edinburgh,* **73**, 135–149.

——, ——, DAUTEL, D., SONET, J. & ZIMMERMAN, J. L. 1987. Granites of western Karakoram and northern Kohistan (Pakistan): a composite Mid-Cretaceous to Upper Cenozoic magmatism. *Lithos,* **20**, 19–40.

DePaolo, D.J. 1979. Implications of correlated Nd- and Sr-isotopic variations for the chemical evolution of the crust and mantle. *Earth and Planetary Science Letters,* **43**, 210–211.

Desio, A. 1979. Geologic evolution of the Karakorum. *In*: Farah, A. & Kees, A. (eds) *Geodynamics of Pakistan*. Geological Survey of Pakistan, Quetta, 111–124.

—— & Martina, E. 1972. Geology of the Upper Hunza Valley, Karakorum, West Pakistan. *Bolletino della Societa Geologica Italiana,* **91**, 283–314.

—— & Zanettin, B. 1970. *Geology of the Baltoro Basin*. E. J. Brill, Leiden.

Dewey, J.F., Shackleton, R.M., Chengfa, C. & Yiyin, S. 1988. The tectonic evolution of the Tibetan Plateau. *Philosophical Transactions of the Royal Society of London,* **A327**, 379–413.

Dietrich, V. & Gansser, A. 1981. The leucogranites of the Bhutan Himalaya (crustal anatexis versus mantle melting). *Schweizerische Mineralogische und Petrographische Mitteilungen,* **61**, 177–202.

England, P.C. & Thompson, A. 1986. Some thermal and tectonic models for crustal melting in continental collision zones. *In*: Coward, M.P. & Ries, A.C. (eds) *Collision Tectonics*. Geological Society, London, Special Publications, **19**, 83–94.

Gansser, A. 1964. *Geology of the Himalayas*. Interscience Publishing, Wiley, London.

Graham, C.M. & Powell, R. 1984. A garnet-hornblende geothermometer: calibration, testing and application to the Pelona Schist, southern California. *Journal of Metamorphic Geology,* **2**, 13–31.

Green, T.H. 1977. Garnet in silicic liquids and its possible use as a P-T indicator. *Contributions to Mineralogy and Petrology,* **65**, 59–67.

Hanson, G.N. 1978. The application of trace elements to the petrogenesis of igneous rocks of granitic composition. *Earth and Planetary Science Letters,* **38**, 26–43.

Harris, N.B.W. & Inger, S. 1992. Trace element modelling of pelite-derived granites. *Contributions to Mineralogy and Petrology,* **110**, 46–56.

Hawkesworth, C.J. & Norry, M.J. (eds) 1983. *Continental Basalts and Mantle Xenoliths*. Shiva Publishing Ltd., Cheshire, UK.

Henderson, P. 1982. *Inorganic Geochemistry*. Pergamon Press.

—— 1984. General geochemical properties and abundances of rare earth elements. *In*: Henderson, P. (ed.) *Rare earth element geochemistry*. Elsevier, 1–29.

Herren, E. 1987. Zanskar shear zone: northeast-southwest extension within the Higher Himalaya (Ladakh, India). *Geology,* **15**, 409–413.

Hodges, K.V. & Crowley, P.C. 1985. Error estimation and empirical geothermobarometry for pelitic systems. *American Mineralogist,* **70**, 702–709.

—— & Spear, F.S. 1982. Geothermometry, geobarometry and the Al_2SiO_5 triple point at Mt. Moosilauke, New Hampshire. *American Mineralogist,* **67**, 1118–1134.

Huppert, H.E. & Sparks, R.S.J. 1988. The generation of granitic magmas by intrusion of basalt into continental crust. *Journal of Petrology,* **29**, 599–624.

Jacobssen, S.B. & Wasserburg, G.J. 1980. Sm-Nd isotopic evolution of chondrites. *Earth and Planetary Science Letters,* **50**, 149–155.

Jaupart, C. & Provost, A. 1985. Heat focussing, granite genesis and inverted metamorphic gradients in continental collision zones. *Earth and Planetary Science Letters,* **73**, 385–397.

Le Fort, P. 1981. Manaslu leucogranite: a collision signature of the Himalaya. A model for its genesis and emplacement. *Journal of Geophysical Research,* **86**, 10545–10568.

——, Debon, F. & Sonet, J. 1983. Petrography, geochemistry and geochronology of some samples from the Karakorum Batholith (N. Pakistan). *In*: Shams, F.A. (ed.) *Granites of Himalayas, Karakoram and Hindu Kush*. Punjab University, Lahore, 377–387.

——, Cuney, M., Deniel, C., France-Lenord, C., Sheppard, S.M.F. Upreti, B.N. & Vidal, Ph. 1987. Crustal generation of the Himalayan leucogranites. *Tectonophysics,* **134**, 39–57.

Malinconico, L.L. 1986. The structure of the Kohistan-arc terrane in northern Pakistan as inferred from gravity data. *Tectonophysics,* **124**, 297–307.

Marsh, N.G., Saunders, A.D., Tarney, J. & Dick, H.J.B. 1980. Geochemistry of basalts from the Shikoku and Daito basins, Deep Sea Drilling Project Leg 58. *In*: De Vries Klein, G., Kobayashi, K. *et al.* (eds) *Initial Reports of the Deep Sea Drilling Project*, Washington D.C., **58**, 805–842.

Miller, C.F. 1985. Are strongly peraluminous magmas derived from pelitic sedimentary sources? *Journal of Geology,* **93**, 673–689.

—— & Stoddard, E.F. 1981. The role of manganese in the paragenesis of magmatic garnet: an example from the Old Woman – Puite Range, California. *Journal of Geology,* **89**, 233–246.

Parrish, R.R. & Tirrul, R. 1989. U-Pb age of the Baltoro granite, northwest Himalaya, and implications for monazite U-Pb systematics. *Geology,* **17**, 1076–1079.

Pichavant, M. 1987. Effects of B and H_2O on the liquidus phase relationships in the haplogranitic system at 1 kbar. *American Mineralogist,* **72**, 1056–1070.

—— & Manning, D.A.C. 1984. Petrogenesis of tourmaline granites and topaz granites: the contribution of experimental data. *Physics of the Earth and Planetary Interiors,* **35**, 31–50.

Pinet, C. & Jaupart, C. 1987. A thermal model for the distribution in space and time of the Himalayan granites. *Earth and Planetary Science Letters,* **84**, 87–99.

Pognante, U. 1990. Shoshonitic and ultra-potassic post-collisional dykes from northern Karakorum, (Sinkiang, China). *Lithos,* **26**, 305–316.

Powell, C.McA. & Vernon, R.H. 1979. Growth and

rotation history of garnet porphyroblasts with inclusion spirals in a Karakorum schist. *Tectonophysics*, **54**, 25–43.

PRIOR, D.J. 1987. Syntectonic porphyroblast growth in phyllites: textures and processes. *Journal of Metamorphic Geology*, **5**, 27–39.

REX, A.J., SEARLE, M.P., TIRRUL, R., CRAWFORD, M.B., PRIOR, D.J., REX, D.C. & BARNICOAT, A. 1988. The geochemical and tectonic evolution of the central Karakoram, N. Pakistan. *Philosophical Transactions of the Royal Society of London*, **A326**, 229–255.

SCHNEIDER, H.J. 1957. Tektonik und Magmatismus in NW Karakorum. *Geologische Rundschau*, **46**, 426–476.

SEARLE, M.P. 1986. Structural evolution and sequences of thrusting in the High Himalayan, Tibetan-Tethys and Indus Suture zones of Zanskar and Ladakh, Western Himalaya. *Journal of Structural Geology*, **8**, 923–936.

—— 1991. *Geology and Tectonics of the Karakoram Mountains*. J. Wiley & Sons, Chichester.

—— & FRYER, B.J. 1986. Garnet, tourmaline and muscovite-bearing leucogranites, gneisses and migmatites of the Higher Himalaya from Zanskar, Kulu, Lahoul, and Kashmir. *In*: COWARD, M.P. & RIES, A.C. (eds) *Collision Tectonics*. Geological Society, London, Special Publications, **19**, 185–201.

——, REX, A.J., TIRRUL, R., WINDLEY, B.F., ST ONGE, M. & HOFFMAN, P. 1986. A geological profile across the Baltoro Karakoram range, N. Pakistan. *Geological Bulletin of the University of Peshawar*, **19**, 1–12.

——, WINDLEY, B.F., COWARD, M.P., COOPER, D.J.W., REX, A.J., REX, D.C., LI TINGDONG, XIAO XUCHANG, JAN, M.Q., THAKUR, V.C. & KUMAR, S. 1987. The closing of Tethys and the tectonics of the Himalayas. *Geological Society of America Bulletin*, **98**, 678–701.

——, COOPER, D.J.W. & REX, A.J. 1988. Collision tectonics of the Ladakh-Zanskar Himalaya. *Philosophical Transactions of the Royal Society of London*, **A326**, 117–150.

——, REX, A.J., TIRRUL, R., REX, D.C., BARNICOAT, A. & WINDLEY, B.F. 1989. Metamorphic, magmatic and tectonic evolution of the Central Karakoram in the Biafo – Baltoro – Hushe regions of N. Pakistan. *Geological Society of America Special Paper*, **232**, 47–73.

——, CRAWFORD, M.B. & REX, A.J. 1992. Field relations, geochemistry, origin and emplacement of the Baltoro granite, central Karakoram. *Transactions of the Royal Society of Edinburgh*, **83**, 519–538.

——, METCALFE, R.P., REX, A.J. & NORRY, M.J. 1993. Field relations and emplacement of the Bhagirathi leucogranite, Garhwal Himalaya. *This volume*.

TARNEY, J., SAUNDERS, A.D., WEAVER, S.D., DONELLAN, N.C.B. & HENDRY, G.L. 1979. Minor element chemistry of Leg 49, North Atlantic Ocean. *In*: LUYENDYK, J.R., CANN, J. *et al.* (eds) *Initial Reports of the Deep Sea Drilling Project*, Washington D.C., **49**, 657–691.

TAYLOR, S.R. & MCLENNAN, S.M. 1981. The composition and evolution of the continental crust: rare-earth evidence from sedimentary rocks. *Philosophical Transactions of the Royal Society of London*, **A301**, 381.

—— & —— 1985. *The Continental Crust: its Composition and Evolution*. Blackwell Scientific Publications, Oxford.

THOMPSON, A.B. 1982. Dehydration melting of pelitic rocks and the generation of H_2O-undersaturated granitic liquids. *American Journal of Science*, **282**, 1567–1595.

TINDLE, A.G. & PEARCE, J.A. 1981. Petrogenetic modelling of in situ fractional crystallisation in the zoned Loch Doon pluton, Scotland. *Contributions to Mineralogy and Petrology*, **78**, 196–207.

TRELOAR, P.J., REX, D.C., GUISE, P.G., COWARD, M.P., SEARLE, M.P., WINDLEY, B.F., PETTERSON, M.G., JAN, M.Q. & LUFF, I.W. 1989. K-Ar and Ar-Ar geochronology of the Himalayan collision in NW Pakistan: constraints on the timing of suturing, deformation, metamorphism and uplift. *Tectonics*, **8**, 881–909.

VIDAL, PH., COCHERIE, A. & LE FORT, P. 1982. Geochemical investigations of the origin of the Manaslu leucogranite (Himalaya, Nepal). *Geochimica et Cosmochimica Acta*, **46**, 2279–2292.

VIELZEUF, D. & HOLLOWAY, J.R. 1988. Experimental determination of the fluid-absent melting relations in the pelitic system. *Contributions to Mineralogy and Petrology*, **98**, 257–276.

WALSH, J.N., BUCKLEY, F. & BARKER, J. 1981. The simultaneous determination of the rare-earth elements in rocks using inductively coupled plasma source spectrometry. *Chemical Geology*, **33**, 141–153.

WATSON, E.B. & HARRISON, T.M. 1983. Zircon saturation revisited: temperature and composition effects in a variety of crustal magma types. *Earth and Planetary Science Letters*, **64**, 295–304.

—— &—— 1984. Accessory minerals and the geochemical evolution of crustal magmatic systems: a summary and prospectus of experimental approaches. *Physics of the Earth and Planetary Interiors*, **35**, 19–30.

WICKHAM, S.M. 1987. The segregation and emplacement of granitic magmas. *Journal of the Geological Society, London*, **144**, 281–297.

WYLLIE, P.J. 1983. Experimental studies on biotite- and muscovite-granites and some crustal magmatic sources. *In*: ATHERTON, M.P. & GRIBBLE, C.D. (eds) *Migmatites, melting and metamorphism*. Shiva Publishing Ltd., Cheshire, UK, 12–26.

——, HUANG, W.L., STERN, C.R. & MAALØE, S. 1976. Granitic magmas: possible and impossible sources, water contents and crystallisation sequences. *Canadian Journal of Earth Sciences*, **13**, 1007–1019.

ZEITLER, P.K. 1985. Cooling history of the NW Himalaya, Pakistan. *Tectonics*, **1**, 127–151.

Tectonics of the Himalayan collision between the Indian Plate and the Afghan Block: a synthesis

PETER J. TRELOAR[1] & CHRISTOPHER N. IZATT[2]

[1] *School of Geological Sciences, Kingston University, Kingston upon Thames, Surrey KT1 2EE, UK*

[2] *British Gas Exploration and Production, 100 Thames Valley Park, Reading RG6 1PT, UK*

Abstract: A synthesis of the available tectonic, structural, magmatic and geochronologic data from Afghanistan and western Pakistan suggests that collision between the Indian Plate and the Afghan Block (a collage of blocks assembled during the Cimmeride orogeny), took place during the Pliocene. Prior to this, the thinned western margin of the Indian Plate, together with the transitional to oceanic crust which separated the Indian Plate from the Afghan Block, was overlain by a wedge of Tertiary clastic sediments deposited in a large flexural basin that included what is now the Katawaz basin as well as the Sulaiman foredeep. At least three phases of collision can thus be recognised in the northwest Himalaya: the first, at about 100 Ma involved the suturing of the Kohistan–Ladakh–East Nuristan island arc to southern Asia; the second, at about 55 Ma involved India–Asia collision and the third, during the Pliocene, involved India–Afghanistan collision. During the Eocene and Miocene, the Afghan block was displaced southwards in response to Indian indentation into Asia. Simultaneous counterclockwise rotation of the Afghan Block resulted in intracontinental extension causing alkaline magmatism and right lateral displacements along the Herat Fault. As the Indian Plate moved northward with respect to Afghanistan, some of the wedge of clastic sediments deposited on the Indian transitional crust were accreted to the leading edge of the Afghan block and displaced sinistrally with respect to the rest of the sediments in the wedge. These displaced sediments now form the Makran sequence. Collisional structures on both sides of the India–Afghan collision zone in the Helmand block and the Sulaiman–Kirthar Ranges probably represent the inversion of continental margin basins rather than thrust belts with large shortening amounts. After initial inversion, the dominant feature of this highly oblique collision has been transpressional tectonics. Prior to its final docking with the Afghan Block, the Kabul block was a continental slice, completely surrounded by oceanic crust, and located northwest of continental India. Palaeocene collision between India and the Kabul block could have led to thickening of the Indian Plate prior to initiation of India–Asia collision.

Afghanistan lies at the western end of the Himalayan orogenic belt. Although most of the country lies within the confines of the Asian Plate, the western end of the Indus Tsangpo suture outcrops in eastern Afghanistan, with Indian Plate rocks outcropping in the southeastern part of the country (Fig. 1). The collision zone is a complicated one involving island/volcanic arcs, sedimentary basins, ophiolites and strike slip faults. Although much of the tectonic evolution of Afghanistan is dominated by the effects of the Cimmeride orogeny (Sengor 1984) a wide variety of Himalayan tectonic and magmatic features are recognised, an understanding of which is critical to our understanding of the evolution of the western end of the Himalayan orogenic belt.

Geologically, Afghanistan is comprised of a number of crustal blocks separated by fault zones, many of which have Cenozoic, commonly strike slip, displacement histories (Fig. 2). Everything to the north and west of the Herat and Central Badakhstan faults (the North Afghanistan, or Tadzhik, platform) has been part of the Asian Plate since before the Permo-Trias. Those blocks which lie to the south of the Herat–Central Badakhstan fault system and to the north and west of the Chaman and Konar faults were once part of Gondwanaland but separated from that supercontinent before India separated from it, and accreted to the southern margin of Eurasia during the Cimmeride orogeny (Sengor 1984; Boulin 1988, 1991). It is only those rocks to the south and east of the Chaman and Konar faults that form part of the Indian Plate or were developed on oceanic crust prior to collision between India and Eurasia.

Since the Soviet intervention in Afghanistan

From Treloar, P.J. & Searle, M.P. (eds), *Himalayan Tectonics*
Geological Society Special Publication, No. 74, pp. 69–87.

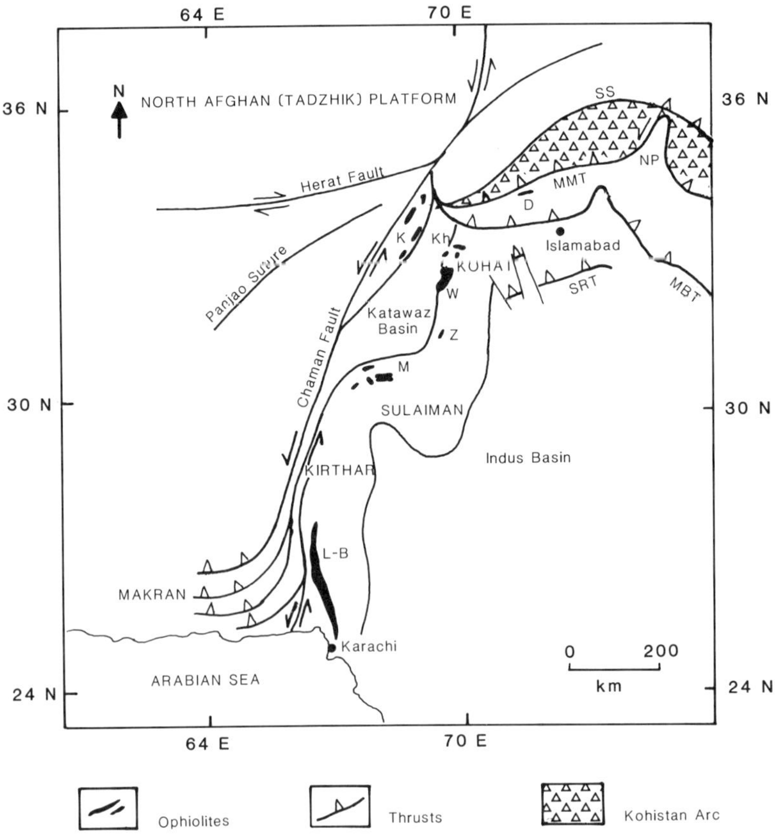

Fig. 1. Schematic map of the western end of the Himalayan orogenic belt in eastern Afghanistan and western Pakistan showing the location of the major ophiolites, thrusts and strike slip faults developed at the western end of the Himalayan belt. Note the way that the Chaman fault bifurcates around the Kabul block, and the way that it acts as a lateral ramp to the major south-verging thrusts within the Makran prism. NP Nanga Parbat. Ophiolites: D, Dargai; K, Kabul; Kh, Khost; L-B, Las Bela; M, Muslimbagh; W, Waziristan; Z, Zhob. Faults: MBT, Main Boundary Thrust; MMT, Main Mantle Thrust; SRT, Salt Range Thrust; SS, Shyok Suture.

in December 1979, the country has been closed to western geologists. At that time, due to the then recent mapping by German (Wittekindt & Weippert 1973; Wolfart & Wittekindt 1980), Soviet (Kafarskiy *et al.* 1975) and French teams (e.g. Boulin & Bouyx 1977; Boulin 1981; Tapponnier *et al.* 1981), Afghan geology (Shareq 1981) was probably known in more detail than that of any other segment of the Himalaya. Since then, although considerable advances have been made in our knowledge of the rest of the Himalayan chain, our understanding of Afghan geological evolution has become somewhat frozen. However, the data available in 1979 are amenable to interpretation in the context of what we now know about adjoining regions, especially in northern and western Pakistan.

In order to understand the timing and tectonic evolution of the western end of the Himalayan orogen, we need to unravel the tectonics of Afghanistan. This paper, partly review and partly provocative interpretation, aims to do that within the context of what is now known about the western Himalaya. In particular the aim is to ascertain: the timing of Afghan–Pakistan (Indian Plate) collision; the geometry of collision related structures within Afghanistan; the sequence of events in the evolution of the westernmost Himalaya; and the significance of strike slip faulting within Afghanistan especially as it relates to models of crustal extrusion.

Geological framework of Afghanistan prior to Himalayan collision

Prior to the Himalayan collision the tectonic framework of Afghanistan was dominated by

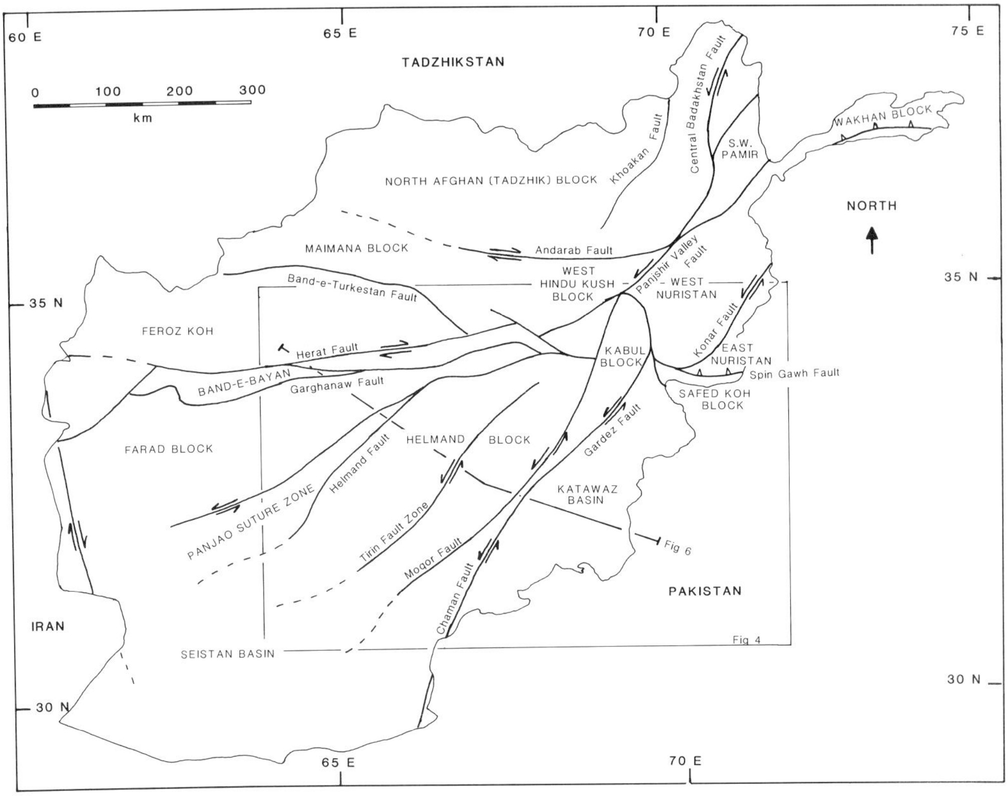

Fig. 2. Map of the major crustal scale fault blocks within Afghanistan (after Schreiber *et al.* 1972), and the strike slip faults which separate them. Also shown are the location of the map shown in Fig. 4, and of the section plotted in Fig. 6.

structures related to the Cimmeride suturing of exotic blocks, to the southern margin of Asia (Sengor *et al.* 1988; Boulin 1988, 1991). Two ophiolite-bearing suture zones are recognized. The northerly suture follows the line of the Herat fault and its eastern continuation into the Panjshir Valley (Sengor 1984). The southerly suture, the Panjao suture, separates the Helmand and Farad blocks of central Afghanistan. On the basis of the relationship between fabrics within the suture zones and the age of overlying sediments, timing of suturing within the northern and southern sutures is constrained as being pre-Early and pre-Mid-Cretaceous respectively (Tapponnier *et al.* 1981; Sengor 1984; Montenant *et al.* 1986; Girardeau *et al.* 1988). This post-dates the emplacement of a series of Permo-Triassic subduction related granitoids into rocks of the southernmost part of the Tadzhik block (Fig. 3a; Debon *et al.* 1987).

Subsequent to Cimmeride suturing, the accreted crustal blocks were overlain by an extensive sequence of Mesozoic shelf sediments (Schreiber *et al.* 1972; Wolfart & Wittekindt 1980). In northern Afghanistan, the Tadzhik block was overlain by a Jurassic clastic sequence that changes upwards to a carbonate-rich platform sequence of Cretaceous age. On the Farad block, the Palaeozoic basement sequence is overlain by Upper Jurassic – Cretaceous sediments with up to 5000 m of Cretaceous rocks, of which 3000 m are massively bedded rudist limestone. On the Helmand block, the base of the Cretaceous is marked by an erosional contact overlain by red coarse clastic sediments themselves overlain by a thick (>1000 m) sequence of clastic marls and limestones. Within the Panjao suture, the steeply dipping suture related rocks are unconformably overlain by thick bioclastic limestones of Barremian and younger age (Tapponnier *et al.* 1981).

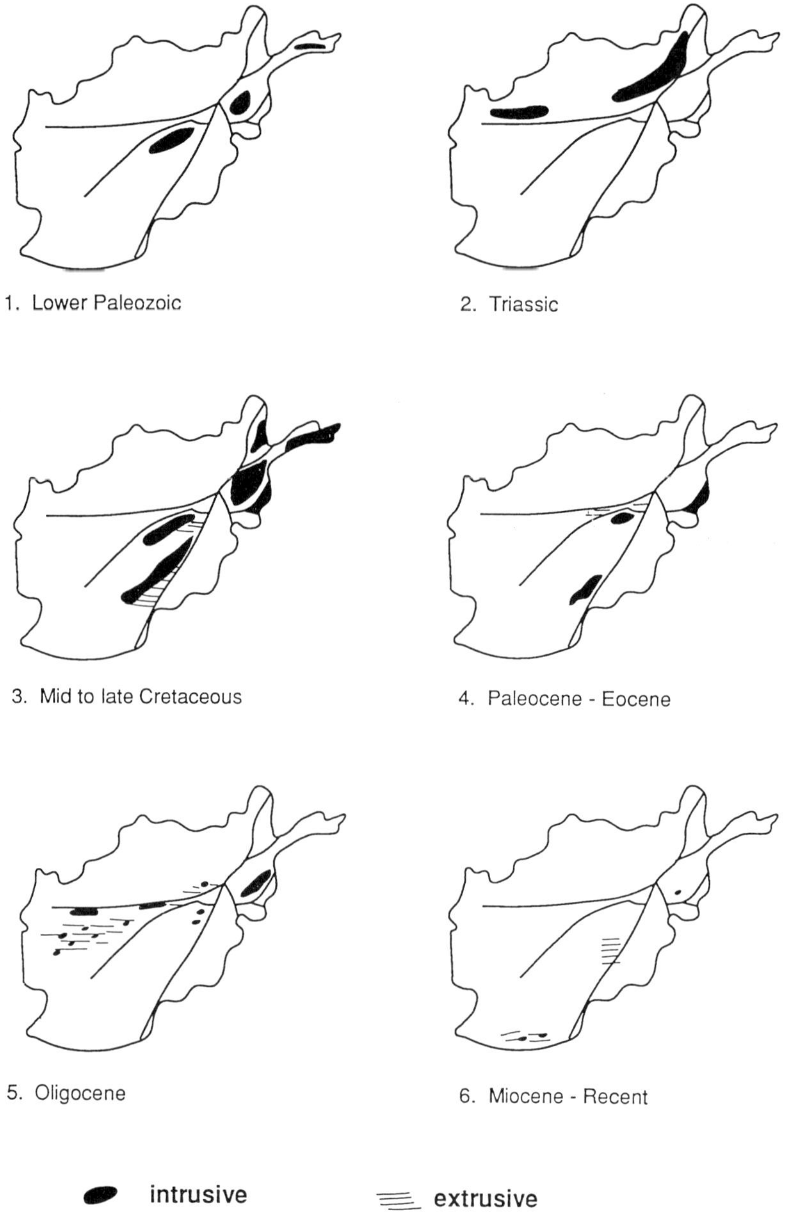

Fig. 3. Map of Afghanistan showing the location of centres of magmatic activity from the Palaeozoic to the present (modified from Bellon *et al.* 1979).

Effects of the Himalayan orogeny on the Afghan Block

The term 'Afghan Block' will be used to describe the collage of crustal blocks assembled, in present-day Afghanistan, during the Cimmeride orogeny, and located west and north of the Chaman and Konar faults. This Cimmeride basement, and its Mesozoic and younger sedimentary cover, shows a variety of tectonic and magmatic effects that result from the Himalayan collision or from the subduction of Tethyan oceanic material beneath the Afghan margin prior to collision.

Magmatic events

Firstly a series of subduction related 'I-type' granitoids were emplaced into the Helmand and West Nuristan blocks. This subduction related magmatism is dated by Rb-Sr methods as Cretaceous in age (Debon *et al.* 1987) with activity terminating by the earliest Tertiary (Fig. 3c, d). The calc-alkaline plutons represent a volcanic arc built on the southern margin of Eurasian continental crust. They postdate the extrusion, between the Upper Jurassic and Middle Cretaceous, of the Kandahar volcanics (Tapponnier *et al.* 1981) which represent early stages of the development of the calc-alkaline volcanic arc.

Secondly, a number of Oligocene-aged alkaline intrusions and basaltic extrusives were emplaced/extruded into the Farad block, the Band-e-Bayan belt to the immediate south of the Herat fault zone and into sedimentary basins within the Herat fault zone (Fig. 3e; Debon *et al.* 1987). The alkaline to peralkaline chemistry of these rocks would be consistent with magmas derived from the mantle beneath a zone of continental extension. In addition Oligo-Miocene plutons (Fig. 3e) were intruded into the thickened Asian Plate crust north of the East Nuristan arc and to the east of the Central Badakhstan fault (Debon *et al.* 1987).

Sedimentary basins

At the end of the Mesozoic, widespread carbonate shelf sedimentation became restricted to deposition of marine sediments within localised basins on the Tadzhik platform (Wolfart & Wittekindt 1980). By the Oligo-Miocene the main sedimentation on the Tadzhik platform was restricted to the deposition of fluvial red bed molassic deposits in localised basins. Similar sediments extend south of the Herat fault onto the Farad block. Deposition of these sediments represents beginning of regional uplift above sea level of the southern margin of Asia following early stages of India–Asia collision.

The Herat fault zone, originally active as a Cimmeride suture zone, was re-activated during the Tertiary as a dextral strike slip fault system (Tapponnier *et al.* 1981). A series of elongate Oligo-Miocene basins were developed within the fault zone. The best studied of these basins, (the Bamyan, Yakawlang and Ghorband basins) contain Oligo-Miocene fluvial-aeolian sediments with localised carbonate horizons (Lang 1977). In places sedimentary geometries are consistent with deposition within pull-apart basins (Tapponnier *et al.* 1981). Oligocene alkaline volcanism is recorded within some of the basins (Lang 1977) and elsewhere along the fault zone (Debon *et al.* 1987). Dextral displacements along the fault zone, pull apart basins and alkaline volcanism are all consistent with a dextral transtensional setting for the Herat fault zone during the Oligo-Miocene.

Deformation

Most of the internal block boundaries, including Cimmeride suture zones, within the Afghan Block have been reactivated during the Tertiary as fault zones with strike slip displacement senses (Fig. 2). The best studied of these are the dextral Herat and sinistral Chaman faults. That post-Miocene deposits are not clearly offset along the Herat fault led Tapponnier *et al.* (1981) to conclude that it has not been active since the Miocene. By contrast the Chaman fault is demonstrably active to the present (Yeats *et al.* 1979; Lawrence *et al.* 1981). A number of other major faults are still active or have been until recently (Wellman 1966). These include the dextral Andarab fault within the North Afghan Tadzhik platform; the sinistral Tirin fault within the Helmand block; and the sinistral Central Badakhstan fault.

In addition to strike slip features a variety of Himalayan compressional structures are recognised on the Afghan Block. In that they are post-Cimmeride and deform Palaeogene to Neogene(?) rocks they must be Himalayan in age.

On the Farad block, Oligocene basaltic to andesitic volcanics, with K-Ar ages of as young as 31.5 Ma (Debon *et al.* 1987), interbedded with red clastic flysch deposits, were deformed during a major phase of folding with axes which are east–west trending in the northern part of the block, but which have a southwesterly trend toward the east and south of the block (Wolfart & Wittekindt, 1980). The folds generally have steep axial planes, although towards the southeast anticlines and associated thrusts become southeast verging.

Within the Helmand block, Mesozoic shelf sediments have been deformed into a series of northeast-trending folds with generally steep axial planes. Around Kandahar, in the southeastern part of the block, folds face towards the southeast. By contrast, within the northwestern and central part of the block the folds form part of a major northwest verging fold and thrust belt (Fig. 4) which imbricates the Mesozoic shelf sequence with Palaeozoic basement. To the northwest, similar thrusting shortened and imbricated the Panjao suture sequence within which the Cretaceous shelf carbonates are

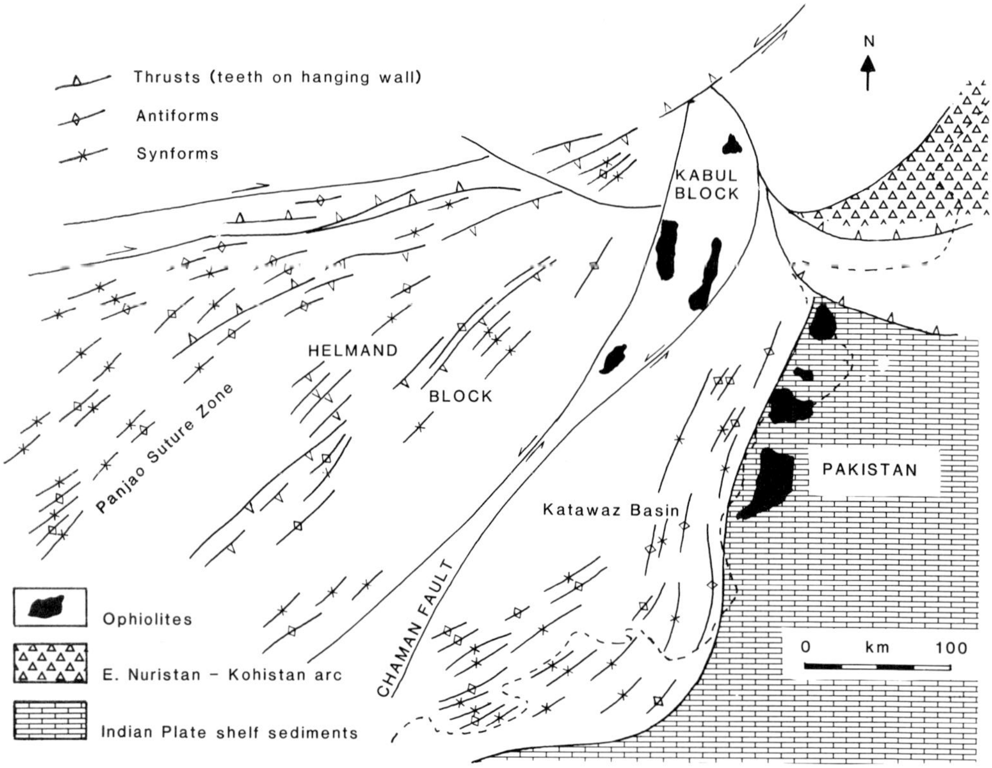

Fig. 4. Structural map showing the main thrust faults and fold belts of central Afghanistan (after Wittekindt & Weippert 1973).

imbricated with units of the ophiolitic sequence (Tapponnier *et al.* 1981).

Along the northern margin of the Helmand block thrusts have an ENE-strike (Fig. 5). Here north-vergent structures invert the southern margin of the Bamyan basin (Boulin & Bouyx 1977). Figure 5, an amalgam of the structural sections drawn across the Southwest Pamir and Herat fault zone by Boulin & Bouyx (1977), shows these structures. These postdate the Oligo-Miocene sedimentation and are described as Mio-Pliocene thrusts by Boulin & Bouyx (1977). The northwest-vergent thrusts elsewhere within the Helmand block are interpreted here as being synchronous with those which invert the Bamyan basin.

Figure 6 is a northwest–southeast section drawn across southern Afghanistan and constructed here on the basis of mapping published by Wittekindt & Weippert (1973) and Girardeau *et al.* 1988). Significant features to note are the sequence of northwest-vergent thrusts in the Helmand block – Panjao suture – Farad block region and the change in vergence direction across the Helmand block to southeast-vergent in the southeast of the block. The change in thrust vergence across the Helmand block (Fig. 6), suggests that the Tertiary tectonics of the whole block may be viewed in terms of a crustal scale pop-up structure inverting an earlier transtensional basin. The section drawn in Fig. 6, based on models of Ziegler (1986, Fig. 10), reflects our contention that Himalayan deformation within the Helmand block essentially involved the re-activation of earlier extensional faults during the inversion of a Mesozoic continental margin basin.

Tectonic units to the east of the Afghan Block

The zone that marks the boundary, within Afghanistan, between the Afghan Block and the Indian Plate is a complex collage of tectonic units. These include: the East Nuristan island arc complex; the Kabul block; the Katawaz basin; and the Kabul and Khost ophiolites (Fig. 1).

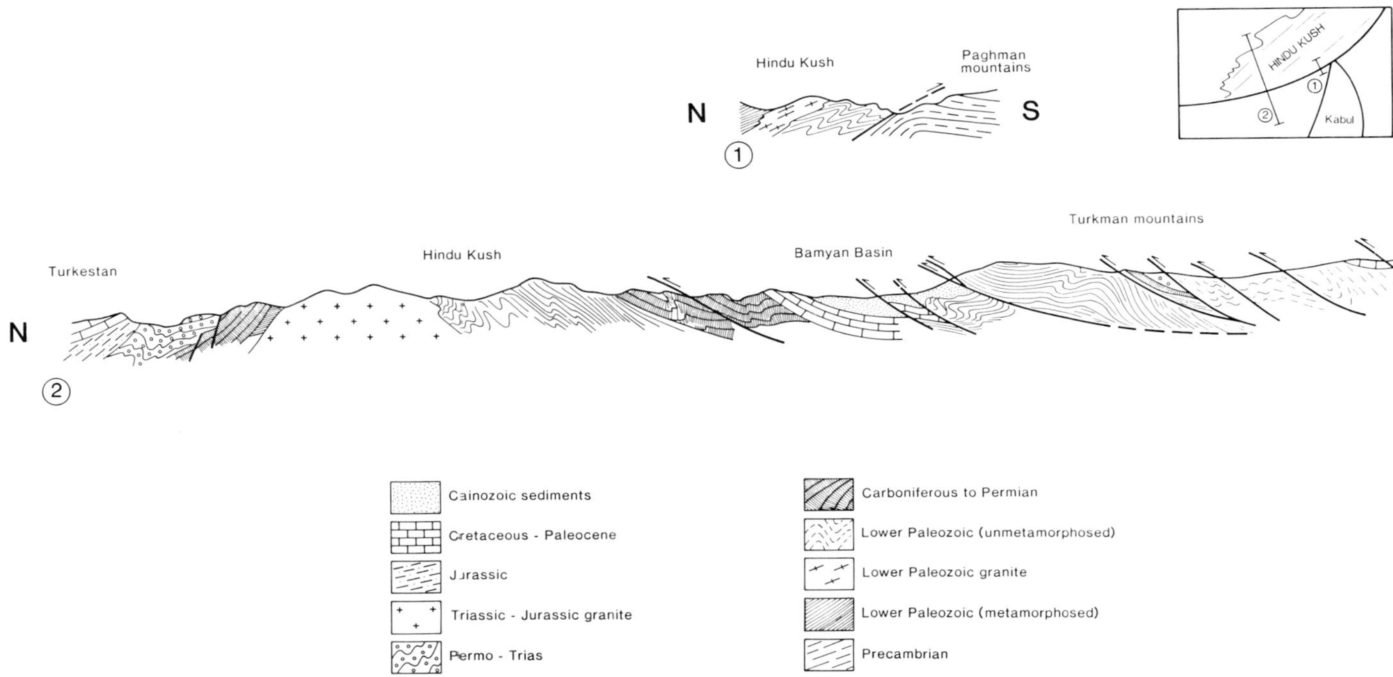

Fig 5. Structural sections (an amalgam of those constructed by Boulin & Bouyx 1977) across the Herat Fault Zone/suture zone showing the major, dominantly north verging thrusts and folds, developed across the zone.

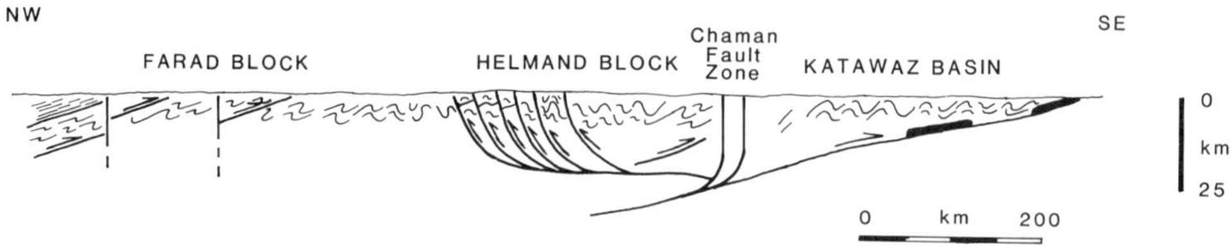

Fig. 6. Sketch section across southern Afghanistan showing the major Himalayan-aged structures of the region. Note the way that thrusts within the Helmand block define a pop-up structure developed during inversion of the Mesozoic continental margin basin on the hanging wall of the southeast-verging thrust that emplaces the Afghan Plate onto the Katawaz flysch basin.

The East Nuristan island arc

The island arc is the westernmost continuation of the Kohistan–Ladakh island arc complex exposed in North Pakistan and Northwest India. Like the better studied and documented parts of the complex the East Nuristan arc is composed of basic to acid plutons which intrude a range of basic volcanics and volcaniclastic sediments (Debon *et al.* 1987). By analogy with Kohistan, volcanism within East Nuristan will have commenced during the Mid-Cretaceous. The arc sutured to the southern Asian margin at about 100 Ma ago with the main phase of subduction related volcanism continuing into the earliest Eocene (Treloar *et al.* 1989). In Pakistan the early Eocene cessation of subduction related volcanism is part of the evidence that dates India–Asia collision as pre-50 Ma.

The Kabul ophiolite

Two Eo-Himalayan suture zones, represented by the presence of ophiolitic material, are present in eastern Afghanistan. The westerly of the two is the Kabul ophiolite (Mattauer *et al.* 1978; Tapponnier *et al.* 1981). Ophiolitic peridotites outcrop as klippen which tectonically overlie intensely folded shelf sediments of the Kabul block. Emplacement, which involved imbrication of Jurassic and Maastrichtian limestones with basic volcanics and cherts, was from the northwest to southeast (Tapponnier *et al.* 1981). The ophiolitic sequence is overlain by Palaeocene to Eocene sediments which date the timing of ophiolite emplacement as end-Cretaceous–early-Palaeocene.

The Khost ophiolite

The Khost ophiolite is distinct from the Kabul ophiolite, and is the direct northward continuation of the Palaeocene Waziristan, Muslimbagh and Zhob ophiolites (Fig. 1) of western Pakistan (Asrarullah & Abbas 1979). Here mafic and ultramafic rocks have been thrust southwestward onto lower Palaeozoic to Jurassic calcareous shelf sediments of the western margin of the Indian Plate. At Muslimbagh and Zhob, the ophiolites are overlain by Lower to Mid-Eocene shallow water foraminiferal limestones, interpreted as evidence for a Palaeocene age of emplacement (Alleman 1979).

The Kabul block

The Kabul block is an apparent enigma. It is formed of crystalline and intensely deformed rocks of Lower Palaeozoic age overlain by Permian to Mesozoic shelf limestones imbricated, during emplacement of the Kabul ophiolite, with Jurassic to Cretaceous deep water shales, pillow lavas and cherts (Mennessier 1977; Mattauer *et al.* 1978; Tapponnier *et al.* 1981). The original tectonic setting of the block is uncertain. It could be part of the southern Asian Cimmeride collage displaced during early stages of Himalayan collision; a fragment of the Indian Plate displaced by strike slip motions; or a separate 'exotic' fragment. Part of the key to understanding the Kabul block is that prior to the Palaeocene the block was completely surrounded by oceanic crust, preserved now in the form of the Kabul ophiolite to the west and the Khost ophiolite to the east. This relationship is interpreted here as implying that prior to the Palaeocene, the Kabul block represented a narrow sliver of continental crust of unknown length, located at an unknown distance offshore of the northwestern margin of the Indian Plate. Modern analogies for this slice of continental crust, may be the Rockall Shelf offshore of northwest Scotland and the off Bombay High of the western Indian margin (Kolla & Coumes 1990).

The Katawaz basin

The Katawaz basin outcrops to the east of the Chaman fault and extends east into Pakistan.

Although the lowermost parts of the sedimentary sequence may be pelagic (Tapponnier *et al.* 1981), the majority appear to be shallow water. It contains a thick sequence of dominantly clastic deposits (Ganss 1970; Schreiber *et al.* 1972), the base of which is dated by nummulitic limestones of Eocene age which represent the westernmost extent of the Indian Plate calcareous shelf sequence. The top is represented by Pliocene units of the Oligocene to Pliocene aged Shaigalu Formation (Hunting Survey Corporation 1960). The lower parts of the clastic sequence are represented by the >3000 m thick Eocene to Early Oligocene Siahan Shales. At their base these consist of sandy and silty shales, shales and mudstones with lenses of sandstone. The sequence coarsens upwards with increasing contents of thickly bedded sandstones intercalated with conglomerates, siltstones and clay rich marls (Hunting Survey Corporation 1960; Schreiber *et al.* 1972). The Siahan Shales unit passes up into the >6000 m Shaigalu Sandstone unit (Hunting Survey Corporation 1970), which consists of intercalated crossbedded and ripple marked sandstones and shales, often red in colour, with horizons of conglomerate and limestone. Although the lowermost parts of the Siahan Shales contain evidence for marine deposition that was remarkably devoid of coarse clastic sediments there is a clear upward change, through the Oligocene, from shale dominated to sandstone and conglomerate dominated lithologies with the presence of nummulites indicating a shallow to intertidal depositional setting. The Hunting Survey Corporation Report (1960) implies both an upward and a northward transition from shallow marine to estuarine or fluviatile deposition conditions for the Shaigalu Sandstone unit.

The basement of the sequence is interpreted here as transitional continental crust for the following reasons. Firstly, to the east, the clastic sediments overlie the Indian Plate Mesozoic shelf sediments of western Pakistan and ophiolitic rocks of the Khost–Waziristan–Muslimbagh complexes thrust onto them. In the north they overlie thinly bedded Jurassic and Cretaceous limestones of the Kabul block. Secondly, gravity data (Marussi 1976; Kurshid 1991 *in* Jadoon 1992) is consistent with the clastic sediments of the Katawaz basin overlying transitional crust that straddles the transition from Indian Plate continental crust, overlain by Mesozoic shelf carbonates, and oceanic crust (Izatt 1990; Jadoon 1992).

The rocks of the Katawaz Basin have often been loosely described as 'flysch' (Schreiber *et al.* 1972; Tapponnier *et al.* 1981), a description that has generated the view that they are trench deposits (Sengor *et al.* 1988). However, the term "flysch" has been of somewhat nebulous use world wide, is of diverse definition and of disputed worth. The Hunting Survey Corporation Report (1960) does not describe turbidites and specifically refrained (p. 20–21) from attributing a flysch-type genetic origin to these deposits. At best, therefore, a trench setting is unproven. Indeed the fauna and sedimentology of the Shaigalu Formation is more typically that of shallow water marine to fluviatile environments.

Rocks of the Shaigalu Formation are lithologically similar to those of the Oligocene–Pliocene aged Sibi Supergroup of the Indus Basin/Sulaiman foredeep west of the Sulaiman Ranges (Hunting Survey Corporation 1960) which also show an upward transition from shallow marine to estuarine/fluviatile sediments and with which they can be correlated. A better analogue for the base of the Sibi Supergroup – Katawaz basin sediments may be the present day Indus Fan. Both Auden (1974) and the Hunting Survey Corporation report (1960) demonstrate strong lithological similarities between the Katawaz basin sediments and those within the Makran region to the south (Fig. 1).

Thus the Katawaz basin can be interpreted as part of one large flexural basin, located on the western margin of the Indian Plate (Fig. 7a). Subsidence, synchronous with sedimentation, of the shelf margin in the early Eocene is indicated by the deposition of clay rich sediments on nummulites bearing limestones. The thickness of Tertiary sediment preserved within the Katawaz basin, where it is probably in excess of 10 000 m as well as in the Sulaiman foredeep and the Makran, is indicative of the magnitude of the basin subsidence. However, the dominance of shallow marine to fluviatile sediments in the upper half or more of the sedimentary succession is indicative that, although subsidence continued as the basin filled during the Oligo-Pliocene, sedimentation was not in a trench environment.

This large flexural basin has been shortened and inverted during Late Tertiary collision between India and Afghanistan. The dominant structural elements within the Katawaz basin, and which affect the entire stratigraphic sequence, are tight folds characterised by relatively open synforms and basins separated by tight antiforms and domes (Ganss 1970; Wittekindt & Weippert 1973). A well-pronounced slaty cleavage is developed in suitable lithologies throughout the sequence. The structural grain of the basin swings from an average SW trend

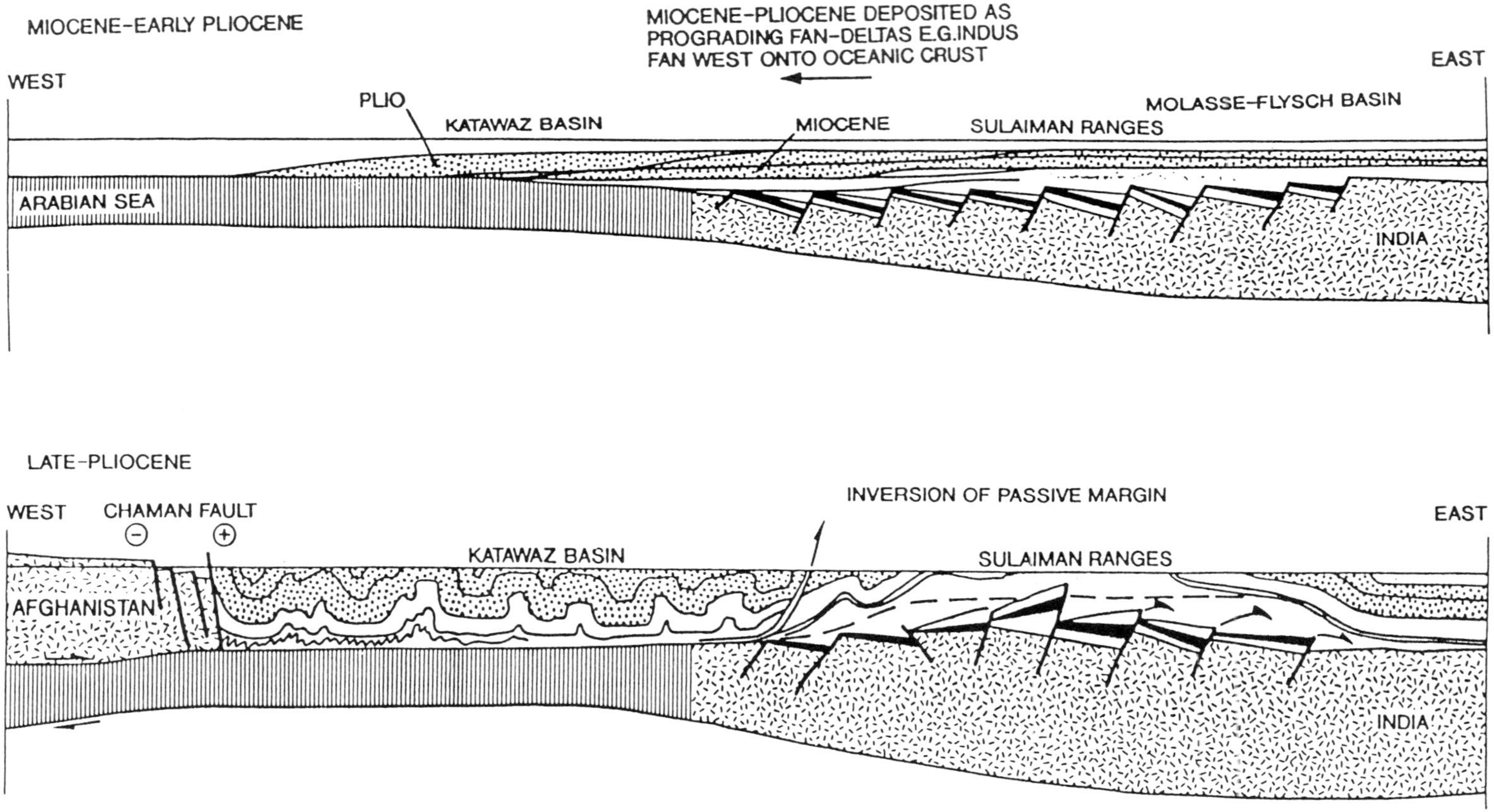

Fig. 7. Schematic east–west sections across the India-Afghanistan collision zone during the Mio-Pliocene and Late Pliocene showing deposition of clastic sediments of the Katawaz and Indus basins on extended Indian Plate continental and transitional crust that continued until the Early Pliocene. Late Pliocene deformation involved inversion of the Indian Plate passive margin with the resultant pop-up of the Sulaiman thrust duplex; shortening across the Katawaz basin; and the over-riding of the transitional to oceanic crust by the Afghan Plate, the eastern margin of which is marked by the Chaman fault.

(locally WSW) in the south to NNE (locally N) in the north. The apparent correlation between the Shaigalu Formation of the Katawaz basin and the Sibi Supergroup of the Sulaiman foredeep has implications for deformation style within the Sulaiman Ranges. Rather than being a thrust type deformation with shortening measured in hundreds of kilometres (Jadoon *et al.* 1992), this deformation should maybe be related to inversion of the Indian Plate margin which involved shortening and uplift of the Sulaiman Ranges from beneath the overlying Neogene cover (Fig. 7b) but with little total shortening (M.P. Coward, pers. comm. 1992).

Within the Katawaz basin, folding took place prior to the deposition, on an erosional unconformity, of the Pleistocene Bostan Formation (Hunting Survey Corporation 1970). Although we cannot exclude the possibility that some of the lower parts of the sequence underwent some deformation during the Miocene or earlier, a Late Pliocene age appears likely for the age of the main deformation. This age is not tightly constrained, although appears likely as the uppermost members of the Shaigalu Formation, which scanty palaeontological evidence suggests to be lowest Pliocene, are folded together with the rest of the sequence.

Synthesis and implications

Timing of regional 'Himalayan deformation' in southern and central Afghanistan and western Pakistan

The timing of the regional deformation must be inferred on the basis of the age of the youngest rocks deformed. Within the Farad block, the major phase of folding which deformed Oligocene basaltic to andesitic volcanics dated at 31.5 Ma (Debon *et al.* 1987), and the interbedded red clastic flysch deposits, must postdate 30 Ma.

Within the northern part of the Helmand block and the southern part of the Hindu Kush, ENE–WSW-striking thrusts deform the basement sequences and invert Miocene sediments within the Bamyan basin. These thrusts must be late-Miocene at the earliest, and are interpreted as Mio-Pliocene by Boulin & Bouyx (1977). They curve into, and are correlated here with, the northwest striking thrusts within the Helmand block and the Panjao suture (Fig. 4), that accommodate shortening of the Helmand block and which imbricate Cretaceous shelf sediments with their Paleozoic basement. This deformation is correlated here with that within the Farad block.

An Oligocene aged syn-depositional deformation within the Bamyan basin (Lang 1977), is interpreted here as being related to shortening episodes within a generally transtensional strike slip zone.

Within the Katawaz basin, deformation postdates the deposition of the majority of the sedimentary sequence (Hunting Survey Corporation 1960). As deposition continued into at least the Upper Miocene or Lower Pliocene this places the main shortening as commencing during the Pliocene, but before the deposition of the undeformed Pleistocene Bostam Formation. A similar age is inferred, on the basis of relationships between thrusts and the Pliocene Dhok Pathan Formation, for the main stage deformation in the Kirthar and Sulaiman Ranges of Pakistan where it is post-Miocene or Late Pliocene (Izatt 1990) in age. Similarly, transpressive deformation in the Kohat Plateau, which postdates an early Miocene to Pliocene south-vergent thrusting, appears to have reached its peak during the late Pliocene to Pleistocene (Pivnik & Sercombe this volume).

Thus it appears that much of the regional shortening deformation within southern and central Afghanistan, in the Katawaz basin and the Hill Ranges of western Pakistan is probably post-Miocene in age. Significant transpressive movements are Plio-Pleistocene and younger in age. Movement along the majority of the strike slip faults continues to the present day, as documented by earthquake data (Haghipour *et al.* 1984) and recent river offsets (Wellman 1966). An important exception to this is the Herat fault, dextral movement along which occurred during the Oligo-Miocene, although geomorphic data show little evidence for post-Miocene movement (Tapponnier *et al.* 1981).

The timing of collision between the Indian plate and the Afghan Block

Timing of collision between the Indian Plate and the southern margin of the Asian Block is poorly constrained. On the basis of palaeomagnetic data, Klootwijk *et al.* (1991) constrain this at about 55 Ma, although with the possibility that collision was earlier in the northwest than further east. By contrast, on grounds of pre-collisional plate shapes and rates of plate movement Dewey *et al.* (1989) and Le Pichon *et al.* (1992) date initial collision as late as 45 Ma. In North Pakistan and Northeast Afghanistan this collision essentially dates that between the

Indian Plate and the East Nuristan–Kohistan–Ladakh island arc complex which had sutured to Asia 100 Ma ago.

The early Tertiary age does not, however, date collision across the Chaman fault zone between the West Pakistan margin of the Indian Plate and the Afghan Plate. The evidence outlined above shows that the majority of Himalayan age deformation that affected Afghanistan and West Pakistan from the Herat fault zone southward to east of the Kirthar and Sulaiman Ranges is post-Miocene in age. The age data are consistent with two possible scenarios. Firstly, deformation migrated from northwest to southeast across the collision zone. Secondly, all the deformation is Pliocene in age. The second hypothesis, preferred here, would be consistent with India–Afghan collision taking place during the early to mid-Pliocene. In particular, the fact that the Katawaz sediments were not deformed until the Late Pliocene must imply that collision between India and Afghanistan cannot have occurred until then.

The timing of metamorphism on the Indian Plate of North Pakistan

Collision between the Indian Plate and the Kohistan island arc involved subduction of the former beneath the latter. On its northern margin the Kabul block abuts against the East Nuristan terrain, and by implication any northward extension of the Kabul Block must have been subducted beneath the East Nuristan arc before the leading edge of continental India. There is now sufficient data to demonstrate that, in the Indian plate rocks of northwestern Pakistan, the metamorphic peak was attained at about 50 Ma ago (Treloar & Rex 1990; Chamberlain *et al.* 1991). Although this could be consistent with India–Kohistan collision at about 55 Ma ago (Treloar & Coward 1991; Klootwijk *et al.* 1991) it still represents a very short time interval from collision to metamorphic peak, not obviously compatible with conductive models of metamorphism after crustal thickening. However, this apparent dichotomy could be explained if metamorphism of the Indian plate of northwestern Pakistan represented crustal thickening following Palaeocene collision between the leading edge of continental India and an exotic block such as the Kabul block. The thickened Indian continental crust could then have been subsequently overthrust by Kohistan with little new deformation and metamorphism. Such a collision could also explain the Palaeocene ophiolite emplacement event following sudden changes in plate movements.

Implications for plate reconstructions

There are some implications for plate reconstruction models here, in that these should have to provide for the separation of Northwest India and Afghanistan throughout the Tertiary until some time in the Pliocene, although by how much is uncertain. In addition, it may not be appropriate to estimate the southern margin of pre-collisional Asia as being determined at its western end by the present day position of Afghanistan (cf. Tapponnier *et al.* 1981; Dewey *et al.* 1989). Models of the behaviour of Asian crust since collision have tended to imply that the southern margin of Asia was essentially a great circle segment into which India indented (Le Pichon *et al.* 1992). That this was not so may be indicated by the separation of India and Afghanistan until the Pliocene. A plausible case can be made for significant translation of the Afghan block following initial collision between India and Asia prior to closing of the oceanic gap between India and Afghanistan. Klootwijk & Radhakrishnamurthy (1981) document counterclockwise rotations within Afghanistan during the Tertiary, and it is possible that part of the response of Asian crust to initial collision with India was the southward translation with associated counterclockwise rotation of the entire Afghan Block.

The role of strike slip faulting in accommodating crustal extrusion

Since Tapponnier *et al.* (1986) suggested that a process of lateral extrusion of thickened crust by strike slip faulting, could accommodate some of the shortening resultant from India–Asia collision, a number of the Afghanistan strike slip faults have been modelled in terms of lateral extrusion (Fig. 8). In particular, the dextral Herat fault and the sinistral Chaman fault have featured on maps of the southern Asia strike slip fault system (cf. Dewey *et al.* 1989; Searle 1991) with the, possibly unintentional, implication that they were active together. However, the Herat fault, although active during the Oligo-Miocene, shows little evidence for displacements subsequent to that. By contrast, the Chaman fault, which marks the eastern margin of the Afghan block (Izatt 1990), is seismically active (Yeats *et al.* 1979; Haghipour *et al.* 1984), although cannot have been so before India–Afghan collision during the Pliocene. Thus the two faults cannot have acted together as structures that controlled crustal extrusion. Some limited southwest directed extrusion of Afghanistan may be accommodated by continu-

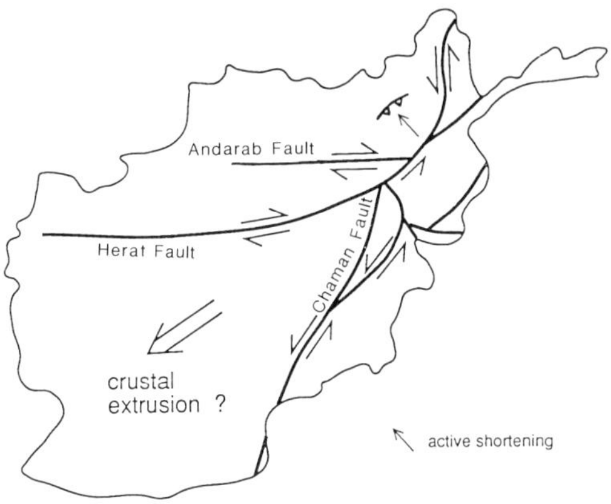

Fig. 8. A simplified, and incorrect, model of crustal extrusion within Afghanistan. The dextral Herat and sinistral Chaman faults were never active together and can therefore not have controlled crustal extrusion. However, it is possible that some extrusion can be accommodated by the Chaman fault and the recently operative dextral faults to the north of the Herat Fault such as the Andarab fault. Otherwise, thrusts at the base of the Western Pamir may accommodate much of the recent crustal shortening in northern Pakistan and Afghanistan.

ing movements along the Chaman Fault and the Andarab and Talemazar faults of northern Afghanistan. Historical and recorded earthquakes along the Andarab and Talemazar faults have focal depths of >50 km and magnitudes of >6 (Haghipour *et al.* 1984). Displacements are, though, apparently not large (Wellman 1966; Tapponnier *et al.* 1981) and it is unlikely that major crustal extrusion is taking place along them.

A geological history of the India–Afghan collision

Pre-Palaeocene (Fig. 9a)

Prior to the Palaeocene, as the Indian Plate moved northward towards Asia, subduction of the oceanic part of the plate beneath southern Asia generated the volcanic arc complexes which extend from the Gangdese batholith in the east, through the East Nuristan–Kohistan–Ladakh arc which had sutured to Asia at about 100 Ma ago, to the calc-alkaline complexes of the Helmand block on the eastern edge of the Afghan Block. Cretaceous volcanism in the Chagai arc of southwestern Pakistan and southeastern Iran, south of the present southern border of Afghanistan, represented volcanism related to subduction of the Tethyan oceanic plate beneath southernmost Afghanistan (Arthurton *et al.* 1979, 1982). Included within the confines of the overall Indian Plate but surrounded by, and separated from continental India, oceanic crust was the Kabul block continental micro-plate.

Palaeocene (Fig. 9a)

The Palaeocene was marked by ophiolite emplacement events along the northwestern margin of the Indian Plate from Bela through Muslimbagh-Khost to Spontang in northwestern India (Asruralluh & Abbas 1979; Searle 1986). Synchronously, the Kabul ophiolite was emplaced onto the Kabul block (Mattauer *et al.* 1978). Although magmatism in the Kohistan–East Nuristan arc continued into the early Eocene (Treloar *et al.* 1989) and was continuous throughout the period in the Chagai arc, it ceased in the Helmand block in the late Cretaceous. This is probably because the eastern margin of the Afghan Plate was parallel or sub-parallel to the movement vector of the Indian Plate. The margin was thus either a true conservative one, or one of highly oblique subduction with little or no associated magmatism. As Indian Plate movement was veering in a counterclockwise direction, this change from a subduction type margin to a conservative one must have involved a counterclockwise rotation of the Afghan Plate, possibly in response to suturing of the Kohistan–East Nuristan arc to the southern margin of Asia.

Eocene–Miocene (Fig. 9b)

The Indian Plate continued to move northwards with direct collision occurring between India and the island arc complex that marked the southern margin of the Asian Plate. Crustal thickening and eventual crustal melting commenced within the Asian Plate to the north of the Kohistan–East Nuristan arc in the Pamir, Hindu Kush and Karakorum (Debon *et al.* 1983, 1987; Crawford & Searle 1992). Continued indentation of the Indian Plate resulted in sinistral strike slip movements along major faults in northeastern Afghanistan, such as the central Badakhstan fault, and along the eastern end of the Shyok suture. As a result of this indentation a considerable re-adjustment of the south Asian margin took place, with the Afghan Block being displaced southward and rotating in a counterclockwise direction (Klootwijk & Radhakhrishanamurthy 1981). Internal stresses associated with this rotation could have led both

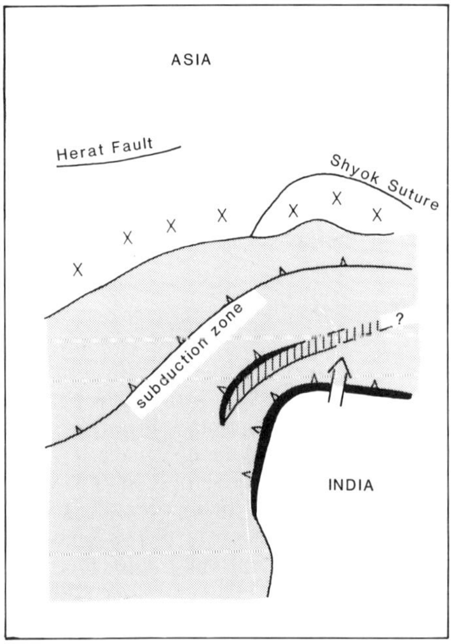

a) Late Cretaceous - Paleocene

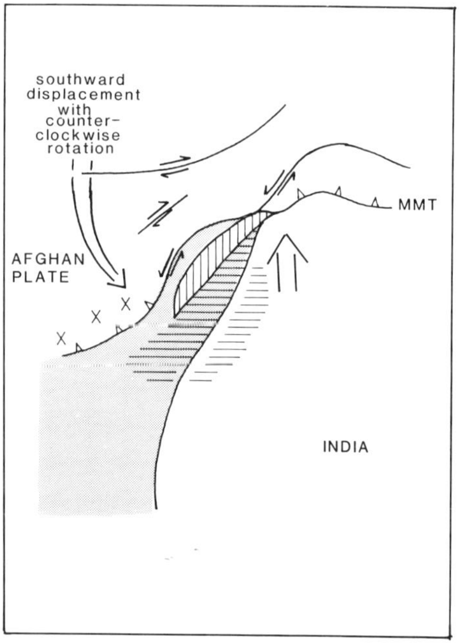

b) Eocene - Miocene

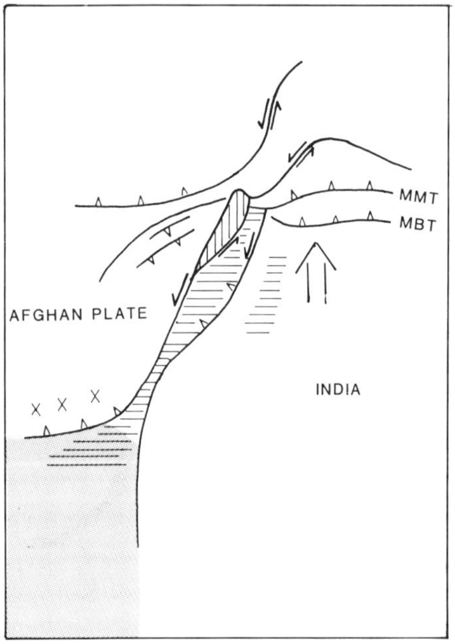

c) Pliocene

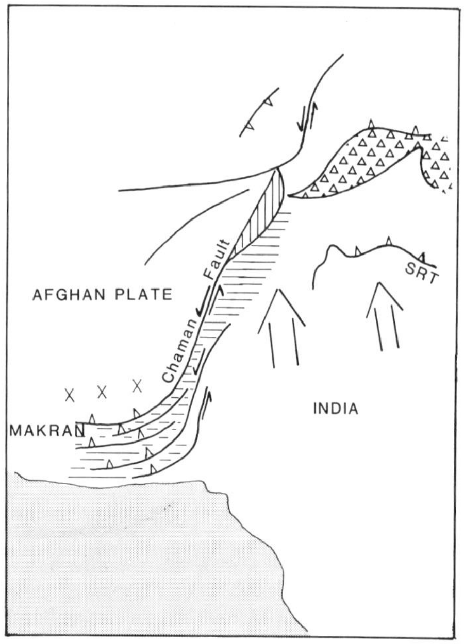

d) Recent

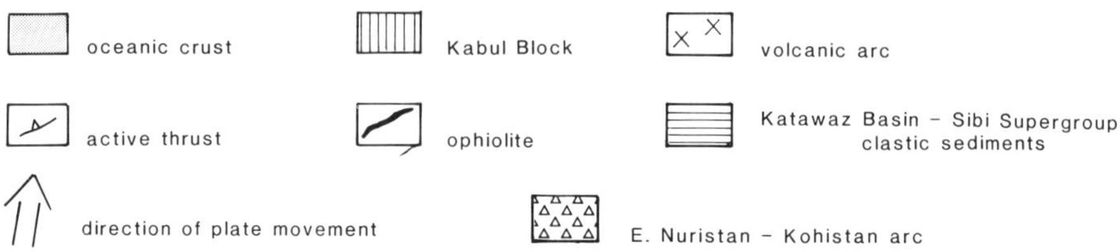

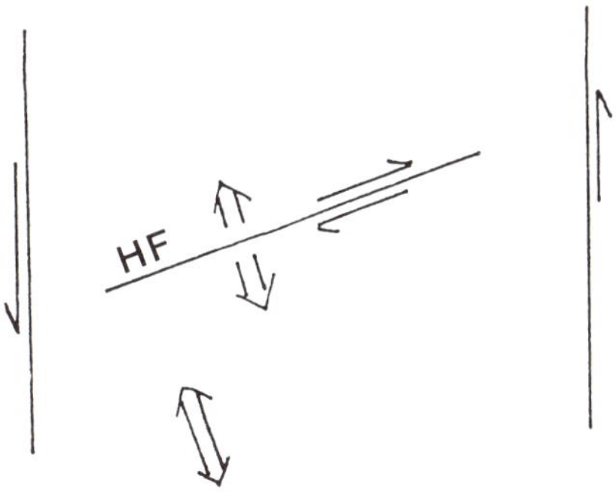

Fig. 10. Simplified model explaining both crustal extension in central Afghanistan and right lateral movement along the Heraut fault (HF) within a region of counter-clockwise rotation, modelled here as a block bounded by two large strike slip faults.

to the re-activation of the Cimmeride Herat–Panjshir suture as the dextral Herat fault (Fig. 10) and to intraplate extension leading to alkaline volcanism in the Herat fault zone, as well as in the Band-e-Bayan and Farad blocks.

As India continued to move northward, clastic sedimentation, in what are now the Katawaz basin and the Sulaiman foredeep, commenced on the extended and thinned crust of westernmost Indian Plate and on the transitional oceanic crust separating India from Afghanistan (Fig. 7a). The Indian Plate oceanic crust must have bent downwards beneath the Afghan Block (Izatt 1990), linking beneath southern Afghanistan with crust being subducted northwards under Afghanistan. The deeper water sediments at the base of the Katawaz basin sedimentary pile may reflect early stages of this downbending of the transitional crust.

Pliocene (Fig. 9c)

All the evidence that relates to timing of deformation across the region from the Farad block in western Afghanistan to the Sulaiman and Kirthar Ranges in western Pakistan, is consistent with shortening across the India–Afghanistan collision zone as having commenced during the Pliocene. This must mark the major collision between continental India and Afghanistan. The lack of ophiolites along the suture and of Tertiary subduction related volcanism in the Helmand block suggests that the final collision was highly oblique. This would be consistent with an overall northward vector of movement for the Indian plate sub-parallel to the trend of the eastern Afghanistan margin. The Pliocene collision must have involved thrusting of Afghanistan onto the sediments deposited in the Katawaz basin upon transitional and thinned crust of the Indian Plate margin (Fig. 7). Inversion of earlier continental margin basins characterises the deformation of both the Indian and Afghanistan margins. Change in structural vergence across the Helmand block is consistent with the reactivation of earlier extensional structures during a pop-up style inversion of a Mesozoic continental margins. Southeast vergent folds developed within the southeastern part of the Helmand Block may, however, be related to southeast directed thrusting of the block onto the sediments of the Katawaz basin, the folds detaching on the thrust located at the base of the Afghan continental crust. On the western margin of the Indian Plate, the collision involved shortening and inversion of the marginal basin uplifting Mesozoic sediments from beneath the Katawaz Basin/Sulaiman foredeep clastic sequence within the thrust duplexes of the Kirthar and Sulaiman Ranges (Fig. 7b).

Pliocene to Recent (Fig. 9d)

With the continued northward movement of India past Afghanistan, the margin between India and Afghanistan has been marked by transpressional tectonics. In the Kohat Plateau of western Pakistan, south vergent Miocene to Pliocene structures have been overprinted by north-trending Plio-Pleistocene transpressional ones (Pivnik & Sercombe this volume). The sinistral Chaman strike slip fault essentially marks the eastern boundary of the Afghan Plate (Izatt 1990). At its southern end the Chaman fault and associated faults to the east (Ornach Nal and Ghazaband) act as lateral ramps to south-vergent thrusts within the Makran prism (Fig. 1). If the Eocene to Miocene sediments of the Makran are laterally equivalent to those of the Katawaz basin (Hunting Survey Corporation Report 1960), and were originally deposited on transitional to oceanic Indian crust, then they must have been displaced left laterally from the rest of the clastic sequence accreting to the leading edge of the Afghan Block as India

Fig. 9. Plate tectonic reconstructions of the Indian Plate–Afghan Plate collision zone. Note that final collision did not take place until the Pliocene. MBT, Main Boundary Thrust; MMT, Main Mantle Thrust; SRT, Salt Range Thrust.

moved northwards passed it. Assuming a maximum relative displacement between the Indian plate and Afghanistan of 5 mm a^{-1} from the Early Pliocene onwards, the maximum total displacement on the Chaman fault zone cannot exceed 300 km. However, the left lateral displacement of sediments within the Makran with respect to those of the Katawaz basin may be greater than this if the leading edge of the Afghan Block collided with the clastic wedge offshore of continental India during the Miocene. All of the original tectonic relationships along the eastern margin of the Afghan Plate have been obscured by Pliocene and post-Pliocene strike slip faulting with the exotic Kabul block, for instance, now bounded on both sides by left lateral faults. Within Afghanistan the left lateral Chaman fault and right lateral Andarab fault between them may accommodate some limited southwest directed lateral extrusion of the Central Afghan crust. However, a more significant structure may be the thrust system developed along the northwest side of the Pamirs, along which that mountain chain is currently being thrust out over the molasse filled basins of the Tadzhik Platform (Abers *et al.* 1988).

The only remaining volcanism within the sector is that now located across the Iran-Pakistan border (Arthurton *et al.* 1979, 1982; Dykstra & Birnie 1979; Jacob & Quittmeyer 1979) and in southernmost Afghanistan, where volcanoes such as the Bazman, Taftan and Koh-I-Sultan are currently active. These represent melting of the Indian Plate as it subducts beneath the Makran prism.

Conclusions

Collision between India and Asia along the northern part of the Indian Plate took place in the early Tertiary. However, timing of deformation on structures as far apart as the Kohat plateau of western Pakistan and the Farad block of west central Afghanistan demonstrate that collision between the western part of the Indian Plate and Afghanistan did not occur until the Pliocene. India and Afghanistan were thus separated until the Pliocene by oceanic and extended continental margin crust of the Indian Plate, on which were deposited the marginal marine to fluviatile clastic deposits of the Katawaz basin.

A number of implications arise from this: the southern margin of the Asian Plate prior to collision cannot have been a straight line, or great circle segment, extending eastward from the present day position of southern Afghanistan; the Chaman fault, which marks the western margin of continental Afghanistan where it over-rides the Katawaz basin, cannot have existed prior to the Pliocene and hence cannot have accommodated more than 300 km of left lateral displacement; and, as movements along the right lateral Herat fault had ceased by the late Miocene, it is not possible for the Herat and Chaman faults to have acted together in accommodating southwest directed extrusion of Afghanistan crust.

The earliest Himalayan event on the Asian Plate was subduction related magmatism along the southern Asian and eastern Afghanistan margins which, in Afghanistan, ended in the late Cretaceous. On the Indian Plate the earliest event is the Palaeocene ophiolite emplacement recorded at Spontang in India, and Khost–Waziristan–Muslimbagh in Pakistan and Afghanistan. That this was contemporaneous with ophiolite emplacement onto the Kabul block has an important implication in that the Kabul block, composed of continental crust, must have been fully surrounded by oceanic crust before the Palaeocene. The block may therefore have formed a slice of continental crust offshore of continental India. Recognition of this may solve an apparent enigma that relates to the timing of collision and metamorphism in the northwest Himalaya. Here peak metamorphism occurred at or before 50 Ma ago, although independent estimates of the initiation of India-Asia collision range from 55 to 45 Ma. Although it would not help constrain the timing of initial collision between India and Asia, a Palaeocene collision between continental India and the Kabul block could have resulted in thickening and metamorphism of the leading edge of the Indian Plate before the Indian Plate collided with and was subducted beneath the East Nuristan–Kohistan–Ladakh island arc.

During much of the Tertiary, after initial collision between India and Asia but before India–Afghanistan collision, a re-arrangement of blocks within the South Asian collage resulted in a displacement of the Afghan Plate probably involving an overall southward displacement associated with an anticlockwise rotation. Resultant stresses within the rotating plate caused local crustal extension within central Afghanistan leading to Oligocene alkaline magmatism, and the initiation of dextral displacements along the Herat fault zone, thus reactivating structures in an older Cimmeride suture zone.

The western margin of the continental Indian Plate was marked, throughout the Tertiary, by the development of a large marginal basin located across the transition between thinned continental crust and oceanic crust. A thick sequence of clastic sediments deposited within this basin is exposed now in the Katawaz basin, the Sulaiman foredeep and the Makran. Although the Makran part of this clastic wedge was accreted to the leading edge of the Afghan Plate and displaced left laterally from the rest of the sedimentary sequence during the Mio-Pliocene, the remainder was shortened and inverted during India–Afghan collision. The Pliocene collision between India and Afghanistan was highly oblique. Structures within both the Helmand block of the eastern Afghanistan margin and the Sulaiman and Kirthar Ranges of western Pakistan can be interpreted as resulting from inversion of earlier continental margin structures, possibly with no great amounts of shortening across them, but increasingly transpressive features are characteristic of the Plio-Pleistocene deformation associated with India–Afghanistan convergence.

We acknowledge reviews from M.P. Searle and A.M.C. Sengor, the latter a particularly detailed one. Discussions over many years with M.P. Coward have strongly influenced our thinking. We are grateful to I.A.K. Jadoon both for discussion and for providing preprints of his relevant papers.

References

ABERS, G., BRYAN, C., ROECKER, S. & MCCAFFREY, R. 1988. Thrusting of the Hindu Kush over the southeastern Tafzhik Basin, Afghanistan: evidence from two large earthquakes. *Tectonics*, **7**, 41–56.

ALLEMAN, F. 1979. Time of emplacement of the Zhob Valley ophiolites and Bela ophiolites, Baluchistan. *In*: FARAH, A. & DEJONG, K.A. (eds) *Geodynamics of Pakistan*. Geological Survey of Pakistan, Quetta, 215–242.

ARTHURTON, R.S., ALAM, G.S., AHMAD, S.A. & IQBAL, S. 1979. Geological history of the Alamreg-Mashki Chah area, Chagai district, Baluchistan. *In*: FARAH, A. & DEJONG, K.A. (eds) *Geodynamics of Pakistan*. Geological Survey of Pakistan, Quetta, 325–331.

——, FARAH, A. & AHMED, W. 1982. The Late Cretaceous-Cenozoic history of western Baluchistan, Pakistan – the northern edge of the Makran subduction complex. *In*: LEGGETT, J.K. (ed.) *Trench Forearc Geology: sedimentation and tectonics on modern and ancient active plate margins*. Geological Society, London, Special Publications, **10**, 373–385.

ASRARULLAH, Z.A. & ABBAS, S.G. 1979. Ophiolites in Pakistan: an introduction. *In*: FARAH, A. & DEJONG, K.A. (eds) *Geodynamics of Pakistan*. Geological Survey of Pakistan, Quetta, 325–331.

AUDEN, J.B. 1974. Afghanistan-West Pakistan. *In*: SPENCER, A.M. (ed.) *Mesozoic–Cenozoic orogenic belts*. Geological Society, London, Special Publications, **4**, 235–253.

BELLON, H., BORDET, P. & MONTENANT, C. 1979. Histoire magmatique de l'Afghanistan central: nouvelles donnees chronometriques K-Ar. *Comptes Rendus de l'Academy des Sciences, Paris, Series D*, **289**, 1113–1116.

BOULIN, J. 1981. Afghanistan structure, greater India concept, and eastern Tethys evolution. *Tectonophysics*, **72**, 261–287.

—— 1988. Hercynian and Eocimmerian events in Afghanistan and adjoining regions. *Tectonophysics*, **148**, 253–278.

—— 1991. Structures in southwest Asia and adjoining regions. *Tectonophysics*, **196**, 211–268.

—— & BOUYX, E. 1977. Introduction a la geologie de l'Hindu Kouch oriental. *Memoir hors-series de la Societe Geologique de France*, **8**, 87–106.

CHAMBERLAIN, P.C., ZEITLER, P.K. & ERICKSON, E. 1991. Constraints on the tectonic evolution of the northwestern Himalaya from geochronologic and petrologic studies of Babusar Pass. *Journal of Geology*, **99**, 829–849.

CRAWFORD, M.B. & SEARLE, M.P. 1992. Field relationships and geochemistry of pre-collisional (India-Asia) granitoid magmatism in the central Karakoram, northern Pakistan. *Tectonophysics*, **206**, 171–192.

DEBON, F., AFZALI, H., LE FORT, P. & SONET, J. 1987. Major intrusive stages in Afghanistan: typology, age and geodynamic setting. *Geologisches Rundschau*, **76**, 245–264.

——, LE FORT, P., DAUTEL, D., SONET, J. & ZIMMERMAN, J-L. 1983. Chemical-mineraological typology, chronology and geodynamic setting of the western Hindu-Kush-Badakshan plutonic belt (Afghanistan). *In*: SHAMS, F.A. (ed.) *Granites of the Himalayas, Karakorum and Hindu Kush*. Punjab University, Lahore, 407–423.

DEWEY, J. F., CANDE, S. & PITMAN, W.C. 1989. Tectonic evolution of the India/Eurasia collision zone. *Eclogae Geologicae Helvetia*, **82**, 717–734.

DYKSTRA, J.D. & BIRNIE, R.W. 1979. Segmentation of the Quaternary subduction zone under the Baluchistan region of Pakistan and Iran. *In*: FARAH, A. & DEJONG, K.A. (eds) *Geodynamics of Pakistan*. Geological Survey of Pakistan, Quetta, 319–323.

GANSS, O. 1970. *Zur geologi von Sudost Afghanistan*. Geologische Jahrbruch Beihefte, **84.**

GIRARDEAU, J., MARCOUX, J. & MONTENANT, C. 1988. The Neo-Cimmerian ophiolite belt in Afghanistan and Tibet: comparison and evolution. *In*: SENGOR, A.M.C. (ed.) *Tectonic evolution of the Tethyan region*. Kluwer Academic Publishers, 477–504.

HAGHIPOUR, A., GHORASHI, M. & KADJAR, M. H. 1984. *Explanatory tect of the seismotectonic map of Iran, Afghanistan and Pakistan*. Geological Survey of Iran, Tehran.

HUNTING SURVEY CORPORATION 1960. *Reconnaisance geology of part of West Pakistan. A Colombo Plan co-operative project*. Toronto.

IZATT, C.N. 1990. *Variations in thrust front geometry*

across the Potwar Plateau and Hazara/Kalachitta Hill Ranges, northern Pakistan. PhD thesis. Imperial College, London.

JACOB, K.H. & QUITTMEYER, R.L. 1979. The Makran region of Pakistan and Iran: trench-arc system with active plate subduction. *In*: FARAH, A. & DEJONG, K.A. (eds) *Geodynamics of Pakistan*. Geological Survey of Pakistan, Quetta, 305–317.

JADOON, I.A.K. 1992. Ocean/continent transitional crust underneath the Sulaiman thrust lobe and an evolutionary tectonic model for the Indian/Afghan collision zone. *Pakistan Journal of Hydrocarbon Research*, **4**, 33–44.

——, LAWRENCE, R.D. & LILLIE, R.J. 1992. Balanced and retrodeformed geological cross-section from the frontal Sulaiman lobe, Palistan: duplex development in thick strata along the western margin of the Indian Plate. *In*: MCCLAY, K.R. *Thrust Tectonics*. Chapman & Hall, London, 343–356.

KAFARSKIY, A.K., GHMYRIOV, V.M., DRONOV, V.I., STAZHILO-ALEKSEEV, K.F., ABDULLAH, J. & SAIKOVSKIY, V.S. 1975. Tectonic map of Afghanistan. *Himalayan Geology Seminar*. Geological Survey of India, Miscellaneous Publications, **41**.

KHURSHID, A. 1991. *Crustal structure of the Sulaiman Range*. MSc thesis. Oregon State University.

KOLLA, V. & COUMES, F. 1990. Extension of structural elements and tectonic trends from the Indian sub-continent into the Eastern Arabian Sea. *Marine and Petroleum Geology*, **7**, 188–196.

KLOOTWIJK, C.T. & RADHAKRISHANAMURTHY, C. 1981. Phanerozoic paleomagnetism of the Indian plate and the India-Asia collision. *In*: *Paleoreconstruction of Continents*. American Geophysical Union. Geodynamics Series, **2**, 93–105.

——, GEE, J.S., PEIRCE, J.W. & SMITH, G.M. 1991. Constraints on the India-Asia convergence: Paleomagnetic results from Ninetyeast Ridge. *Proceedings of the Ocean Drilling Program, Scientific results*, **121**, 777–882.

LANG, J. 1977. Evolution geodynamic de bassins intramontagneux tertiares d'Afghanistan central (Bamyan, Yakawlung et vallee du Ghorband). *Memoir hors-series de la Societe Geologique de France*, **8**, 107–116.

LAWRENCE, R.D., KHAN, S.H., DEJONG, K.A., FARAH, A. & YEATS, R.S. 1981. Thrust and strike slip fault interaction along the Chaman transform Zone, Pakistan. *In*: MCCLAY, K. & PRICE, N.J. (eds) *Thrust and Nappe Tectonics*. Geological Society, London, Special Publications, **9**, 363–368.

LE PICHON, X., FOURNIER, M. & JOLIAT, L. 1992. Kinematics, topography, shortening and extrusion in the India-Eurasia collision. *Tectonics*, **11**, 1085–1098.

MARUSSI, A. 1976. Introductory report of geophysics. *In*: *Geotectonica Della zone orogeniche del Kashmir Himalaya, Karakorum-Hindu Kush-Pamir*. Accademia dei Lincei, **21**, 237–242.

MATTAUER, M., PROUST, F., TAPPONNIER, P. & CASSAIGNEAU, C. 1978. Ophiolites, obductions et tectonique globale dans l'est de l'Afghanistan. *Comptes Rendus Academy de Sciences, Paris, Series D*, **287**, 983–985.

MENNESSIER, G. 1977. Stratigraphie, tectonique et evolution du fosse de Kaboul (Afghanistan). *Memoir hors-series de la Societe Geologique de France*, **8**, 153–168.

MONTENANT, C., GIRARDEAU, J. & MARCOUX, J. 1986. La ceinture ophiolitique Neo-Cimmerian au Tibet, dans les Pamirs et en Afghanistan.: evolution geodynamic comparative. *Science de la Terre memoire*, **47**, 229–252.

PIVNIK, D.A. & SERCOMBE, W.J. 1993. Compression and transpression related deformation in the Kohat Plateau, NW Pakistan. *This volume*.

SCHREIBER, A., WEIPPERT, D., WITTEKINDT, H-P. & WOLFART, R. 1972. Geology and petroleum potentials of central and south Afghanistan. *Bulletin of the American Association of Petroleum Geologists*, **56**, 1494–1519.

SEARLE, M.P. 1986. Structural evolution and sequence of thrusting in the High Himalayan, Tibetan Tethys and Indus Suture zones of Zanskar and Ladakh, western Himalaya. *Journal of Structural Geology*, **8**, 923–936.

SEARLE, M.P. 1991. *Geology and Tectonics of the Karakorum Mountains*. Wiley, London.

SENGOR, A.M.C. 1984. *The Cimmeride orogenic system and the tectonics of Eurasia*. Geological Society of America, Special Papers, **195**.

——, ALTIMER, D., CIN, A., USTOMER, T. & HSU, K.J. 1988. Origin, assembly of the Tethyside orogenic collage at the expense of Gondwanaland. *In*: AUDLEY-CHARLES, M.G. & HALLAM, A. (eds) *Gondwana and Tethys*. Geological Society, London, Special Publications, **37**, 119–181.

SHAREQ, A. 1981. Geological observations and geophysical investigations carried out in Afghanistan over the period of 1972–1979. *In*: GUPTA, H. & DELANY, F. (eds) *Zagros–Hindu Kush–Himalaya Geodynamic Evolution*. American Geophysical Union Series, **3**, 75–86.

TAPPONNIER, P., MATTAUER, M., PROUST, F. & CASSAIBNEAU, C. 1981. Mesozoic ophiolites, sutures and large scale tectonic movements in Afghanistan. *Earth and Planetary Science Letters*, **52**, 355–371.

TAPPONNIER, P., PELTZER, G. & ARMIJO, R. 1986. On the mechanics of the collision between India and Asia. *In*: COWARD, M.P. & RIES, A.C. (eds) *Collision Tectonics*. Geological Society, London, Special Publications, **19**, 115–157.

TRELOAR, P.J. & COWARD, M.P. 1991. Indian plate motion and shape: constraints on the geometry of the Himalayan orogen. *Tectonophysics*, **191**, 189–198.

—— & REX, D.C. 1990. Cooling and uplift histories of the crystalline thrust stack of the Indian Plate internal zones west of Nanga Parbat, Pakistan Himalaya. *Tectonophysics*, **180**, 323–349.

——, REX, D.C., GUISE, P.G., COWARD, M.P., SEARLE, M.P., WINDLEY, B.F., PETTERSON, M.G., JAN, M.Q. & LUFF, I.W. 1989. K-Ar and Ar-Ar geochronology of the Himalayan collision in NW Pakistan: constraints on the timing of suturing, deformation, metamorphism and uplift. *Tectonics*, **8**, 881–909.

WELLMAN, H.W. 1966. Active wrench faults in Iran, Afghanistan and Pakistan. *Geologisches Rundschau* **55**, 716–735.

WITTEKINDT, H. & WEIPPERT, D. 1973. *Geologisches karte von Zentral und Sud-Afghanistan. 1:500000*, 4 sheets. Bundesanst fur Bodenforsch, Hannover.

WOLFART, R. & WITTEKINDT, H. 1980. *Geologie von Afghanistan*. Beitrage zur regionalen geolgie der Erde, **14**.

YEATS, R.S., LAWRENCE, R.D., JAMIDUDDIN, S. & KHAN, S.H. 1979. Surface effects of the 16 March 1978 earthquake, Pakistan–Afghanistan border. *In*: FARAH, A. & DEJONG, K.A. (eds) *Geodynamics of Pakistan*. Geological Survey of Pakistan, Quetta, 359–361.

ZIEGLER, P.A. 19879. Geodynamic model for Alpine intra-plate compressional deformation in western and central Europe. *In*: COOPER, M.A. & WILLIAMS, G.D. (eds) *Inversion Tectonics*. Geological Society, London, Special Publication, **44**, 63–85.

Leucocratic rocks from the Bela ophiolite, Khuzdar District, Pakistan

ZULFIQAR AHMED

National Centre of Excellence in Mineralogy, University of Balochistan, GPO Box 43, Quetta, Pakistan

Abstract: This is the first detailed report of the well-developed leucocratic rocks, dominantly plagiogranites, from the Bela ophiolite, the most complete and largest ophiolitic outcrop of Pakistan. A wide range of leucocratic rock types occur within the ophiolite including leucogabbro, gabbroic pegmatite, diorite, quartz diorite, trondhjemite, epidote trondhjemite, granite, keratophyre and acidic pegmatite. Two suites of acidic leucocratic rocks are recognized: a very low potash and high soda suite containing magma-fractionated plagiogranite characteristic of most ophiolites, is accompanied by an anatectic suite with higher potash content. Each suite possesses distinct trace element and rare earth contents, together with distinctly different Sr and Nd systematics. These differences indicate differences in tectonic setting for each leucocratic suite. U-Pb isotopic measurements on zircons separated from the acidic rocks yield a crystallization age of 65 ± 1 Ma for both the rock suites. On the basis of geochemical data a suprasubduction zone origin dominated by a marginal basin is proposed for the Bela ophiolite. The plagiogranitic rock suite could have been derived from basic magma by fractional crystallization. The potash-richer granitic rocks may have been generated by crustal anatexis forming a melt that locally invaded the ophiolite.

Plagiogranites form conspicuous outcrops of relatively small dimensions in most ophiolites. Based on the specific geological features of each occurrence, plagiogranitic rocks have been shown to form in a variety of different ways (e.g., Coleman & Peterman 1975; Gerlach *et al.* 1981; Pedersen & Malpas 1984; Flagler & Spray 1991). The primary aim of this paper is to characterize the leucocratic rocks of the Bela ophiolite (Ahmed 1986, 1990) and to discuss the nature and possible origins of the K-rich varieties which are an unusual feature of an ophiolite. The rock suites described here are hosted by ophiolitic rocks that are well-exposed in a large stretch of land spanning the geographic districts of Khuzdar, Bela, Uthal and Kalat. The leucocratic rock exposures, however, occur only in the Khuzdar District (Fig. 1). Plagiogranites of such large size and diversity are not known from anywhere else in Pakistani ophiolites. The rock types identified from the sodic suite conform to the definition of oceanic plagiogranite of Coleman & Peterman (1975). The potassic granitic rocks which dominate the second rock suite are comparable to the anatectic granites reported from certain ophiolites (e.g., Saunders *et al.* 1979; Pearce 1989).

This paper presents new data on major, trace and rare earth element abundances in these rocks obtained by DCP, XRF and ICP-AES techniques. Radiogenic isotopic data on two samples, one from each rock suite, are also included. These data are used to characterize the genesis, age and tectonic setting of the Bela ophiolite.

Field relations and petrography

The Bela ophiolite (Ahmed 1986) has a north–south-trending outcrop parallel to and close to the western margin of the Indian plate. This plate boundary is known to be defined by two major sinistral transform faults, the Ornach–Nal and Chaman–Noshki faults (Kazmi 1979). The present study concurs with the idea that the Bela ophiolite itself marks the western boundary of the continental part of the Indian plate in this sector. The minimum age of the Bela ophiolite based upon overlying fossiliferous strata is late Eocene (Allemann 1979). The ophiolite is widely distributed as shown in the outcrop map (Fig. 1). Ophiolitic exposures just west of Karachi city (Gansser 1979) continue to appear northwards as far as Karku in the Khuzdar District. In addition, tectonically emplaced minor outcrops occur several kilometres away from the main outcrop, such as the basalt body at Pir Umar (Fig. 1), located 24 km south of Khuzdar and 20 km east of the main ophiolitic outcrop.

All the components of the stratigraphic sequence of a typical ophiolite are exposed and

From Treloar, P.J. & Searle, M.P. (eds), *Himalayan Tectonics*
Geological Society Special Publication, No. 74, pp. 89–100.

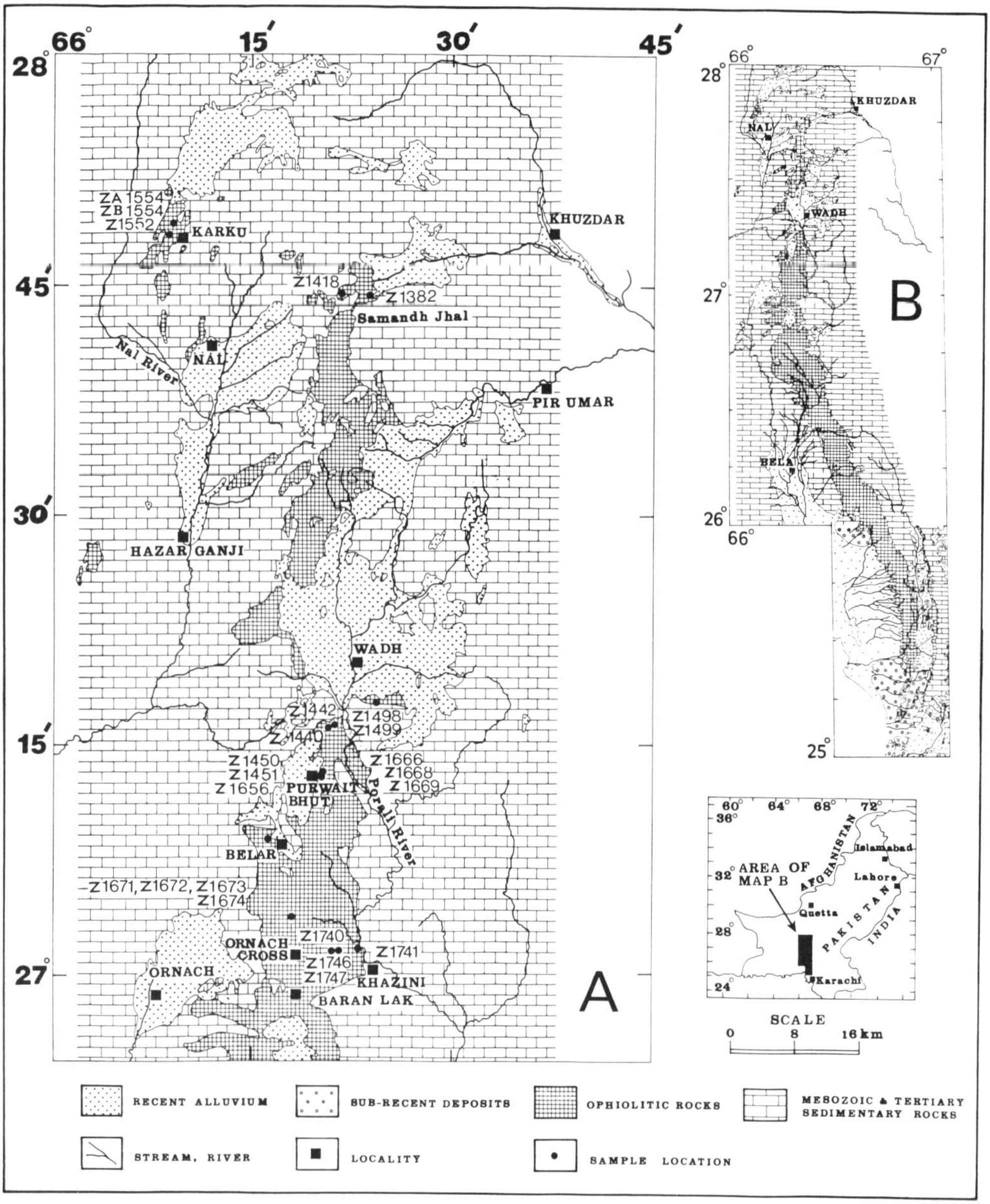

Fig. 1. Map showing the rock outcrops of the Bela ophiolite and the locations of samples included in this study. (**A**) Enlarged view of the upper part of map (**B**) which covers the area shown on the map of Pakistan.

well developed in the Bela ophiolite. These include lherzolite, harzburgite, wehrlite, dunite, chromitite, clinopyroxenite, layered gabbro, isotropic gabbro, sheeted diabase dykes, basaltic pillow lavas with interbedded chert, shale and limestone, and the leucocratic rocks described in this paper. The pyroxenites and dolerites also occur as satellite dykes in the ultramafic rocks. The mantle sequence forms only a small part of the ophiolite. The mafic volcanic rocks of the ophiolite indicate a composite source dominated by basalts geochemically akin to marginal basin basalts (Ahmed 1991).

Fig. 2. Field photographs of the leucocratic rocks of Khuzdar District. (**a**) The anatectic granite at Purwait Bhut contains angular mafic fragments of diorite. (**b**) Foliations in diorite near the acidic rock outcrop at Purwait Bhut are weakly folded due to deformation at high temperature. (**c**) Leucocratic rock outcrop at Laya Garr is veined by late stage mafic dyke splinters. (**d**) Anatectic granite interdigitating with diabase near Laya Garr. (**e**) Vein of granite-breccia enclosed in the biotite-granite near Karku Jhal. Latitudes and longitudes of localities are given in text. (**f**) A mafic rock lense enclosed within diorite (DR) overlain by trondhjemite (PG) outcrops in the Mamir area.

Leucocratic rocks form discrete major bodies as well as minor bodies resembling dykes, veins and lenses intruded into mafic or intermediate rocks. The outcrops are confined to the Khuzdar District which forms the northern part (north of 27° latitude) of the >400 km north–south length of the ophiolite (lat. 25° N to 27°32′ N). The Bela region covering the southern half of the Bela ophiolite, lacks plagiogranite altogether (Fig. 1). The leucocratic rocks of the Bela ophiolite cover a broad spectrum of lithologies described below separately for the two rock suites distinguished in this study.

The potassic acidic suite

Rocks of the potassic acidic rock suite comprise bodies of granites with or without pegmatites, granophyres, trondhjemites and albite granites. These occur at the following five main localities:

(1) Purwait Bhut (lat. 27°12′32″ to 27°12′54″ N; long. 66°20′20″ to 66°20′24″ E). A conspicuous white hillock and its surroundings are made of albite granite (samples Z1450, Z1451, Z1656, Z1666) intrusive into the underlying diorite. The contact zone shows diorite xenoliths trapped in acidic rock (Fig. 2), apophyses of which intrude the host diorite. However, these features are restricted to the contact zone only and are not pervasive within the outcrops of each rock type. At places, the diorite is foliated and folded (Fig. 2b) due to high temperature deformation.

(2) Laya Garr (lat. 27°15′54″ N; long. 66°20′40″ E). Minor granophyric (sample Z1442) tectonic slices occur in microgabbro. In the western part of this section a cluster of dykes and irregular minor bodies of acidic rock and keratophyre (sample Z1440) are intruded by diabase dykes (Fig. 2c, d). The keratophyre shows porphyritic texture with feldspar and quartz phenocrysts.

(3) Belar (lat. 27°8′51″ N; long. 66°16′40″ E). Keratophyre (sample Z1405) outcrops and hosts dykes of porphyritic basalt. In addition, keratophyre and acid porphyries occur in a few km long outcrop at Goth Shafi (27°32′ N; 66°24′ E) north of Wadh town.

(4) Karku Jhal (27°45′30″ N; 66°8′14″ E). A greyish biotite granite (samples Z1552, ZB1554) forms a discrete outcrop in contact with pillowed basalts and Mn-bearing sediments. The granite is cut by veins of quartz breccia (Fig. 2c), pinkish pegmatite (sample ZA1554) and thin mafic bands of dolerite. It is sufficiently calcic to contain andesine in the mode.

(5) North Khazini area (lat. 27°1′57″ N; long. 66°24′18″ E). A discrete outcrop of pink granite (sample Z1741) is present adjacent to basalts at the eastern top end of the ophiolite.

The plagiogranitic suite

The plagiogranitic rock suite spans dioritic to sodic acidic rocks that occur at the following four main localities.

(1) Mamir – Khushhal Garr. A large zone of plagiogranite extends from Mamir (lat. 27°3′41″ N; long. 66°17′44″ E) in the NW to Khushhal Garr (lat. 27°1′27″ N; long. 66°23′9″ E) and beyond towards the SE. Nearer the latter locality, the zone overlies isotropic gabbro and dolerite. This sequence includes lherzolite and harzburgite outcrops near Lak Baran, succeeded northwards and eastwards by cumulate dunite with mineable lenticular layers of chromitite in the Pahar Khan Bidrang area. Further up-section, successive outcrops of wehrlite, clinopyroxenite, layered gabbro, isotropic gabbro, diorite and acidic plagiogranite are present. The area is represented by samples Z1746 (calcic trondhjemite), Z1747 (granophyric albite granite) and Z1740 (diorite). In the Mamir area, trondhjemite (sample Z1673) and green (epidote)-spotted trondhjemite (sample Z1674) occur with an intermediary outcrop of epidote-free trondhjemite (sample Z1672) that is veined by epidote-trondhjemite (sample Z1671).

(2) Lohi Jhal (lat. 27°18′17″ N; long. 66°24′5″ E). Minor outcrops of lensoid and vein shaped trondhjemite (samples Z1498, Z1499) occur.

(3) Samandh Jhal (lat. 27°44′13″ N; long. 66°20′26″ E). This area is dominated by a tectonic ophiolitic melange. A >3 km long granophyric quartz diorite body (sample Z1418) is overlain unconformably by Nal Limestone Formation of Eocene age. The isotropic gabbro in the country rock contains minor leucocratic veins, upto a metre thick (sample Z1382).

(4) Purwait Bhut. The acidic rocks are hosted by a large diorite (sample Z1668) which also hosts dykes of dolerite and gabbroic pegmatite (sample Z1669).

Mineralogy

The leucocratic rocks of Khuzdar District contain the mineral assemblage quartz, albite, plagioclase (zoned from andesine to albite), orthoclase, chlorite, zircon, ilmenite, apatite, rutile, biotite, sphene, epidote and magnetite. One granitic occurrence is biotite rich (sample no. Z1552, Table 1). Granophyric and symplec-

Table 1. *Bulk-rock analyses of potassic (anatectic) acidic rocks*

Sample no.	ZA1554	Z1741	Z1442	Z1450	ZB1554	Z1552	Z1440*	Z1405*	Z1666	Z1451
SiO_2	71.30	70.19	71.20	77.23	68.82	69.23	74.62	72.25	75.44	77.47
TiO_2	0.09	0.50	0.31	0.18	0.53	0.64	0.32	0.34	0.12	0.13
Al_2O_3	14.97	14.30	13.77	12.74	16.44	15.74	12.62	13.36	12.52	13.00
Fe_2O_3	0.57	3.18	2.90	1.01	3.11	3.47	3.09	2.13	2.62	1.57
MnO	0.01	0.05	0.06	0.12	0.03	0.04	0.14	0.13	0.03	0.03
MgO	0.26	0.69	1.01	0.28	1.39	1.48	1.04	0.85	0.33	0.29
CaO	0.77	0.99	2.74	1.19	3.08	3.48	0.76	1.41	1.15	0.87
Na_2O	1.88	4.43	3.24	4.37	3.77	3.74	4.93	5.40	4.79	5.68
K_2O	8.67	5.07	3.14	2.41	2.41	1.94	2.18	1.93	1.29	0.34
P_2O_5	0.04	0.11	0.09	0.02	0.19	0.18	0.06	0.09	0.02	0.03
LOI	0.74	1.31	1.60	0.71	1.35	0.76	1.06	2.38	0.82	0.77
Total	99.30	100.82	100.06	100.16	101.12	100.70	100.82	100.27	99.13	100.18
Trace elements (ppm)										
S	–	–	–	0	–	–	20	120	–	–
Cl	–	–	–	0	–	–	30	60	–	–
Ba	1137	356	986	460	924	729	1084	828	273	96
Rb	115	102	48	49	64	72	31	28	18	–
Th	–	–	–	4.8	–	–	7.2	7.1	–	–
Nb	–	–	–	14	–	–	8.6	7.4	–	–
Ta	–	–	–	2.6	–	–	0.5	0.6	–	–
La	–	–	–	21.4	–	–	19.6	21.1	–	–
Sr	578	65	108	125	528	524	69	84	139	90
Zr	25	655	106	125	172	172	140	154	110	84
Hf	–	–	–	3.7	–	–	3.7	4	–	–
Zn	11	120	41	6	74	92	43	29	6	7
U	–	–	–	1.1	–	–	2.3	2.7	–	–
Y	3	58	–	30	8	8	34	31	17	24
Ga	–	–	–	12	–	–	10	12	–	–
Pb	–	–	–	3	–	–	2.4	5	–	–
Cu	5	7	7	11	5	–	29	42	10	–
Cr	3	8	8	6	30	11	44	12	1	5
Ni	3	2	3	2	17	4	8	5	1	3
V	–	–	–	24	–	–	29	29	–	–
Sc	2	6	10	–	8	8	–	–	3	7

* XRF analysis; other analyses are by DCP

tic intergrowths between quartz and feldspar are prevalent in both the sodic and potassic acidic rocks.

Albite is, in general, the most abundant feldspar. More calcic laths commonly exhibit saussuritization. K-feldspar occurs in anatectic rock samples such as Z1442 and Z1656. Plagioclase is often strongly zoned with more sodic rims and calcic cores. In Z1656, for example, K-feldspar coexists with plagioclase zoned from $An_{1.5}$ to An_{29}. The biotite granite Z1552 contains plagioclase of composition An_{30} which indicates an extraneous source of Ca.

In plagiogranitic acidic rocks, albite is the only feldspar. Epidote trondhjemite contains 10 to 20 modal percent coarse, green magmatic epidote that forms large, euhedral prismatic crystals and occurs mainly at one locality, represented by sample Z1671 in Fig. 1. Within epidote, pistacite content ranges from 12 to 28%. Chlorite shows spherulitic as well as flaky habits and is a ripidolite. Biotite from sample Z1552 has TiO_2 content around 3% and Fe/(Fe + Mg) around 58%. Sphene and zircon lack REE. Ilmenite and magnetite are common opaque minerals; rutile is present in sample Z1656.

Geochemistry

The leucocratic rock samples were analysed for their major and trace elements. Those analysed by XRF are marked with an asterisk in Tables 1 and 2; the remainder being analysed by DCP. XRF analyses were performed at the Department of Geological Sciences, University of Southern California, Los Angeles, USA. DCP analyses were performed at the California Institute of Technology (Caltech), USA,

Table 2. *Bulk-rock analyses of plagiogranites (magma-fractionated)*

Sample no.	Z1672	Z1673*	Z1671*	Z1674	Z1747	Z1498	Z1499	Z1746	Z1418	Z1382*	Z1669	Z1668	Z1740
SiO_2	73.34	74.04	75.78	78.78	75.48	71.24	69.52	69.75	62.72	60.71	55.34	50.79	50.32
TiO_2	0.20	0.22	0.18	0.15	0.22	0.33	0.37	0.62	0.76	0.11	0.19	0.63	0.44
Al_2O_3	13.10	12.91	12.55	12.07	12.45	13.78	15.46	13.93	14.68	20.95	21.08	15.16	15.90
Fe_2O_3	3.94	4.64	2.06	1.08	2.87	3.47	4.85	3.86	9.00	1.07	1.68	8.02	6.54
MnO	0.01	0.13	0.13	0.01	0.03	0.05	0.06	0.05	0.14	0.12	0.03	0.14	0.12
MgO	0.32	0.20	0.14	0.11	0.26	1.39	2.11	0.90	2.57	1.16	1.09	9.39	9.45
CaO	1.25	2.16	3.50	2.40	2.75	4.04	2.77	7.37	4.42	5.96	12.79	13.03	16.96
Na_2O	6.40	5.90	5.01	5.20	5.03	4.97	5.67	2.56	3.38	8.65	5.81	2.48	1.20
K_2O	0.14	0.16	0.19	0.11	0.01	0.03	0.04	0.01	3.03	0.25	0.01	0.16	0.02
P_2O_5	0.05	0.03	0.02	0.02	0.05	0.00	–	0.22	0.14	0.07	0.01	0.05	0.02
LOI	0.68	0.65	0.77	0.51	0.86	1.60	–	2.02	0.66	1.79	2.76	1.70	–
Total	99.14	101.04	100.33	100.44	100.01	100.90	100.85	101.33	101.50	100.84	100.79	101.55	100.97
Trace elements (ppm)													
S	–	0	0	–	–	–	–	–	–	20	–	–	–
Cl	–	0	10	–	–	–	–	–	–	60	–	–	–
Ba	33	19	0	13	16	0	17	5	140	70	4	−1	–
Rb	4	1.3	2.2	–	–	–	–	–	–	4.8	3	–	–
Th	–	0.3	0.3	–	–	–	–	–	–	0	–	–	–
Nb	–	7.1	4.3	–	–	–	–	–	–	2.7	–	–	–
Ta	–	1.09	2.42	–	–	–	–	–	–	1.3	–	–	–
La	–	4.3	1.7	–	–	–	–	–	–	4.7	–	–	–
Sr	75	117	157	106	97	125	109	255	1277	205	127	224	187
Zr	145	159	185	144	232	71	88	154	76	48	137	38	16
Hf	–	4.1	4.8	–	–	–	–	–	–	2	–	–	–
Zn	4	6	2	–	6	30	49	–	72	7.2	3	–	–
U	–	0.05	0.77	–	–	–	–	–	–	0.09	–	–	–
Y	46	42	58	43	62	15	16	49	27	21.3	18	54	12
Ga	–	15	14	–	–	–	–	–	–	7.8	–	–	–
Pb	–	0.9	0.04	–	–	–	–	–	–	1.4	–	–	–
Cu	2	4	7	–	–	–	–	–	–	4.3	27	–	–
Cr	2	54	6	–	11	10	14	–	–	20	10	414	111
Ni	2	14	3	–	8	8	11	8	–	21	5	124	71
V	–	26	27	–	–	–	–	–	–	28	–	–	–
Sc	15	13	8	11	8	13	15	15	–	6	5	61	81

* XRF analysis; other analyses are by DCP

using rock standards analysed by B. Chappell, and supplied by J.B. Saleeby. The precision of the XRF analyses is better than ±3% whereas that of DCP analyses is ±6%. Rare earth element (REE) analyses of three samples were performed by ICP-AES at the RHB New College, University of London, UK, using standards KC10, KC11, KC12 and KC13. The precision of these analyses is estimated at ±5%.

For the geochronological measurements, each of two rock samples of about 3 kg weight were crushed and zircon separated using standard procedures with a Frantz isodynamic separator and heavy liquids. Final fractions were picked by needle under the microscope in order to collect clear zircon euhedra. The fractions were abraded and weighed into teflon minibombs, dissolved and subjected to chemical separation of U and Pb before their analysis with a VG Instruments mass spectrometer at the laboratory of J.B. Saleeby at Caltech. In the same laboratory, Sm/Nd and Rb/Sr isotopic determinations were made after chemical separation by the isotope dilution technique.

Representative compositions of a number of the granitic rock types are given in Tables 1 (potassic) and 2 (plagiogranitic). The plagiogranitic rocks (Table 2) show very low K_2O and high Na_2O. Although the K_2O contents of the samples are generally low, some of the rock samples deviate from typical oceanic plagiogranite in possessing relatively high K_2O. The upper limit for trondhjemites at 2.5% K_2O is exceeded by the potassic granites and pegmatites (Table 1).

It is generally agreed that the high-field-strength elements (HFSE: Ti, Zr, Y, Nb, Ta, Hf), transition metals (TM: Sc, V, Cr, Ni) and REE are essentially immobile during all but the most severe sea-floor hydrothermal alteration; and that the low-field-strength elements (LFSE: Cs, Rb, Ba, Sr) and elements such as Si, K, Na are mobile in hydrous fluids (e.g., Middleburg *et al.* 1988; Pearce 1975; Humphris & Thompson 1978; Wood *et al.* 1979). Because of this, these trace elements store significant information relating to the origin and evolution of igneous rocks.

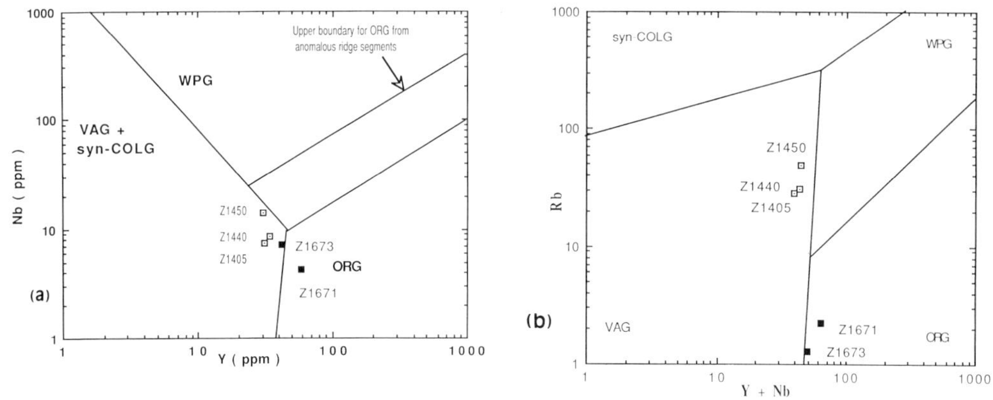

Fig. 3. Trace element data for the leucocratic rocks of Khuzdar District to discriminate between various tectonic settings: (**a**) Nb versus Y; (**b**) Rb versus Y + Nb. Anatectic potassic rocks (Z1673, Z1671) rocks are shown by squares with dots; and magma-fractionated plagiogranitic rocks (Z1440, Z1450) by solid squares.

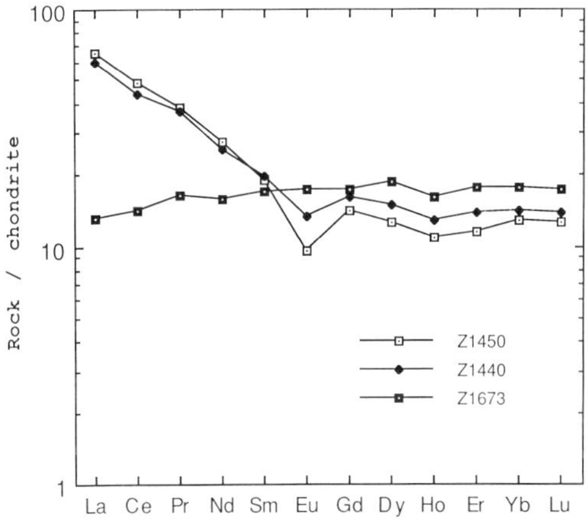

Fig. 4. Chondrite – normalized REE plot for the samples representative of the magma-fractionated (Z1440 and 1450) and anatectic leucocratic rocks (Z1673) of the Khuzdar District. Normalizing values after Nakamura (1974).

The potassic granitic samples are richer in the incompatible trace elements (Rb, Ba, K, Nb, Zr, Th, La, Ce) than the plagiogranitic samples. The whole rock trace element data can be used to compare these rocks with the classification of granite settings established by Pearce *et al.* (1984). Y values (Fig. 3) for all but one of the anatectic K-rich granites are consistent with their being volcanic arc granites (VAG). For half of the plagiogranites, Y values are consistent with an orogenic ridge granite (ORG), although the rest have lower Y values more consistent with a VAG origin. Trace element plots (Fig. 3a & b) also show this distribution, with the plagiogranitic rocks plotting in the ORG field or on the ORG-VAG boundary line, and with the potassic granites clearly plotting in the VAG field.

Plagiogranitic acidic rocks typically plot in the ORG field on the discrimination plot of Nb versus Y and Rb versus Y + Nb drawn after Pearce *et al.* (1984).

The REE geochemistry of acidic rock samples shows the presence of two distinct types of chondrite-normalized patterns (Fig. 4). The anatectic potassic granites (e.g., sample nos. Z1440, Z1450) show LREE-enriched patterns with distinct negative Eu anomalies. These features resemble those of the Fournier sodic plagiogranite shown to have formed by anatexis of amphibolite (Flagler & Spray 1991, Fig. 3A). However, the HREE part of the potassic Bela samples is flatter with higher (11 to 16 times chondritic) abundances than those of the Fournier plagiogranite. The patterns of the potassic Bela samples are also unlike those of the amphibolite-anatectic plagiogranites of East Karmoy which display positive Eu anomalies and lower REE abundances (Pedersen & Malpas 1984). The pattern of the potassic Bela samples resembles that of the sample OM2 of the anatectic granite from the Oman ophiolite thrust sheet (Pearce 1989, Fig. 4a) which is interpreted as having formed by melting of crustal sediments.

A second type of REE pattern from Bela is shown by the plagiogranitic trondhjemite sample (no. Z1673, Fig. 4). It has an almost flat pattern at 10 to 18 times chondritic abundances. There is only slight depletion of LREE relative to HREE without a Eu anomaly. The flatness in this pattern resembles that of Visnes type plagiogranite of Pedersen & Malpas (1984) which,

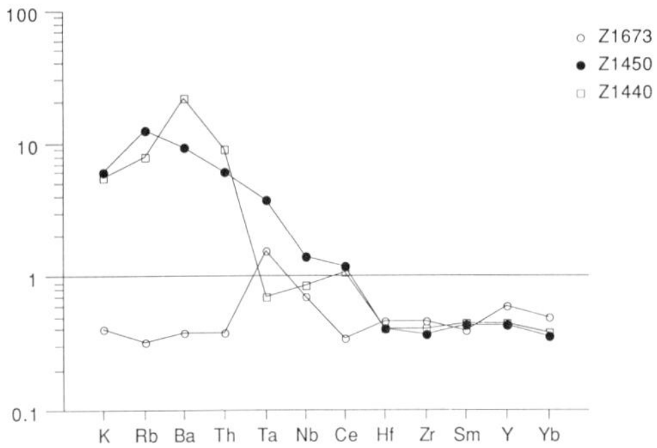

Fig. 5. Data for the leucocratic rocks of the Khuzdar District normalized to hypothetical ocean-ridge granite from Pearce *et al.* (1984). Sample numbers as in Fig. 4.

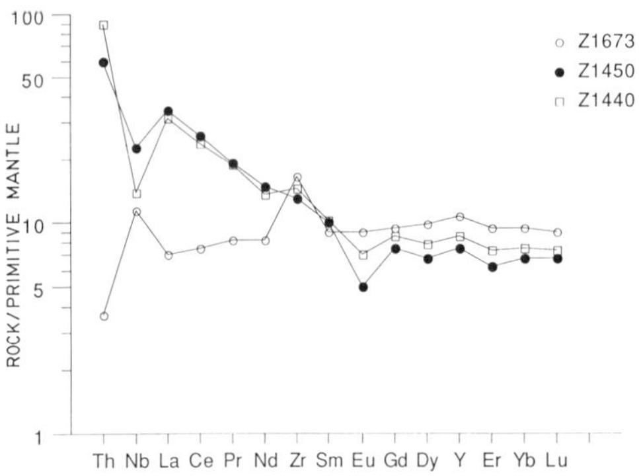

Fig. 6. Primitive mantle normalized abundances of Th, REE, and HFSE in the leucocratic rocks of Khuzdar region. Normalizing values from Hoffmann (1988). Sample numbers as in Fig. 4.

however, differs in possessing higher REE and a distinct negative Eu anomaly. The plagiogranites from the Ibra area of the Samail ophiolite, Oman (Pallister & Knight 1981) also differ in their higher REE abundances, negative Eu anomalies and stronger LREE depletions.

When plotted on multi-element spidergram diagrams, normalized against the hypothetical ocean ridge granite of Pearce *et al.* (1984), trace element patterns from the two rock suites are distinctly different. The plagiogranitic trondhjemite (Z1673) pattern is relatively flat being depleted in Nb through Yb as well as K, Rb, Ba and Th relative to ORG. The low K and Rb may be due to their loss in a volatile phase or to alteration. The pattern differs from that of the suprasubduction zone ridge granite of the Troodos Massif which contains higher Th and Ba and lower K, Rb and Th relative to Ta. The pattern parallels the line of ORG values, except for its apparently higher Ta value. The depletion in K, Rb and Ba is a typical ORG feature (Pearce *et al.* 1984). Comparison with Fig. 1a and b of Pearce *et al.* (1984) indicates a slight suprasubduction zone influence.

The potassic granites (Z1450, Z1440, Fig. 5) show sub-ORG concentrations in the Hf to Yb part of the plot and are enriched in Th relative to Nb; indicating a suprasubduction zone environment (Jenner *et al.* 1991). They are also rich in K, Rb, Ba and Th, a trend typical of volcanic arc granites. Z1450 is richer in Th, Ta, Nb and Ce, like that of the granite from the anomalous ridge segment (MAR 45° N) but differs in its high Rb and K. The patterns of anatectic samples tally with the abundances of crustal anatectic granites from the Oman ophiolite (Pearce 1989).

A trace element variation plot of the three samples incorporating REE, Th, Nb, Zr and Y and normalized against primitive mantle using values given by Hofmann (1988) is shown in Fig. 6. Depletion of Nb relative to Th is featured by the potassic samples, whereas the plagiogranite samples show the opposite behaviour. Th is strongly enriched relative to La in the potassic samples within which a negative Eu anomaly is clearly present.

This anomaly is not present in the pattern for sample Z1673, which shows a marked enrichment in Zr relative to the REE. Nb depletion relative to Th on the primitive mantle normalized plot is often taken as a sign of a suprasubduction zone environment (e.g., Jenner *et al.* 1991). This feature is clearly shown by the potassic granitic samples (e.g., Z1450, Z1440, Z1405; Table 1); but not by those of the plagiogranitic samples (e.g., Z1671; Table 2).

Radiogenic isotopic data

Sm-Nd and Rb-Sr isotopic data for samples representing each of the two rock suites are given in Table 3. The isotopic abundance ratios are normalized as follows: $^{146}Nd/^{144}Nd = 0.7219$ (for Nd); $^{149}Sm/^{152}Sm = 0.51686$ and $^{86}Sr/^{88}Sr = 0.1194$ (for Sr). The data reveal significant differences between the two types of acidic rocks within the Bela ophiolite. The samples do not plot on a whole-rock isochron. The samples differ in their initial ε_{Nd} as well as the present time deviation of the $^{143}Nd/^{144}Nd$ value from that of CHUR (chondritic uniform reservoir) (De Paolo & Wasserburg 1976), represented by $\varepsilon_{Nd} = (0)$. This suggests that each type has a different magma source composition.

The ε_{Nd} value for Z1673 at 7.49 is relatively closer to that of typical zero-age MORB (= +10 with standard deviation of 1.5; De Paolo 1988),

Table 3. *Rb/Sr and Sm/Nd data for a magma-fractionated trondhjemite sample (Z1673) and an anatectic granite sample (Z1450) from Khuzdar District*

Sample no.	Z1673	Z1450
Rb (ppm)	0.965	42.832
Sr (ppm)	113.707	122.942
Rb/Sr	0.0085	0.3484
$^{87}Rb/^{86}Sr$	0.02439	1.001586
$^{87}Sr/^{86}Sr$ (0 Ma)	0.704607	0.705397
$^{87}Sr/^{86}Sr$ (65 Ma)	0.704585	0.704471
εSr (65 Ma)	2.284	0.682
Sm (ppm)	3.614	3.967
Nd (ppm)	11.060	19.018
Sm/Nd	0.3268	0.2086
$^{147}Sm/^{144}Nd$	0.1976	0.1261
$^{143}Nd/^{144}Nd$ (0)	0.513022	0.51284
$^{143}Nd/^{144}Nd$ (65 Ma)	0.512938	0.512796
ε_{Nd} (0 Ma)	7.49	3.93
ε_{Nd} (65 Ma)	7.49	4.53

and farther from that of chondrites. The value for the plagiogranite from Ibra area, Oman (McCulloch *et al.* 1980) is 8.2 and that for the Bay of Islands ophiolite (Jacobsen & Wasserburg 1979) is 7.0 ± 0.5. This indicates derivation of sample Z1673 from a light REE-depleted source region, possibly with an oceanic affinity (McCulloch *et al.* 1980).

The initial ε_{Nd} value of sample Z1450 at 4.53 is unlike island arc, marginal basin or MORB values (McCulloch *et al.* 1980) and falls either in the range of ocean island rocks (+4 to +8), or in the range encompassed by rocks contaminated by older continental material that carries lower ε_{Nd}.

Sample Z1673 has a much lower, and sample Z1450 has a much higher Rb/Sr value compared to that of undifferentiated mantle taken as 0.029 ± 0.003 (DePaolo 1988). The samples deviate from the oceanic mantle correlation trend (DePaolo & Wasserburg 1976; O'Nions *et al.* 1977) in terms of the Nd and Sr isotopic data. The $^{87}Sr/^{86}Sr$ values are higher than expected for the observed ε_{Nd} values. This indicates special petrogenetic conditions for the Khuzdar rocks.

Geochronology

Zircons analysed for geochronology are clear and colourless with minor inclusions and cracks. The crystals are faceted and euhedral with sharp terminations; interpreted to possess an igneous morphology. The U-Pb data on the zircons separated from two samples, Z1673 (sodic type) and Z1450 (potassic type) are given in Table 4. The age of crystallization derived from this data is 65 ± 1 Ma for both the samples. The similarity in ages between the two samples implies a temporal relationship between the plagiogranitic and anatectic rocks.

Discussion

Geochemical data from leucocratic rocks of the Khuzdar District show a distinct grouping into two rock suites (e.g., Table 3), a Na-rich suite of trondhjemitic plagiogranites and a K-rich suite of anatectic plagiogranites.

The trace element discrimination plots (Fig. 3) drawn after Pearce *et al.* (1984) show that the soda-rich plagiogranite samples (e.g., Z1673, Z1671) plot in the ocean ridge granite (ORG) field or close to the VAG–ORG boundary. More potassic samples (e.g., Z1450, Z1440) form a distinct rock suite and plot partly in the volcanic arc granite (VAG) field where suprasubduction zone granites also plot. A parallel result is also obvious when Rb is plotted against Y + Nb (Fig. 3). Therefore, the sodic samples of the plagiogranitic suite are interpreted as representing a suprasubduction zone, marginal basin ridge environment, whereas the potassic group of samples which plots in the 'volcanic arc granites' field may be anatectic and derived from the ophiolite aureole.

Table 4. *U-Pb geochronological data on zircons obtained from samples of the two rock suites of leucocratic rocks of Khuzdar District.*

Sample no.	U (ppm)	Pb (ppm)	Fractions	Radiogenic atomic ratios			Age (Ma)		
				$^{206}Pb/^{238}U$	$^{207}Pb/^{235}U$	$^{207}Pb/^{206}Pb$	$t^{206}Pb/^{238}U$	$t^{207}Pb/^{235}U$	$t^{207}Pb/^{206}Pb$
Z1673	0.05	0.94	<45 μm	0.01036	0.06765	0.04738	66.4 ± 1	66.5 ± 1	68 ± 5
			45–62 μm	0.01007	0.06570	0.04734	64.6 ± 1	64.6 ± 1	66 ± 4
Z1450	1.14	3.05	<62 μm	0.01023	0.06679	0.04737	65.6 ± 1	65.7 ± 1	67 ± 6

Z1673 is a trondhjemitic sample; Z1450 is a potassic granite.

Plots of rare earth and trace element concentrations from samples of the two suites display distinctly different patterns (Figs 4, 5 & 6). Whereas the potassic granites are markedly enriched in the light rare earths and the incompatible elements with respect to primitive mantle and ocean ridge granites, the plagiogranitic trondhjemites are less enriched in those elements and are depleted in them with respect to ocean ridge granites (Figs 4, 5 & 6). The relative enrichment in the incompatible elements and the depletion of Nb relative to Th in the potassic granites is compatible with their evolution in a suprasubduction zone environment.

The isotopic data provide significant information that can be used to separate the two plagiogranitic types. For basalts, ε_{Nd} values range from +7 to +13 in the mid-ocean ridges, +8 to +9 in marginal basins, +7 to +10 in island arcs, +4 to +8 in oceanic islands and approximately zero on continents. In the Samail ophiolite in Oman, there is no marked shift in ε_{Nd} values between basalts and other rock units including plagiogranite (McCulloch *et al.* 1980). Thus, Nd in the Khuzdar samples shows an oceanic rather than continental affinity. The Nd-Sm isotopes for sample Z1673 indicate a depleted mantle source whereas values for sample Z1450, e.g., lower ε_{Nd}, indicate a possible source involving recycled continental material.

In magmas, the Rb/Sr ratio generally increases with progressive crystallization. Earlier-formed granites with high Rb/Sr develop higher $^{87}Sr/^{86}Sr$ than does the upper mantle which has lower Rb/Sr. In the continental crust, the isotopic composition of Sr is markedly heterogeneous. Crustal mixing with mantle materials often results in higher and widely varying $^{87}Sr/^{86}Sr$ values (e.g., Wilson 1988; Arakawa 1990). Both the samples in Table 3 show initial $^{87}Sr/^{86}Sr$ close to the value for bulk Earth or unfractionated mantle at 0.7045 (e.g., Wilson 1988). These samples do not represent a depleted source region. The higher $^{87}Sr/^{86}Sr$ for sample Z 1673 may be due to exchange with seawater; whereas that of sample Z1450 could be due to crustal contamination.

The trace element and isotopic data is consistent with the sodic trondhjemitic plagiogranites representing a late stage fractionate of a basic magma. By contrast, the potassic granites could be the product of anatexis associated with ophiolite emplacement and involving at least some contribution from continental crustal material. Amongst the mechanisms proposed for the formation of plagiogranite, those involving the processes of Na-metasomatism, or liquid immiscibility seem unlikely for the leucocratic rocks of Bela ophiolite. Na-metasomatism of a K-rich acid precursor (Gilluly 1933) does not seem viable for the Bela rocks, as both sodic and potassic rock types are texturally similar and form discrete magmatic bodies lacking metasomatic texture and appearance.

It is equally unlikely that these rocks may represent crystallisation from immiscible liquids in equilibrium with conjugate mafic melts. Experimental production of two immiscible liquids of plagiogranitic and Fe-rich basaltic compositions by Dixon & Rutherford (1979) was found to be inapplicable to formation of natural plagiogranite at the Canyon Mountain ophiolite by Gerlach *et al.* (1981). At Bela, the large size of the sodic plagiogranite outcrops, the existence of rocks of diorite and granodiorite compositions and the overlap in chemistry of the two rock suites do not favour the model of liquid immiscibility.

Anatectic plagiogranite may form during partial melting of basic rocks under hydrous conditions (e.g., Pedersen & Malpas 1984; Flagler & Spray 1991; Spray & Dunning 1991; Payne & Strong 1979; Gerlach *et al.* 1981). High temperature, shear-zone-related partial melting of hydrated gabbro has been proposed. The shear zones may be low angle asthenospheric shears extending up into the still plastic crust near an active spreading ridge possibly adjacent to a magma chamber. The shears may also be due to off-axis intrusion of related mafic bodies into the still plastic crust. Were the acidic rocks to have been formed by such a process, they may bear no temporal relationship to the ophiolite itself. The plagiogranite of the Canyon Mountain ophiolite is considered to have formed by the partial melting of basic rocks in the presence of water at shallow depths with total pressure below 5 kbar. This plagiogranite includes varieties from diorite to trondhjemite which occur mostly as sills (Gerlach *et al.* 1981). Pedersen & Malpas (1984) demonstrated the coexistence of two suites of plagiogranite in the Karmoy ophiolite: one associated with high temperature, pre-basic dyke shear zones, formed by anatexis of amphibolites, and the other formed by filter pressing of a differentiated interstitial liquid combined with autometasomatism. In addition it is possible that anatectic granites from sedimentary protoliths may form by the residual heat from the ophiolite (Pearce 1989).

Fractional crystallization of basic magma has been invoked to explain plagiogranite formation by many workers (e.g., Menzies *et al.* 1980; Pallister & Knight 1981; Lippard *et al.* 1986).

Coleman & Peterman (1975) held that oceanic plagiogranite forms by the low pressure differentiation of a subalkaline, tholeiitic basaltic magma at slower spreading centres. Intrusive relationships in the Bela ophiolite are consistent with such a model. The outcrop of plagiogranite at Purwait Bhut has gabbro in the west overlain to the east by diorite and then by trondhjemite which also occurs intruding the diorite and not the adjacent gabbro. That the diorite had at least partly solidified before the trondhjemite was emplaced is shown by the intrusive nature of the contact of trondhjemite against diorite. The compositional variation within individual zoned feldspar crystals in certain plagiogranite samples is gradual and spans a large range and this supports a fractional crystallization model.

Conclusions

The Bela ophiolite is the largest and most complete ophiolite in Pakistan, exceeding 400 km in its north–south outcrop length. The northern half of the ophiolite displays a variety of leucocratic rock types that belong to two distinct rock suites: a magma-fractionated plagiogranitic rock suite and an anatectic granitic rock suite. Rock types in both these suites contrast in their major, trace and rare earth element concentrations and in their Rb/Sr and Nd/Sm isotopic ratios. However, their crystallization is related in time and space. Both crystallized about 65 Ma ago and both are components of the Bela ophiolite. The geochemical signatures and the age and field constraints suggest that the plagiogranites may have formed in a marginal basin by the fractional crystallization of a basic magma. Their source region was depleted in light REE; and their $^{87}Sr/^{86}Sr$ increased due to seawater exchange. The associated more potassic acidic rocks could have developed by crustal anatexis due to heat supplied by the ophiolite in a suprasubduction zone environment.

The Bela ophiolite may be the location of the western margin of the Indian plate, which is usually marked along a N–S-trending transform fault zone. A mid-ocean origin for the Bela ophiolite is not likely as the similar crystallization ages for sodic and potassic acidic rocks fall near the minimum age of emplacement of the Bela ophiolite.

The author thanks the following for extending laboratory facilities and assistance: J.B. Saleeby, D. Pickett and M.A. Chaudhry of Caltech, USA for the isotopic work; Lawford-Anderson, E. Bender and V. Wilson of the University of Southern California, for the XRF work; J.N. Walsh and C. Lowe of RHB New College, University of London, UK, for the REE analyses. P. LeDell of San Marino, USA extended assistance throughout the author's period of work in USA. The funds for the project were provided by the USAID, Pakistan Science Foundation, Centre of Excellence in Mineralogy, Quetta, and the University Grants Commission, Pakistan. B.J. Murton, J.G. Spray and P.J. Treloar are thanked for their reviews which have greatly improved the initial manuscript.

References

Ahmed, Z. 1986. Ophiolites and chromite deposits of Pakistan. *In*: Petraschek, W., Karamata, S., Karavchenko, G.G., Johan, Z., Economou, M. & Engin, T. (eds) *Chromites: UNESCO's IGCP 197 Project: Metallogeny of ophiolites*. Theophrastus Publications, Athens, 241–262.

—— 1990. An overview of Pakistani ophiolites. *Science, Technology & Development*. Pakistan Council for Science & Technology, Islamabad, **9**, 31–48.

—— 1991. Basalt geochemistry and the suprasubduction zone origin of the Bela ophiolite, Pakistan. *Acta Mineralogica Pakistanica*, **5**, 83–98.

Alleman, F. 1979. Time of emplacement of the Zhob Valley ophiolites and Bela ophiolites. *In*: Farah, A. & DeJong, K.A. (eds) *Geodynamics of Pakistan*. Geological Survey of Pakistan, Quetta, Pakistan, 215–242.

Arakawa, Y. 1990. Strontium isotopic compositions of Mesozoic granitic rocks in the Hida belt, central Japan: diversities of magma sources and of processes of magma evolution in a continental margin area. *Lithos*, **24**, 261–273.

Coleman, R.G. & Donato, M.M. 1979. Oceanic plagiogranite revisited. *In*: Barker, F. (ed.) *Trondhjemites, Dacites and Related Rocks*. Elsevier, Amsterdam, 149–165.

—— & Peterman, Z.E. 1975. Oceanic plagiogranite. *Journal of Geophysical Research*, **80**, 1099–1108.

De Paolo, D.J. 1988. *Neodymium Isotope Geochemistry: An Introduction*. Springer-Verlag, Berlin.

—— & Wasserburg, G.J. 1976. Neodymium isotopic variations and petrogenetic models. *Geophysical Research Letters*, **3**, 249–252.

Dixon, S. & Rutherford, M.T. 1979. Plagiogranites and late stage immiscible liquids in ophiolite and mid-ocean ridge suites: An experimental study. *Earth and Planetary Science Letters*, **45**, 45–60.

Flagler, P.A. & Spray, J.G. 1991. Generation of plagiogranite by amphibolite anatexis in oceanic shear zones. *Geology*, **19**, 70–73.

Gansser, A. 1979. Reconnaissance visit to the ophiolites in Baluchistan and the Himalayas. *In*: Farah, A. & DeJong, K.A. (eds) *Geodynamics of Pakistan*. Geological Survey of Pakistan, Quetta, Pakistan, 193–214.

Gerlach, D.C., Leeman, W.P. & Ave Lallemant, H.G. 1981. Petrology and geochemistry of plagiogranite in the Canyon Mountain ophiolite, Oregon. *Contributions to Mineralogy and Petrology,* **77**, 82–92.

GILLULY, J. 1933. Replacement origin of the albite granite near Sparta, Oregon. *United States Geological Survey Professional Paper,* **175C**, 63–81.

HOFMANN, A.W. 1988. Chemical differentiation of the Earth: the relationship between mantle, continental crust and oceanic crust. *Earth and Planetary Science Letters,* **90**, 297–314.

HUMPHRIS, S.E. & THOMPSON, G. 1978. Hydrothermal alteration of oceanic basalts by seawater. *Geochimica et Cosmochimica Acta*, **42**, 107–125.

JACOBSEN, S.B. & WASSERBURG, G.J. 1979. Nd and Sr isotopic study of the Bay of Islands ophiolite complex and the evolution of the source of midocean ridge basalts. *Journal of Geophysical Research,* **84 (B13)**, 7429–7445.

JENNER, G.A., DUNNING, G.R., MALPAS, J., BROWN, M. & BRACE, T. 1991. Bay of Islands and Little Port Complexes, revisited: age, geochemical and isotopic evidence confirm suprasubduction zone origin. *Canadian Journal of Earth Science,* **28**, 1635–1652.

KAZMI, A.H. 1979. Active fault systems in Pakistan. *In*: FARAH, A. & DEJONG, K.A. (eds) *Geodynamics of Pakistan.* Geological Survey of Pakistan, Quetta, Pakistan, 285–294.

LIPPARD, S.J., SHELTON, A.W. & GASS, I.G. 1986. *The ophiolite of northern Oman.* Geological Society, London, *Memoirs*, **11**.

MCCULLOCH, M.T., GREGORY, R.T., WASSERBURG, G.J. & TAYLOR, H.P. 1980. A neodymium, strontium and oxygen isotopic study of the Cretaceous Samail ophiolite and implications for the petrogenesis and seawater hydrothermal alteration of oceanic crust. *Earth and Planetary Science Letters,* **46**, 201–211.

MENZIES, M., BLANCHARD, D. & XENOPHONTOS, C. 1980. Genesis of the Smartville arc-ophiolite, Sierra Nevada foothills, California. *American Journal of Science,* **280-A**, 329–344.

MIDDLEBURG, J.J., VAN DER WEIJDEN, C.H. & WOITTIEZ, J.R.W. 1988. Chemical processes affecting the mobility of major, minor and trace elements during weathering of granitic rocks. *Chemical Geology,* **68**, 253–273.

NAKAMURA, N. 1974. Determination of REE, Ba, Fe, Mg, Na and K in carbonaceous and ordinary chondrites. *Geochimica et Cosmochimica Acta,* **38**, 757–775.

O'NIONS, R.K., HAMILTON, P.J. & EVENSEN, N.M. 1977. Variations in Nd/Nd and Sr/Sr ratios in oceanic basalts. *Earth and Planetary Science Letters,* **34**, 13–22.

PALLISTER, J.S. & KNIGHT, R.Y. 1981. Rare-earth element geochemistry of the Samail ophiolite near Ibra, Oman. *Journal of Geophysical Research,* **86**, 2673–2697.

PAYNE, J.G. & STRONG, D.F. 1979. Origin of the Twillingate trondhjemite, north-central Newfoundland: Partial melting in the roots of an island arc. *In*: BARKER, F. (ed.) *Trondhjemites, Dacites and Related Rocks*, Elsevier, Amsterdam, 489–516.

PEARCE, J.A. 1975. Basalt geochemistry used to investigate past tectonic environments on Cyprus. *Tectonophysics,* **25**, 41–67.

—— 1989. High T/P metamorphism and granite genesis beneath ophiolite thrust sheets. *Ofioliti,* **14**, 195–211.

——, HARRIS, N.B.W. & TINDLE, A.G. 1984. Trace element discrimination diagrams for the tectonic interpretation of igneous rocks. *Journal of Petrology,* **25**, 956–983.

PEDERSEN, R.B. & MALPAS, J. 1984. The origin of oceanic plagiogranites from the Karmoy ophiolite, Western Norway. *Contributions to Mineralogy and Petrology,* **88**, 36–52.

SPRAY, J.G. & DUNNING, G.P. 1991. A U/Pb age for the Shetland Islands oceanic fragment, Scottish Caledonides: evidence from anatectic plagiogranites in 'layer 3' shear zones. *Geological Magazine,* **128**, 667–671.

SAUNDERS, A.D., TARNEY, J., STERN, C.R. & DALZIEL, I.W.D. 1979. Geochemistry of Mesozoic marginal basin floor igneous rocks from southern Chile. *Geological Society of America Bulletin,* **90**, 237–258.

WILSON, M. 1988. *Igneous Petrogenesis: A global tectonic approach.* Unwin Hyman, London.

WOOD, D.A., JORON, J.L. & TREUIL, M. 1979. A re-appraisal of the use of trace elements to classify and discriminate between magma series erupted in different tectonic settings. *Earth and Planetary Science Letters,* **50**, 325–336.

Chemistry of chromite and associated phases from the Shangla ultramafic body in the Indus suture zone of Pakistan

MOHAMMAD ARIF[1] & M. QASIM JAN

NCE in Geology, University of Peshawar, Peshawar, Pakistan

[1] *Present address: Department of Geology, University of Leicester, Leicester LE1 7RH, UK*

Abstract: The partly to wholly serpentinized ultramafic rocks of the Shangla area constitute the northern part of the Mingora ophiolitic melange which lies along the Main Mantle Thrust (MMT) zone between the Kohistan island arc and the Indo-Pakistan plate. These rocks contain disseminated grains and massive bodies of segregated chromite. Field, textural, and chemical characteristics of the segregated chromite closely resemble those of podiform chromite deposits. The chemical composition of the chromite varies considerably from sample to sample with modal proportion, from grain to grain in the domain of a thin section, and from core to margin within individual grains. In general, the segregated chromites are richer in Cr and Mg, but poorer in Fe, Al and Mn than the accessory chromites. The inter- and intragranular chemical variations are much greater and more common in the case of accessory chromite due probably to a greater degree of subsolidus re-equilibration with the associated abundant silicate matrix. The intragranular variation, probably a result of alteration, mostly involves enrichment in Fe, Mn (and Ti), and depletion in Al, Mg and Cr, leading to the development of ferritchromit and, rarely, magnetite along fractures and margins of individual grains.

The studied rocks from the Shangla area display cumulus textures but contain abnormally magnesian olivine ($Fo_{96.0-98.5}$). The 100Cr/(Cr + Al) ratios of their chromite suggest a complex origin. These rocks may have formed under conditions transitional between those of arc and oceanic settings. The Shangla rocks underwent a low- to ?medium-grade metamorphism as indicated by the development of talc and antigorite. This prograde metamorphic process probably also accounted for the extreme chemical zoning of the Shangla chromite.

Mafic-ultramafic rock associations occur at a number of localities in Pakistan, some of which are considered to be ocean-ridge ophiolites (cf. Gansser 1979; Asrarullah *et al.* 1979). The major six mafic–ultramafic sequences are : (1) The Bela ophiolites (DeJong & Subhani 1979); (2) The Zhob valley ophiolites (Bilgrami & Ingamells 1960; Moores *et al.* 1980); (3) The Waziristan ophiolites (Jan *et al.* 1985); (4) The Skhakot–Qila ophiolites (Ahmed 1978); (5) The Shangla ophiolites (Gansser 1979); and (6) The Jijal ophiolites (Jan & Windley 1990).

Chromite occurs in the ultramafic rocks of all these complexes both as an accessory constituent and as segregated masses of varying size. In view of their petrological, tectonic and economic significance, the chromite occurrences of the Zhob valley, Waziristan, Skhakot–Qila, and Jijal have been studied in detail in recent years. This paper presents the field, textural, and chemical characteristics of chromite and its associated phases from Shangla. Since the ophiolitic rocks of Shangla and adjoining areas constitute large bodies located along the well-defined suture zone in northern Pakistan, an understanding of their origin should throw light on the petrogenesis of mafic–ultramafic complexes trapped in collisional orogenic terrains. Not all the ophiolite complexes around Pakistan are remnants of oceanic crust and subjacent mantle. For example, the Jijal complex is regarded by Jan & Windley (1990) to be a series of arc cumulates.

General geology

The Shangla and adjoining areas in the Swat valley, NW Pakistan, provide one of the best exposures of ophiolitic rocks within the Main Mantle Thrust (MMT) zone which marks collision of the Indo-Pakistan plate with the Kohistan island arc. Petrographic details of these rocks have been presented by Jan & Symes (1977), Shams (1980), and Jan (1985). Kazmi *et al.* (1984) grouped the rocks into blueschist, greenschist, and ophiolitic melanges.

The ophiolitic melange consists of a series of lensoidal bodies of mafic–ultramafic rocks incorporated within the metasediments of the Indian

From TRELOAR, P.J. & SEARLE, M.P. (eds), *Himalayan Tectonics*
Geological Society Special Publication, No. 74, pp. 101–112.

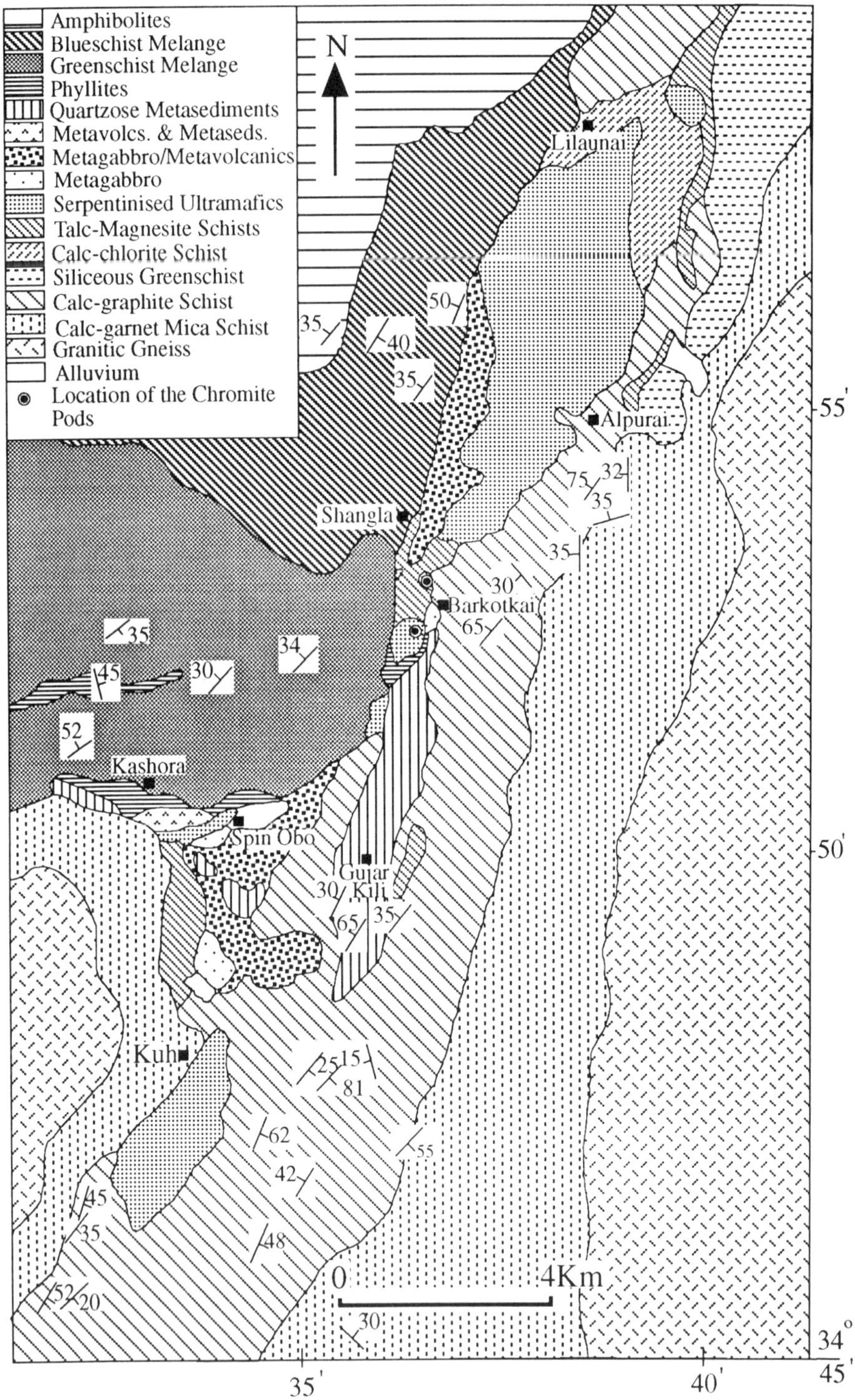

Fig. 1. Geological map of the Shangla area, Swat (modified after Kazmi *et al.* 1984).

plate. The component lithologies, including ultramafic rocks, gabbros, basalts and pillow lavas as well as metamorphosed manganiferous quartzose sediments and cherts, are representative of a typical ophiolite complex. The ultramafic rocks are invariably serpentinized, carbonated and/or steatitized to varying degrees. The mafic rocks are altered and metamorphosed under greenschist to lower amphibolite facies conditions such that their original mineralogy is completely obliterated.

The occurrence of chromite

Petrographic details and whole rock analyses of the serpentinized ultramafic rocks of the ophiolitic melange were carried out by Chaudhry & Ashraf (1986). These authors also reported the occurrence of lensoidal bodies of chromitite in the Shangla area. The contacts of these bodies with the enclosing rocks are sharp and they are not economic because of their extremely small (1.0 × 2.0 m to 1.5 × 2.0 m) sizes. The present work is based on mineralogical studies on two of these bodies and the enclosing serpentinized ultramafic (dunite and harzburgite) rocks (Fig. 1).

The chromite bodies of the Shangla area display a well developed nodular texture. The individual nodules vary both in shape and size (Fig. 2). Most of them are, however, spherical and 3 to 15 mm across. The Shangla chromites also exhibit occluded silicate and pull-apart textures. These are considered to be critical features of podiform chromite deposits (cf. Thayer 1970).

The Shangla chromitites contain variable amounts (80 to more than 98 volume %) of chromite, the rest being silicate(s). The chromite is mostly medium grained, subhedral, translucent and brownish red in refracted light. Along cracks and margins of the individual grains, the chromite is opaque 'ferritchromit' and/or magnetite due to alteration. The associated silicate is mostly chlorite and, rarely, serpentine and talc. It may be interesting to note that the modal proportion of chromite is markedly low in the nodular chromitite compared to the other textural types. From a study of the Japanese ores, Bamba (1978) concluded that nodular chromite was associated with chlorite only but the nodules of the Shangla chromite are accompanied by chlorite and serpentine as well as talc.

The Shangla chromitites are hosted by serpentinites that locally contain patches of partly serpentinized coarse-grained dunite. The serpentinites are green to yellowish green and locally jointed and sheared. They commonly consist of abundant (70–90%) fine-grained serpentine and accessory to subordinate amounts of chromite and magnetite with or without olivine, talc, chlorite, and carbonate(s). They also contain bastite textures, mostly after orthopyroxene. This suggests that the original rocks of the Shangla serpentinites were dominantly harzburgites. An outstanding feature of these rocks is the occurrence of 1–3 mm thick veins of olivine

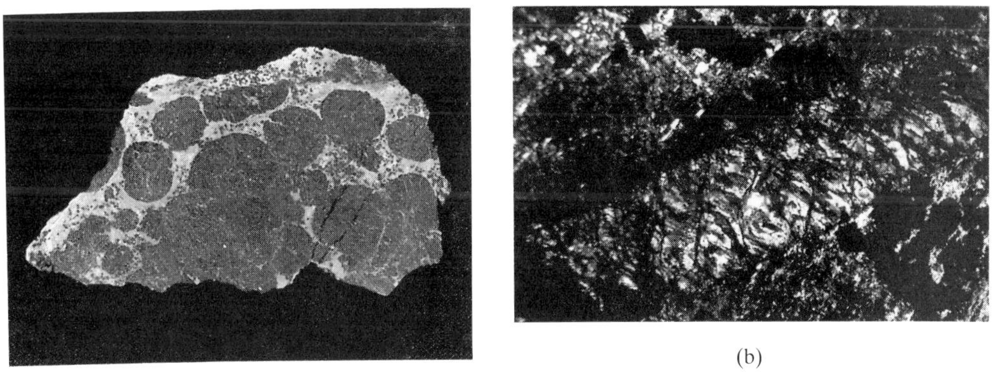

(a) (b)

Fig. 2 (**a**) Shangla chromite displaying nodular texture: length of the photograph is 8.33 cm; (**b**) a vein of olivine cutting across the serpentine matrix of a serpentinised harzburgite. Actual thickness of the vein is about 2 mm.

cutting across the main serpentine matrix (Fig. 2). Moreover, thin microscopic veins of olivine also traverse the bastites.

Texturally, the Shangla serpentinites are dominated by the non-pseudomorphic textures of Wicks & O'Hanley (1988), e.g., antigorite interpenetrating texture, antigorite interlocking texture, antigorite-lizardite interlocking texture, and chrysotile serrate veins. The only pseudomorphic texture in these rocks are the bastites (e.g., the 'shaggy' and 'indistinct' types of Wicks & O'Hanley 1988) seemingly after orthopyroxene. Wicks & O'Hanley (1988) defined five types of bastite which, in the order of increasing degree of crystallization, are uniform, dominal, patchy, shaggy and indistinct.

Mineral chemistry

A total of 15 polished thin sections (six chromitites and nine host silicate rocks) were studied in detail. An average of ten to twelve points, representing three to five chromite grains, and two spots per silicate phase were analysed in each section. Many spot analyses were performed on chromite and associated phases for checking inter-granular variation. In order to generate data about zonation and alteration pattern, spot compositions of both the core and margin of individual chromite grains were determined in most cases. Most of the analyses were repeated two, and in some cases three, times and almost similar results obtained.

Chromite

Representative analyses of the Shangla chromite and chrome spinels are listed in Table 1. Analyses representing central portions (cores) of chromite grains contain 30–62 wt% Cr_2O_3 10–27 wt% Al_2O_3, 4.8–15.0 wt% MgO, and 13–65 wt% FeO*. A computer programme based on spinel stoichiometry was used to split the total iron (FeO*) into Fe^{2+} and Fe^{3+}. The MnO content of the Shangla chrome spinel ranges from 0.01 to 5.42 wt%. Spinels with such a high and highly variable MnO have also been reported in metamorphosed ultramafic rocks of Kusa, Bergslagen, Sweden (Zakrzweski 1989). Except for three analyses, the TiO_2 content of the Shangla chromites does not exceed 0.3 wt%. This is a typical feature of residual ultramafic rocks (cf. Ozawa 1983). The data compiled by Herbert (1982), however, show that not only residual but also a large number of oceanic cumulate rocks can have spinels with low (<0.3 wt%) TiO_2 values.

Table 1. *Selected analyses of chrome spinels in chromitites and host silicate rocks*

No.*	SC1	SF1	SC2	SM2	SC3	SM3	AC4	AM4	AR4	AC5	AR5
SiO_2	0.06	0.00	0.03	0.00	0.06	0.21	0.09	0.00	0.16	0.05	0.10
TiO_2	0.16	0.16	0.14	0.62	0.07	0.14	0.05	0.03	0.01	0.07	0.00
Cr_2O_3	59.85	22.14	54.41	51.95	53.90	52.37	39.39	36.75	1.68	50.71	6.40
Al_2O_3	11.58	6.60	12.79	0.45	8.24	5.23	26.94	1.49	0.03	14.71	0.10
Fe_2O_3	1.78	41.45	0.89	16.56	8.56	12.78	5.56	32.01	67.46	5.45	64.01
FeO	15.23	24.35	22.67	22.62	21.96	22.21	17.18	14.77	27.80	19.35	27.95
MnO	0.07	0.06	0.68	0.88	0.47	0.47	0.19	5.63	0.06	0.35	0.42
MgO	12.68	5.57	6.97	5.35	7.49	6.85	13.18	7.21	2.17	9.97	2.27
Total	101.41	100.33	98.58	98.43	100.75	100.26	102.58	97.89	99.37	100.66	101.25
Number of ions on the basis of 32 (O)											
Si	0.02	0.00	0.01	0.00	0.02	0.06	0.02	0.00	0.05	0.01	0.03
Ti	0.03	0.03	0.03	0.14	0.01	0.03	0.01	0.01	0.00	0.01	0.00
Cr	12.06	4.93	11.63	11.98	11.53	11.46	7.38	8.43	0.40	10.36	1.51
Al	3.48	2.19	4.08	0.15	2.63	1.71	7.50	0.51	0.01	4.48	0.04
Fe^{3+}	0.34	8.78	0.18	3.64	1.74	2.66	0.99	6.99	15.46	1.06	14.36
Fe^{2+}	3.25	5.73	5.13	5.52	4.97	5.14	3.40	3.59	7.08	4.18	6.97
Mn	0.02	0.01	0.16	0.22	0.11	0.11	0.04	1.38	0.02	0.08	0.11
Mg	4.82	2.34	2.81	2.33	3.02	2.83	4.65	3.12	0.99	3.84	1.01
Cr No.	77.61	69.22	74.04	98.72	81.43	87.04	49.51	94.30	97.41	69.80	97.72
Mg No.	59.74	28.96	35.39	29.65	37.80	35.47	57.75	46.52	12.21	47.87	12.64
Fe^{3+} No.	2.14	55.22	1.13	23.08	10.94	16.80	6.24	43.88	97.42	6.67	90.26

* S, segregated chromite analysis; A, analysis of accessory chromite; C, core of the analysed grain; M, margin of the analysed grain; R, rim around the grain; F, fracture within the grain; analyses marked with the same numeral represent different positions within one grain.

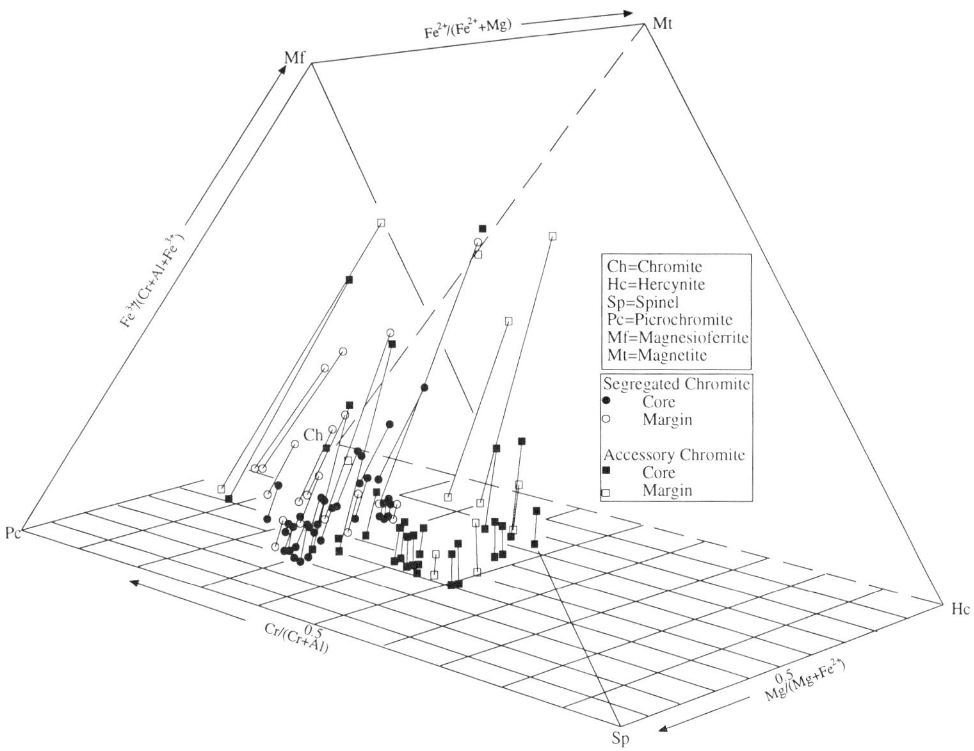

Fig. 3. Plots of the Shangla chromite in the three-dimensional compositional prism of Irvine (1965).

The Shangla chromites have been plotted in the spinel compositional prism (Fig. 3). They have also been compared with chromites from different geological settings in Fig. 4a and b which represent the Ch-Hc-Sp-Pc and Mf-Mt-Ch-Pc faces of the spinel prism (Irvine 1965). The chemical relationship between Cr, Al, and $Fe^{3+} + 2Ti$ of the Shangla chromite is shown in Fig. 5. In most of these diagrams, the least altered (core compositions of the segregated grains) of the Shangla chromites mostly fall into the area overlapped by the fields of chromites from alpine-type and stratiform complexes. The wt% TiO_2 v. wt% Cr_2O_3 plots of the chromites and their comparison with chromites from different types of oceanic rocks have been presented in Fig. 6. The relation of wt% TiO_2 with Fe^{2+}/Mg ratios in Shangla chromite was also studied and compared with that observed in podiform and stratiform chromites (Fig. 7). On this particular diagram, almost all of the Shangla chromite analyses fall into the field of podiform chromites. It should, however, be noted that the core–margin tie lines and some of the analyses, representing the rims of magnetite and chrome magnetite around chromite and ferritchromit and shown plotted on the other diagrams, have not been presented on this plot for the sake of clarity.

The composition of Shangla chromite and chrome spinels shows a substantial variation with their modal proportion in different rocks. A chemical comparison, based on the core compositions of the analysed grains, between the accessory and segregated chromites shows that the former displays a wider range of wt% Cr_2O_3, and an overall impoverishment in wt% Cr_2O_3 and MgO but enrichment in wt% FeO*, MnO and Al_2O_3.

The Shangla chromites also exhibit chemical variation from one grain to another within the domain of a thin section. Moreover, there is a drastic variation from core to margin within the individual chromite grains. These variations are most significant in the accessory chromites but also occur in segregated chromites (Figs 3 to 7).

The chemical variation from core to margin of individual chromite grains in Shangla rocks mostly involves increase in Fe, Mn (and Ti) and decrease in Al, Mg, and Cr. A similar trend, leading to the development of highly reflecting true ferritchromit (Fig. 8), has also been

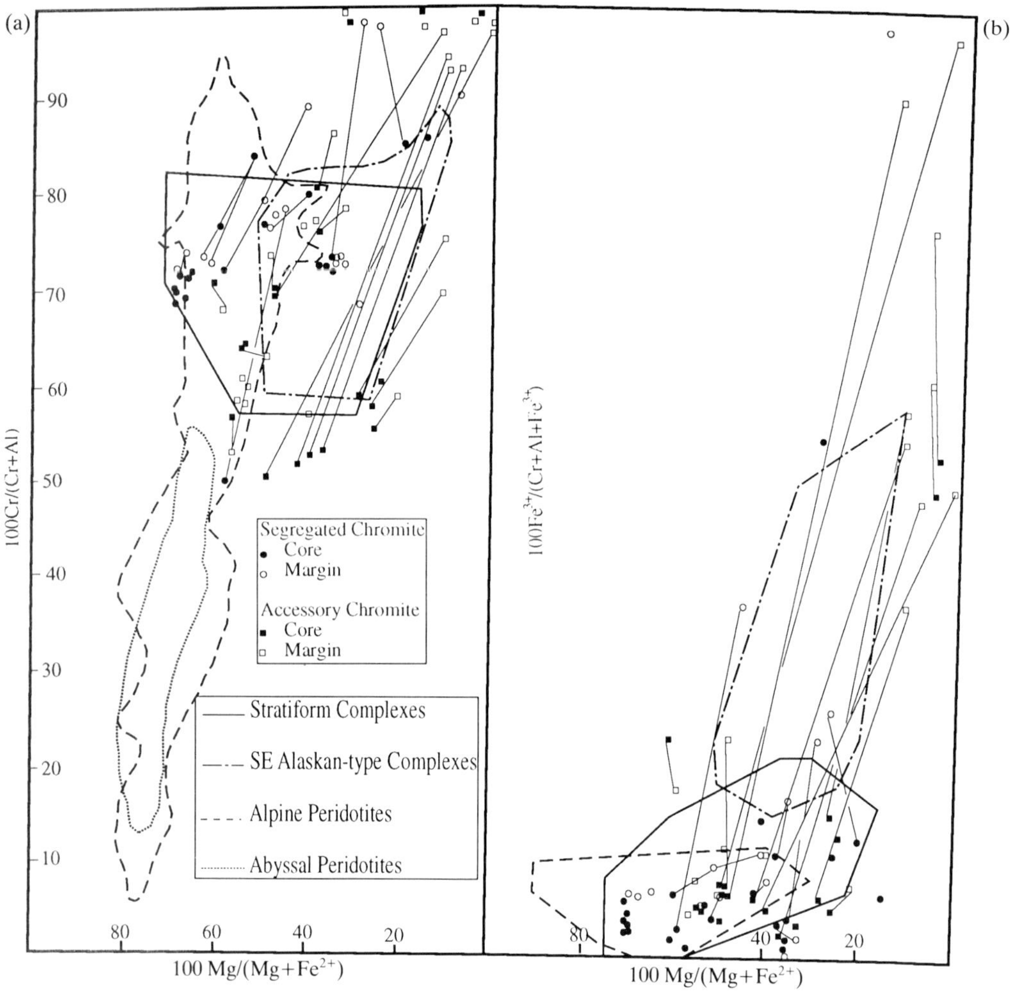

Fig. 4. (**a**) Variation of Cr and Mg Nos in the chromites of Shangla. The various fields are after Dick & Bullen (1984). Tie lines connect core and margin of zoned grains. Many of the tie lines were abandoned for clarity. (**b**) Fe^{3+} and Mg No. variation in Shangla chromites. Fields are from Jan & Windley (1990).

re-imported from other areas (cf. Ahmed & Hall 1981). Ferritchromit is occasionally also seen between the chromite cores and the magnetite outer rims in some of the Shangla serpentinite and dunite.

Ferritchromit is unanimously believed to be the product of alteration of chromite. However, very little agreement exists concerning the actual process of alteration. It has been related to (1) serpentinization (Bliss & Maclean 1975), (2) chloritization (Beeson & Jackson 1969; Lipin 1984; Jan & Windley 1990), and (3) low-grade (upper greenschist to lower amphibolite facies) metamorphism (Bliss & Maclean 1975; Srikantappa & Friend 1982).

The distinctly higher Fe^{3+}/Fe^{2+} ratios of the Shangla ferritchromit than their parent (original) chromite suggest an increase in oxygen fugacity (see also Beeson & Jackson 1969). Moreover, the conversion of chromite to ferritchromit of the Shangla area involves increase in Fe^{3+}/Cr and Fe^{3+}/R^{3+} ratios. It may have resulted from the oxidation of some of the Fe^{2+} of the parent chromite at the margins of the individual grains.

Associated phases

Besides chromite, its associated phases, i.e., olivine, serpentine, chlorite, talc, and car-

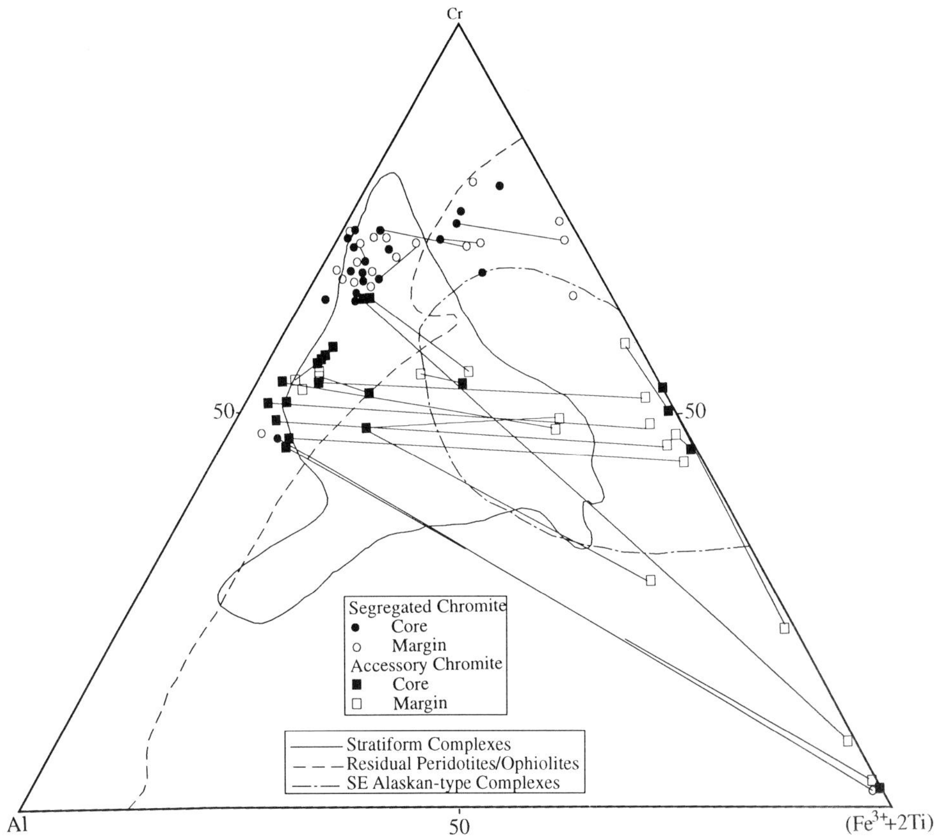

Fig. 5. Cr, Al, and Fe^{3+} + 2Ti relations in the chromites of Shangla. Other details as in Fig. 4b.

bonate, were also analysed in the Shangla rocks.

Representative analyses of olivine are listed in Table 2. They contain notable amounts of NiO (up to 0.40 wt%) and Cr_2O_3 (up to 0.06 wt%). The olivine occurs in the form of large grains distributed in the serpentine matrix. As pointed out earlier, olivine also occurs as veins that cut across the main serpentine matrix and traverse the orthopyroxene bastites. The composition of the vein olivine is more or less similar to that of the matrix olivine. Both types are characterized by high to very high (96.0–98.5) Mg Nos [= 100 Mg/(Mg + Fe^{2+})]. SEM studies of some of the olivine grains revealed the occurrence of abundant preferentially oriented, submicroscopic lamellae of magnetite (Fig. 9).

Examples of very highly magnesian olivine are: Fo_{95-100} in meteorites of the enstatite chondrite type (Binns 1967); $Fo_{96.-98.9}$ in metamorphosed ultramafites (cf. Deer *et al.* 1982); $Fo_{91.4-98.0}$ in chromites (Sinton 1977; Ahmed 1982; Jan & Windley 1990); Fo_{96} as inclusions in chromite grains (Prichard & Neary 1981); and $Fo_{96.2}$ in a metamorphosed limestone (cf. Deer *et al.* 1962). It should, however, be noted that the highly magnesian character of the Shangla olivines cannot be ascribed to an Mg = Fe exchange with chromite because none of the analysed grains occurs either adjacent to or as an inclusion in chromite. Moreover, the significantly high amounts of NiO and Cr_2O_3 make them different from those with otherwise similar composition noted in some ophicalcites.

It follows, therefore, that metamorphosed ultramafic rocks are the only terrestrial silicate rocks whose olivine (Mg No. 96.0–98.9 with significant Ni and Cr; cf. Deer *et al.* 1982) is comparable in composition to that of the Shangla rocks. Typical examples are those described from East Dover, south-central Vermont (Hoffman & Walker 1978) and Toulumne River, Sonora, California (Morgan 1978). A lower to medium-grained (upper greenschist to lower amphibolite facies) regional metamorphism has been suggested for the development of olivine in

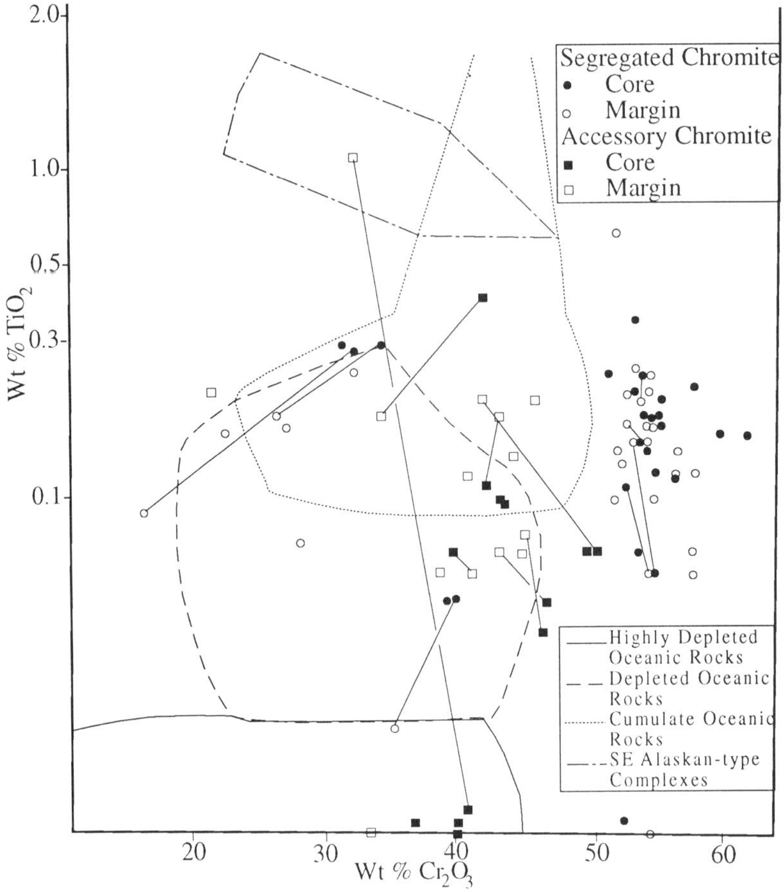

Fig. 6. Relationship between wt% TiO_2 and Cr_2O_3 in the Shangla chromites. The fields for different types of oceanic rocks are from Herbert (1982). The field for the SE Alaskan-type complex is after Jan & Windley (1990).

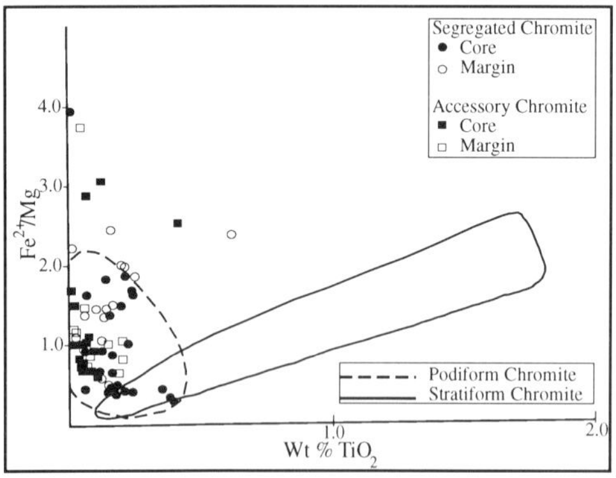

Fig. 7. Wt% TiO_2 and Fe^{2+}/Mg variation in the chromites of Shangla. Some of the analyses approaching magnetite in composition have been excluded from this diagram. The fields for stratiform and podiform chromites are from Prichard & Neary (1981).

Fig. 8. Alteration of chromite to ferritchromit along grain margin and fractures in a serpentinized ultramafic rock of Shangla.

Table 2. *Selected analyses of the silicate phases associated with chrome spinel*

No.*	1	2	3	4	5	6	7	8	9	10
SiO_2	41.33	40.44	41.64	43.38	41.45	43.72	29.12	33.23	61.10	60.38
TiO_2	0.00	0.01	0.02	0.01	0.02	0.01	0.03	0.00	0.01	0.00
Al_2O_3	0.02	0.07	0.01	0.01	0.47	0.10	13.65	14.34	0.02	0.19
FeO*	2.92	3.74	2.20	1.60	3.19	1.99	0.89	3.86	0.83	0.51
MnO	0.88	0.86	0.23	0.60	0.01	0.04	0.08	0.03	0.01	0.00
MgO	53.74	52.79	55.50	53.50	40.28	40.99	31.99	34.18	29.76	28.85
CaO	0.03	0.05	0.00	na	na	na	0.03	na	0.08	na
Na_2O	0.10	0.15	na	na	na	na	0.54	na	0.08	na
NiO	na	na	na	0.36	na	na	0.70	na	na	na
Cr_2O_3	0.03	0.01	0.04	0.03	0.49	0.14	8.41	1.07	0.00	0.20
Total	99.05	98.12	99.64	99.49	85.91	86.99	85.44	86.71	91.89	90.13
*Number of ions**										
Si	0.99	0.99	0.99	1.01	1.97	2.03	5.74	6.32	8.75	8.81
Ti	0.00	0.00	0.00	0.00	0.00	0.00	0.00	0.00	0.00	0.00
Al	0.00	0.00	0.00	0.00	0.03	0.00	3.17	3.21	0.00	0.03
Fe^{2+}	0.06	0.08	0.04	0.03	0.13	0.08	0.15	0.61	0.10	0.06
Mn	0.02	0.02	0.00	0.01	0.00	0.00	0.01	0.01	0.00	0.00
Mg	1.93	1.92	1.97	1.93	2.86	2.84	9.40	9.68	6.36	6.23
Ca	0.00	0.00	0.00	na	na	na	0.01	na	0.01	na
Na	0.00	0.01	na	na	na	na	0.21	na	0.02	na
Ni	na	na	na	0.01	na	na	0.11	na	na	na
Cr	0.00	0.00	0.00	0.00	0.02	0.00	1.31	0.16	0.00	0.01

* Analyses 1–4 are olivine and the number of ions are based on 4(O); analyses 5 and 6 are serpentine and the number of ions are based on 7(O, OH); analyses 7 and 8 are chlorite in chromitite and serpentinite, respectively, and the number of ions are based on 28(O, OH); analyses 9 and 10 are talc in serpentinite and nodular chromitite, respectively, and their recalculation is based on 24(O, OH).

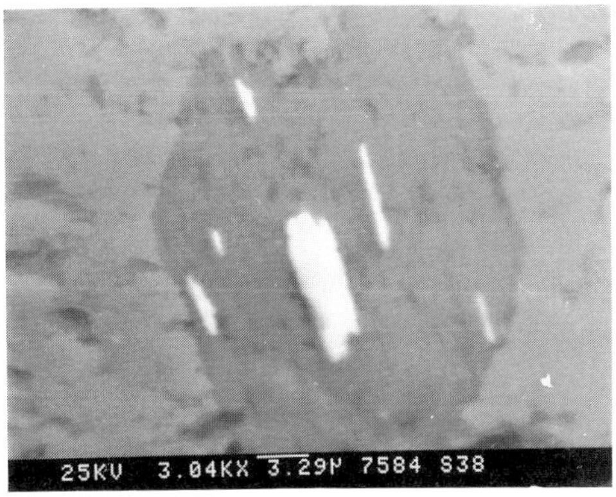

Fig. 9. Lamellae of magnetite in a grain of olivine.

rocks of both these areas. Hoffman & Walker (1978) further noted a progressive increase in the Fo content of olivine with the degree of recrystallization. Morgan (1978) suggests that magnetite produced during serpentinisation of the primary Fo_{92} olivine was not reduced or involved in the reaction that presumably occurred under conditions of relatively high oxygen fugacity.

Representative analyses of serpentine are listed in Table 2. Their Al_2O_3 content is mostly low, although the distinctly high (1.3–1.75 wt%) Al_2O_3 of some of the analyses may be due to a reaction with chromite during serpentinization. Minor amounts (mostly 0.1–0.5 wt%) of Cr_2O_3, drawn presumably from the adjacent chromite grains, were also noted in the analysed grains. The SiO_2/MgO ratio of the Shangla serpentines is more than one and they, therefore, resemble the antigorites of Page (1968). Their FeO* contents, however, suggest that some (the Fe-poorer analyses) are antigorite whilst others (the Fe-richer compositions) could be lizardite. The petrographic studies and X-ray diffraction patterns, obtained for six of the partly to wholly serpentinised rocks, also revealed the occurrence of antigorite as the major serpentine type; accompanied by only small amounts of lizardite and chrysotile in the studied samples.

Antigorite seems to be stable at higher temperatures than lizardite and chrysotile and its presence indicates that the serpentinite has undergone a prograde (upper greenschist to lower amphibolite facies) metamorphism or that the rock was serpentinised in a higher *P–T* regime (Evans & Frost 1975; Moody 1976). That the Shangla rocks have undergone a prograde recrystallization is also shown by the development in some of them of talc.

Selected chlorite analyses are presented in Table 2. All the analyses are chromian with wt% Cr_2O_3 ranging from 1.1 to 8.4. Although highly variable, the composition of the chlorite grains (especially their Al_2O_3, Cr_2O_3 and FeO* contents) is dependent upon their associated chromite grains. Thus the formation of the chromian chlorite of Shangla is connected with the alteration of chromite. A similar conclusion was also reached by Bamba (1978) and Jan & Windley (1990).

Table 2 shows representative analyses of talc from two rocks: a serpentinite and a nodular chromitite. Their MgO and FeO* contents are low and relatively uniform (27.9–28.8 wt% and 0.5–0.6 wt%, respectively) in the chromitite but high and highly variable (28.2–34.9 wt% and 0.8–7.7 wt%, respectively) in the serpentinite. The trace amount of Cr_2O_3 in the talc analyses can also be related to bulk composition. Hence wt% Cr_2O_3 (0.11–0.20) of talc in chromitite is higher than that (0.00–0.11) in the serpentinite. Small amounts of Na_2O (0.09 wt% average) and CaO (0.02–0.11 wt%) were also recorded in the Shangla talc.

Carbonate, analysed in one of the samples, contains 0.10 wt% FeO*, 0.54 wt% MnO, 21.55 wt% MgO, and 31.32 wt% CaO. The recalculated analysis $[Ca_{1.014}(Mg_{0.970}Mn_{0.074}Fe_{0.002})]$ is similar to the ideal dolomite.

Discussion

The tectonic setting, extensive serpentinisation, and the field and petrographic features of their contained chromite bodies suggest that the ultramafic rocks of Shangla are similar to those of the alpine-type peridotite–gabbro complexes. Based upon the Cr No. [= 100Cr/(Cr + Al)] of their chrome spinels, the peridotites and associated volcanic rocks of the alpine-type complexes have been assigned to three different tectonic settings (Dick & Bullen 1984): (1) mid-ocean ridges, (2) island arcs, and (3) environments transitional between (1) and (2). The Cr Nos (50–100) of chrome spinels in Shangla rocks group them with the type (3) rocks which, according to Dick & Bullen (1984), contain spinels with Cr No. covering the entire range from <60 to >60 and reflect composite origins, such as the formation of a young island arc on an older oceanic crust. The Shangla rocks lie along the Main Mantle Thrust (MMT) in NW Pakistan which is considered to be the site of a major oceanic basin and are located at the base of the Cretaceous Kohistan island arc. The available data on chrome spinels show that neither oceanic (cf. Herbert 1982) nor arc-related ultramafic rocks, e.g., the Goodnews Bay ultramafic complex, Alaska (Bird & Clark 1976), can contain spinels with Cr_2O_3 content exceeding 50 wt%. In contrast, the Cr_2O_3 content of chrome spinels in Shangla rocks is mostly high and ranges up to 62 wt%. Hence the range of Cr No. of chrome spinel, the geological setting and the Cr_2O_3 content of their chromite all favour the assumption that the Shangla rocks may have formed under conditions transitional between those of arc and oceanic settings.

The low (mostly <0.3 wt%) TiO_2 content of chromite and the abnormally high Mg No. (96.0–98.5) of their olivine suggest that the Shangla rocks could be residual (highly depleted mantle) in origin. The presence of cumulus texture, the lack of a typical tectonite fabric at least in the studied samples (cf. Herbert *et al.* 1983), and the very wide range (83–96) (based on the analyses presented by Chaudhry & Ashraf 1986) of their Mg No. as compared with that (around 90) of the ultramafic tectonites (Dick 1977; Himmelberg & Loney 1980), however, argue against a residual origin for the Shangla rocks.

The abnormally high (96.0–98.5) Mg No. of the Shangla olivine may be due to one of the following reasons.

(1) The Shangla rocks crystallized from a highly magnesian (e.g., komatiite or boninite) magma.
(2) The Shangla olivines represent one of the earliest products of crystallisation from a magma of ordinary basaltic composition. They, however, retained their initial highly magnesian character because of failure to re-equilibrate with changing magma composition as crystallization proceeded (cf. DeBari *et al.* 1987).
(3) The crystallization of Shangla olivines took place under highly oxidizing conditions.
(4) The Shangla olivines are of metamorphic origin.

The last mentioned possibility is favoured on the following grounds: (i) the only terrestrial rocks to contain olivine comparable in composition to that in the Shangla rocks are the metamorphosed ultramafites (cf. Deer *et al.* 1982), such as those of the East Dover, south-central Vermont (Hoffman & Walker 1978) and the Tuolumne River, Sonora, California (Morgan 1978); (ii) the occurrence of preferentially orientated magnetite lamellae in the Shangla olivine; (iii) the occurrence of Shangla olivine also as veins; and (iv) the presence of talc in the

Shangla rocks is consistent with the well documented occurrence of talc + forsterite assemblages in metamorphosed ultramafic terrains (cf. Evans & Frost 1975). The metamorphism of Shangla rocks was, however, static in character so that the original (cumulus) textures were not completely obliterated.

The financial and laboratory facilities were provided by the NCE in Geology, University of Peshawar, Pakistan. Mr Barkatullah and Mr Jehanzeb are thanked for assisting in the performance of probe analyses and polishing the thin sections, respectively. R.N. Wilson is appreciated for his help in doing some of the repeat analyses through the Jeol Superprobe at the Department of Geology, University of Leicester. We are thankful to M.A. Khan for reading the earlier draft and H.M. Prichard for a critical review.

References

AHMED, Z. 1978. Chromite from Skhakot-Qila area, Malakand Agency, Pakistan. *Mineralogical Magazine,* **42**, 155–157.

—— 1982. Porphyritic-nodular, nodular and orbicular chrome ores from Skhakot–Qila complex, Pakistan, and their chemical variations. *Mineralogical Magazine,* **45**, 167–178.

—— & HALL, A. 1981. Alteration of chromite from the Skhakot-Qila complex, Malakand Agency, Pakistan. *Chemical Erde,* **40**, 209–239.

ASRARULLAH, AHMAD, Z. & ABBAS, S.G. 1979. Ophiolites in Pakistan: an introduction. *In*: FARAH, A. & DEJONG, K.A. (eds) *Geodynamics of Pakistan*. Geological Survey of Pakistan, Pakistan, 181–192.

BAMBA, T. 1978. The alpine chrome ores from Japan. *Hokkaido University Faculty of Science Journal Series,* **4(18)**, 167–179.

BEESON, M.H. & JACKSON, E.D. 1969. Chemical composition of altered chromite from the Stillwater complex, Montana. *American Mineralogist,* **54**, 1584–1610.

BILGRAMI, S.A. & INGAMELLS, C.O. 1960. Chemical composition of the Zhob valley chromite. *American Mineralogist,* **45**, 576–590.

BINNS, R.A. 1967. Olivine in enstatite chondrite. *American Mineralogist,* **52**, 1549–1554.

BIRD, M.L. & CLARK, A.L. 1976. Microprobe studies of olivine chromitites of the Goodnews Bay ultramafic complex, Alaska, and the occurrence of Platinum. *Journal of Research, Geological Survey, United States,* **4**, 717–725.

BLISS, N.W. & MACLEAN, W.H. 1975. The paragenesis of zoned chromite from central Manitoba. *Geochimica et Cosmochimica Acta,* **39**, 973–990.

CHAUDHRY, M.N. & ASHRAF, M. 1986. Petrology of ultramafics from Shangla–Alpurai–Malan Jabba area, Swat *Kashmir Journal of Geology*, **4**, 15–32.

DEBAR, S., KAY, S.M. & KAY, R.W. 1987. Ultramafic xenoliths from Adagdak volcano, Adak, Aleutian Islands, Alaska: deformed igneous cumulates from the Moho of an island arc. *Journal of Geology,* **95**, 329–341.

DEER, W.A., HOWIE, R.A. & ZUSSMAN, J. 1982. *Rock-forming minerals*. Volume **1A**. Longman, London.

DEJONG, K.A. & SUBHANI, A.M. 1979. Note on the Bela ophiolites, with special reference to the Kanar area. *In*: FARAH, A. & DEJONG, K.A. (eds) *Geodynamics of Pakistan*. Geological Survey of Pakistan, Pakistan, 263–269.

DICK, H.J.B. 1977. Partial melting in the Josephine peridotite. I, The effect on mineral composition and geothermometry. *American Journal of Science,* **277**, 804–832.

—— & BULLEN, T. 1984. Chromian spinel as a petrogenetic indicator in abyssal and alpine-type peridotites and spatially associated lavas. *Contributions to Mineralogy and Petrology,* **86**, 54–76.

EVANS, B.W. & FROST, B.R. 1975. Chromian spinel in progressive metamorphism: a preliminary analysis. *Geochimica et Cosmochimica Acta,* **39**, 959–972.

GANSSER, A. 1979. Reconnaissance visit to the ophiolites in Baluchistan and Himalayas. *In*: FARAH, A. & DEJONG, K.A. (eds) *Geodynamics of Pakistan*. Geological Survey of Pakistan, Pakistan, 193–213.

HERBERT, R. 1982. Petrography and mineralogy of oceanic peridotites and gabbros: Some comparison with ophiolite complexes. *Ofioliti,* **7 (2/3)**, 299–324.

HIMMELBERG, G.R. & LONEY, R.A. 1980. Petrology of the ultramafic and gabbroic rocks of Canyon mountain ophiolites, Oregon. *American Journal of Science,* **280-A**, 232–268.

HOFFMAN, M.A. & WALKER, D. 1978. Textural and chemical variations of olivine and chrome spinel in the East Dover ultramafic bodies, Vermont. *Geological Society of America Bulletin,* **89**, 699–710.

IRVINE, T.N. 1965. Chrome spinel as a petrogenetic indicator. Part 1, Theory. *Canadian Journal of Earth Sciences,* **2**, 648–671.

Jan, M.Q. 1985. High-P rocks along the suture zones around Indo-Pakistan plate and phase chemistry of blueschists from eastern Ladakh. *Geological Bulletin University of Peshawar* **18**, 1–40.

—— & SYMES, R.F. 1977. Piemontite schist from upper Swat, northwest Pakistan. *Mineralogical Magazine,* **41**, 537–540.

—— & WINDLEY, B.F. 1990. Cr-spinel-silicate chemistry in ultramafic rocks of the Jijal complex, NW Pakistan. *Journal of Petrology,* **31**, 667–715.

——, —— & KHAN, A. 1985. The Waziristan ophiolite, Pakistan: general geology and chemistry of chromite and associated phases. *Economic Geology,* **80**, 294–306.

KAZMI, A.H., LAWRENCE, R.D., DAWOOD, H., SNEE, L.W. & HUSSAIN, S. 1984. Geology of the Indus suture in the Mingora-Shangla area of Swat, N. Pakistan. *Geological Bulletin University of Peshawar,* **17**, 127–144.

LIPIN, B.R. 1984. Chromite from the Blue Ridge

Province of North Carolina. *American Journal of Science,* **284**, 507–529.

MOODY, J.B. 1976. Serpentinisation: a review. *Lithos,* **9**, 125–138.

MOORES, E., ROEDER, D.H., ABBAS, S.G. & AHMAD, Z. 1980. Geology and emplacement of the Muslimbagh ophiolite complex. *In*: PANAYIOTOU, A. (ed.) *Ophiolites symposium*. Geological Survey Department, Ministry of Agriculture and Natural Resources, Cyprus, 424–429.

MORGAN, B.A. 1978. Metamorphic forsterite and diopside from the ultramafic complex at the Toulumne River, California. *Journal of Research, Geological Survey, United States,* **6**, 73–80.

OZAWA, K. 1983. Relationships between tectonite and cumulate in ophiolites in Miyamori ultramafic complex, Kitakami Mountains, northeast Japan. *Lithos,* **16**, 1–16.

PAGE, N.J. 1968. Chemical differences among the serpentine 'polymorphs'. *American Mineralogist,* **53**, 201–215.

PRICHARD, H. & NEARY, C.R. 1981. Chromite in the Shetland ophiolite complex. *UNESCO international syposium on metallogeny of mafic and ultramafic complexes, Athens project 169,* **3**, 343–360.

SHAMS, F.A. 1980. Origin of the Shangla blueschist, Swat Himalaya, Pakistan. *Geological Bulletin University of Peshawar,* **13**, 67–70.

SINTON, J.M. 1977. Equilibration history of the alpine-type peridotite, Red Mountain, New Zealand. *Journal of Petrology,* 216–246.

SRIKANTAPPA, C. & FRIEND, C.R.L. 1982. Chromian edenite from the Siduvalli ultramafic body, Karnataka, south India. *Journal of the Geological Society of India,* **23**, 539–545.

THAYER, T.P. 1970. Chromite segregations as petrogenetic indicators. *Geological Society of South Africa, Special Publication,* **1**, 380–390.

WICKS, F.J. & O'HANLEY, D.S. 1988. Serpentine minerals: structure and petrology. *Reviews in Mineralogy, Mineralogical Society of America,* **19**, 91–169.

ZAKRZWESKI, M.A. 1989. Chromian spinel from Kusa, Bergslagen, Sweden. *American Mineralogist,* **74**, 448–455.

The Sapat mafic–ultramafic complex, Kohistan arc, North Pakistan

M. QASIM JAN, M. ASIF KHAN & M. SUFYAN QAZI
National Centre of Excellence in Geology, University of Peshawar, Pakistan

Abstract: The Sapat layered complex, occupying the tectonic base of the Cretaceous Kohistan island arc, occurs in the immediate hanging wall of the Indus suture zone north of the Kaghan valley. Previously undescribed, much of the complex consists of gabbroic rocks, but in the basal part there are ultramafic cumulates ($15 \times < 2$ km) in direct tectonic contact with the Indian plate. In the central part, the complex can be divided into five zones. From the base these are: (1) homogeneous serpentinized dunites (350 m thick), (2) dunites with thin chromite layers (100 m), (3) thinly layered dunites–pyroxenites (30 m), (4) layered dunites–pyroxenites–gabbros (170 m) with chromite layers in the lower part, and (5) gabbros. The gabbroic rocks are amphibolitized and consist of a variety of lithologies ranging from troctolites to anorthosites, with local pyroxenite (± olivine) layers and horizons of olivine-rich ultramafics. These rocks are locally cut by pyroxene-rich pegmatites containing up to 45 cm long pyroxene crystals. Preliminary microprobe studies on the ultramafic rocks show that the clinopyroxene is Mg-Ca-rich diopside, and olivine has a forsterite content of 92 to 78, but in rare cases > 95. The chromite displays high Cr No. (75–67) and low Fe^{3+}/R^{3+} (<10). These data, especially the chromite analyses, are akin to those of ultramafic rocks of island arcs, including the Jijal complex, 80 km to the west. We propose that the Sapat complex formed the basal part of the Kohistan terrain and is related to arc magmatism. It has been metamorphosed under epidote amphibolite facies conditions.

The Indus suture in northern Pakistan (locally known as the Main Mantle Thrust) demarcates the Indian plate to the south from the Cretaceous Kohistan island arc to the north (Tahirkheli *et al.* 1979; Coward *et al.* 1986). A number of mafic–ultramafic plutonic complexes have been described from the suture zone (Jan & Jabeen 1990). Of these, the following six are more prominent.

(1) The Dargai ophiolite, a 30 km × 3.5 km sized body which consists of tectonized peridotite, ultramafic cumulates, and subordinate mafic cumulates (Rossman & Abbas 1970; Ahmed 1984).

(2) The Tora Tigga complex, western Dir. This 6 km^2 body consists of cumulate peridotites, pyroxenites, and abundant hornblendites and is surrounded by amphibolitized gabbros (Jan & Tahirkheli 1990).

(3) The Shangla ophiolitic melange in upper Swat, with lenses of serpentinized ultramafic rocks, the largest of which measures some 7 × 1 km (Kazmi *et al.* 1984; Ashraf *et al.* 1989; Arif 1990). Similar ultramafic bodies occur in the melange zone of Allai, 40 km to the east (Shah 1986; Baig 1989).

(4) The Jijal complex, a tectonic wedge covering about 200 km^2 area along the Indus river. This consists of dunite, peridotites and pyroxenites in the south, and garnet granulites in the north (Jan & Windley 1990; Miller *et al.* 1991).

(5) The Babusar amphibolites (metagabbros) and associated ultramafic rocks (Ahmed & Chaudhry 1976).

(6) The Chilas complex, which is perhaps the largest of its type, stretching for > 300 km with a maximum width of 40 km (Fig. 1). This consists of gabbronorites with subordinate ultramafic and anorthositic rocks (Jan *et al.* 1984).

Despite considerable field and laboratory data, the origin and tectonic setting of some of these complexes is debated. A clear understanding of these is pertinent to the magmatic evolution of island arcs in general and Kohistan in particular. The Tora Tigga, Jijal, and Babusar bodies are clearly related in space and time to the growth of the Kohistan magmatic arc. Others, such as the Shangla body, occur in tectonic melanges of the suture zone and presumably represent sub-arc/oceanic lower crust/upper mantle. The Dargai ophiolite (Ahmed 1988), like that of Spongtang in western Ladakh, may be a klippe that overrides the Indian plate some 30 km to the south of the suture. The Chilas complex, forming an immense belt in the southern part of the Kohistan arc, presumably represents mafic melts derived from a mantle diapir during the early spliting of the Kohistan arc or back arc basin (Khan *et al.* 1989).

From TRELOAR, P.J. & SEARLE, M.P. (eds), *Himalayan Tectonics*
Geological Society Special Publication, No. 74, pp. 113–121.

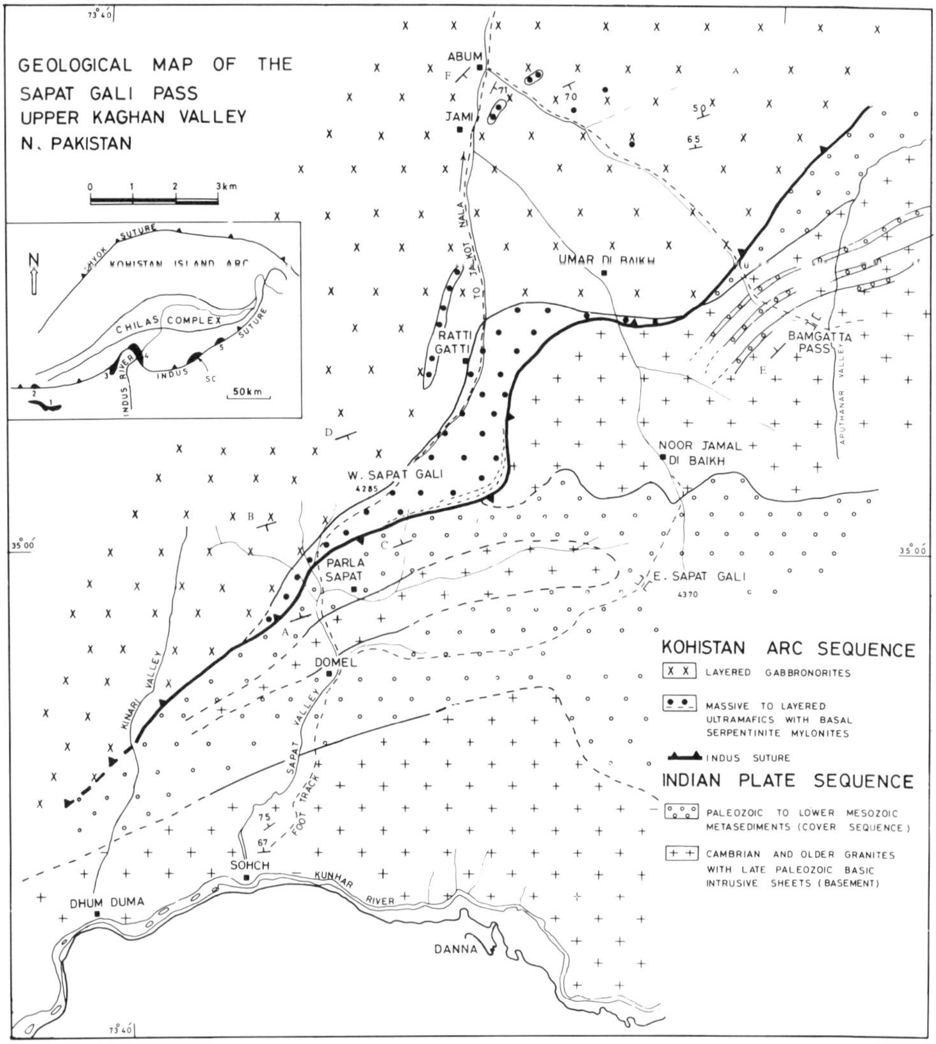

Fig. 1. Geological map of the Indus suture zone near Sapat, NE Kaghan. Inset shows location of mafic–ultramafic plutonic complexes (1–6) briefly described in the text. SC: Sapat complex.

The suture zone at Sapat

Our recent studies have revealed additional occurrences of mafic–ultramafic plutonic complexes in the southern part of the Kohistan arc. Here we present preliminary data on the previously undescribed Sapat mafic-ultramafic complex, which occurs along the drainage divide between the Kaghan valley and Kohistan, some 40 km west of Babusar. The Sapat complex consists of ultramafic rocks which are hosted by layered metagabbros and gabbronorites in the southern amphibolite belt of the Kohistan arc (Jan 1988). There are several ultramafic bodies, of which the principal one extends NE-SW immediately to the north of the suture and covers more than 15 km^2 area (Fig. 1).

The Indian plate rocks to the south of the suture, in the western part of the studied area, consist of quartz–mica ± garnet schists and graphitic schists. These are followed southwards by a series of micaceous to pure marbles and calc-silicate rocks. The latter include bands up to 25 m thick which are rich in garnet up to 2 cm across. There are small granitic bodies, pegmatites and quartz boudins within this sequence of Phanerozoic cover. Further south, strongly sheared and lineated granitic gneisses outcrop

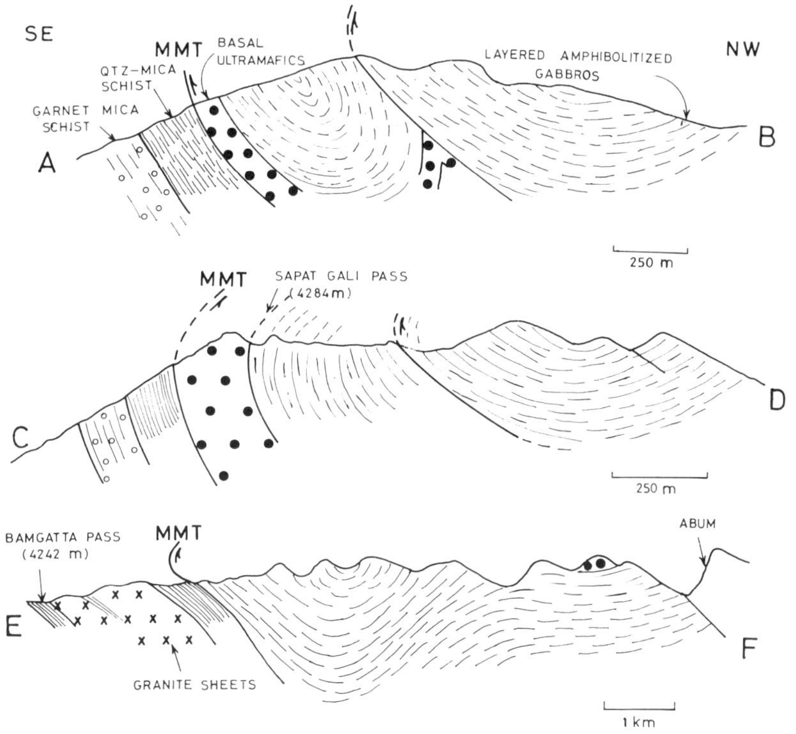

Fig. 2. Cross sections across the Indus suture (MMT, Main Mantle Thrust) showing hanging-wall structures in southern Kohistan. Top and middle near the Sapat Gali Pass and bottom near the Bamgatta Pass.

which belong to the Precambrian/Cambrian basement. They contain schistose and foliated amphibolite dykes with or without garnet. Both the basement and the cover sequence are intruded by pegmatite dykes and sills (± muscovite, ± tourmaline, ± garnet) which are folded, but commonly unsheared, and cut the internal fabric in the host rocks. In the northeastern part of the area, the Sapat complex is directly in contact with the granitic gneisses (Fig. 1).

The suture in the area is sharp and without the development of a melange zone. However, the rocks on both sides are intensely sheared and mylonitized. The quartz–mica schists are banded and contain centimetre-scale isoclinal folds. They are retrogressed near the fault and are devoid of garnet and staurolite, which are essential components of the assemblage elsewhere. The plunge of lineations suggests an oblique movement towards the southwest associated with thrusting in this part of the suture zone. Recumbent-type folding of the fault along a horizontal axis is well displayed in vertical profiles exposed in deeply disected valleys and steep cliffs (Fig. 2).

The Sapat mafic–ultramafic complex

The complex is made up of a variety of lithologies ranging from chromitite and dunite to gabbro and anorthosite. The rocks are generally well-layered, the layering apparently younging to the northwest. Ultramafic rocks occur principally along the southeastern edge of the complex where it is sutured to the Indian plate. These represent basal cumulates underlying gabbros. However, since the complex is synclinally folded, the ultramafic rocks are repeated in the northwestern part of the studied area (Fig. 2). There are also isolated bodies of ultramafic rocks which may represent lenses of deformed cumulates, and small intrusions of ultramafic pegmatites.

In the central part of the area the ultramafic rocks are more extensive. The stratigraphy observed here in the lower part of the complex is summarized in Fig. 3. This consists of five divisions with the following general features from bottom to top.

(1) At the base, a homogeneous dunite, generally devoid of layering, makes up the basal

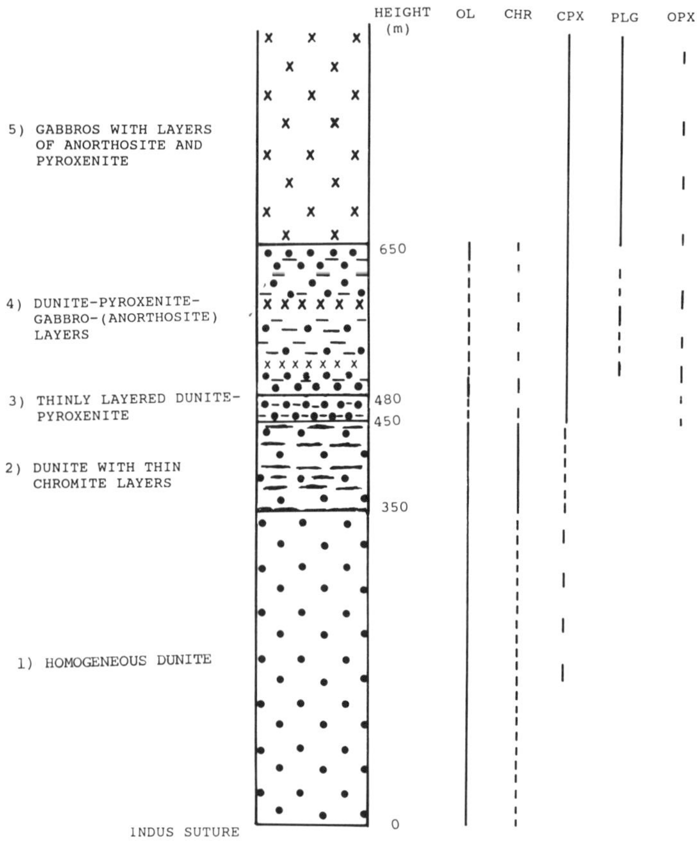

Fig. 3. Stratigraphic column of the lower central part of the Sapat layered complex, 2.5 km south of Ratti Gatti.

350 m cumulates. It contains sparse chromite grains, tending to increase modally upward, and very locally forming oval clots measuring up to 14 × 10 cm. The lower 10 m part of this zone is strongly sheared and serpentinized, but serpentinization is common throughout, especially along the abundant shear zones. There are rare and thin (< 5 cm) dykelets and veins of pyroxenite containing coarse crystals, up to 3 cm long.

(2) Overlying this is a 100 m thick zone of dunite with abundant layers of chromite. The chromite layers range from streaks to 3 cm in thickness and may disappear laterally within a few metres. The layers display (?slump) folding, faulting and some are brecciated. Solid chromitite layers (>80% chromite) are rare and most chromite layers have abundant serpentine after olivine. No deposits of commercial interest have been found. Veins of serpentine, <1 to 30 cm thick, are common and post-date the faulting in the layers. Again, rare thin dykes of pyroxenite occur in this zone.

(3) The overlying 30 m thick zone consists of thinly layered dunite, pyroxenite and rare wehrlite. The layers, which display typical rhythmic disposition, range from mere streaks to some centimetres in thickness and may phase out in short distances. In one place, 25 layers were counted within a 20 cm thick horizon and in another 50 layers in 63 cm.

(4) Upward in the sequence is a complex zone of dunite, pyroxenite, gabbro, and minor peridotite and anorthosite. The upper and lower parts (*c.* 20 m) of this zone are made up of dunite with some pyroxenite layers, whereas in the middle part the pyroxenites are as abundant as the dunites forming layers up to 50 cm thick or more. The rocks are generally medium grained as elsewhere, but some pyroxenites are coarse grained with grains up to 2 cm long. The pyroxenite layers in the upper part show a pinch and

swell geometry, and at places contain olivine clots. Above the lowest dunite of this zone, completely epidotized plagioclase makes its first appearance in a 2 m thick horizon containing gabbro layers. Some 125 m from the base, there is a 25 m thick sequence containing abundant gabbro layers up to 35 cm thick, and rare anorthosite layers alternating with ultramafic rocks. The following rhythmic disposition was noted in one place: 1 m dunite – 3.5 cm pyroxenite – 5 cm gabbro – 3.5 cm pyroxenite – 30 cm dunite – 2.5 cm pyroxenite – 8 cm gabbro – 2.5 cm pyroxenite – thick dunite.

(5) Overlying these rocks are homogeneous to layered gabbros with some pyroxenite and anorthosite layers. Elsewhere, these are associated with some peridotites and troctolites.

Petrography

Ultramafic rocks. These consist of dunites, pyroxenites, their gradational lithologies, chromitites and pyroxene pegmatites. These are generally medium grained, but some pyroxenites are coarse grained and the pegmatites contain up to 45 cm long crystals. Near the suture and along many shear zones the rocks are mylonitized. Layering is common and ranges from millimetre to metre-scale (Fig. 4). Layering types include size-grading, mineral-grading and phase layering, slump-folds, syndepositional faults and breccias. The layers may pinch and swell, they sometimes wedge-out abruptly or phase out laterally over a few metres. The pyroxenites contain clots and enclaves of dunite, which at places are aligned and appear to be broken layers. Some of the pegmatites are interlayered with pyroxenites, but others look like intrusive masses containing enclaves of pyroxenite and dunite. Intrusive into these are rare, fine-grained dunitic veins up to 5 cm thick. The chromite-rich rocks commonly occur in well-defined layers, locally broken up, but there are also rare millimetre-scale veins and centimetre-scale clots in dunites. Such field features have been reported in the Chilas (Jan *et al.* 1984; Khan *et al.* 1989) and several other mafic–ultramafic complexes (e.g. Sierra Nevada: James 1971) and suggest turbulent magmatic conditions, slumping, deformation and remobilization.

The dunites contain less than a percent or two of chromite, but in the chromite-rich zones there are gradational lithologies to chromitite. Some dunites contain clinopyroxene and pass through wehrlite (rare) and olivine clinopyroxenite to pure clinopyroxenite. Orthopyroxene is a rare mineral in the ultramafic rocks and websterites occur only sparingly. A few pyroxenites contain appreciable amounts of brown hornblende, but hornblendites have not been noticed in the area. Both the pyroxenes may display exsolution and the clinopyroxene and hornblende commonly form ophitic/poikilitic grains up to 2 cm in length. The pegmatites, consisting of two pyroxenes ± olivine, are an interesting group of rocks, because of their coarse-grained texture, ultramafic composition and general lack of plagioclase. These rocks form small bodies which are more common in the southwestern part of the complex.

The olivine-rich rocks commonly display varying degrees of alteration to serpentine and magnetite, accompanied by Mg-rich chlorite (clinochlore) around chromite; some also contain talc and a carbonate mineral. Serpentinization is particularly intense near the suture,

Fig. 4. (**a**) Dunite with folded chromite layers. Bands in the lower part are due to serpentinization (1.5 cm coin for scale). (**b**) Layered dunite–pyroxenite with ultramafic vein (left of photograph). Note shearing in some of the layers.

along shear zones and in chromite-rich rocks. The pyroxene-rich rocks contain secondary (metamorphic) amphibole (in some abundantly) ± chlorite, probably formed during late stage movement and hydration of the complex. In the pegmatites the pyroxene may be sieved and shot through by abundant amphibole.

Gabbroic rocks. Forming the bulk of the complex, these range from homogeneous to very well-layered varieties. The layers are from centimetre to metre-scale in thickness and display similar features to those noted in the ultramafic rocks. Most of these rocks are medium-grained gabbros with sharp to gradational contacts with the adjacent anorthosites, pyroxenite and rare peridotite layers. The anorthosites typically occur in thin layers, but the ultramafic rocks may form layers several metres thick. In the northern part of the complex there are gabbronorites similar to those of the Chilas complex. Here also outcrop some sheared, well-layered metagabbro–anorthosite–pyroxenite sequences which resemble the banded amphibolites of Kamila, in the Indus Valley (Treloar *et al.* 1990).

The gabbros have been metamorphosed in epidote amphibolite facies conditions and consist of epidote, amphibole, and ripidolite chlorite with or without quartz and Fe-oxide. Relicts of pyroxene and brown hornblende occur locally. Some rocks have escaped high to medium-temperature hydration and contain only minor amounts of hydrous minerals. A few of the more sheared rocks contain patches and veins of calcic-amphibole. Actinolitic amphibole and chlorite may also occur abundantly in the clinopyroxenite (rarely websterite) layers within the gabbros, whereas the anorthosite layers are now made up of zoisite with minor quartz and white mica (? paragonite). In the ultramafic body to the northwest of Ratti Gatti, which is made up of dunite, pyroxenite, wehrlite and pyroxene pegmatite, there are coarse- to medium-grained olivine gabbros, troctolite, and allivalite. In some of these, thin coronas of amphibole/clinopyroxene–spinel have grown due to plagioclase–olivine reaction during cooling and uplift.

Mineral chemistry

The geochemistry of the complex is under investigation and only preliminary data are included here. Silicate phases have been analysed in some ultramafic rocks and a metagabbro. Detailed data have been collected on chromite in many rocks. The salient features of the mineral chemistry are summarized below.

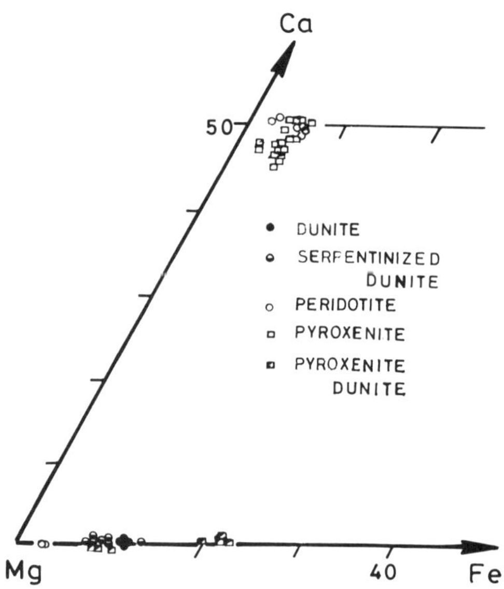

Fig. 5. Plots of olivine and clinopyroxene analyses from the ultramafic rocks on a Mg–Fe–Ca triangle.

The olivine analyses contain 0.2 to 0.5 wt% NiO and usually have a forsterite content of 92.5 to 86.0%, although one peridotite contains Fo_{78}. Olivines in the chromitites are highly magnesian (Fo_{98}–Fo_{96}) probably as a result of subsolidus exchange of Mg–Fe with associated chromite (Irvine 1967). The clinopyroxene in the ultramafic rocks is diopsidic, rich in Mg and Ca (Fig. 5). It is depleted in TiO_2 (<0.2%) and Na_2O (<0.1%), with Al_2O_3 below 3.1%. Metamorphic amphiboles range from Al-poor hornblende to actinolite.

The chromite analyses display considerable compositional variations, but all are characterized by high Cr No. (100Cr/(Cr + Al)) and low Fe^{3+} No. (100 Fe^{3+}/R^{3+}). Plots of Cr-Al-Fe^{3+}, and Cr No. and Fe^{3+} No. against 100 Mg/(Mg + Fe^{2+}) are displayed in Figs 6 and 7. Also shown in these figures are the fields of Cr-spinels from the Jijal and Chilas complexes of the Kohistan arc. The general similarity of the Sapat complex and Jijal chromites is obvious. The high Cr No. of the Sapat complex chromites are akin to those of island arc-related environments; chromites formed in oceanic environments contain Cr No. below 60 (Dick & Bullen 1984; Jan & Windley 1990). In serpentinized rocks the chromites have been altered completely or along margins and cracks to ferritchromite. These are enriched in iron content, have higher Cr No.

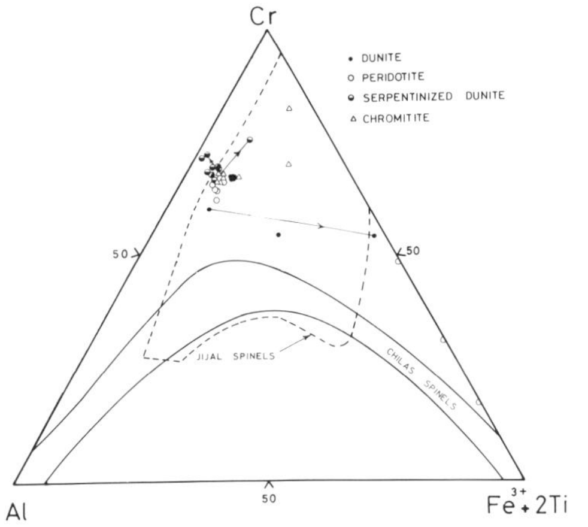

Fig. 6. Cr–Al–Fe^{3+} + 2Ti plots of Cr-Spinels from the Sapat complex. Fields of spinels from the Jijal complex (Jan & Windley 1990) and Chilas complex (Jan *et al.* 1992) are also shown for comparison. Arrows point to margins of zoned grains.

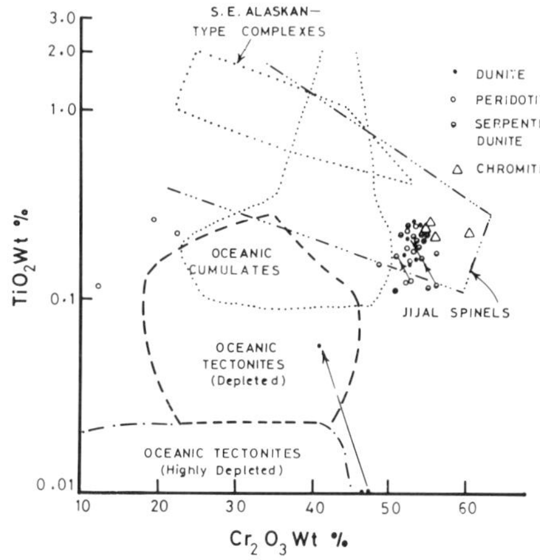

Fig. 8. TiO_2–Cr_2O_3 wt% relations in the spinels. Field for spinels from oceanic rocks are after Herbert (1982) and those for Alaskan-type complexes and Jijal are from Jan & Windley (1990).

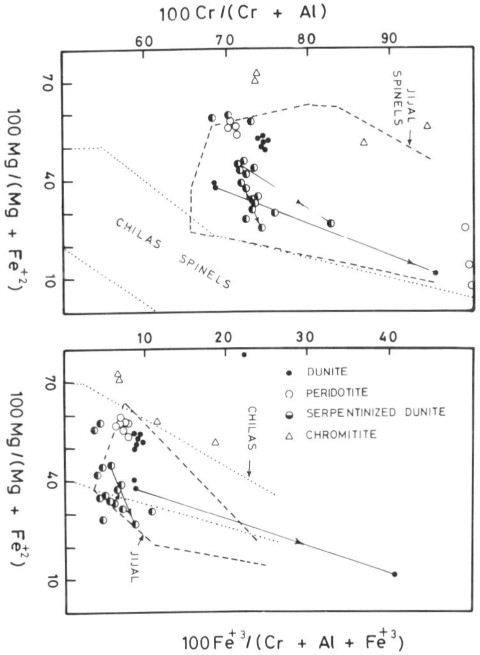

Fig. 7. Variation in the 100Cr/(Cr + Al) and 100 Fe^{3+}/R^{3+} against 100 Mg/(Mg + Fe^{2+}) in Cr-spinels. The Jijal and Chilas spinels fields are shown for comparison. Other details as in Fig. 6.

than cores, and are accompanied by a halo of clinochlore in many cases. In Fig. 7 all the altered chromites plot distinctly above the main population due to their high Cr No. and low Mg No.

Figure 8 shows the plots of TiO_2 against Cr_2O_3 wt% in the chromite analyses. Also shown are the fields of chromites from Alaskan-type complexes, oceanic rocks, and the Jijal complex. As in Jijal the Sapat chromites do not plot in the area of oceanic rocks which strengthens the probability that the complex is related to an island arc setting.

Discussion

The Indus suture traces the boundary of the Indian plate with 'Asiatic' continental blocks or island arcs (Powell 1979). The suture zone is characterized by the occurrence of mafic–ultramafic complexes which have been loosely termed as (oceanic) ophiolites (Gansser 1980). In northern Pakistan, the mafic–ultramafic plutonic complexes located along the Indus suture can be grouped broadly into two genetic types. One group, which is associated with volcanic and sedimentary rocks (e.g. Dargai, Shangla, ? Allai), occurs in tectonic melanges and probably represents dismembered oceanic ophiolites (Ahmed 1988). The other overrides the Indian plate in the immediate hanging wall of the suture. Complexes of this group (Tora Tigga, Jijal, ? Babusar) have been considered as cumulates to magma(s) of the Cretaceous Kohistan island arc (Jan & Jabeen 1990; Jan & Windley 1990). Our preliminary studies on the Sapat complex, which occurs in the southeastern part of the arc, provide further insight into this problem.

The Sapat complex is made up of up to 2 km broad basal ultramafic cumulates ranging from dunites to pyroxenites which are overlain by gabbros containing layers of peridotites, pyroxenites and anorthosites. These pass upward into generally isotropic gabbros. Several aspects of the complex preclude the possibility that it represents a segment of oceanic crust: (1) it is closely associated in space and time with island arc-related rocks without a tectonic break, (2) it is too thick when compared with oceanic mafic–ultramafic cumulates and isotropic gabbros, (3) it is devoid of sheeted dykes, pillow lavas and pelagic sediments (cf. Coleman 1977), and (4) the composition of its chromite, especially the high Cr No., is akin to arc-related rocks (Dick & Bullen 1984). We, therefore, propose that the Sapat represents a basal complex related to the Kohistan arc magmatism.

The Sapat complex displays broad similarities with some other arc-related equivalents such as the Border Ranges ultramafic–mafic complex in Alaska (Burns 1985; DeBari & Coleman 1989). Within the Kohistan terrain, the fault-bounded Jijal complex can be roughly equated with the Sapat complex. Although the Jijal complex was re-equilibrated at 10–12 kbar and its metagabbros are garnet-rich, the basal ultramafic cumulates are similar in many respects to those of Sapat. Both occur in the immediate hanging wall of the suture and seemingly form the basal part of the Kohistan magmatic arc, and both form thick sequences of similar rock types and mineral (especially chromite) chemistry (cf. Jan & Windley 1990).

In terms of layering and rock-types, the Sapat complex displays similarities with the Chilas complex also. However, the tectonic locale and spinel composition of the Chilas complex are distinctly different than those of Jijal and Sapat. Unaltered spinel in the Chilas complex is invariably low in Cr No. (Jan *et al.* 1992), and similar to oceanic peridotites of Dick & Bullen (1984). The Chilas complex forms a 300 km long, E–W-trending body which divides the Kohistan arc effectively into two parts. The northern part is made up of tholeiitic to calc-alkaline volcanics, the Kohistan batholith, and minor sediments (Petterson & Windley 1985; Petterson *et al.* 1990), and the southern part mainly of metamorphosed basaltic rocks, gabbros, diorites and ultramafic cumulates (Jan 1988). Geochemical investigation of the Chilas complex suggests that it may have formed either in a back arc setting or through early rifting of a mature island arc (Khan *et al.* 1989). The Jijal and Sapat complexes, on the other hand, are cumulates to early arc tholeiitic magma (Khan & Thirlwall 1988; Jan & Windley 1990), implying that they cannot be equated with the Chilas complex.

Metamorphic mineral assemblages in the Sapat complex are suggestive of epidote amphibolite facies conditions: hornblende + epidote ± chlorite in gabbros, amphibole ± chlorite in pyroxenites and pyroxene pegmatites, and serpentine ± talc ± amphibole ± chlorite in peridotites and dunites. These assemblages are locally overprinted by greenschist facies minerals (chlorite ± actinolite ± albite). The hydration of the complex took place due to shearing related to the thrusting of the Kohistan arc onto the Indian plate during the Early Tertiary (Treloar *et al.* 1989).

We acknowledge an NSRDB grant (No. E.Sc.24) for financial support, and thank M. Barkatullah for advice during microprobe analyses. Peter Treloar and Mike Searle are thanked for inviting us to present this paper in the 1992 Himalayan Workshop at Oxford. Peter Treloar and an anonymous referee reviewed the paper and offered useful suggestions for improvement.

References

Ahmed, Z. 1984. Stratigraphic and textural variation in the chromite composition of ophiolitic Sakhakot-Qila complex, Pakistan. *Economic Geology,* **79**, 1334–1359.

—— 1988. Bulk chemistry and petrography of the Sakhakot-Qila ophiolite, Pakistan. *Acta Mineralogica Pakistanica,* **4**, 4–29.

—— & Chaudhry, M.N. 1976. Geology of Babusar area. Diamir district, Gilgit, Pakistan. *Geological Bulletin, Punjab University,* **12**, 67–78.

Arif, M. 1990. *Mineral chemistry of chromite occurrences from Shangla (Swat) N.W.F.P. and Bela, Baluchistan.* MPhil Thesis, University of Peshawar.

Ashraf, M., Loucks, R.R. & Awan, M.A. 1989. Serpentinization of cumulate ultramafites and development of heazlewoodite–pentlandite–awaruite–magnetite and pentlandite–chalcopyrite–pyrrhotite–pyrite association in Alpurai and Kishora, Swat, Pakistan. *Kashmir Journal of Geology,* **6–7**, 1–17.

Baig, M.S. 1989. New occurrences of blueschist from Shin-kamar and Marin areas of Allai-Kohistan, northwest Himalaya, Pakistan. *Kashmir Journal of Geology,* **6 & 7**, 103–108.

Burns, L.E. 1985. The Border Ranges ultramafic and mafic complex, south-central Alaska: Cumulate fractionates and island arc volcanics. *Canadian Journal of Earth Sciences,* **22**, 1028–1038.

Coleman, R.G. 1977. *Ophiolites.* Springer-Verlag, New York.

COWARD, M.P., WINDLEY, B.F., BROUGHTON, R.D., LUFF, I.W., PETTERSON, M.G., PUDSEY, C.J., REX, D.C. & KHAN, M.A. 1986. Collision tectonics in the NW Himalaya. *In*: COWARD, M.P. & RIES, A. (eds) *Collision Tectonics*. Geological Society, London, Special Publications, **19**, 203–219.

DEBARI, S.M. & COLEMAN, R.G. 1989. Examination of the deep levels of an island arc: Evidence from the Tonsina ultramafic-mafic assemblage, Tonsina, Alaska. *Journal of Geophysical Research*, **94**, 4373–4391.

DICK, H.J.B. & BULLEN, T. 1984. Chromian spinel as a petrogenetic indicator in abyssal and alpine-type peridotites and spacially associated lavas. *Contributions to Mineralogy and Petrology*, **86**, 54–76.

GANSSER, A. 1980. The peri-Indian suture zone. *In*: AUBOUIN, J., DEBELMAS, J. & LATREILLE, M. (Coord) *Geology of the Alpine chains born of the Tethys*. Bureau de Recherches Geologique et Minieres memoires, **115**, 140–148.

HERBERT, R. 1982. Petrology and mineralogy of oceanic peridotites and gabbros: some comparisons with ophiolite examples. *Ofioliti*, **7**, 299–324.

IRVINE, T.N. 1967. Chrome spinel as a petrogenetic indicator. Part-II. Petrologic applications. *Canadian Journal of Earth Sciences*, **4**, 71–103.

JAMES, O.B. 1971. Origin and emplacement of the ultramafic rocks of the Emigrant Gap area, California. *Journal of Petrology*, **12**, 523–560.

JAN, M.Q. 1988. Geochemistry of amphibolites from the southern part of Kohistan arc, N. Pakistan. *Mineralogical Magazine*, **25**, 147–159.

—— & JABEEN, N. 1990. A review of mafic–ultramafic plutonic complexes in the Indus Suture Zone of Pakistan. *Physics and Chemistry of the Earth*, **17**, 93–113.

—— & TAHIRKHELI, A.Z. 1990. The Tora Tigga Complex, southern Dir, NW Pakistan; An example of mafic–ultramafic rocks in the bottom of an island arc. *Geological Bulletin, University of Peshawar*, **23**, 231–251.

—— & WINDLEY, B.F. 1990. Chromian spinel-silicate chemistry in ultramafic rocks of the Jijal complex, northwest Pakistan. *Journal of Petrology*, **31**, 667–715.

——, KHAN, M.A. & WINDLEY, B.F. 1992. Exsolution in Al-Cr-Fe^{3+}-rich spinels from the Chilas mafic–ultramafic complex, Pakistan. *American Mineralogist*, **77**, 1074–1079.

——, KHATTAK, M.U.K., PARVEZ, M.K. & WINDLEY, B.F. 1984. The Chilas stratiform complex; field and mineralogical aspects. *Geological Bulletin, University of Peshawar*, **17**, 153–169.

KAZMI, A.H., LAWRENCE, R.D., DAWOOD, H., SNEE, L.W. & HUSSAIN, S.S. 1984. Geology of the Indus Suture Zone in the Mingora-Shangla area of Swat, N. Pakistan. *Geological Bulletin, University of Peshawar*, **17**, 127–143.

KHAN, M.A., JAN, M.Q., WINDLEY, B.F., TARNEY, J. & THIRLWALL, M. 1989. The Chilas mafic–ultramafic igneous complex; the root of the Kohistan island arc in the Himalaya of northern Pakistan. *Geological Society of America, Special paper*, **232**, 75–94.

—— & THIRLWALL, M.F. 1988. Babusar amphibolites: arc tholeiites from the southern Kohistan arc, N. Pakistan. *Geological Bulletin, University of Peshawar*, **21**, 147–158.

MILLER, D.J., LOUCKS, R.R. & ASHRAF, M. 1991. Platinum group elements mineralization in the Jijal layered mafic–ultramafic complex, Pakistani Himalayas. *Economic Geology*, **86**, 1093–1102.

PETTERSON, M.G. & WINDLEY, B.F. 1985. Rb-Sr dating of the Kohistan arc-batholith in the trans Himalaya of N. Pakistan and its tectonic implications. *Earth and Planetary Science Letters*, **74**, 45–57.

——, —— & LUFF, I.W. 1990. The Chalt volcanics, Kohistan, Pakistan: high-Mg tholeiitic and low-Mg calc-alkaline volcanism in a Cretaceous island arc. *Physics and Chemistry of the Earth*, **17**, 19–30.

POWELL, C.MCA. 1979. Geodynamics of Pakistan: an introduction. *In*: FARAH, A. & DEJONG, K. (eds) *Geodynamics of Pakistan*. Geological Survey of Pakistan, Quetta, 5–24.

ROSSMAN, D.L. & ABBAS, S.G. 1970. *Geology and Economic potential for chromite in the ultramafic rock complex near Dargai, Peshawar division*. Pakistan Geological Survey and US Geological Survey Report (unpublished).

SHAH, M.T. 1986. *Petrology of the rocks from Shergarh Sar area, Allai Kohistan, northern Pakistan*. MPhil Thesis, University of Peshawar.

TAHIRKHELI, R.A.K., MATTAUER, M., PROUST, F. & TAPPONNIER, P. 1979. The India-Eurasia suture zone in northern Pakistan: Synthesis and interpretation of data at plate scale. *In*: FARAH, A. & DEJONG, K.A. (eds) *Geodynamics of Pakistan*. Geological Survey of Pakistan, Quetta, 125–130.

TRELOAR, P.J., BRODIE, K.H., COWARD, M.P., JAN, M.Q., KHAN, M.A., KNIPE, R.J., REX, D.C. & WILLIAMS, M.P. 1990. The evolution of the Kamila shear zone, Kohistan, Pakistan. *In*: SALISBURY, M.H. & FOUNTAIN, D.M. (eds) *Exposed Cross-Sections of the Continental Crust*. Kluwer Academic Press, Amsterdam, 175–214.

——, REX, D.C., GUISE, P.G., COWARD, M.P., SEARLE, M.P., WINDLEY, B.F., PETTERSON, M.G., JAN, M.Q. & LUFF, I.W. 1989. K-Ar and Ar-Ar geochronology of the Himalayan collision in NW Pakistan: Constraints on the timing of suturing, deformation, metamorphism and uplift. *Tectonics*, **8**, 881–909.

Evolution of the lower arc crust in Kohistan, N. Pakistan: temporal arc magmatism through early, mature and intra-arc rift stages

M. ASIF KHAN[1], M. QASIM JAN[1] & B.L. WEAVER[2]
[1] *NCE in Geology, University of Peshawar, Pakistan*
[2] *School of Geology and Geophysics, University of Oklahoma, Norman, Oklahoma 73019, USA*

Abstract: The middle to lower island-arc crust in southern Kohistan represents a 40 Ma life-span of subduction-related magmatism in an intraoceanic setting. The Kamila amphibolite belt comprises two varieties of metavolcanic amphibolites. One enriched in high-field strength (HFS) and heavy rare-earth (HRE) elements is transitional in character between N- and E-MORB with a minor subduction-related component, and represents the earliest arc basement in Kohistan. The other variety of the metavolcanic amphibolites, together with deformed and amphibolitised intrusive basic plutons, has a transitional tholeiitic to calc-alkaline nature marked by depleted HFS and HRE elements and a distinct negative anomaly for Nb, suggesting emplacement in the early to mature stages of arc growth. The subsequent magmatism in southern Kohistan took place in two stages. The extensive calc-alkaline gabbronorites of the Chilas complex were derived from partial melting of a mantle diapir emplaced into a mature arc or, more probably, during the initial stages of sub-arc splitting. This intra-arc rifting, at its advanced stages, generated tholeiitic picrite to high-Mg basalts which crystallized the ultramafic–mafic–anorthosite (UMA) association and basic dykes of the Chilas complex. The Late Cretaceous accretion of the Kohistan island arc with the Karakoram plate in the north ceased the intraoceanic history of magmatism in southern Kohistan.

The Kohistan island arc of northern Pakistan is tectonically sandwiched between the Karakoram and Indian continental plates (Tahirkheli *et al.* 1979; Coward *et al.* 1986). This Cretaceous arc developed in the neo-Tethys due to northward subduction of oceanic lithosphere at the leading edge of the Indian plate. Closure of the marginal basin to the north along the Late Cretaceous Shyok suture welded the arc to the Karakoram plate (Tahirkheli 1983; Petterson & Windley 1985; Pudsey *et al.* 1985; Coward *et al.* 1986; Searle *et al.* 1987). From then on the Kohistan arc became an Andean-type continental margin (Petterson & Windley 1985). Continued subduction resulted in complete consumption of the leading oceanic edge of the Indian plate, resulting in Eocene collision between the arc and the Indian continent at the site of the Indus suture (Treloar *et al.* 1989). The deformation associated with the two collision events folded and tilted the arc so that the basal arc crust is now exposed in southern Kohistan in the hangingwall of the Indus suture, while the upper arc crust occurs in the vicinity of the Shyok suture (Fig. 1).

The phases of magmatism associated with the two stages of arc growth are compositionally distinct. The arc sequence developed during the intraoceanic stage comprises mainly ultramafic–mafic igneous rocks, while magmatism in the continental-margin stage was intermediate to felsic in composition. Much of the Kohistan batholith and Dir volcanic group developed during the latter stage.

The magmatic stratigraphy of the intraoceanic stage of the Kohistan arc is schematically shown in Fig. 2. The exposed lower crust consists of stratiform ultramafic and gabbroic rocks emplaced into amphibolites derived from basaltic to andesitic volcanics. The rocks of the Jijal and Chilas complexes (Jan & Howie 1981; Khan *et al.* 1989) and the Kamila amphibolite belt (Jan 1988) are representative lithologies of this sequence (discussed later). The intermediate to upper arc crust comprises a sequence of supracrustal sedimentary and volcanic rocks. This sequence, in its lower levels, contains high-grade metasediments which are intercalated with amphibolites and called the Gilgit paragneisses (Tehseen Khan, pers. comm. 1991). A sequence of volcanic rocks of basic to intermediate composition, locally containing pillow lavas (the Chalt volcanic group of Petterson & Windley 1985), overlies the Gilgit paragneisses. The Yasin Group is the uppermost lithological unit in the Kohistan sequence. It

From Treloar, P.J. & Searle, M.P. (eds), *Himalayan Tectonics*
Geological Society Special Publication, No. 74, pp. 123–138.

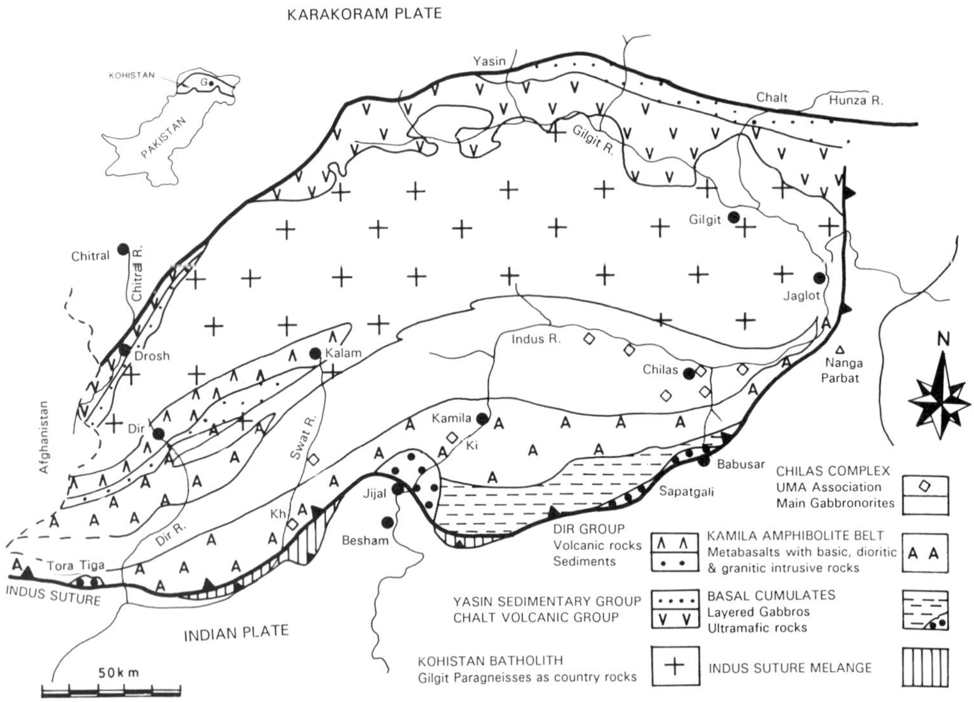

Fig. 1. Simplified geological map of the Kohistan arc (modified after Tahirkheli & Jan 1979; Coward *et al.* 1986). Kh, Khawaza Khela; Ki, Kiru; UMA, ultramafic–mafic-anorthosite.

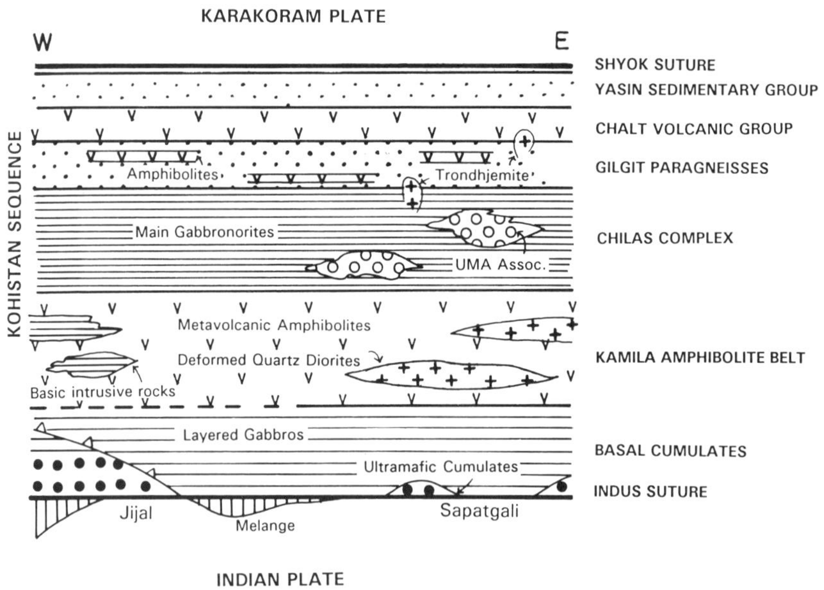

Fig. 2. A schematic profile across the Kohistan arc (modified after Coward *et al.* 1986; Treloar *et al.* 1990) indicating magmatic stratigraphy developed during the intra-oceanic stage of arc growth (prior to Late Cretaceous Kohistan–Karakoram collision).

comprises quartzite, limestone, sandstone and shale (Tahirkheli 1982; Pudsey *et al.* 1985), and contains Albian–Aptian fossils (Pudsey 1986).

Recent data from the Kohistan batholith and Chalt volcanic group from northern Kohistan provide well-constrained models for the magmatic evolution of the Kohistan arc during its continental-arc stage of development (Petterson & Windley 1985, 1991; Petterson *et al.* 1990; Sullivan *et al.* 1993). However, the pre-collision history of the arc is still poorly understood. In this paper, we explore variations in the geochemical composition of successive phases of magmatic activity in the Kamila amphibolites and the Chilas mafic–ultramafic complex from southern Kohistan. These were emplaced prior to the 80–90 Ma deformation related to the Kohistan–Karakoram collision, and thus represent part of the intraoceanic stage of arc growth (Coward *et al.* 1986, 1987; Khan *et al.* 1989; Khan & Coward 1990). The data and models presented in this paper have implications for a better understanding of the lower crust in island arcs.

Geological setting

Southern Kohistan (Fig. 2) comprises three main tectonic elements (i) basal ultramafic–mafic cumulates, (ii) the Chilas complex, and (iii) the Kamila amphibolite belt. The basal cumulates outcrop on the immediate hangingwall of the Indus suture and include, from east to west, the Babusar, Sapatgali, Jijal and Tora Tiga complexes. These are made up of dunite, peridotite, pyroxenite and, locally, chromitite overlain by layered gabbros, gabbronorites and anorthosites. In the Jijal complex these lithologies are, however, metamorphosed to higher grades resulting in the development of garnet granulites (Jan & Howie 1981; Jan & Windley 1990). The Jijal complex has a tectonic contact with the basal cumulates towards the east (Miller *et al.* 1991) and may represent a slice of even deeper arc crust tectonically transported to its present position by obduction associated with the Indus suture. The Chilas complex occupies the medial axis of the Kohistan arc along its entire east-west extension. Rocks of the Kamila amphibolite belt outcrop between the Chilas complex in the north and the ultramafic–mafic plutons to the south. The following discussion is based on data collected from parts of the Chilas complex and the Kamila amphibolite belt exposed in the Indus and Swat valleys and in the area south of Chilas (Fig. 1). The ultramafic–mafic plutons to the south will be dealt with separately in a subsequent publication (Jan & Weaver, in prep.).

The Kamila amphibolite belt

The Kamila amphibolite belt is a composite mass, predominantly comprising amphibolites, with subordinate hornblendite, hornblende gneisses, diorites, plagiogranites, and minor metasediments. Jan (1988) and Treloar *et al.* (1990) distinguished two types of amphibolites; (1) fine to medium grained, either homogeneous or banded, and (2) medium to coarse grained and homogeneous. The former are interpreted to be mainly metamorphosed basic to intermediate volcanic and volcanoclastic rocks, while the latter are considered to be derived from intrusive gabbros, gabbronorites and diorites. Much of the Kamila belt has been strongly sheared resulting in the obliteration of intrusive relations between the metavolcanic lithologies and intrusions. The metavolcanic and metaplutonic lithologies, can however, be distinguished on the basis of relict structures. The coarse-grained amphibolites contain lenses of the protolith and relict igneous textures such as tabular to prismatic plagioclase crystals, amphibole pseudomorphs after prismatic pyroxenes, and local igneous layering and lamination. The fine-grained homogeneous amphibolites in the Indus valley do not contain any relict structures indicative of their metavolcanic nature, although, in the valleys to the south of Chilas there are relict pillow structures. Similarly, in the Swat valley, the fine-grained amphibolites of the Kamila belt contain a 200 m thick succession comprising stretched pillow lavas.

The relative abundance of the metavolcanic and metaplutonic amphibolites is variable along the strike. That part of the Kamila amphibolite belt exposed in the valleys to the south of Chilas is exclusively composed of fine to medium grained amphibolites of volcanic origin (Khan 1988). Westward in the Indus and Swat valleys, more than two thirds of the amphibolite belt comprises metamorphosed basic plutonic rocks initially intrusive into the metavolcanic amphibolites (Fig. 2).

The Chilas complex

The Chilas complex is a 300 km long and up to 40 km broad mafic-ultramafic body intrusive into a basement comprising Kamila amphibolites in the south and Gilgit paragneisses in the north. The complex is remarkably coherent and extends from Nanga Parbat in the east to Dir in the west. More than 85% of the Chilas complex is made up of gabbronorites (hereafter called the main gabbronorites), which locally grade into pyroxenites, anorthosites, and two-pyroxene diorites. Ductile shearing is common (Khan & Coward 1990), resulting in local transformation of the gabbronorites into granulites (Jan & Howie 1980). The main gabbronorites, particularly in the vicinity of Chilas, contain several bodies of ultramafic rocks (commonly less than 5 km^2 in area) with associated troctolites, olivine gabbros, gabbronorites and anorthosites. Khan *et al.* (1989) proposed the name 'UMA association' for these ultramafic, mafic and anorthositic rocks, and outlined differences with the main gabbronorites. The rocks of the UMA association are characterized by highly calcic plagioclase ($An_{98.5}$–An_{83}), and magnesian orthopyroxene (En_{92-65}), clinopyroxene (mg_{87-68}) and amphibole (mg_{84-62}) compared to the main gabbronorites (An_{64-40}, En_{76-48}, mg_{75-55}, mg_{65-57} respectively). Unlike the weakly layered main gabbronorites, sedimentary structures such as modal layering, cross and/or graded bedding, slump folding, syn-sedimentary faulting and scour channels are a common feature of the UMA association (Jan *et al.* 1984).

The mutual relationship between the main gabbronorites and the UMA association has been discussed by Jan *et al.* (1984) and Khan *et al.* (1989). Although there is a possibility that the ultramafic rocks are cumulates complementary to the main gabbronorites, Khan *et al.* (1985, 1989) favoured a relatively later phase of magma generation for the UMA association. The characteristic mode of emplacement of the latter as small bodies within the main gabbronorites, existence of appophyses and dykes originating from the UMA bodies and intruding the main gabbronorites, and substantial petrographic, mineralogical and compositional contrasts between the two associations support this conclusion. The rocks of the two associations are intruded by a suite of basic dykes representing the youngest igneous activity in the Chilas complex. At places these dykes cross cut the sheared gabbronorites, suggesting that at least some of them were emplaced after the main phase of deformation in the Kohistan arc which is related to its collision with the Karakoram plate (Coward *et al.* 1986; Petterson & Windley 1985; Treloar *et al.* 1989). It is notable that a suite of basic dykes contemporaneous with those in the Chilas complex occurs in northern Kohistan (the Jutal dykes of Petterson & Windley 1992).

Sequence of magmatic activity

The mutual field relations between the Chilas complex and the Kamila amphibolite belt are a key to the sequence of magmatic activity in southern Kohistan. Unfortunately, in the area to the south of Chilas, the contact between these two units is obliterated by intense shearing associated with the Jal shear zone (Khan & Coward 1990). Westward in the Indus and Swat valleys, however, shearing is less intense in the contact zone and plutons with composition identical to the main gabbronorites of the Chilas complex intrude the Kamila amphibolite belt.

The field relations within the Kamila amphibolite belt suggest that the fine to medium grained amphibolites of metavolcanic origin are probably the earliest lithology formed in the Kohistan arc. The plutons in this belt are clearly younger than the metavolcanic component and bear close similarity in terms of field relations, lithologies, and mineral composition to both the UMA association and the main gabbronorites of the Chilas complex. More than two-thirds of the Kamila belt is made up of gabbronorites, gabbros, and two-pyroxene diorites identical to those of the Chilas complex, while the local troctolite–olivine gabbro–anorthosite bodies at Khawaza Khela and Kiru (Fig. 1) are close analogues of the UMA association. All the lithologies in the Kamila amphibolite belt are cross cut by basic dykes similar to those in the Chilas complex.

The relationship between the Chilas complex and the Kamila amphibolite belt can also be constrained on the basis of metasedimentary xenoliths in the Chilas complex. The fine- to medium-grained amphibolites in the Kamila amphibolite belt are locally associated with minor marble and calc silicates. These lithologies are recognised in xenoliths at more than five occurrences in the Chilas complex (Khan 1988). The southern marginal parts of the Chilas complex, in the valleys to the south of Chilas, commonly contain xenoliths of basic composition. A great majority of these xenoliths are derived from lithologies within the Chilas complex but some fine-grained basic varieties have close petrographic resemblance to the metavolcanic amphibolites in the Kamila belt exposed just south of the contact. Fine-grained amphibolites, closely similar to those of the Kamila belt have also been recognized at the northern contact of the Chilas complex. These amphibolites are intercalated with the paragneisses, and are found as abundant xenoliths within the main gabbronorites to the north of Chilas. Eastward, in the vicinity of Jaglot, similar paragneisses are cross cut by small plutons comprising layered gabbronorites closely resembling the Chilas complex.

Geochemistry and tectonic setting of magma generation

Analytical techniques

Representative samples from the Chilas complex and part of the Kamila amphibolites exposed in the valleys to the south of Chilas were analysed using a Philips PW1400 XRF at Royal Holloway and Bedford New College (RHBNC), Egham. The Indus and Swat valley samples of the Kamila amphibolites were analysed at the University of Oklahoma, Norman, using a Rigaku XRF. In both cases fused discs were analysed for major elements using a Rh anode X-ray tube. Pressed pellets were used for trace elements; Ni, Cu, Zn, Rb, Sr, Th, Pb, Zr, Nb and Y were analysed using a Rh anode X-ray tube, while Cr, V, Ba, La, Ce, and Nd were analysed with a W anode X-ray tube. In the case of the RHBNC analyses, matrix corrections were calculated from major elements, and estimated errors for reproducibility are reported in Khan *et al.* (1989). The set of trace elements analysed by the Rh tube at Oklahoma was corrected for mass absorption using the intensity of Rh Kα Compton scatter peak (Harvey & Atkin 1982), while data analysed by the W tube were corrected using the intensities of the W Lα Rayleigh scatter and Fe Kα peaks to cross the Fe K absorption edge (Nesbit *et al.* 1976). Both major- and trace-element calibrations were based on a wide range of international standards. The detection limits (2 sigma) by Rigaku XRF at Oklahoma are approximately 1 ppm for Cr, V, Zn, Rb, Sr, Zr, Nb, and Y, 2 ppm for Ni, Cu, Th, and Pb, and 3 ppm for Ba, La, Ce, and Nd.

Rare earth elements (REE) for a set of selected samples from the Chilas complex and part of the Kamila amphibolite exposed in the Thak valley were analysed using an inductively coupled plasma emission spectrometer (Philips PV 8210) at RHBNC, Egham, following the method of Walsh *et al.* (1981). Preconcentration of the REE, following dissolution was carried out at Imperial College, London, using the method of Watkins & Nolan (1992). Errors are generally *c.* 5%, but are 20–50% for samples with concentrations near detection limits. A few samples from part of the Kamila amphibolite belt in the Indus valley were analysed for REE by INNA. Irradiations were performed at Georgia Institute of Technology under the Department of Energy Reactor Sharing Program. The analytical technique and counting scheme employed generally followed that of Potts *et al.* (1981, 1985). Analyses for representative samples from the various magmatic phases discussed in this paper are presented in Table 1.

The Kamila amphibolite

A preliminary account of the geochemistry of the Kamila amphibolites was presented by Jan (1988). Additional geochemical data for more than 50 samples have recently been obtained. For this paper, we restrict ourselves to amphibolites and hornblende gneisses derived from basalt, gabbro or gabbronorite precursors. Only those samples are used which contain reasonably high proportions of incompatible trace elements and thus approach non-cumulus compositions. High-field strength elements (HFSE) have a reasonably good mutual correlation, suggesting little effect of remobilization during metamorphism (cf. Weaver *et al.* 1981). Although large-ion lithophile elements (LILE) are included in the following interpretation, their abundances are suspect due to alteration during amphibolite-facies metamorphism and subsequent retrogression.

The fine-grained metavolcanic amphibolites contain SiO_2 ranging from 44 to 53.6% (average 50.8%), MgO from 2.9 to 7.8% (average 5.9%), and alkalies (total) from 1.56 to 4.6% (average 2.8%). According to the classification scheme of De La Roche *et al.* (1980) and Le Bas *et al.* (1986) they range in composition from basalts, through basaltic andesites, to andesites. No high-Mg andesite or boninite (cf. Hickey & Frey 1982) has been found, although such rocks are recorded in the Chalt volcanic group from northern Kohistan (Petterson & Windley 1991). A transitional tholeiite to calc- alkaline affinity is indicated by Fe/Mg ratios which increase moderately with fractionation as monitored by SiO_2 and Zr. This is supported by their position in the AFM plot of Irvine & Baragar (1971). The order of crystallization is dominated by plagioclase in the early stages of fractionation, reflected in Al_2O_3 and CaO depletion with decreasing MgO (major-element plots not presented).

Evaluation based on incompatible trace elements suggests a two-fold division of the metavolcanic amphibolites; (a) HFSE-enriched (or E-type) and (b) HFSE-depleted (or D-type). Zr, Nb, and Y are generally twice the abundance in the E-type amphibolites as in the D-type. Similarly TiO_2 and P_2O_5 are distinctly more abundant in the E-type. These differences are highlighted by the mantle-normalized trace-element patterns (Fig. 3a). E-type amphibolites have a trace-element pattern characterized by enriched HFSE (*c.* 10 × primordial mantle abundance). The LILE are either depleted relative to the HFSE or have similar abundances, resulting in patterns which are either flat or inclined towards the left. In comparison the trace-element patterns for the D-type slope towards the right, with relatively higher LILE and lower HFSE. A limited number of samples from the metavolcanic amphibolites have

Table 1. *Representative whole-rock analyses from southern Kohistan including lithologies from the Chilas complex and the Kamila amphibolite belt*

	CH221 1	CH401 2	CH31b 3	CH224 4	CH44b 5	CH34 6	CH806 7	CH802 8	CH30 9	IK1027 10	IK1048 11	IK1039 12	IK1054 13	IK1061 14	IK1068 15
Oxides (wt%)															
SiO_2	52.53	55.16	57.26	40.60	45.25	50.28	42.75	44.75	46.08	55.74	58.52	50.19	57.35	43.87	52.18
TiO_2	0.45	0.98	0.76	0.10	0.07	0.20	1.37	0.99	0.98	0.73	0.36	1.88	1.77	1.18	0.72
Al_2O_3	19.62	17.11	16.43	4.69	25.32	20.83	20.19	19.72	17.34	17.21	16.95	14.79	14.87	19.99	17.41
Fe_2O_3*	7.24	8.43	7.86	15.59	3.63	6.11	11.61	11.36	12.11	9.46	5.57	11.72	11.48	14.11	10.55
MnO	0.14	0.15	0.12	0.22	0.06	0.11	0.14	0.15	0.23	0.20	0.11	0.16	0.12	0.19	0.15
MgO	6.47	4.74	4.56	32.77	5.77	6.74	8.91	8.34	8.75	4.34	5.87	6.42	2.89	6.51	5.50
CaO	9.86	8.23	7.35	5.45	15.73	12.04	12.59	12.62	11.39	7.60	10.53	12.59	8.60	13.22	10.08
Na_2O	3.06	3.43	3.23	0.25	1.37	2.48	2.08	1.98	2.32	3.01	1.50	1.39	2.59	1.90	2.20
K_2O	0.29	1.20	1.34	0.01	0.04	0.10	0.13	0.09	0.16	0.34	0.34	0.16	0.03	0.23	0.44
P_2O_5	0.07	0.23	0.17	0.01	0.03	0.05	0.03	0.04	0.08	0.12	0.06	0.18	0.43	0.19	0.10
LOI	0.25	0.11	0.85	0.00	1.85	1.73	1.21	1.43	0.93	0.49	1.31	0.58	0.64	0.04	0.08
Total	100.0	99.8	99.9	99.7	99.1	100.7	101.0	101.5	100.4	99.2	101.1	100.1	100.8	101.4	99.4
Trace elements (ppm)															
Sc	na	na	25	na	na	29	58	54	na	na	na	na	na	na	na
Cr	102	78	84	178	245	120	69	113	316	41	48	94	9	44	49
Ni	56	49	53	72	57	46	60	48	82	14	53	43	10	11	17
Cu	18	71	81	20	20	5	160	135	96	51	36	43	26	9	34
Zn	59	71	78	54	6	53	66	67	89	89	39	36	24	108	69
V	129	208	193	114	57	90	437	382	313	232	187	373	157	338	225
Ga	21	20	18	22	0	18	17	15	20	na	na	na	na	na	na
Rb	1.7	27.3	17.0	2.2	0.0	1.8	1.0	0.6	1.3	5.0	6.0	2.0	1.0	2.0	3.0
Sr	401.2	340.4	375.0	367.8	381.8	407.1	368.4	330.7	280.6	371.0	220.0	150.0	208.0	368.0	281.0
Ba	84	210	247	63	26	48	43	43	25	59	61	46	44	49	164
Th	0.7	6.9	0.5	0.1	0.0	0.3	0.1	0.0	2.7	2.0	2.0	2.0	2.0	2.0	2.0
Pb	4	9	7	5	0	3	2	0	0	9	4	3	2	4	6
Y	8.4	32.2	19.0	8.6	0.0	8.1	19.2	15.4	21.0	25.0	7.0	37.0	70.0	23.0	21.0
Zr	7.8	167.6	100.0	2.5	24.2	7.8	21.8	22.1	27.6	58.0	22.0	99.0	231.0	33.0	36.0
Nb	0.8	5.6	4.5	0.7	0.0	0.1	1.4	0.8	1.4	2.2	1.3	4.7	9.6	1.4	1.2
La	7	22	13	5	11	2	2	1	4	3	3	4	9	3	5
Ce	17.8	36.6	27.0	0.0	0.0	5.2	3.7	4.2	0.0	10.0	5.0	17.0	30.0	14.0	15.0
Nd	6.6	24.3	14.0	2.3	0.0	2.5	4.2	4.8	2.4	8.0	3.0	8.0	21.0	7.0	8.0

Table 1 *Continued*

	CH221 1	CH401 2	CH31b 3	CH224 4	CH44b 5	CH34 6	CH806 7	CH802 8	CH30 9	IK1027 10	IK1048 11	IK1039 12	IK1054 13	IK1061 14	IK1068 15
Rare earth elements (ppm)															
La	4.5	14.4	5.8	2.3	4.7	2.1	1.0	1.1	5.0	7.4	na	6.3	15.3	3.0	7.1
Ce	7.3	30.9	11.9	0.8	2.9	4.0	3.4	3.5	7.2	18.6	na	17.2	38.7	10.5	16.7
Pr	1.1	4.4	1.7	0.1	0.4	0.5	1.0	0.9	1.3	na	na	na	na	na	na
Nd	4.6	17.5	7.2	0.6	1.9	3.0	5.6	4.7	6.9	11.9	na	13.3	30.0	10.2	11.4
Sm	1.17	3.85	1.79	0.18	0.49	0.79	2.10	1.63	2.15	3.22	na	4.14	8.85	3.41	2.79
Eu	0.57	1.30	0.85	0.06	0.23	0.46	0.90	0.69	0.70	1.07	na	1.75	3.04	1.25	1.06
Tb	na	na	na	na	na	na	na	na	na	0.61	na	0.98	1.93	0.65	0.59
Gd	1.46	4.21	2.16	0.25	0.52	0.88	2.85	2.17	2.81	na	na	na	na	na	na
Dy	1.56	4.11	2.29	0.00	0.58	1.17	3.26	2.48	3.43	na	na	na	na	na	na
Ho	0.33	0.88	0.54	0.30	0.12	0.25	0.67	0.54	0.74	na	na	na	na	na	na
Er	0.94	2.20	1.33	0.06	0.41	0.62	1.70	1.33	2.10	na	na	na	na	na	na
Yb	0.92	2.44	1.33	0.18	0.35	0.87	1.67	1.35	2.29	2.94	na	3.61	7.22	2.33	2.11
Lu	0.20	0.30	0.20	0.16	0.10	0.10	0.20	0.20	0.30	0.48	na	0.57	1.13	0.33	0.37

* All iron calculated as Fe_2O_3. na not analysed. CH samples from the Chilas complex (analysed using XRF & ICP; RHBNC, Egham, UK), IK from the Kamila amphibolite belt (XRF and INNA, Oklahoma, USA).
1–3, Main gabbronorites; 4–6, UMA association; 7–9, basic dykes; 10–11, HFSE depleted metavolcanic amphibolites; 12–13, HFSE enriched metavolcanic amphibolites; 14–15, metaplutonic amphibolites.

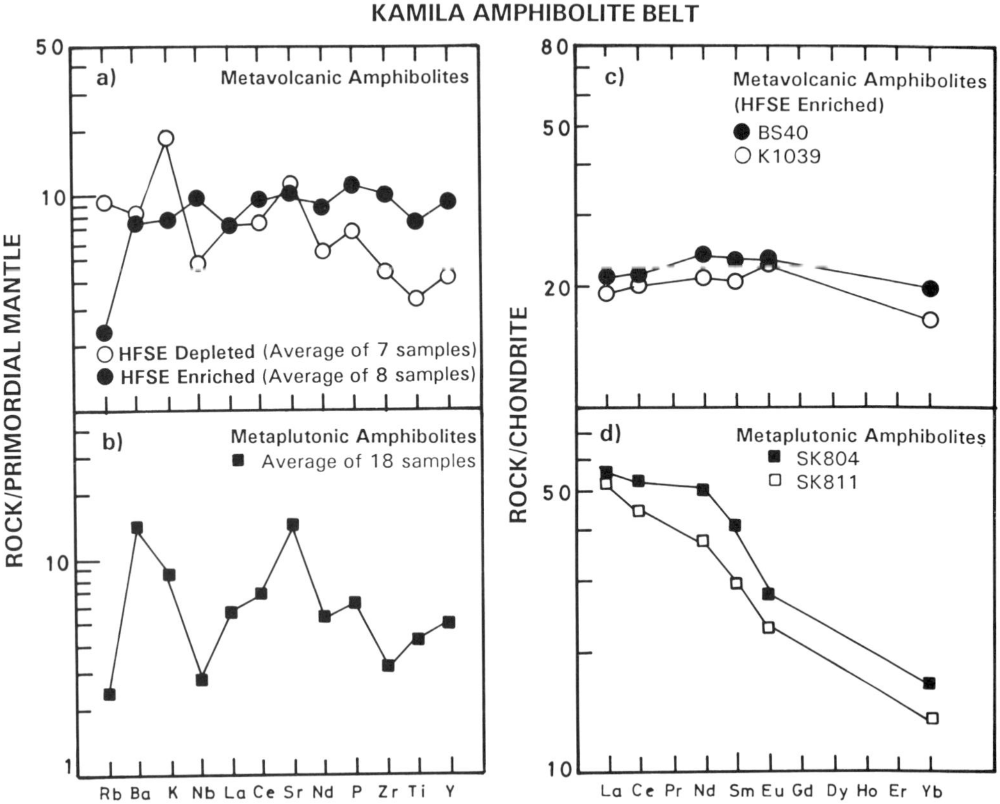

Fig. 3. Multi-element trace- and rare-earth element patterns for the basic rocks in the Kamila amphibolite belt. Normalizing trace-element data for the primordial mantle are from Wood *et al.* (1979) and rare-earth data for the chondrite are from Nakamura (1974).

trace-element characteristics which are transitional between the E- and D-types.

The metaplutonic amphibolites are characterized by a wider variety of rock assemblage, including anorthosites, olivine gabbros, gabbros, gabbronorites and diorites. All the rock types may be related to each other through fractional crystallization involving successive fractionation of olivine, clinopyroxene, magnetite, and plagioclase (Jan 1988). A calc-alkaline nature is indicated by moderate FeO/MgO ratios, and lack of an iron-enrichment trend on the AFM plot (not presented). It is noted that there is a geochemical similarity between the D-type metavolcanic and the metaplutonic amphibolites in terms of their predominantly calc-alkaline character; the two may belong to the same phase of magmatism. The E-type metavolcanic amphibolites are tholeiitic and belong to an earlier phase of magmatism.

When compared using incompatible element abundances, the meta-plutonic amphibolites are markedly different from the E-type metavolcanic amphibolites, but have a resemblance to the D-type (Fig. 3b). The metaplutonic amphibolites have a mantle-normalized trace-element pattern characterized by an inclination towards the right resulting from slight enrichment of LILE over HFSE [$(\mathrm{LILE})_N/(\mathrm{HFSE})_N > 1$, as in the D-type metavolcanic amphibolites]. Concentration of HFSE is distinctly lower than that in the E-type metavolcanic amphibolites. The compositional difference between the metaplutonic amphibolites and E-type metavolcanic amphibolites is particularly well displayed by the chondrite-normalized REE patterns (Fig. 3c & d). The metaplutonic amphibolites are distinctly more enriched in LREE compared to E-type metavolcanic amphibolites (i.e., 40 × v. 10–12 × chondrite abundances).

The mantle-normalized trace-element patterns for the metaplutonic amphibolites and D-type metavolcanic amphibolites (Fig. 3a & b) are marked by positive spikes for P, Sr, and K, and negative spikes for Nb and Zr. These, combined with $(LILE)_N/(HFSE)_N$ ratios in excess of 1, point to a subduction-related origin for the metaplutonic amphibolites (Pearce 1982; Holm 1985). The E-type metavolcanic amphibolites (Fig. 3a), on the other hand, do not reveal an unambiguous tectonic regime of magma generation. The lack of spikes and positive slope (inclined towards the left) in their mantle-normalized patterns appears similar to N-type MORB. The positive slope in the E-type metavolcanic amphibolites is, however, less pronounced than in N-MORB. This difference is highlighted when the E-type metavolcanic amphibolites are normalised against an N-MORB (Fig. 4a) (cf. Pearce 1983). All the LILE (except

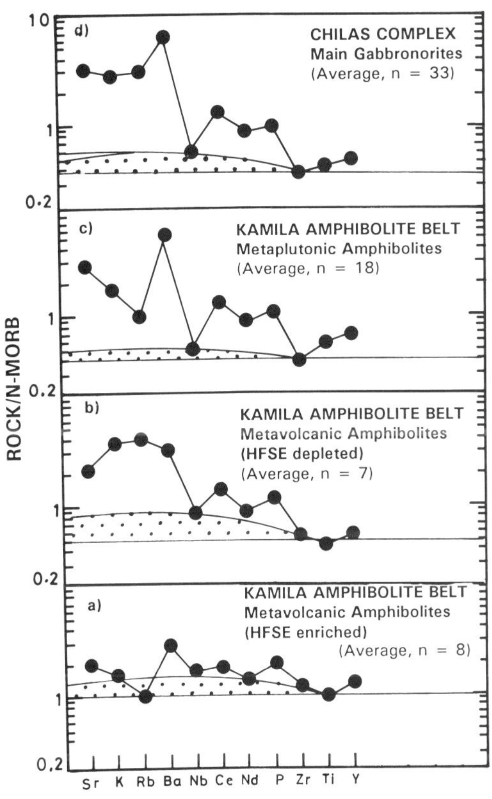

Fig. 4. N-MORB-normalized trace-element patterns for average compositions in four successive phases of magmatic activity in the Kamila amphibolite belt and Chilas complex, southern Kohistan. Tholeiitic metavolcanic amphibolites (HFSE-enriched) considered to represent earliest phase of magmatic activity, are most primitive compared to the later phases. All the phases show an enrichment in some of the HFSE elements inherited from the source mantle (dotted field). N-MORB normalizing values from Pearce (1983).

Rb) are enriched in the E-type metavolcanic amphibolites relative to N-MORB. Besides there is a slight but distinct enrichment of some of the HFSE (Nb, P, Zr) in the metavolcanic amphibolites over N-MORB. Normally such LILE enrichment is attributed to metasomatism of the mantle source induced by subduction, but in the present case subduction-related enrichment cannot be distinguished from remobilization during metamorphism. However, the HFSE enrichment points to an involvement of an E-MORB source in the generation of the E-type metavolcanic amphibolites.

The contributions from an E-MORB source in the generation of the three varieties of the Kamila amphibolites are estimated in Fig. 4. All the three varieties are derived from a mantle source previously enriched in incompatible trace elements. There is a distinctly greater contribution from an E-type mantle source in the case of the E-type metavolcanic amphibolites, compared to the D-type metavolcanic amphibolites and the metaplutonic amphibolites, which may have a higher contribution from a subduction component. It is interesting to note that a primarily enriched source is also envisaged for the Kohistan batholith in northern Kohistan (M.G. Petterson, pers. comm. 1992).

The Chilas Complex

The main gabbronorites. The geochemistry of the main gabbronorites has been evaluated in detail by Khan *et al.* (1989). These rocks are characterized by a calc-alkaline trend on the AFM diagram, comparable to island-arc non-cumulates of Beard (1986). On a mantle-normalized multi-element plot (Fig. 5a), they show (1) positive spikes for Ba, K, Sr, and P, and a marked negative spike for Nb; (2) low Rb, and a high $(K/Rb)_N$ ratio (up to 5), and (3) a slope of the trace-element pattern inclined towards the right characterized by enriched LILE and depleted HFSE. It may be mentioned here that LILE in the Chilas complex are near their primary abundances as reflected by their good correlation with the HFSE (Khan 1988). The REE patterns for the Chilas main gabbronorites (Fig. 5b) are characterized by moderate enrichment in the light REE $[(Ce/Yb)_N = 2$ to $3.5)]$ and a concave-upward shape resulting from a lesser enrichment of the middle REE compared to the heavy REE $[(Sm/Yb)_N = 1.5$ to $1.8)]$. Khan *et al.* (1989) demonstrated the calc-alkaline nature of the Chilas main gabbronorites, and suggested that they formed in a mature or intra-arc rifting stage of the arc growth in Kohistan.

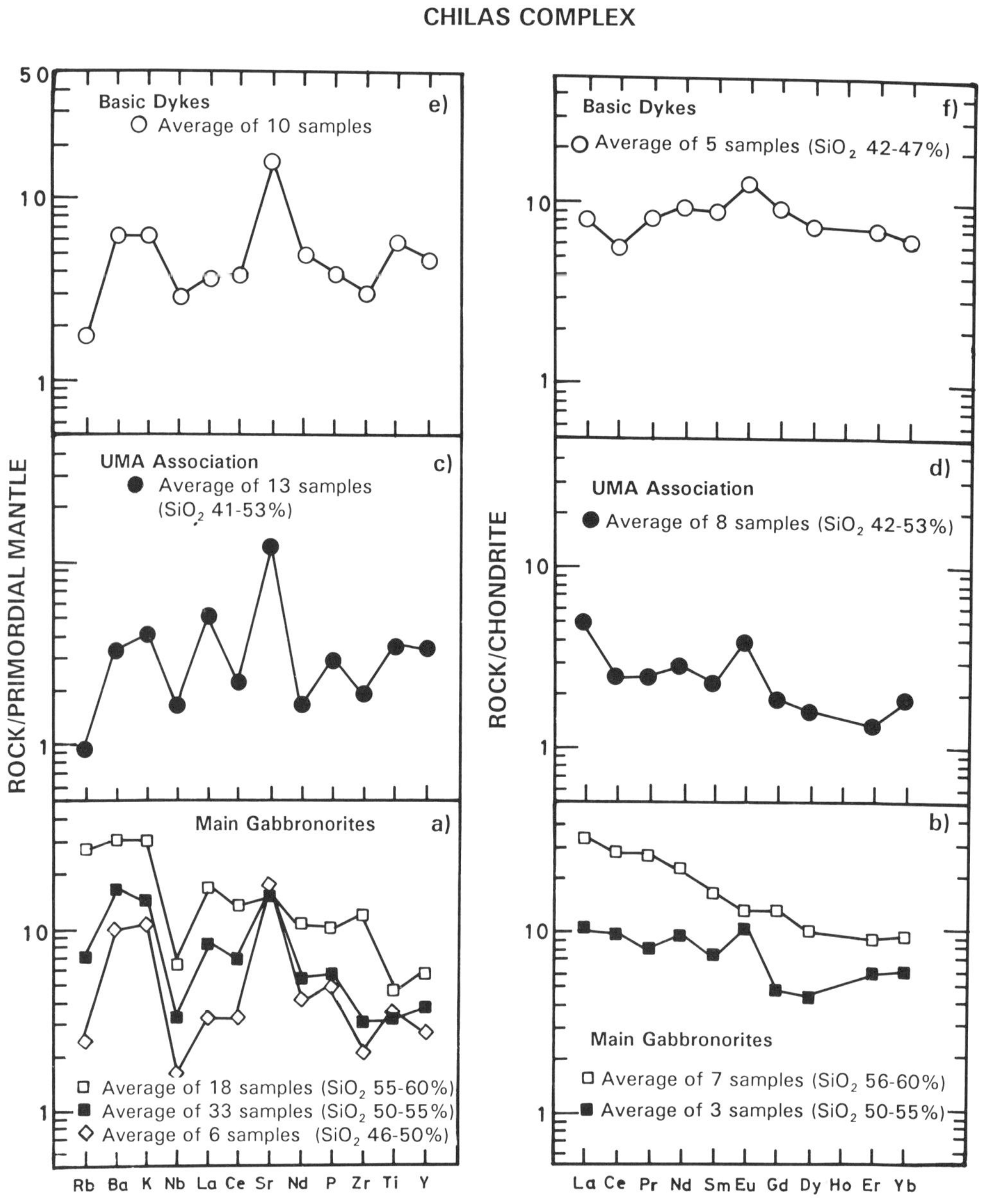

Fig. 5. Multi-element trace- and rare-earth patterns for the three phases of magmatism in the Chilas complex. Same normalizing values are used as in Fig. 3.

The UMA association. The predominantly cumulate character of the UMA association is reflected in unusually high abundances of compatible major and trace elements (Table 1). The geochemical evaluation may be possible only for rocks with a high content of intercumulus material. In such rocks, despite lower absolute abundances, the ratios between incompatible trace elements may approach those of the liquid (Weaver *et al.* 1981). A mantle normalized average pattern for rocks with high abundance of Zr (>15 ppm) is shown in Fig. (5c). In general, the trace-element characteristics of the UMA association are compatible with the main gabbronorites, with high $(Ba/Rb)_N$ and $(K/Rb)_N$ ratios, a negative spike for Nb and a positive spike for Sr. The difference is mainly in the slope of the pattern, which is slightly inclined towards the left in the case of the UMA association [$(Y/Ba)_N$ ratio < 1] and towards the right in the main gabbronorites (Fig. 5a). The REE pattern for an average of eight samples from the UMA association (Fig. 5d) has general shape characteristics similar to the main gabbronorites, such as a slope towards the right [$(Ce/Y)_N \approx 1.4$ and a concave upward shape [$(Sm/Yb)_N = 1.5$]. The positive Eu anomaly reflects the presence of cumulus plagioclase as in the case of the mafic members (SiO_2 50–55%) of the main gabbronorites (Fig. 5b). However, the abundances of the REE in the UMA are distinctly lower than in the main gabbronorites. It is possible that the UMA association represents cumulates complementary to the main gabbronorites, however, the field evidence such as localized occurrence and intrusive relations discount this. Khan *et al.* (1989) suggested that the main gabbronorites and the UMA association were both derived through the melting of a sub-arc mantle source. The UMA association may represent either a higher degree of partial melting or melts of a more refractory source previously impoverished in trace elements through melt extraction.

The basic dykes. The majority of the basic dykes in the Chilas complex are characterized by low SiO_2 and high Al_2O_3, CaO, MgO, Cr, and Ni (Table 1). They have major element characteristics intermediate between alkali-olivine basalts and picrites.

The mantle-normalized trace-element pattern for an average of ten samples (Fig. 5e) shows characteristics like low Rb, high Ba, Sr, Ti and distinctive negative Nb anomaly similar to the UMA association and the main gabbronorites. There are, however, some important differences. The patterns for the main gabbronorites are characterized by a negative overall slope indicative of a calc-alkaline nature. In comparison the pattern for the basic dykes is horizontal to slightly positive, approaching that found in tholeiitic rocks. This feature is also reflected in the chondrite-normalized REE pattern (Fig. 5f), where the middle to heavy REE are distinctly enriched compared to light REE [$(Ce/Sm)_N \approx 0.5$, $(Ce/Yb)_N \approx 0.8$). It is to be noted that the normalised trace- and rare-earth element patterns for the Chilas basic dykes closely resemble the tholeiitic basic dykes intrusive into the Kohistan batholith (the Jutal dykes of Petterson & Windley 1992).

Discussion

Southern Kohistan is commonly considered to represent lower to intermediate levels of an island-arc crust. The uplift and exposure of this part of the arc crust is attributed to (a) imbrication and large-scale folding related to Late Cretaceous Kohistan–Karakoram collision at the site of the Shyok suture, and (b) Eocene obduction of the Kohistan arc onto the Indian plate along the Indus suture (Tahirkheli *et al.* 1979; Coward *et al.* 1986). Several interpretations have been made about the nature and evolution of this intermediate to lower arc crust in Kohistan:

(1) Tahirkheli *et al.* (1979), Bard *et al.* (1980), and Bard (1983) considered the Kamila amphibolites to be a metamorphosed oceanic basement intruded by subduction-related Jijal and Chilas mafic–ultramafic complexes;
(2) according to Miller *et al.* (1991) the Kamila amphibolites are exclusively plutonic, comprising a stack of mafic–ultramafic complexes similar to the Chilas and Jijal complexes;
(3) Jan (1988), and Treloar *et al.* (1990) emphasized the composite nature of the Kamila amphibolites, comprising both metamorphosed plutonic and volcanic basic rocks of island arc affinities.

The existence of metavolcanic rocks in the Kamila amphibolite belt cannot be denied on the basis of the occurrence of metamorphosed pillow lavas associated with massive to thinly banded, fine grained amphibolite. A large part of the Kamila amphibolite belt, particularly in the Indus valley, however, is derived from basic plutons through metamorphism accompanying shearing (Treloar *et al.* 1990). A volcanic basement intruded by basic plutons appears to be a logical interpretation for the Kamila amphibolites. This leads us to establish a sequence of magmatic activity in southern Kohistan starting with metavolcanic amphibolites in the Kamila amphibolite belt (Table 2). The main gabbronorites of the Chilas complex are essentially younger than the metavolcanic basement in the Kamila amphibolites, and may be equivalent to or even younger than the basic plutons in the Kamila amphibolites. The UMA association of the Chilas complex marks the next stage of magmatism, while the basic dykes in the Chilas complex and the Kamila amphibolites represent the last magmatic phase in southern Kohistan in the waning stages of collision of the arc with the Karakoram plate in the north.

Magmatism in southern Kohistan occurred in

Table 2. *A summary of magmatic events in southern Kohistan*

Stage in arc growth	Type of magmatism	Petrographic aspects	Age and correlation in Kohistan
Kamila amphibolite belt			
Early	E-type metavolcanic amphibolites: tholeiitic, HFSE-enriched, LILE-depleted; resemble MORB with some subduction component	Fine–medium grained, homogeneous to banded, local relicts of pillows	80–90 Ma* cooling ages; intruded by pre-collision granitoids (102 Ma Matum Das type). Equivalents of Chalt volcanics and basal cumulates sitting on the Indus suture of which Jijal may be >120 Ma.
Mature	D-type metavolcanic/ metaplutonic amphibolites: tholeiitic to calc-alkaline, HFSE-depleted, significant subduction component	(i) Metavolcanic amphibolites indistinguishable from above (ii) Metaplutonic amphibolites contain relict igneous textures, mineralogy and relict gabbroic lenses	Amphibolitized plutons younger than or coeval with the metavolcanic amphibolites. Early phases of gabbronorites in the Chilas complex may be equivalents.
Chilas complex			
Mature, intra-arc rifting	Main gabbronorites: typical calc-alkaline, enriched in LILE and LREE, depleted in HFSE and HREE. Negative Nb and positive Sr anomalies	Immense volume of coherent gabbronorites, massive to foliated, locally layered	Minimum Ar-Ar cooling age of 80 Ma. Locally intruded by granitoids similar to 102 Ma old Matum Das. Many basic plutonic relicts in the Kamila belt may be equivalents.
Intra-arc rifting	UMA association: tholeiitic, depleted incompatible trace elements, ? higher degree of melting of sub-arc mantle	Mainly cumulate. Includes ultramafic, mafic and anorthositic rocks, massive to strongly layered, apophyses and dykes intrude the main gabbronorites	Local plutonic relicts in the Kamila amphibolite belt (e.g., Khawaza Khela) may be equivalents.
Intra-arc rifting, last stages	Basic dykes: tholeiitic to picritic, LILE depleted relative to HFSE	Fine–medium grained amphibolitic, post-date all the above lithologies, some cut deformed gabbronorites	Ar-Ar cooling ages of 80 Ma. May be equivalent to the Jutal dykes (*c.* 75 Ma) in the Kohistan batholith.

* See Treloar *et al.* (1989) and Petterson and Windley (1985) for the radiometric dates.

successive phases (Table 2) each of which displays distinct geochemical characters, which can be used to characterize changes in tectonic setting of magma generation within the intra-oceanic life span of the Kohistan island arc (Fig. 6).

(1) The metavolcanic rocks of the Kamila amphibolites have an incompatible trace-element chemistry transitional between N- and E-MORB, with a minor subduction component. Their geochemistry suggests a primitive tholeiitic nature in early arc history. There is a possibility that the southern mafic-ultramafic plutons (Tora Tiga, Jijal, Sapatgali, and Babusar complexes), which are predominantly tholeiitic in composition (Jan & Windley 1990; Jan *et al.* this volume), are the plutonic equivalents of the metavolcanic component of the Kamila amphibolites (Fig. 6a). Some of the tholeiitic volcanics recorded in the Chalt volcanic group (Petterson & Windley 1991) in northern Kohistan may be later pulses of the same phase of magmatism.

(2) The transitional tholeiite to calc-alkaline composition of the bulk of the plutons in the Kamila amphibolites, with a distinct subduction component reflected by depleted Nb suggests magma generation in early to mature stages of arc growth (Fig. 6b). They have trace-element compositions resembling the main gabbronorites of the Chilas complex, though the latter are slightly more enriched in LILE and LREE. The Chilas main gabbronorites have a calc-alkaline geochemistry compatible with their origin in the mature stages of arc growth (Fig. 6b). Khan *et al.* (1989) pointed out the thermal problem associated with the generation of the immense volume of magma crystallized into the coherent mass of gabbronorites in the Chilas complex. They suggested that the Chilas magma

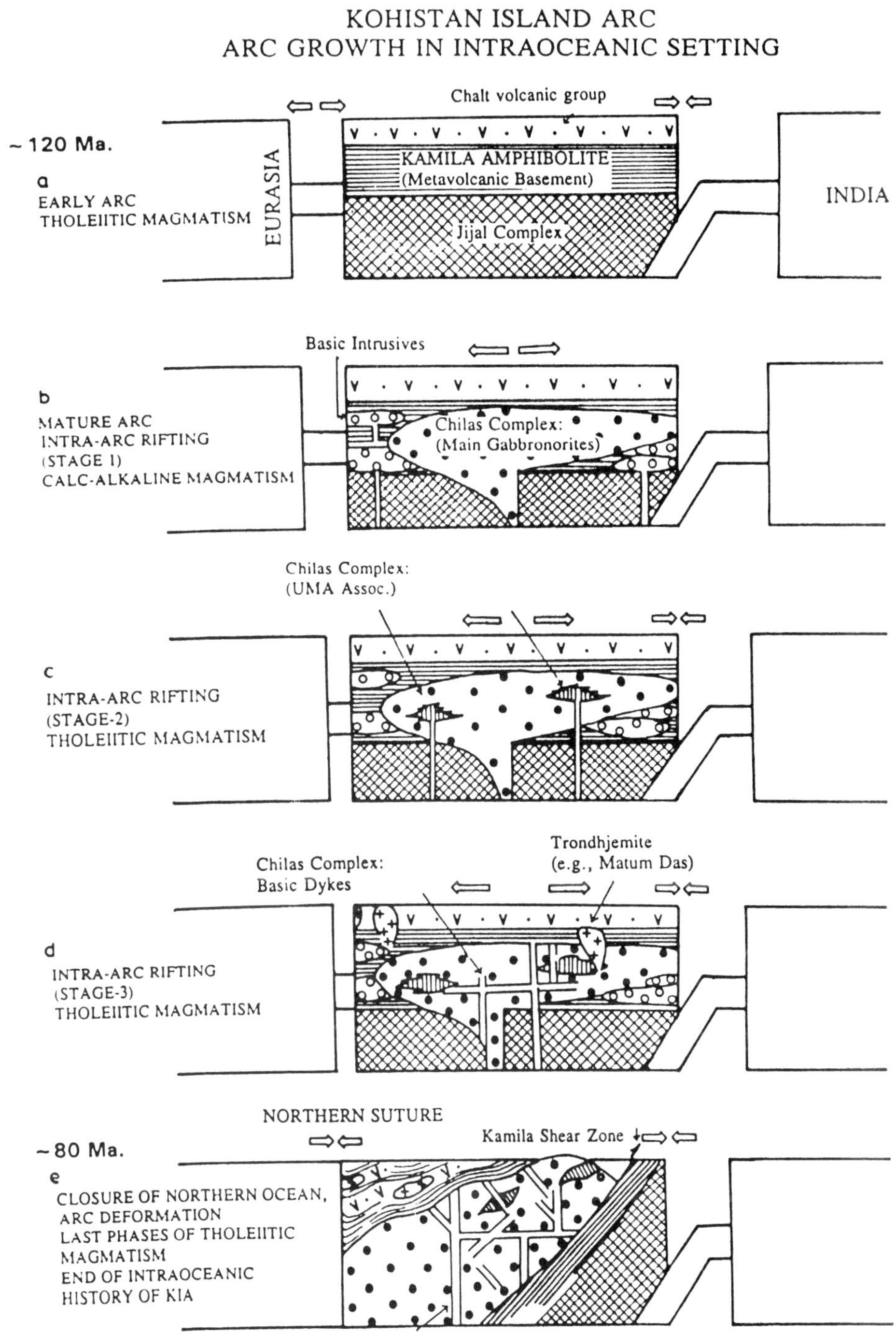

Fig. 6. A schematic model showing various stages in the magmatic evolution of the Kohistan island arc in its intraoceanic life span between 120 and 80 Ma. KIA is Kohistan island arc.

was generated by mantle diapirism during the initial stages of intra- or back-arc splitting of the mature Kohistan island arc.

(3) The UMA association in the Chilas complex marks a reversal to tholeiite composition in the magmatic history of the Kohistan island arc (Fig. 6c). The magma responsible for this association of rocks is picritic to high magnesian basalt, which is distinctly depleted in incompatible trace elements compared to all the previous phases of magmatic activity. A greater degree of partial melting of a depleted but metasomatised mantle source in the relatively advanced stages of intra- or back-arc rifting might have contributed to the magmas of the UMA association. The basic dykes, representing the last phase of magmatic activity in southern Kohistan, are a continuum of the tholeiitic, incompatible trace-element depleted magmatism associated with the last stages of intra-arc rifting (Fig. 6d). The marginal ocean between the Kohistan arc and Karakoram closed by about 80 Ma, resulting in the internal deformation of the arc (depicted in Fig. 6e).

In recent years suggestions have been made that Kohistan is a composite arc system, comprising a northern or Chilas arc and a southern or Jijal arc, now accreted along the Kamila shear zone (Coward *et al.* 1987; Khan & Coward 1990; Khan & Thirlwall 1988; Treloar *et al.* 1990; Ghazanfar *et al.* 1991). Though the Kamila shear zone is a laterally continuous regional structure, its status as a suture zone between two separate arcs could not be established because of the absence of oceanic sediments and ophiolites.

The model presented in this paper favours a composite nature for the Kohistan arc system but rules out any ocean closure at the site of the Kamila shear zone. The Chilas complex is a thick, coherent and laterally continuous mass of closely related igneous rocks which virtually divides the Kohistan arc into two entities. There is a possibility that sections of the arc crust exposed to the north and south of the Chilas complex represent rifted parts of initially one arc. In this scenario, the basal ultramafic–mafic plutons in the hangingwall of the Indus suture and the overlying Kamila amphibolites in southern Kohistan would represent the fore arc, while the Gilgit paragneisses, Chalt volcanic group and the Yasin group of sediments in northern Kohistan the back arc. The Chilas complex was emplaced along the central spine of the Kohistan while undergoing an episode of intra-arc rifting similar to that documented from modern arcs like Mariana, Western Pacific (Karig *et al.* 1978; Stern *et al.* 1990) and New Hebrides, Fiji (Karig 1971; Marcelot *et al.* 1983). The subsequent compression related with collision at the site of the Shyok suture (Petterson & Windley 1985; Coward *et al.* 1986, 1987) uplifted the Chilas complex, and emplaced it tectonically over the fore-arc region comprising the Kamila amphibolite. There is no evidence that the postulated intra-arc rift in Kohistan attained the status of a full-fledge oceanic basin. This can explain the absence of ophiolitic rocks in the Kamila shear zone. Treloar *et al.* (1990) suggested a Late Cretaceous age (Ar-Ar geochronology; see Treloar *et al.* 1989) for deformation within the Kamila shear zone and interpreted it to be related to southward propagation of shortening across the Kohistan arc following suturing with the Karakoram.

The existing data on absolute ages of rocks from southern Kohistan is extremely limited. The minimum age for the intraoceanic period of the Kohistan island arc can be placed at 80 Ma on the basis of Ar-Ar spectra obtained from the hornblende in the basic dykes from the Chilas complex (Khan 1988; Treloar *et al.* 1989). The main gabbronorites and the UMA association in the Chilas complex are locally intruded by trondhjemites compositionally similar to those in the Kohistan batholith dated at 102 Ma on the basis of Rb-Sr whole-rock isochrons (Petterson & Windley 1985). The maximum age from southern Kohistan is from the Jijal complex based on an L-shaped Ar-Ar spectra with a plateaux at 120 Ma (Treloar *et al.* 1989). These data, despite being limited in number and handicapped by problems with excess argon, suggest that the Kohistan island arc had an intraoceanic life span of *c.* 40 Ma. prior to its collision with the Karakoram plate to become an Andean-type continental margin.

M.A.K. acknowledges a Ministry of Education, Pakistan studentship at Imperial College, London, and M.Q.J. a Senior Fulbright Fellowship at the University of Oklahoma. The field work in Kohistan by the first two authors was supported by NCE in Geology, University of Peshawar, and recently by NSRDB project No. ESc. 24. Field assistance and discussions with B.F. Windley, M.P. Coward, M.G. Petterson, P.J. Treloar and T. Khan are gratefully acknowledged. P. Watkins at the Imperial College, M.F. Thirlwall, G. Marriner, N. Walsh at RHBNC, Egham, and D. Powell at Oklahoma are thanked for help with analyses. R. Durrani assisted with drafting. P.J. Treloar and M.P. Searle are thanked for invitation to the first two authors for participation in the Himalayan workshop, Oxford (April 1992) and Geological Society, London, for financing the travel. Extremely detailed reviews by M.G. Petterson, M.A. Sullivan and P.J. Treloar have greatly improved the paper.

References

Bard, J.P. 1983. Metamorphism of an obducted island arc: example of the Kohistan sequence (Pakistan) in the Himalayan collided range. *Earth and Planetary Science Letters,* **65**, 133–144.

——, Maluski, H., Matte, P. & Proust, F. 1980. The Kohistan sequence: crust and mantle of an obducted island arc. *Geological Bulletin, University of Peshawar,* **13**, 87–93.

Beard, J.S. 1986. Characteristic mineralogy of arc related cumulate gabbros; Implications for the tectonic setting of gabbroic plutons and for andesite genesis. *Geology,* **14**, 848–851.

Coward, M.P., Butler, R.W.H., Khan, M.A. & Knipe, R.J. 1987. The tectonic history of Kohistan and its implications for Himalayan structure. *Journal of Geological Society, London,* **144**, 377–391.

——, Windley, B.F., Broughton, R.D., Luff, I.W., Petterson, M.G., Pudsey, C.J., Rex, D.C. & Khan, M.A. 1986. Collision Tectonics in the NW Himalaya. *In*: Coward, M.P. & Ries, A. (eds) *Collision Tectonics*. Geological Society, London, Special Publications, **19**, 203–219.

De La Roche, H., Letterier, J., Grandclaude, P. & Marchal, M. 1980. A classification of volcanic and plutonic rocks using R1R2-diagram and major element analyses-its relationships with current nomenclature. *Chemical Geology,* **29**, 183–210.

Ghazanfar, M., Chaudhry, M.N. & Hussain, M.S. 1991. Geology and petrotectonics of Southwest Kohistan, Northwestern Himalaya, Pakistan. *Kashmir Journal of Geology,* **8 & 9**, 67–98.

Harvey, P.K. & Atkins, B. 1982. The estimation of mass absorption coefficients by Compton scattering: Extensions to the use of RhKα Compton radiation and intensity ratios. *American Mineralogist,* **67**, 534–537.

Hickey, R.L. & Frey, F.A. 1982. Geochemical characteristics of boninite series volcanics: implications for their source. *Geochimica et Cosmochimica Acta,* **46**, 2099–2115.

Holm, P.E. 1985. The chemical fingerprints of different tectonomagmatic environments using hygromagmatophile element abundance of tholeiite basalts and basaltic andesites. *Chemical Geology,* **51**, 303–321.

Irvine, T.N. & Barager, W.R.A. 1971. A guide to the chemical classification of common volcanic rocks. *Canadian Journal of Earth Sciences,* **8**, 523–548.

Jan, M.Q. 1988. Geochemistry of amphibolites from the southern part of Kohistan arc, N. Pakistan. *Mineralogical Magazine,* **52**, 147–159.

—— & Howie, R.A. 1980. Ortho- and clinopyroxenes from pyroxene granulites of Swat Kohistan, northern Pakistan. *Mineralogical Magazine,* **43**, 715–728.

—— & —— 1981. The mineralogy and geochemistry of the metamorphosed basic and ultrabasic rocks of the Jijal complex, Kohistan, NW Pakistan. *Journal of Petrology,* **22**, 85–126.

—— & Windley, B.F. 1990. Chromian spinel-silicate chemistry in ultramafic rocks of the Jijal complex, northwest Pakistan. *Journal of Petrology,* **31**, 667–715.

——, Khan, M.A. & Qazi, M.S. 1993. The Sapat mafic–ultramafic complex, Kohistan arc, North Pakistan. *This volume*.

——, Khattak, M.U.K., Parvez, M.K. & Windley, B.F. 1984. The Chilas stratiform complex: field and mineralogical aspects. *Geological Bulletin, University of Peshawar,* **17**, 153–169.

Karig, D.E. 1971. Origin and development of marginal basins in the western Pacific. *Journal of Geophysical Research,* **74**, 2543–2579.

——, Anderson, R.N. & Bibee, L.D. 1978. Characteristics of back arc spreading in the Mariana trough. *Journal of Geophysical Research,* **83**, 1213–1226.

Khan, M.A. 1988. *Petrology and structure of the Chilas mafic-ultramafic complex, Kohistan, NW Himalayas, Pakistan*. PhD Thesis, Imperial College of Science, Technology, and Medicine, University of London.

—— & Coward, M.P. 1990. Entrapment of an island arc in collision tectonics: A review of the structural history of the Kohistan arc complex, N.W. Himalayas. *Physics and Chemistry of the Earth,* **17**, 1–18.

—— & Thirlwall, M.F. 1988. Babusar amphibolites: arc tholeiites from the southern Kohistan arc, N. Pakistan. *Geological Bulletin, University of Peshawar,* **21**, 147–158.

——, Habib, M. & Jan, M.Q. 1985. Ultramafic and mafic rocks of Thurly Gah and their relationship to the Chilas complex, N. Pakistan. *Geological Bulletin, University of Peshawar,* **18**, 83–102.

——, Jan, M.Q., Windley, B.F., Tarney, J. & Thirlwall, M.F. 1989. The Chilas mafic-ultramafic igneous complex; the root of the Kohistan island arc in the Himalaya of northern Pakistan. *Geological Society of America, Special Paper,* **232**, 75–94.

Le Bas, M.J., Le Maitre, R.W., Streckeisen, A. & Zanettin, B. 1986. A chemical classification of volcanic rocks based on the total alkali-silica diagrams. *Journal of Petrology,* **27**, 745–750.

Marcelot, G., Dupuy, C., Girod, M. & Maury, R.C. 1983. Petrology of Futuna island lavas (New Hebrides): An example of calc-alkaline magmatism associated with the initial stages of back-arc spreading. *Chemical Geology,* **38**, 23–37.

Miller, D.J., Loucks, R.R. & Ashraf, M. 1991. Platinum group elements mineralization in the Jijal layered mafic-ultramafic complex, Pakistani Himalayas. *Economic Geology,* **86**, 1093–1102.

Nakamura, K. 1974. Determination of REE, Ba, Fe, Mg, Na and K in Carbonaceous and ordinary chondrites. *Geochimica et Cosmochimica Acta,* **38**, 757–775.

Nesbitt, R.W., Mastons, H., Stolz, G.W. & Bruce, D.R. 1976. Matrix corrections in trace element analysis by x-ray fluorescence: An extension of the Compton scattering technique to long wavelengths. *Chemical Geology,* **18**, 203–213

Pearce, J.A. 1982. Trace element characteristics of

lavas from destructive plate boundaries. *In*: Thorpe, R.S. (ed.) *Andesites: orogenic andesites and related rocks*. Wiley, Chichester, 525–548.

—— 1983. Role of sub-continental lithosphere in magma genesis at active continental margins. *In*: Hawkesworth, C.J. & Norry, M.J. (eds) *Continental Basalts and Mantle Xenoliths*. Shiva, Nantwich, 230–249.

Petterson, M.G. & Windley, B.F. 1985. Rb-Sr dating of the Kohistan arc-batholith in the Trans-Himalaya of north Pakistan, and tectonic implications. *Earth and Planetary Science Letters*, **74**, 45–57.

—— & —— 1991. Changing source regions of magmas and crustal growth in the Trans-Himalayas: Evidence from the Chalt volcanics and Kohistan batholith, Kohistan, N. Pakistan. *Earth and Planetary Science Letters*, **102**, 326–341.

—— & —— 1992. Field relations, geochemistry and petrogenesis of the Cretaceous basaltic Jutal dykes, Kohistan, northern Pakistan. *Journal of Geological Society, London*, **149**, 299–308.

——, —— & Luff, I.W. 1990. The Chalt volcanics, Kohistan, N. Pakistan: high-Mg tholeiitic and low-Mg calc-alkaline volcanism in a Cretaceous island arc. *In*: Sharma, K.K. (ed.) *Geology and dynamic evolution of the Himalayan collision zone*. Pergamon Press, Oxford, 19–30.

Potts, P.J., Thorpe, O.W. & Watson, J.S. 1981. Determination of rare- earth element abundances in 29 international rock standards by instrumental neutron activation analysis: A critical appraisal of calibration errors. *Chemical Geology*, **34**, 331–352.

——, ——, Issacs, M.C. & Wright, D.W. 1985. High precision instrumental neutron activation analyses of geological samples employing simultaneous counting with both planar and coaxial detectors. *Chemical Geology*, **48**, 145–155.

Pudsey, C.J. 1986. The northern suture, Pakistan: margin of a Cretaceous island arc. *Geological Magazine*, **123**, 405–423.

——, Coward, M.P., Luff, I.W., Shackleton, R.M., Windley, B.F. & Jan, M.Q. 1985. Collision zone between the Kohistan arc and the Asian plate in NW Pakistan. *Transactions of the Royal Society of Edinburgh, Earth Sciences*, **76**, 463–479.

Searle, M.P., Windley, B.F., Coward, M.P., Cooper, D.J.W., Rex, A.J., Rex, D.C., Li Tingdong., Xiao Xuchang., Jan, M.Q., Thakur, V.C. & Kumar, S. 1987. The closing of Tethys and tectonics of the Himalaya. *Geological Society of America Bulletin*, **98**, 678–701.

Stern, R.J., Ping-Nan, L., Morris, J.D., Jackson, M.C., Fryer, P., Bloomer, S.H. & Ito, E. 1990. Enriched back-arc basin basalts from the northern Mariana trough: implications for the magmatic evolution of back arc basins. *Earth and Planetary Science Letters*, **100**, 210–225.

Sullivan, M.A., Windley, B.F., Saunders, A.D., Haynes, J.R. & Rex, D.C. 1993. A palaeogeographic reconstruction of the Dir group: evidence for magmatic arc migration within the NW Himalayas. *This volume*.

Tahirkheli, R.A.K. 1982. Geology of the Himalaya, Karakoram and Hindukush. *Geological Bulletin, University of Peshawar*, **13**, 1–51.

—— 1983. Geological evolution of Kohistan island arc on the southern flank of the Karakorum-Hindukush in Pakistan. *Bollettino Geofisica Terica ed Applicata*, **25**, 351–364.

—— & Jan, M.Q. 1979. Geology of Kohistan, Karakorum Himalaya, northern Pakistan. *Geological Bulletin, University of Peshawar*, **11**, (map in pocket).

——, Mattauer, M., Proust, F. & Tapponnier, P. 1979. The India–Eurasia suture zone in northern Pakistan: Synthesis and interpretation of data at plate scale. *In*: Farah, A. & DeJong, K.A. (eds) *Geodynamics of Pakistan*. Geological Survey of Pakistan, Quetta, 125–130.

Treloar, P.J., Brodie, K.H., Coward, M.P., Jan, M.Q., Khan, M.A., Knipe, R.J., Rex, D.C. & Williams, M.P. 1990. The evolution of the Kamila shear zone, Kohistan, Pakistan. *In*: Salisbury, M.H. & Fountain, D.M. (eds) *Exposed Cross-Sections of the Continental Crust*. Kluwer Academic Press, Amsterdam, 175–214.

——, Rex, D.C., Guise, P.G., Coward, M.P., Searle, M.P., Windley, B.F., Petterson, M.G., Jan, M.Q. & Luff, I.W. 1989. K-Ar and Ar-Ar geochronology of the Himalayan collision in NW Pakistan: Constraints on the timing of suturing, deformation, metamorphism and uplift. *Tectonics*, **8**, 881–909.

Walsh, J.M., Buckley, F. & Baker, J. 1981. The simultaneous determination of REEs in rocks using the inductively coupled spectrometer. *Chemical Geology*, **33**, 141–153.

Watkins, P. & Nolan, J. 1992. Determination of rare-earth elements, yttrium, scandium and hafnium using cation-exchange separation and inductively coupled plasma-atomic emission spectrometry. *Chemical Geology*, **95**, 131–139.

Weaver, B.L., Tarney, J. & Windley, B.F. 1981. Geochemistry and petrogenesis of the Fiskenaesset anorthosite complex, southern west Greenland: Nature of the parent magma. *Geochimica et Cosmochimica Acta*, **45**, 711–725.

Wood, D.A., Tarney, J., Varet, J., Saunders, A.D., Bougault, H., Joron, J.L., Teruil, M. & Cann, J.R. 1979. Geochemistry of basalt drilled in the north Atlantic by IPOD leg 49: implication for mantle heterogeneity. *Earth and Planetary Science Letters*, **42**, 77–97.

A palaeogeographic reconstruction of the Dir Group: evidence for magmatic arc migration within Kohistan, N. Pakistan

M.A. SULLIVAN[1,4], B.F. WINDLEY[1], A.D. SAUNDERS[1], J.R. HAYNES[2] & D.C. REX[3]

[1] *Department of Geology, The University, Leicester LE1 7RH, UK*

[2] *Institute of Earth Studies, University College of Wales, Aberystwyth SY23 3DB, UK*

[3] *Department of Earth Sciences, The University, Leeds LS2 9JT, UK*

[4] *Present address: British Geological Survey, Keyworth, Nottingham NG12 5GG, UK*

Abstract: The Late Palaeocene Dir Group records the resumption of volcanic activity along the southern margin of Eurasia after a prolonged period of uplift and erosion. The group forms an integral part of the Kohistan Batholith and is readily divisible into two distinctly contrasting volcanic successions.

The Baraul Banda Slate Formation comprises 2700 m of fore-arc sandstones and siltstones. The basin formed during the collapse of the Kohistan continental margin and was filled initially by subaerial debris and mass-flow deposits. Subsidence was rapid such that >90% of the sedimentary record consists of thin-bedded sheet turbidites. These maintain a remarkable lateral continuity and indicate deposition in a restricted deep-water environment. Rare interbedded limestones have yielded a Thanetian (60.2–54.9 Ma) marine fauna of *Miscellanea miscella* and *Actinosiphon tibeticus*.

In direct contrast the Utror Volcanic Formation comprises 3000 m of volcaniclastic sedimentary material, lava flows and ignimbrites. Eruptive style was predominantly explosive with fragmental material dominating the volcanic record. Lavas range in composition from 53% to 79% SiO_2 with rhyolite the predominant rock type. A basaltic andesite has an ^{40}Ar-^{39}Ar age of 55 ± 2 Ma. Although the volcanic stratigraphy is complex preliminary investigations suggest accumulation in a predominantly subaerial ring-plain or flanking facies distal to the main focus of volcanic activity. Sedimentary thickness and the abundance of silicic lavas and pyroclastic flows imply extensional tectonic control even though the arc developed at a convergent margin.

Juxtaposition of the two facies along the Dir Thrust excludes any record of shore-line or coastal processes from the volcanic stratigraphy. The inferred original geographical separation of the two environments coupled with their relationships to isotopically dated plutonic suites reveals that between Late Cretaceous and Early Eocene times the locus of magmatic activity within the Kohistan arc migrated progressively northward.

The Shamran volcanic rocks although originally correlated with the island-arc Chalt Volcanic Group are now identified on the basis of a 58 ± 1 Ma (^{40}Ar-^{39}Ar) age as a northern continuation of the Late Palaeocene volcanic arc.

Early Cenozoic arc volcanism along the southern margin of Eurasia records the northward subduction of Tethyan oceanic lithosphere. In northwestern Pakistan, the Dir Group (Tahirkheli 1979) forms a remnant of this volcanic arc and represents the extrusive equivalents of the later stages of the Kohistan batholith. Despite its relatively restricted outcrop, the group provides an important record of volcanic and tectonic activity in the region, and places important constraints on both the temporal as well as spatial evolution of the Kohistan arc-system prior to continent–continent collision.

The aim of this contribution is to introduce the essential characteristics of the Dir Group. Combined sedimentological and volcanological studies will be utilized to interpret the local geology and present a preliminary integrated palaeotectonic and palaeogeographical model. Attention is drawn to the style of volcanic activity and its regional extent, whilst emphasis is placed on volcaniclastic rocks because of their importance in reconstructing palaeoenvironments. We will also present new palaeontological data which constrain both the timing of forearc development in the region and, in a wider context, the timing of the India–Eurasia collision. New evidence is also discussed that establishes a temporal link between the Dir Group and the northern Shamran volcanic rocks (Pudsey *et al.* 1985).

From Treloar, P.J. & Searle, M.P. (eds), *Himalayan Tectonics*
Geological Society Special Publication, No. 74, pp. 139–160.

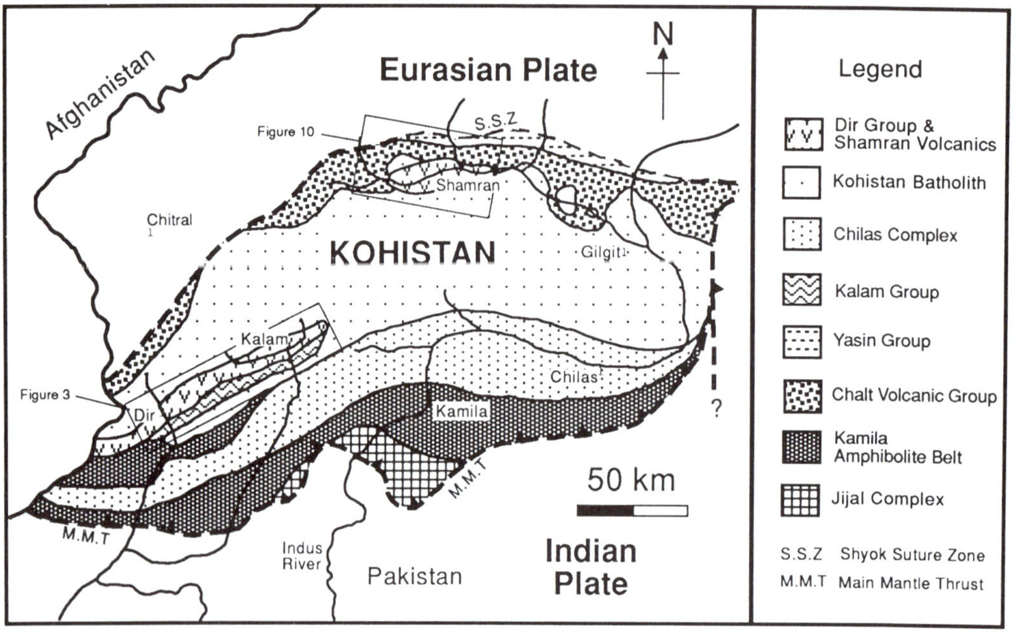

Fig. 1. Simplified geological sketch map of the Kohistan terrane, NW Pakistan. Areas of study are outlined by boxes.

Geological setting

It is beyond the scope of this paper to discuss in detail the geology of the Kohistan terrane and the reader is refered to Tahirkheli *et al.* (1979), Coward *et al.* (1986, 1987) and Treloar *et al.* (1989, 1990). The Kohistan terrane consists of a series of east–west-trending geological units (Fig. 1) which are : (1) Jijal Ultramafic Complex (Jan & Windley 1990), (2) Kamila Amphibolite Belt (Jan 1988; Treloar *et al.* 1990), (3) Chilas Complex (Khan *et al.* 1989), (4) Chalt Volcanic Group (Coward *et al.* 1986; Petterson *et al.* 1990*a*), (5) Yasin Group (Pudsey *et al.* 1985) and (6) Kohistan Batholith (Petterson & Windley 1985, 1991; Petterson *et al.* 1990*b*). They are interpreted as the remains of an upturned intra-oceanic arc that was accreted along the Eurasian margin between 102 and 85 Ma (Petterson & Windley 1985; Treloar *et al.* 1989). Subduction continued after collision with the majority of the Kohistan batholith emplaced into a newly formed Andean margin. Two phases of post-collisional plutonism can be distinguished in western Kohistan (Fig. 2); Stage 1 at *c.* 78–75 Ma and Stage 2 at *c.* 48–45 Ma (Zeitler 1985; Treloar *et al.* 1989). Separating the two is a screen of volcano-sedimentary cover (Fig. 3), the Dir Group, and it is this which forms the focus of our study.

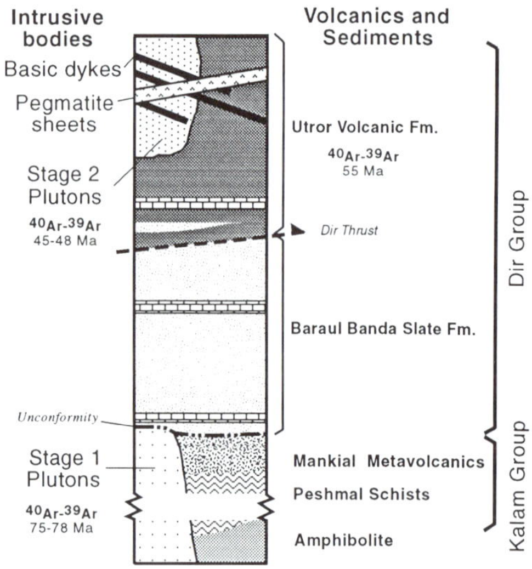

Fig. 2. Simplified lithostratigraphy of the western part of the Kohistan Batholith. Diagram is compiled from field observations made in Dir and Upper Swat and existing radiometric ages (Zeitler 1985; Treloar *et al.* 1989).

The Dir Group is confined to the western half of the Kohistan arc terrane and forms a 10–15 km wide (Fig. 3), moderate to steeply

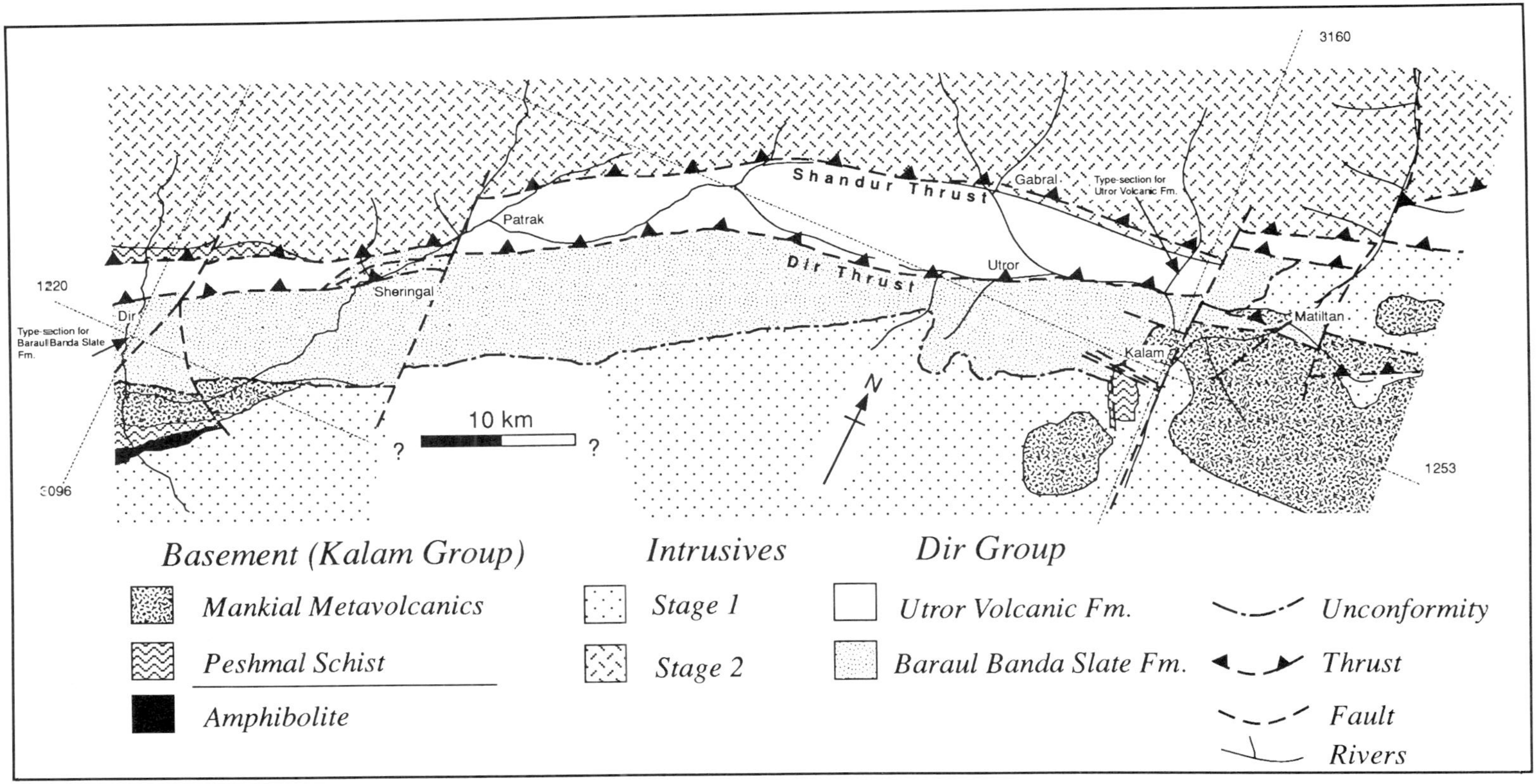

Fig. 3. Geological map of the Dir and Upper Swat districts of NW Pakistan. Basemaps are 1 : 50 000 Government of Pakistan topographic sheets.

dipping, gently folded belt that stretches approximately 120 km from the Upper Swat valley southwestward into Dir. On the basis of existing palaeontological and radiometric evidence the rocks are assigned a Late Palaeocene to Early Eocene age (Kakar *et al.* 1971; Khan 1979; Treloar *et al.* 1989).

Tahirkheli (1979) divided the Dir Group on an informal basis into two, the Baraul Banda Slates and the Utror Volcanics, and for the purposes of this study these same divisions have been adopted and assigned formation status. Unfortunately later plutonism and erosion have obscured and obliterated the majority of the volcanic record and what remains has, in addition, undergone tectonic dismemberment such that lavas and coarse volcaniclastic material are overthrust southwestward onto fine grained forearc sediments.

The Dir Group: lithological description and interpretation

Each formation preserves a record of contrasting styles of volcanism and environments of deposition, and accordingly they will be described and interpreted separately.

Baraul Banda Slate Formation

In lithostratigraphic terms the Baraul Banda Slate Fm. forms the lowest part of the Dir Group (Fig. 2) and unconformably overlies an erosion surface cut deep into Stage 1 granitoid intrusions (*c.* 78–75 Ma) of the batholith and the metavolcanic rocks of the Kalam Group which they intrude. The Baraul Banda Slate Fm. displays a remarkable uniformity in character, both laterally as well as vertically, with steeply dipping strata providing sections through the stratigraphy. Along the type-section south of Dir a great deal of internal coherence is maintained with the rocks dipping at 50°–60° towards the northwest. No fold closures are observed and occasional limestones and intraformational breccias display no obvious imbrication or repetition. Evidence exists for slip along bedding planes, with lineations plunging towards the northwest. However, the absence of any 'ramp and flat' structures indicates that original sedimentary thicknesses were maintained. The present day thickness of the Baraul Banda Slate Fm., as measured from its depositional contact with the Kalam Group in Dir to its faulted contact with the Utror Volcanic Fm., is approximately 2700 m. Although this is clearly a minimum estimate, it corresponds well with similar thicknesses of sediment recorded from continental shelf forearc basins off the coast of Peru (Thornburg & Kulm 1981) and Java (Moore *et al.* 1982).

Three facies are identified from recent fieldwork (Fig. 4): (1) basal conglomerate and breccia; (2) stratified and graded pebbly sandstone; (3) thinly bedded and graded sandstone, siltstone and mudstone. Limestones and intraformational breccias, although important are restricted to the thinly bedded sandstone facies and represent a very small part of the whole succession (< 1%).

Facies 1: basal conglomerate and breccia. A basal unconformity is of regional extent. It is overlain by a disorganized ungraded polymict breccia and conglomerate which displays rapid lateral thickness variations (1 m to 30–40 m) implying deposition over a buried topography. It can vary from clast-supported to matrix-supported over short distances. Clasts are subangular to subrounded and range in size from <1 cm to 100 cm. The larger clasts are restricted to the basal section, whereas smaller cobbles occur in occasional stratified and unstratified beds and in small channel sequences. The matrix varies in colour from green in Upper Swat to red–brown in Dir, and consists of a poorly sorted mixture of plutonic and reworked volcanic lithic fragments, broken feldspar and quartz crystals, opaque minerals and an unidentifiable fine sand and silt component. The argillaceous component of the matrix has undergone limited recrystallization with the development of white micas.

Clast compositions in Upper Swat include, in order of increasing abundance, granitoids, limestone, schist, metavolcanic rocks, dolerite, vein quartz, dacite and volcaniclastic sandstone and siltstone. Plutonic material is undeformed and has petrographic characteristics identical to underlying Stage 1 diorites and granitoids. Kalam Group lithologies are also recorded together with minor amphibolite. What remains problematical, however, is the prominent abundance of bioclastic limestone detritus which can constitute up to 70% of the clast mode in some exposures. Although extensive searches were undertaken below the unconformity an in situ source of this limestone was never located.

Limestone clasts are well rounded and attain a maximum diameter of 100 cm. Acid extraction techniques have yielded a mixed fauna of foraminifera, ostracods and broken shelly material. Poor preservation hampers identification, but *Miscellanea miscella* previously described from the Ranikot Fm. of the Salt Range and Thal Fm. (Davis & Pinfold 1935) provides a reliable Late Palaeocene (Thanetian, 60.2–54.9 Ma) age for the clasts.

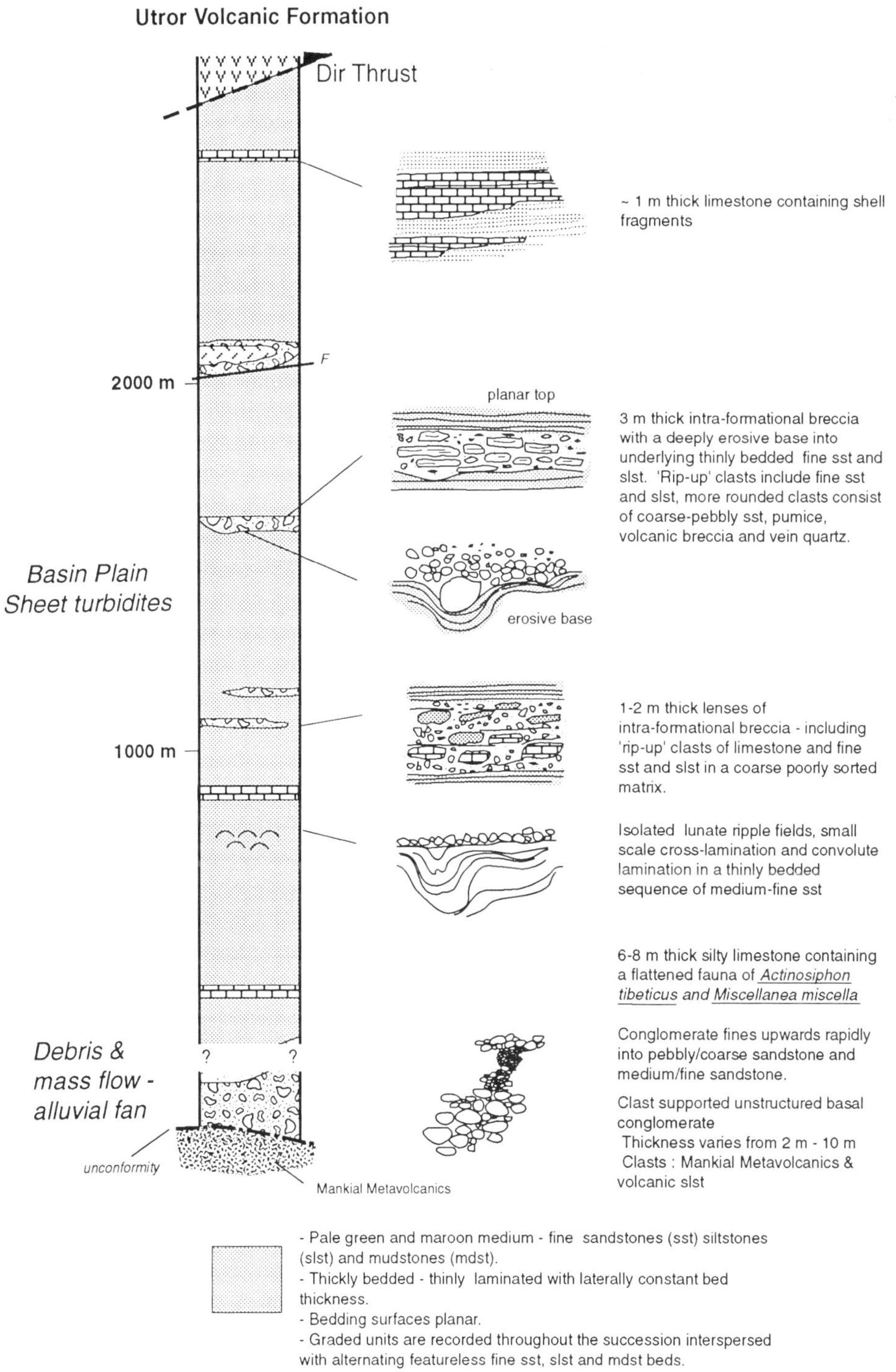

Fig. 4. Schematic stratigraphic section based on reconnaissance mapping of the Baraul Banda Slate Fm. Section is a compilation of observations made along the road section south of Dir (type-section) and additional sections along strike in the vicinity of Sheringal and Zhandrai Pass. Total thickness estimations assume no thrust thickening (see text).

Further to the southwest in Dir the conglomerate framework is monolithic in character. Mankial Metavolcanic (Kalam Group) lithologies dominate, with an absence of any limestone, schist or granitoid detritus. Stratified beds of coarse–very coarse-grained sandstone vary in thickness from 20 cm to 1 m. They contain isolated boulders and cobble sized detritus with tabular cross-bedding indicating current flow from the NNE.

Facies 2: pebbly sandstone. Regionally the conglomerates and breccias fine upwards after approximately 30–70 m into stratified and unstratified, poorly sorted, pebbly and coarse-grained sandstones. Two different styles of bedding are observed. The first involves alternations of pebble- and sand-rich beds with stringers and irregular patches of larger clasts. Thicknesses vary from 1 to 2 m and bedding surfaces are essentially planar with few large-scale erosive features. Occasional lenticular conglomerate lenses composed of limestone and granodiorite are interbedded within coarser sandstone units. The second type is positively graded with a grain size variation from coarse sandstone (locally pebbly) to fine sandstone. Beds range in thickness from 0.5 to 1 m and alternate with stratified pebbly sandstones. Flame-structures occur locally at the interface of individual units. There is little evidence of channelisation, and bedding surfaces are essentially parallel and planar suggesting a sheet-like geometry. Erosive bases are recorded in the thickest graded beds and scour-marks together with sparsely developed low-angle tabular cross-bedding indicate current flow from the north and northwest. Detritus has the same compositional variation as the basal conglomerate facies.

Facies 3: sheet sandstone, siltstone and mudstone. The pebbly sandstone facies passes upwards conformably into a thick succession (approximately 2500 m) of thinly bedded and laminated sandstones, siltstones and mudstones. These lithologies dominate the sedimentary record of the basin such that thinly bedded pale green sandstones and siltstones represent approximately 90% of the present-day outcrop (Fig. 4). They comprise a monotonous sequence of positively graded beds that alternate with ungraded beds. Grain-size varies from medium sand to very fine silt. Bed thicknesses vary from 1 cm to 50 cm with exceptional 2 m thick units recorded northwest of Kalam. No thinning of strata is recorded, and instead the beds display impressive lateral consistencies of thickness. All thickness. All surfaces are even and parallel and there is a general absence of wave or current-generated structures. Interbedded with the repetitive graded units are thick to thinly laminated fine sandstones and siltstones. Soft-sediment deformation structures, flame structures and slumps occur locally but are usually restricted to beds no thicker than 10 cm.

Detrital grains have a subrounded to subangular morphology and this, together with isolated ripple fields, indicates a significant period of reworking and abrasion probably by low-concentration turbidity currents (Cas 1979). This is further substantiated by occasional groove or scour marks observed at the base of thicker graded units.

Interbedded limestones are rare and attain a maximum thickness of 1.5 m. An exceptional 8 m thick silty limestone in Dir has yielded an abundant fauna of *Actinosiphon tibeticus* and *Miscellanea miscella*. This is of crucial importance as it provides not only an Late Palaeocene age for the Baraul Banda Slate Fm. but also implies contemporaneity with limestone detritus recorded in the basal conglomerate in Upper Swat. Other limestone horizons have suffered extensive silicification and although broken shell fragments are observed on bedding surfaces no faunas are identified. The development of bedding at all scales indicates a lack of bioturbation and corresponds well with the absence of any trace-fossils and the restricted faunal variation with interbedded limestones.

In the stratigraphically higher part of the succession occasional intraformational breccias provide a direct indication of volcanogenic input into the basin. They form beds no thicker than 2 to 4 m (Fig. 4) and have erosive bases and planar upper surfaces containing 'rip-up' clasts derived from underlying sandstones and siltstones. Two of the breccias are noteworthy in that they contain, in addition to volcanically derived detritus, angular limestone fragments. Unfortunately silicification has obliterated any palaeontological record.

The most striking feature of this facies and the succession as a whole is its remarkable uniformity in character over the entire outcrop, with a conservative volume estimate of between 1500 and 1700 km^3.

Sandstone provenance

The overlap between the new palaeontological age of the Baraul Banda Slate Fm. (60.2–54.9 Ma) and the isotopic age of the Utror Volcanic Fm. (55 ± 2 Ma, Treloar *et al.* 1989) suggests a relatively straight forward volcanic arc model for the Dir Group. However, the current age resolution is such that volcanic activity may not have been contemporaneous with basin development and this is indeed what the general absence of coarse volcanic detritus suggests. Furthermore, as the contact between the two formations is tectonic rather than stratigraphic a direct relationship needs to be proven rather than assumed. Apart from the abundant limestone clasts in the basal conglomerate and pebbly sandstones facies, detrital characteristics can be matched directly to the underlying batholith lithologies and reflect provenance from a dissected arc source. In contrast, the stratigraphically higher intraformational breccias are dominated by volcanogenic detritus, implying that there was perhaps some temporal variability in the source of sediment entering the basin.

To place tighter constraints on sediment provenance five sandstone samples from facies 3 were selected for petrographic modal analysis. The framework consists of quartz and feldspar grains (plagioclase dominant over alkali feldspar), opaque material and very minor plutonic and volcanic fragments. The matrix is composed of unidentifiable fine sand, silt or clay minerals. Detrital proportions were calculated on the basis of 300 points, for quartz, feldspar and lithic fragments. The results are displayed on a standard QFL diagram (Fig. 5) of Dickinson & Suczek (1979). They clearly show derivation from a dissected magmatic arc and as such agree with conclusions drawn from the detritus in the basal conglomerate.

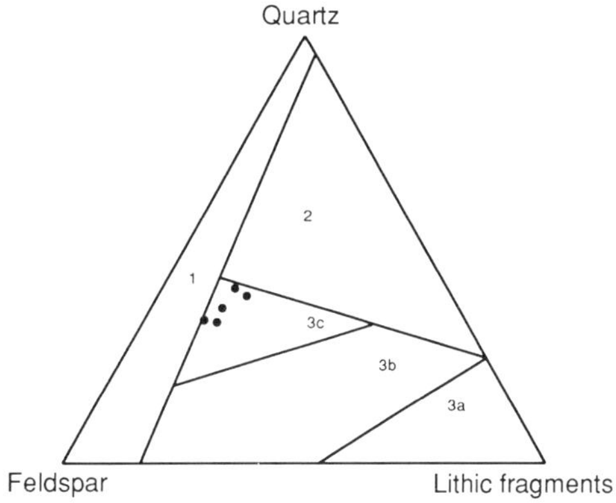

Fig. 5. Detrital modes for sandstones from the Baraul Banda Slate Fm plotted on a standard QFL diagram (Dickinson & Suczek 1979). 1, Continental block field; 2, recycled orogen field; 3a, b, c, undissected, transitional and dissected magmatic arc fields.

If volcanism was contemporaneous with forearc sedimentation then the paucity of volcanic lithic fragments is somewhat surprising especially considering the dispersive nature of explosive eruptions. This raises the question as to whether the measured modal proportions reflect the true provenance or record later

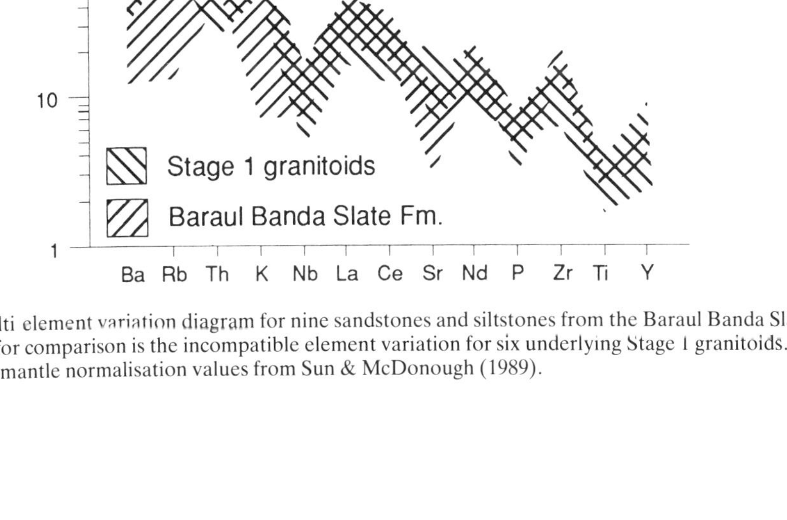

Fig. 6. Multi element variation diagram for nine sandstones and siltstones from the Baraul Banda Slate Fm. Displayed for comparison is the incompatible element variation for six underlying Stage 1 granitoids. Primordial mantle normalisation values from Sun & McDonough (1989).

modification during burial and compaction. During diagenesis and low grade metamorphism lithic fragments (especially pumice) are commonly converted to matrix. One possible alternative therefore is to examine the rock chemistry (especially high-field strength elements), which under such circumstances should remain constant (Taylor & McLennan 1985).

Trace element variation for nine sandstones and siltstones is shown on a multi-element variation diagram (Fig. 6). Patterns are typical of silicic rocks derived from a subduction-related magmatic arc. Large ion lithophile elements are enriched relative to the high-field strength elements (although there is evidence for Ba, Rb and K mobility), light rare earth elements are enriched relative to heavy rare earth elements and La/Nb ratios are greater than 1. Furthermore depletions in Sr, P and Ti are consistent with derivation from an intermediate–acid source which evolved in part by fractionation of plagioclase, apatite and titanomagnetite. Displayed for comparison are patterns for underlying Stage 1 granitoids. The similarity in both patterns (cross-hatched region) leaves little doubt that the sediments have an arc provenance but does not distinguish between an active and dissected arc.

Values for compatible trace elements such as V, Cr and Ni values (Fig. 7) display surprisingly high concentration levels compared with other forearc sands of similar age in the Himalayas (Table 1). Garzanti & Van Haver (1988) studied sandstone chemistry within the Indus basin in Ladakh and attributed high V, Cr and Ni levels to the inclusion of ophiolitic or accretionary complex detritus. The Eocene Nimu and Nurla Formations record erosion of suture-zone material either during or after the India–Eurasia collision. However, in the case of the Baraul Banda Slate Fm. age considerations suggest that a more appropriate explanation for such high values is that they reflect the inclusion of detritus derived from exposed Kamila Amphibolite and/or Chalt Volcanic Group lithologies. It is

Table 1. *Comparison of average V, Cr and Ni values (in ppm) between clastic sediments from the Baraul Banda Slate Fm. and Indus Group (Garzanti & Van Haver 1988)*

Lithology	V	Cr	Ni
Baraul Banda Slate Fm.	88	124	47
Nimu Fm.[a]	67	62	47
Nurla Fm.[a]	131	41	12
Temesgam Fm.[b]	91	11	7
Basgo Fm.[b]	55	7	5

Indus Group: [a] post-collisional continental sandstones; [b] fore-arc sandstones.

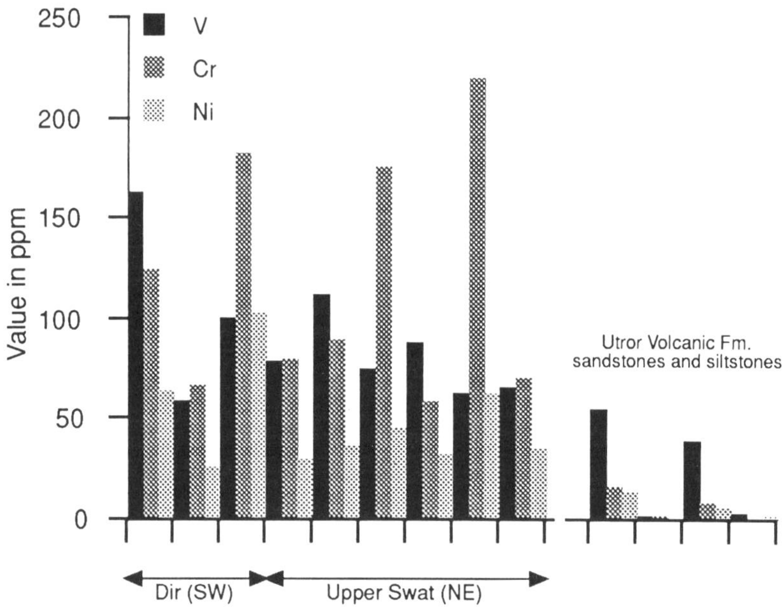

Fig. 7. Histogram for V, Cr, and Ni, illustrating the variation (in ppm) within the Baraul Banda Slate Fm. Also shown for comparison are values from volcanic sandstones and siltstones from the Utror Volcanic Fm. (notice the very low abundances).

important also to note that Cr and Ni (and to some extent V) contents for four volcanic sandstones and siltstones from the Utror Volcanic Fm. (Fig. 7) are very low and as such emphasise the limited volcanogenic input into the basin.

Depositional model for the Baraul Banda Slate forearc basin

In this section we attempt to present a sedimentological facies model for the Baraul Banda Slate Fm. Broad lithological correspondence between Dir and Kalam allows interpretations made from both localities to be applicable over a much wider area.

Facies analysis of the Baraul Banda Slate Fm. provides a local record of forearc development and sedimentation after a prolonged period of uplift and erosion along the continental margin of Eurasia. During the Late Palaeocene a phase of faulting occurred along the continental margin in the region that is now western Kohistan. This resulted in differential uplift, collapse of the forearc region into the trench and produced an elongate basin or trough that was founded not on oceanic crust but accreted island-arc crust and continental margin plutonic rocks. The basin was located within the 'arc–trench gap' (Dickinson 1973) and corresponds to an intra-massif forearc basin of Dickinson & Seely (1979).

In spite of the absence of red beds, the lack of any sediments below the regional basal unconformity favours a subaerial rather than subaqueous setting. Polymict breccias and conglomerates display lateral variation in clast sizing and bed morphology suggests that they represent debris-flows and mass-flows deposited by semi-arid alluvial fans (Reading 1986). Rapid uplift was necessary to supply the large quantities of coarse clasts whilst sharp topographic contrast caused rapid deposition. Sparse palaeocurrent data reflect transverse drainage from the northeast across the continental borderland.

Clast lithologies can be matched directly to the underlying basement and this observation, coupled with the lack of any exposed Stage 1 plutons below the unconformity in Dir, suggests that unroofing of the batholith occurred earlier in the east of Kohistan than in the west, prior to Late Palaeocene times. The abundant limestone detritus in Upper Swat remains problematical and models to explain its presence include faulting close to an embayed shore-line or collapse of a fringing reef during fault block movement. However, it is important to emphasize that within the present resolution of the palaeontological ages appreciable reworking of detritus is also possible.

Evidence shows that subaerial conditions were rapidly replaced by subaqueous ones as subsidence proceeded. Pebbly sandstones display few channel features and their sheet-like geometry, rather than indicating stream flow, suggest deposition by high-concentration turbidity currents. Available data do not permit a detailed investigation of this facies but it does not appear to be a typical near-shore facies. Such a rapid transition from subaerial alluvial-fan to deep-water sediments with an absence of any near-shore facies is somewhat surprising. However, similar characteristics have been observed elsewhere along tectonically active margins such as the Californian borderlands (Link & Nilsen 1980).

The final stage in the evolution of the basin is characterized by the fining-up into a thick repetitive sequence of thin graded beds. The paucity of features associated with oscillatory bottom currents together with the lateral consistencies in bed thickness suggests that the majority of sedimentation in the Baraul Banda Slate Fm. occurred below wave base (approximately 100–300 m) in a deep-water environment. The absence of distributary channel sequences and progradational sandstone lobes coupled with the planar bed forms points initially to deposition in either an outer fan (fan-fringe) or basinal environment (Normark 1970; Mutti & Ricci Lucci 1972). However, the absence of any upper- or mid-fan facies somewhat questions a fan-style of deposition. Both the sandstones and siltstone and the underlying pebbly-sands are characterized by a conspicuous absence of channel features and instead display a sheet-like geometry. In a similar way many active margin forearc basins do not show typical submarine fan facies associations (Pickering *et al.* 1989), but rather contain sheet-like, deep-water deposits (Mutti 1977; Thornburg & Kulm 1981). Therefore the monotonous thinly bedded nature of the sandstones can be more easily interpreted as low-concentration turbidites deposited by sheet distributary systems in a deep-water restricted basinal environment.

In terms of the lithostratigraphic record >90% of the Baraul Banda Slate Fm records quiet deep-water deposition well away from the influence of volcanic centres. Such conditions prevailed for the bulk of the history of the basin and there are very few recorded instances of the progradation of coarse volcanogenic sandstone bodies. The limited development of limestone indicates that the basin was dominated by clastic input and that carbonate accumulation only took

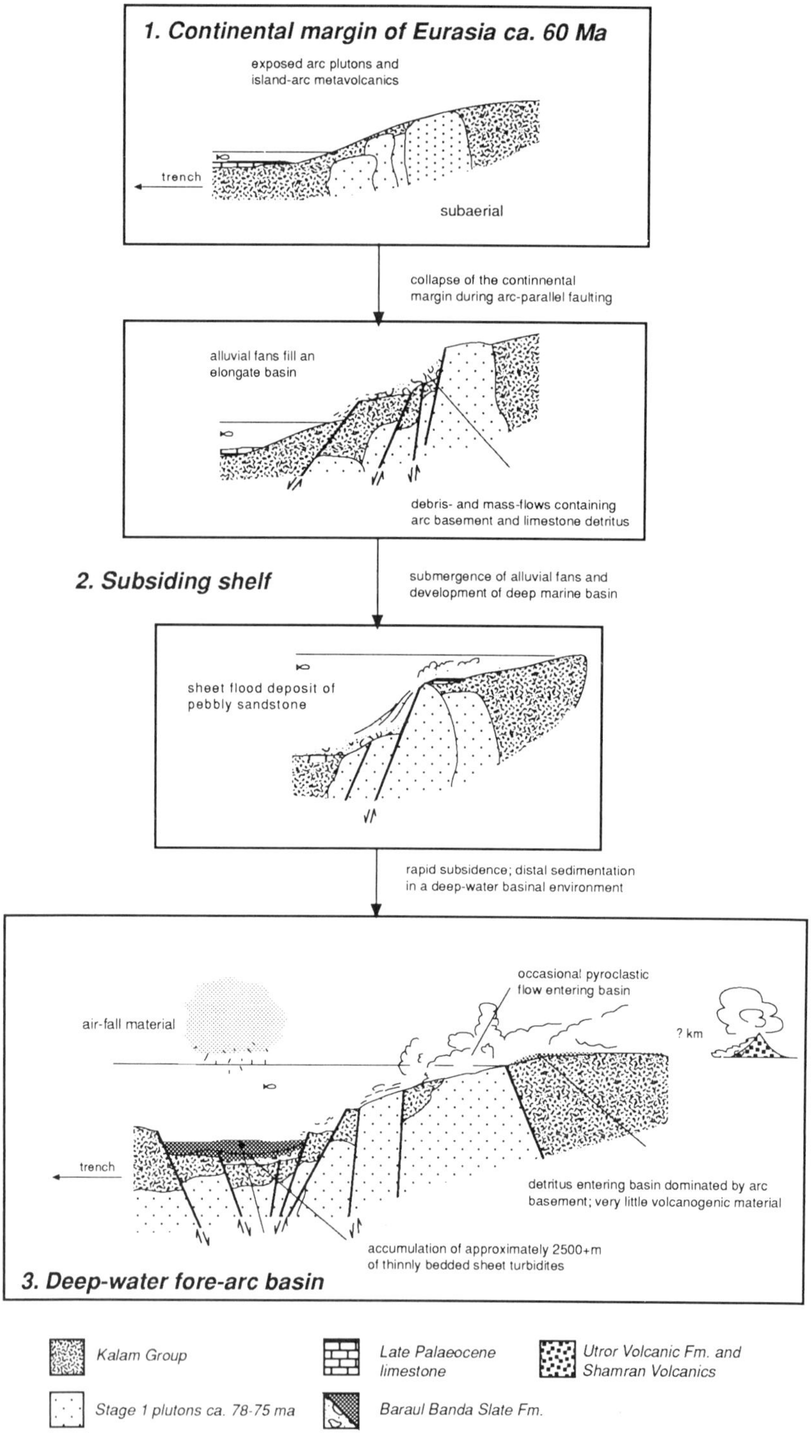

Fig. 8. Conceptual diagram illustrating the proposed model for the development of the Baraul Banda forearc basin.

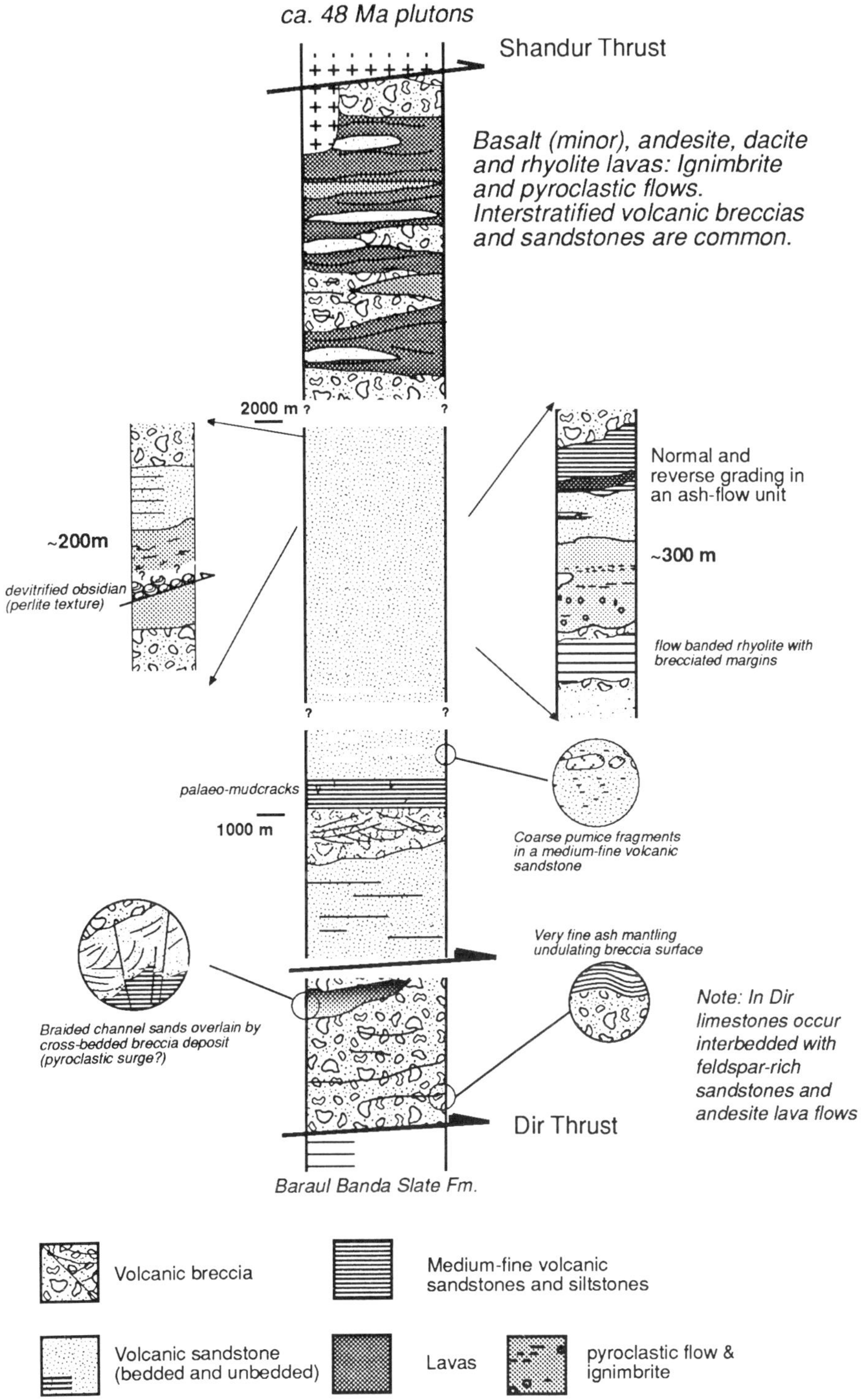

Fig. 9. Schematic lithostratigraphic section through the Utror Volcanic Fm., Anakar Gol, Upper Swat.

place during pauses in sediment supply. Rare, thick graded beds probably involved slope re-working during episodic seismic activity. The absence of slump structures and disrupted bedding indicates that, although rapid, subsidence was continuous rather than episodic and balanced detrital input. The depth of the basin is difficult to constrain because included faunas lack sufficient diversity and were probably transported into position. Figure 8 illustrates schematically the proposed model for the evolution of the Baraul Banda Slate forearc basin.

Utror Volcanic Formation

The Utror Volcanic Fm. comprises a structurally complex and lithologically diverse succession of volcaniclastic rocks and lava flows that crop out in the hanging wall of the Dir Thrust (Figs 2 & 3). It is intruded by *c.* 48–45 Ma calc-alkaline granitoid plutons (Stage 2). The northern boundary is defined by the Shandur Thrust which has locally reworked intrusive contacts. A ^{40}Ar-^{39}Ar age of 55 ± 2 Ma. (Treloar *et al.* 1989) makes the Utror Volcanic Fm. broadly contemporaneous with the Baraul Banda Slate Fm.

The Utror Volcanic Fm. displays a marked variation along strike in structural style and complexity. In the Dir area ductile deformation fabrics tend to obscure primary characteristics, whereas to the northeast in Upper Swat the rocks are more coherent and display brittle structures. As in many volcanic terranes dominated by fragmental material, great spatial and lithological diversity exists and the restoration of tectonically thickened sequences faces serious problems (Cas & Wright 1987). The present-day thickness from the Dir Thrust to the Shandur Thrust is approximately 2500–3000 m. During this study no estimates were obtained as to the magnitude of thrust thickening. Nevertheless, a preliminary lithostratigraphy is presented (Fig. 9) which is based on the section measured along Anakar Gol (5 km NW of Kalam), and as such forms the basis for the formal recognition of the succession. Four main lithological types exist and are described below.

Volcanic breccias. Polymict volcanic breccias are the predominant rock type. They form deposits of variable thickness and can have either an open or closed framework. Clasts range in size from approximately 3 m to less than 5 cm, are angular to subrounded and include a wide range of litho-types including andesite, dacite, rhyolite, ignimbrite, volcanic breccia, sandstone and siltstone. The matrix is generally poorly sorted and consists of very coarse to fine sand-sized volcanic detritus, lithic fragments, broken crystals (feldspar and quartz), devitrified glass-shards, spherulite and perlite fragments, opaque minerals and fine silt and clay grade material. The abundance of glass-shards testifies to explosive fragmentation during volcanic eruptions.

Breccias are up to 20 m thick and vary in form from thickly bedded to massive, often within the same exposure. Occasionally a crude alignment of volcanic lithic clasts defines bedding. Both breccias and pebbly volcanic sandstones display normal and reverse grading. Individual beds are tabular or lens-shaped; the angular to sub-angular nature of the fragments indicates little re-working. Spindle-shaped clasts occur in one exposure east of Dir, but their inclusion in a granular open framework suggests a secondary rather than a primary origin. Some fragments have irregular chilled margins implying that they were hot at the time of incorporation.

Planar bed surfaces are generally followed by a rapid fining-upwards into pebbly sandstones and coarse sandstones, whereas erosive features commonly herald the onset of very thick beds that show variable internal stratification. In such deposits there is occasional large-scale cross-stratification but in most cases it is not clear whether this is attributable to pyroclastic surge processes or simply channelisation during re-working.

Volcanic sandstones and siltstones. These include any deposit with a dominant grain size below 2 mm. They range from well-sorted, lithic-poor sandstones to poorly-sorted pebbly sandstones with a variety of clast compositions. Detritus includes juvenile volcanic material (quartz, feldspar, pumice and glass-shards), opaque phases and lithic fragments. Colour varies considerably from pale green, buff, brown and red.

Bed form is usually planar and laterally continuous. Nevertheless, erosive channels are common, with scours and lag breccias indicating fluvial reworking of volcanogenic detritus. Cross-bedding is usually restricted to tabular units indicating current flow from the north-west–northeast. Normal grading is also common, although on a smaller scale (5–50 cm) than in the breccias. Many beds fine up into poorly sorted siltstones which are rich in fine euhedral feldspar laths. Where the overlying bed consists of coarser material, load-casts and flame-structures are common and provide important 'way-up' criteria.

Lithic fragments display a wide variation in composition and include pre-existing volcaniclastic breccias, sandstones and lavas. Occasional fragments display coherent crystalline textures and indicate fragmentation of lavas and welded pyroclastic flows. Accretionary lapilli are absent. Crystal-rich sandstones are common and contain, in order of increasing abundance feldspar, quartz, chlorite (after hornblende), lithic fragments and pyroxene. Although it cannot be proven conclusively that they have a primary origin, the common retention of embayed quartz margins indicates little reworking. In exposures north of the town of Utror the common occurrence of finer grained tops to crystal-rich sandstones suggests co-pyroclastic ash-fall deposits.

Volcanic siltstones are well bedded and ripple cross-lamination indicates current flow from the northwest. Rare desiccation cracks point to temporary emergence during deposition. In a few localities small pipes are filled by coarser material from the overlying beds. Whether these structures are infilled gas-escape tubes or burrows is unclear. Interbedded limestones are rare and have undergone recrystallization and silicification such that no faunas are preserved.

Lava flows. They consist of a variably altered suite of tholeiitic and calc-alkaline lavas, that range in composition from basaltic andesite (53% SiO_2) to high-silica rhyolite (79% SiO_2), with a bimodal distribution of andesite and rhyolite. They are restricted to the upper parts of the lithostratigraphic section, although a few dacites are interbedded lower down with volcaniclastic sandstones and breccias (Fig. 9).

Basaltic andesites are lenticular in habit and tend to pinch out laterally. Andesites range in thickness from 10–50 m and have rubbly margins. They are sparsely porphyritic with euhedral augite, hornblende and plagioclase the most common phenocrystic assemblages. Ferromagnesian minerals all have Mg-rich compositions with no tendency to show iron enrichment. Most are massive and many grade both vertically and laterally into laterally discontinuous monomict lava breccias. Dacites and rhyolites have a varied flow morphology although no dome or intrusive contacts are recorded. Flow-banding occurs locally in a few rhyolites; however, most are massive and irregular and lack any observable features apart from conspicuous feldspar and quartz phenocrysts in a microcrystalline groundmass of plagioclase, quartz, orthoclase and finely divided opaque minerals. They are all flinty and break with a conchoidal fracture. Devitrification textures are common and flattened spherules are observed in the basal part of a few rhyolitic bodies. The presence of obsidian provides evidence for an originally glassy nature to some flows. Brecciated margins are common along the Anakar Gol section and are interpreted to be the result of autobrecciation processes. Intercalated volcaniclastic material is particularly abundant when associated with the silica-rich lavas.

In terms of their geochemical variation the lavas comprise a diversity of major and trace element compositions. Analyses of six-representative samples are presented in Table 2. They belong to a basaltic andesite to rhyolite, sub-alkalic magma series and as such display both tholeiitic and calc-alkaline characteristics. Values for major and minor element oxides broadly correspond with those for tholeiitic and calc-alkaline arc lavas from other continental margin settings having high Al_2O_3 and low TiO_2 contents. Geochemical characteristics (Fig. 10) show a selective enrichment of large ion lithophile elements (Ba, Rb, K, Sr, Th) relative to high-field strength elements (Nb, P, Ti, Y, Yb) with Ba/Nb ratios in the range 4.7–101. Similarly light rare earth elements are enriched relative to heavy rare earth elements with $(Ce/Yb)_n$ ratios

Table 2. *Whole-rock geochemistry for selected analyses from the Utror Volcanic Fm. lavas*

	D56–29	1–13a	81–26	D91–4	54–4	46–4
SiO_2	52.8	56.6	62.3	67.9	70.2	75
TiO_2	0.8	0.9	0.5	0.9	0.4	0.2
Al_2O_3	18.3	22.9	16.9	14.9	15.1	13.3
FeO^t	9.5	4.1	5.5	5.2	4.4	1.2
MnO	0.2	0.1	0.1	0.2	0.1	0.1
MgO	5.6	3.1	3.5	1.4	0.4	0.4
CaO	9.4	4.3	3.3	2.1	1.4	1.4
Na_2O	3.6	6.5	4.6	5.8	4.7	3.9
K_2O	0.5	2.2	2.6	1.6	3.2	3.6
P_2O_5	0.2	0.3	0.1	0.3	0.1	0.1
Total	100.9	101.0	99.5	100.3	100.0	99.1
LOI	1.0	2.0	1.0	1.3	0.5	0.7
V	226	78	127	43	11	6
Cr	115	7	50	0	1	0
Ni	64	5	44	4	5	3
Zn	98	92	72	94	87	27
Ga	19	21	19	20	21	12
Rb	18	84	126	26	94	101
Sr	381	319	253	133	162	70
Y	23	28	22	52	64	20
Zr	89	207	131	334	390	123
Nb	3	8	4	10	12	5
Ba	118	235	299	286	319	432
La	8	16	12	21	26	17
Ce	20	40	27	51	60	32
Nd	13	21	15	31	35	14
Th	1	5	9	4	6	7

Major elements in wt% oxide: trace elements in ppm. LOI, loss on ignition.

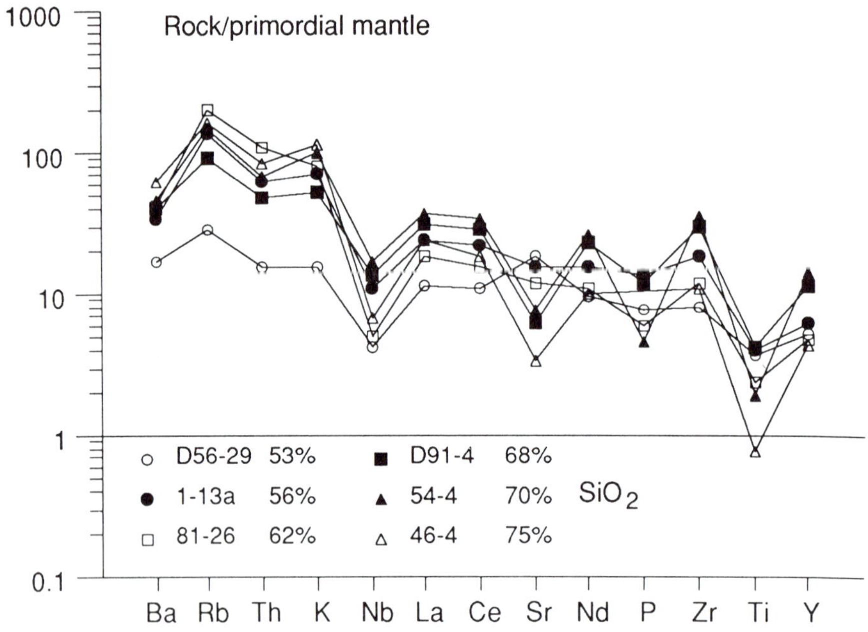

Fig. 10. Multi-element variation diagram for six lavas from the Utror Volcanic Fm. Primordial mantle normalization values from Sun & McDonough (1989).

in the range 2.3–9.9. Such characteristics are typical of subduction-related volcanic arc provinces. A detailed and comprehensive account of their petrographic and geochemical variation is beyond the scope of this paper and will be presented elsewhere.

Alteration products indicate that hydration and oxidation are the most important secondary processes taking place. Mafic silicates such as pyroxene, hornblende and biotite are converted to Mg- and Fe-rich chlorite and epidote, white mica replaces plagioclase, and mixed clays and carbonate occur in the groundmass. Zoisite and epidote are commonly associated with white micas during the hydrothermal alteration of calcic plagioclase.

Pyroclastic flows and ignimbrites. Pyroclastic flows are present along the type-section at Anakar Gol (Fig. 9) as well as along the road between Kalam and Utror. Flattened and deformed pumice is the most diagnostic feature of this lithology although this has been variably replaced by secondary minerals. Many outcrops display complex stratigraphical relationships and this hinders along-strike correlation and thickness determination of flow units. In general thicknesses vary considerably from 10 m to greater than 100 m.

Along the type-section fluvially deposited coarse sandstones and breccia lenses are interbedded with welded pyroclastic flows. One of these flows shows the typical association of a fine basal layer rich in pumice and lithic fragments, a welded zone, a structureless monolithic upper zone (with reverse grading of pumice) and reworked top. Indications of a primary pyroclastic origin come from contorted fiamé and crystal-rich pipe structures (fossil fumarole) that, although inclined at present, indicate post-depositional gas elutriation of a pyroclastic flow.

Dacitic and rhyolitic ignimbrites are identified in a few localities and have chemical compositions similar to the lavas. Welding is a ubiquitous feature with pumice fiamé displaying eutaxitic and parataxitic textures. In places such textures are contorted, indicating rheomorphism (i.e. post-depositional flowage). Shard morphology is still recognizable in a few exposures although devitrification has taken place resulting in replacement by a fine mosaic of quartz, feldspar and epidote. Lithic- and crystal-rich zones are indentified within individual flows. At Anakar Gol, a >50 m thick (Fig. 9) breccia displays normal grading of volcanic lithic fragments and reverse grading higher up of pumice. Such grading is typical of the main body of an ignimbrite (Sparks *et al.* 1973).

Facies model for the Utror Volcanic Formation

In contrast to the Baraul Banda Slate Fm. forearc succession, the Utror Volcanic Fm. is characterized by a structurally complex and lithologically diverse assemblage of volcanic rocks. Nevertheless a few points are diagnostic and allow at least a preliminary analysis concerning the depositional setting and volcanic style.

Heterolithic volcaniclastic breccias and sandstones are a persistent feature throughout the succession. They suggest a secondary rather than primary origin for the majority of the volcaniclastic rocks and represent the reworking of unstable slopes and volcanic edifices by mass-flows and sheet-floods during volcanically quiescent periods. In thin section they display evidence of varying levels of abrasion with a full spectrum of sediment maturity from shard-bearing sandstones (probably tuffs) to heterolithic sandstones with a clastic framework and subrounded grain morphology. Evidence of marine volcanism and deposition is rare and it is suggested that the vast majority of the rocks accumulated within an ensialic, subaerial setting. The predominance of sedimentary material suggests accumulation away from active vents as reworking would seldom preserve such material unless ponded in valleys. Furthermore, bed form, although variable, tends to maintain lateral consistency and this implies accumulation in a tectonically active ring-plain facies flanking the volcanic front (Hacket & Houghton 1989).

The presence of pyroclastic flows and ignimbrites demonstrate that volcanism was explosive in nature. However, the absence of any sizable volcanic conduit or megabreccias together with the conformable contacts between most lavas and volcaniclastic rocks indicates that the rocks accumulated in flanking region away

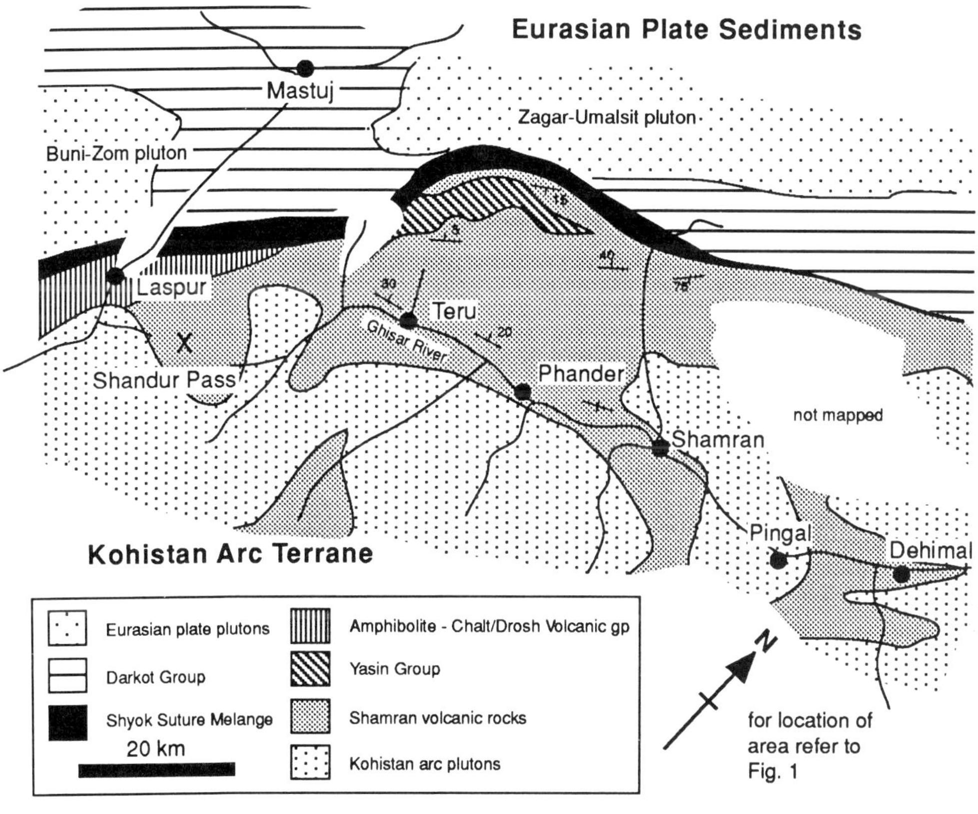

Fig. 11. Geological map of the northern margin of the Kohistan terrane between Shandur Pass and Dehimal (from Pudsey *et al.* 1985). The Shamran volcanic rocks, although previously correlated with the Chalt Volcanic Group (*c.* 110 Ma), have recently yielded a ^{40}Ar-^{39}Ar age of 58 ± 1 Ma from a basaltic andesite in the vicinity of Phander village (Treloar *et al.* 1989).

from volcanic centres. Although the porphyritic nature of the rhyolite lavas implies a high viscosity the absence of any dome or intrusive morphology further enforces the suggestion of a distal or ring-plain volcanic facies rather than a proximal vent facies.

Shamran volcanic rocks

Geological maps of the northern margin of the Kohistan terrane (Figs 1 & 11) show the Chalt Volcanic Group separated from Eurasian Plate metasediments by the Shyok Suture. They represent part of the Cretaceous island-arc lava pile that was deformed and metamorphosed during the arc collision (Petterson & Windley 1985). Pudsey *et al.* (1985) described arc lithologies along the northern margin of Kohistan in some detail and made brief mention of a succession of unmetamorphosed volcanic rocks in the vicinity of Shamran (Fig. 11). On the basis of *Orbitolina* sp.- and *Rudist* sp.- bearing limestones in the overlying Yasin Group, the Shamran volcanic rocks were correlated with the Chalt Volcanic Group, and assigned a mid-Cretaceous age (Pudsey *et al.* 1986).

However, recent isotopic dating of a hornblende-bearing basaltic andesite near Phander village (Fig. 11) has provided a high quality ^{40}Ar-^{39}Ar plateau age of 58 ± 1 Ma (Treloar *et al.* 1989). This spectrum shows no evidence of resetting during later thermal events and can be viewed as a reliable cooling age for the lava flow.

Lithologically the rocks share many of the characteristics of the Utror Volcanic Fm. Coherent lavas range in composition from basalt to rhyolite; silicic compositions are dominant. Abundant fragmental rocks (volcaniclastic breccias and sandstones) testify to an explosive eruptive setting and this is confirmed by the identification of pyroclastic flow and ignimbrite deposits. Unlike the Chalt Volcanic Group strata are essentially unmetamorphosed and undeformed with dips (30°–40°) to the north and northwest.

Thus the currently available evidence suggests that the Shamran volcanic rocks form a remnant of the continental margin volcanic cover of the Kohistan arc rather than part of the island-arc basement. Petterson & Windley (1991) describe the petrology and geochemistry of the Chalt Volcanic Group in the Hunza and Gilgit regions and identify two discrete volcanic suites. The Western Volcanic rocks (from the Gilgit, Yasin and Ishkuman valleys) display petrographic and geochemical characteristics similar to those of

Table 3. *Summary of lithological characteristics and proposed facies interpretations for the Dir Group and Shamran volcanic rocks*

	Age	Thickness	Lithology	Interpretation
Dir Group				
Utror Volcanic Fm. type-section Anakar Gol (Upper Swat)	Late Palaeocene – lava flow; 55 ± 2 Ma ^{40}Ar-^{39}Ar (Hbl) basaltic andesite (Treloar *et al.* 1989).	*c.* 3000 m	Basement unexposed. Heterolithic volcanic breccias, sandstones and siltstones. Lava flows, pyroclastic flows and ignimbrites. Bimodal distribution of andesite and rhyolite. Few limestones.	**Distal or ring-plain facies.** Deposition and eruption in a subsiding/rifted subaerial environment; fluvial reworking dominant. Explosive eruptive style Welded zones indicative of pyroclastic flows and ignimbrite forming eruptions. Occasional lacustrine or shallow marine incursions.
Baraul Banda Slate Fm. type-section Dir Khwar (Dir)	Late Palaeocene – interbedded limestones; *Miscellanea miscella & Actinosiphon tibeticus*. Early Eocene – interbedded shales? *Discocyclina/ Aklinocyclina sp.* (Khan 1979). Eocene – limestone; *Lockhartia sp.* (Kakar *et al.* 1971).	*c.* 2700 m	Basal conglomerate overlies arc basement. Fines upwards rapidly through pebbly sandstone into thinly bedded, medium- to fine-graded sandstones and siltstones. Turbidites with incomplete Bouma sequences; planar bed surfaces; laterally persistent bed thicknesses. Very few marine limestones and intraformational breccias. Dissected arc provenance.	**Deep-water forearc facies.** Deposition in a fault controlled elongate basin (120 km). Alluvial fans submerged rapidly; replaced by sheet turbidites. Absence of any channel, wave or current generated features indicates accumulation at depth > 100–300 m in quiet deep marine basinal environment. Subsidence balanced sedimentation. Sedimentation commenced slightly earlier than volcanism.
Shamran Volcanic rocks	Late Palaeocene – lava flow; 58 ± 1 Ma ^{40}Ar–^{39}Ar (Hbl) hornblende andesite (Treloar *et al.* 1989).	?	Basalt-rhyolite lava flows. Flow banding and welded textures common in more silicic material. Fragmental rocks dominant.	**Volcanic core facies** Founded on Chalt Volcanic Group basement?

both the Shamran and Utror volcanics (Sullivan 1993) and preliminary indications suggest that this suite may represent a further extension of the Palaeocene/Eocene continental margin volcanic cover.

Discussion: early Cenozoic volcanism along the Kohistan continental margin

Cenozoic volcanic rocks of the Dir Group form a remnant of the more extensive Trans-Himalayan volcanic arc which formed in response to the northward subduction of Tethyan oceanic lithosphere prior to the India-Eurasia collision. Much of the volcanic record has been obliterated by the combined effects of intrusion, deformation and erosion. However, the Dir Group and its northerly continuation retains valuable sedimentary data that establish important constraints on the tectonic history of the Kohistan arc system after island-arc accretion. Table 3 presents a lithological summary for both the Dir Group and Shamran volcanic rocks and includes preliminary facies interpretations.

Dismembered volcanic arc

Although geographically separate, the Dir Group and Shamran material indicate that during the Late Palaeocene a subaerial volcanic arc lay along the southern margin of Eurasia, in what is now the Kohistan region of northern Pakistan. The two main lines of evidence for a continental nature are : (1) the abundance of coarse grained, poorly sorted volcanic breccias, sandstones and red-beds with an absence of wave- or current-generated structures (2) the occurrence of rhyolitic lavas and welded pyroclastic flows and ignimbrites.

Reworked material clearly dominates the volcanic record and this, coupled with the restricted incidence of basic lava compositions, indicates that volcanism was predominantly explosive in character. Breccias contain a high proportion of angular dacite and rhyolite clasts and these must have been derived from penecontemporaneous eruptions. Basaltic andesite cooling ages show that volcanism spanned a period of approximately 6 Ma between 58 ± 1 and 55 ± 2 Ma. The Utror Volcanic Fm. is interpreted as a complex association of lavas and volcaniclastic material that accumulated within a tectonically active ring-plain or flanking facies distal to the main focus of volcanic activity. Unfortunately data from the Shamran region are too restricted at this stage to identify with confidence the palaeovolcanic environment. However, its northerly location within Kohistan and the presence of great thicknesses of coarse unstructured breccia and silicic lavas implies in a simplistic sense that it may mark the position of the Palaeocene volcanic front.

In direct contrast to these two subaerial facies, the Baraul Banda Slate Fm. (60–55 Ma) records forearc basin formation within a subsiding continental margin environment. Initial deposition was by alluvial fans filling an elongate fault-bound trough. Debris- and mass-flows deposited a myriad of detritus reflecting erosion of the deformed island-arc metavolcanic basement and earliest post-collisional Andean-type plutons (Stage 1). The basin is characterized by an early phase of rapid subsidence such that a shore-line and coastal facies is very poorly developed, providing further evidence for continued tectonic instability in the region. The majority of deposition, however, took place in a quiet basinal environment and was characterized by thin-bedded sheet-turbidite distributary systems. Once established, deep-water conditions prevailed with only very minor incursions of prograding sandstone bodies.

Timing of basin formation and volcanism

Currently available age data place all continental margin sedimentation and volcanic activity in Kohistan within a 7 Ma period. However, provenance studies have clearly demonstrated that although contemporaneous with the Utror Volcanic Fm. and the Shamran volcanic material, the forearc region received very little in the way of volcanogenic input. Pyroclastic dominated volcanism such as that indicated by the Utror and Shamran material would have erupted ash and coeval ignimbrite fall-out tens to hundreds of kilometres away from the source. It is very difficult to conceive of any physical barrier which can totally prevent ejecta from falling into a continental based forearc region. Thus the implication is that deposition of the Baraul Banda Slate Fm. almost certainly predated the major explosive volcanic phase, even if only by a few million years.

Great variability exists in the dispersal of the volcanic fragments and is reflected in both the magnitude of eruptions and direction of prevailing winds. Pumice tends to travel farther than lithic fragments of the same dimensions (Walker & Croasdale 1971), and this fact coupled with its floatational capacity allows it to be transported great distances from volcanic centres. It has been noted already that intraformational volcanic breccias are observed in the upper parts of the forearc basin illustrating at least some

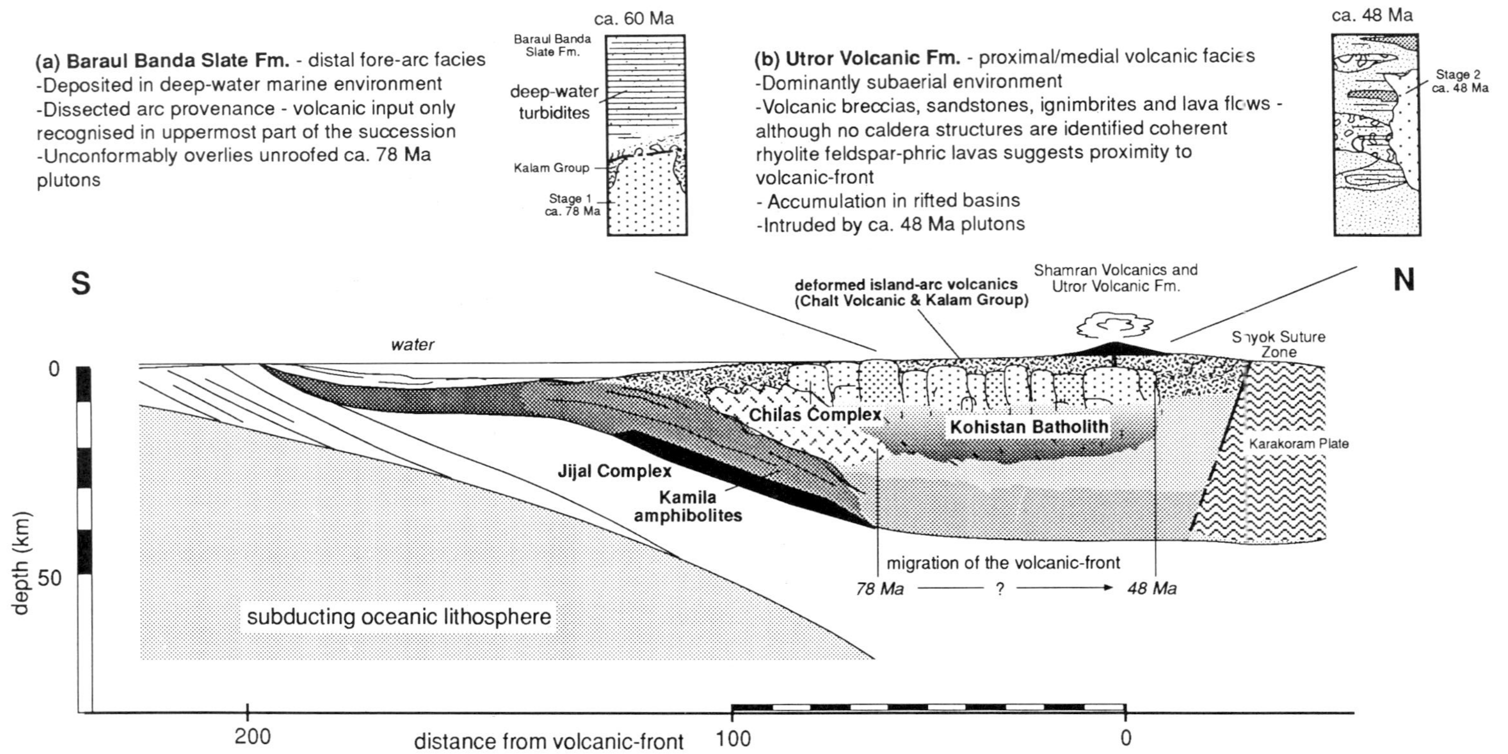

Fig. 12. Schematic cross-section of the Kohistan continental margin subduction system (*c.* 60–55 Ma) during eruption of the Utror Volcanic Fm. Although no record remains of the accretionary prism and residual forearc they are illustrated for completeness. Model is based on Sumatra (after Hamilton 1979).

overlap between the onset of volcanism and basin development. Moreover, the sparsity of pumice and the restricted fragment size implies a significant geographical separation between the location of active vents and the basin depocentre. This intervening region is excluded from the current geological record due to out-of-sequence thrusting along the Dir Thrust during imbrication of the Kohistan terrane after collision with the Indian Plate.

Magmatic arc migration

Temporal migration of magmatism is a common feature within arc provinces and relates to fluctuations in the subduction process. In southern Chile Lopez-Escobar *et al.* (1979) report an eastward decrease in pluton age and within the Peruvian Batholith there has been a similar hinterland migration of magmas from Upper Cretaceous to Miocene times (Pitcher 1985). In a similar way, there was a temporal northward migration in the focus of magmatic activity in the Kohistan arc. Evidence to suggest this is based on the relationship of the individual Dir Group facies with dated plutonic suites.

The Baraul Banda forearc basin is founded on a basement composed of granitoids dated at *c.* 78–76 Ma. These sub-volcanic roots clearly mark the approximate location of the magmatic front at this time. However, by Late Palaeocene times this region had unroofed and subsided sufficiently that it formed the continental basement to a deep-water marine basin. As the limited volcanogenic input demonstrates, this basin was geographically some distance from the focus of volcanic activity. This implies therefore that between *c.* 78–76 Ma and *c.* 55–58 Ma magmatism must have progressively migrated northwards (Fig. 12). Further confirmation of this migratory magma model comes from the intrusive relationships of the Stage 2 plutons. They mark the approximate location of the magmatic front at *c.* 48–45 Ma and intrude the Utror Volcanic Fm. along its entire length but are entirely absent from the Baraul Banda Slate Fm. even though the two are presently juxtaposed. Unfortunately the magnitude of this migration is impossible to establish at the moment without further detailed mapping of the batholith interior.

Tectonic history of the Kohistan arc accretion to Eurasia

After collision of the arc with Eurasia and intrusion of the Stage 1 plutons at *c.* 78–75 Ma, erosion and uplift dominated the continental margin such that the batholith was unroofed. This tectonism implies that a change took place in the stress state of the overriding plate from one of extension to one of compression. At the present-day the Peruvian Coastal Batholith is unroofing during a phase of upper plate compression. This uplift is accompanied by cessation in volcanic activity and development of a wide forearc and retro-arc basin and is a consequence of the subduction of the buoyant aseismic Nazca ridge (Nur & Ben-Avraham 1981). In a similar way subduction of either a buoyant feature, such as a ridge or young oceanic crust, beneath the Kohistan margin between *c.* 78–55 may have been responsible not only for the unroofing of the batholith but also the northward migration of the magmatic front. It is also worth speculating that the Reshun Formation in Chitral (Calkins *et al.* 1981; Pudsey *et al.* 1985) may represent a retro-arc basin that formed during this time.

During the Late Palaeocene the stress configuration in the overriding plate changed. A tensional regime took over along the continental margin with the batholith breaking the surface and voluminous silicic volcanism ensuing. Intra-arc extension is implied both by the collapse of the margin during forearc basin formation and by the accumulation of between 2000–3000 m of sediment and coarse volcanic material. It is interesting to note that in NW Mexico during the Tertiary the Sierra Madre Occidental arc moved from the Cretaceous margin into the continental interior and that this event was also accompanied by vast outpourings of rhyolitic magma (Cameron *et al.* 1980).

Geographically related volcanic successions of similar age such as the Khardung Volcanics (Srimal *et al.* 1987) and Dras II (Reuber 1988) of Ladakh and the Lingzizong of Tibet (Coulon *et al.* 1986) show a similar predominance of silicic volcanic material (rhyolites and pyroclastic flows). Thus implying that the entire northern Tethyan margin was undergoing a similar extensional tectonic phase.

Post-collisional thickening within the Kohistan terrane

Of crucial importance concerning the structural history of western Kohistan is the recognition that the Dir Thrust juxtaposes two contrasting volcanic facies. This implies that considerable shortening has taken place within Kohistan since the docking of India with Eurasia. Independent confirmation of this comes from fission-track cooling ages within western Kohistan. Zeitler (1985) compiled mineral data from the Kohistan region and identified a number of discontinuities in the zircon cooling age contours. One particular unexplained discontinuity occurs in the vicinity of Dir and Upper Swat. It has a NE–SW trend and corresponds to the position of the Dir Thrust as mapped during this study. This illustrates therefore that the Dir Group is in fact an assembly of two similarly aged volcanic facies that underwent slightly different cooling histories on account of their original geographical separation.

Timing of the India–Eurasia collision

The actual timing of the India–Eurasia collision is uncertain and has been inferred from marine magnetic to have been at either *c.* 54 Ma (Patriat & Achache 1984) or *c.* 45 Ma (Dewey *et al.* 1989). Recent work by Klootwijk *et al.* (1992) confirms these findings and identifies a distinct reduction in India's northward movement rate at *c.* 55 Ma which is interpreted as the completion of suturing between India and Eurasia. The identification of the Late Palaeocene (60–55 Ma) forearc succession in western Kohistan agrees with the palaeomagnetic data and illustrates that marine conditions were still in existence prior to the beginning of the Eocene. The newly identified *Miscellanea miscella* fauna is unreported elsewhere along the northern margin. However, Allemann (1979) identifies it in the Murree Brewery Limestone of the Tethyan Shelf sediments near Quetta and this provides an important link with marine sedimentation taking place on the southern India margin prior to ocean closure.

Conclusions

(1) The Late Palaeocene Dir Group forms the dissected remnants of a volcanic arc belonging to the Kohistan Batholith. It acts as a screen separating two post-collisional intrusive suites.

(2) The Baraul Banda Slate Fm. is a marine forearc sequence deposited following collapse of the western Tethyan continental margin. The basin, although initially subaerial, subsided rapidly and is characterized by over 2000 m of thin-bedded sheet turbidites indicative of restricted deep-water basinal environment. Detritus has dissected arc provenance with only a minor volcanogenic signature.

(3) The Utror Volcanic Fm. comprises a lithologically diverse assemblage of coarse volcaniclastic material, basaltic andesite to rhyolite lava flows and pyroclastic material. The majority of material accumulated in a subaerial environment within a subsiding ring-plain facies distal to the focus of volcanic activity.

(4) The Shamran volcanic rocks, although originally identified as part of the island-arc lava pile, have been shown to be Late Palaeocene in age (58 ± 1 Ma) and therefore contemporaneous with the Dir Group.

(5) Soon after accretion of the Kohistan arc to Eurasia the continental margin underwent a prolonged period of uplift and erosion. The *c.* 78–75 Ma component of the batholith was unroofed and formed the basement above which the forearc basin developed.

(6) Between Late Cretaceous and Palaeocene times the focus of magmatic activity within the Kohistan arc migrated progressively northwards.

We are particularly grateful to the Department of Geology and National Centre of Excellence in Geology Peshawar University for assistance during fieldwork, especially M. Farooq. Additional logistical support was provided by Nawaz Khan and Noroz Khan of the Sahad Development Authority, Dir. The Natural Environment Research Council provided fieldwork funding and a studentship to M.A.S. during the period of this research. M.G. Petterson and M.A. Khan are gratefully acknowledged for their exhaustive reviews of this paper. K. Pickering and S. Davis are thanked for infotmative discussions concerning forearc basin sedimentation.

References

Allemann, F. 1979. Time of emplacement of the Zhob Valley Ophiolites and Bela Ophiolites, Baluchistan (Preliminary Report). *In*: Farah, A. & DeJong, K.E. (eds) *Geodynamics of Pakistan*. Geological Survey of Pakistan, Quetta, 215–242.

Calkins, J.A., Jamiluddin, S., Bhuyan, K. & Hussain, A. 1981. *Geology and mineral resources of the Chitral-Partsan area, Hindu Kush range, N. Pakistan*. US Geological Survey Professional Papers, **716-G**.

Cameron, M., Bagby, W.C. & Cameron, K.L. 1980. Petrogenesis of voluminous Mid-Tertiary ignimbrites of the Sierra Madre Occidental, Chihuahua, Mexico. *Contributions to Mineralogy and Petrology*, **74**, 271–284.

Cas, R.A.F. 1979. Mass-flow arenites from the Paleozoic interarc basin, New South Wales, Australia: mode and environment of emplacement. *Journal of Sedimentary Petrology*, **49**, 29–44.

—— & Wright, J.V. 1987. *Volcanic Successions – modern and ancient*. Unwin Hyman Ltd, London.

Coulon, C., Maluski, H., Bollinger, C. & Wang, S. 1986. Mesozoic and Cenozoic volcanic rocks from central and southern Tibet: ^{40}Ar-^{39}Ar dating, petrological characteristics and geodynamic significance. *Earth and Planetary Science Letters*, **79**, 281–302.

Coward, M.P., Windley, B.F., Broughton, R.D.,

LUFF, I.W., PETTERSON, M.G., PUDSEY, C.J. & KHAN, M.A. 1986. Collision tectonics in the NW Himalaya. *In*: COWARD, M.P. & RIES, A.C. (eds) *Collision Tectonics*. Geological Society, London, Special Publications, **19**, 203–219.

COWARD, M.P., BUTLER, R.W.H., KHAN, M.A. & KNIPE, R.J. 1987. The tectonic history of Kohistan and its implications for Himalayan structure. *Journal of the Geological Society, London,* **144**, 377–391.

DAVIS, L.M. & PINFOLD, E.S. 1937. The Eocene beds of the Punjab Salt range. *Memoirs of the Geological Survey of India,* **24**, 1–79.

DEWEY, J.F., CANDE, S. & PITMAN, W.C. 1989. Tectonic evolution of the India/Eurasia collision zone. *Eclogae Geologicae Helvetiae,* **82**, 717–734.

DICKINSON, W.R. 1973. Widths of modern arc-trench gaps proportional to past duration of igneous activity in associated magmatic arcs. *Journal of Geophysical Research,* **78**, 3376–3389.

DICKINSON, W.R. & SEELY, D.R. 1979. Structure and stratigraphy of forearc regions. *The American Association of Petroleum Geologists Bulletin,* **63**, 2–31.

—— & SUCZEK, C.A. 1979. Plate tectonics and sandstone composition. *The American Association of Petroleum Geologists Bulletin,* **63**, 2164–2172.

GARZANTI, E. & VAN HAVER, T. 1988. The Indus Clastics: forearc basin sedimentation in the Ladakh Himalaya. *Sedimentary Geology,* **59**, 237–249.

HACKETT, W.R. & HOUGHTON, B.F. 1989. A facies model for a Quaternary andesitic composite volcano: Ruapehu, New Zealand. *Bulletin of Volcanology,* **51**, 51–68.

HAMILTON, W. 1979. *Tectonics of the Indonesian region*. United States Geological Survey Professional Papers, **1078.**

JAN, M.Q. 1988. Geology of amphibolites from the southern part of the Kohistan Arc, N. Pakistan. *Mineralogical Magazine,* **52**, 147–159.

—— & WINDLEY, B.F. 1990. Chromian spinel chemistry in ultramafic rocks of the Jijal Complex northwestern Pakistan. *Journal of Petrology,* **22**, 667–715.

KAKAR, S.K., MIAN, S.B. & KHAN, J. 1971. The geology of Jandul valley, western Dir. *Geological Bulletin of the University of Peshawar,* **6**, 54–73.

KHAN, M. A. 1979. Geology of the Baraul Valley, Dir. *Geological Bulletin of the University of Peshawar,* **11**, 153–162.

——, JAN, M.Q., WINDLEY, B.F., TARNEY, J. & THIRWALL, M.F. 1989. The Chilas mafic-ultramafic igneous complex; the root zone of the Kohistan island arc in the Himalaya of northern Pakistan. *Geological Society of America Special Papers,* **232**, 75–94.

KLOOTWIJK, C.T., GEE, J.S., PEIRCE, J.W., SMITH, G.M. & MCFADDEN, P.L. 1992. An early India-Eurasia contact; Palaeomagnetic constraints from the Ninetyeast Ridge, ODP leg 121. *Geology,* **20**, 395–398.

LINK, M.H. & NILSEN, T.H. 1980. The Rocks Sandstone, an Eocene sand-rich deepsea fan deposit, northern Santa Lucia Range, California. *Journal of Sedimentary Petrology,* **50**, 583–601.

LOPEZ-ESCOBAR, L., FREY, F.A. & OYARZUN, J. 1979. Geochemical characteristics of Central Chile (33°–34° S) granitoids. *Contributions to Mineralogy and Petrology,* **70**, 439–450.

MOORE, G.F., CURRAY, J.R. & EMMEL, F.J. 1982. Sedimentation in the Sunda Trench and forearc region. *In*: LEGGETT, J.K. (ed.) *Trench-forearc geology*. Geological Society, London, Special Publications, **10**, 245–258.

MUTTI, E. 1977. Distinctive thin-bedded turbidite facies and related depositional environments in the Eocene Hecho Group (South-Central Pyrenees, Spain). *Sedimentology,* **24**, 107–131.

MUTTI, E. & RICCI LUCCHI, F. 1972. Le torbiditi dell'Apennino settentrionale: introduzione all' analisi di facies. *Memoires of the Geological Society of Italy,* **11**, 161–199.

NORMARK, W.R. 1970. Growth patterns of deep-sea fans. *The American Association of Petroleum Geologists Bulletin,* **54**, 2170–2195.

NUR, A. & BEN-AVRAHAM, Z. 1981. Volcanic gaps and the consumption of aseismic ridges in South America. *Geological Society of America, Memoirs,* **154**, 729–740.

PATRIAT, P. & ACHACHE, J. 1984. India-Eurasia collision chronology and its implications for crustal shortening and driving mechanisms of plates. *Nature,* **311**, 615–621.

PETTERSON, M.G. & WINDLEY, B.F. 1985. Rb-Sr dating of the Kohistan arc-batholith in the Trans-Himalaya of north Pakistan, and tectonic implications. *Earth and Planetary Science Letters,* **74**, 45–57.

—— & —— 1991. Changing source regions of magmas and crustal growth in the Trans-Himalayas: evidence from the Chalt volcanics and Kohistan batholith, Kohistan, northern Pakistan. *Earth and Planetary Science Letters,* **102**, 326–341.

——, —— & LUFF, I.W. 1990a. The Chalt volcanics, Kohistan, N. Pakistan: high-Mg tholeiitic and low-Mg calc-alkaline volcanism in a Cretaceous island-arc. *Physics and Chemistry of the Earth,* **17**, 19–30.

——, —— & SULLIVAN, M.A. 1990b. A petrological, chronological, structural and geochemical review of the Kohistan Batholith and its relationship to regional tectonics. *Physics and Chemistry of the Earth,* **17**, 47–70.

PICKERING, K.T., HISCOTT, R.N. & HEIN, F.J. 1989. *Deep-marine environments: clastic sedimentation and tectonics.* Unwin Hyman, London.

PITCHER, W.S. 1985. A multiple and composite batholith. *In*: PITCHER, W.S., ATHERTON, M.P., COBBING, E.J. & BECKINSALE, R.D. (eds) *Magmatism at a plate edge – the Peruvian Andes*. Blackie, London, 93–101.

PUDSEY, C.J. 1986. The Northern Suture, Pakistan: margin of a Cretaceous island-arc. *Geological Magazine,* **123**, 405–423.

——, SCHROEDER, R. & SKELTON, P.W. 1986. Cretaceous (Aptian-Albian) age for the island-arc volcanics, Kohistan. *In*: GUPTA, V.J. (ed.) *Recent researches in geology. Palaeontology, stratigraphy and structure of the Western Himalayas.* Hindustani Publishing Co, Delhi, **3**, 150–168.

——, COWARD, M.P., LUFF, I.W., SHAKLETON, R.M., WINDLEY, B.F. & JAN, M.Q. 1985. Collision zone between the Kohistan arc and the Asian plate in NW Pakistan. *Transactions of the Royal Society of Edinburgh,* **76**, 463–479.

READING, H.G. 1986. *Sedimentary environments and facies.* Blackwell Scientific Publications, Oxford.

REUBER, I. 1989. The Dras arc: two successive volcanic events on eroded oceanic crust. *Tectonophysics,* **161**, 93–106.

SPARKS, R.S.J., SELF, S. & WALKER, G.P.L. 1973. Products of ignimbrite eruptions. *Geology,* **1**, 115–118.

SULLIVAN, M.A. 1993. *The geology of the roof-zone of the Kohistan Batholith.* PhD thesis, University of Leicester.

SRIMAL, N., BASU, A.R. & KYSER, T.K. 1987. Tectonic inferences from oxygen isotopes in volcano-plutonic complexes of the India-Asia collision zone, NW India. *Tectonics,* **6**, 261–273.

SUN, S.-S. & MCDONOUGH, W.F. 1989. Chemical and isotopic systematics of oceanic basalts: implications for mantle composition and processes. *In*: SAUNDERS, A.D. & NORRY, M.J. (eds) *Magmatism in the Ocean Basins.* Geological Society, London, Special Publications, **42**, 313–345.

TAHIRKHELI, R.A.K. 1979. Geology of Kohistan, and ajoining Eurasian and Indo-Pakistan continents, Pakistan. *Geological Bulletin of the University of Peshawar,* **11**, 1–30.

——, MATTAUER, M., PROUST, F. & TAPPONNIER, P. 1979. The India-Eurasia suture zone in northern Pakistan: synthesis and interpretation of recent data at plate scale. *In*: FARAH, A. & DEJONG, K.A. (eds) *Geodynamics of Pakistan.* Geological Survey of Pakistan, Quetta, 125–130.

TAYLOR, S.R. & MCLENNAN, S.M. 1985. *The continental crust: its composition and evolution.* Blackwell Scientific Publications, Oxford.

THORNBURG, T. & KULM, L.D. 1981. Sedimentary basins of the Peru continental margin: structure, stratigraphy, and Cenozoic tectonics from 6° S to 16° S latitude. *Geological Society of America Memoirs,* **154**, 393–422.

TRELOAR, P.J., BRODIE, K.H., COWARD, M.P., JAN, M.Q., KHAN, M.A., KNIPE, R.J., REX, D.C. & WILLIAMS, M.P. 1990. Evolution of the Kamila Shear-Zone, Kohistan, Pakistan. *In*: SALISBURY, M.H. & FOUNTAIN, D.M. (eds) *Exposed cross-sections of the continental crust.* Kluwer Academic Publishers, 175–214.

TRELOAR, P.J., REX, D.C., GUISE, P.G., COWARD, M.P., SEARLE, M.P., WINDLEY, B.F., PETTERSON, M.G., JAN, M.Q. & LUFF, I.W. 1989. K-Ar and Ar-Ar geochronology of the Himalayan collision in NW Pakistan: constraints on the timing of suturing, deformation, metamorphism and uplift. *Tectonics,* **8**, 881–909.

WALKER, G.P.L. & CROASDALE, R. 1971. Two Plinian-type eruptions in the Azores. *Journal of the Geological Society, London,* **127**, 17–55.

YAMADA, E. 1984. Subaqueous pyroclastic flows: their development and deposits. *In*: KOKELAAR, B.P. & HOWELLS, M.F. (eds) *Marginal Basin Geology.* Geological Society, London, Special Publications, **16**, 29–35.

ZEITLER, P.K. 1985. Cooling history of the NW Himalaya, Pakistan. *Tectonics,* **4**, 127–151.

High-pressure metamorphism in the High Himalayan Crystallines of the Stak valley, northeastern Nanga Parbat–Haramosh syntaxis, Pakistan Himalaya

UGO POGNANTE[1]*, PIERA BENNA[1] & PATRICK LE FORT[2]

[1] *Dipartimento di Scienze Mineralogiche e Petrologiche and C. S. Geodinamica Catene Collisionali (CNR), Via Valperga Caluso, 37, 10125 Torino, Italy*

[2] *Laboratoire de Géodynamique des Chaînes Alpines, Institut Dolomieu, 15 rue Maurice Gignoux, 38031 Grenoble, France*

Abstract: The High Himalayan Crystallines (HHC) from the little-known NE termination of the Nanga Parbat–Haramosh syntaxis (Stak and upper Turmik valleys, Pakistan) consist of kyanite-bearing gneiss with minor garnet-granulite and garnet-amphibolite. The HHC underwent a Himalayan metamorphism with a peak at high pressure (8–13 kbar) and high temperature (650–700° C). During exhumation the HHC rocks followed a rapid exhumation path at high temperature with little or no medium to low pressure re-equilibration. These lines of evidence, combined with geochronological and petrological data from Eocene eclogites recently found in the Kaghan nappe, indicate that, after subduction and high to very high pressure metamorphism, part of the HHC from northern Pakistan underwent very rapid cooling and exhumation. By contrast, exhumation along paths of increasing temperature are recorded by the HHC from regions located east of Pakistan (e.g. Nepal). These differences along strike in the HHC suggest that, during the Eocene collision between India and Eurasia, subduction and exhumation occurred at higher rates in northwestern than in central-eastern Himalaya.

The Nanga Parbat– Haramosh syntaxis (northern Pakistan) is one of the most peculiar and interesting geological features of the collided Indian plate in the northwestern part of the Himalayan belt (e.g. Bossart *et al.* 1988; Coward *et al.* 1988; Chamberlain *et al.* 1989; Madin *et al.* 1989; Rehman & Majid 1989; Treloar *et al.* 1991). The syntaxis suffered rapid uplift (Zeitler 1985) and according to Treloar *et al.* (1991) it represents a structural interference between southwest-verging and southeast-verging thrusts. According to part of the papers cited above, asymmetric uplift occurred in the syntaxis with northwest-verging thrusting on the western margin.

In spite of the studies cited above, very little is known about the northern termination of the syntaxis and the relations with the surrounding rocks of the Shyok Suture Zone, the Kohistan–Ladakh Unit and the Karakorum Metamorphic Complex.

This paper presents preliminary field and petrological data collected during an expedition to the northeastern termination of the Haramosh syntaxis in the Stak and Turmik valleys (NW Skardu, Pakistan; Fig. 1). Previous studies concerning this area include the pioneering work of Zanettin (1964) and, for the southernmost part of the Stak valley, Verplanck's thesis (1986). Zanettin (1964) compiled the first geological sketch map of the Stak and Turmik valleys and provided a preliminary petrographic study. Verplanck's thesis (1986) chiefly deals with the structural relationships between the Haramosh syntaxis and the Ladakh Unit.

Geology and petrography

Structural setting of the NE termination of the Haramosh syntaxis

The Stak valley is located between the Indus river, to the south, and the Haramosh massif (7397 m), to the north, and is carved into the northeastern termination of the Nanga Parbat–Haramosh syntaxis (Fig. 1). The rocks of the syntaxis are part of the High Himalayan Crystallines (HHC) which along the Indus Gorge section are dominated by the north-plunging Bulache antiform (Madin *et al.* 1989). Folding along the Bulache antiform post-dates an early

* Following the untimely death of Ugo Pognante in a skiing accident this paper has been revised by his co-authors with the assistance of Bruno Lombardo.

From Treloar, P.J. & Searle, M.P. (eds), *Himalayan Tectonics*
Geological Society Special Publication, No. 74, pp. 161–172.

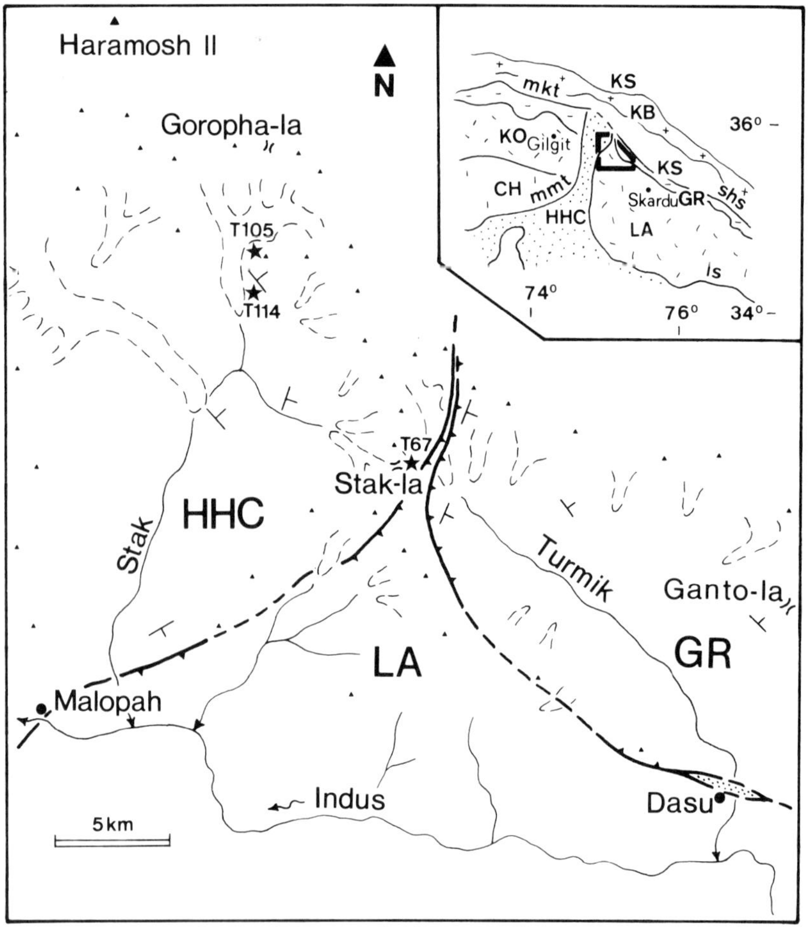

Fig. 1. Tectonic sketch map of the boundary between the High Himalayan Crystallines (HHC), the Ladakh Unit (LA) and the Greenstone Complex (GR) in the Stak and Turmik valleys. Inset shows the location of the studied area in the NE termination of the Nanga Parbat–Haramosh syntaxis. Other symbols: KS, Karakoram metasediments; KB, Karakoram batholith; KO, Kohistan Unit; CH, Chilas Complex; mkt, Main Karakoram Thrust; mmt, Main Mantle Thrust; is, Indus suture; shs, Shyok suture. The lenticular body (dots) north of Dasu consists of serpentinized peridotite. Stars indicate the locations of the samples analysed in Table 1. Dashed lines indicate the main glaciers of the Stak and Turmik valleys.

imbrication between the Iskere Gneiss and the Shengus Gneiss with imbrication defined by sillimanite-bearing mylonites (Treloar *et al.* 1991).

In the studied area the HHC is tectonically bounded, to the east of Stak-la and in the upper Turmik valley, by amphibolites of the Ladakh Unit and by low-grade rocks attributed to the Greenstone Complex of Tahirkheli & Jan (1979) and Hanson (1986, 1989). The Greenstone Complex consists of sedimentary, metabasic and minor ultrabasic rocks. We have found kyanite, chloritoid and pseudomorphs of clinozoisite/epidote, possibly replacing lawsonite, in the rocks of the Greenstone Complex.

The Ladakh Unit and the Greenstone Complex overlie the HHC and in the upper Turmik valley, just to the east of Stak-la, the boundary is defined by a shear zone dipping 25–40° toward the ESE (Fig. 2). The mylonitic zone is a few tens of metres thick and shear-sense indicators (i.e. asymmetrical folds and minor S–C fabrics) give a prevalent dextral strike-slip regime with shear of the Ladakh Unit to the south. This movement is contemporaneous with retrograde metamorphism of the topmost HHC. Near the

Fig. 2. Tectonic boundary between the HHC and the overlying Ladakh Unit and Greenstone Complex. The upper part of the HHC consists of prevalent metabasic rocks and pegmatitic dykes with minor paragneiss and orthogneiss. The face is about 900 m high. View looking NE on the upper left bank of the Turmik valley.

confluence of the Stak valley with the Indus valley, the contact between the HHC and amphibolites and gabbros of the Ladakh Unit is a normal fault zone with mylonitization and brecciation (Verplanck 1986). From south to north thinning of the Ladakh Unit is obvious, and strike of the foliations and of the main tectonic boundary changes from NE–SW to N–S (Fig. 1).

Field relations and microstructures

The HHC rocks of the Stak valley are part of the Shengus Gneiss which is a series of paragneisses imbricated with orthogneisses derived from early Palaeozoic granitoids (Zeitler *et al.* 1989), metabasic rocks and rare calcite-diopside-bearing marbles. Garnet-bearing leucogranites forming small layers or dykelets are very rare while deformed pegmatitic dykes are common near the upper shear zone (Fig. 2).

Though the paragneisses are rich in micas and garnet and poor in alkali feldspar with respect to the orthogneisses, these two rock-types have a rather homogeneous mineralogy and the most common assemblage is quartz–plagioclase–biotite–muscovite–garnet–kyanite ± rutile. Except for the uppermost structural levels (upper Turmik valley), this assemblage is well preserved and shows very little re-equilibration. K-feldspar often occurs as porphyroclasts that do not seem to pertain to the paragenesis. The gneisses show a planar fabric defined by biotite, muscovite and fine-grained kyanite. In some paragneisses the foliation wraps around centimetre-sized garnet porphyroblasts (Fig. 3a). Kyanite also occurs within syn-tectonic garnet porphyroblasts to define a gentle S inclusion fabric (S_i) often discordant to the external fabric (S_e) (Fig. 3a). Kyanite also shows evidence of polyphase recrystallization and is always well preserved external to garnets where it defines S_e (Fig. 3b). All this suggests that kyanite was stable during different stages of recrystallization of the HHC. South of Gorophala-la, at structural levels lower than in the upper Turmik valley (hereafter indicated as 'middle HHC'), the gneisses show an incipient migmatization with formation of quartzofeldspathic leucosomes. Even in these rocks kyanite is part of the stable assemblage and sillimanite is lacking. The stability of kyanite in the migmatitic rocks was emphasized also by Zanettin (1964).

The metabasites form metre-thick layers which in places define isoclinal folds. They derive from original dykes (Fig. 4) and occur both in the paragneiss and in the orthogneiss. The metabasites are fine-grained rocks characterized by the assemblage clinopyroxene–plagioclase–garnet–quartz–rutile ± hornblende ± biotite ± sphene ± calcite. Pale green clinopyroxene includes very small plagioclase grains and is rimmed by and partly transformed to

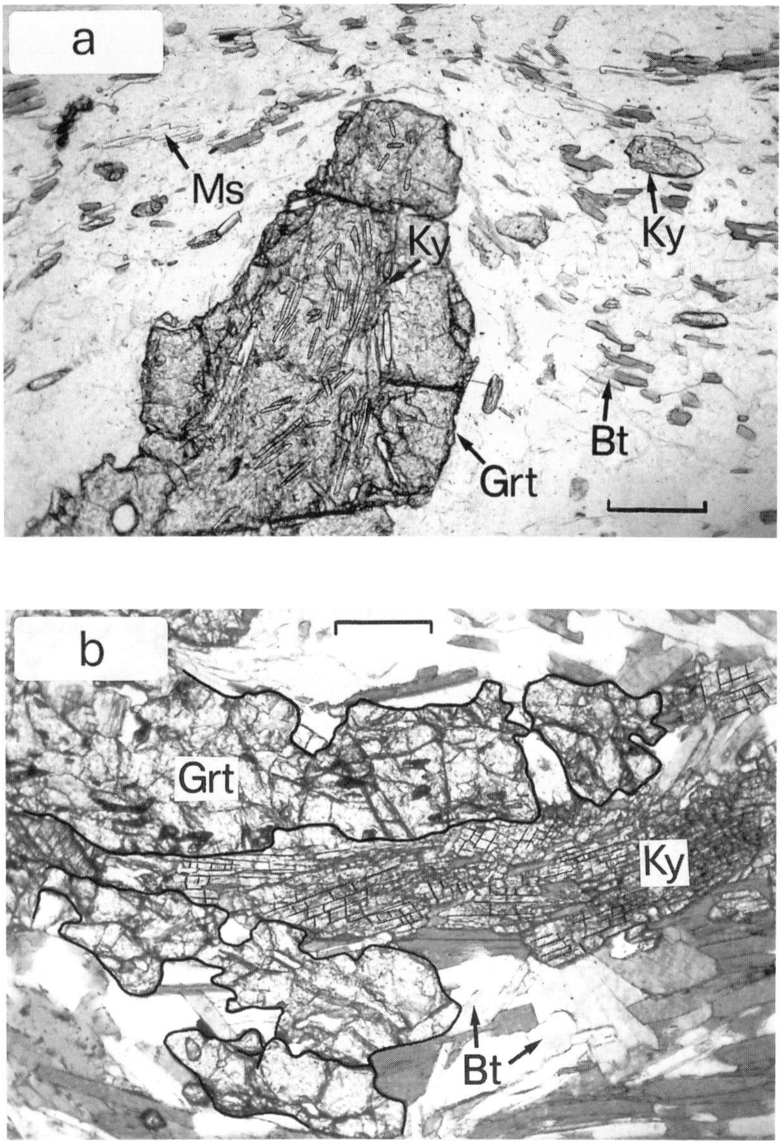

Fig. 3. (**a**) Paragneiss from Stak-la (4600 m, T77). Garnet with sub-vertical S_i defined by small kyanite needles and discordant to sub-horizontal S_e defined by muscovite, biotite and larger kyanite blasts. The bar is 0.2 mm long. Abbreviations are after Kretz (1983). (**b**) Paragneiss from west of Stak-la (4400 m, T88). Syntectonic and partly recrystallized kyanite coexists with biotite and garnet (rich in small rutile inclusions). The bar is 0.2 mm long.

brown amphiboles (Fig. 5). In a few samples and along centimetre-thick metamorphic veins clinopyroxene is completely replaced by brown or green amphibole, biotite ± calcite, while rutile is replaced by sphene and ilmenite.

In the uppermost levels, for a few tens of metres below the main contact with the Ladakh Unit, a pervasive greenschist-facies recrystallization occurs in the HHC rocks. The following assemblages have been observed in the upper shear zone: quartz–albite–biotite–muscovite–clinozoisite with relicts of garnet in the gneiss;

Fig. 4. Metamorphic basic dyke transformed to garnet-granulite and included in kyanite-bearing gneiss (upper Stak valley, south of Goropha-la).

green or blue–green hornblende–albite–clinozoisite/epidote–chlorite–sphene with relicts of biotite, garnet and brown hornblende in the metabasic rocks; and quartz–albite/plagioclase–muscovite–biotite–clinozoisite/epidote in the pegmatitic dykes. Clinozoisite forms poikiloblasts rimmed by Fe-epidote and overgrows the main foliation, albite borders plagioclase whilst biotite and amphibole transform to chlorite.

Mineral chemistry

Minerals have been analysed using an EDS/SEM microprobe and representative compositions are reported in Table 1. The composition of garnets, clinopyroxenes and amphiboles are plotted in the (Fe + Mn)–Mg–Ca diagram of Fig. 6.

In the metabasic rocks, garnets are rich in almandine (X_{Alm} = 0.43–0.56) and have significant amounts of grossularite (X_{Grs} = 0.13–0.31) and pyrope (X_{Prp} = 0.20–0.27). They are zoned with rims enriched in almandine and depleted in grossularite relative to cores. In the gneissic rocks garnets are very rich in almandine (X_{Alm} = 0.65–0.85) with lower amounts of pyrope (X_{Prp} = 0.10–0.21), spessartine (X_{Sps} = 0.01–0.10) and grossularite (X_{Grs} = 0.01–0.11). These garnets have rims enriched in almandine and grossularite and depleted in pyrope relative to cores.

Biotites from gneiss have a compositional range (X_{Mg}[Mg/Mg + Fe_{tot}] = 0.43–0.69) which reflects the different bulk compositions and the slight irregular zoning. In the metabasic rocks biotites usually show high Mg contents (X_{Mg} = 0.61–0.63) and are rich in Ti (≤0.296 atoms per 12 oxygens). Muscovites have relatively low celadonite contents (Si ≤ 3.132 atoms per 12 oxygens) and moderate to high X_{Mg}(0.51–0.73).

The analysed plagioclases are oligoclase in the gneiss (X_{An} = 0.13–0.26) and in the metabasic rocks (X_{An} = 0.19–0.30). In a few samples they show slight inverse zoning.

Clinopyroxenes are diopsides and augites with low but significant amounts of jadeite (X_{Jd} ≤ 0.14). They have high X_{Mg} (0.72–0.78), low acmite contents (X_{Acm} ≤ 0.15) and low Ti contents (Ti ≤ 0.011 atoms per 6 oxygens). Cores are usually richer in jadeite than rims.

Amphiboles are tschermakitic, edenitic or ferroan hornblendes (nomenclature according to Leake 1978) with X_{Mg} = 0.52–0.63 and low to moderate Ti contents (0.050–0.176 atoms per 24 oxygens). Al^{IV} and Al^{VI} are 1.455–1.814 and 0.490–0.629 atoms per 24 oxygens, respectively.

Geothermobarometry

P–T estimates determined using different thermobarometers are reported in Table 2 and the main results are summarized in Fig. 7.

Metamorphic temperatures of the gneissic rocks were determined with the Grt-Bt thermometers of Hodges & Spear (1982), Perchuk & Lavrent'eva (1983) and Indares & Martignole (1985). Estimates give a wide range with temperatures from 550 ± 40° C to 680 ± 20° C for rocks of the middle structural levels (2 samples, 15 pairs) and are higher for garnet cores and included biotites than for garnet rims and external biotites (Table 2; Fig. 7). Temperatures between 450 ± 50° C and 590 ± 40° C were obtained in the gneiss of the upper levels (2 samples, 12 pairs). Among the three calibrations used for thermometry, the estimates obtained for cores (650–700° C) using the Hodges & Spear (1982) and the Perchuk & Lavrent'eva (1983) calibrations may broadly correspond to the temperature peak. In fact they are consistent with incipient migmatization and with data concerning the metabasic rocks. By contrast, the

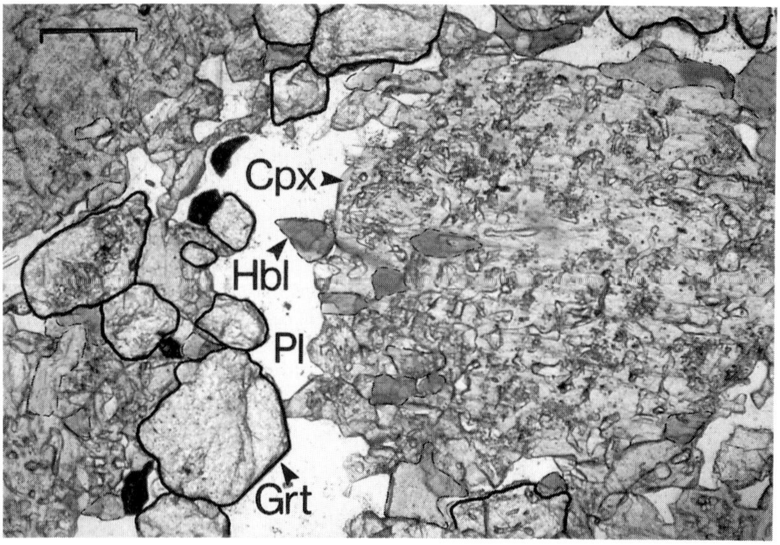

Fig. 5. Garnet-granulite from the upper Stak valley, south of Goropha-la (4000 m, T102). Clinopyroxene rich in plagioclase and opaque inclusions, is partly transformed to very fine-grained brown hornblende. The bar is 0.2 mm long.

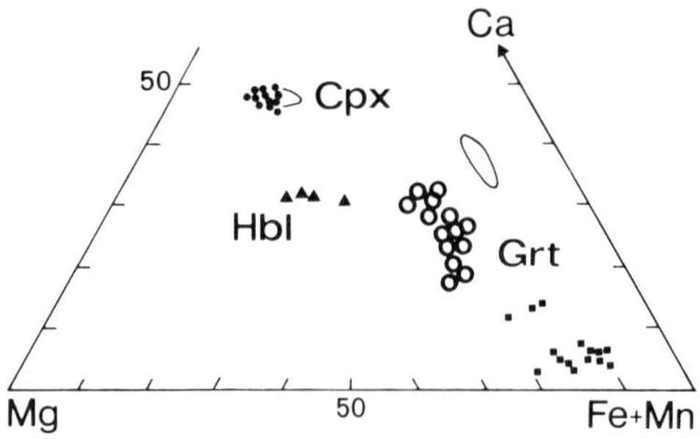

Fig. 6. (Fe + Mn)–Mg–Ca diagram showing the compositions of clinopyroxenes (dots) and garnets (circles) from the Stak metabasites. Small squares represent garnets of the gneissic rocks while triangles represent the composition of amphiboles from the metabasites. The compositional fields of clinopyroxenes and garnets from the Zanskar garnet-granulites (Pognante & Lombardo 1989) are shown for comparison.

Indares & Martignole calibration (1985) gives too low temperatures.

For the gneiss, pressures were determined using the garnet–plagioclase–kyanite (Newton & Haselton 1981; Koziol & Newton (1988), and the garnet–plagioclase–biotite–muscovite and garnet–plagioclase–muscovite (Hodges & Crowley 1985) geobarometers. Simultaneous solution of the garnet–plagioclase–kyanite geobarometer and the garnet–biotite thermometer of Perchuk & Lavrent'eva (1983) gives higher pressures at middle than at higher structural levels (7.0–11.5 kbar versus 5.0–9.5 kbar). In addition to that, higher pressures are obtained using garnet cores and included biotites, while the Newton & Haselton (1981) calibration gives values 0.5–1.0 kbar lower than the Koziol & Newton (1988) calibration. Pressures a few kilobars lower are obtained using the garnet–plagioclase–biotite–muscovite and garnet–plagioclase–muscovite barometers (Hodges & Crowley 1985).

In the metabasic rocks temperatures from 600 ± 80° C to 770 ± 20° C (2 samples) are calculated using the garnet–clinopyroxene (Powell 1985; Krogh 1988), garnet–biotite (Hodges & Spear 1982; Perchuk & Lavrent'eva 1983; Indares & Martignole 1985) and garnet–hornblende (Graham & Powell 1984) geothermometers. For both calibrations of the garnet–clinopyroxene thermometer, values of around 700° C were calculated for 15 pairs of core compositions, and these probably approach the temperature peak. These values are 50–100° C higher than those obtained using rim compositions. However, values obtained with cores are valid only if we assume equilibrium between cores of garnet and pyroxene while rims should give closure temperatures during cooling. For the garnet–biotite (5 pairs) and garnet–hornblende (6 pairs) thermometers, only rim compositions were used for thermobarometry

Table 1. *Representative microprobe analyses of minerals of metabasic and gneissic rocks from different structural levels of the HHC*

	Middle HHC									Upper HHC				
	T114					T105				T67				
	Grt core	Cpx core	Hbl	Bt rim	Pl rim	Grt core	Bt core	Ms rim	Pl	Grt core	Bt core	Ms	Pl	Or rim
SiO_2	38.33	51.83	43.09	36.46	61.26	36.46	35.88	45.90	64.13	36.81	35.79	46.16	63.28	64.05
TiO_2	0.00	0.00	0.45	4.30	0.00	0.00	3.39	1.60	0.00	0.00	2.84	1.54	0.00	0.00
Al_2O_3	22.11	7.05	12.83	15.19	23.48	20.48	17.45	33.06	21.67	20.90	18.16	32.73	22.20	18.56
Fe_2O_3	0.58	1.90	5.40	n.c.	0.52	1.38	n.c.	n.c.	0.37	1.29	n.c.	n.c.	0.00	0.00
FeO	24.96	6.14	11.94	15.29	0.00	31.56	19.04	1.41	0.07	33.18	17.79	1.29	0.00	0.00
MnO	0.73	0.00	0.00	0.00	0.00	4.08	0.00	0.00	0.00	0.83	0.00	0.00	0.00	0.00
MgO	6.95	11.05	10.13	13.50	0.00	3.23	9.65	1.31	0.00	5.25	10.10	1.14	0.00	0.00
CaO	6.57	18.62	11.73	0.00	5.64	2.17	0.00	0.00	2.66	1.09	0.00	0.00	3.55	0.00
Na_2O	0.00	2.65	1.75	0.00	8.75	0.00	0.00	0.42	9.88	0.00	0.48	0.00	9.88	1.68
K_2O	0.00	0.00	0.28	9.73	0.27	0.00	9.60	10.93	0.18	0.00	9.67	10.73	0.20	14.74
Total	100.23	99.24	97.60	94.47	99.92	99.36	95.01	94.62	98.97	99.35	94.83	93.59	99.11	99.03
Si	2.957	1.915	6.388	2.820	2.719	2.958	2.813	3.095	2.857	2.949	2.806	3.132	2.809	2.966
Al^{IV}	0.043	0.085	1.613	1.180	0.281	0.042	1.187	0.906	0.143	0.051	1.194	0.868	0.191	0.034
Al^{VI}	1.967	0.222	0.629	0.204	0.947	1.916	0.425	1.721	0.995	1.922	0.485	1.750	0.970	0.979
Ti	0.000	0.000	0.050	0.250	0.000	0.000	0.200	0.081	0.000	0.000	0.168	0.079	0.000	0.000
Fe^{3+}	0.034	0.053	0.602	n.c.	0.017	0.084	n.c.	n.c.	0.013	0.078	n.c.	n.c.	0.000	0.000
Fe^{2+}	1.610	0.190	1.481	0.989	0.000	2.141	1.248	0.079	0.003	2.223	1.166	0.074	0.000	0.000
Mn	0.047	0.000	0.000	0.000	0.000	0.280	0.000	0.000	0.000	0.056	0.000	0.000	0.000	0.000
Mg	0.799	0.609	2.239	1.557	0.000	0.391	1.128	0.132	0.000	0.627	1.181	0.115	0.000	0.000
Ca	0.543	0.737	1.863	0.000	0.268	0.189	0.000	0.000	0.127	0.094	0.000	0.000	0.169	0.000
Na	0.000	0.190	0.503	0.000	0.753	0.000	0.000	0.056	0.853	0.000	0.073	0.000	0.850	0.151
K	0.000	0.000	0.053	0.960	0.015	0.000	0.960	0.940	0.010	0.000	0.968	0.929	0.011	0.870

T114: mafic dyke metamorphosed to quartz–rutile-bearing garnet-granulite (South of Goropha-la, 3800 m, upper Stak valley). T105: kyanite-bearing gneiss showing incipient migmatization (South of Goropha-la, 3800 m, Stak valley). T67: kyanite-bearing gneiss (Stak-la, 4600 m). Location of the samples is given in Fig. 1. Structural formulae are calculated on the basis of 8 cations and 12 oxygens for garnet, 4 cations and 6 oxygens for pyroxene, 13 cations (N_{cat}-Ca-Na-K = 13) and 24 oxygens for amphibole, 7 cations (N_{cat}-Ca-Na-K = 7) and 12 oxygens for biotite, 6 cations (N_{cat}-Ca-Na-K = 6) and 12 oxygens for muscovite and 5 cations and 8 oxygens for feldspars; n.c. = not calculated.

Table 2. *Thermobarometric estimates for gneissic (Gn) and metabasic (Mb) rocks*

Thermobarometer, calibration, rock-type		Temperature (° C)		Pressure (kbar)	
		cores	rims	cores	rims
Grt–Bt					
HS	(Gn)	680 ± 20	650 ± 40		
IM	(Gn)	580 ± 40	550 ± 40		
PL	(Gn)	650 ± 30	590 ± 30		
HS	(Gn)*	580 ± 30*	560 ± 50*		
IM	(Gn)*	490 ± 30*	450 ± 50*		
PL	(Gn)*	590 ± 40*	580 ± 25*		
HS	(Mb)		770 ± 20		
IM	(Mb)		670 ± 30		
PL	(Mb)		690 ± 15		
Grt–Cpx					
K	(Mb)	700 ± 50	600 ± 80		
P	(Mb)	700 ± 25	630 ± 45		
Grt–Hbl					
GP	(Mb)		680 ± 35		
Grt–Pl–Ky					
KN	(Gn)			11.0 ± 1.5	8.0 ± 1.5
NH	(Gn)			10.5 ± 1.0	7.5 ± 1.0
KN	(Gn)*			9.0 ± 0.5*	7.5 ± 2.0*
NH	(Gn)*			8.0 ± 1.0*	7.0 ± 2.0*
Grt–Pl–Bt–Ms					
HG	(Gn)			9.0 ± 1.0	7.5 ± 1.0
HG	(Gn)*			6.0 ± 1.5*	7.0 ± 2.0*
Grt–Pl–Cpx					
NP	(Mb)			10.5 ± 2.0	9.0 ± 2.0
PH	(Mb)			11.0 ± 1.0	10.0 ± 1.5
Grt–Pl–Hbl					
KS	(Mb)				8.5 ± 1.5
Ab–Jd–Qtz					
GL-HO	(Mb)			≈12.0 (at 650° C)	

Calibration: HS, Hodges & Spear (1982); IM, Indares & Martignole (1985); PL, Perchuck & Lavrent'eva (1983); K, Krogh (1988); P, Powell (1985); GP, Graham & Powell (1984); KN, Koziol & Newton (1988); NH, Newton & Haselton (1981); HG, Hodges & Crowley (1985); NP, Newton & Perkins (1982); PH, Powell & Holland (1988); KS, Khon & Spear (1990); GL, Gasparik & Lindsley (1980); HO, Holland (1980).

* Rocks from the upper structural levels.

Pressures have been calculated by simultaneous solution of the barometer and the Grt-Bt thermometer (PL calibration) for gneiss, and of the barometer and the Grt–Cpx thermometer (K calibration) for the metabasic rocks.

because biotites and amphiboles usually postdate growth of pyroxenes and garnets (Fig. 5). Among the geothermometers we have used, the garnet–biotite thermometer of Hodges & Spear (1982) gives values, probably somewhat too high, in excess of 750° C.

Pressures of the garnet-granulites of the middle structural levels were determined using the garnet–plagioclase–clinopyroxene–quartz barometer (Newton & Perkins 1982; Powell & Holland 1988). Estimates range from *c.* 7 to *c.* 13 kbar and are more or less comparable to those of the associated gneiss. Pressure values are 1.0–1.5 kbar higher using cores rather than rims. Using the clinopyroxene richest in jadeite ($X_{Jd} = 0.14$) that coexists with plagioclase ($X_{An} = 0.29$), the calibrations of Gasparik & Lindsley (1980) and of Holland (1980) of the albite–jadeite–quartz barometer give P of about 12 kbar at T of 650° C. The garnet–plagioclase–hornblende–quartz barometer (Khon & Spear 1990) was also applied to the Stak metabasites

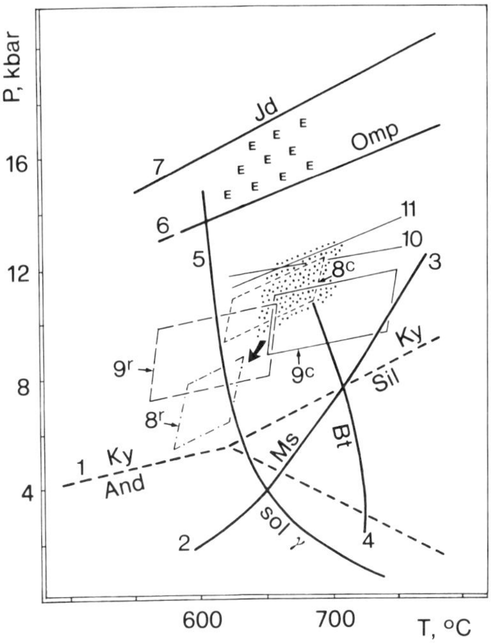

Fig. 7. *P–T* grid showing the conditions of the metamorphic peak of the middle structural levels of the HHC from the Stak valley (dotted area). The arrow defines the possible retrograde path suggested by study of garnet zoning profiles and other considerations (see text). The conditions of recrystallization of the Kaghan eclogites (E) are shown for comparison (Pognante & Spencer 1991). Legend. **1**, kyanite–andalusite–sillimanite stability fields (Richardson *et al.* 1969); **2**, muscovite + quartz = andalusite + orthoclase + H_2O (Kerrick 1972); **3**, muscovite + albite + quartz = orthoclase + sillimanite + liquid (Petö 1976); **4**, biotite (X_{Mg} = 0.4) + sillimanite + plagioclase + quartz = garnet + orthoclase + liquid (Clemens 1984); **5**, beginning of melting in the albite–orthoclase–quartz–H_2O system (Merrill *et al.* 1970); **6**, albite = jadeite$_{50}$ + quartz (Holland 1980); **7**, albite = jadeite + quartz (Holland 1980); **8**, *P–T* conditions of recrystallization of the gneissic rocks defined by garnet–biotite thermometry and garnet–plagioclase–kyanite barometry for core (c) and rim (r) compositions; **9**, P–T conditions of recrystallization of the metabasic rocks defined by garnet–clinopyroxene thermometry and garnet–plagioclase–clinopyroxene barometry for core (c) and rim (r) compositions; **10**, and **11**, curves calculated with the Holland (1980) and Gasparik & Lindsley (1980) calibrations respectively of the albite–jadeite–quartz barometer using the clinopyroxene richest in jadeite (X_{jd} = 0.14) coexisting with plagioclase (X_{An} = 0.29). Mineral abbreviations on the figure are after Kretz (1983).

and gives pressures a few kilobars lower than the barometers discussed above.

In conclusion, the thermobarometric data obtained suggest that the rocks of the Stak valley suffered a metamorphic peak at T = 650–700° C and P = 8–13 kbar. These values are similar to those derived from the Alpurai schists in the Swat Valley on the immediate footwall of the Main Mantle Thrust 300 km southwest of Nanga Parbat (Treloar *et al.* 1989*a*; DiPietro 1991.) Comparable *P–T* estimates have also been reported for garnet-granulites from the HHC of Zanskar (northern India, Pognante & Lombardo 1989). However, in Zanskar the assemblages of the pressure peak occur as rare relicts in rocks showing a pervasive high-temperature recrystallization during exhumation.

Metamorphic Zonation and Evolution

Metamorphic assemblages are the same throughout the HHC of the Stak valley but in the upper structural levels thermobarometry indicates temperatures 50–70° C lower and pressures 1–2 kbar lower than in the lower structural units. In addition, within the upper shear zone the syn-tectonic formation of green/green-blue amphibole, clinozoisite, albite and chlorite indicates a greenschist-facies retrograde recrystallization linked to southward shearing of the Ladakh Unit.

Away from the upper shear zones the post-peak transformations are rare and are restricted to discrete zones. In the metabasites these transformations chiefly occur along millimetre- to decimetre-thick veins and their formation is controlled by fluid-infiltration along fractures. Clinopyroxenes are zoned with decreasing jadeite contents in the rims while green amphiboles form in veins and reaction rims bordering clinopyroxene and brown hornblende. This is consistent with a retrograde decompressional evolution.

Comparable post-peak transformations are uncommon in the gneiss which often show evidence of polyphase recrystallization of the peak metamorphic assemblage. For these rocks a possible *P–T* path was determined using chemical zoning in garnets following the 'Gibbs method' discussed by Spear & Selverstone (1983). In the system KNCFMASH, the gneisses contain the divariant assemblage garnet–biotite–plagioclase–quartz–kyanite–muscovite–orthoclase and the *P–T* path was calculated considering the inverse zonation within plagioclase. The calculated path suggests a decompressional path at decreasing temperature in the kyanite stability field (arrow in Fig. 7) which is consistent with the discussion reported above.

Discussion and conclusions

After the recent discovery of early Eocene eclogites formed at $T = 650 \pm 50°$ C and $P = 13$–18 kbar (Fig. 7) in the upper Kaghan valley (Spencer *et al.* 1990; Pognante & Spencer 1991; Tonarini *et al.* 1993) and the documentation of metamorphic pressures of >10 kbar in Swat Valley (Treloar *et al.* 1989a; DiPietro 1991), the study of metamorphism in the HHC from northern Pakistan attracts increasing interest. During summer 1991, eclogites have been recognized in other parts of the HHC from northern Pakistan (Neelum valley; D. Fontan, M. Schouppe & G. Martinotti, pers. com. 1991) and post-eclogitic glaucophane has been found in the structurally upper part of the Kaghan nappe (U.P. pers. obs). The present discovery of high-pressure, garnet-granulites equilibrated at $T \approx 650$–700° C and $P \approx 8$–13 kbar (Fig. 7) in the Shengus Gneiss forming the northeastern termination of the Nanga Parbat–Haramosh syntaxis, suggests that the entire HHC in Pakistan may have been subjected to high-pressure metamorphism.

In the Stak valley, the high-pressure assemblages suffered polyphase recrystallization but show little reequilibration during exhumation. Only at the uppermost structural levels does the HHC show a pervasive greenschist-facies recrystallization caused by shearing and conductive cooling along the contact with the Ladakh Unit. The microstructural and petrological data discussed in this paper demonstrate that, by comparison with many other Himalayan regions to the east of Pakistan (e.g. Nepal: Caby *et al.* 1983; Pêcher 1989; Pognante & Benna, this volume), exhumation of the Stak rocks occurred at decreasing temperatures in the kyanite stability field instead of increasing temperatures in the sillimanite field. Exhumation of the HHC from the NE termination of the Haramosh syntaxis thus most likely occurred at relatively high rates which allowed rapid cooling and prevented pervasive low-pressure reequilibrations. Rapid cooling of the HHC in northern Pakistan has been proposed by Treloar *et al.* (1989*b*) and Treloar & Rex (1990) and rapid cooling rates of 15–20° C Ma^{-1} have been obtained for the early exhumation of Eocene age of the Kaghan eclogites (Tonarini *et al.* 1993).

Based on the previous lines of evidence, the question arises as to whether high exhumation rates in the Pakistan Himalaya depend on plate-scale tectonic processes or merely reflect variations due to local factors. In our opinion, the characteristics of the HHC from many parts of Pakistan, notably the occasional formation of eclogites and garnet-granulites and the lack of a heating stage during rapid exhumation, are not accidental and have important implications for the evolution of the collided Indian margin.

India–Eurasia collision must have begun during the early Eocene between the NW Indian spur and Kohistan arc and, after the initial collision, the northward drift of India continued at lower rates (Klootwijk *et al.* 1985; Garzanti *et al.* 1987; Dewey *et al.* 1989; Gaetani & Garzanti 1991). As recently discussed by Pognante (1992), it is proposed here that high rates of convergence in the zone of the first collision, now represented by northern Pakistan, caused higher rates of subduction and exhumation (and hence lower thermal gradients and only incipient migmatization) relative to zones of the northern Indian margin located to the east. During eastward progressive suturing, subduction and exhumation of central and eastern Himalaya likely occurred at lower rates allowing the isotherms to rise and giving a relatively warm thickened crust. During exhumation this crust suffered extensive anatexis with formation of numerous plutons of leucogranite.

Structural and geochronological evidence (Zeitler 1985; Treloar *et al.* 1991) indicates that, in northern Pakistan, very rapid uplift (>6 mm a^{-1}) and erosion resumed in the zone of the Nanga Parbat–Haramosh syntaxis during the last million years.

We thank M. Gaetani for encouraging the expedition in the Turmik and Stak valleys. Reviews by A.C. Barnicoat, an anonymous referee and P.J. Treloar, and discussions with B. Lombardo greatly improved this paper. Fieldwork has been financed by European Community Contract (CEC) NoCl1*-CT90-0852 (Geology of the Karakorum). The laboratory work of U.P and P.B. was financially supported by C.S. Geodinamica Catene Collisionali, CNR, Torino. Minerals have been analysed using the EDS/SEM Cambridge microprobe installed in the Dipartimento di Scienze Mineralogiche e Petrologiche of Torino.

References

Bossart, P., Dietrich, D., Greco, A., Ottiger, R. & Ramsay, J.G. 1988. The tectonic structure of the Hazara–Kashmir Syntaxis, southern Himalayas, Pakistan. *Tectonics*, **7**, 273–297.

Caby, R.A., Pêcher, A. & Le Fort, P. 1983. Le grand chevauchement central himalayen: nouvelles données sur le métamorphisme inverse à la base de la dalle du Tibet. *Revue Géologie Dynamique Géographie Physique*, **24**, 89–100.

Chamberlain, C.P., Zeitler, P.K. & Jan, M.Q. 1989. The dynamics of the suture between the Kohistan island arc and the Indian plate, northwest Himalaya. *Geological Society of America Special Papers,* **232**, 23–32.

CLEMENS, J.D. 1984. Water contents of silicic to intermediate magmas. *Lithos,* **17**, 273–287.

COWARD, M.P., BUTLER, R.W.H., CHAMBERS, A.F., GRAHAM, R.H., IZATT, C.N., KHAN, M.A., KNIPE, R.J., PRIOR, D.J., TRELOAR, P.J. & WILLIAMS, M.P. 1988. Folding and imbrication of the Indian crust during Himalayan collision. *Philosophical Transactions of the Royal Society London,* **A326**, 89–116.

DEWEY, J.F., CANDE, S. & PITMAN, W.C. 1989. Tectonic evolution of the India/Eurasia Collision Zone. *Eclogae Geologicae Helvetiae,* **82**, 717–734.

DIPIETRO, J.A. 1991. Metamorphic pressure-temperature conditions of Indian Plate rocks south of the Main Mantle Thrust, Lower Swat, Pakistan. *Tectonics,* **10**, 742–757.

GAETANI, M. & GARZANTI, E. 1991. Multicyclic history of the northern India continental margin (northwestern Himalaya). *The American Association of Petroleum Geologists Bulletin,* **75**, 1427–1446.

GARZANTI, E., BAUD, A. & MASCLE, G. 1987. Sedimentary record of the northward flight of India and its collision with Eurasia. *Geodinamica Acta,* **1**, 87–102.

GASPARIK, T. & LINDSLEY, D.H. 1980. Phase equilibria at high pressure of pyroxenes containing monovalent and trivalent ions. *In*: PREWITT, C.T. (ed.) *Pyroxenes.* Mineralogical Society of America, Reviews in Mineralogy, **7**, 309–339.

GRAHAM, C.M. & POWELL, R. 1984. A garnet-hornblende geothermometer: calibration, testing, and application to the Pelona Schist, Southern California. *Journal of Metamorphic Geology,* **2**, 13–31.

HANSON, C.R. 1986. *Bedrock geology of the Shigar valley area, Skardu, Northern Pakistan.* MSc Thesis, Dartmouth College, New Hampshire.

—— 1989. The northern suture in the Shigar valley, Baltistan, northern Pakistan. *Geological Society of America Special Papers,* **232**, 203–215.

HODGES, K.V. & CROWLEY, P.D. 1985. Error estimation and empirical geothermobarometry for pelitic systems. *American Mineralogist,* **70**, 702–709.

—— & SPEAR, F.S. 1982. Geothermometry, geobarometry and the Al_2SiO_5 triple point at Mt. Moosilauke, New Hampshire, *American Mineralogist,* **67**, 1118–1134.

HOLLAND, T.J.B. 1980. The reaction albite = jadeite + quartz determined experimentally in the range 600–1200° C. *American Mineralogist,* **65**, 129–134.

INDARES, A. & MARTIGNOLE, J. 1985. Biotite-garnet geothermometry in the granulite facies: the influence of Ti and Al in biotite. *American Mineralogist,* **70**, 272–278.

KERRICK, D.M. 1972. Experimental determination of muscovite + quartz stability with $P_{H2O} < P_{tot}$. *American Journal of Science,* **272**, 946–958.

KHON, M.J. & SPEAR, F.S. 1990. Two new geobarometers for garnet amphibolites, with applications to southeastern Vermont. *American Mineralogist,* **75**, 89–96.

KLOOTWIJK, C.T., CONAGHAN, P.J. & POWELL, C.MCA. 1985. The Himalayan arc: large scale continental subduction, oroclinal bending and backarc spreading. *Earth and Planetary Science Letters,* **75**, 167–183.

KOZIOL, A.M. & NEWTON, R.C. 1988. Redetermination of the anorthite breakdown reaction and improvement of the plagioclase-garnet-Al_2SiO_5-quartz geobarometer. *American Mineralogist,* **73**, 216–223.

KRETZ, R. 1983. Symbols for rock-forming minerals. *American Mineralogist,* **68**, 277–279.

KROGH, E.J. 1988. The garnet-clinopyroxene Fe-Mg geothermometer- a reinterpretation of existing experimental data. *Contributions to Mineralogy and Petrology,* **99**, 44–48.

LEAKE, B.E. 1978. Nomenclature of amphiboles. *American Mineralogist,* **63**, 1023–1052.

MADIN, I.P., LAWRENCE, R.D. & REHMAN, S.S. 1989. The northwestern Nanga Parbat–Haramosh massif: evidence for crustal uplift at the northwestern corner of the Indian craton. *Geological Society of America Special Papers,* **232**, 169–182.

MERRILL, R.B., ROBERTSON, J. K. & WYLLIE, P.J. 1970. Melting reactions in the system $NaAlSi_3O_8$-SiO_2-H_2O to 20 kilobars compared with results for other feldspar-quartz-H_2O and rock-H_2O systems. *Journal of Geology,* **93**, 673–689.

NEWTON, R.C. & HASELTON, H.T. 1981. Thermodynamics of the garnet-plagioclase-Al_2SiO_5-quartz geobarometer. *In*: NEWTON, R.C., NAVROTSKY, A. & WOOD, B.J. (eds) *Thermodynamics of Minerals and Melts* Springer-Verlag, New York. 129–145.

—— & PERKINS, D. 1982. Thermodynamic calibration of geobarometers based on the assemblages garnet–plagioclase–orthopyroxene (clinopyroxene)–quartz. *American Mineralogist,* **67**, 203–222.

PÊCHER, A. 1989. The metamorphism in the Central Himalaya. *Journal of Metamorphic Geology,* **7**, 31–41.

PERCHUK, L.L. & LAVRENT'EVA, I.V. 1983. Experimental investigation of exchange equilibria in the system cordierite-garnet-biotite. *In*: SAXENA, S.K. (ed.) *Kinetics and Equilibrium in Mineral Reactions.* Springer-Verlag, New York. 199–239.

PETÖ, P. 1976. An experimental investigation of melting reactions involving muscovite and paragonite in the silica-saturated portion of the system K_2O–Na_2O–Al_2O_3–SiO_2–H_2O to 15 kbar total pressure. *Progress in Experimental Petrology, NERC, London, 3rd report,* 41–45.

POGNANTE, U. 1992. Different P-T-t paths and leucogranite occurrences along the High Himalayan Crystallines: implications for subduction and collision along the northern Indian margin. *Geodinamica Acta,* **6**, 5–17.

—— & BENNA, P. 1993. Metamorphic zonation, migmatization and leucogranites along the Everest transect (Eastern Nepal and Tibet): record of an exhumation history. *This Volume.*

—— & LOMBARDO, B. 1989. Metamorphic evolution of the High Himalayan Crystallines in SE Zanskar, India. *Journal of Metamorphic Geology*, **7**, 9–17.

—— & SPENCER, D.A. 1991. First report of eclogites from the Himalayan belt, Kaghan valley (northern Pakistan). *European Journal of Mineralogy*, **3**, 613–618.

POWELL, R. 1985. Regression diagnostics and robust regression in geothermometer/geobarometer calibration: the garnet-clinopyroxene geothermometer revisited. *Journal of Metamorphic Geology*, **3**, 231–243.

—— & HOLLAND, T.J.B. 1988. An internally consistent thermodynamic dataset with uncertainties and correlations: 3. Applications to geobarometry, worked examples and a computer program. *Journal of Metamorphic Geology*, **6**, 173–204.

REHMAN, S.S. & MAJID, M. 1989. Petrology of a part of the western limb of the Nanga Parbat-Haramosh loop, northern Pakistan. *Geological Bulletin University Peshawar*, **22**, 181–195.

RICHARDSON, S.W., GILBERT, M.C. & BELL, P.M. 1969. Experimental determination of kyanite-andalusite and andalusite-sillimanite equilibria; the aluminium silicate triple point. *American Journal of Science*, **267**, 259–272.

SPEAR, F.S. & SELVERSTONE, J. 1983. Quantitative P-T paths from zoned minerals: theory and tectonic applications. *Contributions to Mineralogy and Petrology*, **83**, 348–357.

SPENCER, D.A., RAMSAY, J.G., SPENCER-CERVATO, C., POGNANTE, U., CHAUDHRY, M.N. & GHAZANFAR, M. 1990. High pressure (eclogite facies) metamorphism in the Indian plate, NW Himalaya, Pakistan. *Geological Bulletin of University Peshawar, Special issue, Proceedings of the II Pakistan Geological Congress*, **23**.

TAHIRKHELI, R.A.K. & JAN, M.Q. 1979. Geology of Kohistan, Karakoram Himalaya, northern Pakistan. *Geological Bulletin of University of Peshawar Special Issue*, **11**, 1–189.

TONARINI, S., VILLA, I.M., OBERLI, F., MEIER, M., SPENCER, D.A., POGNANTE, U. & RAMSAY, J.G. 1993. Eocene age of eclogite metamorphism in Pakistan Himalaya. *Terra Nova*, **5**, 13–20.

TRELOAR, P.J. & REX, D.C. 1990. Cooling and uplift histories of the crystalline thrust stack of the Indian Plate internal zones west of Nanga Parbat, Pakistan Himalaya. *Tectonophysics*, **180**, 323–349.

——, COWARD, M.P., WILLIAMS, M.P. & KHAN, M.A. 1989*a*. Basement-cover imbrication south of the Main Mantle Thrust, North Pakistan. *Geological Society of America Special Papers*, **232**, 167–184.

——, POTTS, G.J., WHEELER, J. & REX, D.C. 1991. Structural evolution and asymmetric uplift of the Nanga Parbat syntaxis, Pakistan Himalaya. *Geologische Rundschau*, **80**, 411–428.

——, REX, D.C., GUISE, P.G., COWARD, M.P., SEARLE, M.P., PETTERSON, M.G., WINDLEY, B.F., JAN, M.Q. & LUFF, I. W. 1989*b*. K-Ar and Ar-Ar geochronology of the Himalayan collision in NW Pakistan: constraints on the timing of collision, deformation, metamorphism and uplift. *Tectonics*, **8**, 881–909.

VERPLANCK, P.L. 1986. *A field and geochemical study of the boundary between the Nanga Parbat–Haramosh massif and the Ladakh arc terrane, northern Pakistan*. MSc Thesis, Oregon State University.

ZANETTIN, B. 1964. Geology and Petrology of Haramosh–Mango Gusor Area. *Scientific Reports, Italian Expeditions to the Karakorum (K^2) and Hindu Kush*. Series 3, Vol. 1. Brill, Leiden.

ZEITLER, P.K. 1985. Cooling history of the NW Himalaya. *Tectonics*, **4**, 127–151.

——, SUTTER, J.F., WILLIAMS, I.S., ZARTMAN, R. & TAHIRKHELI, R.A.K. 1989. Geochronology and temperature history of the Nanga Parbat–Haramosh massif, Pakistan. *Geological Society of America Special Papers*, **232**, 1–22.

The tectonic implications of contrasting granite magmatism between the Kohistan island arc and the Nanga Parbat–Haramosh Massif, Pakistan Himalaya

MARK T. GEORGE, NIGEL B.W. HARRIS & ROBERT W.H. BUTLER[1]

Department of Earth Sciences, Open University, Milton Keynes, MK7 6AA, UK

[1] *Department of Earth Sciences, University of Leeds, Leeds LS2 9JT, UK*

Abstract: In northern Pakistan, an integrated structural and geochemical study across the contact between the Cretaceous–Tertiary Kohistan island arc and Indian crust of the Nanga Parbat–Haramosh massif (NPHM) has identified strongly contrasting source regions contributing to post-50 Ma magmatism, which constrains the tectonic evolution of the region. Nd-model ages of Tertiary intrusives increase dramatically across the contact from *c*. 700 Ma in the Kohistan terrane to *c*. 2500 Ma in the NPHM.

Following collision between the Indian continent and the Kohistan island arc terrane at *c*. 50 Ma, biotite granite sheets (Confluence granites) were emplaced into the Kohistan batholith at 50–30 Ma, followed by muscovite granite sheets (Parri granites) at *c*. 26 Ma. These are geochemically distinct, with the Confluence granites comprising a range of granitic compositions with high Sr and Ba abundances and the Parri granites forming granite sheets enriched in Rb. Undeformed granites in both suites have $(^{87}Sr/^{86}Sr)_i$ in the range 0.7045–0.7054 and $\varepsilon_{Nd}(T)$ of +0.1 to +2.7, suggesting that both groups may be derived from juvenile arc sources. Detailed structural studies verify that the Kohistan granite sheets postdate the initial collision of the Kohistan terrane with the Indian continent, but are pre- or syntectonic with respect to continued underthrusting of the Indian continent below the Kohistan arc. These results, together with isotopic evidence from the undeformed granite sheets, suggest that significant underthrusting of northern Kohistan by the Indian continental crust could not have occurred until after 26 Ma.

Along the western margin of the NPHM, deformed granite sheets show a marked increase in $(^{87}Sr/^{86}Sr)_i$ (0.7075–0.7784), with decreased $\varepsilon_{Nd}(T)$ (−13 to −26). These trends are thought to be due to a combination of sub-solidus fluid infiltration and assimilation of crustal material, with fluids or material derived from the adjacent, isotopically evolved NPHM crust.

The NPHM (Indian continent) has been intruded by a series of tourmaline leucogranite dykes and plutons at 2–12 Ma. These intrusives have trace-element signatures consistent with generation by vapour-absent melting of a pelitic source. Sr-Nd isotope systematics indicate derivation from metasediments such as are currently exposed in the basement rocks of the massif, although melting at the present exposure level is generally precluded by subsolidus metamorphic grades in the country rock. Extreme heterogeneities in radiogenic Sr($(^{87}Sr/^{86}Sr)_i > 0.9$) are observed in leucogranite dykes from the Liachar thrust zone, probably indicating subsolidus fluid infiltration. In general the NPHM leucogranites result from rapid exhumation of metasediments characterised by unusually high heat productivity.

Collision orogenesis commonly involves the stacking of lithospheric levels through large-scale thrusting. Charting the juxtaposition of these different materials during orogenesis is of fundamental importance in understanding how collision belts evolve. Where available, geophysical methods, particularly deep seismic reflection profiling, provide insights on the modern geometry of mountain belts. An alternative line of evidence comes from the chemistry and chronology of syn-orogenic magmatism. The variation of Sr or Nd isotopic ratios in granitoids across terrane boundaries can provide critical information on the nature of the crustal stack at depth (Geist *et al*. 1988).

In this contribution we examine the Tertiary magmatism on the western flank of the Nanga Parbat massif, along the contact between the Kohistan island arc and the Indian subcontinent (Fig. 1). Granites which post-date the Eocene collision between India and Eurasia occur on either side of this zone, and provide contrasting geochemical and isotopic signatures that can be related to the varying source regions available in the two terranes during post collisional tectonics. Consequently, an evaluation of the timing and

From Treloar, P.J. & Searle, M.P. (eds), *Himalayan Tectonics*
Geological Society Special Publication, No. 74, pp. 173–191.

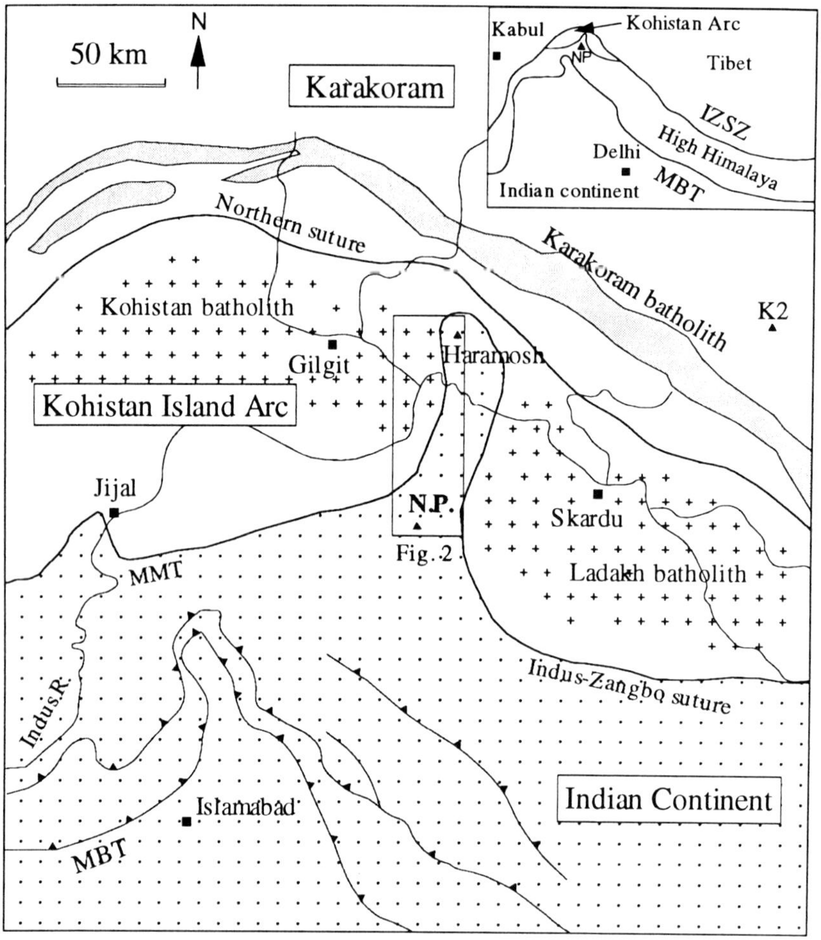

Fig. 1. Sketch map of the NW Himalayas, showing positions of the major tectonic terranes. MMT, Main Mantle Thrust; MBT, Main Boundary Thrust; N.P., Nanga Parbat. Inset shows location of the Kohistan arc in the Himalayan chain (IZSZ, Indus Zangpo suture zone).

geochemistry of magmatism combined with detailed structural studies of the relationship between intrusion and deformation, can be used as evidence in reconstructing the tectonometamorphic evolution of the suture zone.

Regional structure of the Tertiary suture zone

In the Himalayas the principal Tertiary (Indus–Zangbo) suture between the Indian subcontinent in the south and Eurasia in the north appears readily traceable as a major tectonic break (Fig. 1), decorated by small remnants of ophiolitic material (e.g. Le Fort 1989). Plate-tectonic reconstructions of the opening of the Indian Ocean and palaeomagnetic data led Patriat & Achache (1984) to conclude that collision occurred between *c.* 55 and 52 Ma. Besse *et al.* (1984) conclude from palaeomagnetic data in the eastern Himalayas that suturing occurred at 53–47 Ma, whereas in the western Himalayas collision has been dated by a change in sedimentation from marine to continental environments in early–mid-Eocene times between 44 and 55 Ma (Le Fort 1989).

The tectonic evolution of the western Himalayas is complicated by the presence of the Cretaceous–Tertiary Kohistan–Ladakh island arc terrane (Tahirkheli *et al.* 1979; Honegger *et al.* 1982), thought to have sutured to Eurasia (Karakoram block) in the late Cretaceous (Klootwijk *et al.* 1979; Pudsey 1986; Treloar *et al.* 1989). The northern part of the arc is dominated by a range of calcalkaline intrusives dated at *c.* 100–40 Ma, related to the northward subduction of Tethyan oceanic lithosphere and representing the westward continuation of the Trans-Himalayan batholith (Petterson & Windley 1985; Honegger *et al.* 1982). Along the southern margin of the Kohistan arc, the Indus–Zangbo suture zone has been traced as the Main Mantle Thrust (MMT). This suture, which carries peridotites and high-pressure granulites of the Kohistan island arc terrane in its hangingwall, was defined by Tahirkheli *et al.* (1979) along the Indus valley in the Hazara area, near the town of Jijal (Fig. 1). The adjacent footwall rocks derived from the Indian continent have been stacked up into a post-collisional thrust belt (Coward *et al.* 1988). These metamorphic rocks had cooled through the blocking temperature for Ar in hornblende (*c.* 500° C) by 39 Ma (Treloar *et al.* 1989). These and other isotopic data are consistent with suturing between the Kohistan terrane and the Indian continent at Hazara, with displacements occurring on the MMT, at *c.* 50 Ma.

Since the pioneering work of Tahirkheli and colleagues (Tahirkheli & Jan 1979) the MMT has been traced to the NE, from its type area at Hazara, to define the western flank of the NPHM. This represents the northernmost outcrops of Indian continental rocks. The eastern boundary of the massif can be traced in Ladakh to link with the main trace of the Indus–Zangbo suture (Fig. 1; Tahirkheli *et al.* 1979). The northern contact between the Nanga Parbat–Haramosh Massif (NPHM) and the Kohistan terrane has now been identified (Butler *et al.* 1992) thus linking the map traces of the MMT with the Indus–Zangbo suture. On both sides of the massif the primary contact is a high strain zone of ductile deformation preserving structures that developed under amphibolite-facies conditions (Treloar *et al.* 1991). It is tempting to link these ductile structures in time to those at the MMT at Hazara. There are however, strong indications that this temporal correlation is not appropriate.

Butler & Prior (1988*a*) studied the kinematics of the contact between the Kohistan arc and the NPHM in the Sassi area (Fig. 2). They found evidence for an early, penetrative deformation marked by top-to-the-SSE thrusting under amphibolite facies conditions. As this is consistent with the relative motions across the suture predicted by plate reconstructions, they inferred this ductile structure to be the MMT. However, Butler & Prior (1988*a*) recognize that, as upper arc rocks are juxtaposed against the Indian continent, this segment of the 'MMT' cannot be the primary suture, but instead may be a form of break-back thrust. Regardless of the nature of the early ductile contact, it has certainly been modified by later structures developed under decreasing metamorphic conditions. Detailed structual mapping on the western margin of the massif at Sassi has indicated that younger movements are dominated by dextral-transtension under greenschist facies conditions, suggesting uplift of the NPHM relative to the Kohistan arc (see also Butler *et al.* 1989). The present structure of the contact between the Kohistan arc and the NPHM is thus very different to the type-section of the MMT seen in the Hazara area.

Further south in the Nanga Parbat area, the Liachar shear zone (Fig. 2) carries rocks of the NPHM northwestwards, back over the Kohistan arc. This recently active shear zone operated from amphibolite-facies conditions to the present surface, and probably accommodated substantial amounts of uplift of the NPHM (Butler & Prior 1988*b*).

Granite sheets of the Kohistan arc

Field relations and petrography

Late stage magmatism within the Kohistan batholith has been studied near to the eastern edge of the Kohistan arc (see Fig. 2), where swarms of granite sheets, each 0.5–10 m thick, locally make up to 30% of exposure. Away from the contact between the Kohistan arc and the NPHM, the sheets are undeformed and sharply cross-cut the country rocks, which consist of semi-pelites and metagabbros at epidote-amphibolite grade. In detail, igneous relations are complex and at least three generations of granite sheets can be observed. On the basis of mineralogy and geochemistry, Petterson & Windley (1985) divided the granites into the 'Indus Confluence Acid Sheets' and the 'Parri Acid Sheets', the former having biotite as the dominant mica, whereas for Parri granites muscovite predominates. This study has enlarged the geochemical and isotopic database of the granites by sampling a greater number of different sheets, particularly necessary for the Parri granites which had, until now, only been studied at one locality.

The Confluence granites constitute about 70% of the granite exposure in the studied area. They form medium grained, two-mica granites occurring both as dykes and more massive plugs up to 50 m thick. The mineral assemblage consists of quartz (25–50%), plagioclase (10–30%), alkali feldspar (20–40%), biotite (2–20%), ± muscovite (<5%), ± epidote or clinozoisite

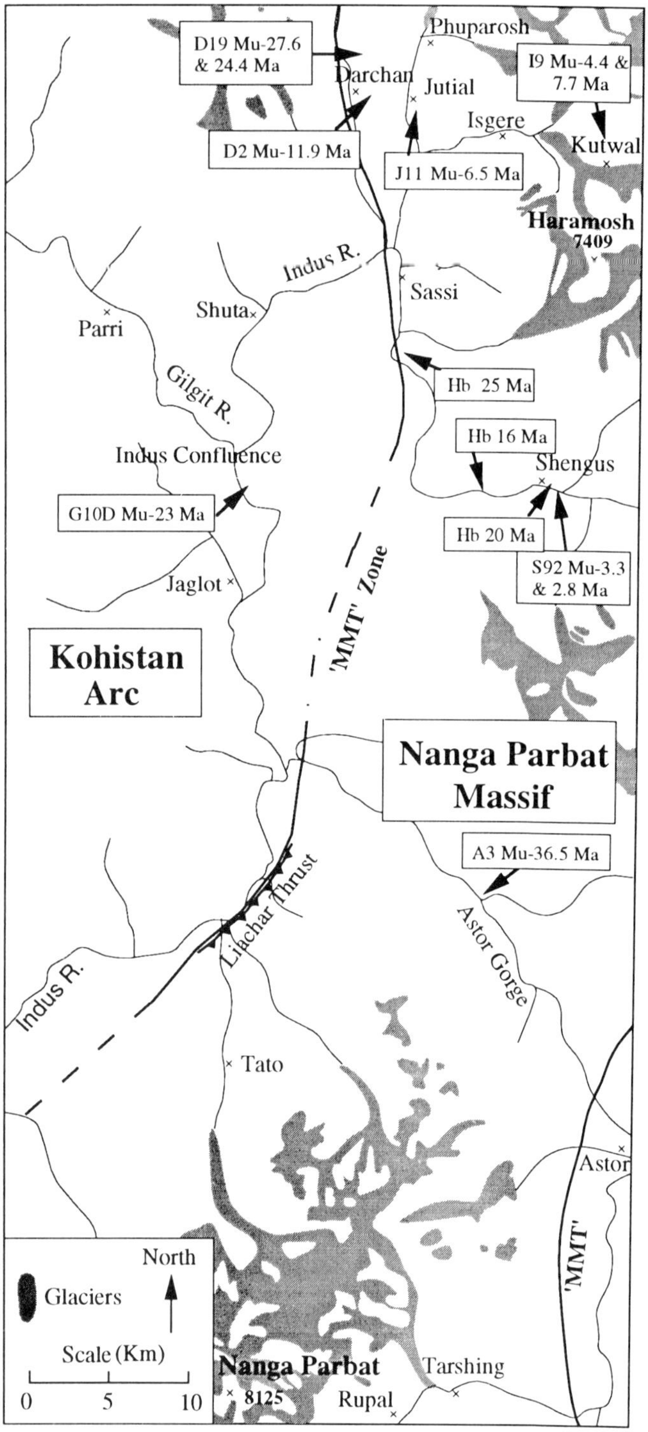

Fig. 2. Location map of the western margin of the Nanga Parbat–Haramosh Massif, showing sample localities for Rb-Sr muscovite (Mu) (Table 3) and ^{40}Ar-^{39}Ar hornblende (Hb) dating (Zeitler *et al.* 1989).

(1–8%), sometimes enclosed by biotite. Myrmekitic textures are common. Accessory minerals include apatite, magnetite and occasionally sphene, with rare garnet and zircon.

The Parri granites are mineralogically more variable, and tend to cross-cut the Confluence granites, whilst both types are cross-cut by coarse pegmatites. The diagnostic mineral assemblage of the Parri granites is quartz (20–50%), plagioclase (10–30%), alkali feldspar (25–50%), muscovite (6–20%), ± biotite (1–4%) ± garnet (1–4%), with subordinate and similar accessory mineral contents to the Confluence granites. The Parri granites are leucogranites, but to avoid confusion with those granites intruding the NPHM, are referred to by their type locality.

Major and trace element geochemistry

The main geochemical features of the Confluence and Parri sheets have been detailed by Petterson (1984) and are summarized below. The Confluence granites are peraluminous with SiO_2 varying between 64–75%: antipathetic trends between Mg, Fe, Ti, Mn and Ca and silica can be linked to fractionation of micas and feldspars during crystallization. The Parri granites are also peraluminous but are uniformly siliceous ($SiO_2 > 71\%$), although Petterson (1984) noted some fractionation trends within a single intrusion.

Average trace-element abundances (Table 1) are plotted for both suites and are found to be distinct (Fig. 3a). The Confluence granites have low Rb and high Ba compared to the Parri granites. High-field-strength-element (HFSE) abundances are similar, although the Confluence granites display a strong depletion in Ta and Nb and are more depleted in Y and HREE. The two suites of granite sheets can be most easily distinguished on a Rb versus Sr plot (Fig. 4), where Confluence granites are characterised by low Rb/Sr (0.01–0.4) compared to Parri granites that show higher, and more variable Rb/Sr ratios (generally 1–2).

Table 1. *Average whole-rock major and trace element abundances*

	Confluence granites $n = 24$	Parri granites $n = 14$	Leucogranite Dykes $n = 14$	Jutial granite $n = 11$
wt%				
SiO_2	72.17	74.02	74.14	73.63
TiO_2	0.17	0.06	0.05	0.10
Al_2O_3	15.69	15.14	14.88	14.79
Fe_2O_3	1.24	0.60	0.73	1.00
MnO	0.02	0.04	0.03	0.01
MgO	0.37	0.11	0.10	0.14
CaO	2.55	1.44	0.94	1.20
Na_2O	4.55	4.34	4.16	3.54
K_2O	2.92	4.03	4.43	5.16
P_2O_5	0.07	0.03	0.07	0.05
LOI	0.43	0.56	0.55	0.63
Total	100.15	100.36	100.07	100.26
ppm				
Rb	67	184	319	294
Sr	440	132	46	83
Y	5	11	30	12
Zr	84	40	38	47
Nb	5	11	13	11
Ba	815	224	121	221
Pb	29	68	58	76
Th	6	5	14	22
U	2	5	13	20
Ta	0.17	2.4	–	1.3
Hf	3.1	2.9	–	1.7
Ce	46.9	40.9	–	55.4
Sm	2.3	3.0	–	4.0
Yb	0.3	1.0	–	0.9

Ta, Hf, Ce, Sm and Yb were analysed by instrumental neutron activation; all other elements by XRF. n is number of analyses.

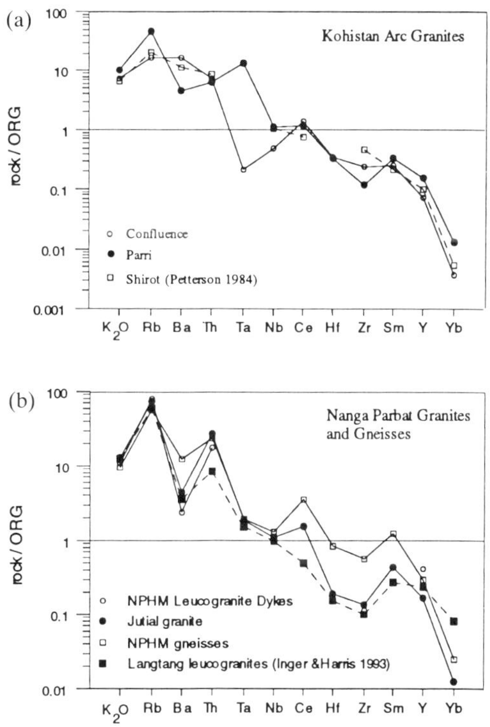

Fig. 3. Variations in trace-element abundances, normalized against Ocean Ridge Granite (Pearce *et al.* 1984) for (**a**) Kohistan granites and (**b**) Leucogranites and gneisses of the Nanga Parbat–Haramosh massif.

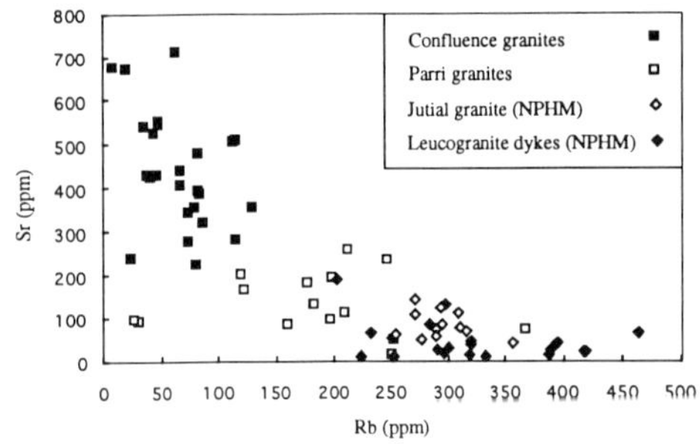

Fig. 4. Covariation between Rb and Sr abundances observed in granites in the Kohistan arc and the Nanga Parbat–Haramosh Massif.

The Confluence granites have geochemical compositions broadly similar to relatively large, evolved plutons, such as the main Shirot granitoid, which intruded the Kohistan batholith between 40 and 54 Ma (Petterson & Windley 1985). The Shirot granitoid shows a wide range in geochemical composition, and was probably generated by the fractionation of a magma derived from a variably-metasomatised mantle source (Petterson & Windley 1991). Detailed petrogenetic modelling is beyond the scope of this study, but fractionation trends suggest that the Confluence granites may share a similar petrogenesis to plutons such as the Shirot granitoid. Alternatively, as suggested by Petterson & Windley (1991), the Confluence granites may have been generated by partial melting of juvenile, possibly crystalline, arc crust. If this is the case, the high Ba (up to 2000 ppm) of the Confluence granites would imply that temperatures in the source region were sufficiently high to exhaust biotite, whilst, following the reasoning of Zen & Hammarstrom (1984), the presence of magmatic epidote would suggest that pressures in the source region were also moderately high. However, these geochemical features are generally shared by the Shirot granitoid, and there is no definitive way of distinguishing between the fractionation of a mantle-derived magma and the partial melting of a lower arc

Table 2. Whole-rock Sr-isotopic data for the Kohistan granites

Sample	Location	Rb/Sr	$^{87}Rb/^{86}Sr$	$^{87}Sr/^{86}Sr$	2σ	$(^{87}Sr/^{86}Sr)_{30}$
Confluence Granites						
G10-A[1]	Jaglot	0.21	0.62	0.70535	2	0.70509
G10-D[1]	Jaglot	0.19	0.55	0.70521	2	0.70498
G10-E[1]	Jaglot	0.08	0.24	0.70501	2	0.70491
G10-F[1]	Jaglot	0.21	0.61	0.70512	6	0.70486
G10-G[1]	Jaglot	0.17	0.48	0.70511	5	0.70491
G10-H[1]	Jaglot	0.09	0.27	0.70501	2	0.70503
S95	Darchan	0.23	0.65	0.70488	3	0.70460
S6G	Shuta	0.22	0.64	0.70480	2	0.70453
S81*	Sassi	0.27	0.77	0.70786	2	0.70753
D17*	Darchan	0.09	0.26	0.70812	2	0.70801
S67*	Sassi	0.06	0.19	0.70846	2	0.70838
S75*[2]	Sassi	0.01	0.03	0.70969	2	0.70967
S76*[2]	Sassi	0.03	0.08	0.70820	2	0.70816
Parri Granites						
G2	Parri	2.03	5.87	0.70752	2	0.70502
G3	Confluence	19.31	55.80	0.72832	2	0.70455
G14	Shuta	0.33	0.96	0.70581	2	0.70540
G23	Shuta	5.04	14.57	0.71076	2	0.70456
S97*	Darchan	0.82	2.38	0.73546	2	0.73445
G43*	Sassi	1.86	5.37	0.76032	2	0.75803
G35*[3]	Sassi	1.02	2.95	0.72505	2	0.72379
G36*[3]	Sassi	0.74	2.14	0.70838	9	0.70747
G39*	Sassi	1.92	5.55	0.78072	4	0.77836
G48*	Sassi	0.98	2.82	0.75640	5	0.75520
S66*	Sassi	1.89	5.45	0.75842	5	0.75610
D5*	Darchan	0.60	1.74	0.72350	2	0.72276
G30*	Sassi	0.37	1.06	0.74051	2	0.74006

Rb and Sr data from XRF. $^{87}Sr/^{86}Sr$ ratio determinations were made on a Finnigan MAT 261 multi-collector
$^{87}Sr/^{86}Sr_i$ calculated for an age of 30 Ma.
* Deformed granites.
[1,2,3] Samples from the same granite sheets.

source that has been recently derived from the mantle.

The high Rb/Sr ratios of the Parri granites (Fig. 4) suggest either extreme fractionation or small-degree partial melting of a mica-rich source (Harris & Inger 1992). It is probable that the Parri granites were derived at shallower, cooler crustal levels, such that mica was retained in the protolith, thus limiting the Ba abundance in the melt. Although the metasediments exposed in the Gilgit valley would be a potential source for the Parri granites, their epidote-amphibolite grade precludes partial melting of these rocks at the present exposure level.

Age of intrusion

Field evidence shows that the Parri granites are younger than the Confluence granites. Petterson & Windley (1985) obtained poorly constrained, whole-rock errorchrons for a Confluence and a Parri granite yielding ages of 34 ± 14 Ma (MSWD = 0.74, $(^{87}Sr/^{86}Sr)_i$ = 0.7045) and 29 ± 8 Ma (MSWD = 0.2, $(^{87}Sr/^{86}Sr)_i$ = 0.7052)

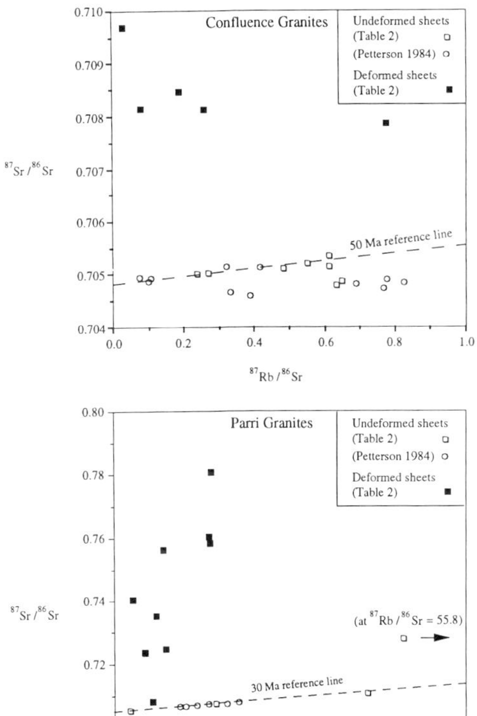

Fig. 5. Rb-Sr isochron diagram for granite sheets from the Kohistan Arc. (**a**) Confluence granites; (**b**) Parri granites.

respectively. Whole-rock data from undeformed and deformed granites together (this study, Table 2, Fig. 5) do not define an isochron for either Confluence or Parri suites. However, the data from the four undeformed Parri granites, collected from widely separated localities, do define an errorchron yielding an age of 26.8 ± 2.6 Ma (MSWD = 15, $(^{87}Sr/^{86}Sr)_i$ = 0.7054). These data can be combined with the seven isotopic ratios from the Parri granite analysed by Petterson (1984), which yields an improved errorchron indicating an age of 26.2 ± 1.2 Ma (MSWD = 5.9, $(^{87}Sr/^{86}Sr)_i$ = 0.7054).

Petterson (1984) was unable to date a granite exposed 3 km south of the Indus Confluence (the Sainallah leucogranite) due to the restricted range in Rb/Sr ratios for this granite. Sr-isotope data from an adjacent sheet at Jaglot (this study) can be combined with the Sainallah data to define a ten-point errorchron with an age of 49.1 ± 11 Ma (MSWD = 11; $(^{87}Sr/^{86}Sr)_i$ = 0.70485 ± 4). Whilst the age has significant uncertainties, it does suggest that proximal Confluence granites may be coeval. On the basis of present whole-rock data we can conclude that the Confluence sheets were emplaced at 50–30 Ma, whereas the Parri sheets were intruded at *c.* 26 Ma.

The time of granite emplacement is further constrained by a range of cooling ages from granites and metamorphic rocks in the arc (Treloar *et al.* 1989; Zeitler *et al.* 1989). K-Ar and ^{40}Ar-^{39}Ar studies indicate closure for Ar at 26–38 Ma for hornblende, 18–36 Ma for biotite and 18–21 Ma for muscovite. A muscovite Rb-Sr age from the Jaglot granite of 23 ± 1 Ma (Table 3) places a lower limit on the time of granite emplacement, and the coincidence of ages in the granites and their country rocks indicates that the granites were intruded into relatively warm country rocks which had not yet cooled below the blocking temperature (500–550° C) for Sr in muscovite (Purdy & Jäger 1976; Cliff 1985).

Sr and Nd isotopes systematics

Sr-isotope data from 13 Confluence granite samples, representing 7 separate sheets, and 13 Parri granite samples, from 12 separate sheets, are shown in Table 2. The majority of the undeformed samples were collected from the Gilgit valley–Indus confluence area, whilst all the deformed samples come from within the ductile shear zone bordering the NPHM (MMT of Butler & Prior 1988*a*). The isotopic composition of each suite is variable, with $(^{87}Sr/^{86}Sr)_{30}$ in the range 0.704–0.710 for the Confluence granites and 0.705–0.778 for the Parri granites.

Table 3. Sr-isotopic data of muscovites

Sample and lithology	Rb (ppm)	Sr (ppm)	Rb/Sr	$^{87}Rb/^{86}Sr$	$^{87}Sr/^{86}Sr$	2σ	Age (Ma)
Kohistan Arc							
Jaglot Granite							
G10-D WR	92	482	0.19	0.55	0.70521	2	
G10-D-M	250	106	2.35	6.79	0.70725	2	23.0 ± 1.0
NPHM							
Jutial Granite							
J11 WR	309	110	2.81	8.12	0.88438	3	
J11-M	603	9.43	64.0	184.9	0.90010	6	6.5 ± 0.04
Leucogranite dyke							
S92 WR	389	25	15.6	44.97	0.83015	2	
S92-M	865	3.14	275	796.5	0.86564	6	3.3 ± 0.02
S92-M-dup	928	3.37	275	795.2	0.86019	5	2.8 ± 0.02
Orthogneiss							
D2 WR	227	81.5	2.79	8.05	0.96392	3	
D2-M	409	4.56	90.0	260.0	1.00792	9	11.9 ± 0.08
Paragneiss							
19 WR	109	151	0.72	2.08	0.76613	4	
19-M	249	17.0	14.6	42.26	0.76868	7	4.4 ± 0.78
19-M-dup	311	21.9	14.2	41.10	0.77043	3	7.7 ± 0.08
Paragneiss							
A3 WR	375	77.4	4.84	14.00	0.83808	2	
A3-M	967	23.3	41.4	119.7	0.89377	12	36.3 ± 0.3
Paragneiss							
D19 WR	175	181	0.97	2.79	0.77988	2	
D19-M	344	86.7	3.98	11.49	0.78333	3	27.6 ± 0.3
D19-M dup	356	85.3	4.17	12.05	0.78313	11	24.4 ± 0.5

Rb and Sr concentrations by I.D, using a VG Isomass 54E single-collector mass spectrometer at the Open University. All errors on the ages calculated using $^{87}Rb/^{86}Sr \pm 2\%$. WR, whole-rock; M, muscovite; dup, duplicate dissolution and analysis.

The lack of isotopic homogenization within whole-rock samples of peraluminous granites is well documented and is generally attributed either to isotopically distinct source regions, coupled with lack of homogenisation in viscous intrusions (Deniel *et al.* 1987) or to subsolidus fluid infiltration (Cliff *et al.* 1991). It is clear from Fig. 5 that the deformed Confluence and Parri granites show an enrichment in radiogenic Sr compared with their undeformed counterparts outside the shear zone. For the Parri granites for example, undeformed samples are characterized by $(^{87}Sr/^{86}Sr)_{30} < 0.705$, but this ratio rises up to 0.778 in deformed sheets near Sassi. Individual Parri and Confluence granite sheets in the shear zone also show marked isotopic disequilibrium on the whole-rock scale. This is demonstrated by samples such as S75 and S76, which were collected about 20 m apart from separate, 2 m thick boudins of the same sheet.

Undeformed Parri granites from widely separated localities define an isochron with $(^{87}Sr/^{86}Sr)_i = 0.7054$, as described above, showing that these granite sheets are coeval. Whole-rock samples of undeformed Confluence granites show slight isotopic heterogeneities, both between granite sheets and within the same intrusion (e.g. G10). Such fine-scale heterogeneities may reflect processes occurring just prior to crystallization, such as wall-rock contamination or subsolidus fluid interaction. The Confluence granites yield $(^{87}Sr/^{86}Sr)_i$ of 0.7045 and 0.7048. The initial-Sr isotopic ratios of the undeformed Confluence and Parri granites are low compared with Miocene leucogranites from the Central Himalayas, derived from metapelites (Deniel *et al.* 1987; Inger & Harris 1993), indicating less radiogenic source regions for the former.

Further constraints on the nature of the source of the Kohistan granites are provided by Nd isotopic data obtained from two samples from the Jaglot sheet (a typical Confluence granite) and five Parri granites (Table 4). The Jaglot sheet has $\varepsilon_{Nd}(T) \approx +1.0$, and two undeformed Parri granites have $\varepsilon_{Nd}(T) \approx +1.0$ and +2.7. In contrast, three deformed Parri granites from the western margin of the NPHM have $\varepsilon_{Nd}(T) \approx -13$ to -26. This depletion in $^{143}Nd/^{144}Nd$ for granites within the shear zone is clearly coupled with an enrichment in $^{87}Sr/^{86}Sr$, as can be seen

Table 4. Whole-rock Nd-isotope data for the Kohistan Arc and the NPHM

	Sm/Nd	$^{147}Sm/^{144}Nd$	$^{143}Nd/^{144}Nd$	2σ	$^{143}Nd/^{144}Nd_i$	$\varepsilon_{Sr}(T)$	$\varepsilon_{Nd}(T)$	T_{DM}(Ga)
Kohistan								
Confluence								
G10-D Jaglot	0.145	0.087	0.51267	1	0.51265	4.5	1.0	0.7
G10-H Jaglot	0.147	0.089	0.51266	1	0.51265	3.3	0.9	0.7
Parri								
G2 Parri	0.207	0.124	0.51276	3	0.51274	5.1	2.7	0.5
G3 Parri	n/d	–	0.51263	4	0.51261	−1.8	0.1	0.7
G30* Sassi	0.171	0.103	0.51190	2	0.51188	505	−14	1.7
G39* Sassi	n/d	–	0.51127	4	0.51125	1045	−26	2.6
G43* Sassi	0.211	0.127	0.51196	2	0.51193	757	−13	1.7
NPHM								
Jutial granite								
J19	0.094	0.057	0.51132	2	0.51131	2496	−25.7	2.5
J1G	0.153	0.092	0.51130	1	0.51129	2607	−26.1	2.6
J1i	0.137	0.083	0.51127	2	0.51126	2568	−26.6	2.6
J1B	0.127	0.077	0.51129	1	0.51128	2609	−26.3	2.6
J7 dyke	n/d	–	0.51138	2	0.51138	1730	−24.3	2.4[1]
Gneisses								
I6 Iskere	0.224	0.135	0.51162	1	0.51161	1902	−19.8	2.1
A13 Rupal	0.184	0.111	0.51130	1	0.51129	2498	−26.1	2.6
L3 Liachar	n/d	–	0.51117	2	0.51117	1153	−28.4	2.7[1]
I9 Iskere	n/d	–	0.51100	3	0.51100	868	−31.8	3.0[1]
J14 Jutial	n/d	–	0.51125	2	0.51125	544	−26.8	2.6[1]

[1] Average crust $^{147}Sm/^{144}Nd$ value used
$^{143}Nd/^{144}Nd$ ratios normalized to $^{146}Nd/^{144}Nd = 0.7219$
ε values relative to CHUR at time T($T = 30$ Ma for Kohistan & $T = 10$ Ma for NPHM samples).
T_{DM} relative to depleted mantle from DePaolo (1981).
* Deformed granites.

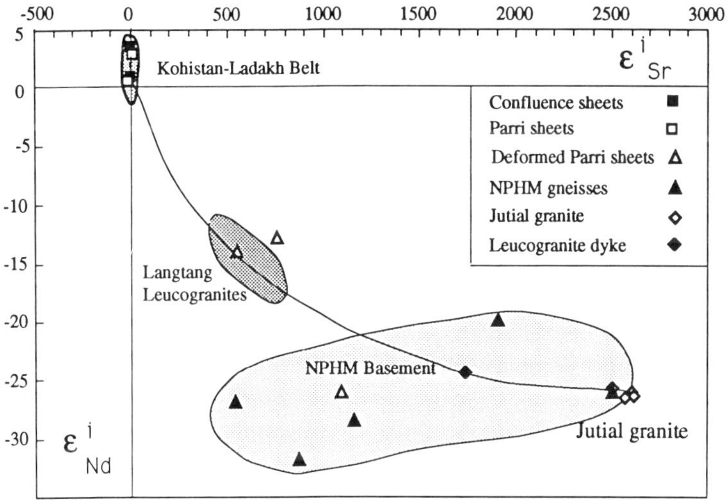

Fig. 6. $\varepsilon_{Nd}(T)$–$\varepsilon_{Sr}(T)$ plot for samples of granites and gneisses from Kohistan and the NPHM. $T = 30$ Ma for Kohistan granites and $T = 10$ Ma for NPHM granites and gneisses. Fields plotted using data for Langtang from Inger & Harris (1993), for Ladakh from Allègre & Ben Othman (1980) and for Kohistan from Petterson *et al.* (1993). Curved line is a mixing line between undeformed Parri granites and leucogranites derived from the NPHM basement at 30 Ma.

from an $\varepsilon_{Nd}(T)$–$\varepsilon_{Sr}(T)$ plot (Fig. 6), in which these data are expressed relative to the isotope ratios of bulk earth at the time of formation. For the sake of comparison we have assumed formation at 30 Ma for this plot. The undeformed Confluence and Parri granites plot within the field defined by arc-derived magmas from Kohistan and Ladakh. Both Nd and Sr data are consistent with a juvenile source for the undeformed Kohistan granites, similar to the source of the Kohistan–Ladakh batholith.

The age of igneous protoliths can be constrained by Nd model ages (Table 4) that represent the time passed since a sample, or its crustal protolith, was extracted from a mantle reservoir. The underlying assumption for such ages is that Sm/Nd has not fractionated during the crustal residency of the source regions. The interpretation of such ages has been widely discussed in the literature (e.g. Arndt & Goldstein 1987; Harris *et al.* 1988; Cliff *et al.* 1991) but for this study the low Sm/Nd ratios of many granites suggest that accessory phase behaviour has resulted in depletion of Sm/Nd during crustal anatexis. In order to circumvent problems of intracrustal Sm/Nd fractionation, Nd model ages have been calculated by assuming an average crustal value for Sm/Nd (0.19) between the time of extraction from a depleted mantle and crustal anatexis. For undeformed granites of both types, Nd model ages lie in the range 500–700 Ma (Table 4; Petterson *et al.* 1993). They are similar to the Nd model ages obtained from the large plutons of the Kohistan-Ladakh batholith, which lie in the range 300–800 Ma (Allègre & Ben Othman 1980; Petterson *et al.* 1992). All of these intrusives have model ages which are considerably older than their age of emplacement, indicating that limited continental material has contributed to their protolith. This may reflect the subduction of continental-derived sediments during the evolution of the island arc.

The cause of the enrichment in $^{87}Sr/^{86}Sr$ observed in those Confluence and Parri granites which have been deformed along the western margin of the NPHM is equivocal. One explanation for such a marked disturbance to the Sr isotopic system, coupled with the absence of any consistent trace element variation between undeformed and deformed granites, is the infiltration of fluids through the shear zone. This interpretation is supported, firstly by isotopic heterogeneites within individual deformed sheets (Table 2), and secondly by analyses of amphibolite lithologies within the shear zone, which are also characterized by similarly high initial-Sr isotope ratios ($(^{87}Sr/^{86}Sr)_{30} = 0.74$–$0.76$), suggesting that the wallrocks to the granites may have also been affected by fluids. The enrichment trend is less marked in the Confluence sheets, perhaps because higher elemental abundances of Sr in these sheets makes them less sensitive to isotopic shifts induced by fluid infiltration. The elevation of $(^{87}Sr/^{86}Sr)_{30}$ in the deformed granites indicates that if fluids were involved, they must have been relatively radiogenic, suggesting a derivation from the Indian crust. However, significant mobility of REE by fluid infiltration is unusual, and consequently the depleted, crustal Nd signatures seen in the deformed Parri granites are unlikely to have been caused by subsolidus fluids. Furthermore, REE data available for two of the deformed samples contain no geochemical anomalies which might be expected if the REE had been mobilized by fluids.

A second explanation for the evolved isotopic signatures of the deformed granite sheets, is that sources in both the arc and the NPHM have contributed to the melts. In this case, data from such magmas should plot on a mixing line between endmember compositions from each source. It can be seen from Fig. 6 that only one of the deformed sheets (G30) lies close to such a line, if we assume that the NPHM source was similar to that of the Jutial granite. The wide variation in the Sr and Nd isotopic ratios from the three deformed sheets may only be the result of magma mixing if the contributing melts from the NPHM were characterized by highly variable isotopic characteristics. However, there is currently no evidence to suggest that melting in the arc and in the NPHM was synchronous. This, coupled with the close association between deformation and isotopic ratios, makes a mechanism involving magma mixing unlikely.

A third explanation is that fragments of material from the Indian plate have been assimilated by melts derived from the arc, within the shear zone. Given the radiogenic Sr and Nd isotopic composition of the NPHM crust (Table 5), assimilation of such material might have caused the isotopic characteristics of the deformed granites. A study of the garnets from two deformed Parri granites (G43 and G30) provides some evidence for wallrock contamination during emplacement of the magma. Garnets in these granites have similar morphology and geochemical composition to garnets in adjacent amphibolite-grade metasedimentary rocks, suggesting that they are xenocrysts. The grains are highly fractured and corroded, and contain quartz, mica and sillimanite inclusions. Electron microprobe studies show that these garnets have the average composition $Alm_{74}Py_{18}Gr_6Sp_2$. In contrast, garnets within undeformed Parri granites are generally small, undeformed, euhedral and inclusion-free grains. These garnets have relatively high spessartine contents ($Alm_{62}Py_2Gr_6Sp_{30}$), indicative of an igneous paragenesis (e.g. Green 1977; Harris *et al.* 1992). Given the

Table 5. Whole-rock Sr-isotopic data for NPHM gneisses

Sample	Location	Rb/Sr	$^{87}Rb/^{86}Sr$	$^{87}Sr/^{86}Sr$	2σ	$(^{87}Sr/^{86}Sr)_{10}$
I6	Iskere	1.63	4.71	0.83941	2	0.83874
J6	Phuparosh	0.60	1.75	0.73173	3	0.73148
J13	Jutial	2.13	6.15	0.87565	3	0.87478
J14	Jutial	5.16	14.93	0.87851	7	0.87639
J21	Jutial	1.30	3.74	0.80397	6	0.80344
T25	Tato track	1.79	5.16	0.83801	3	0.83728
I9	Haramosh	0.72	2.07	0.76613	4	0.76584
I10	Haramosh	0.52	1.52	0.76878	9	0.76857
D2	Darchan	2.79	8.05	0.96392	3	0.96278
D19	Darchan	0.97	2.79	0.77989	2	0.77950
G31	Sassi	1.12	3.24	0.75943	2	0.75897
S91	Shengus	1.85	5.35	0.76962	2	0.76887
A3	Astor gorge	4.84	14.00	0.83808	2	0.83609
A7	Tarshing	0.55	1.60	0.71708	2	0.71685
A13	Rupal	2.91	8.40	0.88193	2	0.88074
L3	Liachar	0.98	2.83	0.78633	2	0.78593
M10	Tato	6.64	19.19	0.97311	2	0.97038
M3-L	Tato	0.78	2.26	0.8438	2	0.84347
M3-R	Tato	1.42	4.10	0.85084	3	0.85026

Details as for Table 2. $^{87}Sr/^{86}Sr_i$ calculated for an age of 10 Ma.

high REE contents of garnets, it is probable that inclusion of garnet xenocrysts into some granites has strongly altered the whole-rock Nd isotopic compositions of these samples, although confirmation of this theory would require an analysis of the Nd isotopic compositions of the garnets themselves. Assimilation of Indian crust by arc-derived granites also suggests that at the time of intrusion, considerable imbrication between arc and continental lithologies had already occurred along the western margin of the NPHM. However, it is unlikely that the shift in Nd isotopes results entirely from the assimilation of garnet, since another deformed Parri granite (G39) shows an even greater depletion in radiogenic Nd($\varepsilon_{Nd}(T) \approx -27$) but contains only trace amounts of garnet. It is probable that other phases and possibly other processes, such as fluid transport, were involved.

Despite the limited Nd data-set, the correlation between the enrichement in radiogenic Sr and the depletion in radiogenic Nd for deformed Parri granites from the shear zone indicates the involvement of ancient crust with an evolved isotopic signature. This may have been incorporated into the melt geochemistry through fluid infiltration, magma hybridisation or crustal assimilation, or indeed a combination of these processes.

Field relationships at Sassi

In the Indus gorge at Sassi (Fig. 2), the contact between the Kohistan arc and the NPHM is marked by a 2.5 km wide, steeply inclined ductile shear zone containing intercalated lithologies typical of the upper part of the Kohistan island arc (metabasic volcanics and intrusives) together with marbles, para and granitic orthogneisses thought to be derived from the Indian continent (Butler & Prior 1988*a*). Butler *et al.* (1992) mapped this zone of intercalation (their 'layered unit') northwards around the antiform defining the northern termination of the Nanga Parbat massif.

Both Confluence and Parri granites can be traced eastwards along the Indus gorge into the shear zone at Sassi where they can be discriminated in the field on the basis of mineralogy. This distinction is confirmed by trace element analyses on collected samples. Butler & Prior (1988*a*) recognized that in this zone, Kohistan arc granites become sheared and that the kinematics of the shearing fabrics that deform the sheets are top-to-the SSE. The ductile fabrics are at amphibolite facies and are found both within and outside the granites. They concluded that the age of the arc granites is pre to syn-peak shearing along the 'MMT', a conclusion supported by the apparent restriction of these sheets to Kohistan-derived units.

The majority of Kohistan granites in the shear zone are highly deformed and concordant with the banding in the rocks. The deformation fabrics within the granites vary from an intense penetrative foliation to a more poorly defined foliation with recrystallized quartz around large alkali feldspar porphyroblasts. Detailed work associated with this study has found rare localities where Confluence-type granites cross-cut deformation fabrics, and intrude some of the intercalations of Indian continental material. Since the geochemical and isotopic evidence suggests that the Confluence granites are a relatively coherent group with presumably a restricted intrusion age, these relationships clearly indicate that the arc-derived granites post-date initial collision of the arc with the Indian continent. Furthermore, the granites were emplaced into an actively deforming shear zone, which resulted in the subsequent deformation of many granite sheets. These observations have important tectonic implications, discussed further below. The apparent lack of any significant penetration of the arc granites into the Indian plate may be due to the thermal contrast between the two terranes at the time of granite intrusion, or to continued displacements in the shear zone.

Nanga Parbat–Haramosh Massif leucogranites

Field relations and petrography

Dykes and small plutons of leucogranite have a widespread but scattered distribution across the NPHM. The dykes are heterogeneous and coarse grained, and generally 1–3 m wide. They generally have sharp margins, and individual dykes can commonly be traced for 10–100 m. Whilst the majority of leucogranites cross-cut the regional metamorphic fabrics of the NPHM, locally they are deformed by both cataclastic and ductile shears. Deformation of leucogranites is particularly marked in the recently active Liachar shear zone (Butler & Prior 1988*b*).

Leucogranite dykes were sampled in the Nanga Parbat area at Rupal, Tato and Liachar and in the Haramosh area from the Indus gorge northwards into the Iskere and Phuparosh valleys (Fig. 2). The leucogranites are composed of quartz (15–50%), plagioclase (10–40%), alkali-feldspar (25–45%), muscovite (5–15%), ± biotite (<4%), ± tourmaline (<10%). Garnets are rarely observed (<1%), and have relatively high spessartine contents ($Alm_{69}Py_2Gr_1Sp_{28}$), indicative of an igneous paragenesis. The tourmaline is believed to be magmatic in origin since it occasionally forms inclusions in garnet. Zircon and apatite are common accessory phases. The usual presence of tourmaline in these sheets distinguishes them in the field from the Parri granites of the Kohistan arc. The Jutial granite is a larger, medium-grained pluton composed of biotite (5–10%) and muscovite (2–10%) with rare tourmaline, and can be traced for 2.5 km along the Phuparosh valley (Fig. 2). Numerous leucogranite dykes and pegmatites cross-cut the margins of the Jutial granite. Large roof pendants of paragneiss within the granite maintain a constant orientation with respect to the country rocks, suggesting a fairly passive intrusion of magma.

The country rocks of the leucogranites are a range of partially migmatised orthogneisses intercalated with metasediments, and paragneisses (termed Iskere and Shengus gneisses respectively by Madin *et al.* 1989). The pelitic lithologies contain kyanite and occasionally sillimanite, and are at upper amphibolite grade. All of the sampled leucogranites cross-cut these rocks with abrupt contacts, indicating that the melts have been able to effectively leave their sources.

Major and trace element geochemistry

The NPHM leucogranites are peraluminous with a high, restricted silica range (71–75%), high Rb/Sr ratios (3–23 in the dykes, 2–8 in the Jutial pluton), and depleted HFSE abundances. They have similar trace element chemistries to the Miocene leucogranites from the central Himalaya (Fig. 3b), except for the relatively high Th and LREE concentrations of the NPHM leucogranites (Table 1). Compared with the Parri sheets of Kohistan, NPHM leucogranites are characterized by even higher Rb/Sr ratios (Fig. 4).

A range of metamorphic lithologies sampled from the NPHM basement include pelites (biotite + muscovite + garnet ± kyanite/sillimanite), semi-pelites (biotite ± muscovite) and orthogneisses. Compared with the leucogranites, the gneisses are enriched in both Ba and Sr (Fig. 3b), with Rb/Sr ratios in the range 0.5–5. The relative concentrations of LILE, together with the high Rb/Sr ratios in the leucogranites, are consistent with vapour-absent melting of metapelite source by incongruent melting of muscovite (Harris & Inger 1992). For example, an Rb/Sr ratio of 8 can result from a 0.15 melt fraction from a pelitic source under vapour-absent conditions (Harris *et al.* this volume). The limited Rb enrichment in the leucogranites compared with the gneisses may indicate that muscovite, as opposed to biotite, was the dominant mica involved in the melting reactions. Depletion in HFSE in the granites, such as Zr, Hf, Y and REE, is consistent with their retention in accessory phases that were largely residual during melting.

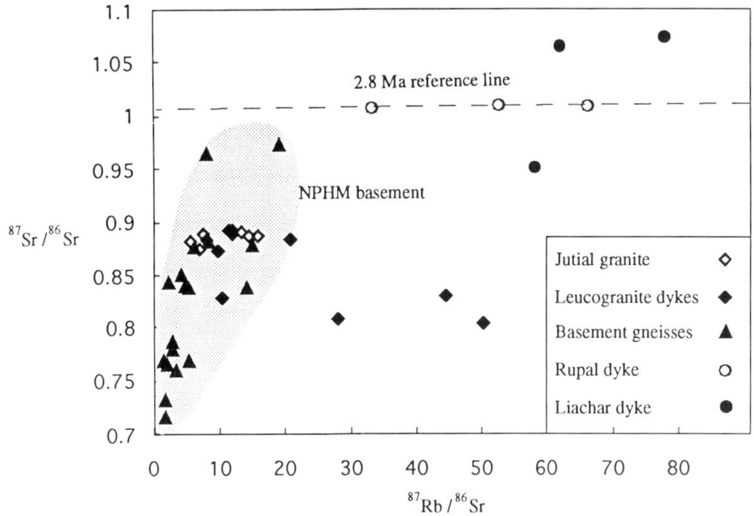

Fig. 7. Rb-Sr isochron diagram for leucogranites and gneisses from the Nanga Parbat–Haramosh massif. Dashed line is a 2.8 Ma reference line for the Rupal granites; shaded region delineates field of NPHM crust.

Age of intrusion

The Rb-Sr isotope ratios for ten widely-spaced samples from the Jutial granite do not define an isochron (Fig. 7). However, work by Zeitler & Chamberlain (1991) on young zircon overgrowths from three leucogranite dykes has yielded concordant spreads in ages of 2.2–4.3 Ma; 5–8 Ma; and 7–12 Ma. In each case, the youngest age is interpreted by these authors as the leucogranite intrusion age. Interestingly, three whole-rock samples of a ten-metre thick leucogranite from Rupal define an approximate age of 2.8 ± 0.1 Ma (($^{87}Sr/^{86}Sr)_i$ = 1.0056), although the small number of samples on such a young age makes the calculated uncertainty highly unrealistic.

A further constraint on the age of the leucogranite magmatism comes from Rb-Sr muscovite ages, interpreted as the time of cooling through a 500–550° C blocking temperature (Purdy & Jäger 1976; Cliff 1985). The Jutial granite has a muscovite age of 6.5 Ma, compared with muscovite ages of 2.8 and 3.3 Ma for a leucogranite dyke in the Haramosh area (Table 3).

Muscovite Rb-Sr ages from metamorphic rocks in the NPHM are extremely variable and lie in the range 4–36 Ma (Fig. 2; Table 3), indicating contrasting thermal histories for different parts of the NPHM. The blocking temperature for Ar in hornblende is similar to the blocking temperature for Sr in muscovite (≈500° C, Harrison 1981), although both are affected by cooling rate. Thus the older muscovite Rb-Sr ages noted above correspond broadly to hornblende ^{40}Ar-^{39}Ar ages of 16–25 Ma from the NPHM (Zeitler *et al.* 1989), also shown in Fig. 2. Interestingly, the Rb-Sr muscovite ages for the NPHM are locally very young (4.4–7.7 Ma). These are similar to the young monazite ages (4–11 Ma) recorded by Smith *et al.* (1992), suggesting that at least some parts of the NPHM experienced a much younger metamorphic event, or even more recent cooling. However, in general, the younger muscovite ages from the leucogranites suggest that the leucogranites intruded after their country rocks had cooled below about 500° C. For example, from Shengus (Fig. 2), Zeitler *et al.* (1989) obtained a metamorphic hornblende ^{40}Ar-^{39}Ar age of 20 Ma, which contrasts with Rb-Sr muscovite ages of 3.3–2.8 Ma for a leucogranite from this locality. Magmatic cooling rates will be affected by volume and depth of intrusion, but in general post-magmatic closure of Sr in muscovite occurs shortly after crystallization. Following from this, the variable leucogranite muscovite cooling ages above may indicate a slightly older age for the Jutial granite than for the leucogranite sheets at Haramosh and Rupal, although detailed U-Pb zircon dating would be required to confirm this.

In summary, the timing of leucogranite magmatism is rather poorly constrained, but probably took the form of separate magma batches intruded between 2 and 12 Ma.

Sr and Nd isotope systematics

Initial Sr-isotope ratios from all leucogranites in the NPHM are >0.80, well in excess of values observed in the Kohistan granites. Whereas widely spaced samples of the Jutial granite have $(^{87}Sr/^{86}Sr)_{10}$ ratios in the range 0.87–0.89, leucogranite dykes near the Jutial granite have more variable ratios of 0.80–0.87, possibly reflecting isotopically distinct sources for the leucogranite dykes and the Jutial granite. Extreme Sr-isotope ratios (0.98–1.06) have been identified in leucogranite dykes from the Liachar thrust zone and at Rupal in the NPHM. Nineteen samples of Indian crust material from across the western NPHM have varied $(^{87}Sr/^{86}Sr)_{10}$ in the range 0.716–0.970. There is no obvious correlation between metamorphic lithology or geographic location and the Sr-isotope composition. The Sr-isotope compositions of the analysed gneisses encompass the range of all the leucogranites except the extremely radiogenic bodies mentioned above (Fig. 7). Even in these cases, it is possible that further analyses of basement gneisses could yield potential source materials. Alternatively, the very high Sr-isotope compositions locally observed in the NPHM leucogranites may be the result of disequilibrium melting of a mica-rich protolith, where the radiogenic Sr of the reactant mica is incorporated in the melt, and the melt extracted before isotopic equilibration is reached between melt and restite phases.

The significance of heterogeneous Sr isotopic compositions, observed not only between different granites but also within individual intrusions such as the Jutial granite, is equivocal, though it must reflect the characteristics of the source region, possibly modified by wall-rock assimilation or subsolidus fluid infiltration. In the light of the variable isotopic compositions of the NPHM gneisses, together with the geochronological evidence which indicates emplacement of magmas as separate pulses over several million years, the isotopic variation existing between and within different granites can be accounted for by batch melting of heterogeneous lithologies within the ancient Indian crust, similar to those now exposed in the NPHM basement.

A leucogranite dyke from Liachar, deformed in the hanging wall of the Liachar thrust, provides an example of extreme isotopic heterogeneity within a single body (L17, Table 6). In an isotopic study of another leucogranite at Lia-

Table 6. Whole-rock Sr-isotopic data for NPHM leucogranites

Sample	Location	Rb/Sr	$^{87}Rb/^{86}Sr$	$^{87}Sr/^{86}Sr$	2σ	$(^{87}Sr/^{86}Sr)_{10}$
Jutial Granite						
J1A	Jutial	5.41	15.64	0.88723	2	0.88501
J1B	Jutial	4.58	13.24	0.89041	3	0.88853
J1C	Jutial	4.08	11.79	0.89199	3	0.89031
J1E	Jutial	4.15	11.99	0.88839	3	0.88668
J1F	Jutial	3.91	11.29	0.89171	3	0.89011
J1G	Jutial	2.55	7.36	0.88947	2	0.88842
J1i	Jutial	4.97	14.35	0.88769	2	0.88565
J1D	Jutial	2.40	6.94	0.87465	2	0.87366
J19	Jutial	1.88	5.44	0.88132	2	0.88055
J11	Jutial	2.80	8.09	0.88438	3	0.88323
Leucogranite dykes						
J7	Phuparosh	3.64	10.52	0.82808	2	0.82659
I11	Haramosh	9.67	27.95	0.80804	2	0.80407
I8	Iskere	17.41	50.32	0.80485	10	0.79770
S87	Shengus	3.32	9.59	0.87362	3	0.87225
S92	Shengus	15.50	44.80	0.83015	2	0.82379
M-18	Tato	7.22	20.86	0.88122	2	0.87826
L17-F[1]	Liachar	21.20	61.27	1.06583	3	1.05712
L17-E[1]	Liachar	26.77	77.35	1.07385	5	1.06286
L17-G[1]	Liachar	19.95	57.66	0.9499	3	0.94172
A14[2]	Rupal	11.60	33.52	1.00696	3	1.00220
A10[2]	Rupal	18.67	53.96	1.00780	2	1.00014
A9[2]	Rupal	22.91	66.21	1.00831	6	0.99891

Details as for Table 2. $^{87}Sr/^{86}Sr_i$ calculated for an age of 10 Ma.
[1,2] Samples from the same granite dykes.

char, Cliff *et al.* (1991) noted isotopic disequilibrium between adjacent phases, even at a millimetre scale, which was ascribed to the action of subsolidus fluids. Further evidence for the action of fluids in the Liachar thrust comes from oxygen-isotope disequilibrium between minerals in pegmatites (Barnett *et al.* 1991). Thus it seems that fluid infiltration has resulted in Sr mobility within those NPHM leucogranites which have been deformed in Liachar shear zone.

Samples of the Jutial granite have uniform Nd-isotope compositions with $\varepsilon_{Nd}(T) \approx -26$ (Fig. 6). A leucogranite dyke at Phuparosh has $\varepsilon_{Nd}(T) = -24.4$, comparable to a deformed dyke at Liachar which has $\varepsilon_{Nd}(T) = -24.3$ (Cliff *et al.* 1991). Five samples of NPHM gneiss with a wide geographical spread have $\varepsilon_{Nd}(T)$ in the range −20 to −32. On the basis of Sr and Nd isotope systematics, the leucogranites could have been derived from the NPHM at 2–12 Ma ago.

The mineralogy, peraluminous composition, high Rb/Sr ratio and radiogenic Sr and Nd isotopic compositions of NPHM leucogranites indicate a derivation by partial melting of an ancient, highly evolved crustal source. Model Nd ages for the leucogranites lie between 2.4 and 2.6 Ga, within the range provided by the sampled gneisses (2.1–2.7 Ga). These compare with somewhat younger model ages (1.6–2.0 Ga) for High Himalayan leucogranites (e.g. Deniel *et al.* 1987; Inger & Harris 1993) indicative of the relative ages of the metasedimentary protoliths from these two regions of the Himalayan orogen.

A crustal source for the NPHM leucogranites is also suggested by the common inheritance of radiogenic Pb seen in zircons from these intrusives (Zeitler & Chamberlain 1991). Conventional and ion microprobe zircon U-Pb ages of *c.* 1.9 Ga (Iskere gneisses) and 400–500 Ma (Shengus gneisses) have been obtained (Zeitler *et al.* 1989) from country rocks to the leucogranites, providing suitably ancient potential sources for the granites. Furthermore, xenocrystic zircons from leucogranites intruding both the Iskere and Shengus gneisses have the same age as the Iskere gneisses. This suggests that the Iskere gneisses are more likely to have been the source of the leucogranites (Zeitler & Chamberlain 1990; Zeitler *et al.* 1989), although this work has shown that, in general, the melting zone is not presently exposed. Unfortunately, Sr-isotope studies cannot resolve any significant difference in isotopic composition between the Iskere gneisses and the Shengus gneisses.

In most cases, the Sr-isotope ratios of leucogranites do not correspond to the ratios observed in their immediate country rocks. For example, the Rupal dyke has $(^{87}Sr/^{86}Sr)_i = 1.005$ compared with an adjacent gneiss (sample A13) which is characterized by $(^{87}Sr/^{86}Sr) = 0.882$. These observations, together with the cross-cutting nature of the leucogranites, indicate that the magma was sufficiently mobile to leave its source. This is consistent with the younger muscovite cooling ages observed in the leucogranites, and the widspread presence of stable muscovite in the country rocks. Thus there is little evidence for widespread, Himalayan age, in situ melting in the massif at present exposure levels, although locally sillimanite-grade assemblages and migmatites are observed.

Discussion

Field, petrographic, geochemical and isotopic data enable us to distinguish between three suites of granites adjacent to the western margin of the NPHM. As expected, the most significant isotopic and geochemical contrasts exist between granites intruding the arc and the relatively young leucogranites intruding the NPHM, due to primary differences between the source characteristics of the two terranes. This contrast is exemplified by Nd model ages for intrusives which increase dramatically across the contact from *c.* 700 Ma in the Kohistan terrane to *c.* 2500 Ma in the Indian Plate.

Two suites of Kohistan granite sheets can be traced into the shear zone, showing that for each suite, magmatic ages are similar both in and out of the shear zone. The majority of the Confluence and Parri granites which have been deformed in the shear zone show elevated whole-rock $(^{87}Sr/^{86}Sr)$ ratios relative to their undeformed counterparts in the arc, whereas major and trace elements remain unaffected. Compared with the undeformed granites in the Kohistan arc, granites within the shear zone also show increased scatter in $(^{87}Sr/^{86}Sr)$ ratios between different granites, and extreme heterogeneity has even been preserved within single granite sheets. These variations are consistent with infiltration of subsolidus fluids through the shear zone. However, an evaluation of the Nd isotopic data suggest that incorporation of ancient crustal material into the melts during ascent has also played a role in the isotopic evolution of the granites within the shear zone.

Deformed leucogranites in the Liachar thrust zone also provide evidence of fluid infiltration effecting Sr-isotope ratios (Cliff *et al.* 1991; this contribution). Interestingly, the similarity in

$\varepsilon_{Nd}(T)$ between these leucogranites and undeformed dykes within the NPHM could be taken as confirmation that Sr, but not Nd, can be mobilised by fault-related fluids. The absence of melting associated with fluid infiltration suggests that advection through the shear zone occurred after amphibolite-grade ductile deformation. The fluids may be associated with crystallization of leucogranite bodies within the NPHM basement <12 Ma ago.

Ion probe U-Pb depth-profiling of basement zircons reveals young overgrowths with ages of 2–10 Ma, possibly reflecting post-metamorphic fluid flow at temperatures of 200–500° C (Zeitler & Williams 1988). These ages are comparable to the ages of NPHM leucogranite magmatic zircons (Zeitler & Chamberlain 1991), and coincide with the time of young, rapid uplift of the NPHM (Zeitler 1985), suggesting that leucogranite magmatism within the NPHM, pervasive fluid flow and rapid uplift are contemporaneous and possibly related.

Petterson & Windley (1991) suggested that frictional heating and dehydration of the Indian continental crust during underthrusting below the arc caused anatexis of lower arc crust and generation of the undeformed Confluence and Parri granites, whilst Petterson *et al.* (1993) proposed that fluids derived from the Indian crust had enriched the Sr isotopic composition of the source of the Parri granites. The data presented in this study are consistent with an arc source for post-50 Ma magmatism in Kohistan. However, if fluids from the underthrusting Indian continental crust had caused melting or metasomatism in the arc, the isotope systematics of the 26–50 Ma, undeformed arc granites should record a continental input, which is not observed. The absence of Archean signatures in such young granites in the arc is perhaps surprising, because the present crustal levels of the Indian continent exposed in the NPHM include thick wedges of metasediments with evolved, radiogenic isotopic compositions (Table 6). During underthrusting, such hydrous assemblages would be expected to undergo partial melting (Patino-Douce *et al.* 1990), generating melts with the isotopic and geochemical characteristics of the leucogranites currently exposed in the NPHM. The presence of isotopically juvenile melts in northern Kohistan, rather than NPHM-type leucogranites, therefore suggests that significant underthrusting of northern Kohistan by Indian continental crust did not occur before 26 Ma. We should emphasise, however, that this study does not constrain the time at which either collision or crustal thickening was initiated, but places a maximum constraint of 26 Ma on the time of large-scale underthrusting of the Indian crust in northern Kohistan.

The monazites of the Miocene Baltoro plutonic unit (BPU) in the Karakoram (Searle *et al.* 1989) share similar geochemical characteristics to the Confluence granites. The BPU is characterized by low $(^{87}Sr/^{86}Sr)_i$ ratios thought to reflect large scale melting of the lower Karakoram crust during homogeneous crustal thickening, and enrichment of LILE possibly caused by melts derived from the mantle wedge (Searle *et al.* 1992). Significantly, there is no evidence for the involvement of ancient Indian crust (Rex *et al.* 1988; Searle & Tirrul 1991). Crawford & Windley (1990) identified the BPU as a 'type b' collision-related leucogranite, with melting at the base of the crust being triggered by an input of heat or material from the mantle wedge. If northern Kohistan was still underlain by a Tethyan mantle wedge at 50–30 Ma, as would be the case if underthrusting had not occurred until then, then the Confluence granites may share a similar origin to the BPU. However, if melting in the arc had been triggered by a mantle heat source after underthrusting, fertile Indian continental material would be expected to melt before juvenile arc crust, although this would also be dependent on the initial crustal temperatures. Again, the absence of mature crustal melts intruding the Kohistan arc therefore suggests, if the mantle was indeed the heat source, that underthrusting of continental material beneath the northern arc may not have occurred until after *c.* 30 Ma. Interestingly, the BPU has been dated at 21–25 Ma (U-Pb zircon, Parrish & Tirrul 1989; Scharer *et al.* 1990), slightly younger than the Confluence and Parri granites of Kohistan to the south. This may therefore reflect crustal thickening and melting progressively migrating northwards from the collision zone, as predicted by England & Searle (1986).

Cross-cutting igneous relations at Sassi show that the Kohistan granites post-date the closure of the collisional suture, observations which are supported by geochronology. Kinematic work shows that the same granites have been deformed by a top-to-the-SSE shearing, associated with underthrusting of the Indian continental crust below Kohistan (Butler & Prior 1988*a*). If underthrusting had commenced during the initial collision, then either this deformation phase must have lasted for at least 20 Ma, or there have been two separate deformation phases characterized by top-to-the-SSE shearing. Alternatively, either the initial collision in northeastern Kohistan occurred later than in the Hazara area (for which there is no evidence), or significant underthrusting of Kohistan by the

Indian craton did not occur until after 26 Ma. Both structural and isotopic evidence therefore suggest that before 26 Ma, continental underthrusting below northern Kohistan was limited.

The trace element contents of the NPHM leucogranites are consistent with vapour-absent melting of a metapelite source (Harris & Inger 1992). The high Rb/Sr of the leucogranites and Sr and Nd isotope systematics suggest that the sources of the leucogranites were similar to the NPHM gneisses currently exposed, although there is little evidence at the surface for Himalayan-age in situ melting. Low volume, batch melting of these ancient, evolved and isotopically heterogeneous gneisses occurred over a period of about 2–12 Ma. The association of rapid, young uplift in the NPHM (Zeitler 1985) with young leucogranite magmatism suggests an origin by decompression melting (Zeitler & Chamberlain 1991). For a crustal exhumation from 20 to 10 km depth, a melt fraction of $F = 0.1$ can be produced from subsolidus temperatures if exhumation is sufficiently rapid for adiabatic ascent (Harris *et al.* this volume). Melting during decompression would be enhanced by the high concentrations of U and Th in pelites within the NPHM, which result in extremely high heat production (5–7 $\mu W\ m^{-3}$). Thus the unusually young leucogranites intruding the NPHM were derived from crustal rocks with high heat productivities, buried beneath an insulating layer of Kohistan rocks before extremely rapid uplift and exhumation.

This work was supported by a NERC studentship (M.T.G.) and fieldwork expenses (R.W.H.B. & N.B.W.H.) were funded by a Royal Society study visit to Pakistan and a NERC research grant (GR3/6113). We are indebted to M. Ghazanfar and N. Chaudary of the Punjab University, Lahore, for hospitality and logistical support in the field. We acknowledge the advice given by P. Treloar and two anonymous reviewers of an earlier draft, and M.T.G. thanks P. van Calsteren and M. Johnston for support in the laboratory.

References

ALLÈGRE, C.J. & BEN OTHMAN, D. 1980. Nd-Sr isotopic relationships in granitoid rocks and continental crust: a chemical approach to orogenesis. *Nature*, **296**, 335–342.

ARNDT, N.T. & GOLSTEIN, S.L. 1987. Use and abuse of crust-forming ages. *Geology*, **15**, 893–895.

BARNETT, D.E., CHAMBERLAIN, C.P. & ZEITLER, P.K. 1991. Oxygen Isotope Constraints on the Genesis of Young (<5 Ma) Leucogranites from the Nanga Parbat Haramosh Massif. *EOS Transactions, AGU*, **72**.

BESSE, J., COURTILLOT, V., POZZI, J.P., WESTPHAL, M. & ZHOU, Y.X. 1984. Palaeomagnetic estimates of crustal shortening in the Himalayan thrusts and Zangbo suture. *Nature*, **311**, 621–626.

BUTLER, R.W.H. & PRIOR, D.J. 1988a. Anatomy of a continental subduction zone: the Main Mantle Thrust in northern Pakistan. *Geologische Rundschau*, **77**, 239–255.

—— & —— 1988*b*. Tectonic controls on the uplift of the Nanga Parbat Massif, Pakistan Himalayas. *Nature*, **333**, 247–250.

——, —— & KNIPE, R.J. 1989. Neotectonics of the Nanga Parbat Syntaxis, Pakistan, and crustal stacking in the N.W. Himalayas. *Earth and Planetary Science Letters*, **94**, 329–343.

——, GEORGE, M.T., HARRIS, N.B.W., JONES, C., PRIOR, D.J., TRELOAR, P.J. & WHEELER, J. 1992. Geology of the Northern part of the Nanga Parbat massif, Northern Pakistan, and its implications for Himalayan tectonics. *Journal of the Geologial Society, London*, **149**, 557–567.

CLIFF, R.A. 1985. Isotopic Dating in Metamorphic Belts. *Journal of the Geological Society, London*, **142**, 97–110.

——, BARNICOAT, A.C. & TRELOAR, P.J. 1991. Generation and Migration of Isotopically Extreme Partial Melts in Active Fault Zones. *Terra Abstracts*, **3**, 489.

COWARD, M.P., BUTLER, R.W.H., CHAMBERS, A.F., GRAHAM, R.H., IZATT, C.N., KHAN, M.A., KNIPE, R.J., PRIOR, D.J., TRELOAR, P.J. & WILLIAMS, M.P. 1988. Folding and imbrication of the Indian crust during Himalayan collision. *Philosophical Transactions of the Royal Society, London*, **A326**, 89–116.

CRAWFORD, M.B. & WINDLEY, B.F. 1990. Leucogranites of the Himalaya/Karakoram: implications for magmatic evolution within collisional belts and the study of collisional-related leucogranite petrogenesis. *Journal of Volcanology and Geothermal Research*, **44**, 1–19.

DENIEL, C., VIDAL, P., FERNANDEZ, A., LE FORT, P. & PEUCAT, J.J. 1987. Isotopic study of the Manaslu granite (Himalaya, Nepal): inferences of the age and source of Himalayan leucogranites. *Contributions to Mineralogy and Petrology*, **96**, 78–92.

DEPAOLO, D.J. 1981. Nd isotope systematics in the Colorado Front Range and crustal mantle evolution in the Proterozoic. *Nature*, **291**, 193–196.

ENGLAND, P.C. & SEARLE, M.P. 1986. The Cretaceous-Tertiary deformation of the Lhasa block and its implications for crustal thickening in Tibet. *Tectonics*, **5**, 1–14.

GEIST, D.J., FROST, C.D., KOLKER, A. & FROST, B.R. 1988. A geochemical study of magmatism across a major terrane boundary: Sr and Nd isotopes in Proterozoic granitoids of the southern Laramie Range, Wyoming. *Journal of Geology*, **97**, 331–342.

GREEN, T.H. 1977. Garnet in silicic liquids and its possible use as a P-T indicator. *Contributions to Mineralogy and Petrology*, **65**, 59–67.

HARRIS, N.B.W. & INGER, S. 1992. Trace Element modelling of Pelite derived Granites. *Contributions to Mineralogy and Petrology*, **110**, 46–56.

——, Xu Ronghua, Lewis, C.L., Hawkesworth, C.J. & Zhang Yuquan 1988. Isotope geochemistry of the 1985 Tibet Geotraverse, Lhasa to Golmud. *Philosophical Transactions of the Royal Society*, **A327**, 263–285.

——, Gravestock, P. & Inger, S. 1992. Ion-microprobe determinations of trace-element concentrations in garnets from anatectic assemblages. *Chemical Geology*, **100**, 41–49.

——, Inger, S. & Massey, J.A. 1993. The role of fluids in the formation of High Himalayan leucogranites. *This volume.*

Harrison, T.M. 1981. Diffusion of argon in hornblende. *Contributions to Mineralogy and Petrology*, **78**, 324–331.

Honegger, K., Dietrich, V., Frank, W., Gansser, A., Thoni, M. & Trommsdorff, V. 1982. Magmatism and metamorphism in the Ladakh Himalaya (the Indus-Tsangpo suture zone). *Earth and Planetary Science Letters*, **60**, 253–292.

Inger, S. & Harris, N. 1993. Geochemical constraints on leucogranite magmatism in the Langtang Valley, Nepal Himalaya. *Journal of Petrology*, in press.

Klootwijk, C., Sharma, M.L., Gergan, J., Tirkey, B., Shah, S.K. & Agarwal, V. 1979. The extent of greater India, 2. Palaeomagnetic data from the Ladakh intrusives at Kargil, NW Himalayas. *Earth and Planetary Science Letters*, **44**, 47–64.

Le Fort, P. 1989. The Himalayan Orogenic Segment. *In*: Sengor, A.M.C. (ed.) *Tectonic Evolution of the Tethyan Region.* NATO ASI Series C, **259**, 289–386.

Madin, I.P., Lawrence, R.D. & Ur-Rahman, S. 1989. The northwestern Nanga-Parbat Haramosh Massif, evidence for crustal uplift at the N.W. corner of the Indian Craton. *Geological Society of America Special Paper*, **232**, 169–182.

Parrish, R. & Tirrul, R. 1989. U-Pb age of the Baltoro granite, northwest Himalaya, and implications for monazite U-Pb systematics. *Geology*, **17**, 1076–1079.

Patino-Douce, A.E., Humphreys, E.D. & Johnston, A.D. 1990. Anatexis and metamorphism in tectonically thickened crust exemplified by the Sevier hinterland, western North America. *Earth and Planetary Science Letters*, **97**, 290–315.

Patriat, P. & Achache, J. 1984. India-Eurasia collision chronology has implications for Crustal Shortening and Driving mechanism of Plates. *Nature*, **311**, 615–621.

Pearce, J.A., Harris, N.B.W. & Tindle, A.G. 1984. Trace element discrimination diagrams for the tectonic interpretation of granitic rocks. *Journal of Petrology*, **25**, 956–983.

Petterson, M.G. 1984. *The structure, petrology and geochemistry of the Kohistan batholith, Gilgit, Kashmir, N. Pakistan.* PhD thesis, Leicester University.

—— & Windley, B.F. 1985. Rb-Sr dating of the Kohistan arc-batholith in the Trans-Himalaya of north Pakistan, and tectonic implications. *Earth and Planetary Science Letters*, **74**, 45–57.

—— & —— 1991. Changing source regions of magmas and crustal growth in the trans-Himalayas: evidence from the Chalt batholith, Kohistan, northern Pakistan. *Earth and Planetary Science Letters*, **102**, 326–341.

——, Crawford, M.B. & Windley, B.F. 1993. Petrogenetic implications of neodymium isotope data from the Kohistan batholith, North Pakistan. *Journal of the Geological Society, London*, **150**, 125–129.

Pudsey, C.J. 1986. The northern suture, Pakistan: margin of a Cretaceous island arc. *Geological Magazine*, **123**, 405–423.

Purdy, J.W. & Jäger, E. 1976. K-Ar Ages on Rock-Forming Minerals from the Central Alps. *Memorie degli Instituti di Geologia e Mineralogia dell'Universita di Padova*, **30**, 1–31.

Rex, A.J., Searle, M.P., Tirrul, R., Crawford, M.B., Prior, D.J., Rex, D.C. & Barnicoat, A. 1988. The geochemical and tectonic evolution of the central Karakoram, northern Pakistan. *Philosophical Transactions of the Royal Society*, **A326**, 229–255.

Scharer, U., Copeland, P.Y., Harrison, T.M. & Searle, M. 1990. Age, cooling history and origin of post-collisional leucogranites in the Karakoram batholith; a multi-system isotope study. *Journal of Geology*, **98**, 233–251.

Searle, M.P. & Tirrul, R. 1991. Structural and Thermal evolution of the Karakoram crust. *Journal of the Geological Society, London*, **148**, 65–82.

——, Crawford, M.B. & Rex, A.J. 1992. Field relations, geochemistry, origin and emplacement of the Baltoro granite, Central Karakoram. *Transactions of the Royal Society of Edinburgh: Earth Sciences*, **83**, 519–538.

——, Rex, A.J., Tirrul, R., Rex, D.C., Barnicoat, A. & Windley, B.F. 1989. Metamorphic, magmatic and tectonic evolution of the central Karakoram in the Biafo–Baltoro–Hushe regions of northern Pakistan. *Geological Society of America Special Paper*, **232**, 47–73.

Smith, H.A., Chamberlain, C.P. & Zeitler, P.K. 1992. Documentation of Neogene regional metamorphism in the Himalayas of Pakistan using U-Pb in monazite. *Earth and Planetary Science Letters*, **113**, 93–105.

Tahirkheli, R.A.K. & Jan, M.Q. 1979. Geology of Kohistan, Karakoram Himalaya, northern Pakistan. *Special Issue Geological Bulletin University Peshawar*, **11**, 1–30.

——, Mattauer, M., Proust, F. & Tapponnier, P. 1979. The India Eurasia suture zone in northern Pakistan: synthesis and interpretation of recent data at plate scale *In*: Farah, A. & DeJong, K.A. (eds) *Geodynamics of Pakistan.* Geological survey of Pakistan, Quetta, 125–130.

Treloar, P.J., Rex, D.C., Guise, P.G., Coward, M.P., Searle, M.P., Windley, B.F., Petterson, M.G., Jan, M.Q. & Luff, I.W. 1989. K-Ar and Ar-Ar geochronology of the Himalayan collision in N.W. Pakistan: Constraints on the timing of suturing, deformation, metamorphism and uplift. *Tectonics*, **8**, 881–909.

——, Potts, G.J., Wheeler, J. & Rex, D.C. 1991.

Structural evolution and asymmetric uplift of the Nanga Parbat Syntaxis, Pakistan Himalaya. *Geologische Rundschau,* **80**, 411–428.

ZEN, E-AN & HAMMARSTROM, J.M. 1983. Magmatic epidote and its petrological significance. *Geology,* **12**, 515–518.

ZEITLER, P.K. 1985. Cooling history of the N.W. Himalaya, Pakistan. *Tectonics,* **4**, 127–151.

—— & CHAMBERLAIN, C.P. 1991. Petrogenetic and Tectonic Significance of Young Leucogranites from the Northwestern Himalaya, Pakistan. *Tectonics,* **10**, 729–741.

—— & WILLIAMS, I.S. 1988. U-Pb Dating of metamorphic Zircon Overgrowths by means of Depth Profiling with an Ion Microprobe. *EOS Transactions, AGU,* **69**, 464.

——, SUTTER, J.F., WILLIAMS, I.S., ZARTMAN, R. & TAHIRKHELI, R.A.K. 1989. Geochronology and temperature history of the Nanga Parbat–Haramosh massif, Pakistan. *Geological Society of America Special Paper,* **232**, 1–22.

Preliminary Landsat lineament analysis of the northern Nanga Parbat–Haramosh Massif, northwest Himalaya

V.S. CRONIN,[1] G.J. SCHURTER[1,2] & K.A. SVERDRUP[1]

[1] *Geosciences Department, University of Wisconsin-Milwaukee, PO Box 413, Milwaukee, Wisconsin 53201, USA*

[2] *Amoco Production Company, PO Box 3092, Houston, Texas 77253, USA*

Abstract: Analysis of a mosaic of Landsat 2 MSS band 7 images covering the northern Nanga Parbat-Haramosh Massif (NPHM) has yielded approximately 540 lineaments, defined by sets of colinear drainage segments and aligned linear graytone boundaries. We recognized segments of the few well-mapped faults along the NPHM as lineaments. The paucity of field data concerning faults, joints, lithological fabric and contacts prevents us from interpreting most of the lineaments at this time. However, some of the prominent lineament trends probably mark the trace of previously unmapped active faults that accommodate the recent uplift of the NPHM. The origins of these lineaments must be established through structural geological field work.

The Nanga Parbat–Haramosh Massif (NPHM) of northern Pakistan is a prominent salient of allochthonous Indian continental crust that is exposed within the Kohistan–Ladakh magmatic arc (Fig. 1). Earth's sixth tallest mountain, Nanga Parbat (8126 m), has the greatest local relief and one of the most rapid rates of sustained uplift yet observed in continental crust, at between 5 and 10 mm per year (Zeitler *et al.* 1982*a*, 1989; Zeitler 1985; Treloar *et al.* 1989). The northerly trend of the NPHM is nearly perpendicular to the active thrust faults of the Himalayan foreland, and crosses the course of the Indus River. Uplift of the NPHM may have resulted in the ponding of the Indus River at Skardu, where the Bunthang Sequence of glacial, lacustrine, alluvial, and fluvial strata began to accumulate by *c.* 1.1 to 1.3 Ma (Cronin *et al.* 1989; Cronin & Johnson in press). The reasons for the sustained rapid uplift and anomalous trend of the NPHM are not yet fully understood. A better understanding of the NPHM awaits collection of a more complete suite of primary structural, seismic, topographic, and lithological data throughout the area.

We conducted a Landsat lineament analysis as a reconnaissance tool to assist in field mapping of active structures along the NPHM (Figs 2 & 3). The study area lies within a 1° by 1° geographic box around the northern part of the NPHM, between latitudes 35°15′ N and 36°15′ N, and longitudes 74°15′ E and 75°15′ E. The boundaries of the study area were chosen for several reasons. First, the area has documented activity in terms of uplift (Zeitler *et al.* 1982*a*, 1989; Zeitler 1985), neotectonic faulting (Lawrence & Ghauri 1983; Madin 1986; Shroder *et al.* 1989), and seismicity (Khurshid *et al.* 1984; Yielding *et al.* 1984; Sverdrup *et al.* 1990; Schurter 1991). Second, the traces of the Raikot, Baroluma, and Stak Faults, which are thought to accommodate part of the uplift of the NPHM, lie within the study area (Lawrence & Ghauri 1983; Verplanck *et al.* 1985; Madin 1986; Verplanck 1987; Madin *et al.* 1989; Shroder *et al.* 1989). Third, previous mapping in this area is more extensive than in other areas of the NPHM, although

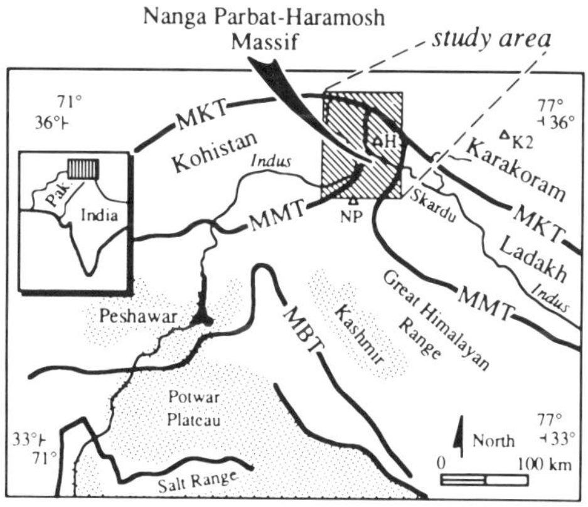

Fig. 1. Location of the study area within the structural context of the northwest Himalaya. Abbreviations are: MKT, Main Karakoram Thrust; MMT, Main Mantle Thrust; MBT, Main Boundary Thrust; H, Haramosh; NP, Nanga Parbat. Stippled pattern is molasse basin fill.

From Treloar, P.J. & Searle, M.P. (eds), *Himalayan Tectonics*
Geological Society Special Publication, No. 74, pp. 193–206.

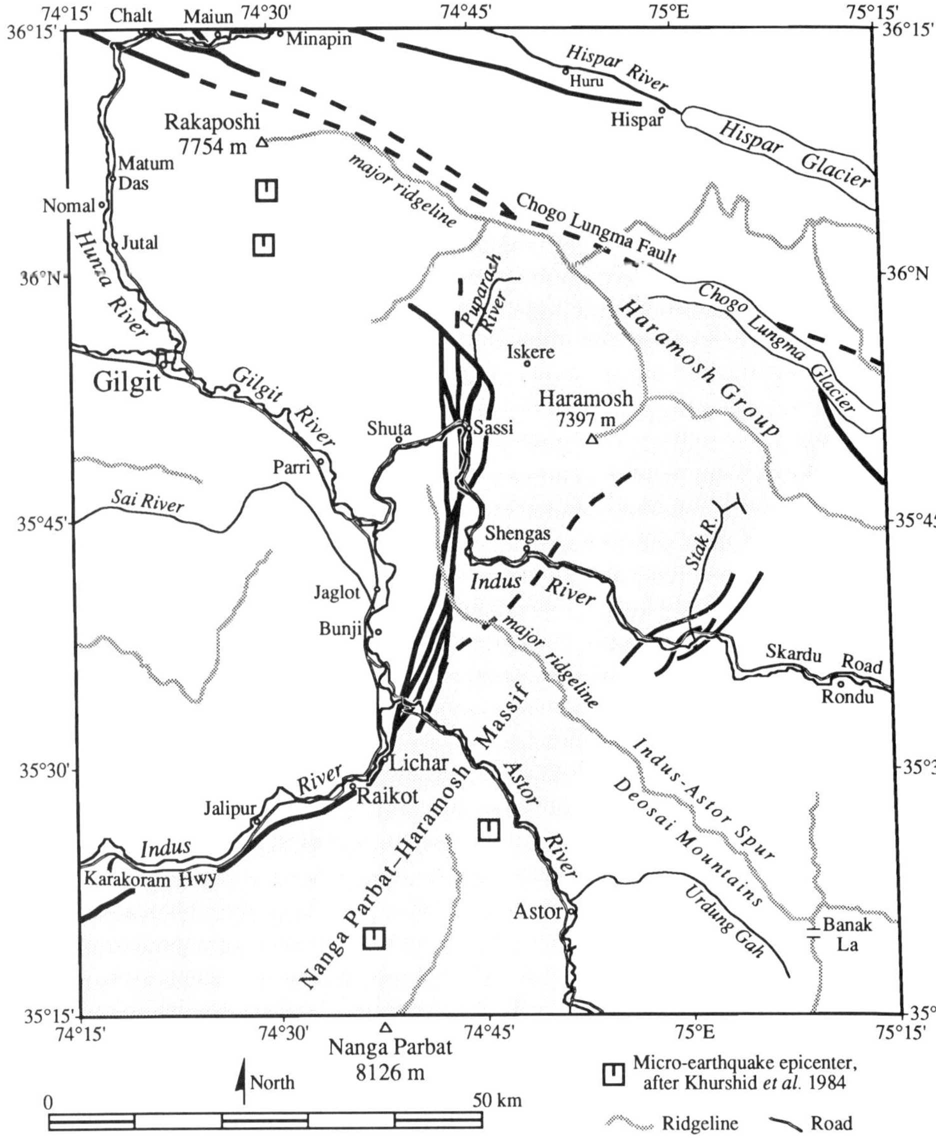

Fig. 2. Map of the study area, including the principal cultural features (roads, settlements), rivers, topographic features (ridgelines, mountain peaks), and the relatively well-mapped fault zones. Bold curves are faults, dashed where location is in doubt. Light grey area is the NPHM. The trace of the Main Mantle Thrust is traditionally depicted around the edge of the NPHM, but is not mapped here because it is locally truncated by the Younger Stak and Raikot Faults. Microearthquake epicentres from local network data recorded by Khurshid *et al.* (1984) during a 2 month period. The same area and the major rivers are mapped in Figs 3, 5–6 & 8–9. Map data are after Verplanck (1987) Madin *et al.* (1989), Shroder *et al.* (1989), and Searle (1991); geography registered to the Landsat mosaic in Fig. 3.

published maps are often incomplete and locally incompatible with one another (cf. Wadia 1932; Misch 1949; Desio 1964; Gansser 1964; Tahirkheli 1982; Searle 1991). Most of the published field data have been collected from the areas immediately adjacent to the roadways that follow the Indus, Gilgit, Hunza, and Astor Rivers, leaving geologic mapping of intervening areas largely unconstrained by data. Though incomplete, the existing base of field data permits

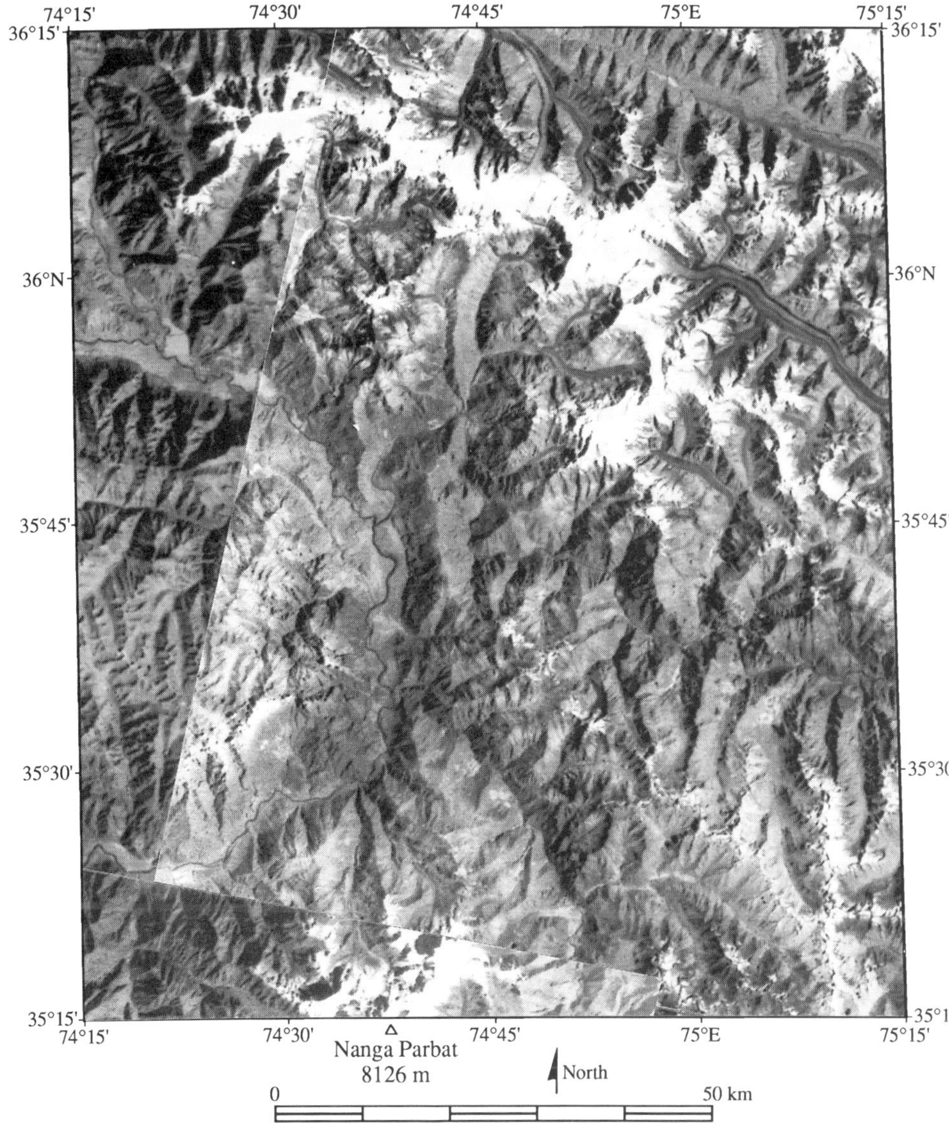

Fig. 3. Mosaic of Landsat 2 MSS band 7 images, corresponding to the area shown in Fig. 2. Serial numbers for the images used are listed in Table 1.

preliminary interpretation of the structural significance of some lineaments within the study area. Finally, this area has historically been of interest to many geologists, and is generally accessible via good roads along the major river valleys. We intend that geologists who visit the NPHM area in the future will use some of their field time to help evaluate the nature of some of the lineaments identified in this study.

The purpose of this study is to use Landsat images of the NPHM to identify lineaments that may be related to active structures exposed at the surface that accommodate uplift of the NPHM. Primary interest is in locating the brittle structures that accommodate uplift through the upper crust. The recognition of lineaments is the functional equivalent of the generation of a family of hypotheses that must be tested during structural geological field work. It is not the intent of this paper to propose structural interpretations of lineaments in the absence of corresponding field data; however, many of the lineaments identified in this study are thought to have developed along structural discontinuities (e.g., faults, joints, spatially persistent foliation, and lithological variations that result in differential weathering.) Our preliminary lineament mapping should enhance both the focus and economy of future structural reconnaissance mapping in areas along the NPHM that are difficult to reach due to glaciation, extreme topographic relief, and a variety of natural and logistical obstacles inherent in Himalayan field work.

Lineaments

Aerial photographic images have been used in regional geological studies since the reconnaissance geological mapping of Libya in 1913 (Lillesand & Kiefer 1987). Hobbs (1904) recognized that geomorphic lineaments can often be correlated with structural discontinuities that may not be obvious from other data observable at the surface. In this paper, the term *lineament* is used to describe a long (generally ≥5 km) colinear or slightly curving array of stream drainage segments or tonal boundaries within the Landsat mosaic that does not appear to be related to human construction or other activities.

Lineament recognition is a partially, though not totally, subjective process. Subjectivity in lineament recognition has been minimized in this study by examining multiple images of the same area, by performing duplicate analyses, and by using a computer-assisted drainage-lineament analysis. After an array of lineaments is recognized, the interpretation of the origin and significance of each lineament becomes the focus of study, along with the temporal, spatial, and possible mechanical relationships among the structures that are marked by lineaments. Lineaments may be related to (1) human activities and human-made structures, (2) surficial-deposit morphology that is not controlled by subsurface structure or morphology, as with some glacial features, (3) regional topographic controls on drainage courses, as across broad pediments, (4) differential erosion on tilted beds, (5) planar structures such as joints, faults, and persistent foliation, (6) coincidental alignment of planar or linear features, or (7) artifacts in the images, such as scan lines, missing or garbled data, or flaws related to photographic processing. Reliable interpretation of the origin and significance of lineaments generally requires field work. Given the scale difference between a Landsat image and human observations in the field, it may be very difficult to relate some Landsat lineaments to any particular origin, even with field work. For example, a fault along which a stream is localized may have no obvious surface expression other than as a lineament, because its trace is buried under sediment in the lower reaches of the valley and by snow and ice at the head of the valley.

Stream courses may be easily established along structural discontinuities or zones of structurally controlled differential weathering, which may be more easily eroded than adjacent rock masses. For example, parallel or colinear sets of drainages can be established along trends defined by inclined metamorphic foliation (e.g., Turner 1952; Zumberge 1955) developed as a result of ductile/plastic faulting or distributed flow in the lower crust. Identification of drainage lineaments (i.e., lineaments defined by colinearity of drainage segments) is potentially useful in locating high-angle joints, faults, foliation, and planar lithological contacts, as well as the trace of active thrust faults. Active thrust faults often have regular linear or bow-shaped traces, whereas less active or fossil thrust faults have a more sinuous trace developed as erosion modifies the leading edge of the thrust sheet.

Prominent lineaments were recognized in the Himalaya as soon as the first orbital images became available (e.g., Casnedi 1976; Ebblin 1976; Lawrence 1981). Structural interpretation of Landsat images has been used to particular advantage in reconnaissance structural mapping in and around the Tibetan Plateau, where major faults like the Altyn Tagh, Kunlun, and Karakoram Faults, and many more minor faults, were recognized as lineaments and are still known almost exclusively from the analysis of orbital images (e.g., Molnar & Tapponnier 1975, 1978; Tapponnier & Molnar 1977; Ni & York 1978; Srimal 1983, 1986; Rothery & Drury 1984; Ni & Barazangi 1985; Dewey *et al.* 1988; Searle 1991).

Kazmi (1979; Kazmi & Rana 1982) used Landsat lineaments to assist in locating active faults in Pakistan, but did not extend the analysis as far north as the NPHM. Gornitz & Seeber (1981) mapped Landsat lineaments southwest of the NPHM (approximately latitude 34° N–35° N and longitude 71.5° E–74.5° E), and suggested that the lineaments are defined along faults and joints, some of which developed above lower-plate structures such as the Indus–Kohistan Seismic Zone (Seeber & Jacob 1977; Armbruster *et al.* 1978; Seeber & Armbruster 1979). Most commonly, Landsat images have been used as a mapping base during geological field work, as well as the basis for standard photogeological analyses (e.g., Olson 1982; Cronin 1982, 1989; Yeats *et al.* 1984; Madin 1986; Molnar *et al.* 1987; Madin *et al.* 1989; Yeats & Hussain 1989, Searle 1991; Searle *et al.* 1989).

Methods

We have used two complementary approaches to identify prominent lineaments along the northern NPHM. First, we analysed the trends of stream drainages discernible on the Landsat images to identify lineaments defined by anomalous drainage systems. The drainage analysis was partially automated to minimize interpreter biases. Second, we made a visual assessment of the images to identify persistent linear trends marked by drainages as well as by linear gray-tone boundaries, using traditional methods of photogeology. Finally, the results of these analyses were combined to yield the final lineament map.

A mosaic of Landsat 2 multi-spectral scanner (MSS) band 7 images was created from photographic prints made from first-generation 70 mm film negatives obtained from the American Geographic Society's Landsat collection (Fig. 3). Landsat MSS images were chosen because of their low cost and wide availability. The mosaic was printed at a scale of *c.* 1:500000 for this analysis. Other Landsat images that were not included in the mosaic were used for comparison and reference during the lineament analysis, taking advantage of varying snow cover and sun angles among different images. Table 1 contains relevant data concerning the Landsat images used in this study.

Satellite images constitute the best mapping base for studies along the NPHM because accurate topographic maps of the area are not widely available. All maps in this paper that are derived from Landsat images are registered to the images rather than to a standard map projection. The pixel size for Landsat 2 MSS images is 79 × 58 m (Slater 1979). The native resolution of the Landsat digital data is degraded somewhat during the production of photographic negatives and the printing of film positives. In spite of the relatively low resolution of a photographic mosaic of Landsat 2 MSS images (e.g., Rothery 1985), we were able to recognize the traces of the several known faults and hundreds of prominent lineaments in the study area. The superior 10 m resolution of the SPOT images and the 30 m resolution of more recent Landsat TM images are more suitable as bases for larger-scale field mapping. It was perhaps two orders of magnitude less costly to produce the mosaic using Landsat 2 images, because we already owned the negatives, than it would have been if we had acquired comparable coverage in SPOT images.

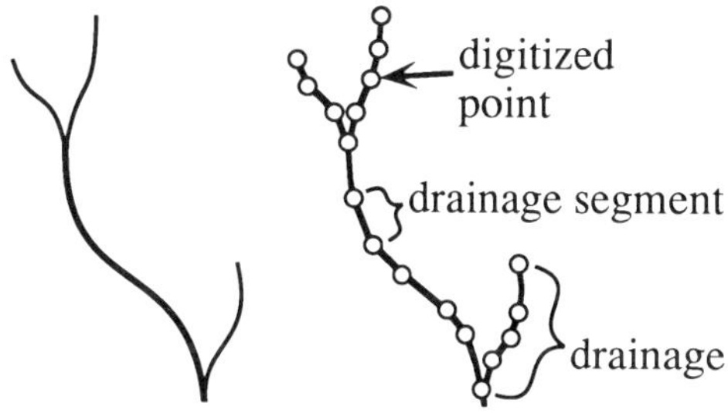

Fig. 4. The trace of a typical drainage network observed in the Landsat mosaic is shown at left. The drainage system is digitized into a series of polylines, shown at right. Spacing of digitized points varies with the curvature of the drainage course. Drainage segments are lines between adjacent digitized points.

Table 1. *Landsat images used in this study*

Image number	Band no.	Centre Location	Acquisition date	Sun		Cloud cover	Use in study
				Elevation	Azimuth		
E-2978-04342	7	N36°6′, E74°11′	26 Sept 1977	40°	131°	0%	Mosaic
E-2978-04345	7	N34°40′, E73°45′	26 Sept 1977	41°	129°	<5%	Mosaic
E-2923-04315	7	N36°, E75°35′	2 Aug 1977	51°	108°	<5%	Mosaic
E-2564-04524	7	N36°5′, E74°13′	8 Aug 1976	53°	114°	<5%	Reference
E-2923-04321	7	N34°34′, E75°8′	2 Aug 1977	52°	106°	<20%	Reference
E-2941-04305	7	N36° 2′, E75°35′	20 Aug 1977	48°	115°	<20%	Reference
E-30118-04545	7	N35°54′, E75°43′	1 July 1978	59	106	<20%	Reference
E-30136-04551	7	N35°55′, E75°42′	19 July 1978	58	109	<10%	Reference
E-30155-05011	7	N35°55′, E74°15′	7 Aug 1978	55	116	<20%	Reference
E-30155-05013	7	N34°29′, E73°48′	7 Aug 1978	56	113	<30%	Reference

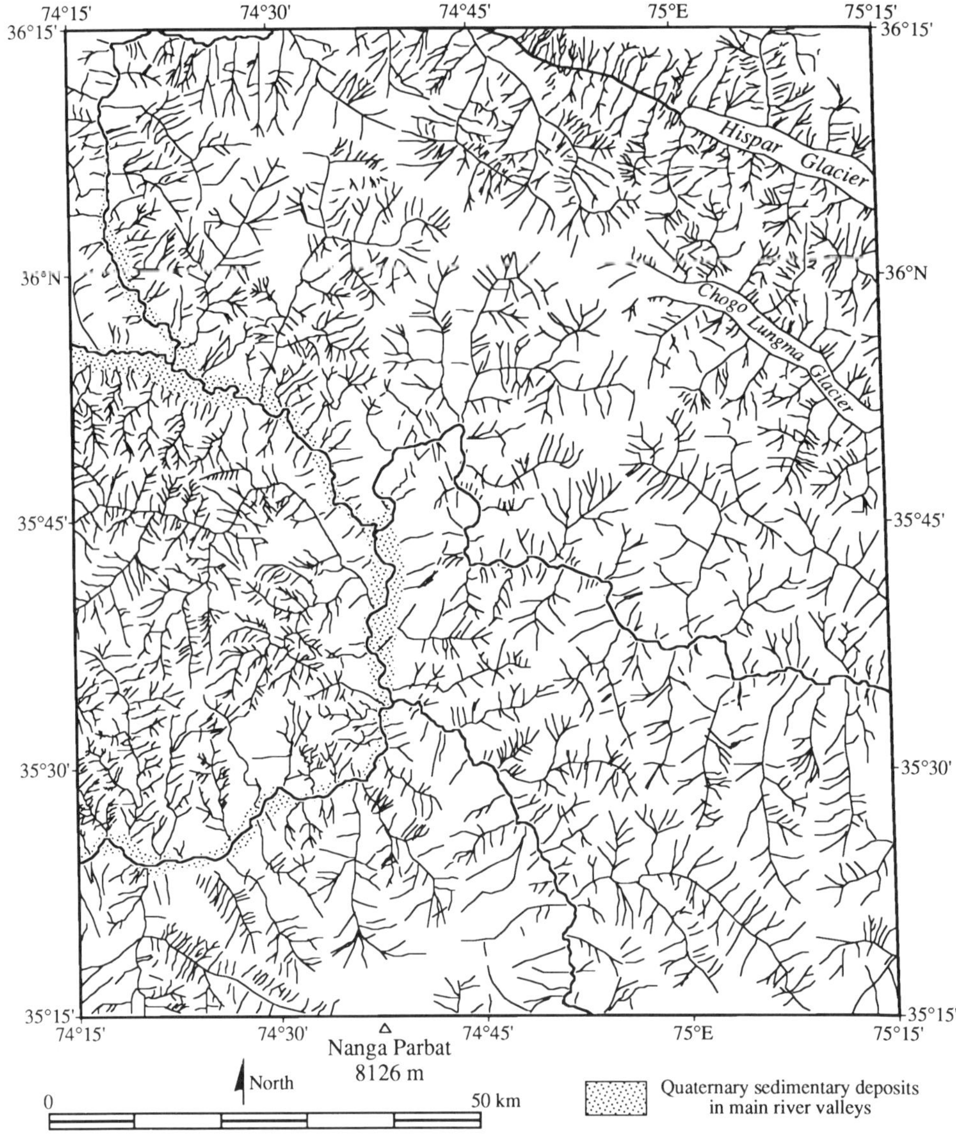

Fig. 5. Map of all drainage segments digitized for this study. Stippled area corresponds to sedimentary valley fill along the Gilgit, Hunza and Indus Rivers.

Drainage lineament analysis

A GTCO Digi-pad L-series digitizer with a resolution of 1200 dpi was used to digitize the trace of drainages discernible in the Landsat mosaic. The trace of some drainages is marked by visible streams, while other drainage courses are marked solely by greytone boundaries related to ground surface morphology. All obvious drainage courses were digitized. Where a valley is occupied by a glacier, the midline of the glacier was digitized. Each drainage was digitized as a polyline of short, connected line segments that are called *drainage segments* (Fig. 4). Drainage segments tend to be longer where the drainage is straighter, and shorter where the drainage course is more tightly curved. Approximately 3100 drainages were digitized, including *c*. 15000 drainage segments (Fig. 5).

The digitizer was driven using a Macintosh microcomputer by the commercial application DIGITZE (Rockware 1988), which generated a

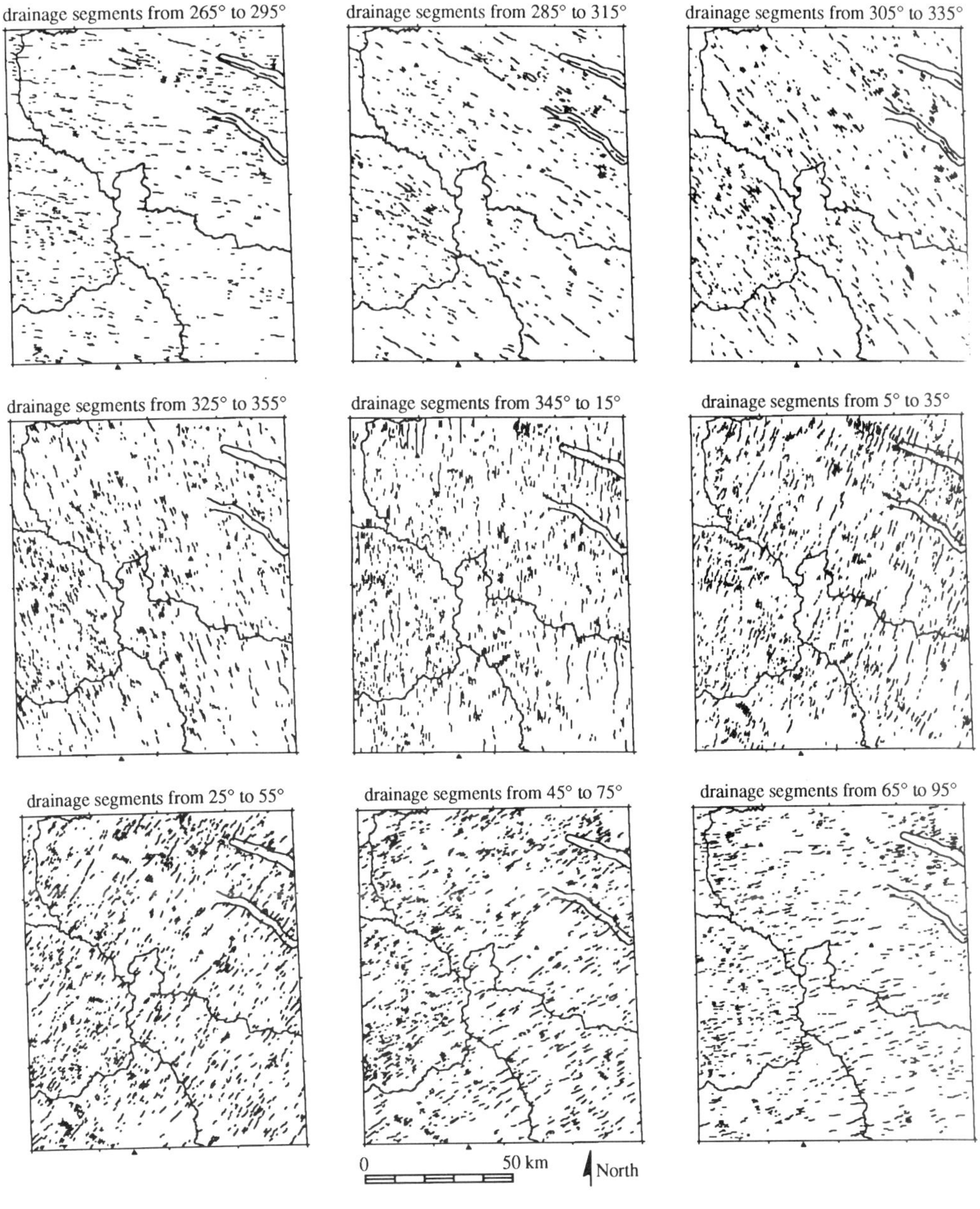

Fig. 6. Drainage segments mapped by azimuth in 9 overlapping azimuth classes. Azimuthal filtering assists in the recognition of aligned drainage segments. Principal drainages and valley glaciers are shown for geographic reference, as in Fig. 2.

PICT file readable by the commercial graphics application CANVAS 3.02 (Deneba Software 1991) as well as a flat file readable by the original program *Azimuthal Analysis of Lineaments* written in TrueBASIC 2.1 (TrueBASIC, Inc. 1991). *Azimuthal Analysis of Lineaments* deter mines the length and azimuth of each drainage segment, sorts drainage segments by azimuth, and creates PICT maps of all drainage segments whose azimuth lies between user-defined boundaries.

Most of the drainages in the study area are in bedrock channels containing a minor volume of unconsolidated sediments. Only the Gilgit,

Hunza, and Indus River Valleys within the study area contain a volume of sedimentary fill that is significant at the resolution of our Landsat mosaic. Valley glaciers and extensive snow mantles are common along the higher ridges in the area. Vegetative and soil cover is generally insignificant above the valley floors within the study area, so bedrock visibility in satellite images is generally good.

Azimuthal analyses are useful in identifying linear fabrics in aerial images. Ronchi filters, which are coarse diffraction gratings that accentuate linear features oriented perpendicular to the grating and suppress features with other orientations, have been used to recognize lineaments on aerial images (Pohn 1978). In this study, drainage segments that are orientated within a specified azimuthal range were identified automatically, as described above, and plotted by a laser printer on acetate, at the same scale as the Landsat mosaic. For example, drainages established along a long north-trending fault would contribute a number of colinear drainage segments that are easily recognizable on an azimuthal plot of all drainage segments oriented north $\pm n°$, where n is an azimuth range selected by the investigator. A suite of azimuthally filtered maps of drainage segments was made to assist in the recognition of drainage lineaments (Fig. 6). Azimuth ranges of from 10° to 30° were used in producing overlapping series of drainage-segment maps. The length of each drainage segment could be increased by a user-defined factor, and the artifically elongated segment plotted over the midpoint of the original segment, to further accentuate linear trends.

Drainage lineaments were identified based upon several criteria for recognizing anomalies that may indicate structural control of drainage courses (Fig. 7). Drainage-course anomalies included the following: (a) lower-order drainages that intersect higher-order drainages in an upstream direction, rather than in the expected downstream direction; (b) two lower-order drainages that are colinear across a higher-order drainage into which both flow; (c) two drainages that are colinear on opposite sides of a drainage divide; (d) two or more segments of the same drainage that are colinear; and (e) atypically long, straight segments of drainages. Most of the lineaments identified in this study met more than one of these criteria. Acetate overlays of the azimuthally filtered drainage-segment maps were placed over the Landsat mosaic to assist in the visual identification of lineaments.

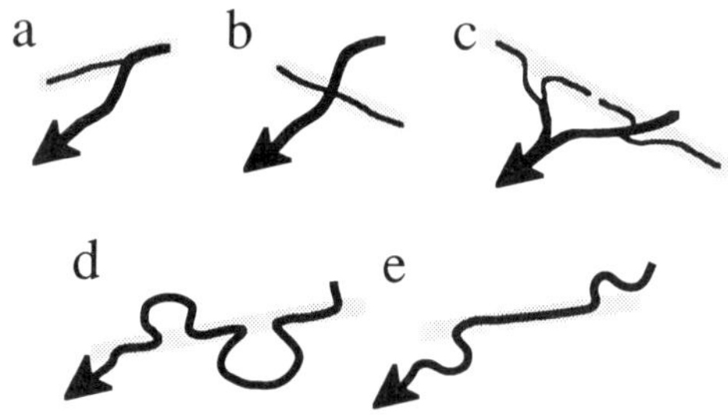

Fig. 7. Criteria for recognizing drainage lineaments: (**a**) lower order stream intersecting higher order stream in an upstream direction; collinear stream segments (**b**) across a higher-order stream, (**c**) across drainage divides, and (**d**) at different places along a single stream; and (**e**) anomalously long, straight stream segment. Lineaments are along halftone strips. Arrowheads indicate direction of stream flow.

General lineament analysis

Published maps of the NPHM area by Madin *et al.* (1989), Shroder *et al.* (1989), and Searle (1991) indicate that segments of some faults cross slopes, rather than exclusively following the bottom of drainage courses. Similarly, linear features that might be interpreted as structural discontinuities were commonly observed along slopes and across ridges between colinear drainage segments. These features included linear tonal boundaries that are apparently unrelated to topography, thin strips of lighter or darker tone that are apparently unrelated to topography, large cliff faces along peaks and ridge lines, and notches or offsets in ridgelines that are aligned with other linear features. Our final lineament maps are a composite of drainage-defined lineaments and lineaments defined in whole or in part by linear features that are unrelated to drainages.

Limitations of lineament analyses

The two principal ridges in the study area (the Haramosh Group north of the Indus Gorge and the Deosai Mountains south of the Indus Gorge) trend northwest, roughly parallel to the direction of solar illumination for the Landsat images. Stream valleys that drain from the northwest-trending ridges into northwest-trending segments of the Indus, Gilgit, and Astor Rivers are particularly well expressed in the mosaic, because of their orientation at a high angle to the solar illumination. Lower-order drainages that traverse southeast-facing slopes tend to be less distinct because they are washed with sunlight, while deep shadows may obscure lower-order drainages that traverse northwest-facing slopes.

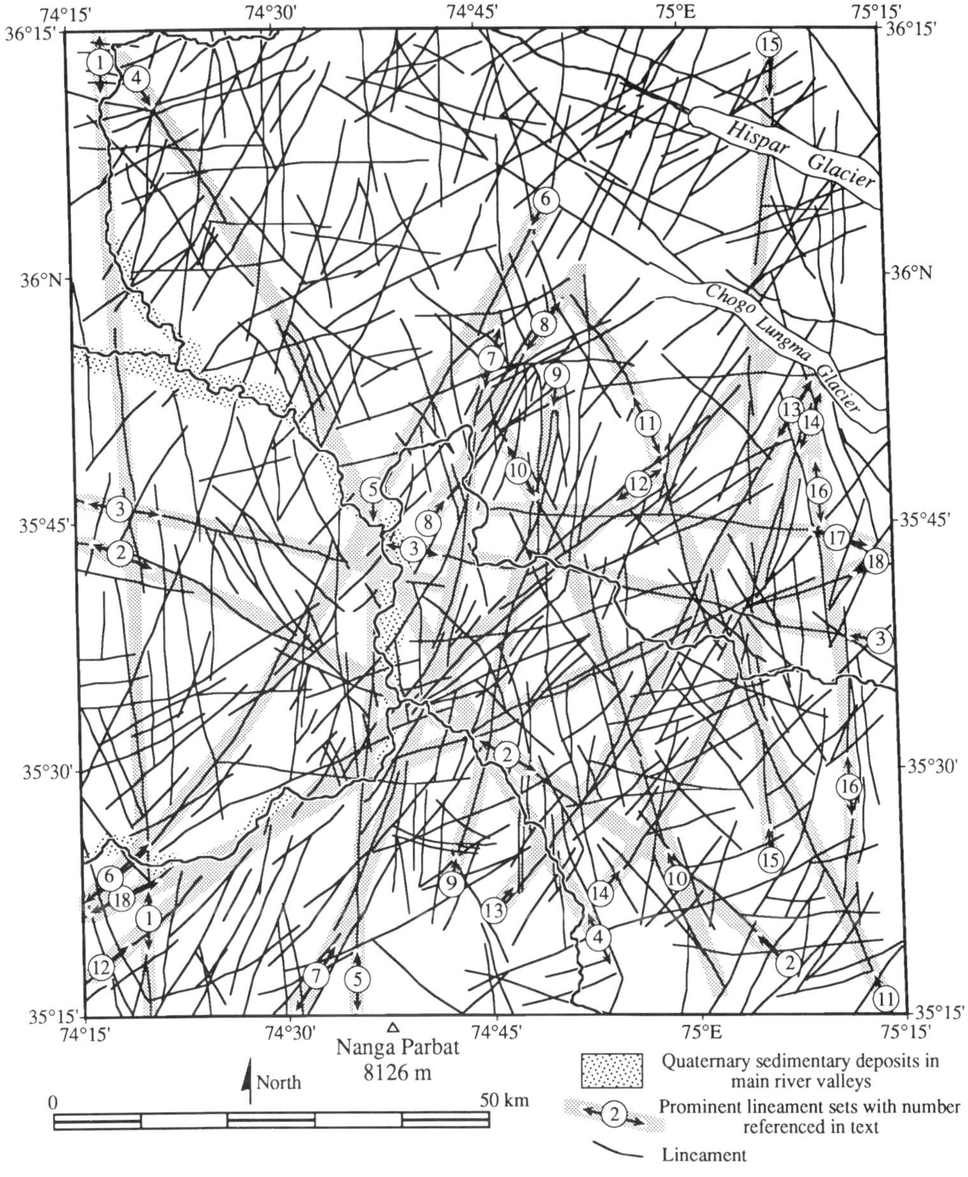

Fig. 8. Map of *c*. 540 lineaments, including drainage lineaments and lineaments defined along graytone boundaries. Circled numbers and halftone strips identify prominent lineament sets discussed in the text.

Topographic data were also attenuated along some ridge lines due to the very high albedo of snow and glacial ice. We sought to minimize biases related to the direction of solar illumination through (1) the use of multiple Landsat images of the same scene, (2) the continuous tracing of drainages as their orientations varied from high-contrast to low-contrast trends, and (3) the computer-assisted compilation of drainage-segment maps plotted by drainage segment azimuth. Faint tonal artifacts produced by the sensor scan lines, which trend west-northwest, can also mislead interpreters. Interpreter biases were partially mitigated by duplicate lineament analyses by different interpreters.

Results

Approximately 540 lineaments were identified in this study, ranging in length from <5 km to nearly 100 km (Fig. 8). Some lineaments could be interpreted as being continuous with other lineaments; however, we have attempted to be conservative in making such connections prior to evaluation of lineaments in the field. For the moment, the lineaments shown in Fig. 8 represent suggestions about where a geologist should look for faults, joints, foliation trends, lithologic boundaries, and other laterally extensive linear features that could be expressed as Landsat lineaments.

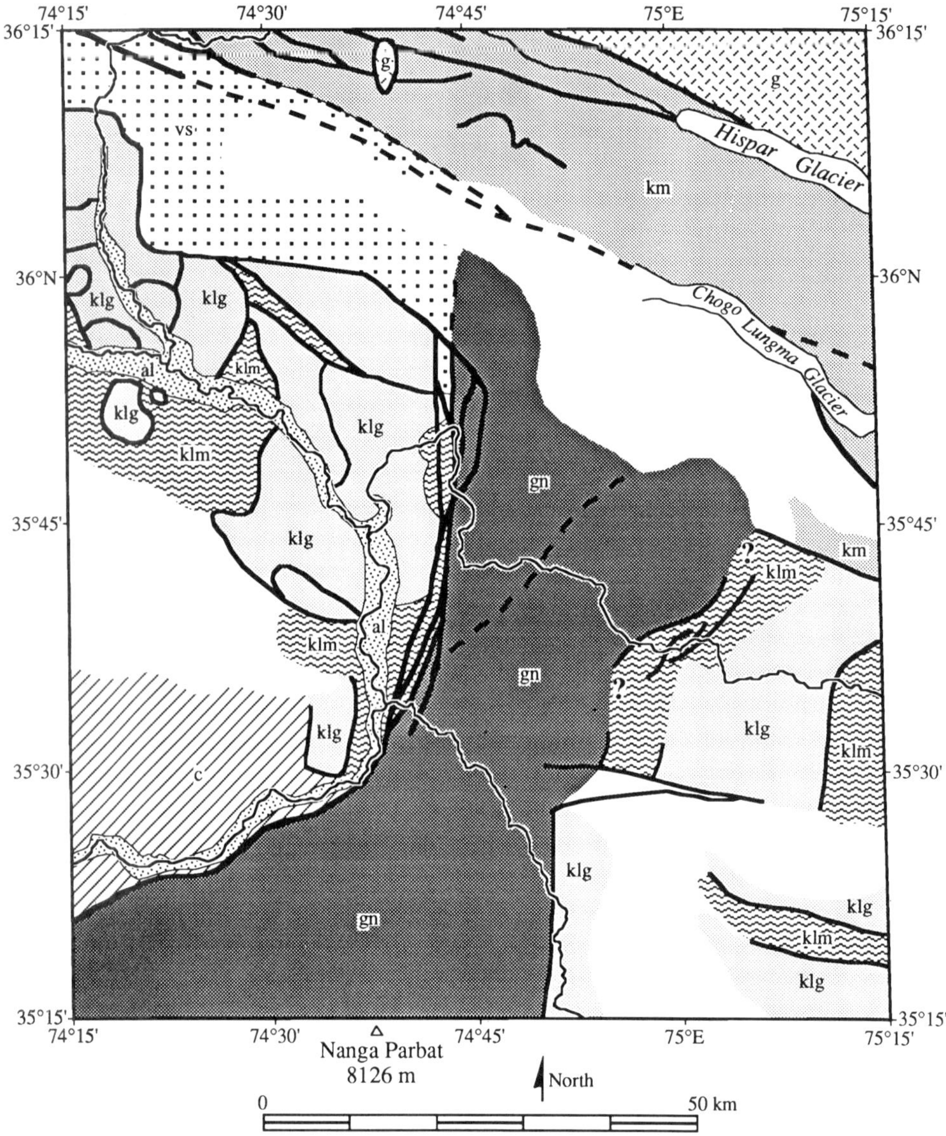

Fig. 9. Geological sketch map of the study area, simplified from Madin *et al.* (1989), Shroder *et al.* (1989), and Searle (1991). Bold curves are faults, dashed where location is in doubt; thin curves are lithological contacts. White areas are unmapped. Abbreviations are: al, Quaternary alluvium; g, Karakoram igneous rock units; km, Karakoram metamorphic complex; vs, Chalt & Yasin Groups; klg, igneous rock units of the Kohistan–Ladakh magmatic arc; klm, metasedimentary rock units of the Kohistan–Ladakh magmatic arc; c, Kamila Amphibolite and Chilas Complex; gn, Precambrian gneisses of the NPHM.

Do any of the lineaments correspond to known faults or joint sets? Figure 8 includes lineaments that correlate with segments of the Raikot, Baroluma, and Stak Faults (Madin 1986; Verplanck 1987; Madin *et al.* 1989; Shroder *et al.* 1989), as well as other unnamed faults in published maps (Fig. 9). Some of the faults included on published maps do not correspond to strong Landsat lineaments along all or part of their traces. For example, the strand of the Raikot Fault that extends southwest from the Astor River (Shroder *et al.* 1989) is not obvious in the Landsat images. No adequate data has been published concerning joint orientations in the study area.

Do any of the lineaments correspond to known lithological contacts or rock-fabric trends? Many of the lithological contacts in the study area that are likely to be observable in Landsat images are faulted contacts. Intrusive contacts in the area are either non-planar or not laterally extensive enough to result in obvious Landsat lineaments. Virtually all of the published foliation orientation data were collected along the roadways in the Astor, Hunza, Gilgit, and Indus River Valleys, so correlations between lineaments and foliation trends are not yet possible for most of the study area. Foliation in the Nanga Parbat gneisses, as measured along the Indus and Astor Rivers, generally strikes north $\pm 10°$ (Searle 1991), so some of the north-trending lineament fabric within the NPHM is probably related to foliation. Many of the prominent lineaments discussed earlier in this paper are long enough to extend across the known lithological boundaries in the area. The mode of origin for most of the lineaments identified in this study is not definable using current published geological maps.

We are reluctant to interpret lineaments without reference to greater geological control than is currently available; however, several trends along the NPHM seem especially noteworthy (see labels in Fig. 8). The following discussion does not include the significant lineament trends north of the Main Karakoram Thrust (MKT; Mattauer pers. comm. 1979 *in* Tahirkheli & Jan 1984; Bard *et al.* 1980), and is not meant to be an exhaustive survey of significant trends in the mosaic south of the MKT. The 18 trends mentioned below seem of particular geological interest and are generally accessible for field study.

Trend 1 is one of a family of north-northwest-trending drainage lineaments that are particularly common in the area south of Gilgit and west/north of the Indus River. The Hunza Valley between Jutal and Chalt is coincident with trend 1, which extends beyond the mosaic boundaries to the north and south. Trend 2 is most evident west of the Indus River, and coincides with the lower *c.* 10 km of the Astor River Valley and the upper reaches of the Urdung Gah, in the southeast corner of the mosaic (Casnedi & Ebblin 1977). Trend 3 extends across the entire study area, approximately through Shengas. The Indus flows along trend 3 for *c.* 10 km east of Shengas. In the eastern half of the area, trend 3 is typically marked by tonal boundaries. The three main segments of trend 3 (east of Shengas, Shengas to near the Indus–Gilgit River confluence, west of the Indus) have a left-stepping relationship to one another.

Trend 4 is a *c.* 10–15 km wide zone of aligned or parallel drainages extending from Astor northwest to between Nomal and Chalt on the Hunza River. This trend includes *c.* 20 km of the Astor Valley immediately northwest from Astor, as well as the lower *c.* 20 km of the Sai, Gilgit, and Hunza River Valleys. Cronin (1989) named this trend the Nomal–Astor Line. Trend 5 coincides with the Indus Valley from the confluence of the Indus and Gilgit Rivers to Raikot, south of which the trend coincides with the Raikot Valley. Trend 6 extends from the Indus Valley *c.* 15 km west of Jalipur northwest through Shuta, along part of the Puparash Valley, and across the Haramosh Range ridgeline towards the Hispar Valley.

Trend 7 extends from the west side of Nanga Parbat northeast to near the mouth of the Astor River, where it coincides with the northern Raikot Fault. Trend 8 extends in a northeast direction from just south of Sassi to near Iskere. Trend 9 is approximately parallel to the Raikot Fault, and passes through Shengas along the Indus River. Trend 10 extends from the Urdung Gah Valley over the Indus–Astor spur of the Deosai Mountains, down a prominent glacial valley, across the Indus River just east of Shengas, to the Phuparash River *c.* 5 km north of Sassi. Trend 11 is defined by a set of north-northwest-trending lineaments that extend from the southeast corner of the mosaic through the lower Stak River Valley. The Indus River flows along a 5 km right step that coincides with trend 11 near the Stak River confluence.

Lineaments generally aligned with the Baroluma Fault (Madin 1986; Madin *et al.* 1989), trend 12, can be interpreted extending from the Hispar Valley southwest to the Indus Valley. The course of the Indus Valley follows trend 12 from the confluence of the Indus and Astor Rivers to just south of Jalipur. Lineaments generally aligned with the Stak Fault (Verplanck 1987) are along trends 13 and 14, which extend northeast from the Astor Valley to the Chogo Lungma Glacier. Trends 15 and 16 parallel a family of north-trending drainage lineaments that are common in the mosaic east of 75° E longitude. This family of lineaments can be interpreted as extending from the Deosai Mountains northward beyond the edge of the mosaic. Trend 17 is orientated west-northwest, and is observable approximately 5 km north of Shengas extending from the Indus River to the east edge of the mosaic. Trend 18 extends east-northeast across the entire mosaic, coinciding with the Indus Valley near Jalipur and Raikot, and crossing the Indus again near the mouth of the Stak River. The western part of this trend coincides with the southwestern part of the Raikot Fault.

It is reasonable to speculate that structural activity may be accommodated on more than the few active faults that have been mapped in this area, given that the NPHM is in one of the most structurally active areas on Earth. Many of the lineaments that do not correlate with known faults, joints, lithological contacts, or foliation trends are located in areas that have not yet been mapped adequately, or at all. Some of the lineaments may correspond to previously unmapped active faults. Other lineaments can be interpreted as extensions of previously mapped faults. Evaluation of the various hypotheses concerning the origin and significance of these lineaments will require field work.

Conclusions and comments

Approximately 540 lineaments have been identified in Landsat 2 MSS band 7 images of the NPHM. Segments of the relatively well known Raikot, Baroluma, and Stak Fault Zones were recognized as lineaments. Lineaments associated with these faults that do not coincide with the mapped fault traces can be interpreted as unmapped extensions of the known faults. Not all faults mapped in the literature were observable as lineaments, partly because of inaccuracies and inconsistencies in published fault maps of the NPHM area. Some lineaments may be controlled by metamorphic foliation or primary stratification; however, the paucity of rock fabric data across much of the study area makes it impossible to correlate lineaments with bedding or foliation at this time. At least some of the lineaments whose origins are not yet constrained by map data are likely to have developed along active faults. Structural field work is necessary to evaluate the origin of the lineaments along the NPHM.

The NPHM is one of Earth's great natural laboratories for the study of mountain-building processes at all crustal levels; the opportunities for cooperative multi-disciplinary research are many. The principal logistical problem in conducting geological field work along the NPHM, as elsewhere in the Himalaya, involves accurately locating field data and registering those locations to geographically accurate base maps. Lacking accurate site locations and base maps, compilation of geological data from disparate sources is tenuous at best.

The anticipated use of portable global positioning systems, geographic information systems for data management and analysis, and space images with improved spatial and spectral resolution will dramatically improve the accuracy and precision of geological mapping along the NPHM. Field work assisted by these technologies remains necessary to collect the primary geological data required to constrain models for the evolution of the NPHM. Data sets that are essential for future NPHM work include: (1) better topographic control via accurate digital elevation/terrain models; (2) basic geological mapping to establish the distribution of lithologies and to accurately locate the surface trace of contacts, faults, joints, foliation, and bedding; (3) basic structural petrofabric work on the NPHM to establish the kinematics of deformation within the cataclastic regime of the upper crust as well as the ductile/plastic regime of the lower crust; (4) long-term geodetic network surveys across suspected active structures; and (5) long-term microearthquake studies using a permanent digital seismograph network in the northwest Himalaya.

Computer resources used in this study were provided through faculty startup grants from the University of Wisconsin-Milwaukee to V.S. Cronin. This study evolved from earlier work funded by NSF grants to G. Johnson and N. Johnson at Dartmouth College. The costs of this pilot study were largely borne by C.E. and K.E. Cronin. C. Baruth of the American Geographic Society Collection at UW-Milwaukee assisted with our acquisition of Landsat images. We thank J. Risch, D. Rothery and other reviewers who suggested improvements to the manuscript, as well as R. Lawrence of Oregon State University, J. Shroder of the University of Nebraska-Omaha, and G. Johnson of Dartmouth College for their continued assistance and encouragement in this work.

References

Armbruster, J., Seeber, L. & Jacob, K.H. 1978. The northwestern termination of the Himalayan mountain front: active tectonics from microearthquakes. *Journal of Geophysical Research*, **83**, 269–282.

BARD, J.P., MALUSKI, H., MATTE, P. & PROUST, F. 1980. The Kohistan sequence – crust and mantle of an obducted island arc. *University of Peshawar Geological Bulletin,* **13**, 87–94.

CASNEDI, R. 1976. Geological notes on the junction between the Haramosh-Nanga Parbat structure and the Karakorum Range. *Accademia Nazionale dei Lincei, Rendiconti,* **61**, 631–633.

—— & EBBLIN, C. 1977. Geological notes on the area between Astor and Skardu, Kashmir. *Accademia Nazionale dei Lincei, Rendiconti,* **62**, 662–668.

CRONIN, V.S. 1982. *The physical and magnetic polarity stratigraphy of the Skardu Basin, Baltistan, northern Pakistan.* AM thesis Dartmouth College, New Hampshire.

—— 1989. Structural setting of the Skardu intermontane basin, Karakoram Himalaya, Pakistan. *In*: MALINCONICO, L.L. & LILLIE, R.J. (eds) *Tectonics and Geophysics of the Western Himalaya.* Geological Society of America, Special Papers, **232**, 183–201.

—— & JOHNSON, G.D. 1992. Revised chronostratigraphy of the late Cenozoic Bunthang sequence of Skardu intermontane basin, Karakoram Himalaya, Pakistan. *In*: SHRODER, J.F., JR. (ed.) *Himalaya to the Sea: Geology, Geomorphology, and the Quaternary.* Routledge, London, in press.

——, JOHNSON, W.P., JOHNSON, N.M. & JOHNSON, G.D. 1989. Chronostratigraphy of the upper Cenozoic Bunthang sequence and possible mechanisms controlling base level in Skardu intermontane basin, Karakoram Himalaya, Pakistan. *In*: MALINCONICO, L.L. & LILLIE, R.J. (eds) *Tectonics and Geophysics of the Western Himalaya.* Geological Society of America, Special Papers, **232**, 295–309.

DENEBA SOFTWARE 1991. *Canvas v.3.01.* 3305 N.W. 7th Ave., Miami, Florida, USA.

DESIO, A. 1964. *Geological tentative map of the western Karakorum, scale 1:500,000.* Instituto di Geologia, Universita di Milano, Milano.

DEWEY, J.F., SHACKLETON, R.M., CHANG CHENGFA & SUN YIYIN 1988. The tectonic evolution of the Tibetan Plateau. *Philosophical Transactions of the Royal Society of London,* **A327**, 379–413.

EBBLIN, C. 1976. Tectonic lineaments in Karakorum, Pamir, and Hindu Kush from ERTS imageries. *Accademia Nazionale dei Lincei, Rendiconti,* **60**, 245–253.

GANSSER, A. 1964. *Geology of the Himalayas.* New York, Interscience.

GORNITZ, V. & SEEBER, L. 1981. Morphotectonic analysis of the Hazara Arc region of the Himalayas, north Pakistan and northwest India. *Tectonophysics,* **74**, 263–282.

HOBBS, W.H. 1904. Lineaments of the Atlantic border region. *Geological Society of America Bulletin,* **15**, 483–506.

KAZMI, A.H. 1979. Active fault systems in Pakistan. *In*: FARAH, A. & DEJONG, K.A. (eds) *Geodynamics of Pakistan.* Geological Survey of Pakistan, Quetta, 285–294.

—— & RANA, R.A. 1982. *Tectonic map of Pakistan.* Scale 1:2 000 000. Geological Survey of Pakistan, Quetta.

KHURSHID, A., YIELDING, G., AHMAD, S., DAVISON, I., JACKSON, J.A., KING, G.C.P. & LIN, B.Z. 1984. The seismicity of northernmost Pakistan. *Tectonophysics,* **109**, 209–226.

LAWRENCE, R.D. 1981. *Landsat mosaic of Pakistan.* Scale 1:2 000 000. Oregon State University.

—— & GHAURI, A.A.K. 1983. Evidence of active faulting in Chilas district, northern Pakistan. *University of Peshawar Geological Bulletin,* **16**, 185–186.

LILLESAND, T.M. & KIEFER, R.W. 1987. *Remote sensing and image interpretation* [2nd ed.]. John Wiley & Sons, New York.

MADIN, I.P. 1986. *Structure and neotectonics of the northwestern Nanga Parbat–Haramosh massif.* MS thesis Oregon State University.

MADIN, I.P., LAWRENCE, R.D. & UR-REHMAN, S. 1989. The northwestern Nanga Parbat–Haramosh Massif; evidence for crustal uplift at the northwestern corner of the Indian craton. *In*: MALINCONICO, L.L. & LILLIE, R.J. (eds) *Tectonics and Geophysics of the Western Himalaya.* Geological Society of America, Special Papers, **232**, 169–182.

MISCH, P. 1949. Metasomatic granitization of batholithic dimensions, part 1. *American Journal of Science,* **247**, 209–249.

MOLNAR, P. & TAPPONNIER, P. 1975. Cenozoic tectonics of Asia: effects of a continental collision. *Science,* **189**, 419–426.

—— & —— 1978. Active tectonics of Tibet. *Journal of Geophysical Research,* **83**, 5361–5375.

——, BURCHFIEL, B.C., LIANG, K. & ZHAO, Z. 1987. Geomorphic evidence for active faulting in the Altyn Tagh and northern Tibet and qualitative estimates of its contribution to the covergence of India and Eurasia. *Geology,* **15**, 249–253.

NI, J. & BARAZANGI, M. 1985. Active tectonics of the western Tethyan Himalaya above the underthrusting Indian plate: the upper Sutlej River Basin as a pull-apart structure. *Tectonophysics,* **112**, 277–295.

—— & YORK, J. 1978. Late Cenozoic tectonics of Tibetan Plateau. *Journal of Geophysical Research,* **83**, 5375–5384.

OLSON, T.M. 1982. *Sedimentary tectonics of the Jalipur sequence, northwest Himalaya, Pakistan.* AM thesis Dartmouth College, New Hampshire.

POHN, H.A. 1978. The use of the Ronchi rulings in geologic lineament analysis. *In*: O'LEARY, D.W. & EARLE, J.L. (eds) *Proceedings of the Third International Conference on Basement Tectonics.* Basement Tectonics Committee, Denver, 89.

ROCKWARE, INC. 1988. *Digitize.* 4251 Kipling Street, Wheat Ridge, Colorado, USA.

ROTHERY, D.A. 1985. Interactive processing of satellite images for geological interpretation – a case study. *Geological Magazine,* **122**, 57–63.

—— & DRURY, S.A. 1984. The neotectonics of the Tibetan Plateau. *Tectonics,* **3**, 19–26.

SCHURTER, G.J. 1991. *A seismotectonic and lineament analysis of the Nanga Parbat–Haramosh Massif, northwest Himalaya, Pakistan*. MS thesis University of Wisconsin-Milwaukee.

SEARLE, M.P. 1991. *Geology and tectonics of the Karakoram Mountains*. John Wiley & Sons, Chichester.

——, REX, A.J., TIRRUL, R., REX, D.C., BARNICOAT, A. & WINDLEY, B.F. 1989. Metamorphic, magmatic, and tectonic evolution of the central Karakoram in the Biafo-Baltoro-Hushe regions of northern Pakistan. *In*: MALINCONICO, L.L. & LILLE, R.J. (eds) *Tectonics and Geophysics of the Western Himalaya*. Geological Society of America, Special Papers, **232**, 47–73.

SEEBER, L. & ARMBRUSTER, J. 1979. Seismicity of the Hazara arc in northern Pakistan: decollement vs. basement faulting. *In*: FARAH, A. & DEJONG, K.A. (eds) *Geodynamics of Pakistan*. Geological Survey of Pakistan, Quetta, 131–142.

—— & JACOB, K.H. 1977. Microearthquake survey of northern Pakistan: preliminary results and tectonic implications. *Proceedings of the C.N.R.S. Symposium on the Geology and Ecology of the Himalayas, Paris,* 347–360.

SHRODER, J.F., KHAN, M.S., LAWRENCE, R.D., MADIN, I.P. & HIGGINS, S.M. 1989. Quaternary glacial chronology and neotectonics in the Himalaya of northern Pakistan. *In:* MALINCONICO, L.L. & LILLIE, R.J. (eds) *Tectonics and Geophysics of the Western Himalaya.* Geological Society of America, Special Papers, **232**, 275–294.

SLATER, P.N. 1979. A re-examination of the Landsat MSS. *Photogrammetric Engineering and Remote Sensing,* **45**, 1479–1485.

SRIMAL, N. 1983. Geology and tectonics of the eastern Karakoram: implications for India-Asia collision. *Geological Society of America, Abstracts with Programs,* **15**, 694.

—— 1986. India-Asia collision: implications from the geology of the eastern Karakoram. *Geology,* **14**, 523–527.

SVERDRUP, K.A., SCHURTER, G.J. & CRONIN, V.S. 1990. Relocation of earthquakes near the Nanga Parbat–Haramosh Massif in the Karakoram-Himalaya Mountains, Pakistan (abs.). *EOS Transactions of the American Geophysical Union,* **71**, 562.

TAHIRKHELI, R.A.K. 1982. Geology of the Himalaya, Karakoram, and Hindukush in Pakistan. *University of Peshawar Geological Bulletin,* **15**, 1–51.

—— & JAN, M.Q. 1984. The geographical and geological domains of the Karakoram. *In*: MILLER, K.J. (ed.) *Proceedings of the International Karakoram Project.* Royal Geographic Society, London, **2**, 57–70.

TAPPONNIER, P. & MOLNAR, P. 1977. Active faulting and tectonics in China. *Journal of Geophysical Research,* **82**, 2905–2930.

TRELOAR, P.J., REX, D.C., GUISE, P.G., COWARD, M.P., SEARLE, M.P., WINDLEY, B.F., PETTERSON, M.G., JAN, M.Q. & LUFF, I.W. 1989. K-Ar and Ar-Ar geochronology of the Himalayan collision in NW Pakistan: constraints on the timing of suturing, deformation, metamorphism and uplift. *Tectonics,* **8**, 881–909.

TRUEBASIC, INC. 1991. *TrueBASIC v.2.1.* 12 Commerce Avenue, West Lebanon, New Hampshire, USA.

TURNER, F.J. 1952. "Gefugerelief" illustrated by "schist tor" topography in central Otago, New Zealand. *American Journal of Science,* **250**, 802–807.

VERPLANCK, P.L. 1987. *A field and geochemical study of the boundary between the Nanga Parbat–Haramosh massif and the Ladakh arc terrane, northern Pakistan.* MS thesis Oregon State University.

——, SNEE, L.W. & LUND, K. 1985. The boundary between the Nanga Parbat Massif and Ladakh island arc terrane, northern Pakistan: A cross fault on the Main Mantle Thrust (abs.). *EOS Transactions of the American Geophysical Union,* **66**, 1074.

WADIA, D.N. 1932. Note on the geology of Nanga Parbat, Mt. Diamir, and adjoining parts of Chilas, Gilgit district, Kashmir. *Geological Survey of India Records,* **66**, 212–234.

YEATS, R.S. & HUSSAIN, A. 1989. Zone of Late Quaternary deformation in the southern Peshawar Basin, Pakistan. *In*: MALINCONICO, L.L. & LILLIE, R.J. (eds) *Tectonics and Geophysics of the Western Himalaya.* Geological Society of America, Special Papers, **232**, 265–274.

—— & LAWRENCE, R.D. 1983. Tectonics of the Himalayan thrust belt in northern Pakistan. *In*: HAQ, B.U. & MILLIMAN, J.D. (eds) *Marine geology and oceanography of Arabian Sea and coastal Pakistan.* Van Nostrand Reinhold, New York, 177–198.

——, KHAN, S.H. & AKHTAR, M. 1984. Late Quaternary offset of the Salt Range of Pakistan. *Geological Society of America Bulletin,* **95**, 958–966.

YIELDING, G., AHMAD, S., DAVISON, I., JACKSON, J.A., KHATTAK, R., KHURSHID, A., KING, G.C.P. & LIN, B.Z. 1984. A microearthquake study in the Karakoram. *In*: MILLER, K.J. (ed.) *Proceedings of the International Karakoram Project.* Royal Geographic Society, London, **2**, 150–169.

ZEITLER, P.K. 1985. Cooling history of the NW Himalaya, Pakistan. *Tectonics,* **4**, 127–151.

——, JOHNSON, N.M., NAESER, C.W. & TAHIRKHELI, R.A.K. 1982*a* Fission-track evidence for Quaternary uplift of the Nanga Parbat region, Pakistan. *Nature,* **298**, 255–257.

——, TAHIRKHELI, R.A.K., NAESER, C.W. & JOHNSON, N.M. 1982*b*. Unroofing history of a suture zone in the Himalaya of Pakistan by means of fission-track annealing ages. *Earth and Planetary Science Letters,* **57**, 227–240.

——, SUTTER, J.F., WILLIAMS, I.S., ZARTMAN, R.E. & TAHIRKHELI, R.A.K. 1989. Geochronology and temperature history of the Nanga Parbat–Haramosh Massif, Pakistan. *In:* MALINCONICO, L.L. & LILLIE, R.J. (eds) *Tectonics and Geophysics of the Western Himalaya.* Geological Society of America, Special Papers, **232**, 1–22.

ZUMBERGE, J.H. 1955. Glacial erosion in tilted rock layers. *Journal of Geology,* **63**, 149–158.

Stratigraphy south of the Main Mantle Thrust, Lower Swat, Pakistan

JOSEPH A. DIPIETRO[1], KEVIN R. POGUE[2], ROBERT D. LAWRENCE[3], MIRZA S. BAIG[4], AHMAD HUSSAIN[5] & IRSHAD AHMAD[6]

[1] *Department of Geology, University of Southern Indiana, 8600 University Boulevard, Evansville, Indiana 47712, USA*

[2] *Department of Geology, Whitman College, Walla Walla, Washington 99362, USA*

[3] *Department of Geosciences, Oregon State University, Corvallis, Oregon 97331, USA*

[4] *Institute of Geology, University of Azad Jammu and Kashmir, Muzaffarabad, Azad Kashmir, Pakistan*

[5] *Geological Survey of Pakistan, 86-C New Defense Officer's Colony, Shami Road, Peshawar, Pakistan*

[6] *National Center of Excellence in Geology, University of Peshawar, Peshawar, Pakistan*

Abstract: The metamorphic sequence in Lower Swat is described and placed into the geological framework of Pakistan. The stratigraphic sequence consists of the Precambrian–Cambrian(?) Manglaur formation unconformably overlain by the Alpurai group which is subdivided into the Carboniferous or younger Marghazar formation, and the Triassic or younger Kashala, Saidu, and Nikanai Ghar formations. A third unit, the Jobra formation of unknown age, is present as discontinuous lenses unconformably below the Alpurai group. Type sections are indicated for each rock unit. Comparison with the stratigraphy in the Peshawar Basin indicates that the Lower Swat area existed as a highland with active normal faulting during and perhaps before deposition of the Marghazar formation. The Marghazar formation appears to be a rift facies that correlates with the Panjal Traps of western India. The Kashala and Nikanai Ghar formations represent a transition to a stable shelf environment. These units may correlate, in part, with the Zanskar Supergroup of western India. The Saidu formation may represent drowning of the shelf as it was pulled down and overridden by the Main Mantle Thrust suture melange. It may correlate with the Lamayuru Formation of western India. The Alpurai group thus records a depositional history from Late Paleozoic breakup of Gondwana, to development of a passive Mesozoic shelf, to drowning of the shelf at the onset of Himalayan orogeny.

Lower Swat is located in the foothills of the Himalayan mountains where the Peshawar Basin, at an altitude of 400 m gives way to 5500 m mountains in Kohistan (Fig. 1). The southern part of Swat lies on the Indian plate where Precambrian to Mesozoic argillites, quartzites, and limestones record a history of shelf deposition interrupted by numerous erosional unconformities (Pogue *et al.* 1992*a*). This rock sequence can be followed northward into progressively higher grade rock that culminates in a structural dome where rocks are at kyanite grade (DiPietro & Lawrence 1991). The northern flank of this dome is terminated by a major melange-bearing suture-zone called the Main Mantle Thrust (MMT). The MMT separates the Indian plate rocks from amphibolites that are considered to be part of the accreted Mesozoic Kohistan island arc sequence (Tahirkheli *et al.* 1979). Although much is known about the structural history of the Indian plate in Pakistan, few detailed studies are available on the metamorphic stratigraphy adjacent to the MMT. Lower Swat is an exceptional region in which to address this concern because there is no documented disruption of the stratigraphy by thrust faults such as are present in the Hazara region of Pakistan (Calkins *et al.* 1975 Bossart *et al.* 1988; Treloar *et al.* 1989*a*). In order to understand properly the tectonic history of the Indian plate it is fundamental to establish the stratigraphy of the region. In this paper, the stratigraphy of the metamorphic rocks along the northern edge of the Indian plate is described and correlated with other parts of Pakistan. Type localities are presented as well as age constraints and depositional environments.

From TRELOAR, P.J. & SEARLE, M.P. (eds), *Himalayan Tectonics*
Geological Society Special Publication, No. 74, pp. 207–220.

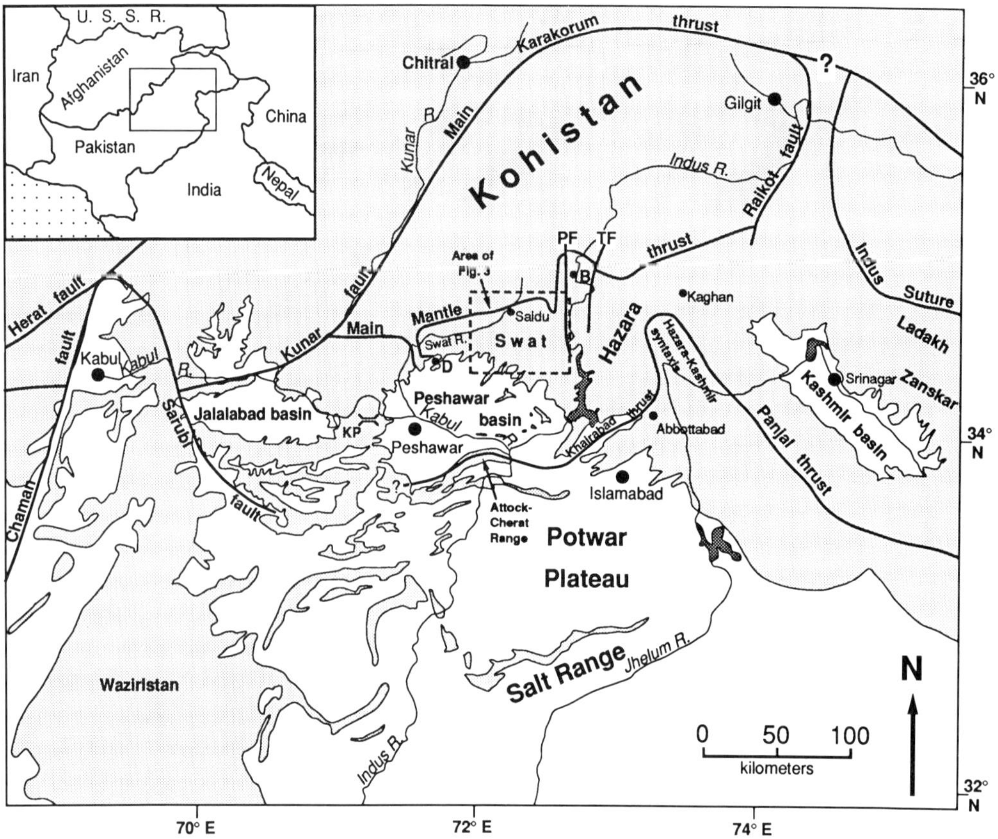

Fig. 1. Map of Northern Pakistan showing location of Fig. 3. Shaded area represents pre-Quaternary rock. B, Besham; D, Dargai; KP, Khyber Pass; PF, Puran Fault; TF, Thakot Fault.

Previous work

The stratigraphy of Lower Swat was originally established by Martin *et al.* (1962) and by King (1964). They described a sequence of quartz-schists, amphibolites, calc-schists, and phyllites, which they named the Lower Swat–Buner schistose group (Fig. 2). The base of this sequence is intruded by augen and tourmaline granite gneisses which they called the Swat granites and granitic gneisses. Martin *et al.* (1962) and King (1964) considered at least the upper part of the Lower Swat–Buner schistose group to be Upper Paleozoic or younger on the basis of poorly preserved fossils at Nikanai Ghar (King 1961). Stauffer (1967, 1968) re-examined the fossils and assigned a probable Silurian–Devonian age to all of the metasedimentary rocks in the region.

In the Hazara region of Pakistan, Le Fort *et al.* (1980) obtained a whole rock Rb/Sr age of 516 ± 16 Ma on the Mansehra granite. This granite was later considered to be contemporaneous with the granitic gneisses in Lower Swat (Jan *et al.* 1981; Le Fort *et al.* 1983). In order to explain the apparent discrepancy of Upper Palaeozoic metasedimentary rocks intruded by Lower Palaeozoic granitic rocks, Lawrence & Ghauri (1984), and Kazmi *et al.* (1984, 1986) proposed that the granitic rocks intrude only the lower part of the metasedimentary sequence and that an unconformity separates these rocks from the upper part of the metasedimentary sequence. Based on the presence of an unconformity, Kazmi *et al.* (1984) discarded the designation Lower Swat–Buner schistose group and introduced the name Manglaur schist for rocks intruded by granitic gneisses, and the names Alpurai schist and Saidu schist for rocks that unconformably overlie the Manglaur schist and granitic gneisses (Fig. 2).

Palmer-Rosenberg (1985), Ahmad (1986) and Ahmad *et al.* (1987) found that the tourmaline granite-gneisses intruded the lower part of the

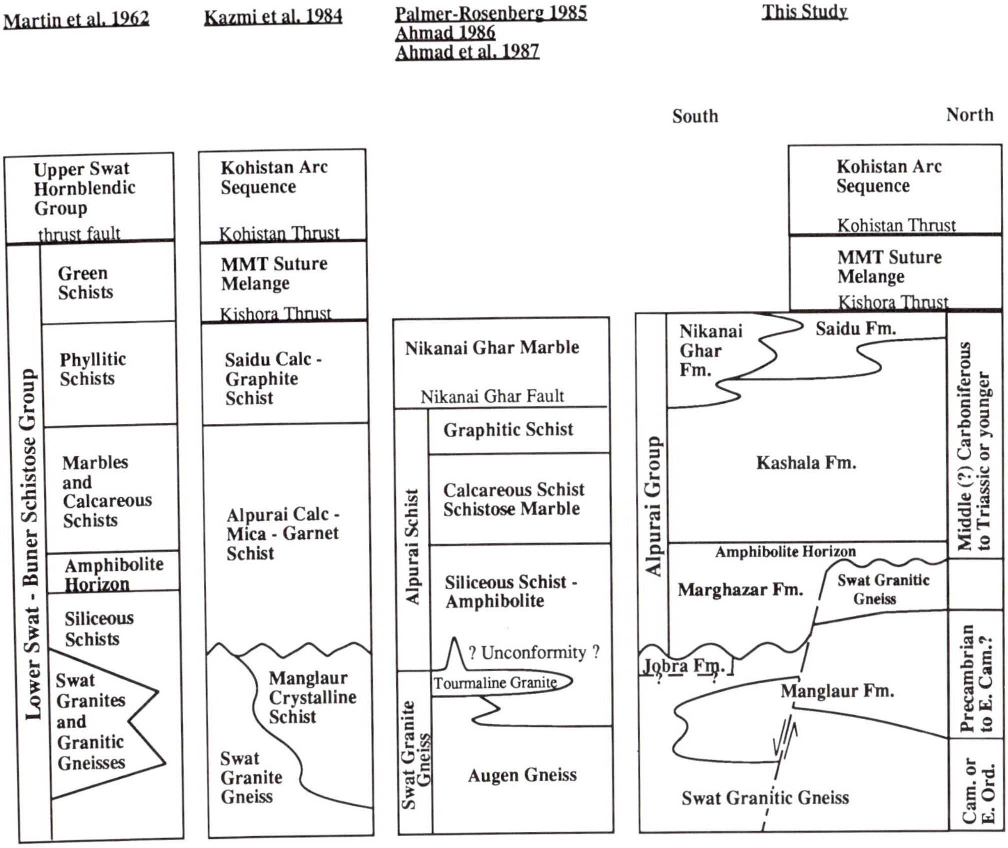

Fig. 2. Comparative structural/stratigraphic columns for Lower Swat. The column for this study does not show the tourmaline granite gneiss which intrudes the Marghazar formation.

Alpurai schist and therefore could not confirm the existence of an unconformity. These authors also introduced the name Nikanai Ghar marble for a thick bedded sequence of marble in the southern part of Lower Swat that they considered to be in fault contact with the underlying Alpurai schists (Fig. 2).

DiPietro (1990) re-examined the Swat area and, as a result, assigned the Manglaur schist and a new unit, the Jobra calc-silicate, to formation status, and the Alpurai schist to group status. The Alpurai group was further subdivided into the Marghazar, Kashala, Saidu, and Nikanai Ghar formations. DiPietro (1990) found no evidence for a fault below the Nikanai Ghar formation, although an unconformity between the Manglaur formation and Alpurai group was documented on the basis of both field and fossil evidence (DiPietro 1990; Pogue *et al.* 1992*a*). This paper follows the nomenclature of DiPietro (1990) who refers to the Manglaur–Jobra–Alpurai sequence along with the surrounding intrusive rock as the Lower Swat rock sequence (Fig. 2).

More than 95% of the intrusive rock consists of strongly deformed augen–flaser granite and granodiorite gneiss. This unit, traditionally referred to as the Swat granitic gneiss, intrudes the Manglaur formation but unconformably underlies the Alpurai group. The four largest bodies of Swat granitic gneiss are given local names as outlined in Fig. 3. The Swat granitic gneiss is believed to be Late Cambrian to Early Ordovician in age (Le Fort *et al.* 1980; Jan *et al.* 1981). The only other intrusive rock of volumetric importance is a leucocratic tourmaline granite gneiss that occurs principally along the margins of the Swat gneiss (Fig. 4). The tourmaline granite-gneiss intrudes the Alpurai group and therefore must be younger than the Swat gneiss. It is believed to be Palaeogene in age (DiPietro 1990).

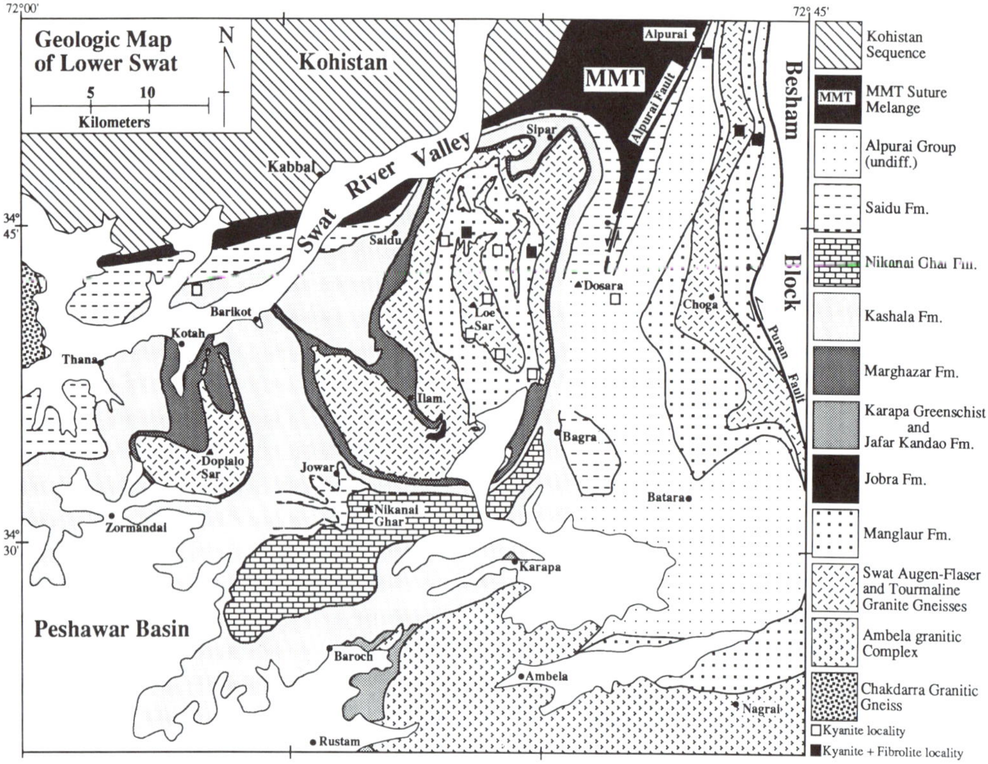

Fig. 3. Geological map of Lower Swat. Local names for the main bodies of Swat augen–flaser granitic gneiss are, from west to east: Dopialo Sar gneiss, Ilam gneiss, Loe Sar gneiss, and Choga gneiss. All four bodies contain minor amounts of younger tourmaline granite-gneiss particularly at their margins. The Besham block area consists of Precambrian gneiss, schist and amphibolite. Other nonpatterned areas represent Quaternary sediments.

General geology

The major structure in the Lower Swat rock sequence is a composite dome with two anticlinal culminations occupied by the Dopialo Sar and Loe Sar gneisses (Fig. 3). These culminations are separated by a synclinal trough that is occupied by the Kashala formation. The dome developed late in the structural and metamorphic history and is the result of uplift coupled with the superposition of E–W-trending open folds on earlier N–S-trending open folds (DiPietro & Lawrence 1991).

The earliest folds are west-verging basement involved recumbent fold nappes with NNW–SSE axial traces (DiPietro & Lawrence 1991). These folds have been recognized along the unconformity between the Ilam gneiss and Marghazar formation on the western margins of the dome north of Jowar and south of Ilam (Fig. 3). The folds die out in the Alpurai group above the unconformity and do not extend beyond the southern margin of the Ilam gneiss.

The early fold nappes developed synchronously with amphibolite facies metamorphism dated by $^{40}Ar/^{39}Ar$ techniques as Eocene–Oligocene in age (Maluski & Matte 1984; L. Snee *in* Palmer-Rosenberg 1985; Treloar *et al.* 1989*b*; Baig 1990). Metamorphic intensity decreases southward toward Baroch and Rustam where the rocks are at greenschist facies (Fig. 3). Garnet disappears south of the latitude of Jowar but is almost ubiquitous in schistose rock to the north. Aluminosilicate minerals are rare throughout the region. Figure 3 shows the known occurrences of kyanite and kyanite plus fibrolite. The Manglaur formation, Loe Sar gneiss, and Choga gneiss contain only trace amounts of aluminosilicate minerals. Those in the Alpurai group are restricted to the Kashala formation where typically they occur as coarse aggregates within quartz veins. Geothermobar-

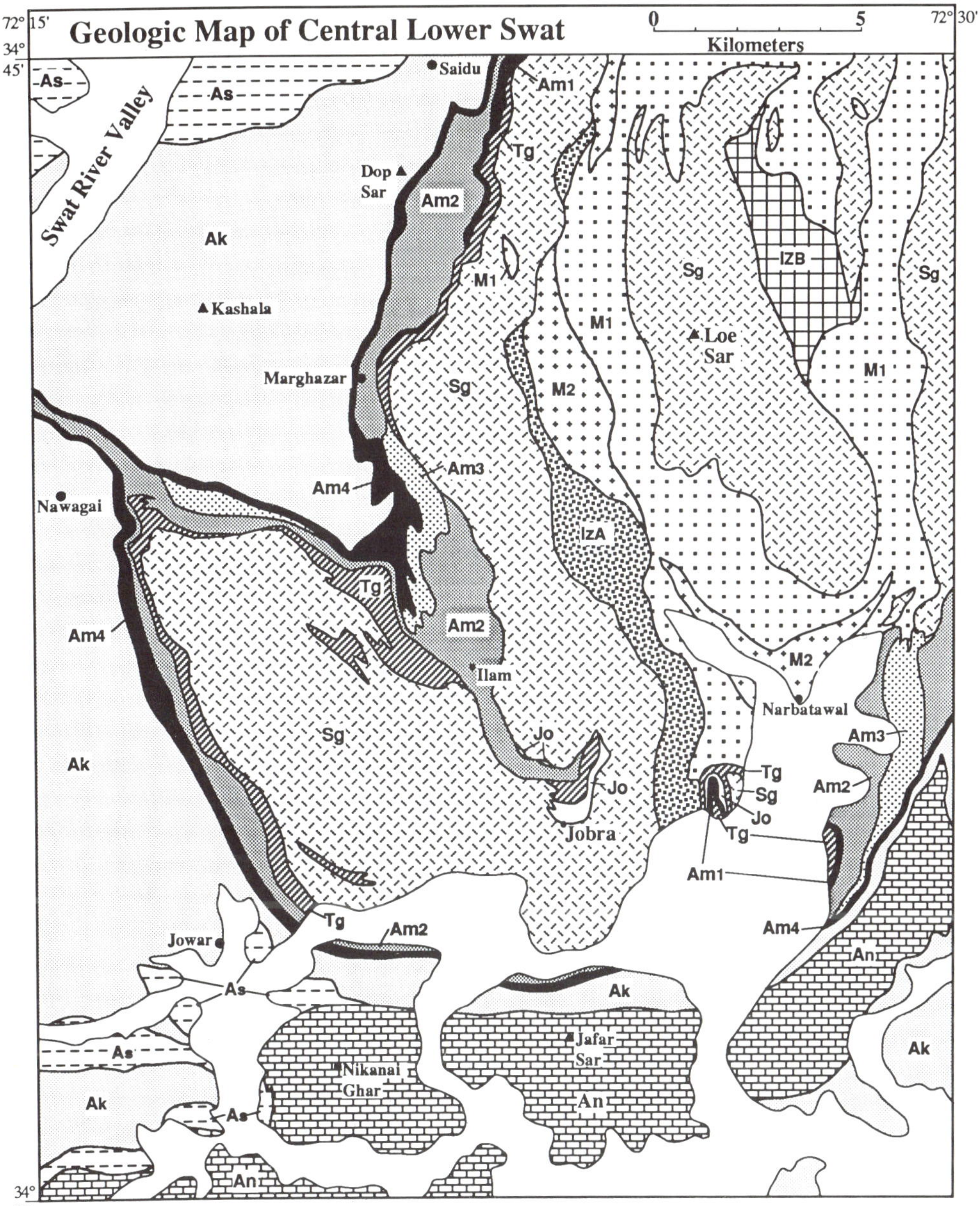

Fig. 4. Geological map of central Lower Swat showing type locations and lithological facies relationships in the Manglaur formation and Alpurai group. Location is the central part of Fig. 3. As, Saidu formation; An, Nikanai Ghar formation; Ak, Kashala formation; Am1, Marghazar phlogopite-bearing marble; Am2, undifferentiated Marghazar formation; Am3, Marghazar psammitic and amphibolitic schists; Am4, Marghazar amphibolite horizon; Jo, Jobra formation; M1, undifferentiated Manglaur formation; M2, Manglaur quartzite and tremolite marble; Sg, Swat granitic gneiss; Tg, Tourmaline granite-gneiss; IZA, intruded zone A; IZB, intruded zone B. Non-patterned areas represent Quaternary sediments (after DiPietro 1990).

ometry on rocks in the Manglaur and Marghazar formations between Barikot and Dosara indicate peak metamorphic conditions between 600–700° C and 9–11 kbar (DiPietro 1991). Farther south near Baroch the colour alteration index of conodonts from the Kashala formation is between 5.5 and 7 (Pogue *et al.* 1992*a*) suggesting peak metamorphic temperatures between 300 and 400° C (Epstein *et al.* 1977).

The Lower Swat rock sequence extends from the Peshawar Basin northward to where it truncates against the MMT. This suture zone is considered to be the western continuation of the 2500 km long Indus–Tsangpo suture zone and therefore marks the northern boundary of the exposed part of the Indian plate (Gansser 1964, 1980; Tahirkheli 1979*a*, *b*; Yeats and Lawrence 1984). In the Swat region the MMT consists of serpentinite, talc–carbonate schist, greenschist, and blueschist blocks in a sheared, greenschist facies matrix. It separates the Lower Swat rock sequence from amphibolites of the Kohistan island arc sequence (Fig. 3).

Toward the east, the Lower Swat rock sequence is truncated against the Puran fault which is an oblique, high-angle, left-lateral shear zone, downthrown on the west side (Baig *et al.* 1989; Baig 1990). Rocks in the Besham area, east of the Puran fault and west of the Thakot fault (Fig. 1), are entirely different from those in Lower Swat. The Besham basement complex consists of Precambrian paragneiss, amphibolite, and granitic gneiss that is unconformably overlain by Precambrian metaconglomerate, psammitic schist, pelitic schist, and graphitic schist of the Karora group (Jan & Tahirkheli 1969; Gansser 1979; Ashraf *et al.* 1980; Baig *et al.* 1989; Baig 1990). These rocks were metamorphosed and strongly deformed in the Precambrian but were little affected by Himalayan metamorphism (Treloar *et al.* 1989*b*; Baig 1990). Rocks of similar age and of similar depositional and deformational history as those in the Swat region reappear east of the Thakot fault in Hazara (Greco *et al.* 1989; Treloar *et al.* 1989*a*). Nick points in side streams that cross the Puran and Thakot faults suggest that both faults are active features (Baig 1990). Baig (1990) and DiPietro & Lawrence (1991) explain the lack of Himalayan metamorphism in the Besham rocks by suggesting that the Swat and Hazara rocks were well north of Besham during the Himalayan metamorphism. The Swat and Hazara rocks were later carried southward over the

Table 1. *Representative mineral assemblages*

Formation	Mineral assemblage	Rock type
Nikanai Ghar	Cal–Do	Marble
	Do–Cal	Dolomitic marble
Saidu	Ms–Qtz–Gr–Clz–Chl–Ilm	Graphitic phyllite
	Cal–Qtz–Pl–Ms	Marble
Kashala	Ms–Clz–Cal–Qtz–Grt–Ilm–Rt	Calcareous schist
	Cal–Clz–Ms–Qtz	Schistose marble
	Cal–Qtz–Ms–Pl–Grt–Ilm	Schistose marble
	Cal-Qtz-Ms	Marble
	Do–Cal	Dolomitic marble
Marghazar	Qtz–Pl–Ms–Bi–Ilm–Grt	Garnetiferous schist
	Bi–Clz–Qtz–Pl	Biotite schist
	Pl–Qtz–Bi–Chl–Clz–Grt	Psammitic schist
	Qtz–Pl–Kfs–Bi–Ms	Psammitic schist
	Qtz–Pl–Ms	Feldspathic quartzite
	Cal–Pl–Qtz–Clz–Phl–Ilm	Marble
	Hbl–Qtz–Bi–Pl–Grt–Rt	Hornblende schist
	Hbl–Qtz–Ms–Grt–Rt	Hornblende schist
	Hbl–Pl–Qtz–Clz–Cal–Rt–Ilm	Amphibolite
Jobra	Cal–Qtz–Wo–Grs–Vs–Di–Kfs–Spn	Calc-silicate
	Cal–Do–Tr–Chl	Tremolite marble
	Hb–Pl–Cal–Qtz–Clz	Amphibolite
Manglaur	Ms–Qtz–Pl–Grt–Bi–Ilm	Garnetiferous schist
	Qtz–Ms–Grt	Micaceous quartzite
	Qtz–Pl–Bi–Ky–Ms–Grt–Ilm	Feldspathic quartzite
	Tr–Cal–Do	Tremolite marble

Each assemblage represents a single thin section with the most abundant mineral listed first. Trace minerals are not shown. Mineral abbreviations: Bi, biotite; Cal, calcite; Chl, chlorite; Clz, clinozoisite; Di, diopside; Dol, dolomite; Gr, graphite; Grs, grossular; Grt, garnet; Hbl, hornblende; Ilm, ilmenite; Kfs, K-feldspar; Ky, kyanite; Ms, muscovite; Pl, plagioclase; Qtz, quartz; Rt, rutile; Spn, sphene; Tr, tremolite; Vs, vesuvianite; Wo, wollastonite.

Besham rocks along one or more foreland thrust faults. Recent uplift along the Puran and Thakot faults then exhumed the Besham block from below the thrust faults.

Toward the west, the Lower Swat rock sequence extends beyond Zormandai possibly as far as Khyber, however, these areas have not been mapped in detail (Lawrence *et al.* 1989).

Stratigraphy

Manglaur Formation (M1 and M2)

The Manglaur formation proper is restricted to a domal area near Loe Sar where it is sandwiched between layers of Swat granitic gneiss (Fig. 3). The proposed type locality is along the ridge north of Narbatawal (Fig. 4). At its type locality, the Manglaur formation is subdivided into two lithological units. Undifferentiated Manglaur formation (M1) is composed mainly of garnetiferous schist and micaceous quartzite. The garnetiferous schist alternates with fine-grained, garnet-poor, psammitic layers at a scale of a few centimetres. Minor quartzite and rare interlayers of graphitic schist, amphibolite, and marble also occur (Table 1). The second lithological unit (M2) is distinguished on the basis of thick interlayers of quartzite, fine-grained feldspathic quartzite and/or tremolite marble. These rocks occur as lenses near the top of the exposed part of the Manglaur formation. The observed maximum exposed thickness, uncorrected for strain and folding, is 1500 m.

The Manglaur formation name has been applied to a similar sequence of rock located along the margins of the Choga granitic gneiss (Baig 1990, Fig. 3). Both lithological units are recognized in this area with tremolite-bearing marbles near the top of the formation. Descriptions by Martin *et al.* (1962), suggest that this belt of Manglaur formation extends southward to the Ambela complex (Fig. 3).

Jobra Formation (Jo)

The Jobra formation is present only as a series of small, isolated lenses that overlie the Ilam gneiss and underlie the Marghazar formation. It is mapped separately from the Alpurai group and Manglaur formation because it is lithologically distinct and because its age and contact relationships with the other units are not certain. The largest exposure and proposed type locality occurs in the village of Jobra (Fig. 4). The distinguishing rock is a fine-grained (<0.5 mm) wollastonite-bearing calc-silicate rock that occurs interlayered with quartzite, tremolite marble, garnetiferous schist and amphibolite (Table 1).

The Jobra formation is not in contact with the Manglaur formation. The lower contact with the Ilam gneiss is sharp with no interlayering. Shams (1963) mapped a xenolith of similar calc-silicate rock within the Swat gneiss northeast of Saidu. No xenoliths were found near the type locality, therefore, it is not clear if the Jobra formation is intruded by the Swat gneiss or unconformably overlies the Swat gneiss. Truncation of marker units in the Jobra formation indicate that the upper contact with the Marghazar formation is an unconformity (DiPietro 1990).

Alpurai Group

The Alpurai schist was introduced by Kazmi *et al.* (1984) for a sequence of largely calcareous schists that unconformably overlie the Manglaur formation and Swat gneiss. DiPietro (1990) assigned the Alpurai to group status and subdivided the group into the Marghazar, Kashala, Saidu, and Nikanai Ghar formations. Each formation is characterized by facies changes and thickness variations. The Marghazar formation everywhere is stratigraphically below the other three formations. In places it lithologically resembles the Kashala formation but is distinguished as being the only formation in the Alpurai group that contains amphibolites. The Kashala, Saidu and Nikanai Ghar formations show close lithological similarities. There is, in addition, at least partial chronostratigraphic equivalence between these units.

Marghazar Formation (Am1, Am2, Am3, Am4). In central Lower Swat, the Marghazar formation unconformably overlies the Ilam gneiss and Manglaur formation (Figs 3 and 4). The proposed type locality is along the stream valley north and south of Marghazar village (Fig. 4). In this area, and throughout central Lower Swat, four mappable lithological units are recognized. A phlogopite-bearing marble (Am1) forms discontinuous layers at the base of the Marghazar formation where it can be mapped as a separate marker horizon. In a few areas the marble makes up more than half of the formation and, near Sipar, marble and calc-schist occur throughout the Marghazar formation. Much of the marble is identical to marbles in the Kashala formation. A persistent 20–50 m thick layer of dark greenish grey, fine-grained, amphibolite (Am4) is present everywhere at the top of the Marghazar formation. This is the amphibolite horizon of Martin *et al.* (1962) and it marks the stratigraphically highest position of any amphibolitic rock in Lower Swat. Between the laterally discontinuous marble and the laterally continuous amphibolite horizon, the Marghazar formation is characterized by great horizontal variability and thickness changes. Undifferentiated Marghazar formation (Am2) consists of garnetiferous schist, biotite schist, and hornblende schist interlayered with lesser psammitic schist, feldspathic quartzite, marble, and amphibolite (Table 1). In several areas, thick sequences of psammitic schist, hornblende schist, and amphibolite form a separate mappable unit (Am3) that locally makes up nearly the entire Marghazar formation (Fig. 4).

Extreme changes in thickness and lithology in the Marghazar formation are well documented in the area east of Narbatawal (DiPietro 1990). The maximum observed thickness, uncorrected for strain and folding is at least 1000 m in the area east of Narbatawal where the Marghazar formation overlies the Manglaur formation (Fig. 4). Included within the Marghazar formation is a psammitic schist unit (Am3) that contains scattered pebbles of Swat gneiss. The psammitic schist can be followed northward to where it ends abruptly against the Swat gneiss. At this point the Marghazar formation unconformably overlies the Swat gneiss and its thickness is reduced by at least 50% (Fig. 4). The Marghazar formation continues to thin northward where locally it is less than 100 m thick and consists almost entirely of the amphibolite horizon (Am4). The Marghazar formation is never more than about 500 m thick where it overlies the Swat gneiss.

The Marghazar formation name is applied to a similar sequence of rock north of, and surrounding, the Dopialo Sar granitic gneiss (Fig. 3). In the area southeast of Kotah the Marghazar formation both overlies and structurally underlies the granitic gneiss. This area is interpreted as an early, syn-metamorphic, west-verging fold nappe that is similar in style to the early west-verging recumbent fold nappes described by DiPietro & Lawrence (1991) in the areas south of Ilam and north of Jowar (Fig. 5). The two 'fingers' of granitic gneiss that extend northward toward the Swat River Valley represent the anticlinal core of the nappe. The western finger is breached forming erosional outliers of granitic gneiss structurally above the Marghazar formation (Figs 3 and 5). Along the eastern finger and southern margin of the Dopialo Sar gneiss, the stratigraphically right-side-up section of the Marghazar formation is 30–50 m thick and consists almost entirely of the amphibolite horizon. This section thickens to between 50 and 100 m at the northern tip of the western finger with the addition of marble, psammitic schist, garnetiferous schist, and hornblende schist below the amphibolite horizon. The stratigraphically inverted section between the two fingers of granitic gneiss is over 300 m thick and consists of a nearly horizontal sequence of psammitic schist, hornblende schist, amphibolite, and marble.

The facies changes and thickness variations in the Marghazar formation across the fold are not surprising considering the rapid lateral changes documented in the area surrounding the Ilam gneiss. The changes seen at Dopialo Sar are similar to those seen in the Ilam gneiss east of Narbatawal.

Marghazar correlatives are also present east and south of Alpurai village and near Batara where they surround the Manglaur formation (shown as undifferentiated Alpurai group in Fig.

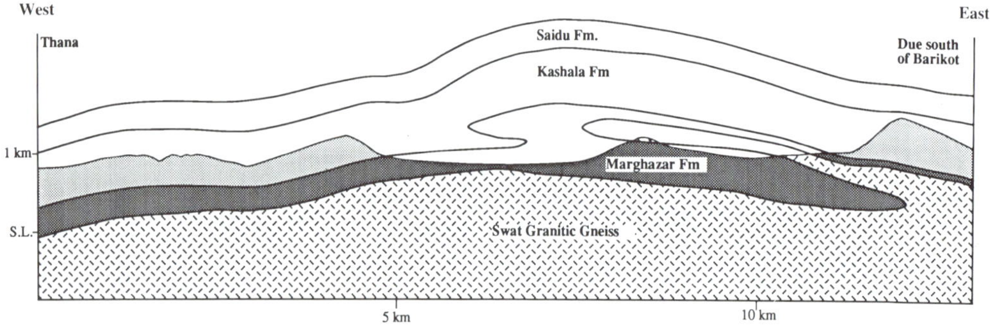

Fig. 5. Cross section from Thana along a line due east to the longitude of Barikot. Patterned below surface level only.

3). In these areas the Marghazar formation typically is less than 150 m thick and composed of psammitic schist, hornblende schist, and amphibolite, with discontinuous layers of phlogopite-bearing marble at the base. The amphibolite horizon, which is present everywhere at the top of the Marghazar formation in central Lower Swat, is at least locally absent in this area.

Kashala formation (Ak). The Kashala formation underlies a large area of Lower Swat from the MMT to the Peshawar Basin. It takes its name from Kashala mountain about 13 km SSW of Saidu (Fig. 4). Because of significant north to south compositional and metamorphic changes, two type localities are proposed. The first type locality is along road cuts on Dop Sar in an area of high-grade metamorphism (Fig. 4). Here, the Kashala formation consists of garnet and/or epidote-bearing calcareous schist that grades into schistose marble and finally into grey or brown-weathering calcite marble that locally is coarsely crystalline (Table 1). A few pelitic layers lack calcite entirely but psammitic layers are absent. North to south changes include the disappearance of garnet and the appearance of fine-grained dolomitic marble south of the latitude of Jowar (Fig. 3). The second type locality is within an area of low-grade metamorphism along stream cuts north of Baroch (Fig. 3). Here, the Kashala formation consists principally of brown-weathering marble and dolomitic marble with minor interlayers of calcareous phyllite and phyllite. The observed thickness of the Kashala is less than 700 m north of Jowar but more than 1200 m near Baroch. The lower contact with the amphibolite horizon of the Marghazar formation is sharp.

The Kashala formation can be followed eastward to the vicinity of Sipar and Bagra where interlayers of graphitic phyllite are common. Farther east, much of the undifferentiated Alpurai group shown in Fig. 3 is Kashala formation. In this area the marble tends to be less schistose with thick intervals of grey or brown-weathering marble alternating with calcareous schist and graphitic phyllite. The extent of the Kashala formation west of Zormandai is not known.

Saidu Formation (As). The Saidu formation proper occurs as a continuous belt adjacent to the MMT where it overlies the Kashala formation (Fig. 3). The type location is along road cuts one kilometer north of Saidu (Martin *et al.* 1962; Kazmi *et al.* 1984). Here the rock consists mainly of graphitic phyllite with minor lenses of grey or brown-weathering marble that is identical to marble in the Kashala formation (Table 1). Laminated silty layers are present but psammitic layers are absent. Garnet is present locally in the graphitic phyllite but is uncommon and does not characterize the rock. The maximum observed thickness, uncorrected for strain and folding, is 1000 m. The lower contact with marble of the Kashala formation is sharp near Saidu but is gradational and difficult to place near Sipar and Bagra where the Kashala formation contains abundant interlayers of graphitic phyllite. The upper contact with distinctly green phyllites, talc carbonates, and greenschists of the MMT is interlayered due to thrust imbrication.

The Saidu formation name is applied to a similar sequence of rock near Jowar. Here, the Saidu formation consists of two belts of graphitic phyllite and lesser marble that converge toward the east such that the Saidu formation overlies the Kashala formation and underlies the Nikanai Ghar formation. Farther east, the Saidu pinches out and the Kashala formation becomes thicker. These relationships suggest stratigraphic equivalence between the lower part of the Saidu formation and the upper part of the Kashala formation. The western continuation of this belt has not been mapped but reconnaissance suggests that the two belts pinch out in the Kashala formation.

Thick intervals of graphitic phyllite, correlated with the Saidu formation, are also present east of Bagra (Fig. 3). Reconnaissance mapping suggests that this belt may extend northward and connect with the Saidu formation near Dosara. The Alpurai fault, which cuts the Saidu formation near Dosara, is considered to be a reactivated fault between the MMT and the Saidu formation. The reactivated fault is an oblique, down on the west side, high-angle fault that post-dates metamorphism and the final phase of folding. The fault has not been traced to the south and presumably dies out.

Nikanai Ghar Formation (An). The Nikanai Ghar formation is restricted to a large lens shaped body in the south-central part of Lower Swat (Figs 3 and 4). The proposed type locality is along stream valleys on the south side of Jafar Sar (Fig. 4). Here the Nikanai Ghar formation consists of a thick-bedded sequence of dark grey to white marble and dolomitic marble with rare interlayers of graphitic phyllite and feldspathic quartzite. The marble is similar to that in the Kashala formation but coarsely crystalline varieties are more common. The major lithological difference between the two formations is the almost complete lack of micaceous rock in the

Nikanai Ghar formation. The contact between the Nikanai Ghar and Kashala formations is gradational and is placed above the last appearance of micaceous rock in the Kashala formation. This coincides roughly with the first appearance of coarsely crystalline marble in the Nikanai Ghar formation. The contact with graphitic phyllites of the Saidu formation is sharp.

The Nikanai Ghar formation is at least 1200 m thick in its southern part north of Baroch where it occupies the high mountain peaks and overlies the surrounding Kashala formation. It thins to zero thickness toward the northeast due to facies changes with the Kashala formation. At its northeastern termination near Bagra, crystalline marble is interlayered with schistose marble and garnet-bearing calcareous schist. Beyond this point, the Nikanai Ghar formation is not recognizable as a distinct lithologic unit. In this area, it appears that the Nikanai Ghar formation is interlayered with the Kashala formation and underlies Saidu correlatives at Bagra. These relationships suggest at least partial stratigraphic equivalence between the top of the Kashala formation and the Nikanai Ghar formation. Partial stratigraphic equivalence is also indicated between the Nikanai Ghar and Saidu formations by their similar position above the Kashala formation in different parts of Lower Swat.

Discussion

Age correlation and depositional environment of the Manglaur and Jobra Formations

There are no fossils in the Manglaur Formation. Lithologically, it resembles the Tanawal Formation which, in the eastern Peshawar Basin, unconformably overlies the Precambrian Salkhala Formation and unconformably underlies the Cambrian(?) Ambar formation (Pogue & Hussain 1988; Pogue *et al.* 1992*a*). In Hazara, the Tanawal Formation is intruded by the 516 ± 16 Ma Mansehra granite. Baig (1990) interprets the Tanawal Formation as a Late Precambrian molasse that was deposited following the *c.* 650 Ma Hazaran orogeny (Baig & Lawrence 1987; Baig *et al.* 1989). A Tanawal–Manglaur correlation would imply a Late Precambrian to Early Cambrian age for the Manglaur formation.

The age of the Jobra formation is problematic as its contact relationship with the Swat gneiss is not clear. It is believed to represent part of an early to middle Palaeozoic shelf sequence that was eroded prior to deposition of the Marghazar formation. Its position, unconformably below the Marghazar formation, suggests that it may correlate with the Devonian Nowshera formation which is exposed in the Peshawar Basin.

Age, correlation and depositional environment of the Alpurai Group

There are no reported fossils in the Marghazar formation. Pogue *et al.* (1992*a*) mapped a persistent 20–50 m thick layer of greenschist directly below the Kashala formation in the vicinity of Baroch and named it the Karapa greenschist. This unit is correlated with the higher-grade Marghazar amphibolite horizon based on its identical stratigraphic position. Below the Karapa greenschist, Pogue & Hussain (1988) and Pogue *et al.* (1992*a*) describe a laterally variable, 1100 m thick sequence of argillite, sandy limestone, conglomerate, quartzite, and calcareous quartzite, which they named the Jafar Kandao formation. This unit contains Late Devonian (late Famennian) to Early Carboniferous (Tournaisian) conodonts near its base and Mid-Carboniferous (Westphalian B) conodonts 20 m below the Karapa greenschist. These ages indicate that the Karapa greenschist, and the correlative Marghazar amphibolite horizon, is Westphalian or younger. The entire Marghazar formation must be Westphalian or younger where only the amphibolite horizon is present. The Jafar Kandao formation lacks interlayered greenschist, whereas amphibolite is abundant throughout the Marghazar formation. Although it is possible that the lower part of the Marghazar formation is as old as Late Devonian, we suggest, based on the lack of interlayered greenschist in the Jafar Kandao formation, that the entire Marghazar formation is Mid-Carboniferous or younger. In any case, it is clear that a major erosional unconformity separates the Marghazar formation from the Precambrian to Early Cambrian Manglaur formation.

Late Triassic (late Carnian) conodonts are reported from the middle of the Kashala formation near Baroch (Pogue *et al.* 1992*a*). This indicates a Triassic age for at least part of the Kashala formation and a Triassic or younger age for the Saidu and Nikanai Ghar formations. No Permian fossils have been recovered from the Jafar Kandao formation or the Alpurai group. However, the lateral continuity of the amphibolite horizon, coupled with lithologic and thickness variations in the lower part of the Marghazar formation, suggests that an erosional unconformity exists in the Marghazar formation directly below the amphibolite horizon. The continuity of the amphibolite horizon/Karapa greenschist argues against an extended period of erosion immediately following deposition of the amphibolite horizon. It is possible, therefore, that the amphibolite horizon, the Karapa greenschist, and/or the lower part of the Kashala formation are Permian in age.

The stratigraphy in the Peshawar Basin is markedly different from that in Lower Swat. In the Peshawar Basin, the Swat gneiss is absent and the Jafar Kandao formation is underlain by a thick sequence of Early and Mid-Palaeozoic argillite, quartzite, and limestone that, with the possible exception of the Jobra formation, has no counterpart to the north (Fig. 6). Numerous erosional unconformities in this section suggest periodic uplift throughout the Palaeozoic. Pogue *et al.* (1992*b*) interpret the stratigraphy as an epicontinental sea well inland from the continental margin that received detritus from the craton to the south and, periodically, from highlands to the north. Part of the Peshawar Basin stratigraphy is younger than the supposed Late Cambrian–Early Ordovician age of the Swat gneiss. This suggests that the Lower Swat stratigraphy is separated from the Peshawar Basin stratigraphy by a major basement fault or by a series of basement faults that preferentially elevated and eroded the Lower Swat area prior to deposition of the Alpurai group. The location of one of these basement faults is inferred to be below the Nikanai Ghar formation because this is where the stratigraphy changes. A north to south cross section depicting this relationship is shown in Fig. 7.

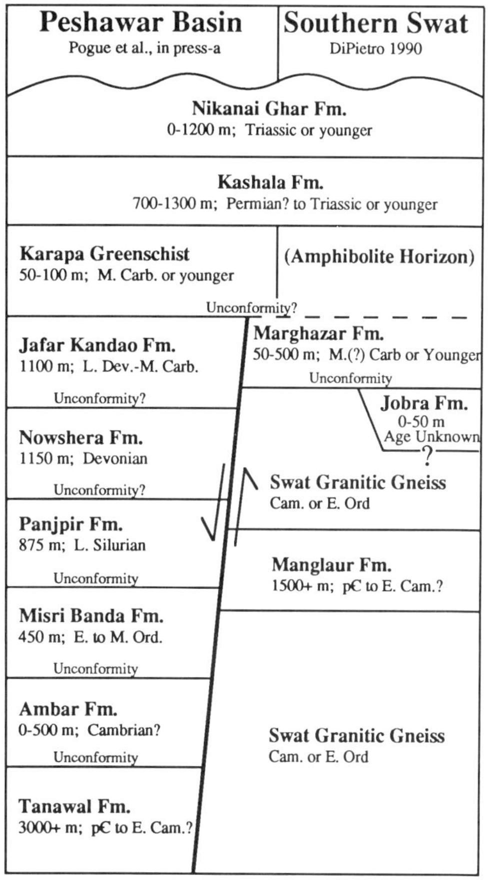

Fig. 6. Comparative stratigraphic columns and inferred relationships between the Peshawar Basin and southern Swat.

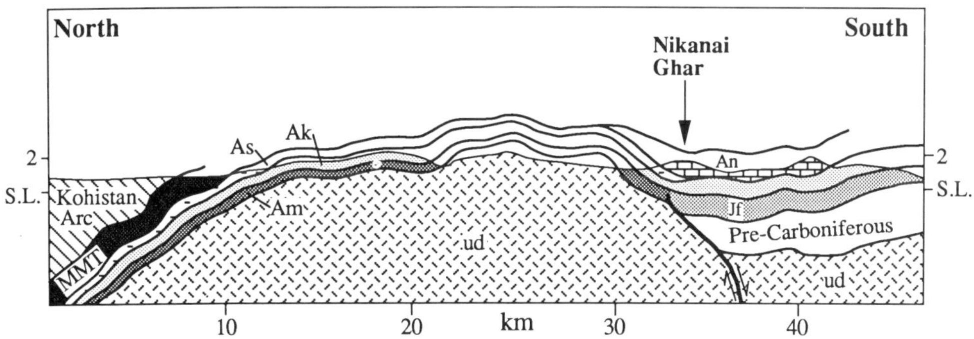

Fig. 7. North to south cross section through Nikanai Ghar. As, Saidu formation; An, Nikanai Ghar formation; Ak, Kashala formation; Am, Marghazar formation; Jf, Karapa greenschist/Jafar Kandao formation; ud, undifferentiated granite, granitic gneiss, and Precambrian metasedimentary rock. The Pre-Carboniferous strata are represented by the Nowshera, Panjpir, Misri Banda, and Ambar formations which are exposed in the Peshawar Basin. Patterned below surface level only. After DiPietro & Lawrence (1991).

The oldest Palaeozoic unit common to both the Peshawar Basin and the Lower Swat area is the amphibolite horizon/Karapa greenschist suggesting that faulting predates deposition of these units. The age at which faulting began is not well constrained. Field relationships suggest active faulting in the Carboniferous during deposition of the lower part of the Marghazar formation. The best evidence for this is east of Narbatawal where the Ilam gneiss abruptly terminates against the Marghazar formation (Fig. 4). The field relationships in this area have already been described. The relationships suggest that the Marghazar formation was deposited in a half-graben adjacent to an upthrown block of Swat gneiss (Fig. 8). The area was then eroded to a lowland by post-Westphalian time prior to deposition of the basalt protolith of the amphibolite horizon/Karapa greenschist. The age, lithology, and tectonic setting of the Marghazar formation suggests that it is a rift facies associated with the breakup of Gondwana (Pogue *et al.* 1992*b*). It correlates broadly with the Panjal Traps of western India.

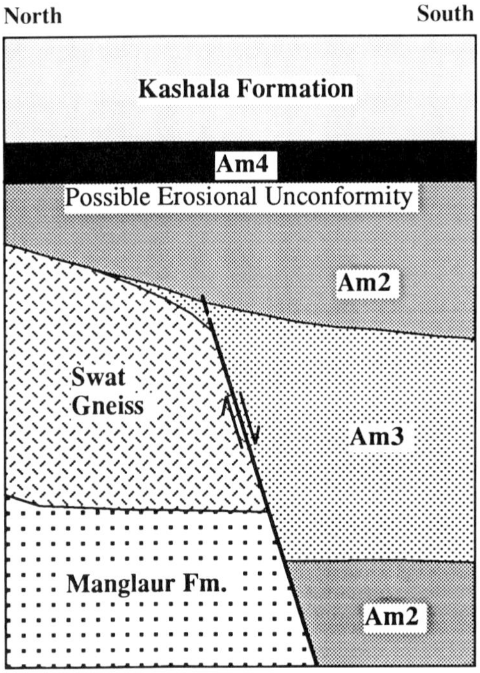

Fig. 8. Interpretive cross section of the area east of Narbatawal prior to Himalayan deformation. The Marghazar formation is deposited in a half-graben and thins dramatically where it onlaps the upthrown block of Swat gneiss. Faulting ceased and the area was eroded to a lowland prior to deposition of the amphibolite horizon. Am2, undifferentiated Marghazar formation; Am3, psammitic and amphibolitic schist unit of the Marghazar formation; Am4, amphibolite horizon of the Marghazar formation. The psammitic schist contains scattered pebbles of Swat gneiss. After Pogue *et al.* (1992*b*).

The largely calcareous Triassic or younger Kashala and Nikanai Ghar formations are interpreted as shelf deposits. This implies a return to epicontinental shelf sedimentation across the entire Peshawar Basin–Lower Swat area. The Nikanai Ghar formation is a lens-shaped body within and above the Kashala formation. It may represent a reef complex or a large bioherm. These rocks may correlate, in part, with the Upper Permian to Eocene Zanskar Supergroup of western India (Searle 1983). The graphitic phyllites in the Saidu formation may represent a transition to deeper water oceanic conditions. It is significant that the Saidu formation is always found structurally below the MMT. It may mark initial drowning of the shelf as it was pulled down and overridden by the MMT. Thus, in a tectonic sense, it correlates with the Triassic to Cretaceous(?) Lamayuru formation of western India (Frank *et al.* 1977; Gansser 1980, 1981).

The Alpurai group as a whole has been lithologically correlated with the Banna group which occurs in northern Hazara, just east of the Thakot fault and south of the MMT (Treloar *et al.* 1989*c*; Baig 1990). Farther east, Greco *et al.* (1989) describe a sequence of amphibolite and schistose marble (Burawai formation) from just south of the MMT near Kaghan that resemble the Marghazar–Kashala rocks (Fig. 1).

In conclusion, the Alpurai group records a depositional history from Late Palaeozoic rifting associated with the breakup of Gondwana; to development of a passive carbonate shelf in the Mesozoic; to drowning of the shelf as it was overridden by the MMT at the onset of the Himalayan orogeny. Lithologically similar, and possibly age equivalent, crystalline rock occur throughout the Pakistani Himalaya just south of the MMT.

Field work in Pakistan would not have been possible without the assistance and support of R.A.K. Tahirkheli, M. Q. Jan, A.A.K. Ghauri, I. Ahmad and M. Riaz of Peshawar University; and A.H. Kazmi of the Geological Survey of Pakistan. We thank D. Spencer and an anonymous referee for critical review of the paper. Financial support was provided from grants National Science Foundation grants INT 86–09914 and EAR 86–17543.

References

AHMAD, I. 1986. *Geology of Jowar area, Karakar Pass, Swat District, N.W.F.P., Pakistan*. MPhil Thesis, University of Peshawar.

——, Rosenberg, R.S., Lawrence, R.D., Ghauri, A.A.K. & Majid, M. 1987. Lithostratigraphy of the Karakar Pass section, south of the Main Mantle thrust, Swat, N.W. Pakistan. *Geological Bulletin, University of Peshawar,* **20**, 199–208.

Ashraf, M., Chaudhry, M.N. & Hussain, S.S. 1980. General geology and economic significance of the Lahor granite and rocks of the southern ophiolite belt in Allai-Kohistan area. *Geological Bulletin, University of Peshawar,* **13**, 207–213.

Baig, M.S. 1990. *Structure and geochronology of pre-Himalayan and Himalayan orogenic events in the northwest Himalaya, Pakistan, with special reference to the Besham area.* PhD Thesis, Oregon State University.

—— & Lawrence, R.D. 1987. Precambrian to Early Paleozoic orogenesis in the Himalaya. *Kashmir Journal of Geology,* **5**, 1–22.

——, Snee, L.W. & Lafortune, R.J. 1989. Timing of pre-Himalayan orogenic events in the northwest Himalaya: $^{40}Ar/^{39}Ar$ constraints. *Kashmir Journal of Geology,* **6&7**, 29–39.

Bossart, P., Dietrich, D., Greco, A., Ottiger, R. & Ramsay, J.G. 1988. The tectonic structure of the Hazara-Kashmir syntaxis, Southern Himalayas, Pakistan. *Tectonics,* **7**, 273–297.

Calkins, J.A., Offield, T.W., Abdullah, S.K.M. & Tayyab Ali, S. 1975. *Geology of the southern Himalaya in Hazara, Pakistan, and adjacent areas.* United States Geological Survey Professional Paper **716-C**.

DiPietro, J.A. 1990. *Stratigraphy, structure and metamorphism near Saidu Sharif, Lower Swat, Pakistan.* PhD Thesis, Oregon State University.

—— 1991. Metamorphic pressure-temperature conditions of Indian plate rocks south of the Main Mantle thrust, Lower Swat, Pakistan. *Tectonics,* **10**, 742–757.

—— & Lawrence, R.D. 1991. Himalayan structure and metamorphism south of the Main Mantle thrust, Lower Swat, Pakistan. *Journal of Metamorphic Geology,* **9**, 481–495.

Epstein, A.G., Epstein, J.B. & Harris, L.D. 1977. *Conodont color alteration – an index to organic metamorphism.* United States Geological Survey Professional Paper **915**.

Frank, W., Gansser, A. & Trommsdorff, V. 1977. Geological observations in the Ladakh area (Himalayas) a preliminary report. *Schweizerische Mineralogische Petrographische Mitteilungen,* **57**, 89–113.

Gansser, A. 1964. *Geology of the Himalayas.* Interscience Publishers, London.

—— 1980. The significance of the Himalayan suture zone. *Tectonophysics,* **62**, 37–52.

—— 1979. A reconnaissance visit to the ophiolites in Baluchistan and the Himalaya. *In*: Farah, A. & DeJong, K.A. (eds) *Geodynamics of Pakistan.* Geological Survey of Pakistan, Quetta, 193–213.

—— 1981. The geodynamic history of the Himalaya. *In*: Gupta, H.K. & Delany, F.M. (eds) *Zagros–Hindu Kush–Himalaya geodynamic evolution.* American Geophysical Union Geodynamic Series, **3**, 111–121.

Greco, A., Martinotti, G., Papritz, K., Ramsay, J.G., & Rey, R. 1989. The crystalline rocks of the Kaghan Valley (NE-Pakistan). *Eclogae Geologicae Helvetiae,* **82/3**, 629–653.

Jan, M.Q. & Tahirkheli, T. 1969. The geology of the lower part of Indus Kohistan (Swat), West Pakistan. *Geological Bulletin, University of Peshawar,* **4**, 1–13.

——, Asif, M., Tahirkheli, T. & Kamal, M. 1981. Tectonic subdivision of granitic rocks of north Pakistan. *Geological Bulletin, University of Peshawar,* **14**, 159–182.

Kazmi, A.H., Lawrence, R.D., Dawood, H., Snee, L.W. & Hussain, S. 1984. Geology of the Indus suture zone in the Mingora-Shangla area of Swat, N. Pakistan. *Geological Bulletin, University of Peshawar,* **17**, 127–144.

——, ——, Anwar, J., Snee, J.W. & Hussain, S.S. 1986. Mingora emerald deposits (Pakistan): suture associated gem mineralization. *Economic Geology*, **81**, 2022–2028.

King, B.H. 1961. A new fossil locality in Swat, West Pakistan. *Geological Bulletin, Punjab University,* **1**, 65.

—— 1964. *The structure and petrology of part of Lower Swat, West Pakistan, with special reference to the origin of the granitic gneisses.* PhD Thesis, University of London.

Lawrence, R.D. & Ghauri, A.A.K. 1984. Tectonics of the western Indus suture in Pakistan. *EOS (American Geophysical Union Transactions)*, **65**, 1094.

——, Kazmi, A.H. & Snee, L.W. 1989. Geological setting of the emerald deposits. *In*: Kazmi, A.H. & Snee, L.W. (eds) *Emeralds of Pakistan: Geology, gemology, and genesis.* Van Nostrand Reinhold, New York, 13–38.

Le Fort, P., Debon, F. & Sonet, J. 1980. The "Lesser Himalayan" cordierite granite belt, typology and age of the pluton of Manserah, Pakistan. *Geological Bulletin, University of Peshawar,* **13**, 51–62.

——, ——& —— 1983. The Lower Paleozoic "Lesser Himalayan" granite belt: emphasis on the Simchar pluton of central Nepal. *In*: Shams, F.A. (ed.) *Granites of the Himalayas Karakoram and Hindu Kush.* Institute of Geology, Punjab University, Lahore, 235–256.

Maluski, H. & Matte, P. 1984. Ages of Alpine tectono-metamorphic events in the northwestern Himalaya (northern Pakistan) by $^{39}Ar/^{40}Ar$ methods. *Tectonics,* **3**, 1–18.

Martin, N.R., Siddiqui, S.F.A. & King, B.H. 1962. A geological reconnaissance of the region between the lower Swat and Indus Rivers of Pakistan. *Geological Bulletin, Punjab University,* **2**, 1–14.

Palmer-Rosenberg, P.S. 1985. *Himalayan deformation and metamorphism of rocks south of the Main Mantle thrust zone, Karakar Pass area, southern Swat, Pakistan.* MS Thesis, Oregon State University.

Pogue, K.R. & Hussain, A. 1988. Paleozoic stratigraphy of the Peshawar Basin, Pakistan. *Geological*

Society of America Abstracts with Programs, **20**, 266–267.

——, WARDLAW, B.R., HARRIS, A.G. & HUSSAIN, A. 1992*a*. Paleozoic and Mesozoic stratigraphy of the Peshawar Basin, Pakistan: Correlations and Implications. *Geological Society of America, Bulletin*, **104**, 915–927.

——, DIPIETRO, J.A., RAHIM, S., HUGHES, S., DILLES, J.H. & LAWRENCE, R.D. 1992*b*. Late Paleozoic rifting in northern Pakistan. *Tectonics*, **11**, 871–883.

SEARLE, M.P. 1983. Stratigraphy, structure and evolution of the Tibetan–Tethys zone in Zanskar and the Indus suture zone in the Ladakh Himalaya. *Transactions of the Royal Society of Edinburgh: Earth Sciences*, **73**, 205–219.

SHAMS, F.A. 1963. Reactions in and around a calcareous xenolith lying within the granite-gneiss of Manglaur, Swat State, West Pakistan. *Geological Bulletin, Punjab University*, **3**, 7–18.

STAUFFER, K.W. 1967. Devonian in India and Pakistan. *International Symposium on the Devonian System, Calgary, Canada*, 545–556.

—— 1968. Silurian-Devonian reef complex near Nowshera, West Pakistan. *Geological Society of America Bulletin*, **79**, 1331–1350.

TAHIRKHELI, R.A.K. 1979*a*. Geology of Kohistan and adjoining Eurasian Indo-Pakistan continents, Pakistan. *Geological Bulletin, University of Peshawar*, **11**, 1–30.

—— 1979*b*. Geotectonic evolution of Kohistan. *Geological Bulletin, University of Peshawar*, **11**, 113–130.

——, MATTAUER, M., PROUST, F. & TAPPONIER, P. 1979. The India-Eurasia suture zone in northern Pakistan; synthesis and interpretation of recent data at plate scale. *In*: FARAH, A. & DEJONG, K.A. (eds) *Geodynamics of Pakistan*. Geological Survey of Pakistan, Quetta, 125–130.

TRELOAR, P.J., BROUGHTON, R.D., WILLIAMS, M.P., COWARD, M.P. & WINDLEY, B.F. 1989*a*. Deformation, metamorphism and imbrication of the Indian plate, south of the Main Mantle thrust, north Pakistan. *Journal of Metamorphic Geology*, **7**, 111–125.

——, REX, D.C., GUISE, P.G., COWARD, M.P., SEARLE, M.P., WINDLEY, B.F., PETTERSON, M.G., JAN, M.Q. & LUFF, I.A. 1989*b*. K-Ar and Ar-Ar geochronology of the Himalayan collision in NW Pakistan: constraints on the timing of suturing, deformation, metamorphism and uplift. *Tectonics*, **8**, 881–909.

——, WILLIAMS, M.P. & COWARD, M.P. 1989*c*. Metamorphism and crustal stacking in the north Indian plate, north Pakistan. *Tectonophysics*, **165**, 167–184.

YEATS, R.S. & LAWRENCE, R.D. 1984. Tectonics of the Himalayan thrust belt in northern Pakistan: *In*: HAQ, B.U. & MILLIMAN, J.D. (eds) *Marine geology and oceanography of Arabian Sea and coastal Pakistan*. Van Nostrand Reinhold, New York, 177–198.

A section through the Indian Plate, Kaghan Valley, NW Himalaya, Pakistan

ANTONIO GRECO[1,2] & DAVID A. SPENCER[1]

[1] *Institute of Geology, ETH, Zürich, CH 8092, Switzerland*

[2] *Kreuzackerstrasse 1, 5012 Schonenwerd, Switzerland*

Abstract: A map is presented which summarizes the work of the ETH Himalayan Group in the Kaghan Valley and adjacent areas of the NW Himalaya in Pakistan. The stratigraphical, structural and metamorphic features of the lithologies belonging to a virtually complete section through the Indian plate are described. Evidence is presented which shows a stratigraphic range in ages from the Precambrian Salkhala Formation to the Eocene Murree Formation which have been metamorphosed at various metamorphic grades ranging from very low grades/unmetamorphosed to eclogite facies. An inverse metamorphism, associated with the Main Central Thrust is also described. A large-scale tectonic subdivision of the area into the classical Sub-, Lesser and Higher Himalayan elements can be correlated with the Kashmir Himalaya to the east. A tectonic model is presented which suggests that an anticlockwise rotation of the transport direction caused the refolding and stacking of the nappe piles in the Paleogene.

In the past ten years, geologists from the Institute of Geology, ETH-Zürich have intensively investigated parts of the Himalayan ranges in NE-Pakistan (Kaghan Valley and Azad Kashmir). The results have been reported in several PhD theses (Bossart 1986; Ottiger 1986; Greco 1989; Spencer 1993), diploma theses (Papritz 1989; Rey 1989; Wahrenberger 1992; Meier 1992) and publications (Bossart *et al.* 1988; Greco *et al.* 1989; Pognante & Spencer 1991). The aim of these works has been to investigate stratigraphical aspects, to correlate large- and small-scale tectonic structures, to study the complex structural geometry and the relationships between the deformational and metamorphic histories of the investigated areas.

This paper aims to summarize the stratigraphic, metamorphic and tectonic features of the Hazara Kashmir Syntaxis (Sub- and Lesser Himalaya) exposed in the Kaghan Valley and the accessible parts of Azad Kashmir. A colour compilation map of the research from the past decade is presented. The adjacent Higher Himalaya with its metamorphosed Tethyan covers, as well as the suture zone with the adjacent Island Arc Sequences of Kohistan (western Himalayan syntaxial area), have also recently been remapped.

Description and age of the rock sequences

The composition, age and distribution of the rocks of the investigated area is complicated by the evolution of several structural domains which contain repeatedly deformed and metamorphosed parts of the stratigraphical sequence. Figure 1 shows the stratigraphy and Fig. 2 the tectonic subdivision of the lithologies which we define in the map (Fig. 3).

The Salkhala Formation

The oldest rocks exposed in the investigated area are low grade metamorphic, unfossiliferous, crystalline pelitic schists alternated with carbonates and basic and granitic intrusions. In the present work, the name Salkhala Formation has been used to group all the graphitic, chloritic, porphyroblastic and quartzitic micaschists, often interbedded with marbles (i.e., groups together the Salkhala Series and Dogra slates series of Wadia 1931), that are sandwiched between the Panjal and the Main Central Thrust, and which extend from the axial zone of the syntaxis towards the southeast. The most common rock type is a fine-grained micaschist showing alternations of quartz-rich and mica-rich layers. This layering is best interpreted as relict sedimentary banding. Despite the severe deformation, primary sedimentary structures such as cross-bedding and current lamination are preserved in quartz-rich layers in the Middle Kaghan Valley (Greco *et al.* 1989).

As already suggested by Tahirkheli & Jan (1979), a Cambrian age is assigned to the granitic intrusions and porphyroblastic, two mica granitic gneisses because of their lithological

From Treloar, P.J. & Searle, M.P. (eds), *Himalayan Tectonics*
Geological Society Special Publication, No. 74, pp. 221–236.

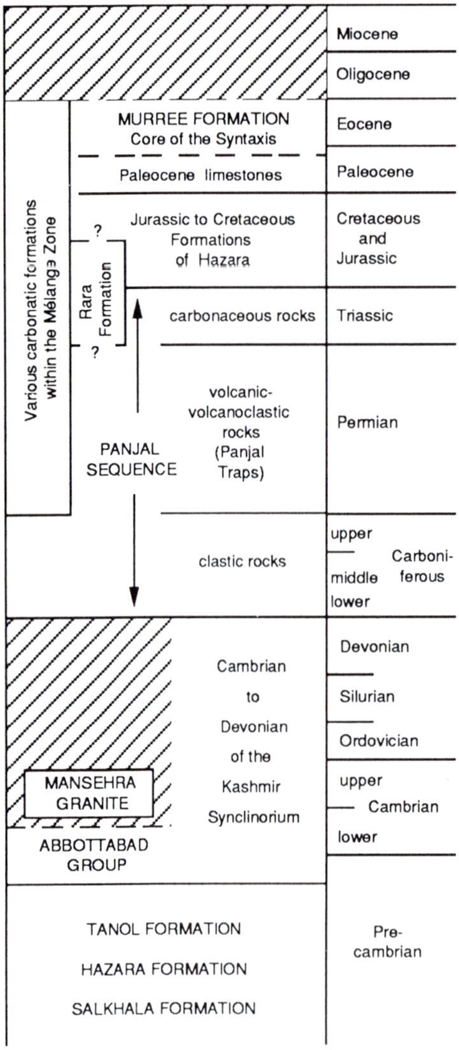

Fig. 1. Stratigraphy of the investigated area.

similarity with the Mansehra Granite (see section below). They should therefore be separated from the Salkhala Formation. The basic intercalations can also be separated from the Salkhala schists because petrographic observations and geochemical analysis indicate their magmatic origin as hypabyssal equivalents of the Permian Panjal Traps (Papritz & Rey 1989).

The Hazara Formation

In the Hazara district of NE-Pakistan, the argillaceous, non-metamorphic equivalent of the Dogra Slates have been called Hazara Slates by Waagen & Wynne (1872) and by Wadia (1934), and more recently Hazara Slates Formation by Marks & Ali (1961). In the area between Nathiagali and Abbottabad, Gardezi (1968) was able to distinguish several lithological zones within the formation ('Pelitic Zone', 'Psammitic Zone', 'Miranjani and Langrial Limestones'). Later, the whole sequence was renamed by Latif (1970) as the Hazara Group or the Hazara Formation by Calkins *et al.* (1975). A Late Precambrian age has been assigned (Gardezi 1968; Latif 1974; Crawford & Davies 1975).

In the investigated area, the Hazara Formation occurs along the western limb of the Hazara–Kashmir Syntaxis (Muzaffarabad area). It is predominantly composed of slightly metamorphosed, but strongly deformed, argillaceous rocks (slates) accompanied by two distinct intercalations of carbonaceous and sulphate rocks respectively. These non-argillaceous intercalations are considered to belong stratigraphically to the Hazara Formation and extend southwestward into the Miranjani and Langrial Limestones. The slates are very monotonous, with a characteristic very fine-grained, graded lamination which represent the bedding trace; their gradation indicates the stratigraphic younging direction.

The sedimentary cycles of the Hazara Formation are best interpreted as Bouma cycles, with a lack of the coarser subdivisions and have poorly developed tractive-current structures. These features, accompanied by primary and compactional structures are typical of turbidite-like sedimentation. By contrast, the sedimentological features of the oolitic and stromatolitic Miranjani Limestone (Gardezi 1968) suggest a shallow sea with important tidal and wave activity. Stromatolites and the cherty limestones which pass laterally into gypsum and dolomite indicate an evaporitic environment, accompanied by terrigenous input. A primary sedimentary contact between the limestone and the pelitic facies, exposed in the Nathiajan area, is indicated by the transition from the completely-submerged shale facies to the very shallow to intertidal sedimentation of the stromatolitic carbonates. Sediments at the top of the section indicate a return to completely-submerged shale facies.

On the basis of the present sedimentological data, a depositional environment similar to that of the homotaxial Simla Group of NW-India (Kumar & Brookfield 1987) can be proposed (Greco 1989). The lower parts of the Hazara formation could have been built up in a muddy, poorly channelized, large deltaic system on the northern margin of the Indian Shield. Shortly

MAIN SUBDIVISIONS (7)	INDUS KOHISTAN - HAZARA and POTWAR PLATEAU	HAZARA-KASHMIR SYNTAXIS and UPPER KAGHAN VALLEY	KASHMIR - LADAKH and ZANSKAR
	Kohistan Sequence (12)	**Kohistan Sequence** (12)	**Indus - Tsangpo Suture Zone** (9)
IS	*Indus Suture*	*Indus Suture*	*Indus Suture*
HIGHER HIMALAYA	**Basement and Cover Imbrications** (14) Tertiary granitoids (14) Paleozoic to Mesozoic (2) Cambrian: Mansehra-type granitoids and Karora Group (5) Proterozoic:gneiss (14), Besham and Pazang Gps. (5)	**Crystalline Unit** (8) Cover of Permian and Mesozoic with "Panjal affinities" Pre-Permian basement of detritic Lower to Middle Paleozoic, Lower Paleozoic granite (Mansehra Type) and Precambrian paragneisses	**Tibetan Zone** Tethys platform sediments of Zanskar-Spiti area (6,9) **Upper Crystalline Nappe** (6) Suru: Zanskar Crystalline (Precambrian to Lower Paleozoic) and Paleozoic to Lower Mesozoic metasedimentary cover (9)
MCT	*Oghi Shear Thrust* (4)	*Mylonite Zone* (1)	Kashmir - Chamba Syncline: Late PC to Trias (13) / **Lower Crystalline Nappe** Precambrian schists around the Kistwar Window (6)
internal unit	**Tanol Unit** (1) Cambrian: Mansehra Granite Upper Precambrian to Lower Cambrian: Abottabad Group. Upper Precambrian: Tanol Fm.	**Salkhala Unit** (1,8) Permian (?): dolerites Lower Paleozoic: granite gneiss Possible remains of Upper Precambrian: Tanol Formation Precambrian: metasedimentary rocks (Salkhala Series)	**Chail Nappe** Late Precambrian to Middle Paleozoic (6)
	Mansehra Thrust (3)	*Panjal Thrust* (15)	*Panjal Thrust* (15)
LESSER HIMALAYA	**Hazara Unit** (1) Paleo-Eocene: Gali Groups. Mesozoic: Thandiani and Hotla Groups. Upper Precamrian to Lower Cambrian: Abottabad Group. Precambrian Hazara Fm.(10)	**Panjal Unit** Panjal Imbricate Zone Triassic (Jurassic ?): Limestone and Dolomite Permian: Panjal Volcanics Upper Carboniferous: tilloides (1)	**Parautochtonous Unit** (6) "Schuppen" of Late Precambrian to Cambrian, Paleomesozoic and Eocene Formations (13)
external units	*Nathiagali Thrust* (11) **Main Boundary Thrust Zone** (3) Paleo-Eocene: Gali Gp. Mesozoic: Thandiani and Hotla Groups (10)	Mélange Zone Permian to Eocene (1)	
MBT	*Murree Thrust* (15)	*Murree Thrust* (15)	*Murree Thrust* (15)
SUB-HIMALAYA	**Murree Unit** **Siwalik Unit** Cambrian to Pleistocene (16)	**Murree Unit** Late Paleocene to Middle Eocene: Murree Fm. Paleocene formations Upper Precamrian to Lower Cambrian: Abottabad Group.	**Tertiary Zone** (6) or **Outer Himalayan Zone** (13) Cambrian, Paleocene to Pleistocene
MFT	*Salt Range Thrust* (16)		
ACTUAL FORELAND	Holocene of the Indus and Jhelum alluvial plains (16)		

Fig. 2. Table of the tectonic subdivision of the Northwestern Himalayas. References: (1) Bossart *et al.* 1988; (2) Coward *et al.* 1982; (3) Coward & Butler 1985; (4) Coward *et al.* 1986; (5) Fletcher *et al.* 1986; (6) Fuchs 1975; (7) Gansser 1964, 1981; (8) Greco *et al.* 1989; (9) Honegger *et al.* 1982; (10) Latif 1970; (11) Latif 1984; (12) Tahirkheli *et al.* 1979; (13) Thakur & Gupta 1983; (14) Treloar *et al.* 1989*a*; (15) Wadia 1931; (16) Yeats *et al.* 1984.

after deposition, these fine-grained turbidites were partially overlain by carbonates and evaporites of the Salt Range facies. Subsequently, the depositional environment developed into a stable slope-basin system with more psammitic facies receiving the more coarse-grained sediments of the Tanol Formation.

The Tanol Formation

The Tanol Formation (or Tanawal Formation of Wynne 1879) consists essentially of arenaceous beds with subordinate argillaceous, conglomeratic and carbonitic rocks and can be considered to be entirely of Upper Precambrian age (Latif 1974). In the present work, the Tanol Formation lies to the west of the mapped area where low metamorphic grade metapelitic schists of this formation are considered to bound the Hazara Formation to the northeast along the Mansehra Thrust.

The Abbottabad Group

In the Northwestern Himalaya, the Upper Precambrian is characterized by turbidite-like sedimentation which passes from an initial carbonaceous and argillaceous composition

(Salkhala and Hazara Formations) to a more arenaceous one (Tanol Formation). In the Hazara area, an important unconformity in the Upper Precambrian is represented by the deposition of the glacially derived Tanakki conglomerate (Tanakki Boulder Beds of Wadia 1931 or Tanakki Member of Latif 1970) which forms the base of the Abbottabad Group (Latif 1974). The rest of the sequence consists of arenaceous and carbonitic rocks, the latter with characteristic cherty and stromatolitic dolomites. The occurrence of Cambrian *Hyliolites* in the overlying Hazira Formation (Latif 1974; confirmed by Ottiger 1986) assigns an Upper Precambrian to Lower Cambrian age to the Abbottabad Group.

In the investigated area, the Abbottabad Group forms the core of the Garhi Habibullah Syncline and of the Muzaffarabad Anticline.

The Mansehra Granite

Granitic rocks, associated with the Pan-African orogenic event, were intruded during the Cambrian time into the entire region of the future Himalayan chain (Le Fort *et al.* 1980). In the investigated area, these plutonic intrusions remained restricted to the Tanol and Salkhala Formations and did not affect the Hazara Formation and the Abbottabad Group.

The well-known plutonic association of the Mansehra Granite (Shams 1961) became the type model for such intrusions in the Northwest Himalayas. Deformed and metamorphic equivalents are also described in the Alpurai and Swat area (Fletcher *et al.* 1986; Treloar *et al.* 1989*a*) and in the Kaghan Valley, where they contain metamorphic kyanite and garnet (Greco *et al.* 1989; Spencer 1993).

As already discussed, rocks which can be correlated with the Mansehra granite are exposed within the Salkhala Formation in the eastern limb of the Syntaxis (Neelum and Jhelum Valleys). These gneisses are coarse-grained with augen structure. They are quartzofeldspathic in composition and contain muscovite, biotite, rutile, sphene, epidote and tourmaline as secondary minerals. Apatite and zircon are important mineral accessories. Together with xenolith inclusions and micaschists, decimetre-thick tourmaline-rich quartz ribbons (deformed tourmaline-pegmatite?) have been observed in strongly deformed gneisses.

The Panjal Sequence

The Panjal Sequence is a term originally used by Calkins *et al.* (1975) to include a continuous stratigraphic sequence of Carboniferous to Triassic age with clastic, volcanic and carbonitic rocks. This sequence is exposed in the investigated area between the molasse core and the crystalline rocks of the outer northern and eastern rim of the Syntaxis and in the Higher Himalayan nappes of Kaghan Valley. Ottiger (1986) and Bossart (1986) have identified the sedimentological origin and stratigraphic position of this rock sequence in the Lesser Himalaya of the Kaghan Valley. Detailed mapping (Greco 1989) indicates that the lithotypes extend into the Neelum and Jhelum Valleys as well as the Higher Himalaya (Greco *et al.* 1989; Spencer 1993). Further subdivisions of this sequence are also proposed.

The clastic rocks of the Panjal Formation. The observed lithologies are: pyrite-bearing graphitic phyllites, sericite–chlorite schists, chlorite–sericite quartzite, and metapsammitic to metapelitic rocks with a fine-grained matrix of sericite, chlorite and graphite. The microscopic, as well as the macroscopic detrital components, indicate a metamorphic and granitic origin for the source rocks. No signs of volcanic components have been observed. The pebbles of the metaconglomerates are mainly fragments of quartzitic and granitic rocks with less carbonates and slates. Typically, the metaconglomerates are poorly packed and the components 'swim' in a dark matrix. These so-called Panjal Mixtites have been interpreted as tillites of glacial origin (Ottiger 1986). It is impossible to correlate these clastic rocks directly with the Carboniferous Agglomeratic Slates series of Kashmir because, contrary to the observations of Wadia (1931) and Calkins *et al.* (1975), no sign of volcanic origin has been identified. However, because of its close association with the volcanics, a (Silurian to) Upper Carboniferous age for these pelitic–psammitic sequence is suggested.

The volcanic rocks of the Panjal Formation. The volcanic rocks outcropping in the investigated area are directly comparable with the Upper Carboniferous to Permian volcanic sequence of the Panjal Traps of the Kashmir Himalayas (Wadia 1931). In the investigated area the Panjal volcanics can be subdivided at the outcrop scale into decimetre to metre thick basaltic lava flows,

Fig. 3. Geological map of the Himalayas of NE Pakistan (Hazara–Kashmir Syntaxis).

alternating with decimetre-thick (locally reaching several tens of metres) tuffaceous layers.

The basalts are metamorphosed into an epidote- and chlorite-rich, weakly schistose groundmass, containing amygdales infilled with epidote, quartz and plagioclase aggregates. Microscopically, the presence of rare pyroxene and amphibole minerals as well as the idiomorphic texture of the saussuritic plagioclase indicate a magmatic origin for these layers. Locally, isolated sub-spherical occurrences of massive, epidote-rich rocks, half-metre in size, with quartz-filled amygdales and surrounded by very schistose rocks can be interpreted as deformed and transformed pillow structures (Ghazanfar & Chaudhry 1984). Other observed magmatic features include flow banding and degassing structures.

Fine-grained, very schistose rocks of green, violet and red colours are associated with the massive, amygdaloidal lavas and represent the volcanoclastic (tuffaceous) layers. These schistose rocks are predominantly composed of chlorite, actinolite and epidote, but lack the mineral aggregates described above. A spotted appearance is given by the concentration of iron-rich chlorite in very flattened vesicles. These structures may represent chips of glassy material which was deposited in the pyroclastic sediment (ignimbrite?) and subsequently flattened during the collapse of the porous rock under the load of the overlying sediments. The tectonic overprint has further increased their oblate shape.

Geochemical analysis by Papritz & Rey (1989) indicate a tholeiitic-subalkalic composition of the Panjal Volcanics of the Hazara–Kashmir Syntaxis. Consequently, these rocks could be considered as equivalents of the Panjal Trap of Kashmir (Bhat & Zainuddin 1979; Honegger *et al.* 1982). Macroscopic features (pillows, amygdales, interlayered tuffaceous levels) as well as conformable alternations with crinoidal limestone exclude an oceanic origin (as suggested by Ghazanfar & Chaudhry 1984) and indicate a continental, marine depositional environment of alternating shallow water and subareal episodes.

The carbonitic rocks of the Panjal Formation. A wide range of (meta)carbonates and quartzites are associated with the volcanic and clastic rocks. These carbonates have undergone a more or less pronounced recrystallization. However, fossil remains have been found. Poorly preserved segments of crinoids, as well as remains of undeterminable echinoderms, shells, gastropods and bryozoa have been observed in sparitic, cross-bedded limestone as well as in microsparitic to micritic laminated limestone. Above Balgiran (Khetar, Neelum Valley), a strong convolute lamination of macroscopic rusty brown colour is observed in medium-grained dolomite suggesting a stromatolitic origin for these laminated structures. The stratigraphic top indicated by the stromatolitic shape is consistent with the local stratigraphic setting.

Wadia (1934) discussed the ‘contemporary interstratification’ of the Panjal volcanics with fossiliferous (Triassic) and unfossiliferous (Infra-Trias) carbonates. Although some of these intercalations are of tectonic origin, the close relationships between Upper Carboniferous to Lower Triassic Panjal volcanics and the crinoidal limestones suggest an Upper Paleozoic to Lower Mesozoic age for the carbonates.

The Panjal Sequence of the Higher Himalaya. Greco *et al.* (1989) correlated the cover sequences of the Higher Himalayan crystalline with the Panjal Sequence. This correlation is extensively discussed by Spencer (1993) who essentially agrees with Greco *et al.* (1989). The evidence is based on petrographic observations (amphibolites intercalated with marble horizons and subsequently overlain by mica-schists, before passing up into an extensive carbonate unit of marbles and dolomites which is the same as that in the Zanskar sedimentary shelf as documented by Herren (1987) and Gaetani *et al.* (1986)); geochemical evidence (see Papritz & Rey 1989 and Spencer 1993); field evidence (unconformable relationship of the Panjal Sequence (called the Burawai Formation in Greco *et al.* 1989) to the underlying Naran Formation (Greco *et al.* 1989) and the basement rocks suggests that a transgressive sequence of progradation, similar to that for the Panjal Traps is necessary); structural evidence (small-scale listric faulting occurs in between the amphibolite horizons in the marble layers, suggesting an extensional or rift-related environment, similar to that known for the Panjal Traps); fossil evidence (stromatolites were found which may, although not necessarily, imply an environment and age of formation that is similar to that of Tethys) and geochronological evidence (whole rock regression line Sm/Nd ages of amphibolites and eclogites which pertain to the extrusion age are Upper Carboniferous; S. Tonarini, pers. comm.).

The Rara Formation

South of Muzaffarabad, along the Murree Fault, a rock sequence composed of dolomites, quartzites and limestones is sandwiched between the Murree Formation and the Hazara Formation. The section exposed along the northern bank of the Kunhar River, between the bridge on the Jhelum River and the village of Ril, revealed the presence of fossiliferous limestone associated with dolomites, quartzites and carbonaceous slates. Consequently, this rock sequence of about 100 mm in thickness, can be distinguished from the carbonates of the Hazara Formation and is provisionally called the Rara Formation (Greco 1989).

The lithological association of the Rara Formation (peloidal, bioclastic limestones, quartzites, shales and dolomites) and its fossil content suggest a very shallow, high energy marine depositional environment, typical of several Triassic to Jurassic formations exposed in the southeastern parts of the Hazara district: the Lower Triassic Mianwali Formation (Tahirkheli 1982), the Liassic Maira Formation (Latif 1970), the Early Jurassic Datta Formation (Latif 1970), and the Upper Jurassic Samana Suk Limestone (Calkins *et al.* 1975). On the basis of its tectonic position, the Rara Formation correlates better with the Thandiani Group, which according to the map of Latif (1970), outcrops between the Hazara and the Murree Formations southwest of the Kohala bridge, and forms the stratigraphical base of a Jurassic to Eocene rock sequence.

The Tertiary (Paleocene and Murree Formation)

In the investigated area, Tertiary rocks are exposed in the core of the Hazara Kashmir Syntaxis as Palaeocene limestones and marls gradationally overlain by the red clastic rocks of the Murree Formation (late Palaeocene to Eocene).

On the northeastern limb of the Muzaffarabad Anticline, a good section across the Lower Tertiary is exposed along the road of the Neelum Valley. A similar section, exposed east of Balakot (Beran Katha), has been studied in detail by Bossart (1986). In the Neelum Valley, the stratigraphical gap between the Lower Cambrian Abbottabad Group and the Palaeocene is filled by 4 m of black, graphitic, coal beds accompanied by bauxite soils (Jurassic?). The Palaeocene is represented by about 200 m of alternating silty marls and limestones, the latter occurring as bedded and nodular strata. The rich macrofossil fauna (foraminifera, lamellibranchia, gastropoda, corallia, bryozoa etc.) allowed Bossart (1986) to date these reefal sequences as Illerdian (upper Palaeocene) and to compare these rocks with the southern exposed Lockart Limestone and Patala Formation of the Kala Chitta Zone (Tahirkheli 1982). Towards the top (northeast), the Palaeocene reefal sequence grades into the basal conglomerates and sandstone of the Murree Formation, as evident by an increase of detrital material in the Palaeocene marls and by the presence of calcareous clasts in the Murree conglomerates.

The Murree Formation is characterized by a cyclic sedimentation of sandstone and conglomerate grading into siltstone, with a few intercalations of marls and limestones containing a fauna of Late Palaeocene to Mid-Eocene, subtropical foraminifera (Bossart & Ottiger 1989). These authors suggest that the Murree Formation was deposited in an intertidal environment (clastic rocks), episodically transgressed by shallow marine conditions (nummulitic limestone), and filling the southwards migrating Tertiary foreland basin of the Himalayan chain (see also Lillie *et al.* 1987). This hypothesis is based on sedimentological features (e.g., cyclicity), on the detritus type and on the age of the formation. The age of the Murree Formation becomes younger away from the axis of the Muzaffarabad Anticline towards the Northeast, where it reaches the Mid-Eocene along the Main Boundary Thrust and towards the South (Potwar Plateau), where it reaches an Early Miocene age. We therefore speculate on the basis of this sedimentological evidence, that the collision was not synchronous, but diachronous in the Northwest Himalayan region, with a Palaeocene/Eocene impingement in the Hazara–Kashmir and Kaghan area.

Structural style and tectonic subdivisions

The investigated area is tectonically divisible into three main elements: the Subhimalaya, the Lesser Himalaya and the Higher Himalaya. In detail, the rock distribution in the Kaghan Valley and in Azad Kashmir indicates that each element is composed of one or more tectonic units, each having a particular stratigraphic sequence and a local name. The lateral continuity of such units, which could extend for several hundred kilometres, is one of the most impressive features of the Himalayan Range. A correlation with the tectonic equivalents in the Indian Kashmir and in the Hazara, Kohistan and Potwar areas of North Pakistan, based on data from the literature is schematically suggested in Fig. 2.

In general, the Himalayan structures emphasize the age and facies differences between the different elements. The thrusted units become older in stratigraphic age towards the uppermost tectonic elements, because the thrust planes reached deeper levels of the Indian crust. The deepest crustal level is represented by the lithologies which overlie the Main Central Thrust: the Higher Himalayan crystalline basement, with its Phanerozoic, predominantly carbonitic cover, thrust over the Lesser Himalayan detrital formations. The deformational style is very ductile at the base of the Higher Himalaya. Polyphase folding, associated penetrative schistosity and ductile shearing affect the nappe structure of basement and cover rocks. Characteristic of the Lesser Himalaya is the imbrication and folding of the sedimentary sequence, which is, in part or completely, detached from its basement substratum. The number of brittle thrust planes, increase downwards into the Lesser Himalaya. Folding and internal thrusting is also developed in the Subhimalaya.

The Subhimalaya

From the Indus to the Chenab River, the Murree Formation outcrops continuously along the southern margin of the Lesser Himalayan units, forming almost all of the Subhimalayn Unit. In the core of the Hazara–Kashmir Syntaxis, folding and subsequent doming are the principal structural features, resulting in the exposure of Lower Cambrian and Palaeocene rocks along the hinges of the Muzaffarabad and Kotli anticlines. Around the former anticline, metamorphism reaches prehnite–pumpellyite facies.

To the north, the Main Boundary Thrust, locally called the Murree Fault, borders the Subhimalaya. Geometrically, the fault plane cuts the steep dipping molasse beds on its footwall, and carries northeast dipping Lesser Himalayan schists and gneisses on its hanging wall.

The Lesser Himalaya

The Lesser Himalaya is composed of folded and thrusted rock formations, ranging in age from Precambrian to Eocene. It is delimited to the south by the Tertiary molasse along the Main Boundary Thrust, and to the north by the Higher Himalayan crystalline rocks along the Main Central Thrust. The metamorphism in this element is of very low- to low-grade. A further subdivision of the Lesser Himalaya into separate tectonic units is necessary because of the presence of several thrust slabs, of slightly different stratigraphical sequence and metamorphic grade.

West of the Hazara–Kashmir Syntaxis, the Main Boundary Thrust Zone (Coward & Butler 1985 or Islamabad Zone of Fuchs 1975, or Kala Chitta Zone of Tahirkheli 1982) forms a complex mosaic of thrusts and folds consisting of Mesozoic and Palaeogene formations. The northern boundary is represented by the Nathiagali Thrust (Latif 1984), which branches from the Main Boundary Thrust (or Murree Thrust) 5 km south of Muzaffarabad, where the, probably Jurassic, limestone of the Rara Formation has been found.

North of the Nathiagali Thrust, the Hazara Unit contains a thick sequence of the Upper Precambrian Hazara Formation at the base followed by the Upper Precambrian to Lower Cambrian Abbottabad Group. Toward the south the Cambrian disappears and Mesozoic rocks directly overlie the Hazara slates. The structures developed in the Precambrian slate formation can be inferred to be of Himalayan age because the associated folds contain Jurassic and Palaeogene formations. In the Hazara unit the metamorphism reaches anchi- to low epizonal facies (Greco 1989).

The northern boundary of the Hazara unit is represented by the Mansehra Thrust (Coward & Butler 1985), which is supposed to branch with the Murree Thrust south of Balakot under the alluvium of the Kunhar river. The interpretation of the Mansehra Thrust as a continuation of the Panjal Thrust (Calkins *et al.* 1975) suggests an equivalence between the Panjal- and the Hazara Units. In contrast, the age, metamorphism and deformational character of the formations involved indicate that it is preferable to consider the Hazara Unit as an individual sheet, which is missing in the apex of the Hazara–Kashmir Syntaxis. Therefore, it is suggested here that the term Panjal Thrust be abandoned to the west of the syntaxial bend and the name Mansehra Thrust be used instead. We note here, however, that Coward *et al.* (1988) show that the Mansehra Thrust is located to the north of the Panjal Thrust west of the syntaxial bend. They suggest that the Tanawal Formation extends well north of the Oghi Shear and that the rocks intruded by the Manshera granite extend north up through Hazara and are collectively described as the Tanawal Formation by Calkins *et al.* (1975). This would therefore cross our division of the Lesser and Higher Himalaya.

In the Hazara area the Tanol Unit, consisting of the Tanol Formation of Upper Precambrian age and of the Cambrian Mansehra Granite, represents the highest structural level of the Lesser Himalaya. The metamorphic grade of this Tanol Unit is appreciably higher than the underlying Hazara units. The northern tectonic limit of the Tanol Unit in Hazara is represented by the Oghi Shear (Coward *et al.* 1986), which should correspond to the Mylonite Zone of the Kaghan Valley and to the Main Central Thrust.

The Panjal Unit is exposed in the apex and eastern limb of the Hazara–Kashmir Syntaxis. In the Kaghan Valley, this unit is composed of four different imbricate slices. The lowermost slice is formed by a tectonic mélange of Permian to Eocene rocks (the Mélange Zone), whereas the remaining three (the Panjal Imbricate Zone) represent an Upper Palaeozoic to Lower Mesozoic stratigraphic sequence of Carboniferous tilloides, Permian Panjal Trap volcanics, and Triassic Limestone (Bossart *et al.* 1988). The metamorphic grade reached greenschist facies. Geometrically this unit shows a typical 'duplex' configuration with the Main Boundary Thrust as floor-thrust and the Panjal Thrust as roof-thrust. With small changes in their geometry, this unit can be followed eastward into the Indian Kashmir–Chamba area.

In the apex and eastern limb of the Hazara–Kashmir Syntaxis, the uppermost Lesser Himalayan unit is mainly represented by the Precambrian metamorphic rocks of the Salkhala Formation. The Salkhala Unit is mainly composed of metapelitic schists with characteristic graphite, quartzite and carbonate sediments and is intruded by pegmatite dykes and two mica augen gneisses, which represent the deformed equivalent of the Lower Palaeozoic Mansehra Granites. The widespread basic dykes should belong to the Permian magmatic event because of their geochemical similarity with the Panjal volcanics (Papritz & Rey 1989). The Salkhala sequences are folded, well foliated and thrusted. Similar to the Tanol unit, pre-Himalayan metamorphism could exist but is overprinted by the strong Himalayan event. The northern limit of the Salkhala Unit is the Mylonite Zone or Main Central Thrust.

The Main Central Thrust

The Main Central Thrust occurs in the Kaghan Valley (Naran area) as a granite-derived mylonitic zone which divides the low metamorphic grade Salkhala schists from amphibolite-facies metamorphic carbonaceous schists belonging to the cover rock sequence of the Higher Himalaya (Ghazanfar & Chaudhry 1986; Chaudhry & Ghazanfar 1990; Greco *et al.* 1989). The same discontinuity was first observed in the western part of the Syntaxis near Balakot by Bossart (1986), who called it 'Mylonite Zone' or 'Balakot Discontinuity'. In the eastern limb of the Syntaxis, the Main Central Thrust could be mapped in the Kalasar area (Neelum Valley; Greco 1989), where it forms an up to 500 m thick pile of metapelitic mylonites containing fibrolitic sillimanite, and which passes upward into paragneisses of the Higher Himalayan basement. Therefore, this main intracrustal discontinuity cuts across the basement-cover sequence of the Higher Himalayan crystalline unit.

The Higher Himalaya

The Higher Himalayan Crystalline Unit is exposed in the uppermost parts of the investigated area in Upper Kaghan. Recent investigations in the Kaghan Valley (Greco *et al.* 1989; Spencer, 1993), allowed a basement and a cover sequence in the Higher Himalayan Crystalline Unit to be distinguished. The basement is composed of granitoid rocks, lithologically similar to the Cambrian Mansehra Granite, intruded into older paragneisses perhaps corresponding to the Precambrian Salkhala Formation. A distinct, second, lithotype of the basement is the metapelitic–metagreywacke gneiss, probably corresponding to Lower Palaeozoic sediments of the Kashmir Synclinorium (Cambrian to Ordovician in Fig. 3) and of the Zanskar area (see Garzanti *et al.* 1986). In Upper Kaghan, this has been termed the Naran Formation by Greco *et al.* (1989). All these rocks are the metamorphic, magmatic and sedimentary products of a Late Pan-African orogenic event and, therefore, form the (pre-Permian) basement on which rocks of the Alpine cycle were deposited. In fact, the cover sequence begins with thick amphibolitic layers, which together with the doleritic dykes of the basement, and on the base of the petrographic, geochemical and field evidence, have been correlated with the Permian Panjal Trap. Towards the top, the cover is carbonitic and pelitic (Lower Mesozoic?), and is always accompanied by interlayered basaltic flows. In Upper Kaghan, this has been termed the Burawai Formation by Greco *et al.* (1989). The tectonic limits of this sequence are represented by the mylonites of the Main Central Thrust Zone at the base, and by the mafic-ultramafic rocks of the Kohistan Sequence at the top. We suggest that the previously named Sharda Group (Ghazanfar & Chaudhry 1987), which described the complete Higher Himalaya section of Upper Kaghan between the Main Central Thrust and the Indus Suture, be abandoned.

Recumbent folding and ductile thrusting formed several nappes with a sheet-like structure. They were subsequently refolded by a late-stage doming event.

Grade and type of metamorphism

Except for the Subhimalayan Formations and the Higher Himalaya, the boundaries between the different metamorphic facies almost coincide with the main thrust faults. Because these structures do no represent reaction isograds, the distribution of the metamorphic grade is therefore suggested to be a direct consequence of the tectonic evolution of the area.

Pre-Himalayan metamorphism

Evidence of pre-Himalayan metamorphism in the Lesser- and Higher Himalayan realms are scarce. Microstructural evidence indicates that the main metamorphic recrystallization, in the Precambrian rocks of both the Lesser and Higher Himalaya, is associated with the Himalayan deformation (Bossart *et al.* 1988; Greco 1989). Because clear evidence for the presence of Precambrian events is at present lacking in the Hazara–Kashmir and Kaghan region, it is reasonable to consider the metamorphism as mainly associated with Himalayan tectonism (although a Precambrian metamorphism has been clearly documented in the areas to the west of Hazara–Kashmir; see Baig *et al.* 1989; Williams *et al.* 1989). Nevertheless, the presence of Cambrian intrusives (Mansehra Granite) and some sedimentary evidence in the Northwest Himalayas (Garzanti *et al.* 1986), as well as the Higher Himalaya of Kaghan (Spencer 1993), support the existence of a Late-Pan African event in the future Himalayan realm. In addition, records of a Precambrian orogenic event have been found in the basement of the Indian Plate south of the Indus Suture in the Besham area (Northern Pakistan, Treloar *et al.* 1989*b*; Treloar & Rex 1990) and in the Nanga Parbat area (Zeitler *et al.* 1989).

Higher Himalaya

The highest tectonic element of the Indian Plate, the Higher Himalaya, consists of a basement with an originally overlying sedimentary cover sequence. In the Kaghan Valley, both basement and cover have suffered an intense common syn- to post-deformational medium grade of metamorphism with localised pre- and syn-deformational high grade eclogite facies metamorphism of 650° C ± 50° C at 15–17 kbar Pognante & Spencer 1991). In addition, within this element a decrease in metamorphic grade, showing a Barrovian type gradient, is observed towards the sedimentary cover of the Kashmir Syncline. As the regional metamorphism involves Mesozoic rocks, it is considered of Himalayan age. Isotopic ages point to a Miocene age for the metamorphic peak in the Central Himalaya (Le Fort 1986) but are Eocene in age in Upper Kaghan (Tonarini *et al.* 1993) with cooling generally to below 300° C by the Micoene (Zeitler 1985; Treloar & Rex 1990; Spencer 1993).

Main Central Thrust

The southwest-directed transport of the Higher Himalayan basement–cover sequence took place along an important intracrustal shear zone, the Main Central Thrust. Evidence that deformation along the Main Central Thrust started to develop under hanging-wall metamorphic conditions are found in the mylonitic metapelites at the Kalasar area (Neelum Valley, see also Greco 1989). These rocks are a product of shear deformation occurring during amphibolite-facies metamorphism, and have subsequently been overprinted by post-kinematic mineral growth within a change in metamorphic conditions. Because fibrolitic sillimanite is stable during the shear deformation, we conclude that the mylonitization developed within the temperature-dominated metamorphism of this portion of the Higher Himalaya (see Greco 1989).

These features are rare, whereas rocks of the Mylonite Zone, south of Naran (Greco *et al.* 1989) and around Balakot (Bossart *et al.* 1988) indicate deformation occurring during upper greenschist-facies conditions, similar to those of the upper Salkhala Unit. In these mylonitized rocks, hornblende remains stable and only garnet has grown after the deformation (Greco 1989).

Therefore, the Main Central Thrust did not simply develop after the metamorphism but represents a narrow transition zone between two metamorphic realms. A transition therefore exists from the very ductile deformation (which developed from the upper amphibolite metamorphic conditions in the hangingwall rocks) to the lesser metamorphic grade (greenschist facies) of the footwall sequences. After deformation, and far from the Main Central Thrust, the two tectonic units maintained their original thermal conditions long enough to allow post-deformational mineral growth (see Greco 1989). A syn- to post-kinematic metamorphic re-equilibrium between the two units has only been reached along the Mylonite Zone. The Main Central Thrust at the Kalasar area records the entire tectonic throw. From the petrological features observed, it is estimated that deformation along this thrust occurred between

550° C and 650° C. In this type of metamorphic gradient, these temperatures correspond to a lithostatic pressure of 6–9 kbar (depth of 20–30 km) and record a vertical displacement between the Higher and the Lesser Himalaya of at least 10 km (Greco 1989).

Lesser Himalaya

In the Lesser Himalaya, basement and cover have in part been completely detached, leading to the separation of different nappe units or imbricated sequences. The older thrust units lie more internally (and tectonically higher) in the tectonic system, and show higher metamorphic grades than the younger, tectonically lower units. The series of metamorphic facies observed in the Lesser Himalaya belong to a medium pressure, regional Barrovian metamorphism, as evident by the presence of chloritoid- and garnet-bearing mineral assemblages (Greco 1989).

Inverse metamorphism

As inverted distribution of the metamorphic facies is particularly well developed in the Lesser Himalaya, and its origin has been investigated with microstructural analysis in these rock units (Greco 1989). Microtextures in both the Panjal and the Salkhala Units indicate that peak metamorphic conditions had been reached during the main, thrust-related deformational phase. In addition, the microtextures related to the development of the Panjal Thrust indicate that the rocks at the base of the lower Salkhala Unit have suffered a retrograde mineral transformation at lower greenschist conditions. The metamorphism across the Panjal Thrust also tended to re-equilibrate. This last observation leads us to conclude that the inverted distribution of metamorphic grade of the Lesser Himalaya had been caused by continuous development of the thrust planes (Greco 1989). By contrast to the west of Kaghan, the thrusting is seen to be post-metamorphic in the 'Higher Himalayas' (Treloar *et al.* 1989*b*, *c*; Treloar & Rex, 1990).

Main Boundary Thrust

The Main Boundary Thrust cuts across several Lesser Himalayan Units with different metamorphic grades, and also cuts across the stilpnomelane isograd in the Subhimalayan Murree Formation (Bossart 1986). As brittle–ductile deformation features are present, and because postkinematic mineral growth is only rarely observed in the Panjal and Hazara Units, the Main Boundary Thrust must have developed after the peak of regional metamorphism. An important vertical displacement of less than 10 km, is recorded across this tectonic discontinuity.

Subhimalaya

As already emphasized, in the Subhimalayan rocks of the apex of the Hazara–Kashmir Syntaxis, the metamorphic conditions reached the beginning of the epizone (lower greenschist facies) with the formation of stilpnomelane in the arenaceous beds. As suggested by Bossart (1986), the sedimentary thickening of the molasse formation caused the low to very low grade mineral assemblages in the stratigraphic base of these clastic rocks, which have subsequently been deformed by the thrust tectonics without metamorphic consequences. Therefore, the metamorphism in the foreland is of a different character from that seen in the Himalayan nappes.

Interpretation of tectonic evolution

A structural evolution of the investigated area is based on detailed small-scale structures, rock distribution and textural studies. Summarising the structural data elaborated by Bossart *et al.* (1984, 1988), Greco *et al.* (1989) and Spencer (1993), two superimposed deformational phases can be observed in each tectonic unit of the Hazara–Kashmir Syntaxis area on a macroscopic as well as on a microscopic scale (Figs 4 and 5).

In all tectonic units, a first deformational phase produced the major recumbent fold nappes of the Higher Himalaya as well as the small scale isoclinal folds and imbrications in the Lesser Himalaya. In the Subhimalaya, this phase is emphasized by the Muzaffrabad Anticline and by the tilting of the Murree beds before the development of the Main Boundary Thrust. All these structures are not necessary contemporaneous, but are the first to develop in each tectonic unit.

The well-established NE–SW-directed thrust movements, recognized by the stretching lineations, are probably related to the initiation of folding and continued later deformation. The Main Central Thrust cuts across these nappes, as different rock types are found in its hanging wall. The overlapping of the mylonitization on the first phase structures is particularly clear in the western limb of the Syntaxis, as the associated deformation produced the curved bend of the Panjal imbricate structures (Bossart 1986). The

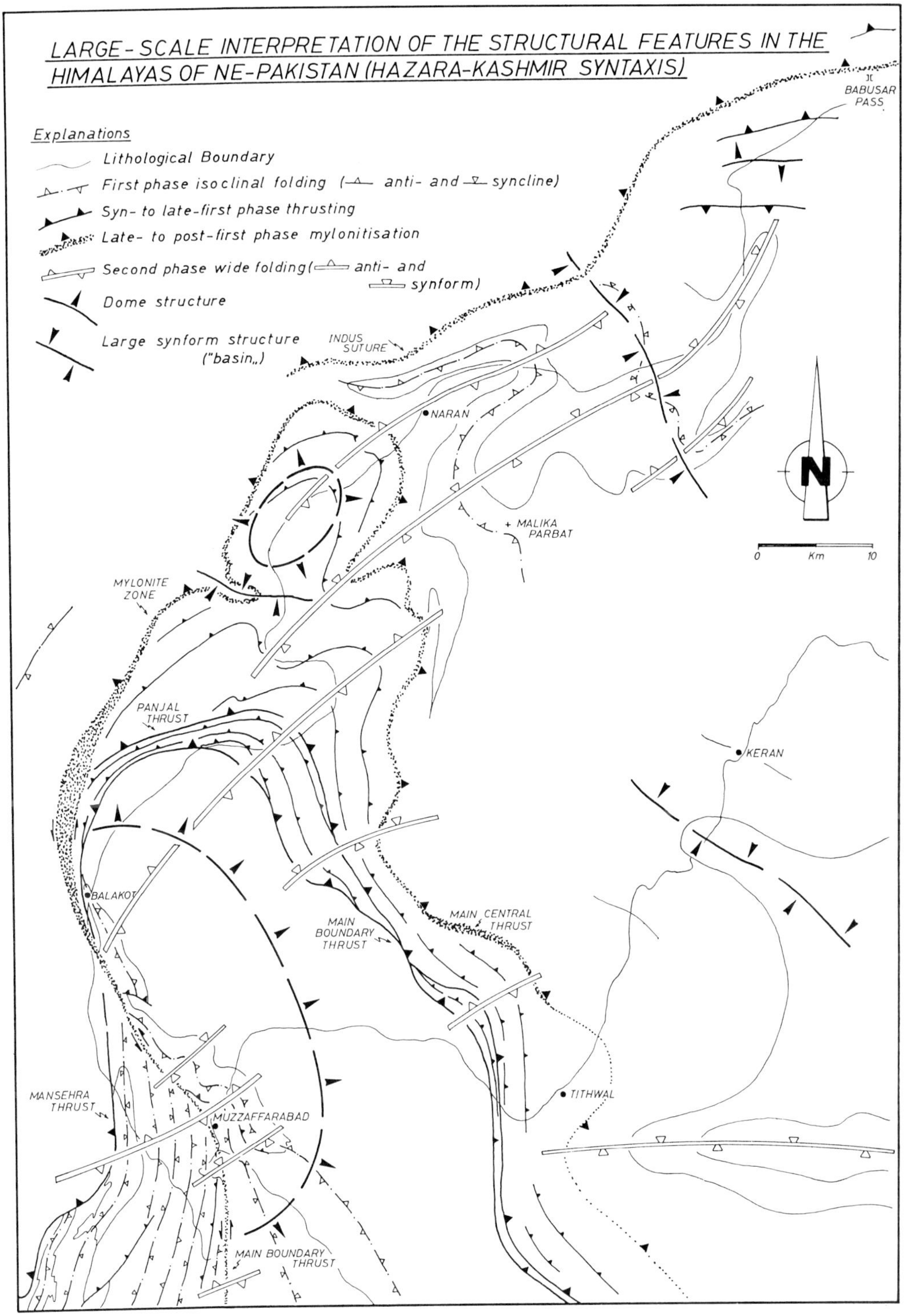

Fig. 4. Large-scale interpretation of the main structures exposed in the western Himalayan Syntaxis area.

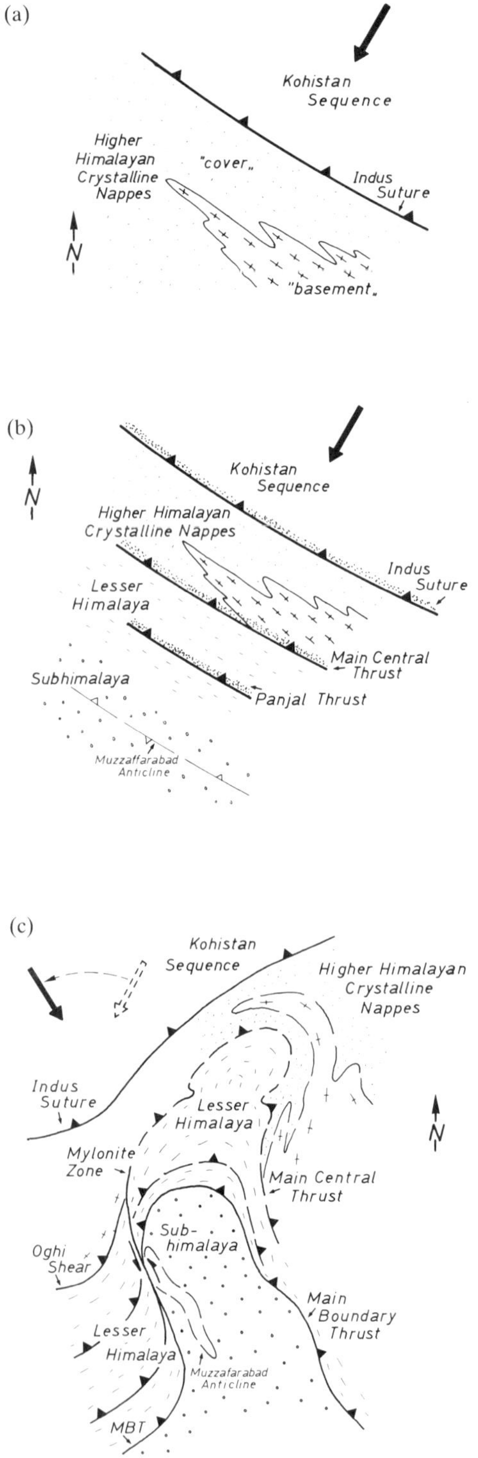

Fig. 5. Model for the tectonic evolution of the western Himalayan Syntaxis area (not to scale).

Muzaffarabad anticline and the Murree sediments are clearly cut by the Main Boundary Thrust in the lower Kaghan Valley, which indicates that movements on the most frontal thrust plane took place after the first deformational phase. The most likely explanation for these geometric features is that the deformation developed first in the higher tectonic units and propagated towards the Subhimalaya. The north–south trend of the Mylonite Zone near Balakot and the stretching lineations which become progressively north–south orientated in the lowermost tectonic elements (Greco 1989), are the results of rotation of the initially formed northeast-southwest thrust during the later southeasterly-directed overthrusting.

A second, regionally identified, phase of folds is superimposed on both isoclinal folds and thrusts. It is possible to follow these structures through the different tectonic elements, hence we can suggest that they developed contemporaneously (Greco *et al.* 1989). The S2 crenulation cleavage remains fairly constantly orientated and cuts the previously formed penetrative S1 schistosity and mylonitic foliation. Regionally, this deformation can be correlated with the northeast–southwest orientated structures which develop from the Hazara to the Indus Kohistan (Coward & Butler 1985; Coward *et al.* 1986; Bossart *et al.* 1988). In the Kaghan Valley it is clear that this second set of structures is superimposed on the older set, producing an anticlockwise rotation of the earlier thrust lineations.

A doming of the earlier structures, the bowing up of the Murree Formation into the core of the Hazara–Kashmir Syntaxis, and the domal arrangement of the dark phyllitic schists in the Salkhalas south of Naran has also been observed. Between the domes, 'basins' and regionally important synformal depressions have been developed. This doming causes the flexure of the Mylonite Zone around Kaghan village, while a major axial depression has led to the extensive area of surface outcrop of the Upper Palaeozoic to Lower Mesozoic cover in the upper Kaghan Valley. In the Sub- and Lesser Himalaya discrete small-scale tectonic features related to the doming (minor folds, schistosity or shear structures) have not been observed. The domes are recognized only by the flexuring of the preexisting thrust planes, foliations and lineations. Consequently, it is suggested that the domes of the Subhimalayan and Lesser Himalayan units do not belong to a discrete third deformation, but are the product of the non-coaxial type 1 interference pattern (–>2) (Ramsay 1967) of the two previous phases. In the Higher Himalaya, the extreme ductility of the rocks caused the rotation of the first phase fold axis into a direction parallel to that of the transport direction and the superimposed deformation produced a coaxial interference pattern of type 3.

Model of tectonic evolution

The observation that the deformational regime and the metamorphic development of the area are intimately associated allows the use of petrologic data to suggest a model for both the tectonic and metamorphic evolution. The evolution begins with the emplacement of the Kohistan Sequence over the Indian Plate, although the exact age of this emplacement is questionable; presumably it would have occurred before the peak metamorphic age of the eclogite facies metamorphism (49 Ma by Sm/Nd; age data in Tonarini *et al.* 1993). This also agrees with the age of the Murree formation in the Hazara–Kashmir area which is Upper Palaeocene in the lower part of the formation, implying a late Cretaceous–early Palaeocene collision in this region (Spencer 1993). Movements along this thrust plane ended after the penetrative deformation, in amphibolite facies, of the Higher Himalaya (Greco *et al.* 1989; Spencer 1993), and therefore correspond to nappe formation in the Higher Himalaya, which developed under amphibolite and eclogite facies conditions (Fig. 5a).

In a subsequent phase, the Main Central Thrust developed, and the inverse metamorphic zoning pattern started to form (Fig. 5b). Rocks at the base of the Higher Himalaya cooled during the exhumation and reached upper greenschist facies conditions (the Mylonite Zone of the Kaghan Valley). Contemporaneously, the Lesser Himalaya became tectonically loaded and heated by the overriding Higher Himalaya synchronously with the development of the imbricated structure. Consequently, rocks of the Salkhala and Panjal Units underwent syndeformational greenschist grade metamorphism. The internal thrusting within the Lesser Himalaya caused cooling at the base of the Salkhala Unit (Panjal Thrust), where we observe the retrograde mineral transformations. The end of this tectono-metamorphic event, which also caused the folding of the Subhimalayan Unit, is characterised by the end of the movement on the Indus Suture (suggested to be at 20 Ma; Chamberlain *et al.* 1991) which produced the greenschist facies overprint of the granulites of the Kaghan Valley (Greco *et al.* 1989).

In order to prevent the complete thermal re-equilibration of the tectonic pile, which would have obliterated the inverse metamorphic pattern, a rapid uplift of the area is required (Fig. 5c). Zeitler (1985) has shown that this uplift occurred in the investigated area in the last 15 Ma. However, the isostatic re-equilibration of the thickened Indian crust could not solely have produced the rate and the amount of requested uplift. Consequently, the uplift must have had other important tectonic components, such as extension along the Indus Suture (Treloar *et al.* 1991; Chamberlain *et al.* 1991). In the present model, uplift had begun before the development of the Main Boundary Thrust in the apex and eastern limb of the Hazara–Kashmir Syntaxis, which is confirmed by all the cooling ages in the Higher Himalaya having passed through 300° C by Miocene times. This was achieved by thrusting (probably on the Main Central Thrust or on one of the lower thrusts such as the Panjal Thrust), coupled with pre Early Miocene extension along the Indus Suture (Chamberlain *et al.* 1991; Treloar *et al.* 1991; Spencer 1993) as well as folding (Spencer 1993). The anticlockwise rotation of the transport direction and therefore the development of the second deformational phase signals the shift of deformation to the western limb of the syntaxis. Finally, the progressive interference of the two transport directions produced the doming of the Western Himalayan Syntaxis region.

The ETH Himalayan group has been supervised by J.G. Ramsay. We would like to thank P. Bossart, R. Ottiger, D. Dietrich, R. Rey, K. Papritz, E. Herren, G. Martinotti and C. Spencer-Cervato for discussions on numerous occasions. In particular, we also wish to thank our Pakistan colleagues R.H. Malik, S.T. Ali, M. Ashraf, M. Ghazanfar and N. Chaudhry. A.G. was financially supported by the ETH Grant No. 0330.084.48/4. D.S. was financially supported by the ETH Grant No. 0.20.367.89, as well as an ETH–Imperial College Scholarship 0.330.028.02/8. We would also like to acknowledge financial support from the Staub Fund of the Institute of Geology, ETH, for the costs of the map reproduction. A. Uhr (ETH) and M. Collins (Geological Society, Publishing House) are thanked for the help in the map production. This paper benefited from a thorough review from P. Treloar.

References

Baig, M.S., Lawrence, R.D. & Snee, L.W. 1989. Evidence for late Precambrian to early Cambrian orogeny in northwest Himalaya, Pakistan. *Geological Magazine*, **125**, 83–86.

Bhat, M.I. & Zainuddin, S.M. 1979. Origin and evolution of the Panjal volcanics. *Himalayan Geology*, **9/2**, 421–461.

Bossart, P. 1986. *Eine Neue interpretation der Tektonik der Hazara–Kashmir Syntaxis (Pakistan)*. Dissertation ETH, **8032**, Zürich.

—— & Ottiger, R. 1989. Rocks of the Murree formation: indicators of a descending foreland basin of late Palaeocene to middle Eocene age? *Eclogae Geologicae Helvetiae,* **82/1**, 133–165.

——, Dietrich, D., Greco, A., Ottiger, R. & Ramsay, J.G. 1984. A new structural interpretation of the Hazara–Kashmir Syntaxis (Southern Himalaya) Pakistan. *Kashmir Journal of Geology,* **2**, 19–36.

——, ——, ——, —— & —— 1988. The tectonic structure of the Hazara–Kashmir Syntaxis, Southern Himalayas, Pakistan. *Tectonics,* **7/2**, 273–297.

Calkins, J.A., Offield, T.W., Abdullah, S.K.M. & Tayyab, Ali, S. 1975. *Geology of the Southern Himalaya in Hazara, Pakistan, and adjacent areas.* USGS Professional Papers, **716C**.

Chamberlain, C.P., Zeitler, P.K. & Erickson, E. 1991. Constraints on the tectonic evolution of the northwestern Himalaya from geochronologic and petrologic studies of Babusar Pass, Pakistan. *Journal of Geology,* **99**, 829–849.

Chaudhry, M.N. & Ghazanfar, M. 1990. Position of the Main Central Thrust in the tectonic framework of the western Himalaya. *Tectonophysics,* **174**, 321–329.

Coward, M.P. 1985. A section through the Nanga Parbat Syntaxis, Indus Valley, Kohistan. *Geological Bulletin of the University of Peshawar,* **18**, 147–152.

—— & Butler, R.W.H. 1985. Thrust tectonics and deep structure of the Pakistan Himalaya. *Geology,* **13**, 417–420.

——, Jan, M.Q., Rex, D., Tarney, J., Thirlwall, M. & Windley, B.F. 1982. Geotectonic framework of the Himalaya of N. Pakistan. *Journal of the Geological Society, London,* **139**, 299–308.

——, Windley, B.F., Broughton, R.D., Luff, I.W., Petterson, M.G., Pudsey, C.J., Rex, D.C. & Khan, M.A. 1986. Collision tectonics in the NW Himalayas. *In:* Coward, M.P. & Ries, A.C. (eds) *Collision Tectonics* Special Publications Geological Society, London, **19**, 205–221.

——, Butler, R.W.H., Chambers, A.F., Graham, R.H., Izatt, C.N., Khan, M.A., Knipe, R.J., Prior, D.J., Treloar, P.J. & Williams, M.P. 1988. Folding and imbrication of the Indian crust during Himalayan collision. *Philosophical Transactions of the Royal Society of London,* **A326**, 89–116.

Crawford, A.R. & Davies, R.G. 1975. Ages of Pre-Mesozoic formations of the Lesser Himalaya, Hazara District, Northern Pakistan. *Geological Magazine,* **112/5**, 509–514.

Fletcher, C.J.N., Leake, R.C. & Haslam, H.W. 1986. Tectonic setting, mineralogy and chemistry of a metamorphosed stratiform base metal deposit within the Himalayas of Pakistan. *Journal of the Geological Society, London,* **143**, 521–536.

Fuchs, G. 1975. Contributions to the Geology of the North-Western Himalayas. *Abhandlungen der Geologischen Bundesanstalt, Wien,* **32**, 1–59.

Gaetani, M., Casnedi, R., Fois, E., Garzanti, E., Jadul, F., Nicora, A. & Tintori, A. 1986. Stratigraphy of the Tethys Himalaya in Zanskar, Ladakh, Initial report. *Rivista Italiana di Paleontologia Stratigrafia,* **91/4**, 443–478.

Gansser, A. 1964. *Geology of the Himalayas.* Wiley-Interscience, London.

—— 1981. The geodynamic history of the Himalaya. *In*: Gupta, H.K. & Delany, F.M. (eds) *Zargos, Hindukush, Himalaya Geodynamic Evolution*. American Geophysical Union Geodynamic Series, **3**, 11–121.

Gardezi, A.H. 1968. Note on the geology of area around Nathiagali, District Hazara, West Pakistan. *Geological Bulletin of the University of the Punjab,* **7**, 71–78.

Garzanti, E., Casnedi, R. & Jadul, F. 1986. Sedimentary evidence of a Cambro-Ordovician orogenic event in the Northwestern Himalaya. *Sedimentary Geology,* **48**, 237–265.

Ghazanfar, M. & Chaudhry, M.N. 1984. A Palaeozoic ophiolite and island arc sequence of Hazara-Kashmir Syntaxis, District Manshera. *Kashmir Journal of Geology,* **2/1**, 37–38.

—— & —— 1986. Reporting M.C.T. in Northwest Himalaya, Pakistan. *Geological Bulletin of the University of the Punjab,* **21**, 10–18.

—— & —— 1987. Geology, structure and geomorphology of Upper Kaghan Valley, northwest Himalaya, Pakistan. *Geological Bulletin of the University of the Punjab,* **22**, 13–56.

——, Baig, M.S. & Chaudhry, M.N. 1983. Geology of Tithwal Kel area Neelum Valley, Azad Jammu and Kashmir. *Kashmir Journal of Geology,* **1**, 1–10.

Greco, A. 1989. *Tectonics and Metamorphism in the Western Himalayan Syntaxis Area (Azad Kashmir, NE-Pakistan)*. Mitteilungen aus dem Geologischen Institut der Eidgenössische Technische Hochschule und der Universität Zürich, Neue Folge, **274**.

——, Martinotti, G., Papritz, K., Ramsay, J.G. & Rey, R. 1989. The Himalayan crystalline nappes of the Kaghan Valley (NE-Pakistan). *Eclogae Geologicae Helvetiae,* **82/2**, 692–653.

Herren, E. 1987. *Structure, deformation and metamorphism of the Zanskar area.* Dissertation ETH.

Honegger, K., Dietrich, V., Frank, W., Gansser, A., Thöni, M. & Trommsdorff, V. 1982. Magmatism and metamorphism in the Ladakh Himalayas (The Indus – Tsangpo Suture Zone). *Earth and Planetary Science Letters,* **60**, 253–292.

Kumar, R. & Brookfield, M.E. 1987. Sedimentary environments of the Simla Group (Upper Precambrian), Lesser Himalaya, and their Palaeotectonic significance. *Sedimentary Geology,* **52**, 27–43.

Latif, M.A. 1970. Explanatory notes on the Geology of South Eastern Hazara, to accompany the revised Geological Map. *Jahr buch fur Geologischen Bundesanstalt, Wien, Sonderband,* **15**, 5–19.

—— 1973. Partial Extension of the Evaporite Facies of the Salt Range to Hazara, Pakistan. *Nature Physical Science,* **244**, 124–125.

—— 1974. A Cambrian age for the Abbottabad Group of Hazara, Pakistan. *Geological Bulletin of the University of the Punjab*, **10**, 1–20.

—— 1984. Age of the Salt Range Formation in the light of broader setting of the Himalayan Geology. *Kashmir Journal of Geology*, **2**, 39–44.

LE FORT, P. 1986. Metamorphism and magmatism during the Himalayan collision. *In*: COWARD, M.P. & RIES, A.C. (eds) *Collision Tectonics*. Special Publications, Geological Society, London, **19**, 159–172.

——, DEBON, F. & SONET, J. 1980. The 'Lesser Himalayan' Cordierite Granite Belt. Typology and Age of the Pluton of Mansehra (Pakistan). *Special Issue, Geological Bulletin of the University of Peshawar*, **3**, 51–61.

LILLIE, R.J., JOHNSON, G.D., YOUSUF, M., AGHA SHER HAMID ZAMIN, & YEATS, R.S. 1987. Structural development within the Himalayan Foreland fold-and-thrust belt of Pakistan. *In*: BEAUMONT, C. & TANKARD, A.J. (eds) *Sedimentary Basins and Basin-Forming Mechanisms*, Canadian Society of Petroleum Geologist, Memoir, **12**, 379–392.

MARKS, P. & ALI, C.M. 1961. The geology of the Abbottabad area, with special reference to the Infra-Trias. *Geological Bulletin of the University of the Punjab*, **1**, 47–55.

MEIER, A. 1992. *The geology of the Higher Himalaya in Upper Kaghan Valley, NE-Pakistan: Burawai–Jalkhad section. Aspects of tectonics*. Diploma thesis, ETH-Zürich.

OTTIGER, R. 1986. *Einige Aspekte der Geologie der Hazara–Kashmir Syntaxis (Pakistan)*. Dissertation ETH, **8083**, Zürich.

PAPRITZ, K. 1989. *The geology of Kaghan Valley, NE-Pakistan: aspects of tectonics and geochemistry*. Diploma thesis. ETH-Zürich.

—— & REY, R. 1989. Evidence for the occurrence of Panjal Trap Basalts in the Lesser- and Higher Himalaya of the Western Syntaxis area, NE Pakistan. *Eclogae Geologicae Helvetiae*, **82/2**, 603–627.

POGNANTE, U. & SPENCER, D.A. 1991. First record of eclogites from the Himalayan belt, Kaghan Valley, Northern Pakistan. *European Journal of Mineralogy*, **3/3**, 613–618.

RAMSAY, J.G. 1967. *Folding and fracturing of rocks*. McGraw-Hill, New York.

REY, R. 1989. *Einibe aspekte über die Geologie des Kaghan-Tales, NE-Pakistan*. Diploma thesis. ETH-Zürich.

SHAMS, F.A. 1961. A preliminary account on the geology of the Mansehra area, District Hazara, West Pakistan. *Geological Bulletin of the University of the Punjab*, **1**, 57–62.

SPENCER, D.A. 1993. *Tectonics of the Higher- and Tethyan Himalaya, Upper Kaghan Valley, NW Himalaya, Pakistan: Implications of an early collisional, high pressure (eclogite facies) metamorphism to the Himalayan belt*. Dissertation ETH, Zürich.

TAHIRKHELI, R.A.K. 1982. Geology of the Himalaya, Karakoram and Hindukush in Pakistan. *Special Issue, Geological Bulletin of the University of Peshawar*, **15**, 1–51.

—— & JAN, M.Q. 1979. Geology of Kohistan Karakoram Himalaya Northern Pakistan. *Special Issue of the Geological Bulletin of the University of Peshawar*, **2/1**, 1–30.

THAKUR, V.C. & GUPTA, V.J. 1983. Regional stratigraphy, paleontology and structure of Kashmir and Ladakh Himalayas. *Contributions to Himalayan Geology*, **2**, 1–32.

TONARINI, S., VILLA, I., OBERLI, F., MEIER, M., SPENCER, D.A., POGNANTE, U. & RAMSAY, J.G. 1993. Eocene age of eclogite metamorphism in Pakistan Himalaya: implications for India-Eurasian collision. *Terra Nova*, **5**, 13–20.

TRELOAR, P.J. & REX, D.C. 1990. Post-metamorphic cooling history of the Indian plate crystalline thrust stack, Pakistan Himalaya. *Journal of the Geological Society of London*, **147**, 735–738.

TRELOAR, P.J., REX, D.C. & WILLIAMS, M.P. 1991. The role of erosion and extension in unroofing the Indian Plate thrust stack, Pakistan Himalaya. *Geological Magazine*, **128**, 465–478.

——, COWARD, M.P., WILLIAMS, M.P. & KHAN, M.A. 1989*a*. Basement-cover imbrication south of the Main Mantle Thrust, North Pakistan. *Geological Society of America, Special Papers*, **232**, 137–152.

——, —— & —— 1989*b*. Metamorphism and crustal stacking in the North Indian Plate, North Pakistan. *Tectonophysics*, **165**, 167–184.

——, BROUGHTON, R.D., COWARD, M.P., WILLIAMS, M.P. & WINDLEY, B.F. 1989*c*. Deformation, metamorphism and imbrication of the Indian Plate, south of the MMT, North Pakistan. *Journal of Metamorphic Geology*, **7**, 111–127.

WAAGEN, W. & WYNNE, A.B. 1872. The geology of Mount Sirban in the Upper Punjab. *Geological Survey of India, Memoir*, **9/2**, 331–350.

WADIA, D.N. 1931. The Syntaxis of the North-West Himalaya: Its Rocks, Tectonics and Orogeny. *Records of the Geological Survey of India*, **65/2**, 189–220.

—— 1934. The Cambrian-Trias Sequence of North-Western Kashmir (Parts of Muzaffarabad and Bramula District). *Records of the Geological Survey of India*, **66**, 212–234.

WAHRENBERGER, C. 1992. *The geology of the Higher Himalaya of Upper Kaghan, NE-Pakistan: aspects of metamorphism*. Diploma thesis, ETH-Zürich.

WILLIAMS, M.P., TRELOAR, P.J. & COWARD, M.P. 1989. More evidence of pre-Himalayan orogenesis in north Pakistan. *Geological Magazine*, **125**, 651–652.

WYNNE, A.B. 1879. Further notes on the geology of the Upper Punjab. *Records of the Geological Survey of India*, **12/2**, 114–133.

YEATS, R.S., HASAN KHAN, S. & AKHTAR, M. 1984. Late Quaternary deformation of the Salt Range of Pakistan. *Geological Society of America Bulletin*, **95**, 958–966.

ZEITLER, P.K. 1985. Cooling history of the NW Himalaya, Pakistan. *Tectonics*, **4/1**, 127–151.

——, SUTTER, J.F., WILLIAMS, I.S., ZARTMAN, R. &

TAHIRKHELI, R.A.K. 1989. Geochronology and temperature history of the Nanga Parbat–Haramosh Massif, Pakistan. *Geological Society of America, Special Papers*, **232**, 1–22.

Geochemistry of the continental basalts within the Tethyan Himalaya of Lahul–Spiti and SE Zanskar, northwest India

J.C. VANNAY & L. SPRING

Institut de Minéralogie et de Pétrographie, Université de Lausanne, 1015 Lausanne, Switzerland

Abstract: The intrusive continental basalts of the Tethyan Himalaya are not necessarily comagmatic feeder dykes or the hypabyssal equivalents of the middle Permian Panjal Traps flows associated with the Permo-Mesozoic rifting of the Tethys, as is generally considered. A comparative study of basic rocks from SE Zanskar, NW Spiti and Upper Lahul shows the coexistence of geochemically different basalts in this region. The Panjal Traps of SE Zanskar and NW Spiti are tholeiitic continental flood basalts (CFB). They originated from an 'enriched' P-MORB-type magma which underwent a limited evolution by fractional crystallization and probable crustal contamination. In the adjacent Upper Lahul area, the Baralacha La Dyke Swarm shows a close spatial association with Lower Carboniferous transtensional synsedimentary faults. These dykes consist of tholeiitic to alkalic basalts, belonging to a comagmatic suite that evolved mainly by fractional crystallization. Geochemical data indicate that the Panjal Traps and the dykes are not comagmatically related. These characteristics imply either the coexistence, in the studied area, of contemporaneous but chemically different magmas or temporally unrelated magmatic events. The Baralacha La Dyke Swarm could be associated with a Lower Carboniferous transtensional tectonic event, related to an early phase of Permian rifting of Neo-Tethys.

Palaeozoic continental basalts of the NW Himalayas mainly consist of voluminous middle Permian flows called Panjal Traps. Recent studies have interpreted these extrusions as continental flood basalts (CFB) associated with Permian rifting of Neo-Tethys (Honegger *et al.* 1982; Papritz & Rey 1989). The Panjal Traps have been mainly studied in Kashmir and in NE Pakistan and no other Palaeozoic basaltic event has been described in the Tethyan Zone. The basaltic dykes cropping out in units older than the middle Permian are generally considered to be the comagmatic feeder dykes or hypabyssal equivalents of the Panjal Traps.

This paper presents the results of a geochemical study of the basalts from the SE Zanskar, NW Spiti and Upper Lahul regions. This study indicates that the Panjal Traps of Zanskar–Spiti are geochemically similar to their Kashmir equivalents and it confirms that they correspond to CFB-type basalts. Moreover, the basaltic dykes and the Panjal Traps occurring in the studied area were found not to be comagmatically related. The plan of the paper is as follows: geochemical analyses of the studied basalts are presented after a description of their geological occurrences and petrography; then the geochemical data and the tectonomagmatic signification of these basalts are discussed; and finally the conclusions derived from this work are presented.

Geological setting and petrography of the basic rocks

The investigated area is situated in the Tethyan Zone of the Higher Himalayas of NW India (Fig. 1). Geologically, it consists of a complete stratigraphic sequence of Precambrian to Mesozoic sediments, which underwent a complex tectonic and metamorphic history during the Tertiary Himalayan orogeny (Spring & Crespo-Blanc 1992). The basic rocks cropping out in this region have been metamorphosed and only rarely are original mineralogies and textures preserved (Table 1). The basalts display various types of occurrence which have been divided into the three groups described in the following sections.

The Panjal Traps of SE Zanskar and NW Spiti (PT group)

The Panjal Traps are voluminous basaltic flows in the NW Himalayas (Lydekker 1878). In the Tethyan Zone of Kashmir, the Panjal Traps basaltic extrusions mark the climax of volcanic activity that began with the production of intermediate to acid pyroclastic rocks, present in the Agglomeratic Slates. Together, these rocks form the Panjal Volcanic Series which reaches a maximum thickness of 2500 m in Kashmir (Pareek 1976). The Panjal Traps of Kashmir are

From Treloar, P.J. & Searle, M.P. (eds), *Himalayan Tectonics*
Geological Society Special Publication, No. 74, pp. 237–249.

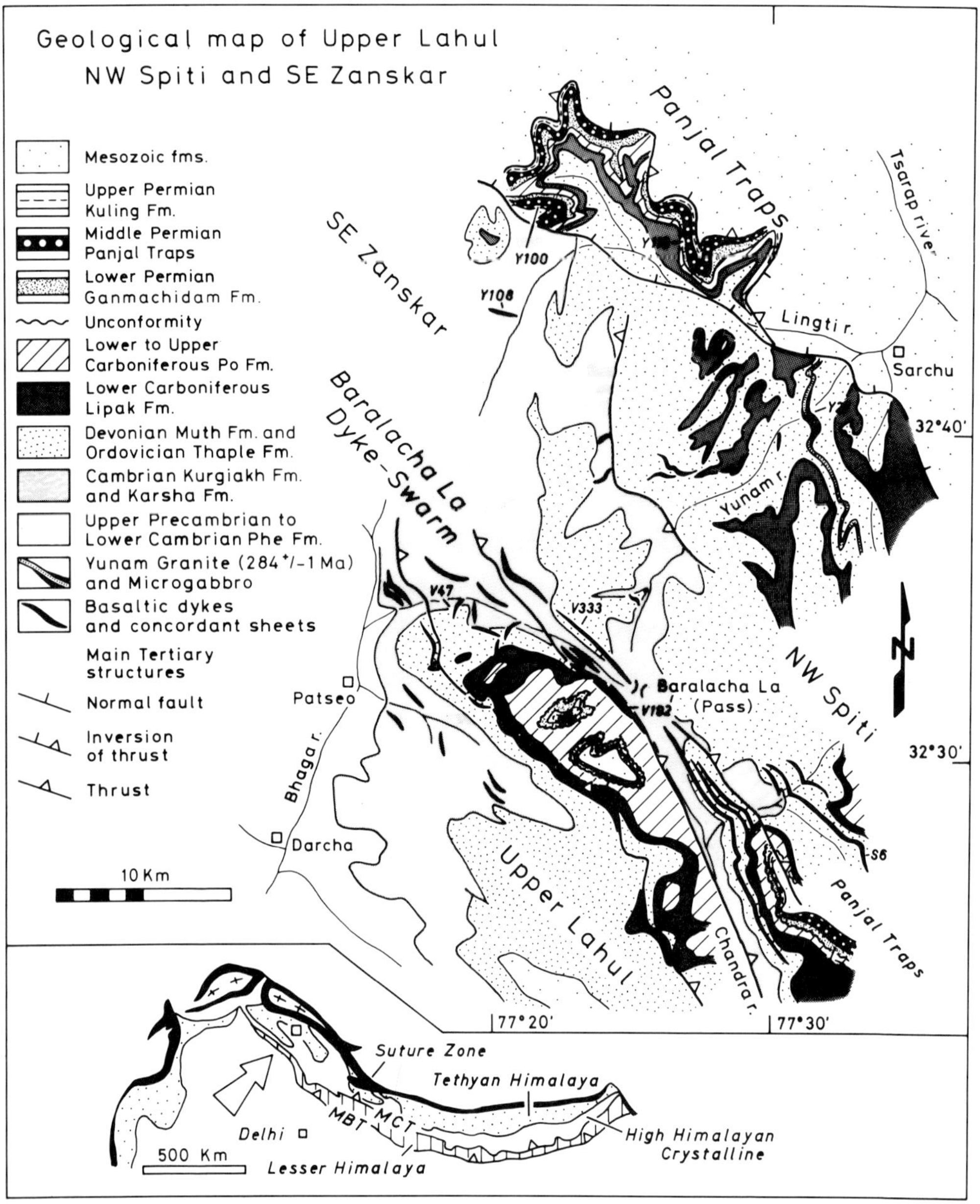

Fig. 1. Geographical and geological setting of the studied area in the Indian NW Himalayas, with selected sample locations. The map has been drawn after the authors' original 1/50 000 scale geological survey. Topographic base: Spot satellite images IDN 0137411K and 0137412L. Stratigraphic formation names and ages have been compiled after Hayden (1904), Fuchs (1982, 1987), Baud *et al.* (1984), Gaetani *et al.* (1986, 1990) and authors' observations.

massive aphyric basaltic flows, showing mainly tholeiitic and sometimes slightly alkalic compositions (Nakazawa & Kapoor 1973; Singh *et al.* 1976; Bhat & Zainuddin 1979; Honegger *et al.* 1982; Gupta *et al.* 1983; Pareek 1983). Rare intermediate to acid differentiation products associated with the Panjal Traps have been identified (Pareek 1976). The occasional pres-

Table 1. *Petrography of the basic rocks from Lahul–Spiti and SE Zanskar*

Group	Mineralogy	Texture	Metamorphic facies
Panjal Traps of SE Zanskar and NW Spiti (PT)	ab, chl, act, ep, wm, stilp, bi, ti, (cc)	Fine-grained and/or vesicular	Lower greenschist
Baralacha La Dyke-Swarm (BDS)	*magmatic*: cpx, plag, hbl, ap, ox	Sub-ophitic to intergranular	
	metamorphic: ab, chl, act, ep, pump, stilp, (cc)	Fine-grained and/or foliated	Pumpellyite-actinolite
	ab, chl, act, ep, wm, stilp, bi, ti, (cc)		Greenschist
Yunam microgabbro (MG)	hbl, olig, ep, chl, bi, wm, ti, (cc)	Relict microgabbroic	Lower amphibolite

Abbreviations

ab:	albite	cpx:	augitic clinopyroxene	plag:	calcic plagioclase
act:	actinolite	clz:	clinozoisite	pump:	pumpellyite
ap:	apatite	ep:	epidote	stilp:	stilpnomelane
bi:	biotite	hbl:	hornblende	tour:	tourmaline
cc:	calcite	olig-and:	oligoclase-andesine	wm:	white mica
chl:	chlorite	ox:	oxides	ti:	titanite

ence of spilitic pillow lavas, and the nature of the under- and overlying sediments, indicate that these basalts were extruded in coastal aerial to subaqueous environments (Nakazawa & Kapoor 1973; Nakazawa *et al.* 1975). Stratigraphic relations indicate an age between the Artinskian (270–275 Ma, Kapoor 1977) and the Kungurian to Kazanian periods (258–270 Ma, Nakazawa *et al.* 1975) for these basaltic extrusions.

The Panjal Traps have been mainly studied in Kashmir but they also occur from NE Pakistan to NW Spiti. To the NW of Kashmir, Panjal Traps have recently been identified and described in the Lesser Himalayas of the Western Syntaxis Area in NE Pakistan (Papritz & Rey 1989). To the SE of Kashmir, the Panjal Traps can be followed to SE Zanskar and NW Spiti, where they wedge out. These flows form the PT group of this study. In SE Zanskar, the Panjal Traps unit corresponds to 30–150 m thick massive basalts whereas in NW Spiti, it crops out locally as 30 m thick basalts, occasionally showing pillow lavas and hyaloclastic textures. According to Gaetani *et al.* (1990) for the SE Zanskar Panjal Traps, stratigraphic arguments define an age between the late Sakmarian or early Artinskian and the Kungurian–Kazanian (middle Permian).

Succeeding publications have provided different interpretations for the Panjal Traps basalts. This late Palaeozoic volcanism was assigned to the Hercynian orogeny by Wakhaloo (1972). Later, Nakazawa & Kapoor (1973) demonstrated the geochemical similarity between the Panjal Traps and the oceanic basalts. Since then, the interpretation relating the Panjal Traps to a 'geosynclinal' context has been adopted by many authors (Bhat & Zainuddin 1979; Wakhaloo 1979). Nevertheless, Rao (1981) considered these basalts as the product of a Hercynian volcanism related to subduction processes between India and Asia. More recently, the Panjal Traps of Kashmir have been interpreted as continental flood basalts (CFB) associated with middle Permian rifting processes and related to the opening of the Permo-Mesozoic Neo-Tethyan ocean between India and the Cimmerian continent (Honegger *et al.* 1982). The present study will demonstrate that the Panjal Traps of SE Zanskar and NW Spiti are geochemically comparable with their Kashmir equivalents and supports their identification as CFB.

The Baralacha La Dyke Swarm (BDS group)

In the Baralacha La area (Upper Lahul), numerous basaltic dykes cut the stratigraphical units of the Tethyan Himalaya from the Cambrian Karsha Formation up to the Lower Carboniferous Lipak Fm. (Fig. 1). Most of these intrusions are grouped in a parallel swarm. These dykes have a fairly constant orientation (NW–SE strike) and dip toward the NE. The intrusions are 1–20 m thick and some may be followed for several kilometres. Associated with these dykes, rare local concordant basaltic sheets were observed in the evaporites of the Lower Carboniferous Lipak Fm. and in the lowest part of the

Lower to Upper Carboniferous Po Fm. In SE Zanskar, similar concordant basaltic lavas and tuffs have been described at the base of the evaporites of the Lipak Fm. (Fuchs 1987).

Detailed geological mapping shows that the dykes never cut lithologies younger than the Lower Carboniferous Lipak Fm. (Lower Tournaisian), and that they are frequently injected along normal faults dipping toward the NE. Some of these faults proved to be synsedimentary transtensional structures, contemporaneous with the sedimentation of the Lipak Fm. (Fig. 2). The dykes could therefore be as old as Early Carboniferous. In SE Spiti, where the Panjal Traps extrusions have never been observed, basaltic dykes occasionally intruding along faults have been described in Carboniferous, Ordovician and Devonian formations (Hayden 1904; Fuchs 1982). These basic dykes in the Spiti area could mark the SE continuation of the Baralacha La Dyke Swarm. In the Baralacha La area, some dykes have been observed along NE-dipping Tertiary reverse faults, but they are affected by the Tertiary metamorphism and by the main regional schistosity. All these data are interpreted as a consequence of a tectonic inversion of the Carboniferous normal faults, with their associated dykes, during the Himalayan orogeny.

Radiochronological dating by the $^{40}Ar/^{39}Ar$ method failed to establish the age of this dyke swarm. In Lahul–Spiti and in Zanskar, the basaltic intrusions are generally considered to be the feeder dykes or hypabyssal equivalents of the middle Permian Panjal Traps (Fuchs 1982, 1987; Gaetani *et al.* 1986, 1990). In contrast, this geochemical study will indicate that the Panjal Traps of SE Zanskar-NW Spiti and the Baralacha La Dyke Swarm are not comagmatically related.

The Yunam microgabbro (MG group)

Cropping out on the banks of the Yunam river south of Sarchu, a composite discordant intrusion composed of a granitic dyke (Yunam Granite) and a parallel microgabbroic dyke (Yunam Microgabbro) cuts into the Ordovician Thaple Fm. (Fig. 1). These 2–15 m thick dykes are closely associated but the relative chronological relationship of the granite and the microgabbro has not been established. A geochronological and geochemical study indicates that the Yunam granite has a Lower Permian age and that it is most probably related to an anorogenic or extensional context (Spring *et al.* 1993). This intrusion is temporally associated with a contemporaneous tectonic event, documents, in this part of the Tethyan Himalaya by an angular unconformity between the Lower Permian Ganmachidam Fm. and the underlying sediments. In the Baralacha La area, we mapped such an unconformity (Fig. 1), which is comparable to the one described in Spiti by Fuchs (1982). This tectonomagmatic activity probably

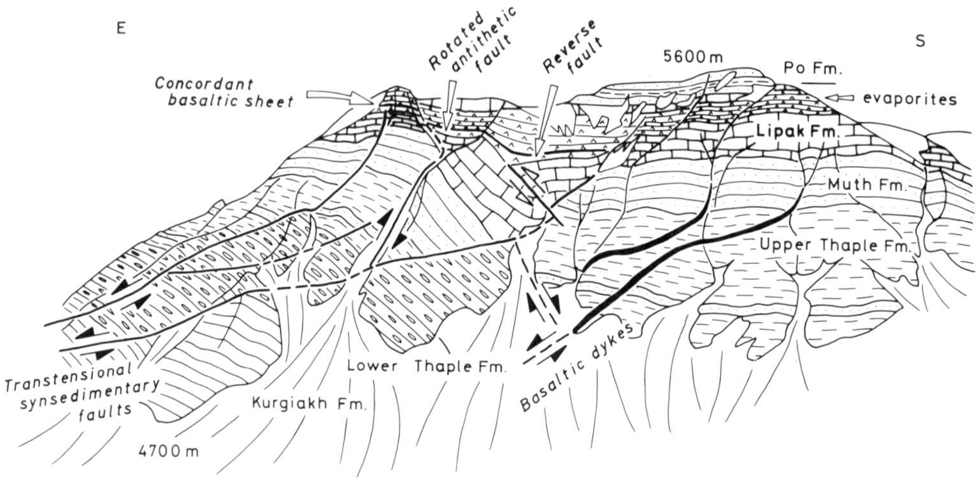

Fig. 2. Association of basaltic dykes with Lower Carboniferous synsedimentary transtensional faults (Zingzingbar, upper Bhaga Valley, Lahul). The evaporites of the Lower Carboniferous Lipak Fm. represent the syn-deformational sequence. The presence of an associated reverse fault (compressive component) indicates a transtensional 'negative flower structure' rather than simple extensional normal faults. For ages of the stratigraphic formations see Fig. 1.

represents an early phase of the Permo-Mesozoic Tethys rifting. The geochemical signature of the Yunam Microgabbro indicates that it is most probably comagmatically related to the BDS group of intrusions.

Geochemistry

In order to characterize and compare the different types of basalts cropping out in the investigated area, sixty-seven samples were

Table 2. *Chemical composition of selected basalts from Lahul–Spiti and SE Zanskar*

	Panjal Traps (PT)			Baralacha La Dyke Swarm (BDS)					MG
	Y100	Y118	V408A	V47	V192	V333	S6	Y108	Y21
Major elements (wt%)									
SiO_2	54.0	50.0	48.4	44.8	45.8	43.0	45.9	45.6	48.4
TiO_2	1.54	1.18	1.13	2.55	1.33	3.33	2.86	3.78	2.02
Al_2O_3	14.7	15.0	14.5	16.6	17.0	13.9	12.3	14.0	14.0
Fe_2O_3	6.79	4.64	4.49	4.53	1.37	4.10	10.22	3.51	4.94
FeO	3.5	5.2	4.4	9.5	10.6	10.7	5.4	12.7	8.6
MnO	0.22	0.18	0.17	0.16	0.17	0.23	0.24	0.25	0.22
MgO	5.6	8.6	5.7	6.5	6.8	5.4	3.6	4.5	6.2
CaO	7.2	7.3	11.0	3.9	9.1	6.8	7.0	7.5	10.4
Na_2O	2.3	2.9	1.3	4.4	2.5	3.0	3.0	2.8	2.3
K_2O	bd	bd	0.48	0.07	0.27	0.86	1.50	0.84	0.57
P_2O_5	0.13	0.09	0.08	0.32	0.17	0.64	0.81	0.83	0.29
H_2O	3.5	4.1	3.8	4.9	4.3	4.3	2.3	3.0	1.6
CO_2	0.4	0.6	4.3	1.6	0.2	3.5	4.7	0.2	0.2
Total	99.88	99.79	99.75	99.83	99.61	99.76	99.83	99.51	99.74
Trace elements (ppm)									
Ba	bd	4	264	37	377	129	355	304	124
Rb	3	3	16	bd	9	57	49	52	24
Sr	667	358	319	194	404	203	185	265	245
Th	1	1	bd	bd	bd	1	7	6	1
Nb	7	bd	bd	9	bd	20	27	30	11
Y	24	20	15	40	31	41	64	60	31
Zr	127	108	93	168	92	222	319	392	160
V	296	255	240	245	177	268	248	314	399
Cr	196	386	359	121	63	78	47	98	122
Ni	53	93	81	114	138	68	20	41	70
Co	59	56	50	75	73	63	51	70	68
Cu	bd	bd	25	37	76	80	115	bd	210
Zn	92	79	76	117	89	127	168	172	107
Ga	17	15	8	20	17	20	25	25	20
Hf	bd	bd	bd	5	3	3	6	7	3
S	bd	bd	27	649	427	bd	360	108	44
REE (ppm)									
La	13	11	9	17	12	25	37	35	14
Ce	29	30	26	39	29	53	76	70	34
Nd	17	14	17	20	13	32	45	42	19
Sm	4.8	3.9	na	5.9	3.9	8.6	11.6	10.3	5.1
Eu	1.4	1.1	na	1.9	1.3	2.8	3.3	3.5	1.7
Gd	5.4	4.0	na	6.5	4.4	8.4	11.7	11.3	6.0
Dy	4.3	3.7	na	6.5	4.78	7.4	11.3	10.3	5.6
Er	2.6	2.1	na	4.0	2.9	3.5	5.3	4.9	3.2
Yb	2.3	1.8	na	3.3	2.6	3.5	5.4	5.4	2.7
Lu	0.42	0.31	na	0.46	0.37	0.62	0.93	0.91	0.47
$(La/Yb)_N$	3.9	4.2	na	3.4	3.2	4.9	4.5	4.3	3.4
Mg′	0.53	0.64	0.57	0.48	0.53	0.42	0.32	0.36	0.48

MG: Yunam microgabbro. XRF analyses for Y, La, Ce, and Nd are presented when the complete REE and Y determinations by ICP are not available. $(La/Yb)_N$ are normalized ratios (normalization values as in Fig. 6). bd, below detection limit; na, not analysed. Uncertainties (2σ% rel.). major elements max. 4% (except FeO – 7%), trace elements max. 10% (except S = 17%) and REE max. 10% (except Nd = 20%).

systematically collected and were analysed for major and trace elements. Rare earth element contents were determined for nine of the samples (Table 2).

Sampling and analytical method

The sampling was carried out in order to minimize the influence of alteration on the results. Analyses were performed by X-ray fluorescence using glass beads for the major elements and powder pellets for the trace elements. A PW 1400 Philips spectrometer with Sc-Mo tube was used (Mineral Analysis Center of the University of Lausanne, analysts H.R. Pfeifer and J.C. Lavanchy). FeO was measured by colorimetry, CO_2 by coulometry and H_2O was calculated from the loss on ignition, taking into account the FeO, CO_2 and S contents. The spectrometer calibration was based on international standards from the US Geological Survey, NIM-South Africa and ANRT-France. Finally, the rare earth elements (REE) were analysed using the ICP method (CRPG laboratory at Vandoeuvre-les-Nancy, France; analyst J. Morel).

Metasomatic changes

As the analysed rocks were observed to have been metamorphosed under various conditions ranging from pumpellyite–actinolite facies to lower amphibolite facies (Table 1), it is necessary to discuss the possible mobility of some elements during metamorphism and/or alteration. K_2O, Na_2O, CaO, Sr, Rb and Ba show little correlation with Zr, which is known for its low mobility. Moreover, there is a correlation between abnormal contents of some of these elements (e.g. K_2O and Ba) and the intensity of the regional metamorphism or the degree of alteration of some samples. These variations strongly suggest that these elements have indeed undergone variable remobilization. The mobility of these 'large ion lithophile elements' (LILE) during metamorphism and alteration is a well-documented process, but the variations could also be the consequence of crustal contamination (Wilson 1989). However, without isotopic data it is not possible to quantify precisely the influence of crustal contamination on the evolution of the studied rocks.

Apart from the elements mentioned above, the other major and trace elements generally show discernable to good correlation with Zr. Consequently, these 'high field strength elements' (Saunders *et al.* 1980) are concluded to have undergone limited to no remobilization during secondary processes and are thus more likely to mirror the original chemical composition of the studied basalts.

Major element characteristics

The Panjal Traps of SE Zanskar and Spiti (PT group) have rather primitive compositions, characterized by low contents of incompatible elements such as TiO_2, P_2O_5 and FeO^{tot}. (Fig. 3) and relatively high Mg numbers ($Mg' = Mg/(Mg + Fe^{2+})$). The Mg numbers (mean $Mg' = 0.58 \pm 0.07$) also indicate that these basalts do not correspond to primary magmas in equilibrium with mantle compositions ($Mg' = 0.68$–0.75, Wilson 1989). In respect to their P_2O_5 contents, these basalts display a tholeiitic character. A limited evolutionary trend is visible in the variation diagrams.

The Baralacha La Dyke Swarm basalts (BDS group) have alkalic to tholeiitic compositions, in respect to their P_2O_5 contents (Fig. 3). It can be noted that in a ($Na_2O + K_2O$) versus SiO_2 diagram (not represented), despite the sensitivity of the elements to remobilization, the PT group plots in the tholeiitic to subalkalic field while the BDS group extends from the alkalic to the tholeiitic field. In the major elements variation diagrams, the BDS group displays coherent evolutionary trends towards high contents of the incompatible elements (Fig. 3). Together with the decrease in Mg′, these trends indicate a magmatic evolution by fractional crystallization. These trends can be subdivided into two segments. The first at below 200 ppm Zr, corresponds to linear trends in the TiO_2 and P_2O_5 diagrams. These well defined trends prove that the analysed rocks belong to a comagmatic suite. In the FeO^{tot}. diagram and, to a lesser extent, in the Mg′ diagram, the more scattered arrays probably account for a higher sensitivity of the elements to remobilization. Above 200 ppm Zr, the second segment is characterized by different rates of incompatible element enrichments and Mg′ depletion, compared to the first part of the trends. These modifications probably indicate a change in the crystallizing mineralogical assemblage during the magmatic evolution. The larger scattering of the analyses along the second segment could be a consequence of a larger influence of the contamination on the most evolved rocks (Lightfoot & Naldrett 1989). In the TiO_2 and P_2O_5 variation diagrams, the PT and BDS groups evolutionary trends are characterized by different Zr versus incompatible element ratios, reflecting the distinct evolutions of two magmas of different initial compositions. In terms of major element features, the Yunam microgabbro (MG group) is close to the BDS group.

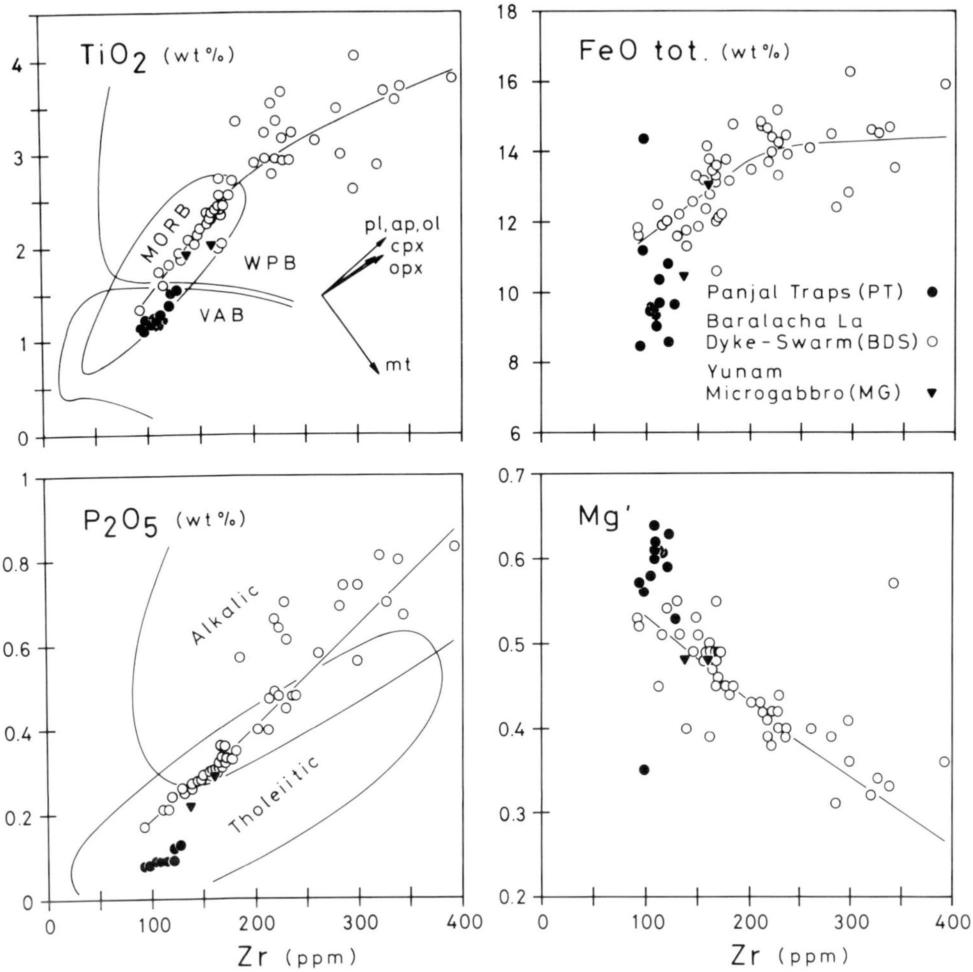

Fig. 3. Variation diagrams for some representative major elements. MORB (mid-ocean ridge basalts), VAB (volcanic arc basalts) and WPB (within plate basalts) compositional fields and fractionation vectors after Pearce (1982). Abbreviations; pl, plagioclase; ol, olivine; ap, apatite; cpx, clinopyroxene; opx, orthopyroxene; mt, magnetite. Alkalic and tholeiitic fields in the P_2O_5 versus Zr diagram after Floyd & Winchester (1975). Mg′ ('Mg number') = $Mg/(Mg + Fe^{2+})$, calculated with the assumption that $Fe_2O_3/FeO = 0.1$ (mean ratio for unaltered basalts). Smoothed weighted curve fits for the BDS group analyses are drawn. Uncertainties for the selected elements (2σ% relative): TiO_2 = 2%, FeO^{tot} = 9%, Mg′ = 11%, P_2O_5 = 4% and Zr = 5%.

Trace element characteristics

The PT group shows comparatively low concentrations of incompatible trace elements such as Y and high concentrations of compatible Cr (Fig. 4), characteristic of relatively primitive magmas that have not undergone strong fractionation of olivine, Cr-spinel and/or pyroxenes (Pearce 1982). However, the relatively low Ni concentrations, typical of many CFB (Wilson 1989), indicate that the PT group magmas do not correspond to primary liquids.

Compared to the PT group, the BDS group is characterized by progressive coherent enrichments of incompatible trace elements and by a depletion of compatible elements (Fig. 4). The markedly lower Cr contents and the trend toward low Ni concentrations indicate that the BDS group underwent significant fractionation of olivine, Cr-spinel and/or pyroxenes. Y is a good index of fractionation since this incompatible element is relatively insensitive to crustal contamination (Thompson *et al.* 1983). In the Y versus Zr diagram (Fig. 4), the clear trend

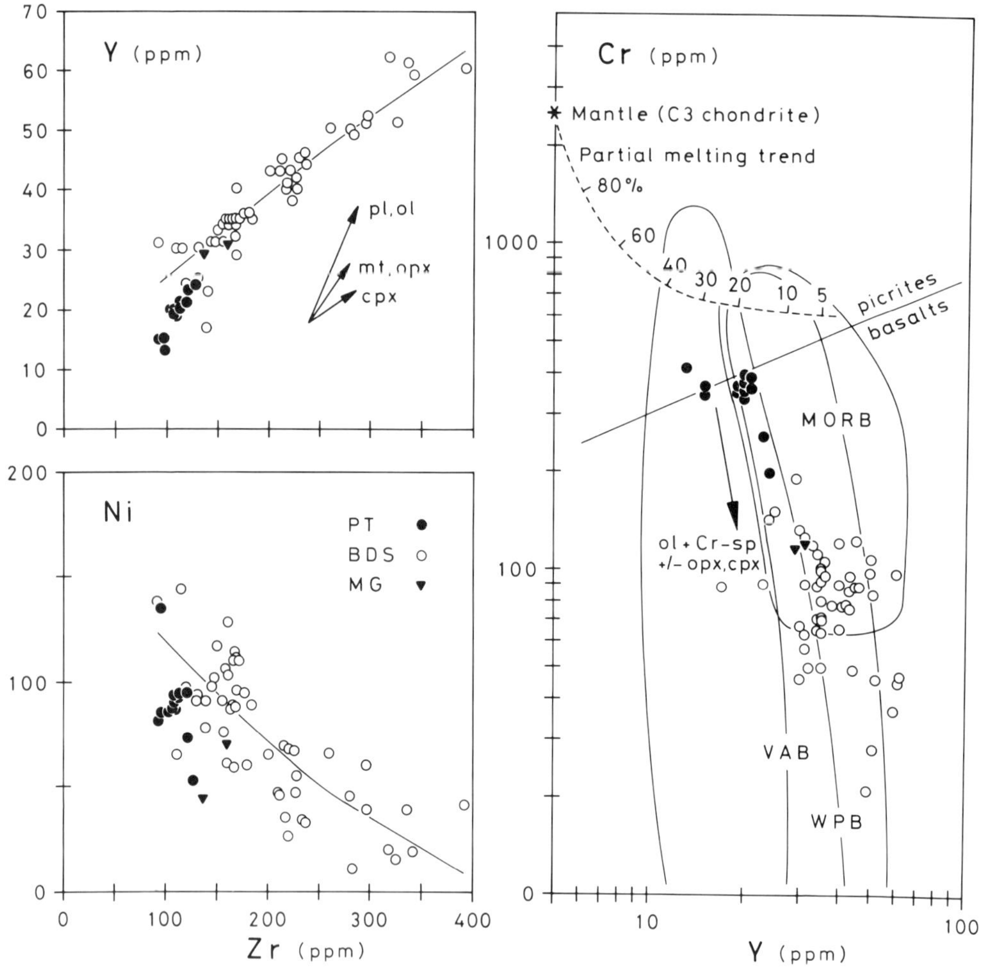

Fig. 4. Variation diagrams for selected trace elements. Fractionation vectors in the Y versus Zr diagram after Pearce & Norry (1979). Cr versus Y diagram after Pearce (1982). Uncertainties (2σ% relative): Y = 4%, Cr = 7% and Ni = 7%.

toward high concentrations demonstrates the importance of fractional crystallization in the magmatic evolution of the BDS rocks. If crustal contamination affected this evolution, its influence was never strong enough to mask the net evolutionary trend. Nevertheless, its is possible that a moderate contamination acted jointly with fractional crystallization to produce a coherent evolutionary trend (Lightfoot & Naldrett 1989). The MG and BDS groups have similar trace elements characteristics indicating that these rocks are most probably comagmatically related.

Mantle-normalized diagrams

The PT group displays mantle-normalized contents which are similar to the Kashmir Panjal Traps (Fig. 5A and C). The Panjal Traps patterns show many similarities to tholeiitic CFB and to 'enriched' mid-ocean ridge basalts (P-MORB). The PT group profiles have a marked trough for Nb and P and the Kashmir Panjal Traps also show a Nb negative anomaly. Some CFBs such as the Bushe Fm. of the Deccan Traps (Fig. 5C) display comparable geochemical profiles, characterized by Nb and P negative

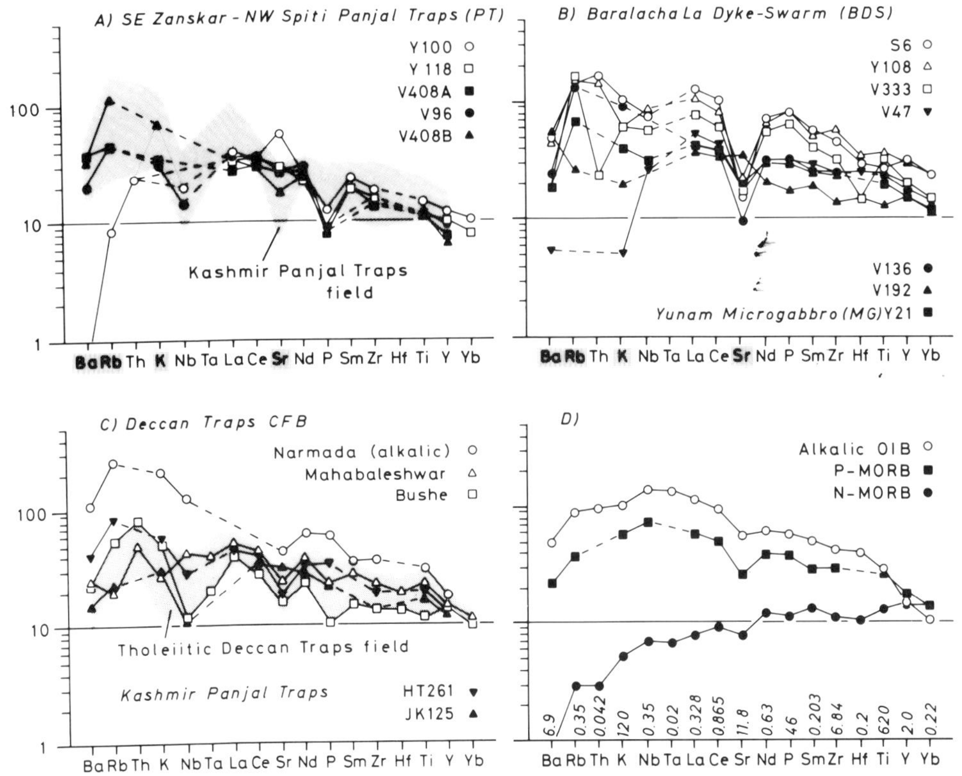

Fig. 5. Mantle-normalized 'spider-diagrams' for representative SE Zanskar, NW Spiti and Upper Lahul continental basalts (A and B). Shaded LIL elements (K, Rb, Ba and Sr) must be considered cautiously as they are susceptible to show variations due to remobilization. (C) Representative tholeiitic and alkalic Deccan Traps CFB after Lightfoot *et al.* (1990) and Mahoney *et al.* (1985) respectively. Selected Kashmir Panjal Traps after Honegger *et al.* (1982). (D) Typical N-MORB and alkalic OIB after Sun & McDonough (1989), P-MORB after Le Roex *et al.* (1983). Normalization values after Thompson *et al.* (1984).

anomalies. These features are generally interpreted as a sign of crustal contamination of a MORB- or ocean island basalt-type (OIB) magma (Thompson *et al.* 1983).

The mantle-normalized patterns of the less enriched BDS group analyses display close similarities to tholeiitic 'enriched' P-MORB and CFB. Samples characterized by high enrichments of incompatible elements show profiles comparable to alkalic OIB and alkalic continental basalts (Fig. 5B, C and D). In these patterns, the negative Nb anomaly compared to La and Ce is not as important as in the PT group and additionally there is no P anomaly. These features could reflect either a lower influence of contamination and/or different initial magma compositions for the BDS group compared to the PT group. The marked Sr negative anomaly characteristic of the BDS group profiles is coherent with low pressure plagioclase fractionation but could also be accounted for by Sr depletion during secondary processes. The close similarities of the MG and BDS group patterns indicate that these rocks are most probably comagmatically related.

Rare earth elements (REE)

The PT group shows REE chondrite-normalized patterns which coincide with tholeiitic P-MORB (Fig. 6A and C). This type of profile is frequently observed in continental tholeiites (Dupuy & Dostal 1984) and is comparable to the tholeiitic Deccan Traps patterns. A slight Eu anomaly suggests a plagioclase fractionation.

The BDS and MG groups display subparallel REE normalized profiles (Fig. 6B), most probably related to fractional crystallization

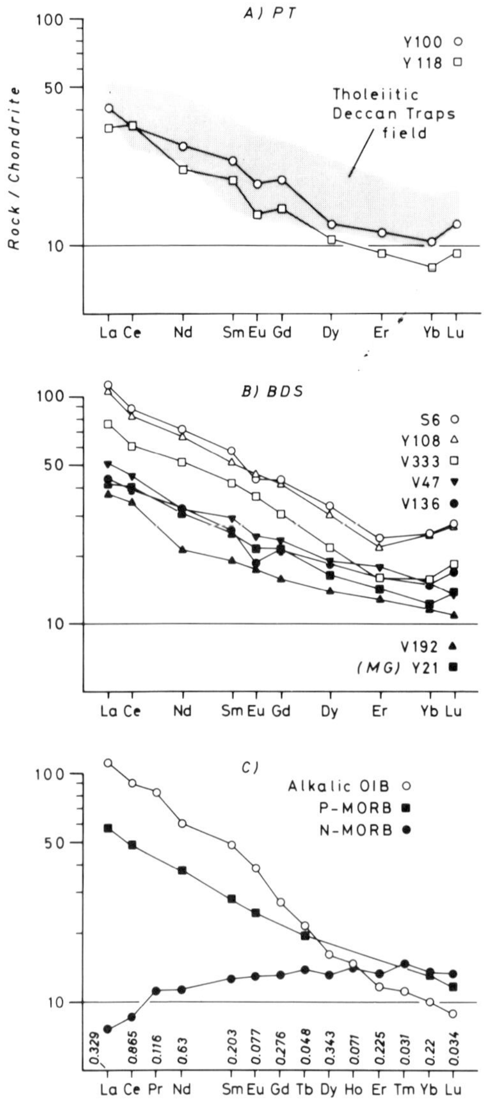

Fig. 6. Chondrite-normalized rare earth element analyses for representative SE Zanskar, NW Spiti and Upper Lahul continental basalts (A and B). Tholeiitic Deccan Traps field in (A) and MORB and alkalic OIB in (C) after the same references as in Fig. 5. Normalization values after Nakamura (1974).

processes in a comagmatic suite (Wilson 1989). REE enrichments in the selected samples are consistent with their trace element contents. The less enriched samples show REE patterns comparable with tholeiitic P-MORB and the most enriched samples have light REE (LREE) contents similar to alkalic basalts such as alkalic OIB (Fig. 6B and C).

Discussion

Compositions of continental basalts are generally considered to reflect the interaction between more or less enriched MORB-type magmas and continental lithosphere (Thompson *et al.* 1984; Hawkesworth *et al.* 1990). Major, trace and rare earth element data indicate that the Panjal Traps of SE Zanskar and NW Spiti (PT group) display relatively primitive tholeiitic compositions. These rocks originated from an 'enriched' P-MORB-type magma which underwent a limited magmatic evolution by fractional crystallization and probable crustal contamination. The PT group compositions compare well to the Kashmir Panjal Traps. This feature suggests the relative chemical homogeneity of this magmatic event over large distances (more than 200 km.). The geochemical signature of the PT group also displays close similarities to some of the tholeiitic Deccan Traps basalts. These characteristics indicate that the PT group corresponds to continental flood basalts (CFB), in agreement with the conclusions of some geochemical studies of Kashmir and NE Pakistan Panjal Traps (Honegger *et al.* 1982; Papritz & Rey 1989). The scarcity of associated intermediate and acidic melts (Pareek 1976) and the geochemical signature of the Panjal Traps indicate that they do not correspond to active continental margin basalts, generally characterized by low contents of Nb, Ce, Sm, Zr, Ti and Y compared to N-MORB (Pearce 1982). Consequently, they cannot be related to subduction processes, as was proposed by Rao (1981). The PT group magmas were not primary liquids as indicated by their Mg′ and relatively low Ni contents. However, their relatively primitive compositions (low contents of incompatible elements and high Cr concentrations) indicate that they have not been affected by important differentiation and suggests that these magmas probably rose rapidly through the crust. This conclusion is supported by stratigraphic arguments which indicate that the duration of the volcanic activity may have been very short. According to Gaetani *et al.* (1990), the time interval between the dated stratigraphic horizons below and above the Panjal Traps of SE Zanskar suggests a duration of 2–3 Ma for these extrusions.

CFB are generally related to continental break up. In the Tethyan Himalaya, the middle Permian Panjal Traps extrusions precede a passive margin evolution from Triassic to Eocene (Baud *et al.* 1984; Gaetani *et al.* 1986, 1990;

Fuchs 1987). Therefore, the Panjal Traps CFB extrusions are mostly probably related to the Permo-Mesozoic opening of the Tethys ocean (Honegger *et al.* 1982; Stampfli *et al.* 1991).

The BDS group compositions range from tholeiitic P-MORB-type basalts to more evolved alkalic OIB-type basalts. Coherent major and trace element variations indicate that these rocks belong to a comagmatic suite which evolved mainly by fractional crystallization. The basaltic dykes in the Tethyan Himalaya are widely considered to represent comagmatic equivalents of the Panjal Traps. In CFB provinces, the flows are linked to their magma reservoir by swarms of feeder dykes (Thompson *et al.* 1983). However, continental dyke-swarms related to rifting environments are not necessarily associated with CFB (Oliveira *et al.* 1990). Incompatible element ratios indicate that BDS and PT groups basalts cannot be related by fractional crystallization processes and that they evolved from magmas characterized by different initial compositions. Consequently, the BDS group intrusions are not the feeder dykes or hypabyssal equivalents of the PT group. Dating of the PT and BDS group samples by $^{40}Ar/^{39}Ar$ method on mineral separates gave only Tertiary metamorphic ages. In the absence of radiochronological data for the original age of the BDS group rocks, we propose two possible interpretations for these intrusions (Fig. 7).

(1) The BDS and PT group basalts could correspond to contemporaneous but non-comagmatic magmas, characterized by different initial compositions and subsequent evolutions. This hypothesis implies that these basaltic magmas were related either to a geochemically heterogeneous source or to different rates of partial melting of a common source. It can be noted that these two geochemically distinct basalt groups coexist in a relatively restricted area of the Himalayas. This fact contrasts with the apparent geochemical homogeneity of the Panjal Traps over large distances. In our opinion, this consideration does not favour this first interpretation.

(2) The BDS intrusions could represent a distinct magmatic event, temporally unrelated to the PT group extrusions. Several observations in the studied region prove the existence of magmatic and tectonic activities preceding the middle Permian PT group magmatic phase. The Yunam granite is a marker of an Early Permian magmatic event, associated with an anorogenic tectonic setting (Spring *et al.* this volume). The BDS dykes never cut rocks younger than Lower Carboniferous and show a close spatial association with synsedimentary transtensional faults of the same age. These structures are the first evidence for an extensional tectonic phase within the Tethyan Himalaya during this period. They precede by *c.* 70 Ma the middle Permian Panjal Traps extrusions associated with the Permo-Mesozoic Tethys rifting. Some recent models show that rifting processes start with a transtensional phase, associated with syn-rift

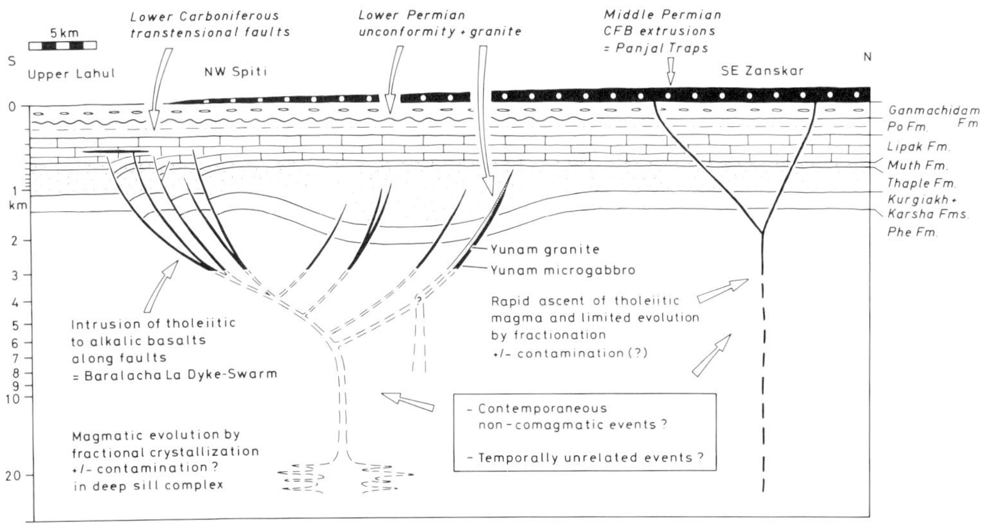

Fig. 7. Spatial and temporal relationship of the continental basalts from Lahul-Spiti and SE Zanskar with the Permo-Carboniferous tectonomagmatic events related to the Permo-Mesozoic Tethys rifting. Stratigraphy as in Fig. 1.

volcanism. This transtensional stage precedes a simple shear extensional phase, contemporaneous with 'plateau basalts' extrusions (Coleman & McGuire 1988; Stampfli *et al.* 1991). We interpret the Lower Carboniferous faults of the Baralacha La area as markers of an early phase of the Neo-Tethys rifting. The Baralacha La Dyke Swarm could be related to this tectonic event.

Conclusions

The middle Permian Panjal Traps of SE Zanskar and NW Spiti are tholeiitic continental flood basalts. They originated from an "enriched" P-MORB-type magma, which underwent a limited evolution by fractional crystallization and probable crustal contamination. The basaltic dykes of the Baralacha La area have tholeiitic to alkalic compositions and they correspond to a comagmatic suite which evolved mainly by fractional crystallization processes. These dykes are not comagmatically related to the Panjal Traps and are derived from a geochemically different initial magma. Consequently, the Baralacha La Dyke-Swarm basalts do not represent the feeder dykes or hypabyssal equivalents of the Panjal Traps. Dating using the $^{40}Ar/^{39}Ar$ method failed to establish the original age of the BDS group rocks. The lack of data for the age of these intrusions renders their precise relationship with the Panjal Traps magmatic event rather difficult to determine. However, the close spatial association of these dykes with Lower Carboniferous transtensional structures suggests that they could be related to an early phase of the Permo-Mesozoic Tethys rifting.

We wish to thank all the people who helped in this study. This work was taken from two PhD theses in preparation at the University of Lausanne. R. Allemann, H. Bücher, A. Crespo, L. Calmbach, R. Marchant, G. Mascle, G. Müller, B. Reymond and J.P. Schärer assisted the authors with field work. The XRF analyses was carried out with the assistance of J.C. Lavanchy. Critical reviews by A. Baud, M. Cosca, B.P. Duval, J. Hernandez, D. Laduron, J.D. Macdougall, H.R. Pfeifer, A. Saunders, G. Stampfli, A. Steck and an anonymous reviewer greatly improved the manuscript. This work received financial support from the University of Lausanne and the Swiss National Research Fund (grants FNR nr. 21–25 573.88 and 20–30 297.90).

References

BAUD, A., GAETANI, M., GARZANTI, E., FOIS, E., NICORA, A. & TINTORI, A. 1984. Geological observations in southeastern Zanskar and adjacent Lahul area (northwestern Himalaya). *Eclogae Geologicae Helvetiae*, **77**, 171–197.

BHAT, M.I. & ZAINUDDIN, S.M. 1979. Origin and evolution of Panjal Volcanics. *Himalayan Geology*, Dehra Dun, **9**, 421–461.

COLEMAN, R.G. & MCGUIRE, A.V. 1988. Magma systems related to the Red Sea opening. *Tectonophysics*, **150**, 77–100.

DUPUY, C. & DOSTAL, J. 1984. Trace element geochemistry of some continental tholeiites. *Earth and Planetary Science Letters*, **67**, 61–69.

FLOYD, P.A. & WINCHESTER, J.A. 1975. Magma type and tectonic setting discrimination using immobile elements. *Earth and Planetary Science Letters*, **27**, 211–218.

FUCHS, G. 1982. The geology of the Pin valley in Spiti (H.P., India). *Jahrbuch der Geologischen Bundesanstalt*, Wien, **124**, 35–359.

—— 1987. The geology of southern Zanskar (Ladakh) – Evidence for the autochtony of the Tethys Zone of the Himalaya. *Jahrbuch der Geologischen Bundesanstalt*, Wien, **130**, 465–491.

GAETANI, M., CASNEDI, R., FOIS, E., GARZANTI, E., JADOUL, F., NICORA, A. & TINTORI, A. 1986. Stratigraphy of the Tethys Himalaya in Zanskar, Ladakh. *Rivista Italiana di Paleontologia e Stratigrafia*, **91**, 443–478.

——, GARZANTI, E. & TINTORI, A. 1990. Permo-Carboniferous stratigraphy in SE Zanskar and NW Lahul (NW Himalaya, India). *Eclogae Geologicae Helvetiae*, **83**, 143–161.

GUPTA, K.R., GERGAN, J.T. & KUMAR, S. 1983. Geochemistry of the volcanic rocks of Northwestern Himalaya and its bearing on tectonics-a review. *In*: SINHA, A.K. (ed.) *Contemporary Geoscientific Researches in Himalaya*, **2**, Dehra Dun, 9–18.

HAWKESWORTH, C.J., KEMPTON, P.D., ROGERS, N.W., ELLAM, R.M. & VAN CALSTEREN, P.W. 1990. Continental mantle lithosphere and shallow level enrichment processes in the Earth's mantle. *Earth and Planetary Science Letters*, **96**, 256–268.

HAYDEN, H.H. 1904. *The geology of Spiti with parts of Bashahr and Rupshu*. Memoirs of the Geological Survey of India, **36**.

HONEGGER, K., DIETRICH, V., FRANK, W., GANSSER, A., THOENI, M. & TROMMSDORFF, V. 1982. Magmatism and metamorphism in the Ladakh Himalayas (The Indus-Tsangpo Suture Zone). *Earth and Planetary Science Letters*, **60**, 253–292.

KAPOOR, H.M. 1977. Lower Gondwana of Nishatbagh, Kashmir and its significance. *Geophytology*, **7**, 188–196.

LE ROEX, A.P., DICK, H.J.B., ERLANK, A.J., REID, A.M., FREY, F.A. & HART, S.R. 1983. Geochemistry, mineralogy and petrogenesis of lavas erupted along the Southwest Indian Ridge between the Bouvet triple junction and 11 degrees east. *Journal of Petrology*, **24**, 267–318.

LIGHTFOOT, P.C. & NALDRETT, A.J. 1989. Assimilation and crystallization in basic magma chambers: trace element and Nd-isotopic variations in the Kerns sill, Nipissing diabase province, Ontario. *Canadian Journal of Earth Science*, **26**, 737–754.

——, HAWKESWORTH, C.J., DEWEY, C.W., ROGERS, N.W. & VAN CALSTEREN, P.W. 1990. Source and differentiation of Deccan Traps lavas: Implications of geochemical and mineral chemical variations. *Journal of Petrology*, **31**, 1165–1200.

LYDEKKER, R. 1878. Geology of Kashmir, Kisthwar and Panghi. *Records of the Geological Survey of India*, **11**, 31–64.

MAHONEY, J.J., MACDOUGALL, J.D., LUGMAIR, G.W., GOPALAN, K. & KRISHNAMURTHY, P. 1985. Origin of contemporaneous tholeiitic and K-rich alkalic lavas: a case study from the northern Deccan Plateau, India. *Earth and Planetary Science Letters*, **72**, 39–53.

NAKAZAWA, K. & KAPOOR, H.M. 1973. Spilitic pillow lava in Panjal Trap of Kashmir, India. *Memoirs of the Faculty of Science, Kyoto University, Series of Geology & Mineralogy*, **39**, 83–98.

——, ——, ISHII, K., BANDO, Y., OKIMURA, Y. & TUKUOKA, T. 1975. The Upper Permian and the Lower Triassic in Kashmir, India. *Memoirs of the Faculty of Science, Kyoto University, Series of Geology and Mineralogy*, **42**, 1–106.

NAKAMURA, N. 1974. Determination of REE, Ba, Fe, Mg, Na and K in carbonaceous and ordinary chondrites, *Geochimica et Cosmochimica Acta*, **38**, 757–775.

OLIVEIRA, E.P., TARNEY, J. & JOAO, X.J. 1990. Geochemistry of the Mesozoic Amapa and Jari Dyke Swarms, northern Brazil: Plume-related magmatism during the opening of the central Atlantic. *In*: PARKER, A.J., RICKWOOD, P.C. & TUCKER, D.H. (eds) *Mafic dykes and emplacement mechanisms*. Balkema Publishers, Rotterdam, 173–184.

PAPRITZ, K. & REY, R. 1989. Evidence for the occurrence of Permian Panjal Trap basalts in the Lesser and Higher Himalayas of the Western Syntaxis Area, NE Pakistan. *Eclogae Geologicae Helvetiae*, **82**, 603–627.

PAREEK, H.S. 1976. On studies of the Agglomeratic Slate and Panjal Trap in the Jhelum, Liddar and Sind Valleys, Kashmir. *Records of the Geological Survey of India*, **107**, 12–37.

—— 1983. The Himachal and Panjal Traps; a geochemical appraisal. *In*: SINHA, A.K. (ed.) *Contemporary Geoscientific Researches in Himalaya*, **2**, Dehra Dun, 1–8.

PEARCE, J.A. 1982. Trace element characteristics of lavas from destructive plate boundaries. *In*: THORPE, R.S. (ed.) *Andesites*. Wiley, Chichester, 525–548.

—— & NORRY, M.J. 1979. Petrogenetic implications of Ti, Zr, Y and Nb variations in volcanic rocks. *Contributions to Mineralogy and Petrology*, **69**, 33–47.

RAO, V.D. 1981. Basic igneous episodes in the Himalaya and their tectonic significance. *In*: GUPTA, H.K. & DELANY, F.M. (eds) *Zagros–Hindu Kush–Himalaya geodynamic evolution*. Geodynamic Series **3**, 205–214.

SAUNDERS, A.D., TARNEY, J., MARSH, N.G. & WOOD, D.A. 1980. Ophiolites as ocean crust and marginal basin crust: a geochemical approach. *In*: PANAYIOTOU, A. (ed.) *Proceedings of the International Ophiolites Symposium*, Cyprus Geological Survey, 193–204.

SINGH, M.P., NANDA, M.M. & SINHA, P.K. 1976. The Ralakung volcanics of the Zanskar Valley (Ladakh), its geological setting, petrography, petrochemistry and a comparative study with the Panjal volcanics of the NW Himalaya. *Geological Survey of India Miscellaneous Publications*, **41**, 218–228.

SPRING, L. & CRESPO-BLANC, A. 1992. Nappe tectonics, extension and metamorphic evolution in the Indian Tethys Himalaya (Higher Himalaya, SE Zanskar and NW Lahul). *Tectonics*, **11**, 978–989.

——, BUSSY, F., VANNAY, J.C., HUON, S. & COSCA, M.A. 1993. Early Permian granitic dykes of alkaline affinity in the Indian High Himalaya of Upper Lahul, SE Zanskar: geochemical characterization and geotectonic implications. *This volume*.

STAMPFLI, G., MARCOUX, J. & BAUD, A. 1991. Tethyan margins in space and time. *Palaeogeography, Palaeoclimatology, Palaeoecology*, **87**, 373–409.

SUN, S.S. & MCDONOUGH, W.F. 1989. Chemical and isotopic systematics of oceanic basalts: implications for mantle compositions and processes. *In*: SAUNDERS, A.D. & NORRY, M.J. (eds) *Magmatism in the Ocean Basins*. Geological Society, London, Special Publications, **42**, 313–345.

THOMPSON, R.N., MORRISON, M.A., DICKIN, A.P. & HENDRY, G.L. 1983. Continental flood basalts. Arachnids rule OK? *In*: HAWKESWORTH, C.J. & NORRY, M.J. (eds) *Continental basalts and mantle xenoliths*. Shiva, Nantwich, 158–185.

——, ——, HENDRY, G.L. & PARRY, S.J. 1984. An assessment of the relative roles of crust and mantle in magma genesis: an elemental approach. *Philosophical Transactions of the Royal Society of London*, **A310**, 549–590.

WAKHALOO, S.M. 1972. Hercynian volcanicity in Kashmir. *Revue of the University of Jammu*, **3**, 44–50.

—— 1979. The Panjal volcanics of Kashmir. *In*: GUPTA, V.J. (ed.) *Contribution to Himalayan Geology*, **1**, Hindustan Publishing Corporation, Delhi, 189–194.

WILSON, M. 1989. *Igneous petrogenesis, a global tectonic approach*. Unwin Hyman, London.

Early Permian granitic dykes of alkaline affinity in the Indian High Himalaya of Upper Lahul and SE Zanskar: geochemical characterization and geotectonic implications

L. SPRING[1], F. BUSSY[1,2], J.-C. VANNAY[1], S. HUON[1] & M.A. COSCA[1]

[1] *Instituts de Géologie et de Minéralogie, Université de Lausanne, 1015 Lausanne, Switzerland*

[2] *Department of Geology, Royal Ontario Museum, 100 Queen's Park Toronto, Ontario M5S 2C6, Canada*

Abstract: In the Tethyan sedimentary series and the High-Himalayan 'crystalline' units of NW India, two generations of granite have previously been described. These rocks include large Cambro-Ordovician batholiths and small Tertiary bodies. A third type, the Yunam granite, has been discovered in the SE Zanskar – Upper Lahul region. It crops out as discordant dykes in the Ordovician Thaple Fm. Radiometric (U-Pb) measurements on zircons give an Early Permian crystallization age of 284 ± 1 Ma. Tertiary metamorphic cooling ages of 33.5, 20.8 ± 0.4, 22.8 ± 0.5 and 19.3 ± 0.2 Ma have been obtained by $^{40}Ar/^{39}Ar$ and K-Ar methods on amphiboles, muscovites and biotites of the granite and associated rocks. In contrast to the crustal affinity of the Cambro-Ordovician and Tertiary Himalayan granites, the Yunam granite is associated with an anorogenic, possibly thinned lithosphere context. This magmatic event is most probably related to the Permo-Mesozoic rifting of Neo-Tethys.

The granites of Upper Lahul and SE Zanskar mainly correspond to the two Cambro-Ordovician and Tertiary granitic events that have been described in many other parts of the Himalayas. In this study we report the first evidence for an Early Permian granitic magmatic event in the Indian Tibetan zone. The geochronological and geochemical characteristics of this granite are presented and compared to those of other granites cropping out in the region, and the implications of this granitic event on the late Palaeozoic tectonomagmatic evolution of the North Indian plate are discussed.

Geological setting

The northwestern part of the Indian Himalaya can be divided into the following tectonic domains, from north to south (Fig. 1):

(1) the Indus–Yarlung suture zone, comprising ultramafic to mafic rocks, oceanic sediments, flysch, tectonic melanges, and molassic conglomerates;
(2) the Tibetan zone, formed by sediments ranging from Late Precambrian to Eocene in a continuous series;
(3) the 'crystalline' series of the High Himalaya (HHC), composed of metasediments, ortho- and paragneiss and magmatic intrusions of Palaeozoic and Tertiary age;
(4) the Lesser Himalayan series, separated from the HHC by the Main Central Thrust (MCT), and including the transition from the Gondwanan facies to the Tethyan series;
(5) the Sub Himalayan molassic sediments, separated from the preceding domain by the Main Boundary Thrust (MBT).

The study has been carried out in the Tibetan zone of Upper Lahul and SE Zanskar, where a number of granitic bodies crop out (Fig. 1). Cambro-Ordovician granites (Mehta 1977; Frank *et al.* 1977; Le Fort *et al.* 1986; Stutz & Thöni 1987; Pognante *et al.* 1990) are widely represented, and record the whole deformational and metamorphic history related to the Tertiary Himalayan orogeny. Tertiary granites are represented by the small Gumburajun (Ferrara *et al.* 1987) and Oji Bihal (Vannay unpub.) plutons (Fig. 1, no. 3–4), which preserve original textures and parageneses.

In addition to these two generations of granite, a third type has been discovered in the investigated area. On the banks of the Yunam River in SE Zanskar (Fig. 2), a group of granitic dykes intrudes the Ordovician sediments of the Thaple Fm. These dykes have a general WNW–ESE orientation and a southerly dip of 10–60°. Kanwar & Bhandari (1979) observed them for the first time and interpreted them as evidence of

From TRELOAR, P.J. & SEARLE, M.P. (eds), *Himalayan Tectonics*
Geological Society Special Publication, No. 74, pp. 251–264.

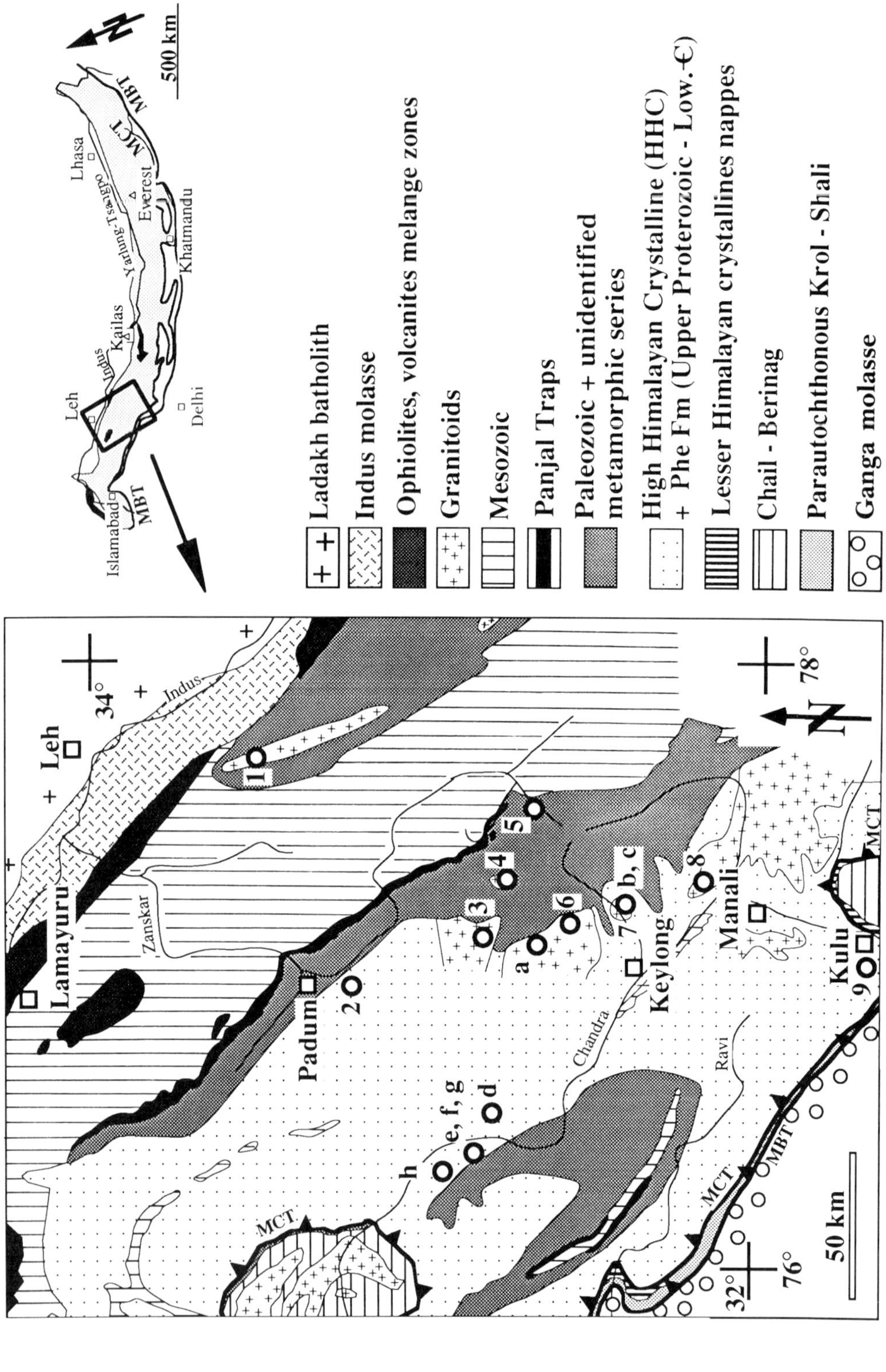
Ladakh batholith
Indus molasse
Ophiolites, volcanites melange zones
Granitoids
Mesozoic
Panjal Traps
Paleozoic + unidentified metamorphic series
High Himalayan Crystalline (HHC) + Phe Fm (Upper Proterozoic - Low.-€)
Lesser Himalayan crystallines nappes
Chail - Berinag
Parautochthonous Krol - Shali
Ganga molasse
500 km
50 km
Leh
Lamayuru
Padum
Keylong
Manali
Kulu
Zanskar
Indus
Chandra
Ravi
MCT
MBT
34°
32°
76°
78°
Islamabad
Delhi
Kailas
Yarlung-Tsangpo
Everest
Khatmandu
Lhasa

'Caledonian' magmatic activity on the basis of their stratigraphic position. Fuchs (1987) indicated their presence on his map. We propose the name of Yunam granite for these intrusions. The granite is locally closely associated with parallel microgabbroic dykes. Both types of intrusion are 2–15 m wide. No interpenetrations, mobile margins, or enclaves of either rock into the other have been found. Their spatial coexistence has been observed for several hundreds of metres to one kilometre (Fig. 2), whereas the granite alone can be followed for more than 7 km. No evidence of a chronological or genetic relationship between the granite and the microgabbro has been established, due to the poor exposure of their contact. The microgabbroic dykes are most probably cogenetically related to a basaltic dyke swarm cropping out to the south of the studied field, in the Baralacha La area (Vannay & Spring, this volume).

Petrography

The Yunam granite crops out in terranes that underwent a complex tectonic and metamorphic history during the Tertiary Himalayan orogeny (Spring & Crespo 1992).

It has been metamorphosed in the amphibolite facies and is characterized by the following metamorphic paragenesis: quartz–oligoclase–microcline – biotite – muscovite – hornblende – sphene–apatite. However, the rock preserves an original porphyritic microgranitic texture. Relicts of the primary magmatic assemblage are represented by blue quartz, plagioclase and porphyroclasts of K-feldspar. The absence of zoning in the plagioclases and the presence of 'alkaline-type' perthites in the potassic feldspars suggest an alkaline to hyperalkaline rock (J. Hernandez pers. comm.). Dark enclaves of centimetric to decimetric size, containing quartz ocelli, are locally present. In areas less affected by deformation, a chilled margin and a flux structure are clearly visible. A narrow contact metamorphic zone in the adjacent Ordovician sandstones has been observed.

Geochemical characterization of the granite

The 'large ion lithophile elements' (LILE: CaO, K_2O, Na_2O, Sr, Rb and Ba), widely used in characterization diagrams for granites, are susceptible to remobilization during metamorphism. As the Yunam granite has been metamorphosed in the amphibolite facies, these elements must be considered cautiously. However, the 'high field strength elements' (HFS: Ti, Zr, Hf, Nb, Y, Th and REE) generally display lower sensitivity to remobilization processes (Fig. 3A–C).

Whole-rock analyses (Table 1) indicate that the Yunam granite is a peraluminous acid rock (Shand 1927) (A/CNK = 1.1–1.24), with a normative corundum value between 3.5 and 4.8%. The Fe/(Fe + Mg) ratios are high, between 0.71 and 0.87. In contrast, the granite plots in the alkaline granite field of the R1/R2 diagram (de La Roche *et al.* 1980, not presented), but its $(K_2O + Na_2O)/SiO_2$ ratio is relatively moderate. In terms of its log $(CaO/(Na_2O + K_2O))$ features, the granite is alkali-calcic (Peacock 1931; Brown 1981, not presented). These discrepancies may indicate that the LIL elements have actually been remobilized. Consequently, a precise typology based on these elements is not possible.

In terms of 'immobile' elements (HFS), the Yunam granite displays marked differences from the Cambro-Ordovician and Tertiary granites of Lahul–Ladakh (Fig. 3A–C), or from peraluminous granites of mesocrustal origin in general. The Yunam granite is significantly

Fig. 1. Geological map of the Indian Higher Himalaya in Ladakh-Lahul. Compiled from Frank *et al.* (1987), Fuchs (1982*b*), Stutz (1988), and Spring & Crespo (1992). MBT, Main Boundary Thrust; MCT, Main Central Thrust.
Higher Himalayan granites. 1: Nyimaling granite (Stutz & Thöni 1987): 460 ± 8 Ma (Rb-Sr WR). 2: HHC orthogneiss (Pognante *et al.* 1990): 472 ± 9 Ma (U-Pb zircon). 3: Oji Bihal Tertiary granite (Vannay, unpub. data). 4: Gumburajun Tertiary granite (Ferrara *et al.* 1987): 18.8 ± 0.3 Ma (Rb-Sr biotite), 20.7 ± 0.3 Ma (Rb-Sr muscovite); (Villa & Oddone 1988): 19 Ma ($^{40}Ar/^{39}Ar$ muscovite). 5: Yunam granite (this paper) 284 ± 1 Ma (U-Pb zircon), 19.3 ± 0.2 Ma ($^{40}Ar/^{39}Ar$ biotite), 22.8 ± 0.5 (K-Ar muscovite), 20.8 ± 0.5 Ma (K-Ar biotite), 33.5 Ma ($^{40}Ar/^{39}Ar$ amphibole). 6: Lahul granites (Frank *et al.* 1977) 495 ± 16 Ma (Rb-Sr WR), >400 Ma (Rb-Sr muscovite), 16 Ma (Rb-Sr biotite); Kade granite (Ferrara *et al.* 1987): 621 ± 41 Ma (Rb-Sr WR); (Pognante *et al.* 1990): 549 ± 70 Ma (Rb-Sr WR), 20 ± 0.3 Ma (Rb-Sr biotite). 7: Jispa granite (Pande & Kumar 1974): 20–50 Ma (K-Ar). 8: Rohtang granite (Mehta 1977): 581 ± 9 Ma (Rb-Sr WR). 9: Kulu granite (Mehta 1977): 500 ± 8 Ma (Rb-Sr WR).
a–h: Location of analysed High Himalayan granites, from Searle & Fryer (1986).

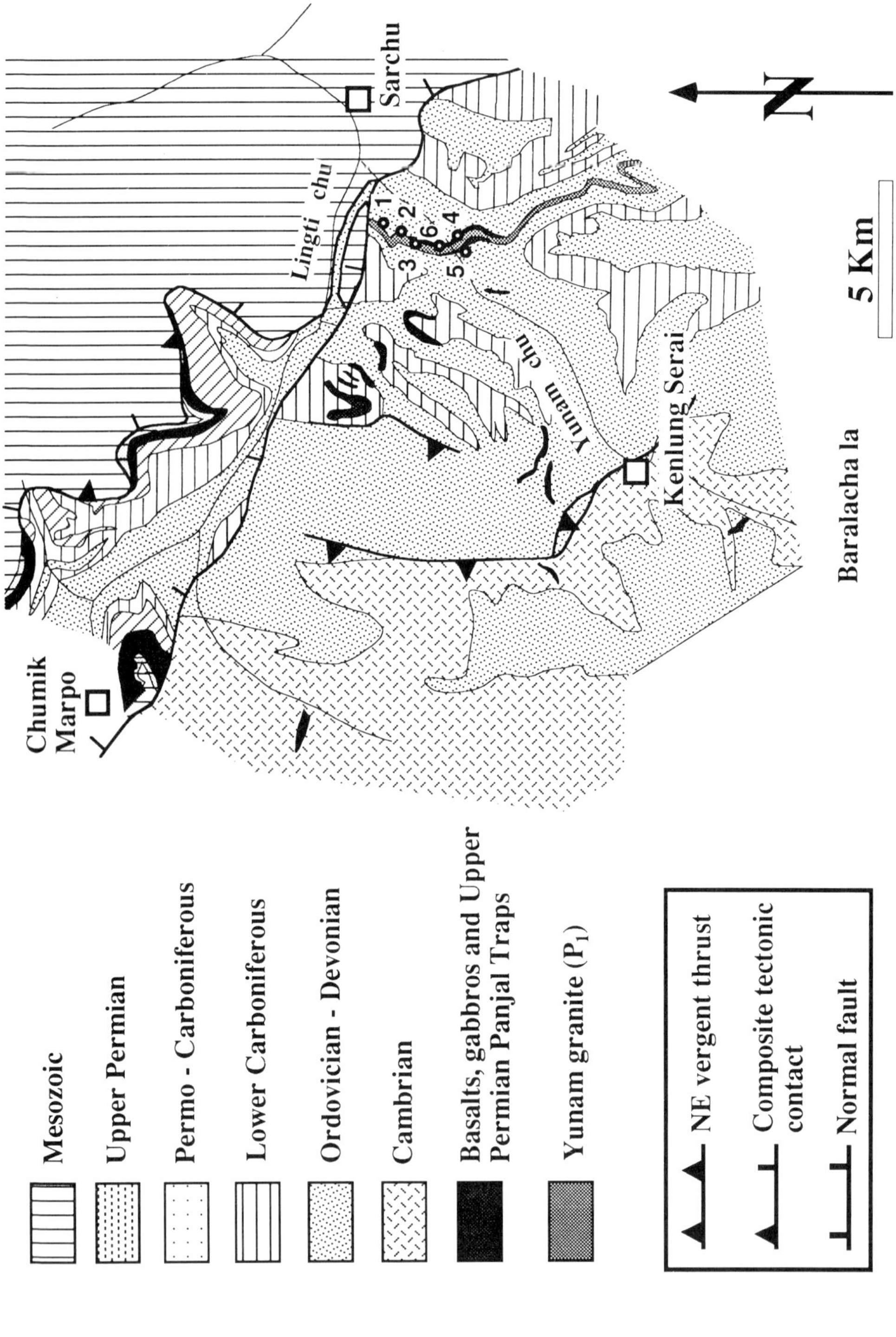
Sarchu
Lingti chu
1
2
3
4
5
6
Yunam chu
Kenlung Serai
Baralacha la
Chumik Marpo
N
5 Km
Mesozoic
Upper Permian
Permo - Carboniferous
Lower Carboniferous
Ordovician - Devonian
Cambrian
Basalts, gabbros and Upper Permian Panjal Traps
Yunam granite (P_1)
NE vergent thrust
Composite tectonic contact
Normal fault

Table 1. *Whole-rock major elements, trace elements and REE contents of the Yunam granite and the microgabbro*

	Yunam granite						Microgabbro	
Sample	FG	R1	FG2	HM25	V106	48931	Y1	Y21
Major elements (wt%)								
SiO_2	74.25	72.24	74.40	74.58	73.20	75.26	49.83	48.37
TiO_2	0.27	0.31	0.26	0.33	0.17	0.15	1.92	2.02
Al_2O_3	12.58	13.71	12.73	12.67	15.50	13.44	15.46	13.99
$Fe_2O_3^T$	2.55	3.23	2.71	3.21	1.38	1.53	10.82	13.50
MnO	0.03	0.03	0.04	0.03	0.03	0.04	0.18	0.22
MgO	0.26	0.50	0.20	0.25	0.28	0.25	4.90	6.15
CaO	0.74	0.58	0.70	0.56	0.73	0.65	9.92	10.40
Na_2O	2.44	1.87	2.15	2.36	4.39	2.68	1.25	2.32
K_2O	5.61	6.22	5.47	4.92	3.21	4.90	2.46	0.57
P_2O_5	0.09	0.09	0.08	0.09	0.09	0.17	0.22	0.29
H_2O	0.80	0.62	0.76	0.79	bd	bd	2.14	1.57
CO_2	0.47	0.36	0.42	0.12	0.64	0.71	0.30	0.18
Total	100.09	99.76	99.92	99.91	99.62	99.78	99.40	99.58
Trace elements (ppm)								
Ba	1012	1091	1064	913	333	104	570	124
Rb	198	204	195	196	188	396	136	24
Sr	72	95	72	78	285	29	419	245
Pb	47	79	36	31	152	28	16	3
Th	30	38	33	35	3	9	2	bd
U	bd	bd	bd	bd	4	9	bd	bd
Nb	16	23	22	22	bd	10	9	11
La	101	101	100	110	14	13	15	18
Ce	188	203	191	209	21	26	40	43
Nd	96	99	92	104	9	11	19	21
Y	76	82	73	80	10	25	29	35
Zr	313	292	301	296	65	71	137	160
V	21	25	16	29	11	7	354	399
Cr	12	bd	bd	bd	6	bd	117	122
Ni	19	4	3	4	bd	6	44	70
Co	22	33	16	29	38	28	58	68
Cu	7	bd	bd	bd	bd	bd	140	210
Zn	71	74	61	60	56	36	82	107
Ga	21	22	19	19	19	19	18	20
Hf	9	3	4	5	bd	bd	bd	3
S	59	56	82	102	bd	bd	39	44
REE (ppm)								
Eu	1.76						1.65	
Dy	12.59						5.64	
Yb	5.83						2.68	
Sm	16.47						5.11	
Gd	14.35						5.95	
Er	6.58						3.22	
Lu	0.96						0.47	

The location of the samples is shown on Fig. 2. Nyimaling granite: 48931; Tertiary Oji Bihal granite (Vannay, unpub): V106.
The major and trace elements analyses have been carried out by XRF on a PW Philips spectrometer at the Mineral Analysis Centre of Lausanne University (analysts H.R. Pfeifer and J-C Lavanchy). The REE have been analysed by ICP at the CNRS laboratory of Vandoeuvre-lès-Nancy.

Fig. 2. Geological map of SE Zanskar–Upper Lahul and localization of the magmatic rocks Location of samples: (Samples 1–4: Yunam granite; samples 5–6: microgabbro) 1, FG, geochemistry + U-Pb + $^{40}Ar/^{39}Ar$ + K-Ar dates; 2, FG2, geochemistry; 3, HM 25, geochemistry; 4, R1, geochemistry; 5, Y1, geochemistry; 6, Y21, $^{40}Ar/^{39}Ar$ date + geochemistry.

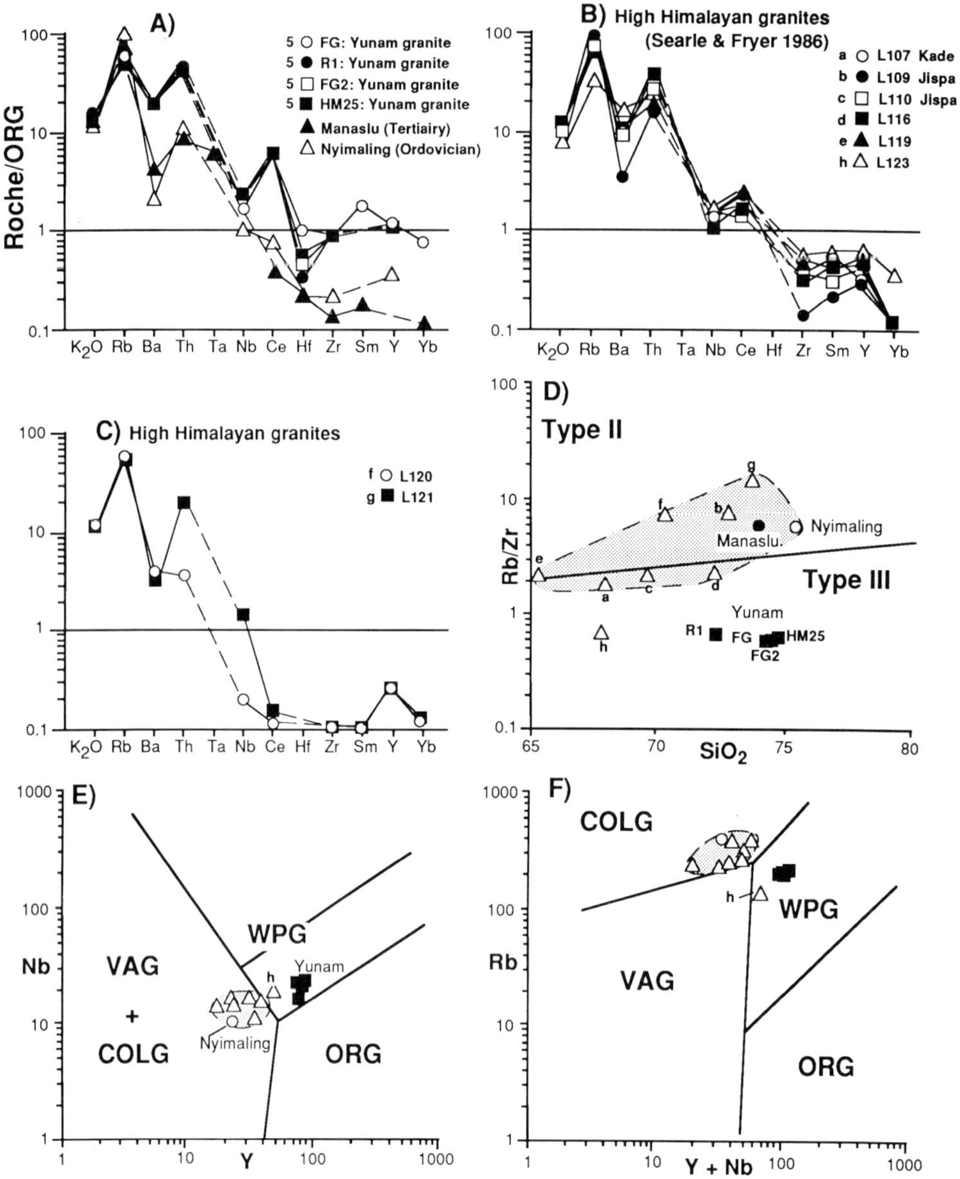

Fig. 3. Ocean ridge granite (ORG)-normalized (Pearce *et al.* 1984) geochemical pattern for the Yunam granite (**A**) and the High Himalayan granites (**B** and **C**). The high abundances of Rb, Th and Nb are a characteristic of WPG, which can be explained in terms of derivation from incompatible elements-enriched mantle. (**D**) Tectono-discriminant diagram after Harris *et al.* (1986). The field III indicates a mantle derived component. (**E**) Nb-Y and (**F**) Rb-(Y + Nb) discriminant diagrams for syn-collisional granites (COLG), volcanic arc granites (VAG), within plate granites (WPG) and ocean ridge granites (ORG) after Pearce *et al.* (1984). The Yunam granite (black squares) plots in the WPG field in both diagrams, while the Cambro-Ordovician and Tertiary Himalayan granites from Upper Lahul and Ladakh (Searle & Fryer 1986 and own unpublished data) plot in the syn-collisional field.

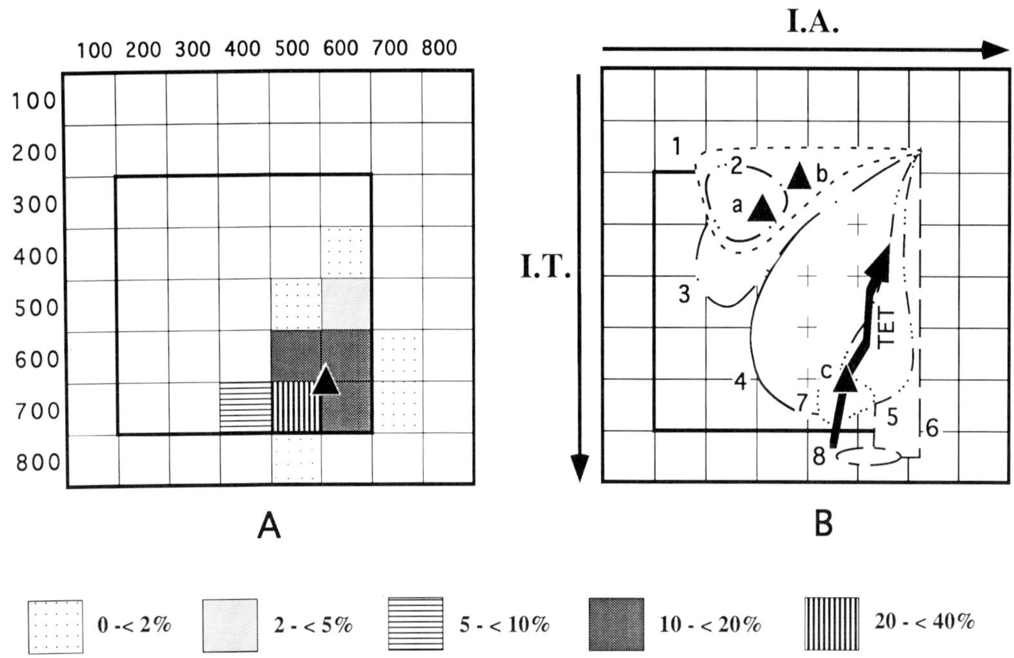

Fig. 4. Location of the zircon populations studied in the typological diagram of Pupin (1988). The A index (IA), or index of alkalinity, describes the relative development of the {211} (= low IA) and the {101} (= high IA) pyramidal faces. The T index (IT), or index of temperature, describes the relative development of the {110} (= low IT) and the {100} (= high IT) prismatic faces. (**A**) Early Permian Yunam granite. Mean point values: IA = 532, IT = 654. (**B**) Typological evolutionary trend (TET) of the Yunam granite. Also shown are the mean points of the Ordovician Nyimaling granite (a) (Stutz & Thöni 1987, IA: 315, IT: 325), the Tertiary Manaslu granite (b) (Brouand *et al.* 1990; IA: 430, IT: 270) and the Yunam granite (c). The different fields correspond to zircons from: 1, aluminous granites; 2, (sub)autochthonous monzogranites–granodiorites; 3, intrusive aluminous monzogranites-granodiorites; 4, calc-alkaline and K-calc-alkaline series granites; 5, subalkaline series granites; 6, alkaline series granites; 7, continental tholeiitic granites; 8, oceanic tholeiitic series granites.

depleted in Al_2O_3, and has higher values of Th, Nb, REE, Y and Zr (Fig. 3A–F). Moreover, its geochemical characteristics fit several of the Whalen *et al.* (1987) criteria for A-type granites: for a given SiO_2 value, it has relatively low concentrations of Al_2O_3, MgO and high concentrations of FeO, Zr (Fig. 3D) and REE.

The zircon typology method (Pupin 1980, 1988), provides additional arguments for the characterization of the studied rock, since its zircons have not been affected by Tertiary recrystallization (see zircon characteristics section and U-Pb data). The Yunam granite defines a Typological Evolutional Trend (TET) that plots in the subalkaline granite field (Fig. 4b) from high to lower crystallization temperatures. In its zircon typology, the Yunam granite clearly differs from the mesocrustal peraluminous granites, such as the Ordovician Nyimaling and the Tertiary Manaslu granites (Fig. 4B). Therefore, this method confirms the alkaline to subalkaline nature of the Yunam granite, in contrast to the peraluminous Himalayan granites.

According to Bonin (1986, 1990) the alkaline granites can be subdivided into two groups: (1) post-orogenic granites, rich in Ba and Sr, usually red, containing Mg-rich mafic minerals, and with a high Mn content. (2) early anorogenic hypersolvus granites, poor in Ba and Sr, white-green in colour, with a low Mn content. The petrography and the geochemistry of the Yunam granite do not fully fit either of these categories, and there is no field evidence of an emplacement related to an orogeny.

In conclusion, the Yunam granite shows an alkaline affinity, sometimes blurred by a late metamorphic overprint, but it does not fit any classical alkaline type.

It is beyond the scope of this paper to discuss the ultimate origin of alkaline granites. However

Table 2. *U-Pb analytical result*

Sample	Mass	Concentrations			Atomic ratios				Apparent ages (Ma)		
(No of mineral fraction, grain characteristics, and amount of selected zircons)	(mg)	U (ppm)[a]	Pb_{rad} (ppm)[a]	$^{208}Pb_{rad}$ (mol.%)[b]	$^{206}Pb/^{204}Pb$[c]	$^{206}Pb/^{238}U$[d]	$^{206}Pb/^{235}U$[d]	$^{207}Pb/^{206}Pb$[d]	$^{206}Pb/^{238}U$	$^{207}Pb/^{235}U$	$^{207}Pb/^{206}Pb$
1 Prism tips (35)	0.080	203	10	17.9	15099	0.04504 ± 21[e]	0.3231 ± 17	0.05202 ± 9	284.0	284.3	286.2
2 Long prisms (37)	0.071	127	6	19.0	8146	0.04499 ± 21	0.3223 ± 17	0.05196 ± 9	283.7	283.7	283.7
3 Short prisms (15)	0.038	66	3	29.9	2320	0.04501 ± 21	0.3227 ± 19	0.05200 ± 14	283.8	284.0	285.5
4 Prism tips (32)	0.229	107	5	18.2	11903	0.04522 ± 21	0.3243 ± 17	0.05201 ± 10	285.1	285.2	286.0
5 Long prisms (31)	0.117	148	7	13.7	23155	0.04663 ± 22	0.3415 ± 18	0.05311 ± 9	293.8	298.8	337.0
6 Short prisms (61)	0.128	194	10	17.3	32521	0.04528 ± 22	0.3268 ± 18	0.05234 ± 8	285.5	287.1	300.4

[a] Concentrations are known to about 1–2% for samples weights between 1.0 and 0.2 mg, about 5% for sample weights of about 0.1 mg and about 10% for sample weights of less than 0.05 mg.
[b] Relative to total radiogenic Pb.
[c] Corrected for fractionation and for spike.
[d] Corrected for fractionation, spike and blank; no initial common lead is present.
[e] Error estimates refer to the last significant digits of the isotopic ratios and reflect reproducibility of standards and measurement errors.

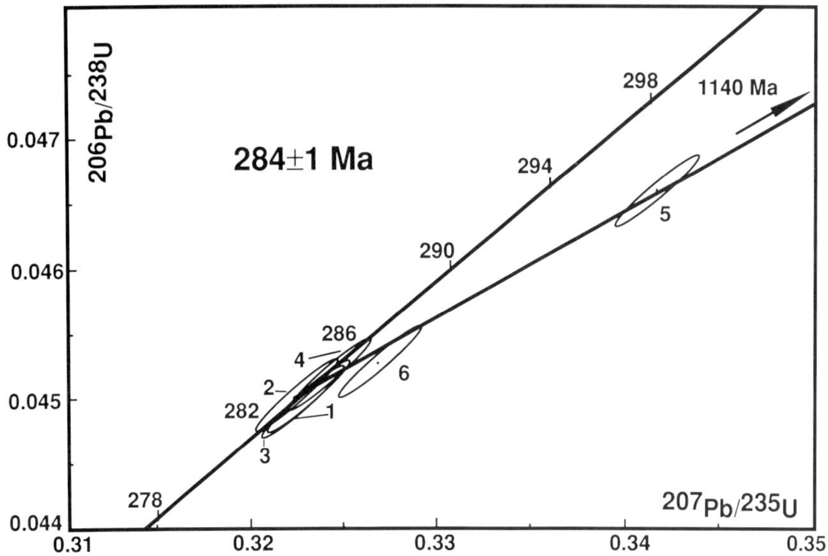

Fig. 5. Concordia diagram for zircon fractions of the Yunam granite. Ellipses around data points indicate 2σ analytical error. The 284 ± 1 Ma age is a mean value of the U-Pb apparent ages of fractions (1) to (4), the lower intercept of the discordia line is at 283.6 + 0.9/−1.0 Ma.

this type of magmatism is usually considered to be related to anorogenic or extensional tectonic environments (Collins *et al.* 1982; Pitcher 1983; Pearce *et al.* 1984; Harris *et al.* 1986; Whalen *et al.* 1987).

U-Pb geochronology

Analytical procedure

A single 4 kg sample was pulverized, and its zircons were isolated using conventional separation techniques. Air abrasion was applied extensively in order to increase the degree of concordancy of the ages (Krogh 1982). Chemistry was carried out using the standard techniques developed at the Royal Ontario Museum (Krogh 1973; Davis *et al.* 1982). Isotopic measurements were performed on a VG354 mass spectrometer in single collector mode. Pb and U were loaded together with silica gel and H_3PO_4 on Re-filaments. A conversion factor of 0.4%/AMU was applied to Daly data. Pb and U analyses were corrected for a fractionation of +0.1%/AMU (Corfu & Grunsky 1987), and for blanks of 2 to 5 picograms (Pb-blank isotopic composition: ^{208}Pb: ^{207}Pb: ^{206}Pb: ^{204}Pb = 37.62: 15.56: 18.3: 1). Regression lines were computed using the Davis method (1982), and decay constants were those recommended by Steiger & Jäger (1977). Error calculations were carried out according to the procedure described in Corfu & Ayres (1991) and Corfu & Grunsky (1987). Quoted errors are given at the 95% confidence level.

Zircon characteristics and U-Pb data

Zircons from the Yunam granite define a fairly homogeneous population (Fig. 4). They are colourless to slightly pinkish and transparent. The crystals are euhedral and short-prismatic with largely dominant {100} and {101} faces. According to Pupin (1980, 1988), this typological distribution pattern is typical of zircons formed at high temperatures (around 850° C) in a rather alkaline environment. There is no evidence of a resorption and/or an overgrowth event. Most, if not all zircons contain various types of inclusions: rounded to elongated melt inclusions, apatite needles, opaque material. No inherited zircon cores have been observed.

Six zircon fractions were selected (Table 2). Only crystals free of cracks and traces of turbidity were considered, but melt inclusions were still present in some of the abraded grains. Fractions (1) and (4) consist of elongated prism tips, selected to minimize a possible inherited component potentially located in the central part of the crystals. Fractions (2) and (5) include long-prismatic zircons up to 0.5 mm in length with a length/width ratio between 3 and 8. Fractions (3) and (6) are short-prismatic grains up to 0.2 mm in length with a l/w ratio <3.

Isotopic data are presented in Table 2 and Fig. 5. The fractions do not contain common Pb after correction for blank. Analyses (1) to (4) give no indication of inherited lead; all are concordant and overlap within error; the mean of all four fractions for $^{206}Pb/^{238}U$, $^{207}Pb/^{235}U$ and $^{207}Pb/^{206}Pb$ ages are 284.2 ± 0.7, 284.3 ± 0.7 and 285.4 ± 1.1 Ma respectively. Relying preferentially on U-Pb data, an age of 284 ± 1 Ma will be considered in the following discussion. Since neither metamorphic resorption nor overgrowths are detectable within the zircons, and considering their very simple euhedral shape, the 284 Ma event is interpreted as the crystallization age of the Yunam granite.

Fractions (5) and (6) are significantly discordant; they certainly include inherited lead. A best fit line may be drawn through all six fractions, with concordia intercepts at 283.6 + 0.9/−1.0 Ma and 1140 + 162/−133 Ma. This points to the presence of a single inherited component of Precambrian age within the magmatic system. Nevertheless, inheritance seems to be of limited importance, since several zircon fractions containing more than 30 crystals are totally free of detectable inherited lead. Possible sources for old lead are zircon relics from the melted material which generated the granitic magma and/or wall-rock zircons incorporated within the magma during its ascent.

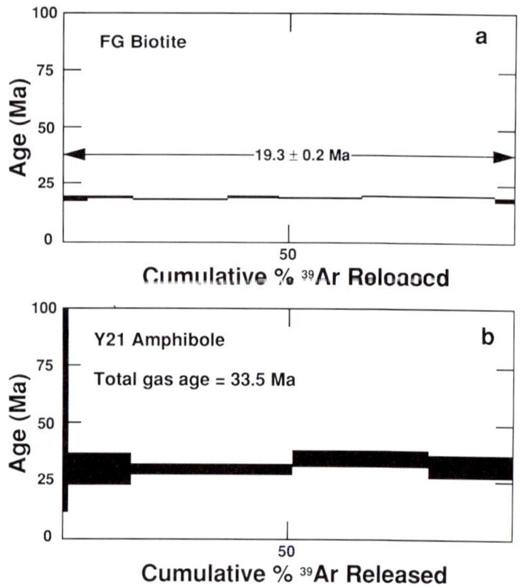

Fig. 6. $^{40}Ar/^{39}Ar$ release spectra for the Yunam granite biotite (**a**) and the microgabbro hornblende (**b**).

$^{40}Ar/^{39}Ar$ and K-Ar geochronology

The region comprising the granite outcrops has been metamorphosed in the amphibolite facies. $^{40}Ar/^{39}Ar$ and K-Ar geochronology has been performed on the metamorphic micas of the Yunam granitic dykes and on the amphiboles of the microgabbro at the isotope laboratories of Lausanne and Geneva Universities. The micas

Table 3. *$^{40}Ar/^{39}Ar$ data for the Yunam granite biotite and the microgabbro hornblende*

T(° C)	$^{40}Ar/^{39}Ar$	$^{36}Ar/^{39}Ar$	$^{39}Ar_k$ (% of total)	$^{40}Ar_R$ (% of total)	$^{39}Ar_k$ ($\times 10^{-14}$ mol)	Apparent age (Ma)
FG Biotite; J = 0.00457; wt = 0.04543 g						
550	2.29	0.02	5.54	33.1	0.27	18.8 ± 0.9
650	2.29	<0.01	9.88	74.6	0.48	18.8 ± 0.5
750	2.24	<0.01	21.11	84.5	1.03	18.4 ± 0.3
850	2.35	<0.01	11.38	89.1	0.55	19.3 ± 0.4
950	2.31	<0.01	18.25	91.1	0.89	19.0 ± 0.3
1050	2.47	<0.01	29.55	99.9	1.44	20.3 ± 0.2
Fuse	2.27	<0.01	4.29	85.9	0.20	18.6 ± 1.1
Plateau age (550° C-Fuse) = 19.3 ± 0.2 Ma						

T(° C)	$^{40}Ar/^{39}Ar$	$^{37}Ar/^{39}Ar$	$^{36}Ar/^{39}Ar$	$^{39}Ar_k$ (% of total)	$^{40}Ar_k$ (% of total)	$^{39}Ar_k$ ($\times 10^{-15}$ mol)	Apparent K/Ca, mol/mol	Apparent Age (Ma)
Y21 Amphibole; J = 0.00457; wt = 0.0467 g								
650	19.64	1.53	0.81	1.02	7.6	0.02	0.320	155 ± 146.6
950	3.62	10.63	0.02	14.23	41.9	0.32	0.046	29.6 ± 7.1
1050	3.71	11.64	0.01	35.87	60.8	0.81	0.042	30.4 ± 2.8
1150	4.34	12.56	<0.01	30.16	69.0	0.68	0.039	35.4 ± 3.4
Fuse	3.99	12.52	0.01	18.73	65.4	0.42	0.039	32.6 ± 5.4
Total gas age = 33.5 Ma								

Table 4. *K/Ar data for the Yunam granite biotite and muscovite (sample FG)*

	%K	[^{40}Ar]	%^{40}Ar*	Age
Biotite	5.95	2.16238*^{10}−10 moles/g	86.32	20.8±0.4Ma (±1σ)
Muscovite	3.57	1.42103*^{10}−10 moles/g	65.73	22.8±0.5Ma (±1σ)

%K: is potassium content in the mineral; [^{40}Ar]: is concentration of radiogenic argon; %^{40}Ar*: is content of radiogenic argon relative to atmospheric argon; ±1σ: is age error. The βγ values are after Steiger & Jäger (1977).

Table 5. *Dates from the Peshawar alkaline province*

Rock unit	Method	Age	Reference
Makaland granite	U-Pb zircons	294Ma	Zeitler 1988
Ambela syenite	Rb-Sr whole rock	315±15Ma	Le Bas *et al.* 1987
Ambela ijolite	Rb-Sr whole rock	297 ± 4 Ma	Le Bas *et al.* 1987
Ambela syenite	U-Pb zircons	Carboniferous	Zeitler 1988
Tarbela albititite	K-Ar	350 ± 15 Ma	Kempe 1986

were separated simultaneously with the zircons. ^{40}Ar/^{39}Ar analyses have been performed according to the procedure of Cosca *et al.* (1991). The biotite (sample FG) yields an ^{40}Ar/^{39}Ar plateau age of 19.3 ± 0.2 Ma and the amphibole (sample Y21) yields an ^{40}Ar/^{39}Ar total gas age of 33.5 Ma (Fig. 6a & b, Table 3). The amphibole had little potassium as indicated by the data, and yielded large errors on the individual heating steps. The amphibole and biotite ages are interpreted as cooling ages following Tertiary metamorphism. Given argon closure temperatures in amphibole and biotite are approximatively 500 and 300° C, respectively. A metamorphic cooling rate of 14° C/Ma has been calculated for this temperature interval from these data. The K-Ar data for the same biotite (sample FG, Table 4) have a slightly older age than the ^{40}Ar/^{39}Ar results, but clearly, these data suggest a complete resetting of the Ar system, whereas the U-Pb system in zircon remained intact.

The Lower Oligocene and Miocene cooling ages record a Tertiary regional metamorphic phase, affecting the area of Sarchu. Several geothermometers have been used to quantify the *P*/*T* conditions of the metamorphism south of Sarchu (Spring & Crespo 1992), yielding values of about 550° C and 5 kbar. These conditions were consequently insufficient to rejuvenate the granite zircons, although they sometimes present a slightly pinkish coloration, characteristic of a regional metamorphism.

Discussion

As far as currently known, the Yunam granite is the first report of an alkaline to subalkaline granitic event of Lower Permian age in the Indian High Himalaya. It is most probably equivalent to the alkaline province of Peshawar in Pakistan (Coulson 1936; Ahmad *et al.* 1969; Kempe & Jan 1970; Kempe 1973; Ashraf & Chaudary 1977; Jan *et al.* 1981; Kempe 1983), 600 kilometres to the west, where Middle to Upper Carboniferous alkaline granites, gabbros and carbonatites have recently been dated (Table 5).

Some granites of Upper Palaeozoic age have been described in the Indian High Himalaya. Mehta (1977) reports a Rb-Sr age of 311 ± 6 Ma for the Mandi leucogranite, which is spatially associated with a granite dated at 545 ± 12 Ma. Bhanot *et al.* (1975) give a Rb-Sr age of 350 ± 50 Ma for a pegmatite intruding the 456 ± 50 Ma Dalhousie granite. Unfortunately, there is no geochemical data for these rocks. The most interesting age is undoubtedly that of Trivedi *et al.* (1982) for the Permo-Triassic Gaik granite (235 ± 13 Ma, Rb-Sr), situated north of the Indus–Yarlung Suture, in the Ladakh–Rupshu region. No geochemical data are available for this granite, but its rather low initial ratio of 0.7081 ± 0.0004 ^{87}Sr/^{86}Sr is compatible with an alkaline type granite. This granite intruded the North Gondwanan margin during or just after the Permo-Mesozoic formation of the Tethys ocean.

The Yunam granite of alkaline to subalkaline affinity clearly differs from the peraluminous granites of this part of the Himalayas. In the tectono-discriminant diagram of Pearce *et al.* (1984), the Yunam granite analyses plot in the 'within plate granite' (WPG) field, usually observed in anorogenic or thinned lithospheric context (Fig. 3E and F). In contrast, the Cambro-Ordovician and Tertiary granites are clearly associated with a crustal environment. Moreover, the very small inherited Pb component of the studied rock is quite unusual in common orogenic granites. It is more typical of alkaline post- to anorogenic granitic magmas, generated either by an efficient high temperature melting of the lower crust with almost complete dissolution of old zircons, or by a mantle source free of this mineral. All these characteristics strongly suggest that the Yunam granite has been associated with an anorogenic or a thinned lithosphere context.

The Yunam granite intrusion precedes, by approximately 10 Ma, the Middle Permian basaltic extrusions (Panjal Traps), associated with the rifting of the Permo-Mesozoic Tethys (Honegger *et al.* 1982). The Panjal Traps are contemporaneous with tectonic movements recorded in this part of the Himalaya by the sedimentation of Lower Permian conglomerates (Chumik Fm.), which unconformably overlie older sediments. Such tectonic activity has been interpreted as evidence of a Hercynian orogenic phase (Fuchs 1982*a*, 1987) or of early movements related to the Permo-Mesozoic Tethys rifting (Gaetani *et al.* 1990). In Lahul-Zanskar, there is no evidence for Hercynian orogenesis. In particular, the huge masses of granitic material so typical of the Hercynian belt are completely lacking. The presence of a lower Carboniferous extensional phase (Vannay & Spring this volume) and the proximity of the Neo-Tethys rifting suggest an extensional environment for the emplacement of the Yunam granite. Consequently we interpret this rock as the marker of an alkaline to subalkaline granitic event related to an early tectono-magmatic phase of the Permo-Mesozoic rifting. During this extensional phase, a thinning of the crust may have led to the upwelling of the upper mantle and melting of the lower crust. The melts produced could have subsequently intruded the upper crust.

Conclusions

The Yunam granite (284 ± 1 Ma, U-Pb zircon) provides the first evidence for a granitic event during the Early Permian in the Tibetan zone of India. Its geochemical signature contrasts with the characteristics of the Cambro-Ordovician and Tertiary syn-collisional peraluminous Himalayan granites. The geochemistry and the zircon typology indicate that the Yunam granite is of alkaline type, emplaced in a thinned lithosphere. The intrusion is contemporaneous with tectonic movements which created an unconformity in the Lower Permian sediments. This tectonomagmatic event took place in a succession of transtensional to extensional events ranging from the Early Carboniferous to the Mid-Permian, linked to the rifting and opening of the Permo-Mesozoic Neo-Tethys.

This study was supported by the Swiss National Science Foundation grants Nr. 21–25 573.88 and 20–30297 (Géologie de l'Himalaya du Ladakh, coupe Manali – Leh, H. Masson and A. Steck, head of the project). We thank F. Debon and M.Q. Jan for their critical review, E.J. Essene, J. Hernandez and Ph. Thélin who provided helpful comments of an early version of the manuscript, A. Crespo, R. Marchant, G. Müller and M. Sartori for their help in the field work, J-P. Pupin (Nice) for the determination of the zircon population, H.R. Pfeifer and J.C. Lavanchy for the XRF analysis, and A. Spring for the English translation. F.B. is greatly indebted to the SNSF (grant Nr 8220–028439) and to the Natural Sciences and Engineering Research Council of Canada for their financial support, and to F. Corfu, who guided his first steps at the ROM through this study.

References

Ahmad, M. Ali, K.S.S., Khan, B., Shah, M.A. & Ullah, I. 1969. The geology of the Warsak area, Peshawar, West Pakistan. *Geological Bulletin of University of Peshawar,* **4**, 44–78.

Ashraf, M. & Chaudary, M.N. 1977. A discovery of carbonatite from Malakand. *Geological Bulletin of Punjab University,* **14**, 91–94.

Bhanot, V.B., Goel, A.K., Singh, V.P. & Kwatra, S.K. 1975. Radiometric studies for Dalhousie and Rohtang areas, Himachal Pradesh. *Current Science,* **44**, 219.

Bonin, B. 1986. *Ring complex granites and anorogenic magmatism.* North Oxford Academic Publishers.

—— 1990. From orogenic to anorogenic settings: evolution of granitoid suites after a major orogenesis. *Geological Journal,* **25**, 261–270.

Brouand, M., Banzet, G. & Barbey, P. 1990. Zircon behaviour during crustal anatexis. Evidence from the Tibetan Slab migmatites (Nepal). *Journal of Volcanology and Geothermal Research,* **44**, 143–161.

Brown, G.C. 1981. Space and time in granite plutonism. *Philosophical Transactions of the Royal Society of London,* **A301**, 321–336.

Collins, W.J., Bearns, S.D., White, A.J.R. & Chappell, B.W. 1982. Nature and origin of A-type granites with particular references to

Southeastern Australia. *Contributions to Mineralogy and Petrology*, **10**, 189–200.

CORFU, F. & AYRES, L.D. 1991. Unscrambling the stratigraphy of an Archean greenstone belt: a U-Pb geochronological study of the Favourable Lake belt, northwestern Ontario, Canada. *Precambrian Research*, **50**, 201–220.

—— & GRUNSKY, E.C. 1987. Igneous and tectonic evolution of the Batchawana greenstone belt, Superior Province: a U-Pb zircon and titanite study. *Journal of Geology*, **95**, 87–105.

COSCA, M.A., SUTTER, J.F. & ESSENE, E.J. 1991. Cooling and inferred uplift/erosion history of the Grenville Orogen, Ontario: constraints from $^{40}Ar/^{39}Ar$ thermochronology. *Tectonics*, **5**, 959–977.

COULSON, A.L. 1936. A soda-granite suite in the North-West Frontier Province. *Proceedings of the Natural Institute of Sciences of India*, **2**, 103–111.

DAVIS, D.W. 1982. Optimum linear regression and error estimation applied to U-Pb data. *Canadian Journal of Earth Sciences*, **19**, 2141–2149.

DAVIS, D.W., BLACKBURN, C.E. & KROGH, T.E. 1982. Zircon U-Pb ages from the Wabigoon–Manitou lakes region, Wabigoo Subprovince, northwest Ontario. *Canadian Journal of Earth Sciences*, **19**, 254–266.

DE LA ROCHE, H., LETERRIER, J., GRANDCLAUDE, P. & MARCHAL, M. 1980. A classification of volcanic and plutonic rocks using R1R2 – diagram and major elements analyses – its relationships with current nomenclature. *Chemical Geology*, **29**, 183–210.

FERRARA, G., LOMBARDO, B., TONARINI, S. & TURI, B. 1987. New Rb-Sr data on granitoids from Gumburajun and Kade-Chu (High Himalaya). *II Himalaya- Karakorum Workshop*, Vandoeuvre-les-Nancy. Abstract vol. 33.

FRANK, W., THÖNI, M. & PURTSCHELLER, F. 1977. Geology and petrography of Kullu-South Lahul area. *Colloque international du CNRS* (Paris), **268**, 147–172.

——, BAUD, A., HONEGGER, K. & TROMMSDORFF, V. 1987. Comparative studies on profiles across the northwest Himalayas. *In*: SCHAER, J-P., & RODGERS, J. (eds) *The anatomy of mountain ranges*. Princeton University Press, 261–275.

FUCHS, G. 1982*a*. Geology of the Pin Valley in Spiti, H.P., India. *Jahrbuch der Geologischen Bundesanstalt* (Wien), **124**, 325–359.

—— 1982*b*. Explanations of the geologic-tectonic map of the Himalaya. *Jahresbericht von Geologisches Bundesanstalt* (Wien), **124**.

—— 1987. The geology of southern Zanskar (Ladakh). Evidence for the autochtony of the Tethys zone of the Himalaya. *Jahrbuch der Geologischen Bundesanstalt* (Wien), **130**, 465–491.

GAETANI, M., GARZANTI, E. & TINTORI, A. 1990. Permo-Carboniferous stratigraphy in SE Zanskar and NW Lahul (NW Himalaya, India). *Eclogae Geologicae Helvetiae*, **83**, 143–161.

HARRIS, N.B.W., PEARCE, J.A. & TINDLE, A.G. 1986. Geochemical characteristics of collision-zone magmatism. *In*: COWARD, M.P. & RIES, A.C. (eds) *Collision Tectonics*. Geological Society, London, Special Publications, **19**, 67–81.

HONEGGER, K., DIETRICH, V., FRANK, W., GANSSER, A., THONI, M. & TROMMSDORFF, V. 1982. Magmatism and metamorphism in the Ladakh Himalaya. *Earth and Planetary Science Letters*, **60**, 253–292.

JAN, M.Q., ASIF, M. & TAHIRKELI, T. 1981. The geology and petrography of the Tarbela 'Alkaline' complex. *Geological Bulletin of University of Peshawar*, **14**, 1–28.

KANWAR, S.S. & BHANDARI, A.K. 1979. Stratigraphy, structure and sedimentation of part of Lahaul and Spiti District, Himachal Pradesh. *In: Himalayan geology seminar Delhi 1976*, Geological Survey of India Miscellaneous Publications, **41**, 169–184.

KEMPE, D.R.C. 1973. The petrology of the Warsak alkaline granites, Pakistan, and their relationships to other alkaline rocks of the region. *Geological Magazine*, **110**, 385–404.

—— 1983. Alkaline granites, syenites and associated rocks of the Peshawar Plain Alkaline Igneous Province, NW Pakistan. *In*: SHAMS, F.A. (ed.) *Granites of Himalaya, Karakoram and Hindukush*. Institute of Geology Punjab University, Lahore, 143–169.

—— 1986. A note on the ages of the alkaline rocks of the Peshawar plain alkaline igneous province, NW Pakistan. *Geological Bulletin of University of Peshawar*, **19**, 113–119.

—— & JAN, M.Q. 1970. An alkaline igneous province in the North-West Frontier province, West Pakistan. *Geological Magazine*, **107**, 395–398.

KROGH, T.E. 1973. A low-contamination method for hydrothermal decomposition of zircon and extraction of U and Pb for isotopic age determinations. *Geochimica et Cosmochimica Acta*, **37**, 485–494.

—— 1982. Improved accuracy of U-Pb zircon ages by the creation of more concordant systems using an air abrasion technique. *Geochimica et Cosmochimica Acta*, **46**, 637–649.

LE BAS, M.J., MIAN, I. & REX, D.C. 1987. Age and nature of carbonatite emplacement in North Pakistan. *Geologisches Rundschau*, **76**, 317–323.

LE FORT, P., DEBON, F., PECHER, A., SONET, J. & VIDAL, PH. 1986. The 500 Ma magmatic event in alpine southern Asia, a thermal episode at Gondwana scale. *Mémoires des Sciences de la Terre*, **47**, 191–209.

MEHTA, P.K. 1977. Rb-Sr geochronology of the Kulu–Mandi Belt. Its implications for the Himalayan tectonogenesis. *Geologisches Rundschau*, **66**, 156–175.

PANDE, I.C. & KUMAR, S. 1974. Absolute age determinations of the crystalline rocks of Manali-Jispa region, northwestern Himalaya. *Geologisches Rundschau*, **63**, 539–548.

PEACOCK, M.A. 1931. Classification of igneous rock series. *Journal of Geology*, **39**, 54–67.

PEARCE, J.A., HARRIS, N.B.W. & TINDLE, A.G. 1984. Trace element discrimination diagrams for the tectonic interpretation of granitic rock. *Journal of Petrology*, **25**, 956–983.

PITCHER, W.S. 1983. Granite type and tectonic

environment. *In*: R. Hsü(ed.) *Mountain Building processes.* London, Academic Press, 19–40.

Pognante, U., Castelli, D., Benna, P., Genovese, G., Oberli, F., Meyer, M. & Tonarini, S. 1990. The crystalline units of the High Himalayas in the Lahul-Zanskar region (North West India): metamorphic-tectonic history and geochronology of the collided and imbricated Indian plate. *Geological Magazine,* **127**, 101–116.

Pupin, J-P. 1980. Zircon and granite petrology. *Contributions to Mineralogy and Petrology,* **73**, 207–220.

—— 1988. Granites as indicators in paleogeodynamics. *Rendiconti della Società Italiana di Mineralogia e Petrologia,* **43**, 237–262.

Searle, M.P. & Fryer, B.J. 1986. Garnet, tourmaline and muscovite-bearing leucogranites, gneisses and migmatites of the Higher Himalayas from Zanskar, Kulu, Lahoul and Kashmir. *In*: Coward, M.P. & Ries, A.C. (eds) *Collision Tectonics.* Geological Society, London, Special Publications, **19**, 185–201.

Shand, S.J. 1927. *Eruptive rocks. Their genesis, composition classification and their relations to ore deposits.* Murby editor, London.

Spring, L. & Crespo, A. 1992. Nappe tectonics, extension and metamorphic evolution in the Indian Tethys Himalaya (SE Zanskar-NW Lahul). *Tectonics,* **11**, 978–989.

Steiger, R.H. & Jäger, E. 1977. Subcommission on geochronology: convention on the use of decay constants in geo- and cosmochronology. *Earth and Planetary Science Letters,* **36**, 359–362.

Stutz, E. 1988. *Géologie de la chaîne de Nyimaling aux confins du Ladakh et du Rupshu (NW Himalaya, Inde) – évolution paléogéographique et tectonique d'un segment de la marge nord-indienne.* Mémoires de géologie de l'Université de Lausanne, **3**.

—— & Thöni, M. 1987. The Lower Palaeozoic Nyimaling granite in the Indian Himalaya (Ladakh): new Rb-Sr data versus zircon typology. *Geologisches Rundschau,* **76**, 307–315.

Trivedi, J.R., Gopalan, K., Sharma, K., Gupta, K.R. & Choubey, V.M. 1982. Rb-Sr age of Gaik granite, Ladakh batholith, northwest Himalaya. *Proceedings of the Indian Academy of sciences (earth and planetary Sciences),* **91**, 65–73.

Vannay, J-C. & Spring, L. 1993. Geochemistry of the continental basalts within the Tethyan Himalaya of Lahul–Spiti and SE Zanskar, northwest India. *This volume.*

Villa, I.M. & Oddone, M. 1988. $^{39}Ar/^{40}Ar$ ages of Himalayan leucogranites decrease eastward. *Abstract vol., 3rd Himalaya-Karakorom-Tibet workshop,* Lausanne, 16.

Whalen, J.B., Currie, K.L. & Chappell, B.W. 1987. A-type granites: geochemical characteristics, discrimination and petrogenesis. *Contributions to Mineralogy and Petrology,* **95**, 407–419.

Zeitler, P.K. 1988. Ion microprobe dating of zircon from the Malakand granite, NW Himalaya, Pakistan. A constraint on the timing of Tertiary metamorphism in the region. *Geological Society of America Abstract with programs,* **20**, 323.

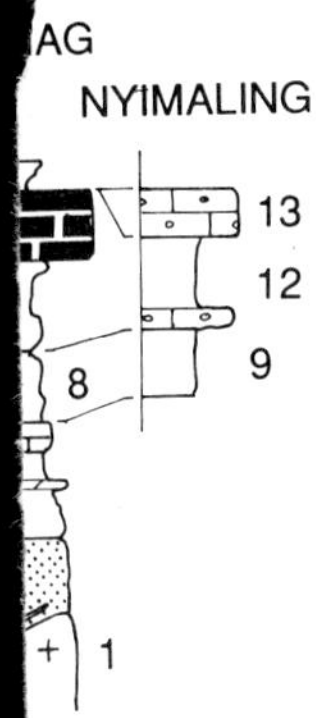

15 Mesozoic limestones (TANDI LIMESTONES)
14 Upper Jurassic-Tertiary shales, sandstones and limestones (SPITI, GIUMAL, CHIKKIM)
13 Middle Jurassic calciturbidites (LALUNG LA)
12 Triassic-Liassic shales (DOLTO)
11 Upper Triassic - Middle Jurassic limestones (KIOTO)
10 Triassic shales, limestones and sandstones (LILANG GR.)
9 Permian shales (KULING)
8 Permian shales, limestones and sandstones (KULING)
7 PANJAL TRAPS
6 Lower Permian polygenic conglomerates (CHUMIK)
5 Devonian sandstones (MUTH) and Carboniferous shales, limestones and sandstones (LIPAK, PO)
4 Ordovician-Silurian sandstones and conglomerates (THAPLE)
3 Cambrian slates sandstones and dolomites (KARSHA + KURGIAKH)
2 Precambrian-Cambrian slates and sandstones (PHE)
1 Granitic gneisses (protolith of Cambro-Ordovician age)

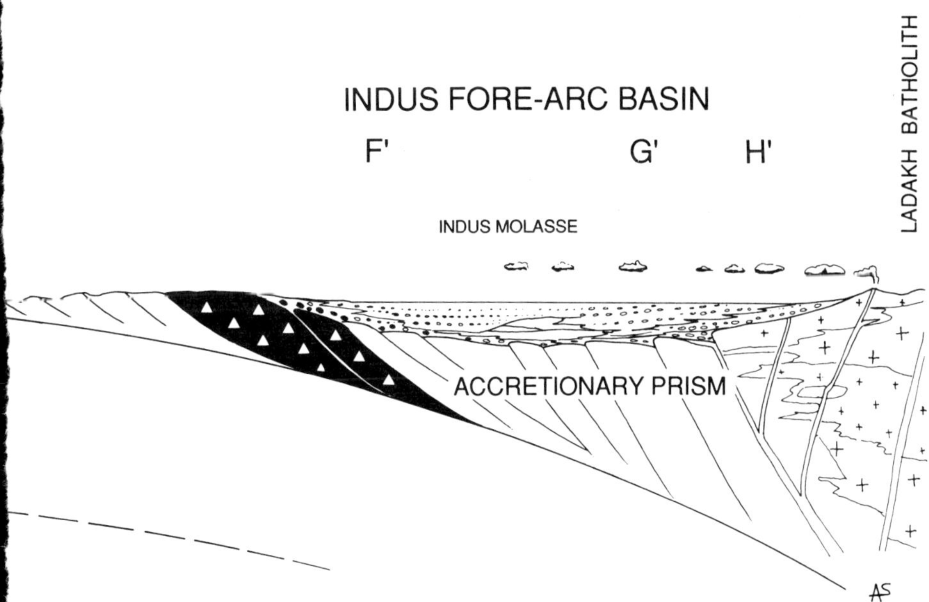

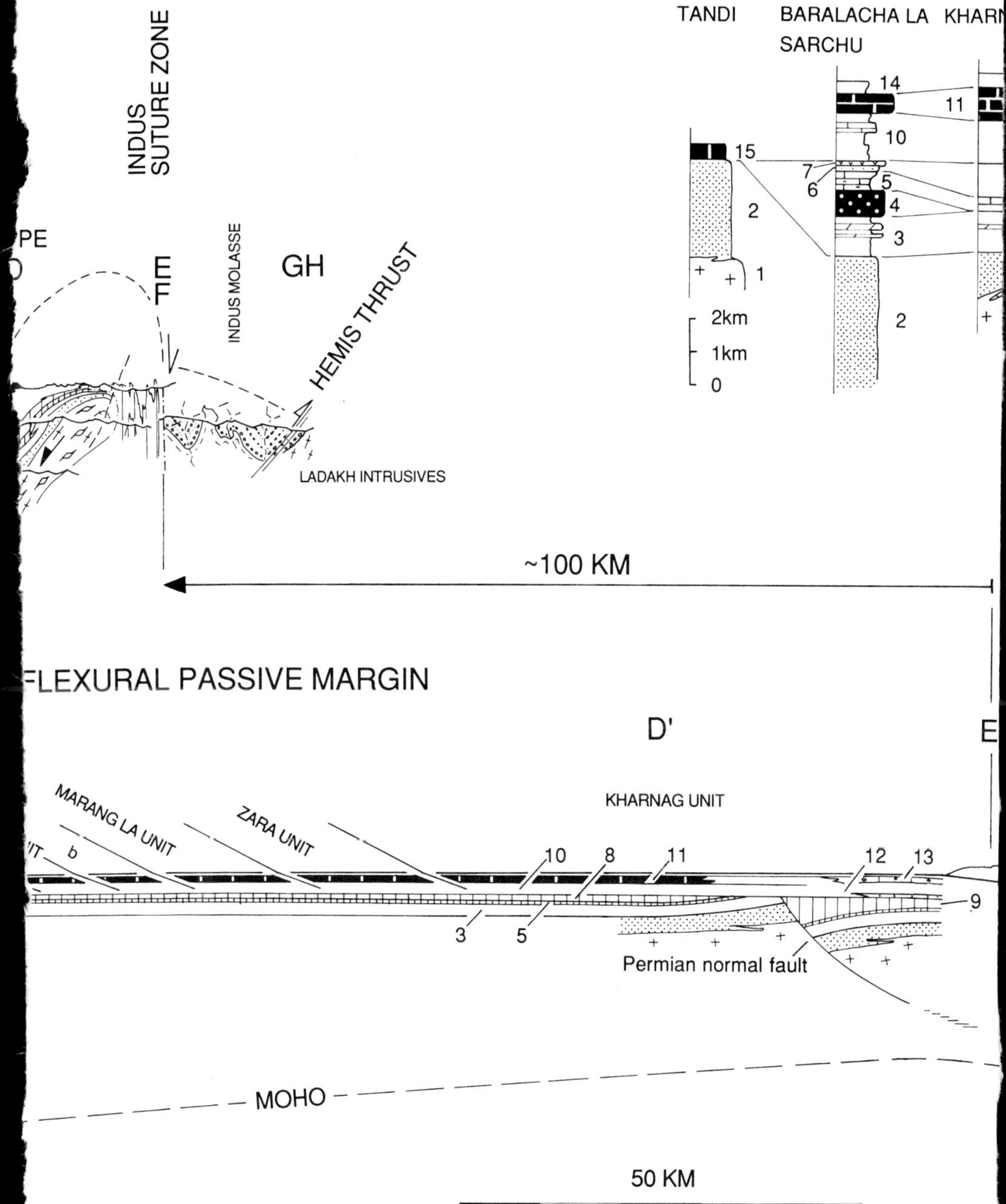

INDUS SUTURE ZONE
PE
INDUS MOLASSE
GH
HEMIS THRUST
LADAKH INTRUSIVES
TANDI
BARALACHA LA
SARCHU
KHARN
2km
1km
0
~100 KM
FLEXURAL PASSIVE MARGIN
D'
E
MARANG LA UNIT
ZARA UNIT
KHARNAG UNIT
Permian normal fault
MOHO
50 KM

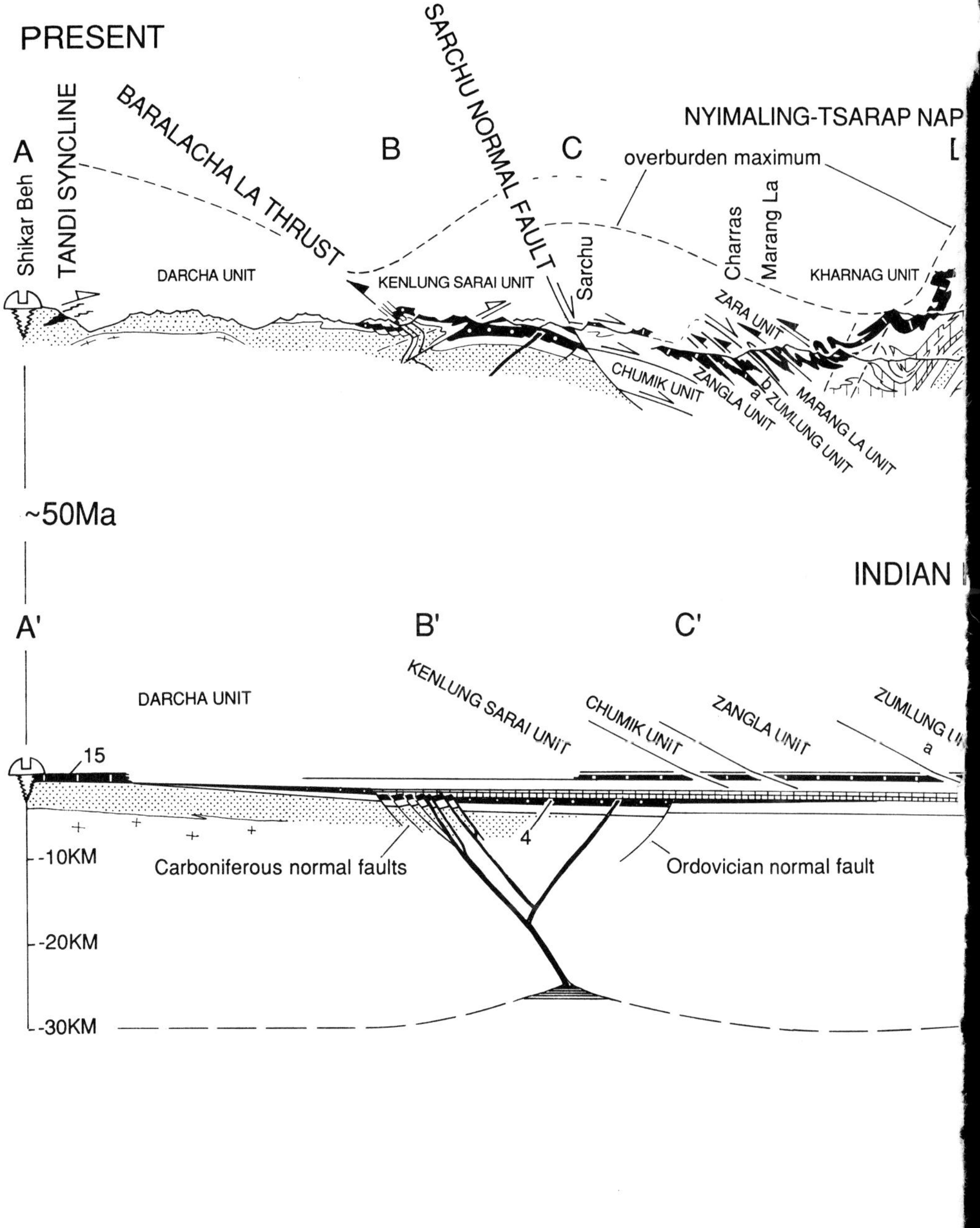

PRESENT
A
Shikar Beh
TANDI SYNCLINE
BARALACHA LA THRUST
B
SARCHU NORMAL FAULT
C
NYIMALING-TSARAP NAP
overburden maximum
Charras
Marang La
Sarchu
DARCHA UNIT
KENLUNG SARAI UNIT
KHARNAG UNIT
ZARA UNIT
CHUMIK UNIT
ZANGLA UNIT
a
b
ZUMLUNG UNIT
MARANG LA UNIT
~50Ma
INDIAN
A'
B'
C'
DARCHA UNIT
KENLUNG SARAI UNIT
CHUMIK UNIT
ZANGLA UNIT
ZUMLUNG U
a
15
4
-10KM
-20KM
-30KM
Carboniferous normal faults
Ordovician normal fault

The tectonic evolution of the Northwestern Himalaya in eastern Ladakh and Lahul, India

A. STECK, L. SPRING, J.-C. VANNAY, H. MASSON, H. BUCHER, E. STUTZ, R. MARCHANT & J.-C. TIECHE

Section des Sciences de la Terre, Université de Lausanne, BFSH 2, CH-1015 Lausanne, Switzerland

Abstract: Geological studies along a transect across the Himalaya in eastern Ladakh and Lahul provide new insights into the Tertiary structural evolution of this region. The initiation of the Nyimaling–Tsarap Nappe is related to an early phase of underthrusting of India below Asia. In Lahul, an opposite vergent intra-continental underthrusting develops immediately after continental collision (NE-vergent Tandi Syncline and Shikar Beh Nappe). This NE-vergent nappe stack is responsible for the amphibolite-facies regional metamorphism of the lower Chandra Valley. The subsequent phase corresponds to the main thrusting of the SW-vergent Nyimaling–Tsarap Nappe, developed by ductile shear (87 km Eocene shortening). This nappe pile is responsible for the regional metamorphism of SE Zanskar (kyanite–staurolite near Sarchu). The root zone and the frontal part of the Nyimaling–Tsarap Nappe are subsequently overprinted by two NW–SE-orientated dextral transpressional shear zones. To the south of the investigated area, the Main Central Thrust has been developed as a shear zone in the regional metamorphic ductile crustal rocks below the older nappes to the north. In the Sarchu and Nyimaling regions, the following tectonic phase corresponds to NE-vergent 'backfolding' (Miocene). Normal faults in the Sarchu area record a late extension of approximately 14 km.

Numerous geological observations made during several scientific expeditions in the eastern Ladakh and Lahul regions of the Himalaya, have unravelled the structural evolution of this mountain chain. A detailed geological map, a cross-section and a palinspastic section of this part of the northern Indian plate margin are presented in Figs 1, 2 and 3 respectively.

The Ladakh Batholith, which marks the northern limit of the study area, is interpreted as the product of partial melting of the subducted Mesozoic ocean floor of the Himalayan Tethys beneath the Asian Plate (Frank *et al.* 1977; Honegger *et al.* 1982). During continental collision, two main compressional structures characterized the tectonic evolution of the eastern Ladakh and Lahul regions, to the south the NE-vergent Tandi Syncline and the associated Shikar Beh Nappe, and to the north the SW-vergent Nyimaling–Tsarap Nappe (Figs 3 and 4). This tectonic evolution is summarized in Table 1 and briefly described in the following sections. A more detailed account of the regional geology of eastern Ladakh and Lahul is presented in Steck *et al.* (1993).

The Tandi Syncline and the NE-vergent Shikar Beh Nappe

The Tandi Syncline is an enigmatic structure composed of Mesozoic limestones in-folded with the Upper Precambrian to Lower Cambrian meta-sandstones of the Phe Fm. (Figs 3 and 5), in the lower part of the Chandra Valley (Upper Lahul). The basal stratigraphic contact corresponds to an unconformity and the Tandi Unit represents a normal stratigraphic sequence. Powell & Conaghan (1973) and Picket *et al.* (1975) interpret this structure as a SW-vergent antiformal syncline related to the first Tertiary tectonic phase. Frank *et al.* (1973) and Srikantia & Bhargava (1982) described it as a NE-vergent synform. Our mapping and structural evidence show that the Tandi Syncline is a NE-vergent synform, created during the earliest Tertiary deformation recognizable in the studied area (D_{Ta1}). In the lower part of the Chandra Valley, this structure is associated with a zone of medium pressure and temperature regional metamorphism (Frank *et al.* 1973, 1975). Kyanite–staurolite–garnet mineral assemblages observed near Khoksar imply a lithostatic overburden of approximately 20 km (Fig. 3). Toward the NE, the regional metamorphism decreases to actinolite–pumpellyite facies south of the Baralacha La. Therefore, the high grade regional metamorphism of the lower part of the Chandra Valley is not related to the SW-vergent Nyimaling–Tsarap Nappe observed to the north, in SE Zanskar. We interpret this metamorphism as a consequence of a NE-vergent nappe stacking, associated with the tectonic phase

From Treloar, P.J. & Searle, M.P. (eds), *Himalayan Tectonics*
Geological Society Special Publication, No. 74, pp. 265–276.

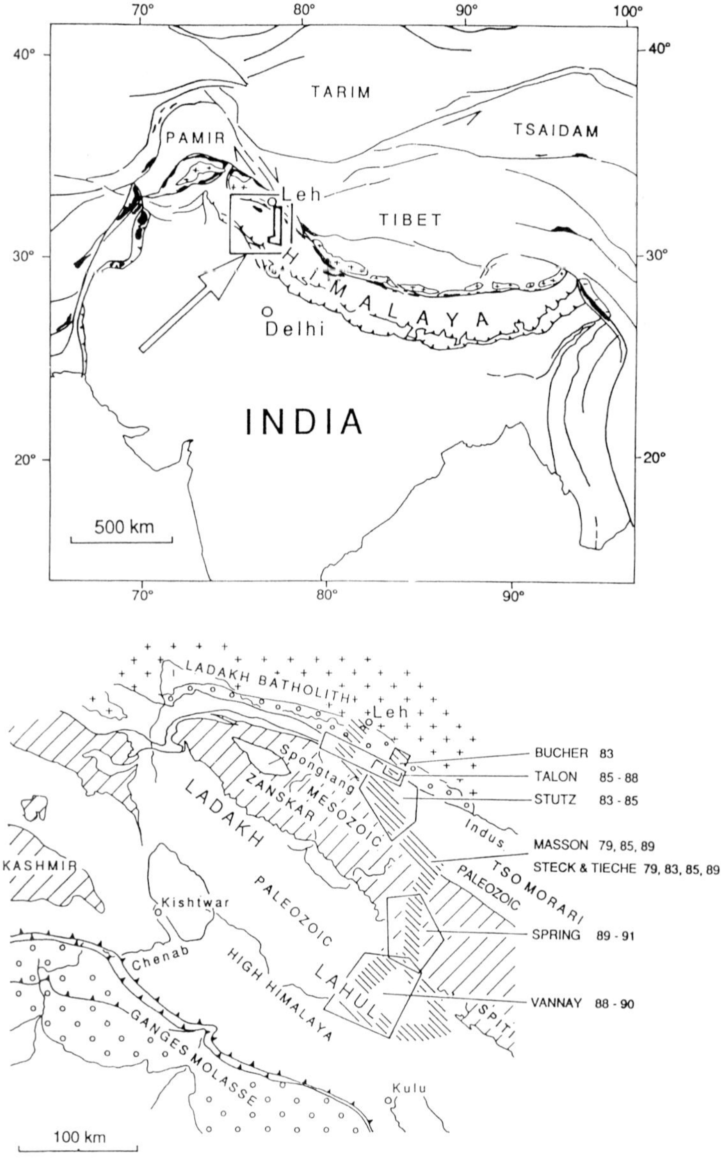

Fig. 1. Index map of the investigated area in Ladakh and Lahul.

responsible for the NE-vergent Tandi Syncline. We propose the name Shikar Beh Nappe for the lower unit of this presently eroded nappe stack. This structure is probably the result of a south-dipping intracontinental zone of underthrusting, caused by a mechanical destabilisation in the Indian crust during the progressive underthrusting of India below Asia. Deep seismic profiles suggest the presence of such a S dipping underthrusting zone below the High Himalaya of Nepal (Hirn *et al.* 1984). This conclusion contrasts with classical models of Himalayan tectonic evolution, which postulate that the high grade metamorphism of the 'High Himalayan Crystalline' is due exclusively to crustal thickening related to SW-vergent structures.

Fig. 2. Palinspastic reconstruction of the Rohtang La–Leh section.

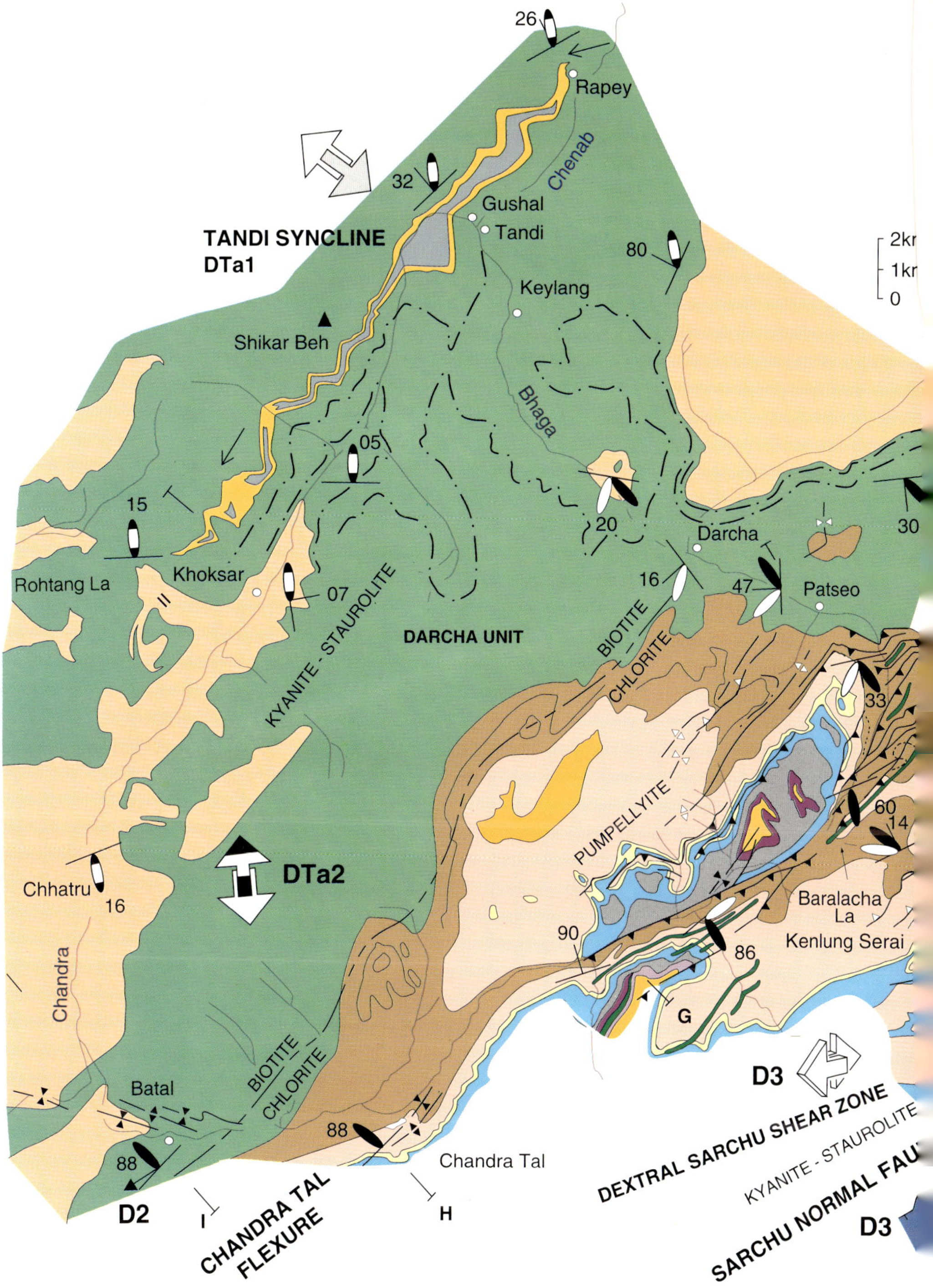

GEOLOGICAL TRANSECT OF WESTERN HIMALAYA IN LADAKH AND LAHUL

by A. STECK, L. SPRING, J.-C. VANNAY, H. MASSON, H. BUCHER, R. MARCHANT, E. STUTZ and J.-C. TIECHE

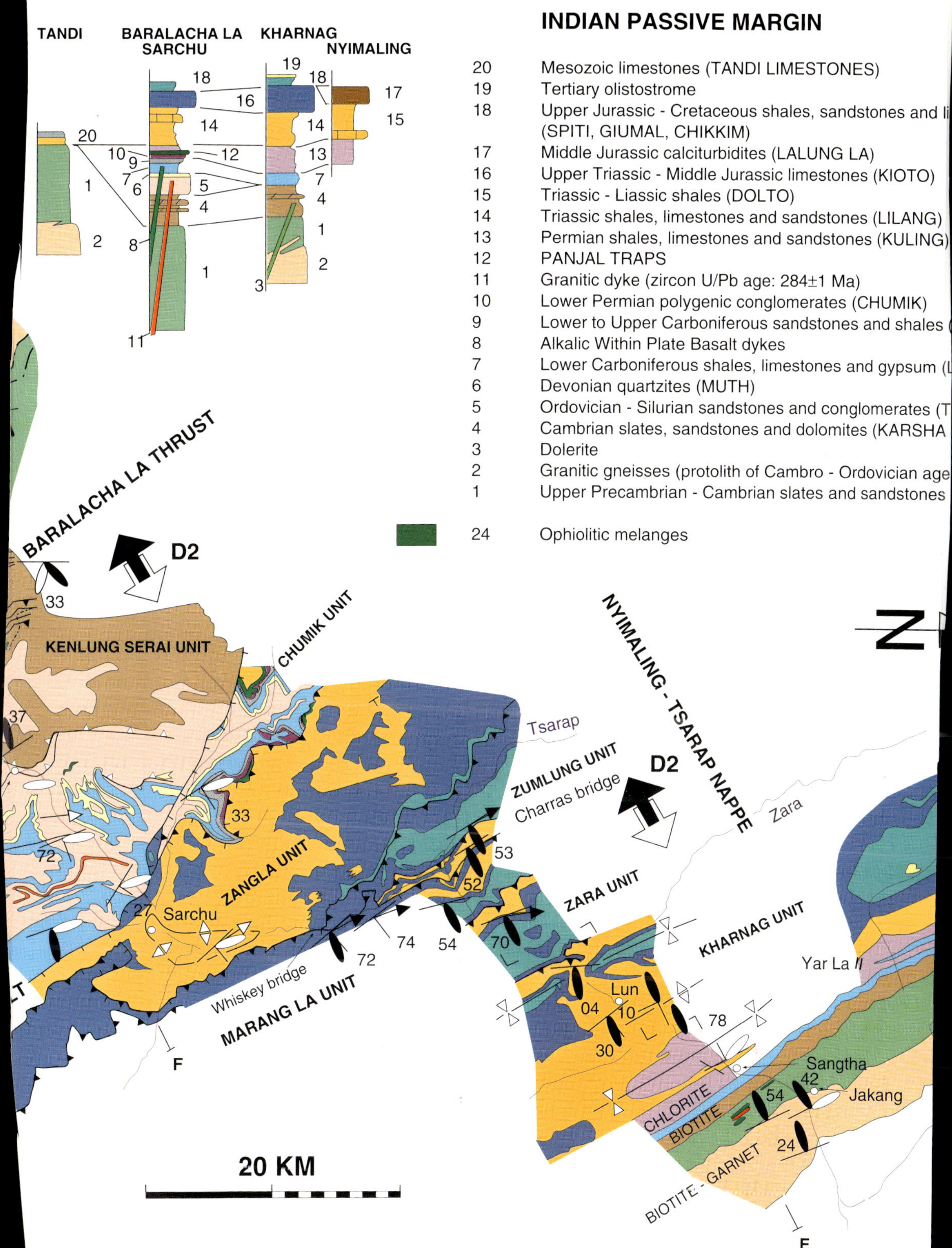
TANDI
BARALACHA LA SARCHU
KHARNAG
NYIMALING
INDIAN PASSIVE MARGIN
20 Mesozoic limestones (TANDI LIMESTONES)
19 Tertiary olistostrome
18 Upper Jurassic - Cretaceous shales, sandstones and li
(SPITI, GIUMAL, CHIKKIM)
17 Middle Jurassic calciturbidites (LALUNG LA)
16 Upper Triassic - Middle Jurassic limestones (KIOTO)
15 Triassic - Liassic shales (DOLTO)
14 Triassic shales, limestones and sandstones (LILANG)
13 Permian shales, limestones and sandstones (KULING)
12 PANJAL TRAPS
11 Granitic dyke (zircon U/Pb age: 284±1 Ma)
10 Lower Permian polygenic conglomerates (CHUMIK)
9 Lower to Upper Carboniferous sandstones and shales
8 Alkalic Within Plate Basalt dykes
7 Lower Carboniferous shales, limestones and gypsum
6 Devonian quartzites (MUTH)
5 Ordovician - Silurian sandstones and conglomerates (T
4 Cambrian slates, sandstones and dolomites (KARSHA
3 Dolerite
2 Granitic gneisses (protolith of Cambro - Ordovician age
1 Upper Precambrian - Cambrian slates and sandstones
24 Ophiolitic melanges
BARALACHA LA THRUST
D2
KENLUNG SERAI UNIT
CHUMIK UNIT
NYIMALING - TSARAP NAPPE
Tsarap
ZUMLUNG UNIT
Charras bridge
Zara
ZANGLA UNIT
ZARA UNIT
KHARNAG UNIT
Sarchu
Whiskey bridge
MARANG LA UNIT
Lun
Yar La
Sangtha
Jakang
CHLORITE
BIOTITE
BIOTITE - GARNET
F
E
20 KM

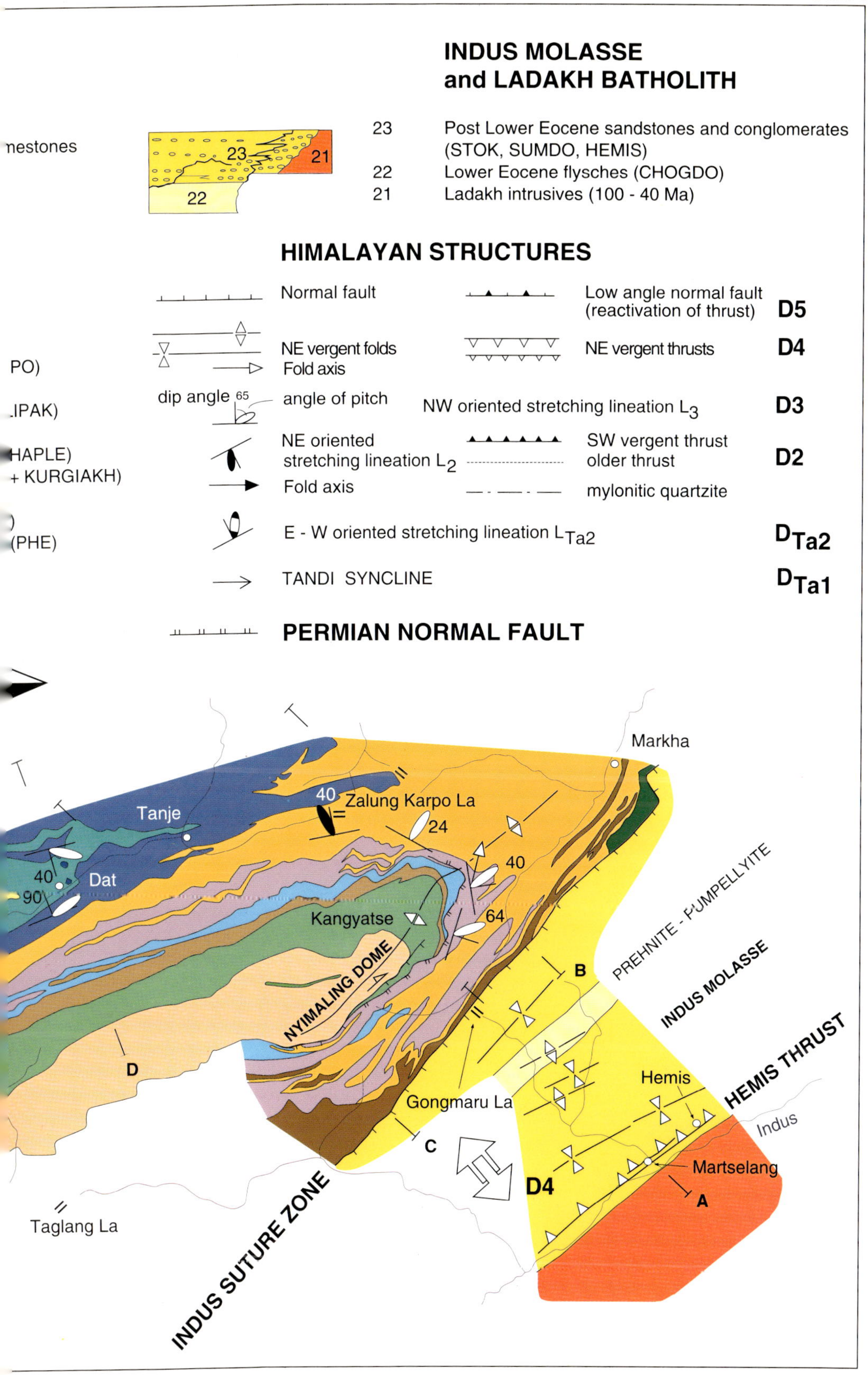
INDUS MOLASSE and LADAKH BATHOLITH
23 Post Lower Eocene sandstones and conglomerates (STOK, SUMDO, HEMIS)
22 Lower Eocene flysches (CHOGDO)
21 Ladakh intrusives (100 - 40 Ma)
nestones
PO)
.IPAK)
HAPLE)
+ KURGIAKH)
(PHE)
HIMALAYAN STRUCTURES
Normal fault
Low angle normal fault (reactivation of thrust)
D5
NE vergent folds
Fold axis
NE vergent thrusts
D4
dip angle
65
angle of pitch
NW oriented stretching lineation L_3
D3
NE oriented stretching lineation L_2
Fold axis
SW vergent thrust
older thrust
mylonitic quartzite
D2
E - W oriented stretching lineation L_{Ta2}
D_{Ta2}
TANDI SYNCLINE
D_{Ta1}
PERMIAN NORMAL FAULT
Markha
Zalung Karpo La
Tanje
Dat
Kangyatse
NYIMALING DOME
PREHNITE - PUMPELLYITE
INDUS MOLASSE
HEMIS THRUST
Hemis
Indus
Martselang
Gongmaru La
D4
Taglang La
INDUS SUTURE ZONE
A
B
C
D
40
24
40
64
90
40

Table 1. *Chronology of the Tertiary Himalayan structures in eastern Ladakh and Lahul*

Continental collision and beginning of underthrusting of India below Asia (*c.* 54 Ma)		
(A)	Early initiation of SW-vergent Nyimaling–Tsarap Nappe	D1
Early NE-vergent movements		
(B)	NE-vergent folds (F_{Ta1}), axial surface schistosity (S_{Ta1}) and stretching lineation (L_{Ta1}): Tandi Syncline, Shikar Beh Nappe	DTa1
Main SW-vergent deformations and subsequent dextral transpression and backfolding		
(C)	SW-vergent folds (F_{Ta2}), axial surface schistosity (S_{Ta2}) and E–ENE-orientated stretching lineation (L_{Ta2}): SW-vergent deformation of Tandi Syncline	DTa2
(D)	SW-vergent thrusts and folds (F2), axial schistosity (S2) and NE-orientated stretching lineation (L2): main development of the Nyimaling–Tsarap Nappe, Baralacha La Thrust and Chandra Tal Flexure	D2
(E)	Dextral shear and underthrusting (transpression) parallel to NW-orientated stretching lineation (L3): Nyimaling and Sarchu Shear Zones	D3
(F)	NE-vergent 'back-folds' F4, doming and dextral transpression: Nyimaling Dome	D4
(G)	Late dextral shear in Nyimaling–Markha Valley Steep Belt (S_5)	
Late extensional structures of Sarchu region		
(H)	Low-angle normal faults (reactivation of pre-existing D2 thrust planes)	D5
(I)	Steep Sarchu Normal Fault	
Late compression		
(J)	NE–SW compression and vertical extension: symmetric box-folds in the Nyimaling Dome W-vergent folds in the Kenlung Sarai Unit near Sarchu	D6

The Tandi Syncline has been overprinted by the subsequent SW-vergent deformations. Near Batal, in the Chandra Valley, the main S2 schistosity of the Chandra Tal Flexure and Baralacha La Thrust crosscut the older NE vergent F_{Ta1} folds associated with the Tandi Syncline. The S2 schistosity is related to the main SW-vergent thrusting (D2) of the Nyimaling–Tsarap Nappe, farther to the north. This local chronological relationship in the front of these two opposite-vergent structures does not exclude the possibility that the Nyimaling–Tsarap Nappe initiated at the same time or even earlier than the Tandi structure (D_{Ta1}).

The Nyimaling–Tsarap Nappe

The initiation of the SW-vergent thrusting of the Nyimaling–Tsarap Nappe, away from the North Indian continental border, probably corresponds to the incipient stage of crustal thickening related to the continental collision of India and Asia. This stage is probably older than the NE vergent Tandi Syncline deformation (D_{Ta1}). However, as mentioned above, the main deformation phase of the Nyimaling–Tsarap Nappe (D2) is younger than the Tandi structure.

The Nyimaling–Tsarap Nappe developed by ductile shear of the upper Indian crust during underthrusting of the Indian continental border below the rigid margin of the Asian Plate. The lower and frontal limit of the ductile shear zone propagated progressively through the Indian upper crust toward the SW. The Nyimaling–Tsarap Nappe is characterized by a set of high structural level discrete thrust surfaces, which grade with depth into a large ductile shear zone with associated isoclinal folds (Figs 2 & 6). The geometry of the Nyimaling–Tsarap Nappe before backfolding and late extension is shown in Fig. 7. The Nyimaling–Tsarap Nappe represents an imbricate structure ('structure imbriquée') as defined and drawn by De Margerie & Heim (1888, fig. 105), more than 100 years ago. This imbricate structure differs from the 'imbricate fan' defined by Boyer & Elliot (1982) as the frontal thrust slices are not related to a unique 'sole thrust' but to a large ductile shear zone associated with fold structures and situated in the upper and middle crust. The model of Boyer & Elliott (1982) is only suitable for thrust systems developed at shallow tectonic levels, as for instance the Jura Thrust system in the Alps (Buxtorf 1907; Laubscher 1965). The preservation of the normal stratigraphic sequence and the continuity between the thrust front and the autochthonous root zone of the Nyimaling–Tsarap Nappe indicate that simple shear is the

Fig. 3. (Plate 1). Geological transect of the Northwestern Himalaya in eastern Ladakh and Lahul.

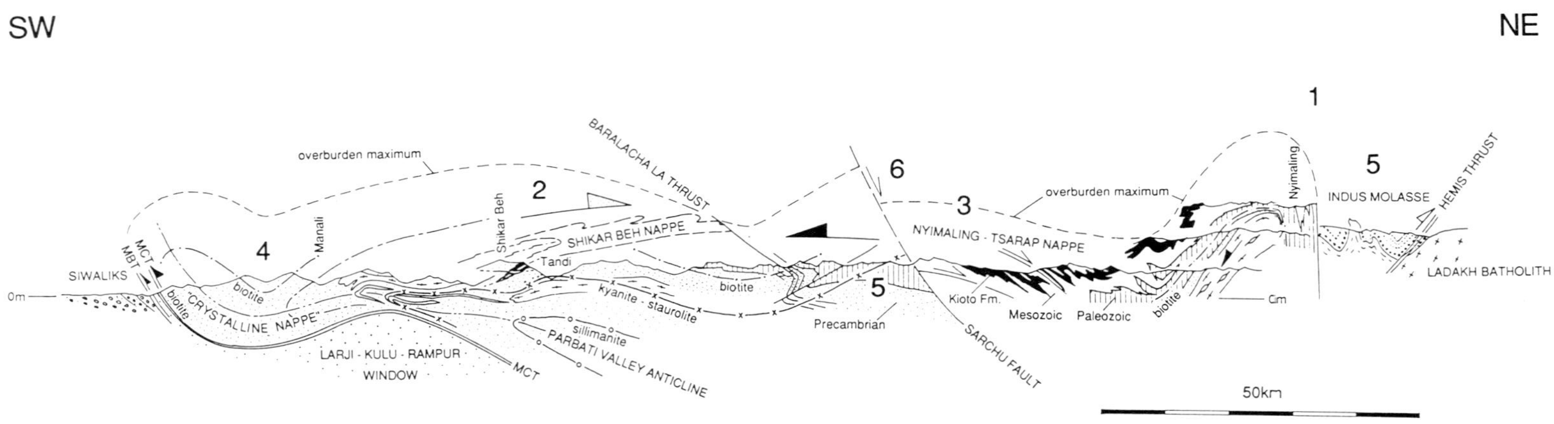

Chronology of the Himalayan structures

1-	Continental collision, underthrusting of India below Asia and initiation of the Nymaling-Tsarap Nappe	(Eocene)
2-	NE vergent Shikar Beh Nappe	(Eocene)
3-	SW vergent Nyimaling-Tsarap Nappe	(Eocene)
4-	SW vergent "Crystalline Nappe" (MCT and MBT)	(Oligocene-Miocene)
5-	NE vergent folding	(Miocene)
6-	Late extension	(Miocene)

Fig. 4. Geological cross section of the Himalaya in Ladakh and Lahul. The geology of the region situated south of the Rohtang La–Shikar Beh area has been compiled after Frank *et al.* (1987).

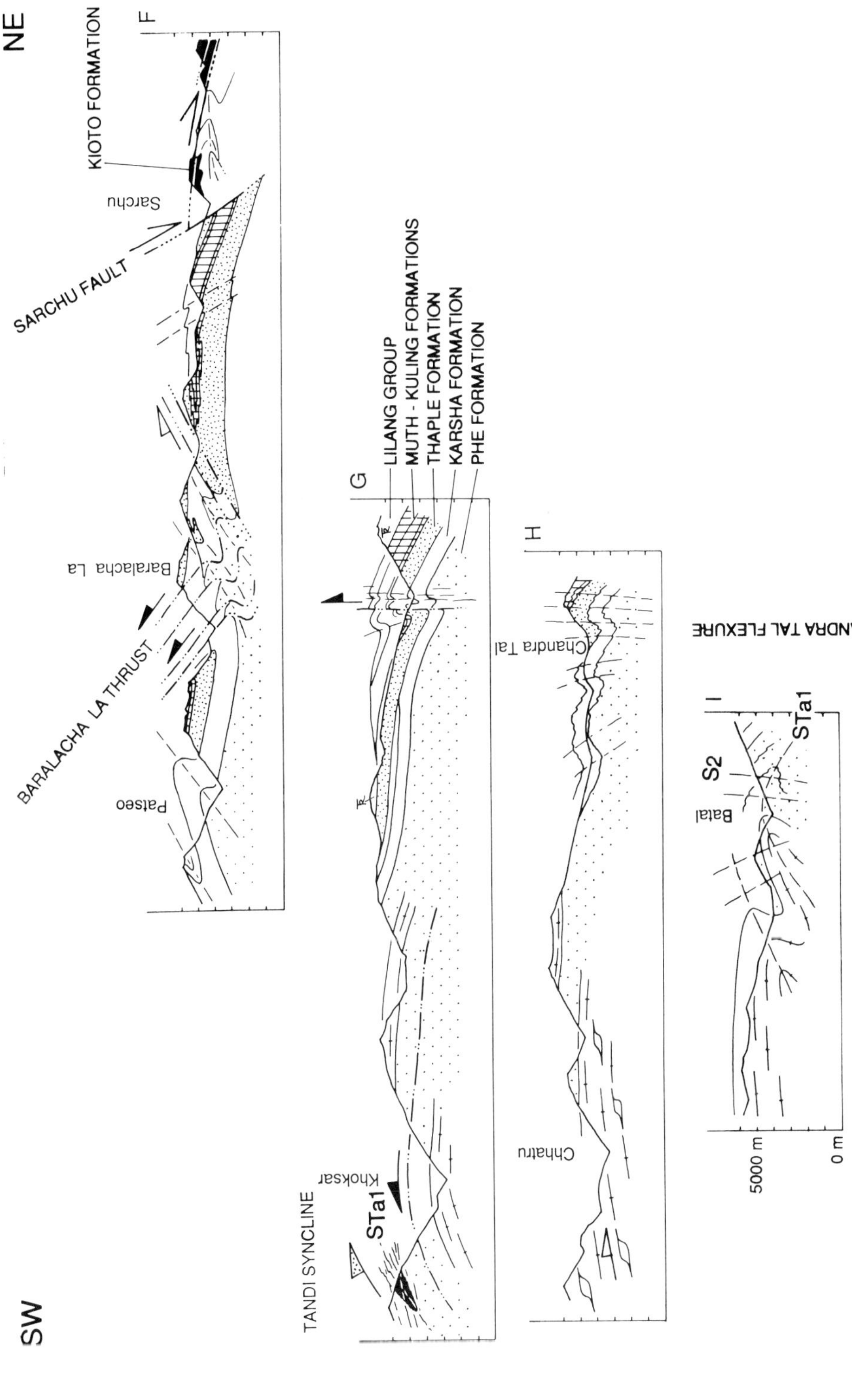

Fig. 5. Geological sections of the region between Rohtang La, Tandi and Sarchu (locations of the sections F–I on Fig. 3).

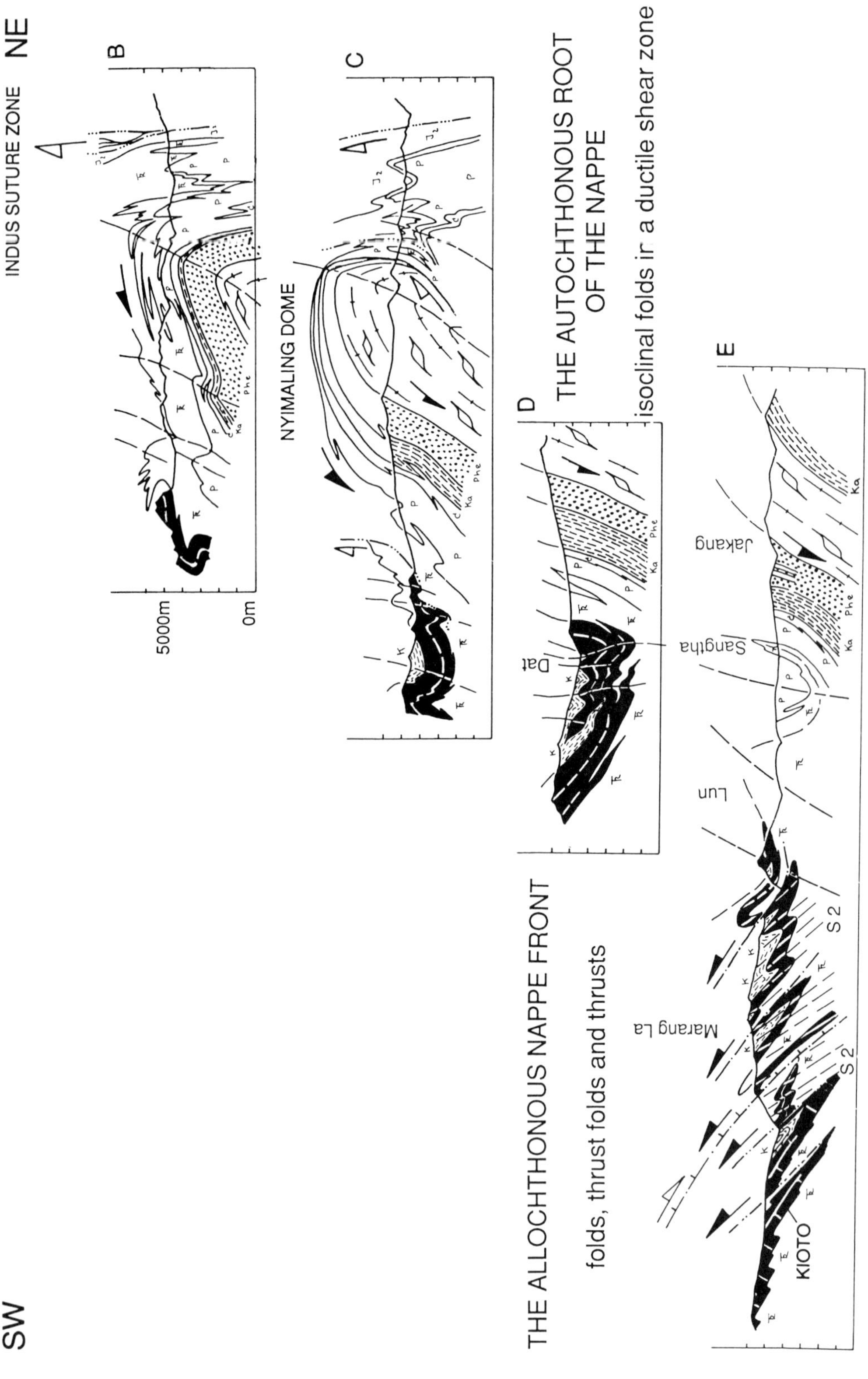

Fig. 6. Geological sections of the Nyimaling–Tsarap Nappe (locations of the sections B–E on Fig. 3).

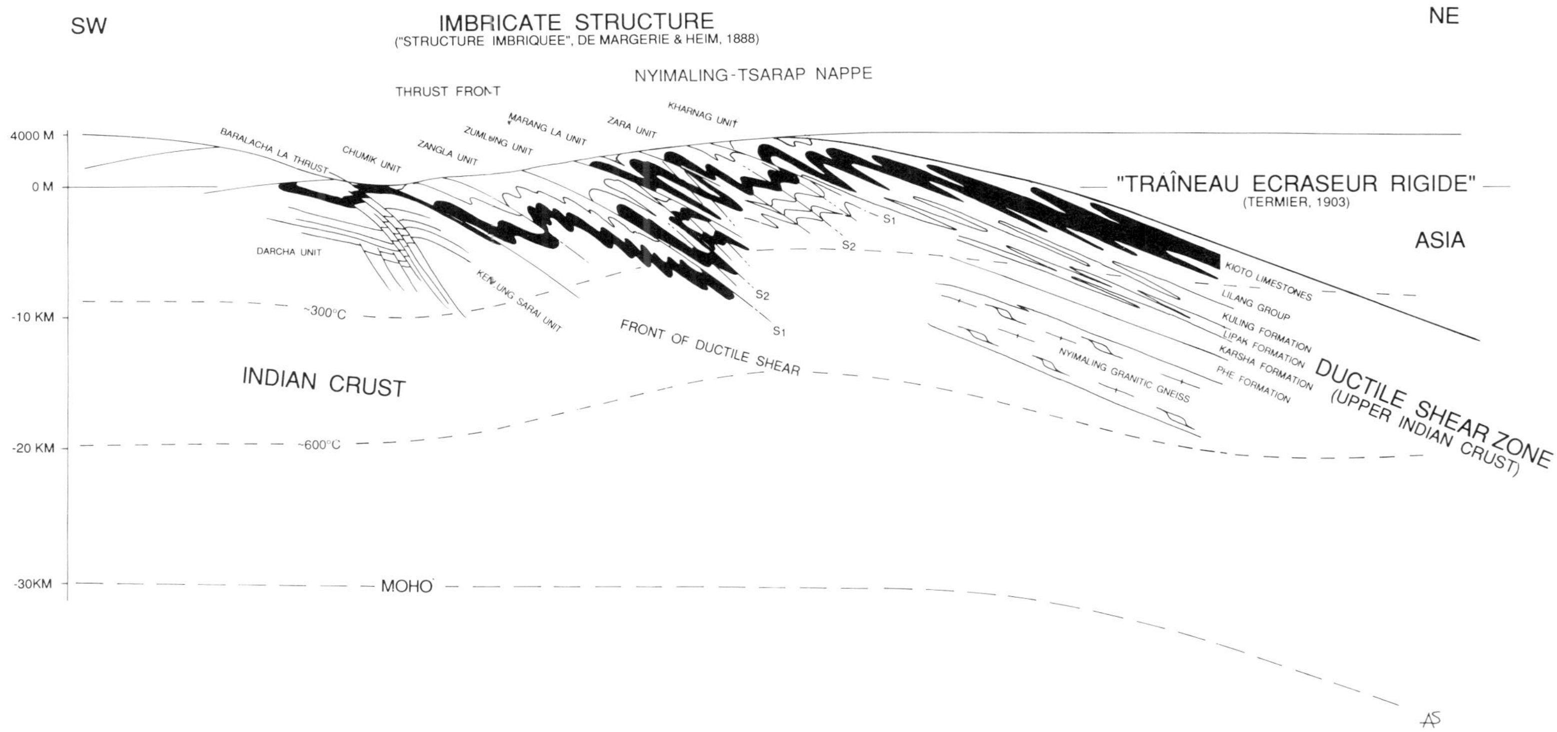

Fig. 7. Structural model of the Nyimaling–Tsarap Nappe (situation before backfolding, extension, uplift and erosion). An imbricate structure with discrete thrust surfaces characterizes the upper tectonic level. This structure grades with depth into a large ductile shear zone, associated with large recumbent folds.

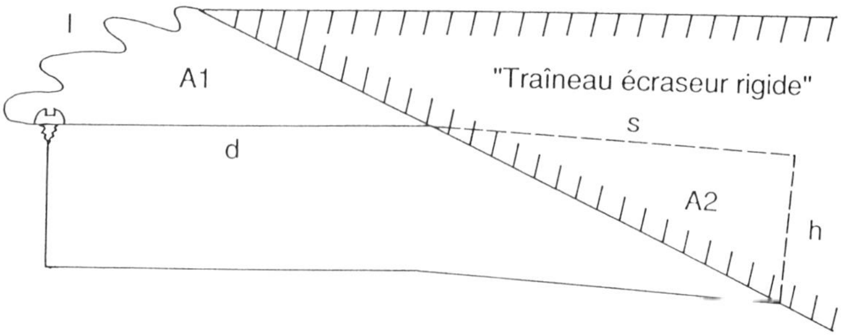

BOUNDARY CONDITIONS

Constant area: $A1 = A2$

Constant length of strata: $s = l - d$

Simple shear: $h = \frac{A2 \times 2}{s} = \frac{A1 \times 2}{l - d}$

Fig. 8. Shear model.

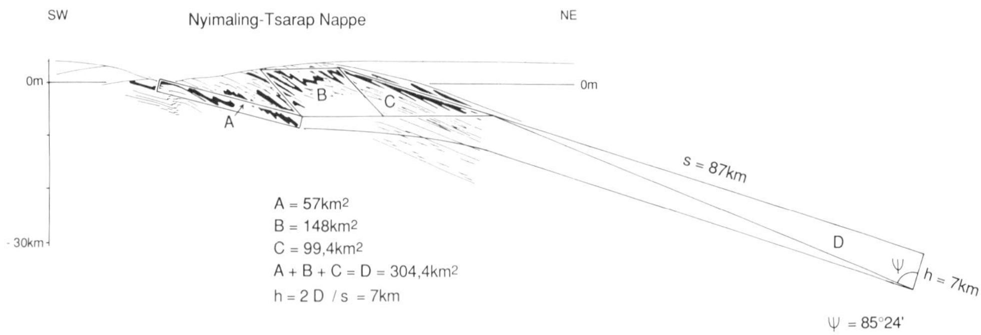

Fig. 9. Application of the shear model (Fig. 8) to the Nyimaling–Tsarap Nappe.

main mechanism of nappe translation. A simple shear model, assuming the boundary conditions shown in Fig. 8, allows a quantitative estimation of the deformation related to this nappe. This model indicates a crustal shortening of approximately 87 km (Fig. 9). The structural evolution of the Himalaya along the eastern Ladakh–Upper Lahul studied transect contrasts with the tectonic style proposed in some models for the Zanskar nappes, about 50 km to the west (Searle 1986; McElroy *et al.* 1990).

The Nyimaling and Sarchu dextral shear zones

The SW-vergent thrusting of the Nyimaling–Tsarap Nappe (D2) is followed by a N to NW-orientated dextral transpression (oblique underthrusting, D3), associated with a stretching lineation L3. Two important shear zones are associated with this tectonic movement: the Dextral Sarchu Shear Zone in the frontal part of the Nyimaling–Tsarap Nappe, and the Dextral Nyimaling Shear Zone near the root zone of the nappe (Figs 3 & 4). The transition from SW-vergent thrusting (D2) to N–NNW orientated dextral transpression (D3) can be correlated with the anticlockwise rotation of the underthrusting direction of India with respect to Asia which took place about 50, 45 and 36 Ma ago (Mid-Eocene to Early Oligocene) according to plate tectonic models (Molnar & Tapponnier

1975; Pierce 1978; Patriat & Achache 1984; Klootwijk *et al.* 1985; Dewey *et al.* 1986).

If the proposed correlation between plate movements and deformation structures is correct, the tectonic development of the Nyimaling–Tsarap Nappe was completed in the Eocene or Lower Oligocene (by 35 Ma). The observed dextral transpressional movements agree with the lateral strike-slip escape of the Tibetan block toward the SE with respect to India (Fig. 1; e.g. Tapponnier *et al.* 1982, 1986).

NE-vergent backfolds and late dextral shear

In the Sarchu and Nyimaling regions, the D4 tectonic phase corresponds to a gradual superposition of NE vergent 'backfolds' on the zones of dextral transpression. Between Sarchu and Kenlung Sarai, a spectacular isoclinal recumbent syncline composed of Devonian to Carboniferous sediments and illustrated in Baud *et al.* (1984, fig. 7), is related to this deformation phase. In the Nyimaling region, these 'backfolds' were subsequently modified by late dextral shear. Backfolding and associated dextral transpression are probably responsible for the tectonic unroofing of the Nyimaling–Tso Morari Crystalline (Stutz & Steck 1986; Stutz 1988). NE-vergent folds and thrusts are the only Tertiary structures observed in the Indus Molasse and late dextral shear does not affect this tectonic unit, situated to the north of the Indus Suture (Fig. 3).

Late extension

During a late period of the Himalayan orogeny, an important extension affected the frontal part of the Nyimaling–Tsarap Nappe. In the Sarchu region, two generations of extensional structures have been observed (Figs 2, 3 and 4). A first set of low angle normal faults developed parallel to pre-existing thrust surfaces of the Nyimaling–Tsarap Nappe. These low angle faults are crosscut by the steep Sarchu Normal Fault, which places amphibolite facies metamorphic rocks to the south in contact with lower greenschist facies rocks to the north (Spring & Crespo 1992). The tectonic unroofing of the amphibolite facies rocks south of Sarchu is probably a consequence of this extension. ^{40}Ar/^{39}Ar cooling ages of 33.5 Ma for amphibole (total gas age) and of 19.3 ± 0.2 Ma for biotite (plateau) are probably related to this unroofing (Spring *et al.* this volume). The relationship between the Sarchu Normal Fault and the Zanskar Normal Fault farther to the west (Searle 1986; Herren 1987) is unknown. However, it is quite possible that the discrete extensional faults of the Sarchu region represent a high-structural level tectonic feature of the ductile extensional shear zone developed in the deeper tectonic level that outcrops in Zanskar.

The late extensional structures characterized by normal faulting and backfolding are frequently interpreted as related to the so-called phenomenon of 'gravitational collapse' (e.g., Burg & Cheng 1984; Burchfiel & Royden 1985; Herren 1987; Searle *et al.* 1987). In our opinion this model cannot explain the structures in the Sarchu region, as the normal faults are younger than the NE-vergent backfolds. We interpret these structures as the consequence of a dextral transtensional tectonic phase that followed dextral transpressional movements. In the Himalaya of Nepal, Pêcher & Le Fort (1989) and Pêcher (1991) proposed a similar transtensional phase to explain comparable structures.

The age of continental collision and nappe stacking

The youngest marine sediments incorporated in the Himalayan nappes are marls and limestones containing reworked Palaeocene nummulites and were deposited along the North Indian margin. These sediments are probably of Early Eocene age. They crop out in the Lingshed and Zangla nappes and beneath the Spongtang ophiolite klippe (Garzanti *et al.* 1987). In the Kharnag region, we observed such reworked Palaeocene nummulites in a Tertiary olistostrome within the Nyimaling–Tsarap nappe. Reworked Palaeocene nummulites are also present in the Indus Molasse sediments of the active Asiatic margin (Frank *et al.* 1977*a*; Baud *et al.* 1982; Van Haver *et al.* 1984; Bucher & Steck 1987). These sediments are also probably of Early Eocene age. The Early Eocene age of the youngest marine sediments of the Indian Tethys and of the oldest continental sediments of the Indus Molasse, is coeval with a rapid decrease in the plate motion of India between 54 and 35 Ma (Molnar & Tapponnier 1975; Pierce 1978; Patriat & Achache 1984; Scotese *et al.* 1988; Dewey *et al.* 1989). These data indicate that the continental collision of India and Asia occurred during the Early Eocene and that the nappe thrusting started at this time. As discussed earlier, formation of the Nyimaling–Tsarap Nappe was completed in the Eocene or Lower Oligocene (by 35 Ma).

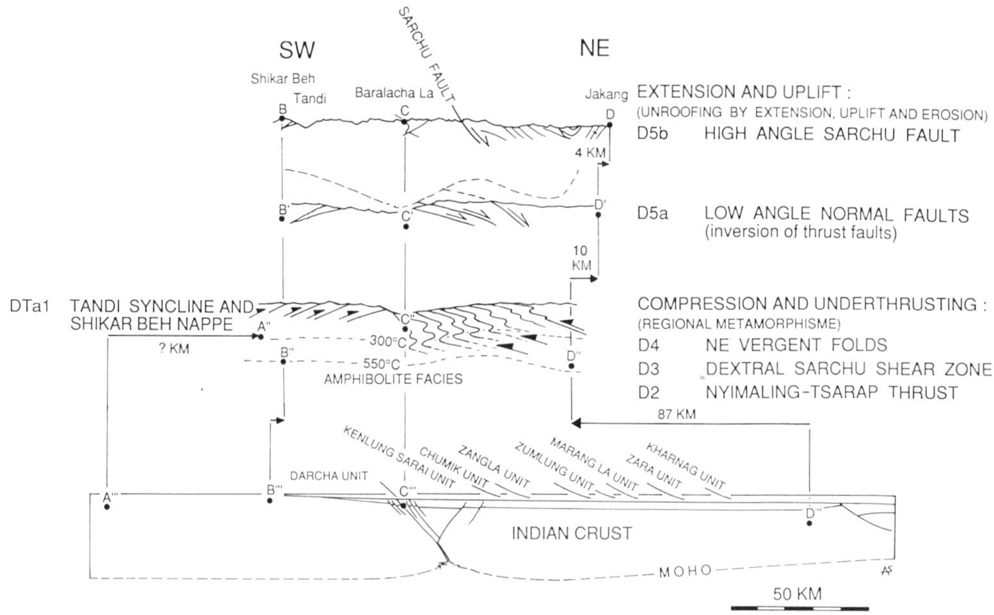

Fig. 10. Tectonic model for the Rohtang La–Leh section.

Conclusions

The tectonic evolution of the Himalayan chain along the investigated transect is summarized in the updated geological cross section presented in Figs 4 and 10. This profile includes a compilation of the results of Frank *et al.* (1977*b*, 1987) for the region situated between the Shikar Beh and the Siwaliks, to the south of the studied area. The following deformational events characterize the model we propose. The initiation of the SW-vergent Nyimaling–Tsarap Nappe corresponds to the first stage of the continental collision. The NE-vergent Tandi Syncline and Shikar Beh Nappe developed during an early phase of the collision, as a consequence of intra continental underthrusting developed within the Indian Plate. Note that seismic data suggest the presence of a south-dipping underthrusting zone below the High Himalaya of Nepal (Hirn *et al.* 1984). This structure might have been generated by a 'flake tectonics' mechanism, as described by Oxburgh (1972). These early intracontinental nappe structures were subsequently overprinted by SW-vergent deformation associated with the main thrusting of the Nyimaling–Tsarap Nappe to the north and with the Main Central Thrust (MCT) to the south. Our observations contrast with the classical models for the Himalayan chain, characterized mainly by SW–vergent nappe structures. However, more regional investigations are needed to test the overall significance of these early NE-vergent nappe structures in the Himalaya.

The SW-vergent Nyimaling–Tsarap Nappe is a consequence of the ductile shearing of the upper Indian crust during its underthrusting below Asia. The Nyimaling–Tsarap Nappe is characterized by a change in the tectonic style associated with different structural levels. The discrete thrust surfaces of the imbricate structure observed in the nappe front grade into a large ductile shear zone at deeper crustal levels. The structural features of the Nyimaling–Tsarap Nappe are consistent with the orogenic lid model ('traineau écraseur rigide', Fig. 7) described by Termier (1903).

The superposition of NW–SE-orientated dextral transpression zones on the root zone and the frontal part of the SW-vergent Nyimaling–Tsarap Nappe, indicates an anticlockwise rotation of the underthrusting direction of India with respect to Asia. This conclusion agrees with the plate tectonic models, which show an anticlockwise change in the motion direction of India between Mid-Eocene and Early Oligocene. During a late phase of the Himalayan orogeny, an important extensional event affected the frontal part of the Nyimaling–Tsarap Nappe.

We thank M.A. Cosca, E.J. Essene, J.C. Hunziker, D. Kirschner, D. Marshall, M.P. Searle and G. Stampfli for critical reviews and helpful discussions.

Support from Swiss National Science Foundation, project No. 20–30297.90 are gratefully acknowledged. Our field expeditions in 1979, 1983 and 1989 and the printing of the coloured geological map have been generously supported by the Herbette Foundation, University of Lausanne.

References

BAUD, A., ARN, R., BUGNON, P., CRISINEL, A., DOLIVO, E., ESCHER, A., HAMMERSCHLAG, J.G., MARTHALER, M., MASSON, H., STECK, A. & TIECHE, J.-C. 1982. Le contact Gondwana-péri-Gondwana dans le Zanskar oriental (Ladakh, Himalaya). *Bulletin de la Société géologique de la France*, **24**, 341–361.

——, GAETANI, M., GARZANTI, E., FOIS, E., NICORA, A. & TINTORI, A. 1984. Geological observations in southeastern Zanskar and adjacent Lahul area (northwestern Himalaya). *Eclogae Geologicae Helvetiae*, **77**, 171–197.

BOYER, S.E. & ELLIOTT, D. 1982. Thrust Systems. *Bulletin of the American Association of Petroleum Geologists*, **66**, 1196–1230.

BUCHER, H. & STECK, A. 1987. Stratigraphy and tectonics of the Indus basin: the Martselang-Gongmaru La section (Ladakh). *Terra Cognita*, **7**, 112.

BURCHFIEL, B.C. & ROYDEN, L.H. 1985. North–south extension within the convergent Himalayan region. *Geology*, **13**, 679–682.

BURG, J.P. & CHENG, G.M. 1984. Tectonics and structural zonation of southern Tibet, China. *Nature*, **311**, 219–223.

BUXTORF, A. 1907. Zur Tektonik des Kettenjura. *Bericht über die 40. Versammlung des Oberrheinischen geologischen Vereins zu Lindau*, 1907, 1–10.

DEWEY, J.F., CANDE, S. & PITMAN III, W.C. 1989. Tectonic evolution of the India/Eurasia Collision Zone. *Eclogae Geologicae Helvetiae*, **82**, 717–734.

FRANK, W., BAUD, A., HONEGGER, K. & TROMMSDORFF, V. 1987. Comparative studies on profiles across the northwest Himalayas. *In*: SCHAER, J.P. & RODGERS, J. (eds) *The anatomy of mountain ranges*. Princeton University Press, 261–275.

——, GANSSER, A. & TROMMSDORFF, V. 1977*a*. Geological observations in the Ladakh area (Himalayas). A preliminary report. *Schweizerische Mineralogische und Petrographische Mitteilungen*, **57**, 89–113.

——, HOINKES, G., MILLER, C., PURTSCHELLER, F., RICHTER, W. & THÖNI, M. 1973. Relations between metamorphism and orogeny in a typical section of the Indian Himalayas. *Tschermaks Mineralogische und Petrographische Mitteilungen*, **20**, 303–332.

——, THÖNI & PURTSCHELLER, F. 1977*b*. Geology and petrography of Kulu-South Lahul area. *Colloques internationaux du C.N.R.S.* **268/2**, 147–172.

GARZANTI, E. & VAN HAVER, T. 1988. The Indus clastics: forearc basin sedimentation in the Ladakh Himalaya (India). *Sedimentary Geology*, **59**, 237–249.

——, BAUD, E. & MASCLE, G. 1987. Sedimentary record of the northward flight of India and its collision with Eurasia (Ladakh Himalaya, India). *Geodynamica Acta*, 1, 297–312.

HERREN, E. 1987. Northeast–southwest extension within the Higher Himalayas (Ladakh, India). *Geology*, **15**, 409–413.

HIRN, A., LEPINE, J.-C., JOBERT, G., SAPIN, M., WITTLINGER, G., XIN, X.Z., YUAN, G.A., JING, W.X., WEN, T.J., BAI, X.S., PANDEY, M.R. & TATER, J.M. 1984. Crustal structure and variability of the Himalayan border of Tibet. *Nature*, **307**, 23–27.

HONEGGER, K., DIETRICH, V., FRANK, W., GANSSER, A., THÖNI, M. & TROMMSDORFF, V. 1982. Magmatism and metamorphism in the Ladakh Himalayas (the Indus-Tsangpo suture zone). *Earth and Planetary Science Letters*, **60**, 253–292.

KLOOTWIJK, C.T., CONAGHAN, P.J. & MCA. POWELL, C. 1985. The Himalayan Arc: large-scale continental subduction, oroclinal bending and back-arc spreading. *Earth and Planetary Science Letters*, **75**, 167–183.

LAUBSCHER, H. 1965. Ein kinematisches Modell der Jurafaltung. *Eclogae Geologicae Helvetiae*, **58**, 232–318.

MARGERIE, DE E. & HEIM, A. 1988. *Les dislocations de l'écorce terrestre*. J. Wurster, Zürich.

MCELROY, R., CATER, J., ROBERTS, I., PECKHAM, A. & BOND, M. 1990: The structure and stratigraphy of SE Zanskar, Ladakh Himalaya. *Journal of the Geological Society, London*, **147**, 989–997.

MOLNAR, P. & TAPPONNIER, P. 1975. Cenozoic tectonics of Asia: effects of a continental collision. *Science*, **189**, 419–426.

OXBURGH, E.R. 1972. Flake tectonics and continental collision. *Nature*, **239**, 202–204.

PATRIAT, P. & ACHACHE, J. 1984. India-Eurasia collision chronology has implications for crustal shortening and driving mechanism of plates. *Nature*, **311**, 615–621.

PÊCHER, A. 1991. The contact between the higher Himalaya Crystallines and the Tibetan sedimentary series: Miocene large-scale dextral shearing. *Tectonics*, **10**, 587–598.

—— & LE FORT, P. 1989. Stretching lineation trajectories in the crystalline pile of central Nepal, *5th Himalaya–Tibet–Karakorum Workshop, Milano*, Abstract vol., 39.

PIERCE, J. 1978. The Northward motion of India since the Late Cretaceous, *Geophysical Journal of the Royal Astronomical Society*, **52**, 277–311.

PICKETT, J.W., JELL, J.S., CONAGHAN, P.J. & POWELL, C.MCA. 1975. Jurassic invertebrates from the Himalayan Central Gneiss. *Alcheringa*, **1**, 71–85.

POWELL, C.MCA. & CONAGHAN, P.J. 1973. Polyphase deformation in Phanerozoic rocks of the Central Himalayan gneiss, Northwest India. *Journal of Geology*, **81**, 127–143.

RAMSAY, J.G. 1980. Shear zone geometry: a review. *Journal of Structural Geology*, **2**, 83–99.

SCOTESE, C.R., GAHAGAN, L.M. & LARSON, R.L.

1988. Plate tectonic reconstructions of the Cretaceous and Cenozoic ocean basins. *Tectonophysics,* **155**, 27–48.

SEARLE, M.P. 1986. Structural evolution and sequence of thrusting in the High Himalayan, Tibetan-Tethys and Indus suture zones of Zanskar and Ladakh, Western Himalaya. *Journal of Structural Geology,* **8**, 923–936.

——, M.P., WINDLEY, B.F., COWARD, M.P., COOPER, D.J.W., REX, A.J., REX, D., TINDONG, L., XUCHANG, X., JAN, M.Q., THAKUR, V.C. & KUMAR, S. 1987. The closing of Tethys and the tectonics of the Himalaya. *Geological Society of America Bulletin,* **98**, 687–701.

SPRING, L. & CRESPO, A. 1992. Nappe tectonics, extension and metamorphic evolution in the Indian Tethys Himalaya. *Tectonics,* **11/5**, 978–989.

——, BUSSY, F., VANNAY, J.C., HUON, S. & COSCA, M.A. 1993. Early Permian granite magmatism of alkaline affinity in the Indian High Himalaya (Upper Lahul–SE Zanskar): geochemical characterization and geotectonic implications. *This volume.*

——, STUTZ, E., THELIN, P., MARCHANT, R., MASSON, H. & STECK, A. 1992. Inverse metamorphic zonation in very low-grade Tibetan series of SE Zanskar and its tectonic consequences (Himalaya, NW India). *Schweizerische Mineralogische und Petrographische Mitteilungen,* **73**, 85–95.

SRIKANTIA, S.V. & BHARGAVA, O.N. 1982. An outline of the structure of the area between the Rohtang pass in Lahaul and the Indus Valley in Ladakh. *Geological Survey of India, Miscellaneous Publications,* **41/3**, 193–204, Calcutta.

STECK, A., SPRING, L., VANNAY, J.-C., MASSON, H., BUCHER, H., STUTZ, E., MARCHANT, R. & TIECHE, J.C. 1993. Geological transect across the Northwestern Himalaya in eastern Ladakh and Lahul. *Eclogae Geologicae Helvetiae,* **86**, in press.

STUTZ, E.A., 1988. Géologie de la chaîne de Nyimaling aux confins du Ladakh et du Rupshu (NW-Himalaya, Inde). *Memoires de Geologie (Lausanne),* 3.

—— & STECK, A. 1986. La terminaison occidentale due Cristallin du Tso Morari (Haut-Himalaya; Ladakh méridional, Inde). *Eclogae Geologicae Helvetiae,* **79**, 253–269.

TAPPONNIER, P., PELTZER, G., LE DAIN, A.Y., ARMIJO, R. & COBBOLD, P. 1982. Propagation extrusion tectonics in Asia: New insights from simple experiments with plasticine. *Geology,* **10**, 611–616.

——, —— & ARMIJO, R. 1986. On the mechanics of the collision between India and Asia. *Geological Society of London Special Publications,* **19**, 115–157.

TERMIER, P. 1903. Quatre coupes à travers les Alpes franco-italiennes. *Bulletin de la Société Géologique de France,* **28**, 411–433.

VAN HAVER, T., BASSOULET, J.P., BLONDEAU, A. & MASCLE, G. 1984. Les séries détritiques du bassin de l'Indus au Ladakh: nouvelles données stratigraphiques et structurales. *Rivista Italiana di Paleontologia e Stratigrafia,* 90, 87–102.

Sedimentary evolution and drowning of a passive margin shelf (Giumal Group; Zanskar Tethys Himalaya, India): palaeoenvironmental changes during final break-up of Gondwanaland

EDUARDO GARZANTI

Dipartimento di Scienze della Terra, Via Mangiagalli 34, 20133, Milano, Italy

Abstract: The Lower Cretaceous terrigenous shelf succession of the Zanskar Range (uppermost Spiti Shale and Giumal Group) testifies to major tectono-eustatic events which affected the Tethys Himalaya passive margin during final fragmentation of Gondwanaland.

The sudden supply of sand-sized terrigenous detritus in the Lower Takh Formation, deposited in coastal to inner shelf environments probably around late Neocomian to Barremian times, documents a first stage of tectonic uplift and rejuvenation of the Indian continental block.

Next, the overall deepening-upward trend recorded by the inner to middle shelf sediments of the largely Aptian Upper Takh Formation indicates relative sea-level rise, culminated with flooding, starvation and deposition of the glauconitic Labar La arenite.

This tectono-eustatic event occurred probably around the Aptian–Albian boundary, and was followed abruptly by influx of abundant 'trachytic' volcanic detritus, recorded at the base of the Pingdon La Formation.

Eruption of alkalic lavas derived from deep-seated mantle sources was confined within the Albian, when volcanic arenites continued to be deposited in inner to middle shelf settings. The local occurrence of up to very coarse-grained, coastal quartz arenites points to episodes of differential uplift and renewed erosion of the Indian Craton.

Suddenly, the Zanskar clastic shelf was drowned. Widespread deposition of the Nerak glauco-phosphorite on the sediment-starved outer shelf in the early Late Albian was followed by very discontinuous sedimentation and major stratigraphic gaps at latest Albian to Cenomanian times. Largely palimpsest and condensed arenites (Oma Chu glauco-phosphorite) were periodically reworked on the outermost shelf by intruding oceanic currents, which prevented deposition of micrite. Only some decimetres of pelagic oozes (base of the Fatu La Formation) could thus accumulate on the uppermost slope until the Early Turonian, when all across the Zanskar margin sufficient depth was reached to allow rapid sedimentation of foraminiferal mudstones on the upper slope (Chikkim and Fatu La Formations).

A complex interplay of tectonic, eustatic and oceanographic processes concurred in the development of 'drowning unconformities' at the top of the Giumal clastic shelf. Rapid deepening and waning of both volcanic and quartzo-feldspathic terrigenous detritus are mainly ascribed to the global mid-Cretaceous sea-level rise and to rapid thermo-tectonic subsidence at the end of the short-lived Albian magmatic event, possibly related to a mantle plume rising beneath northern India. Intensification of the east-bound oceanic current off the north margin of India after the final break-up of Gondwanaland was responsible for continuous resuspension and minimal accumulation rates around the shelf-break. The associated coastal upwelling favoured impingement of the oxygen-minimum zone on the outer shelf, with glauco-phosphorite deposition coinciding in time with peak global transgressions and 'anoxic events' in the world oceans.

The Zanskar Range, lying between the High Himalaya Crystalline to the SW and the Indus–Tsangpo suture to the NE, is formed by Upper Precambrian to Lower Eocene sedimentary rocks belonging to the Tethys Himalaya Zone (Fig. 1). After initial opening of the Neo-Tethys in the Permian, the Mesozoic succession was deposited on the N passive continental margin of the Indian sub-continent (Gaetani & Garzanti 1991).

Reconnaissance stratigraphic work on Jurassic to Cretaceous units cropping out in this area was carried out in the late 1970s and 1980s by several scientific parties (Raina & Bhattacharyya 1977; Kanwar & Bhandari 1979; Srikantia *et al.* 1980; Fuchs 1982, 1984, 1986, 1987;

From Treloar, P.J. & Searle, M.P. (eds), *Himalayan Tectonics*
Geological Society Special Publication, No. 74, pp. 277–298.

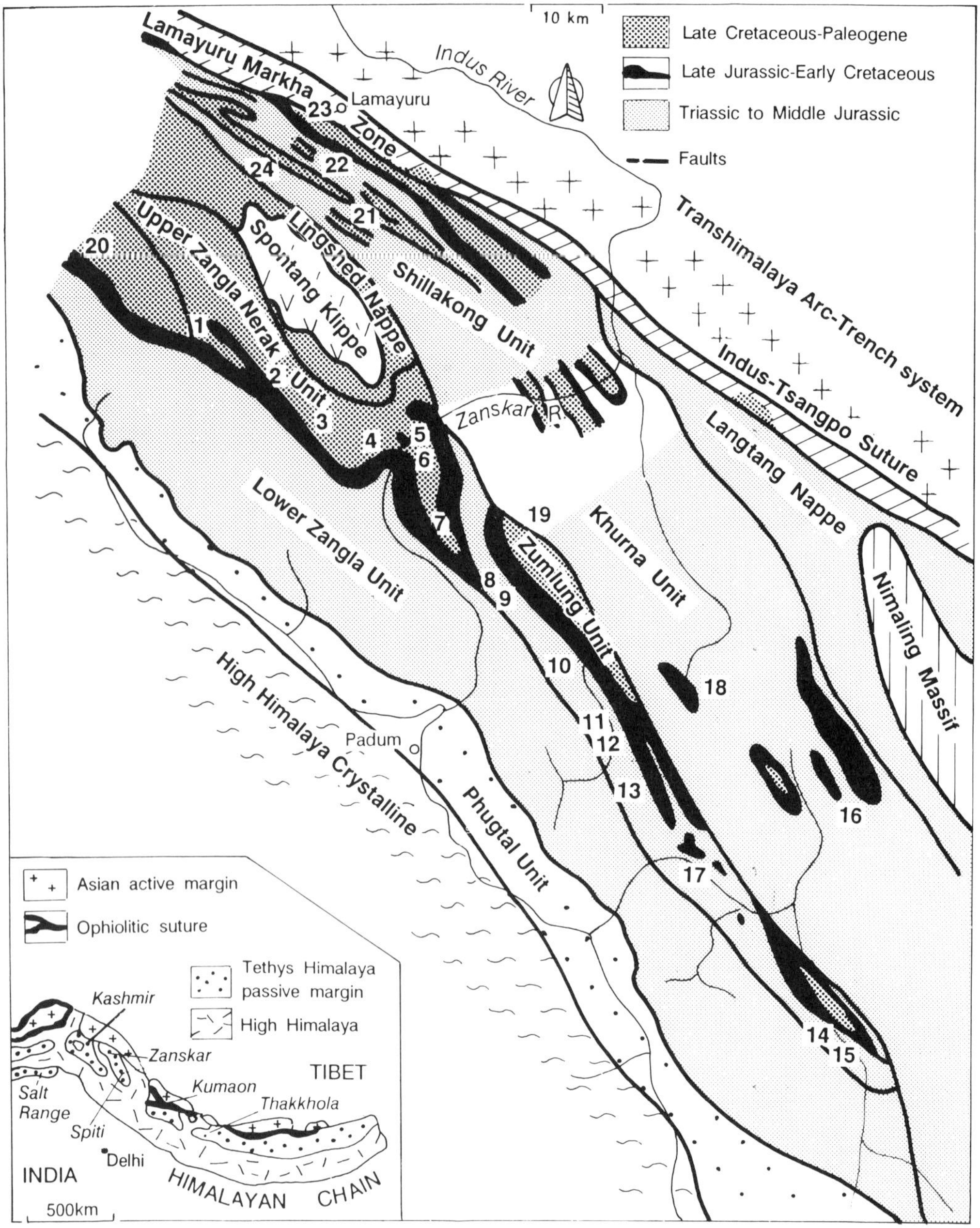

Fig. 1. Geological sketch map of the Zanskar Range (compiled from several sources cited in the text), with location of the studied stratigraphic sections. Shown in inset is the areal distribution of Tethyan passive margin sediments in the Himalayan Chain. Numbers of measured sections from 1 to 15 as in Figs 3 and 4. (16) Khurnak syncline; (17) Gotunda La; (18) Shapodak La; (19) Chirche valley; (20) Spanboth valley; (21) Honupatta valley; (22) Shillakong valley; (23) Fatu La; (24) Kangi.

Baud *et al.* 1982, 1984; Bassoullett *et al.* 1983, 1984; Kelemen & Sonnenfeld 1983; Fuchs & Willems 1990; Mathur 1990).

After closure of Neo-Tethys and initial collision between India and Asia around the Palaeocene–Eocene boundary (Garzanti *et al.*

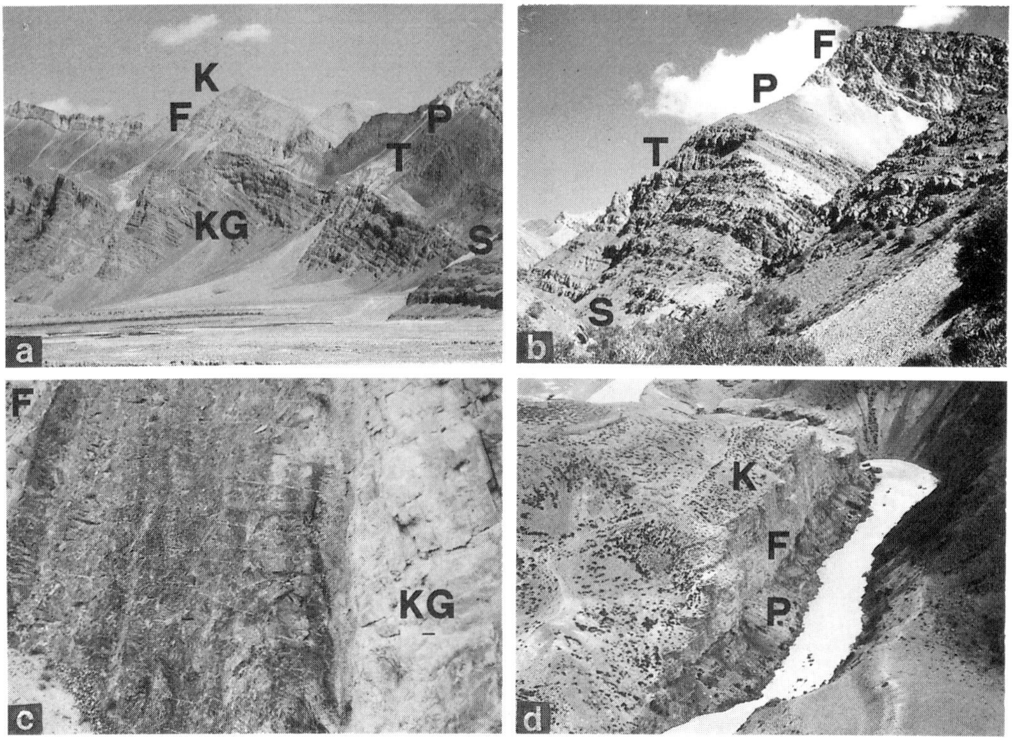

Fig. 2. The Jurassic to Cretaceous succession of the Zanskar Range. (a) Zangla Unit at Zangla; (b) Zumlung Unit at W Sumdo; (c) Shillakong Unit in the Honupatta gorge; (d) Zumlung Unit at Takh. KG, Kioto Limestone Group; S, Spiti Shale; T, Takh Fm.; P, Pingdon La Fm.; F, Fatu La Fm.; K, Kangi La Marls.

1987), the Zanskar continental terrace was dismembered into several thrust sheets, which were tectonically transported to the SW beneath the Spongtang oceanic allochton (Fig. 1). The Zangla thrust sheet, which is subdivided into a lower and an upper unit by a decollement surface following the incompetent Spiti Shale along the Oma Chu valley, is the southernmost one containing Cretaceous sedimentary rocks (Fig. 2a). To the NE, Cretaceous formations crop out in footwall synclines belonging to the Zumlung and Khurna Units (Fig. 2b,d). In the Nerak area, the Upper Zangla and Zumlung Units wedge out, and further to the north a very reduced Upper Jurassic–Lower Cretaceous section is exposed in the Shillakong Unit (Fig. 2c).

The aim of the present paper, which follows preliminary reports of the 1981 (Baud *et al.* 1984) and 1984 (Gaetani *et al.* 1986) expeditions, is to describe in detail the sedimentary evolution of the Zanskar passive margin during the Early Cretaceous. A number of companion publications deal specifically with the Giumal Group stratigraphy (Garzanti 1992), with the mid- to Late Cretaceous stratigraphy and sedimentary evolution (Premoli Silva *et al.* 1992), and with sandstone petrography and provenance of terrigenous detritus (Garzanti in prep).

Stratigraphy

The Upper Jurassic to lowest Cretaceous succession is represented by cephalopod-bearing black pelites (*Spiti Shale*), overlain by the Giumal Group shelf clastics. The latter have been recently subdivided into two formations, both capped by condensed glauconitic horizons representing laterally-continuous stratigraphic markers (Fig. 3 and 4; Garzanti 1992). The Upper Neocomian? to Aptian *Takh Formation* (Fig. 5a,b), nearly 200 m thick and characterized by quartzo-feldspathic sandstones, is capped by the *Labar La Arenite* (Fig. 5c). The Albian *Pingdon La Formation*, 214 to 240 m thick in SW Zanskar and only 92 to 130 m thick in NE Zanskar, is characterized by widespread volcanic arenites and capped by the *Nerak* and *Oma*

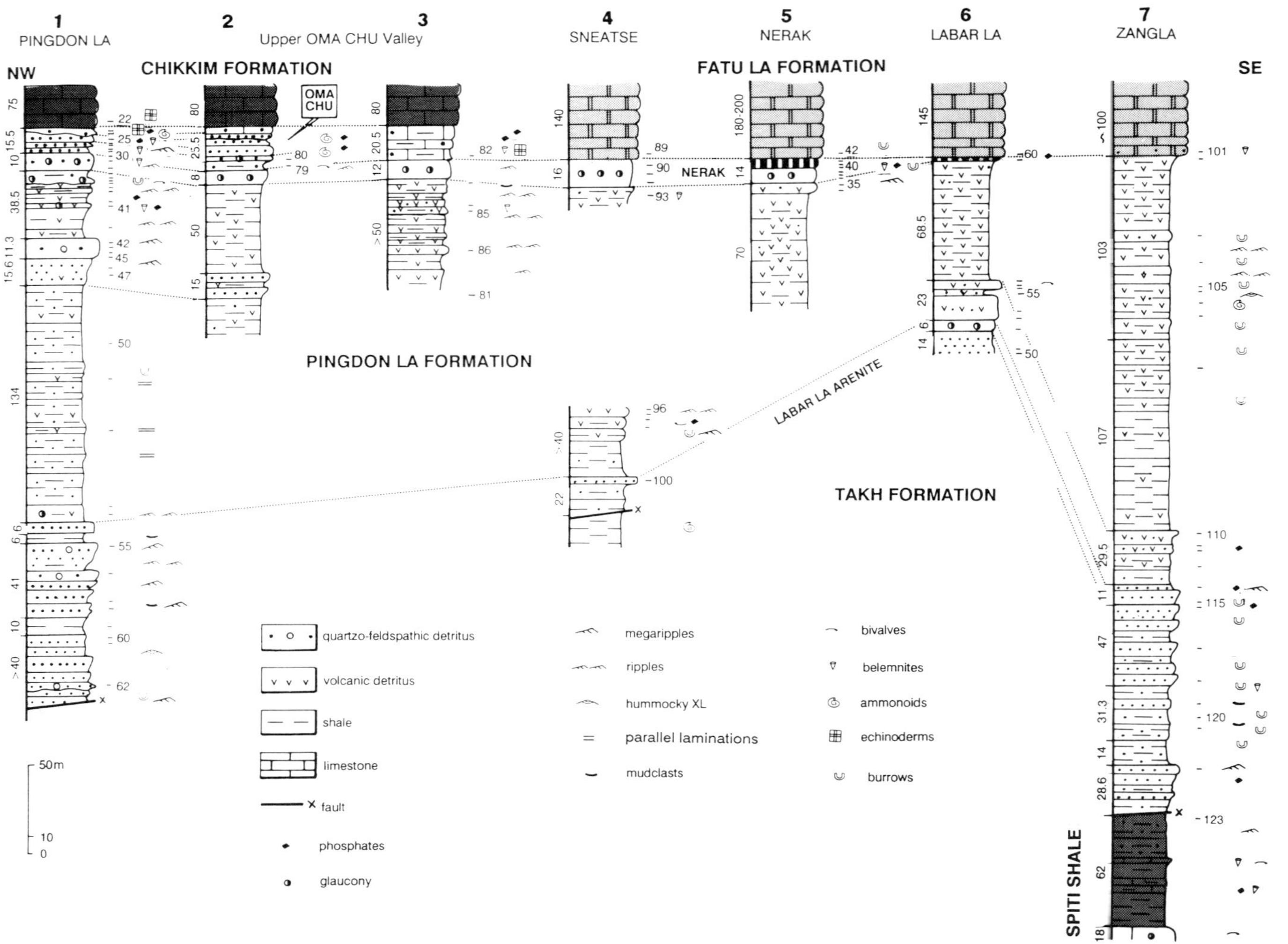

Fig. 3. Measured stratigraphic sections in the Zangla-Nerak Unit. Thickness of each interval in metres. Dark shading, Spiti Shale at the base and Chikkim Fm. at the top; light shading, Fatu La Fm.

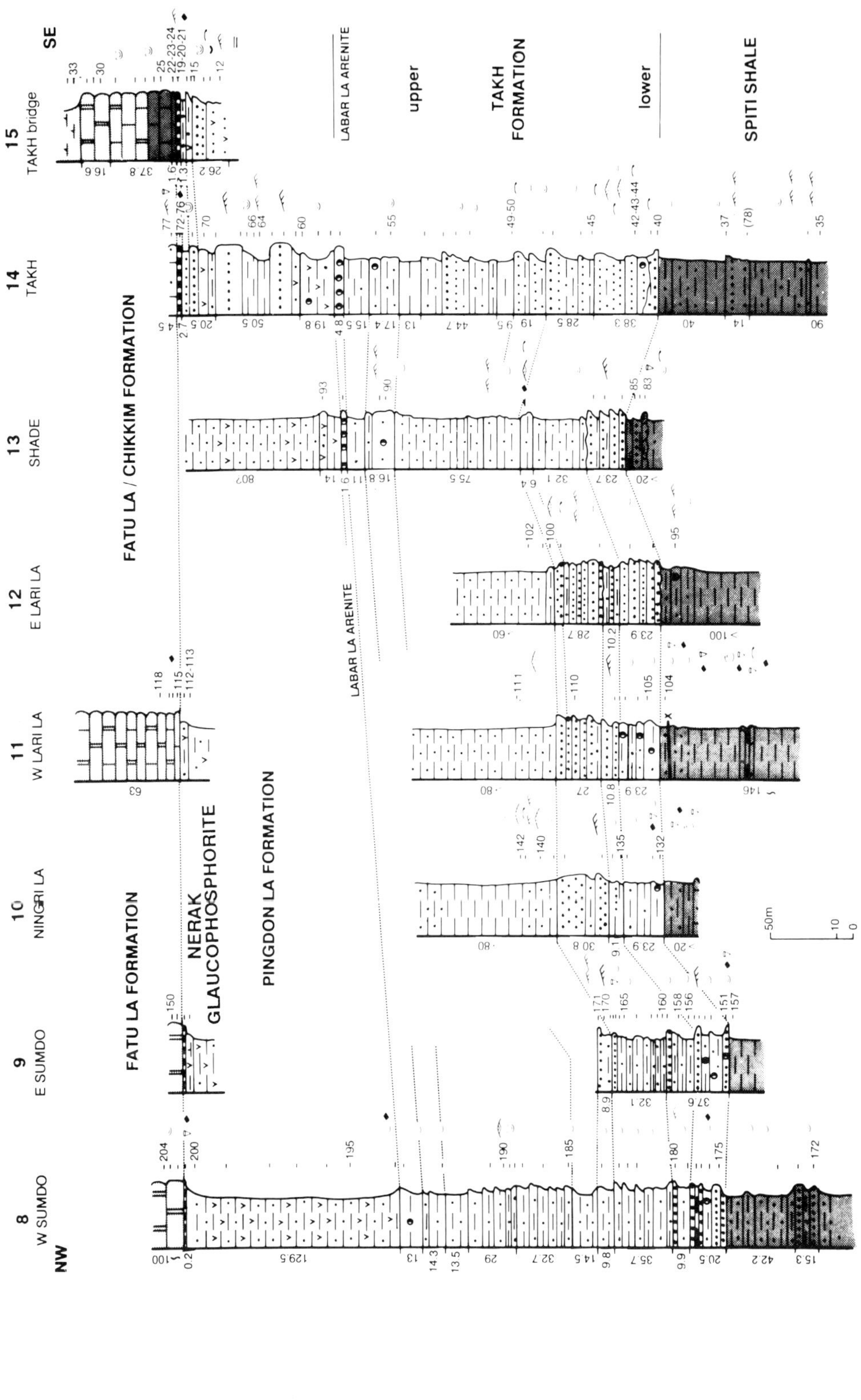

Fig. 4. Measured stratigraphic sections in the Zumlung Unit. Thickness in metres. Dark shading: Spiti Shale at the base and Chikkim-like facies in the Fatu La Fm. at the top; light shading: Fatu La Fm. Symbols as in Fig. 3.

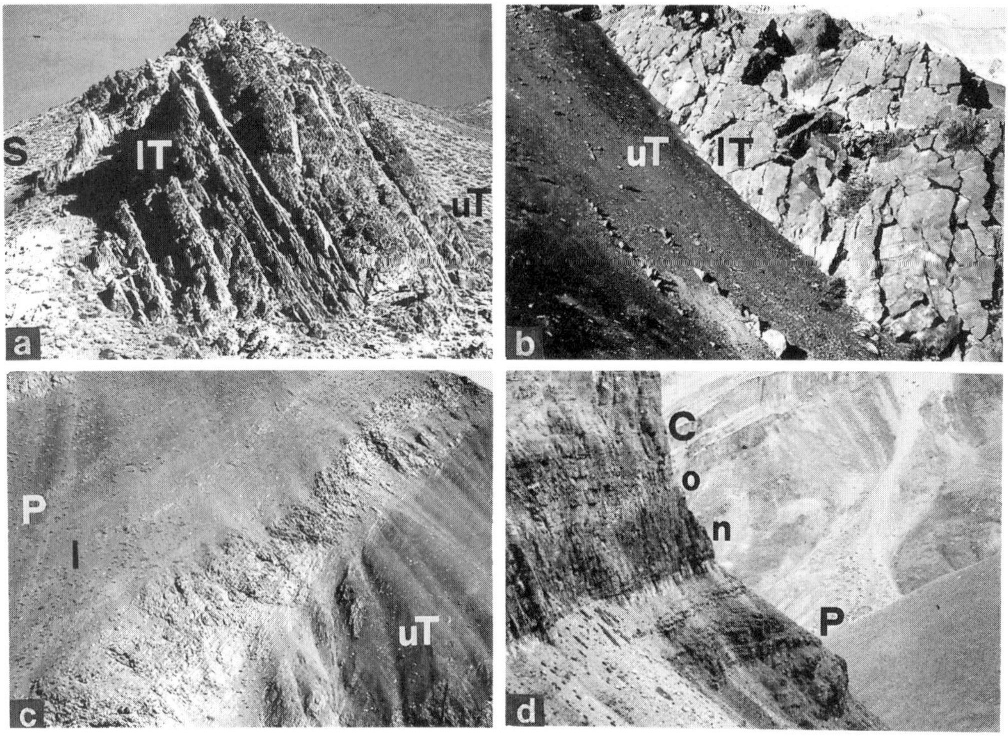

Fig. 5. Stratigraphy of the Giumal Group. (a) Transition between Spiti Shale and coarsening-upward Lower Takh Fm. (W Lari La); (b) sharp boundary between lower and Upper Takh Fm. (Ningri La); (c) transition between Takh and Pingdon La Fms. (Shade); (d) transition between Giumal Group and Chikkim Fm. (Pingdon La). S. Spiti Shale; T, Takh Fm. (l, lower; u, upper); l, Labar La Arenite; P, Pingdon La Fm.; n, Nerak Glauco-phosphorite; o, Oma Chu Glauco-phosphorite; C, Chikkim Fm.

Chu Glauco-phosphorites (Fig. 5d; Garzanti *et al.* 1989).

The Giumal Group is sharply followed by pelagic 'Scaglia-like' foraminiferal mudstones and wackestones, beginning from the Late Albian *R. ticinensis* or *R. appenninica* Zones (*Fatu La Formation*) or from the Early Turonian *H. helvetica* Zone (*Chikkim Formation*) and reaching up to the Campanian/Maastrichtian boundary (*G. calcarata*/*G. havanensis* Zones; Premoli Silva *et al.* 1992). The Cretaceous succession is completed by shelfal silty marls and shallow-water limestones of Maastrichtian age (*Kangi La* and *Marpo Formations*), sharply overlain by the Stumpata Quartzarenite (Gaetani *et al.* 1986; Nicora *et al.* 1987).

Fig. 6. Facies associations in the Late Jurassic to mid-Cretaceous succession of the Zanskar Range. (a) Bipolar cross-lamination in very coarse-grained quartzarenites (facies A; middle Pingdon La Fm., Pingdon La); (b) interbedded pelites and fine to lower medium-grained rippled volcanic arenites (facies C; upper Pingdon La Fm., upper Oma Chu valley); (c) thickening-upward cycle of clean and moderately well-sorted upper fine-grained quartz-rich subarkoses (facies D), isolated in black pelites (facies H; Upper Spiti Shale, E Shade); (d) beds with scoured base (arrow) of lower very fine-grained bioclastic sandstones passing laterally to glaucony-bearing sandstones (facies E, F; basal Lower Takh Fm., Takh); (e) coarse-grained graded bed with belemnite lag at the top (arrow; Upper Spiti Shale, Shade); (f) cross-laminated, lower fine-grained, hematitic, glauconitic arenites (facies I; Nerak Glauco-phosphorite, Nerak); (g) pararuditic phosphorite (facies K; Oma Chu Glauco-phosphorite, Pingdon La); (h) grey condensed deep-water limestones with bacterial mats (arrow; facies M; base of the Chikkim Fm., upper Oma Chu valley).

a
b
c
d
e
f
g
h

Sedimentology

The Lower Cretaceous succession consists of shelfal sediments, as testified by the occurrence of marine fossils in all stratigraphic units (Spiti Shale, Takh and Pingdon La Formations). Fifteen principal facies associations were recognized, according to sedimentary features, texture, mineralogical composition and faunal content. They are listed below, in order of inferred distance from shore and increasing water depth.

(A) Very coarse-grained, well- to moderately well-sorted and quartz-cemented quartzarenites with large-scale cross-bedding and herringbone cross-lamination (Fig. 6a).

(B) Upper fine- to very coarse-grained, mainly well-sorted, quartz-cemented quartzarenites with megaripple cross-lamination. Large burrows may occur on stratal surfaces.

(C) Fine- to medium-grained, well to moderately-sorted sandstones (volcanic arenites, subarkoses, sublitharenites, rarely quartzarenites) with common ripple marks, low-angle, planar or hummocky lamination (Fig. 6b). Megaripple cross-lamination, graded-bedding, 'Bouma-type' Tb-c structures and concentration of mud rip-up clasts were also recorded. Interbedded dark pelites or calcareous arenites with bivalves are subordinate. Bioturbation is common, but sedimentary structures are generally preserved.

(D) Up to 30 m thick coarsening-upward sequences, beginning with dark pelites and capped by upper, fine-grained, well and bimodally-sorted, mainly quartz-cemented, cross-laminated quartzose sandstones (subarkoses, sublitharenites or quartzarenites; Fig. 6c). Planar, hummocky, ripple and megaripple cross-lamination were observed. Bioclasts are very subordinate; bioturbation may be important.

(E) Up to medium-grained, moderately well- to moderately-sorted sandstones (subarkoses, volcanic arenites, rarely quartzarenites), either quartz-cemented, matrix-rich or calcite-rich. Megaripple cross-lamination or 'Bouma-type' structures and common bioclasts (belemnites, bivalves) were recorded. Phosphates, authigenic ferroan dolomite and particularly greenish illitic or silicate peloids, possibly of fecal origin and generally representing over 10% of framework grains, are characteristic.

(F) Largely burrowed, up to lower fine-grained, mainly medium-bedded and moderately well-sorted sandstones (mostly subarkoses in the Takh Fm. and volcanic arenites to feldspathic volcanic arenites in the Pingdon La Fm.) interbedded with subordinate pelites. Sandstones may show amalgamated hummocky cross-lamination, low-angle to planar lamination, ripple marks, channellized bodies with scoured base, graded-bedding and 'Bouma-type' structures. Small fragile bivalve shells are commonly found, along with echinoderms, belemnites or foraminifera. Silicate peloids, greenish illitic clasts, immature glaucony and small phosphate grains may be present (Fig. 6d).

(G) Burrowed dark pelites with subordinate up to lower fine-grained, thin- to medium-bedded, moderately well to moderately-sorted sandstones (mostly subarkoses in the Takh Fm. and feldspathic volcanic arenites to lithic arkoses in the Pingdon La Fm.), showing planar to hummocky lamination and ripple marks. Bivalves, echinoderms, ostracodes, belemnites, planktonic foraminifera and siliceous sponge spicules may be present; ammonoids are very rare.

(H) Black shales with open marine fauna (belemnites, ammonoids, echinoderms, bivalves, gastropods, radiolaria, foraminifera). Medium to thick graded beds with 'Bouma-type' structures display scoured base overlain by lags of moderately-sorted intraformational hybrid rudites containing bioclasts (largely cephalopods) or pelitic to calcareous intraclasts (Fig. 6e; see also Gaetani *et al.* 1986, fig. 14). Belemnite lags and thin-bedded bioclastic arenites may occur. Pyrite is common.

(I) Very fine- to coarse-grained, moderately well to poorly-sorted, pyritic to hematitic, either quartzose or volcanic lithic-rich glauconitic arenites, commonly showing large-scale cross-lamination (Fig. 6f) or ripple marks. Belemnites, echinoderms, bivalves (including inoceramids), ostracodes and foraminifera may be observed. Fully mature glaucony peloids tend to decrease upward in abundance with respect to siliciclastics and phosphates. Dolomite and ferriferous carbonates commonly occur.

(J) Medium- to coarse-grained, moderately well- to very poorly-sorted, quartzose, phosphatic and bioclastic arenites, yielding belemnites, echinoderms and foraminifera. Channellized bodies with scoured base, large-scale cross-lamination and coarse bioclastic lags occur. Multiple generations of early diagenetic calcite cements may be observed (Garzanti *et al.* 1989, fig. 8). Micrite is absent.

(K) Moderately- to poorly-sorted reworked phosphorites, commonly ruditic and with scoured base (Fig. 6g). These layers yield ammonoids, belemnites, echinoderms, radiolaria, calcispheres, foraminifera, and contain micrite. Quartzose siliciclastic detritus reaches coarse sand size.

Table 1. *Distribution of bedding sequences in the Late Jurassic to early Late Cretaceous succession of the Zanskar Range*

Sources		1 Pingdon La	1 Oma Chu	1 Sneatse	1 Nerak	1 Labar La	1 Zangla	1 Sumdo	1 Ningri/ Lari La	1 Shade/ Tantak	1 Takh	1 Khurna South	2 & 3 Khurna North	1 & 4 Shillak. South	4 Shillak. North
Chikkim/Fatu La *H. helvetica*		M->N	M->N	N-O	O	O	O	O	N-O	N	N	O	O	O	O
Oma Chu/Fatu La *R.tic. - W. arch.*		J->K->L	J->K->L	M	M	L->M	L->M	M	L	---	M	---	---	L->M	---
Nerak Glauco-Ph *R. subticinensis*		I	G-I	I	I->K	I	X-----X	I->J->K	G->K	---	J->K				
Upper	**Pingdon La Fm.**	E-G	C->E	E-G	G-H	G-H	G	E-G	G		C->G				
Middle		E->A	B-C	---			F-G		---	G-H	B-C-E	G	G	H	
Lower		G	---	---	G	G	G	G	---		G->E				
Basal			---	G->E	---	E	G->E		---	G->F					
Labar La Arenite	*glaucony*			E		I	(J)			I	I				
Upper	arenite	G	---	G	---	I	E	G	---	G	G				
	pelite	D		C		E	F	G		E-F	E-F				
Upper Takh Fm.		G		G			F	G			G				
		A-B-C	---	F		F	F	F	G->F	G	D-F				
Basal	pelite	H		X-----X	X-----X	G	G	G-H	H		G-H				
Topmost								E			F				
Lower Takh Fm.		A-B-C					D	D-F	D	D	D-F				
Basal		X-----X	X-----X	X-----X	X-----X	X-----X	X-----X	B-C-E	E-F	E-F	E-F	E-F			
Upper		X-----X	X-----X	X-----X	X-----X	X-----X	X-----X	D-G	G-H	D-G-H	D-H	X-----X			
Middle	**Spiti Shale**	X-----X	X-----X	H	X-----X	X-----X	X-----X	---	H	X-----X	D-H				
Lower		----	---	D-G-H	D-H	H	G-H	---		H		H			

Sources of information: (1) own data; (2) Fuchs 1984, 1986; (3) Stutz 1988; (4) Bassoulett *et al.* 1983.

(L) Grey pelagic limestones with phosphate nodules, alternating with lenticular beds of up to medium-grained, moderately to poorly-sorted, bioclastic quartzose arenites to quartzose echinoderm grainstones. Also bivalves, ammonoids, belemnites and glaucony may be found.

(M) Grey mudstones/wackestones with 'pseudo-stromatolitic' laminae, or even winnowed bioclastic grainstones, yielding belemnites, echinoderms, calcispheres and planktonic foraminifera. Quartzose siliciclastic detritus is very subordinate, but maximum diameter may still reach upper medium sand size; along with phosphates it wanes upsection. Biostratigraphic evidence indicates deposition at very low accumulation rates and major time gaps (Fig. 6h).

(N) Grey foraminiferal mudstones/wackestones with sporadic benthic fauna and devoid of siliciclastics.

(O) Light grey to greenish and reddish foraminiferal marly mudstones/wackestones devoid of silt and sand. Benthic fauna is uncommon. Wave ripple marks exceptionally occur (Premoli Silva *et al.* 1992, fig. 21).

Depositional environments

Sedimentary evolution can be reconstructed from lateral and vertical distribution of depositional environments, as documented by sedimentary structures and faunal content (Table 1), and by comparison with modern and ancient storm-dominated shelves (Frey & Pinet 1978; Specht & Bremner 1979; Handford 1986).

The unfossiliferous and coarse-grained quartzose sandstones observed in the Takh Formation of W Zanskar and in the middle part of the Pingdon La Formation (Pingdon La, Takh), are interpreted as desposited at very shallow water depths (few metres), in estuarine (*facies association A*) to upper shoreface (*facies association B*) settings. Estuarine sands accumulated under the action of tidal currents within valleys incised while sea-level was low, at the onset of the subsequent transgression (Fig. 6a; 'bay mouth sand facies' developed in the distal part of transgressive estuarine wedges; Nichols *et al.* 1991). Lateral transition within a few kilometres at most to volcanic arenites of facies

C indicates strong enrichment in quartz with respect to volcanic rock fragments in high-energy coastal environments.

Facies association C occurs in the Takh Formation (Pingdon La, Sneatse), but is particularly common in the middle and upper Pingdon La Fm. (Fig. 6b; Oma Chu valley, Takh), where it is associated or can be physically traced shoreward to facies A and B. These sediments are interpreted as deposited in distal tidal bars to lower shoreface settings mainly influenced by action of waves and storms, at water depths of several metres.

Facies association D, occurring commonly in the Upper Spiti Shale (Fig. 6c) and Lower Takh Fm. of the Zumlung Unit, and rarely in the Upper Takh Fm. (Pingdon La, Takh), documents prograding sandstone bars on a storm-dominated shelf (Campbell 1971; Winn *et al.* 1984; Ramli 1986). These coarsening-upward sequences are thought to have formed preferentially at lowstand stages (Plint 1988). Water depth was around storm wave-base for the pelitic lower part, whereas clean sands at the top indicate fair-weather wave reworking.

Facies association E, occurring at the base and top of the Lower Takh Fm. in the Zumlung Unit, at the top of the Upper Takh Fm. and at the base and top of the Pingdon La Formation, was deposited in high energy open marine conditions. Relatively coarse grain size, abundance of bioclasts, occurrence of open marine cephalopods, and enrichment in non-carbonate intrabasinal grains suggest sedimentation during sea-level rise (shelf margin wedge and transgressive stages; Garzanti 1991).

Storm-dominated, thin to medium-bedded and highly burrowed facies F and G are both very common in the Lower Cretaceous succession of the Zanskar Range. *Facies association F*, most common in the Takh Formation, was deposited on the inner shelf, as indicated by abundance of storm generated sedimentary structures and by poorly diversified fauna (Flügel 1982, p. 36). Water depth was between fair-weather and average storm wave-base (indicatively 10 to 30 m). Scour-and-fill structures, associated with amalgamated hummocky cross-lamination in the basal part of the Lower Takh Fm. at Takh (Fig. 6d), are ascribed to offshore directed bottom flows induced by storm surge, in the transition zone between the shoreface and the shelf (Niedoroda *et al.* 1985; Massari 1988, p. 154). Facies F passed laterally seaward to *facies association G*, which is widespread both in the Upper Takh and Pingdon La Formations. These sediments were deposited under the action of sporadic storms, largely between average and major storm wave base. Occurrence of rare planktonic foraminifera and siliceous sponge spicules indicates water depths of some tens of metres (Maxwell & Swinchatt 1970).

Facies association H, widespread in the Spiti Shale, and occurring at major transgressive stages in the Upper Takh Fm. and at the top of the Pingdon La Fm., was deposited on an almost undisturbed middle shelf. Belemnite beds possibly formed at times of flooding, reduced fine terrigenous supply and condensation. The seafloor was stirred only during exceptional events, which produced thick-bedded, graded, intraclastic or belemnite-rich rudites. These deposits may be related either to sediment-gravity flows and associated turbidity currents triggered by penecontemporaneous tectonic activity, or to 'megastorms' (tropical cyclones?, tsunamis?), which caused resuspension of shelfal sediments, winnowing and concentration of intrabasinal mudclasts or coarse shell debris (Fig. 6e). Alternatively, they may have originated by storm reworking due to lowered storm wave base at stages of relative sea-level fall (Plint 1988). Similar 'event beds', becoming thicker and coarser-grained in deeper waters (Bayer *et al.* 1984; Gagan *et al.* 1988, 1990), are commonly described from other Upper Jurassic to Lower Cretaceous Gondwanian passive margin shelves (Hallam & Maynard 1987, p. 111; Koutsoukos *et al.* 1991, p. 488).

Poor oxygenation and preservation of organic matter are largely ascribed to palaeoceanographic conditions prone to anoxia in the Late Jurassic and Early Cretaceous. Sedimentation took place mostly below storm wave-base at water depths indicatively between 40 and 100 metres, consistent with occurrence of shelfal ammonoid genera in the Middle Spiti Shale of the Zangla Unit (Oloriz & Tintori 1991), and of a few shallow-dwelling planktonic foraminifers (Hedbergellids and Ticinellids; Leckie 1987) in the Upper Takh and Pingdon La Formations.

Facies association I, characterizing the Labar La Arenite of NE Zanskar and the Nerak Glauco-phosphorite of west and central Zanskar, was deposited during major transgressive stages, when siliciclastic detritus was trapped in flooded river valleys and estuarine mouths (Garzanti 1991). Pelagic faunal content indicates water depths around 100 m. Coarse grain-size and cross-lamination are ascribed to action of strong oceanic currents (Fig. 6f; Flemming 1978, 1980), which repeatedly reworked palimpsest quartzose detritus (Swift *et al.* 1971). These arenite intervals become very thin and may almost lack siliciclastics far from fluvial entry points of terrigenous detritus, where the outer shelf remained starved of sand and currents were strong enough to prevent settling of mud.

Facies association J ('outer shelf facies' of Stanley & Wear 1978) is best displayed in the lower part of the Oma Chu Glauco-phosphorite and locally at the top of the Takh Formation. It comprises largely palimpsest, coarse and clean arenites reworked by oceanic currents, which caused continuous resuspension of micrite and prevented deposition of pelagic mud in the outermost shelf. Water depths were in the 100 to 150 m range, as indicated by occurrence of still predominant shallow-dwelling planktonic foraminifera (Hedbergellids, Ticinellids and *Globigerinelloides*; Premoli Silva *et al.* 1992). Multiphase early cementation, scours and other high-energy features testify to bottom current velocities of some to several tens of centimetres per second (Hollister & Hezeen 1972; Quine & Bosence 1991).

Facies association K, characterizing the Nerak Glauco-phosphorite in the Zumlung Unit and the upper Oma Chu Glauco-phosphorite in west Zanskar (Fig. 6g), was deposited close to the shelfbreak at times of oceanic anoxia and increased environmental stress, when upwelling was intensified and the expanded oxygen-minimum zone impinged on the shelf margin (Jenkyns 1980; Cruickshank & Rowland 1983). At these transgressive stages, scouring and reworking by oceanic currents was still intense, but the water column was deep enough (well over 100 m, as indicated by occurrence of Rotaliporids; Leckie 1987) to allow sporadic accumulation of micrite.

Facies association L ('transitional facies' of Stanley & Wear 1978) is well displayed at the top of the Oma Chu Glauco-phosphorite, and marks the transition from high-energy outermost shelf deposits (palimpsest quartzose arenites and encrinites) to low-energy pelagic mudstones).

Facies association M, found at the base of the Chikkim (Fig. 6h) and Fatu La Formations, was deposited on the uppermost slope. Upper bathyal conditions are testified by abundance of typical Tethyan, deep-dwelling keeled planktonic foraminifera. At water depths approaching 200 m, oceanic currents still affected the sea-bottom, causing resuspension of foraminiferal tests and preventing sediment accumulation during most of the time in question. 'Pseudostromatolitic' lamination is ascribed to bacterial mats, similar to those forming today in oxygen-deficient marine basins and coastal upwelling regimes, at non-photic water depths of many tens to a few hundred metres (Williams & Reimers 1983; Williams 1984).

Facies associations N and O ('slope facies' of Stanley & Wear 1978) were deposited on the upper slope, below a type II 'mudline' (Stanley *et al.* 1983). The grey facies N, typical of the Chikkim Formation, indicates deposition in still poorly-oxygenated environments under the direct influence of the oxygen-minimum zone. The multicoloured facies O, characteristic of the Fatu La Formation, points to deeper (well over 200 m) and more oxygenated conditions oceanward of upwelling zones. From time to time the sea-bottom was still under the action of oceanic currents and exceptional storm or internal waves, but deposition of both micrite and clay occurred continuously at high accumulation rates.

Palaeocurrent directions

Palaeocurrent indicators in the Upper Spiti Shale and Giumal Group (53 measurements) are mainly directed between the north and east (30 measures). The palaeotopographic gradient of the Zanskar shelf was thus towards the NE, where the Neo-Tethys Ocean lay. A similar distribution was found in the underlying Upper Triassic to Jurassic succession (Garzanti 1986; Jadoul *et al.* 1985, 1990), suggesting that palaeogeography of the N India margin did not radically change during the Early Cretaceous.

In proximal deposits (facies associations A, B and C), chiefly north-northwestward to eastward palaeocurrent directions (22 measured dips of cross-laminae) indicate that sediment dispersal was mainly downslope (perpendicular and oblique to shore). The estuarine sandstones of the middle Pingdon La Fm. (facies A) are characterized by a complex multidirectional pattern (330° to 150° N, with modal values between 70° and 90° N); northwestward to southeastward bipolar structures (7 measures; Fig. 6a) point to roughly shore-parallel, reversing tidal currents.

Disperse patterns are found in cross-laminated offshore bars (facies associations D and E), including directions oblique and sub-parallel to coast (southeastward: 2 measures; north-northeastward: 2 measures; east-north-eastward: 1 measure), to landward-oriented (southwestward: 3 measures), which may suggest a modified current regime particularly during transgressive stages.

Storm beds (facies association F) are composed of up to lower fine-grained sand transported towards the NE, from shoreface to inner shelf settings (4 measures on sporadic current ripples; Leckie & Krystinik 1989; Duke *et al.* 1991). Rippled offshore arenites (facies association G) indicate a shore-parallel (northwestward: 3 measures) transport direction, possibly related to activity of inshore countercurrents (Flemming 1978, fig. 13C). Elongation of belemnite rostra found in the Spiti Shale ruditic lags (facies association H) is mostly SW–NE (4 measures).

Large-scale cross-lamination in the condensed and palimpsest arenites at the top of the Giumal Group (Fig. 6f; facies associations I and J), pointing to southeastward palaeocurrents (3 measures), is consistent with deposition under the action of a SE-bound oceanic current. The existence of such currents off the north margin of India is inferred from modern reconstructions of mid-Cretaceous palaeoceanographic circulation (Barron & Peterson 1990; Baumgartner *et al.* 1991). Axes of scours and channels in these glauco-phosphoritic horizons (facies associations J and K) are directed S–N and SW–NE (2 measures), roughly perpendicular to bathymetry.

Starved or straight-crested ripple marks, occasionally observed in Late Cretaceous pelagic mudstones (facies associations L, M and N) are symmetrical sedimentary structures presumably generated by hurricane, tsunami or internal waves.

Palaeocurrent data thus indicate that sediment transport was largely perpendicular to the coast in shoreface to inner shelf settings (Leckie & Krystinik 1989), and became sub-parallel to shore in the middle and outer shelf, with geostrophic veering taking place at water depths of some to several tens of metres (Snedden & Swift 1991).

Sedimentary evolution

Early Cretaceous sedimentation on the Zanskar passive margin shelf was controlled by the interaction of extensional tectonics, related magmatic activity and eustatic sea-level fluctuations (Cloetingh 1988; Fulthorpe 1991). Major changes in palaeoenvironments resulted from the complex interplay of tectono-eustasy with climate, oceanic circulation, chemistry of ocean waters, and biological processes (Fig. 7; Fischer & Arthur 1977; Schlanger & Arthur 1990; Jansa 1991; Schlager 1991).

Table 2. *Measured thickness (in metres) of recognized stratigraphic intervals*

5 My	6 My	Sources	1 Pingdon La	1 Oma Chu	1 Sneatse	1 Nerak	1 Labar La	1 Zangla	1 Sumdo	1 Ningri/ Lari La	1 Shade/ Tantak	1 Takh	1 Khurna South	2 & 3 Khurna North	1 & 4 Shillak. South	4 Shillak. North
89	90	**Chikkim/Fatu La** *H. helvetica*	67	---	≥35	---	>30	>10	>10	50	---	41	---	---	---	---
90	91	**Oma Chu/Fatu La** *R.tic. - W. arch.*	15.5	20÷26	1.5	≥ 0.5	≥0.5	≥0.5	0÷1.5	0.5	---	1.6	---	---	---	---
99.5	98	**Nerak Glauco-Ph.** *R. subticinensis*	10	8÷12	16	14	0.3÷0.8	X-----X	0.2÷0.4	0.5	---	1.3÷2.7				
101	99	**Pingdon La Fm.** Upper	38.5	50						---		20÷26				
102.5	100.5	Middle	26.9	15		70	68.5	103		---	≥ 80	50.5		30÷60	2÷10	0
105.5	103	Lower	123	---	>100			107	129.5	---		19.8				
109.5	106	Basal		---	≥20	---	23	29.5		---	14					
110.5	107	**Labar La Arenite** Upper *glaucony*			2.2		6				1.6	4.8				
		arenite	11	---	20	---		11	13		11	15.5				
112	108	pelite	6		4		14	15	14.3		16.8	17.4				
			6		9			6	13.5			13				
		Upper Takh Fm.	41	---	>9		65	56.3	61.7	>70	75.5	44.7		0?		
120	111	Basal pelite	10		X-----X	X-----X	12	14	14.5	10		9.5				
122	112	Topmost							9÷10			19				
		Lower Takh Fm.	>40					>29	45÷49	38÷40	38.5	46.5				
		Basal	X-----X	X-----X	X-----X	X-----X	X-----X	X-----X	20.5	23.9	23.7	20.3	≥100			
135.5	121.5	**Spiti Shale** Upper	X-----X	X-----X	X-----X	X-----X	X-----X	X-----X	≥ 60		≥ 57	≥ 54	X-----X			
148	136	Middle	X-----X	X-----X	≥100	X-----X	X-----X	X-----X	---	≥146	X-----X	≥150		0?		
152	140	Lower	---	---		30÷60	50÷60	62	---		---		6			
156	150.5															

Note fairly constant thickness in the Takh Formation, whereas in the Pingdon La Formation accumulation rates became double in south-western Zanskar (Zangla Unit) with respect to northeastern Zanskar (Nerak area and Zumlung Unit).

Sources of information: (1) own data; Premoli Silva *et al.* 1992; (2) Fuchs 1984, 1986; (3) Stutz 1988; (4) Bassoullet *et al.* 1983; (5) Harland *et al.* 1989; (6) Haq *et al.* 1988.

Age

Since age-diagnostic fossils occur only at the top of the Giumal Group, ages of various stratigraphic intervals could only be inferred tentatively. Sporadic occurrence of nannofossils and silicified planktonic foraminifera from the middle part of the lower Takh Fm. onwards, recognition of laterally-continuous condensed horizons and petrographic markers, and information available from adjacent areas (Brookfield & Westermann 1982; Krishna *et al.* 1982; Krishna 1983; Hallam & Maynard 1987), suggest the following chronostratigraphic scheme (Garzanti 1992): (a) the Spiti Shale extends into the Valanginian (supercycles LZA-4 to lower part of LZB-2; Haq *et al.* 1988); (b) the Lower Takh Formation is largely Hauterivian to Barremian in age (upper part of supercycle LZB-2 to LZB-3); (c) the Upper Takh Formation is chiefly ascribed to the Aptian (supercycle LZB-4); (d) the Labar La Arenite was deposited at earliest Albian times (transgressive tract of cycle UZA-1.1); (e) the volcanic arenites of the lower Pingdon La Fm. are Early Albian in age (cycles UZA-1.1 and 1.2); (f) the middle part of the Pingdon La Fm., characterized by coastal coarse-grained quartzose arenites in proximal sections, is Mid-Albian in age (cycle UZA-1.3); (g) the upper part of the Pingdon La Fm. straddles the boundary between the Mid- and the Late Albian (cycle UZA-1.4); (h) the Nerak Glauco-phosphorite is well-dated as Late Albian (*R. subticinensis* Subzone; cycle UZA-1.5); (i) the Oma Chu Glauco-phosphorite reaches up to the Late Cenomanian (*W. archaeocretacea* Zone; lower and middle parts of supercycle UZA-2).

Accumulation rates

If the inferred age framework is assumed as correct, indicative accumulation rates can be calculated from the data set presented in Table 2, according to the time scales of Haq *et al.* (1988) or Harland *et al.* (1989).

In the Upper Jurassic to lowest Cretaceous Spiti Shale, average accumulation rates were minimal in the Khurna and Shillakong Units (less than 0.5 m Ma^{-1}), between 0.5 and 1 m Ma^{-1} in the Spanboth valley, and between 4 and 10 m Ma^{-1} in central Zanskar and in the Zumlung Unit.

In the Lower Takh Fm. rates seemingly remained uniformly low (between 5 and 7 m Ma^{-1}). In the Zumlung Unit the transition between the Spiti Shale and the Giumal Group is locally sharp, but evidence of a significant time gap is lacking. Progressively increasing terrigenous supply possibly contributed to the overall coarsening-upward trend (Fig. 5a). Metric cyclothems, which are locally bounded by sharp unconformable surfaces, are likely to represent distinct depositional sequences, each spanning about 1 Ma.

In the Upper Takh Fm. rates increased sharply to about 20 m Ma^{-1} according to the Haq *et al.* (1988) time scale, or only slightly to about 10 m Ma^{-1} according to the Harland *et al.* (1989) time scale, which indicates a much longer duration for the Aptian stage. Black shale tongues in this part of the succession were probably deposited during peak global transgressions (at 111 and 108 Ma, according to the Haq *et al.* chart).

If the Labar La Arenite, a glauconitic marker horizon characterized by unique petrographic features and followed by sudden influx of volcaniclastics in all of the studied sections, is indeed isochronous, average accumulation rates of the Pingdon La Formation were double in SW Zanskar (20 to 30 m Ma^{-1}) than in NE Zanskar (10 to 15 m Ma^{-1}).

During deposition of the Nerak Glauco-phosphorite, rates were not much reduced in central and W Zanskar (between 10 and 15 m Ma^{-1} if this horizon represents about 1 Ma), whereas east of Sneatse and Nerak they dropped abruptly to less than 1 m Ma^{-1}, with slightly higher values recorded only at Takh.

During the latest Albian to Cenomanian, average accumulation rates dropped to about 2 m Ma^{-1} for the Oma Chu Glauco-phosphorite of W Zanskar, whereas in central and E Zanskar they were only 0.2 m Ma^{-1} or less.

In the Early Turonian, when sedimentation took place in deeper environments below the 'mudline', rates sharply increased again to over 30 m Ma^{-1}.

The tectono-eustatic signal

The Spiti Shale, which represents the upper part of the Middle Jurassic to lowest Cretaceous supersequence IV (Gaetani & Garzanti 1991), was deposited in middle shelf settings. Deepening in the Late Jurassic is indicated by upward transition from the mainly belemnitiferous *Belemnopsis gerardi* Beds (facies D, G and H) to the ammonoid-bearing shales of the Chidamu Beds (facies H). The black pelites rich in late Early Tithonian ammonoids found at Sneatse (Oloriz & Tintori 1991) may correspond to the global condensed section marked at 137 Ma on the Haq *et al.* curve. In the earliest Cretaceous, a regressive trend is documented by the common occurrence of arenaceous offshore bars (facies D) in the Lochambel Beds of the Zumlung Unit.

The Giumal Group represents the lower part of a long and complex Cretaceous sedimentary cycle (supersequence V of Gaetani & Garzanti 1991). Nine flooding events, marked by offshore shales (facies G to H) or by arenites enriched in non-carbonate intrabasinal grains (facies E, or I to K), are well documented: (1) basal part of the Lower Takh Fm.; (2) pelitic tongue in the basal part of the Upper Takh Fm.; (3) pelitic tongue in the upper part of the Upper Takh Fm.; (4) Labar La Arenite; (5) basal part of the lower Pingdon La Formation; (6) basal part of the middle Pingdon La Fm.; (7) upper part of the Pingdon La Formation; (8) Nerak Glauco-phosphorite; (9) Oma Chu Glauco-phosphorite.

These flooding episodes were typically associated with reducing conditions at the sea bottom, probably related to stages of expanded oxygen-minimum layer in the Neo-Tethys Ocean. The Nerak and Oma Chu Glauco-phosphorites (Fig. 5d) were deposited during 'oceanic anoxic events' OAE 1c and OAE 2 respectively, whereas the black pelitic tongue in the basal part of the Upper Takh Fm. (Fig. 5b) and the Labar La Arenite (Fig. 5c) might correspond to OAE 1a and OAE 1b (Arthur *et al.* 1990).

Where transgressive intervals contain age-diagnostic fossils, they can be invariably correlated with global events of peak eustatic rise and drowning (Haq *et al.* 1988; Schlager 1991, fig. 16). Some however, and surely major ones, are not purely eustatic in origin (Hubbard 1988; Miall 1991). The flooding event testified by the Labar La Arenite, suddenly followed by volcaniclastic influx, was associated with onset of alkalic volcanism, whereas final drowning of the Tethys Himalayan shelf coincided with the last pulse of magmatism all along the N India continental margin (Premoli Silva *et al.* 1992).

Drowning of the Giumal clastic shelf

At the top of the Giumal Group, during a period of rapid relative sea-level rise and starvation, water depth increased within some million years from less than 100 m to more than 200 m. The shelf break, which previously lay some tens of kilometres to the north as documented by very prolonged hiatuses spanning from the mid-Jurassic to the end of the Early Cretaceous in the Shillakong and Khurna tectonic units (Bassoullet *et al.* 1983; Fuchs 1986; Stutz 1988), was displaced just oceanward of NE Zanskar (Nerak, Takh) during the Late Albian *R. subticinensis* Subzone. In the Cenomanian, continuous transgression led the shelfbreak very close to the proximal W Zanskar area (upper Oma Chu), whereas in central and E Zanskar condensed sedimentation was taking place on the uppermost slope. The position and morphology of the shelfbreak was largely controlled by the previous sea-floor topography, influenced in turn by proximity of entry points of terrigenous detritus and occurrence of coarser sandstones in the Pingdon La Formation (Garzanti *et al.* 1989, p. 240).

Under the action of SE-bound oceanic currents, up to coarse-grained quartz and glaucony grains were transported in large bedforms at water depths of 100 to 150 m. Siliciclastic detritus was largely provided by erosion of the shelf and coastal plain during lowstand stages, as testified by occurrence of abundant soil material (Garzanti *et al.* 1989, p. 237). Sediments were reworked by oceanic currents intruding onto the outer shelf at major transgressive stages, when a few centimetres of pelagic mud could be accumulated on the uppermost slope. During the intervening lowstands, instead, intensified action of oceanic currents was strong enough not only to cause continuous resuspension and prevent deposition of micrite and planktonic tests, but also to scour the sea-floor and rework the phosphatic hardgrounds deposited at times of flooding (Garzanti *et al.* 1989; Quine & Bosence 1991).

Low oxygenation, indicated by occurrence of bacterial mats and pyrite growth, is ascribed to direct influence of the oxygen-minimum layer, expanded during peak transgressions and 'global oceanic events' (OAE 1c and OAE 2; Jenkyns 1980; Arthur *et al.* 1990), when phosphates were deposited under the direct influence of strong upwelling. In central Zanskar (Sneatse, Nerak), extensive hematitization at subsequent stages points to better oxygenation in deeper environments, farther from the narrow zone of coastal upwelling. Very similar condensed phosphorites were deposited on the N margin of Alpine Tethys under the action of west-bound oceanic currents during the global mid-Cretaceous transgression (Föllmi & Ouwehand 1987; Föllmi 1989; Delamette 1988, 1990).

In the Early Turonian, sea-level rose to its maximum (Haq *et al.* 1988) and, due to continuing strong thermo-tectonic subsidence, the shelfbreak was displaced south of the studied area. On the whole Zanskar margin, sediments with planktonic faunas were thus deposited on the upper slope at water depths of a few hundred metres.

Depositional supersequences

The Tethys Himalaya continental terrace, as well as other passive margin successions (Hubbard *et al.* 1985; Boote & Kirk 1989), is formed by superposed long-term transgressive–regressive sequences, displaying cyclic repetition of sedimentary patterns primarily controlled by regional extensional tectonics and global tectono-eustasy (Gaetani & Garzanti 1991; Garzanti & Pagni Frette 1991). These 'Sloss-type' supersequences, generally representing each from 35 to 60 Ma, extend toward the low-frequency part of the spectrum the hierarchy of multifold cyclicity recognized in sedimentary successions (Mitchum & Van Wagoner 1991). They are built by nested sets of higher-frequency 'Milankovitch-type' (5th to 4th-order parasequences or PACs; Anderson & Goodwin 1985) to 'Vail-type' (3rd to 2nd-order sequences and composite sequences; Haq *et al.* 1987) cyclothems, which are stacked in forestepping, aggrading, backstepping, drowning and infilling stratal packages to form lowstand, transgressive and highstand tracts.

The following scheme is obtained by applying this concept to the Himalayan sedimentary succession sandwiched between the top of the Kioto Group (central part of the Middle Jurassic) and the base of the Stumpata Quartzarenite (Cretaceous/Tertiary boundary) (supersequences IV and V of Gaetani & Garzanti 1991).

Supersequence IV. Top of the Kioto Limestone Group: lower Boundary. Laptal Formation: Lowstand Tract. Ferruginous Oolite Formation: Transgressive Tract. Belemnopsis gerardi and Chidamu Beds: deepening Early Highstand Tract. Lochambel Beds: prograding Late Highstand Tract.

Supersequence V. Spiti/Giumal transition: lower Boundary. Lower Takh Fm.: prograding Lowstand Tract (third-order sequences are recognized). Top of the Lower Takh Fm.: Transgressive Surface. Upper Takh Fm.: 'Keep-up' phase as sedimentation kept up with subsidence in the lower part of the Transgressive Tract. Labar La Arenite: major Flooding Horizon. Pingdon La Formation: 'Give up' phase as sedimentation became outpaced by subsidence in the upper part of the Transgressive Tract (higher-order downward shifts of coastal onlap are recognized in the middle part of the unit). Nerak Glauco-phosphorite: major flooding horizon (lower part of the condensed section). Top of the Giumal Group in central and eastern Zanskar: surface of Maximum Starvation (Baum & Vail 1988). Oma Chu Glauco-phosphorite: major Flooding Horizon (upper part of the Condensed Section). Top of the Giumal Group in western Zanskar: top of the Condensed Section. Chikkim and Fatu La Formations: deepening Early Highstand Tract. Kangi La and Marpo Formations: prograding Late Highstand Tract. Base of the Stumpata Quartzarenite: upper Boundary.

Note that supersequence boundaries correspond in time with major tectonic episodes and break-up pulses which led to final disintegration of Gondwanaland: (a) initial opening of the Indian Ocean between India/Madagascar and Africa in the late Mid-Jurassic; (b) opening of the Indian Ocean between India and Australia/Antarctica in the mid-Neocomian; (c) final detachment of India from the Seychelles microcontinent close to the Cretaceous–Tertiary boundary.

Also note that the 'drowning unconformity' developed at the top of the Giumal clastic shelf corresponds to maximum flooding and did by no means involve subaerial exposure. Vertical succession of bedding sequences indicate that sea-level fluctuations in the virtually ice-free Cretaceous were of some to several tens of metres at most, and never strong enough to cause wide exposure of the shelf, with the possible exception of tectonically-enhanced relative sea-level fall occurring in the middle part of the Pingdon La Formation.

Tectonic and magmatic history

The rapid increase in sand-sized quartzo-feldspathic detritus at the base of the Takh Formation points to erosion of the tectonically rejuvenated Indian continental block during final fragmentation of Gondwanaland. About 15 to 25 Ma later (according to the Haq *et al.* and Harland *et al.* time scales respectively), abundant 'trachytic' detritus at the base of the Pingdon La Formation testifies to sudden outpouring of alkalic magmas onto N India. For another 10 Ma, short-lived pulses of volcanism alternated with stages of renewed uplift of continental blocks to the south, until in the Late Albian both volcanic and quartzo-feldspathic detritus was sharply reduced. Intraplate alkalic magmatism was thus confined to the Early–Mid-Albian, when deep-seated faults propagated through the continental crust and tapped magma sources in the upper mantle (Garzanti & Jansa 1990).

This evolution through time compares well with the classic sequence of tectono-sedimentary episodes during rifting, with initial doming followed by erosion, eruption of alkalic lavas at the climax of tectonic extension, strong subsidence, break-up, drowning, and finally gradual building up of sedimentary strata to form an isostatistically balanced continental terrace (Blatt 1982, p. 17). Uplift and erosion followed by volcanism, enhanced and then slower thermal subsidence is predicted also in the case of uprising hot plumes, when rifting above anomalous mantle generates huge volumes of magma by extensive decompression melting of the asthenosphere as it rises passively beneath the stretched and thinned lithosphere (White & McKenzie 1989; Campbell & Griffiths 1990).

Early Cretaceous volcanic activity and sedimentation on the Zanskar margin occurred during final separation of India from the rest of Gondwanaland, which had already begun with the late Mid-Jurassic opening of the Somali and Mozambique Basins (Segoufin & Patriat 1981; Scotese *et al.* 1988) and was completed at the close of the Cretaceous with rifting from the Seychelles Block and eruption of the Deccan Trap basalts (Norton & Sclater 1979; Cox 1989; Jaeger *et al.* 1989; Courtillot 1990).

In W India, Early Cretaceous rifting is recorded in the Cambay Basin (Biswas 1987), and lithospheric doming is indirectly suggested by widespread siliciclastic deposits, from the Lumshiwal Formation of the W Salt Range to Jaisalmer, Kutch and the Narmada Basin (Das Gupta 1975; Krishna 1987; Acharyya & Lahiri 1991; Akhtar & Ahmad 1991). Volcanic detritus has not been reported in these clastic units, even though Lower Cretaceous volcanics are locally found (Krishna 1987), and basaltic rocks occur in SE Africa, Madagascar and in the Mozambique Ridge (DSDP Site 249; Vallier 1974).

During the Early Cretaceous, India also rotated away from Australia/Antarctica, with seafloor spreading beginning in the Valanginian (Powell *et al.* 1988; Scotese *et al.* 1988; Baumgartner *et al.* 1991). All basins of the E India coast are characterized by Hauterivian to Aptian sandstones (Sastri *et al.* 1981; Alam 1989; Acharyya & Lahiri 1991), pointing to rejuvenation of the craton ascribed to lithospheric doming. A sudden burst of flood-basalt magmatism, linked to the activity of the Kerguelen mantle plume (Mahoney *et al.* 1983), took place at around 115 Ma, as recorded in the Rajmahal–Sylhet–Bengal Trap Province of NE India (Baksi *et al.* 1987; possibly extending to the Abor volcanics of the E Lesser Himalaya; Acharyya & Lahiri 1991, p. 16) and in the Naturaliste and Kerguelen Plateaux (Leclaire *et al.* 1987; Shaming & Rotstein 1990; Weis *et al.* 1991). Magmatic activity was characterized by both alkalic and tholeiitic suites, with alkalic volcanism spanning from about 122 to 100 Ma and occurring along a number of discrete volcanic centres established along deep lineament zones, both during and subsequent to the Rajmahal flood-basalt eruptions (Kent 1991).

Alkalic magmatism affected also the N Tethyan margin of the Indian subcontinent (Garzanti & Jansa 1990). Volcanism and enhanced tectonic subsidence both started significantly later in Zanskar than in central Nepal (Kali Gandaki valley, probably along one such old Precambrian? lineaments; Bordet *et al.* 1971; Sakai 1983; Garzanti & Pagni Frette 1991). Westward migration of rifting at a rate of some tens of kilometres per million years is compatible both with modern analogues (Courtillot 1982; Bonatti 1986; Brozena & White 1990) and with westward-migrating continental break-up north of Australia during the Jurassic to Early Cretaceous (Pigram & Panggabean 1984).

All along the Tethys Himalayan Zone, tectonic extension associated with volcanism and volcaniclastic supply lasted until the latest Albian, as determined by both radiometric (Arita *et al.* 1991; Rb/Sr biotite/feldspar isochron age of 97 Ma obtained on the Aulis volcanics of central Nepal) and palaeontological methods (Premoli Silva *et al.* 1992). Next, rapid opening of the Indian Ocean and consequent northward drift of India was accompanied by onset of subduction of Neo-Tethyan oceanic crust underneath Eurasia. Calc-alkalic magmatism and volcaniclastic sedimentation in the Transhimalayan arc-trench system is in fact recorded to have begun in Late Albian times (Garzanti & Van Haver 1988). During the Late Cretaceous, India continued its rapid northward flight between the opening Indian Ocean to the S and the closing Neo-Tethys Ocean to the north, until it collided with Eurasia close to the Palaeocene/Eocene boundary (Garzanti *et al.* 1987).

The Early Cretaceous evolution of the N India continental margin was thus directly influenced by a complex sequence of tectonic and magmatic episodes closely related to final disintegration of Gondwanaland. About 400 Ma after its assembly (Dalziel 1991), this super-continent was finally weakened and disrupted by multiple ‘plume’ and ‘superplume’ events affecting the world oceans and their margins (Sheridan 1987; Kent 1991; Larson 1991; Garzanti 1993).

Conclusions

After transition from the offshore Spiti Shale to sandier inner shelf sediments at the base of the

Giumal Group, the Early Cretaceous succession of the Zanskar Range displays an overall backstepping character. A long-term transgressive trend is indicated by repeated flooding events, increasing upward in intensity, and by progressively more common nektonic and planktonic fauna.

Sedimentary features of the Giumal Group indicate predominating storm influence, with fluviatile and tidal processes documented only in most proximal areas. Since in the Early Cretaceous the N India margin lay at middle latitudes (between 30° and 45° S; Scotese *et al.* 1988), both tropical hurricanes and intense winter storms are likely to have affected sedimentation (Marsaglia & Klein 1983; Duke 1985). In the inner shelf, sand transport was largely normal or oblique to shore. The finer terrigenous fractions were transported offshore by wave and unidirectional currents, where they accumulated in largely undisturbed middle shelf environments. Episodic resuspension occurred during exceptional events, and action of geostrophically-balanced alongshelf-orientated currents was felt particularly at transgressive stages. Grain size decreases gradually seaward from estuarine mouths to the middle–outer shelf (facies associations A to H), but it increases again in the outermost shelf (facies associations I to L), which was kept free of fines by continuous action of strong oceanic currents. Sand rapidly disappears in the uppermost slope (facies associations M to O), where sedimentation could still be affected episodically by exceptional storm, tsunami or internal waves.

During the Late Albian to Cenomanian, drowning of the Giumal clastic shelf occurred through successive episodes of deepening and starvation, caused by a complex interplay of tectonic, eustatic and oceanographic processes (Fig. 7). At times of global rapid eustatic rise and anoxia, condensed glauco-phosphorites were deposited under the action of oceanic currents impinging onto the outer shelf and shelfbreak. During the Early Turonian, at times of maximum transgression, these hardgrounds were rapidly buried under thick pelagic oozes. Once the sea-bottom fell below the mudline, seaward of the high-velocity core of the currents, accumulation rates increased by over a hundred times, and 95% of the pelagic mid-Cretaceous section was deposited in less than 2 Ma (no more than 10% of total elapsed time).

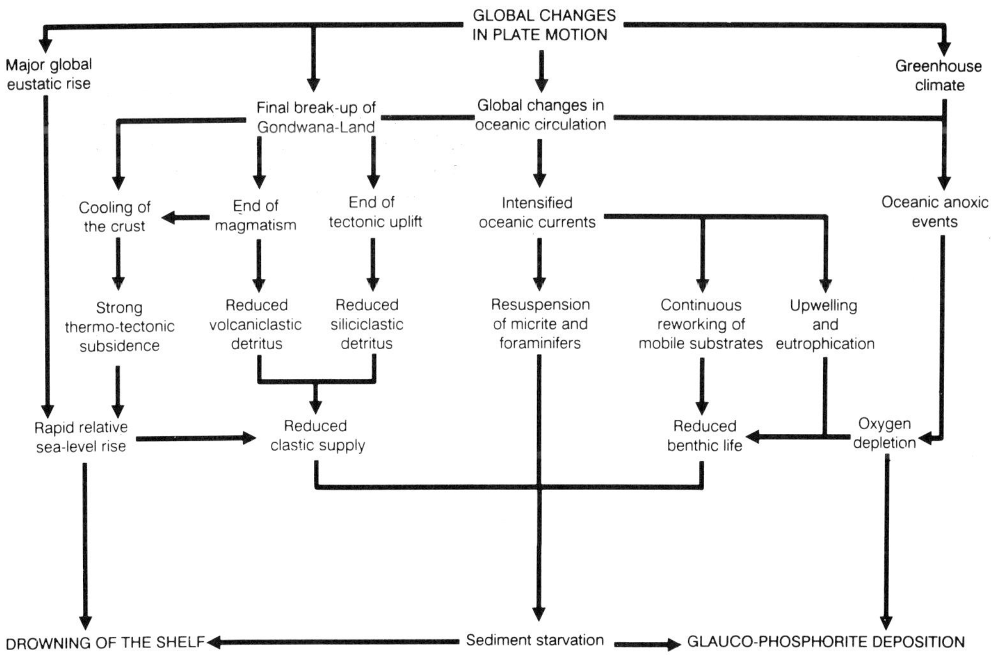

Fig. 7. The complex interplay of factors and processes leading to drowning of the Giumal clastic shelf and glauco-phosphorite deposition in the Late Albian to Cenomanian. Platform drowning at this stage was recorded world-wide (Schlager 1991, fig. 12) and is here ultimately ascribed to global fast plate spreading and increased tectonic activity on the mid-Cretaceous Earth (Sheridan 1987; Larson 1991).

This study began as part of my Doctoral and post-Doctoral Theses under the guidance of M. Gaetani. I. Premoli Silva gave invaluable help in unravelling the Cretaceous biostratigraphy. P. Bruni, D. Castradori and E. Erba gave precious advice at various stages of the work. A. Baud, G. Mascle, R. Casnedi, F. Jadoul, A. Nicora and A. Tintori visited with me many Giumal outcrops during our expeditions to Zanskar. M. Carmine, E. Drigo, L. Ferrario, C. Rosemberg, C. Monesi, E. Zanetello, P. Covelli, N. Chinaglia, A. Arensi, M. Lucchetta, L. Aveta, S. Dossi, D. Sala, A. Doneda, P. Perfumi and G. Borgogna gave great help in point-counting of hundreds of arenite thin sections, which were all produced by C. Malinverno. S. Antico made the drawings. An earlier draft benefitted from reviews by A. Robertson and an anonymous referee.

References

ACHARYYA, S.K. & LAHIRI, T.C. 1991. Cretaceous palaeogeography of the Indian subcontinent; a review. *Cretaceous Research,* **12**, 3–26.

AKHTAR, K. & AHMAD, A.H.M. 1991. Single-cycle cratonic quartzarenites produced by tropical weathering: the Nimar Sandstone (Lower Cretaceous), Narmada Basin, India. *Sedimentary Geology,* **71**, 23–32.

ALAM, M. 1989. Geology and depositional history of Cenozoic sediments of the Bengal Basin of Bangladesh. *Palaeogeography, Palaeoclimatology, Palaeoecology,* **69**, 125–139.

ARITA, K., SAKAI, H. & KOIDE, Y. 1991. Cretaceous alkaline magmatism in the Nepalese Lesser Himalayas. *Géologie Alpine*, Mém. H.S., **16**, 9–10.

ARTHUR, M.A., BRUMSACK, H.-J., JENKYNS, H.C. & SCHLANGER, S.O. 1990. Stratigraphy, geochemistry and paleocéanography of organic carbon-rich Cretaceous sequences. *In*: GINSBURG, R.N. & BEAUDOIN, B. (eds) *Cretaceous resources, events and rhythms.* Kluwer Academic Publishers, 75–119.

BASKI, A.K., BARMAN, T.R., PAUL, D.K. & FARRAR, E. 1987. Widespread Early Cretaceous flood basalt volcanism in eastern India: geochemical data from the Rajmahal-Bengal-Sylhet Traps. *Chemical Geology,* **63**, 133–141.

BARRON, E.J. & PETERSON, W.H. 1990. Mid-Cretaceous ocean circulation: results from model sensitivity studies. *Paleoceanography,* **5/3**, 319–337.

BASSOULLET, J.-P., COLCHEN, M., GILBERT, E., MARCOUX, J., MASCLE, G., SUTRE, E. & VAN HAVER, T. 1984. L'orogène himalayen au Crétacé. *Mémoires Societé Géologique de France*, N.S., **147**, 9–20.

——, ——, JUTEAU, T., MARCOUX, J., MASCLE, G. & REIBEL, G. 1983. Geological studies in the Indus suture zone of Ladakh (Himalayas). *In*: GUPTA, V.J. (ed.) *Stratigraphy and structure of Kashmir and Ladakh Himalayas.* Contribution to Himalayan geology, **2**, Hindustani Publishing Company, 96–124.

BAUD, A., ARN, R., BUGNON, P., CRISINEL, A., DOLIVO, E., ESCHER, A., HAMMERSCHLAG, J.-G., MARTHALER, M., MASSON, H., STECK, A. & TIECKE, J.-C. 1982. Le contact Gondwana – péri-Gondwana dans le Zanskar oriental (Ladakh Himalaya). *Bulletin Societé Géologique de France,* s.7, **24**, 341–361.

——, GAETANI, M., GARZANTI, E., FOIS, E., NICORA, A. & TINTORI, A. 1984. Geological observations in southeastern Zanskar and adjacent Lahul area (northwestern Himalaya). *Eclogae Geologicae Helvetiae,* **77**, 171–197.

BAUM, G.R. & VAIL, P.R. 1988. Sequence stratigraphic concepts applied to Paleogene outcrops, Gulf and Atlantic Basins. *SEPM Special Publication,* **42**, 309–327.

BAUMGARTNER, P.O., BOWN, P., MARCOUX, J., MUTTERLOSE, J., KAMINSKI, M., HAIG, D. & MCMINN, A. 1991. Early Cretaceous biogeographic and oceanographic synthesis, of Leg 123 (off NW-Australia). *In*: GRADSTEIN, F.M. & LUDDEN, J.N. (eds) *Proceedings of the ODP, Scientific Results*, **123**, in press.

BAYER, U., ALTHEIMER, E. & DEUTSCHLE, W. 1984. Environmental evolution in shallow epicontinental seas: sedimentary cycles and bed formation. *In*: BAYER, U. & SEILACHER, A. (eds) *Sedimentary and evolutionary cycle.* Lecture Notes in Earth Sciences. Springer Verlag, 347–381.

BISWAS, S.K. 1987. Regional tectonic framework, structure and evolution of the western marginal basins of India. *Tectonophysics,* **135**, 307–327.

BLATT, H. 1982. *Sedimentary petrology.* Freeman & Co., San Francisco.

BONATTI, E. 1986. Variabilità nelle zone di accrescimentio di litosfera oceanica: rift "lenti" e "veloci", rift che si propagano, rift che si accavallano, rift che si esauriscono. *Memorie della Società Geologica Italiana,* **27** (1984), 23–34.

BOOTE, D.R.D. & KIRK, R.B. 1989. Depositional wedge cycles on evolving plate margin, Western and Northwestern Australia. *American Association of Petroleum Geologists Bulletin,* **73**, 216–243.

BORDET, P., COLCHEN, M., KRUMMENACHER, D., LE FORT, P., MOUTERDE, R. & REMY, M. 1971. Recherches géologiques dans l'Himalaya du Nepal, region de la Thakkhola. *Editions du CNRS.*

BROOKFIELD, M.E. & WESTERMANN, G.E.G. 1982. Mesozoic ammonites from the Spong Valley, Zanskar, N.W. India. *Journal of the Geological Society of India,* **23**, 263–266.

BROZENA, J.M. & WHITE, R.S. 1990. Ridge jumps and propagations in the South Atlantic Ocean. *Nature,* **348**, 149–152.

CAMPBELL, C.V. 1971. Depositional model – Upper Cretaceous Gallup beach shoreline, Ship Rock area, northwestern New Mexico. *Journal Sedimentary Petrology,* **41**, 395–409.

CAMPBELL, I.H. & GRIFFITHS, R.W. 1990. Implications of mantle plume structure for the evolution of flood basalts. *Earth and Planetary Science Letters,* **99**, 79–93.

CLOETINGH, S. 1988. Intraplate stresses: a tectonic

cause for third-order cycles in apparent sea-level? *SEPM Special Publication*, **42**, 19–29.

COURTILLOT, V. 1982. Propagating rifts and continental breakup. *Tectonics*, **1**, 239–250.

—— 1990. Deccan volcanism at the Cretaceous-Tertiary boundary: past climatic crises as a key to the future? *Paleogeography, palaeoclimatology, Palaeoceanography*, **189**, 291–299.

COX, K.G. 1989. The role of mantle plumes in the development of continental drainage patterns. *Nature*, **342**, 873–877.

CRUICKSHANK, M.J. & ROWLAND, T.J. 1983. Mineral deposits at the shelfbreak. *SEPM Special Publication*, **33**, 429–436.

DALZIEL, I.W.D. 1991. Pacific margins of Laurentia and East Antarctica-Australia as a conjugate rift pair: evidence and implications for an Eocambrian supercontinent. *Geology*, **19**, 598–601.

DAS GUPTA, S.K. 1975. A revision of the Mesozoic–Tertiary stratigraphy of the Jaisalmer Basin, Rajasthan. *Indian Journal of Earth Sciences*, **2**, 77–94.

DELAMETTE, M. 1988. Relation between the condensed Albian deposits of the Helvetic domain and the oceanic current-influenced continental margin of the northern Tethys. *Bulletin Societé Géologique de France*, **8/4**, 739–745.

—— 1990. Aptian, Albian and Cenomanian microbialites from the condensed phosphatic deposits of the Helvetic shelf, Western Alps. *Eclogae Geologicae Helvetiae*, **83**, 99–121.

DUKE, W.L. 1985. Hummocky cross-stratification, tropical hurricanes, and intense winter storms. *Sedimentology*, **32**, 167–194.

——, ARNOTT, R.W.C. & CHEEL, R.J. 1991. Shelf sandstones and hummocky cross-stratification: new insights on a stormy debate. *Geology*, **19**, 625–628.

FISCHER, A.G. & ARTHUR, M.A. 1977. Secular variations in the pelagic realm. *SEPM Special Publication*, **25**, 19–50.

FLEMMING, B.W. 1978. Underwater sand dunes along the Southeast African continental margin – observations and implications. *Marine Geology*, **26**, 177–198.

—— 1980. Sand transport and bedform patterns on the continental shelf between Durban and Port Elisabeth (southeast African continental margin). *Sedimentary Geology*, **26**, 179–205.

FLÜGEL, E. 1982. *Microfacies analysis of limestones*. Springer Verlag, Berlin.

FÖLLMI, B.K. 1989. Mid-Cretaceous platform drowning, current-induced condensation and phosphogenesis, and pelagic sedimentation along the eastern Helvetic shelf (Northern Tethys margin). *In*: WIEDMANN, J. (ed.) *Cretaceous of the Western Tethys*. Proceedings 3rd International Cretaceous Symposium, Tübingen 1987. E. Schweizerbart-'sche Verlagsbuchanglung, Stuttgart, 585–606.

—— & OUWEHAND, P.J. 1987. Garschella-Formation und Götzis-Schichten (Aptian-Coniacian): Neue stratigraphische Daten aus dem Helvetikum der Ostschweiz und des Vorarlbergs. *Eclogae Geologicae Helvetiae*, **80**, 141–191.

FREY, R.W. & PINET, P.R. 1978. Calcium-carbonate content of surficial sands seaward of Altamaha and Doboy Sounds, Georgia. *Journal Sedimentary Petrology*, **48**, 1249–1256.

FUCHS, G. 1982. The geology of western Zanskar. *Jahrbuch Geologisches Bundesanstalt*, **125**, 1–50.

—— 1984. Note on the geology of the Markha – Nimaling area in Ladakh (India). *Jahrbuch Geologisches Bundesanstalt*, **127**, 5–12.

—— 1986. The geology of the Markha–Khurnak region in Ladakh (India). *Jahrbuch Geologisches Bundesanstalt*, **128**, 403–437.

—— 1987. The geology of southern Zanskar (Ladakh). Evidence for the autochthony of the Tethys Zone of the Himalaya. *Jahrbuch Geologisches Bundesanstalt*, **130**, 465–491.

—— & WILLEMS, H. 1990. The final stages of sedimentation in the Tethyan Zone of Zanskar and their geodynamic significance (Ladakh-Himalaya). *Jahrbuch Geologisches Bundesanstalt*, **133**, 259–273.

FULTHORPE, C.S. 1991. Geological controls on seismic sequence resolution. *Geology*, **19**, 61–65.

GAETANI, M., CASNEDI, R., FOIS, E., GARZANTI, E., JADOUL, F., NICORA, A. & TINTORI, A. 1986. Stratigraphy of the Tethys Himalaya in Zanskar, Ladakh. *Rivista Italiana di Paleontologia e Stratigrafia*, **91**, 443–478.

—— & GARZANTI, E. 1991. Multicyclic history of the northern India continental margin (Northwestern Himalaya). *American Association of Petroleum Geologists Bulletin*, **75**, 1427–1446.

GAGAN, M.K., JOHNSON, D.P. & CARTER, R.M. 1988. The cyclone Winifred storm bed, central Great Barrier Reef shelf, Australia. *Journal Sedimentary Petrology*, **58**, 845–856.

——, CHIVAS, A.R. & HERCZEG, A.L. 1990. Shelf-wide erosion, deposition, and suspended sediment transport during cyclone Winifred, central Great Barrier Reef, Australia. *Journal Sedimentary Petrology*, **60**, 456–470.

GARZANTI, E. 1986. Storia sedimentaria del margine continentale settentrionale della placca indiana (Tethys Himalaya, Ladakh, India). *Tesi di Dottorato inedita*, Univ. di Milano.

—— 1991. Non-carbonate intrabasinal grains in arenites: their recognition, significance and relationship to eustatic cycles and tectonic setting. *Journal Sedimentary Petrology*, **61**, 959–975.

—— 1992. Stratigraphy of the Early Cretaceous Giumal Group (Zanskar Range, Northern India). *Rivista Italiana di Paleontologia e Stratigrafia*, **97**, 485–510.

—— 1993. Himalayan ironstones, "Superplumes" and the breakup of Gondwana. *Geology*, **21**, 105–108.

—— & JANSA, L.F. 1990. Geodynamic significance of Early Cretaceous volcaniclastic sandstones from the northern passive margin of the Indian Plate. *V Himalaya – Tibet – Karakorum Workshop, Milano* 19.

—— & PAGNI FRETTE, M. 1991. The stratigraphic succession of the Thakkhola region (central Nepal) – comparisons with the northwestern

Tethys Himalaya. *Rivista Italiana di Paleontologia e Stratigrafia*, **96**, 3–26.

—— & VAN HAVER, T. 1988. The Indus clastics: forearc basin sedimentation in the Ladakh Himalaya (India). *Sedimentary Geology*, **59**, 237–249.

——, BAUD, A. & MASCLE, G. 1987. Sedimentary record of the northward flight of India and its collision with Eurasia (Ladakh Himalaya, India). *Geodinamica Acta*, **1**, 297–312.

——, HAAS, R. & JADOUL, F. 1989. Ironstones in the Mesozoic passive margin sequence of the Tethys Himalaya (Zanskar, Northern India): sedimentology and metamorphism. *In*: YOUNG, T.P. & TAYLOR, W.E.G. (eds) *Phanerozoic Ironstones* Geological Society, London, Special Publications, **46**, 229–244.

GOODWIN, P.W. & ANDERSON, E.J. 1985. Punctuated aggradational cycles: a general hypothesis of episodic stratigraphic accumulation. *Journal of Geology*, **93**, 515–533.

HALLAM, A. & MAYNARD, J.B. 1987. The iron ores and associated sediments of the Chichali formation (Oxfordian to Valanginian) of the Trans-Indus Salt Range, Pakistan. *Journal of the Geological Society, London*, **144**, 107–114.

HANDFORD, C.R. 1986. Facies and bedding sequences in shelf-storm-deposited carbonates – Fayetteville Shale and Pitkin Limestone (Mississippian), Arkansas. *Journal Sedimentary Petrology*, **56**, 123–137.

HAQ, B.U., HARDENBOL, J. & VAIL, P.R. 1987. The chronology of fluctuating sea level since the Triassic. *Science*, **235**, 1156–1167.

——, —— & —— 1988. Mesozoic and Cenozoic chronostratigraphy and cycles of sea-level change. *In*: WILGUS C.K. *et al.* (eds) Sea-level changes – an integrated approach, *SEPM Special Publication*, 71–108.

HARLAND, W.B., ARMSTRONG, R.L., COX, A.V., CRAIG, L.E., SMITH, A.G. & SMITH, D.G. 1989. *A geologic time scale*. Cambridge University Press, Cambridge.

HOLLISTER, C.D. & HEZEEN, B.C. 1972. Geologic effects of ocean bottom currents. *In*: GORDON, A.L. (ed.) *Studies in physical oceanography – a tribute to G. Wust on his 80th birthday*, Gordon & Breach, New York, 37–66.

HUBBARD, R.J. 1988. Age and significance of sequence boundaries on Jurassic and Early Cretaceous rifted continental margins. *American Association of Petroleum Geologists Bulletin*, **72**, 49–72.

——, PAPE, J. & ROBERTS, D.G. 1985. Depositional sequence mapping to illustrate the evolution of a passive continental margin. *In*: BERG, O.R. & WOOLVERTON, D.G. (eds) *Seismic stratigraphy II – An integrated approach*. American Association of Petroleum Geologists Memoirs, **39**, 93–115.

JADOUL, F., FOIS, E., TINTORI, A. & GARZANTI, E. 1985. Preliminary results on Jurassic stratigraphy in Zanskar (NW Himalaya). *Rendiconti della Società Geologica Italiana*, **8**, 9–13.

——, GARZANTI, E. & FOIS, E. 1990. Upper Triassic – Lower Jurassic stratigraphy and palaeogeographic evolution of the Zanskar Tethys Himalaya (Zangla Unit). *Rivista Italiana di Paleontologia e Stratigrafia*, **95**, 351–396.

JAEGER, J.-J., COURTILLOT, V. & TAPPONNIER, P. 1989. Paleontological view of the ages of the Deccan Traps, the Cretaceous/Tertiary boundary, and the India/Asia collision. *Geology*, **17**, 316–319.

JANSA, L.F. 1991. Processes affecting paleogeography, with examples from the Tethys. *Palaeogeography, Palaeoclimatology, Palaeoecology*, **87**, 345–371.

JENKYNS, H.C. 1980. Cretaceous anoxic events: from continents to oceans. *Journal of the Geological Society, London*, **137**, 171–188.

KANWAR, S.S. & BHANDARI, A.K. 1974. Stratigraphy, structure and sedimentation of part of Lahaul and Spiti District, Himachal Pradesh. *Miscellaneous Publications Geological Survey India*, **41**, 169–178.

KELEMEN, P.B. & SONNENFELD, M.D. 1983. Stratigraphy, structure, petrology and local tectonics, central Ladakh, NW Himalaya. *Schweizerische Mineralogische Petrographische Mitteilungen*, **63**, 267–287.

KENT, R. 1991. Lithospheric uplift in eastern Gondwana: evidence for a long-lived mantle plume system? *Geology*, **19**, 19–23.

KOUTSOUKOS, E.A.M., MELLO, M.R., DE AZAMBUJA FILHO, N.C., HART, M.B. & MAXWELL, J.R. 1991. The Upper Aptian-Albian succession of the Sergipe Basin, Brazil: an integrated paleoenvironmental assessment. *American Association of Petroleum Geologists Bulletin*, **75**, 479–498.

KRISHNA, J. 1983. Reappraisal of the marine and/or "mixed" Lower Cretaceous sedimentary sequences of India: palaeogeography and time boundaries. *In: Cretaceous of India*. Indian Association Palynostratigraphy, 94–119.

—— 1987. An overview of the Mesozoic stratigraphy of Kachchh and Jaisalmer Basins. *Journal Palaeontological Society India*, **32**, 136–149.

——, KUMAR, S. & SINGH, I.B. 1982. Ammonoid stratigraphy of the Spiti Shale (Upper Jurassic), Tethys Himalaya, India. *Neues Jahrbuch Geologische Paläontologische Monatshefte*, 580–592.

LARSON, R.L. 1991. Latest pulse of Earth: evidence for a mid-Cretaceous superplume. *Geology*, **19**, 547–550.

LECKIE, D.A. & KRYSTINIK, L.F. 1989. Is there evidence for geostrophic currents preserved in the sedimentary record of inner to middle-shelf deposits? *Journal Sedimentary Petrology*, **59**, 862–870.

LECKIE, R.M. 1987. Paleoecology of mid-Cretaceous planktonic foraminifera: a comparison of open ocean and epicontinental sea assemblages. *Micropaleontology*, **33**, 164–176.

LECLAIRE, L., BASSIAS, Y., DENIS-CLOCCHIATTI, M., DAVIES, H., GAUTIER, I., GENSOUS, B., GIANNESINI, P.-J., PATRIAT, P., SEGOUFIN, J., TESSON, M. & WANNESSON, J. 1987. Lower Cretaceous basalt and sediments from the Kerguelen Plateau. *Geo-marine Letters*, **7**, 169–176.

MAHONEY, J.J., MACDOUGALL, J.D. & LUGMAIR, G.W. 1983. Kerguelen hotspot source for

Rajmahal Traps and Ninetyeast Ridge? *Nature,* **303**, 385–389.

MARSAGLIA, K.M. & KLEIN DE VRIES, G. 1983. The palaeogeography of Paleozoic and Mesozoic storm depositional systems. *Journal of Geology,* **91**, 117–142.

MASSARI, F. 1988. Le facies paraliche: confronto tra depositi costieri e fluviali. *In*: ORI, G.G. & CREMONA, M. (eds) *I depositi continentali.* Giornale di Geologia, 147–161.

MATHUR, N.S. 1990. Tethyan Cretaceous sediments in the Northwest Himalaya. *Cretaceous Research,* **11**, 289–305.

MAXWELL, W.G.H. & SWINCHATT, J.P. 1970. Great Barrier reef: regional variation in a terrigenous-carbonate province. *Geological Society of America Bulletin,* **81**, 691–724.

MIALL, A.D. 1991. Stratigraphic sequences and their chronostratigraphic correlation. *Journal Sedimentary Petrology,* **61**, 497–505.

MITCHUM, R.M. & VAN WAGONER, J.C. 1991. High-frequency sequences and their stacking patterns: sequence-stratigraphic evidence of high-frequency eustatic cycles. *Sedimentary Geology,* **70**, 131–160.

NICHOLAS, M.M., JOHNSON, G.H. & PEEBLES, P.C. 1991. Modern sediments and facies model for a microtidal coastal plain estuary, the James estuary, Virginia. *Journal Sedimentary Petrology,* **61**, 883–899.

NICORA, A., GARZANTI, E. & FOIS, E. 1987. Evolution of the Tethys Himalaya continental shelf during Maastrichtian to Paleocene (Zanskar, India). *Rivista Italiana di Paleontologia e Stratigrafia,* **92**, 439–496.

NIEDORODA, A.W., SWIFT, D.J.P., FIGUEREIDO, A.G. & FREELAND, G.L. 1985. Barrier island evolution, middle Atlantic shelf, U.S.A. Part II: evidence from the shelf floor. *Marine Geology,* **63**, 363–396.

NORTON, I.O. & SCLATER, J.G. 1979. A model for the evolution of the Indian Ocean and the breakup of Gondwanaland. *Journal of Geophysical Research,* **84**, 6803–6830.

OLORIZ, F. & TINTORI, A. 1991. Upper Jurassic (Tithonian) ammonites from the Spiti shales in western Zanskar (NW Himalayas). *Rivista Italiana di Paleontologia e Stratigrafia,* **96**, 461–486.

PIGRAM, C.J. & PANGGABEAN, H. 1984. Rifting of the northern margin of the Australian continent and the origin of some microcontinents in eastern Indonesia. *Tectonophysics,* **107**, 331–353.

PLINT, A.G. 1988. Sharp-based shoreface sequences and "offshore bars" in the Cardium Formation of Alberta: their relationship to relative changes in sea-level. *SEPM Special Publication,* **42**, 357–370.

POWELL, C.MCA., ROOTS, S.R. & VEEVERS, J.J. 1988. Pre-breakup continental extension in East Gondwanaland and the early opening of the eastern Indian Ocean. *Tectonophysics,* **155**, 261–283.

PREMOLI SILVA, I., GARZANTI, E & GAETANI, M. 1992. Stratigraphy of the Chikkim and Fatu La Formations in the Zangla and Zumlung Units (Zanskar Range, India), with comparisons to the Thakkhola region (central Nepal): mid-Cretaceous evolution of the Indian passive margin. *Rivista Italiana di Paleontologia e Stratigrafia,* **97**, 511–564.

QUINE, M. & BOSENCE, D. 1991. Stratal geometries, facies and sea-floor erosion in Upper Cretaceous Chalk, Normandy, France. *Sedimentology,* **38**, 1113–1152.

RAINA, V.K. & BHATTACHARYYA, D.P. 1977. The geology of a part of the Charap and Sarchu Valleys, Lahaul and Spiti District, Himachal Pradesh. *Golden Jubilee Volume*, Geological Mineralogical and Metallurgical Society of India, 129–142.

RAMLI, N. 1986. Depositional model of a Miocene barred wave- and storm-dominated shoreface and shelf, southeastern Malay Basin, offshore Western Malaysia. *American Association of Petroleum Geologists Bulletin,* **70**, 34–47.

SAKAI, H. 1983. Geology of the Tansen Group of the Lesser Himalaya in Nepal. *Memoirs Faculty of Science Kyushu University,* s. D. Geol., **25**, 27–74.

SASTRI, V.V., VENKATACHALA, B.S. & NARAYANAN, V. 1981. The evolution of the east coast of India. *Palaeogeography, Palaeoclimatology, Palaeoecology,* **36**, 23–54.

SCHLAGER, W. 1991. Depositional bias and environmental change – important factors in sequence stratigraphy. *Sedimentary Geology,* **70**, 109–130.

SCHLANGER, S.O. & ARTHUR, M.A. 1990. Relevance of Cretaceous volcanic activity to CRER. *In*: GINSBURG, R.N. & BEAUDOIN, B. (eds) *Cretaceous resources, events and rhythms.* Kluwer Academic Publisher, 305–308.

SCOTESE, C.R., GAHAGAN, L.M. & LARSON, R.L. 1988. Plate tectonic reconstructions of the Cretaceous and Cenozoic ocean basins. *Tectonophysics,* **155**, 27–48.

SEARLE, M.P. 1986. Structural evolution and sequence of thrusting in the High Himalayan, Tibetan-Tethys and Indus suture zones of Zanskar and Ladakh, western Himalaya. *Journal of Structural Geology*, **8**, 923–936.

SEGOUFIN, J. & PATRIAT, P. 1981. Anomalies magnétiques mésozoiques dans l'Océan Indien occidental. Paléoposition de Madagascar. *Bulletin Societé Géologique de France,* **23/6**, 603–607.

SHAMING, M. & ROTSTEIN, Y. 1990. Basement reflectors in the Kerguelen Plateau, south Indian Ocean: indications for the structure and early history of the plateau. *Geological Society of America Bulletin,* **102**, 580–592.

SHERIDAN, R.E. 1987. Pulsation tectonics as the control of long-term stratigraphic cycles. *Paleoceanography,* **2**, 97–118.

SNEDDEN J.W. & SWIFT, D.J.P. 1991. Is there evidence for geostrophic currents preserved in the sedimentary record of inner to middle-shelf deposits? – Discussion. *Journal Sedimentary Petrology,* **61**, 148–151.

SPECHT, R.W. & BREMNER, R.L. 1979. Storm-wave genesis of bioclastic carbonates in Upper Jurassic epicontinental mudstones, east-central Wyoming. *Journal Sedimentary Petrology,* **49**, 1307–1322.

SRIKANTIA, S.V., GANESAN, T.M., RAO, P.N., SINHA, P.K. & TIRKEY, B. 1980. Geology of Zanskar area, Ladakh Himalaya. *Himalayan Geology,* **8**, 1009–1033.

STANLEY, D.J., ADDY, S.K. & BEHRENS, E.W. 1983. The mudline: variability of its position relative to shelfbreak. *SEPM Special Publication,* **33**, 279–298.

—— & WEAR, C.M. 1978. The "mud-line": an erosion-deposition boundary on the upper continental slope. *Marine Geology,* **28**, M19–M29.

STUTZ, E. 1988. *Géologie de la châine de Nyimaling aux confins du Ladakh et du Rupschu (NW-Himalaya, Inde) – évolution paléogéographique et tectonique d'un segment de la marge nord-indienne.* Mémoires de Géologie, Lausanne.

SWIFT, D.J.P., STANLEY, D.J. & CURRAY, J.R. 1971. Relict sediments on continental shelves: a reconsideration. *Journal of Geology,* **79**, 322–346.

VALLIER, T.L. 1974. Volcanogenic sediments and their relation to landmass volcanism and sea floor-continent movements, western Indian Ocean, Leg 25, DSDP. *Initial Reports of the DSDP,* **25**, 515–542.

WEIS, D., FREY, F.A., SAUNDERS, A., GIBSON, I. & Leg 121 Scientific Shipboard Party 1991. Ninetyeast Ridge (Indian Ocean): a 5000 km record of a Dupal mantle plume. *Geology,* **19**, 99–102.

WHITE, R. & MCKENZIE, D. 1989. Magmatism at rift zones: the generation of volcanic continental margins and flood basalts. *Journal Geophysical Research,* **94**, 7685–7729.

WILLIAMS, L.A. 1984. Subtidal stromatolites in Monterey Formation and other organic-rich rocks as suggested source contributors to petroleum formation. *American Association of Petroleum Geologists Bulletin,* **68**, 1879–1893.

—— & REIMERS, C. 1983. Role of bacterial mats in oxygen-deficient marine basins and coastal upwelling regimes: preliminary report. *Geology,* **11**, 267–269.

WINN, R.D., STONECIPHER, S.A. & BISHOP, M.G. 1984. Sorting and wave abrasion: controls on composition and diagenesis in Lower Frontier Sandstones, Southwestern Wyoming. *American Association of Petroleum Geologists Bulletin,* **68**, 268–284.

Sedimentology and tectonic implications of the Lamayuru Complex: deep-water facies of the Indian passive margin, Indus Suture Zone, Ladakh Himalaya

ALASTAIR H.F. ROBERTSON & PAUL J. DEGNAN

Department of Geology and Geophysics, Grant Institute, West Mains Road, Edinburgh EH9 3JW, UK

Abstract: The Lamayuru Complex, exposed in the Indus Suture Zone of the Ladakh Himalaya, comprises remnants of the Triassic to Upper Cretaceous, north-facing Indian passive margin. This margin experienced pulses of extension and/or transtension in the Late Permian, early Mid-Jurassic, Late Jurassic and Early Cretaceous. Two areas studied preserve mainly relatively intact slope, to base-of-slope settings.

Initial sea-floor spreading in the Late Permian is reflected in early deposition of base-of-slope turbidites in the Lamayuru Complex, and exotic limestones and volcanics in associated melange units. Quartzose turbidites, calciturbidites and shales then accumulated in slope, to base-of-slope settings during the Triassic and Early Jurassic, related to subsidence of the adjacent Zanskar shelf after spreading began. The carbonate platform edge underwent drastic extensional collapse in the early Mid-Jurassic (Aalenian), giving rise to limestone 'exotics' in the west and carbonate-filled submarine channels in the east. Extension related alkaline volcanism, dated as Mid–Late Jurassic by radiolarians, exploited the inferred continent-boundary area in the west. Mixed quartzose clastic, volcaniclastic, radiolarian and pelagic carbonates accumulated in the Cretaceous, but remain poorly dated. Radiolarian cherts reflect siliceous plankton productivity along the passive margin. During the latest Cretaceous, ophiolitic melange was accreted at a trench within the Tethys ocean to the north, but there is, as yet, no definite evidence of emplacement onto the Zanskar shelf until formation of an early Eocene foredeep, followed by emplacement of the Spontang ophiolite and related melange.

The purpose of this paper is to discuss and interpret the Mesozoic deep-water sedimentary rocks of the Indus Suture Zone in the Ladakh Himalayas, in the light of information on the Zanskar shelf succession to the south and comparisons with other Tethyan margins (Figs 1 & 2). Successions in the Indus Suture Zone are pervasively sheared, folded and thrust-bounded and are known as the Lamayuru Complex.

Earlier descriptions of 'Indus Flysch' were summarized by Frank *et al.* (1977) and Thakur (1981). Bassoullet *et al.* (1981) focused on the age, palaeontology and petrography of Triassic–Jurassic limestones and associated sediments in Western Ladakh. These authors established that thick limestone units were interbedded with 'flysch', rather than being 'exotics' related to tectonic emplacement, as suggested by Frank *et al.* (1977). They also interpreted local units of Late Triassic 'platform limestone' ('Mulbeck exotics') as Jurassic sedimentary slide sheets. Brookfield & Andrews-Speed (1984) envisaged a deep-water passive margin setting for the Lamayuru, similar to other Mesozoic Tethyan passive margins, a conclusion that is supported here. Descriptions of the Lamayuru facies in central Ladakh were also given by Fuchs (1979), Bassoullet *et al.* (1983), Cannat & Mascle (1990) and A.K. Sinha (pers. comm.). The petrology of Mesozoic volcanic rocks within the Lamayuru Complex in Western Ladakh has been documented in detail (Honegger *et al.* 1982, 1989; Honegger 1983). Sutre (1991) and Reuber (*in* Sutre 1991) mapped these volcanics and associated deep-water sedimentary rocks ('Sapi-Karamba unit') and interpreted them in terms of magmatic activity along the northern passive margin of the Indian plate in Mid–Late Jurassic time. Lamayuru-type sedimentary rocks have also been described from eastern Ladakh ('Lamayuru Unit', or 'Markha Unit' by Baud *et al.* (1982), Fuchs (1986), Talon (1988) and Stutz (1988). Bassoullet *et al.* (1978, 1980*b*) and Searle *et al.* (1988) also report Lamayuru-type sediments beneath the Spontang ophiolite in the Zanskar Mountains, south of the Indus Suture Zone.

Recently, information from both the Indus Suture Zone and from melanges beneath the Spontang ophiolite (Colchen *et al.* 1987; Reuber *et al.* 1987, 1992) has been used to reconstruct the evolution of the Indian passive margin from

From Treloar, P.J. & Searle, M.P. (eds), *Himalayan Tectonics*
Geological Society Special Publication, No. 74, pp. 299–321.

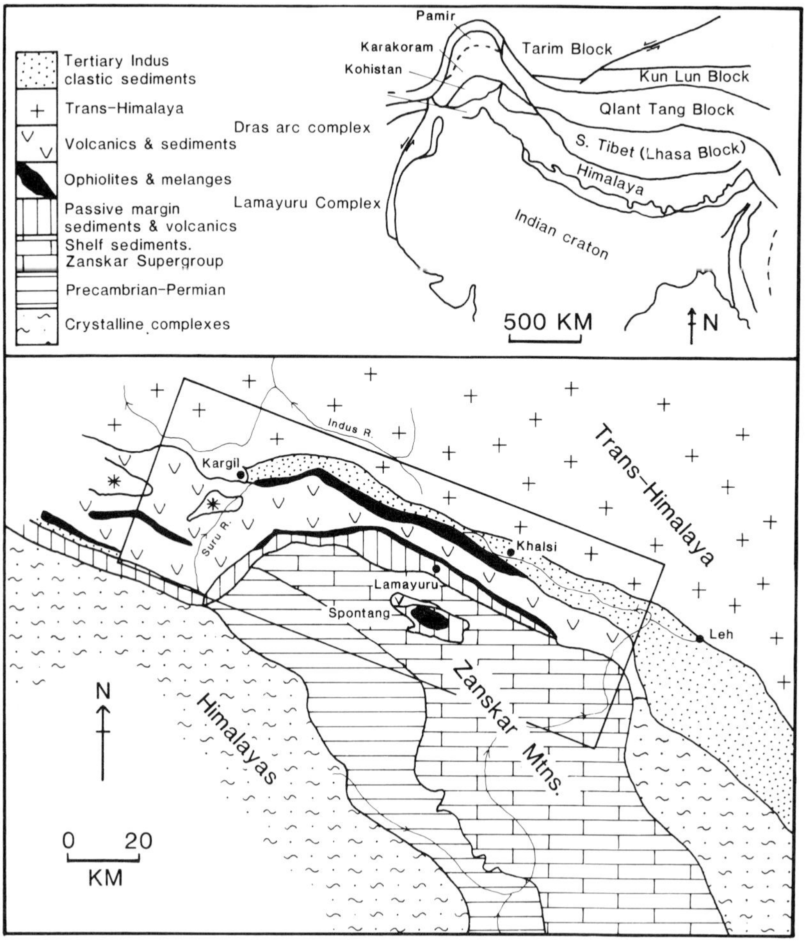

Fig. 1. Geological setting of the Indus Suture Zone in Ladakh. (Study area shown in box below).

the Permian to early Tertiary (Cannat & Mascle 1990; Reuber *et al.* 1992). However, a number of important aspects still remain unclear, or controversial: (i) to what extent was the Lamayuru Complex emplaced as an essentially intact stratigraphic succession, prior to Late Tertiary backthrusting (Sutre 1991), as opposed to having been emplaced as an imbricated stack of thrust sheets (Searle *et al.* 1988)? (ii) Is the Lamayuru Complex as a whole, entirely allochthonous, thrust over the Zanskar shelf from far to the north (Baud *et al.* 1982, 1984; Baud 1989), or is it relatively autochthonous, interfingering with the Zanskar shelf successions to the south, as argued by Fuchs (1982, 1986, 1989) for areas east of the Zanskar River? (iii) By what depositional processes and in what palaeoenvironments did the Lamayuru Complex sediments accumulate? (iv) Were the associated volcanics at least in part, erupted within the terrigenous Lamayuru sediments (Sutre 1991, Robertson & Degnan 1992), or are they slices of Tethyan ocean crust, unrelated to the Lamayuru continental margin successions (Honegger *et al.* 1982, 1989; Fuchs & Sinha 1992)? (v) Was the Lamayuru Complex first deformed and emplaced in the Late Cretaceous (Frank *et al.* 1977; Searle 1983, 1986; Brookfield & Andrews-Speed 1984; Searle *et al.* 1988), or not until the Eocene (e.g. Kelemen & Sonnenfeld 1983; Garzanti *et al.* 1987; Sutre 1991; Reuber *et al.* 1992)?

Structural setting

In Western Ladakh, west of the Zanskar River (Fig. 2), the Lamayuru Complex is structurally

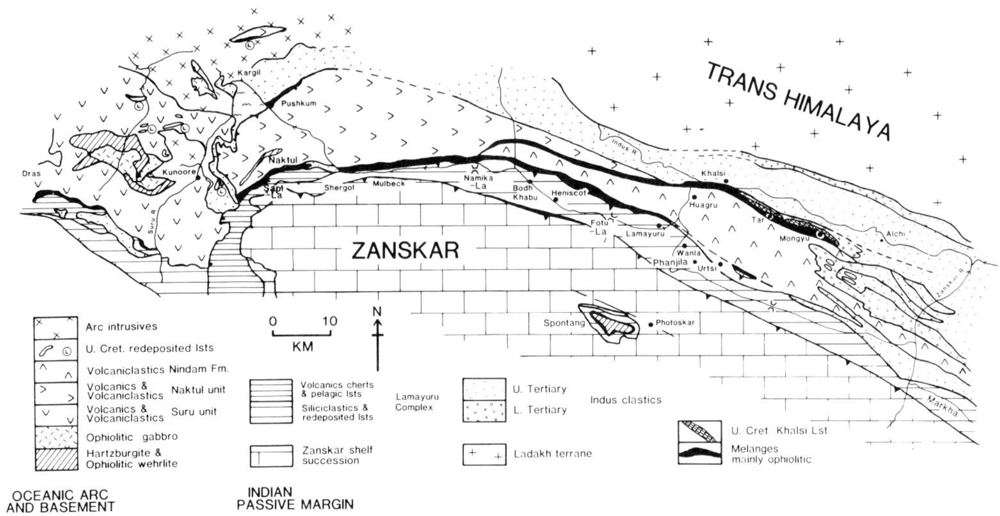

Fig. 2. Outline geological map of Ladakh, including the deep-water passive margin units of the Indian plate (Lamayuru Complex), based on literature cited. In this paper the Lamayuru Complex is divided, for the purposes of description, into *Eastern* and *Western* areas on either side of Namika-La.

overlain by the Mesozoic Zanskar shelf succession along a southward-dipping backthrust (Bassoullet *et al.* 1980*b*; Searle 1983, 1986; Searle *et al.* 1988). In this area, the Lamayuru Complex is thrust, again with a south-dipping thrust contact, over melange units (i.e. Shergol Melange and Urtsi Melange), and then over volcanics and sediments of the Cretaceous Naktal and Suru units of the Dras arc (Sutre 1991). In eastern Ladakh, east of the Zanskar River (Fig. 2), Fuchs (1986) reports that facies of his 'Lamayuru Unit' depositionally interfinger with sediments of the Zanskar shelf succession, but this is challenged by Baud (e.g. 1989) who shows major thrusts between the two units. The Dras volcanic arc unit wedges out eastwards (Nindam Formation) and the Lamayuru Complex is in direct tectonic contact with ophiolitic melanges (Omlung melange; Talon 1988) to the north, in eastern Ladakh.

The Lamayuru complex is strongly sheared, stratally disrupted, folded and locally inverted. Searle *et al.* (1988) envisage the Complex as a southward-directed thrust stack, that was rotated and inverted by Late Tertiary, northeast verging back-thrusting. Colchen *et al.* (1985) report large-scale early south-facing folds with an axial planar cleavage. This was followed by early Tertiary chevron folding, then pervasive northward-directed shearing and folding associated with the development of a second cleavage. We observed mainly bedding-parallel cleavage in most outcrops, locally cut by a spaced fracture cleavage. Numerous outcrop-scale folds mainly verge northwards.

In the east, younging directions are mainly to the south across the outcrop (i.e. east of Namika-La; Fig. 2), while in the west numerous younging direction changes and stratal inversions indicate large-scale isoclinal folding. Most of this folding and shearing relates to early Tertiary backthrusting of the Zanskar shelf over the Indus Suture Zone.

Intact, loggable, stratigraphic successions are locally preserved, particularly within competent units (e.g. thick limestones, volcanics and sills), mainly in the east. Thick units of Jurassic redeposited limestone, which we describe in detail, are mainly steeply-dipping and young southward (Fig. 3). The northward margins of these units are truncated by shear planes, marked by pressure solution cleavage, calcite veining and crushing of chert nodules (e.g. at Namika-La; Fig. 2). These limestones were taken to be detached blocks (i.e. olistoliths) by Frank *et al.* (1977) and Sinha & Upadhyay (pers. comm.), but can clearly be seen at many locations to pass gradationally into other Lamayuru Complex sediments to the south (Fig. 3). Also, in the western area, shallow-water successions in the 'Mulbeck exotics' pass depositionally upwards into deeper-water facies (e.g. shales, redeposited limestones and conglomerates), thus excluding an origin simply as carbonate-derived olistoliths.

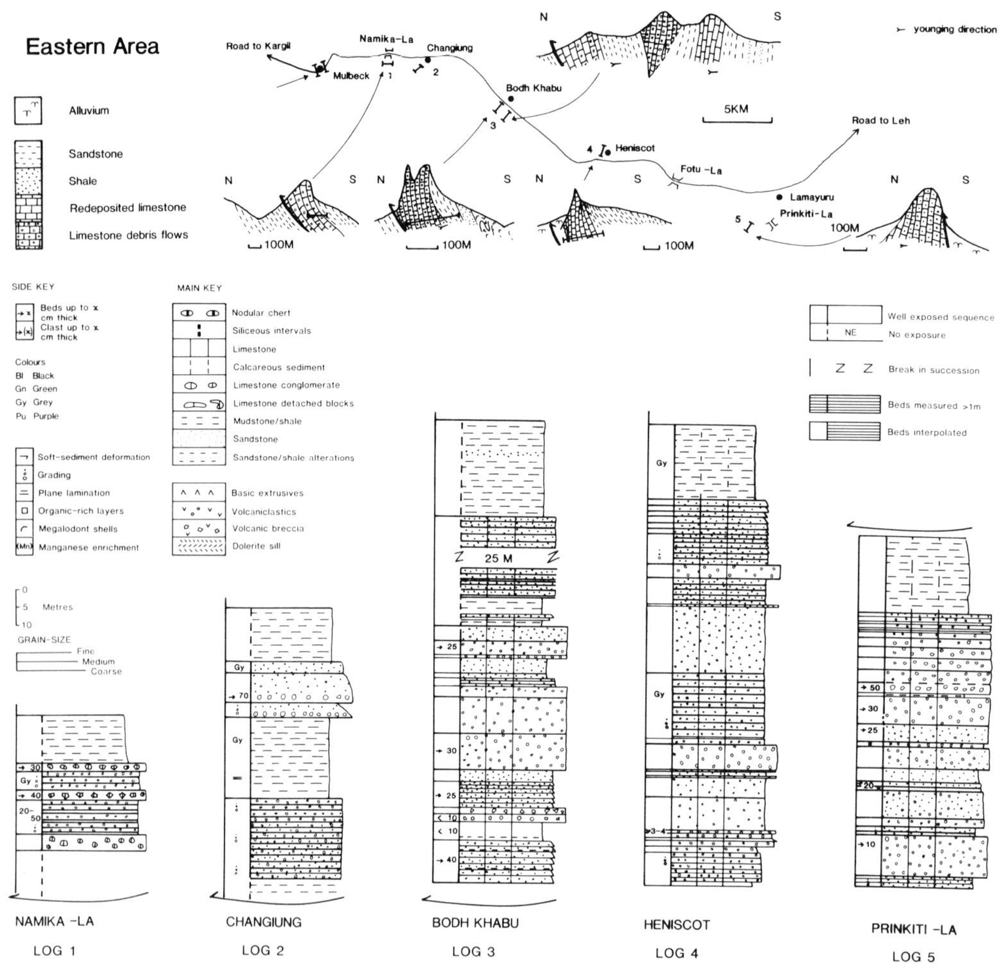

Fig. 3. Measured logs of proximal slope redeposited carbonates and interbedded lithologies in the Late Triassic–Jurassic interval in the eastern area. Locations and younging directions are shown in sketches above.

Lithostratigraphy

No formal stratigraphy is currently available for the Lamayuru Complex. Frank *et al.* (1977) divided the 'Indus Flysch' into: (i) 'Namika-La Flysch', west of Namika-La, comprising mainly siliciclastic sediments with 'limestone blocks', and (ii) 'Lamayuru Flysch', in the vicinity of Lamayuru further east, mainly comprising relatively fine-grained siliciclastics and shales. Honegger (1983) termed volcanics, including blueschists in Western Ladakh, the Karamba Formation; this was renamed the Sapi-Karamba Formation by Sutre (1991), who excluded the blueschists (Shergol, Sapi-La, Suru valley areas; Fig. 2). Fuchs (1986) used the term Lamayuru Formation for all 'dark calcareous flysch' exposed in Eastern Ladakh.

In view of the structural complexity it is not possible to deduce an accurate lithostratigraphy for the Lamayuru Complex and we therefore retain an informal terminology. We describe contrasting lithological assemblages exposed east and west of Namika-La (Fig. 2), as the *Eastern area* and the *Western area*, respectively. Within these two areas we discuss a variety of facies and discrete lithological successions.

Biostratigraphy

The lowest part of the succession in our eastern area is dated early Triassic (lower Scythian) based on the ammonite *Meekoceras sp.* (Bassoullet *et al.* 1980*a*; Thakur 1990). A Mid-Triassic age (mid-Ladinian) is assigned, based on

Daonella (H. Riebel, *in* Frank *et al.* 1977) from a roadcut just north of Lamayuru gompa (Fig. 2). *Daonella indica* and *D. longobardica*, or *D. moussoni* of mid-Ladinian age from the low in the 'Lamayuru Formation' and Ceratitid ammonites, belemnites and brachiopods are also reported southeast of Lamayuru (Frank *et al.* 1977). Limestones at Prinkiti-La are dated as Late Triassic (Rhaetic) by the presence of *Alpinophragmium* (Bassoullet *et al.* 1981; Cannat & Mascle 1990). Clasts in overlying Jurassic limestone conglomerates commonly also contain fossils of Late Triassic age (Bassoullet *et al.* 1981). Higher stratigraphic levels in the eastern area (south of Lamayuru, west of Prinkiti-La, at Heniskot and elsewhere; Fig. 3) are well dated as Aalenian (early Mid-Jurassic) based on diverse assemblages, including *Lucasella* cf. *cayeuxi* in the matrix of redeposited limestones (Bassoullet *et al.* 1981; Cannat & Mascle 1990). In addition, Sinha & Upadhyay (pers. comm.) report Upper Jurassic ammonites from the mid-part of the Lamayuru succession in the Eastern area. Pelagic carbonates, well exposed between Wanla and Phanjila (Fig. 2) are similar to the Campanian pelagic carbonates of the Shillakong unit of the distal Zanskar shelf succession (Bassoullet *et al.* 1978).

The oldest rocks definitely dated in our western area are late Triassic, based on Megalodonts in large detached masses of limestone ('Mulbeck exotics'; Bassoullet *et al.* 1981). An early to mid-Jurassic age is indicated by the foraminifera (*Haurania*) in the matrix of overlying redeposited limestones, within one of the large detached blocks. Further west, at Namika-La, Dasycladacean algae in redeposited limestone indicate an early to mid-Jurassic age. A.K. Sinha (pers. comm.) reports Maastrichtian planktonic foraminifera in pelagic limestones north of Fatu-La, while similar pelagic carbonates with planktonic foraminifera are also present west of Namika-La (Fig. 2). He also reports Palaeocene–Eocene benthic foraminifera locally near the contact between the Lamayuru Complex and ophiolitic melange to the north.

Further west (Shergol, Sapi-La, Suru Valley area), Sutre (1991) dates the base of the succession as Triassic, based mainly on *Daonella*, and as Late Triassic–Early Jurassic on the evidence of redeposited Megalodonts. Overlying basic lavas and volcaniclastics contain radiolarian cherts with radiolaria of Mid–Late Jurassic age. Sutre (1991) believes that contacts are depositional and, therefore the volcanics are mid–late Jurassic. In contrast, Honegger (1983) suggested Late Triassic–Early Jurassic ages, but these are unconfirmed. Sutre (1991) also reports that 'overlying' pelagic limestones contain *Globotruncana* and are, thus Late Cretaceous.

In Eastern Ladakh ('Markha Unit'), Stutz (1988) reports late Permian sandstones and shales (Lutchungse unit) and late Triassic marls and sandstones (Sumlung unit). Baud *et al.* (1982) report a diverse assemblage including *Lucasella cayeuxi* of Aalenian (early middle Jurassic). Pelagic carbonates of late Cretaceous age, determined by Fuchs (1986) and Stutz (1988) are assigned by others to the Zanskar shelf succession (Baud *et al.* 1982, 1984; Baud 1989) and more work is needed in this area to determine exactly which units are related to the Lamayuru Complex and which to the Zanskar shelf succession.

In summary, available fossil evidence indicates a late Permian to Upper Cretaceous (Eocene?) age for the Lamayuru Complex. A passive margin history extending back to the late Permian is confirmed, taking account of the ages of exotic limestones, locally exposed in tectonic contact with the Lamayuru Complex (e.g. Lamayuru exotic; Colchen *et al.* 1978) and beneath the Spontang ophiolite (Reuber *et al.* 1987, 1992; Colchen *et al.* 1987). A further important implication of the biostratigraphical data is that there is no evidence of major age reversals or repititions within the Lamayuru Complex, at least in the better dated eastern area, where successions young southwards. This implies that the Lamayuru Complex is not an imbricate thrust complex. However, this conclusion may not apply to the poorly dated western area, including the volcanics exposed adjacent to the tectonic melange to the north. More work is needed on this aspect.

Sedimentology and facies

Eastern area

We reconstruct the eastern area as a folded, locally sheared, but essentially intact, mainly southward younging succession, *c.* 3 km thick, of early Triassic to late Cretaceous age (Fig. 4a, b). In general, the succession begins with calcareous shales, commonly containing pelagic bivalves, subordinate interbeds of redeposited limestones and mainly fine-grained, relatively thin-bedded quartzose siltstones (early–mid-Triassic). These are then overlain by thicker-bedded, coarser-grained, lenticular, quartzose sandstones, then by major intervals of redeposited limestones (early–late Jurassic). These limestones are lenticular on scales of tens, to hundreds of metres along strike (Fig. 4c). The interior parts of lenses are often dominated by debris flow and other

Fig. 4. Field photographs. (**a**) Vertically dipping redeposited limestones 1 km west of Prinkiti-La; (**b**) vertically dipping redeposited limestones; (**c**) lenticular redeposited calcirudites and calcarenite grainstone interbedded with terrigenous shale; (b–c) Eastern area, near Bodh Khabu (Fig. 3); (**d**) Thick-bedded calcirudites, Namika-La; (**e**) 'Mulbeck exotics'. Late Triassic neritic limestones young to the left (W) passing into mudstones interbedded with limestone debris flows; (**f**) 'Mulbeck exotics', typical neritic limestone; large shell fragments (e.g. Megalodonts) infilled with coarse sparry calcite; (e–f) Western area.

mass flow-type calcirudites, while marginal areas are mainly composed of thin- to medium-bedded calcarenites. Shales, lenticular redeposited limestones (mainly medium-bedded grainstones) and subordinate relatively fine-grained, micaceous and locally ferruginous sandstones stratigraphically overlie the thick Jurassic limestones. Pink micritic limestone intercalations up to 10 m thick are well exposed between Wanla and Phanjila and between Phanjila and Urtsi (Fig. 2).

The following lithofacies are recognized in the Eastern area.

Shales. Shales that are associated with limestone are mainly calcareous, while those interbedded with quartzose sandstone are largely carbonate-free. Pelagic bivalve fragments are common in the Triassic intervals (e.g. near Lamayuru). Where interbedded with limestone debris flows, calcareous shales commonly contain elongate calcilutite rip-up clasts (e.g. Namika-La, Fig. 3, log 1). Bioturbation of deep-water type (e.g. *Nerites*) is often seen. The less calcareous shales are dominantly quartzose, with subordinate muscovite and plagioclase in a fine-grained terrigenous silt and clay matrix. Radiolarians and sponge spicules are locally present. Calcareous shales interbedded with the Jurassic redeposited limestone are micritic, with calcite-replaced radiolarians and small quartz and plagioclase grains set in a muddy matrix.

Siliciclastics. In the stratigraphically lower, Triassic, parts of the succession (e.g. near Lamayuru), mainly fine-grained, medium-bedded quartzose sandstones, showing classical turbidite structures (e.g. grading, bottom structures) dominate. Thicker-bedded units include completely preserved Bouma A–D turbidite divisions (Bouma 1962). Sandstones higher in the succession are lenticular (up to 50 m wide and 10 m thick), medium- to thick-bedded (up to 2.1 m thick) and mainly medium-grained (e.g. Lamayuru River area). Sandstones stratigraphically above Jurassic redeposited limestones include medium- to thick-bedded quartzose sandstones, with a rusty red ferruginous matrix, forming a 25 m thick horizon, traceable 500 m along strike (south of Bodh Khabu, Fig. 3, log 3). In thin section, the siliciclastics comprise quartz, plagioclase (including minor microcline) and small lithoclasts in a fine-grained terrigenous matrix. Quartz grains commonly show undulose extinction and are interpreted mainly as of plutonic and/or vein origin, together with minor polycrystalline quartz of metamorphic origin. Some muscovite laths show strain-slip cleavage. Intraclasts are mainly fine-grained micaceous siltstone.

Calcisiltites and calcilutites. Occurring sporadically throughout the succession in units, locally up to 5–10 m thick, these sediments are thin-bedded, graded, often parallel laminated and commonly interbedded with calcareous shale (e.g. Fig. 3, log 3). Some facies are black and organic-rich (e.g. near Bodh Khabu), while others are pink and ferruginous (e.g. south of Urtsi). Petrographic study shows that the Triassic calcilutites are dominantly micritic, with pelagic bivalve and brachiopod fragments, echinoderm debris and subordinate siliciclastic grains, in a muddy micritic matrix (e.g. near Lamayuru; Fig. 2).

Calcarenites. These are mainly oolitic grainstone/packstones, or (less commonly) lithoclastic limestones, varying from thin- to medium- and thick-bedded. The distribution of calcarenites range from occasional beds within shale in the Triassic parts of the succession, to regular interbeds within the overlying Jurassic limestones. There are also up to 40 m thick, uniformly bedded, lenticular intervals higher in the succession, which include repeated, amalgamated, massive, and/or parallel laminated beds, up to 0.8 m thick, with numerous mudstone, or calcilutite intraclasts (e.g. Prinkiti-La, Fig. 3, log. 5).

The Jurassic calcarenites are mainly grainstones and packstones, comprising rounded micritic grains. There are ooids and pisoliths, echinoderm plates and spines (with micritic envelopes), shell fragments, other bioclastic debris and scattered quartz grains, cemented by sparry calcite (Fig. 5a, b). Large, angular, quartz grains of plutonic, or vein origin are locally present. Grainstone clasts within the calcirudites are mainly oosparite, while micritic clasts contain scattered shells, echinoderm and calcareous algae.

Matrix-supported calcirudites. These form a volumetrically substantial component of the Jurassic limestones. Up to 1–2 m thick units of matrix-supported calcirudites are interbedded with grainstones (e.g. Fig. 4d), or form repeated, amalgamated units, separated by shaly partings (e.g. Heniscot; Fig. 3, log. 4). Clasts are mainly pale, to dark (occasionally black) micritic limestone, ranging from elongate to sub-rounded and angular (up to 40 cm long) (e.g. Prinkiti-La; Fig. 3, log 1). The matrix is typically oolitic grainstone. In places, matrix-supported conglomerates pass upwards, within single

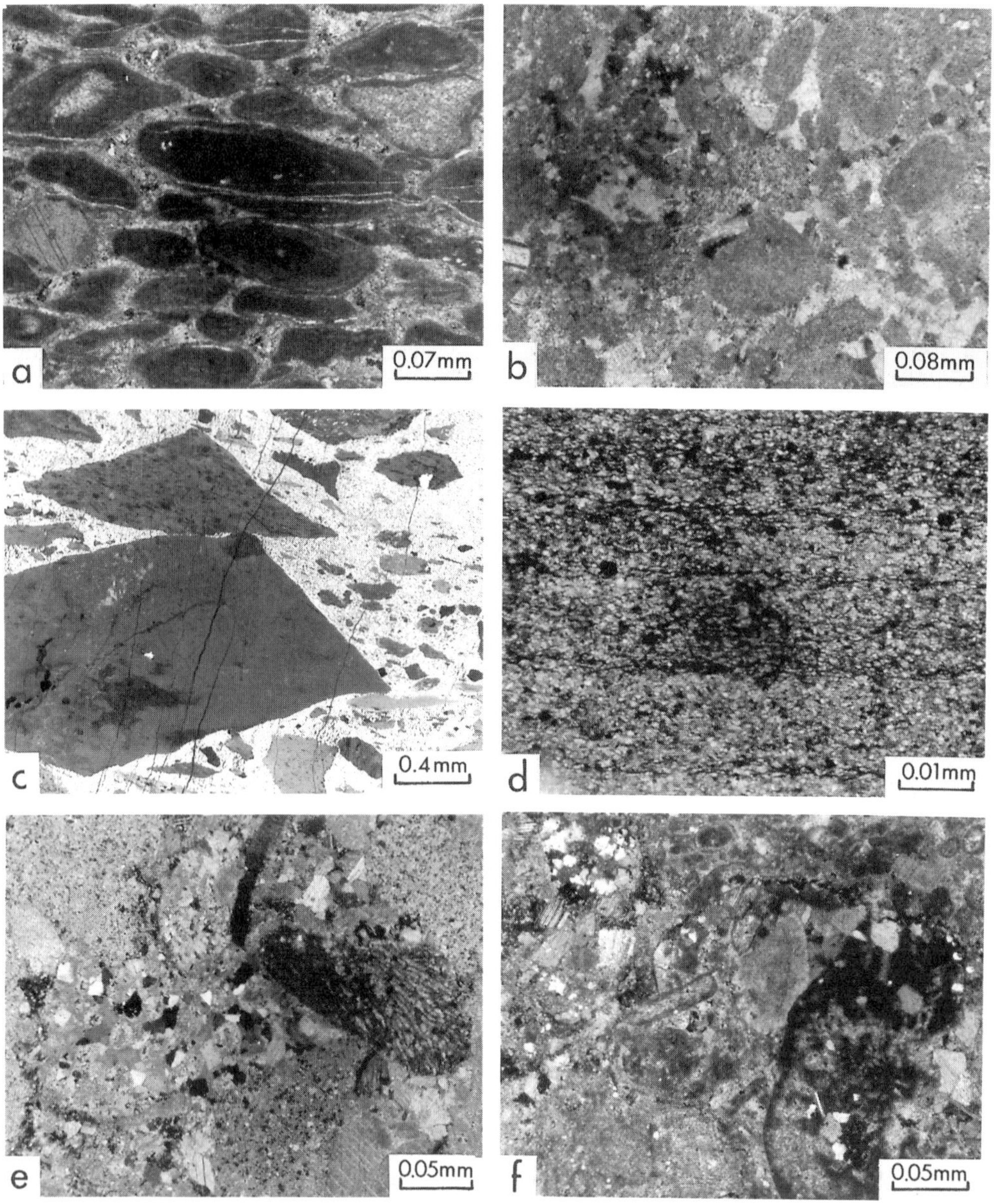

Fig. 5. Photomicrographs. (**a**) Redeposited grainstone, ooids, pisoliths and oolitically coated grains with a sparry calcite matrix, Prinkiti-La (Fig. 3); (**b**) poorly sorted redeposited oosparite, oolitic nature is typically indistinct (Fig. 3); (**c**) angular fragments of neritic limestone deposited in an impure calcareous matrix from debris flow stratigraphically above the Late Triassic 'Mulbeck exotic', negative print from a thin section (Fig. 7); (**d**) finely-laminated turbiditic siliciclastic siltstone; Western area; (**e**) coralline algal fragment (left), with terrigenous and micritic grains); (**f**) redeposited limestone with fragments of echinoderms, detrital quartz grains, shells, ooids, pisoliths and calcareous algae; upper levels, Western area.

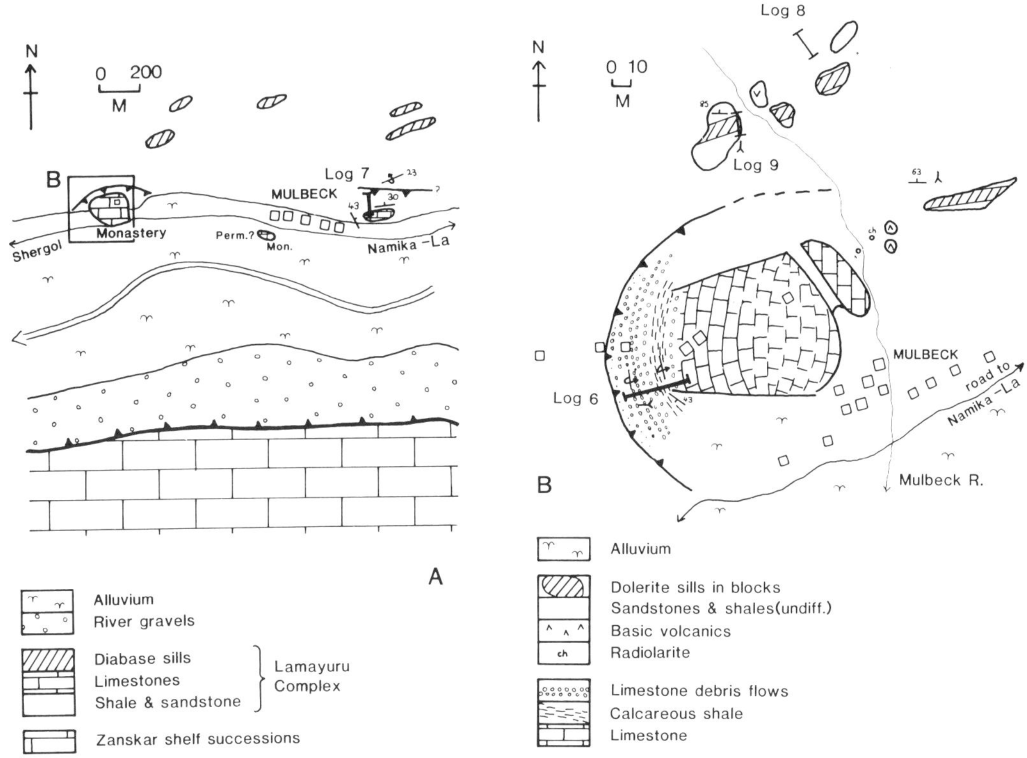

Fig. 6. Sketch maps of the Late Triassic 'Mulbeck exotic' (Western area) in the type area. (**a**) Setting of platform limestone units; (**b**) detail of the largest limestone block.

depositional units, into graded, laminated calciturbidites (e.g. near Namika-La; Fig. 3, Log. 1). Elsewhere (e.g. Heniskot; Fig. 3, log. 3), individual calcirudites merge gradationally upwards, or downwards into redeposited oolitic grainstones.

Clast-supported calcirudites. These are locally associated with the thickest and coarsest limestone conglomerate units (Fig. 3). Clasts are mainly very angular (up to 30–50 cm in size), dark, to pale coloured grainstones. Individual beds, up to 5 m thick, are laterally discontinuous over 5–10 m and are interbedded with medium- to thick-bedded oolitic grainstones.

Western area

Successions in the western area are much more sheared, folded and thrust faulted, and are commonly inverted. In contrast to the eastern area, the western area includes swarms of dolerite sills, individually up to *c.* 10 m thick, that have often undergone intense shearing to form elongate blocks.

Three relatively structurally intact lithostratigraphic units can be recognised, separated by sheared contacts: (i) A lower, *Limestone–conglomerate–shale unit* ('Mulbeck exotics'), dated Late Triassic–Jurassic; (ii) an overlying, *Shale–sandstone–limestone unit* (undated); and (iii) a *Heterogeneous volcanogenic unit* (partly Cretaceous). Although separated by thrust faults, facies in the uppermost levels of the limestone–conglomerate–shale unit are similar to some of those in the shale–sandstone–limestone unit. This latter unit also contains volcanic-derived detritus, suggesting original proximity to the heterogeneous volcanogenic unit.

(i) *Limestone–conglomerate–shale unit.* A succession east of Mulbeck (Figs 4e, 6 & 7) exhibits an important sedimentary transition from Late Triassic shallow-water limestones ('Mulbeck exotics') to deeper-water shales and siliciclastics. Thick-bedded, sub-vertically dipping (but not inverted), recrystallised, neritic limestones (Figs 4f & 7, log 7), with poorly preserved Megalodonts, pass upwards into a 60 m thick

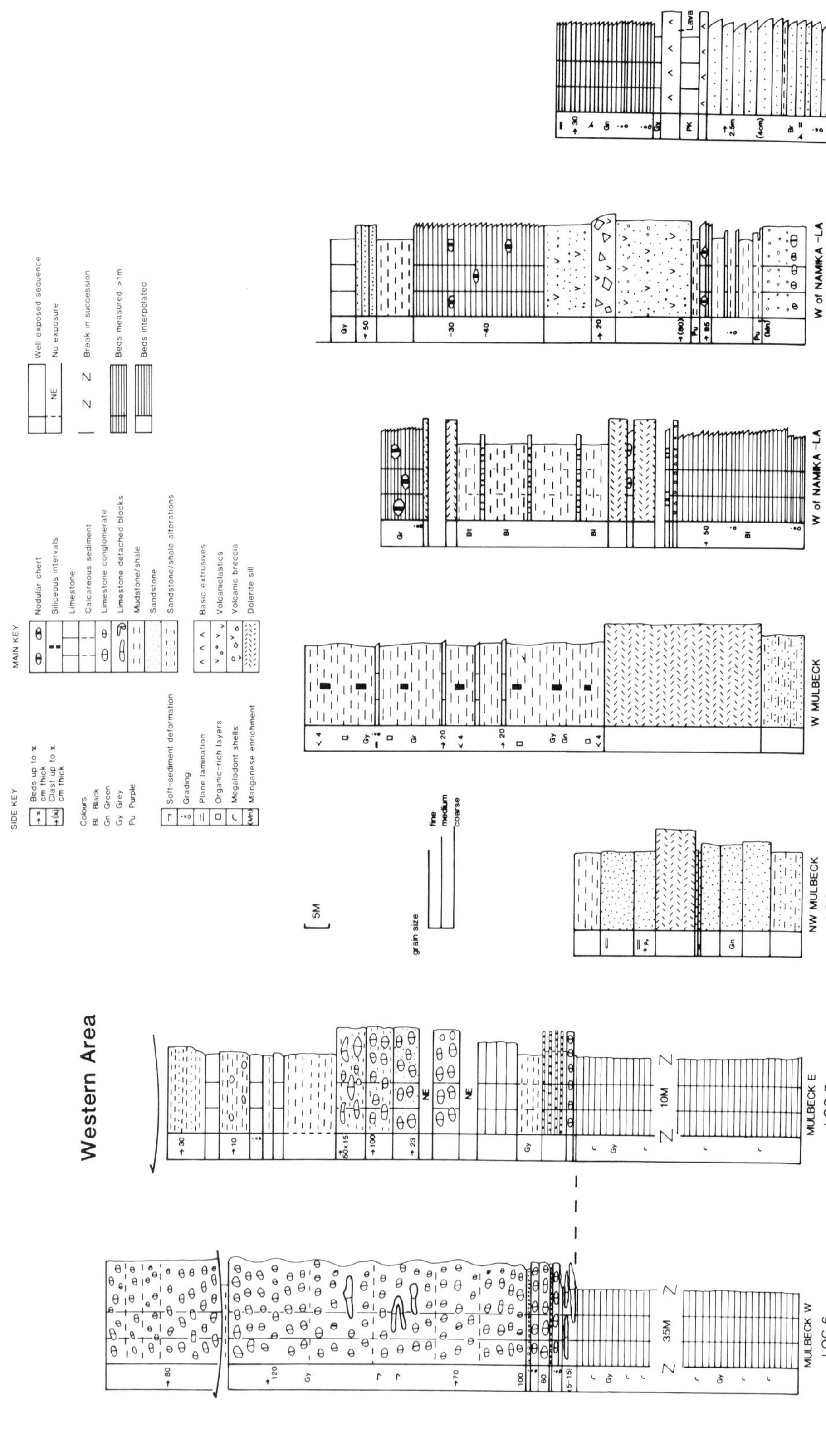

Fig. 7. Measured logs of successions in the Western area. See Figs 2, 3 and 6 for locations.

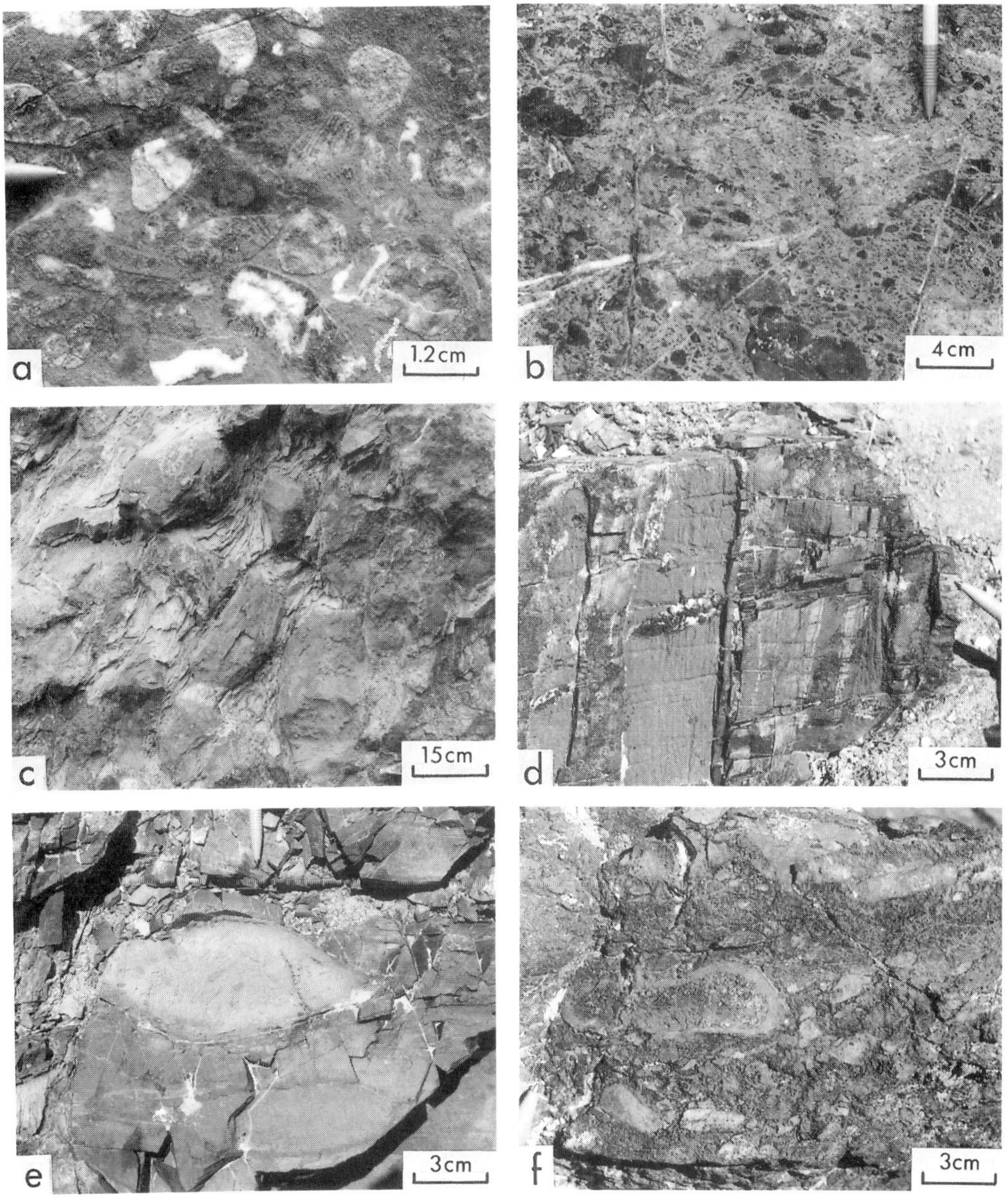

Fig. 8. Field photographs. (**a**) Redeposited limestone debris flow stratigraphically overlying the Late Triassic 'Mulbeck exotic' (Western area); (**b**) debris flows of clasts of neritic and hemipelagic carbonate with a muddy, terrigenous matrix, stratigraphically above (a); (**c**) debris-flows with large clasts of neritic limestones in a shaly matrix, dated Early to Mid-Jurassic, stratigraphically above (b); see Fig. 6 for locations; (**d**) calciturbidites interbedded with black shale, Mulbeck R.; (Fig. 6b); (**e**) calcareous concretion with hemipelagic black shale, Mulbeck R. (Fig. 6b); (**f**) Lapilli tuff from upper structural levels, Western area (see Fig. 7).

succession of thick-bedded, redeposited limestones. These limestones are overlain by crudely stratified clast-supported, to matrix-supported, limestone conglomerates and breccias, interbedded with sheared shale and grainstones (Fig. 8a, b). Petrographic study shows that the redeposited carbonates are micritic, with scattered dolomite rhombs, broken, volcanic derived, plagioclase crystals, shell fragments, echinoderm spines, ferruginous clay grains, small angular quartz grains and patches of diagenetic replacement quartz. Micritic clasts include calcite replaced radiolarians and sponge spicules. Clasts in the overlying conglomerates are mainly dark grey micrite and range from angular (Fig. 5c) to sub-rounded. Limestone conglomerates higher in the succession include micritic limestone blocks (up to 1 m long) in a shaly matrix (Fig. 8c). Limestone and micaceous shale alternations continue above this and then pass depositionally upwards into thin- to medium-bedded quartzitic sandstones, which become increasingly sheared and disrupted upwards.

Further west, similar limestones (such as that capped by Mulbeck gompa; Fig. 6) are shown to be inverted by geopetal structures and grading in depositionally overlying sediments. The exposed succession begins with metre-thick beds of massive, bioclastic limestone (Fig. 7 log 6), with an in situ biota (e.g. Megalodonts), in a buff micritic matrix. Aragonitic shells are replaced by sparry calcite. In thin section, this limestone is composed of micrite, partly recrystallized to microspar, with fossils, including coral, polyzoans, echinoderms (with micritic envelopes), benthic foraminifera, calcareous algae and shell fragments. Syndepositional fractures cut this limestone and are filled with ferruginous and calcareous silt, bioclastic fragments, calcareous silt and scattered quartz grains, including polycrystalline quartz.

The shallow-water limestone is depositionally overlain by matrix-supported, monomict, conglomerate, with elongate clasts (5 cm× 15 cm) of biomicrite in a buff to reddish brown, micritic matrix. Many clasts were deformed prior to complete lithification. The stratigraphically highest limestone beds are calcirudites composed of sub-rounded, to irregularly shaped, micritic clasts, together with bioclastic debris (up to 1 cm sized) in a pale grey micritic matrix. Diffuse edges indicate that many of the clasts were soft when redeposited.

The succession continues upwards with a 1 m thick interval of matrix-supported calcirudite, containing dark grey, micritic clasts in a pale brown, oolitic grainstone matrix. Many clasts show 'plastic' fabrics, indicating deformation while still soft. Interbeds of micritic limestones (up to 0.6 m thick and 3–11 m long) and oolitic grainstones are also present. Similar grainstones are locally present as rip-up clasts within interbedded calcirudites. In thin section, the grainstones are mainly oolitic, with scattered calcite-replaced radiolarians, small shell fragments, echinoderm plates, rare quartz grains and coarsely crystalline limestone lithoclasts, set in a micritic matrix. The clasts commonly show a streaky texture, indicating deformation while still soft. Some clasts retain prismatic calcite overgrowths. The matrix includes muscovite and minor quartz (mainly polycrystalline). The carbonate matrix is strongly dolomitized, in contrast to the allochems.

An overlying 60–80 m thick interval of mainly clast-supported limestone conglomerate is dated as early to mid-Jurassic. Most clasts are sub-rounded, to angular; some have merged due to plastic deformation before lithification. In thin section, the clasts are mainly oolitic grainstone, with minor detrital quartz. Thick veins are infilled with diagenetic quartz–calcite intergrowths. The matrix is dark micrite which has been almost entirely recrystallised to small dolomite rhombs.

The conglomerates pass transitionally upward into alternations of micaceous shale and lenticular, mainly matrix-supported limestone conglomerates, then the succession is obscured by intense shearing.

(ii) *Shale–sandstone–limestone unit.* Facies of this unit dominate the outcrop of the western area (Fig. 7, logs 8–10). Only short successions are preserved intact, however, owing to intense shearing and minor thrust faulting. The following lithofacies are present.

Black shales. Non-calcareous, fissile, micaceous shales volumetrically dominate the background sediment. Unoxidized exposures are black and organic-rich with abundant pyrite. Some thin layers are siliceous and cherty, or contain small (less than 5 cm sized) elliptical, to lenticular carbonate concretions (up to 0.6 m long × 0.18 m wide; Fig. 8e). Bedding surfaces are commonly coated with secondary manganese oxide.

Calcarenites and calcilutites. These are orange weathering, mainly recrystallized, up to 0.35 m thick and show subtle grading and parallel lamination, mainly confined to the basal 2 cm of individual beds (Fig. 8d).

Siliciclastics. These are mainly thin-bedded siltstones, showing subtle grading, parallel lamination and, locally a streaky fine lamination. Common sandstones are mainly fine- to medium-grained (Fig. 5d) and up to medium-bedded. Shale rip-up clasts are locally abundant. Thin section examination reveals compositions similar to siliciclastics in the eastern area (described above), except that basic volcanic rock fragments are present only in the western area.

(iii) *Heterogeneous volcanogenic unit.* Volcanic-related successions crop out between Mulbeck and Namika-La (north of the road; Fig. 2) and also as detached blocks near Mulbeck (in the Mulbeck River, Fig. 6b). Several intact, inverted successions are locally present, bounded by thrusts.

One short succession (Fig. 7, log 12) begins with massive thick-bedded coarse-grained, quartzose sandstones (with beds up to 2.5 m thick), interbedded with minor debris flows (with shale clasts up to 4 cm in size). The sandstones are *depositionally* overlain by vesicular, feldspar-phyric lava, interbedded with green and purple coarse-grained volcaniclastic sandstones and carbonate cemented hyaloclastite breccia. These volcaniclastics are, in turn, interbedded and depositionally overlain by pink pelagic carbonates, sheared greenish shales and up to medium-bedded calciturbidites. These observations are important, as according to some authors (Fuchs 1979, 1982; Fuchs & Sinha 1992; Honegger 1983) the volcanics represent entirely exotic thrust sheets unrelated to the Lamayuru quartzose clastics successions.

Another local succession (Fig. 7, log 10) begins with thin- to medium-bedded packstone/grainstone calciturbidites, interbedded with finely-laminated, black, micritic limestones. These then pass stratigraphically upwards into debris flows, mainly composed of small limestone lithoclasts, medium-bedded calcilutites (with replacement chert) and interbedded siliceous shales. Numerous sills intrude and thermally metamorphose this part of the succession. Above a shear zone (of uncertain significance) come alternations (Fig. 7, log 11) of limestone debris flows, purple manganiferous shales, calcilutites, medium-bedded calciturbidites (with partly silicified shaly partings), purple/green calciturbidites (Figs. 5e,f) and volcaniclastics. Lithofacies include an 0.8 m thick lapilli tuff horizon (Fig. 8f), rich in black devitrified glass and coarse volcanic breccia, with angular, green volcanic clasts grading upward into a finer-grained volcaniclastic upper interval. These volcaniclastics are depositionally overlain by medium- to locally thick-bedded grainstone calciturbidites, manganese-coated micaceous sandstone, grey and purple shale and reddish purple calcilutite. In thin section, the reddish calcilutites contain calcified radiolarians and rare planktonic foraminifera, while the interbedded limestone is poorly recrystallised grainstone, with quartz grains, shell fragments, large muscovite laths and algal micrite. This local succession again confirms that the volcanics are interbedded with relatively proximal redeposited limestones and quartzose sandstones derived from the Zanskar shelf.

Petrography of volcaniclastics. The volcaniclastics include both pyroclastic and epiclastic sediment. Typical epiclastics are made up of small angular, basaltic and silicic lithoclasts in a chloritic and calcareous matrix, mixed with detrital carbonate grains, including dolomitised micrite clasts and echinoderm spines (Fig. 9a,b). Detached blocks at Mulbeck (Fig. 6b) are petrographically diverse. Extrusive lithoclasts are mainly altered basalt and devitrified glass, trachyte with flow-aligned plagioclase microphenocrysts and rhyolite (Fig. 9b,c). Individual detrital phenocryst fragments include broken, zoned plagioclase, altered olivine, titaniferous augite and rare clear, angular volcanic quartz crystals, with occasional embayed margins. Large sub-rounded grains of strained quartz, of probable plutonic origin, and rarer grains of polycrystalline quartz are also present. Sedimentary lithoclasts are sub-rounded grains of partly recrystallised fossiliferous micrite; angular limestone clasts with small angular quartz grains; radiolarite, with radiolarians preserved as calcite and/or microcrystalline quartz in a fine-grained terrigenous matrix; and small siltstone grains with small angular single crystal grains (partly replaced by calcite). The matrix contains abundant chlorite, with extensive alteration to calcite. The petrography thus provides additional evidence that mafic volcanics were associated with shallow-water carbonate and coarse sandstone, as well as radiolarite.

Comparison with adjacent areas

The Lamayuru Complex also crops out east and west of the areas studied and, reportedly, also in the Zanskar Mountains.

Western Ladakh. In the Shergol, Sapi-La and Suru valley areas (Fig. 2) Sutre (1991) reports the following succession, cut by sills: The Triassic consists of terrigenous 'flysch', interbedded with lithoclastic and bioclastic limestones (with Conodonts and foraminifera), and hemipelagic limestones containing *Daonella*. Local limestone breccias contain Megalodonts. The Mid–Late Jurassic succession is dominated by alka-

Fig. 9. Photomicrographs. (**a**) Large quartz grains (left) mixed with basalt (right) and other volcanic derived and terrigenous grains, east of Mulbeck, Western area (Fig. 7); (**b**) epiclastic volcaniclastic sandstone; note the angular grains of basalt, plagioclase and quartz and other reworked grains; Western area; (**c**) as (b), but also containing clasts of radiolarian chert and terrigenous siltstone, (a–c) from the Western area; (**d**) shell, echinoderm, micritic and terrigenous grains from a calciturbidite interbedded with mafic extrusives, Volcanic heterogeneous unit, near Shergol (Fig. 2).

line volcanics, fragmental extrusives and volcaniclastics, interbedded with redeposited limestones, thin-bedded red silts and locally, radiolarites, together with local detached blocks of Jurassic limestone. This thick volcanic interval is, in turn overlain by redeposited oolitic limestones, red siltstones and radiolarites of Late Jurassic? age. Early Cretaceous lithofacies are not confirmed, while the Late Cretaceous is represented by pelagic limestones containing *Globotruncana*, interbedded with limestones, red shales, radiolarites and volcaniclastics. It is, however, by no means clear that the succession is not repeated by early, emplacement related thrusting, especially close to the Shergol ophiolitic melange to the north.

Our own reconnaissance near the Shergol Melange contact (Fig. 2) revealed the following units: tens-of-metre-thick, lithologically variable volcanic-sedimentary units are repeated by thrusts. Above a local basal thust, partly silicified grainstone calciturbidites, with black replacement cherts are overlain by 60–80 m of aphyric and feldspar-phyric basalt flows (including pillow basalts), interbedded with volcaniclastics and reddish shales containing vesicular lava clasts (up to 15 cm in size), and red radiolarite. Structurally overlying (right way up) sequences include shale, calcilutites, packstone calciturbidites, and thick-bedded lithoclastic limestones, with volcanic and micritic grains (Fig. 9d). Structurally higher sheets include thick basaltic extrusives, with up to 40 m thick limestone intercalations. Although limited, these observations again show that redeposited limestones are interbedded with mafic volcanics and are not exotic, oceanic derived thrust slices.

Eastern Ladakh. In the Markha valley area, east of the Zanskar River (Fig. 2), Fuchs (1986) describes a 'Lamayuru Formation' which he took to be depositionally intergradational with the Zanskar shelf succession to the south and with ophiolitic melanges (Omlung Melange, Talon 1988) to the north. Other workers (Baud *et al.* 1982, 1984), however, recognize an important thrust within the Lamayuru-type units, separating true equivalents of the Lamayuru Complex ('Markha Unit') from similar facies ('schistes lustres') of the Zanskar shelf succession (Lantang Group). Further sedimentological and structural work is needed to resolve this controversy. However, all authors agree (Baud *et al.* 1982; Fuchs 1982, 1986) that Lamayuru Complex equivalents are present in the Markha area. These are dominantly black schists, phyllites, calcschists, siltstones, marls and sandstones. These are commonly interbedded with limestone units up to 100 m thick and traceable laterally for up to several kilometres. Typically, these limestones are calcilutites, pelletal limestones, calcarenites and 'microbreccias'. Up to centimetre-sized grains include micrite, calcarenite, oolite, dark mudstone, finely crystalline dolomite and shallow-water derived bioclastic debris. Stutz (1988) argues that facies as old as late Permian in this area accumulated near the base of the rifted Zanskar carbonate platform slope.

Zanskar Mountains. Possible equivalents of the Lamayuru Complex have been reported in the vicinity of the Spontang ophiolite (e.g. Bassoullet *et al.* 1978, 1980*b*; Searle *et al.* 1988). Individual sub-vertical, mixed sandstone–limestone successions up to 500 m thick and 1–5 km long in this area are erosionally resistant and have weathered out along strike as trains of pinnacles. Successions we examined near Sirsir-La, west of Photaksar (Fig. 2) are dominantly cleaved micaceous shales with alternations of mainly thin- to medium-bedded, fine- to medium-grained, mixed carbonate clastic sediments, showing grading, locally well developed basal flute casts, parallel and cross-lamination and scattered carbonate concretions. Thin sections reveal poorly preserved *Globotruncana*.

The succession described by Sutre (1991) in western Ladakh is broadly similar to that in our western area, but probably more complete. The eastern Ladakh succession reported by Fuchs (1986) is similar to that in our eastern area, particularly the thick, lenticular redeposited limestones, although it may in part be older (late Permian). It appears that the successions near Sirsir-La should be correlated with the late Cretaceous facies of the Zanskar shelf succession (Kangi-La Formation), rather than the Lamayuru Complex (Kelemen & Sonnenfeld 1983; Garzanti *et al.* 1987; Cannat & Mascle 1990). However, more work is needed to determine if both thrust imbricated Zanskar shelf and Lamayuru Complex sediments are, in fact present beneath the Spontang ophiolite.

Discussion and interpretation

Depositional processes

The mudstone facies throughout the succession are interpreted as background hemipelagic sediments, of mainly terrigenous provenance. The presence of locally developed 'streaky' fine lamination points to bottom current activity. The subdued colours, with common pyrite, indicate sub-surface reducing conditions, probably developed in response to high input of organic matter and/or accumulation within an expanded oxygen minimum zone (e.g. the black shales north of Mulbeck). The thin-bedded siliciclastics are interpreted as low-density turbidites, again terrigenous. Coarser-grained, thicker-bedded, lenticular sandstones are considered to be channelized, high-density turbidites. Fine-grained carbonates, which contain radiolarians, are periplatform ooze deposits, derived from the Zanskar shelf. The calcarenites are calciturbidites, with both lenticular and laterally extensive tabular morphologies, indicating both channelled and unchannelled transport. Most are oolitic, while some are lithoclastic; all were derived from the Zanskar shelf to the south. Both clast- and matrix-supported conglomerates of the eastern area are interpreted as gravity flows within major channels, carrying Triassic and Jurassic aged carbonate detritus.

The Late Triassic shallow-water limestones in the western area ('Mulbeck exotics') show evidence of gravitational instability and break-up, allowing carbonate and terrigenous sediment to percolate downwards into fissures. The highest stratigraphic levels of the shallow-water succession exhibit soft-sediment deformation and gravity reworking, indicating tilting and subsidence. Overlying oolitic grainstones were derived from a still extant carbonate platform, while interbedded hemipelagic shales indicate continued deepening. Overlying limestone debris flows accumulated in the Jurassic by mass wasting of carbonate platform lithologies.

Bassoullet *et al.* (1981) envisaged the 'Mulbeck exotics' as sedimentary slide blocks from the adjacent carbonate platform, while Searle (1983) interpreted them as tectonically emplaced oceanic carbonate build-ups, similar to those in Oman (Robertson & Searle 1990). A platform margin, rather than oceanic setting is, however established by the existence of a depositional cover, including terrigenous sediments and platform-derived sediment (i.e. oolites) derived from the Zanskar shelf. Age-equivalent, Jurassic shallow-water carbonates are notably absent from exotics associated with the Spontang ophiolite (Colchen *et al.* 1987; Reuber *et al.* 1987, 1992). These exotics are envisaged as true seamounts which formed within Tethys well to the north of the Zanskar shelf. Accepting an Indian margin origin for the 'Mulbeck exotics', the question arises as to whether they represent the product of relatively far-travelled gravity slides (i.e. olistoliths), as envisaged by Bassoullet *et al.* (1981), or if, instead they are fault blocks which rotated and subsided near the platform edge. The presence of the intact, effectively conformable deep-water Jurassic succession above the Late Triassic platform carbonates and the absence of voluminous limestone detritus in the adjacent Lamayuru succession (i.e. olistostromes) favour the second possibility: that the 'Mulbeck exotics' formed by extensional faulting and break-up of the platform edge.

The interbedded nature of the alkaline volcanics with quartzose and calcareous turbidites indicates eruption in deep-water, presumably in a slope setting, judging from the spalled lava breccias and hyaloclastites. The associated heterogeneous epiclastic sediments reflect extensive erosion and reworking, along with sandstone, limestone and radiolarite, which probably resulted from uplift, although there is no evidence of subaerial exposure. The local interbedding of radiolarites and pelagic carbonates in the Cretaceous is suggestive of development of the radiolarites in response to upwelling and high productivity along the continental margin (De Wever 1989), rather than simply accumulation below the carbonate compensation depth.

Reconstruction of the passive margin

The Lamayuru Complex preserves slope, to base-of-slope depositional settings of a deep-water passive margin. The thick, channelized, Jurassic redeposited limestones of the eastern area are similar, for example to the carbonate platform slope units of the Sumeini Group in Oman (Watts & Garrison 1986) and equivalents in Antalya, SW Turkey (Robertson & Woodcock 1982). The shale–siliciclastic facies are also similar to inferred slope to base-of slope successions of other Tethyan deep-water successions, including the Hamrat Duru Group in Oman (e.g. for a recent review see Robertson & Searle 1990), the Ayios Photios Group in SW Cyprus and the Pindos of W Greece (Robertson *et al.* 1991). If the Late Triassic 'Mulbeck exotics' are correctly interpreted as platform edge-derived fault blocks, rather than far-travelled olistoliths, then associated units of the Lamayuru Complex in the Western area formed in an upper slope setting. The relatively thin-bedded and fine-grained nature of many of the sandstone and limestone turbidites in this area may thus, reflect sediment by-passing, at least in Late Triassic–Mid-Jurassic.

Reuber *et al.* (1987, 1992) infer a semi-enclosed continental margin basin origin for the Lamayuru Complex, bordered to the north (i.e oceanwards) by large peripheral platforms, that are now represented by exotics below the Spontang ophiolite. We think it more likely, however that these exotics formed seamounts scattered within Tethys, well to the north of the Zanskar shelf. We have not found any evidence of input from the exotic successions into the Lamayuru Complex as this alternative reconstruction would imply. We therefore interpret the Lamayuru Complex as a passive margin slope-rise succession, bordering the Tethys ocean to the north.

The successions in the eastern area and western area are broadly lateral equivalents. However, at least some of those in the western area are probably relatively proximal (upper–mid-slope), in the light of the origin of the 'Mulbeck exotics', discussed above. Successions in the eastern area accumulated mainly in a lower slope–continental rise setting. The volcanics in the Western area are interbedded with siliciclastic and calcareous turbidites and thus, represent eruptions near the rifted Zanskar shelf, rather than out in the Tethys ocean. Sutre (1991) proposed that volcanic reactivation of an already established passive margin took place in the Mid–Late Jurassic. Emplaced truly oceanic crust appears instead, to be restricted to structurally adjacent melange units within the Indus Suture Zone (e.g. Urtsi melange; Sutre 1991; Omlung melange; Talon 1988), together with the Spontang ophiolite and associated melange, emplaced over the Zanskar shelf.

The Zanskar shelf succession for comparison

As summarized by Gaetani & Garzanti (1991), rifting of the north Indian plate began with widespread basaltic volcanism (Panjal Traps) in early–late Permian, followed by marine transgression, in response to thermal subsidence during the early–mid-Triassic. A thick carbonate platform succession (Hanse unit) developed in the late Triassic, followed, in the Norian, by terrigenous clastics on a storm-dominated shelf (Quartzite unit). After a hiatus with local subaerial exposure, carbonate platform accumulation resumed (Kioto unit) during the early Jurassic (Garzanti *et al.* 1987), followed by a regional unconformity and 10 Ma hiatus. Above this comes shales (Laptal unit) in distal units and ferruginous oolites, shales and sandstones of early Callovian age in proximal settings. Mafic volcanic detritus is present above the ferruginous oolite unit. Muds come in above in the latest Jurassic (Spiti unit) and are, in turn overlain by prograding deltaic, terrigenous clastics in the Early Cretaceous (Guimal unit) (Gaetani *et al.* 1983, 1986). There is a localized influx of alkaline volcanic detritus in the Albian (Baud *et al.* 1984; Garzanti this volume). Late Albian time is marked by the presence of glauconitic sandstone, overlain by pelagic carbonate deposition in outer margin areas (Fatu-La unit), spreading inboard in Cenomanian–Turonian time. Early Tertiary began with continental sandstones (Stumpata unit), deepening upwards into late Palaeocene open shelf carbonates (Dibling unit). This was overlain first, by open shelf pelagic carbonates (Shingo-La unit), then by late Palaeocene nummulitic limestones (Kesi unit) and greenish sandstones (Kong unit), representing the latest marine sediments before final emplacement of the Spontang ophiolitic units.

Correlation of the Zanskar shelf and Lamayuru slope-rise units

The Late Permian proximal, base-of-slope turbidites in eastern Ladakh (Zutchungse unit), reported by Stutz (1988), have counterparts in the Zanskar shelf transgressive limestones. The Triassic muddy Lamayuru sequences, with silt and calcareous turbidites (e.g. Samlung unit of Stutz 1988) reflect progradation of the Zanskar carbonate platform (Hanse unit). Undated, overlying thick-bedded, channelized sandstone turbidites in the Lamayuru (e.g. south of Lamayuru) can probably be correlated with the Norian clastics on the Zanskar shelf (Fig. 10).

The widespread, lenticular, largely oolitic, redeposited carbonates in the eastern area (and in eastern Ladakh) can be broadly correlated with the early Jurassic platform carbonate succession (Kioto unit). The hiatus and unconformity which ensued could have resulted from

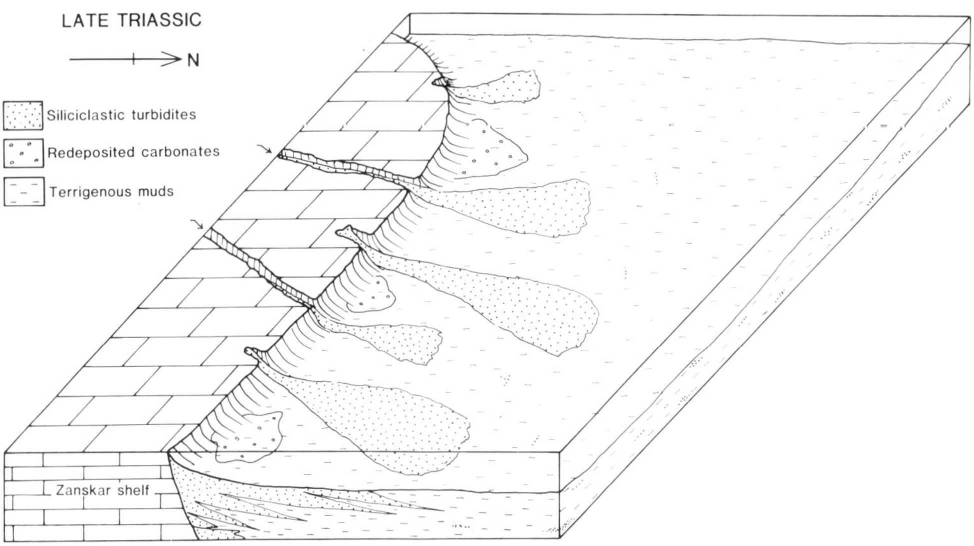

Fig. 10. Facies reconstruction of the Lamayuru Complex in the late Triassic. Platform carbonates were redeposited in channels away from the Zanskar shelf.

regional crustal extension and flexural uplift of the Zanskar shelf in early Mid-Jurassic time. This was possibly related to regional plate reorganisation, including opening of the Somali and Mozambique basins (Patriat, *et al.* 1982; Gaetani & Garzanti 1991). Localized ferruginous sandstones overlying the redeposited limestones (south of Bodh Khabu) in the Lamayuru Complex might correspond to the ensuing late Jurassic hiatus, with condensed ferruginous deposition on the Zanskar shelf.

The thick, lenticular, Jurassic redeposited limestones in the Lamayuru Complex (eastern area) could, in principle, relate either to steady state early Jurassic progradation and/or to rapid break-up and subsidence of the platform edge associated with regional crustal extension at the end of the early Jurassic. The presence of abundant lithoclastic detritus, including Triassic limestone clasts, indicates that large-scale, tectonically induced, mass wasting of the platform edge took place and this, together with reported Aalenian (early mid-Jurassic) ages, supports an origin related to catastrophic platform edge break-up, rather than steady-state progradation. Also, the collapse succession overlying the 'Mulbeck exotics' is dated only as early to mid-Jurassic and this also can be related to the inferred early mid-Jurassic crustal extensional event.

The localized alkaline volcanic detritus within the Callovian Zanskar shelf succession may relate to the widespread alkaline volcanism in the western area of the Lamayuru Complex, dated as mid–late Jurassic based on radiolarians. This magmatism also developed in response to crustal extension, which exploited segments of the Permian continent–ocean boundary north of the Zanskar shelf.

Further rifting took place in the early Cretaceous, associated with a terrigenous clastic influx (Guimal unit) and additional mafic volcanism. The shelf clastics can be tentatively correlated with sandstone turbidites in the upper levels of the Lamayuru Complex (e.g. near Wanla and north of Mulbeck). Sutre (1991) infers that basic volcanism took place within the Lamayuru Complex in western Ladakh during the Cretaceous and this could correspond to the Albian alkaline volcanic detritus on the Zanskar shelf.

The widespread late Cretaceous pelagic carbonates on the Zanskar shelf (e.g. Kangi-La and Fotu-La units) have counterparts in the pink

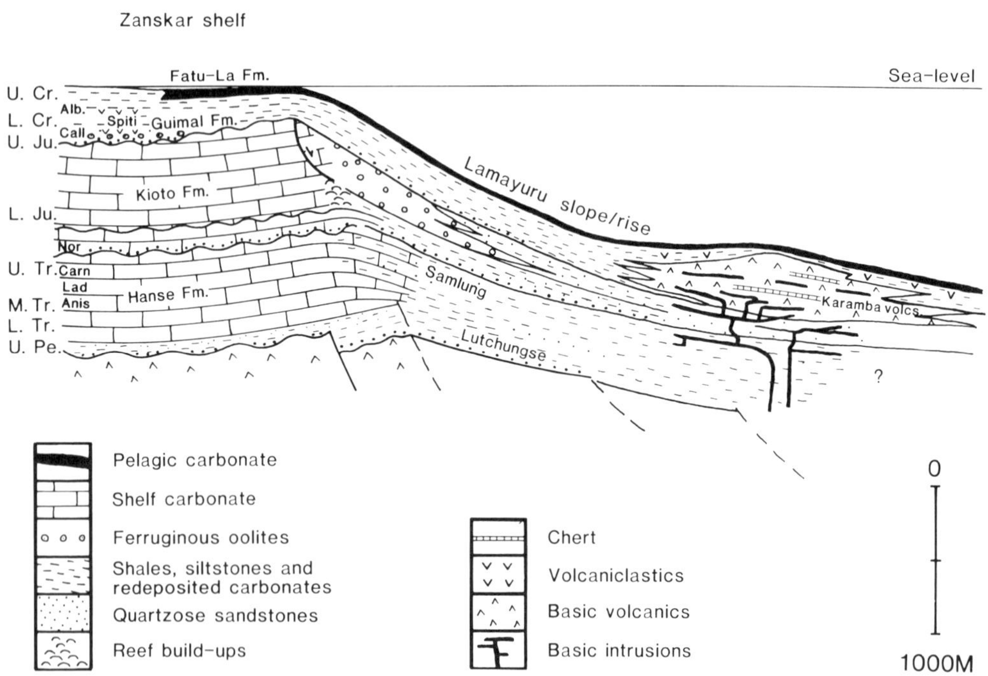

Fig. 11. Reconstruction of the Indian passive margin based on evidence from the Zanskar Mountains and the Indus Suture Zone. Literature sources cited in the text.

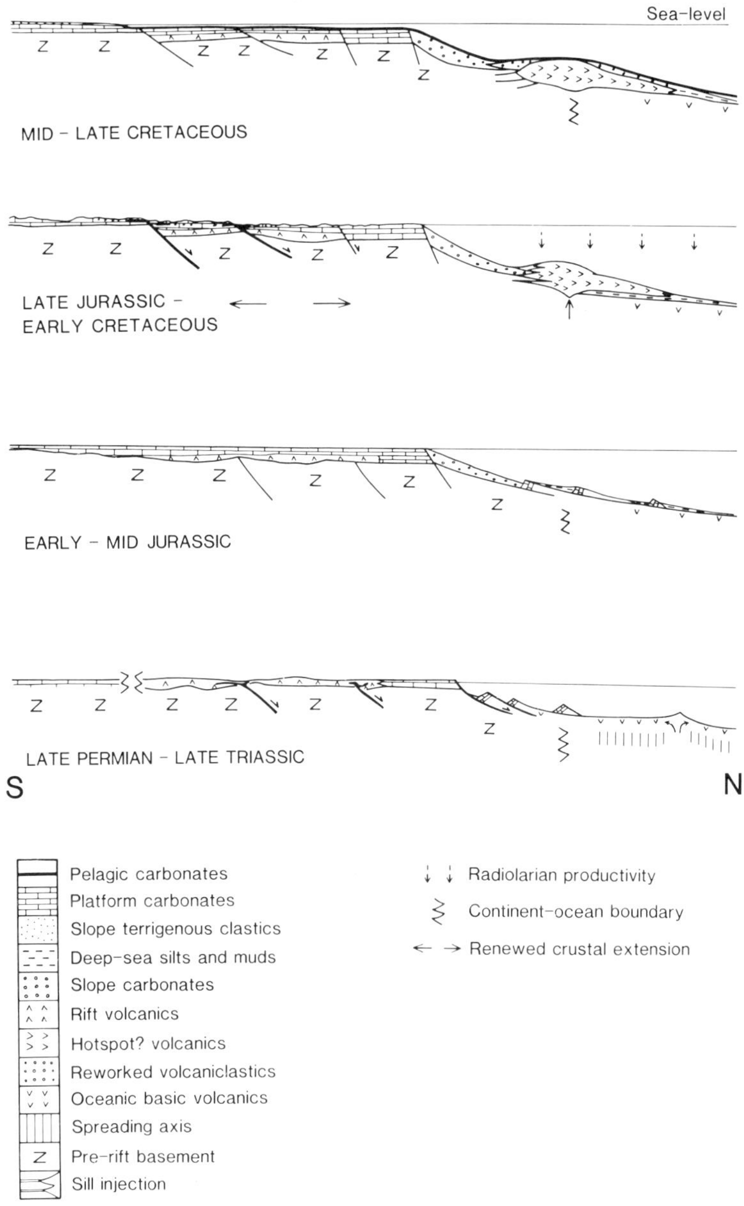

Fig. 12. Summary of the passive margin evolution recorded in the Lamayuru Complex. See text for discussion.

pelagic carbonates in the higher levels of the Lamayuru Complex (e.g. west of Namika-La, between Phanjila and Wanla and in Eastern Ladakh) (Bassoullet 1978; Nicora *et al.* 1987). The pelagic carbonates accumulated during a time of subsidence along the northern Indian passive margin and the northward migration of India (Gaetani & Garzanti 1991). Lastly, the Palaeocene–Eocene benthonic foraminifera reported from the Lamayuru succession by Sinha & Upadhyay (pers. comm.) could possibly relate to late Eocene nummulitic carbonate deposition on the Zanskar shelf (Kesi unit) but more work is needed on this aspect. Overall correlations are summarized in Fig. 11.

Conclusions: summary of deep-water passive margin evolution

The following tectonic/sedimentary evolution can be inferred for the deep-water passive margin units of the Lamayuru Complex.

Permian. Initial spreading took place in the late Permian (Searle *et al.* 1987, 1988; Garzanti *et al.* 1987; Gaetani & Garzanti 1991; Reuber *et al.* 1992). Late Permian proximal, base-of-slope redeposited facies are preserved in eastern Ladakh (Stutz 1988). Elsewhere, Permian shallow-water carbonates were associated with alkaline volcanics. These are interpreted as oceanic seamounts, now preserved as rare exotic blocks within melange in the Indus Suture Zone (at Lamayuru; Tewari & Pande 1970; Colchen *et al.* 1978), and, more extensively, as melange below the Spontang ophiolite (Colchen *et al.* 1987; Reuber *et al.* 1987; Cannat & Mascle 1990).

Triassic. During the early to mid-Triassic terrigenous shales, calciturbidites and siliciclastic carbonates accumulated as small channelised complexes, interbedded with deep-sea muds on the passive margin slope. Thick-bedded, channelised quartzose sandstones apparently were deposited in the late Triassic (Norian), based on correlation with the Zanskar shelf succession.

Jurassic. During the Early Jurassic, relatively thin-bedded calciturbidites, rich in ooids, were introduced from the prograding Zanskar carbonate platform. Then, during the early Mid-Jurassic (Aalenian) the carbonate platform edge distintegrated, in response to regional crustal extension, giving rise to the 'Mulbeck exotics' (western area). At the same time, shallow-water carbonate detritus, including ooids and lithoclastic carbonates were reworked from the platform, giving rise to large channelised units on the lower slope (eastern area). Crustal extension also initiated alkaline volcanism near the inferred continent–ocean boundary in the western area. Swarms of alkaline sills in the western area are assumed to have fed overlying alkaline volcanics. Post-spreading alkaline volcanism in Northern Oman, similarly has been interpreted in terms of reactivation of the continental margin, in response to transform tectonics (Robertson & Searle 1990). MORB crust formed within Neotethys to the north of the Indian plate during the late Jurassic, dated by associated radiolarites, and is now preserved as fragments in associated melange units (e.g. Urtsi Melange; Sutre 1991).

Cretaceous. Volcanism, volcaniclastic, radiolarian and pelagic carbonate deposition took place in deep-water during the Cretaceous, but the record is obscure, because of poor dating and the highly deformed state of the upper parts of the Lamayuru succession in many areas (e.g. western area). Deposition may have continued into the early Tertiary, based on a report of benthic foraminifera (Sinha & Upadhyay pers. comm.). Post-Cretaceous sediments are also present beneath the Spontang ophiolite.

During the late Cretaceous northward drift of India, ocean crust north of the Zanskar/Lamayuru passive margin was subducted, giving rise to oceanic arc volcanism (Reuber 1989), but fragments remained, now represented by blueschists and melanges in the Indus Suture Zone (e.g. Searle 1986; Honegger *et al.* 1989, Spencer 1992) and by melanges below the Spontang ophiolite (Reuber *et al.* 1987, 1992; Colchen *et al.* 1987). However, these units should not be considered as part of the Lamayuru Complex. Indeed, one important conclusion of this work is that only the more proximal slope and base-of-slope units were extensively preserved in the Indus Suture Zone following emplacement, explaining the paucity of pelagic sediments (e.g. radiolarian cherts) relative to other emplaced South Tethyan margins (e.g. Oman; Glennie *et al.* 1973). More distal units were largely subducted and/or emplaced over the Zanskar shelf and later eroded.

The timing of the initial deformation of the Zanskar/Lamayuru passive margin remains controversial. Well-discussed points (see above) opposing late Cretaceous emplacement include the absence of deformation and the entirely terrigenous nature of the Late Cretaceous Zanskar shelf succession south of the Spontang ophiolite. Also, platform margin collapse units and foredeep facies (Muti Formation) that accompanied the late Cretaceous ophiolite emplacement in Oman (Robertson 1987) have not been observed in Ladakh. One alternative is that ophiolitic units were emplaced only onto the distal, leading edge of the Zanskar/Lamayuru passive margin in the latest Cretaceous and then re-thrust to their present position in the post-early Eocene (Searle 1986).

Ophiolitic melange below the Spontang ophiolite contains matrices of both late Cretaceous (upper Campanian) and early Eocene ages in different units. Reuber *et al.* (1992) interpret the late Cretaceous melanges in terms of gravity collapse of the Zanskar/Lamayuru passive margin. However, it is more probable that the late Cretaceous melange formed in a subduction trench setting. The presence of terrigenous clastics within the melange suggests that the latest Cretaceous trench/foredeep was located near the Zanskar/Lamayuru passive margin, rather than far out in the Tethys ocean.

In summary, northward subduction gave rise to a trench/accretion complex in the late Cretaceous, which migrated to a position close to the Zanskar/Lamayuru passive margin, but there is, as yet, no direct evidence that actual collision took place until post-early Eocene time, when a new foredeep migrated southward to its present position on the Zanskar shelf, associated with final emplacement of the ophiolitic melanges and Spontang ophiolite.

This work was funded by a Royal Society expedition grant. We thank M.P. Searle for introducing us to the geology of the Indus Suture Zone. E. Sutre kindly made available Volume 1 of his thesis soon after publication. Thanks go to G. Fuchs, M. Searle and A. Sinha for discussion. E. Bull assisted with drafting and Y. Cooper with photography. Helpful comments on the manuscript were received from E. Garzanti and M. Searle.

References

BASSOULLET, J.-P. 1978. Découverte de Crétacé supérieur calcaire pélagique dans le Zanskar (Himalaya du Ladakh). *Bulletin de la Societé Géologique de France,* (7), **20**, 961–964.

——, BOULIN, J., COLCHEN, M., MARCOUX, G. & MONTENANT, C. 1980*a*. L'évolution des domaine téthysiens du partour du bouchien indien du carbonifère du crétacé. *Report of the 26th International Congress,* **115**, 180–198.

——, COLCHEN, M., JUTEAU, TH., MARCOUX, J. & MASCLE, G. 1980*b*. L'édifice de nappes du Zanskar (Ladakh, Himalaya). *Comptes Rendus de l'Académie des Sciences,* Série D., **290**, 389–392.

——, ——, ——, ——, —— & RIEBEL, G. 1983. Geological studies in the Indus Suture Zone of Ladakh (Himalayas). *In*: GUPTA, V.J. (ed. *Contributions to Himalayan Geology,* **2**, Hindustan Publishing Corporation, Delhi, 96–124.

——, ——, MARCOUX, J. & MASCLE, G. 1978. Une traversale de la zone de l'Indus de Khalsi a Phothaksar, Himalaya du Ladakh. *Compte Rendus de l'Académie Sciences de Paris,* Série D, **286**, 563–566.

——, ——, —— & —— 1981. Les masses calcaires du flysch Triasico-Jurassique le Lamayuru (zone de la suture de L'Indus, Himalaya du Ladakh): klippes sédimentaires et élements de plate-forme rémanies. *Rivista Italiano di Paleontologica e Stratigraphica,* **86**, 825–844.

BAUD, A. 1989. L'autochtonie en pays de nappes: responses à G. Fuchs. *Eclogae Geologicae Helvetiae,* **82**, 693–697.

——, ARN, R., BUGNON, P., CRISINEL, A., DOLINO, E., ESHER, HAMMERSCHLANG, J.-G., MARTHALER, M., MASSON, H., STECK, A. & TIECHE, J.-C. 1982. Le contact Gondwana-peri-Gondwana dans le Zanskar oriental (Ladakh, Himalaya). *Bullétin de la Société Géologique de France,* **24**, 341–361.

——, GAETANI, M., GARZANTI, E., FOIS, E., NICORA, A. & TINTORI, A. 1984. Geological observations in southeast Zanskar and adjacent Lahul area (northwestern Himalaya). *Eclogae Geologicae Helvetiae,* **77**, 171–197.

BROOKFIELD, M.E. & ANDREWS-SPEED, C.P. 1984. Sedimentology, petrography and tectonic significance of the shelf, flysch and molasse clastic deposits across the Indus suture zone, Ladakh, NW India. *Sedimentary Geology,* **40**, 249–286.

BOUMA, A.H. 1962. *Sedimentology of some flysch deposits*. Elsevier, Amsterdam.

CANNAT, M. & MASCLE, G. 1990. Réunian extraodinaire de la Société Géologique de France en Himalaya du Ladakh. *Bullétin de la Société Géologique de France,* (8), **6**, No 4, 53–582.

COLCHEN, M., GUEX, J., LYS, M., MARCOUX, J. & MASCLE, M. 1978. Permien terminal néritique, Scythien pélagique et volcanique sous-marin, indices de la suture de l'Indus (Himalaya du Ladakh). *Comptes Rendus de l'Académie des Sciences, Paris*, Série D, **287**, 675–678.

——, MASCLE, G. & VAN HAVER, T. 1985. Some aspects of collision tectonics in the Indus Suture Zone, Ladakh. *In*: COWARD, M.P. & RIES, A. (eds) *Collision Tectonics* Geological Society, London, Special Publications, **19**, 175–186.

——, REUBER, I., BASSOULLET, J.P., BELLIER, J.P., BLONDEAU, A., LYS, M. & DE WEVER, P. 1987. Donneés biostratigraphique sur les mélanges du Zanskar, Himalaya du Ladakh. *Comptes Rendus de l'Académie des Sciences, Paris* Série 2, **305**, 305–306.

DE WEVER, P. 1989. Radiolarians, radiolarites, and Mesozoic palaeogeography of the Circum-Mediterranean Alpine Belts. *In*: HEIM, J.R. & OBRADOVIĆ, J. (eds) *Silceous Deposits of the Tethys and Pacific Regions*. Springer, 31–50.

FRANK, W., GANSSER, A. & TROMMSDORF, V. 1977. Geological observations in the Ladakh Himalaya. *Schweizerische Mineralogische und Petrographisches Mitteilungen,* **57**, 89–113.

FUCHS, G. 1979. On the geology of western Ladakh. *Geologisches Jahrbuch,* **122**, 513–540.

—— 1982. The geology of the Western Zanskar. *Geologisches Jahrbuch,* **122**, 513–540.

—— 1986. The geology of the Markha-Khurnak Region in Ladakh (India). *Geologisches Jahrbuch,* **128**, 403–437.

—— 1989. Arguments for the autochoneity of the Tibetan Zone. *Eclogae Geologicae Helvetiae,* **82**, 685–692.

—— & SINHA, A.K. 1992. A terrane map of the Himalayas (abstract). *7th Himalaya–Karakoram–Tibet Workshop, University of Oxford, U.K. April, 1992*, p. 27.

GAETANI, M. & GARZANTI, E. 1991. Multicycle history of the Northern India continental margin (Northwestern Himalayas). *American Association of Petroleum Geologists Bulletin,* **75**, 1427–1446.

——, CASNEDI, R., FOIS. E., GARZANTI, E., JADOUL, F., NICORA, A. & TINTORI, A. 1986. Stratigraphy of the Tethys Himalaya in Zanskar, Ladakh. *Rivista Italiano di Paleontologica e Stratigraphica,* **91**, 443–478.

——, NICORA, A., SILVA, I.P., FOIS, E., GARZANTI, E. & TINTORI, A. 1983. Upper Cretaceous and Paleocene in Zanskar range (NW Himalaya). *Rivista Italiano di Paleontologica e Stratigraphica,* **89**, 81–118.

GARZANTI, E. 1993. Sedimentary evolution and drowning of a passive margin shelf (Giumal Group, Zanskar Tethys Himalaya, India): Palaeoenvironmental changes during final break-up of Gondwanaland. *This volume.*

——, BAUD, A. & MASCLE, G. 1987. Sedimentary record of the northward flight of India and its collision with Eurasia (Ladakh Himalaya, India), *Geodinamica Acta (Paris),* **1**, 297–312.

GLENNIE, K.W., BOEUF, M.G.A., HUGHES CLARKE, M.W., MOODY-STUART, M., PILAAR, W.F.H. & REINHARDT, B.M. 1973. Late Cretaceous nappes in the Oman Mountains and their geological significance. *American Association of Petroleum Geologists Bulletin,* **57**, 5–27.

HONEGGER, K. 1983. *Strukturen and Metamorphose im Zanskar Kristallin (Ladakh, Kashmir, Indien).* PhD thesis, ETH Zurich, Switzerland.

——, DIETRICH, V., FRANCK, W., GANSSER, A., THONI, M. & TROMMSDORF, V. 1982. Magmatism and metamorphism of the Ladakh Himalayas (the Indus-Tsanpo suture zone). *Earth and Planetary Science Letters,* **60**, 253–292.

——, LE FORT, P., MASCLE, G. & ZIMMERMAN, J.L. 1989. The blueschists along the Indus suture zone in Ladakh, NW Himalaya. *Journal of Metamorphic Geology,* **7**, 57–72.

KELEMEN, P.B. & SONNENFELD, M.D. 1983. Stratigraphy, structure, petrology and local tectonics, Central Ladakh, NW Himalaya. *Scweitzerische Mineralagische und Petgraphische Mitteilungen,* **63**, 276–287.

NICORA, A., GARZANTI, E. & FOIS, E. 1987. Evolution of the Tethys Himalaya continental shelf during Maastrichtian to Paleocene (Zanskar Himalaya), *Rivista Italiano di Paleontologica e Stratigraphica,* **92**, 439–496.

PATRIAT, P.J., SEGOUFIN, R., SCHLICH, R., GOSLIN, J., AUZENFDE, J.M., BEUZART, P., BONNIN, J. & OLIVET, J.L. 1982. Les mouvements relatifs de l'Inde, de l'Afrique et de l'Eurasie. *Bulletin de la Societé Géologique de France,* **24**, 363–373.

REUBER, I. 1989. The Drac arc: two successive volcanic events on eroded oceanic crust. *Tectonophysics,* **161**, 93–106.

——, COLCHEN, M. & MEVEL, C. 1987. The geodynamic evolution of the South Tethyan margin in Zanskar, NW Himalaya, as revealed by the Spontang ophiolitic melanges. *Geodynamic Acta (Paris),* **1**, 283–296.

——, —— & —— 1992. The Spontang Ophiolite and ophiolitic melanges of the Zanskar, NW Himalaya, Tracing the evolution of the closing Tethys in the upper Cretaceous to the early Tertiary. *In*: SINHA, A.K. (ed.) *Himalayan Orogen and Global Tectonics.* Oxford and IBH Publishing Co., New Delhi, 235–26.

ROBERTSON, A.H.F. & DEGNAN, P.J. 1992. Mesozoic-Tertiary tectonic evolution of the Indus suture zone: Ladakh Himalaya. *7th Himalaya-Karakorum–Tibet Workshop, Oxford University, April, 1992*, 76–77.

—— & SEARLE, M.P. 1990. The Northern Oman Tethyan continental margin: stratigraphy, structure, concepts and controversies. *In*: ROBERTSON, A.H.F., SEARLE, M.P. & RIES, A.C. (eds) *The Geology and Tectonics of the Oman Region.* Geological Society, London, Special Publications, **49**, 3–25.

—— & WOODCOCK, N.H. 1982. Sedimentary history of the southwestern segment–Tertiary Antalya continental margin, southwestern Turkey. *Eclogae Geologicae Helvetiae,* **75**, 517–562.

——, CLIFT, P.D. DEGNAN, P.J. & JONES, G. 1991. Palaeogeographic and palaeotectonic evolution of the Eastern Mediterranean Neotethys. *Palaeogeography, Palaeoclimatology, Palaeoecology,* **87**, 289–343.

SEARLE, M.P. 1983. Stratigraphy, structure and evolution of the Tibetan-Tethys zone in Zanskar and the Indus suture zone in the Ladakh Himalaya. *Royal Society of Edinburgh Transactions, Earth Sciences,* **73**, 205–219.

—— 1986. Structural evolution and sequence of thrusting in the High Himalayan Tibetan-Tethys and Indus suture zones of Zanskar and Ladakh, western Himalaya. *Journal of Structural Geology,* **8**, 923–938.

——, COOPER, D.W.J. & REX, A.J. 1988. Collision tectonics of the Ladakh–Zanskar Himalaya. *Philosophical Transactions of the Royal Society, London,* **A326**, 117–150.

——, WINDLEY, B.F., COWARD, M.P., COOPER, D.W.J., REX, A.J., LI TINGDONG, D.C., XUCHANG, X., JAN, M.Q., THAKUR, V.C. & KUMAR, S. 1987. The closing of Tethys and the tectonics of the Himalayas. *Geological Society of America Bulletin,* **98**, 678–701.

SINHA, A.K. & UPADHYAY, R. 1992. Geodynamic and significance of algal oncolite structures recorded in the Dras and Lamayuru area, western Ladakh Himalaya, India. *Current Science,* **62**, 530–532.

SPENCER, D. 1992. Evidence of the deepest and earliest derived nappe of the Indian plate, NW Himalaya, Pakistan. *Symposium on Himalayan geology, Shimane '92*, Japan, Abstract, 51–55.

STUTZ, E.1988. *Géologie de la chaine de Nyimaling aux confins du Ladakh et du Rupshu (NW Himalaya, Inde), évolution palaeogéographique et téctonique d'un segment de la marge nord-indienne.* Memoires de Géologie, Imprivité, Lausanne, **3**

SUTRE, E. 1991. *Les formations de la marge nord-Neotethsienne et les mélanges ophiolitique de la zone de suture de l'Indus en Himalaya du Ladakh.* Thèse, Docteur de L'Universite de Poitiers, Tome 1.

TALON, CH. 1988. L'unité des melanges d'Omlung

(Abstract). *Himalaya–Karakorum–Tibet Workshop Meeting*, **36**, Lausanne.

TEWARI, B.S. & PANDE, I.C. 1970. Permian fossiliferous limestone from Lamayuru, Ladakh. *Publication of the Centre of Advanced Study in Geology*, Panjab University, Chandigarh, **7**, 188–190.

THAKUR, V.C. 1981. Regional framework and geodynamic evolution of the Indus–Tsangpo suture zone in the Ladakh Himalayas. *Transactions of the Royal Society of Edinburgh, Earth Sciences*, **72**, 89–97.

—— 1990. Indus Tsangpo suture in Ladakh–biostratigraphy and tectonics. *Proceedings of the Indian Academy of Science*, **99**, 169–185.

WATTS, K.W. & GARRISON, R.E. 1986. Sumeini Group, Oman–Evolution of a Mesozoic carbonate slope on a South Tethyan continental margin. *Sedimentary Geology*, **48**, 107–168.

Metamorphic zonation, migmatization and leucogranites along the Everest transect of Eastern Nepal and Tibet: record of an exhumation history

UGO POGNANTE* & PIERA BENNA

Dipartimento di Scienze Mineralogiche e Petrologiche and C. S. Geodinamica Catene Collisionali (CNR), Via Valperga Caluso 37, 10125 Torino, Italy

Abstract: A polyphase evolution and a thermal and mineral zonation characterize metamorphism of the High Himalayan Crystallines (HHC) in the Everest transect. After an early Barrovian-type metamorphism (M1: T = 550–680° C; P = 8–10 kbar) caused by subduction and crustal thickening during India–Eurasia collision, exhumation of the HHC occurred at rates which allowed the isotherms to rise and the thickened crust to suffer partial melting. In the upper structural levels, the thermal conductivity contrast between crystalline and upper sedimentary rocks together with the infiltration of metamorphic fluids released during exhumation, favoured migmatization and formation of leucogranite magmas through partial melting of muscovite-bearing metasediments chiefly at medium (M2: T = 650–750° C; P = 4–7 kbar) but also at low-pressure (M3: T = 600–700° C; P = 2–4 kbar). The leucogranite melts segregated as dykes and kilometre-sized bodies into extensional shear zones situated at the uppermost levels of the HHC and at the base of the overlying Tibetan Series. During exhumation, a metamorphic zonation was imposed on the HHC in response to the complex interplay between rise of the isotherms in the thickened crust, emplacement of leucogranites in the uppermost levels of the HHC and conductive cooling near the 'cold' Lesser Himalaya and Tibetan Series coupled with thrust imbrication.

The High Himalayan Crystallines (HHC), a tectonic unit which is part of the collided and thickened Indian plate, has been studied along the north–south transect of Mt Everest (Chomolungma; 8848 m) in the Khumbu region, the Arun and Barun valleys (Nepal) and the Rongbuk valley (Tibet). A complex metamorphic evolution and a metamorphic and thermal zonation are recorded within the HHC and the adjacent tectonic units in the cross-strike Everest transect. As in other regions of the central and eastern Himalayas (e.g. Garhwal, Central Nepal, Bhutan; Gansser 1991) the upper structural levels of the HHC show extensive migmatization and are intruded by Tertiary leucogranites. Kilometre-sized plutons of leucogranite are very well exposed in the 3000 m high south faces of Cho Oyu (8153 m), Nuptse (7879 m) and Makalu (8475 m). Based on new field, petrographical and petrological data, this paper provides a metamorphic map of the Everest transect, describes the metamorphic history of the HHC units and discusses the relations between metamorphism, tectonism and Tertiary leucogranites. A companion paper (Lombardo *et al.* this volume) describes the geology of the Everest transect and discusses the tectonomagmatic evolution.

Geological setting and previous studies

The HHC is tectonically bounded at the top by the low-grade metasediments of the Tibetan Series and at the bottom, beneath the Main Central Thrust zone (MCT), by the Lesser Himalaya (Fig. 1).

High Himalayan Crystallines and Main Central Thrust zone

After the pioneering works of Odell (1948), Lombard (1958), Bordet (1961), Wager (1965), Krummenacher (1967) and Hagen (1969), the geology and petrography of the Everest region have been reinvestigated by Bortolami *et al.* (1974, 1976), Yin & Kuo (1978) and Brunel (1983). In addition to providing much needed geothermobarometric data, recent studies by Brunel and Kienast (1986) and Hubbard (1989) discussed the relationship between tectonic history and metamorphic evolution, a

* Following the death of Ugo Pognante in a ski-ing accident this manuscript has been revised on his behalf by Bruno Lombardo and Piera Benna

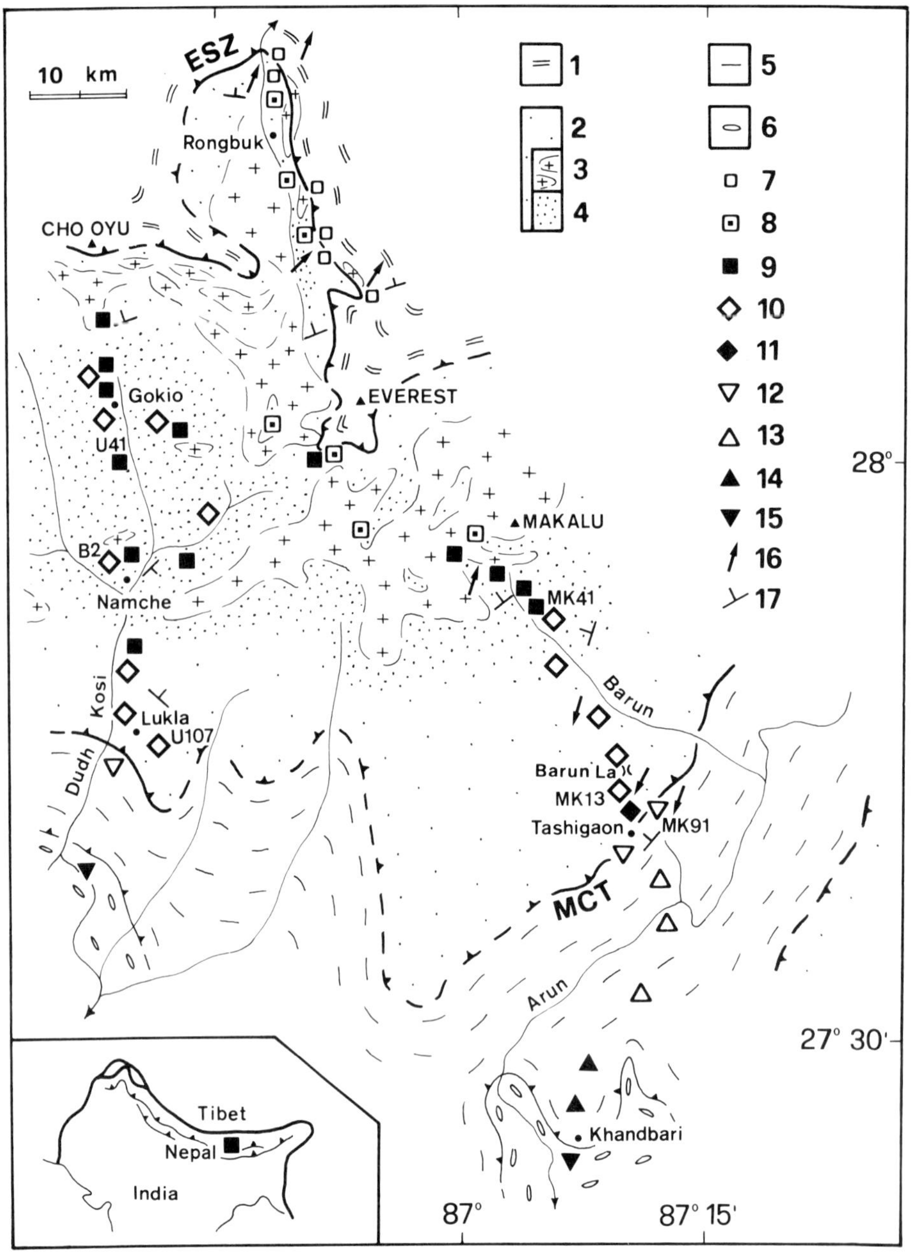

Fig. 1. Tectonic sketch map of the Everest–Makalu region (eastern Nepal and Tibet) with distribution of the main metamorphic assemblages of gneiss and micaschist. Legend. 1, Tibetan Series; 2, HHC; 3, HHC leucogranites; 4, zones of extensive migmatization in the HHC; 5, Lesser Himalaya orthogneiss; 6, Lesser Himalaya metasediments. Location of the samples analyzed in Table 1 is indicated. Main assemblages are: 7, quartz–muscovite–biotite–albite/plagioclase; 8, quartz–plagioclase–biotite–muscovite–sillimanite ± orthoclase ± cordierite; 9, quartz–plagioclase–biotite–orthoclase–cordierite ± andalusite ± sillimanite; 10, quartz–plagioclase–biotite–orthoclase–garnet–sillimanite ± rutile; 11, quartz–plagioclase–biotite–muscovite–kyanite–garnet ± sillimanite ± orthoclase ± rutile; 12, quartz–biotite–muscovite–plagioclase–kyanite ± garnet ± staurolite; 13, quartz–plagioclase–biotite–muscovite ± orthoclase and quartz–phlogopite–kyanite–muscovite; 14, quartz–biotite–muscovite–garnet ± plagioclase/albite; 15, quartz–muscovite–garnet ± chloritoid and quartz–muscovite–chlorite ± garnet.; 16, main lineation trajectories defining NNE extension and SSW compression; 17, representative dips of foliation planes and of the ductile tectonic discontinuities. ESZ, upper extensional shear zones; MCT, Main Central Thrust zone. The location of the MCT between Arun and Dudh Kosi is after Bordet (1961), that of the ESZ south of Mt Everest is after Bortolami *et al.* (1976).

hotly-debated topic in the Himalayan mountain belt, and a major point of disagreement between them. For Brunel & Kienast the Himalayan metamorphism in the HHC is polyphase with an early Barrovian stage followed by heating during decompression with the metamorphic zonation (the well known 'inverted metamorphism') essentially resulting from post-metamorphic thrusting along the MCT. For Hubbard (1989) the Himalayan metamorphism is essentially a single-stage event, and the inverted metamorphism results from synmetamorphic thrusting along the MCT as proposed by Le Fort (1975).

The HHC of the Everest region has been subdivided into: a complex of migmatized paragneiss with minor metabasites, calc-silicate rocks and marbles ('Gneiss du Barun': Bordet 1961; 'Gneiss inférieur du Khumbu': Krummenacher 1967); a complex of migmatized orthogneiss ('Migmatites de Namche Bazar': Lombard 1958; 'Migmatites du Khumbu': Krummenacher 1967); and an upper complex of black paragneiss ('Gneiss Noirs': Bordet 1961; 'Gneiss supérieurs du Khumbu': Krummenacher 1967).

In agreement with the radiometric ages of Ferrara *et al.* (1983, 1992), our field observations suggest that the orthogneiss is a pre-Himalayan (Early Palaeozoic) granitoid intruded into a sequence of metasedimentary rocks, an hypothesis originally proposed by Bortolami *et al.* (1976). During the Himalayan history the granitoids have been altered to migmatized orthogneiss while the metasedimentary host rocks have been further metamorphosed into a migmatized paragneiss which should thus be considered a polymetamorphic rock. For all these rock types, isotopic data derived from separates of the metamorphic minerals give prevalent Miocene cooling ages (Bassett *et al.* 1978; Ferrara *et al.* 1983; Hubbard & Harrison 1989).

Layers of metabasic rocks a few metres thick occur within the ortho- and paragneiss and probably derive from pre-Himalayan dykes. Magmatic relicts of plagioclase and orthopyroxene occur in a few metamorphosed porphyritic dykes from the Barun valley.

The upper complex of black paragneiss has been considered to be either a monometamorphic sedimentary cover of the other complexes (Bortolami *et al.* 1976) or, chiefly on the base of isotope data, a polymetamorphic complex (Ferrara *et al.* 1983). In black paragneiss from the south face of Makalu, we have found metaconglomeratic levels with rounded quartz pebbles and small cobbles of orthogneiss. If the orthogneiss has been derived from Palaeozoic granitoids which suffered only the Himalayan metamorphism, the black paragneiss (or at least part of it) should represent the monometamorphic cover of the basement rocks.

The Tertiary leucogranites are concentrated in the upper part of the HHC, notably in the upper black paragneiss (Fig. 2a). They are rich in metasedimentary xenoliths and form concordant layers, dykes and lens-shaped bodies (up to a few kilometres in size) such as are exposed on the south faces of Cho Oyu, Nuptse and Makalu. A few leucogranite bodies show a preferred orientation of micas and form veins and layers which have been affected by ductile deformation along extensional shear zones. The floors of many bodies are not well defined as they grade into anatectic leucosomes and migmatites (Fig. 2a) while the roofs only locally (e.g. Nuptse: Bortolami *et al.* 1976) have a rounded or mushroom shape, suggestive of diapiric uprise. At the uppermost levels of the HHC syn-tectonic pegmatitic dykes and layers are numerous (Fig. 2b) and are crosscut by fine-grained leucogranites. The age of the leucogranites is constrained by U-Th-Pb data on the Makalu and Rongbuk plutons (Schärer 1984; Hodges *et al.* 1992) and by Rb-Sr and Ar-Ar data on magmatic micas of the Nuptse–Lhotse leucogranites (Ferrara *et al.* 1983; Villa 1990). These data indicate formation and cooling during the Miocene, specifically between 24 and 15 Ma.

The Main Central Thrust (MCT) is a complex mylonitic zone a few hundred metres thick which generally dips 15–30° to the north and separates the HHC from the underlying Lesser Himalaya. It has been studied previously in the Arun valley where the prevalent monoclinal structure dipping to the north is complicated by later open megafolds (i.e. the 'transanticlinal de l'Arun': Bordet 1961). In recent times, the MCT has been studied by Brunel (1983) in the Arun valley and by Hubbard (1989) in the region between the Dudh Kosi and Hongu rivers. The MCT is defined by tectonic sheets of micaschist, gneiss, quartzite and amphibolite (Bordet 1961). S–C fabrics and stretching lineations oriented around NNE-SSW are very common in the MCT zone and in the lower HHC and indicate SSW thrusting of the HHC.

Tibetan Series

The contact between the HHC and the overlying Tibetan Series is well exposed in the Rongbuk valley, where it dips 10–15° to the north (Fig. 1). Here the gneisses of the HHC structurally underlie a series of fine-grained schists and impure marbles of the, possibly Sinian–Cambrian, North Col Formation of Yin & Kuo (1978) (Fig. 2b). A mylonitic zone several tens of

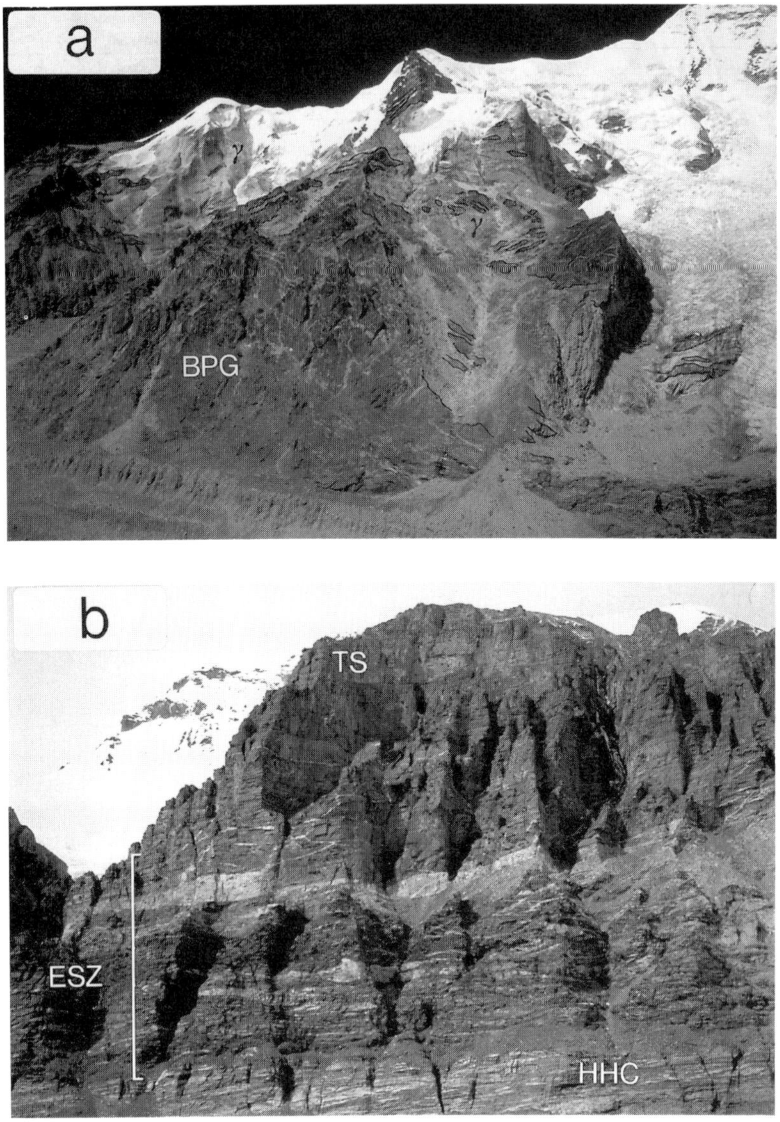

Fig. 2. (**a**) Migmatized black paragneiss (BPG) with different generations of dykes, layers and lens-shaped bodies of Miocene leucogranites (γ) including abundant metasedimentary xenoliths. Upper structural levels of the HHC south of Makalu (east face of Peak 4, upper Barun valley, Nepal). The floors of the leucogranite bodies are not well defined and grade into the migmatized paragneiss. The face is about 1000 m high and the altitude of its base is 5200 m. (**b**) Uppermost part of the HHC and lower part of the Tibetan Series (TS) (south ridge of Lixin Peak, East Rongbuk Glacier, Tibet). Abundant syn-tectonic pegmatitic layers and minor leucogranites occur in the extensional shear zone (ESZ) along the tectonic boundary between HHC and TS. The face is about 600 m high and the altitude of its base is 6000 m.

metres thick occurs along the contact. In the mylonitic rocks of the lowermost Tibetan Series and of the uppermost HHC, shear-sense indicators (chiefly S–C fabrics) indicate an extensional regime with NNE-directed shear of the Tibetan Series. In agreement with the observations of Burg *et al.* (1987) at the top of the HHC in southern Tibet and particularly with the kinematic interpretation proposed for the Everest area by Burchfiel *et al.* (1992) on the base of mapping and observations along the east side of Rongbuk valley, this contact is interpreted here as an extensional ductile shear zone (ESZ).

Hectometre-sized bodies of orthogneiss and deformed leucogranite occur in the ESZ (Fig. 2b). Though the orthogneiss could derive from pre-Himalayan granite as suggested by some Rb-Sr data (Yin & Kuo 1978), the deformed leucogranite is petrographically similar to the Tertiary leucogranites and shows the same S–C fabrics and stretching lineations orientated NNE–SSW as are observed in the host schist (observations with B. Lombardo and P. Pertusati).

Lesser Himalaya

The upper part of the Lesser Himalaya is a thick mylonitic unit interposed between the MCT and the structurally lower metasediments. It corresponds to the 'Migmatites du Bas Himalaya' of Bordet (1961) and consists dominantly of orthogneiss, probably derived from Palaeozoic granitoids (Kay 1981). Kyanite-bearing phlogopite schist forming metre-thick layers, shows a gradual transition to the mylonitic orthogneiss and probably formed along syn-metamorphic shear zones. As in the MCT, orthogneiss and micaschist are characterized by S–C fabrics and stretching lineations which indicate SSW-directed thrusting.

The lower part of the Lesser Himalaya consists of phyllite and quartzite which, near Khandbari, are overthrust by the mylonitic orthogneiss (Fig. 1).

Distribution of the metamorphic assemblages and microstructural relations

The distribution of the main metamorphic assemblages along the Everest transect is shown in the metamorphic map of Fig. 1. A complex metamorphic evolution with three texturally different stages (M1, M2, M3) and a metamorphic zonation are recorded in the HHC.

High Himalayan Crystallines and Main Central Thrust zone

Mylonitic and S–C fabrics which indicate SSW-directed thrusting above the Lesser Himalaya are very common in the rocks of the MCT zone. These rocks show the following assemblages: quartz – biotite – muscovite – kyanite – plagioclase ± garnet ± staurolite and quartz–biotite–muscovite–staurolite–plagioclase in the gneisses and micaschists; and hornblende–plagioclase–sphene ± biotite ± garnet in the metabasics. Micas, kyanite and amphiboles define the planar fabric; helicitic inclusion trails are common in syn-tectonic garnet porphyroblasts and kyanite partly replaces staurolite.

At the lower to middle structural levels of the HHC paragneiss and orthogneiss have planar or S–C fabrics that record a SSW-directed sense of shear. Starting from the MCT zone (i.e. above Tashigaon in the Arun valley and south of Lukla in the Khumbu region) and extending up to the middle levels (i.e. the Barun valley up to Nehe Kharka-Yanle and the Dudh Kosi valley up to Gokyo), the following representative assemblages have been observed: quartz–plagioclase–biotite–muscovite–garnet–kyanite ± rutile, quartz – plagioclase – biotite – muscovite – garnet – sillimanite and quartz–plagioclase–biotite–sillimanite–garnet–orthoclase ± rutile in the gneisses; and plagioclase–hornblende–sphene–biotite ± quartz ± garnet in the metabasic rocks. The sillimanite-bearing gneiss is the most widely distributed rock type and shows increasing degrees of migmatization toward the middle levels. In these rocks the planar fabric is defined by biotite and prismatic or fibrolitic sillimanite wrapping around garnet porphyroblasts. Prismatic sillimanite (Fig. 3a) defines lineations orientated SSW–NNE or SW–NE, while in many samples corroded relicts of kyanite are included within garnet (Fig. 3b). Consequently, besides the metamorphic zonation, the rocks of the lower and middle levels record the evolution from an early kyanite-bearing stage M1 (completely preserved only at the lowermost levels) to an orthoclase-sillimanite bearing stage M2 that is muscovite absent. Migmatization and formation of concordant garnet–orthoclase–sillimanite-bearing leucosomes are related to M2.

At upper levels of the HHC (upper Barun valley, Dudh Kosi valley north of Gokyo) migmatization is more extensive than at lower levels and migmatized gneiss including varieties with centimetre-sized aggregates of sillimanite (fibrolite) and quartz is rather common. Except for very rare corroded grains of staurolite and kyanite, relicts of M1 are lacking at middle structural levels and the most common M2 assemblages are: quartz–plagioclase–orthoclase–biotite–garnet–sillimanite in the gneisses; plagioclase–hornblende–biotite–quartz–sphene ± garnet in the metabasic rocks; and plagioclase–diopside–biotite–quartz–calcite–sphene ± hornblende ± orthoclase ± garnet ± scapolite in the calc-silicate units. M2 is associated with the formation of numerous garnet–sillimanite–orthoclase-bearing leucosomes (Fig. 4a) and was

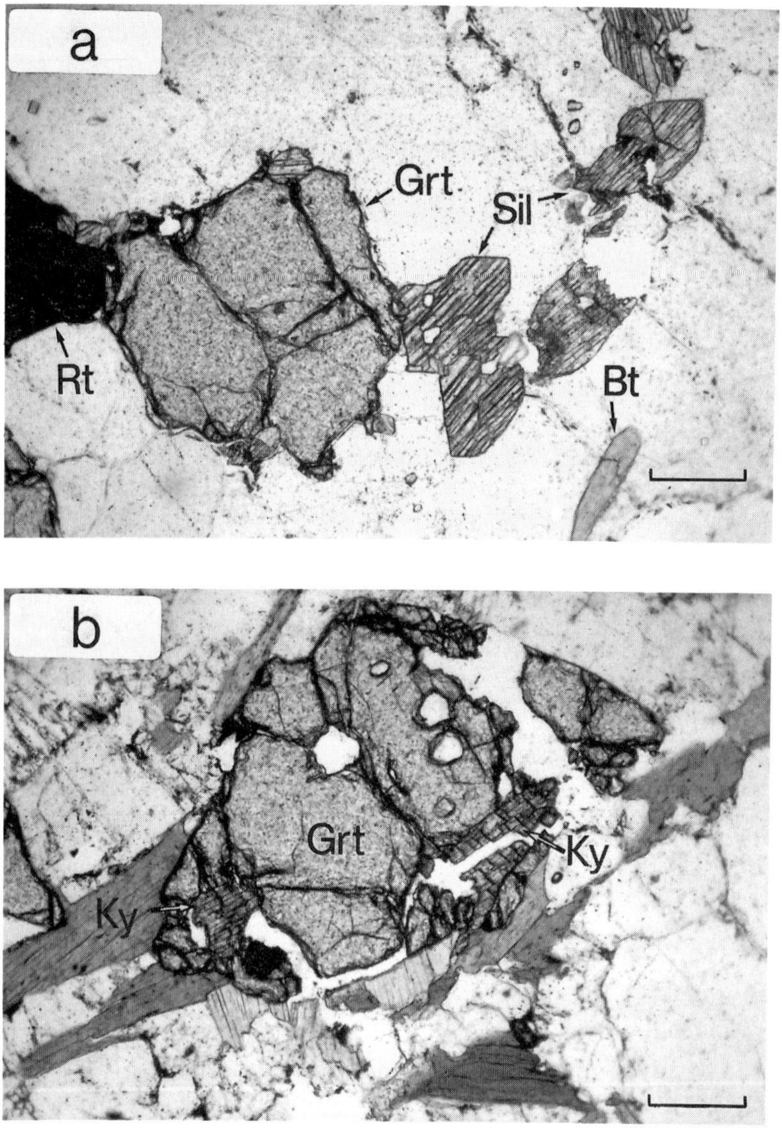

Fig. 3. (**a**) Garnet–biotite-bearing gneiss with prismatic sillimanite and rutile (sample MK83). Lower structural levels of the HHC (south of Barun La, Arun Valley, 4100 m; Nepal). The bar is 0.26 mm long. (**b**) Relicts of kyanite preserved within garnet (sample MK83). The bar is 0.2 mm long.

followed by a further period of migmatization (M3). M3 is characterized by the formation of andalusite-cordierite-orthoclase-bearing discordant leucosomes which cross-cut M2 leucosomes (Fig. 4b) and by the growth of cordierite which borders and partly replaces syn-tectonic garnet and sillimanite (Fig. 5a). M3 is very well recorded in the upper black paragneiss where the cordierite-andalusite-bearing leucosomes are common (Fig. 5b) and the stable assemblage is quartz–plagioclase–biotite–orthoclase–cordierite ± andalusite with abundant relicts of fibrolite and rare garnet.

The leucogranites concentrated in the upper structural levels are characterized by the following magmatic assemblages: quartz–orthoclase–albite–biotite–muscovite ± garnet ± sillimanite or quartz–orthoclase–albite–biotite–muscovite–

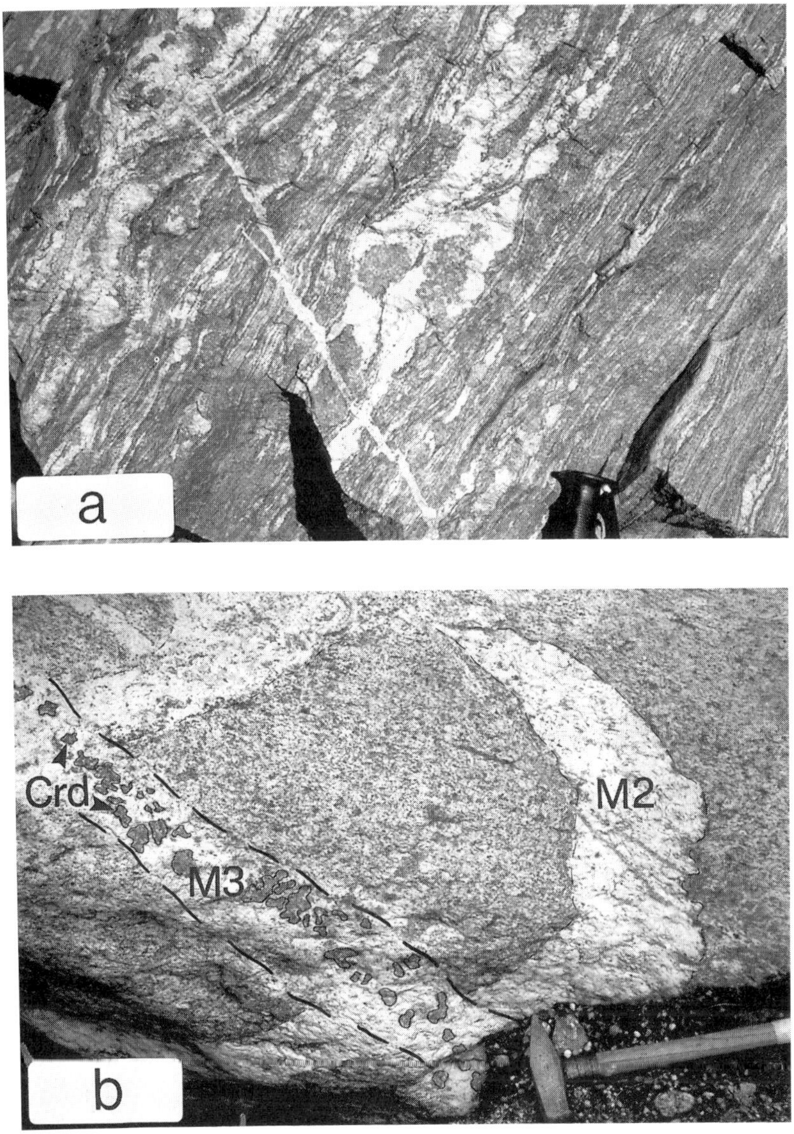

Fig. 4. (**a**) Concordant garnet–sillimanite-bearing leucosome M2 in migmatized paragneiss from the middle structural levels of the HHC (Mera camping ground, Barun valley, 4500 m; Nepal). (**b**) Discordant cordierite-bearing leucosome M3 crosscuts a folded sillimanite-bearing leucosome M2 in migmatized orthogneiss (north of Namche Bazar, 4200 m; Nepal).

cordierite–andalusite. Large crystals of cordierite and andalusite are very common in the leucogranites associated with the upper black paragneiss. Cordierite also occurs in centimetre-thick veinlets and is partly to completely transformed to pinite. Tourmaline is an abundant accessory mineral.

At the uppermost levels of the HHC (upper part of the south faces of Cho Oyu, Lhotse–Nuptse and Makalu; Rongbuk valley) and in the extensional shear zone at the top of the HHC the gneiss has planar or mylonitic fabrics. S–C fabrics that record NNE-directed extension are common in these rocks. The following assemblages have been recorded: quartz–plagioclase biotite–muscovite–sillimanite (fibrolite) ±

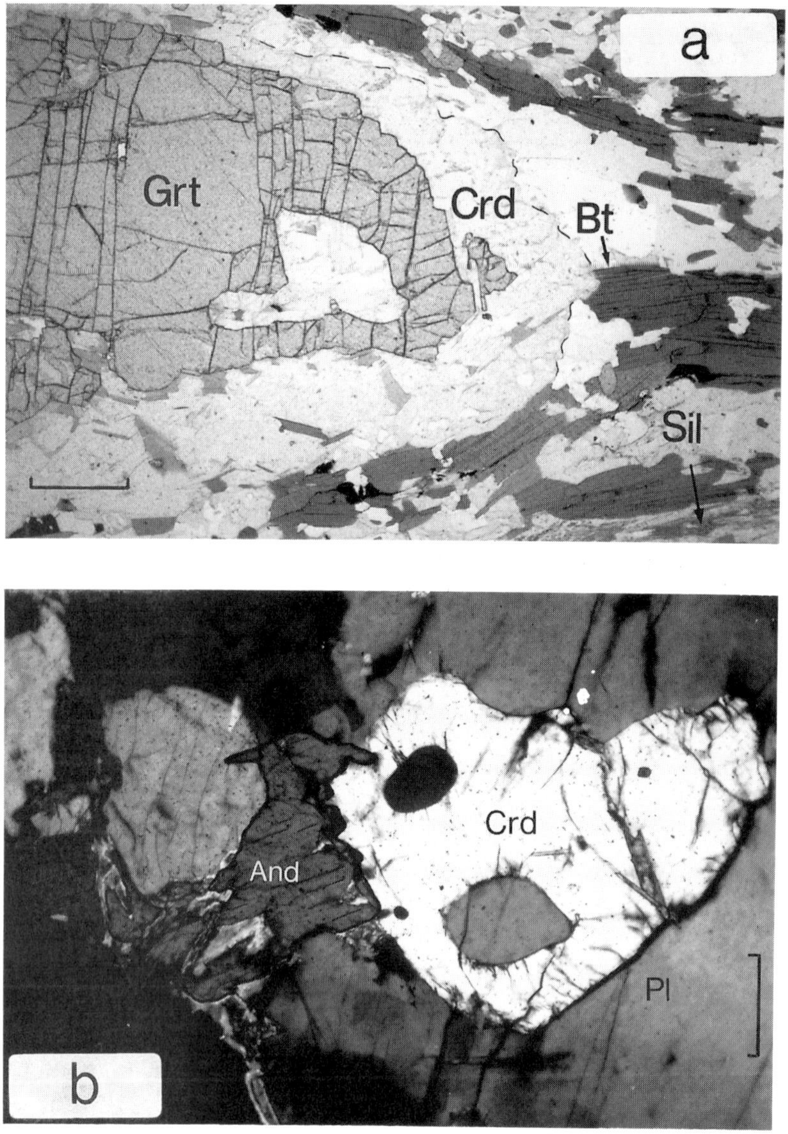

Fig. 5. (**a**) Biotite–sillimanite-bearing gneiss with corroded garnet partly replaced by cordierite (sample U54). Upper structural levels of the HHC (south ridge of Kangchung, east of Gokyo, 4900 m; Nepal). The bar is 0.54 mm long. (**b**) Andalusite–cordierite-bearing leucosome M3 (sample MK52). Upper structural levels of the HHC (northeast face of Peak 6380 m, upper Barun valley, 5200 m; Nepal). The bar is 0.2 mm long.

orthoclase with rare relicts of garnet in the paragneisses; quartz–plagioclase–hornblende–biotite–sphene ± orthoclase in amphibolitic gneisses often associated with leucogranite and calc-silicate rocks; quartz-diopside–plagioclase–hornblende–sphene ± scapolite ± epidote ± vesuvianite ± calcite in the calc-silicate rocks. These assemblages and the numerous syn-tectonic pegmatitic dykes are hypothetically ascribed to M2. Kyanite, a possible relict of the M1 stage, was reported from the upper East Rongbuk Glacier by Wager (1965) and corroded relicts of staurolite (M1?) partly replaced by cordierite (M3?) have been reported by Bortolami *et al.* (1976) from the south faces of Nuptse and Lhotse. The mylonites from the ESZ are characterized by the growth of a second generation of very fine-grained micas (MsII and BtII) which replace sillimanite and, in rocks showing S–C fabrics, define the C shear planes. This incipient recrystallization, hereafter defined M3*, is more pervasive towards the contact with the Tibetan Series. On paragenetic and textural grounds M3* appears to be younger than the M3 migmatization described in the upper and middle levels of the HHC.

Tibetan Series

The metasediments of the North Col Formation show pervasive metamorphic recrystallization and have mylonitic and S–C fabrics restricted to the ESZ. The fine-grained metamorphic minerals define the planar fabric and often show evidence of polyphase recrystallization. This is consistent with a syn-kinematic growth. Except for the lack of sillimanite, the assemblages are similar to those of the uppermost HHC: quartz–muscovite–biotite–plagioclase/albite in schists and diopside–epidote–vesuvianite–garnet in impure marbles. The metamorphic recrystallization of the North Col Formation rapidly decreases from the ESZ upwards.

Lesser Himalaya

In the mylonitic rocks of the upper structural levels of the Lesser Himalaya the main assemblages are: quartz–plagioclase–biotite–muscovite ± orthoclase in the orthogneiss, and phlogopite–quartz–kyanite–muscovite–rutile in the associated micaschists. Mg-chlorite and fine-grained muscovite partly replace kyanite and phlogopite in the micaschist. Above the thrust contact with the underlying metasediments (north of Khandbari), only muscovite-biotite-bearing assemblages often including albite and garnet have been observed in the orthogneiss and micaschist.

In the fine-grained quartzite and phyllite occurring below the Khandbari thrust, metamorphic recrystallization probably decreases toward structurally lower levels. In fact quartz–muscovite–garnet–chloritoid assemblages grade downward to quartz–muscovite ± garnet ± chlorite assemblages.

Mineral chemistry and *P–T* estimates

Minerals have been analysed using an EDS/SEM microprobe and representative compositions are reported in Table 1. Results of geothermobarometry are presented in Table 2. The main metamorphic assemblages of gneiss and micaschist and the *P–T* conditions of metamorphism derived using microstructural relations, geothermobarometry and garnet-zoning profiles are given in Table 3 and are discussed below. For HHC and MCT zone, the topologies of gneiss and micaschist are shown by the AFM diagrams of Fig. 6 while the *P–T* estimates and the metamorphic history are summarized in Fig. 7.

High Himalayan Crystallines and Main Central Thrust zone

Mineral compositions and zoning. The samples analysed are (Table 1): MK91, garnet–kyanite micaschist from the MCT zone; MK13 and U107, two garnet–biotite gneisses from the lower HHC; B2, a biotite orthogneiss from the middle HHC; U41 and MK41, two biotite–cordierite gneisses from the upper HHC. In the HHC, garnet is rich in Fe (X_{Alm} = 0.62–0.83) and Mg, with minor spessartine and grossular components (X_{Sps} = 0.01–0.13, $X_{Grs} \leq 0.06$). Biotite has a relatively restricted compositional range (X_{Mg}[Mg/Mg + Fe] = 0.34–0.43), but for sample MK13, where biotite is more magnesian (X_{Mg} = 0.48–0.53). Plagioclase is oligoclase or andesine (X_{An} = 0.17–0.39). Cordierite has

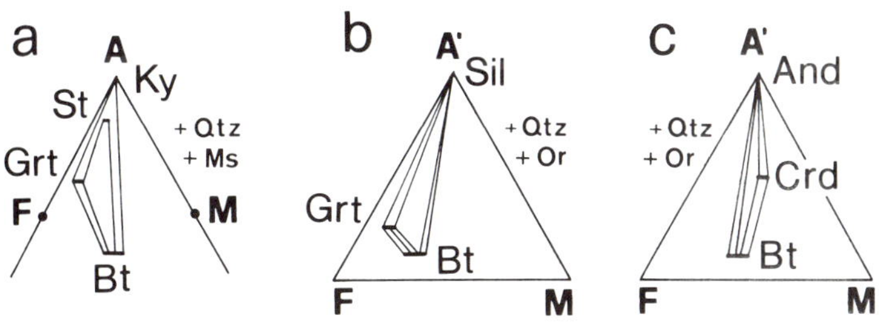

Fig. 6. AFM (projected from muscovite) and A'FM (projected from alkali feldspar) diagrams showing the topologies of mineral assemblages in gneiss and micaschist. (**a**) MCT zone and M1 in HHC, (**b**) M2 in HHC and (**c**) M3 in HHC.

Table 1. *Representative microprobe analyses of minerals of gneiss and micaschist from the MCT zone and different structural levels of the HHC, Everest transect*

	MCT					Lower HHC											
	MK91					MK13						U107					
	Grt core	Grt rim	Bt	Ms	Pl	Grt core	Grt rim	Bt incl	Bt matr	Pl incl	Pl matr	Grt core	Grt rim	Bt core	Bt rim	Pl core	Pl rim
SiO_2	36.96	36.97	36.28	47.24	62.09	37.49	37.74	35.92	35.70	59.27	62.22	36.39	36.91	35.60	35.23	57.82	58.43
TiO_2	0.00	0.00	1.49	0.55	0.00	0.00	0.00	3.87	5.17	0.00	0.00	0.00	0.00	4.94	4.76	0.00	0.00
Al_2O_3	20.34	20.66	18.06	34.00	23.70	21.62	21.46	18.34	17.49	25.93	24.42	20.16	20.59	18.43	18.89	26.02	26.35
Fe_2O_3	2.28	1.80	n.c.	n.c.	0.00	0.67	0.42	n.c.	n.c	0.00	0.00	2.89	1.59	n.c.	n.c.	0.00	0.00
FeO	32.33	30.48	17.70	2.07	0.00	31.39	33.33	17.23	18.45	0.00	0.00	27.80	34.39	19.24	19.28	0.00	0.00
MnO	1.34	4.41	0.00	0.00	0.00	0.61	0.82	0.00	0.00	0.00	0.00	5.48	2.26	0.00	0.00	0.00	0.00
MgO	4.41	3.70	12.00	0.63	0.00	5.56	5.04	10.71	9.53	0.00	0.00	2.75	3.06	8.17	8.28	0.00	0.00
CaO	2.66	2.68	0.00	0.00	5.34	2.83	1.80	0.00	0.00	7.72	5.43	5.04	2.05	0.00	0.00	8.39	8.18
Na_2O	0.00	0.00	0.36	1.38	8.21	0.00	0.00	0.00	0.00	7.15	8.61	0.00	0.00	0.00	0.46	7.14	7.08
K_2O	0.00	0.00	9.04	9.30	0.00	0.00	0.00	9.17	9.40	0.50	0.00	0.00	0.00	9.94	9.88	0.20	0.15
Total	100.32	100.71	94.93	95.17	99.34	100.17	100.61	95.24	95.74	100.57	100.67	100.52	100.85	96.32	96.77	99.57	100.19
Si	2.950	2.949	2.780	3.143	2.781	2.953	2.979	2.769	2.782	2.632	2.741	2.920	2.959	2.791	2.754	2.591	2.605
Al^{IV}	0.050	0.051	1.220	0.857	0.220	0.047	0.021	1.231	1.218	0.368	0.259	0.080	0.041	1.210	1.246	0.409	0.395
Al^{VI}	1.863	1.892	0.410	1.808	1.031	1.960	1.975	0.435	0.388	0.989	1.009	1.826	1.904	0.492	0.495	0.965	0.989
Ti	0.000	0.000	0.086	0.028	0.000	0.000	0.000	0.224	0.303	0.000	0.000	0.000	0.000	0.291	0.280	0.000	0.000
Fe^{3+}	0.137	0.108	n.c.	n.c.	0.000	0.040	0.025	n.c.	n.c.	0.000	0.000	0.174	0.096	n.c.	n.c.	0.000	0.000
Fe^{2+}	2.157	2.033	1.134	0.115	0.000	2.067	2.200	1.110	1.202	0.000	0.000	1.865	2.305	1.261	1.260	0.000	0.000
Mn	0.091	0.298	0.000	0.000	0.000	0.041	0.055	0.000	0.000	0.000	0.000	0.372	0.154	0.000	0.000	0.000	0.000
Mg	0.525	0.440	1.370	0.063	0.000	0.653	0.593	1.231	1.107	0.000	0.000	0.329	0.366	0.955	0.965	0.000	0.000
Ca	0.227	0.229	0.000	0.000	0.256	0.239	0.152	0.000	0.000	0.367	0.256	0.433	0.176	0.000	0.000	0.403	0.391
Na	0.000	0.000	0.054	0.179	0.713	0.000	0.000	0.000	0.000	0.615	0.735	0.000	0.000	0.000	0.069	0.620	0.612
K	0.000	0.000	0.884	0.789	0.000	0.000	0.000	0.901	0.934	0.028	0.000	0.000	0.000	0.994	0.985	0.011	0.008

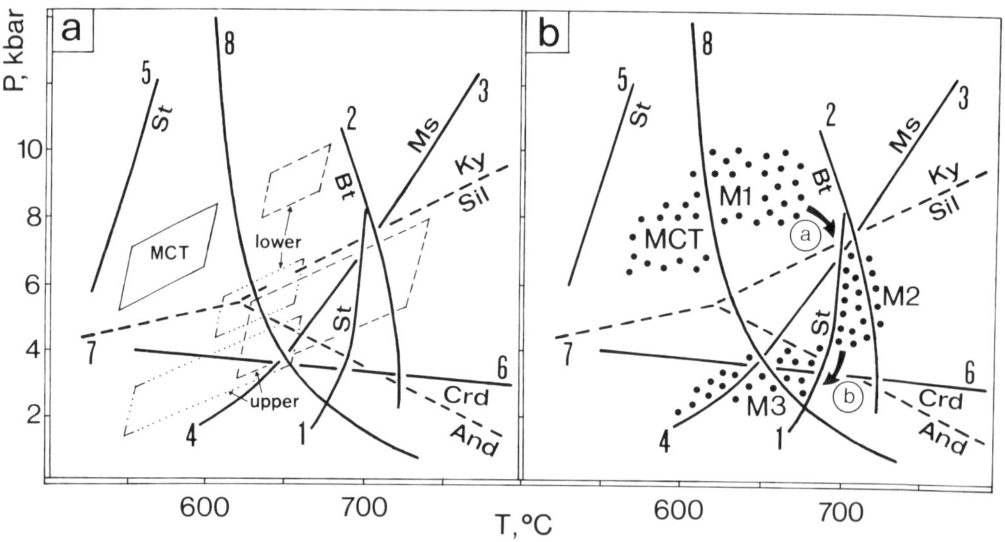

Fig. 7. *P–T* grids showing: (**a**) the conditions estimated using Grt–Bt thermometry (Perchuk & Lavrent'eva calibration 1983) and Grt–Pl–Ky/Sil–Qtz barometry for the MCT zone, lower and upper structural levels of HHC (the areas defined by dashed and dotted lines have been obtained using core and rim compositions, respectively); (**b**) the conditions for M1, M2 and M3 in the HHC inferred by thermobarometry and phase relations. Curves: 1, staurolite + quartz = almandine + sillimanite + H_2O (Richardson 1968); 2, biotite (X_{Mg} = 0.4) + sillimanite + plagioclase + quartz = garnet + orthoclase + liquid (Clemens 1984); 3, muscovite + albite + quartz = orthoclase + sillimanite + liquid (Petö 1976); 4, muscovite + quartz = andalusite + orthoclase + H_2O (Kerrick 1972); 5, chloritoid + kyanite = staurolite + almandine + H_2O (Rao & Johannes 1979); 6, almandine + andalusite/sillimanite + quartz + H_2O = cordierite (Thompson 1976); 7, kyanite–andalusite–sillimanite stability fields (Richardson *et al.* 1969); 8, beginning of melting in the albite–orthoclase–quartz–H_2O system (Merrill *et al.* 1970). Arrows show *P–T* paths indicated by relative thermobarometry for the lower HHC (**a**) and upper HHC (**b**). Mineral abbreviations on the figure are after Kretz (1983).

Middle HHC				Upper HHC													
B2				U41								MK41					
Grt core	Grt rim	Bt	Pl	Grt core	Grt rim	Bt core	Bt rim	Pl core	Pl rim	Or	Crd rim	Grt core	Grt rim	Bt	Pl core	Pl rim	Crd rim
37.14	37.03	34.18	62.43	36.52	36.32	34.57	34.59	63.11	63.13	65.31	47.03	37.58	36.64	33.99	61.24	61.67	48.73
0.00	0.00	4.00	0.00	0.00	0.00	2.42	2.24	0.00	0.00	0.23	0.00	0.00	0.00	3.54	0.00	0.00	0.00
20.12	20.57	18.58	23.92	20.61	20.16	19.05	19.16	22.30	22.38	18.70	31.46	20.80	20.47	18.27	24.83	24.74	32.38
1.37	1.08	n.c.	0.00	0.99	1.99	n.c.	n.c.	0.39	0.00	0.00	0.00	1.28	1.32	n.c.	0.00	0.00	0.00
34.97	35.26	21.10	0.00	36.29	35.96	23.81	22.48	0.00	0.00	0.00	12.17	33.00	33.49	21.08	0.00	0.00	9.80
2.46	2.73	0.40	0.00	3.02	3.16	0.00	0.00	0.00	0.00	0.00	0.31	2.00	5.47	0.00	0.00	0.00	0.35
3.21	2.98	7.73	0.00	2.29	1.91	6.86	7.12	0.00	0.00	0.00	6.09	4.50	2.06	8.02	0.00	0.00	7.34
0.90	0.95	0.00	5.05	0.57	0.59	0.00	0.00	3.91	3.81	0.00	0.00	1.72	1.26	0.00	6.83	6.41	0.00
0.00	0.00	0.58	8.92	0.00	0.37	0.30	0.49	9.44	9.87	3.04	0.34	0.00	0.00	0.31	7.87	7.96	0.00
0.00	0.00	9.27	0.00	0.00	0.00	8.99	9.20	0.28	0.14	12.27	0.00	0.00	0.00	9.13	0.15	0.00	0.00
100.17	100.61	95.85	100.31	100.29	100.46	96.00	95.29	99.45	99.33	99.55	97.40	100.88	100.69	94.34	100.94	100.78	98.60
3.002	2.982	2.700	2.755	2.967	2.949	2.720	2.743	2.804	2.796	2.994	4.994	2.979	2.966	2.711	2.703	2.726	5.041
0.000	0.018	1.300	0.245	0.033	0.051	1.280	1.257	0.196	0.204	0.006	1.006	0.021	0.034	1.289	0.297	0.274	0.959
1.917	1.934	0.430	0.999	1.940	1.879	0.486	0.534	0.971	0.964	1.005	2.931	1.924	1.920	0.428	0.995	1.014	2.990
0.000	0.000	0.238	0.000	0.000	0.000	0.143	0.134	0.000	0.000	0.008	0.000	0.000	0.000	0.212	0.000	0.000	0.000
0.083	0.065	n.c.	0.000	0.060	0.121	n.c.	n.c.	0.013	0.000	0.000	0.000	0.076	0.080	n.c.	0.000	0.000	0.000
2.364	2.374	1.394	0.000	2.465	2.442	1.566	1.490	0.000	0.000	0.000	1.080	2.188	2.267	1.406	0.000	0.000	0.848
0.169	0.186	0.027	0.000	0.208	0.217	0.000	0.000	0.000	0.000	0.000	0.028	0.134	0.375	0.000	0.000	0.000	0.031
0.387	0.357	0.911	0.000	0.277	0.231	0.804	0.842	0.000	0.000	0.000	0.964	0.533	0.249	0.954	0.000	0.000	1.132
0.078	0.082	0.000	0.239	0.050	0.051	0.000	0.000	0.186	0.181	0.000	0.000	0.145	0.109	0.000	0.323	0.303	0.000
0.000	0.000	0.089	0.763	0.000	0.058	0.045	0.076	0.813	0.847	0.270	0.070	0.000	0.000	0.049	0.674	0.682	0.000
0.000	0.000	0.934	0.000	0.000	0.000	0.902	0.931	0.016	0.008	0.717	0.000	0.000	0.000	0.928	0.008	0.000	0.000

MK91: garnet–kyanite–biotite–muscovite micaschist (Khaswa Khola north of Tashigaon, 1800 m; Arun valley).
MK13: sillimanite-bearing garnet–biotite–kyanite gneiss (south of Barun La, 3850 m; Arun valley).
U107: sillimanite-bearing garnet–biotite gneiss with relicts of kyanite (Zatr Og, 3800 m; Dudh Kosi valley east of Lukla). Structural formulae are calculated on the basis of 8 cations and 12 oxygens for garnet, 7 cations (N_{cat}–Ca–Na–K = 7) and 12 oxygens for biotite, 6 cations (N_{cat}–Ca–Na–K = 6) and 12 oxygens for muscovite, 5 cations and 8 oxygens for feldspar, 11 cations and 18 oxygens for cordierite; n.c., not calculated; incl, included in garnet; matr, matrix.
B2: garnet–sillimanite-bearing biotite orthogneiss (southern face of Peak 5200 m north of Khunde; Dudh Kosi valley).
U41: andalusite–fibrolite-bearing biotite–cordierite gneiss with relicts of garnet (east ridge of Peak 5977 m southwest of Gokyo, 5200 m; upper Dudh Kosi valley).
MK41; cordierite–andalusite bearing garnet–biotite–sillimanite gneiss (ridge north of Mera camping ground, 5000 m, upper Barun valley).

$X_{Mg} = 0.58$ in coarse-grained leucosomes of the middle structural levels (sample B5, Khumbui Yul Lha, not in Table 1), X_{Mg} = 0.47–0.57 in biotite–cordierite gneisses from the upper HHC, and X_{Mg} = 0.61 in restitic biotite–cordierite–sillimanite gneiss just below the Makalu granite (sample MK70, not in Table 1). In the MCT zone, garnet has X_{Alm} = 0.68–0.74, plagioclase is oligoclase (X_{An} = 0.20–0.26), biotite is richer in Mg than in the HHC (X_{Mg} = 0.54–0.56) and muscovite has X_{Mg} = 0.33–0.42.

At the lower structural levels of the HHC, garnet shows marked zoning, with rims enriched in Fe and Mg, and depleted in Ca and Mn relative to cores. At middle and upper levels garnet has rather homogeneous compositions, with rims slightly enriched in Fe and depleted in Mg relative to cores. In the MCT zone, garnet often shows irregular zoning with rims either enriched or depleted in Fe relative to cores.

Geothermobarometry. Temperatures of metamorphism have been calculated using geothermometers based on Fe-Mg partitioning in garnet and biotite. Representative results calculated using mineral compositions obtained averaging all the analyzed points are given in Table 2. Except for the extreme lowermost and uppermost structural levels (not considered in Table 2), estimates do not vary much throughout the HHC as also observed along the Darondi and Burhi Gandaki sections in central Nepal by Hodges *et al.* (1988). Among the different garnet–biotite geothermometers, the Perchuk & Lavrent'eva (1983) calibration gives the smallest temperature range, while the Hodges & Spear (1982) and the Indares & Martignole (1985) calibrations give the highest and the lowest estimates, respectively. Temperatures calculated for the 5 HHC samples with the Perchuk & Lavrent'eva (1983) calibration are 660 ± 40° C

Table 2. *Thermobarometric results on gneiss and micashist from the HHC and MCT zone, Everest transect*

Sample	MCT		Lower HHC				Middle HHC		Upper HHC			
	MK91		MK13		U107		B2		U41		MK41	
	a	b	a	b	a	b	a	b	a	b	a	b
Temperature (° C)												
Thermometer												
Grt–Bt												
HS at $P = 4$ kbar	650	608	804	806	732	654	720	684	658	562	924	558
HS at $P = 8$ kbar	664	622	822	825	750	660	736	700	674	578	946	571
PL at $P = 4$ kbar	600	578	672	674	628	606	645	626	616	564	730	559
PL at $P = 8$ kbar	620	594	695	700	649	624	664	646	635	582	751	578
IM at $P = 4$ kbar	596	576	655	609	604	500	578	550	572	488	773	472
IM at $P = 8$ kbar	610	592	674	629	625	516	596	566	590	502	794	486
Pressure (kbar)												
Barometer												
Grt–Pl–Sil–Qtz												
NH at $T = 600°$ C	7.0*	7.0*	6.1*	5.4	7.7*	4.3	3.0	3.1	2.4	3.0	4.3	3.0
NH at $T = 700°$ C	9.1*	9.1*	8.1*	7.4	9.9*	6.2	4.7	5.0	4.1	4.8	6.3	5.0
KN at $T = 600°$ C	7.8*	7.8*	7.0*	6.7*	8.7*	5.2	4.2	4.3	3.3	3.7	5.6	3.8
KN at $T = 700°$ C	9.9*	9.9*	9.0*	8.8*	10.8*	7.0	5.9	6.0	5.0	5.5	7.5	5.6
Grt–Pl–Ms–Bt												
HC at $T = 600°$ C	7.2	7.0										
HC at $T = 700°$ C	9.2	8.9										

Calibrations are HS, Hodges & Spear (1982); PL, Perchuk & Lavrent'eva (1983); IM, Indares & Martignole (1985); NH, Newton & Haselton (1981); KN, Koziol & Newton (1988); HC, Hodges & Crowley (1985).
* Grt–Pl-Ky–Qtz barometer.
Standard deviations of calculated T range from 15° to 50° C, and are on average 30° C; s.d. of calculated P range from 0.5 to 2.0 kbar, and are on average 1.3 kbar.
MK91: a, Grt core + matrix Bt, Pl, Ms; b, Grt rim + matrix Bt, Pl, Ms. MK13: a, Grt core + Bt and Pl included in Grt; b, Grt rim + matrix Bt and Pl. U107 and U41: a, Grt core + Bt and Pl cores; b, Grt rim + Bt and Pl rims. B2: a, Grt core + matrix Bt and Pl; b, Grt rim + matrix Bt and Pl. MK41: a, Grt core + matrix Bt and Pl core; b, Grt rim + matrix Bt and Pl rim.

using garnet cores and biotites included in garnet (10 pairs), and 610 ± 50° C using garnet rims and external biotites (18 pairs). Using the Hodges & Spear (1982) calibration average temperatures are 770° C and 650° C, respectively. The lowest values calculated using rim compositions are probably closure temperatures during cooling and are not consistent with the orthoclase–sillimanite-bearing muscovite-free assemblages of the middle and upper structural levels. In the MCT zone calculated temperatures are lower, at 620 ± 40° C using rim compositions (12 pairs) and the Hodges & Spear calibration (1982). The ternary feldspar solvus geothermometry of Stormer (1975) has also been applied to the studied rocks. The coexistence of plagioclase (X_{An} = 0.20) and alkali feldspar (X_{Ab} = 0.27) gives temperatures of around 650° C (at $P = 5$ kbar) for the HHC gneiss of the middle and upper levels, while elsewhere exsolution in feldspar upon cooling has affected the peak temperatures.

Pressures have been estimated using the garnet–plagioclase–sillimanite–quartz geobarometer (Table 2) for which the Koziol & Newton (1988) calibration gives pressures on average 1.0 kbar higher than the calibration of Newton & Haselton (1981). In the MCT zone pressures calculated for equilibration temperatures intermediate between the values of the Perchuk & Lavrent'eva (1983) and Hodges & Spear (1982) thermometers are 7–8 kbar while in the lower levels of the HHC they are 8–10 kbar and 5–7 kbar using core and rim compositions, respectively. At middle structural levels metamorphic pressures are around 5 kbar. At upper levels pressures are 3–7 kbar and 2–5 kbar for cores and rims, respectively. Using the Hodges & Crowley (1985) calibration of the garnet–plagioclase–biotite–muscovite geobarometer,

Table 3. *Metamorphic assemblages in quartz-rich gneiss and micaschist from the Everest transect*

Structural level	Metamorphic stage	Assemblage	T (° C)	P (kbar)
Tibetan Series				
Lowermost		Ms–Bt–Ab/Pl	450–550	low/very low
HHC				
Uppermost	M1?	relicts of St, Grt, Ky		
	M2–M3?	Pl–Bt–Ms–Sil(F) ± Or ± Crd	550–560	<4–6
	M3*	MsII–BtII–Ab/Pl	450–550	low/very low
Upper	M2	relicts of Grt and Sil		
	M3	Pl–Bt–Or–Crd ± And ± Sil(F)	600–700	2-4
Middle	M1	relicts of Ky and St		
	M2	Pl–Bt–Or–Grt–Sil ± Rt	650–750	4–7
	M3	Pl–Bt–Or–Crd ± Sil(F) ± And	600–700	2–4
Lower	M1	Pl–Bt–Ms–Grt–Ky ± Rt	550–680	8–10
	M2	Pl–Bt–Sil–Grt–Or ± Rt	650–750	5–7
MCT				
		Bt–Ms–Ky–Pl ± Grt ± St	550–620	7–8
		Bt–Ms–Ky–Pl		
Lesser Himalaya				
Upper orthogneiss		Pl–Bt–Ms ± Or	500–550	>4–5
		Phl–Ky–Ms–Rt		
		Ms–Bt–Grt		
Lower metasediments		Ms–Grt–Cld	400–450	medium
		Ms–Chl ± Grt	350–450	medium

Mineral abbreviations are after Kretz (1983), except for *F*, fibrolite.
P and *T* have been estimated from the data of Table 2 for MCT zone and HHC, and by paragenetic constraints for the Lesser Himalaya, the Tibetan Series and the uppermost HHC.

calculated pressures for the muscovite-bearing rocks of the MCT zone are 7.0–7.5 kbar at $T = 610–620°$ C.

In the cordierite-bearing gneiss (samples U41 and MK41) of the upper HHC the Aranovich & Podlesskii (1983) calibration of the garnet–cordierite–andalusite geothermometer, applied to six garnet–cordierite pairs, gives $T = 550–600°$ C and $P = 2–3$ kbar.

P–T evolution. Constraints on the metamorphic history of the HHC and MCT zone are given by the following.

(a) The microstructural relationships between the aluminium silicates (Fig. 3). The kyanite to sillimanite transformation indicates a temperature increase and/or a pressure decrease from M1 to M2. The sillimanite to andalusite transformation at middle and upper levels indicates a pressure drop possibly associated with a temperature decrease from M2 to M3.

(b) The abundance of cordierite in the migmatized gneiss of the upper HHC and the existence of cordierite–andalusite–orthoclase-bearing leucosomes in these rocks. Cordierite formation by reactions involving sillimanite and garnet such as: garnet + sillimanite + quartz + H_2O to cordierite (Fig. 7) is consistent with a pressure drop at high temperature from M2 to M3.

(c) The association of muscovite-free, sillimanite–garnet-orthoclase-biotite-assemblages (texturally pertaining to the M2 event) with anatectic leucosomes. This indicates that temperatures of M2 were higher than those of dehydration melting of muscovite and 'approximately' lower than temperatures defined by the curve of dehydration melting of biotite used in Fig. 7 ($T = 650–750°$ C).

(d) The stability of staurolite in the MCT zone and the existence of rare relicts of this mineral in the high-temperature rocks of the HHC. This indicates temperatures lower than 680° C for the MCT zone and for M1 in the HHC (Fig. 7).

(e) The association of diopside with calcite and quartz in the calc-silicate rocks. For intermediate values of X_{CO_2}, this indicates T = 650–700° C for P = 5 kbar and $T \approx 550°$ C for P = 1 kbar (Winkler 1979).

(f) Relative thermobarometry using the Gibbs method based on garnet zoning profiles (Spear & Selverstone 1983; Spear 1989) which indicates (Fig. 7) decompression under increasing T for the lower HHC (sample U107) and near-isobaric cooling at low P for the upper HHC (samples U41 and MK41). In the calculation we have used the model system KNCF-MASH for the trivariant assemblage garnet–biotite–plagioclase–quartz–aluminium silicate–muscovite/orthoclase–H_2O. In the MCT zone (sample MK91), relative thermobarometry indicates a prograde path under increasing pressure.

Tibetan Series

In the phyllites of the lowermost part of the North Col Formation, low-grade conditions ($T \approx 450$–$500°$ C) are implied by quartz–muscovite–biotite–albite-bearing assemblages. At the contact with the HHC and the bodies of deformed leucogranite, higher temperatures are suggested by the presence of oligoclase instead of albite in the phyllites, and of diopside–calcite–quartz-bearing assemblages in the impure marbles. At intermediate values of X_{CO_2} and at $P = 1$ kbar, the assemblage of the impure marbles indicates $T \approx 550°$ C (Winkler 1979).

This evidence and the fact that the metasediments of the North Col Formation show a rapid upward transition to unmetamorphosed calcareous sediments (the Early Ordovician Mt Jolmo Lungma Formation), suggest that the Tibetan Series suffered a low-pressure metamorphism linked to conductive heating at the contact with the 'hot' HHC and the Tertiary leucogranites.

Lesser Himalaya

In the rocks of the Lesser Himalaya peculiar mineral compositions have been obtained in the kyanite-bearing micaschists associated with the orthogneiss. The micaschists contain muscovite (X_{Mg} = 0.67–0.73) and phlogopite (X_{Mg} = 0.72–0.74) that are significantly richer in Mg than micas of all the other analysed rock-types.

As to the P–T conditions of metamorphism, relatively high-pressures for low to medium grade metamorphism ($T \approx 500$–$550°$ C; $P > 4$–5 kbar) are suggested for the middle and upper parts of the orthogneiss by the coexistence of muscovite and biotite with plagioclase and by the stability of kyanite. Low metamorphic grade is suggested for the lower part of the orthogneiss by the absence of kyanite and plagioclase. Low to very low metamorphic grade is indicated for the underlying metasediments by the absence of biotite and by the occurrence of assemblages including albite ± chloritoid ± chlorite (Table 3).

Discussion and conclusions

Main characteristics of the Everest transect

The data presented in this paper provide detailed information on the metamorphic history and on leucogranite formation along the Everest transect. The main results are summarized in Table 3 and are the following.

(1) *Inverted metamorphism in the Lesser Himalaya.* The mylonitic orthogneiss of the Lesser Himalaya suffered a low to medium-grade metamorphism at relatively high-pressures ($T = 500$–$550°$ C; $P > 4$–5 kbar). From the MCT zone downward, these rocks are characterized by a temperature decrease and they overthrust a low-grade metasedimentary unit. The existence, below the MCT, of Lesser Himalayan orthogneiss with a pervasive recrystallization of presumed Himalayan age, is an important character of the Everest transect. By contrast, sharper lithologic and metamorphic jumps from low-grade meta-sediments of the Lesser Himalaya to medium-grade gneisses of the HHC are exposed along the MCT zone in other parts of the Himalayan belt (e.g. Gansser 1991).

(2) *Metamorphic zonation of the HHC* (Figs 1 & 7a). Metamorphic assemblages are of uniformly high grade ($T = 650$–$750°$ C) throughout the HHC. A decrease in metamorphic grade has been observed only in the lowermost HHC at the contact with the Lesser Himalaya and near the ESZ at the contact with the overlying Tibetan Series. Metamorphic recrystallization is associated with SSW-directed thrusting in the lower part of the HHC and with NNE-directed extension in the upper part of the HHC. Pressures decrease toward structurally upper levels from 7–8 kbar to 2–4 kbar in the uppermost levels. Except for the new data concerning the low-pressure migmatization at the middle-upper levels and the zonation near the ESZ, our P–T estimates are comparable to the results of Brunel & Kienast (1986) and Hubbard (1989). Brunel & Kienast found equilibration temperatures in the Barun valley of around 700° C (for $P \approx 5$ kbar), using the garnet–biotite thermometer as calibrated by Holdaway & Lee (1977), for garnet–biotite–sillimanite gneiss with relicts of kyanite occurring at the lower structural levels of the HHC near Yanle, and temperatures around 680° C (for $P \approx 4$ kbar) using the cordierite–garnet thermometer as calibrated

by Holdaway & Lee (1977) for the biotite–cordierite–sillimanite gneiss occurring below the Makalu granite. Hubbard (1989), using the garnet–biotite thermometer as calibrated by Ferry & Spear (1978), found temperatures in the HHC of the Dudh Kosi valley ranging from *c*. 640° to 880° C in rocks from the lower structural levels, 540° to 710° C for the middle levels and 680° to 750° C for the upper levels. Calculated pressures decrease on average up section with a gradient of 0.28 kbar km^{-1} resembling a lithostatic gradient.

(3) *Polyphase metamorphism of the HHC* (Figs 6 & 7b). Besides the metamorphic zonation, three main metamorphic stages have been differentiated on paragenetic and textural grounds in the HHC. They record an exhumation (decompressional) path at increasing temperatures from M1 to M2 ($T = 550$–680° C to $T = 650$–750° C; $P = 8$–10 kbar to $P = 4$–7 kbar) and at more or less decreasing temperatures from M2 to M3 ($T = 600$–700° C; $P = 2$–4 kbar). The metamorphic stages show a heterogeneous distribution which depends on the structural level of the HHC. M1 is well preserved only in the lower structural levels and elsewhere was partially to completely overprinted during the subsequent thermal evolution. The M2 assemblages are widespread at lower and middle levels, whilst M3 assemblages are widespread at middle and upper levels.

(4) *Migmatization in the HHC*. Partial melting in the high-grade rocks of the middle structural levels produced pods and dykes of garnet–sillimanite-leucosome. Such leucosomes may be crosscut by later cordierite–andalusite-bearing leucosomes (Fig. 4b) suggesting that anatexis of the HHC gneiss occurred during exhumation; it began during M2 in the sillimanite stability field and continued at low-pressure during M3 in the cordierite–andalusite stability field (Figs 4 & 5).

(5) *Leucogranites*. In the upper levels of the HHC crustally-derived peraluminous leucogranites of Tertiary age (Ferrara *et al.* 1983; Schärer 1984; Schärer *et al.* 1986) are associated with and locally grade into anatectic leucosomes and migmatized gneiss which show comparable assemblages including sillimanite, andalusite and cordierite (Fig. 2a). Consequently the leucogranites: (a) formed at medium to low-pressure during M2 and M3 by 'in situ' anatexis of the upper levels of the HHC; (b) suffered limited vertical ascent. After generation, the leucogranites often segregated into extensional shear zones and eventually collected in kilometer-sized bodies (e.g. Makalu, Baruntse). Melt emplacement was locally accompanied by diapiric uprise into the base of the Tibetan Series (North Col Formation) and by 'in situ' ballooning (e.g. Nuptse: Bortolami *et al.* 1976).

(6) *Vapour-absent versus vapour-present anatexis*. *P–T* estimates indicate that during M2 and, in part also, during M3, the conditions were reached in the upper HHC for vapour-absent anatexis of pelitic and quartzofeldspathic rocks in response to dehydration melting involving muscovite (Fig. 7). Such a process produced the migmatites and the 'in situ' leucogranites recently described in the HHC of the Zanskar region (NW Himalayas: Pognante *et al.* 1990, Pognante 1992). However, it appears to be inadequate in order to form the voluminous leucogranite plutons of the Everest transect, as vapour-absent melting will give too small an amount of melt unless temperatures are high enough for biotite to undergo dehydration melting (Clemens & Vielzeuf 1987). As there is no evidence in the HHC of the Everest transect for such high temperatures (*c*. 850° C) and as the bulk compositions of the leucogranites are very close to minimum melts in the system Ab–An–Or–Qtz–H_2O, we conclude that infiltration of an aqueous fluid (i.e. wet anatexis) is required in order to obtain higher degrees of melting. We accordingly propose a two-stage process for generating leucogranite melts from the muscovite-bearing rocks occurring at the top of the HHC. In a first stage, dehydration melting under vapour-absent conditions produces strongly hygroscopic water-undersaturated leucogranite magmas which are able to leave the site of generation and climb to higher structural levels. Upon crystallization such magmas boil off, liberating hydrous fluids which infiltrate country rocks, promoting water-excess partial melting of muscovite-bearing layers and generation of large amounts of leucogranite melts. In the Everest region high degrees of melting are favoured by the low-pressure conditions of anatexis; in fact, as argued by Le Breton & Thompson (1988) at comparable temperatures and water contents, and starting from comparable protoliths, larger amounts of melt are produced at low pressure than at high pressure.

(7) *Contact metamorphism in the Tibetan Series*. The sediments of the Tibetan Series suffered a low-pressure metamorphism restricted to the ESZ at the contact with the 'hot' HHC and the leucogranite bodies emplaced in the shear zone (Fig. 2). The syn-kinematic growth of the metamorphic minerals suggests that recrystallization

is coeval with leucogranite emplacement and extensional NNE-directed shear of the Tibetan Series.

A possible model of exhumation of the HHC along the Everest transect

Figure 8 shows schematically a possible Himalayan metamorphic-tectonic history of the HHC and the surrounding tectonic units along the Everest transect. After Barrovian metamorphism caused by subduction and crustal thickening in the collided Indian plate (Fig. 8a) (see Mattauer 1986; Dewey *et al.* 1989 for general models), the HHC of the Everest transect underwent exhumation at increasing temperatures which caused extensive anatexis at medium- and low-pressures (Figs 7b & 8b). During exhumation a metamorphic zonation formed within the HHC in response to the complex interplay between rise of the isotherms in the thickened crust (Thompson & England 1984) and thermal perturbations near the adjacent units (Le Fort 1975).

Conductive cooling during exhumation and overthrusting above the relatively 'cold' Lesser Himalaya is apparently the most efficient mechanism to maintain lower temperatures at the base of the HHC. Cooling at the uppermost levels of the HHC and along the ESZ was associated with NNE-directed shear and probably resulted from a combination of conduction and advective heat transfer caused by fluid migration along the shear zone and into the Tibetan Series, according to the mechanism advocated by Reynolds & Lister (1987). Thermal interaction along the main tectonic contacts is consistent with the metamorphic gradients observed not only in the marginal parts of the HHC but also in the adjacent tectonic units. The period of heating experienced by the middle and upper levels of the HHC during exhumation is readily achieved in a thickened collided crust (Thompson & England 1984) and was probably favoured by the thermal conductivity contrast between crystalline (i.e. HHC) and sedimentary (i.e. Tibetan Series) rocks (Pinet & Jaupart 1987).

As to the mechanisms which promoted leucogranite emplacement in extensional shear zones, it is proposed here that relative movement of hanging wall and footwall of the ESZ favoured extension and melt segregation. In other words, SSW-directed thrusting in the lower HHC and NNE-directed extensional shear in the upper HHC combined with natural magma buoyancy forces to create space for leucogranite emplacement in the upper levels (Fig. 8b).

In conclusion, different stages of the exhumation history post-dating the India–Eurasia collision are recorded along the Everest transect. In the HHC, the exhumation occurred at rates which allowed the isotherms to rise through the thickened crust and the gneissic rocks to suffer extensive anatexis particularly at upper structural levels, where partial melting of muscovite-bearing metapelites produced in the middle Miocene voluminous leucogranite plutons. By contrast, other regions of the HHC (i.e. part of NW Himalayas in northern Pakistan, e.g. Pognante 1993) were rapidly exhumed along a cooling path and still record the subduction history at deep crustal levels.

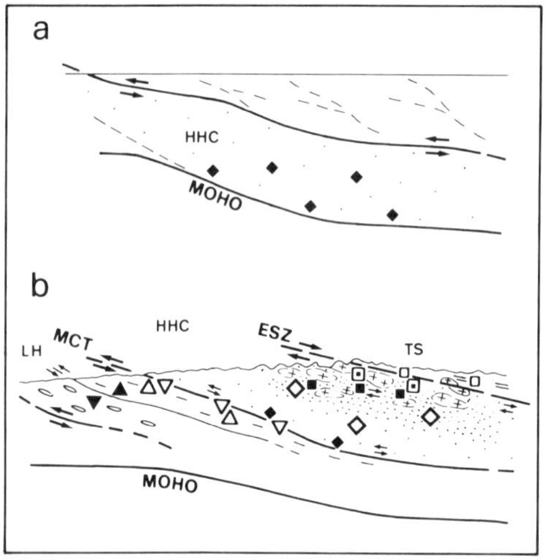

Fig. 8. Cartoon showing the possible tectonic evolution of Himalayan age and the metamorphic setting of HHC, Tibetan Series (TS) and Lesser Himalaya (LH) along the Everest transect. (**a**) Northwards subduction stage M1 in the HHC (Eocene?); (**b**) exhumation stages M2 and M3 in the HHC and prograde metamorphism in TS and LH (Oligocene–Miocene). Symbols as in Fig. 1. The probable thrust imbrication of the isograds near the MCT zone and ESZ is also shown.

We thank A. Desio and M. Gaetani for encouraging the expeditions in Nepal and Tibet. Reviews by K.V. Hodges, an anonymous referee and P.J. Treloar are much appreciated. P.B. would like to thank B. Lombardo, A. Borghi and F. Colombo for their help in the revision. Field work was funded by the National Research Council of Italy (CNR) through the EV-K2-Project. The laboratory work was financially supported by C.S. Geodinamica Catene Collisionali, CNR, Torino. Minerals have been analysed using the EDS/SEM Cambridge microprobe installed in the Dipartimento di Scienze Mineralogiche e Petrologiche of Torino, Italy.

References

ARANOVICH, L. YA. & PODLESSKII, K.K. 1983. The cordierite-garnet-sillimanite-quartz equilibrium: experiments and applications. *In*: SAXENA, S.K. (ed.) *Kinetics and Equilibrium in Mineral Reactions*. Springer-Verlag, New York, 173–198.

BASSETT, A.M., KINGERY, F.A., KRUMMENACHER, D. & LAYNE, H.F. 1978. Petrology, metamorphism and K-Ar age determinations in eastern Nepal. *In*: SAKLANI, P.S. (ed.) *Tectonic geology of the Himalaya*. Today & Tomorrow's, New Delhi, 151–166.

BORDET, P. 1961. *Recherches géologiques dans l'Himalaya du Népal, région du Makalu*. Editions du CNRS, Paris.

BORTOLAMI, C., LOMBARDO, B. & POLINO, R. 1976. The Higher Himalaya and the Tibetan Series in the Lhotse area (eastern Nepal). *Bollettino della Società Geologica Italiana*, **95**, 489–499.

——, SORZANA, F., FADDA, M. & NERVO, R. 1974. Missione scientifica "Cho Oyu – Everest region 1973". Osservazioni geografico-geologiche sulla dorsale Taboche-Jobo Lhaptshan e sulla testata del Khumbu Glacier (Khumbu Himal-Nepal orientale). *Memorie Società Geologica Italiana*, **13**, 409–424.

BRUNEL, M. 1983. *Étude pétro-structurale des chevauchements ductiles en Himalaya (Népal oriental et Himalaya du nord-ouest)*. Thèse doctorat ès sciences, University Paris VII.

—— & KIENAST, J.R. 1986. Étude pétro-structurale des chevauchements ductiles himalayens sur la transversale de l'Everest-Makalu (Népal oriental). *Canadian Journal of Earth Sciences*, **23**, 1117–1137.

BURCHFIEL, B.C., CHEN, Z., HODGES, K.V., LIU, Y., ROYDEN, L.H., DENG, C. & XU, J. 1992. The South Tibetan Detachment System, Himalayan orogen: Extension contemporaneous with and parallel to shortening in a collisional mountain belt. *Geological Society of America, Special Paper*, **269**, 1–41.

BURG, J.P., LEYRELOUP, A., GIRARDEAU, J. & CHEN, G.M. 1987. Structure and metamorphism of a tectonically thickened continental crust: the Yalu Tsangpo suture zone (Tibet). *Philosophical Transactions of the Royal Society London*, **A321**, 67–86.

CLEMENS, J.D. 1984. Water contents of silicic to intermediate magmas. *Lithos*, **17**, 273–287.

—— & VIELZEUF, D. 1987. Constraints on melting and magma production in the crust. *Earth and Planetary Science Letters*, **86**, 287–306.

DEWEY, J.F., CANDE, S & PITMAN III, W.C. 1989. Tectonic evolution of the India/Eurasia Collision Zone. *Eclogae Geologicae Helvetiae*, **82**, 717–734.

FERRARA, G., LOMBARDO, B., MARCASSA, P. & TONARINI, S. 1992. A geochemical and isotopic study of partial melting in the Namche migmatite orthogneiss of Nepal Himalaya. *Abstracts of 7th Himalaya–Karakorum–Tibet Workshop*, Dept. Earth Sciences, Oxford University, 110–111.

——, —— & TONARINI, S. 1983. Rb/Sr geochronology of granites and gneisses from the Mount Everest Region, Nepal Himalaya. *Geologisches Rundschau*, **72**, 119–136.

FERRY, J.M. & SPEAR, F.S. 1978. Experimental calibration of the partitioning of Fe and Mg between biotite and garnet. *Contributions to Mineralogy and Petrology*, **66**, 113–117.

GANSSER, A. 1991. Facts and theories on the Himalayas. *Eclogae Geologicae Helvetiae*, **84**, 33–59.

HAGEN, T. 1969. *Report on the geological survey of Nepal. Volume I: preliminary reconnaissance*. Mémoires de la Societé Helvetique des Sciences Naturelles, **86**.

HODGES, K.V. & CROWLEY, P.D. 1985. Error estimation and empirical geothermobarometry for pelitic systems. *American Mineralogist*, **70**, 702–709.

—— & SPEAR, F.S. 1982. Geothermometry, geobarometry and the Al_2SiO_5 triple point at Mt. Moosilauke, New Hampshire. *American Mineralogist*, **67**, 1118–1134.

——, LE FORT, P. & PÊCHER, A. 1988. Possible thermal buffering by crustal anatexis in collisional orogens: thermobarometric evidence from the Nepalese Himalaya. *Geology*, **16**, 707–710.

——, BURCHFIEL, B.C., CHEN, Z., HOUST, T., LUX, D., PARRISH, R. & ROYDEN, L.H. 1992. Rapid Early Miocene tectonic unroofing of the metamorphic core of the Himalaya: Evidence from the Qomolangma (Everest) region, Tibet. *Abstracts of 7th Himalaya–Karakorum–Tibet Workshop*, Dept. Earth Sciences, Oxford University, 39–40.

HOLDAWAY, M.J. & LEE, S.M. 1977. Fe-Mg cordierite stability in high grade pelitic rocks based on experimental and natural observations. *Contributions to Mineralogy and Petrology*, **63**, 175–198.

HUBBARD, M.S. 1989. Thermobarometric constraints on the thermal history of the Main Central Thrust zone and Tibetan Slab, eastern Nepal Himalaya. *Journal of Metamorphic Geology*, **7**, 19–30.

—— & HARRISON, T.M. 1989. $^{40}Ar/^{39}Ar$ age constraints on deformation and metamorphism in the Main Central Thrust Zone and Tibetan Slab, eastern Nepal Himalaya. *Tectonics*, **8**, 865–880.

INDARES, A. & MARTIGNOLE, J. 1985. Biotite-garnet geothermometry in the granulite facies: the influence of Ti and Al in biotite. *American Mineralogist*, **70**, 272–278.

KAY, K. 1981. Rb-Sr geochronology of the rocks of the Himalayas, Eastern Nepal. Part I. The metamorphic age of the Himalayan Gneiss. *Memoirs of the Faculty of Sciences, Kyoto University, Series of Geology & Mineralogy*, **47**, 135–148.

KERRICK, D.M. 1972. Experimental determination of muscovite + quartz stability with $P_{H_2O} < P_{tot}$. *American Journal of Science*, **272**, 946–958.

KOZIOL, A.M. & NEWTON, R.C. 1988. Redetermination of the anorthite breakdown reaction and improvement of the plagioclase-garnet-Al_2SiO_5-quartz geobarometer. *American Mineralogist*, **73**, 216–223.

KRETZ, R. 1983. Symbols for rock-forming minerals. *American Mineralogist*, **68**, 277–279.

KRUMMENACHER, D. 1967. Géologie et pétrographie des nappes de Kathmandu et de Nawakot, et de la série de Khumbu dans l'est du Népal. *Schweizerische Mineralogische und Petrographische Mitteilungen,* **47**, 855–871.

LE BRETON, N. & THOMPSON, A.B. 1988. Fluid-absent (dehydration) melting of biotite in metapelites in the early stages of crustal anatexis. *Contributions to Mineralogy and Petrology,* **99**, 226–237.

LE FORT, P. 1975. Himalayas: the collided range. Present knowledge of the continental arc. *American Journal of Science,* **275**, 1 44.

LOMBARD, A. 1958. *Un itinéraire géologique dans l'est du Népal (Massif du Mont Everest).* Mémoires de la Societé Helvétique des Sciences Naturelles, **82**.

LOMBARDO, B., PERTUSATI, P. & BORGHI, A. 1993. Geology and tectonomagmatic evolution of the eastern Himalaya along the Chomolungma–Makalu transect. *This volume.*

MATTAUER, M. 1986. Intracontinental subduction, crust-mantle décollement and crustal-stacking wedge in the Himalayas and other collision belts. *In*: COWARD, M.P. & RIES, A.C. (eds) *Collision Tectonics.* Geological Society, London, Special Publication, **19**, 37–50.

MERRILL, R.B., ROBERTSON, J.K. & WYLLIE, P.J. 1970. Melting reactions in the system $NaAlSi_3O_8$–SiO_2–H_2O to 20 kilobars compared with results for other feldspar–quartz–H_2O and rock–H_2O systems. *Journal of Geology,* **93**, 673–689.

NEWTON, R.C. & HASELTON, H.T. 1981. Thermodynamics of the garnet–plagioclase–Al_2SiO_5–quartz geobarometer. *In*: NEWTON, R.C., NAVROTSKY, A. & WOOD, B.J. (eds) *Thermodynamics of Minerals and Melts.* Springer-Verlag, New York, 129–145.

ODELL, N.E. 1948. Geological and some other observations in the Mount Everest region. *In*: TILMAN, H.W. (ed.) *Mount Everest 1938.* Cambridge Univ. Press, London, 143–154.

PERCHUK, L.L. & LAVRENT'EVA, I.V. 1983. Experimental investigation of exchange equilibria in the system cordierite-garnet-biotite. *In*: SAXENA, S.K. (ed.) *Kinetics and Equilibrium in Mineral Reactions.* Springer-Verlag, New York, 199–239.

PËTO, P. 1976. An experimental investigation of melting reactions involving muscovite and paragonite in the silica-saturated portion of the system K_2O–Na_2O–Al_2O_3–SiO_2–H_2O to 15 kbar total pressure. *Progress in Experimental Petrology, NERC, London, 3rd report,* 41–45.

PINET, C. & JAUPART, C. 1987. A thermal model for the distribution in space and time of the Himalayan granites. *Earth and Planetary Science Letters,* **84**, 87–99.

POGNANTE, U. 1992. Migmatites and leucogranites of Tertiary age from the High Himalayan Crystallines of Zanskar (NW India): a case history of anatexis of Palaeozoic Orthogneisses. *Mineralogy and Petrology,* **46**, 291–313.

—— 1993. Different P-T-t paths and leucogranite occurrences along the High Himalayan Crystallines: implications for subduction and collision along the northern Indian margin. *Geodinamica Acta* **6**, 5–17.

——, CASTELLI, D., BENNA, P., GENOVESE, B., OBERLI, F., MEIER, M. & TONARINI, S. 1990. The crystalline units of the High Himalayas in the Lahul-Zanskar region (northwest India): metamorphic-tectonic history and geochronology of the collided and imbricated Indian plate. *Geological Magazine,* **127**, 101–116.

RAO, B.B. & JOHANNES, W. 1979. Further data on the stability of staurolite + quartz and related assemblages. *Neues Jahrbuch für Mineralogie, Abhandlungen,* **10**, 437–447.

REYNOLDS, S.J. & LISTER, G.S. 1987. Structural aspects of fluid–rock interactions in detachment zones. *Geology,* **15**, 362–366.

RICHARDSON, S.W. 1968. Staurolite stability in a part of the system Fe–Al–Si–O–H. *Journal of Petrology,* **9**, 468–488.

——, GILBERT, M.C. & BELL, P.M. 1969 Experimental determination of kyanite–andalusite and andalusite–sillimanite equilibria; the aluminium silicate triple point. *American Journal of Science,* **267**, 259–272.

SCHÄRER, U. 1984. The effect of initial ^{230}Th disequilibrium on young U-Pb ages: the Makalu case, Himalaya. *Earth and Planetary Science Letters,* **67**, 191–204.

——, XU, R.A. & ALLEGRE, C.J. 1986. U-(Th)-Pb systematics and ages of Himalayan leucogranites, South Tibet. *Earth and Planetary Science Letters,* **77**, 35–48.

SPEAR, F.S. 1989. Relative thermobarometry and metamorphic P-T paths. *In*: DALY, J.S., CLIFF, R.A. & YARDLEY, B.W.D. (eds) *Evolution of Metamorphic Belts.* Geological Society, London, Special Publications, **43**, 63–81.

—— & SELVERSTONE, J. 1983. Quantitative P-T paths from zoned minerals: theory and tectonic applications. *Contributions to Mineralogy and Petrology,* **83**, 348–357.

STORMER, J.C. 1975. A practical two feldspar thermometer. *American Mineralogist,* **60**, 667–674.

THOMPSON, A.B. 1976. Mineral reactions in pelitic rocks: II. Calculations of P-T-X (Fe-Mg) phase relations. *American Journal of Science,* **276**, 425–454.

—— & ENGLAND, P.C. 1984. Pressure-Temperature-Time Paths of Regional Metamorphism II. Their Inference and Interpretation using Mineral Assemblages in Metamorphic Rocks. *Journal of Petrology,* **25**, 929–955.

VILLA, I.M. 1990. Geochronology and excess Ar geochemistry of the Lhotse Nup leucogranite, Nepal Himalaya. *Journal of Volcanology and Geothermal Research,* **44**, 89–103.

WAGER, L.R. 1965. Injected granite sheets of the Rongbuk valley and the north face of Mt. Everest. *In: D. N. Wadia commemorative volume.* India Mining, Geology and Metallurgy Institute, 358–379.

WINKLER, H.G.F. 1979. *Petrogenesis of metamorphic rocks.* 5th Edition, Springer-Verlag, New York.

YIN, C.H. & KUO, S.T. 1978. Stratigraphy of the Mount Jolmo Lungma and its North slope. *Scientia Sinica,* **21**, 629–644.

Geology and tectonomagmatic evolution of the eastern Himalaya along the Chomolungma–Makalu transect

BRUNO LOMBARDO[1], PIERO PERTUSATI[2] & SANDRO BORGHI[3]

[1] *CNR, CS Geodinamica Catene Collisionali, c/o DSMP, Università di Torino, Via Valperga Caluso 37, 10125 Torino, Italy*

[2] *Dipartimento di Scienze della Terra, Università di Pisa, Via S. Maria 53, 56100, Pisa, Italy*

[3] *DSMP, Università di Torino, Via Valperga Caluso 37, I–10125 Torino*

Abstract: The Lesser Himalayan units, the base of the Tethyan sedimentary sequence and the intervening High Himalayan Crystallines have been studied along a transect in the eastern Nepal Himalaya and southern Tibet.

The lowest Lesser Himalayan tectonic unit exposed along the transect is the par-autochthonous Tumlingtar Unit, a thick sequence of phyllite and quartzite of presumed Late Proterozoic age. Metamorphic grade is low (chlorite and biotite zones) and deformation much weaker than in the overlying unit, the Lesser Himalayan Crystallines, a unit including large bodies of granitic orthogneiss of probable Early Palaeozoic age. Metamorphic grade of the Lesser Himalayan Crystallines is higher than in the Tumlingtar Unit, being characterized by garnet–biotite–muscovite–kyanite–staurolite assemblages in metapelites and biotite–muscovite–kyanite in blastomylonitic orthogneiss.

A belt of metasedimentary and metagranitic thrust sheets separate the Lesser Himalayan Crystallines from the overlying High Himalayan Crystallines (HHC). Its position corresponds to that of the Main Central Thrust (MCT) as defined in central Nepal. The HHC are a gneiss sequence of 6–7 km visible thickness in which metasediments of upper amphibolite facies (Barun Gneiss) are associated with large tabular bodies of granitic orthogneiss that occur especially in the middle part of the sequence and which display evidence of water-deficient partial melting. The age of the granite protolith from which the orthogneiss derives is Early Paleozoic, possibly Cambrian.

The top of the HHC consists of biotite paragneiss and micaschist (Black Gneiss) hosting networks and lens-shaped bodies of tourmaline leucogranite which may attain a thickness of 1–2 km (Makalu, Baruntse) and evolve into cylindrical diapirs with gently north-plunging long axes (Nuptse). At the regional scale the top of the HHC coincides with the muscovite-in isograd and appears as a major intracrustal detachment zone which brought the HHC into contact with low-grade metasediments of the Tethyan Zone (North Col Fm) during the Miocene uplift of the Himalayan nappe pile, with attendant muscovite dehydration melting of the metasediments occurring at the top of the HHC.

The top of the HHC, the overlying North Col Fm. and the leucogranite sills hosted in both, are affected by a ductile extensional tectonics which appears broadly synchronous with granite emplacement and is characterized by extension lineations trending 20° to 40°. Extensional structures occur as far south as Makalu, whereas further south ductile deformations have the same orientation but are compressional. In the area south of Makalu the ductile deformation involves some of the Miocene leucogranite bodies, suggesting that Miocene compression in the lower and middle HHC may have been synchronous with ductile extension in the upper HHC.

Relations between metamorphism and deformation are inconsistent with a wholly post-metamorphic thrusting of the HHC over the Lesser Himalayan units and rather suggest that the Miocene exhumation of the HHC may be better explained by changes in the internal dynamics of the orogenic wedge produced by the Early Himalayan collision between northward-drifting India and Eurasia.

This paper reviews geological knowledge of the eastern Himalaya along the Everest–Makalu transect and describes its tectonomagmatic evolution during the Himalayan orogeny. Information presented here is derived from literature sources and from observations made along traverses from Lukla to Ngojumba Glacier in 1989, Khandbari to the Barun valley in 1990 and from Shegar to East Rongbuk Glacier in 1991. The area considered is the Arun valley from Tumlingtar to Sedua, the Barun valley, most of Khumbu Himal, the Rongbuk valley and the

From TRELOAR, P.J. & SEARLE, M.P. (eds), *Himalayan Tectonics*
Geological Society Special Publication, No. 74, pp. 341–355.

upper Dzakar Chu area. A strip map covering a 120 km-long transect (Fig. 1) and a geological cross section (Fig. 2) that summarize the main lithological and structural features of the area are included. After briefly describing the regional tectonics and metamorphic evolution we discuss relations between deformation and metamorphism, and explore the tectonic implications of Miocene partial melting in the High Himalayan Crystallines. A companion paper (Pognante & Benna, this volume) describes in detail the metamorphic history and presents a new model for the generation of the Miocene leucogranites of the Everest–Makalu area.

Geological setting: the eastern Nepal Himalaya

As elsewhere in the Himalaya (Gansser 1964) a simple morpho-tectonic zonation characterizes the eastern Nepal Himalaya and southern Tibetan plateau (Fig. 2).

(1) In the south, bordering the Ganges plain, the Siwalik Hills, a belt of folded Neogene molasse sediments.

(2) The geologically complex Lesser Himalayan zone, comprising the Mahabharat Range and the Midland zone, bounded to the south by the Main Boundary Thrust (MBT) and to the north by the Main Central Thrust (MCT). The Lesser Himalaya consist of a thick sequence of weakly metamorphosed sediments, the Nawakot Nappes of Hagen (1969) and of higher-grade metamorphic rocks, the Lesser Himalayan Crystallines, lying on top of the low-grade metasediments.

Around the Kathmandu basin a large WNW–ESE-trending synclinorium in the Lesser Himalayan metasediments has preserved from erosion a higher thrust sheet, the Kathmandu Complex (Kathmandu Nappe of Hagen) consisting of the crystalline Bimphedi Group and of the Early Palaeozoic Pulchauki Group (Stöcklin 1980).

Interference between late, open megafolds trending N–S and WNW–ESE, respectively, has produced dome- and basin-like structures of

MIOCENE INTRUSIVES

Tourmaline leucogranite,biotite-muscovite leucogranite and banded "muscovite granitic gneiss".

TIBETAN SEDIMENTARY SERIES

Shale, sandstone, limestone, conglomerate. (Permian-Carboniferous).

?unconformity..

Mt. Jolmo Lungma Fm. : limestone and crystalline limestone. (Lower Ordovician).
Yellow band : phyllitic quartz marble, calcareous quartz phyllite. ?(Lower Ordovician).

low angle normal fault
....

North Col Fm. : biotite quartz schist and phyllite, banded biotite epidote hornfels and calc-silicate rocks. ?(Cambrian-Sinian).

extensional ductile fault
....

HIGHER HIMALAYAN CRYSTALLINES

Rongbuk Fm. : biotite gneiss, biotite-muscovite-sillimanite +/- cordierite gneiss with relics of staurolite, biotite, quartz schist, marble.
Black Gneisses : biotite gneiss, biotite-sillimanite gneiss, biotite-sillimanite-cordierite gneiss, biotite-sillimanite gneiss with quartz-sillimanite nodules.

Namche Migmatite Orthogneiss : granite-granodiorite orthogneiss, augen gneiss, with minor biotite and accessory garnet, sillimanite and/or cordierite.

Barun Gneiss : garnet-kyanite-biotite gneiss, biotite-sillimanite-garnet gneiss, amphibolite, marble.

Main Central Thrust
MCT..

LESSER HIMALAYA

HIGHER THRUST SHEETS: quartzite, biotite micaschist, marble and calc-silicate rocks, garnet amphibolite, granite orthogneiss.

NUM ORTHOGNEISS : granite augen gneiss, with bands of kyanite-phlogopite schists (Migmatites du Bas Himalaya, BORDET ; Lesser Himalayan Crystallines, STOCKLIN).

LOWER THRUST SHEETS : quartzite, micaschist, granite orthogneiss.

TUMLINGTAR UNIT quartzite, phyllite and garnet micaschist, metaconglomerate (Seti Fm., Geological Map of Eastern Nepal, 1:250,000, Dept. Mines & Geol. of Nepal).

Fig. 1. Tectonic map of the Eastern Himalaya along the Chomolungma–Makalu transect. The map is based on personal observations and on the regional maps of Bordet & Latreille (1958*a*), of Academia Sinica (1979) and of the Nepal Department of Mines & Geology (1984). In our compilation we also used the more detailed geological maps of the Mt Everest region of Bordet & Latreille (1958*b*), Yin & Kuo (1978) and Polino (1983). The key is shown to the left.

(4) Further north, between the main peaks and the Yarlung Zangpo suture zone, the ranges of the Tibetan plateau consist entirely of Tethyan sediments of the north Indian margin. A major south-verging thrust divides the Tethyan sediments into a southern and a northern belt. The south Tethyan sediments comprise a fossiliferous sequence ranging in age from Early Ordovician to Eocene (Mu *et al.* 1973), as well as the North Col Formation, a sequence of low-grade metasediments (phyllite, quartzite, marble). The fossiliferous sediments are non-metamorphic and have suffered only one main phase of deformation (Burg & Chen 1984). At the base of the north Tethyan sequence, Lower Ordovician orthogneisses underly staurolite–kyanite schist and recrystallized Permian limestone. The Mesozoic includes a thick and monotonous calcareous flysch of up to Turonian age that underwent polyphase deformation (Burg & Chen 1984). The north Tethyan sediments are intruded by high-level plutons of two-mica granite emplaced during Late Miocene times, the North Himalayan (or Lhagoi Kangri) granite belt (Debon *et al.* 1986).

Geology of the Chomolungma–Makalu transect

Lesser Himalaya

The lowest (and southernmost) lithotectonic unit exposed along the transect is the parautochthonous Tumlingtar Unit, a sequence of low-grade metasediments (phyllite and quartzite) of presumed Late Proterozoic age, which crops out in the core of the Arun tectonic window. The most elaborate stratigraphical subdivision of the Tumlingtar Unit is by Andrews (1985) who in the Sabhaya Khola area (NE sector of the Arun tectonic window) distinguished six formations. Simpler subdivisions have been proposed by Bordet (1961), Akiba *et al.* (1973) and Brunel (1983). The whole sequence is grouped within the Seti Formation of the Midland Group in the Geological Map of eastern Nepal (Nepal Department of Mines & Geology 1984).

A thrust surface separates the Tumlingtar Unit from an overlying belt of metagranitic and metasedimentary thrust sheets (Zone des Ecailles, Bordet 1961), the Chainpur Group of Andrews (1985). The Chainpur Group includes the metagranitic Barabise Gneiss, a metasedimentary sequence of phyllite, quartzite and limestone and two sheets of micaschist (Simbuwa Schist and Basthala Schist, respectively) that sandwich the Barabise Gneiss. Both the Simbuwa Schist and Basthala Schist are biotite–muscovite–garnet micaschists of possibly meta-igneous origin (Bordet 1961, p. 143). The Simbuwa Schist apparently corresponds to Unit C of Meier & Hiltner (this volume) from which they report the common occurrence of kyanite and staurolite.

Tubular bodies of chlorite–quartz–garnet schist in which garnets are up to four cm in diameter occur within the upper part of the Barabise Gneiss close to the contact with the overlying Basthala micaschists. The bodies of chlorite schist are parallel to N–S trending mineral and stretching lineations within the Barabise Gneiss (Andrews 1985).

The Lesser Himalayan Crystallines are a unit including very large bodies of granitic orthogneiss and of metagranite (Num Orthogneiss in the north, Sikdim Orthogneiss in the Sabhaya Khola area). The granite protolith of the orthogneiss is probably of Early Palaeozoic age (Kai 1981). The Lesser Himalayan Crystallines correspond to the Irkhua Crystalline Nappe of Akiba *et al.* (1973) and to the northern part of Bordet's (1961) 'Nappe de Tinjure'.

The Main Central Thrust zone

A belt of metasedimentary and metagranitic thrust sheets separates the Lesser Himalayan Crystallines from the overlying HHC (Fig. 2). Its top corresponds to the Main Central Thrust (MCT) as defined in central Nepal (Bordet 1977). The position of the MCT is well defined in the area between the Dudh Kosi valley and the Hongu Khola, and in the Arun valley (Bordet & Latreille 1958; Bordet 1961; Brunel 1983), but somewhat uncertain in the area between the Hongu Khola and the western side of the Arun valley, where in placing it on the map we have followed the sketch map of Akiba *et al.* (1973, fig. 4).

In the Dudh Kosi valley the MCT is placed by most authors south of Kharte (Bordet 1961; Krummenacher 1967; Brunel 1983, p. 50), where the garnet–kyanite gneisses of the HHC are separated by a metasedimentary thrust sheet from the underlying augen gneisses of the Lesser Himalayan Crystallines (Solu Khola Gneiss of Brunel 1983). Metasedimentary rocks of the upper thrust sheets are (from top to bottom): garnet–kyanite–staurolite micaschist and green quartzite, black phyllite and calc-schist with marble layers, white quartzite. Garnet amphibolite occurs at the top of the metasedimentary sequence. Chlorite phyllonites occur both at the top and the bottom of the sequence and the top of the augen gneisses is marked by strongly schistose granitic blastomylonite (Brunel 1983). A peculiar rock type met with only in the Khasuwa Khola north of Tashigaon (Arun valley) is zoisite amphibolite with zoisite crystals up to several centimetres in diameter.

Higher Himalayan Crystallines

The Higher Himalayan Crystallines are a gneiss sequence of 6–7 km visible thickness in which a lower section of upper amphibolite facies gneiss (Barun Gneiss: Bordet 1961) is associated with large bodies of granitic orthogneiss occurring especially in the middle part of the sequence (Namche Migmatite Orthogneiss: Bordet 1961). The age of the granite protolith from which the orthogneiss derives is Early Palaeozoic, possibly Cambrian (Kai 1981; Ferrara *et al.* 1983), and Nd model ages suggest it was generated within continental crust not older than 1.9–2.5 Ga (Ferrara *et al.* 1992). In the Barun valley the Migmatite Orthogneiss is much thinner than along the Dudh Kosi, occurring only as a sheet a few hundred metres thick, but its features are the same as further west, in particular the presence of pods and dykes of granite neosome and the ubiquitous occurrence of cordierite (Bordet 1961, p. 162).

The upper part of the HHC is marked by an evident change in lithology, as it consists of biotite paragneiss and micaschist (Black Gneiss: Bordet 1961) which host networks and lens-shaped, sub-concordant bodies of tourmaline leucogranite. The Black Gneiss is biotite–sillimanite gneiss of obvious metasedimentary origin, possibly comprising a sedimentary cover of the Migmatite Orthogneiss, as the occurrence of metaconglomerates and quartzite layers (Island Peak, Barun valley) would suggest. Sillimanite is of the fibrolite variety and also occurs as quartz-sillimanite nodules (Bortolami *et al.* 1976, 1977).

The top of the gneiss belt, hosting the tourmaline leucogranites, coincides with the muscovite-in isograd and is formed by biotite-muscovite gneiss, locally with cordierite and fibrolite (Rongbuk Fm. of Yin & Kuo 1978). Peculiar metaconglomerate-like rocks which probably belong to the Rongbuk Rm. have been found in the moraine material at the base of Lhotse south face. The origin of such rocks is enigmatic, but the high ratio of matrix to pebbles suggest they may be metamorphosed tillite.

Tibetan Allochthon

North Col Formation. In the area around its type locality, North Col Fm (Yin & Kuo 1978) is a thick (1.5 km) sequence of low metamorphic grade (chlorite–biotite phyllite, biotite–calcite metagreywacke and impure quartzite) with some crystalline limestone layers (the 'yellow bands' of Lhotse), whereas in Nuptse and Lhotse it comprises mostly banded biotite–epidote hornfels and calc-silicate rocks. The North Col Fm. may be correlated with the low-grade metasedimentary sequences of Late Proterozoic–Early Cambrian age which are found elsewhere in the Himalaya on top of the HHC (Phe Fm. and Haimantas of Zanskar–Lahul, Martoli and Garbyang Series of Central Himalaya). Another possible correlation is with the middle part of the Kathmandu Complex, specifically with the Kulikhani, Markhu and Tistung Formations (Stöcklin 1980). An important consequence of extrapolating such lithological correlations to geological history is that the North Col Fm. could also have been intruded by Cambro-Ordovician granites and involved in an Early Palaeozoic orogeny, as indicated by sedimentological evidence in NW Himalaya (Garzanti *et al.* 1986) and by the age of granites in the Kathmandu Complex (Schärer & Allègre 1983).

Palaeozoic sedimentary sequence. On Chomolungma and the smaller peaks to the north, and particularly in the low hills south of Dzakar Chu, the North Col Fm. is capped by a limestone sequence, the 'Yellow Band' and the overlying Mt Jolmo Lungma Fm. Fossils in the latter indicate an Early Ordovician age, possibly the Arenigian–Llanvirnian (Mu *et al.* 1973). The contact of the limestone sequence with the underlying North Col Fm. is apparently not tectonic on Chomolungma itself, but north of Changtse the Yellow Band is missing in most outcrops suggesting that the contact between the North Col Fm. and the Ordovician limestone is a north-dipping low-angle normal fault (Burg *et al.* 1984).

A major hiatus divides the Mt Jolmo Lungma Fm. from the overlying Permo-Carboniferous sequence of Gondwanian affinity in the area north of Chomolungma. The existence of such a large time gap in the stratigraphic sequence suggests either that the largely held view of continuous Palaeozoic sedimentation may be locally unwarranted or that the middle section of the Tibetan Series in this area is actually an extensional allochthon. In view of the occurrence of Silurian and Devonian rocks both east and west of the Chomolungma area (Academia Sinica 1979), we consider the latter hypothesis as more likely.

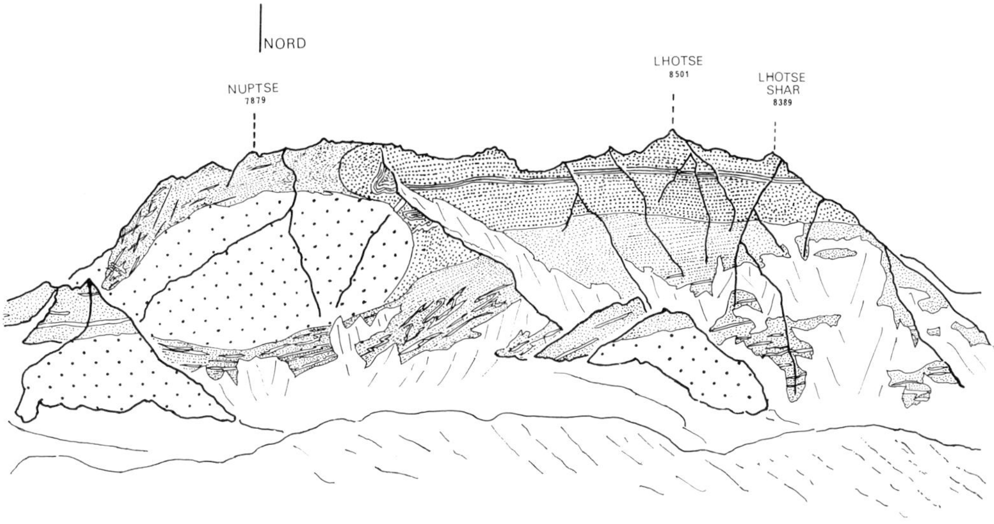

Fig. 4. The south face of the Nuptse–Lhotse ridge looking north from Chhukhung, in the upper Imja Khola (Khumbu Himal). The Nuptse pluton (dots) has pierced the envelope of Black Gneisses and Rongbuk Fm. (light stipple) intruding the overlying North Col Fm. (heavy stipple). Both undeformed and folded dykes (unornamented) can be seen in the hinge of the Nuptse 'fold'.

Miocene leucogranites

The Black Gneiss, Rongbuk Fm. and North Col Fm. host networks of dykes, sheet-like bodies and large, diapiric plutons of tourmaline leucogranite. The granite plutons may attain a thickness of 1–2 km (Makalu–Baruntse) and evolve into cylindrical diapirs with gently north-plunging long axes, which still retain a connection to their root zone as can be seen in the south face of Nuptse, a natural section of nearly three thousand metres (Fig. 4).

The contacts of the granite bodies with their country rocks are sharp and clearly intrusive, with abundant gneiss xenoliths floating in the granite (Bordet 1961; Bortolami *et al.* 1983). As in the other Himalayan leucogranites, the dark microgranular enclaves characteristic of other granite types are notably absent.

Both the Nuptse and Makalu granite are fine- to medium grained two-mica, tourmaline leucogranites, with muscovite usually prevailing over biotite. Garnet-bearing varieties devoid of biotite are less common and apparently restricted to the margins of some granite body (Nuptse). Cordierite-bearing varieties, first reported by Bordet (1961), occur in the upper Barun valley close to the floor of the Makalu pluton, which is also notable for the common presence of sillimanite and andalusite.

Aplitic to occasionally pegmatitic leucogranite sheets are hosted in the Rongbuk Fm. and the overlying North Col Fm. The leucogranites vary from slightly, to strongly foliated (the 'muscovite granitic gneiss' of Yin & Kuo 1978), with S–C fabrics and a stretching lineation that plunges towards between 020 and 040°. Such planar and linear fabrics are believed to result from northward shearing related to low-angle normal faulting during the ascent and emplacement of the leucogranite magma as first suggested by Burg *et al.* (1984).

Tectonics

General tectonic descriptions of the Everest region are given by Bordet & Latreille (1955), Lombard (1958), Hagen (1963, 1969) and Akiba *et al.* (1973). More recent studies, including microstructural analysis, are those of Brunel (1983, 1986), Burg *et al.* (1984), Burg & Chen (1984) and Burchfiel *et al.* (1992).

Three major shear belts are exposed along the Khandbari–Dzakar Chu transect: (1) the thrust zone which separates the parautochthonous Tumlingtar Unit from the overlying Lesser Himalayan Crystallines (Midland Thrust of Akiba *et al.* 1973), (2) the Main Central Thrust between the Lesser Himalayan Crystallines and the overlying Higher Himalayan Crystallines, (3) a north-directed detachment fault between the HHC and the Tethyan Sedimentary Sequence (South Tibetan Detachment: Burchfiel *et al.* 1992).

According to our observations in Rongbuk valley the South Tibetan Detachment actually consists of two major faults, a lower extensional ductile fault which divides the low-grade North Col Fm. from the amphibolite-facies Rongbuk Fm. and an upper low-angle normal fault between the North Col Fm. and the Ordovician limestone. Both faults are shown on the cross sections accompanying the Academia Sinica's (1979) geological map of South Tibet, but the extensional ductile fault is considered as a south-directed thrust, an interpretation incompatible with the observations of Burg *et al.* (1984) on the deformed leucogranites and with our own observations on S–C fabrics in the extensional ductile fault, which all indicate top-to-the-north extension.

The ductile extensional tectonism is characterized by extension lineations that plunge towards the NNE (020–040°) and appears broadly synchronous with the emplacement of the deformed leucogranites, but earlier than the crystallization of the undeformed granite bodies. Macrofolds related to the extensional tectonics are overturned to the NW, as on the north-side of Chienchin valley above the Everest Base Camp.

Extensional S–C fabrics occur as far south as the French 1954 Makalu Base Camp, where folds trend N140 and verge SW. South of Makalu, more precisely of the Mehra camping ground, down to the MCT ductile deformations have the same orientation but are compressional, recording an overall N–S shortening of the nappe pile. According to Brunel & Kienast (1986, table 1) the deformation history in this area begins with a composite deformation producing the main metamorphic foliation, isoclinal folds and a pervasive NNE-plunging, (010–035°) lineation. This phase is synchronous with a first stage of the Himalayan metamorphism (Barrovian event) but is earlier than the second metamorphic stage (*HT–LP* event). The second phase is synchronous with the *HT–LP* event and the generation of leucogranite melts. A third phase, characterized by deformation along retrometamorphic shear zones which cut through the earlier fabrics at all scales, is especially prominent in the zone of high deformation which encompasses the base of the HHC and the Lesser Himalayan units down to the top of Tumlingtar Unit. The thrusting of the HHC over the Lesser Himalayan Crystallines is ascribed by Brunel & Kienast to this phase and considered accordingly to be a late-metamorphic event, a view that we discuss below.

Late structures include large-scale open folds with axes plunging gently towards between 000 and 030° which are the most prominent structures south and west of Mt. Everest. A second group of large-scale folds, with axes trending 280–100° are common in the upper Dudh Kosi valley (Lombard 1958). The first group is earlier, possibly coeval with northward extension during leucogranite emplacement as its orientation would suggest.

Metamorphism

Lesser Himalaya

The metamorphic grade of the Tumlingtar Unit is low- to very low (Meier & Hiltner this volume; Pognante & Benna this volume) with higher-grade assemblages upsection (chlorite–sericite at the bottom, biotite–chlorite–sericite in the middle parts, garnet–biotite–chloritoid at the top). Metamorphic temperatures of 350°–450° C have been estimated by Pognante & Benna. A polyphase evolution is suggested in the garnet–biotite schists by the overprinting relations of a later chlorite-sericite assemblage (Brunel & Kienast 1986).

Metamorphic minerals in metagranitic rocks of the Lesser Himalayan Crystallines include biotite, muscovite, garnet and kyanite, with the same minerals plus staurolite in the metapelites (Brunel & Kienast, 1986; Hubbard 1989; Pognante & Benna this volume; Meier & Hiltner this volume). Magnesian schists within the Num Orthogneiss show the assemblage quartz–kyanite–magnesian phengite–phlogopite (Pognante & Benna this volume).

Metamorphic temperatures ranging from 500° C at the base of the Lesser Himalayan Crystallines to 550°–620° C at the top of the Upper Thrust Sheets have been estimated in the Arun valley by Pognante & Benna (this volume) for pressures of 6–8 kbar in the Upper Thrust Sheets and higher than 4–5 kbar in the Lesser Himalayan Crystallines. Metamorphic pressures of around 7 kbar and temperatures around 600° C have been estimated by Hubbard (1989) for the kyanite rocks of the MCT zone from the Dudh Kosi and Hinku-Hongu transects.

Higher Himalayan Crystallines

The Barun Gneiss is of significantly higher metamorphic grade than the Upper Thrust Sheets, its base being characterized by garnet–kyanite–biotite assemblages without either muscovite or staurolite. The upper portion of the Barun Gneiss shows garnet–sillimanite–biotite assemblages, which also characterize the Namche Migmatite Orthogneiss, together with younger cordierite.

A polyphase metamorphic evolution has been documented by Brunel & Kienast (1986) and Pognante & Benna (this volume), with at least two metamorphic events of Himalayan age, an Early Himalayan and a Late Himalayan event. The former is of Barrovian type, the latter is a nearly-isothermal decompression at high temperatures (*HT–LP* event), followed by a retrograde evolution under low pressures and high temperatures. The Barrovian event is best seen at the base of the HHC, where kyanite and a first generation of biotite are overprinted by garnet–sillimanite–biotite assemblages, but is also recorded in the Rongbuk Fm., at the top of the HHC, where relicts of staurolite, garnet and possibly kyanite are preserved in biotite–muscovite–sillimanite–cordierite assemblages (Wager 1965; Bortolami *et al.* 1977).

Biotite–sillimanite–cordierite assemblages characterize the Black Gneiss, where, in contrast with the overlying Rongbuk Fm., muscovite was apparently not stable. Andalusite, so far reported only in the Makalu granite (Bordet 1961) also occurs in the Black Gneiss (French Makalu Base Camp) and in paraschists within the Migmatite Orthogneiss of the upper Barun valley (U. Pognante, pers. comm., 1992).

Tibetan Allochthon

Both in the upper Rongbuk valley and on Chomolungma metagreywackes and phyllites of the North Col Fm. belong to the biotite zone of the greenschist facies (Yin & Kuo 1978; Pognante & Benna this volume). Among samples collected by Swiss parties climbing Everest by the normal route (Gysin & Lombard 1960) a few are fine-grained metagreywacke and phyllite with post-tectonic biotite porphyroblasts, the typical Budhi Schist facies described in Kumaon by Heim & Gansser (1939, p. 87), which is found all along the Himalaya in the upper Proterozoic sequences at the contact with the HHC. Close to the leucogranite bodies and to the Rongbuk Fm. extensive recrystallization has transformed the impure metagreywackes into banded biotite-epidote-hornblende hornfels and diopside–grossular–epidote rocks which are especially common in the upper parts of Cho Oyu, Nuptse and Lhotse (Gysin & Lombard 1960; Krummenacher 1967; Bortolami *et al.* 1983).

Discussion

Relations between metamorphism and deformation

It is clear from the observations summarized above that, as elsewhere in the Himalayan mountain belt, in the Everest–Makalu transect metamorphic grade increases upwards from the structurally lowest units towards the top of the nappe pile, specifically towards the middle–upper structural levels of the HHC and then decreases in the uppermost HHC and at the base of the Tethyan sequence to die out in the Early Ordovician limestones. Models proposed to explain this paradox have been reviewed by Le Fort (1975) and include post-metamorphic folding of isograds, late- to post-metamorphic imbrication of thrust sheets with different metamorphic grade and syn-tectonic overturning of isotherms resulting from thrusting of the 'hot' hanging wall over the 'cold' footwall of a continental subduction zone.

The last scenario, the model preferred by Le Fort (1975), has been applied to explain the thermal history of the MCT zone and of the HHC in the Everest region by Hubbard (1989). However, as noted by Brunel & Kienast (1986) and Pognante & Benna (this volume), the time relations between metamorphism, deformation and thrusting along the Everest–Makalu transect are complex, and in particular two metamorphic events of Himalayan age are recorded in the HHC, an Early Himalayan event of Barrovian type and a Late Himalayan event of *HT–LP* type. Even if the time span between the two events is not known, and they may be viewed simply as steps in a metamorphic continuum, two observations are not consistent with a simple mechanism of downward heating and recumbent overfolding of earlier isograds. The first observation is that the Lesser Himalayan Tumlingtar Unit, i.e. the deepest tectonic unit exposed along the transect, lacks any definite evidence of *HP–LT* metamorphism. The second observation is that the different metamorphic zones are separated from each other by regional-scale thrust zones, a situation strongly reminiscent of the metamorphic and tectonic pattern described in northwest Himalaya, for example in the Hazara region of northern Pakistan where it is ascribed to late stacking and imbrication resulting from regional breakback thrusting (Treloar *et al.* 1989). On the other hand, to explain the inverted metamorphic zonation as a result only of late- to post-metamorphic thrusting, as proposed by Brunel & Kienast (1986), is inconsistent with the occurrence of the *HT–LP* event throughout the HHC and especially with the subsequent retrograde evolution at high temperatures, which rather suggests shallowing of the HHC through syn-metamorphic thrusting.

Further speculation about tectonic models for the Himalayan metamorphism in the eastern

Himalaya must await precise data on the $P–T–t$ paths followed by individual tectonic units, in particular by those of the Lesser Himalaya. As an interim conclusion we would like to suggest that the internal dynamics of an orogenic wedge (Platt 1986) produced by the Early Himalayan (Eocene) collision between northward-drifting India and Eurasia seems to provide the best way to explain simultaneous compression at the base of the HHC and ductile extension at their top.

Partial melting in the HHC during the Late Himalayan metamorphism

The lower portion of the HHC (i.e. the Barun Gneiss and Namche Migmatite Orthogneiss) displays abundant field evidence of incipient partial melting in the form of granite to pegmatite pods and cross-cutting dykes that are characterized by garnet in the Barun Gneiss and cordierite in the Namche Orthogneiss. In the latter, partial melting produced pods and small dykes of garnet–cordierite granodiorite and granite, first reported from the Khumbu region by Krummenacher (1967).

Two orthogneisses with leucogranite pods have been investigated in detail by Ferrara *et al.* (1992). The orthogneisses have syenomonzogranite and monzogranite composition, with Sr isotope ratios (at 20 Ma) of 0.76169 and 0.75762. Sr isotope compositions of minerals appear to be in equilibrium with those of the host rocks and a biotite–WR isochron gave a precise age for the post-metamorphic cooling of one of the orthogneiss samples of 15.75 ± 0.23 Ma. Nd isotope ratios for the two orthogneiss samples 0.511938 and 0.511945 yield Nd model ages of 1.9 Ga in both samples.

The leucogranite pods have syenogranite and syenomonzogranite composition, with dominant K-feldspar, quartz and plagioclase, minor biotite, accessory garnet and sillimanite or cordierite. Initial Sr and Nd compositions are 0.7582 and 0.7537, and 0.51191 and 0.51186, respectively. Nd model ages are 1.9 Ga in both samples, as in the host orthogneisses. $^{87}Sr/^{86}Sr$ ratios of K-feldspar, plagioclase and biotite are in equilibrium with the whole rock ratio and one of the leucogranite samples gives a biotite–WR age of 15.9 ± 0.24 Ma. Relative to host rocks the leucogranites are depleted in MgO, FeO, transition elements and REE, but enriched in K_2O, Ba, Sr and Eu. Rb/Sr ratios are higher than in host rocks and, significantly, Sr isotope ratios are lower in leucogranites than in host rocks.

These features and the observation that when projected in the Ab–Or–Qtz system both leucogranites plot in the Or field, well away from compositions of minimum melts at 2 and 5 kbar P_{H_2O}, rule out the possibility that they formed by equilibrium partial melting under water-excess conditions and instead suggest closed-system melting of the host orthogneiss under water-deficient conditions as the most likely process involved in their generation. At the high temperatures implied for the Namche Orthogneiss by geothermometry (Hubbard 1989; Pognante & Benna this volume), a possible water-producing reactions is:

biotite + sillimanite + quartz = cordierite + garnet + K-feldspar + H_2O,

which at lower temperatures and pressures is replaced by

biotite + sillimanite + quartz = cordierite + K-feldspar + H_2O (Fig. 5).

Such reactions typically occur at temperatures of 650°–750° C under decreasing pressure, in a field where access of water to the metapelite system results in the production of melt according to the reaction:

biotite + albite (Pl) + sillimanite + quartz + H_2O = cordierite + liquid (Fig. 5).

In granitic compositions lacking sillimanite, biotite is stable and as plagioclase is more calcic than albite, the water-saturated minimum melting is overstepped and melting occurs according to the reaction:

plagioclase + K-feldspar + quartz + H_2O = liquid.

The small amounts of granite melt produced by this mechanism are unable to leave the site of melting (Wickham 1987) and in the Namche Orthogneiss are now exposed as pods and small dykes within much larger portions of unmolten orthogneiss.

The occurrence of water-deficient partial melting producing small amounts of granite melts in both the Barun Gneiss and the granitic Namche Orthogneiss coupled with the lack of total melting in the latter is strong evidence against wholesale H_2O fluxing of the HHC by fluids rising from the underlying low-grade metasediments of the Lesser Himalaya during the overthrusting of the hot Tibetan Slab onto the cold Lesser Himalayan units as proposed for generating the Miocene leucogranites from the base of the Tibetan Slab by Le Fort (1975, 1981). In our opinion, anatexis of the muscovite-bearing metapelites occurring at the upper structural levels of the HHC during the HT–LP event and the subsequent HT retrograde evolution, as proposed by Pognante & Benna (this volume), better explains field relations and petrology of the Miocene leucogranite bodies occurring in the Everest–Makalu region at the top of the High Himalayan Crystallines. In addition to the

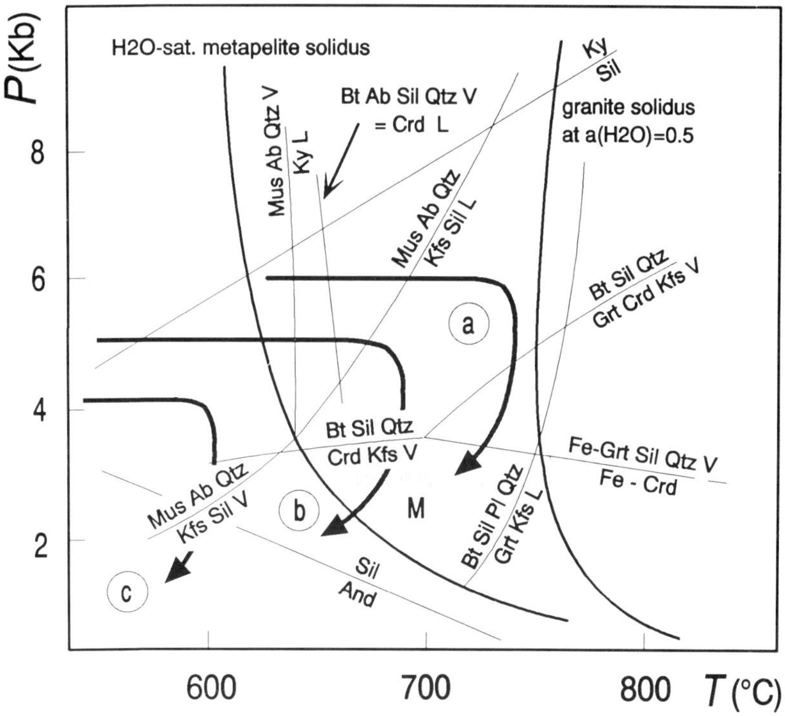

Fig. 5. Pressure–temperature diagram showing a selection of reaction curves relevant to partial melting in Namche Orthogneiss and in muscovite-bearing metapelites of the upper HHC. Reactions shown are:
muscovite + K-feldspar + albite + quartz + H_2O = liquid (water-saturated metapelite solidus, Thompson 1982)
muscovite + albite + quartz + H_2O = liquid + Al-silicate (Storre & Karotke 1972)
muscovite + albite + quartz = K-feldspar + sillimanite + H_2O (Thompson & Algor 1977)
muscovite + albite + quartz = K-feldspar + silimanite + liquid (Thompson & Algor 1977)
biotite + albite + sillimanite + quartz + H_2O = cordierite + liquid (Hoffer 1978)
biotite + sillimanite + quartz = cordierite + K-feldspar + H_2O (Holdaway & Lee 1977)
biotite + sillimanite + quartz = garnet + cordierite + K-feldspar + H_2O (Holdaway & Lee 1977)
Fe-garnet + sillimanite + quartz + H_2O = Fe-cordierite (Holdaway & Lee 1977)
biotite + sillimanite + plagioclase + quartz = garnet + K-feldspar + liquid (Le Breton & Thompson 1988).
The granite solidus at $a_{(H_2O)} = 0.5$ is after Ebadi & Johannes (1991), the kyanite–sillimanite and sillimanite–andalusite reactions after Holdaway (1971). Shaded bar shows the experimentally-determined crystallization interval of the Manaslu leucogranite at $P_{H_2O} = P_{tot} = 3$ kbar (Benard *et al.* 1985). (a), (b) and (c) are *P–T* paths compatible with a combination of assemblage-reaction constraints and thermobarometry for the Namche Orthogneiss (a), the Black Gneiss (b) and Rongbuk Fm (c), respectively. Geothermobarometric data are from Brunel & Kienast (1986), Hubbard (1989) and Pognante & Benna (1993).

evidence discussed above, reasons for preferring the model of Pognante & Benna (this volume) include the observations that the leucogranites crystallized in metamorphic equilibrium with rocks of the upper HHC that host them (Fig. 5) and the mineralogical features displayed by the Black Gneiss, most clearly in the Barun valley, suggesting that parts of it may be restite material from which leucogranite minimum melts have been extracted. Such rocks, the so called Gneiss du Camp de Base (Bordet 1961) are strongly peraluminous, consisting of quartz and feldspar with abundant biotite, cordierite, sillimanite and garnet and may in some way be regarded as a gigantic mafic selvedge of the Makalu granite.

Under water-excess conditions the reactions most likely involved in the generation of the leucogranites from the muscovite-bearing metasediments of the upper HHC are:

muscovite + albite (plagioclase) + quartz + H_2O = liquid + Al-silicate

and

muscovite + K-feldspar + albite (plagioclase) + quartz + H_2O = liquid, the so called water-saturated metapelite solidus (Fig. 5). Large amounts (>30% of melt fraction) of water-saturated melts may be produced through this reaction. Being water-saturated, such melts are clearly rather fluid, but their upwards mobility is very small and they can migrate only laterally, freezing at the same structural level of generation (Wickham 1987).

A major problem of this model is the origin of water necessary for water-saturated melting to proceed since the depth of emplacement of the Nuptse and Makalu bodies (estimated around 10–12 km by Lombardo & Borghi, in prep., on the basis of phase relations in leucogranite) is such that direct access of meteoric water seems problematical and metamorphic fluid flow appears to have been away *from* the HHC and the leucogranites towards the overlying low-grade metasediments of North Col Fm. For such reasons we prefer a model where the leucogranite magma formed by dehydration melting of muscovite-bearing metapelites at the top of the HHC. The most likely reaction is:
albite (plagioclase) + muscovite + quartz = liquid + K-feldspar + sillimanite (kyanite) (Fig. 5).

Melts produced by this reaction are water-undersaturated (Thompson 1982), a condition implied for the Miocene leucogranites by their ability to migrate vertically, and collect in large intrusive bodies (Makalu) and diapirs which have intruded and metamorphosed the North Col Fm. (Nuptse, Rongbuk valley).

This model predicts partial melting of muscovite-bearing metapelites in the upper structural levels of the HHC but not in the Rongbuk Fm. (the uppermost structural level of the HHC), since at pressures below 3 to 4 kbar muscovite cannot be involved in dehydration melting because the muscovite dehydrates by reactions such as:

muscovite + albite + quartz = K-feldspar + sillimanite + H_2O

which occur at even lower temperatures than the inferred water-saturated solidus for metapelites (Thompson 1982).

The time for generation and cooling of the leucogranites is bracketed between 24 and 15 Ma by U-Pb monazite ages and mica Rb-Sr and K-Ar ages, respectively. The oldest monazite ages (24.0 ± 0.2 and 21.9 ± 0.2 Ma) were measured on samples of the Makalu granite from the upper Barun valley (Schärer 1984), the youngest (14.4 ± 0.6 Ma) on a granite body N of Mt Everest (Schärer *et al.* 1986), possibly the same Rongbuk Monastery granite on which Hodges *et al.* (1992) measured a concordant U-Pb xenotime age of 20.6 ± 0.2 Ma. Mica ages range from 18.3 ± 1.2 Ma (muscovite, Rb/Sr) to 15.3 ± 0.4 Ma (biotite, K/Ar) in the Lhotse Nup granite of the upper Imja Khola south of Everest (Villa 1990). In the upper Ngojumba Glacier Ar/Ar plateau ages on both muscovite and biotite (Hubbard & Harrison 1989) cluster around 17.0 ± 0.8 Ma, nearly the same value as 17.6 Ma, the average of K/Ar ages found on biotites from the Khumbu Glacier and the upper Imja Khola by Bassett *et al.* (1970).

Such ages demonstrate very rapid cooling and uplifting of leucogranites (by as much as 100° C Ma^{-1}) and of country rocks alike, suggesting that leucogranite generation and exhumation of the upper HHC by extensional deformation may have been contemporaneous events. As shown by Hollister & Crawford (1986) localized failure within the lithosphere may be promoted by simultaneous melt formation already at melt fractions lower than 20% and the weakening effect may be substantial enough to produce strain localization into shear zones, where shearing produces additional melting, a feedback mechanism resulting in further strain localization in partially molten rocks. Applying this scenario to the Miocene exhumation of the Himalayan nappe pile has two important implications: (1) the extensional ductile fault which separates the HHC from the North Col Fm. is an intracrustal detachment zone formed by magma-induced crustal softening at the top of a thrust wedge and (2) a dramatic increase in the rate of exhumation is to be expected when rising isotherms intersect dehydration reactions in fertile lithologies.

Conclusions

The Lesser Himalayan units along the Chomolungma–Makalu transect are a nappe pile with an apparent inverse metamorphic gradient, where metagranites and metasediments recrystallized at higher pressures and temperatures sit on top of lower-grade (par)autochthonous metasediments. The tectonic contacts between the individual units are south-verging late-to post-metamorphic thrusts.

Two metamorphic events of Himalayan age, an Early Himlayan and a Late Himalayan event are recorded in the High Himalayan Crystallines. The former is of Barrovian type, the latter is of *HT–LP* type and is followed by a retrograde evolution under low pressures and high temperatures. The Barrovian event is best seen at the base of the HHC in the Barun Gneiss but is also recorded in the Rongbuk Fm. at the top of the HHC.

A north-directed detachment fault separates the HHC from the Tethyan Sedimentary Sequence. The latter, far from being the stratigraphical cover of the HHC is an extensional allochton. Ductile extension in the upper HHC is of Miocene age and may have been synchronous with compression in the lower and middle HHC.

Water-deficient anatexis generating cordierite-bearing leucosomes during the exhumation of the HHC is documented in the granitic Namche Migmatite Orthogneiss and is strong evidence against wholesale H_2O fluxing of the HHC by fluids rising from the underlying low-grade metasediments of the Lesser Himalaya during the Miocene overthrusting. This observation, and petrological evidence, suggests that decompression melting in muscovite-bearing, amphibolite facies metasediments of the upper HHC is the most likely mechanism for generating the 'intrusive' Miocene leucogranites occurring at the top of the HHC and in the overlying North Col Fm. Partial melting occurred where rising isotherms intersected dehydration reactions in fertile lithologies, producing a melt-weakened zone in the Early Himalayan orogenic wedge, along which rocks of the hot HHC were exhumed and brought into contact with low-grade and 'cold' metasediments of the Tibetan Zone (North Col Fm.) during the Miocene uplift of the Himalayan nappe pile. Vertical migration of the leucogranite melts generated in the upper HHC during the extensional deformation eventually sutured the North Col Fm. to the HHC. Very rapid cooling of leucogranites (by as much as $100°$ C Ma^{-1}) and of country rocks alike, suggests that leucogranite generation and exhumation of the upper HHC by extensional deformation may have been contemporaneous events, spanning no more than a few million years in the Mid-Miocene.

Fieldwork in the eastern Nepal Himalaya and southern Tibet was financially supported by the National Research Council of Italy through the Everest–K2 Project. The inspiration and companionship in the field provided by Ugo Pognante during the years 1989, 1990 and 1991 will always be remembered with heartfelt appreciation by B.L. and P.P. We thank D. Castelli for help with computer graphics, and P. Treloar and two anonymous reviewers for their comments on an early draft of the manuscript which have considerably improved the paper as did discussions with N.B.W. Harris and S. Inger. Laboratory work was financially supported by CNR, CS Geodinamica Catene Collisionali, Torino.

References

ACADEMIA SINICA, 1979. Geological Map of Lhasa–Nyalam area, Xizang (Tibet), The people's Republic of China. 1:1500000 (Compiled by the Geological Division of the Comprehensive Scientific Expedition of the Xizang (Tibet)–Qingai Plateau, Academia Sinica). *In: A Scientific Guidebook to South Xizang (Tibet)*. Academia Sinica, Beijing.

AKIBA, C., AMMA, S. & OHTA, Y. 1973. Arun River region. *In*: HASHIMOTO, S. (ed.) *Geology of the Nepal Himalayas*. Saikon Publishing Company, Sapporo, 13–31.

ANDREWS, E. 1985. Stratigraphy of the Sabhaya Khola region, Sankhuwa Sabha district, eastern Nepal. *Journal of the Nepal Geological Society,* **2/2**, 12–35.

BASSETT, A.M., KINGERY, F.A., KRUMMENACHER, D. & LAYNE, H.F. 1978. Petrology, metamorphism and K-Ar age determinations in eastern Nepal. *In*: SAKLANI, P.S. (ed.) *Tectonic geology of the Himalaya*. Today & Tomorrow's, New Delhi, 151–166.

BENARD, F., MOUTOU, P. & PICHAVANT, M. 1985. Phase relations of tourmaline leucogranites and significance of tourmaline in silicic magmas. *Journal of Geology,* **93**, 271–291.

BORDET, P. 1961. *Recherches géologiques dans l'Himalaya du Népal, région du Makalu*. Editions du CNRS, Paris.

—— 1977. Géologie de la dalle du Tibet (Himalaya central). *Mémoires hors série de la Societé géologique de France,* **8**, 235–250.

—— & LATREILLE, M. 1955. La géologie de l'Himalaya de l'Arun. *Bulletin de la Societé géologique de France,* série 6, **5**, 529–541.

—— & —— 1958*a*. *Esquisse géologique de l'Himalaya de l'Arun et de la région de l'Everest, 1:250 000*. Editions du CNRS, Paris.

—— & —— 1958*b*. *Esquisse géologique de la région de l'Everest et du Makalu, 1:50 000*. Editions du CNRS, Paris.

BORTOLAMI, G., LOMBARDO, B. & POLINO, R. 1976. The Higher Himalaya and the Tibetan Series in the Lhotse area (Eastern Nepal). *Bollettino della Società Geologica Italiana,* **95**, 489–499.

——, —— & —— 1977. Osservazioni geologiche sull'alta Imja Khola (Gruppo dell'Everest). *In*: CASSIN, R. & NANGERONI, G. (eds) *Lhotse '75*. Club Alpino Italiano, Milano, 95–110.

——, —— & —— 1983. The granites of the upper Imja Khola (Everest region), eastern Nepal. *In*: SHAMS, F.A. (ed.) *Granites of Himalayas, Karakorum and Hindu Kush*. Institute of Geology, Punjab University, Lahore, 257–270.

BRUNEL, M. 1983. *Etude pétro-structurale des chevauchements ductiles en Himalaya (Népal oriental et Himalaya du Nord-Ouest)*. Thèse Doct. ès Sciences, Université de Paris VII.

—— 1986. Ductile thrusting in the Himalaya: shear sense criteria and stretching lineations. *Tectonics,* **5**, 247–265.

—— & KIENAST, J.R. 1986. Etude pétro-structurale des chevauchements ductiles himalayens sur la transversale del l'Everest-Makalu (Népal oriental). *Canadian Journal of Earth Sciences,* **23**, 1117–1137.

BURCHFIEL, B.C., CHEN, Z., HODGES, K.V., LIU, Y., ROYDEN, L.H., DENG, C. & XU, J. 1992. *The South Tibetan Detachment System, Himalayan orogen: Extension contemporaneous with and parallel to shortening in a collisional mountain belt*. Geological Society of America, Special Papers, **269**.

BURG, J.P. & CHEN, G. 1984. Tectonics and structural zonation of southern Tibet, China. *Nature*, **311**, 219–223.

——, BRUNEL, M., GAPAIS, D., CHEN, G. & LIU, G.H. 1984. Deformation of leucogranites of the crystalline Main Central Sheet in southern Tibet (China). *Journal of Structural Geology*, **6**, 535–542.

DEBON, F., LE FORT, P., SHEPPARD, S.M.F. & SONET, J. 1986. The four plutonic belts of the Transhimalaya-Himalaya: a chronological synthesis along a Nepal-Tibet section. *Journal of Petrology*, **27**, 219–250.

EBADI, A. & JOHANNES, W. 1991. Experimental investigation of composition and beginning of melting in the system qz-ab-or-H_2O-CO_2. *Contributions to Mineralogy and Petrology*, **106**, 286–295.

FERRARA, G., LOMBARDO, B. & TONARINI, S. 1983. Rb/Sr geochronology of granites and gneisses from the Mount Everest region, Nepal Himalaya. *Geologisches Rundschau*, **72**, 119–136.

——, ——, MARCASSA, P. & TONARINI, S. 1992. A geochemical and isotopic study of partial melting in the Namche migmatite orthogneiss of Nepal Himalaya. *Abstracts of 7th Himalaya–Karakoram–Tibet Workshop*, Dept. Earth Sciences, Oxford University, 110–111.

GANSSER, A. 1964. *Geology of the Himalayas*. Interscience Publishers, London.

GARZANTI, E., CASNEDI, R. & JADOUL, F. 1986. Sedimentary evidence of a Cambro-Ordovician orogenic event in the northwestern Himalaya. *Sedimentary Geology*, **48**, 237–265.

GYSIN, M. & LOMBARD, A. 1960. Observations complémentaires de Pétrographie et de Géologie dans le Massif du Mont Everest-Lhotse. *Eclogae Geologicae Helvetiae*, **59**, 183–204.

HAGEN, T. 1963. Evolution of the highest mountain in the world. *In*: HAGEN, T., DHYRENFURTH, G.O., v. FUERER-HAIMENDORF, C. & SCHNEIDER, E. (eds). *Mount Everest. Formation, population and exploration of the Everest region*. Oxford University Press, London, 11–112.

—— 1969. *Report on the geological survey of Nepal. Volume I: Preliminary reconnaissance*. Mémoires de la Societé Helvétique des Sciences Naturelles, **86/1**.

HEIM, A. & GANSSER, A. 1939. *Central Himalaya. Geological observations of the Swiss Expedition 1936*. Mémoires de la Societé Helvétique des Sciences Naturelles, **73/1**.

HODGES, K., BURCHFIEL, B.C., CHEN, Z., HOUSH, T., LUX, D., PARRISH, R. & ROYDEN, L.H. 1992. Rapid Early Miocene tectonic unroofing of the metamorphic core of the Himalaya: Evidence from the Qomolangma (Everest) region, Tibet. *Abstracts of 7th Himalaya–Karakoram–Tibet Workshop*, Dept. Earth Sciences, Oxford University, 39–40.

HOFFER, E. 1978. Melting reactions in aluminous metapelites: Stability limits of biotite + sillimanite + quartz in the presence of albite. *Neues Jahrbuch fur Mineralogie, Monatshefte*, **9**, 396–407.

HOLDAWAY, M.J. 1971. Stability of andalusite and the aluminium silicate phase diagram. *American Journal of Science*, **271**, 97–131.

—— & LEE, S.M. 1977. Fe-Mg cordierite stability in high-grade pelitic rocks based on experimental, theoretical and natural observations. *Contributions to Mineralogy and Petrology*, **63**, 175–198.

HOLLISTER, L.S. & CRAWFORD, M.L. 1986. Melt-enhanced deformation: A major tectonic process. *Geology*, **14**, 558–561.

HUBBARD, M.S. 1989. Thermobarometric constraints on the thermal history of the Main Central Thrust Zone and Tibetan Slab, eastern Nepal Himalaya. *Journal of Metamorphic Geology*, **7**, 19–30.

—— & HARRISON, M.T. 1989. $^{39}Ar/^{40}Ar$ age constraints on deformation and metamorphism in the Main Central Thrust Zone and Tibetan Slab, eastern Nepal Himalaya. *Tectonics*, **8**, 65–880.

KAI, K. 1981. Rb-Sr geochronology of the rocks of the Himalayas, Eastern Nepal. Part I. The metamorphic age of the Himalayan Gneiss. *Memoirs of the Faculty of Science, Kyoto University, Series of Geology & Mineralogy*, **47/2**, 135–148.

KRUMMENACHER, D. 1967. Géologie et pétrographie des nappes de Kathmandu et de Nawakot et de la série de Khumbu dans l'est du Népal. *Bulletin Suisse de Minéralogie et Pétrographie*, **47**, 855–871.

LE BRETON, N. & THOMPSON, A.B. 1988. Fluid-absent (dehydration) melting of biotite in metapelites in the early stage of crustal anatexis. *Contributions to Mineralogy and Petrology*, **99**, 226–237.

LE FORT, P. 1975. Himalaya: the collided range. Present knowledge of the continental arc. *American Journal of Science*, **275A**, 1–44.

—— 1981. Manaslu granite: a collision signature of the Himalaya. A model for its genesis and emplacement. *Journal of Geophysical Research*, **86**, 10545–10568.

LOMBARD, A. 1958. Un itinéraire géologique dans l'Est du Népal, Massif du Mont Everest. *Mémoires de la Societé Helvétique des Sciences Naturelles*, **82/1**, 107 p.

MEIER, K. & HILTNER, E. 1993. Deformation and metamorphism within the Main Central Thrust zone, Arun Tectonic Window, Eastern Nepal. *This volume*.

MU, A.T., WEN, S.H., WANG, Y.K., CHANG, P.K. & YIN, C.H. 1973. Stratigraphy of the Mount Jolmo Lungma region in Southern Tibet, China. *Scientia Sinica*, **16**, 96–111.

NEPAL DEPARTMENT OF MINES & GEOLOGY 1984. *Geological Map of Eastern Nepal, 1 : 250 000*. Ministry of Industry, Department of Mines & Geology, Lainchour, Kathmandu.

PLATT, J.P. 1986. Dynamics of orogenic wedges and

the uplift of high-pressure rocks. *Geological Society of America Bulletin*, **97**, 1037–1059.

POGNANTE, U. & BENNA, P. 1993. Metamorphic zonation, migmatization, and leucogranites along the Everest transect of Eastern Nepal and Tibet: record of an exhumation history. *This volume*.

POLINO, R. 1983. Geological map of the upper Imja Khola (Mt. Everest region, eastern Nepal). *In:* SHAMS, F.A. (ed.) *Granites of Himalayas, Karakorum and Hindu Kush*. Institute of Geology, Punjab University, Lahore.

SCHÄRER, U. 1984. The effect of initial ^{230}Th disequilibrium on young U-Pb ages: the Makalu case, Himalaya. *Earth and Planetary Science Letters*, **67**, 191–204.

—— & ALLÈGRE, C.J. 1983. The Palung granite (Himalaya): high-resolution U-Pb systematics in zircon and monazite. *Earth and Planetary Science Letters*, **63**, 423–432.

——, XU, R.H. & ALLÈGRE, C.J. 1986. U-(Th)-Pb systematics and ages of Himalayan leucogranites, South Tibet. *Earth and Planetary Science Letters*, **77**, 35–48.

STÖCKLIN, J. 1980. Geology of Nepal and its regional frame. *Journal of the Geological Society, London*, **137**, 1–34.

STORRE, B. & KAROTKE, E. 1972. Experimental data on melting reactions of muscovite + quartz in the system $K_2O–Al_2O_3–SiO_2–H_2O$ to 20 kb water pressure. *Contributions to Mineralogy and Petrology*, **36**, 343–345.

THOMPSON, A.B. 1982. Dehydration melting of pelitic rocks and the generation of H_2O-unsaturated granitic liquid. *American Journal of Science*, **282**, 1567–1595.

—— & ALGOR, J.R. 1977. Model system for anatexis of pelitic rocks: I Theory of melting reactions in the system $KAlO_2$-$NaAlO_2$-Al_2O_3-SiO_2-H_2O *Contributions to Mineralogy and Petrology*, **63**, 247–269.

TRELOAR, P.J., BROUGHTON, R.D., WILLIAMS, M.P., COWARD, M.P. & WINDLEY, B.F. 1989. Deformation, metamorphism and imbrication of the Indian plate, south of the Main Mantle Thrust, north Pakistan. *Journal of Metamorphic Geology*, **7**, 111–125.

VILLA, I.M. 1990. Geochronology and excess Ar geochemistry of the Lhotse Nup leucogranite, Nepal Himalaya. *Journal of Volcanology and Geothermal Research*, **44**, 89–103.

WAGER, L.R. 1965. Injected granite sheets of the Rongbuk valley and the northern face of Mount Everest. *In*: *D.N. Wadia Commemorative Volume*. India Mining Geology and Metallurgy Institute, 358–379.

WICKHAM, S.M. 1987. The segregation and emplacement of granitic magmas. *Journal of the Geological Society, London*, **144**, 281–297.

YIN, C.H. & KUO, S.T. 1978. Stratigraphy of the Mount Jolmo Lungma and its North slope. *Scientia Sinica*, **21**, 629–644.

The metamorphism of the Tibetan Series from the Manang area, Marsyandi Valley, Central Nepal

CHRISTIAN SCHNEIDER & LUDWIG MASCH

Mineralogisch-Petrographisches Institut, Universität München, Theresienstr. 41, 8000 München 2, Germany

Abstract: In the Manang area a distinctive metamorphic zonation with grade increasing downward to deeper structural levels of the Tibetan Series can be observed. The first appearance of prograde biotite/phlogopite introducing the metamorphic zonation can be located in the Devonian Tilicho Pass Formation. The Silurian Dark Band Formation contains prograde titanite. Tremolite is not observed to the west of the Mutsog region (Nilgiri Limestone). In the lower parts of the Nilgiri Limestone, tremolite appears together with K-feldspar. Diopside only occurs in the lowest parts of the Nilgiri Limestone. Maximum temperatures for the prograde metamorphism, derived from carbonate solvus thermometry, are in the range of 510–530° C for the middle Nilgiri Limestone and 370–390° C for the Lower Carboniferous. The fluid composition during the prograde metamorphism at the base of the Tibetan Series was CO_2 rich (X_{CO_2} between 0.4 and 0.7). A strong retrogressive phase with a fluid composition rich in H_2O is documented by secondary growth of sphene, clinozoisite and amphibole. This retrograde phase is rare in formations above the Nilgiri Limestone and might be contemporaneous with detachment at the base of the Tibetan Series. The prograde mineral phases are related to a first deformation phase. A second deformation phase, most probably related to detachment movements between the Tibetan Series and the High Himalayan Crystalline, resulted in a partly developed crenulation cleavage unrelated to any metamorphic change. The formation of the dominating E–W-orientated syncline took place in a third deformation phase.

The regional metamorphism of the High Himalayan Crystallines, the Main Central Thrust zone and the resulting metamorphic zonation in Central Nepal is well known through several studies that deal with the tectono-metamorphic evolution of this region (e.g. Le Fort 1975, 1981; Colchen *et al.* 1986; Pêcher 1989). By comparison, the metamorphism of the Tibetan Series in Central Nepal has not been well investigated. Of particular interest is the transition zone between the High Himalayan Crystallines and the overlying Tibetan Series. Various authors have described a northward directed displacement along extensional shears that separate the Tibetan Series from their crystalline basement, the shearing being located within this transition zone (Burg *et al.* 1982; Colchen *et al.* 1986; Pêcher & Bouchez 1987; Royden & Burchfiel 1987; Herren 1987). A detailed study of the metamorphism and deformation in this zone is thus essential to the understanding of the tectonic and metamorphic evolution of the Himalayas in Central Nepal.

In this paper we present a short review of our results concerning the regional metamorphism and the relation between crystallization and deformation in the Tibetan Series along the Marsyandi Valley in Central Nepal (Manang area). During three German–Austrian geoscientific expeditions in the Marsyandi area (1987, 1988, 1990) approximately 150 samples were collected and structural data obtained. Several electron microprobe analysis of the solid phases, including biotite/phlogopite, amphibole, clinopyroxene, and scapolite, were used to identify specific metamorphic zones and the confining isograds and reactions. From more than 2500 analyses of coexisting calcite and dolomite, 1600 analyses were selected for evaluation of carbonate solvus geothermometry. Field data together with microprobe analyses and microscopic investigations served as a base for the reconstruction of the metamorphic evolution of the Tibetan Series in the Manang area.

Geological setting

The area north of the Annapurna range in Central Nepal (Manang area, see Fig. 1) consists of meta-sediments of the Tibetan Series, which overlie the upper formations of the High Himalayan Crystallines (formation-III augen-gneisses and formation-II calc-silicate rocks). A detailed description of the lithology of the Tibetan Series can be found in Bordet *et al.* (1975) and Fuchs *et al.* (1988).

Lithologies of the Tibetan Series in the

From Treloar, P.J. & Searle, M.P. (eds), *Himalayan Tectonics*
Geological Society Special Publication, No. 74, pp. 357–374.

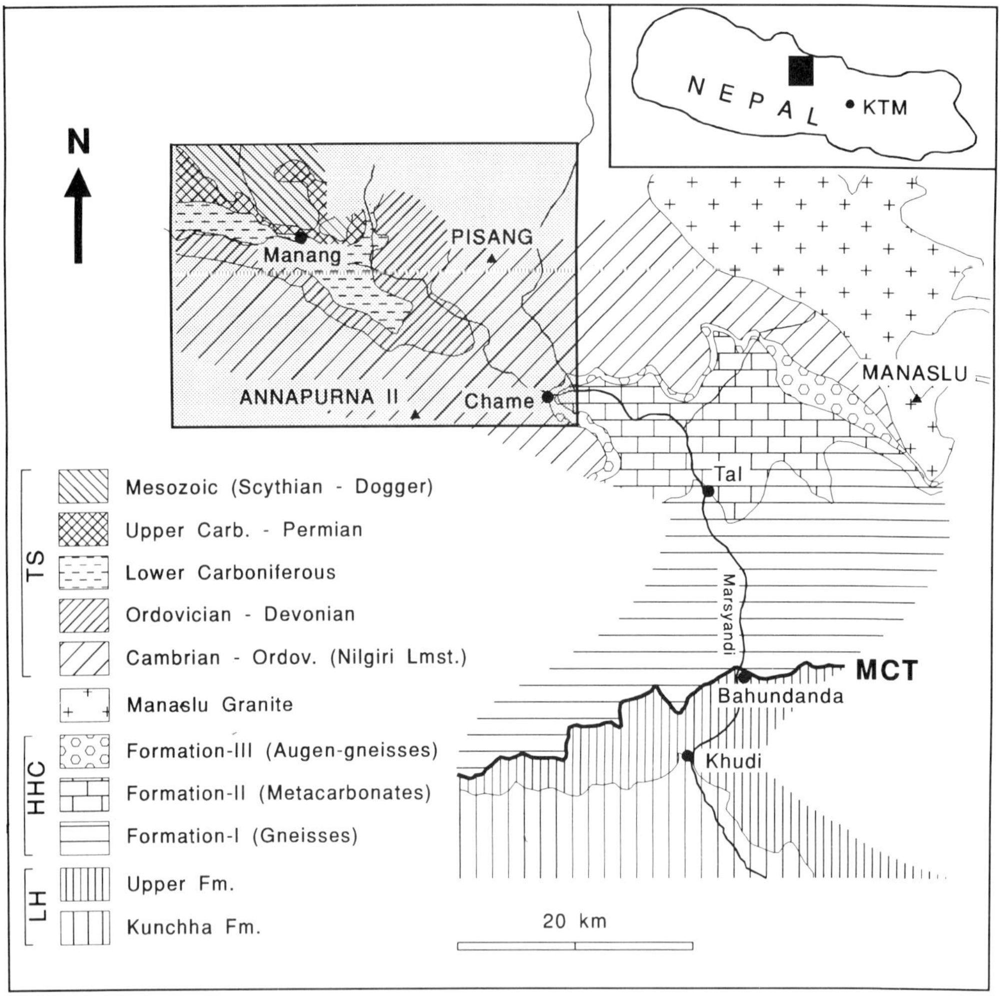

Fig. 1. Simplified geological map of the upper Marsyandi valley (based on Colchen *et al.* 1986 and Fuchs *et al.* 1988) with location of the Manang area. TS, Tibetan Series; HHC, High Himalayan Crystallines; LH, Lower Himalaya; MCT, Main Central Thrust.

Manang area consist of carbonatic and pelitic metasediments from the Cambro-Ordovician Nilgiri Limestone upwards to the lower Jurassic Kioto Limestone (see Fig. 2). As the carbonatic rocks are more abundant than pelitic rocks, the carbonate assemblages were used to investigate the metamorphic evolution.

The lower boundary of the Tibetan Series has been defined, according to Bordet *et al.* (1975) and Fuchs *et al.* (1988), as being located between the Nilgiri Limestone (Formation de Larjung) and the underlying augen-gneisses (Formation III of the High Himalayan Crystallines). Controversy still surrounds the true character of this boundary. On the one hand, the stratigraphy and lithology do not suggest that there is a distinct boundary between the Tibetan Series and the High Himalayan Crystallines. Furthermore, there is a continuous upward decrease in metamorphic grade from the High Himalayan Crystallines to the younger formations of the Tibetan Series. On the other hand, the separation of the two units may be justifiable given the striking differences that exist regarding the style of deformation. The dominant structure of the Tibetan Series in the Manang area is a wide ESE–WNW-orientated syncline. At the base of the Tibetan Series, in the Nilgiri Limestone, the style of deformation changes considerably from open folding to the homoclinal north dipping

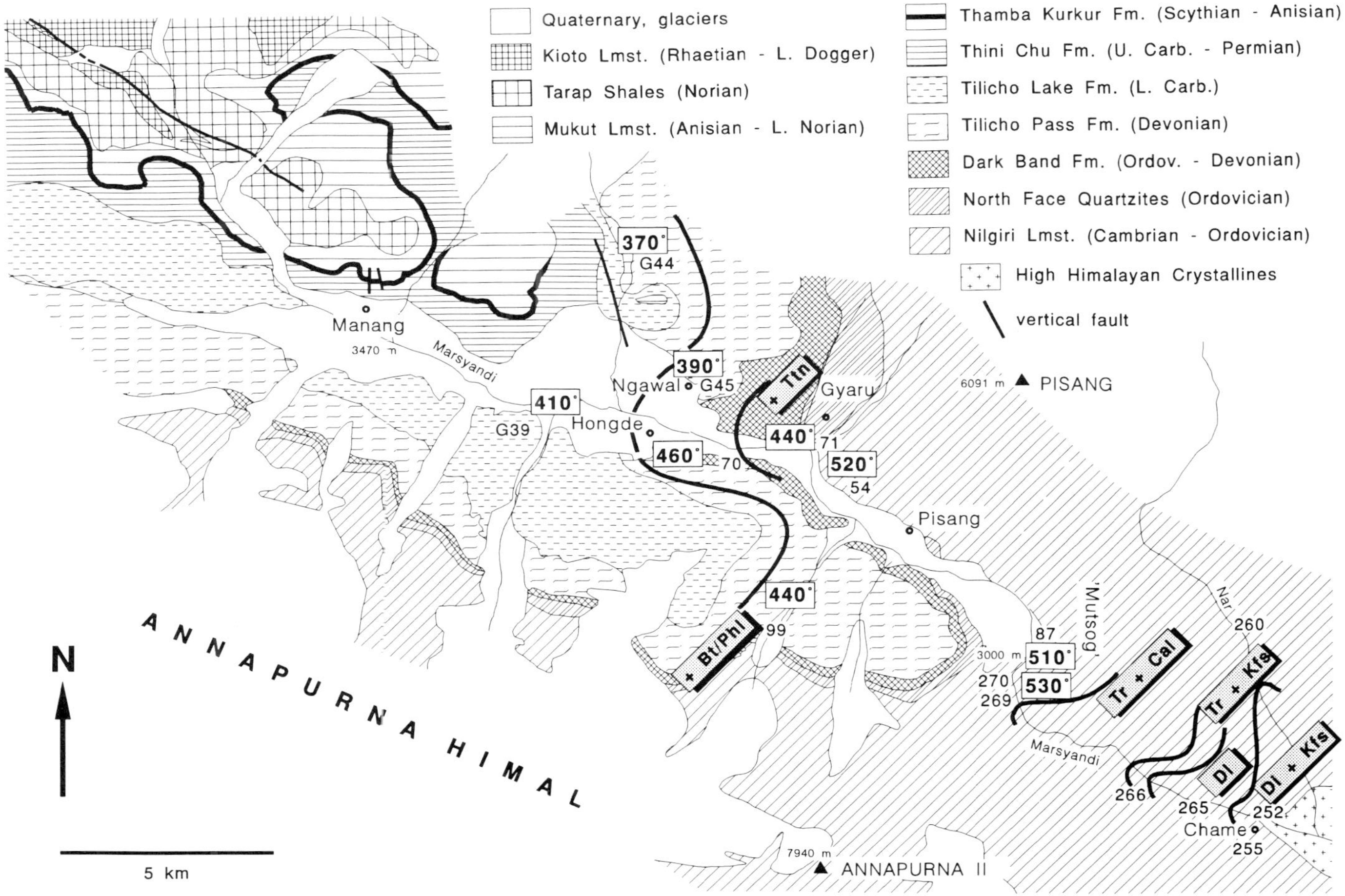

Fig. 2. Geological map of the Manang area (based on Fuchs *et al.* 1988) with orientation of isograds and rounded maximum temperatures of prograde metamorphism derived from carbonate solvus thermometry. The location of samples mentioned in the text or in the tables is indicated. Isograds: Bt/Phl Biotite (Phlogopite); Ttn, Titanite; Tr + Cal, Tremolite + Calcite; Tr + Kfs, Tremolite + K-Feldspar; Di, Diopside; Di + Kfs, Diopside + K-Feldspar.

structure of the underlying High Himalayan Crystallines.

The prograde metamorphism

Field work and investigation of more than 150 thin sections revealed a distinctive prograde metamorphic zonation in the Tibetan Series (see Fig. 2). The mappable zones begin with the appearance of biotite/phlogopite in the Devonian Tilicho Pass Formation. The metamorphic grade increases continuously eastward up to the diopside + K-feldspar zone at the base of the Tibetan Series (Nilgiri Limestone). The assemblages in the underlying meta-carbonates of the High Himalayan Crystallines (formation-II) show that these rocks also belong to the diopside + K-feldspar zone. The expected formation of forsterite was not observed, because of the high Si-content and low Mg-content of these meta-carbonates.

A detailed record of prograde metamorphism for the Tibetan Series older than the Tilicho Lake Formation is achieved by the combination of petrological evidence and geothermometry. An independent temperature estimation was obtained by carbonate solvus geothermometry; the application was restricted to a section from the Lower Carboniferous schists (Tilicho Lake Fm.) down to the Middle Nilgiri Limestone (tremolite + calcite isograd).

There is little prospect of determining the metamorphic conditions for formations younger than the Carboniferous. Besides pure limestones or rare intercalated dolomites, the assemblages calcite–dolomite–quartz–muscovite and calcite–quartz–muscovite can only be found in impure meta-carbonates. These assemblages are insufficient indicators of metamorphic grade, because they are stable within a large pressure and temperature range. The Mg-contents of calcites coexisting with dolomite are too low for use of the calcite–dolomite geothermometer (<1.5 Mol % Mg). The uppermost formations occasionally contain ill-defined chlorite in pelitic assemblages not suitable for mapping as a metamorphic zone. Thus, the determination of the illite crystallinity index (Kubler 1968) is the only possibility of proving the metamorphic imprint to be in the range of 'epimetamorphism' (terminology of Kubler 1968) although reaching the epizone–anchizone transition at the top of the Tibetan Series.

The prograde phase of metamorphism was followed by a strong retrograde stage, developed mainly in the basal parts of the Nilgiri Limestone, often masking the prograde assemblages by the crystallization of new mineral phases (zoisite/clinozoisite, titanite, retrograde amphibole).

Figure 2 shows the mappable zones of prograde metamorphism and the related isograds, based on specific decarbonation reactions, with the exception of the biotite–phlogopite isograd. The mapping of isograds in the field is partly impeded by steep slopes on both sides of the Marsyandi river between Chame village and the Mutsog mountain range. However, the analysis of the relationship between deformation and metamorphism (see Fig. 10) suggests that the isograds are parallel to the lithological layering.

Phlogopite

Phlogopite is observed for the first time in the Devonian Tilicho Lake Formation and appears in all meta-carbonates of structural deeper levels down to the base of Formation II (High Himalayan Crystallines). The isograd represents the first appearance of the index mineral, following the definition of isograd by Tilley (1924).

In carbonate rocks the typical assemblage is calcite–dolomite–phlogopite–quartz. The

Table 1. *Selected microprobe analyses of biotite/phlogopite from the Nilgiri Limestone (for sample location see Fig. 2)*

	266a-1	266a-2	269e-1	54a-1	54a-2
SiO_2	40.43	40.21	40.49	40.91	40.86
TiO_2	1.01	0.92	0.82	1.28	1.03
Al_2O_3	16.41	16.92	17.57	15.61	15.41
FeO^T	7.85	7.77	6.95	8.54	8.67
MnO	0.00	0.04	0.02	0.01	0.00
MgO	19.61	20.08	20.40	19.50	19.57
CaO	0.00	0.00	0.06	0.01	0.01
Na_2O	0.00	0.02	0.05	0.08	0.08
K_2O	9.81	10.11	9.84	9.93	9.89
Total	95.12	96.07	96.20	95.87	95.52
16 cations, 22 oxygens					
Si	5.94	5.84	5.88	5.99	6.00
Al^{IV}	2.06	2.16	2.12	2.01	2.00
Al^{VI}	0.79	0.73	0.89	0.68	0.66
Ti	0.11	0.10	0.09	0.14	0.11
Fe^2	0.97	0.94	0.75	1.05	1.06
Mn	0.00	0.01	0.00	0.00	0.00
Mg	4.30	4.34	4.42	4.25	4.28
Ca	0.00	0.00	0.01	0.00	0.00
Na	0.00	0.01	0.01	0.02	0.02
K	1.84	1.87	1.82	1.85	1.85
X_{Ann}*	0.16	0.15	0.12	0.17	0.17
X_{Phl}	0.70	0.71	0.72	0.70	0.70

* $X_{Ann} = Fe^{2+}/(Fe^{2+} + Mg + Mn + Al^{VI} + Ti)$
$X_{Phl} = Mg/(Fe^{2+} + Mg + Mn + Al^{VI} + Ti)$

chemical composition varies between phlogopite rich and biotite rich phases depending upon the composition of the rock (see Table 1). A possible phlogopite production reaction in carbonatic rocks is:

dolomite + K-feldspar + H_2O
= phlogopite + 3 calcite + $3CO_2$ (1)
(Rice 1977; Puhan 1978).

This reaction may account for only a small amount of the observed phlogopite as there is not enough K-feldspar as reactant in equivalent rocks at lower grades of the biotite/phlogopite zone to form the observed amount of phlogopite.

Titanite

Accessory prograde titanite is found within impure meta-carbonates between the Silurian Dark Band Fm. and the upper parts of the Nilgiri Limestone. Titanite typically appears as very small and colourless rounded grains (<10 μm), similar to the equally prevalent colourless rutile grains (frequently the identification of titanite was possible only by microprobe analysis). The assemblage calcite–quartz–muscovite–biotite–rutile in the calc-schists of the Dark-Band-Fm. suggests the titanite producing reaction:

rutile + calcite + quartz = titanite + CO_2 (2)
(Hunt & Kerrick 1977).

Abundant titanite at the base of the Nilgiri Limestone and in the meta-carbonates of the Formation II (High Himalayan Crystallines) is attributed to a strong retrograde metamorphism.

Table 2. *Selected microprobe analyses of amphiboles from the Nilgiri Limestone (for sample location see Fig. 2)*

	252	255a	255a	260	260	269e	269e
SiO_2	54.60	52.67	51.60	57.44	55.78	56.74	57.59
TiO_2	0.07	0.09	0.17	0.12	0.06	0.06	0.01
Al_2O_3	2.46	2.85	3.77	1.35	1.41	1.75	1.36
FeO^T	8.80	13.47	14.01	4.24	3.89	3.47	3.46
MnO	0.07	0.16	0.14	0.06	0.10	0.03	0.04
MgO	17.79	14.73	14.56	21.39	20.75	21.72	20.79
CaO	13.06	12.55	12.63	13.66	13.89	13.82	13.75
Na_2O	0.22	0.32	0.41	0.10	0.10	0.09	0.03
K_2O	0.14	0.31	0.34	0.19	0.06	0.07	0.02
Total	97.21	97.15	97.63	98.55	96.04	97.75	97.05
TSi	7.76	7.67	7.52	7.88	7.85	7.82	7.97
TAl	0.24	0.33	0.48	0.12	0.15	0.18	0.03
TTi	0.00	0.00	0.00	0.00	0.00	0.00	0.00
CAl	0.17	0.16	0.17	0.09	0.09	0.10	0.19
CTi	0.01	0.01	0.02	0.01	0.01	0.01	0.00
CMg	3.76	3.20	3.16	4.37	4.35	4.46	4.29
CFe^{2+}	1.05	0.63	1.65	0.49	0.45	0.40	0.40
CMn	0.01	0.00	0.00	0.01	0.01	0.00	0.00
CCa	0.00	0.00	0.00	0.03	0.09	0.03	0.12
BMg	0.00	0.00	0.00	0.00	0.00	0.00	0.00
BFe^{2+}	0.00	0.02	0.06	0.00	0.00	0.00	0.00
BMn	0.00	0.02	0.02	0.00	0.00	0.00	0.00
BCa	1.99	1.96	1.92	1.97	2.00	2.00	1.92
BNa	0.01	0.00	0.00	0.03	0.00	0.00	0.01
ACa	0.00	0.00	0.05	0.00	0.01	0.01	0.00
ANa	0.05	0.09	0.12	0.00	0.03	0.03	0.00
AK	0.03	0.06	0.06	0.03	0.01	0.01	0.00
ΣT	8.00	8.00	8.00	8.00	8.00	8.00	8.00
ΣC	5.00	5.00	5.00	5.00	5.00	5.00	5.00
ΣB	2.00	2.00	2.00	2.00	2.00	2.00	1.93
ΣA	0.08	0.15	0.23	0.03	0.05	0.04	0.00

Cations calculated with the computer program 'AMPHIBOL' (Richard & Clarke 1990) (16 cations, 23 oxygens; only Fe^{2+})

Amphiboles

The occurrence of prograde amphiboles may be divided into two zones, depending on the reactions producing the observed assemblages. The growth of amphiboles in zone 1 can be modelled using the system $CaO–MgO–SiO_2–CO_2–H_2O$. The formation of amphibole in zone 2 requires an extension of the system to $CaO–K_2O–MgO–(FeO)–Al_2O_3–SiO_2–CO_2–H_2O$, as the occurrence of K-feldspar in zone 2 has to be considered.

Zone 1 (Tremolite + Calcite). The first appearance of amphiboles is located south of Mutsog in the middle parts of the Nilgiri Limestone. Assemblages observed within this area are: calcite–dolomite–quartz–phlogopite; tremolite–calcite–dolomite–quartz–phlogopite; and tremolite–calcite–quartz–phlogopite.

The amphiboles have a tremolite composition (nomenclature following Leake, 1978) with minor chlorine contents (0.0–0.04 wt%) and no fluorine (see Fig. 3 and Table 2: sample 269e and 260). The tremolites are found either as radial aggregates in a calcitic matrix or as single elongate crystals evenly distributed in the calcitic matrix. Samples of zone 1 not containing tremolite show dolomite in varying amounts (2–30 vol.%); whereas in samples with tremolite the dolomite content never exceeds 2.3 vol.%. In this case dolomite is certainly a relict, which occurs only in tremolite-free (calcite–dolomite–quartz) parts of the thin section, suggesting that here the tremolite-producing reaction has not gone to completion.

The mineral assemblages, as well as inclusions of quartz and calcite within tremolite grains, suggest the reaction (Metz & Trommsdorff

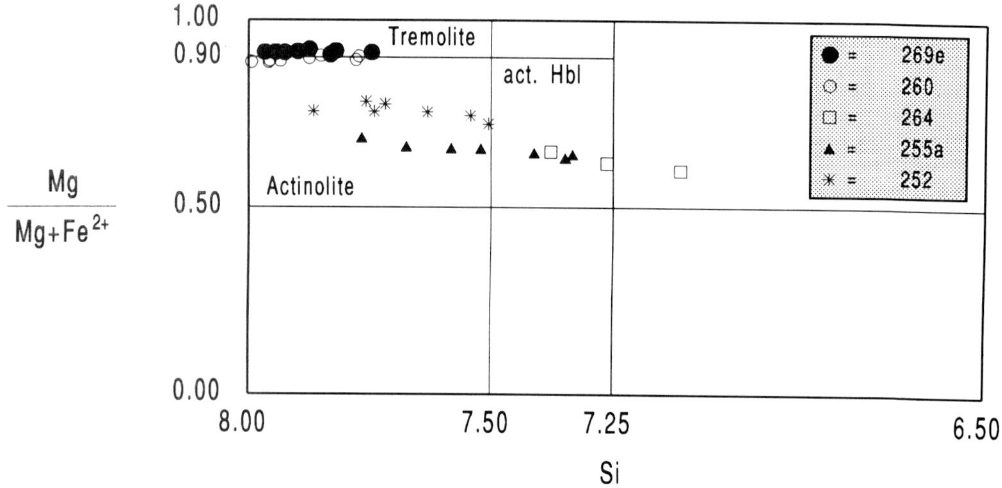

Fig. 3. Chemical composition of prograde amphiboles within the Nilgiri Limestone (diagram according to Leake 1978). For location of samples see Fig. 2. Samples 269e and 260 belong to the amphibole zone 1 (Tr + Cal), samples 264, 255a, and 252 to zone 2 (Tr + Kfs).

1968; Slaughter *et al.* 1975; Eggert & Kerrick 1981; Abercrombie *et al.*, 1987):

$$5 \text{ dolomite} + 8 \text{ quartz} + 1H_2O = 1 \text{ tremolite} + 3 \text{ calcite} + 7CO_2 \quad (3)$$

Reactions involving talc can be excluded in so far as no talc was found in the meta-carbonates of the Manang area.

Zone 2 (tremolite + K-feldspar). In the lower parts of the Nilgiri Limestone the calc-silicate assemblages change. K_2O-containing mineral phases such as K-feldspar or phlogopite are present within the assemblages in larger amounts. The composition of the amphiboles is shifted to more Fe-rich compositions (actinolitic amphiboles, see Fig. 3 and Table 2: samples 264, 255a, 252). Typical assemblages in the calc-silicates of zone 2 are: actinolite–calcite–quartz–phlogopite–K-feldspar and actinolite–calcite–quartz–phlogopite–K-feldspar–plagioclase.

Amphibole-producing reactions in zone 2 are modelled in the system CaO–K_2O–MgO–(FeO)–Al_2O_3–SiO_2–CO_2–H_2O. This system has been investigated by various authors: Hoschek (1973), Hewitt (1973, 1975), Thompson (1975), Rice (1977), Puhan (1978), and Sanford (1980).

The following reaction may be responsible for the prograde growth of amphibole in these assemblages (Hoschek 1973; Hewitt 1975):

$$5 \text{ phlogopite} + 6 \text{ calcite} + 24 \text{ quartz} = 3 \text{ tremolite} + 5\text{K-feldspar} + 6CO_2 + 2H_2O. \quad (4)$$

The only textural proof for this reaction is the presence of phlogopite and quartz inclusions in some prograde amphiboles. Reaction (4) proceeds at slightly higher temperatures than reaction (3) (Hoschek 1973; Gottschalk & Metz 1984). Thus, zone 1 defines a temperature range that is still too low for the participation of phlogopite in the tremolite producing reaction.

Clinopyroxenes

Clinopyroxenes are observed in higher grade meta-carbonates of the Nilgiri Limestone a few hundred metres downward from the amphibole + K-feldspar isograd. Clinopyroxene is present from there on in the assemblages of the underlying metacarbonates down to the base of Formation II within the High Himalayan Crystallines. The composition of clinopyroxenes varies between pure diopside and salite (see Table 3) with Al from 0.38 up to 1.66 wt.% and Na + K not exceeding 0.3 wt.%. The occurrence of clinopyroxenes can be divided into two zones: the separating criterion is, by analogy with the amphiboles, the participation of K_2O-containing phases like phlogopite and K-feldspar in the pyroxene-producing reactions.

Zone 1. Diopside appears as small rounded crystals embedded in a matrix of calcite, quartz and tremolite or as poikiloblasts with inclusions of tremolite and quartz suggesting the formation of diopside in zone 1 according to the reaction

Table 3. *Selected microprobe analyses of clinopyroxenes from the Nilgiri Limestone (for sample location see Fig. 2)*

	252-1	252-2	260-1	260-2	265-1	265-2	266a-1	266a-2
SiO_2	53.73	53.37	54.09	53.51	52.66	53.53	54.14	54.22
TiO_2	0.02	0.00	0.00	0.03	0.02	0.04	0.04	0.00
Al_2O_3	0.30	0.35	0.59	0.78	0.87	0.82	0.54	0.95
FeO^T	6.35	6.71	2.46	2.59	7.98	6.93	4.33	3.56
MnO	0.13	0.04	0.10	0.10	0.07	0.03	0.00	0.01
MgO	13.95	14.12	16.20	16.21	13.16	13.56	15.51	15.15
CaO	24.55	24.43	25.98	26.07	24.27	24.51	25.76	25.59
Na_2O	0.12	0.14	0.12	0.14	0.24	0.24	0.04	0.09
K_2O	0.00	0.00	0.00	0.00	0.02	0.02	0.00	0.01
Cr_2O_3	0.00	0.09	–	–	–	–	–	–
Total	99.15	99.25	99.54	99.43	99.29	99.68	100.36	99.58
4 Cations, 6 oxygens								
Si	2.00	2.00	1.99	1.97	1.98	1.99	1.98	1.99
Al^{IV}	0.00	0.00	0.01	0.03	0.02	0.01	0.02	0.01
Al^{VI}	0.01	0.02	0.02	0.00	0.02	0.03	0.00	0.03
Ti	0.00	0.00	0.00	0.00	0.00	0.00	0.00	0.00
Fe^{2+}	0.20	0.21	0.08	0.08	0.25	0.22	0.13	0.11
Mn	0.00	0.00	0.00	0.00	0.00	0.00	0.00	0.00
Mg	0.78	0.79	0.89	0.89	0.74	0.75	0.85	0.83
Ca	0.98	0.98	1.02	1.03	0.98	0.98	1.01	1.01
Na	0.01	0.01	0.01	0.01	0.02	0.02	0.00	0.01
K	0.00	0.00	0.00	0.00	0.00	0.00	0.00	0.00
Cr	0.00	0.00	–	–	–	–	–	–
Wo*	49.05	48.76	50.52	50.12	48.62	48.49	49.94	49.54
En	40.59	40.50	46.49	47.86	39.28	40.03	43.80	44.58
Fs	10.37	10.75	2.99	2.02	12.10	11.47	6.26	5.88

* Endmembers calculated according to Lindsley (1983).

$$1 \text{ tremolite} + 3 \text{ calcite} + 2 \text{ quartz} = 5 \text{ diopside} + 1H_2O + 3CO_2 \quad (5)$$

(Metz & Trommsdorff 1968; Sanford 1980; Metz 1983; Dachs & Metz 1988).

Zone 2. Calcite and phlogopite inclusions in diopside as well as the intergrowth of diopside with K-feldspar are interpreted as evidence for the reaction:

$$1 \text{ phlogopite} + 3 \text{ calcite} + 6 \text{ quartz} = 3 \text{ diopside} + 1\text{K-feldspar} + 1H_2O + 1CO_2 \quad (6)$$

(Hewitt 1973; Thompson 1975; Sanford 1980).

Retrograde metamorphism

A strong retrograde metamorphism has affected the base of the Nilgiri Limestone only. The crystallization of new solid phases including the hydroxyl group minerals such as zoisite/clinozoisite, amphibole or titanite, has partly overprinted the prograde assemblages.

Retrograde titanite (titanite-II) occurring within the diopside + K-feldspar zone of the Tibetan Series as well as Formation II of the High Himalayan Crystallines, can be a major phase of the assemblages. It crystallized as large idioblastic crystals (sphenoids) overgrowing the prograde phases diopside, amphibole, K-feldspar and calcite.

Together with titanite-II, a second generation of amphiboles (amphibole-II) can be observed in the calc-silicates and marbles of the Nilgiri Limestone near Chame village. Amphibole-II is distinguishable from the prograde amphiboles on the basis of the following criteria:

(1) amphiboles-II are predominantly idiomorphic without embayments due to dissolution or overgrowth at the ends of the elongated crystals;
(2) amphiboles-II are much smaller than amphiboles-I;
(3) amphiboles-II overgrow phlogopite, diopside, K-feldspar and amphibole; frequently they show a preferred orientation in their axis of elongation, the origin of which is not clear;

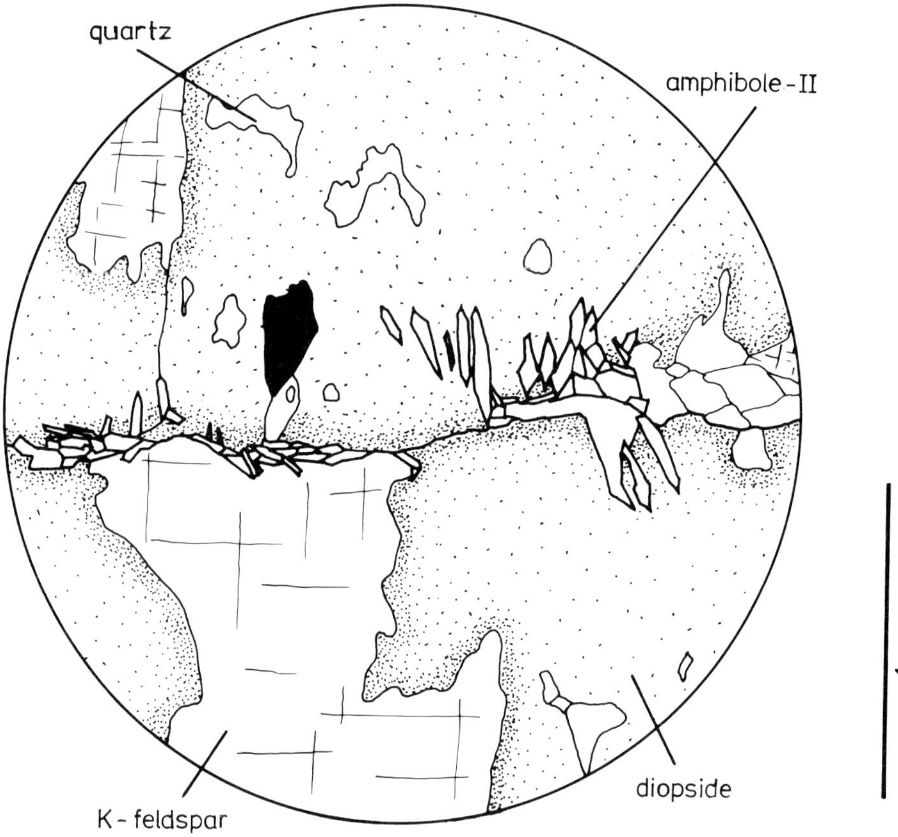

Fig. 4. Retrograde amphiboles-II, crystallized along a crack (sample 252).

(4) amphiboles-II sometimes crystallize along healed micro-cracks (see Fig. 4).

A chemical distinction was not possible. More than 50 microprobe analysis of prograde amphiboles and retrograde amphiboles-II did not reveal any difference in their composition. This lack of difference may be ascribed to the equilibration of prograde amphiboles to the *P–T*-conditions of the retrograde metamorphism. Small overgrowths at the edges of prograde amphiboles with exactly the same composition as the retrograde amphiboles-II support this assumption.

Taking into account that rutile is always present in the assemblages of the meta-carbonates and that titanite occasionally grows directly within a diopside grain, one of the more likely reactions for the formation of both retrograde phases, amphibole and titanite, is given by Hunt & Kerrick (1977):

$$\begin{aligned}&5\ \text{diopside} + 2\ \text{rutile} + CO_2 + H_2O\\&= \text{tremolite} + \text{calcite} + \text{titanite}.\end{aligned} \quad (7)$$

Besides amphibole-II and retrograde titanite, the meta-carbonates of the basal Nilgiri Limestone, as well as the meta-carbonates of Formation II of the High Himalayan Crystallines show clinozoisite/zoisite in their assemblages. Clinozoisite/zoisite rims around scapolite or plagioclase are ubiquitous at the base of the Formation II. At the base of the Tibetan Series (Nilgiri Limestone) clinozoisite/zoisite appears as large poikiloblasts overgrowing diopside, amphibole and K-feldspar. The following reactions may be responsible for the growth of zoisite/clinozoisite: scapolite + H_2O = 2 zoisite + CO_2 (Huckenholz & Seiberl 1989) for the rims around scapolite; and 3 anorthite + calcite + H_2O = 2 zoisite + CO_2 (Storre & Nitsch 1972; Hunt & Kerrick 1977)

Calcite–dolomite geothermometry

The calcite–dolomite solvus provides an appropriate method for an independent temperature estimation in meta-carbonates with coexisting dolomite and calcite. Several investigations of the calcite–dolomite solvus have succeeded in fixing its position in temperature–composition space (see Harker & Tuttle 1955; Graf & Goldsmith 1958; Goldsmith & Newton 1969; Anovitz & Essene 1987). The influence of pressure is negligible; a pressure change of 100 MPa corresponds to a shift of only 4° C in temperature (Gottschalk & Metz 1990 *in* Dachs 1990). But this refers, strictly, only to the $CaCO_3$–$MgCO_3$ system. However, natural calcites often contain considerable amounts of $FeCO_3$. Bickle & Powell (1977) and Anovitz & Essene (1987) developed models for calculating the influence of Fe on the calcite-dolomite solvus, thus extending the geothermometer to the ternary system $CaCO_3$–$MgCO_3$-$FeCO_3$.

The following calculations are based on the model of Anovitz & Essene (1987), which is applicable only between the temperature range 250° C to 800° C, with an increasing error above 500° C. The requirements for application of the geothermometer, coexisting calcite and dolomite with $MgCO_3$ and $FeCO_3$ contents in a range corresponding to a temperature range of 250°–800° C, are fulfilled within the Tibetan Series from the Lower Carboniferous Tilicho Lake Formation down to the Nilgiri Limestone (tremolite + calcite isograd). The base of the Nilgiri Limestone is not suitable for thermometric evaluation because dolomite has completely reacted to form tremolite or diopside.

Calculations based on the correction equations of Bickle & Powell (1977) yielded temperatures more or less identical to those from the Anovitz & Essene (1987) calculation scheme.

All calcite and dolomite analyses were performed using a computer controlled electron microprobe analyser (Camebax SX 50) with 4 spectrometers. Pure calcite and dolomite for Ca and Mg and oxides for Fe and Mn acted as reference standards. The acceleration voltage was 15 kV, the beam current 15 nA. Counting time for each element was 10 s. The beam diameter was set to 10 μm to integrate possible submicroscopic exsolutions. Dolomite exsolution on a microscopic scale was not found. A high number of analysis per grain is essential for evaluating the maximum contents of $MgCO_3$ and $FeCO_3$. Single carbonate grains, as well as domains comprising several carbonate grains were analyzed in intervals varying between 10 μm and 40 μm, yielding between 30 and 200 analyses per grain depending on the grain size.

Mg-distribution patterns in calcites

Microprobe analyses of several calcite grains per sample have shown that the $MgCO_3$-contents may decrease towards grain margins in contact with dolomite. The difference between lowest and highest values within a single grain can be relatively high (up to 2.7 Mol.% $MgCO_3$). This can certainly be attributed to a retrograde equilibration (Wada & Suzuki 1983). Moreover,

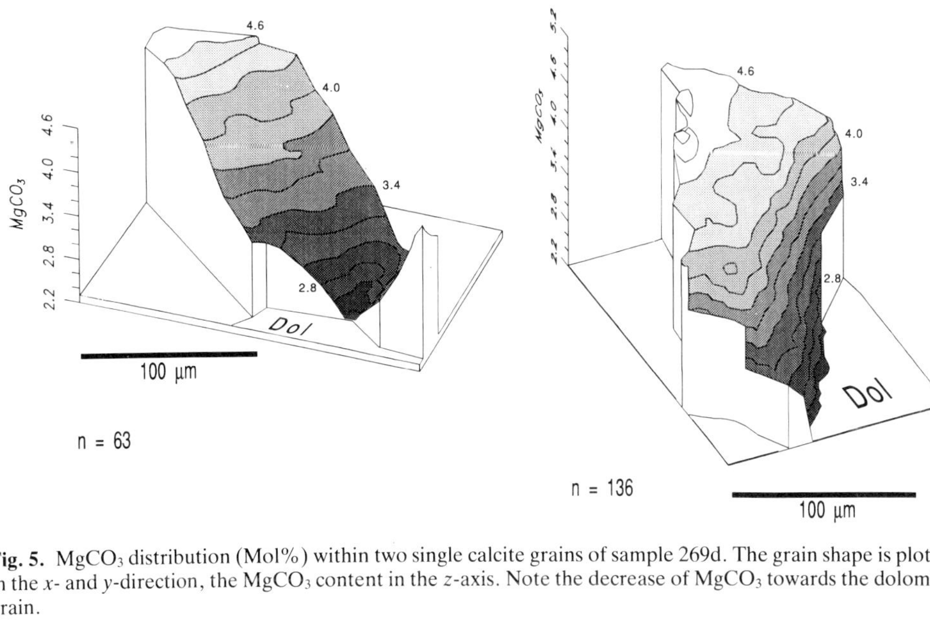

Fig. 5. $MgCO_3$ distribution (Mol%) within two single calcite grains of sample 269d. The grain shape is plotted in the *x*- and *y*-direction, the $MgCO_3$ content in the *z*-axis. Note the decrease of $MgCO_3$ towards the dolomite grain.

calcites coexisting with dolomite are often characterized by irregular distributions of $MgCO_3$ and $FeCO_3$. These distribution patterns are also thought to be caused by retrograde equilibration. Besides the marginal equilibration, mechanisms controlled by crystallographic directions may be taken into account.

The maximum contents of $MgCO_3$ and $FeCO_3$ can be observed in parts of calcite grains, which are furthest from a dolomite grain. Typical examples are calcite grains of sample 269d (see Fig. 5) with highest $MgCO_3$ contents at the calcite/calcite grain boundary and lowest values at the calcite/dolomite grain boundary. It is striking that in some samples the highest $MgCO_3$ and $FeCO_3$ contents occur not in calcite grains directly coexisting with dolomite grains but in calcite grains within a distance of 2–3 grains from the next dolomite grain.

Fe- and Mn-contents of calcites

The Mn content of the analysed calcites are consistently low (<0.5 Mol % $MnCO_3$, in most cases <0.25 Mol % $MnCO_3$).

The Fe content is more variable, maximum values are in the range of 2.5 Mol % $FeCO_3$. The replacement of $MgCO_3$ by $FeCO_3$ results in different $MgCO_3/FeCO_3$ ratios at a fixed temperature. The correlation between both elements with different ratios but the same temperature is negative. The analysed calcites normally show a weak positive correlation, which can be explained by partial retrograde equilibration of the calcites leading to lower $MgCO_3/FeCO_3$ ratios.

Conclusions

The application of the calcite dolomite geothermometer is usually based on a small number of analyses. Our results show, however, that this can cause an underestimation of the prograde maximum temperature. It seems appropriate to choose an empirical approach for interpreting the Mg/Fe data, because it is difficult to determine the maximum value of $MgCO_3$ and $FeCO_3$ with only a few analysis due to the observed irregular distribution patterns.

Table 4 summarizes the statistical data on the

Table 4. *Statistical data for selected calcites*

	n	$MgCO_3$				$FeCO_3$				°C		
		Min	Max	$\bar{x}$	>90	Min	Max	$\bar{x}$	>90	Min	Max	>90
269a	139	1.44	3.08	2.37	2.72	0.66	1.50	1.04	1.26	313	457	432
r1	77	2.01	2.97	2.42	2.72	0.66	1.46	1.03	1.24	371	454	431
r2	18	1.56	2.99	2.23	2.97	0.80	1.42	1.02	1.31	321	456	450
r3	44	1.44	3.08	2.35	2.71	0.76	1.50	1.06	1.32	313	457	432
269d	284	1.28	5.41	4.17	4.72	0.47	1.30	0.87	1.05	260	557	530
r1	63	2.20	4.66	3.84	4.46	0.54	1.17	0.91	1.08	391	529	523
r2	101	3.29	4.97	4.44	4.80	0.50	1.30	0.88	1.07	459	543	532
r3	120	1.28	5.41	4.12	4.73	0.47	1.12	0.85	1.02	260	557	529
270b	64	1.03	3.08	2.66	2.95	1.08	2.08	1.60	1.85	223	466	462
270d	48	2.26	4.85	3.33	3.83	1.38	2.47	2.01	2.33	405	580	522
27012	297	1.30	4.02	3.08	3.51	0.38	1.11	0.78	0.91	(263)	(501)	(474)
87	105	2.17	4.58	3.50	4.06	0.65	1.61	1.09	1.29	393	530	507
r1	24	3.17	4.58	3.63	4.13	0.87	1.39	1.11	1.31	459	530	511
r2	81	2.17	4.32	3.47	4.01	0.65	1.61	1.09	1.27	393	530	502
54a	111	2.06	4.37	3.75	4.15	0.24	1.74	1.31	1.57	378	535	518
70	110	1.32	3.23	2.61	3.00	1.07	2.48	1.69	1.92	297	476	465
r1	55	1.79	3.18	2.70	3.03	1.07	2.03	1.70	1.90	365	476	466
r2	55	1.32	3.23	2.51	2.91	1.31	2.48	1.68	1.94	297	470	470
71b	40	1.12	2.95	2.39	2.80	0.80	1.70	1.32	1.56	231	457	444
99	304	0.85	3.43	2.28	2.92	0.21	1.28	0.65	0.92	200	478	437
r1	170	0.85	3.26	2.06	2.82	0.21	0.84	0.52	0.67	200	455	429
r2	134	1.77	3.43	2.55	2.98	0.41	1.28	0.81	0.97	340	478	444
G39	61	1.25	2.41	2.01	2.34	1.10	1.81	1.46	1.66	284	422	410
G44	17	1.15	1.89	1.54	1.85	1.06	1.71	1.37	1.62	252	375	369
G45	33	1.07	2.37	1.66	1.98	0.23	2.20	1.41	1.99	200	430	387

Fe corrected temperatures are calculated according Anovitz & Essene (1987);
$\bar{x}$, mean value; >90, 90th percentile; r1, r2, r3 refer to different grains within one sample.

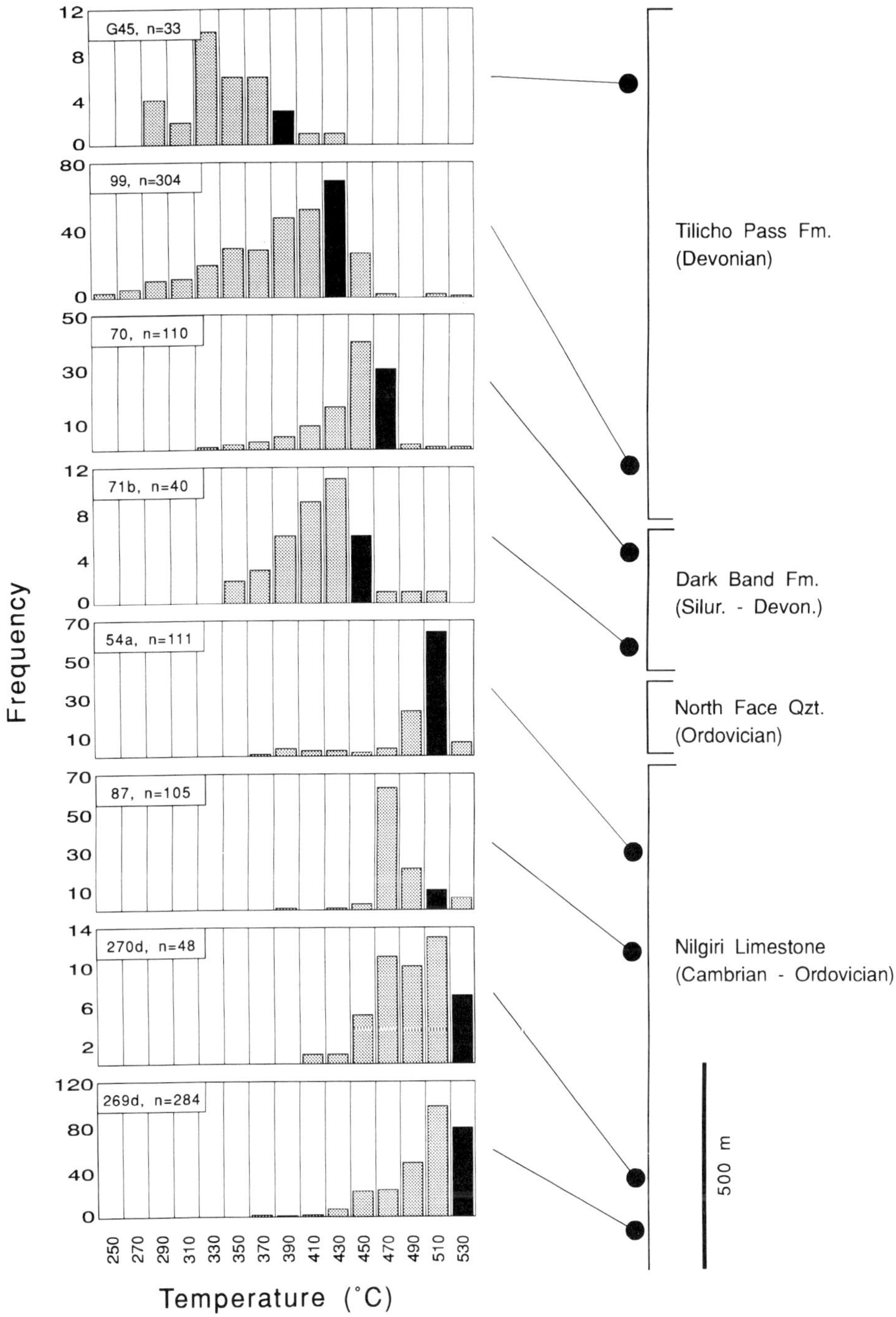

Fig. 6. Frequency distribution of Fe-corrected temperatures calculated according to Anovitz & Essene (1987). The black bar marks the position of the 90th percentile, which was used as the maximum value. For sample locations see Fig. 2.

$MgCO_3$ and $FeCO_3$ contents and the Fe-corrected temperatures calculated after the model of Anovitz & Essene (1987). Figure 6 shows the frequency distribution of the calculated Fe-corrected temperatures within selected samples. The interval width of 20° C corresponds to the error in the pure $CaCO_3$–$MgCO_3$ system and seems to be the best compromise between clearness of the diagram and satisfying 'resolution' so that the essential information is preserved. To avoid the misinterpretation of extreme values caused by analytical errors, we have chosen the 90th percentile as the maximum value, thus excluding the uppermost 10% of data. Extreme values can be caused by analysis contaminated by neighbouring mg-mineral phases like phlogopite. The use of the 90th percentile seems to be arbitrary, but during the study it appeared that maximum values accumulate around the 90th (± 5) percentile. It should be considered, that the maximum $MgCO_3$ and $FeCO_3$ values may be the result of retrograde equilibration, so the derived temperature is a 'minimum' estimation of the prograde maximum temperature. There is no error bar shown, but an error of at least ± 30° C for the Fe-corrected temperatures (Bickle & Powell 1977) has to be taken into account. The histograms clearly show that the frequently used mean value is not appropriate for evaluating the maximum temperature. Retrograde equilibration always causes an accumulation in low temperature regions, so giving an irrelevant mean value (see Table 4 for a comparison of the mean and the 90th percentile). The scatter to higher temperatures ($MgCO_3$) in samples of lower grade (71b, 70, 99, G45) is due to analytical errors at grain margins. The average grain size of these samples is smaller than in metacarbonates of higher grades.

Figure 6 shows the clear correspondence between increasing $MgCO_3$-contents (and temperature) and structural depth equivalent to an increase in metamorphic grade. The derived maximum temperatures are plotted in a temperature versus structural depth diagram (see Fig. 7), so defining a temperature gradient for the time of the thermal maximum of the prograde metamorphism. This gradient is in good agreement with the illite-crystallinity values for the upper Tibetan Series, which are still in the epimetamorphic range. The (Pearson-) correlation between temperature and depth is $r = 0.95$ (excluding sample 269a). Sample 269a yielded much too low a temperature relative to other samples of this structural level. One reason for this may be a more complete retrograde equilibration of the $MgCO_3$ content of the calcites due to the small average grain size of this sample.

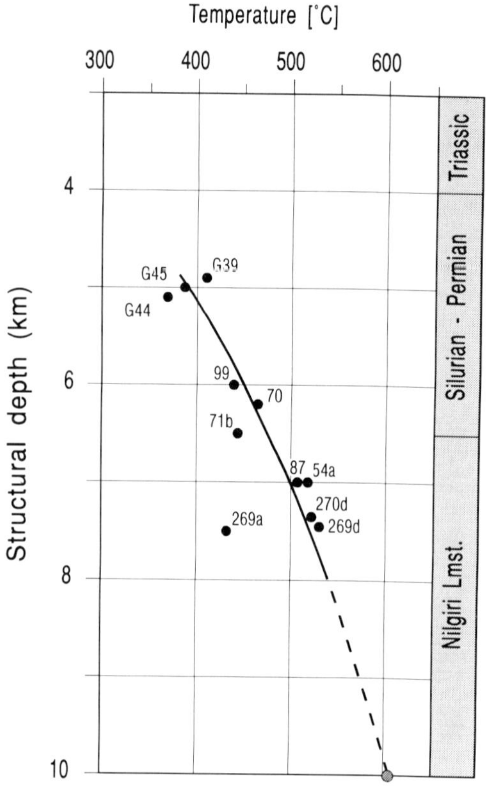

Fig. 7. Maximum temperatures derived by carbonate solvus thermometry plotted against the approximate structural depth. The value at the base of the Tibetan Series represents the temperature value extrapolated from the thermometric data (Schneider 1991) of the High Himalayan Crystallines.

Pressure estimation for the Tibetan Series

A petrological pressure estimation for the time of prograde metamorphism is impeded by the lack of an isobaric invariant point assemblage in the systems $CaO–MgO–SiO_2–CO_2–H_2O$ or $K_2O–Al_2O_3–CaO–MgO–SiO_2–CO_2–H_2O$. Instead, the lithostatic pressure had to be estimated by calculating the overload of the sedimentary pile. The thickness of the Tibetan Series is presumed to be now 7400 m (Fuchs *et al.* 1988). The thickness of already eroded sediments is difficult to estimate, but the Thakkhola graben shows at least 2400 m Jurassic and Cretaceous sediments (Bodenhausen & Egeler 1971; Bordet *et al.* 1975; Colchen *et al.* 1986). In

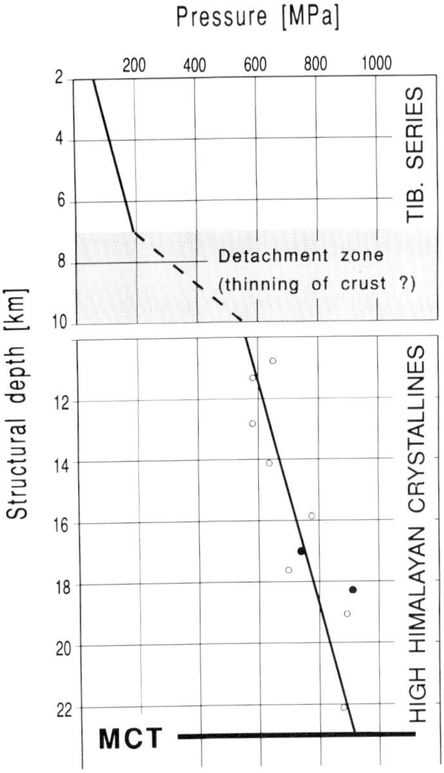

Fig. 8. Pressure gradient for the time of prograde metamorphism. The pressure for the Tibetan Series was calculated by the lithostatic overload, the pressure of the High Himalayan Crystallines is taken from geobarometry results (own values: black circles; values from Hodges *et al.* 1986: open circles).

addition it seems probable that Early Tertiary sediments were present in the Manang area. Other regions of the Himalayas show up to 200 m of Tertiary sediments (e.g. Mt Jolmo-Lungma area in South Tibet; Chang & Cheng 1973). The total thickness of the Tibetan Series in the Manang area at the time of prograde metamorphism can thus be estimated to be approximately 10 km. A serious discrepancy arises when comparing geobarometric data from the High Himalayan Crystallines (Hodges *et al.* 1986; Schneider 1991) extrapolated to the base of the Tibetan Series with the estimated lithostatic pressure at the base of the Tibetan Series (see Fig. 8). The lithostatic pressure is about 300 MPa, the extrapolated geobarometric pressure by comparison is about 560 MPa. A possible explanation for this discrepancy may be a northward detachment of the Tibetan Series at the base of the Nilgiri Limestone causing a thinning of the crust in this region.

X_{CO_2} during the prograde metamorphism

Some indications of the fluid composition (X_{CO_2}) during prograde metamorphism may be deduced from phase diagrams that utilise the carbonate solvus temperature and the pressure estimates. Although the metamorphic temperatures within the Tibetan Series are well documented from the Lower Carboniferous units down to the Nilgiri Limestone (tremolite + calcite isograd), the Nilgiri Limestone includes no usable fluid inclusions, so estimates of fluid composition must be based on petrologic grids that include reactions observed within the sequence. The derived X_{CO_2} values have to be considered with care, because of the following restrictions.

(1) The pressure estimation is subjected to some uncertainties. A shift in pressure to higher values causes a shift of the equilibrium curves to higher temperatures. This means, that the equilibrium X_{CO_2} values for the considered reaction will shift to lower values.

(2) The mineral phases participating in the reactions are solid solutions often with a significant Fe-content and thus differing from the pure endmembers for which the equilibrium curves were determined.

The tremolite–calcite production in the Nilgiri Limestone due to reaction (3) occurred at a temperature of 500–530° C (calcite–dolomite thermometry) and a lithostatic pressure of *c.* 300 MPa (see Fig. 7). Using the petrogenetic grid from Gottschalk & Metz (1984) the equilibrium X_{CO_2} at these conditions is in the range of 0.6–0.8.

At a deeper structural level tremolite + K-feldspar was formed with the participation of phlogopite according to reaction (4). The equilibrium curves for this reaction are located at slightly higher temperatures than for the tremolite–calcite-producing reaction (Hoschek 1973; Hewitt 1975). This reaction proceeds at a slightly higher pressure (*c.* 350 MPa) than the tremolite-calcite producing reaction (3) and thus at a higher temperature (about 550°C). The data from Hoschek (1973) suggest an equilibrium-X_{CO_2} at these conditions of more than 0.4. Shifts of the equilibrium curves caused by fluorine can be neglected as the amphiboles do not contain any fluorine.

The diopside producing reaction (5) of zone 1 at 450 MPa (9.5 km depth) and 560–580° C has a confining equilibrium X_{CO_2} of 0.3 to 0.4 (Metz 1983).

The considerations show, that the fluid phase during the prograde metamorphism had a CO_2-rich composition corresponding to a normal progress of carbonate metamorphism through decarbonisation reactions.

The retrograde metamorphism, in contrast, was controlled by a H_2O-rich fluid phase as is demonstrated by the appearance of OH-containing phases (zoisite/clinozoisite, amphibole-II) and the stoichiometry of the hydration reactions.

Deformation

New mapping of the Tibetan Series in the Manang area by Fuchs *et al.* (1988) and the investigation of microtextures has resulted in a slightly different structural conception compared to previous ideas (Colchen *et al.* 1986). The major structural feature of the Manang area is a wide WNW–ESE-orientated west plunging syncline with Devonian and Lower Carboniferous schists in its core. This syncline resulted from a late deformation phase (D3), which was preceded by two former deformation events (see Fig. 9). A schistosity (S1) was developed parallel or subparallel to bedding during D1. During D2, an S2 crenulation cleavage was developed axial planar to D2 folds. The formation of the Manang Syncline and subordinate N–S directed brittle tectonics is a D3 feature.

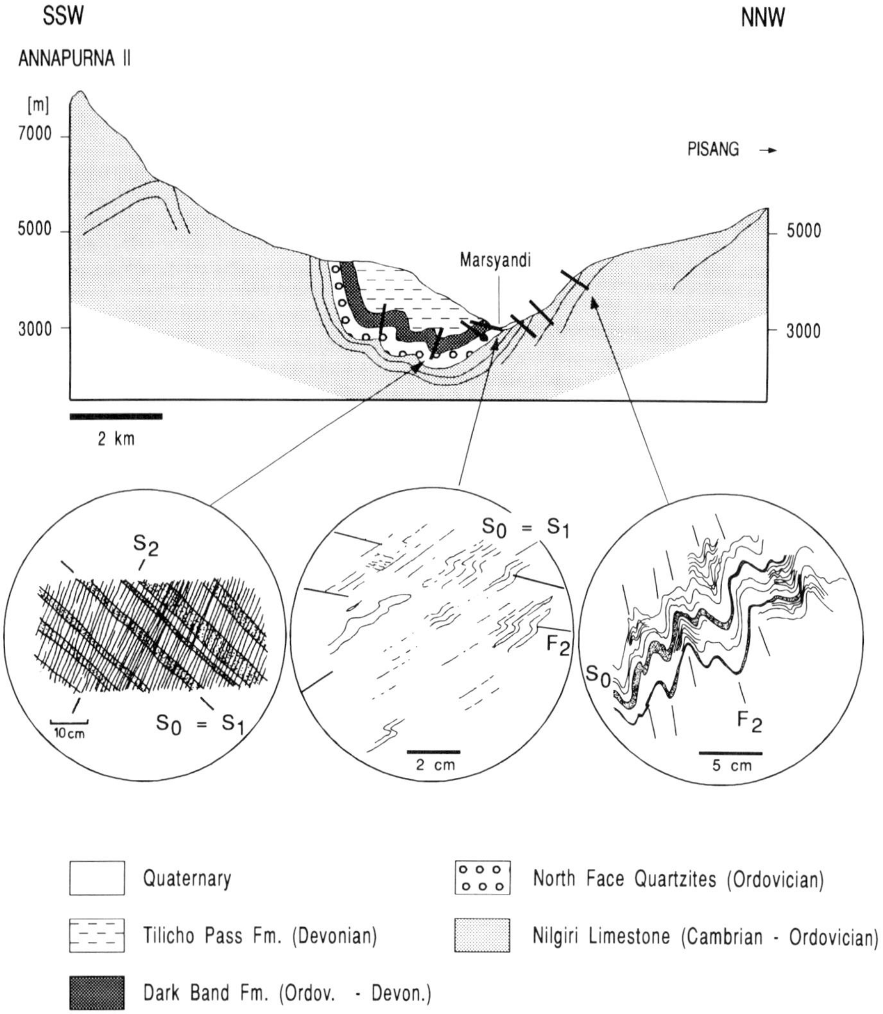

Fig. 9. Cross section through the Tibetan Series of the Manag area showing the orientation of axial planes of D2 small scale folds and the S2 foliation respectively.

First phase D1

From just west of Mutsog (upper Nilgiri Limestone) a planar schistosity, S1, is developed. This schistosity is mostly bedding parallel, although locally may develop at an angle not exceeding 30° to the bedding. S1 is best developed in incompetent layers (pelitic interlayers of the Nilgiri Limestone, phyllites, blackshales and carbonatic shales up to the Lower Carboniferous Tilicho Lake Formation), where it is defined by orientated sheet silicates. In carbonate rich rocks (Cal + Dol > 70 Vol. %) S1 becomes indistinct with few sheet silicates indicating the orientation. In pure metalimestones of the Lower Carboniferous calcite sometimes shows a weak grain flattening, indicating the orientation of S1.

At the base of the Tibetan Series in the Nilgiri Limestone, between Chame and Bhratang, coarse grained tremolite- and diopside-marbles or calc-silicates, partly show a weak foliation parallel to the bedding, that consists of weathered silicate layers. These layers are certainly relicts of the sedimentary layering. A true schistosity S1 has not developed at the base of the Tibetan Series, because of the relative high grade of metamorphic alteration and the coarse recrystallization of diopside, amphibole, and calcite in this region. In addition, the retrograde metamorphism with titanite, zoisite/clinozoisite and amphibole II marks the S1 fabric.

The main stage of prograde metamorphism is correlated with D1. The phases phlogopite, amphibole and diopside grew syn- to postkinematic to S1, but prekinematic to S2. Titanite cannot be attributed definitely to D1 because the small grain size does not allow an identification of the relationship between crystallization and deformation.

Second phase D2

At the base of the Tibetan Series small cracks were formed during the second deformation phase. Those cracks are occasionally healed by retrograde amphibole-II (see above) and calcite. The middle parts of the Nilgiri Limestone show a range of small scale folds from isoclinal to open with a wavelength of some cm that deform S1. Phlogopite, poikiloblastically grown in a late stage of D1 or even postkinematic to D1, is also folded by D2 folds, proving that the folds belong to a second deformation stage. The upper sections of the Nilgiri Limestone show isoclinal to open small scale folds (some centimetres up to a few meters) accompanied by axial plane foliation developed as a crenulation cleavage. This crenulation cleavage is well developed from the upper Nilgiri Limestone up to the Devonian schists (Tilicho Pass Formation). The Lower Carboniferous Tilicho Pass Formation shows only a weakly developed crenulation in pelitic interlayers. Younger Formations do not contain any crenulation cleavage. Newly crystallized mineral phases related to D2 were not found.

Deformation phase D3

A third deformation phase resulted in the formation of the large Manang syncline, with a fold axis plunging at 20° due west, roughly parallel to the small scale D2 folds.

Discussion

This study establishes that there is no major disturbance of the metamorphic gradient within the Tibetan Series. Temperatures at the base certainly did not exceed *c.* 600° C, whilst temperatures in the Lower Carboniferous (Tilicho Lake Formation) were still at about 370–390° C. The carbonate solvus temperatures extrapolated to the base of the Tibetan Series are in good agreement with geothermometry data (Bt-Grt-thermometry) from the High Himalayan Crystallines (Hodges *et al.* 1986). The prograde metamorphism is correlated with a first deformation stage (planar schistosity S1). The second deformation stage, clearly visible between the Lower Carboniferous and the upper Nilgiri Limestone, was not combined with the growth of new mineral phases in this part of the Tibetan Series. A strong retrograde metamorphism is well defined in the basal parts of the Nilgiri Limestone and leads to very extensive crystallization of OH-containing phases like zoisite/clinozoisite, amphibole-II or titanite (see Fig. 10).

Regarding a possible detachment at the base of the Tibetan Series the following facts should be considered. Different directions and mechanisms for a detachment in the transition zone between the Tibetan Series and the High Himalayan Crystallines have been documented. In South Tibet Burg *et al.* (1982) found northward-directed normal faulting at the top of the Crystallines that postdated the granite emplacement. Pêcher & Bouchez (1987) describe a zone of normal faulting combined with a dextral strike slip at the top of the High Himalayan Crystallines resulting in an eastward extrusion of the Tibetan Series during and shortly after the granite emplacement in Central Nepal. Similar late metamorphic normal faulting was also

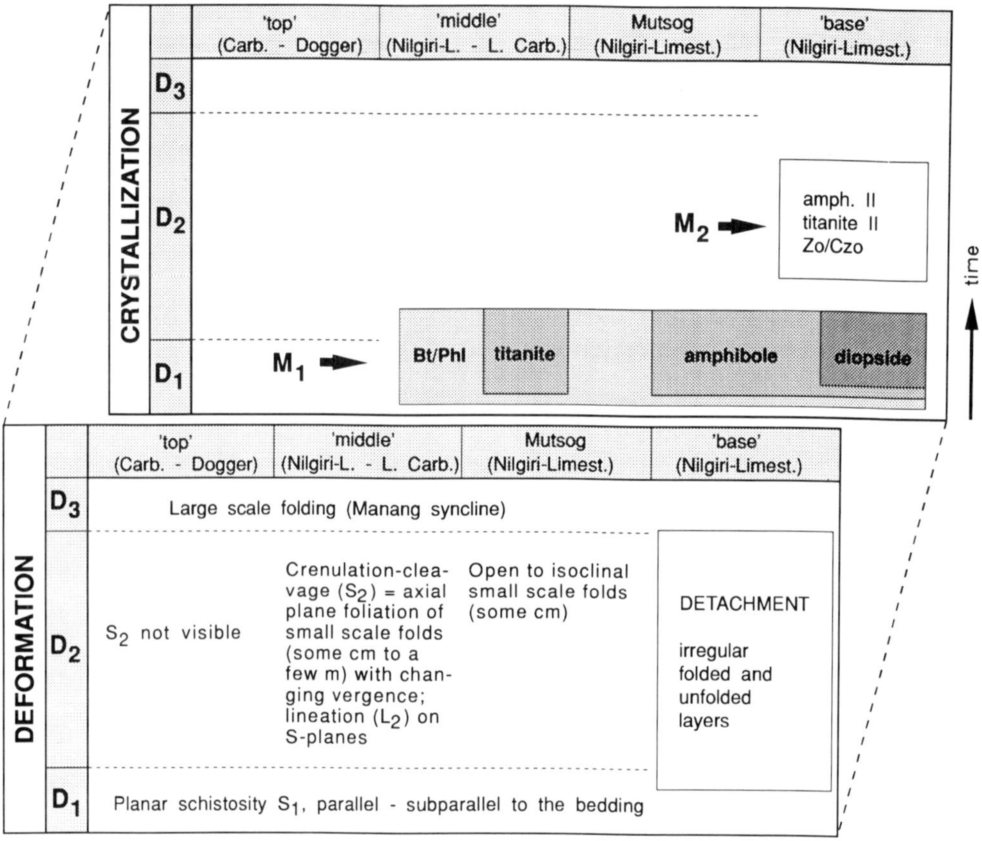

Fig. 10. Schematic illustration for the relation between crystallization and deformation. M1 corresponds to the prograde metamorphism, M2 to the retrograde metamorphism.

described by Herren (1987) in Zanskar, where a vertical displacement of 19 km was calculated from the thickness of the shearzone. Thus, it appears that movements in the transition zone between the Tibetan Series and the High Himalayan Crystallines may be syn- to postmetamorphic to the prograde metamorphism.

For the Marsyandi Valley considerable postmetamorphic displacements at the base of the Tibetan Series can be excluded, because they should disrupt the now observed metamorphic gradient in this zone, whereas the petrologic data clearly show, that the metamorphic gradient is not disturbed.

Thus, on consideration of the following points it is reasonable to assume that there is a syn- to late-metamorphic detachment at the base of the Tibetan Series; firstly, the remarkable change in deformation style between open folding in the Tibetan Series and the homoclinal structure of the High Himalayan Crystallines; secondly, the inhomogeneous structure at the base of the Nilgiri Limestone with sudden changes between irregular folded layers and totally unfolded layers conspicuously visible in the steep walls along the Marsyandi Valley west of Chame; thirdly, some granitic veins, are sheared off at discontinuities parallel to the bedding whereas others cut through undisturbed folded and unfolded layers. As the Manaslu granite intruded syn- to latekinematic to the main S1 schistosity (Le Fort 1981) the movements are contemporaneous to this event; fourthly, outcrops near the village of Chame show cracks, filled with the prograde assemblage diopside and K-feldspar indicating, that the movements began during the prograde metamorphism; fifthly, the asymmetry of shear indicators normally suggest a northward directed displacement of the Tibetan series, and sixthly, a strong retrograde stage of metamorphism is observed only at the base of the Tibetan Series and the underlying High Himalayan Crystallines, but not in the formations younger than the middle Nilgiri Limestone. More research is required to test the theory that this retrograde stage might be related to the granite intrusion and thus contemporaneous to the detachment.

The authors would like to thank S. Heuss-Aßbichler for her helpful comments on the manuscript. The research work was financially supported by the Deutsche Forschungsgemeinschaft. Reviews by M.J. Bickle and G.J.H. Oliver are gratefully acknowledged.

References

ABERCROMBIE, H.J., SKIPPEN, G.B. & MARSHALL, D.D. 1987. F–OH substitution in natural tremolite, talc, and phlogopite. *Contributions to Mineralogy and Petrology,* **97**, 305–312.

ANOVITZ, L.M. & ESSENE, E.J. 1987. Phase equilibria in the system $CaCO_3$–$MgCO_3$–$FeCO_3$. *Journal of Petrology,* **28**, 389–414.

BICKLE, M.J. & POWELL, R. 1977. Calcite-dolomite geothermometry for iron-bearing carbonates: the Glockner Area of the Tauern Window, Austria. *Contributions to Mineralogy and Petrology,* **59**, 281–292.

BODENHAUSEN, J.W.A. & EGELER, C.G. 1971. On the geology of the Upper Kali Gandaki Valley, Nepalese Himalayas I. *Akademie van Wetenschappen, Proceedings. Amsterdam,* **74**, 526–538.

BORDET, P., COLCHEN, M. & LE FORT, P. 1975. *Recherches géologiques dans l'Himalaya du Népal, région du Nyi-Shang.* Edition du Centre National de la Recherche Scientifique, Paris.

BURG, J.P., BRUNEL, M., CHEN, G.M. & LIU, G.H. 1982. Deformation of the leucogranites of the crystalline Main Central Sheet in southern Tibet (China). *Mitteilungen des Geologischen Instituts ETH Zürich und Universität Zürich, Neue Folge,* **239a**, 49–51.

CHANG, C.F. & CHENG, H.L. 1973. Some tectonic features of the Mt. Jolmo Lungma area, Southern Tibet, China. *Scientia Sinica,* **16**, 257–265.

COLCHEN, M., LE FORT, P. & PECHER, A. 1986. *Carte geologique: Annapurna–Manaslu–Ganesh-Himal; Himalaya du Nepal; 1:200.000.* Edition du Centre National de la Recherche Scientifique, Paris.

DACHS, E. 1990. Geothermobarometry in metasediments of the southern Großvenediger area (Tauern Window), Austria. *Journal of Metamorphic Geology,* **8**, 217–230.

—— & METZ, P. 1988. The mechanism of the reaction 1 tremolite + 3 calcite + 2 quartz = 5 diopside + $3CO_2$ + $1H_2O$. *Contributions to Mineralogy and Petrology,* **100**, 542–551.

EGGERT, R.G. & KERRICK, D.M.1981. Metamorphic equilibria in the siliceous dolomite system: 6 kbar experimental data and geologic implications. *Geochimica et Cosmochimica Acta,* **45**, 1039–1049.

FUCHS, G., WIDDER, R. & TULADHAR, R. 1988. Contribution to the Geology of the Manang Area (Annapurna Himal, Nepal). *Jahrbuch der Geologischen Bundesanstalt, Wien,* **131**, 593–607.

GOLDSMITH, J.R. & NEWTON, R.C. 1969. P-T-X relations in the system $CaCO_3$–$MgCO_3$ at high temperatures and pressures. *American Journal of Science,* **267A**, 160–190.

GOTTSCHALK, M. & METZ, P. 1984. Konsistente Gleichgewichtsdaten für Reaktionen in metamorphen kieseligen Dolomiten. *Fortschritte der Mineralogie,* **62/1**, 78–79.

GRAF, D.L. & GOLDSMITH, J.R. 1958. The solid solubility of $MgCO_3$ in $CaCO_3$: A revision. *Geochimica et Cosmochimica Acta,* **13**, 218–219.

HARKER, R.I. & TUTTLE, O.F. 1955. Studies in the system CaO–MgO–CO_2, limits of solid solution along the binary join $CaCO_3$–$MgCO_3$. *American Journal of Science,* **253**,274–282.

HERREN, E. 1987. Zanskar shear zone: Northeast-southwest extension within the Higher Himalayas (Ladakh, India). *Geology,* **15**, 409–413.

HEWITT, D.A. 1973. The metamorphism of micaceous limestones from south-central Connecticut. *American Journal of Science, Cooper Volume,* **273A**, 444–469.

HEWITT, D.A. 1975. Stability of the assemblage phlogopite-calcite-quartz. *American Mineralogist,* **60**, 391–397.

HODGES, K., LE FORT, P. & PECHER, A. 1986. Possible evidence for anatectic buffering of the thermal structure of the Tibetan Slab (Central Nepalese Himalaya). *Geological Society of America Abstracts with Program,* **18**, 638.

HOSCHEK, G. 1973. Die Reaktion Phlogopit + Calcit + Quartz = Tremolit + Kalifeldspat + H_2O + CO_2. *Contributions to Mineralogy and Petrology,* **39**, 231–237.

HUCKENHOLZ, H.G. & SEIBERL, W. 1989. Occurrence of carbonate scapolites and their bearing on geothermometry of rocks of high temperature granulite facies. *28th International Geological Congress, Abstracts,* **2**, 79–80.

HUNT, J.A. & KERRICK, D.M. 1977. The stability of sphene; experimental redetermination and geologic implications. *Geochimica et Cosmochimica Acta,* **41**, 279–288.

KUBLER, B. 1968. Evaluation quantitative du métamorphisme par la cristallinité de l'illite. *Bulletin du Centre de Recherches de Pau – SNPA,* **2**, 385–397.

LE FORT, P. 1975. Himalayas: The collided range. Present knowledge of the continental arc. *American Journal of Science,* **275A**, 1–44.

LE FORT, P. 1981. Manaslu Leucogranite: A collision signature of the Himalaya. A model for its genesis and emplacement. *Journal of Geophysical Research,* **86**, 10545–10568.

LEAKE, B.E. 1978. Nomenclature of amphiboles. *American Mineralogist,* **63**, 1023–1052.

LINDSLEY, D.H. 1983. Pyroxene thermometry. *American Mineralogist,* **68**, 477–493.

METZ, P. 1983. Experimental investigation of the stability conditions of petrologically significant calc-silicate. *In*: MARTIN, H. & EDER, F.W. (eds) *Intracontinental Fold Belts.* Springer, Berlin, 785–793.

—— & TROMMSDORFF, V. 1968. On phase equilibria in

metamorphosed siliceous dolomites. *Contributions to Mineralogy and Petrology,* **18**, 305–309.

PECHER, A. 1989. The metamorphism in the Central Himalayas. *Journal of Metamorphic Geology,* **7**, 31–41.

—— & BOUCHEZ, J.L. 1987. High temperature decoupling between the Higher Himalaya Crystalline and its sedimentary cover. *Terra cognita,* **7**, 110.

PUHAN, D. 1978. Experimental study of the reaction: dolomite + K-feldspar + H_2O = phlogopite + calcite + CO_2 at the total gas pressures of 4000 and 6000 bars. *Neues Jahrbuch für Mineralogie, Monatshefte,* 110–127.

RICE, J.M. 1977. Progressive metamorphism of impure limestone in the Marysville aureole, Montana. *American Journal of Science,* **277**, 1–24.

RICHARD, L.R. & CLARKE, D.B. 1990. AMPHIBOL: A program for calculating structural formulae and for classifying and plotting chemical analysis of amphiboles. *American Mineralogist,* **75**, 421–423.

ROYDEN, L.H. & BURCHFIEL, B.C. 1987. Thin skinned N-S-extension within the convergent Himalayan region: gravitational collapse of a miocene topographic front. *In*: COWARD, M.P., DEWEY, J.F. & HANCOCK, P.L. (eds) *Continental Extensional Tectonics* Geological Society, London, Special Publications, **28**, 611–619.

SANFORD, R.F. 1980. Textures and mechanisms of metamorphic reactions in the Cockeysville Marble near Texas, Maryland. *American Mineralogist,* **65**, 654–669.

SCHNEIDER, C. 1991. *Regionalmetamorphose und Deformation im Hoch-Himalaya Zentral-Nepals (Marsyandi-Profil) unter spezieller Berücksichtigung der Metakarbonate.* PhD thesis, University of Munich, Germany.

SLAUGHTER, J., KERRICK, D.M. & WALL, V.J. 1975. Experimental and thermodynamic study of equilibria in the system $CaO–MgO–SiO_2–H_2O–CO_2$. *American Journal of Science,* **275**, 143–162.

STORRE, B. & NITSCH, K. 1972. Die Reaktion 2Zoisit + $1CO_2$ = 3Anorthit + 1Calcit + H_2O. *Contributions to Mineralogy and Petrology,* **35**, 1–10.

TILLEY, C.E. 1924. The facies classification of metamorphic rocks. *Geological Magazine,* **61**, 167–171.

THOMPSON, A.B. 1975. Mineral reactions in a calc-mica schist from Gassetts, Vermont, USA. *Contributions to Mineralogy and Petrology,* **53**, 105–127.

WADA, H. & SUZUKI, K. 1983. Carbon isotopic thermometry calibrated by dolomite-calcite solvus temperatures. *Geochimica et Cosmochimica Acta,* **47**, 697–706.

Structural evolution of the High Himalayan Gneiss sequence, Langtang Valley, Nepal

STEVEN M. REDDY[1,3], MICHAEL P. SEARLE[2] & JOHN A. MASSEY[1]

[1] *Department of Earth Sciences, The Open University, Milton Keynes, MK7 6AA, UK*

[2] *Department of Earth Sciences, The University of Oxford, Oxford, OX1 3PR, UK*

[3] *Present address: Department of Earth Sciences, Liverpool University, Brownlow Street, Liverpool L69 3BX, UK*

Abstract: Within the High Himalayan Crystalline Sequence (HHCS) of the Langtang Valley of central Nepal, four different lithotectonic units have been identified in the field. The structurally lowest unit, immediately in the hanging wall of the Main Central Thrust (MCT) records an inverted metamorphic sequence with kyanite-grade rocks overlain by rocks containing sillimanite. Pressure–temperature–deformation histories of the different units since Himalayan collision 50 Ma ago, in combination with geochemical data, indicates that the 'apparent' metamorphic inversion is a result of the tectonic juxtaposition along high-temperature shear zones that accommodated top-to-southwest thrusting. It is also argued that the contacts between different sillimanite-grade units are also tectonic in origin and that the HHCS in the Langtang Valley does not represent a structurally undisturbed section through the crust. Internally the different units contain minor extensional and thrust faults which at the same structural level developed at similar metamorphic conditions. Deformation along the high temperature shear zones of the HHCS are coeval with the earliest of these minor fault zones at the lowest structural levels. With time, this deformation migrated upwards and minor faulting within the upper levels of the HHCS took place at chlorite-grade. These minor fault zones are interpreted to be the result of dynamic spreading of the developing orogenic wedge associated with changes in the geometry of this wedge due to underplating and accretion of material to the base of the metamorphic pile.

The continued northward movement of the Indian plate at a rate of approximately 5 cm a year suggests that about 2500 km of post-collisional shortening has taken place since the collision of India with the Asian plate approximately 50 Ma ago (Patriat & Achache 1984). Shortening has been partially accommodated by crustal-scale imbrication of the Indian crust along a series of sub-parallel major thrust faults which resulted in a series of parallel tectonometamorphic units along the length of the Himalayan chain (e.g. Bordet 1961; Gansser 1964). One of these units, the High Himalayan Crystalline Sequence (HHCS) or Tibetan Slab, is a high-grade polymetamorphic unit comprising rocks of possible Precambrian, Palaeozoic and early Mesozoic age, metamorphosed during the late Eocene to early Miocene and intruded by granites of Miocene age.

Despite the significant amount of research undertaken on the HHCS, the thermal-tectonic development of this unit is still relatively poorly understood and as a consequence controversial. One of the major problems pertaining to the HHCS is recognition of the heat source capable of producing the widespread late-stage sillimanite-grade metamorphism and leucogranite formation. Furthermore, despite the need to understand the spatial and temporal interaction among deformation, metamorphism and leucogranite production within the HHCS, very few detailed structural studies have been made to complement the wealth of metamorphic data. In this paper, field and microstructural data from the HHCS exposed within the Langtang Valley of north-central Nepal are presented and used to constrain the tectonic development of the HHCS.

The High Himalayan Crystalline Sequence

The HHCS varies in thickness and lithostratigraphy along the length of the Himalayan chain. Pre-Himalayan deformation and metamorphism within the HHCS has been identified within the Pakistan (Treloar *et al.* 1989) and Zanskar areas (Pognante & Lombardo 1989) but have not been positively identified within more easterly parts of the orogen. Metamorphic studies of the HHCS have identified an 'inverted metamorphic' sequence (Gansser 1964), with kyanite-grade lithologies in the immediate hanging wall of the Main Central Thrust (MCT) passing up to sillimanite-grade units at higher

From TRELOAR, P.J. & SEARLE, M.P. (eds), *Himalayan Tectonics*
Geological Society Special Publication, No. 74, pp. 375–389.

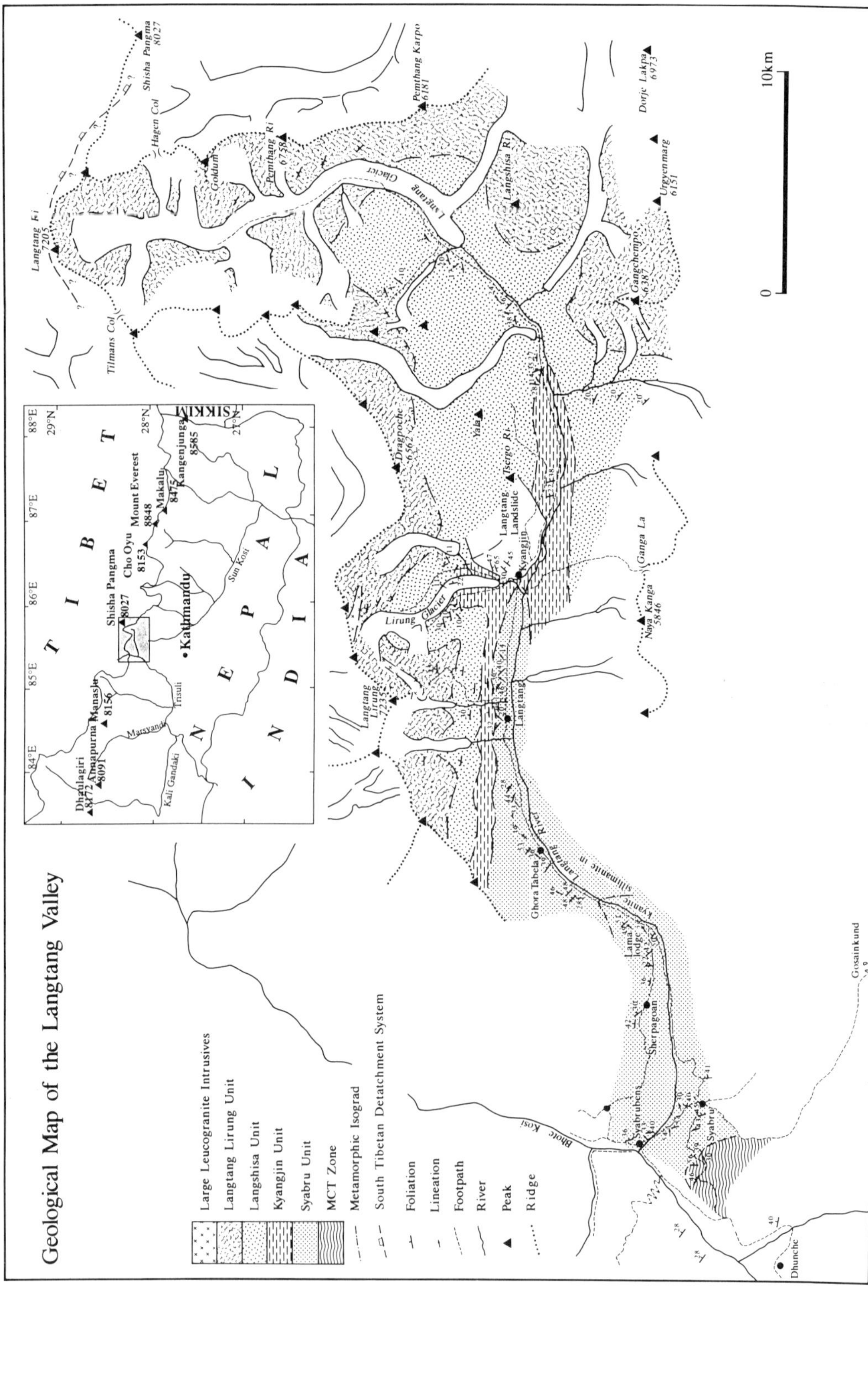

Fig. 1. Geological Map of the Langtang Valley, Nepal based on mapping by the authors and the observations of Inger (1991). Insert shows the geographical location of the valley.

structural levels. A polyphase metamorphic history where sillimanite-grade metamorphism at the top of the Tibetan slab overprints early-Himalayan, kyanite-grade, Barrovian metamorphism has been proposed for central Nepal (Pêcher & Le Fort 1986; Pêcher 1989; Inger & Harris 1992), Garhwal (Hodges & Silverberg 1988), Zanskar–Kistwar (Stäubli 1989; Searle & Rex 1989; Searle *et al.* 1992) and Bhutan (Swapp & Hollister 1991).

To the south, the base of the HHCS is marked by the MCT which emplaces high-grade units of the HHCS onto lower grade units of the Lesser Himalaya Sequence (LHS). To the north, the HHCS is in tectonic contact with a series of low-grade sediments, the Tethyan Sedimentary Series (TSS) which represent Palaeozoic sediments deposited onto the northern continental shelf of the Indian plate. Recent data from southern Tibet (Burg *et al.* 1984; Burchfiel *et al.* 1992), Zanskar (Searle 1986; Herren 1987) and the Annapurna–Manaslu area (Caby *et al.* 1983; Pêcher 1991) show the tectonic contact to be a low-angle, normal-fault zone, named the South Tibetan Detatchment System (STDS) (Burchfiel *et al.* 1992), which downthrows the TTS by up to 30 km to the north.

Geology of the Langtang Valley

The Langtang Valley exposes a 20 km section of the HHCS from the MCT zone in the southwest (Macfarlane 1990), through kyanite and sillimanite-grade metasediments and orthogneiss (Inger & Harris 1992). Field mapping has enabled subdivision of the HHCS in the Langtang area into four major lithological units based upon lithological variations and the percentage of leucogranite melt within the unit (Fig. 1). The rocks throughout the four units of the Langtang section are high-grade metamorphic rocks and nowhere in the Langtang Valley, either in situ or as scree, were low-grade sediments of the overlying TSS observed. These four units are described from the base through to the top of the section.

The Syabru Unit

The lowermost unit of the HHCS in the Langtang area is an approximately 12 km thick sequence containing metapelites and more quartz-rich horizons. The unit contains rocks of kyanite-grade at its base which pass up to sillimanite-grade at higher structural levels (Fig. 1). A detailed petrographic description of these rocks and the chemistry of the mineral phases within the Syabru Unit is given by Inger (1991) and Inger & Harris (1992) who suggested metamorphic conditions of around 740° C and 10 kbar for the kyanite-grade metamorphism (M1) 770° C and 5 kbar for the subsequent sillimanite-grade overprint (M2). Metamorphic data from individual samples suggest a discrete rather than gradual change in metamorphic conditions across the sillimanite–kyanite boundary particularly with respect to pressure data.

Pelitic areas of the lower Syabru Unit are locally migmatitic, contain kyanite and garnet and have a well developed foliation defined by quartz-rich and mica-rich domains that dips at about 40° to the northeast (Fig. 2). Although a well-developed schistosity is present, mineral lineations within these mica-sheets are rare and, unlike exposures of the lower HHCS in other areas of the Nepal Himalaya (eg. Brunel 1986), kyanite laths do not have a preferred orientation within the foliation plane. Shear sense evidence associated with foliation development is poorly preserved.

Both muscovite and biotite are coarse grained (individual grains up to 5 mm) and display little evidence of having undergone significant deformation other than heterogeneously developed undulose extinction and minor localised kink band formation. Kyanite porphyroblasts are usually large (up to 2 mm in length) subhedral to euhedral and often poikiloblastic. Kyanite grains cut across the regionally developed foliation and are therefore interpreted as post-kinematic with respect to the deformation that produced the regionally-developed foliation. In one sample two populations of kyanite are represented by the large kyanite grains described above and by smaller (<0.5 mm), euhedral porphyroblasts which are oblique to the regional foliation and post-date the larger grains. Garnets in the kyanite schists commonly have inclusion-poor rims but inclusion-rich cores. The main inclusion phase is quartz although inclusions of green biotite contrast with the brown biotites now present in the matrix. Inclusion trails are commonly circular although sigmoidal and spiral inclusion trails within the garnet cores suggest syn-kinematic garnet growth. Rims are post-deformation. Quartz is generally coarse-grained and shows features such as undulose extinction and deformation lamellae indicative of low strain deformation but no grain-preferred orientations or fine-grained mylonitic textures are preserved.

Moving up the section, the stable aluminosilicate phase changes to fibrous sillimanite and eventually prismatic sillimanite becomes stable in an upper zone where a muscovite-out isograd has been identified (Inger & Harris 1992).

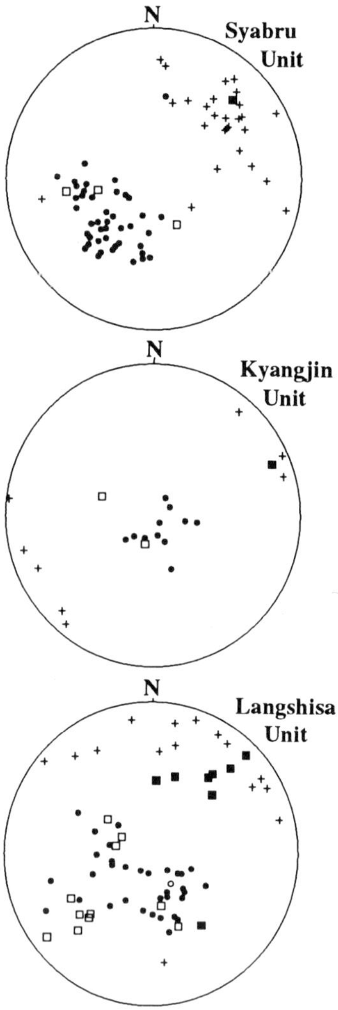

Fig. 2. Equal-area lower hemisphere stereographic projections of structural data from the Syabru Unit, the Kyangjin Unit and the Langshisa Unit. Filled circles, poles to foliation; crosses, mineral lineations on foliation; empty squares, poles to fault zones; filled squares, lineations on fault surfaces.

Fig. 3. (**a**) Sillimanite mineral lineation within the upper levels of the Syabru Unit. The lineation plunges to the northeast and developed during top-to-southwest shearing. Lens cap is 50 mm.
(**b**) Optical micrograph from close to the base of the sillimanite zone (Syabru Unit). Late-muscovite (M) grows around cores of sillimanite (S) and shows an asymmetry suggesting growth during top-to-southwest shearing (top-to-left in the figure). Base of photomicrograph is 2.5 mm.

Associated with the observed increase of metamorphic grade, there is an increase in the degree of migmatisation and leucogranite dykes are developed at high angles to the metamorphic foliation.

Sillimanite commonly forms a mineral lineation (Fig. 3a) that plunges towards the northeast (Fig. 2). At the lowest structural levels of the sillimanite zone, sillimanite appears synkinematic and shear bands associated with this deformation suggest top-to-southwest shearing. No preferred grain-elongation or grain boundary orientation are observed within quartz and feldspar. However, the absence of annealed textures and the presence of large subgrains and sutured boundaries within quartz and feldspar indicate continued low strain rate deformation at temperatures below those required to anneal such microstructures. At higher structural levels, the foliation developed within those rocks, which record top-to-southwest shearing at the hand-specimen scale, is similar to that observed at lower levels and is characterized by recrystallisation.

Although sillimanite is found in the matrix, rare kyanite and staurolite are found as inclusions in garnet, suggesting a polymetamorphic history for the HHCS in the Langtang area (Inger & Harris 1992). Inclusions of kyanite and ilmenite within garnet have a grain elongation that defines a linear inclusion trail showing that garnet must have grown post-kinematically with respect to an early foliation. This contrasts inclusion trails within the kyanite zone garnets. This early fabric is commonly oblique to the external foliation preserved in the matrix but the apparent rotation between the internal and external fabrics is not constant and does not appear to be related to structural level.

The relationship between muscovite and sillimanite is more complicated than has been previously reported in the HHCS and two populations of white mica are present. In some samples, sillimanite is clearly overgrowing muscovite (early muscovite) while in others, muscovite is found surrounding 'cores' of sillimanite (Fig. 3b). In one sample minor folds defined by sillimanite are overgrown by undeformed muscovite. This muscovite is therefore late-stage and overprints the sillimanite-grade assemblage. The spatial distribution of the two muscovite populations is related to structural level (Fig. 4). At the lowest structural level of the sillimanite zone (Lama Lodge), early muscovite is present. Although these rocks contain fibrolite, they have garnets that have spiral inclusion trails which suggest affinites to the kyanite-grade mica schists. Moving up section towards Ghora Tabela, the late-muscovite is well developed (≈10% mode) but decreases with increasing structural height. Garnets within these samples have straight inclusion trails. Where muscovite is clearly seen growing around cores of sillimanite there is an asymmetry consistent with growth during top-to-southwest shearing (Fig. 3b). Around Langtang village very little muscovite is present within pelitic units (<1% mode). However, to the east of Langtang there is a sudden increase in the amount of late muscovite (≈10% mode) which again decreases moving further up the section.

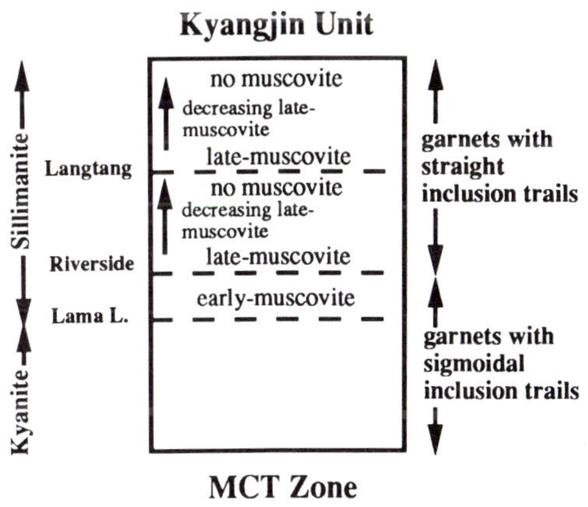

Fig. 4. Schematic diagram illustrating the variation of early- and late-muscovite with structural level within the kyanite- and sillimanite-grade rocks of the Syabru Unit. Also shown is the variation in garnet inclusion trail geometries. Between Lama Lodge and Riverside (see Fig. 1) the growth of sillimanite over early muscovite and the presence of garnets similar to those in the kyanite-grade rocks suggests affinities between these rocks and the kyanite-grade schists.

Locally developed areas of the foliation contain shear criteria that indicate localised northeast-directed extension. These zones are also observed to contain sillimanite which suggests that extension of the metamorphic pile was initiated at relatively high temperatures. Microstructures developed during top-to-the northeast shearing and extension of the metamorphic pile are again largely recrystallized, but shear sense indicators are much more common. Muscovite within these zones appears to be in equilibrium with the observed assemblage.

Kyangjin Unit

The Kyangjin Unit is a *c.* 1 km thick augen orthogneiss which has a K-feldspar–plagioclase–quartz–biotite ± muscovite ± garnet ± sillimanite mineralogy that lies structurally above the Syabru Unit. Two sub-divisions of the augen gneiss can be made based on primary textural features. The structurally lower sub-unit contains feldspar augen of about 2–3 cm in diameter while the upper sub-unit has feldspar augen up to 20 cm in diameter. Unfortunately, the outcrop pattern of the two different gneisses could not be traced in the field due to the nature of the exposure. Both of the subgroups are locally migmatitic. These lithologies are similar to orthogneisses in the Annapurna and Manaslu areas, that have been dated at 517 Ma (Le Fort *et al.* 1983), which suggests by analogy that the Kyangjin unit could represent a Cambro-Ordovician granite which was subsequently deformed and metamorphosed during Himalayan continental collision.

This unit contains a well-developed S–L fabric with foliation and lineations dipping to the northeast and southwest (Fig. 2). The main foliation within the orthogneiss is a spaced anastomosing fabric characterised by the alignment of biotite. Observations made in the field suggest that regional foliation development was associated with top-to-southwest shearing (Fig. 5a). Locally observed within the gneisses are areas of weak foliation development which indicate the heterogeneous nature of deformation within the augen gneiss.

The regional foliation is characterized by

(a)

(b)

(c)

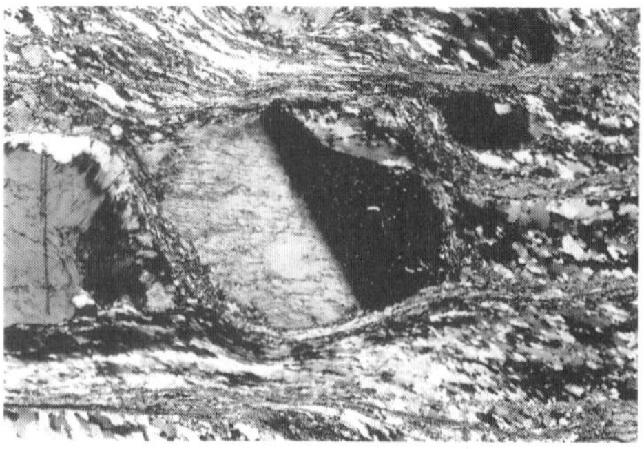

(d)

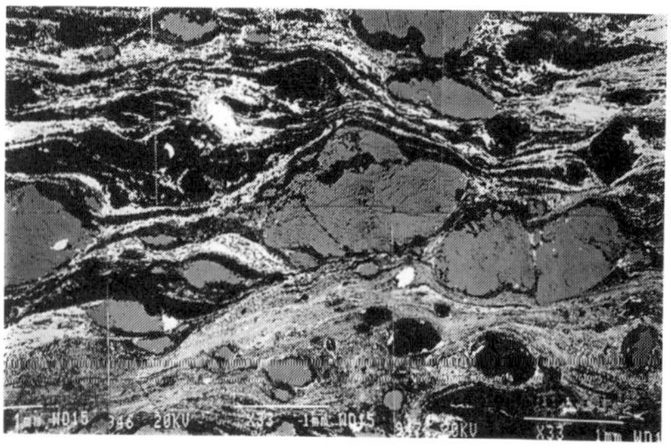

(e)

Fig. 5. (**a**) Foliation developed within augengneiss of the Kyangjin Unit. S-C fabrics and shear bands show deformation was associated with top-to-southwest shearing (i.e. top-to-right in the figure).
(**b**) Discrete, northeast dipping normal fault offsetting a sub-vertical leucogranite dyke seen in the hanging wall. Feldspar augen within the fault zone indicate the shear sense. Northeast is towards the right of the figure.
(**c**) Optical micrograph from a sample from the base of the Kyangjin unit orthogneiss illustrating localised fine-grained shear zones associated with top-to-SW shearing at temperatures significantly below the metamorphic peak. These overprint an earlier foliation with the same kinematics. Base of figure is 2.5 mm.
(**d**) Fabric associated with late top-to-NE (i.e., left) normal fault illustrating the preservation of deformation microstructures and the absence of significant post-deformational recrystallisation. K-feldspar porphyroclasts show recrystallized tails and mantles of plagioclase. Base of figure is 2.5 mm.
(**e**) Backscatter SEM Z-contrast montage showing an early top-to-NE fabric similar to that in figure 5d being overprinted by a localized fine-grained fault zone developed in the same kinematic framework (lower third of figure). Grey and black porphyroclasts are K-feldspar and plagioclase respectively. Biotite-rich and quartz-rich domains that form the matrix are light grey and black respectively. Quartz and plagioclase myrmekite mantles at the edges of K-feldspar porphyroclasts are also black. Scale bar is 1 mm.

microstructures similar to those described for the Syabru Unit. Quartz shows subgrain development and sutured grain boundaries but no grain elongation or grain preferred orientation. Biotite laths define the foliation, reach 1.5 mm in length and show little internal deformation. Sillimanite and white mica are rarely seen within the typical orthogneiss and kinematic indicators are poorly preserved at the thin section scale. However, rarely preserved biotite fish structures confirm the observations made in the field which suggest that foliation development was associated with top-to-southwest non-coaxial deformation.

Within the foliated orthogneiss, higher strain zones up to several metres in width and parallel to the main gneissic foliation are locally developed. Shear criteria observed in the field show these also developed during top-to-southwest overthrusting. Within these zones, feldspars are found as clasts within a finer grained matrix composed of millimetre-scale mica-rich and quartz-rich domains. K-feldspar is often perthitic with albite lamellae being most common in deformed feldspars and filling en-echelon fractures in the host grain. Undulose extinction and localised twinning within the feldspars at grain boundaries and adjacent to albite lamellae is also common. A core-mantle morphology with K-feldspar cores and plagioclase and quartz myrmekitic mantles is commonly observed and embayed feldspar margins often appear to be the result of replacement of feldspar by sillimanite. Occasionally feldspars appear broken and fractures are filled by quartz.

The matrix is composed of quartz-rich and biotite-rich domains. Quartz is commonly recrystallized and forms a polygonal network of new grains and subgrains. Occasionally a quartz grain-elongation fabric is preserved. Mica-rich domains can be subdivided into two types; (i) coarse grained biotite rich bands comprising sillimanite and undeformed biotites (>2 mm in length) and (ii) finer-grained, sillimanite-free bands with a closely spaced (≈0.1 mm), quartz-rich and mica-rich domainal foliation containing larger biotite porphyroclasts (<2 mm). The biotite clasts form fish structures with fine-grained recrystallized margins and tails. The sense of shear inferred from quartz elongation and mica fish suggests deformation during top-to-southwest shearing and confirms shear criteria observed in outcrop. Towards the base of the Kyangjin Unit, fine-grained fault zones are also found. These are microstructurally similar to the fine-grained mica domains described above and overprint an earlier coarse-grained foliation developed during the earlier stages of top-to-southwest shearing (Fig. 5c). These fine-grained zones represent continued but localised post-metamorphic southwest-directed shearing within the augen gneiss.

Discrete shear zones, dipping to the northeast and accommodating top-to-northeast extension, cut both the early augen gneiss foliation and post-foliation leucogranite dykes (Fig. 5b). These fault zones are discrete features characterized by a fine grain-size and strong foliation development and record a range of microstructures. Fault zones generally have a fine-grained matrix with feldspar clasts. Microstructures preserved within the matrix are similar to those described for the fine-grained mica zones associated with late stage southwest-directed thrusting and comprise bands of quartz and mica. However, no sillimanite has been observed within these normal-sense fault zones and large porpyroblastic micas are rare. Chlorite and epidote are also absent. Quartz elongation fabrics are better developed than in the thrusts and quartz ribbon textures are locally preserved (Fig. 5d). Feldspar microstructures are as described for the thrust faults with plagioclase mantles around K-feldspar suggesting that feldspar deformation was not by isochemical recrystallization.

Localized within the top-to-northeast fault zones are areas showing extreme grain size reduction (Fig. 5e). These contain biotite, feldspar, quartz, white mica and fine-grained opaque phases. The only porphyroclastic phase is feldspar but individual grains are generally smaller than feldspars in the major fault zone. Again, K-feldspar cores are surrounded by myrmekitic plagioclase mantles (Fig. 5e). Within these zones are thin dark bands of ultra fine-grained material in which individual grains cannot be resolved by either optical or SEM techniques. These may represent zones of extreme grain-size reduction or may be a result of pressure solution. In one normal-sense shear zone sample, the centre of the fault zone is characterized by one of these fine-grained zones. Kinematic indicators at the microscopic scale suggest that this fine-grained fault zone, with similar microstructures to those described above, developed during top-to-south shearing. The only interpretation of such an observation is that the earlier normal fault has been subsequently reactivated as a thrust.

Langshisa Unit

Above the augen gneisses of the Kyangjin Unit is a 1.5–2 km thick, psammitic, metasedimentary unit with locally developed sub-centimetre pelitic bands containing sillimanite and occasional metre-scale carbonate bands. The psammitic unit contains quartz + biotite + plagioclase + K-feldspar ± sillimanite mineralogy with rare muscovite and garnet. The rock is fine-grained (≈0.2 mm). Within this unit, several different types of leucogranite and pegmatitic intrusives are present and these tend to make up less than 10% of the unit volume. In the basal 50 m of the unit granites are more common and reach 50% of the rock volume. A sequence of dark schists and a stratigraphy similar to that in Langtang is exposed within the HHCS in southern Tibet (Burg *et al.* 1984; Burchfiel *et al.* 1992).

The biotite–quartz-rich gneisses of the Langshisa unit have a well developed foliation which dips to the northeast (Fig. 2). The presence of prismatic sillimanite within pelitic horizons of the Langshisa Unit samples immediately above granitic orthogneisses of the Kyangjin Unit suggests similar metamorphic conditions to those seen in the lower units and indicates that there is not a significant metamorphic break between this and the underlying augen gneisses. Within the scree of the upper Langtang valley, coarse-grained marbles are occasionally seen. The high variance of the marble assemblages and the small amounts of Mg in the mineral phases means that these units are not conducive to reliable thermobarometric determinations. However, available data from garnet-biotite thermometry in the gneisses suggest metamorphic temperatures in excess of 700° C (Massey, unpublished data).

Fabrics preserved within the Langshisa Unit are characterized by recrystallization and commonly have a polygonal texture of closely spaced quartz, feldspar and biotite grains. The gneisses are generally fine grained with grain sizes typically of the order of 0.2 mm. The closely spaced distribution of the different mineral phases appears to have been the major control on the recrystallized grain size. Biotite grains within a single thin section simultaneously go into extinction under crossed polars suggesting dynamic recrystallization of biotite during foliation development. Limited field and microscopic shear criteria are observed. Recrystallised shear bands in thin section show both top-to-southwest and top-to-northeast shear senses. White mica is rarely observed as a primary phase but is found within boudin necks of sillimanite and replacing fibrolite ellipsoids suggesting that it is a retrogressive phase.

Fault zones within the Langshisa Unit are up to a metre wide and cut the earlier foliation and foliation-parallel leucogranites. These fault zones dip to the northeast and show both ductile deformation and cataclasis that developed during top-to-NE extension. North-dipping normal listric faults are exposed within cliff sections to the north of the Lirung Glacier and these cut the

(a)

(b)

(c)

Fig. 6. (**a**) Early two-mica granites at the base of the Langshisa Unit containing the regional foliation and cut by two listric normal faults. The fault on the right is discrete (F) while the one on the left is intruded by tourmaline leucogranite (L). Cliff section in the figure is approximately 50 m high.
(**b**) Cataclastic deformation within the Langshisa Unit shown by a complex network of brecciated psammatic gneiss within a chlorite rich gouge zone. Biotite within the host immediately adjacent to the fault zones has been altered to chlorite. The light coloured band towards the top of the figure is chlorite-free, extremely fine-grained (<1 μm) and may have formed by frictional melting.
(**c**) Foliation-parallel, early two-mica granite (2-M) from the Langtang Lirung Unit cut by a tourmaline leucogranite dyke (T) at high angles to the foliation. The heterogeneous nature of the tourmaline leucogranite is shown by composite banding parallel to its margins. Ice axe is 60 cm.

intercalated granites and metasediments (Fig. 6a). One of these is intruded by a leucogranite. S–C fabrics suggest an extensional framework for mylonitic fabrics containing the northeast-directed stretching lineation.

Mylonites are developed within quartz rich units which also contain feldspar, garnet and tourmaline. This assemblage suggests that the mylonite protolith could have been the late stage leucogranites observed within the area. Bands of up to 2 cm of relatively pure quartz mylonite are also observed. It is possible that these were derived from earlier quartz veins. The general absence of quartz veining away from the fault zone suggests that such veins must have formed at an earlier stage of fault zone development. Subparallel to the mylonites are locally developed, very fine-grained cataclastic zones and smaller areas of pseudotachylyte.

Within the mylonitic areas of the fault zone, quartz commonly shows grain-preferred orientation and grain elongation fabrics. Feldspars show subgrain formation with the development of core–mantle structures and are internally deformed as illustrated by the bending of multiple twin planes. Micas commonly form fish structures in which the mica margins and fish tails are composed of recrystallized fine-grained mica. Fish tails illustrate locally developed high shear strains ($\gamma \approx 200$). Chlorite-grade fault zones are also developed within the Langshisa Unit (Fig. 6b). These faults are found in a variety of orientations and record a microstructurally complex deformation history involving cataclastic deformation of the original gneissic host, the development of fine-grained chlorite-rich gouge and thin bands of pseudotachylyte. Biotite in the gneiss immediately adjacent to these fault zones is altered to chlorite while feldspar undergoes complete to partial alteration to sericite. No garnet has been found in these fault zones. Within the fault zone, large variations in grain size are observed and these variations appear domainal and associated with heterogeneous deformation within the fault. The coarser grained areas are clast supported and comprise coarse sub-angular clasts of the gneiss host (up to 10 mm in size) within an irregular network of fine-grained cataclastic material. The cataclastic zones comprise a fine-grained matrix (1–10 μm) which supports clasts of quartz and feldspar with a complete grain size range from 10 μm up to 0.2 mm in size. These clasts vary from sub-angular to sub-rounded in morphology and there appears to be no relationship between morphology and clast size.

Thin bands of darker material may represent zones of extreme grain size reduction or sites of mineral dissolution although SEM imagery was unable to resolve these possibilities. The margins of these zones are often chlorite-rich and have larger clasts. The chlorite-rich nature of these zones and the coarser clast size suggests that these formed by cataclasis of the gneissic host while the finer grained zones may have developed as a result of localised cataclasis from these already deformed zones. Some of the zones in which chlorite is absent may represent devitrified pseudotachylytes. Carbonate material up to 1 mm is present in irregular patches with irregular margins within the fault zone. The coarse grain size of this carbonate and the fact that it is relatively undeformed within the fault zone suggest that this material formed at a late-stage in the deformation history of the zone.

Langtang Lirung Unit

The Langtang Lirung Unit lies immediately above the Langshisa Unit and differs from it only in the volume of granitic material observed within the paragneiss host. Whereas the lower unit contains <10% of leucogranite, the structurally higher Langtang Lirung Unit contains in the order of 50% by volume leucogranite. The Langtang Lirung Unit is in excess of 5 km in thickness and is best exposed in the faces of the major Langtang peaks (Fig. 1). This unit and the contact between this and the underlying Langshisa Unit is extremely inaccessible and although the contact is a discrete feature that can be easily traced in cliff sections, it is not possible to be specific about the nature of the contact.

Kilometre-scale, south-verging folds with sub-horizontal axial planes affect the foliation and foliation-parallel leucogranite units within the Langtang Lirung Unit and the highest levels of the Langshisa Unit. Hinges associated with this folding appear sub-horizontal and trend approximately E–W. Although this folding is particularly well developed in the Langtang Lirung Unit and affects the Langshisa Unit, it does not appear to affect the underlying Kyangjin unit or the pelitic rocks of the Syabru Unit suggesting decoupling of the south-directed shearing between the lower and upper HHCS.

Granites of the Langtang Valley

Granites within the Langtang valley can be subdivided into two main groups. The earliest group are a series of two-mica granites that are commonly foliation parallel and contain the regional foliation. Biotite is the major mica phase with white mica, where present, being late stage. In outcrop they are similar to the fine-augen gneiss lithology of the Kyangjin Unit. The early granites are cut by later tourmaline-bearing leucogranites at high angles to the regional foliation and are often composite in nature (Fig. 6c). Several mineralogically different granites can be identified within the later suite. No consistent overprinting relationships are seen between these mineralogically distinct groups which illustrates an absence of a consistent temporal relationship and attests to the heterogeneous nature of Himalayan leucogranites. A detailed description of the granites and granite geochemistry is given by Inger (1991) who has shown that (i) both granite types are chemically heterogeneous; (ii) while both types are crustally derived there is still no consensus on the exact source, although, Sr isotopes preclude the sillimanite-grade migmatites of the Syabru Unit from providing a potential source; and (iii) the leucogranites appear to be the result of vapour-absent muscovite breakdown. Preliminary O isotope data from the gneisses of the Langsisa and Langtang Lirung Units (Massey, unpublished data) are compatible with the isotopic signatures required for the source of the Langtang leucogranites.

The leucogranites within the Langshisa and Langtang Lirung Units show a variety of structures related to emplacement and subsequent deformation. As a generalization, the early two-mica granites are foliation parallel and commonly have a weakly developed foliation. The host rock immediately adjacent to these granites is often biotite-rich and may represent restitic material which would imply in situ melting. Early granites are folded, often boudinaged and are cut by late granites. Later tourmaline granites are oblique to, and cross cut, the foliation, are not themselves foliated and are generally much thinner than the earlier granites (Fig. 6c). The late leucogranites cut across fold axial planes and therefore post-date this kilometre-scale folding.

Both early and late granites locally contain thin, anastomosing vein arrays that contain sillimanite. These veins appear to represent one of the youngest high-temperature structural features developed in the granite suggesting that most of the observed deformation associated with the granites and the large-scale folding in the Langshisa and Langtang Lirung Units took place at sillimanite-grades approximately synchronous with leucogranite injection.

Discussion

The regional foliation within the kyanite-grade gneisses of the Syabru Unit is characterized by coarse grain sizes and randomly oriented kyanite which indicates that foliation development predated the kyanite-grade metamorphism. Since the MCT places the HHCS onto lower grade rocks of the Lesser Himalaya then foliation development in the kyanite-grade rocks also predates MCT deformation. In other areas where the base of the HHCS is exposed, kyanite often forms a NE–SW orientated mineral lineation. The best explanation of these contrasting observations is that the lineation is developed by rotation of kyanite towards the shear direction within syn- to post-metamorphic shear zones which may be related to MCT deformation. Although the absence of shear sense indicators within the kyanite-grade unit does not enable the kinematics associated with foliation development to be constrained, the foliation within the lower Syabru unit must therefore represent a pre- or early-Himalayan, pre-MCT fabric which was then carried passively in the hanging wall of the MCT during southwest-directed overthrusting.

Field and microscopic observations within the sillimanite-grade rocks of the Syabru Unit are characterised by structures which indicate that the regional foliation developed as a result of top-to-southwest shearing. The sillimanite-grade rocks of the Syabru Unit therefore represent a wide zone of high temperature non-coaxial deformation. Similarly, the sillimanite-grade gneisses of the Kyangjin Unit contain a regionally developed foliation that also developed during top-to-southwest shearing. Kilometre-scale folds within the Langtang Lirung Unit are south-verging and are interpreted as having developed within the same top-to-south kinematic framework. There is therefore a difference in the deformation histories of the kyanite and sillimanite-grade rocks.

At the base of the sillimanite zone, garnets have an affinity to those in kyanite-grade rocks and also contain early muscovite which is overgrown by sillimanite. This passes up into a zone containing late muscovite and garnets which have affinities to the overlying sillimanite gneisses. Late muscovite post-dates sillimanite growth, has an asymmetry consistent with growth during top-to-SW shearing, and is found in two zones within which its mode decreases up the section. Late muscovite is also found within discrete fault zones at high structural levels. Geochemical evidence shows that the kyanite and sillimanite-grade rocks of the Syabru unit have significantly different initial strontium ratios (<0.76 and >0.80 respectively at 20 Ma ago, Inger 1991) and different stable isotope compositions (Fig. 7). Therefore, the change in

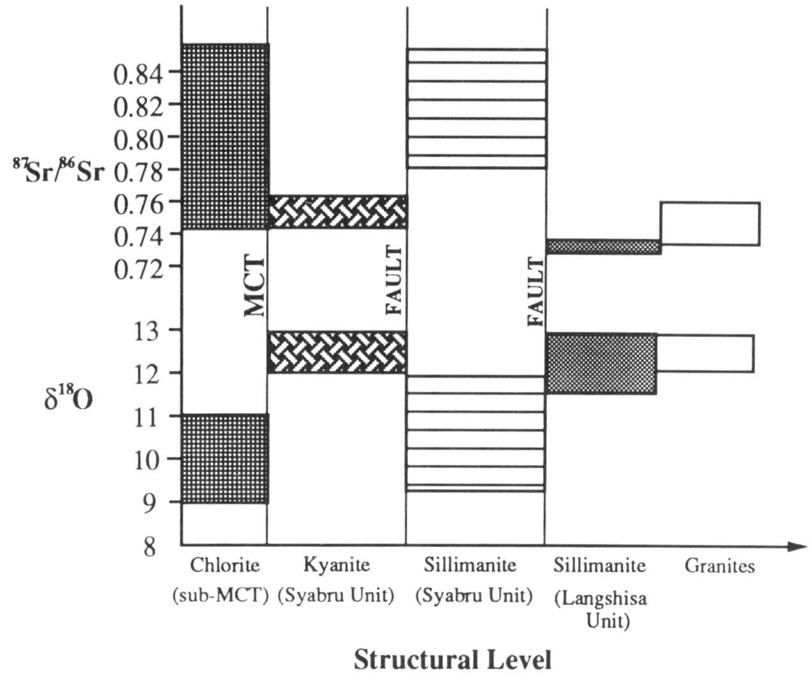

Fig. 7. Diagram showing the variation of $^{87}Sr/^{86}Sr$ and $\delta^{18}O$ within different units of the Langtang Valley. Boxes represent the range of data from the different units. The significant isotopic difference between the sillimanite-grade units of the Syabru Unit and other units of the HHCS contrasts with the very similar isotopic compositions of this unit and samples from the footwall of the MCT.

metamorphic grade corresponds to both changes in deformation style and original geochemistry.

The best interpretation of these data is that the apparent 'inverted metamorphism' is a result of tectonic juxtaposition of sillimanite-grade rocks onto kyanite-grade rocks. Kinematic indicators suggest that this was by south-directed thrusting and not top-to-north extension. The fault zone must have post-dated metamorphism but was still at high grades because both sillimanite and (late-) muscovite were stable. By analogy with the first zone containing late muscovite, the second zone of late muscovite-bearing sillimanite gneisses above Langtang village may represent another high temperature thrust zone. The growth of sillimanite over early muscovite and garnets of kyanite affinity at the very base of the sillimanite zone suggest that the top of the kyanite zone may have undergone deformation and heating during thrusting. Similar models invoking tectonic juxtaposition of metamorphic units have been used to explain the apparent metamorphic inversion of the HHCS in Bhutan (Swapp & Hollister 1991) as well as the apparent inverted metamorphism within the Indian basement south of the MMT in Pakistan (Treloar *et al.* 1989). Although it is not possible to constrain the amounts of displacement associated with this deformation in the Langtang area, it is obvious that the HHCS does not represent a continuous section affected by a static thermal metamorphic overprint but that the sillimanite-grade rocks represent a high temperature zone of non-coaxial deformation. As a consequence, the assumption of a single metamorphic 'event' within a structurally continuous sequence which has placed tight spatial and temporal constraints on thermal-tectonic models of the HHCS is not valid.

For a high-grade shear zone, such as that suggested between the kyanite and sillimanite zones of the Langtang Valley, simple calculations of the rates at which tectonic stresses must be reduced at a given temperature in order to preserve microstructures produced during the deformation (e.g. Knipe 1989) suggest that at temperatures in the order of 700° C such zones are unlikely to preserve syn-deformational microstructures and would be expected to undergo significant post-deformation recrystallization, annealing and rapid microstructural recycling during cooling. As a consequence, the preserved microstructures within the sillimanite-grade rocks are not considered to be those developed during the main phase of ductile deformation associated with foliation development within the HHCS and similarly the fault zone between the sillimanite and kyanite grade units. A quantitative assessment of the strains associated with deformation and foliation development is difficult because of the extensive high temperature recrystallization and microstructural recycling. Fortunately, qualitative features such as shear bands are not necessarily destroyed during recrystallization and this enables the kinematic framework associated with foliation development to be assessed.

The foliation developed within the Langshisa and Langtang Lirung Units is not as regionally consistent as that developed at lower structural levels and has been affected by large-scale folding. Folding of the early, foliated granites suggests a significant component of coaxial deformation associated with south-directed shearing of the HHCS which is consistent with evidence from similar structural levels in southern Tibet (Burchfiel *et al.* 1992). However, the foliation is again recrystallized and there is little consistent shear criteria found within the gneisses at the microscopic scale. The absence of folding within the underlying Kyangjin and Syabru Units suggests decoupling of the different units across the Kyangjin Unit–Langshisa Unit contact with the obvious interpretation being that the contact is tectonic in origin and is associated with south-directed thrusting. Unfortunately, the contact is poorly exposed and this precludes a detailed structural investigation within this area. However, the difference in geochemistry between these units and the underlying sillimanite grade units of the Syabru Unit may support this. If this interpretation is correct, it implies that thrusts propogated down structural section towards the south.

The high-grade gneisses of the HHCS contain a series of post-metamorphic fault zones that developed at various stages during uplift and exhumation of the complex. These faults show a range of microstructures developed by crystal plasticity to cataclastic processes during varying external conditions at different times. The orientation of the late-granites along discrete NW–SE-striking dykes would further suggest a component of northeast–southwest extension during granite intrusion. However, there also appears to be a relationship between the metamorphic grade at which post-metamorphic faulting took place and structural level. Extensional faults within the higher levels of the Syabru Unit are characterized by microstructures indicating significant degrees of secondary recrystallization, contain stable sillimanite and muscovite and therefore developed at temperatures slightly below the metamorphic peak. As such, these extensional faults developed at approximately the same time as the thrusting which placed sillimanite-grade rocks onto kyanite-grade. No fault zones of lower grade were observed within the Syabru Unit.

A wider variation of fault zone microstructures are preserved within the gneisses of the Kyangjin and Langshisa Units. Metre-wide shear zones commonly contain an early high-grade recrystallized foliation that is overprinted by fine-grained zones which record partially recrystallised deformation fabrics. External conditions during the time of deformation are difficult to assess because the limited metamorphic assemblage within these zones precludes the use of mineral equilibria thermobarometers to assess external conditions during deformation. However, by ignoring the strain-rate dependency of microstructural development, minimum temperatures for the deformation may be estimated. Microstructures indicating dislocation creep as an active deformation mechanism suggest deformation took place at temperatures in excess of 300° C while in zones where feldspar plasticity is observed, minimum temperatures of 450° C are likely (Tullis 1983). The preservation of deformation fabrics suggests that temperatures were low enough that significant post-deformational annealing did not take place, that is temperatures during deformation were below 500° C.

The fine-grained fault zones that overprint these earlier fault zone fabrics are discrete features characterised by discontinuous margins and a further grain size reduction in both matrix phases and feldspar clasts. This suggests lower temperature or higher strain-rate deformation. Towards the top of the exposed section, chlorite-grade fault zones developed by cataclastic deformation processes and are characterized by lower temperature rapid strain-rate deformation and by localized metasomatic alteration of biotite and feldspars, suggesting deformation-enhanced fluid infiltration. Stable isotope data indicate that fluids involved with this alteration were meteoric in origin (Massey *et al.* in prep.) which is consistent with their developing late in the exhumation history of the area.

Southwest-directed thrusts and northeast dipping extensional faults show no consistent overprinting relationships and where both thrust and extensional faults are present at the same structural level, microstructures preserved in each are similar. This suggests that thrust and normal

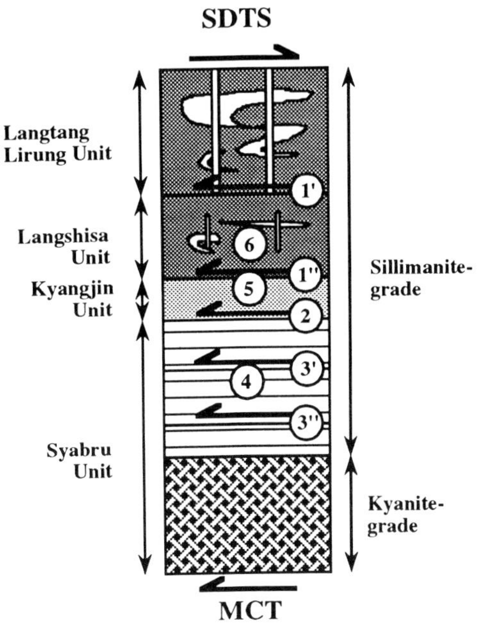

Fig. 8. Schematic model for the structural development of the HHCS of the Langtang Valley. Numbers represent a temporal sequence in the deformation history of the HHCS. South-directed thrusts propagated down structural section with time (1 to 3) and led to the development of a significant thickness of sillimanite-grade rocks which were then emplaced onto the kyanite grade rocks at the base of the section. The whole pile was subsequently thrust onto the Lesser Himalaya along the MCT. Granites are symbolised by unshaded areas in the upper two units. The sillimanite–grade rocks below 3″ are those with affinites to the kyanite-grade rocks below (see Fig. 4). Primed and double-primed numbers represent thrusts that developed at similar metamorphic conditions and therefore approximately coevally. Deformation within the different units generally moves up structural section with time with sillimanite + muscovite-grade fault zones in the Syabru Unit (4), lower temperature deformation within the Kyangjin Unit and base of the Langsisa Unit (5) associated with both top-to-southwest thrusting and northeast-dipping normal faults and chlorite-grade fault zones associated with the infiltration of meteoric water at the higher structural levels of the section (6).

fault development was taking place coevally at the same structural level. However, the fault zones developed at the higher structural levels show mineralogical changes (i.e., the development of chlorite) that suggest that thrust and extensional shear zone development at the top of the HHCS took place at lower temperatures than deformation associated with shear zones at deeper levels (sillimanite and muscovite grade) where no low-grade faulting has been observed. The development of these minor shear zones therefore appears to have migrated to higher structural levels with time. The temporal and spatial distribution of faulting within the HHCS is summarized in Fig. 8.

Theoretical modelling of orogenic wedges has shown that processes which affect the equilibrium geometry of the wedge may lead to extension within convergent orogens and that the equilibrium geometry may be affected by either the underplating of material to the base of the orogenic wedge or by decreasing the basal shear strength of the ductile fault forming the base of the accreted thrust stack (Davis *et al.* 1983; Dahlen *et al.* 1984; Platt 1986). Field evidence has also been used to demonstrate that within the convergent orogens, dynamic spreading of the orogenic wedge may lead to the development of non-coaxial, thrust-parallel extensional strains during foreland-directed thrusting (e.g. Holdsworth & Grant 1990). In Langtang, we argue that the tectonic development of the HHCS is characterized by foreland propogating thrusting at high temperatures which led to the emplacement of sillimanite-grade rocks onto kyanite-grade rocks and finally to the development of the MCT and the Main Boundary Thrust. Such a process would result in the thickening of the orogenic wedge and the development of thrust-parallel extensional strains in the HHCS. Continued convergence would also lead to thrust associated strains developed at the same time as the extensional strains. This fits our observations of syn-convergent extension seen within the sillimanite-grade units of the HHCS. Post-metamorphic deformation within the sillimanite units of the HHCS is therefore considered to be an expression of a changing dynamic orogenic wedge and the attainment of an equilibrium geometry. If such a model is correct for the Langtang area, it would be expected that continued underplating of material due to thrusting along the MCT and MBT would lead to a prolonged post-metamorphic deformation within the HHCS. This is in agreement with observations described earlier that illustrate that shear zone development took place from sillimanite-grade down to chlorite-grade conditions.

Conclusions

(1) Kyanite-grade units at the base of the HHCS of the Langtang Valley record no evidence of post-metamorphic deformation but contain a

strong foliation that developed prior to kyanite-grade conditions. Since the MCT places rocks of the HHCS onto lower-grade rocks of the Lesser Himalaya, the absence of post-metamorphic deformation in the kyanite-grade rocks suggests that this part of the HHCS was not affected by deformation on the MCT, that is the kyanite-grade rocks at the base of the HHCS were passively carried along the MCT.

(2) The regional foliations found within the upper sillimanite units of the Langtang Valley developed as a result of non-coaxial, top-to-south shearing. Towards the base of the sillimanite-grade rocks of the Syabru Unit, sillimanite, which grew during non-coaxial deformation at the metamorphic peak is overgrown by late-muscovite which grew within the same kinematic framework. These observations combined with geochemical, textural and structural arguments suggest that the apparent inverted metamorphism from kyanite to sillimanite grades in the Syabru Unit is the result of thrusting of the sillimanite-grade units over kyanite-grade towards the southwest. By analogy, another zone containing late-muscovite in the area of Langtang Village could be another high temperature shear zone.

(3) Folding developed during top-to-south thrusting within the upper two units of the Langtang Valley does not effect the underlying Kyangjin or Syabru Units. Deformation across the contact between the Kyangjin and Langshisa Unit must have been decoupled and the contact is interpreted as a thrust. Since the folds are a result of non-coaxial deformation and do not affect the underlying units, deformation migrated down structural section with time, that is thrusting was foreland propogating. It is possible to speculate that the contact between the Kyangjin and Syabru Unit and the Langtang Lirung Unit are also tectonic in origin.

(4) Within the different units of the HHCS, minor shear zones with both thrust and normal fault displacements are present. At any particular structural level, both fault types record similar microstructures suggesting they developed coevally. There is an overall decrease in the metamorphic grade in which these zones developed moving up structural section. These faults are considered to have developed as a consequence of the changing geometry of the orogenic wedge during tectonic convergence.

Fieldwork for this research was funded by a Royal Society award to N. Harris and an NERC small grant to S. Reddy. M. Brunel, N. Harris and A. Macfarlane are thanked for critically reviewing an earlier version of the manuscript. An Open University Petrogenesis Research Group Post-Doctoral Fellowship to S. Reddy is gratefully acknowledged.

References

Bordet, P. 1961. *Recherches géologiques dans l'Himalaya du Népal, région de Makalu.* Central National de la Recherche Scientifique, Paris.

Brunel, M. 1986. Ductile thrusting in the Himalayas: Shear sense criteria and stretching lineations. *Tectonics,* **5**, 247–265.

Burchfiel, B.C., Zhiliang, C., Hodges, K.V., Yuping, L., Royden, L.H., Changrong, D. & Jiene, X. 1992. *The South Tibetan Detatchment System, Himalayan orogen: Extension contemporaneous with and parallel to shortening in a collision mountain belt.* Geological Society of America Special Papers, **269**.

Burg, J.P., Brunel, M., Gapais, D., Chen, G.M. & Liu, G.H. 1984. Deformation of leucogranites of the crystalline Main Central Sheet in southern Tibet (China). *Journal of Structural Geology,* **6**, 535–542.

Caby, R., Pêcher, A. & Le Fort, P. 1983. Le grand chevauchement central Himalayen: nouvelles donnees sur le metamorphisme inverse a la base de la Talle du Tibet. *Revue de Geographie Physique et de Geologique Dynamique,* **24**, 89–100.

Dahlen, F.A., Suppe, J. & Davis, D. 1984. Mechanics of fold- and thrust-belts and accretionary wedges: Cohesive Coulomb theory. *Journal of Geophysical Research,* **89**, 10087–10101.

Davis, D., Suppe, J. & Dahlen, F.A. 1983. Mechanics of fold-and-thrust belts and accretionary wedges. *Journal of Geophysical Research,* **88**, 1153–1172.

Gansser, A. 1964. *Geology of the Himalayas.* Wiley-Interscience, London.

Herren, E. 1987. Zanskar shear zone: Northeast-southwest extension within the Higher Himalayas (Ladakh, India). *Geology,* **15**, 409–413.

Hodges, K.V. & Silverberg, D.S. 1988. Thermal evolution of the Greater Himalaya, Garhwal, India. *Tectonics,* **7**, 583–600.

Holdsworth, R.E. & Grant, C.J. 1990. Convergence-related 'dynamic spreading' in a mid-crustal ductile thrust zone: a possible orogenic wedge model. *In*: Knipe, R.J. & Rutter, E.H. (eds) *Deformation Mechanisms, Rheology and Tectonics.* Geological Society, London, Special Publications, **54**, 491–500.

Inger, S. 1991. *Metamorphism and Granite Genesis in the Langtang Region, north-central Nepal.* PhD Thesis (Open University).

—— & Harris, N.B.W. 1992. Tectonothermal evolution of the High Himalayan Crystalline Sequence, Langtang Valley, northern Nepal. *Journal of Metamorphic Geology,* **10**, 439–452.

Knipe, R.J. 1989. Deformation mechanisms: recognition from natural tectonites. *Journal of Structural Geology,* **11**, 127–146.

Le Fort, P., Debon, F. and Sonet, J. 1983. The Lower Palaeozoic 'Lesser Himalayan' granitic belt: emphasis on the Simcharpluton of central Nepal. *In*: SHAMS, F.A. (ed.) *Granites of Himalayas, Karakoram and Hindu Kush*. University of Lahore, Lahore. pp. 277–387.

MACFARLANE, A. 1990. Structural and thermobarometric study of the Main Central Thrust zone, Langtang region, Nepal Himalaya. *Geological Society of America Abstracts with Programs,* **22**, A96–A97.

PATRIAT, P. & ACHACHE, J. 1984. India-Eurasia collision chronology has implications for crustal shortening and driving mechanisms of plates. *Nature,* **311**, 615–621.

PÊCHER, A. 1989. The metamorphism in the Central Himalaya. *Journal of Metamorphic Geology,* **7**, 31–41.

—— 1991. The contact between the Higher Himalaya Crystallines and the Tibetan Sedimentary Series: Miocene large-scale dextral shearing. *Tectonics,* **10**, 587–598.

—— & LE FORT, P. 1986. The metamorphism in central Himalaya, its relation with the thrust tectonic. *Colloques internationaux du CNRS, Paris ecologie et géologie de l'Himalaya,* **286**, 301–318.

PLATT, J.P. 1986. Dynamics of orogenic wedges and the uplift of high-pressure metamorphic rocks. *Geological Society of America Bulletin,* **97**, 1037–1053.

POGNANTE, U. & LOMBARDO, B. 1989. Metamorphic evolution of the High Himalaya crystallines in south east Zanskar, India. *Journal of Metamorphic Geology,* **7**, 9–17.

SEARLE, M.P. 1986. Structural evolution of thrusting in the High Himalayan, Tibetan-Tethys and Indus Suture zones of Zanskar and Ladakh, Western Himalaya. *Journal of Structural Geology,* **8**, 923–936.

—— & REX, A.J. 1989. Thermal model for the Zanskar Himalaya. *Journal of Metamorphic Geology,* **7**, 127–134.

——, WATERS, D.J., REX, D.C. & WILSON, R.N. 1992. Pressure, temperature and time constraints on Himalayan metamorphism from eastern Kashmir and western Zanskar. *Journal of the Geological Society, London,* **149**, 753–774.

STAÜBLI, A. 1989. Polyphase metamorphism and the development of the Main Central Thrust. *Journal of Metamorphic Geology,* **7**, 73–93.

SWAPP, S. & HOLLISTER, L.S. 1991. Inverted metamorphism within the Tibetan slab of Bhutan: evidence for a tectonically transported heat source. *Canadian Mineralogist,* **29**, 1019–1041.

TRELOAR, P.J., BROUGHTON, R.D., WILLIAMS, M.P., COWARD, M.P. & WINDLEY, B.F. 1989. Deformation, metamorphism and imbrication of the Indian plate south of the Main Mantle Thrust, north Pakistan. *Journal of Metamorphic Geology,* **7**, 111–125.

TULLIS, J. 1983. Deformation of feldspars. *In*: RIBBE, P.H. (ed.) *Feldspar Mineralogy*. Reviews in Mineralogy, **2**, 297–323.

The role of fluids in the formation of High Himalayan leucogranites

NIGEL HARRIS, SIMON INGER[1] & JOHN MASSEY

Department of Earth Sciences, Open University, Milton Keynes MK7 6AA

[1] *Present address: Department of Earth Sciences, University of Leeds, Leeds LS2 9JT, UK*

Abstract: During fractional melting of the crust, the availability of fluids is a critical variable that influences the melt fraction obtained, the capacity of the melt to migrate from its source and the relationship between melting and uplift. For the High Himalayan leucogranites, isotope systematics and metamorphic phase equilibria have identified micaceous metasediments as the most likely source, and the incongruent melting of muscovite as the appropriate melting reaction. High-field strength trace elements are controlled by dissolution kinetics and incomplete crystal/liquid segregation of accessory phases during crustal anatexis. In contrast, Rb and Sr are controlled by major phases contributing to the melting reaction. Hence the Rb/Sr ratios in primary melts can be correlated with the availability of fluids during partial melting.

Geochemical modelling of melts resulting from anatexis of a wide range of metasedimentary compositions (plagioclase 5–30%, muscovite 5–30%, biotite 5–20%, quartz 30–85%) indicates that Rb/Sr ratios will vary between 4 and 10 during equilibrium melting and vapour-absent conditions, but are less than 3.5 during vapour-present melting. In either case, Rb/Sr ratios in melts are reduced under disequilibrium conditions. In general Rb/Sr ratios of Himalayan leucogranites vary between 3 and 6. Such high ratios could not have formed under vapour-present conditions unless feldspar fractionation has increased the Rb/Sr ratios of primary melts. Trace-element systematics and petrographic evidence do not support significant fractionation of feldspar. Hence for most, if not all, Himalayan granites, vapour-absent conditions prevailed during fractional melting, irrespective of either source composition or the kinetics of the melting reaction. This is consistent with the observation that the melts have migrated from their source, and allows decompression melting to play a role in their genesis.

The presence or absence of a hydrous fluid during melting of the crust places important constraints on melt formation. Granite melts require five essential components, SiO_2, Al_2O_3, Na_2O, K_2O and H_2O. Three of these, SiO_2, Al_2O_3 and Na_2O, are provided by quartz and plagioclase, which are ubiquitous in a wide range of sedimentary compositions. However, the scarcity of alkali feldspar in sedimentary rocks, and the common absence of a hydrous fluid phase during high-grade metamorphism implies that both potassium and H_2O must be derived from the incongruent melting of micas during the generation of granite melts from metasedimentary protoliths (Holtz & Johannes 1991; Patino-Douce & Johnson 1991).

Early experimental studies on melting in the quartz–albite–orthoclase system (Tuttle & Bowen 1958) were undertaken under hydrous conditions (aH$_2$O = 1). The high solubility of water in granite magmas implies that water saturation of melts will only rarely be achieved in nature (Clemens 1984; Clemens & Vielzeuf 1987). More recent experimental work has explored melting behaviour in the granite ternary system under a range of water-undersaturated conditions (Johannes & Holtz 1990). The presence of water during anatexis will not only depress the solidus temperature and change the minimum melt composition, but also cause a negative dP/dT slope on the granite solidus. For water-undersaturated melting, the positive slope of the solidus means that melts can leave the melt sites and rise through the crust without crystallizing (Cann 1970). A second important consequence of the positive slope of the vapour-absent solidus is that anatexis can be induced for crustal melting, as for partial melting in the mantle, by decompression (Le Breton & Thompson 1988).

In contrast, water-present melting of crustal rocks requires that a large fluid reservoir has access to the protolith throughout the melting event. Such melts are unlikely to leave their source without crystallising, and will thus be retained as leucocratic components in migmatitic complexes. Furthermore, decompression melting will not be possible under water-present conditions since a drop in pressure will drive the melt towards the solidus (Le Breton & Thompson 1988). For water-present conditions therefore, crustal melting cannot be a simple

From TRELOAR, P.J. & SEARLE, M.P. (eds), *Himalayan Tectonics*
Geological Society Special Publication, No. 74, pp. 391–400.

consequence of uplift but must be associated with prograde metamorphism and therefore requires a heat source.

Nevertheless, leucogranites are sometimes cited as examples of crustal melting under hydrous conditions. In the Pyrenees, leucogranites have been associated with shallow melting in the presence of vigorous hydrothermal circulation (Wickham 1987) and, in the Himalayas, leucogranite formation has been associated with movement on major thrust faults with melting induced under hydrous conditions by dehydration of the footwall of the thrust (Vidal *et al.* 1982; Le Fort 1981). Fortunately variations in water activity during melting will generate significant changes in trace-element compositions in the melt (Harris & Inger 1992), and thus the relationship between fluid infiltration and melting can sometimes be constrained through geochemical systematics.

The effects of other fluid species, such as CO_2, N_2 or Ar, on the granite solidus are less well constrained by experiments, but their relatively low solubilities in granite melts suggest that their influence will be considerably less marked than that of H_2O. In this paper we consider the geochemistry of crustal melts formed during the incongruent melting of micas. Since micas provide a ready source of hydrous fluids, we contrast the geochemistry of melts formed under water-rich (vapour-present) and water-deficient (vapour-absent) conditions, although we recognize that there are circumstances in which crustal fluids may not necessarily be water-rich.

Crustal melting in the Himalayas

Peraluminous granites of Miocene age occur in two belts in the Himalayas (Fig. 1); the North Himalaya belt where plutons are intruded into Tethyan sediments (Schärer *et al.* 1986) and the High Himalaya belt emplaced further south (Debon *et al.* 1986). The High Himalaya belt can be traced as discontinuous sheets, sills and minor intrusions for nearly 3000 km along the strike of the High Himalayas from northern Pakistan to Bhutan. These small granitic bodies are intruded near the interface between the high-grade crystalline basement of the High Himalaya, and the overlying low-grade or unmetamorphosed Tibetan–Tethys sediments. Detailed isotopic studies indicate emplacement at about 20 Ma (Deniel *et al.* 1987; Copeland *et al.* 1990) although in the Nanga Parbat Massif of the western Himalayas granite emplacement as young as 5 Ma has been recorded (Zeitler & Chamberlain 1991).

The High Himalayan leucogranites have be-

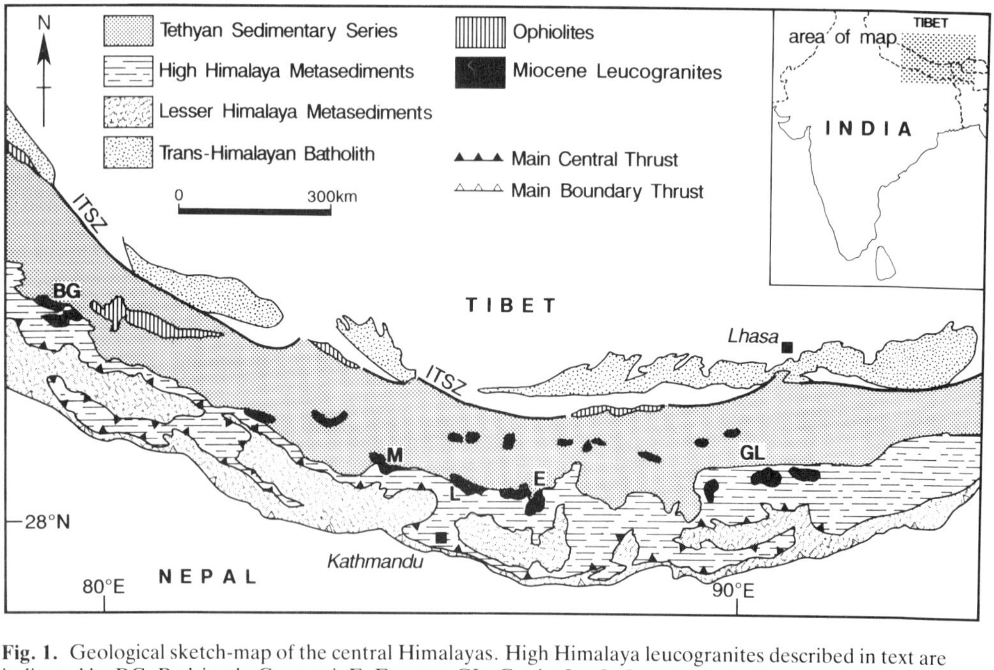

Fig. 1. Geological sketch-map of the central Himalayas. High Himalaya leucogranites described in text are indicated by BG, Badrinath–Gangotri; E, Everest; GL, Gophu La; L, Langtang; M, Manaslu. ITSZ, Indus–Tsangpo Suture Zone.

come classic examples of crustal melts with high concentrations of large-ion-lithophile elements (LILE) and strong depletions of high-field-strength elements (HFSE), indicative of fractional melting from a metasedimentary source (Vidal *et al.* 1982; Harris *et al.* 1986), and Sr and Nd isotopic ratios indicative of a mature crustal source (France-Lanord & Le Fort 1988). Detailed studies of one such intrusion, from Manaslu in central Nepal, has provided initial Sr, Nd and Pb isotopic ratios consistent with derivation from a sequence of metamorphosed pelites and quartzofeldspathic sediments, termed Formation 1 of the crystalline basement (Gariepy *et al.* 1985; Deniel *et al.* 1987; France-Lanord & Le Fort 1988). More recently, Sr and Nd source modelling from the Langtang granite (Fig. 1) has demonstrated that although Formation 1 sediments may represent the granite protolith, melting occurs within metasediments that are isotopically similar to the subsolidus, kyanite-grade pelites from the lower section of the formation, rather than the sillimanite-grade anatectic migmatites (Inger & Harris 1993).

Metamorphic conditions and textural evidence observed in these metasediments (Hodges *et al.* 1988; Inger & Harris 1992) indicate that melts were derived from the incongruent melting of muscovite. This may occur either under vapour-present (1a) or vapour-absent (1b) conditions:

$$\text{Muscovite} + \text{Quartz} + \text{Plagioclase} + \text{Vapour} = \text{Aluminosilicate} + \text{Melt} \quad (1a)$$

$$\text{Muscovite} + \text{Quartz} + \text{Plagioclase} = \text{Aluminosilicate} + \text{Kspar} + \text{Melt} \quad (1b)$$

For reaction (1a), the reaction procedes until one of the three solid reactants is consumed, or until the fluid supply is exhausted. For reaction (1b), sometimes referred to as dehydration melting, the stoichimetric amounts of restitic alkali feldspar are poorly constrained. We have assumed that firstly all muscovite is consumed and secondly that the melt fraction is determined by the muscovite content of the source (see Clemens & Vielzeuf 1987, table 4). Knowing the composition of the granite melt from leucogranite compositions, the amount of restitic alkali feldspar can be determined by mass balance.

In a previous study (Harris & Inger 1992) trace-element concentrations in melts derived from the incongruent melting of muscovite have been modelled for an assumed pelitic composition under conditions of chemical equilibrium between melt and restite. Because of the inhomogeneous nature of Formation I in the crystalline basement, we consider here a wider range of metasedimentary compositions and provide quantitative geochemical models for both equilibrium and disequilibrium melting.

Trace-element behaviour during fractional melting

Quantitative modelling of fractional melting inevitably rests on assumptions about the distribution of trace elements between melt and solid phases. In practice, either chemical equilibrium or complete disequilibrium between melt and restite are generally assumed (Allegre & Minster 1978). For trace elements that reside predominantly in the major phases, such as Rb, Sr and Ba, it is reasonable to infer that these two assumptions underpin end-member models that will constrain geological conditions during crustal melting. However, for trace elements that are predominantly controlled by accessory phase behaviour in granite systems, such as REE, Zr, Th or Nb, their behaviour will be controlled by solution kinetics, solubility dependence on melt composition (Watson & Harrison 1983) and incomplete separation between melt and restite (Copeland *et al.* 1988).

Appropriate trace elements for quantitative modelling in granitic systems are therefore restricted to Rb, Sr and Ba. Of these, Rb and Sr are the most commonly determined trace elements in whole-rock granite analyses. Moreover concentrations of Rb and Sr in minerals, essential for quantitative modelling of disequilibrium melts, are frequently determined for isotopic studies. Hence Rb/Sr ratios can be obtained from the literature for several Himalayan leucogranites, and provide important constraints on magmatic processes.

Rb/Sr ratios in leucogranites

Rb/Sr ratios in evolved magmas are affected not only by the conditions under which fractional melting occurs, but also by any subsequent feldspar fractionation. For some of the well-studied Himalayan leucogranites, such as those from Manaslu (Nepal) and Gophu La (Bhutan) no systematic variation in trace elements is observed that can be ascribed to in situ fractionation. Indeed the variation in Sr and Nd isotopic ratios within individual intrusions is ascribed to varying metasedimentary source compositions (Deniel *et al.* 1987). In contrast geochemical studies of leucogranites from Gangotri (India) and Langtang (Nepal) indicate that tourmaline-bearing granites have markedly higher Rb/Sr ratios than the coexisting biotite granites (Scaillet *et al.* 1990; Inger & Harris 1993). It is unlikely that this variation reflects feldspar fractionation since Rb/Sr ratios are observed to increase with Ba, which can only be modelled by the removal of alkali feldspar. Textural evidence indicates that alkali feldspar is the last feldspar to crystallize, and fractionation of early-formed plagioclase would result in increased Rb/Sr with no concomitant change in Ba. Moreover the lack of density contrast between melt and feldspar, the high viscosity of the melt, and cooling of the magma chamber from above would all inhibit removal of the feldspar from the melt (Marsh 1988). Where feldspar fractionation has been documented in highly siliceous rhyolites (Halliday *et al.* 1991) it results in strongly anomalous Rb/Sr ratios (>1000), which are not observed in the Himalayan granites. We therefore consider that the tourmaline-leucogranite facies, which is characterised by homogeneous minimum-melt compositions, represents a low-degree primary melt, similar to the tourmaline leucogranites of South Dakota (Nabalek & Russ-Nabelek 1991).

Histogram of three well-studied leucogranites (Fig. 2), Manaslu and Langtang in Nepal and Gophu La in Bhutan, indicates that Rb/Sr ratios vary, but for typical leucogranite compositions lie between 2 and 6. An average Rb/Sr value of 4 is obtained from averaging 201 analyses from the Manaslu leucogranite (Le Fort *et al.* 1987). Considerably higher values (up to 12) are obtained from low Sr samples which generally represent aplitic or pegmatitic phases. In the Langtang granite, the biotite-granite facies includes samples with Rb/Sr ratios as low as 1. The Langtang biotite granites are not of minimum-melt composition (Inger & Harris 1993) and field relations indicate they predate tourmaline-granite intrusion. They may represent higher degree partial melts compared to the tourmaline granites which form a well-defined minimum-melt composition with Rb/Sr ratios of 3 to 5.

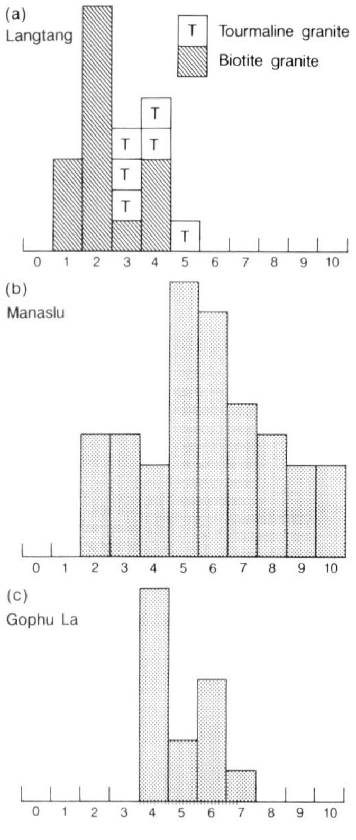

Fig. 2. Histograms of Rb/Sr ratios from Himalayan leucogranites. (**a**) Langtang (Nepal). (**b**) Manaslu (Nepal). (**c**) Gophu La (Bhutan).

Rb/Sr ratios in metasediments

Isotopic constraints and field relations suggest that the High Himalayan leucogranites are derived from a sedimentary protolith, comprised of alternations of metamorphosed greywackes and pelites (Colchen *et al.* 1986). These are characterised by variable Rb/Sr ratios in the range Rb/Sr = 0.7 to 1.4 (Brouand *et al.* 1990; Inger & Harris 1993). During fractional melting the critical minerals controlling melt composition are muscovite, which provides both potassium and a hydrous component, and plagioclase (Patino-Douce & Johnston 1991; Harris & Inger 1992). In this study, we model melts derived from a wide range of metasedimentary source compositions that vary in modal composition from 5 to 30% for both plagioclase and muscovite, covering all likely protolith compositions.

Biotite contents in the metasediments vary between 5 and 20% and a biotite content of 10% is assumed in these calculations. Rb/Sr ratios in melts generated by muscovite breakdown are not sensitive to the biotite content of the source. The relative proportions of other components in the metasediments (quartz, garnet, tourmaline and aluminosilicates) do not affect the Rb/Sr ratios of the resulting melts. Alkali feldspar occurs only in the anatectic migmatites, indicating that it is a product phase of the melt-forming reaction.

In order to predict the concentration of a trace element in the liquid it is necessary to estimate the concentration of that element in the source.

Table 1. *Rb and Sr concentrations (ppm) in minerals from Himalayan mica schist (SSM6)*

	Plagioclase[1]	Muscovite[2]	Biotite[2]
Rb	0	442	755
Sr	760	121	18

[1] Ion probe data (Harris *et al.* 1992)
[2] Isotope dilution (Inger & Harris 1992)

Rb and Sr are almost entirely resident in the three phases plagioclase, muscovite and biotite. The distribution of Rb and Sr between plagioclase, muscovite and biotite are known from mineral analyses of a kyanite-grade schist (Table 1) made up of plagioclase (12%), muscovite (20%), biotite (15%), quartz (38%) and garnet (15%). Rb is distributed between biotite and muscovite in the ratio 1.7:1 and Sr between plagioclase and muscovite in the ratio 6.25:1. Hence Rb and Sr variations in metasedimentary compositions can be estimated from a given modal composition provided that Rb and Sr reside predominantly in the three phases plagioclase, muscovite and biotite. This approach avoids the assumption of a uniform source composition over a wide range of modal compositions, and will generate source compositions (Rb = 100–230 ppm and Sr = 40–230 ppm with Rb/Sr ratios of 0.8 to 1.6) that compare closely with analysed values.

Modelling Rb/Sr ratios during fractional melting

Equilibrium melting

During fractional incongruent melting under equilibrium conditions the ratio of concentrations of a trace element in the liquid and protolith (C_l/C_0) is given by the equation:

$$C_l/C_0 = (1/F)[1 - [1 - PF/(D(1 - F) + PF)]^{1/P}] \qquad (2)$$

assuming continuous removal and collection of the melt of minimum-melt composition (Harris & Inger 1992; equation 5). F is the weight fraction of melt relative to the original parent, D the bulk distribution coefficient for the element between melt and residual mineral phases, and P the weight contribution of the trace element to the liquid. A batch-melting model (which assumes equilibrium between entire melt fraction and restite) generates similar results but is geologically less plausible. Mineral/melt partition coefficients are dependent on both bulk composition of the magma and mineral structure. Plagioclase/melt coefficients for Sr and Ba have been computed from Blundy & Wood (1991). Other coefficients have been taken from values quoted in Nash & Crecraft (1985) for peraluminous granite compositions.

Disequilibrium melting

For disequilibrium melting, we have adapted published models that assume congruent, modal melting (Allegre & Minster 1978; Sawyer 1991) for the incongruent non-modal melts formed by reaction (1). Under these circumstances the ratio (C_l/C_0) is sensitive to the melt fraction generated (F), and may be given by:

$$C_l/C_0 = (1/F)[(x_1 + D_{2/1}x_2 + \ldots\ldots)/(X_1 + D_{2/1}X_2 \ldots\ldots)] \qquad (3)$$

where x_i is the proportion of mineral i entering melt, X_i is the proportion of mineral i in the source rock, and $D_{2/1}$ is the distribution of the trace elements between phases 1 and 2.

D values may be calculated for coexisting biotite, muscovite and plagioclase from a kyanite-grade Himalayan pelite from available trace-element data (Table 1). Sr concentrations in the melt are strongly controlled by the proportion of plagioclase that enters the melt, and Rb by the proportion of muscovite in the source. For vapour-present melting, sillimanite is the only solid phase generated by reaction (1a). In the case of vapour-absent melting, alkali feldspar is created with sillimanite as a restite phase (reaction 1b) and hence assumptions concerning the Rb and Sr concentrations on the restitic feldspar are required. We have considered two possibilities. Firstly that melt extraction was so rapid that both Rb and Sr behaved incompatibly with respect to the nucleating alkali feldspar. Secondly, and more plausibly, that equilibrium was maintained between melt and alkali feldspar during nucleation and growth of the feldspar but that melt was extracted sufficiently rapidly not to equilibrate with pre-existing restite phases (mica and plagioclase), since equilibrium for these phases will be controlled by intra-crystal diffusion.

Results

For the range of protolith compositions modelled in Fig. 3, Rb/Sr ratios may be generated in the melt between <1 to >10, depending on the composition of the source, the activity of H_2O during melting, and on the degree of equilibrium reached during anatexis. Both increased water activity and disequilibrium melting will reduce the Rb/Sr in the melt.

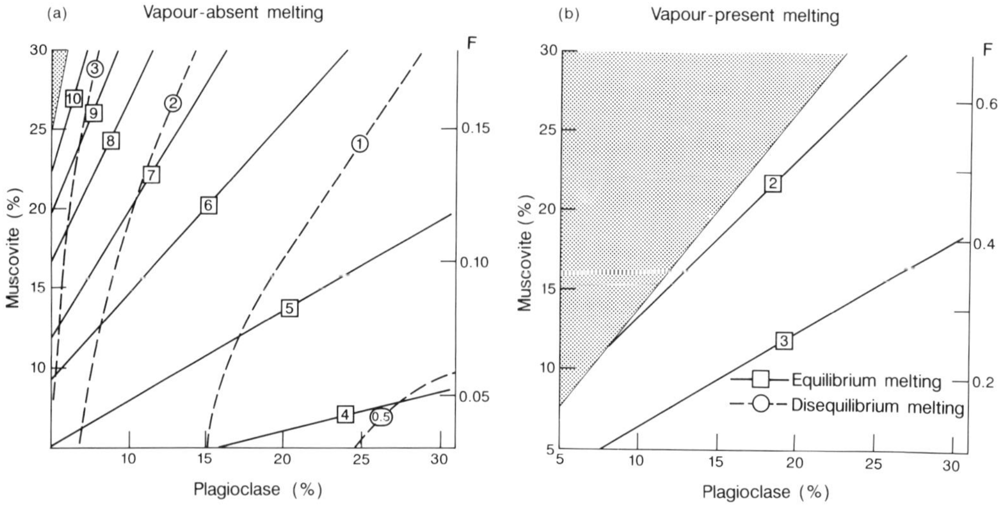

Fig. 3. Modelled Rb/Sr ratios for partial melting of a muscovite-bearing protolith by incongruent melting of muscovite. (**a**) Under vapour-absent conditions; solid lines contour equilibrium melting, dashed lines contour disequilibrium melting. (**b**) Under vapour-present conditions: solid lines contour equilibrium melting. For disequilibrium melting Rb/Sr ≈ 0.6. Shaded region indicates that the protolith cannot generate a minimum-melt composition.

The highest Rb/Sr ratios result from vapour-absent equilibrium melting, since, for most protolith compositions, the low melt fraction ensures some residual plagioclase in the source, and reaction (1b) will generate residual alkali feldspar. Rb/Sr ratios in melts of 3 to 10 can be generated by melting a range of sedimentary compositions under vapour-absent conditions (Fig. 3a), although the low melt fractions obtained in melts of lower Rb/Sr ratios ($F < 0.05$ for Rb/Sr <4) would make it highly unlikely that the melts could be extracted from their source. Nonetheless the range in Rb/Sr ratios compares well with that observed in the High Himalayan leucogranites (Fig. 2). Muscovite-rich, plagioclase-poor sources (pelites) will yield higher Rb/Sr ratios in melts, and higher melt fractions, than plagioclase-rich, muscovite-poor sediments (greywackes).

Under disequilibrium conditions, Rb/Sr ratios are reduced and are less sensitive to source mineralogy. Assuming equilibrium is attained only between melt and restite alkali feldspar, Rb/Sr ratios of 4 to 5 are obtained, irrespective of protolith composition. Under extreme disequilibrium conditions, where restitic alkali feldspar has insignificant Rb and Sr concentrations, melt Rb/Sr ratios are reduced to values of <3 (Fig. 3*a*).

Under vapour-present conditions, Rb/Sr ratios no greater than *c.* 3.5 can be obtained (Fig. 3b). If average partition coefficients for acidic magmas are chosen (Henderson 1982) the maximum Rb/Sr ratio is reduced to less than two. Such low values are generally inconsistent with observed ratios in the Himalayan leucogranites. This conclusion does not preclude the possible presence of fluid during anatexis to generate melts with lower Rb/Sr ratios (1–3). However these ratios could equally represent disequilibrium melting under vapour-absent conditions. For both disequilibrium and vapour-present conditions, a strong increase in Sr in the melt relative to source generates melts of lower Rb/Sr than their source rocks (Rb/Sr ≈ 0.6 in melts from a wide range of source compositions).

It is interesting to note that in general, Rb/Sr ratios will be increased by fractional melting and that a decrease in Rb/Sr of the melt relative to source requires both vapour-present melting and rapid extraction of melt. Hence the assumption, widely-held by isotope geochemists, that intra-crustal melting increases Rb/Sr ratios, is generally supported by the trace-element modelling presented here.

Discussion

During fractional melting in the crust either vapour-present or vapour-absent conditions must prevail. In modelling vapour-absent conditions we have assumed that the melt fraction is internally buffered by muscovite in the protolith. This is not necessarily true for vapour-absent melting. For example, limited fluid infiltration of a melt formed under vapour-absent conditions could result in dissolution of the fluid into the melt without sufficient fluid being available for vapour-present conditions to be established. Equally, some water may be available in the protolith, but insufficient to saturate the melt and cause vapour-present conditions. Under these circumstances, the water activity of the melt remains buffered by the melt but the melt fraction would increase relative to the fraction resulting from simple muscovite breakdown (Powell 1983; Waters 1988), consuming some of the restitic alkali feldspar (reaction 1b). Thus the Rb/Sr ratio in the melt is somewhat reduced, even though melting occurred under vapour-absent conditions. The extent of this reduction depends on the volume of fluid available for dissolution in the melt.

Modelling of equilibrium and disequilibrium conditions also represents a simplification of natural processes because equilibration is a time-dependent process that varies according to the diffusivity of the element concerned and the nature of the medium through which it diffuses. At the temperatures required for crustal melting through the incongruent melting of muscovite, equilibrium of Sr between a biotite crystal (1 mm radius) and melt will take only about 100 years (Giletti 1991), whereas equilibrium of REE between a garnet of similar size and melt will take at least 15 Ma (Harris *et al.* 1992).

These qualifications notwithstanding, the Rb/Sr ratios observed in High Himalayan leucogranites are sufficiently high to preclude an origin by vapour-present melting. The activity of water during melting has a strong control over the solidus temperature. Metamorphic temperatures of *c.* 750° C, calculated from thermobarometric assemblages from Formation 1 migmatites (Hodges *et al.* 1988; Inger & Harris 1992) compare with temperatures of 680–740° C obtained from Zr concentrations in Himalayan leucogranites (Watson & Harrison 1983), assuming melts were saturated in Zr and ignoring the contribution of inherited zircons. Both independant estimates of temperatures during anatexis are well in excess of the haplogranite solidus at 5 kbar under vapour-present conditions (Tuttle & Bowen 1958). This suggests either that a significant temperature overstep occurred or that vapour-absent conditions prevailed. A temperature overstep, under vapour-present conditions, is possible, but influx of fluids would result in rapid, disequilibrium melting (Rubie & Brearley 1990), thereby reducing the Rb/Sr ratio in the melt to less than that of the source.

A possible objection to vapour-absent melting in the Himalayas is that the low melt fraction ($F < 0.15$) would preclude a critical melt fraction being obtained ($F > 0.3$) and hence prevent the melt leaving its source (Wickham 1987). The concept of the critical melt fraction assumes that the melt will migrate through diapirism. In the case of the Himalayan leucogranites, abundant feeder dykes indicate that they were emplaced through brittle fractures, driven either by buoyancy or injection pressure (Clemens & Mawer 1992), and thus diapiric arguments are irrelevant. Under such circumstances it is difficult to place a minimum constraint on the melt fraction that can be extracted from the source.

This study has important implications for the mechanism of leucogranite formation in the High Himalayas. Models that have sought to correlate Himalayan anatexis with fluids derived from footwalls of major thrusts, such as the Main Central Thrust, are attractive in that they seek to integrate the thermal and tectonic evolution of the orogen. However, field observations do not always support such models. For example, in some sections north of the MCT, anatexis occurs at significant stratigraphic distances (5–10 km) above the thrust zone (Inger & Harris 1992). It is interesting that both zones of high-grade metamorphism and of granite emplacement are located close to the northern boundary of the crystalline basement, and that this boundary has been identified as a normal fault both in the eastern and western Himalayas (Burchfiel & Royden 1985; Searle & Rex 1989). The recognition that melting occurred essentially under vapour-absent conditions argues against the correlation of anatexis with regional thrusting on the MCT but is consistent with an association between granite formation and uplift along a normal fault.

It is beyond the scope of this study to identify the heat source required for crustal melting in the Himalayas. However, the geometry of the vapour-absent solidus not only allows melts to leave their source, but also allows decompression melting to occur. Such a mechanism has been suggested for younger (*c.* 5 Ma) leucogranites from the western Himalayas (Zeitler & Chamberlain 1991) and may be equally appropriate for the tectonothermal event that occurred during the early Miocene. During rapid adiabatic uplift of mid-crustal levels, vapour-absent melting will be enhanced because the vapour-absent solidus is depressed at shallower levels. This is illustrated in Fig. 4, where the melt fraction (by volume) resulting from

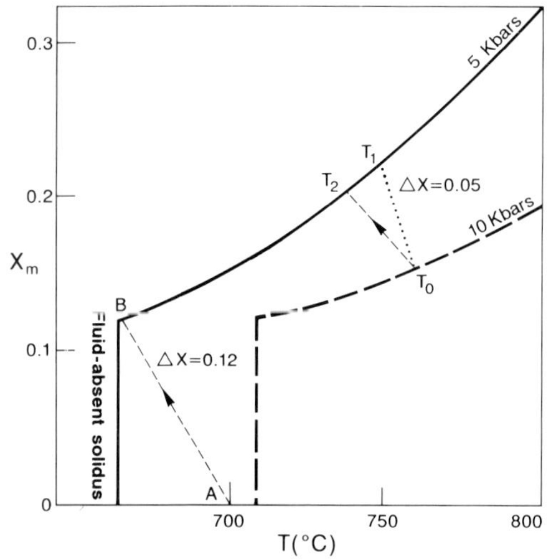

Fig. 4. Decompression melting of a pelite, (H_2O = 1.2%). Melt fraction by volume (X_M) at 5 and 10 kbar derived from Clemens & Vielzeuf (1987).

vapour-absent melting of a pelitic source rock (H_2O = 1.2%) has been determined at 5 and 10 kbar using the model of Clemens & Vielzeuf (1987). During uplift, cooling will occur not only due to adiabatic expansion, but also due to the latent heat of fusion. For example at 760° C (T_0) and pressures of 10 kbar, the melt fraction will be about 0.15. Adiabatic ascent to depths equivalent to 5 kbar will result in cooling to 750° C (T_1), and an increased melt fraction to 0.22, ignoring the effects of enthalpy required for fusion.

However, for enthalpy to be conserved,

$$C_p(T_1 - T_2) = L_s\, \Delta X_m$$

where T_2 is the final temperature, L_s is the heat of fusion, C_p the specific heat of granite and ΔX_m is the change in mass fraction of the melt between T_0 and T_2. Assuming values of 2.93×10^5 J kg^{-1} for L_s and 1.3×10^3 J kg^{-1} K^{-1} for C_p (Huppert & Sparks 1988), T_2 is about 740° C and the melt fraction will increase from 0.15 to 0.20 during decompression. A larger change in melt fraction can be obtained if the melt is uplifted from subsolidus conditions. For example, for a vapour-absent pelite at 700° C and 10 kbar (A on Fig. 4) no melt will be present, but rapid uplift to 5 kbars will result in an overall melt fraction of 0.12 (B on Fig. 4), similar to the minimum melt fraction inferred from melt geochemistry of the Himalayan granites (Fig. 3). Such a mechanism could also supplement other heat sources. For example, thermal models that invoke heat focusing mechanisms resulting from low-conductivity sediments overlying metamorphic basement of high heat production (Jaupart & Provost 1985) are intrinsically linked to decompression melting, if extensional tectonics has downfaulted the sediments onto the basement during uplift. Identification of the ultimate causes of crustal anatexis in the High Himalays however, requires a more detailed knowledge of the thermal properties of the constituent lithologies in the metasedimentary pile than is presently available.

Field work was supported by a NERC (S.I.), NIGL, Keyworth (J.M.) and a Royal Society Overseas Field Research Grant (NH). We gratefully acknowledge the co-operation of the Department of Geology (Tribhuvan University), the Department of Mines and the British Council in Kathmandu. The manuscript benefited from thoughtful reviews by D. Waters and J. Clemens and from useful discussions with M. Bickle.

References

ALLEGRE, C.J. & MINSTER, J.F. 1978. Quantitative models of trace element behaviour in magmatic processes. *Earth and Planetary Science Letters*, **38**, 1–25.

BLUNDY, J.D. & WOOD, B.J. 1991. Crystal-chemical controls on the partitioning of Sr and Ba between plagioclase feldspar, silicate melts, and hydrothermal solutions. *Geochimica et Cosmochimica Acta*, **55**, 193–209.

BROUAND, M., BANZET, G. & BARBEY, P. 1990. Zircon behaviour during crustal anatexis. Evidence from the Tibetan Slab migmatites (Nepal). *Journal of Volcanology and Geothermal Research*, **44**, 143–61.

BURCHFIEL, B.C. & ROYDEN, L.H. 1985. North-south extension within the convergent Himalayan Region. *Geology*, **13**, 679–82.

CANN, J.R. 1970. Upward movement of granite magma. *Geological Magazine*, **107**, 335–40.

CLEMENS, J.D. 1984. Water contents of silicic to intermediate magmas. *Lithos*, **17**, 273–87.

—— & MAWER, C.K. 1992. Granitic magma transport by fracture propagation. *Tectonophysics*, **204**, 339–60.

—— & VIELZEUF, D. 1987. Constraints on melting and magma production in the crust. *Earth and Planetary Science Letters*, **86**, 287–306.

COLCHEN, M., LE FORT, P. & PECHER, A. 1986. *Annapurna–Manaslu–Ganesh Himal; Notice of the 1/200,000 geological map*. Editions CNRS.

COPELAND, P., PARRISH, R.R. & HARRISON, T.M. 1988. Identification of inherited radiogenic Pb in monazite and its implications for U-Pb systematics. *Nature*, **333**, 760–63.

——, HARRISON, T.M. & LE FORT, P. 1990. Age and cooling history of the Manaslu granite: implications for Himalayan tectonics. *Journal of Volcanology and Geothermal Research*, **44**, 33–50.

DEBON, F., LE FORT, P., SHEPPARD, S.M.F. & SONET, J. 1986. The four plutonic belts of the Trans-himalaya-Himalaya: a chemical, mineralogical, isotopic and chronological synthesis along a Tibet-Nepal section. *Journal of Petrology*, **27**, 219–50.

DENIEL, C., VIDAL, P., FERNANDEZ, A., LE FORT, P. & PEUCAT, J.J. 1987. Isotopic study of the Manaslu granite (Himalaya, Nepal): inferences of the age and source of Himalayan leucogranites. *Contributions to Mineralogy and Petrology*, **96**, 78–92.

FRANCE-LANORD, C. & LE FORT, P. 1988. Crustal melting and granite genesis during the Himalayan collision orogenesis. *Transactions of the Royal Society of Edinburgh (Earth Sciences)*, **79**, 183–95.

GARIEPY, C., ALLEGRE, C.J. & RONGHUA, XU 1985. The Pb-isotope geochemistry of granitoids from the Himalaya-Tibet collision zone: implications for crustal evolution. *Earth and Planetary Science Letters*, **74**, 220–34.

GILETTI, B.J. 1991. Rb and Sr diffusion in alkali feldspars, with implications for cooling histories of rocks. *Geochimica et Cosmochimica Acta*, **55**, 1331–43.

HALLIDAY, A.N., DAVIDSON, J.P., HILDRETH, W. & HOLDEN, P. 1991. Modelling the petrogenesis of high Rb/Sr silicic magmas. *Chemical Geology*, **92**, 107–14.

HARRIS, N.B.W., PEARCE, J.A. & TINDLE, A.G. 1986. Geochemical characteristics of collision-zone magmatism. *In*: COWARD, M.P. & RIES, A.C. (eds) *Collision Tectonics*. Geological Society, London, Special Publications, **19**, 67–81.

—— & INGER, S. 1992. Geochemical characteristics of pelite-derived granites. *Contributions to Mineralogy and Petrology*, **110**, 46–56.

——, GRAVESTOCK, P. & INGER, S. 1993. Ion-microprobe determinations of trace-element concentrations in garnets from anatectic assemblages. *Chemical Geology*, **100**, 41–49.

HENDERSON, P. 1982. *Inorganic geochemistry*. Pergamon, Oxford.

HODGES, K.V., HUBBARD, M.S. & SILVERBERG, D.S. 1988. Metamorphic constraints on the thermal evolution of the central Himalayan Orogen. *Philosophical Transactions of the Royal Society*, **A326**, 257–76.

HOLTZ, F. & JOHANNES, W. 1991. Genesis of peraluminous granites I. Experimental investigation of melt compositions at 3 and 5 kb and various H_2O activities. *Journal of Petrology*, **32**, 935–58.

HUPPERT, H.E. & SPARKS, R.S.J. 1988. The generation of granitic magmas by intrusion of basalt into continental crust. *Journal of Petrology*, **29**, 599–624.

INGER, S. & HARRIS, N.B.W. 1992. Tectonothermal evolution of the High Himalayan Crystalline Sequence, Langtang Valley, northern Nepal. *Journal of Metamorphic Geology*, **10**, 438–52.

—— & —— 1993. Geochemical constraints on leucogranite magmatism in the Langtang Valley, Nepal Himalaya. *Journal of Petrology* (in press).

JAUPART, C. & PROVOST, A. 1985. Heat focusssing, granite genesis and inverted metamorphic gradients in continental collision zones. *Earth and Planetary Science Letters*, **73**, 385–97.

JOHANNES, W. & HOLTZ, F. 1990. Formation and composition of H_2O undersaturated granitic melts. *In*: ASHWORTH, J.R. & BROWN, M. (eds) *High temperature metamorphism and crustal anatexis*. Unwin Hyman, 87–104.

LE BRETON, N. & THOMPSON, A.B. 1988. Vapour-absent (dehydration) melting of biotite in metapelites in the early stages of crustal anatexis. *Contributions to Mineralogy and Petrology*, **99**, 226–37.

LE FORT, P. 1981. Manaslu leucogranite: a collision signature of the Himalaya. A model for its genesis and emplacement. *Journal of Geophysical Research*, **86**, 10545–68.

——, CUNEY, M., DENIEL, C., FRANCE-LANORD, C., SHEPPARD, S.M.F., UPRETTI, B.N. & VIDAL, PH. 1987. Crustal generation of the Himalayan leucogranites. *Tectonophysics*, **134**, 39–57.

MARSH, B.D. 1988. Crystal capture, sorting and retention in convecting magma. *Geological Society of America Bulletin*, **100**, 1720–37.

NABELEK, P. & RUSS-NABELEK, C. 1991. Geochemical evidence for the petrogenesis of the Harney Peak leucogranite, Black Hills, USA. *Terra Abstracts*, **3**, 33.

NASH, W.P. & CRECRAFT, H.R. 1985. Partition coefficients for trace elements in silicic magmas. *Geochimica et Cosmochimica Acta*, **49**, 2309–22.

PANTINO-DOUCE, A.E. & JOHNSTON, A.D. 1991. Phase equilibria and melt productivity in the pelitic system: implications for the origin of peraluminous granitoids and aluminous granites. *Contributions to Mineralogy and Petrology*, **107**, 202–18.

POWELL, R. 1983. Fluids and melting under upper amphibolite facies conditions. *Journal of the Geological Society*, **140**, 629–33.

RUBIE, D.C. & BREARLEY, A.J. 1990. A model for rates of disequilibrium melting during metamorphism. *In*: ASHWORTH, J.R. & BROWN, M. (eds) *High temperature metamorphism and crustal anatexis*. Unwin Hyman, 57–86.

SAWYER, E.W. 1991. Disequilibrium melting and the rate of melt-residuum sparation during migmatization of mafic rocks from the Grenville Font, Quebec. *Journal of Petrology*, **32**, 701–38.

SCAILLET, B., FRANCE-LANORD, C. & LE FORT, P. 1990. Badrinath-Gangotri plutons (Garhwal, India): petrological and geochemical evidence for fractionation processes in a High Himalayan leucogranite. *Journal of Volcanology and Geothermal Research*, **44**, 163–88.

SCHARER, U., RONG-HUA, X. & ALLEGRE, C.J. 1986. U-(Th)-Pb systematics and ages of Himalayan leucogranites, South Tibet. *Earth and Planetary Science Letters*, **77**, 35–48.

SEARLE, M.P. & REX, J. 1989. Thermal model for the Zanskar Himalaya. *Journal of Metamorphic Geology*, **7**, 127–34.

TUTTLE, O.F. & BOWEN, N.L. 1958. *Origin of granite in the light of experimental studies in the system*

Ab–Or–Qtz–H_2O. Memoirs of the Geological Society of America, **74**.

VIDAL, PH., COCHERIC, A. & LE FORT, P. 1982. Geochemical investigations of the origin of the Manaslu leucogranite (Himalaya, Nepal). *Geochimica et Cosmochimca Acta,* **46**, 2279–2297.

WATERS, D.J. 1988. Partial melting and the formation of granulite facies assemblages in Namaqualand, South Africa. *Journal of Metamorphic Geology,* **6**, 387–404.

WATSON, E.B. & HARRISON, T.M. 1983. Zircon saturation revisited: temperature and composition effects in a variety of crustal magma types. *Earth and Planetary Science Letters,* **64**, 295–304.

WICKHAM, S. 1987. Crustal anatexis and granite petrogenesis during low pressure regional metamorphism; the Trois Seigneurs Massif, Pyrenees, France. *Journal of Petrology,* **28**, 127–69.

ZEITLER, P.K. & CHAMBERLAIN, C.P. 1991. Petrogenetic and tectonic significance of young leucogranites from the northwestern Himalaya, Pakistan. *Tectonics,* **10**, 729–41.

Cause and effect among thrust and normal faulting, anatectic melting and exhumation in the Himalaya

PHILIP ENGLAND[1] & PETER MOLNAR[2]

[1] *Department of Earth Sciences, Oxford University, Oxford OX1 3PR, UK*

[2] *Laboratoire de Géophysique Interne et Tectonophysique, Boite Postale 53X, Université Joseph Fourier, 38041 Grenoble, France,*

and

Institute for Crustal Studies, University of California, Santa Barbara, CA 93016, USA

Permanent address: Department of Earth, Atmospheric, and Planetary Sciences Massachusetts Institute of Technology, Cambridge, MA 02139, USA

Abstract: The transition from compression to extension in a regime of overall convergence in a narrow mountain belt requires a change with time in at least one of the following: horizontal normal stress applied to the belt by its surroundings; the distribution of shear traction on the base of the belt; the constitutive relation between stress and strain rate of material in the belt itself; the surface height of the belt. The following processes could have caused one or other of these changes in the Himalaya in Miocene time and thus could have triggered the onset of extension by normal faulting on planes parallel to thrust faults that were active at the same time. The shear stress on the thrust fault at the base of the Himalaya could have decreased, for instance because of melting on the fault plane. Alternatively, the creation of a new deeper thrust fault, and hence a new locus of convergence, could have initiated extensional faulting if the distribution of shear stress on the new thrust fault were unable to support the Himalaya above it. The Himalaya could have been weakened by an increase in pore fluid pressure, for instance owing to metamorphic dehydration reactions within the wedge. The Himalaya could have been elevated, for instance following the detachment of a tongue of mantle lithosphere from beneath it, or by the underthrusting of a lens of material beneath its base.

The apparent simultaneity of thrust and normal faulting on nearly parallel planes, of anatexis and of exhumation in the Himalaya in Miocene time can be explained if normal faulting was responsible for most of the exhumation and, through decompression melting, for anatexis. Existing data suggest that the most probable cause for Miocene normal faulting was either uplift of the Himalaya or the cessation of slip on the Main Central Thrust at the base of the orogenic wedge and the creation of a new, deeper thrust fault and a new base to the deforming Himalaya.

Normal faulting is recognized as a common process during mountain building, where the dominant style of deformation is nevertheless thrust faulting. During mountain building, the overall regional finite strain is one of horizontal compression, and normal faulting must contribute only minor regional strain. The mere occurrence of normal faulting, however, suggests that during those periods when, and at those places where, it occurs the deviatoric stress is not one of horizontal compression. Recognition of this apparent anomaly has led to discussions of the stress distributions that might have caused such normal faulting (e.g., Burchfiel & Royden 1985; Burchfiel *et al.* 1992; Dalmayrac & Molnar 1981; England & McKenzie 1982, 1983; Molnar & Tapponnier 1978; Molnar & Lyon-Caen 1988). Most of these studies addressed a static distribution of stress. We suggest that emphasis on a static stress field is misdirected and that the significance of such normal faulting may lie less in the stress distribution at one instant than in the *change with time* in the stress field, and hence in the processes that induce the requisite change in stress (e.g., Dahlen 1984; England & Houseman 1988, 1989).

The region including the Himalaya and the Tibetan Plateau provides evidence of normal faulting in two different environments. Normal faulting and east–west crustal extension characterize the active tectonics of the high plateau (e.g., Armijo *et al.* 1986; Mercier *et al.* 1987; Molnar & Chen 1983; Molnar & Tapponnier 1975, 1978; Molnar & Lyon-Caen 1989; Ni & York 1978). During the underthrusting of India beneath the Himalaya in Miocene time, but not apparently at present, slip on a major, northerly

From Treloar, P.J. & Searle, M.P. (eds), *Himalayan Tectonics*
Geological Society Special Publication, No. 74, pp. 401–411.

dipping normal fault system exhumed the rock in what is now the Greater Himalaya (e.g., Burchfiel & Royden 1985; Burchfiel *et al.* 1992; Burg *et al.* 1983; Herren 1987; Hodges *et al.* 1992; Searle *et al.* 1988; Valdiya & Goel 1983).

It seems likely that the shear tractions on the base of the Tibetan lithosphere are negligible in comparison with the deviatoric stresses within the lithosphere (see England & McKenzie 1982, 1983; Houseman & England 1986 for discussion). Thus, the occurrence of deformation within Tibet and similar high plateaux is determined by the balance between the stresses required to deform the lithosphere and the stresses generated by isostatically balanced density variations within the lithosphere (e.g., Molnar & Tapponnier 1978). On the other hand, the normal faulting in the Himalaya apparently took place synchronously with slip on the Main Central Thrust fault zone, which underlay the normal faults when they were active (Burchfiel *et al.* 1992; Hodges *et al.* 1992, and see below). Analysis of this setting of normal faulting must take into account shear tractions on the base of the deforming region in addition to the stresses required to support the laterally varying mass per unit area of the crustal column and the stresses required to deform it (e.g. Dahlen 1984; Platt 1986).

Neither style of normal faulting is limited to the area of the Himalaya and Tibet. Active normal faulting currently occurs within the high plateau of the Andes (e.g., Dalmayrac 1978; Dalmayrac & Molnar 1981; Mercier *et al.* 1992; Sébrier *et al.* 1985, 1988). This normal faulting, like the active normal faulting in Tibet, may be explained by the balance between the stress applied to the margins of the range and that due to the difference in crustal thickness between the Andes and the neighbouring lowlands (e.g., Artyushkov 1973; Dalmayrac & Molnar 1981; Frank 1972). Normal faulting and crustal extension of the Basin and Range province, which may have begun before the termination of Laramide crustal shortening (e.g., Wernicke *et al.* 1987), can also be treated as analogous with that in Tibet (e.g., Coney & Harms 1984; Molnar & Chen 1983; Molnar & Lyon-Caen 1988; Sonder *et al.* 1987). Moreover, the latest stage of deformation in the Late Palaeozoic, Hercynian belt of Europe can be interpreted as normal faulting in an environment like that of Tibet or the Basin and Range Province (Lorenz & Nicholls 1976; Ménard & Molnar 1988).

Active normal faulting occurs along the western parts of the Apennines (e.g., Anderson & Jackson 1987), while thrust faulting continues along the eastern margin (e.g., Royden *et al.* 1987). Similarly, normal faulting along the west sides of the high mountains of Muztagh-Ata and Kongur, in the easternmost Pamir, seems to occur simultaneously with active overthrusting of the margin of the range onto the Tarim Basin (Brunel *et al.* 1992; Tapponnier & Brunel 1992). These examples of normal faults approximately parallel to and active synchronously with thrust faults may be taken as analogous with the Miocene faulting in the Himalaya.

Thus, normal faulting has been recognized as a ubiquitous aspect of mountain building (e.g., Dewey 1988). Its common presence is, however, easily misunderstood as indicating a common underlying process in the evolution of mountain building. We have purposely distinguished two settings of normal faulting in mountain belts: those occurring within high plateaux, where topographic gradients are small, and those within narrow belts, where strikes of normal and thrust fault planes are nearly parallel and dips may differ by only 0° to 20°. The reason for making this distinction is that the physical processes responsible for the normal faulting need not be, and probably are not, the same for each. Moreover, we do not mean to suggest that these two settings of normal faulting are all-inclusive; other settings and other processes might be important in other belts.

Virtually all normal faulting and crustal extension uses, and therefore could be said to be 'driven by', gravitational potential energy stored within the lithosphere. Crustal thinning lowers mean elevations and therefore extracts potential energy from the elevated terrain and its underlying compensating crust and/or upper mantle. All of the examples of normal faulting cited above could be called 'orogenic collapse' – a term that, by itself, means little. The extension of mountain belts is not unexpected in view of the likelihood that they possess greater gravitational potential energy per unit area than their surroundings (e.g. Artyushkov 1973; Frank 1972; Molnar & Tapponnier 1978; Molnar & Lyon-Caen 1988), and it seems otiose to coin an extra term for the process, or for some manifestations of it. Even for normal faulting occurring within interarc settings where the mean surface elevation lies below sea level, as within the Aegean Sea, the energy source for extension is gravitational potential energy of the (Aegean) lithosphere. The phenomenon named 'slab rollback' is neither an energy source for the extension nor a dynamic process that triggers normal faulting. It is a purely kinematic phenomenon that has its origins in the forces acting on the downgoing slab of lithosphere, including those applied by the overriding plate. Equally, the extension within the overriding plate is powered by potential energy differences between it and its surroundings (e.g. Le Pichon 1983), and the movement of a trench away from a continental interior may as easily be the consequence of extension of the continent as its cause.

The various examples of normal faulting are not distinguished by the source of energy that drives them, and probably not by the basic aspects of their stress distributions; the emphasis of this paper, therefore, lies not upon the extensional faulting itself, but upon the processes preceding the faulting, by which the potential energy stored in the thickened crust becomes available to power normal faulting and crustal extension.

To investigate the processes that initiate extension in plateau-like regions England & McKenzie (1982, 1983) and England & Houseman (1988, 1989) examined the state of stress within a thin viscous sheet. The state of stress within both the viscous sheet and the deforming continental interior can be seen as a balance between stresses of two origins. One is the compressional boundary condition on the margins of the thickened region (as, for instance, is applied by the convergence of India with Asia). The other is the stress difference within the medium owing to gravity acting on lateral variations in density. The common theme of these studies is that a change in this balance is required before extension within the plateau can occur. Stated simply the issue is this: a compressional boundary condition, such as is applied by continental convergence, generates thickened lithosphere in front of the boundary; and horizontal density contrasts within the thickened lithosphere can lead to a stress field that counteracts the compressional stresses applied at the boundary. In the steady state, however, the magnitude of the deviatoric stresses caused by density contrasts within the lithosphere cannot exceed the magnitude of the compressive stresses on the boundary (and so lead to extension), because they are generated solely by those same boundary stresses (England & Houseman 1989).

There are doubtless a large number of complex scenarios that could be imagined that would effect a change in the balance of stresses within a plateau. From the studies cited, however, there appear to be three simple conditions under which a plateau (or part of it) can enter a regime of extension. The most obvious is that the compressional boundary condition is diminished or removed, by the slowing or cessation of convergence. A second possibility is that regions of thickened lithosphere can be carried, by the horizontal flow generated by the boundary condition, to a region distant from the boundary in which compressional stresses are lower (e.g. England & McKenzie 1983, Figs 7–9). England & Houseman (1988, 1989) argued that neither of these conditions can apply to the active extension in the Tibetan plateau. They suggested that convective removal of the lower part of the lithosphere of the plateau occurred, and that this event should have caused a rapid increase of surface height and a change to horizontal extensional deviatoric stress in the plateau.

Change in the components of stress in narrow orogenic belts

For the remainder of this paper we concentrate on the possible causes of extension by normal faulting on planes nearly parallel to thrust faults within narrow mountain belts in which elevation varies across the range. We refer to the processes just discussed, affecting the extension of broad plateaux, only in so far as they bear directly upon this problem. In discussing the general issue of extension during convergence with narrow mountain belts we have specifically in mind the early Miocene, normal faulting and north–south extension within the Himalaya and southern Tibet (Burchfiel *et al.* 1992; Burg *et al.* 1983). The shape of this belt, and of others similar, makes it reasonable to neglect variations of stress along the strike of the belt, compared with variations perpendicular to strike. Therefore we consider the state of stress in only the vertical and one horizontal dimension. For such two dimensional configurations, the stress at a point is specified completely by three independent quantitites: by the magnitudes of the horizontal and vertical normal stresses and of the shear stress on the horizontal (or vertical) plane, or by the magnitudes of the two principal stresses and the orientation (plunge) of one of them. Our goal here is to suggest processes that may cause changes in one of these three components so as to induce normal faulting.

The three independent components of stress are linked to one another via the equations of continuity and by a constitutive relation relating stress to strain or to strain rate and to temperature, pressure, and other intrinsic variables. Thus a change in the stress distribution can be effected by a change in any of these three components of stress or by a change in the constitutive relation. Let us consider how each of these changes might lead to the initiation of normal faulting in what otherwise, and for the most part, is a zone of crustal shortening.

We restrict discussion to aspects of the basic

physical processes. Quantitative analyses appropriate to various constitutive relations can be found in Dahlen (1984), Dahlen *et al.* (1984), Davis *et al.* (1983) and Emerman & Turcotte (1983). Dahlen (1984), Dahlen *et al.* (1984) and Davis *et al.* (1983), investigated in detail the balance of forces within a wedge of material that obeys Coulomb's law and that is bounded by a sloping upper surface and by a rigid lower surface of fixed dip. Emerman & Turcotte (1983) reported similar conclusions for wedges of Newtonian and non-Newtonian viscous fluids.

Obviously one possible change in stress that would lead to normal faulting within a deforming region is a reduction in the horizontal normal stress. Because changes in plate motions are rare and typically not large, they probably do not commonly cause stress distributions to change sufficiently to lead to normal faulting.

As normal faulting implies that the vertical (normal) compressive stress exceeds the horizontal compressive stress, an alternative to lowering the horizontal compressive stress is to increase the vertical stress. Vertical displacement of the deforming region relative to the lithospheric columns bordering it causes such a change. Because the force of gravity contributes to the force balance, an appropriate reference frame for discussing these vertical motions is an equipotential surface of gravity (e.g. sea level). An uplift of the surface of the region with respect to its surroundings might cause normal faulting, just as uplift of a plateau can cause normal faulting within it. Such a process increases the vertical (compressive) normal stress at any given depth below sea level within the region. If no change in surface height occurs in the surroundings of the deforming region, then the state of stress within the deforming region moves towards greater horizontal extensional *deviatoric* stress; if the surface uplift is large enough, extension and normal faulting will occur. Conversely, if the surroundings of the deforming region were to rise rapidly with respect to it, the deforming region would be thrown into a state of enhanced horizontal compression.

There are two simple processes that can cause an uplift in the surface of a mountain belt, or part of it. As noted above, convective processes in the mantle can rapidly remove cold mantle lithosphere and replace it with hot asthenosphere. This process leads, via isostasy, to an uplift of the overlying surface. (In principle, a reduction in density can also be affected by conductive heating of a static lithospheric column, but this process is likely to be too slow to influence the tectonic style of a mountain belt.) The second class of mechanism involves the transfer of material from one part of the deforming belt to another, and its 'underplating' there (Platt 1986). Generalizing about either of these processes is difficult, because the details of each depend too strongly upon the (unknown) distributions of material properties in the deforming belt or its underlying upper mantle. The two processes do, however, share one simple aspect: each causes an increase in mean surface height of part of the belt, which shifts the state of stress there towards decreasing horizontal compression and increasing horizontal extension.

A third process for instigating normal faulting within a deforming wedge is to change the shear stress on the base of the wedge. The mean taper of the wedge is maintained at an angle with respect to the horizontal that depends, among other parameters, on the magnitude of the shear stress at the base of the wedge. A higher basal shear stress supports a steeper wedge. Depending on the taper of the wedge, a lowering of the basal shear stress can cause the wedge to become unstable and to collapse by horizontal extension into a more gently sloping wedge (see, for example, Dahlen 1984, Fig. 13).

Finally, a change in the constitutive relation, or in the rheological properties, can also initiate normal faulting. For instance, an increase in pore fluid pressure within the wedge would lead to a reduction in its strength and, if this reduction is sufficient, to a decrease in its taper (e.g., Dahlen 1984, fig. 2; see also Emerman & Turcotte 1983 who show that the thickness of a fluid wedge is proportional to the one-third power of its viscosity). The result of a reduction in strength of a wedge can, though need not always, be its collapse, via horizontal extension, to one with a gentler taper.

The existence of a slope is a prerequisite for extension of the wedge as a result of its weakening. Consider instead a flat, high plateau along whose margins thrust faulting and crustal shortening occur. This thrust faulting builds the margins of the plateau, so that it grows in areal extent while its interior remains at a constant height. In the interior the horizontal compressional stresses are resisted by the stresses required to deform the plateau and by gravity acting on the lateral contrasts in density associated with its thickened crust. If the strength of the plateau were to decrease, but the compressional stress across it were to remain the same, the plateau would shorten horizontally until a new balance was reached, with a yet higher plateau and a thicker crust. Normal faulting within the plateau should not occur without some other, arbitrarily imposed, condition (England & Houseman 1988).

Possible relations between Miocene normal and thrust faulting in the Himalaya

Our purpose is to use the Miocene thrust and normal faulting to illustrate how various changes in different components of stress, or in the constitutive law, can induce normal faulting in an otherwise convergent setting. We shall use five facts that bear on the thrust tectonics, normal faulting, magmatism, and exhumation of the Greater Himalaya to try to distinguish among several possible causes for the initiation of this normal faulting and extension. Future observations may require different conclusions. Equally, other narrow mountain belts may exhibit different relations among observations that would lead to different conclusions about their evolution.

(1) The Main Central Thrust, the major active thrust fault within the Himalaya, was active in the Everest region at about 21 Ma (Hubbard & Harrison 1989). This fault zone may have been active for a long time before 21 Ma, but we infer that it was not active for long afterwards. The $^{40}Ar/^{39}Ar$ cooling age of hornblende crystals that grew during deformation at geothermometrically determined temperatures of 500–550° C indicates that at 21 Ma shear in a thrust sense persisted on the Main Central Thrust in the Everest region. Hubbard & Harrison's (1989) samples were taken near the base of the thick zone of highly sheared rock in which metamorphic isograds are inverted (Hubbard 1989). A temperature of 500° C near the Main Central Thrust could not have been maintained at a late stage of slip without appreciable dissipative heating in the thrust zone (England *et al.* 1992), and temperatures near a shear zone with such heating rise continuously during slip. Because the closure temperature for the diffusion of argon in hornblende is close to 500° C, we presume that this age marks a period shortly before slip on the Main Central Thrust in this area stopped, for had slip continued later, these samples should have warmed, not cooled. Because the Main Central Thrust zone was a major tectonic boundary along most of the Himalayan chain, it seems likely that while slip was occurring in the Everest region it was also occurring elsewhere along the chain.

(2) Schärer (1984) reported ages of 22–24 Ma for the Makalu granite, but noted the age estimates were problematic because of ^{230}Th disequilibrium. Schärer *et al.* (1986) obtained ages of 16–9 Ma for Himalayan leucogranites but these, too, required corrections for disequilibrium. The most reliable date for the anatectic granites within the Greater Himalaya, above the Main Central Thrust, seems to be the Rb/Sr isochron of 18.1 ± 0.5 Ma for crystallization of the Manaslu granite (Deniel *et al.* 1987). Analyses of the petrology and isotopic composition of this granite imply that melting occurred in the upper block of the Main Central Thrust (Deniel *et al.* 1987; France-Lanord & Le Fort 1988; France-Lanord *et al.* 1988; Vidal *et al.* 1982, 1984).

(3) Very rapid cooling of rock in the Trans-Himalaya near Lhasa at about 18 Ma implies rapid exhumation of these rocks (Copeland *et al.* 1987; Richter *et al.* 1991). Presumably such exhumation occurred either by rapid erosion, as Copeland *et al.* (1987) and Harrison *et al.* (1992) have argued, or by rapid normal faulting.

(4) Major normal faulting in the area between the present high peaks of the Himalaya and the Trans-Himalaya occurred in Neogene time (Burg *et al.* 1983; Herren 1987). The tightest limits on the age of this faulting bracket the major phase between about 21 Ma and 18 Ma (Hodges *et al.* 1992), although there may have been later, and earlier, phases of movement also (Burchfiel *et al.* 1992).

(5) Finally, although perhaps not obviously important, the ratio of $^{87}Sr/^{86}Sr$ deposited in marine limestone increased markedly at about 18 Ma (Hodell *et al.* 1989; Richter *et al.* 1992). As this ratio seems to be controlled largely by the contributions of rivers draining the Himalaya, the Ganga and Brahmaputra (Palmer & Edmond 1989), this increase at 18 Ma is almost surely related to the other processes occurring there at the same time (Edmond 1992; Richter *et al.* 1991). The anatectic granites of the Himalaya have among the highest ratios of $^{87}Sr/^{86}Sr$ in the world (Vidal 1978; Vidal *et al.* 1982, 1984). We, like Edmond (1992), suspect that whatever triggered the anatectic magmatism would have also caused the change in Sr ratios, especially if some of that magma reached the surface by volcanism.

Decompression melting during rapid exhumation as a cause for anatexis

In the following sections, we discuss several processes that could have triggered the Miocene normal faulting in the Himalaya. Whatever its cause, however, the normal faulting and crustal thinning would have rapidly reduced the overburden on the Greater Himalaya, such that decompression melting might have occurred there (Harris *et al.* this volume; P. Kelemen, pers. comm. 1992). If such granites were associated with volcanism, the high $^{87}Sr/^{86}Sr$ ratios of the lavas might have caused the increase in the

$^{87}Sr/^{86}Sr$ ratio of marine carbonates at this time. If however, the magmatic ages for the granites are appreciably older than 20 Ma (as has been suggested by Chen & De Paolo 1992, Deniel *et al.* 1987, and Schärer 1984), then the apparent simultaneity of anatexis and normal faulting may be explained if the granites were emplaced into country rock that was almost as hot as they were (England *et al.* 1992) and so remained close to their emplacement temperatures until they were rapidly exhumed and cooled by the normal faulting.

Reduction in basal shear stress on the Himalaya as the cause of Miocene normal faulting

Suppose that melting along the Main Central Thrust preceded the normal faulting and exhumation of the Trans-Himalayan granite. Such melting near the fault could have led to a major drop in shear stress on the fault (Burchfiel *et al.* 1992). The reduced shear stress at the base of the Himalaya would have been able to support a gentler slope than existed before the reduction in strength and, hence, the potential for normal faulting within the Himalaya would have increased. Moreover, the lower basal shear stress would have led to a lower horizontal compressive stress being applied by the Himalaya to southern Tibet (see, for example, Platt 1986, fig. 4 and equation 15) and the potential for normal faulting would have increased there, also. The normal faulting would have exhumed the Trans-Himalaya.

Change in the configuration of the deforming region as the cause of Miocene normal faulting

Alternatively, suppose that at about 20 Ma, a new thrust fault (presumably though not necessarily the Main Boundary Fault) became active, and slip on the Main Central Thrust ceased. Perhaps continuing convergence had elevated the terrain overlying the Main Central Thrust sufficiently that the increased stress on this fault needed to elevate the Greater Himalaya yet higher would have been greater than that necessary to create a new fault. Whatever its cause, the creation of a new fault and the absorption upon it of the relative motion previously taking place across the Main Central Thrust would have altered both the shape of the deforming region and the stress distribution on its base. Almost certainly these changes would have resulted in a change in the internal stress state of the wedge itself. Without a knowledge of the details of these changes, which is unobtainable, one cannot with certainty say that they would have led to extension within the Himalaya.

Let us imagine, however, a geometrically simple possibility, in which the Main Central Thrust was planar when active, and its successor was also planar, but dipped more steeply. If the shear stresses on the old and new thrust faults were approximately the same, the principal change to the dynamics of the deforming mountain belt would be that the dip of its lower surface increased without a change in the slope of its upper surface. Dahlen (1984, for example in figs 9, 11 and 14) shows that the influence of an increase in basal dip of a critical Coulomb wedge, without a change to the surface slope, moves the state of deviatoric stress in the wedge from horizontal compression towards horizontal extension. Whether net extension is a likely outcome of such a change in a real orogenic belt depends upon its constitutive relation. For Coulomb wedges with low pore pressures the change in basal dip necessary to induce extension are so large as to be implausible. On the other hand, when the pore pressures are high (e.g., Dahlen 1984, fig. 14) transition to a state of extensional deviatoric stress may occur with relatively modest changes in basal dip.

Change in the constitutive law in the Himalaya as the cause of Miocene normal faulting

Suppose that within the stack of thrust sheets, water was released by metamorphic dehydration reactions (P. Le Fort & S. Guillot, pers. comm., 1992). The increase in pore pressure in the stack of thrust sheets would have reduced its strength (e.g. Etheridge *et al.* 1984). A reduction in strength might have caused the Himalayan wedge to collapse to a gentler taper. Such a reduction in taper would occur by extension and normal faulting. Equally, a reduction in strength within the Himalaya caused by changes in other deformation mechanisms would lead to the same result. For example, an increase in temperature within material deforming by one of the mechanisms of plasticity would lead to a reduction in its strength.

Again, because the Himalaya provide the confining southern boundary to the neighbouring Tibetan plateau, a reduction in the strength of the Himalaya would lead to a decrease in the horizontal normal stress applied to the plateau.

Normal faulting could result there too, and the Trans-Himalaya would be exhumed. Moreover, because of decompression, melting would become possible in the Greater Himalaya. Indeed, the flux of water into the wedge might promote melting there, even if the conditions for decompression melting were not achieved (Le Fort 1981).

Changes in surface height as the cause of Miocene normal faulting

A widely cited explanation for the rapid cooling in southern Tibet is that the plateau rose quickly at 18 Ma. Some imagine that this led to accelerated erosion and exhumation of the Trans-Himalaya (Copeland *et al.* 1987; Harrison *et al.* 1992). Although we cannot rule out rapid erosion, it is not obvious how such erosion could either account for, or be the result of, the other nearly simultaneous phenomena discussed above. Furthermore, if the plateau as a whole rose with respect to sea level at this time, this uplift should have increased horizontal compression within the surroundings (including the Greater Himalaya). Thus, unless the Greater Himalaya rose further still than the plateau, extension there ought not to have taken place at that time.

Alternatively, rapid uplift with respect to sea level of the surface of the Greater Himalaya and southern Tibet might have increased potential energy per unit area of this region and then this potential energy was released by normal faulting. Such normal faulting, which is mapped in the Himalaya between the high peaks and the Indus–Tsangpo suture, could have spanned the area as far north as the Trans-Himalaya. If so, it would have given rise to the cooling ages similar to those defining the upper bound on the normal faulting farther south. Again, the pressure drop caused by normal faulting may have been the cause of decompression melting in the Greater Himalaya.

Lyon-Caen & Molnar (1983, 1985) showed that the flexure of the Indian lithosphere requires that there be a load acting on it in addition to the weight of the Himalaya. They suggested that this load could be imposed by a tongue of mantle lithosphere attached to the northern end of the Indian plate, beneath and perhaps north of, the Himalaya. This tongue could represent thickening of the lower lithosphere that is offset northwards from thickening of the crust, which is occurring along the Main Boundary Fault. The form of this tongue is not necessarily that of a rigid slab of lithosphere, but could be a more general down-welling of the mantle such as is envisaged by Molnar (1988, fig. 22). Removal of such a tongue of cold, heavy material might have caused the surface uplift of the overlying region.

Summary

We have offered several possible triggers by which a change in one process or property in Himalayan tectonics could have caused changes in several others. Clearly there are some common themes. For instance, in all cases, the abrupt change at 18 Ma in the $^{87}Sr/^{86}Sr$ ratio recorded by marine carbonates is likely to have been related to the exposure and weathering of anatectic magmas, either by exhumation or by eruption (Edmond 1992). In essentially all cases, the rapid cooling in the Trans-Himalaya would be the result of normal faulting, not of erosion.

Some of the proposed triggers appear, from the available observations, to be less likely than others. Thermal processes govern the timing of two proposed mechanisms: weakening of the Main Central Thrust by melting (e.g., Burchfiel *et al.* 1992) and weakening of the material above it by prograde metamorphism. The characteristic time for thermal processes in thickened continental crust to evolve is typically some tens of millions of years. Yet the present sparse observations suggest that normal faulting developed at nearly the same time along the length of the Himalaya. Therefore, if this suggestion is correct and if the sluggish thermal processes are responsible for triggering extension at widely separated regions, they require a remarkable coincidence. Clearly, however, mechanical processes, such as the propagation of a fault laterally along strike, can occur nearly simultaneously along mountain belts. Moreover, the nearly simultaneous development of a major new thrust system can be the trigger for nearly simultaneous normal faulting and (through decompression) anatexis.

Our inference (above) that slip on the Main Central Thrust ceased at about the same time that normal faulting occurred seems difficult to reconcile with the hypothesis that the fault was weakened at that time. Why would slip have ceased on the Main Central Thrust shortly after it became weakened? This inference, if correct, lends support to the suggestion that the normal faulting was triggered by the initiation of slip on another major fault, presumably though not necessarily the Main Boundary Thrust, which led both to the cessation of slip on the Main Central Thrust and to a change in the stress balance within the Himalaya.

The final possibility that we considered above is that the surface of southernmost Tibet and the Himalaya may have risen following the detachment and sinking of a tongue of dense mantle lithosphere that had been previously attached to the northern end of the Indian plate. If this surface uplift was large enough to throw both the Greater Himalaya and southern Tibet into extension, then the regions both to the south, where the Main Boundary Thrust now lies, and to the north in the Transhimalaya or further north should have been put under greater horizontal compression. Uplift of the region containing the Main Central Thrust might have been great enough to have changed the state of stress there from one of horizontal compression to one of horizontal extension. Thus this mechanism could have led simultaneously to a cessation of motion across the Main Central Thrust and to the absorption of convergence on other structures, either north or south of the region.

Neither the abrupt transfer of convergence from the Main Central Thrust to another major structure, nor the sinking of a dense tongue of lithosphere necessarily occurred simultaneously along the whole Himalayan chain. Once the Main Central Thrust became inactive, however, it did not act again as a major tectonic boundary, though it may have experienced a small thermal perturbation in latest Miocene time (Copeland *et al.* 1991). The transfer of convergence from the Main Central Thrust, therefore, appears to offer the possibility of generating only a single phase of extension at any one locality. In contrast, the descent of lower lithospheric material might well exhibit the episodic nature of convective processes, and so generate repeated episodes of extension. The existence of multiple phases of extension (Burchfiel *et al.* 1992) cannot, at least in the absence of precise dating, distinguish between these two proposed triggers.

On the basis of present evidence we conclude that two mechanisms discussed in the previous paragraph (cessation of slip on the Main Central Thrust by initiation of a new thrust system or rapid uplift of the surface by detachment of the mantle of the Indian lithosphere from beneath the Greater Himalaya and southern Tibet) carry a greater chance of being the trigger for the Miocene extension in the Himalaya than do the others discussed in this paper. To distinguish further between these mechanisms, or to resuscitate suggestions that at present seem less plausible, it is necessary to obtain more precise dates on several of the key processes. Among these processes are the timing of slip on the Main Central Thrust, the dates of initiation of slip on the Main Boundary Fault and on the main normal faults in the northern Himalaya, and the timing of uplift of the surface of the plateau.

Regardless of the present shortcomings in the data, the occurrence of normal faulting during a period when crustal shortening presumably occurred in the Himalaya need not seem puzzling. Instead of a dearth of plausible explanations, we are faced with an excess.

One of us (P.M.) worked at the Laboratoire de Géophysique Interne et Tectonophysique, Université Joseph Fourier when the first draft of this manuscript was written, and at the Institute for Crustal Studies, University of California, Santa Barbara, when the submitted version was finished. N. Kusznir gave a helpful review. We thank J. Platt for thorough and insightful comments that corrected errors in apprehension and presentation.

References

ANDERSON, H. & JACKSON, J. 1987. Active tectonics of the Adriatic region. *Geophysical Journal of the Royal Astronomical Society*, **91**, 937–983.

ARMIJO, R., TAPPONNIER, P., MERCIER, J.L. & HAN, T. 1986. Quaternary extension in southern Tibet: Field observations and tectonic implications. *Journal of Geophysical Research*, **91**, 13803–13872.

ARTYUSHKOV, E.V. 1973. Stresses in the lithosphere caused by crustal thickness inhomogeneities, *Journal of Geophysical Research*, **78**, 7675–7708.

BRUNEL, M., TAPPONNIER, P., BOURJOT, L., ARNAUD, N., SCHÄRER, U., PAN, Y. & WANG, Y. 1992. Tectonics in the eastern Pamirs. *Abstracts, International Symposium on the Karakorum and Kunlun Mountains, June 5–9, 1992, Kashi (sic), China*, 29.

BURCHFIEL, B.C. & ROYDEN, L.H. 1985. North-south extension within the convergent Himalayan region. *Geology*, **13**, 679–682.

—— CHEN, ZHILIANG, HODGES, K.V., LIU YUPING, ROYDEN, L.H., DENG, CHANGRONG, & XU JIENE 1992. *The South Tibetan detachment system, Himalayan orogen: Extension contemporaneous with and parallel to shortening in a collisional mountain belt.* Geological Society of America Special Papers, **269**.

BURG, J.-P., PROUST, F., TAPPONNIER, P. & CHEN, G.M. 1983. Deformation phases and tectonic evolution of the Lhasa block (southern Tibet, China). *Eclogae Geologicae Helvetiae*, **76**, 643–665.

CHEN, C-H. & DEPAOLO, D.J. 1992. Rb-Sr Microchrons in the Manaslu Granite: Implications for Himalayan Thermochronology. *EOS Transactions of the American Geophysical Union*, **73**, 545.

CONEY, P.J. & HARMS, T.A. 1984. Cordilleran metamorphic core complexes: Cenozoic relics of Mesozoic compression. *Geology*, **12**, 550–554.

COPELAND, P., HARRISON, T.M., KIDD, W.S.F., XU RONGHAU & ZHANG YUQUAN 1987. Rapid early

Miocene acceleration of uplift in the Gangdese belt, Xizang (southern Tibet), and its bearing on accommodation mechanisms of the India–Asia collision. *Earth and Planetary Science Letters*, **86**, 240–252.

——, ——, HODGES, K.V., MAUREJOL, P., LE FORT, P. & PÊCHER, A. 1991. An early Pliocene thermal disturbance of the Main Central Thrust, Nepal: Implications for Himalayan Tectonics. *Journal of Geophysical Research*, **96**, 8475–8500.

DAHLEN, F.A. 1984. Noncohesive Critical Coulomb Wedges: an Exact Solution. *Journal of Geophysical Research*, **89**, 10125–10133.

——, SUPPE, J. & DAVIES, D. 1984. Mechanics of fold-and-thrust belts and accretionary wedges: Cohesive Coulomb theory. *Journal of Geophysical Research*, **89**, 10087–10101.

DALMAYRAC, B. 1978. Geologie des Andes peruviennes: geologie de la Cordillere Orientale et de la region de Huanuco: sa place dans un traversale des Andes du Perou centrale (9° S a 10° 30 S). *Trav. Doc. ORSTOM*, **93**.

—— & MOLNAR, P. 1981. Parallel normal and thrust faulting in Peru, and constraints on the state of stress. *Earth and Planetary Science Letters*, **55**, 473–481.

DAVIS, D., SUPPE, J. & DAHLEN, F.A. 1983. Mechanics of fold-and-thrust belts and accretionary wedges. *Journal of Geophysical Research*, **88**, 1153–1172.

DENIEL, C., VIDAL, P., FERNANDEZ, A., LE FORT, P. & PEUCAT, J.-J. 1987. Isotopic study of the Manaslu granite (Himalaya, Nepal): inferences on the age and source of Himalayan leucogranites. *Contributions to Mineralogy and Petrology*, **96**, 78–92.

DEWEY, J.F. 1988. Extensional Collapse of Orogens. *Tectonics*, **7**, 1123–1139.

EDMOND, J.M. 1992. Himalayan tectonics, weathering processes, and the strontium isotope record in marine limestones. *Science*, **258**, 1594–1597.

EMERMAN, S.H. & TURCOTTE, D.L. 1983. A fluid model for the shape of accretionary wedges, *Earth and Planetary Science Letters*, **63**, 379–384.

ENGLAND, P.C. & HOUSEMAN, G.A. 1988. The mechanics of the Tibetan Plateau. *Philosophical Transactions of the Royal Society London*, **A326**, 301–319.

—— & —— 1989. Extension during continental convergence with application to the Tibetan plateau. *Journal of Geophysical Research*, **94**, 17561–17579.

—— & MCKENZIE, D.P. 1982. A thin viscous sheet model for continental deformation. *Geophysical Journal of the Royal Astronomical Society*, **70**, 295–321.

—— & MCKENZIE, D.P. 1983. Correction to: A thin viscous sheet model for continental deformation. *Geophysical Journal of the Royal Astronomical Society*, **73**, 523–532.

——, LE FORT, P., MOLNAR, P. & PÊCHER, A. 1992. Heat sources for Tertiary metamorphism and anatexis in the Annapurna–Manaslu region, central Nepal. *Journal of Geophysical Research*, **97**, 2107–2128.

ETHERIDGE, M.A., WALL, V.J., COX, S.F. & VERNON, R.H. 1984. High fluid pressures during regional metamorphism: implications for mass transfer and deformation mechanisms. *Journal of Geophysical Research*, **89**, 4344–4458.

FRANCE-LANORD, C. & LE FORT, P. 1988. Crustal melting and granite genesis during the Himalayan collision orogenesis. *Transactions of the Royal Society Edinburgh (Earth Sciences)*, **79**, 183–195.

——, SHEPPARD, S.M.F. & LE FORT, P. 1988. Hydrogen and Oxygen isotope variation in the High Himalaya peraluminous Manaslu leucogranite: evidence for heterogeneous source. *Geochimica et Cosmochimica Acta*, **52**, 513–526.

FRANK, F.C. 1972. Plate tectonics, the analogy with glacier flow and isostasy. *In*: HEARD, H.C., BORG, I.Y., CARTER, N.I. & RALEIGH, C.B. (eds) *Flow and fracture of rocks*. American Geophysical Union, Geophysical Monograph Series, **16**, 285–292.

HARRIS, N.B.W., INGER, S. & MASSEY, J.A. 1993. The role of fluids in the formation of High Himalayan leucogranites. *This Volume*.

HARRISON, T.M., COPELAND, P., KIDD, W.S.F & PAN YIN. 1992. Raising Tibet. *Science*, **255**, 1663–1670.

HERREN, E. 1987. The Zanskar shear zone: Northeast-southwest extension within the Higher Himalayas (Ladakh, India). *Geology*, **15**, 409–413.

HODELL, D.A., MUELLER, P.A., MCKENZIE, J.A. & MEAD, G.A. 1989. Strontium isotopes stratigraphy and geochemistry of the late Neogene ocean. *Earth and Planetary Science Letters*, **92**, 165–189.

HODGES, K.V., PARISH, R.R., HOUSH, T.B., LUX, D.R., BURCHFIEL, B.C., ROYDEN, L.H. & CHEN, Z. 1992. Simultaneous Miocene extension and shortening in the Himalayan orogen. *Science*, **258**, 1466–1470.

HOUSEMAN, G.A. & ENGLAND, P.C. 1986. Finite strain calculation of continual deformation 1. Methods and general results for convergent zones. *Journal of Geophysical Research*, **91**, 3651–3663.

HUBBARD, M.S. 1989. Thermo-barometric constraints on the thermal history of the Main Central Thrust zone and the Tibetan slab, Eastern Nepal, Himalaya. *Journal of Metamorphic Geology*, **7**, 19–30.

—— & HARRISON, T.M. 1988. $^{40}Ar/^{39}Ar$ constraints on deformation and metamorphism in the MCT zone and Tibetan slab, eastern Himalaya. *Tectonics*, **8**, 865–880.

LE FORT, P. 1981. Manaslu leucogranite: A collision signature of the Himalaya. A model for its genesis and emplacement. *Journal of Geophysical Research*, **1986**, 10545–10568.

LE PICHON, X. 1983. Land-locked Ocean Basins and Continental Collision: the Eastern Mediterranean as a Case Example. *In*: HSU, K.J. (ed.) *Mountain Building Processes*. Academic Press London, 201–213.

LORENZ, V. & NICHOLLS, I.A. 1976. The Permocarboniferous Basin and Range Province of Europe. An application of plate tectonics. *In*: FALKE, H. (ed.) *The Continental Permian in Central, West and South Europe*. D. Reidel, Dordrecht-Holland, 313–342.

LYON-CAEN, H. & MOLNAR, P. 1983. Constraints on the structure of the Himalaya from an analysis of gravity anomalies and a flexural model of the lithosphere. *Journal of Geophysical Research,* **88**, 8171–8192.

—— & —— 1985. Gravity anomalies, flexure of the Indian plate, and the structure, support and evolution of the Himalaya and Ganges Basin. *Tectonics,* **4**, 513–538.

MÉNARD, G. & MOLNAR, P. 1988. Collapse of a Hercynian Tibetan Plateau into a late Palaeozoic European basin and Range Province. *Nature,* **334**, 235–237.

MERCIER, J.-L., ARMIJO, P., TAPPONNIER, P., CAREY-GAILHARDIS, E. & HAN TONGLIN 1987. Change from Teritary compression to Quarternary extension in southern Tibet during the India–Asia collision. *Tectonics,* **6**, 275–304.

——, SÉBRIER, M., LAVENU, A., CABRERA, J., BELLIER, O., DUMONT, J.-F. & MACHAR, J. 1992. Changes in the tectonic regime above a subduction zone of Andean type: The Andes of Peru and Bolivia during the Pliocene-Pleistocene. *Journal of Geophysical Research,* **97**, 11945–11983.

MOLNAR, P. 1988. A review of the geophysical constraints on the deep structure of the Tibetan plateau, the Himalaya and the Karakoram, and their tectonic implications. *Philosophical Transactions of the Royal Society, London,* **A326**, 33–88.

—— & CHEN, W.-P. 1983. Focal depths and fault plane solutions of earthquakes under the Tibetan plateau. *Journal of Geophysical Research,* **88**, 1180–1196.

—— & LYON-CAEN, H. 1988. Some simple physical aspects of the support, structure and evolution of mountain belts. *In*: CLARK, S.P. JR (ed.) *Processes in continental and lithospheric deformation.* Geological Society of America Special Papers, **218**, 179–207.

—— & —— 1989. Fault plane solutions of earthquakes and active tectonics of the Tibetan plateau and its margins. *Geophysical Journal International,* **99**, 123–153.

—— & TAPPONNIER, P. 1975. Cenozoic tectonics of Asia: Effects of a continental collision. *Science,* **189**, 419–426.

—— & —— 1978. Active tectonics of Tibet. *Journal of Geophysical Research,* **83**, 5361–5375.

NI, J. & YORK, J.E. 1978. Late Cenozic extensional tectonics of the Tibetan plateau. *Journal of Geophysical Research,* **83**, 5377–5387.

PALMER, M.R. & EDMOND, J.M. 1989. The strontium isotope budget of the modern ocean. *Earth and Planetary Science Letters,* **92**, 11–26.

PLATT, J.P. 1986. Dynamics of orogenic wedges and the uplift of high-pressure metamorphic rocks. *Geological Society of America Bulletin,* **97**, 1037–1053.

RICHTER, F.M., LOVERA, O.M., HARRISON, T.M. & COPELAND, P.C. 1991. Tibetan tectonics from a single feldspar sample: An application of the $^{40}Ar/^{39}Ar$ method. *Earth and Planetary Science Letters,* **105**, 266–278.

——, ROWLEY, D.B. & DE PAOLO, D.J. 1992. Strontium isotope evolution of seawater: the role of tectonics. *Earth and Planetary Science Letters,* **109**, 11–23.

ROYDEN, L.H., PATACCA, E. & SCANDONE, P. 1987. Segmentation and configuration of the subducting lithosphere in Italy: An important control on thrust-belt and foredeep-basin evolution. *Geology,* **15**, 714–717.

SCHÄRER, U. 1984. The effect of initial ^{230}Th disequilibrium on young U-Pb ages: the Makalu case, Himalaya. *Earth and Planetary Science Letters,* **67**, 191–204.

——, XU, R.-H. & ALLEGRE, C.J. 1986. U-(Th-)Pb systematics and ages of Himalayan leucogranites, Southern Tibet. *Earth and Planetary Science Letters,* **77**, 35–48.

SEARLE, M.P., COOPER, D.J.W. & REX, A.J. 1988. Collision tectonics of the Ladakh–Zanskar Himalaya. *In*: SHACKLETON, R.M., DEWEY, J.F. & WINDLEY, B.F. (eds) *Tectonic Evolution of the Himalayas and Tibet.* Royal Society London, 117–149.

SÉBRIER, M., MERCIER, J.L., MEGARD, F., LAUBACHER, G. & CAREY-GAILHARDIS, E. 1985. Quaternary normal and reverse faulting and the state of stress in the central Andes of Peru. *Tectonics,* **4**, 739–780.

——, ——, MACHER, J., BONNOT, D., CABRERA, J. & BLANC, J.L. 1988. The state of stress in the overriding plate situated above a flat slab: the Andes of central Peru. *Tectonics,* **7**, 895–928.

SONDER, L.J., ENGLAND, P.C., WERNICKE, B.P. & CHRISTIANSEN, R.L. 1987. A physical model for Cenozoic extension of western North America. *In*: COWARD, M.P., DEWEY, J.F. & HANCOCK, P.L. (eds). *Continental Extensional Tectonics,* Geological Society, London, Special Publications, **28**, 187–201.

TAPPONNIER, P. & BRUNEL, M. 1992. Unpublished appendix to *Introduction to the Integrated Scientific Investigation on Karakorum and Kunlun Mountains*, PAN YUSHENG (ed.) China Meteorological Press.

VALDIYA, K.S. & GOEL, O.P. 1983. Lithological subdivision and petrology of the Great Himalaya Vaikrita Group in Kumaun, India. *Proceedings of the Indian Academy of Sciences (Earth and Planetary Sciences),* **92**, 141–163.

VIDAL, P. 1978. Rb-Sr systematics in granites from Central Nepal (Manaslu): Significance of the Oligocene age and high $^{87}Sr/^{86}Sr$ ratio in the Himalayan orogen (comment). *Geology,* **6**, 196.

——, COCHERIE, A. & LE FORT, P. 1982. Geochemical investigations of the origin of the Manaslu leucogranite (Himalaya, Nepal). *Geochimica et Cosmochimica Acta,* **46**, 2279–2292.

——, BERNARD-GRIFFITHS, J., COCHERIE, A., LE FORT, P., PEUCAT, J.-J. & SHEPPARD, S.M.F. 1984. Geochemical comparison between Himalayan and Hercynian leucogranites. *Physics of the Earth and Planetary Interiors,* **179**, 190.

WERNICKE, B.P., CHRISTIANSEN, R.L., ENGLAND, P.C. & SONDER, L.J. 1987. Tectonomagmatic evolution of Cenozoic extension in the North American Cordillera. *In*: COWARD, M.P., DEWEY, J.F. & HANCOCK, P.L. (eds) *Continental Extensional Tectonics*. Geological Society, London, Special Publications, **28**, 203–221.

The emplacement of the Manaslu granite of Central Nepal: field and magnetic susceptibility constraints

S. GUILLOT[1], A. PÊCHER[1], P. ROCHETTE[2] & P. LE FORT[1]

[1] *Laboratoire de Géodynamique des Chaînes Alpines, Institut Dolomieu, 38031 Grenoble, France*

[2] *Géochimie et Magnétisme des Roches, Faculté St Jérome 13397 Marseille cedex 13, France*

Abstract: This paper investigates the relationship between the emplacement of the Manaslu leucogranite and the regional deformation. Structural data are derived from field measurements and from the anisotropy of magnetic susceptibility of oriented samples. The geometrical relationship between the internal structure of the granite, north vergent folds affecting the Higher Himalayan Sedimentary Series and the E–W dextral shear-zone, separating the Higher Himalayan Crystalline Series and its sedimentary cover, indicate that in the sedimentary series, the collapse structures and E–W dextral shearing are two distinct structural features. Emplacement of the granite post-dates the collapse structures but is related to the E–W dextral shearing. A model for the Manaslu granite emplacement is proposed which requires development of extensional fracturing at 45° to the shear direction, followed by intrusion of granitic magma in successive pulses which results in lateral expansion of the pluton towards the NE.

All along the Himalayan belt, the base of the Higher Himalayan Sedimentary Series (HHSS) is intruded by bodies of leucogranite of the Higher Himalayan leucogranite belt (HHL; Debon *et al.* 1986). These granites are Miocene in age and located at the interface between the Higher Himalayan Crystalline Series (HHCS) and its sedimentary cover. This interface appears to be a major structural break, the North Himalayan Normal Fault (NHNF) (Burg *et al.* 1984; Burchfiel & Royden 1985; Herren 1987; Mattauer & Brunel 1989; Pêcher & Scaillet 1989), which appears to be of Miocene age (Hodges *et al.* 1992). The leucogranites could have been generated in response to thickening of the crust (Le Fort 1975; England *et al.* 1992), after collision between the Indian and Eurasian plates, around 50 Ma. They were produced by melting of the metasedimentary part of the Higher Himalayan Crystalline Series (HHCS) favoured according to Rb/Sr, Sm/Nd, U/Pb and O systematics (Le Fort 1981; Vidal *et al.* 1982; Deniel *et al.* 1987; France-Lanord & Le Fort 1988), by the percolation of fluids as demonstrated by oxygen and hydrogen isotopic studies (France-Lanord & Le Fort 1988), and some frictional heating in the Main Central Thrust zone (England *et al.* 1992).

While petrological models for these granites are now well constrained (review in Le Fort *et al.* 1987), their mechanism of emplacement is still rather poorly understood, despite current interest in the relationship between tectonics and granite emplacement (Brun & Pons 1981; Bateman 1984; Hutton 1988; Lagarde 1989; Paterson *et al.* 1989; Bouchez & Diot 1990). Structural data is available for the leucogranite sheet of Nyalam, north of the main HHL belt (Burg *et al.* 1984; Burchfiel & Royden 1985), the Chhokang arm, an eastward extension of the Manaslu leucogranite, (Pêcher & Bouchez 1987), the leucogranitic sills occurring in the Zanskar normal fault shear zone (Herren 1987), and the Garhwal Badrinath-Gangotri granites (Scaillet 1990). These studies have emphasised subsolidus strain. For example, a strong S–C type fabric has been observed in Zanskar and Manaslu areas (Chhokang arm): in Zanskar, it corresponds to normal faulting, whereas in the Chhokang arm it indicates a dominant dextral shearing component (Pêcher & Bouchez 1987; Pêcher *et al.* 1991). In Garhwal, earlier synmagmatic longitudinal stretching has also been described (Scaillet 1990).

Granite emplacement structure is essentially determined by the measurement of the preferred orientation of early magmatic minerals, through field observations, or petrofabric studies in the laboratory. As granites are often poorly orientated, field and thin section observations must be implemented by a more sensitive technique: the anisotropy of magnetic susceptibility (AMS). Its principle is simple: preferred orientation of anisotropic magnetic

From TRELOAR, P.J. & SEARLE, M.P. (eds), *Himalayan Tectonics*
Geological Society Special Publication, No. 74, pp. 413–428.

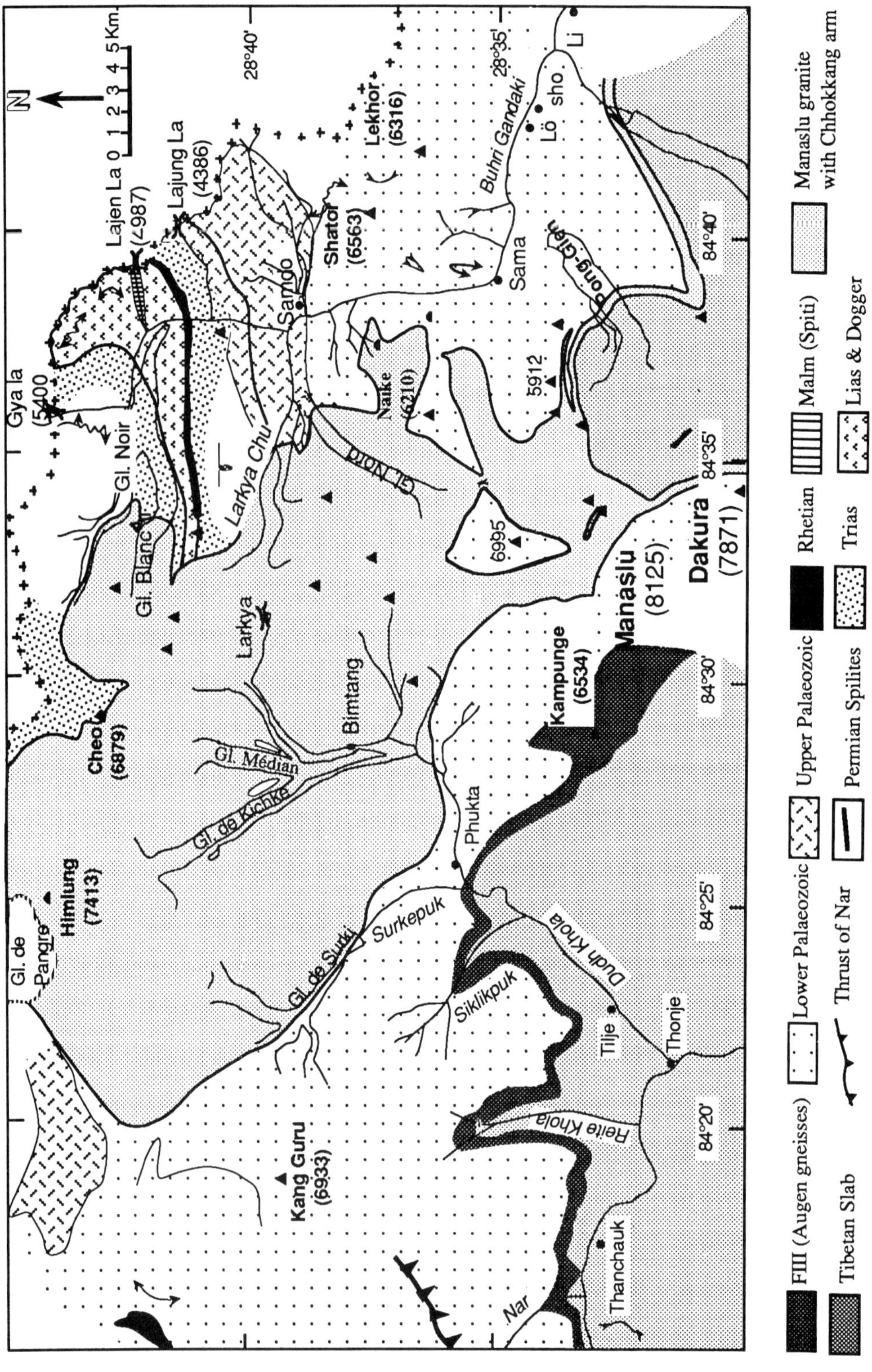

Fig. 1. Geological map of the Manaslu granite.

minerals (such as biotite, tourmaline, iron oxides) results in a detectable magnetic anisotropy of the rock (Hrouda 1982; Rochette *et al.* 1992). AMS appears to be a very rapid and sensitive way to determine the petrofabric once the magnetic minerals responsible for it are identified. For iron oxides, such as magnetite, AMS is due to shape preferred orientations. Previous AMS studies in granites, including syntectonic leucogranites, have shown the importance of this technique for granite emplacement studies (Jover 1986; Bernier *et al.* 1987; Bouchez *et al.* 1987; Gleizes 1992; Rochette 1988; Scaillet 1990).

The aim of this paper is to present structural constraints on Manaslu granite emplacement, through field and magnetic anisotropy data.

Geological setting

The Manaslu granite pluton outcrops over an area of 400 km^2. It is 30 km in length, 13 km in width, elongated in a NW–SE direction. Its base is right above the normal fault/dextral shear zone, at the top of the HHCS. Its top intrudes lower Jurassic limestones (Fig. 1).

The grain size is millimetric to centrimetric, and the mineral assemblage is quartz (±32%), Na-rich plagioclase (±37%), K-feldspar (±21%), muscovite (±7%) and biotite (±3%). The granite is usually tourmaline rich (few %), but nearly completely devoid of accessory minerals and enclaves. Only three types have been recognized (Le Fort 1991): scarce xenoliths of country rocks located in the margins of the pluton, mica-rich enclaves and schlieren, and nodules or 'cocardes' of tourmaline.

Isotopic studies support a Miocene age for the Manaslu granite: Rb/Sr and U/Pb data by Deniel *et al.* (1987) suggest a crystallization age spanning 25 to 18 Ma, whereas $^{40}Ar/^{39}Ar$ systematics of Copeland *et al.* (1990) suggest a shorter time of emplacement and cooling, around 20 Ma.

There is no evidence of thermal disequilibrium between the base of the granite and the surrounding rocks. But at the top of the intrusion there is a contact aureole a few tens of meters thick developed in the sedimentary host rocks. Thermobarometric investigations in the host rock indicate pressures of around 5.5 ± 0.5 kbar and temperature greater than 550° C at the base of the granite, and around 3 kbar and 550–600° C at the upper contact auréole. These values suggest the granite was emplaced at a depth of between 20 and 12 km (Guillot *et al.* 1991*a*).

Field relationships

South of the massif, the eastern cliff of Manaslu displays a 4000 m high section, in which the lower and the upper contacts of the granite are visible (Fig. 2).

The base of the pluton intrudes Lower Palaeozoic limestones of the Annapurna formation, above the orthogneisses (granites around 500 Ma old, Vidal *et al.* 1984) at the top of the HHCS. The contact appears to be as a whole rather flat and parallel to the main metamorphic

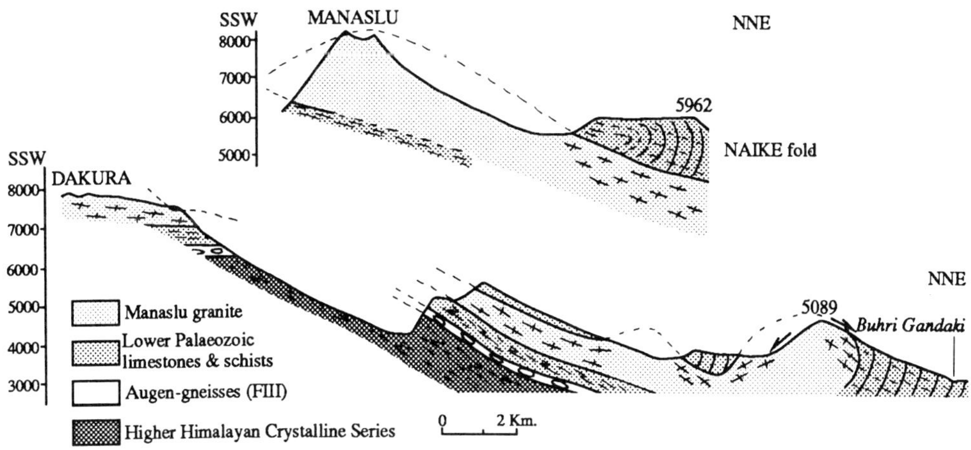

Fig. 2. Geological cross sections through the southern part of the Manaslu granite. They show the northward vergent fold cross-cut by the granite (below the 5962 summit) and the local ballooning effect (below the 5089 summit).

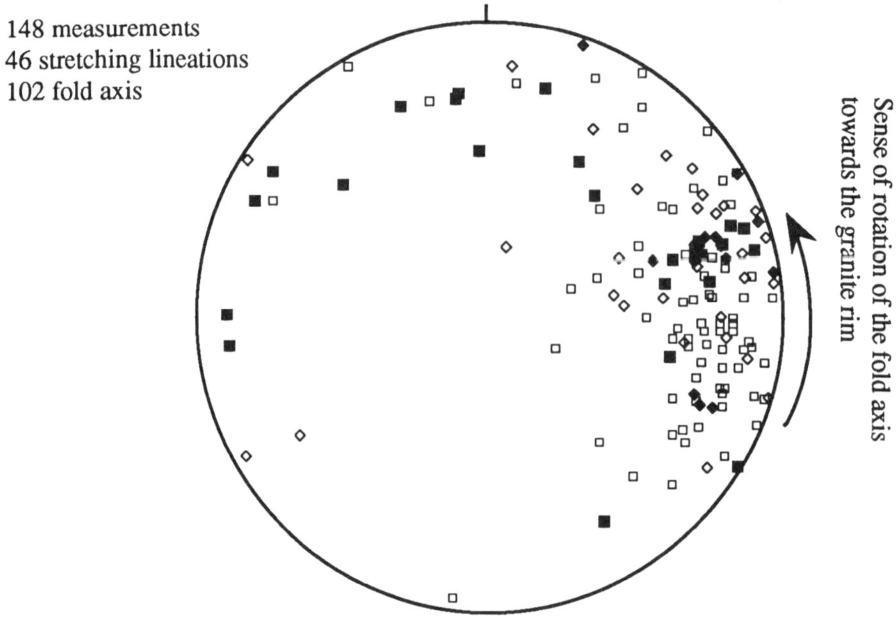

Fig. 3. Stretching lineations far (open rhomb) and close (black rhomb) to the granite. Fold axis far (open square) and close (black square) to the granite in the host rocks. Best axis for the stretching lineations: N74° E, plunge 18° E. The orientation of the stretching lineation is independant of the distance of the granite. Best axis for the fold axis non reorientated by the granite: N100° E, plunge 20° E. Best axis for the fold axis reorientated by the granite: N71° E 21° E. Lower hemisphere and equal angle, Schimdt stereonet (Pêcher 1991).

Fig. 4. Detail of the lateral contact of the granite: cross-cutting of the hinge of the Naike north vergent fold. (NW of Sama, altitude: 4050 m, view towards the south).

cleavage, with a transition zone, a few hundred meters thick, where the gneisses and marbles are cross-cut by a network of sills and dykes.

The top contact, also parallel to the main cleavage, forms a huge domal structure (Fig. 2). Thus the granite forms a large assymetric lens, which disappears southward. Another lens is developed in the Dakura area. The two lenses appear to be connected by a transition zone injected by sills of granite. Such a disposition is similar to the one observed by Scaillet (1990) in the Gangotri–Badrinath leucogranitic lenses.

North of the Manaslu summit, the salient structural feature is the Naike kilometre-sized fold underlain by Palaeozoic series. This regional northward vergent fold, orientated N100°–110° E (Fig. 3), deforms the lithological boundaries S0, and the main metamorphic cleavage Sm. Here, the lateral and upper contact of the granite sharply cut the lower limb of the fold (Figs 2 & 4). Close to the pluton, the fold is reorientated to N70° E, 20° E (Fig. 3), so that the S0–Sm plane becomes parallel to the granite-host rock contact.

Further north, the upper contact of the granite locally defines a new well marked dome. Here the granite displays, close to the contact, a well marked flattening plane, dipping towards the south in the southern part and dipping towards the north in the northern part of the dome and oblique to the contact (Fig. 2). According to Gapais (1989*a*), the obliquity between the contact and the magmatic foliation can be used to determine the local shear-sense during granite emplacement. Following these criteria, the observed relationships, seen better in the southern limb of the dome, would indicate a local syn-emplacement expansion or ballooning of the laccolith (Fig. 2).

In the Glacier Blanc region (Fig. 1), the upper contact is tilted in the inverted limb of the north vergent folds. It dips gently at 20° towards the ENE, then becomes flatter in the Cheo area. East of the Larkya, where the granite intrudes the Triassic and Lower Jurassic Series, the sediments are strongly flattened close to the granite, dykes of granite display pinch and swell structures (axis of boudins from N20° W to N10° E), and are folded in east- to northeast-verging folds (Fig. 5). A new cleavage appears, striking NW–SE and dipping around 40° NE and superimposed by a late N–S strain-slip cleavage. The obliquity between the contact and the reorientated schistosity, and the assymetry of the late small folds, suggest relative upward motion of the granite and the sliding of its cover towards the ENE.

In the Larkya and Glacier Blanc area, the contact metamorphic aureole displays a strong lineation, marked by elongate minerals (biotite or oxides) or by pressure shadows around staurolite or garnet in the pelitic layers and around scapolite and pyroxene in the calc-silicate rocks. Orientation of this lineation is remarkably constant between N70° E and N80° E, plunging 10° E to 20° E (Figs 3 & 6a).

It should be noted that a similar stretching lineation, with the same direction, is also present south of the massif (Naike & Pong-Gien area), even farther from the contact (Fig. 6a), but is only very scarcely observed between the two areas.

Fig. 5. The Triassic pelites around twenty meters above the upper contact of the granite. The pegmatic dyke, with pinch and swell structures is folded by a northeastern vergent fold. The axial plane cleavage of this fold is parallel to the contact (NE of the Larkya pass, altitude: 5400 m, view towards the NW).

Fabrics within the granite

The internal fabric of the granite, better expressed by preferred orientation of ferromagnesian silicates (biotite, muscovite or

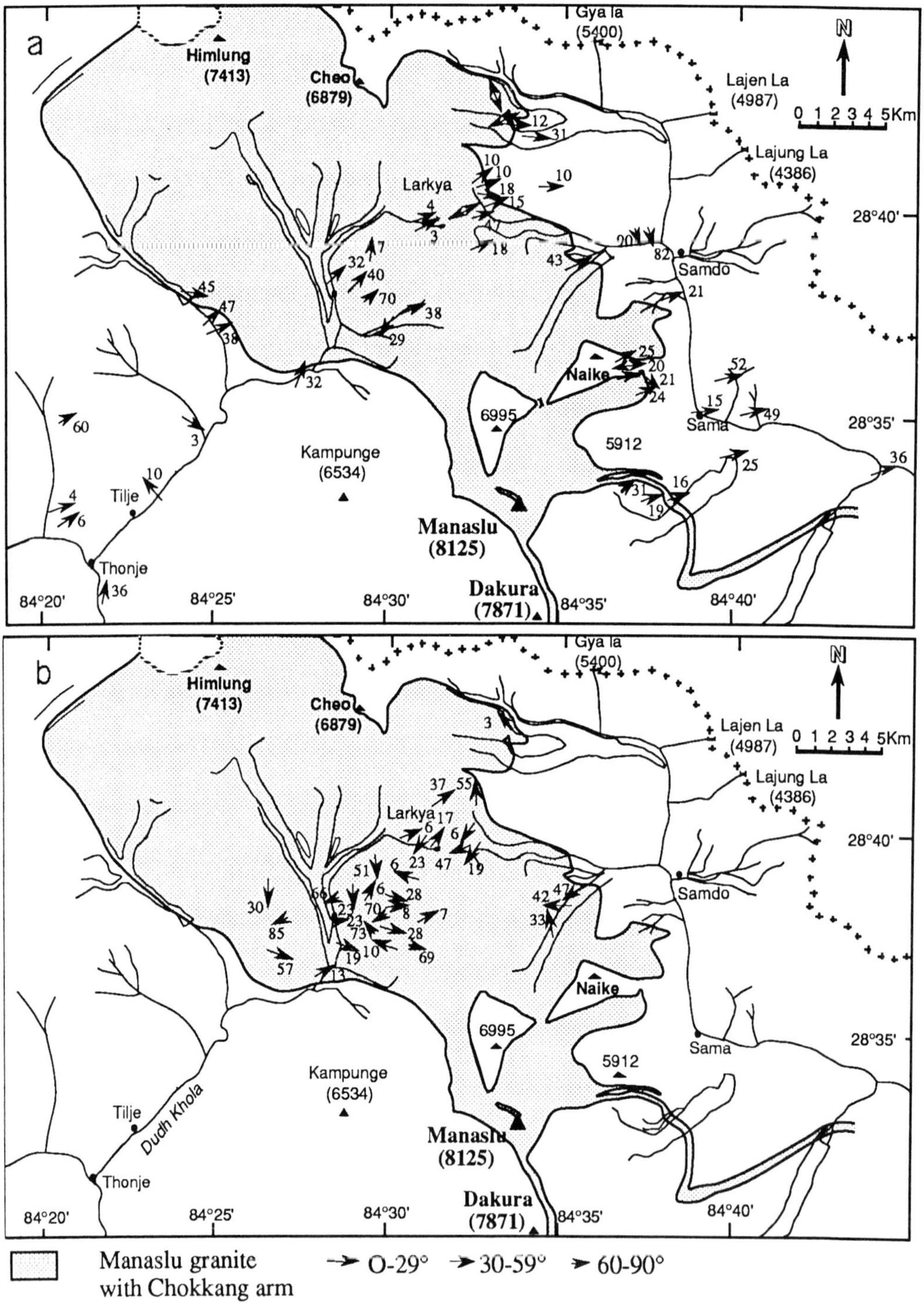

Fig. 6. (**a**) Map of the magmatic lineation in the granite and stretching lineation in the country rocks. (**b**) Map of the magnetic lineations.

tourmaline), is difficult to see in the field. Its development depends on the grain size and the mineralogy of the rock. A planar fabric may be distinguishable in the muscovite and biotite-rich facies, whereas the lineation is easier to detect in the tourmaline-rich facies.

When visible, the foliation is homogeneous at the outcrop scale, but may be locally cross-cut by muscovite rich shear bands with S–C geometries.

In the southern part of the massif, near its base, the strike of the foliation is rather constant at N110° E (±20°), and its dip is usually greater than 40° towards the north. In the core of the massif, the planar foliation has an irregular geometry, with a slight preferred ENE orientation, the dip being usually higher than 50° N. In the northeastern part (Larkya), the foliation gradually becomes N–S striking with a dip of between 20 to 50° to the East.

Due to the scarcity of tourmaline-rich facies, the mineral lineation is often difficult to observe in the field. When present, it is orientated E–W at the base, NE–SW in the core and N60° E to N80° E towards the eastern border (Fig. 6a). Its plunge is around 40° N, in the south and in the core of the massif, but only a few degrees to the NE in the Larkya–Glacier Blanc area.

Microstructural data

Country rocks

Microstructures in the contact aureole are one of the tools available to constrain the relationship between granite emplacement and regional deformation (Hutton 1988; Lagarde 1989; Paterson *et al.* 1989; Vernon *et al.* 1989). Here an outline of microstructural data is summarised.

At the basal contact, in the lower and middle Palaeozoic rocks, cleavage and lineation are defined by high temperature minerals (pyroxene, scapolite, K-feldspar, phlogopite). In pelitic rocks, porphyroblasts of garnet are wrapped by a high temperature foliation defined by fibrous sillimanite (fibrolite) and two micas, which is deformed by the late strain-slip cleavage. Close to the roof of the granite, in the Triassic schists, new fabric related to granite emplacement is defined by biotite, muscovite and tourmaline wrapping the garnet and the staurolite. Staurolite grew during and after the new penetrative cleavage. In the Lower Jurassic limestones, the emplacement related cleavage postdates the minerals of the contact metamorphism (pyroxene and scapolite).

These relationships suggest that:

(i) towards the base of the massif, temperature was high throughout the deformation related to emplacement of the granite, and the N70–80° E lineation is synchronous with the emplacement;

(ii) towards the roof of the granite, the heat required for the high temperature deformation (more than 550° C) was limited to the vicinity of the granite, and decreased strongly in space and time within a few metres of the granite (Guillot *et al.* 1991*b*).

Granitic microstructures: 'magmatic' versus solid-state deformation

The fabric of a granite can be acquired either at 'magmatic stage', that is at an intermediate crystal–melt state, with less than 40% of the melt (Van der Molen & Paterson 1979; Hibbard 1987) or can be due to high-temperature solid-state deformation (Gapais & Barbarin 1986). Microscopic investigations have been carried out on more than 200 thin-sections to assess the magmatic or solid-state character of the fabric.

Fig. 7. Magmatic foliation defined by K-feldspar, plagioclase and magmatic muscovite alignment. Note how plagioclase twins are parallel to the long dimension of crystals. In the left and upper part of the figure, a submagmatic microfracture transects a plagioclase and is sealed by a large monocrystal of quartz. Broken line in quartz indicate prismatic subboundaries perpendicular to the magmatic lineation. (Sample XG 106, Larkya pass, thin-section cut perpendicular to the magnetic foliation and parallel to the magnetic lineation). Size: × 100.

In the core of the Manaslu granite (east and west of Bimtang), the planar fabric is defined by the preferred orientation of micas, as well as by the orientation of euhedral K-feldspar and plagioclase, without evidence of plastic deformation or recrystallization (Fig. 7). Such a fabric is typical of a predominantly 'magmatic' fabric (Shelley 1985; Hutton 1988; Vernon *et al.* 1989; Paterson *et al.* 1989). In places, imbrication of K-feldspar crystals is visible, implying non-coaxial magmatic flow (Blumenfeld 1983; Blumenfeld & Bouchez 1988; Ildefonse 1992). However, the scarcity of this imbrication does not allow a determination of the shear sense. Some K-feldspar crystals contain cracks perpendicular to the lineation, filled with quartz or K-feldspar. According to Bouchez *et al.* (1992), these indicate brittle deformation at the intermediate-state.

Towards the eastern rim of the granite, a late foliation defined by the growth of muscovite cross-cutting the magmatic foliation, indicates a top towards the east shear sense. Late muscovite is similar to magmatic muscovite in the magmatic foliation. Moreover, since there is no quartz recrystallization in the second foliation it can be inferred that this foliation also developed during deformation of a magma in an intermediate-state rather than at solid-state.

Quartz, which readily deforms plastically, is a sensitive indicator of solid-state flow (Vauchez 1980; Bouchez & Pêcher 1981; Marre 1986). In the Manaslu granite, it often occurs as 1–2 cm long crystals with an elongated shape (axial ratio of 1/4 to 1/6 in *xz* section), but displays only weak plastic deformation structures (large slightly disorientated subgrains, without dynamic recrystallization and growth of new grains). Thus its shape fabric should also have been mainly acquired at magmatic stage.

A high temperature solid-state deformation increases towards the northeastern and southwestern rims of the granite. It is marked by the development of sub-grains and new grains at the rim of the elongated magmatic quartz, defining a solid-state foliation parallel to the magmatic one.

Thus deformation mainly occurred at the magmatic stage, although some low-temperature deformation can be observed in the core of the massif. Quartz shows prismatic bands subperpendicular to the foliation (a-slip at temperatures probably below 500° C, Gapais & Barbarin 1986; Gapais 1989*b*), and dynamic recrystallization with grain size reduction (Fig. 8). In the same samples, micas show undulatory extinctions and kink-bands, whereas feldspars are crosscut by millimetre-wide shear-bands containing quartz, chlorite and clay minerals. Quartz–tourmaline-bearing faults associated with this rather low-temperature deformation in the core of the massif (eastern part of Bimtang) strike at around N150° E 70° W.

Fig. 8. Low-temperature solid-state deformation (probably below 500° C) underlined by the stretching of the quartz (in black and white), numerous subgrains in the quartz and kink-bands in the plagioclase. Muscovite recrystallizes into clay minerals at the rim of the K-feldspar (right corner of the figure). (Sample XL25, south-east of Bimtang, thin-section cut perpendicular to the solid-state foliation and parallel to the lineation). Size: X100.

Magnetic measurements

Anisotropy of magnetic susceptibility data (AMS)

In the Manaslu granite, as in other leucogranites, it is often difficult to define the actual magmatic structure in the field or in laboratory due to the weak preferred orientation of the minerals and to the lack of strain markers (enclaves, schlieren planes, dykes). Measurement of the Anisotropy of Magnetic Susceptibility can then be useful in defining structure, even if the magmatic fabric is quite weak (Guillet *et al.* 1983).

To complement our field data, an AMS study has been carried out on 40 blocks, orientated in the field with a magnetic compass, then drilled in the laboratory. An average of seven specimens per sample was possible (total of 279 specimens), each of them having a diameter of 2.5 cm and a length of 2.2 cm. The anisotropy of magnetic susceptibility was measured in the IRIGM (Grenoble) using a Kappabridge KLY-2. The high sensitivity of this instrument (5.10^{-8} SI) is usually necessary for weakly magnetic rocks such as leucogranite.

For each, 15 measurements along different axes were made (following the procedure of Jelinek 1980). From these measurements it is possible to obtain an estimate of the susceptibility ellipsoid, in magnitude and in direction. Principal axes are labelled $K1 > K2 > K3$, and corrected for the diamagnetic term in the matrix (quartz and feldspar) with a value equal to -14.10^{-6} SI (Bernier *et al.* 1987; Rochette 1987).

The mean susceptibility K_m (before diamagnetic correction) is low (5 to 42.10^{-6} SI), but similar to other measurements made from leucogranites (Bernier *et al.* 1987; Rochette 1988). Actually, it is similar to the ones obtained by Rochette *et al.* (1992) in the Everest leucogranite, or by Scaillet (1990) in the Garhwal Badrinath–Gangotri granite which have the same chemical composition as the Manaslu granite. Rochette *et al.* (1992) and Scaillet (1990) have demonstrated, using high field and low temperature magnetic measurements, as well as correlation between K_m and iron amount, that K_m of the HHL is essentially due to diamagnetic minerals (quartz and feldspar) and paramagnetic minerals (ferromagnesian silicates), while ferromagnetic contribution (iron oxides or sulphides) is negligible. Among these minerals, only phyllosilicates and tourmaline have a large intrinsic magnetic anisotropy. Therefore, the AMS will reflect the lattice preferred orientation of either tourmaline or biotite (and only rarely muscovite, due to its low iron content).

As seen above, the fabric of these minerals is of magmatic origin. Consequently, the magnetic fabric reflects the magmatic structure. For tourmaline and biotite crystals, the *c*-axes correspond to minimum susceptibility K_3, with a revolution ellipsoid ($K_1 = K_2$) (Rochette *et al.* 1992). However, in terms of shape preferred orientation it results in a contrasted AMS.

(i) As the *c*-axis is the pole of the phyllosilicate sheet, a biotite- (or muscovite-) bearing sample will yield a K_3 axis parallel to the pole of the foliation and a K_1 axis parallel to the zone axis of the biotite plane. This zone axis is interpreted as a stretching lineation;

(ii) On the contrary, tourmaline occurs as rods, elongated parallel to the *c*-axis: therefore in a tourmaline bearing sample, K_3 is parallel to the mineral lineation and K_1 to the mineral foliation.

Amongst the collected samples, 31 are biotite dominant, 5 are tourmaline dominant and 4 correspond to samples without clear predominance of one mineral over the other. Reliability of AMS directions at the specimen level depends on L (for K_1) and F (for K_3) values. All specimens have values above the detection limit of 1.002 (Table 1). At the site level, mean directions are obtained through the tensorial mean method of Jelinek (1982). On a stereoplot (Fig. 9a), the K_1 and K_3 axes of specimens from the same block are usually well grouped, thus giving a small confidence angle on the average direction.

Four sites have been rejected because they show too high an angular dispersion in the various K_1 and K_3 values, due to difficulties of measurement, heterogeneous distribution of the magnetic grains in the rock or opposition of the contributions made by tourmaline and biotite (Fig. 9b).

For each site, the mean anisotropy parameters (Hrouda 1982; Jover 1986) have been estimated using values corrected for diamagnetic (see Rochette 1987):

the lineation parameter $L = K_1/K_3$
the foliation parameter $F = K_2/K_3$
the anisotropy degree $P = K_1/K_3$
the shape parameter $T = 2(k_2 - k_3)/(k_1 - k_3) - 1$ with $k_i = \mathrm{Log}K_i$

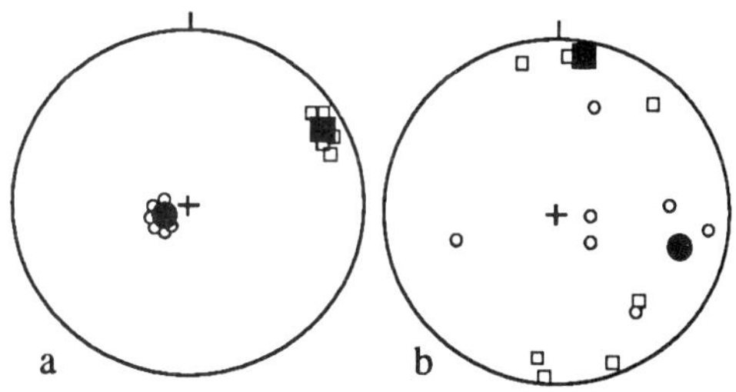

Fig. 9. Stereoplot of AMS data in a biotite-rich sample (**a**). Maximum and minimum susceptibility axes are respectively open squares and circles. The black square and the black circle symbolize respectively the best pole of the maximum and minimum susceptibility. The angular dispersion is weak, then the directions of the lineation and of the foliation are well defined (block KG162, 9 specimens). Stereoplot of AMS data in a mixed sample (**b**) (tourmaline and biotite rich). The opposition of the contributions of the two minerals leads to very high angular dispersion: the sample is rejected. (block XP233, 7 specimens).

Table 1. *Magnetic susceptibility data for the 40 sampling stations*

Sample	CN	Min	K	K1		K3		L	F
			10–5 SI	az.	inc.	az.	inc.		
XG2	5	B	24.1 ± 2.2	347.7 ± 9.6	55.2 ± 3.8	242.9 ± 6.9	10 ± 3.5	1.0308 ± 0.0119	1.0314 ± 0.0049
XG48	6	T + B	18 ± 3.7	8.7 ± 32.7	41.9 ± 13.0	210.3 ± 19.6	46 ± 9.1	1.0121 ± 0.0027	1.0219 ± 0.0091
XG52	6	B	21.6 ± 0.8	60.1 ± 28.4	37.1 ± 5.6	301.1 ± 20.7	32.7 ± 4.0	1.0397 ± 0.0086	1.0284 ± 0.0053
XG102	9	B	26.7 ± 1.7	169.3 ± 6.8	69.2 ± 1.3	55.9 ± 2.8	8.6 ± 1.6	1.0311 ± 0.0024	1.0552 ± 0.0032
XG104	9	B	8.4 ± 0.6	55.7 ± 8.6	6.1 ± 3.2	168 ± 10.8	74.3 ± 4.6	1.0111 ± 0.0009	1.0105 ± 0.0032
XG105	7	B	5.7 ± 2.3	211.6 ± 9.7	20.9 ± 3.9	48 ± 5.7	68.2 ± 3.9	1.0256 ± 0.0083	1.0278 ± 0.0101
XG106	6	B	12.3 ± 2.2	247.5 ± 4.7	23.5 ± 3.8	5.4 ± 13.1	47.1 ± 3.7	1.0347 ± 0.0033	1.0558 ± 0.0063
XG107	9	B	30.7 ± 2.4	237.4 ± 7.7	26 ± 1.8	24.1 ± 9.1	59.8 ± 1.8	1.0421 ± 0.0013	1.0386 ± 0.0018
XG108	10	B	27.9 ± 2	43 ± 2.6	17.1 ± 1.3	226.8 ± 2.6	72.8 ± 1.4	1.0360 ± 0.0020	1.0620 ± 0.0037
XG147	**8**	**T + B**	**22.7 ± 4.7**	**172.2 ± 79.2**	**5.3 ± 20.2**	**273.3 ± 63.0**	**64.3 ± 18.2**	**1.0119 ± 0.0139**	**1.0161 ± 0.005**
XG148	7	T	26.5 ± 1.4	314.5 ± 7.4	2.6 ± 5.5	45.9 ± 6.9	26.8 ± 2.4	1.0120 ± 0.0021	1.0149 ± 0.0040
XG159	7	B	34 ± 3.2	75.7 ± 11.9	23.1 ± 5.7	270 ± 20.5	66.2 ± 5.2	1.0320 ± 0.0025	1.0180 ± 0.0035
XG162	9	B	27.9 ± 1.3	58.7 ± 4.6	12.6 ± 3.0	255.8 ± 5.5	76.9 ± 2.4	1.0183 ± 0.0016	1.0749 ± 0.0049
XL1	10	B	42,1	175.5 ± 23.4	51.2 ± 1.9	303.7 ± 6.9	26.4 ± 4.4	1.0528 ± 0.0023	1.0162 ± 0.0043
XL5	9	B	32 ± 1.1	287.9 ± 7.7	6.4 ± 2.3	26.1 ± 3.4	51.7 ± 1.3	1.0207 ± 0.0034	1.0629 ± 0.0033
XL8	9	B	36.4 ± 0.6	109.5 ± 4.8	28.2 ± 3.4	292.4 ± 12.9	61.7 ± 3.3	1.0243 ± 0.0020	1.0346 ± 0.0023
XL9	8	B	34.4 ± 1.4	87.1 ± 4.7	8.1 ± 1.7	248.7 ± 15.3	81.5 ± 2.0	1.0391 ± 0.0067	1.0313 ± 0.0027
XL11	8	B	33.2 ± 1.8	31.1 ± 7.8	6.4 ± 5.6	125.4 ± 21.9	33.9 ± 2.5	1.0428 ± 0.0035	1.0384 ± 0.0066
XL15	9	B	47.6 ± 11	345.1 ± 10.4	13 ± 6.3	91 ± 14.5	49.8 ± 5.4	1.0354 ± 0.0088	1.0423 ± 0.0120
XL20	8	B + T	38.4 ± 3.7	322.2 ± 19.1	42.2 ± 10.7	23	43.2 ± 10.5	1.0327 ± 0.0140	1.0749 ± 0.0106
XL22	6	T	34.5 ± 2	90.6 ± 23.9	3.5 ± 4.6	181.1 ± 18.8	7.6 ± 13.3	1.0255 ± 0.0043	1.0246 ± 0.0071
XL23	8	B	36.2 ± 2.3	288.3 ± 134	10.2 ± 3.4	38.3 ± 8.9	62.3 ± 5.5	1.0203 ± 0.0034	1.0384 ± 0.0045
XL24	7	B	31.4 ± 0.9	105.3 ± 9.4	68.1 ± 7.8	347.6 ± 27.1	10.6 ± 4.2	1.0157 ± 0.0039	1.0279 ± 0.0068
XL25	5	B	33.5 ± 1.4	115 ± 27.1	27.9 ± 4.8	226.9 ± 5.8	35.2 ± 3.3	1.0114 ± 0.0039	1.0800 ± 0.0096
XL32	5	B	41.6 ± 9.2	48.7 ± 22.3	7.5 ± 7.6	273.5 ± 11.1	79.5 ± 5.7	1.0342 ± 0.0088	1.0525 ± 0.0136
XL37	5	B	23.9 ± 2.6	218.1 ± 15.4	70 ± 1.9	1.8 ± 6.7	16.4 ± 1.3	1.0210 ± 0.0037	1.1146 ± 0.0132
XL45	9	B	26.1 ± 5	246.9 ± 11.6	65.6 ± 7.6	153 ± 11.8	1.8 ± 5.2	1.0251 ± 0.0037	1.0290 ± 0.0072
XP96	8	B	26.1 ± 6	56.2 ± 2.8	20.4 ± 1.8	191.5 ± 1.9	62.4 ± 1.3	1.0456 ± 0.0022	1.1141 ± 0.0109
XP233	**7**	**T**	**21.4 ± 3.7**	**9.0 ± 43.9**	**7.0 ± 15.1**	**102.6 ± 64.9**	**27.3 ± 30.8**	**1.0187 ± 0.0032**	**1.0172 ± 0.0139**
XP265	9	T	26.2 ± 1.5	43.4 ± 6.2	41.6 ± 3.1	237.9 ± 5.1	47.7 ± 3.7	1.0293 ± 0.0060	1.0569 ± 0.0103
XP270	**9**	**B**	**29.9 ± 5.7**	**349.2 ± 32.9**	**32.9 ± 4.4**	**234.9 ± 5.4**	**32.5 ± 3.5**	**1.0125 ± 0.0046**	**1.1060 ± 0.0202**
XP271	5	B	21.6 ± 3.5	271.8 ± 18.7	41.6 ± 12.0	42.3 ± 27.3	36.2 ± 7.1	1.0166 ± 0.0066	1.0623 ± 0.0187
XP307	8	B	27.6 ± 3.2	214.1 ± 7.1	19.5 ± 1.7	353.3 ± 4.7	64.9 ± 1.7	1.0146 ± 0.0047	1.0986 ± 0.0062
XP310	**6**	**B ± T**	**17.2 ± 4.9**	**79.8 ± 58.1**	**23.3 ± 13.1**	**186.5 ± 45.9**	**33.6 ± 13.0**	**1.0170 ± 0.0084**	**1.0323 ± 0.0160**
XP312	7	B	23.4 ± 0.9	243 ± 25.5	46.8 ± 6.8	2 ± 8.6	24.4 ± 3.8	1.0277 ± 0.0068	1.0445 ± 0.0117
DK72	4	B	14.8 ± 1.1	106.8 ± 20.7	19.4 ± 4.4	285.9 ± 12.9	70.6 ± 8.7	1.026 ± 0.005	1.019 ± 0.007
DK188	2	B	14.8 ± 0.1	87.7 ±	29.9 ±	204.6 ±	38.2 ±	1.027 ± 0.002	1.019 ± 0.001
DK310	3	B	16.6 ± 1.7	111.9 ± 30.6	57.5 ± 26.4	269.1 ± 79.3	30.4 ± 23.4	1.016 ± 0.001	1.014 ± 0.005
DK289	3	B	15.1 ± 3.2	185.4 ± 75.2	30.9 ± 27.3	35.7 ± 43.1	55.3 ± 31.1	1.012 ± 0.011	1.035 ± 0.026
DK314	4	T	15.4 ± 10.3	169.6 ± 79.6	68.5 ± 14.4	67 ± 18.3	4.9 ± 10.6	1.010 ± 0.008	1.015 ± 0.015

K, mean susceptibility (before diamagnetic correction); CN, core number; Min, paramagnetic minerals (B, biotite, T, tourmaline); K1, K3 are respectively maximal and minimal susceptibility with K1 = K2. az., azimuth in degrees, inc., plunge. Rejected samples are in bold.

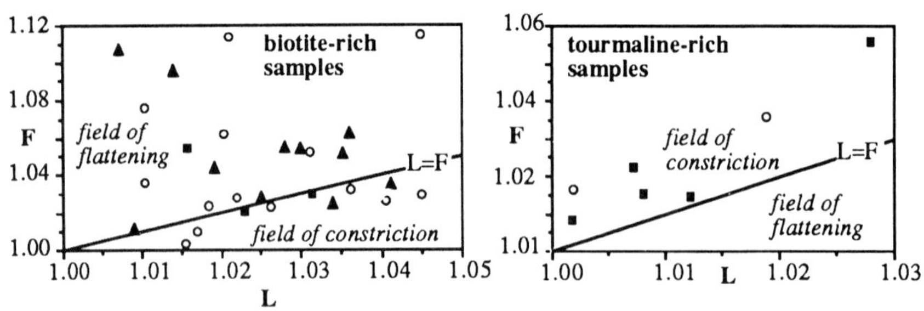

Fig. 10. Shape parameters *L* and *F* of the magnetic ellipsoid. Biotite- and tourmaline-rich samples have been distinguished. The squares refer to the samples from the rim of the granite, triangles from the intermediate levels and circles from the core of the granite.

The type and intensity of magnetic fabric are best demonstrated using an F versus L diagram. These diagrams can be interpreted in terms of petrofabric shape and intensity, taking into account the shape and intensity of the AMS of the magnetic mineral. Therefore, only samples from biotite- or tourmaline-bearing rocks can be compared with each other.

On the F versus L diagram (Fig. 10), most of the samples from the biotite rich rocks fall in the flattening field ($F \geqslant L$). However, in the tourmaline rich samples (Fig. 10), due to the inverse magnetic fabric of tourmaline, the oblate magnetic fabric characterizes a linear petrofabric.

Comparison between AMS and field measurements

In the Everest and Gangotri HHL, the good agreement between magnetic and petrofabric axes have been demonstrated by comparison of AMS and thin-section image analysis data (Rochette *et al*. 1992; Scaillet 1990). In the Manaslu granite, we observe a reasonable agreement between the field fabric and the AMS data (compare Fig. 6a and Fig. 6b and Fig. 11), which emphasises the ability of AMS to depict structures in only slightly deformed plutonic rocks, even if one must be cautious when the susceptibility is carried by two minerals with opposed fabrics (e.g. tourmaline and biotite).

The trends observed in the field are: a foliation plane striking WNW–ESE at the base and in the core of the pluton which trends NW–SE towards the east; a lineation (Figs 6b & 11) constant at around N115° E plunging 10 to 70° E at the south of the massif (i.e. at its base), whereas at the northeast (i.e. at its roof) it is around N50–60° E, plunging 0–30° E. In the core the linear fabric is weak. The lineation seems to be quite dispersed, despite a rough N–S trend.

Discussion

The shape of a granitic body depends mainly on its tectonic environment (Pitcher 1979). Post-tectonic plutons tend to have subcircular contours with concentric internal structures (Pons 1983; Paterson & Tobish 1988). In contrast, syntectonic plutons record in their shape, internal structures and metamorphic aureole, the geometry and the kinetics of the regional deformation (Brun & Pons 1981; Hutton 1988; Lagarde 1989; Lagarde *et al*. 1990; Guineberteau *et al*. 1987; Vernon *et al*. 1989; Castro 1986). The Manaslu granite displays numerous features consistent with emplacement during regional deformation, but the tectonic control of its emplacement is poorly defined, particularly with regard to the North Himalayan normal fault.

Collapse structure versus dextral shearing

The Miocene deformation at the top of the HHCS and within the HHSS is characterized by normal faulting and northward gravity sliding of the HHSS and E–W dextral wrench shear (Caby *et al*. 1983; Burg *et al*. 1984; Brun *et al*. 1985;

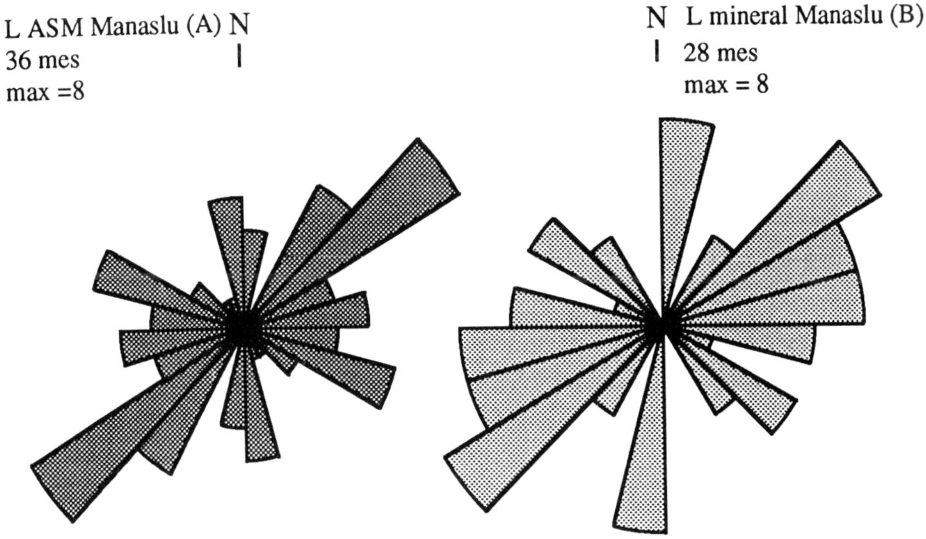

Fig. 11. Rose diagram of the orientation of the magnetic (**A**) and magmatic (**B**) lineations.

Burchfiel & Royden 1985; Herren 1987; Pêcher & Scaillet 1989; Pêcher *et al.* 1991; Hodges *et al.* 1992).

In the upper part of the HHCS, where this deformation is best shown by sillimanite bearing S–C planes, the normal faulting and the dextral shearing appear to be coeval. The transport direction is defined by the same stretching/sliding lineation, but its pitch varies from locality to locality, with predominance either of dip slip or strike slip motion.

Higher in the tectono-metamorphic pile, in the sedimentary series, structures associated with the dextral shearing and the northward collapse are distinct, and the relative timing of the different stages of the deformation is clearer. Dextral shearing is visible in the lower part of the cover, marked by a longitudinal stretching line-

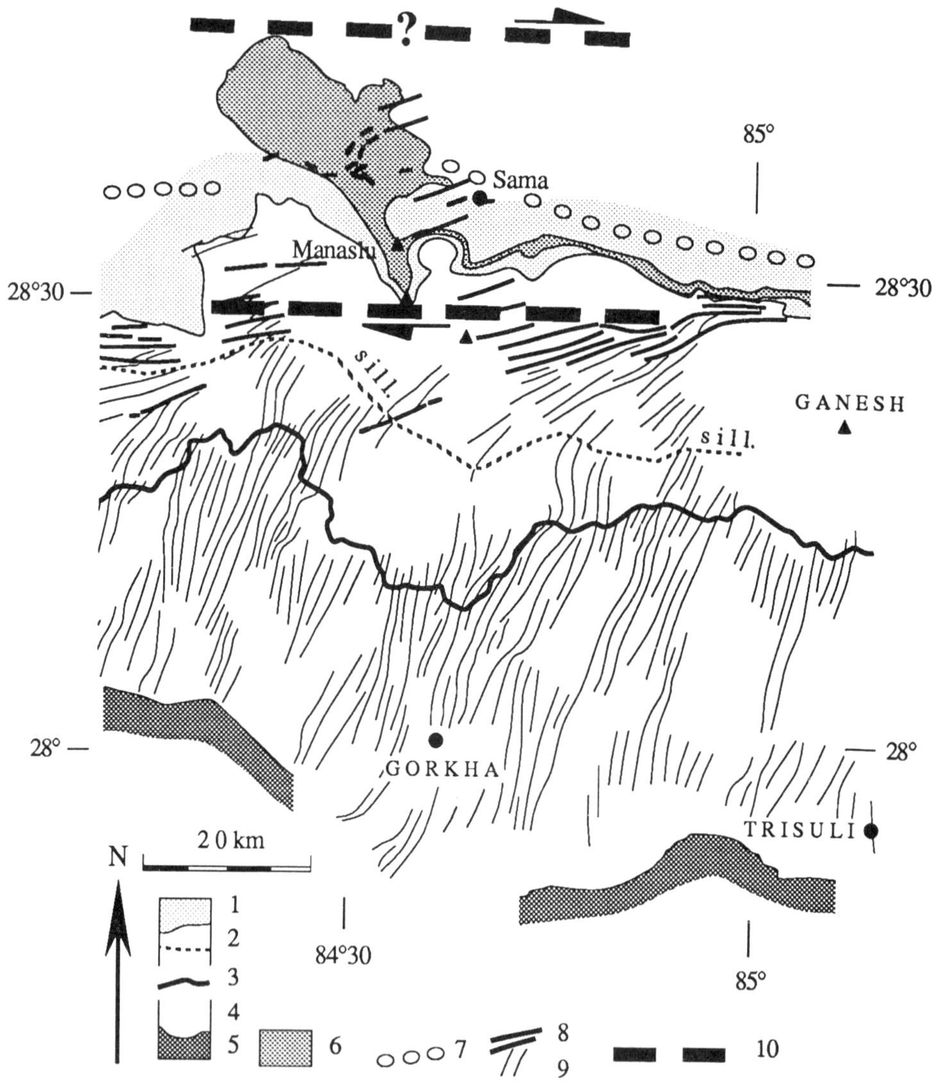

Fig. 12. Stretching lineation patterns in central Nepal. (modified after Pêcher 1991). 1, Higher Himalayan Sedimentary Series; 2, Higher Himalayan Crystalline Series; 3, Main Central Thrust (MCT); 4, Lesser Himalayan nappes; 5, Southern Himalayan sedimentary belt; 6, Manaslu leucogranite and its eastward extension (Chhokkang arm); 7, axial trace of north-vergent Annapurna–Naike fold; 8, late sillimanite lineation and stretching lineation in the sedimentary cover associated to the E–W dextral shears; 9, earlier lineation related to southward MCT thrusting; 10, the two shear-zones boardering the Manaslu granite.

ation, striking around N75°E (Fig. 3). A similar lineation pattern is also observed further to the north, in the Larkya–Gyala area. This suggests the existence of another E–W dextral shear-zone north of the granite (Fig. 12). The normal fault movement is marked by northward facing folds, visible in the whole sedimentary pile, and striking regularly N100° E, 20° E (Fig. 3). It appears that the stretching lineation has a remarkably constant direction around N75° E (Fig. 3), whatever the position in the northward vergent folds, and is not folded about the hinges of the folds. Thus it can be inferred that the longitudinal shear must postdate the folds.

Geometric relationships between the Manaslu granite and the north-vergent folds

As noted above, the north-vergent folds are cut by the granite. The fold axes are reoriented close to the granite contact and the metamorphic contact aureole is characterised by a new flattening plane, parallel to the granite contact and overprinting all the existing structures. This indicates that deformation related to emplacement of the granite postdates the collapse structures. This sequence is similar to that described by Hodges *et al.* (1992) north of Everest, where a HHL clearly crosscuts mylonites developed during normal faulting.

Several authors have suggested that the High Himalayan leucogranites predate the normal faulting (Burg *et al.* 1984; Burchfiel & Royden 1985; Herren 1987; Mattauer & Brunel 1989), or are coeval with the northwards collapse structures (Scaillet 1990). In the case of the Manaslu granite, there is no clear evidence that normal faulting and collapse structures contribute to emplacement of the granite. South of the pluton, in the Dakura and the Manaslu areas, the granite forms kilometric scale lenses similar to the crustal scale 'chocolate tablet' structure observed in Garhwal by Scaillet (1990). Such a structure could reflect the boudinage of the granitic melt during the northward sliding of the sedimentary cover. However, in the Manaslu granite, boudinage is only weakly developed, and is not consistent with the regularity of the roof contact to the north.

Emplacement of the Manaslu granite in a crustal tension gash

The main features defining the structure of the Manaslu granite are its laccolithic shape with an axial ratio (length/width) of about 2.3, and the S shape of the magmatic lineation from the core to the eastern rim of the pluton. Considering that the Manaslu granite is bounded to the south by a large E–W shear zone (NHSZ), and that another similar shear-zone may exist to the north, two models of emplacement can be proposed, both related to opening between two shear-zones at the ductile–brittle transition level: emplacement in a pull-apart system, parallel to the shear-zone, or emplacement in extensional fractures oblique to the shear direction.

The second model proposed by Castro (1986) for the Extremadura batholith in Spain is favoured, as it is more consistent with the obliquity (around 40°) between the bulk elongation of the granite pluton and the shear-zones (Fig. 12), and with the internal magmatic lineation being orientated around N60–80° E in the northeastern part of the granite. Had the granite been emplaced in a pull-apart, the lineation should be broadly parallel to the shear plane (Guineberteau *et al.* 1987).

Conclusion

This study suggests that the Manaslu granite was emplaced in the following manner: initial opening (or rather creation of virtual opening fractures), which could be compared to a tension gash opening, directly related to the E–W dextral shear system at the top of the HHCS. The opening focuses the rise of granitic magma and determines the initial shape of the pluton. A feeder zone may correspond to the central and southwestern part of the massif, where the magmatic lineations and foliations are poorly orientated, and often steeply dipping. Formation of the granite by successive pulses is strongly suggested by the geochemistry (Le Fort *et al.* 1987); Such pulses collect in the centre of the intrusion, which then grows by lateral expansion (a general mechanism proposed by Lagarde *et al.* 1990*b*) towards the NE and is accompanied by local ballooning and sliding of the cover. The base of the pluton (SW of the massif), is still sheared at high temperature during and after emplacement. Finally, N150° E dipping 70–80° W faults crosscut the granite, with an overprinting of the magmatic foliation by low-temperature shear bands.

This study emphasises once more the importance of the regional tectonics and particularly that of strike-slip systems in the control of the emplacement of granitic plutons, and introduces some additional constraints on the relationship between normal faulting, lateral shearing and granite emplacement in the High Himalaya granitic belt.

This paper was presented at the 7th Himalayan–Karakorum–Tibet workshop in Oxford, 1992. The authors wish to thank R.W. England, M.P. Searle, P.J. Treloar and an anonymous reviewer for their constructive comments. The magnetic data are available and can be obtained by writing to S. Guillot or P. Rochette. The financial assistance of CNRS (URA 69) and Commissariat à l'Energie Atomique are gratefully acknowledged.

References

Bateman, R. 1984. On the role of diapirism in the segregation, ascent and final emplacement of granitoid magmas. *Tectonophysics,* **110**, 211–231.

Bernier, S., Bouchez, J.L. & Rochette, P. 1987. Anisotropie de la susceptibilité magnétique dans le granite de Beauvoir (sondage GPF d'Echassières, Massif Central Français). *Compte-rendus de l'Académie des Sciences,* **304**, 986–992.

Blumenfeld, P. 1983. Le 'tuilage des mégacristaux', un critère d'écoulement rotationnel pour les fluidalités des roches magmatiques. Applications au granite de Barbey-Séroux (Vosges, France). *Bulletin de la Société géologique de France,* **3**, 309–318.

—— & Bouchez, J.L. 1988. Shear criteria in granite and migmatite deformed in the magmatic and solid states. *Journal of Structural Geology,* **10**, 361–372.

Bouchez, J. & Diot, H. 1990. Nested granites in question: contrasted emplacement kinematics of independant magmas in the Zaër pluton, Morocco. *Geology,* **18**, 35–38.

—— & Pêcher, A. 1981. Himalaya Main Central Thrust pile and its quartz-rich tectonites in central Nepal. *Tectonophysics,* **78**, 23–50.

——, Bernier, S., Rochette, P. & Guineberteau, B. 1987. Log des susceptibilités magnétiques et anisotropies de susceptibilité dans le granite de Beauvoir: conséquence pour sa mise en place. *Géologie de la France,* **2–3**, 223–232.

——, Delas, C., Gleizes, G., Nédélec, A. & Cuney, M. 1992. Submagmatic microfractures in granites. *Geology,* **20**, 35–38.

Brun, J.P. & Pons, J. 1981. Strain patterns of pluton emplacement in a crust undergoing non coaxial deformation, Sierra Morena, Southern Spain. *Journal of Structural Geology,* **3**, 219–229.

——, Burg, J.P. & Guo, M.C. 1985. Strain trajectories above the Main Central Thrust (Himalaya) in Southern Tibet. *Nature,* **313**, 388–390.

Burchfiel, B.C. & Royden, L.H. 1985. Thin skinned N-S extension within the convergent Himalayan region. *Geology,* **13**, 679–682.

Burg, J.P., Brunel, M., Gapais, D., Chen, G.M. & Liu, G.H. 1984. Deformation of the leucogranites of the crystalline main central sheet in southern Tibet (China). *Journal of Structural Geology,* **6**, 532–542.

Caby, R., Pêcher, A. & Le Fort, P. 1983. Le grand chevauchement central himalayen: nouvelles données sur le métamorphisme inverse à la base de la Dalle du Tibet. *Revue de Géologie dynamique et Géographie Physique,* **24**, 89–100.

Castro, A. 1986. Structural pattern and ascent model in the Central Extremadura batholith, Hercynian belt, Spain. *Journal of Structural Geology,* **8**, 633–645.

Copeland, P., Harrison, T.M. & Le Fort, P. 1990. Age and cooling of the Manaslu granite: implications for Himalayan tectonics. *Journal of Volcanic and Geothermal Research,* **44**, 33–50.

Debon, F., Le Fort, P., Sheppard, S.M.F. & Sonet, J. 1986. The four plutonic belts of the Transhimalaya-Himalaya: a chemical, mineralogical, isotopic and chronogical synthesis along a Tibet-Nepal section. *Journal of Petrology,* **27**, 219–250.

Deniel, C., Vidal, A., Fernandez, A. & Le Fort, P. 1987. Isotopic study of the Manaslu granite (Himalaya, Nepal): Inferences on the age and source of Himalayan leucogranites. *Contributions to Mineralogy and Petrology,* **96**, 78–82.

England, P., Le Fort, P., Molnar, P. & Pêcher, A. 1992. Heat sources for tertiary metamorphism and anatexis in the Annapurna-Manaslu region (Central Nepal). *Journal of Geophysical Research,* **97**, 2107–2128.

Foex, G. 1957. Constantes selectionnées: diamagnétisme et paramagnétisme. *In: Tables de constantes et données numériques,* **7**, Masson, Paris.

France-Lanord, C. & Le Fort, P. 1988. Crustal melting and granite genesis during the Himalayan collision orogenesis. *Transactions of the Royal Society of Edinburgh, Earth Sciences,* **79**, 183–195.

Gapais, D. 1989*a*. *Les orthogneiss: structures, mécanismes de déformation et analyse cinématique.* Mémoires et Documents du CAESS.

—— 1989*b*. Shear structures within deformed granites: mechanical and thermal indicators. *Geology,* **17**, 1144–1147.

—— & Barbarin, B. 1986. Quartz fabric transition in a cooling syntectonic granite (Hermitage massif, France). *Tectonophysics,* **125**, 357–370.

Gleizes, G. 1992. *Structure des granites hercyniens des Pyrénées de Mont-Louis-Andorre à la Maladeta.* M.Sc. thesis, Toulouse.

Guillet, P., Bouchez, J.L. & Wagner, J.J. 1983. Anisotropy of magnetic susceptibility and magmatic structures in the Guérande granite massif (France). *Tectonics,* **2**, 419–429.

Guillot, S., Le Fort, P. & Pêcher, A. 1991*a*. Contact metamorphism of the Manaslu leucogranite: an emplacement within a thickened upper crust. *6th Himalay–Karakorum–Tibet workshop.* Auris. Géologie Alpine, **16**, 47–48.

——, —— & —— 1991*b*. Evolution métamorphique et microstructurale polyphasée de l'auréole du granite du Manaslu. *Bulletin de liaison de la Société Française de Minéralogie,* **2**, 38.

Guineberteau, B., Bouchez, J.L. & Vigneresse, J.L. 1987. The Mortagne granite pluton (France) emplaced by pull-apart along a shear zone: structural and gravimetric arguments and regional implication. *Geological Society of American Bulletin,* **99**, 763–770.

HERREN, E. 1987. Zanskar shear zone: Northeast-southwest extension within the Higher Himalayas (Ladakh, India). *Geology*, **15**, 409–413.

HIBBARD, M.J. 1987. Deformation of incompletely crystallized magma systems: granitic gneisses and their tectonic implications. *Geological Society of American Bulletin*, **90**, 1047–1062.

HODGES, K., BURCHFIEL, B.C., CHEN, Z., HOUSH, T., LUX, D., PARRISH, R. & ROYEN, L.H. 1992. Rapid early Miocene tectonic unroofing of the metamorphic core of the Himalaya: evidence from the Qomolangma region, Tibet. *7th Himalaya–Karakorum–Tibet workshop*. Oxford.

HROUDA, F. 1982. Mathematical model relationship and its application in geology and geophysics. *Geophysical Survey*, **5**, 37–82.

HUTTON, D.W.H. 1988. Granite emplacement mechanisms and tectonic controls: inferences from deformation studies. *The Royal Society of Edinburgh, Earth Sciences*, **79**, 245–255.

ILDEFONSE, B., LAUNEAU, P., BOUCHEZ, J.L. & FERNANDEZ, A. 1992. Effect of mechanical interactions on the development of shape preferred orientations: a two dimensional experimental approach. *Journal of Structural Geology*, **14**, 74–83.

JELINEK, V. 1980. Characterization of magnetic fabric of rocks. *Tectonophysics*, **79**, 63–67.

JOVER, O. 1986. *Les massifs granitiques de Gueret et du Nord-Millevaches: analyse structurale et modèle de mise en place.* PhD. thesis, Grenoble.

LAGARDE, J.L. 1989. Granites tardi-carbonifère et déformation crustale. L'exemple de la Meseta Marocaine. *Mémoires et Documents du CAESS.*

——, BRUN, J.P. & GAPAIS, D. 1990. Formation des plutons granitiques par injection et expansion latérale dans leur site de mise en place: une alternative au diapirisme en domaine épizonal. *Compte-rendus de l'Académie des Sciences*, **310**, 1109–1114.

LAMARCHE, G. 1987. *Analyse microstructurale et fabrique magnétique. L'exemple des calcschistes et des flyshs de la Zone Dauphinoise (Alpes Françaises).* PhD. thesis, Grenoble.

LE FORT, P. 1975. Himalayas: the collided range. Present knowledge of the continental arc. *American Journal of Sciences*, **275A**, 1–44.

—— 1981. Manaslu leucogranite: a collision signature of the Himalaya. A model for its genesis and emplacement. *Journal of Geophysical Research*, **86**, 10545–10568.

—— 1991. Enclaves of the Miocene leucogranites. *In: Enclaves and Granite Petrology*. Dider & Barbarin (eds), Elsevier, 35–46.

——, CUNEY, C., DENIEL, C., FRANCE-LANORD, C., SHEPPARD, S.M.F., UPRETI, B.N. & VIDAL, P. 1987. Crustal generation of the Himalayan leucogranites. *Tectonophysics*, **134**, 39–57.

MARRE, J. 1986. *The structural analysis of granitic rocks*. Elsevier, Amsterdam.

MATTAUER, M. & BRUNEL, M. 1989. La faille normale Nord-Himalayenne (FNNH): conséquence probable d'un diapirisme granitique. *Compte-rendus de l'Académie des Sciences*, **304**, 43–47.

PATERSON, S.R. & TOBISCH, O.T. 1988. Using pluton ages to date regional deformation: problems with commonly used criteria. *Geology*, **16**, 1108–1111.

——, VERNON, R.H. & TOBISCH, O.T. 1989. A review of criteria for the identification of magmatic foliations in granitoids. *Journal of Structural Geology*, **11**, 349–363.

PÊCHER, A. 1989. Schmidt Mac, a program to display and analyse directional data. *Computers and Geosciences*, **15**, 1315–1326.

—— 1991. The contact between the Higher Himalayan crystallines and the Tibetan Sedimentary series: Miocene large-scale dextral shearing. *Tectonics*, **10**, 587–598.

—— & BOUCHEZ, J.L. 1987. High temperature decoupling between the Higher Himalaya Crystallines and its its sedimentary cover. *Terra Cognita* (abstract), **7**.

—— & SCAILLET, B. 1989. La structure du Haut-Himalaya au Garhwal (Indes). *Eclogae geologicae Helvetica*, **82**, 655–668.

——, —— & LE FORT, P. 1991. Micoene dextral shearing between Himalaya and Tibet. *Geology*, **19**, 683–685.

PITCHER, W.S. 1979. The nature, ascent and emplacement of granitic magmas. *Journal of Geological Society, London*, **136**, 627–662.

PONS, J. 1983. Interprétation et utilisation des diverses structures et architectures des plutons acides et basiques. Gabbros et granitoïdes de la Sierra Morena occidentale (Espagne). *Bulletin de la Société Géologique de France*, **5**, 349–355.

ROCHETTE, P. 1987. Magnetic susceptibility of the rock matrix related to magnetic fabric studies. *Journal of Structural Geology*, **9**, 1015–1020.

—— 1988. *La susceptibilité anisotrope des roches faiblement magnétiques: origine et applications.* MSc. thesis, Grenoble.

——, JACKSON, M. & AUBOURG, C. 1992. Rock magnetism and the interpretation of anisotropy of magnety susceptibility. *Reviews of Geophysics*, in press.

SCAILLET, B. 1990. *Structure et Géochimie d'un leucogranite en régime de collision: l'exemple du massif Gangotri Badrinath.* Unpublished thesis, Nancy.

SHELLEY, D.M. 1985. Determining paleo-flow directions from groundmass fabrics in the Lyttleton radial dykes. New Zealand. *Journal of Volcanic and Geothermal Research*, **25**, 69–79.

VAN DER MOLEN, I. & PATERSON, M.S. 1979. Experimental deformation of partially-melted granite. *Contributions to Mineralogy and Petrology*, **72**, 57–72.

VAUCHEZ, A. 1980. Ribbon texture and deformation mechanism of quartz in mylonitized granite of great Kabylie (Algeria). *Tectonophysics*, **67**, 1–12.

VERNON, R.H., PATERSON, S.R. & GEARY, E.E. 1989. Evidence for syntectonic intrusion of plutons in the Bear Mountains fault zone, California. *Geology*, **17**, 723–726.

VIDAL, P., BERNARD-GRIFFITHS, J., COCHERIE, A., LE FORT, P., PEUCAT, J.J. & SHEPPARD, S.M.F. 1984. Geochemical comparison between Himalayan

and Hercynian leucogranites. *Physics of Earth and Planetary Interiors,* **35**, 179–190.
——, Cocherie, A. & Le Fort, P. 1982. Geochemical investigations of the origin of the Manaslu leucogranite (Himalaya, Nepal), *Geochimica et Cosmochimica Acta,* **46**, 2279–2292.

Field relations, petrogenesis and emplacement of the Bhagirathi leucogranite, Garhwal Himalaya

M.P. SEARLE[1], R.P. METCALFE[2*], A.J. REX[2] & M.J. NORRY[2]

[1] *Department of Earth Sciences, Oxford University, Parks Road, Oxford OX1 3PR, UK*

[2] *Department of Geology, Leicester University, Leicester LE1 7RH, UK*

* *Deceased*

Abstract: The Bhagirathi leucogranite forms a series of low-angle en echelon, lensoidal intrusions at the top of the High Himalayan slab in the central Himalaya of Garhwal, northern India. The leucogranite comprises the assemblage: K-feldspar + quartz + plagioclase + tourmaline + muscovite ± biotite ± garnet. Compared to other High Himalayan leucogranites it is particularly rich in tourmaline. The granite is generally compositionally homogeneous although it is magmatically banded in both the upper and lower portions.

The Bhagirathi leucogranite is situated structurally above the kyanite and sillimanite gneisses of the Vaikrita Group which, in turn, overlie the north-dipping Main Central Thrust zone of inverted metamorphic isograds. A pegmatite – aplite leucogranite sill and dyke swarm is present around the margins of the leucogranite. Vaikrita Group gneisses below the leucogranite contain a pronounced northeasterly component to generally randomly orientated mineral stretching lineations. This reflects localized reorientation of early, coaxially constrained, mineral growth by later, non-coaxial deformation. Various shear criteria in the gneisses immediately below the granite document the existence of a major zone of ductile NNE–SSW-directed extension across a northeastward dipping, low-angle normal fault zone. The top of the 1500–1800 m thick leucogranite sheet exposed on the peaks of Bhagirathi, Shivling and Thalay Sagar are intrusive into Martoli Formation metasediments of the Tethyan sequence which locally contain andalusite and staurolite. The roof complex shows numerous stoped blocks, xenoliths and elongate rafts of host rock in the top 100 m together with an extensive zone of layer-parallel leucogranite veining.

The northward-dipping Bhagirathi leucogranite was intruded syn-tectonically during the ductile to brittle transition and has been deformed into linked, en echelon bodies within the extensional shear zone at the interface between the Vaikrita Group gneisses and the Tethyan sedimentary cover. The long axes of the leucogranite lenses lie parallel to the *y–z* plane of the finite strain ellipsoid for this extensional duplex. During extension, sub-orthogonal dilatational forces exceeded sub-horizontal shear stresses thus facilitating the repeated emplacement of sheeted granite melt, a process analogous to low-angle tension gash development. The final emplacement level in the crust must have been ultimately controlled by the density contrast between melt and country rocks, the thermal blanketing of the Tethyan sedimentary cover, and the extensional stress field along the top of the High Himalayan slab.

The High Himalayan crustal melt granites comprise about fifteen major plutons, frequently occurring as gently dipping sheets or lens-shaped bodies, and numerous dykes, sills and irregular migmatitic leucosomes along the length of the Himalaya (LeFort 1975, 1988; LeFort *et al.* 1987). They usually occur at the highest structural levels of the High Himalaya zone or Tibetan slab (Fig. 1). This zone is defined as including the rocks bounded along the base by the Main Central Thrust (MCT) and along the top by the major northeast or north-dipping normal fault, the North Himalayan Fault, also referred to as the South Tibetan detachment system (Burchfiel *et al.* 1992). In the Bhutan Himalaya, the leucogranites have intruded up across the normal fault into the base of the overlying Tethyan sedimentary sequence (Dietrich & Gansser 1981). In the Nepal – south Tibet sector of the Himalaya, the major plutons are represented by the Himalchuli, Manaslu, Shisha Pangma, Cho Oyu, Pumori–Lingtren, Everest–Nuptse, Makalu and Jannu–Kangchenjunga plutons. The bases of these large-scale plutons are usually concordant with the schistosity in the underlying metamorphic rocks in kyanite or sillimanite grade. Hundreds or thousands of leucogranitic sills and dykes intrude the metamorphic rocks beneath the granites and frequently connect up into the larger plutons. In some areas, such as the Langtang region of central Nepal, there is a huge layered intrusion – sheeted sill network but no segregation into large plutons. The upper

From TRELOAR, P.J. & SEARLE, M.P. (eds), *Himalayan Tectonics*
Geological Society Special Publication, No. 74, pp. 429–444.

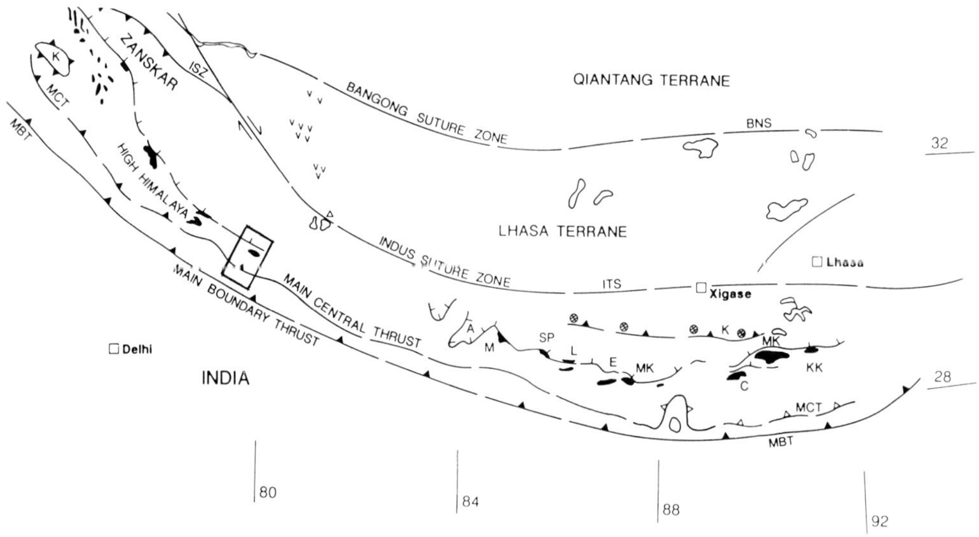

Fig. 1. Geological sketch map of the central Himalaya showing the locations of the main Himalayan leucogranite bodies in black, and the Garhwal Himalaya in particular. K, Kishtwar Window; A, Annapurna; M, Manaslu; SP, Shisha Pangma; L, Langtang; E, Everest; MK, Makalu; C, Chomolhari; MK, Monla Karchung; KK, Kula Kangri. The circles with crosses inside are the North Himalayan granite gneiss domes such as the Kangmar granite (K) cropping out along the hangingwall of a major breakback thrust in southern Tibet south of Xigase. The line with ticks north of the MCT is the North Himalayan Normal Fault, or the South Tibetan Detachment System, which includes the Zanskar Shear Zone in the west and the Jhala Normal Fault in the central Himalaya.

margins of the plutons generally show intrusive contacts into the Palaeozoic or lower Mesozoic sedimentary rocks at the base of the Tethyan shelf sequence. The location of many of the Himalayan granites right at the top of the slab has been attributed to the refractive focusing of heat along the hanging-wall of the normal fault which places sediments with low thermal conductivity on top of metamorphic rocks with high thermal conductivity (Jaupart & Provost 1985; Pinet & Jaupart 1987).

The relationship between metamorphism, migmatisation and melting to produce the peraluminous tourmaline-bearing two-mica granites in the High Himalaya has become one of the major debates in Himalayan geology. Early metamorphism related to crustal thickening after India–Asia collision at about 50 Ma reached kyanite and sillimanite grade, and preceded initiation of movement along the MCT. Later metamorphism, characterized by widespread sillimanite growth, was associated with an increase in temperature, but not pressure. The High Himalayan migmatites and leucogranites are usually associated temporally and spatially with the second metamorphic event. Le Fort (1975), Pêcher (1977) and LeFort *et al.* (1987) argued that melting occurred within the Tibetan Slab gneisses below the Manaslu leucogranite as a result of heat supplied during thrust movement along the Main Central Thrust. England & Molnar (1990) and England *et al.* (1992) argued that dissipative heating caused by frictional shear stresses of about 100 MPa along the Main Central Thrust was great enough to maintain temperatures in the hanging-wall during the later metamorphism. On the other hand, Hodges *et al.* (1988*a,b*) suggested that widespread anatexis could have buffered the temperature in the hanging wall of the MCT near the solidus temperature of the granites. The temperature distribution across the High Himalayan slab is therefore critical to any discussion about the origin of the granites.

Quantitative metamorphic studies involving pressure and temperature distribution in time and space across the High Himalayan zone have been completed in the region south of Everest (Hubbard 1989), the Manaslu region (Hodges *et al.* 1988*a,b*; Pêcher 1989), the Makalu region (Brunel & Kienast 1986) and the Zanskar Himalaya (Stäubli 1989; Searle *et al.* 1992). In the Garhwal Himalaya, Hodges & Silverberg (1988) published a thermobarometric profile from samples collected along the Alaknanda and Dhauli valleys in the east, and Metcalfe (1990, this volume) has produced a P–T profile from samples collected along the Bhagirathi River section in the west. These profiles in general show an inversion of temperature across the MCT zone. Structurally up-section from the MCT, pressures decrease from 6–8 kbar above the MCT zone to 3–4 kbar, 10–12 km up-section, but temperatures remain roughly constant (*c.* 600° C). It appears then that large areas of the High Himalaya have been at uniform high

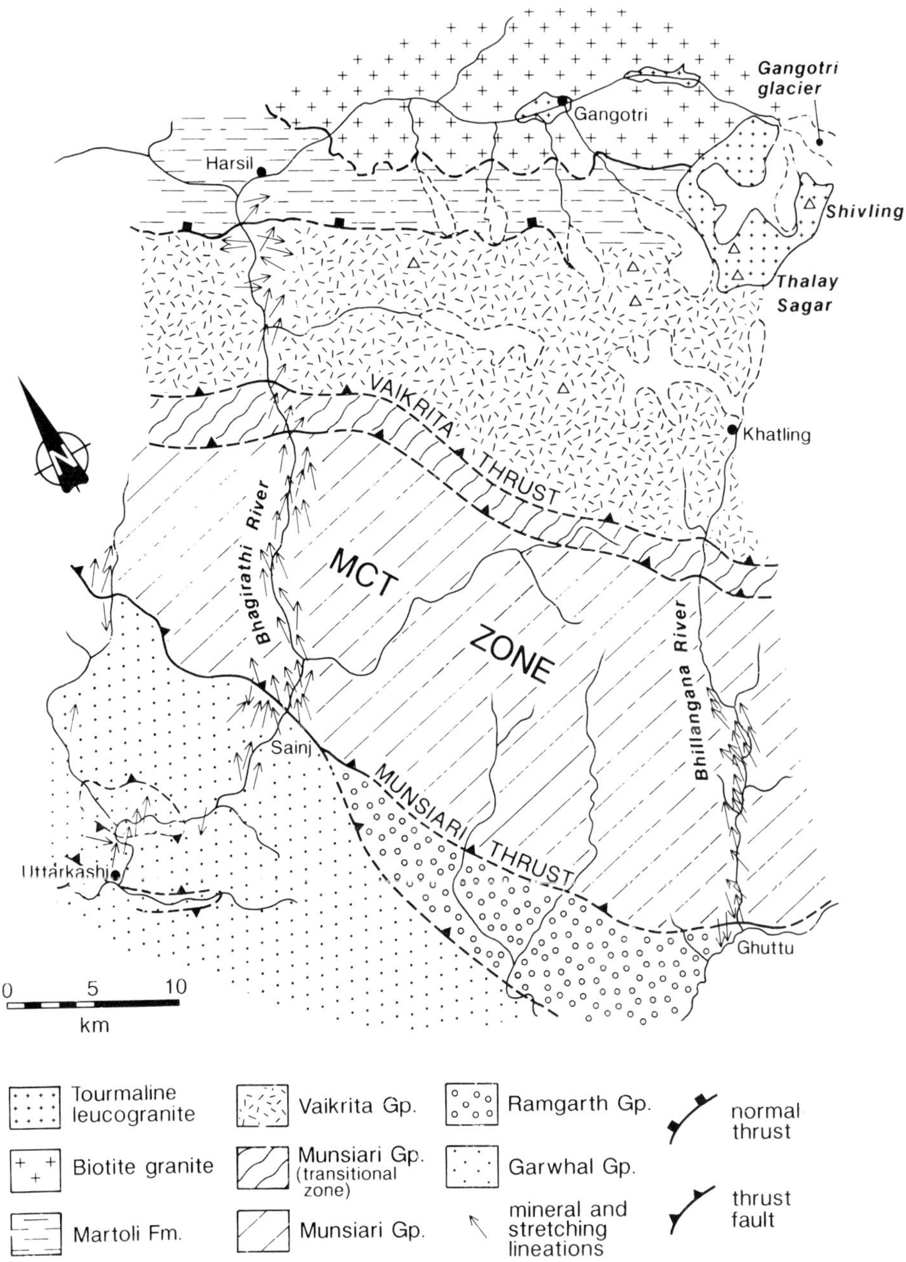

Fig. 2. Lithotectonic map of the Garhwal Himalaya around the Bhagirathi–Bhillangana valleys compiled from maps by Agarwal & Kumar (1973), Pêcher & Scaillet (1989) and the authors, showing stretching lineations along the Bhagirathi and Bhillangana river valleys.

temperatures for horizontal distances up to 50–100 km north of the present outcrop of the MCT. In the Jannu–Kangchenjunga Range of the Darjeeling–Sikkim Himalaya, sillimanite gneisses and Himalayan leucogranites crop out as much as 140 km north of the MCT exposed near Darjeeling (Mohan *et al.* 1989).

This paper describes the field relations of the Bhagirathi leucogranite around the Gangotri glacier in the central Garhwal Himalaya of Kumaun (India) and the mountains of the Bhagirathi Range, Shivling, Meru and Thalay Sagar peaks (Fig. 2). We discuss constraints on melt source and depth of burial, the source of heat and the thermal evolution, and then propose a model for granite emplacement based on the regional structural geology of the Garhwal Himalaya.

Geology of the Garhwal Himalaya

The Indus suture zone and most of the Tethyan sedimentary shelf sequence (the north Indian continental margin) in the central Himalaya lies within southern Tibet. The geology of the Garhwal Himalaya to the south (Fig. 2) is dominated by a series of northward dipping, south-verging thrust sheets. The High Himalayan zone of high-grade metamorphic rocks is bounded to the north by the Jhala normal fault at the top of the slab and to the south by the Main Central Thrust (or Vaikrita thrust) at the base. Along the Bhagirathi and Bhillangana valleys, the Main Central Thrust forms a major SSW-verging shear zone about 10 km thick in which the metamorphic isograds are inverted. Thick K-feldspar augen gneisses, granodiorites, metasediments and amphibolites outcrop within the shear zone, in which numerous kinematic indicators show a SSW-directed sense of shear. The lower boundary of the MCT zone corresponds to the Munsiari thrust, which was first defined as the MCT by Heim & Gansser (1939) further east in the Alaknanda and Dhauli valleys.

Structurally below the Munsiari thrust, to the south, the Lesser Himalaya comprise a series of thrust sheets of low metamorphic grade sediments (Agarwal & Kumar 1973; Valdiya 1980, 1988). These include Precambrian and Palaeozoic quartzites, carbonates and basic volcanics of the Ramgarh and the Garhwal Groups. The next thrust sheets down-section are the Krol and Berinag thrust sheets which structurally overlie the autochthonous foreland of the Indian plate (Valdiya 1980; Johnson 1986). The youngest sediments exposed in the Lesser Himalayan thrust sheets are the Subathu Formation siltstones, mudstones and Nummulitic limestones of late Palaeocene and early Eocene age. These marine sediments were deposited between 62–52 Ma (Johnson & Oliver 1990), immediately prior to collision of India with Asia and closure of the Indus suture zone at about 50 Ma (Searle *et al.* 1988). Johnson (1986) showed that the thrusts of the Kumaun Lesser Himalaya evolved in a foreland propagating sequence, from north to south in the following order: Munsiari–Krol–Berinag–Main Boundary Thrust. The whole thrust pile is folded about steeply-dipping axial planes, and a blind thrust (Himalayan Frontal Thrust) is thought to have acted as the sole thrust which affects the Tertiary Siwalik molasse basin sediments along the footwall of the Main Boundary Thrust. Johnson & Oliver (1990) showed through illite crystallinity studies that individual thrust sheets in the Precambrian and lower Palaeozoic of the Lesser Himalaya have contrasting metamorphic grades, and that thermal effects of the Munsiari Thrust (Heim & Gansser's 1939 MCT, but not the true MCT; see discussion in Searle 1991) are restricted to a section 1300 m below the thrust. Metamorphism in the Lesser Himalaya is therefore dominantly Precambrian or Palaeozoic, whereas the inverted metamorphism across the Vaikrita Thrust (= MCT) and the structurally higher units is Tertiary in age.

Above the Munsiari thrust the metamorphic grade increases up structural section within the MCT zone of inverted metamorphic isograds (Metcalfe 1990; Searle 1991). The MCT zone dips at about 45–50° NNE and the rocks show a mylonitic fabric with abundant evidence (rotated feldspar augen, C–S fabrics) of top-to-the south sense of shear as a result of ductile high-temperature, syn-metamorphic strain. Muscovite K-Ar ages show a southward decrease in cooling ages from the Main Central (Vaikrita) Thrust to the Munsiari Thrust in the south, from 22 to 14 Ma (Metcalfe 1990). Stretching lineations below the Main Central (Vaikrita) Thrust have a general northeasterly component but become more randomly orientated at the top of the slab. Pêcher & Scaillet (1989) and Pêcher (1991) mapped stretching lineation trajectories as becoming progressively rotated towards the north in the Gangotri glacier region, indicative of a dextral sense of shear. However the pattern is not as simple as they show. Early ductile stretching lineations that initially formed parallel to the shortening direction have been folded around later, lower, thrust-related culminations and their orientation as mapped now does not correspond to the regional transport vector or the σ_1 stress direction. The lineation pattern on a local scale, clearly shows that they have been subsequently rotated around the leucogranite lenses in the Bhagirathi and Shivling mountains. The Bhagirathi leucogranite crops out at the top of the High Himalayan slab along the interface between the high-grade gneisses and the sediments of the Martoli Formation above.

Bhagirathi granite

Field relations

Most of the highest peaks around the Gangotri glacier region including the Bhagirathi peaks (6856 m), Shivling (6543 m), Meru (6672 m),

Fig. 3. Geological map of the Gangotri glacier area. T, Tapovan; N, Nandaban base camp sites.

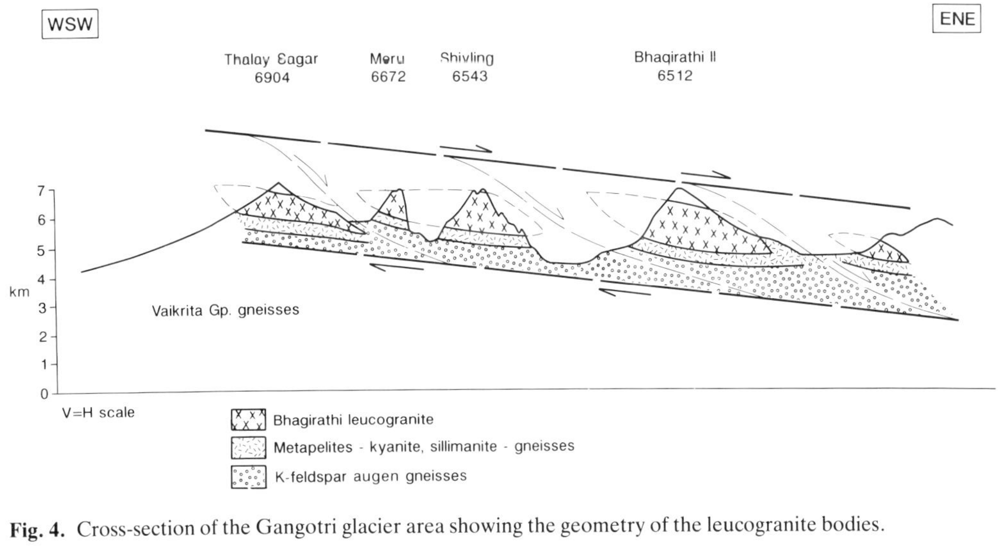

Fig. 4. Cross-section of the Gangotri glacier area showing the geometry of the leucogranite bodies.

(a)

(b)

Fig. 5. (**a**) a photograph and (**b**) a line drawing of the photograph of the southwest faces of Bhagirathi II (6512 m, left) and Bhagirathi III (6454 m, right) showing the geometry of the leucogranite lenses. K-feldspar augen gneiss comprises the pale rock immediately above the Gangotri glacier, with darker pelites above. Note the schistosity in the pelites wrapping around the lateral termination of the leucogranite mega-boudin.

Bhigupanth (6044 m) and Thalay Sagar (6904 m) are composed largely of tourmaline leucogranite that forms massive, gently north-dipping, lens-like sheets (Figs 3 & 4) which extend eastwards to the Badrinath area in eastern Garhwal and further into the Nanda Devi sanctuary region (Changabang). Most of these mountains show the entire thickness of the granite sheet with both the roof and floor exposed in single dramatic vertical faces (Figs 5 & 6a). Smaller outcrops occur around Gangotri village where the tourmaline leucogranite cuts the magmatic foliation in the surrounding, older biotite granite and two-mica porphyritic granite (Fig. 2), the Bhairongathi granite of Pant (1986). These granites lie structurally above the Bhagirathi tourmaline leucogranite and may be related to the 500 Ma Cambro-Ordovician granites along the Lesser Himalaya (Le Fort *et al.* 1986; Stern *et al.* 1989).

The tourmaline leucogranites on the Bhagirathi peaks and Shivling are generally undeformed, although locally they show a weak foliation defined by tourmaline and micas (Scaillet *et al.* 1990). Magmatic layering is also locally apparent where tourmaline ± garnet-rich layers are interbanded with more leucocratic layers. The foliation, where present, is parallel to the

(a)

(b)

Fig. 6. (**a**) The southeast face of Shivling (6543 m) from the junction of the Kirti Bamak and Gangotri glaciers, showing the entire 1800 m thickness of the leucogranite sheet. (**b**) The south face of Meru (6672 m) showing the interleaving due to forceful granite intrusion, of dark metapelites and light-coloured leucogranite along the base of the sheet.

schistosity in the surrounding metamorphic rocks, which wraps around the leucogranite lenses or boudins. This is particularly well exposed along the south and east faces of Bhagirathi III (Fig. 5). Along the base of the leucogranite on Shivling and Meru (Fig. 6a,b) the leucogranite becomes more foliated and has intruded and broken up the country rock by forceful intrusion. Dykes, and more common sills of tourmaline + muscovite ± garnet leucogranite also intrude the underlying metapelites and augen gneisses (Fig. 7a). These sills are parallel to the foliation in the gneisses, contain some magmatic banding and are similar to the 'lit-par-lit' leucogranite intrusions of much of the High Himalaya (e.g. Searle & Fryer 1986, fig. 4d, p. 189).

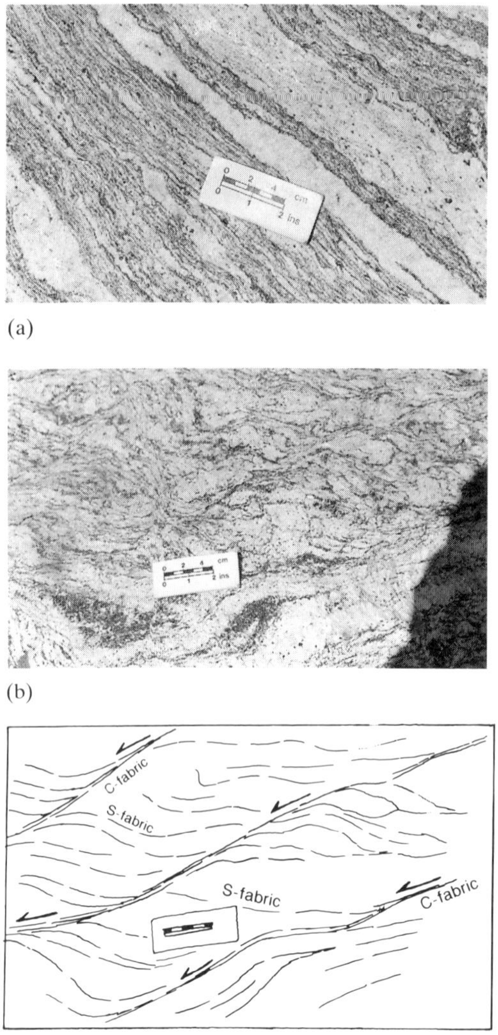

Fig. 7. (**a**) Tourmaline + garnet + muscovite leucogranite pegmatite sill intruding layered K-feldspar augen gneiss from the base of Kedar dome on the Kirti Bamak–Gangotri glacier confluence. (**b**) C–S fabrics in the augen gneisses indicating down-to-the NNE extension, from the Kirti Bamak glacier region. Note the regional schistosity (S fabrics) curving into and out of alignment of intense zones of high shear strain (C fabrics) formed by non-coaxial laminar flow during simple shear. (**c**) Line drawing of (b).

Rotated K-feldspar augen and C–S fabrics in these rocks show strong evidence of down-to-the NNE, normal sense of shear (Fig. 7b,c). These S–C mylonites (Berthé *et al.* 1979) contain two sets of anisotropic foliations, a schistosity (S fabric) related to the accumulation of finite strain, and a shear ('cisaillement' – C fabric) fabric forming localized zones of high shear strain. The regional S fabrics curve into and anastomose with, the narrow zones of high shear strain which form the C fabrics. Both fabrics appear to have developed simultaneously, because they are always present together, and probably developed as a result of non-coaxial laminar flow during simple shear and flattening. They are equivalent to type I S–C mylonites of Lister & Snoke (1984) and can be used as regional kinematic indicators. The localisation of such features close to the granite lead us to believe that the leucogranite was intruded during non-coaxial simple shear in an extensional stress field, although earlier magmatic foliations and compositional layering are also present, particularly in the central part of the leucogranite body. A north-dipping ductile shear zone within the garnet–kyanite-grade metapelites along the base of the granite on Shivling shows early ductile and later brittle fabrics.

The bulk of the tourmaline leucogranite away from the top and base is generally homogeneous and unstrained, varying only in the relative proportion of tourmaline and muscovite. However, the margins do show a weak planar fabric that is parallel to the fabric in the country rock and which is defined predominantly by muscovite (Fig. 8a). Unlike the Manaslu leucogranite, the Bhagirathi leucogranites contain abundant primary tourmaline in the main pluton. Pods or schlieren of tourmaline ± muscovite are also present and probably represent late stage boron-rich fluids crystallizing in vugs or cavities (Fig. 8b). Numerous pegmatite–aplite dykes and sills intrude the country rock around the margins of the leucogranite. These contain the same mineralogy as the main granite body but are usually

(a)

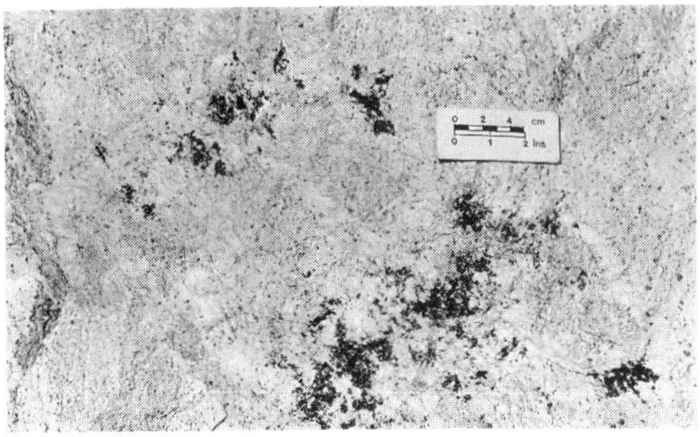

(b)

Fig. 8. (**a**) Garnet-bearing leucogranite that shows a weak planar tectonic fabric, Shivling. (**b**) Pods or schlieren of tourmaline within leucogranite, Shivling. These form the last vestiges of crystallisation of volatile-rich fluid from the Bhagirathi leucogranite magma.

richer in tourmaline, muscovite and garnet and also contain beryl (aquamarine). Early dykes or sills were intruded parallel to the regional schistosity, whereas later dykes were intruded after the formation of the fabric. Along the west bank of the Meru glacier, the upper contact of the Shivling–Meru leucogranite is well exposed. Here, an anastomosing set of leucogranite dykes cross-cuts the regional schistosity in the country rock, showing that these dykes were intruded after the formation of the regional fabric. Along the west bank of the Chaturangi glacier, these same dykes have been folded around upright fold axes trending 133° NW–SE. These late folds, affecting both regional schistosity and leucogranite dykes or sills, may be related to gravitational collapse along the upper contact of the major leucogranite bodies.

The parellelism of the weak fabric within the top of the granite with that in the country rock, and the truncation of the regional fabric by leucogranite dykes implies that the leucogranites were intruded near the end of the ductile phase of extension. The folding of lineation around the granite and its outcrop within discrete en echelon pods (Fig. 4) suggest that granite emplacement, which commenced during the waning stages of ductile extension was completed within the brittle extensional environment during continued uplift and cooling of the gneissic terrain.

The upper contact of the Bhagirathi leucogranite is intrusive into the Martoli Formation

Fig. 9. The roof zone of the Bhagirathi leucogranite on Bhagirathi III showing stoped blocks of dark black slates (Martoli Formation) in the upper part of the leucogranite.

black slates at the base of the Tethyan sedimentary sequence, which locally contain andalusite and staurolite. The uppermost 50–100 m contains numerous stoped blocks and xenoliths of black slates (Fig. 9) and some sills and cross-cutting dykes of tourmaline leucogranite emmanate from the roof complex into the country rock above. Foliation in the low-grade metasediments of the Martoli Formation wraps around the lateral apeces of the leucogranite lenses.

Petrography

The Bhagirathi leucogranite is composed of: quartz + plagioclase + K-feldspar + muscovite + tourmaline ± biotite ± garnet ± beryl and accessory minerals such as zircon and apatite. Tourmaline is always present in the main granite body, unlike some other Himalayan leucogranites where it is only present in the late dykes and pegmatites. It can form up to 15 modal% and occurs as black euhedral–subhedral crystals, scattered throughout the granite or concentrated along layers by magmatic banding. Tourmaline forms an important primary component of the melt which must therefore have been rich in boron. The Bhagirathi and Gangotri leucogranites are therefore more similar to the Everest and Makalu granites (Bordet 1961), which have primary magmatic tourmaline, rather than the Manaslu leucogranite (LeFort 1981). Muscovite is also always present whereas biotite is restricted to the margins of the Bhagirathi leucogranite bodies (Scaillet *et al.* 1990). Both micas and tourmaline are in textural equilibrium and there is no evidence for late metasomatic activity in the main leucogranite body. Garnet is only rarely developed and is restricted to the more tourmaline- and muscovite-rich components. Pale blue beryl (aquamarine) is present as coarse-grained euhedral crystals in the pegmatite dykes and occasionally in the main leucogranite. K-feldspar contains abundant inclusions and probably crystallized late in the sequence. There are no alumino-silicate phases in the Bhagirathi leucogranite, unlike many other Himalayan leucogranites where sillimanite or, more rarely, kyanite (in Kishtwar; Searle & Fryer 1986), or andalusite (in Makalu; Bordet 1961) may be present.

Melt source, fluids and depth of burial

Heterogeneous Sr isotopic values and initial $^{87}Sr/^{86}Sr$ ratios of 0.765–0.785 calculated for 20 Ma (Scaillet *et al.* 1990) imply a melting source of old, heterogeneous metasedimentary rocks in the continental crust. $^{143}Nd/^{144}Nd$ ratios are correspondingly low, less than 0.5119 (Stern *et al.* 1989). These geochemical and isotopic studies, like those of the Manaslu granite (Deniel *et al.* 1986; France-Lanord *et al.* 1988), suggest that there has not been any mantle contribution, or significant magma mixing, during melting or emplacement. The homogeneity of $\delta^{18}O$ values for the Gangotri leucogranite was tentatively ascribed to an homogenization of the source before melting started (Scaillet *et al.* 1990).

LeFort (1975, 1981, 1986) proposed that fluids derived from underthrusted sediments along the footwall of the MCT rose into the hot hanging wall and promoted melting under fluid-saturated minimum melt conditions. Alternatively, fluid-absent melting may result from dehydration reactions when hydrous phases (muscovite, biotite, hornblende) break down. It is generally believed that relatively small volumes of low-temperature (*c.* 650° C), wet granitic melt produced by muscovite breakdown would not be very mobile and would therefore not be capable of intruding far from its source region (Clemens & Vielzeuf 1987). For fluid absent conditions, small quantities of melt can be produced at muscovite breakdown temperatures, but larger volumes can only be produced at higher temperatures (*c.* 850° C) corresponding to biotite breakdown reactions such as:

biotite + aluminium silicate + quartz + plagioclase = K-feldspar + garnet/cordierite + melt.

Peak temperatures reached in the pelites at the top of the High Himalayan slab in the Bhagirathi area are in the range 550–670° C (Metcalfe 1990, this volume) which are not high enough to cross the biotite breakdown reaction. The High Himalayan migmatites therefore probably represent melting of pelites that have not reached high enough temperatures for biotite breakdown, but which have been partially melted by the vapour-absent reaction:

muscovite + quartz + plagioclase = K-feldspar + aluminium silicate ± garnet/biotite/cordierite + melt.

The addition of significant quantities of H_2O-rich fluid into the source region would have the effect of lowering the granite solidus and facilitating melting at lower temperatures. This vapour-present reaction:

muscovite + quartz + plagioclase + H_2O = aluminium silicate + melt

would result in a cool, wet, peraluminous granite melt, which would be relatively immobile and unable to intrude far up into the crust (Clemens & Wall 1981). Migmatites are not widespread beneath the Bhagirathi leucogranite and, although there was an important boron and fluorine-rich fluid phase in the primary magma, there is no strong evidence for extensive metasomatism. Tourmaline and muscovite are both part of the primary magmatic assemblage. Quartzo-feldspathic augen gneisses and amphibolites require higher temperatures for biotite and amphibole breakdown in fluid-absent conditions, and these rock types therefore do not generally show signs of migmatization in the Garhwal area. Along the upper part of the Gangotri glacier area, tourmaline + muscovite ± garnet leucogranite dykes and leucosome components of migmatites are closely associated with biotite + K-feldspar augen gneiss which shows no sign of melting (Fig. 7a,b).

Thermobarometry of metamorphic rocks from the top of the High Himalayan slab along the Bhagirathi River transect suggest pressures in the range 7–10 kbar (Metcalfe 1990, this volume). Using average crustal densities these pressures correspond to depths of peak metamorphism of 25–40 km. The Bhagirathi leucogranite was intruded into these metamorphic rocks at the present top of the slab, and therefore probably originated at melting depths around 40 km presumably further to the north than the present outcrop. These estimated depths of burial are considerably deeper than previous estimates for the Manaslu leucogranite which are around 4 kbar corresponding to depths of burial of only around 14–15 km. Although the viscosity of the magma controls it's ability to intrude, the most important factor must be the stress regime in the crust at the time of crystallization. Although Hollister & Crawford (1986) and Hutton & Ingram (1992) have documented syn-tectonic granite intrusion into contractional thrust-type shear zones, in general terms compression does not aid intrusion of magma and results in crystallization and cooling at depth, whereas tension would facilitate the ascent of magma.

Timing of metamorphism and anatexis

The main regional metamorphism in the High Himalayan slab is constrained in age as following the 50 Ma collision of India and Asia and the subsequent crustal thickening and preceding movement on the Main Central Thrust. Lower Eocene limestones occur at the highest stratigraphic levels of the Lesser Himalaya (Valdiya 1988), south of the MCT, and at the highest stratigraphic levels of the north Indian shelf margin in the Zanskar Range (Searle *et al.* 1988). Internal crustal thickening must have followed collision of the two continental plates and subsequent thrusting in general propagated southwards with time. Hubbard & Harrison (1989) obtained a $^{40}Ar/^{39}Ar$ age on hornblende from within the MCT zone south of Everest of 20.9 ± 0.1 Ma. As the closure temperature for hornblende in the Ar system is around 500° C, this can be regarded as close to the age of metamorphism during movement on the MCT in central Nepal. In the Bhagirathi section, Metcalfe (1990) obtained two $^{40}Ar/^{39}Ar$ hornblende ages from the MCT zone. The structurally lower sample gave a Precambrian age, indicating that the rocks had not been subsequently heated above 500° C since then, while the upper sample gave a Himalayan age of 19.8 ± 2.6 Ma.

As yet there are no accurate (U-Pb zircon, monazite, xenotime) ages of crystallization from the Bhagirathi granite, and we can only assume that anatexis in the Garhwal Himalaya occurred at a similar time to that of the Makalu and Manaslu massifs, which is in the range 25–18 Ma. (Schärer 1984; Deniel *et al.* 1987). For the Bhagirathi leucogranite, Stern *et al.* (1989) published a Rb-Sr mineral isochron of 21.1 ± 0.9 Ma (whole-rock, tourmaline, muscovite, two feldspars) and a K-Ar age on a muscovite separate of 18.9 ± 1.3 Ma from the same rock. These authors obtained an apparent whole rock isochron (?errorchron) from five samples of 64 ± 11 Ma with a larger scatter (MSWD = 32). This geologically unrealistic age probably reflects non-uniform initial Sr isotope ratios and cannot be considered as a true crystallization age.

Petrogenetic and emplacement model

The field relations, mineralogy, geochemistry, isotope chemistry and geochronology of the Miocene High Himalayan leucogranites suggest that they have been derived from partial melting of a metasedimentary protolith as a result of crustal thickening (LeFort *et al.* 1987). The classical model of LeFort (1975) involves intracontinental thrusting along the MCT causing emplacement of a 'hot' hanging wall over a 'cold' footwall resulting in inverted metamorphic isograds. The subduction of wet sedimentary rocks to depths along the MCT zone resulted in liberation of large quantities of fluids released into the hangingwall, which had the effect of triggering melting by lowering the granite solidus temperature. Recent studies of the Langtang leucogranites in central Nepal, however, have questioned the need for fluids in the source region. Harris *et al.* (this volume) suggest that decompression melting during uplift of fluid-absent pelites from 10 to 5 kbar at a starting temperature of around 700° C would generate a 10% melt. These authors also see no correlation between fluid-absent melting and Main Central Thrust movement. In both the Langtang and Garhwal regions it seems that the migmatites may not be related to the melting source of the larger leucogranite bodies. The late sillimanite-grade metamorphism throughout the upper part of the High Himalayan slab is closely associated with the migmatisation and the spatial distribution of the leucogranites. Searle & Rex (1989) proposed a thermal model for the Zanskar Himalaya involving post-metamorphic folding of isograds around large-scale SW-verging recumbent folds during continued compression following peak metamorphism and cooling.

The primary heat source for the melting could have been caused, or buffered, by processes including: (a) frictional (dissipative) heating along the MCT (England & Molnar 1990; England *et al.* 1992), (b) widespread crustal melting (Hodges & Silverberg 1988; Hodges *et al.* 1988*a,b*), (c) ultrametamorphism caused by continued crustal thickening (Searle & Fryer 1986), or (d) exceptionally high heat production rates from a high modal percentage (>1%) of radioactive minerals in the Indian continental crust (Rao *et al.* 1976). The reality probably includes some parts of all these processes. The arguments against frictional heating are that extensive fluid fluxing along the MCT is not compatible with frictional heating, and that the highest temperatures recorded by thermobarometry are not concentrated along the MCT zone at the base of the slab, and that the granites are all concentrated along the top of the slab, 25–40 km structurally above the MCT. The arguments against widespread crustal melting are that the Himalayan leucogranites are the final end-product of regional metamorphism and not the cause of metamorphism. The arguments against exceptionally high heat production rates are that one would expect an even distribution of migmatisation and melting along the entire strike of the Himalaya, rather than the uneven distribution of leucogranites that one sees. Ultrametamorphism, accompanied by renewed crustal thickening as a result of breakback thrusting remains the most suitable process for providing heat for crustal melting.

Calculated *P–T* paths indicate that the later sillimanite-grade metamorphic event involved several kilometers of tectonic burial during the later stages of the thermal history. This later event involved an increase in temperature following the earlier high *T–P* kyanite–sillimanite grade metamorphism (the Eo-Himalayan event). The maintenance of high heat flux was probably caused by continued crustal thickening caused by breakback thrusting, which developed in the hangingwall of active thrusts (e.g. Searle 1986; Searle *et al.* 1992). Although in general terms, major structures and peak metamorphism propagated southwards with time (Fig. 10), much greater crustal thickening occured in the High Himalaya compared to the Lesser Himalaya, and this could only have been accomplished by thrusts developing in the hangingwall of active thrusts.

In this model (Fig. 10) the leucogranite melts form towards the base of the High Himalayan zone, which was being underplated by thrust sheets of the Lesser Himalaya along the footwall of the MCT. Underplating caused extension at high structural levels and the Jhala normal fault was initiated by 'collapse' of the Tethyan sedimentary zone to the north. In reality, the Tethyan sediments remained within 4–5 km of the Earth's surface and almost all the relative motion on the normal fault occurred along the footwall. Erosion rates are considerably greater in the footwall where sillimanite-grade gneisses are now exposed at the Earth's surface.

The low-viscosity, peraluminous, crustal melt leucogranites such as the Bhagirathi leucogranite contain very few restitic enclaves except around the margins, and many of the mineral phases previously regarded as restitic, such as garnet and biotite, are now thought to be magmatic. The main leucogranite body appears to have been fed mainly by dykes and sills feeding magma through by the process of hydraulic fracturing. Muscovite dehydration melting involves a net increase in volume leading to increased pore fluid pressure (Clemens & Vielzeuf 1987). Melt-enhanced deformation (Hollister & Crawford 1986) was undoubtedly a major factor during initial emplacement. The emplacement of the Bhagirathi leucogranite, like most of the other High Himalayan granites was not achieved by a classical, upward, diapiric flow mechanism.

The Bhagirathi leucogranite is located along the interface between the High Himalayan gneisses and the Tethyan sedimentary series. We suggest that the microstructures around the leucogranite lenses show that it was intruded during the ductile to brittle transition of down-to-the-north extensional shear. The geometry of the leucogranite lenses suggest that they form imbricated boudins or large-scale tension gashes within an extensional duplex (Fig. 4). The megaboudins are oriented parallel to the *z–y*

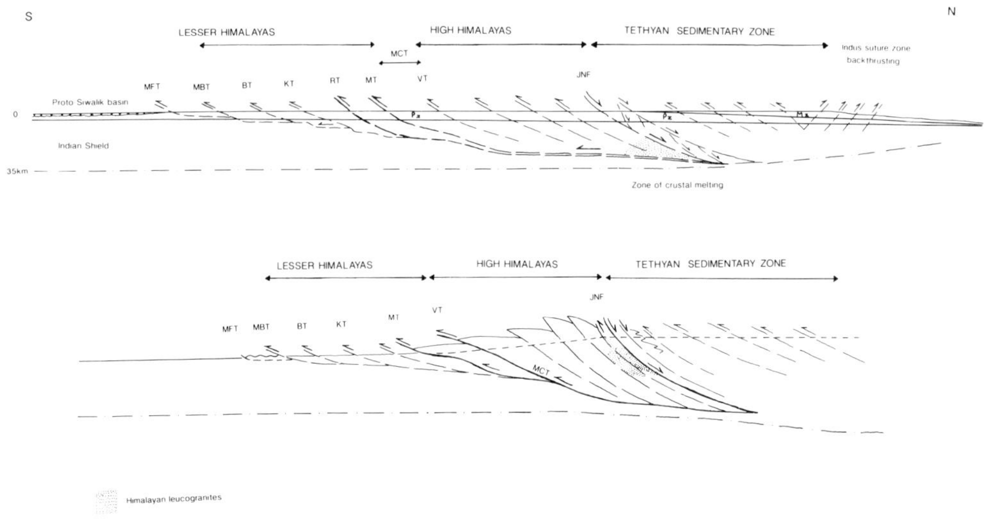

Fig. 10. (**a**) Restored cross-section of the Garhwal Himalaya showing the approximate position of all major thrusts and normal faults prior to compression at about 50 Ma. Thrusting within the Tethyan sedimentary series began during initial collision and in general terms propagated south with time. Thrusting in the High Himalayan zone (Vaikrita duplex) was synchronous with culmination collapse to the north along the Jhala normal fault (JNF). Uplift of the rock on the Main Central Thrust (MCT) was accompanied by simultaneous erosion, so the schematic section in 10b shows the full Vaikrita duplex thickening, ignoring erosion. Note that the melting zone for the leucogranite was at the deeper levels of the Vaikrita duplex and subsequent thrust movement on the lower thrusts within the Vaikrita gneisses and the later MCT has uplifted the zone of melting to higher structural levels. Thrusts propagated in time down-section from the Vaikrita thrust (VT) to the Munsiari thrust (MT) to the Lesser Himalayan Ramgarh (RT), Krol (KT), Berinag (BT) and Main Boundary (MBT) thrusts. The youngest and structurally lowest thrust, the Himalayan Frontal thrust (MFT) is active today and cuts up to the surface in the Tertiary Siwalik Basin along the southern boundary of the Himalaya, where it affects Quaternary Ganga basin gravels. See text for discussion.

plane of the finite strain ellipse for the Jhala normal fault and dip at slightly higher angles to the north than the foliation in the Martoli Formation sediments above. The existence of such a shear regime would normally impose a stress field in which the finite strain ellipse is inclined at *c*. 45° to the shear zone margins. The geometry of the normal fault varies along strike with the angle of dip being greater at Jhala in the west than along the Gangotri glacier in the east. Hutton *et al.* (1990) suggested that rapakivi granites from south Greenland were intruded as large sheets along ductile extensional fault zones which were active during emplacement. However, whereas they suggested that these rapakivi granites were intruded during a phase of regional extension, the Bhagirathi granite was intruded during only localized upper crustal extension caused by excess surface elevation and high erosion-exhumation rates in an overall compressive stress regime.

Exhumation of the mid-crustal Bhagirathi leucogranite was assisted by three simultaneous processes: (a) underplating of Lesser Himalayan thrust slices along the footwall of the MCT zone at deep crustal levels, resulting in uplift of the rock in the hanging wall of the MCT, relative to the geoid, (b) tectonic erosion caused by extension along the footwall of the Jhala normal fault at high crustal levels, and (c) geomorphological erosion resulting in lowering of the surface elevation.

Conclusions

The Bhagirathi leucogranite is one of the High Himalayan crustal melt granites cropping out along the top of the High Himalayan slab, at the interface between the Vaikrita sillimanite-grade gneisses in the footwall and the Tethyan sediments in the hanging wall. It has a primary magmatic mineral assemblage including tourmaline, muscovite ± biotite ± garnet. The leucogranite is compositionally homogeneous although there is some magmatic banding, and has a general lack of restitic material. The uppermost part of the intrusion shows a zone containing xenoliths of Martoli Fm. black slates sometimes containing andalusite or staurolite. Heterogeneous isotopic values for Sr (initial $^{87}Sr/^{86}Sr$ ratios 0.765–0.785) show that the leucogranite was derived from a purely crustal source of Precambrian–Palaeozoic metasediments of the north Indian plate basement with no mantle involvement.

The melting site of the Bhagirathi leucogranite was at the deeper structural level of the High Himalayan (Vaikrita) duplex, and it was intruded into an extensional duplex system along the interface between the High Himalayan gneisses and the low-grade Tethyan sedimentary rocks above. Numerous kinematic indicators, particularly C–S fabrics and rotated K-feldspar augens, show down-to-the-north extensional shear fabrics along the top of the slab. Flattening stresses have resulted in boudinaged lenses of leucogranite within a large-scale, north-dipping extensional duplex. A model for the emplacement of the Bhagirathi leucogranite has been proposed (Fig. 10) which satisfactorily accounts for the macro- and micro-structures as well as the *P–T* conditions of metamorphism for the High Himalayan gneisses. This model may also be applicable to some, but not necessarily all, of the other Himalayan leucogranites, particularly the Manaslu, Langtang–Shisha Pangma, Everest-Nuptse and Kangchenjunga-Jannu leucogranites of Nepal.

This model shows that (a) if frictional heating did cause the excess temperature needed for crustal melting it occurred on the earlier, 'proto-MCT', (Fig. 10a) and not on the present MCT (Fig. 10b), (b) the leucogranites did not intrude far from their source region and their intrusion was along ductile shear zones, (c) melting and emplacement of the leucogranites occurred in a non-coaxial, simple shear environment in which compression at depth and extension at upper structural levels was concomitant, (d) the leucogranites were exhumed by underplating of structurally lower thrust sheets and erosion, and (e) the Jhala normal fault must have been linked at depth to the MCT, and have been active at the same time as the MCT.

This work was funded by NERC grants to B.F. Windley, M.P. Searle and P. Metcalfe, and CNRS grants to P. LeFort. We thank P. LeFort, A. Pêcher, B. Scaillet, J. Dardel and B. Windley for discussions in the field. We also thank Buddhi Singh Rana, Bali and Karan at Mt. Support Trekking in Uttarkashi for logistical help, and B. Scaillet, D. Hutton and P. Treloar for reviews.

References

AGARWAL, N.C. & KUMAR, G. 1973. Geology of the upper Bhagirathi and Yamuna valleys, Uttarkashi district, Kumaun Himalaya. *Himalayan Geology*, **3**, 1–23.

BERTHÉ, D., CHOUKROUNE, P. & JEGOUZO, P. 1979. Orthogneiss, mylonite and non-coaxial deformation of granites: the example of the South Armorican Shear Zone. *Journal of Structural Geology*, **1**, 31–42.

BORDET, P. 1961. *Recherches géologiques dans l'Himalaya du Népal, région du Makalu*. Editions Centre Nationale de Recherches Scientifique, Paris.

BRUNEL, M. & KIENAST, J.R. 1986. Étude pétro-structurale des chevauchements ductile himalayens sur la transversale de L'Everest-Makalu (Népal oriental). *Canadian Journal of Earth Sciences*, **23**, 1117–1137.

BURCHFIEL, B.C., CHEN ZHILIANG, HODGES, K.V., LIU YUPING, ROYDEN, L.H., DENG CHANGRONG & XU JIENE 1992. *The South Tibetan Detachment System, Himalayan Orogen: Extension contemporaneous with and parallel to shortening in a collisional mountain belt.* Geological Society of America, Special Papers, **269**.

CLEMENS, J.D. & WALL, V.J. 1981. Origin and crystallisation of some peraluminous (S-type) granitic magmas. *Canadian Mineralogist*, **19**, 111–131.

—— & VIELZEUF, D. 1987. Constraints on melting and magma production in the crust. *Earth and Planetary Science Letters*, **86**, 287–306.

DENIEL, C., VIDAL, P., FERNANDEZ, A., LE FORT, P. & PEUCAT, J.J. 1987. Isotopic study of the Manaslu granite (Himalaya: Nepal): Inferences on the age and source of Himalayan leucogranites. *Contributions to Mineralogy and Petrology*, **96**, 78–92.

DIETRICH, V. & GANSSER, A. 1981. The leucogranites of the Bhutan Himalaya (crustal anatexis versus mantle melting). *Schweizerische Mineralogische und Petrographische Mittellungen*, **61**, 177–202.

ENGLAND, P.C. & MOLNAR, P. 1990. Temperatures, heat flux, and frictional stress near major thrust faults. *Journal of Geophysical Research*, **95**, 4833–4856.

——, LEFORT, P., MOLNAR, P. & PÊCHER, A. 1992. Heat sources for Tertiary metamorphism and anatexis in the Annapurna – Manaslu region, Central Nepal. *Journal of Geophysical Research*, **97**, 2107–2128.

FRANCE-LANORD, C., SHEPPARD, S.M.F. & LeFort, P. 1988. Hydrogen and oxygen isotope variations in the High Himalayan peraluminous Manaslu leucogranite: evidence for heterogeneous sedimentary source. *Geochimica et Cosmochimica Acta*, **52**, 513–526.

HARRIS, N.B.W., INGER, S. & MASSEY, J.A. 1993. The role of fluids in the formation of High Himalayan leucogranites. *This volume*.

HEIM, A. & GANSSER, A. 1939. *Central Himalaya – Geological observations of Swiss expedition, 1936*. Societe Helvetique Science Naturelle Memoires, **73**.

HODGES, K.V. & SILVERBERG, D.S. 1988. Thermal evolution of the Greater Himalaya, Garhwal, India. *Tectonics*, **7**, 583–600.

——, HUBBARD, M.S. & SILVERBERG, D.S. 1988a. Metamorphic constraints on the thermal evolution of the central Himalayan orogeny. *Philosophical Transactions of the Royal Society, London*, **A326**, 257–280.

——, LE FORT, P. & PÊCHER, A. 1988b. Possible thermal buffering by crustal anatexis in collisional orogens: Thermobarometric evidence from the Nepalese Himalaya. *Geology*, **16**, 707–710.

HOLLISTER, L. & CRAWFORD, M.-L. 1986. Melt enhanced deformation: a major tectonic process. *Geology*, **14**, 555–561.

HUBBARD, M.S. 1989. Thermobarometric constraints on the thermal history of the Main Central Thrust Zone and Tibetan Slab, eastern Nepal Himalaya. *Journal of Metamorphic Geology*, **7**, 19–30.

—— & HARRISON, T.M. 1989. ^{40}Ar-^{39}Ar age constraints on deformation and metamorphism in the MCT zone and Tibetan Slab, eastern Nepal Himalaya. *Tectonics*, **8**, 865–880.

HUTTON, D.H.W. & INGRAM, G.M. 1992. The Great Tonalite Sill of southeastern Alaska and British Columbia: emplacement into an active contractional high angle reverse shear zone (extended Abstract). *Transactions of the Royal Society of Edinburgh, Granite volume*, in press.

——, DEMPSTER, T.J., BROWN, P.E. & BECKER, S.D. 1990. A new mechanism of granite emplacement: intrusion in active extensional shear zones. *Nature*, **343**, 452–5.

JAUPART, C. & PROVOST, A. 1985. Heat focussing, granite genesis and inverted metamorphic gradients in continental collision zones. *Earth and Planetary Science Letters*, **73**, 385–397.

JOHNSON, M.R.W. 1986. The structural evolution of the Kumaun Lesser Himalaya. *Current Trends in Geology*, **9**, 27–39.

—— & OLIVER, G.J.H. 1990. Precollision and postcollision thermal events in the Himalaya. *Geology*, **18**, 753–756.

LEFORT, P. 1975 Himalayas: the collided range, present knowledge of the continental arc. *American Journal of Science*, **275**, 1–44.

—— 1981. Manaslu leucogranite: a collision signature of the Himalaya, a model for its genesis and emplacement. *Journal of Geophysical Research*, **86**, 10545–10568.

—— 1986. Metamorphism and magmatism during Himalayan collision. *In:* COWARD, M.P. & RIES, A.C. (eds) *Collision Tectonics*. Geological Society, London, Special Publications, **19**, 159–172.

—— 1988. Granites in the tectonic evolution of the Himalaya, Karakoram and southern Tibet. *Philosophical Transactions of the Royal Society, London*, **A326**, 281–299.

——, DEBON, F., PÊCHER, A., SONET, J. & VIDAL, P. 1986. The 500 Ma magmatic event in alpine southern Asia, a thermal episode at Gondwana scale. *Science de la Terre*, **47**, 191–209.

——, CUNEY, M., DENIEL, C., FRANCE-LANORD, C., SHEPPARD, S.M.F., UPRETI, B.N. & VIDAL, P. 1987. Generation of the Himalayan leucogranites. Tectonophysics, **134**, 39–57.

LISTER, G.S. & SNOKE, A.W. 1984. S–C Mylonites. *Journal of Structural Geology*, **6**, 617–638.

MOHAN, A., WINDLEY, B.F. & SEARLE, M.P. 1989. Geothermobarometry and development of inverted metamorphism in the Darjeeling-Sikkim region of the eastern Himalaya. *Journal of Metamorphic Geology*, **7**, 95–110.

METCALFE, R.P. 1990. *A thermotectonic evolution for the Main Central Thrust and Higher Himalaya, western Garhwal, India*. PhD thesis, University of Leicester.

—— 1993. Pressure, temperature and time constraints across the Main Central Thrust zone and High Himalayan Slab in the Garhwal Himalaya. *This volume*.

PANT, R. 1986. *Petrochemistry and petrogenesis of the Gangotri granite and associated granitoids, Garhwal Himalaya*. PhD Thesis, University of Roorkee, India.

PÊCHER, A. 1977. Geology of the Nepal Himalaya: deformation and petrography in the Main Central Thrust Zone. *Collogue internationale 268, Ecologie et géologie de l'Himalaya*. Centre Nationale de Recherches de Sciences de la Terre, 301–318.

—— 1989. The metamorphism in the Central Himalaya. *Journal of Metamorphic Geology*, **7**, 31–41.

—— 1991. The contact between the Higher Himalaya crystallines and the Tibetan sedimentary series: Miocene large-scale dextral shearing. *Tectonics*, **10**, 587–598.

—— & SCAILLET, B. 1989. La structure du Haut Himalaya au Garhwal (Indes). *Eclogae Geologicae Helvetiae*, **82**, 655–668.

PINET, C. & JAUPART, C. 1987. A thermal model for the distribution in space and time of the Himalayan granites. *Earth and Planetary Science Letters*, **84**, 87–99.

RAO, R.U.M., RAO, G.V. & NARAIN, H. 1976. Radioactive heat generation and heat flow in the Indian shield. *Earth and Planetary Science Letters*, **30**, 57–64.

SCAILLET, B., FRANCE-LANORD, C. & LE FORT, P. 1990. Badrinath-Gangotri plutons (Garhwal, India): petrological and geochemical evidence for fractionation processes in a high Himalayan leucogranite. *Journal of Volcanology and Geothermal Research*, **44**, 163–188.

SCHÄRER, U. 1984. The effect of initial ^{203}Th equilibrium on young U-Pb ages: the Makalu case, Himalaya. *Earth and Planetary Science Letters*, **67**, 191–204.

SEARLE, M.P. 1986. Structural evolution and sequence of thrusting in the High Himalayan, Tibetan-Tethys and Indus suture zones of Zanskar and Ladakh, western Himalaya. *Journal of Structural Geology*, **8**, 923–936.

—— 1991. Comment on 'Precollision and postcollision thermal events in the Himalaya'. *Geology*, **19**, 666–667.

—— & FRYER, B.J. 1986. Garnet, tourmaline and muscovite-bearing leucogranites, gneisses and migmatites of the Higher Himalaya from

Zanskar, Kulu, Lahoul and Kashmir, *In*. Coward, M.P. & Ries, A.C. (eds) *Collision Tectonics*. Geological Society, London, Special Publications, **19**, 185–201.
—— & Rex, A.J. 1989. Thermal Model for the Zanskar Himalaya. *Journal of Metamorphic Geology*, **7**, 127–134.
——, Windley, B.F., Coward, M.P., Cooper, D.J.W., Rex, A.J., Rex, D.C., Li Tingdong, Xuchang, X., Jan, M.Q., Thakur, V.C. & Kumar, S. 1987. The closing of Tethys and the tectonics of the Himalayas. *Geological Society of America, Bulletin*, **98**, 678–701.
——, Cooper, D.J.W. & Rex, A.J. 1988. Collision Tectonics of the Ladakh-Zanskar Himalaya. *Philosophical Transactions of the Royal Society of London*, **A326**, 117–150.
——, Waters, D.J., Rex, D.C. & Wilson, R.N. 1992. Pressure, temperature and time constraints on Himalayan metamorphism from eastern Kashmir and western Zanskar. *Journal of the Geological Society, London*, **149**, 753–773.
Stäubli, A. 1989. Polyphase metamorphism and the development of the Main Central Thrust. *Journal of Metamorphic Geology*, **7**, 73–93.
Stern, C.R., Kligfield, R., Schelling, D., Virdi, N.S., Futa, K., Peterman, Z.E. & Amini, H. 1989. The Bhagirathi leucogranite of the High Himalaya (Garhwal, India); age, petrogenesis and tectonic implications. *Geological Society of America, Special Publications*, **232**, 33–45.
Valdiya, K.S. 1980. *Geology of the Kumaun Lesser Himalaya*. Wadia Institute of Himalayan Geology, Dehra Dun, India.
—— 1988. Tectonics and evolution of the central sector of the Himalaya. *Philosophical Transactions of the Royal Society, London*, **A326**, 151–175.

Extensional tectonics in the Himalayan orogen, Zanskar, NW India

R.C. PATEL, SANDEEP SINGH, A. ASOKAN, R.M. MANICKAVASAGAM & A.K. JAIN

Department of Earth Sciences, University of Roorkee, Roorkee-247 667, India

Abstract: Large-scale nappes of the deformed Higher Himalayan Crystallines (HHC) were thrust southward over the Lesser Himalayan Proterozoic foreland during the Late Cenozoic. Critical evaluation of shear fabrics, reveals that the HHC underwent an earlier phase of ductile shearing with top-to-southwest overthrust-type sense of movement. On which was superposed layer-parallel NE-trending extension. As a consequence, the Zanskar Shear Zone (ZSZ) along the northern boundary of the HHC is re-interpreted as a complex ductile shear zone with an initial top-to-southwest sense of thrusting followed by the later superposition of top to northeast-directed extension. This period of extension is recognized across the whole of the HHC and is the result of the uplift and exhumation of the HHC by southwestward directed thrusting and ramping along numerous thrusts including the Main Central Thrust (MCT).

The Himalayan orogenic belt displays an unique interplay of large-scale nappe translation and extensional tectonics at different crustal levels. Burg *et al.* (1984), Burchfiel & Royden (1985), Herren (1987) and Royden & Burchfiel (1987) have recognized gravity-controlled collapse of the Himalayan Miocene topographical front along the northern margin of the Higher Himalaya. This contrasts the overall southwest-vergent sense of thrusting within both the Higher Himalayan Crystalline (HHC) belt and the Tethyan sequence in Zanskar.

Late orogenic crustal extension has been well documented in many belts including the Cordilleran Metamorphic Core Complex of western North America (Malavieille 1987), the Alps (Philippot 1990) and the Scandinavian Caledonides (Serguret *et al.* 1989; Andersen *et al.* 1991). In the NW Himalaya, the Zanskar normal fault (Searle 1986) or Zanskar Shear Zone (Herren 1987) is important due to its being an extensional feature developed within a compressional orogenic belt. As models of extension within collisional orogens have been applied to other orogenic belts, as well as the Himalaya, they are, therefore, kinematically-significant.

In the present paper, an extensional tectonic model for the whole of the HHC belt in the NW-Himalaya has been presented together with a critical re-evaluation of the Zanskar Shear Zone.

Geological framework

The India–Asia collision caused crustal-scale thickening, remobilization and thrusting of the Proterozoic basement of the Indian Plate and its cover into an extensively deformed and polymetamorphosed HHC belt (Fig. 1). In the NW Himalaya, this belt is overlain by the Tethyan Sedimentary Zone along the Chenab, Doda and Suru Rivers (Vohra *et al.* 1982; Srikantia *et al.* 1978; Honegger *et al.* 1982; Searle 1986; Herren 1987; Kundig 1989). In Jammu and Kashmir, the HHC belt of the Great Himalayan Range incorporates a vast expanse of meta-pelitic and psammitic rocks together with amphibolite and calc-silicate bands. These are associated with gneissose and granitic bodies of Proterozoic, Early Palaeozoic and Miocene ages (Honegger *et al.* 1982; Searle & Rex 1989). The HHC is thrust southwestward along the folded Main Central Thrust (MCT) over the Lesser Himalayan Shali belt and the Kishtwar–Kulu–Rampur Windows (Fig. 1). Towards the south, in the Kashmir–Bhadarwah basins, the HHC is infolded with unconformably overlying fossiliferous rocks of the Tethyan Sedimentary Zone. However, in the north, the Phanerozoic sedimentary cover is separated from the HHC by the Zanskar Shear Zone (ZSZ) (Fig. 2a).

The 2–3 km wide ZSZ is a major, gently to moderately NE-dipping structure. The zone is characterized by highly strained augen mylonites, sheared bodies of granitoids and tectonized basal Tethyan sedimentary sequences with tectonic fabrics sub-parallel to the basement–cover contact. It contains an extensive tabular tourmaline-bearing leucogranite, deformed into a granite gneiss (Fig. 2b). A steeper NE-dipping extensional shear fabric characterizes the ZSZ (Fig. 2b). Herren (1987) and Searle *et al.* (1988) have the opinion that the ZSZ was active during Miocene–Pliocene times as a normal fault and telescoped the metamorphic isograds within the HHC. The ZSZ is analogous to

From TRELOAR, P.J. & SEARLE, M.P. (eds), *Himalayan Tectonics*
Geological Society Special Publication, No. 74, pp. 445–459.

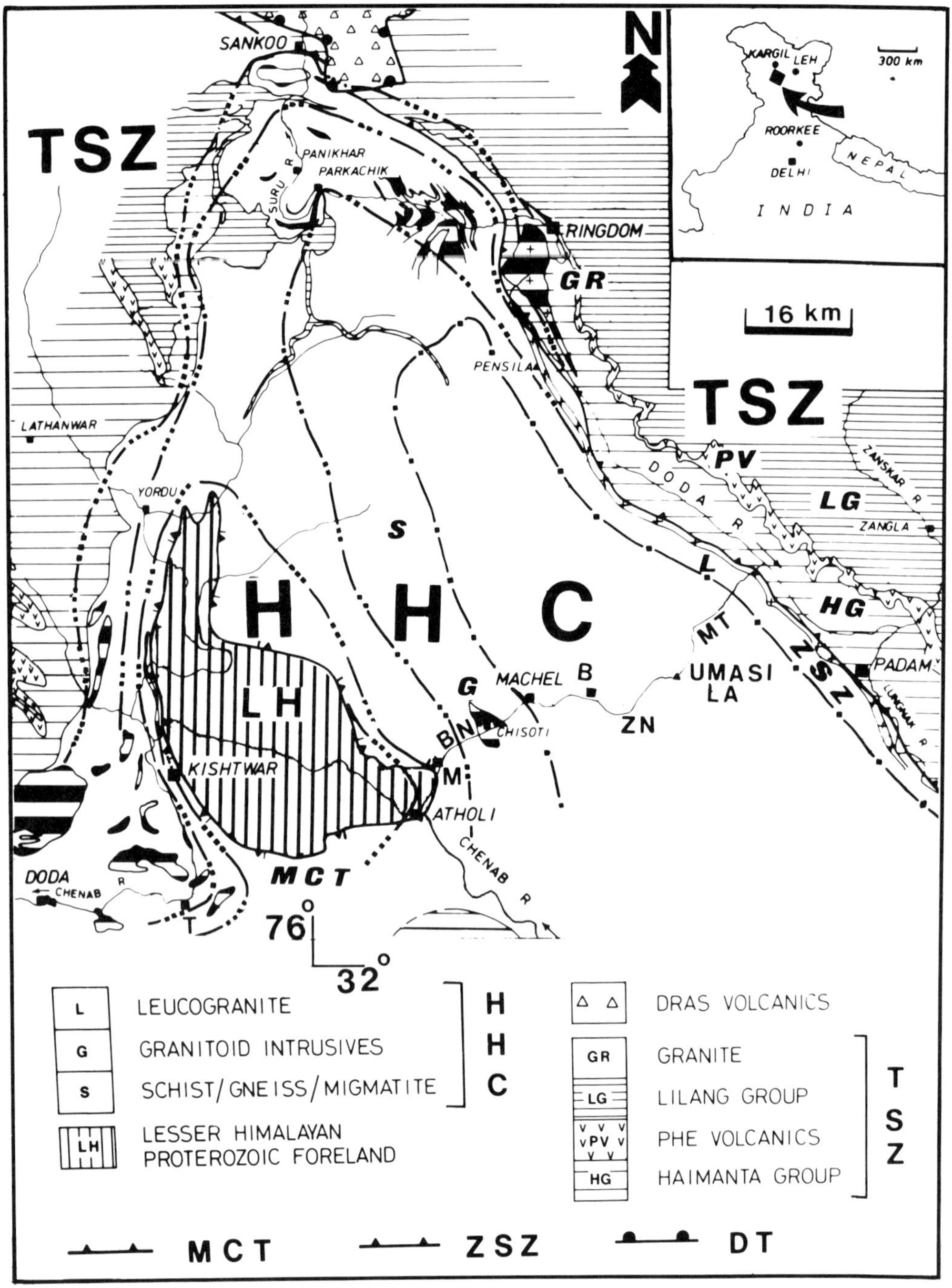

Fig. 1. Simplified tectonic framework of SE Jammu and Kashmir along the Chenab, Suru and Doda Rivers. Abbreviations are BN-Bhot Nala, MT-Mulung Tokpo. Metamorphic isograd boundaries are drawn by thick lines with dots. (— · · · · —) garnet, (— · · ·—) staurolite–kyanite, (— · · —) sillimanite–muscovite and (— . —) sillimanite–K-feldspar grades. Compiled after Vohra *et al.* (1982); Honegger *et al.* (1982); Searle (1986); Herren (1987); Searle *et al.* (1988); Kundig (1989), Searle & Rex (1989) and our observations.

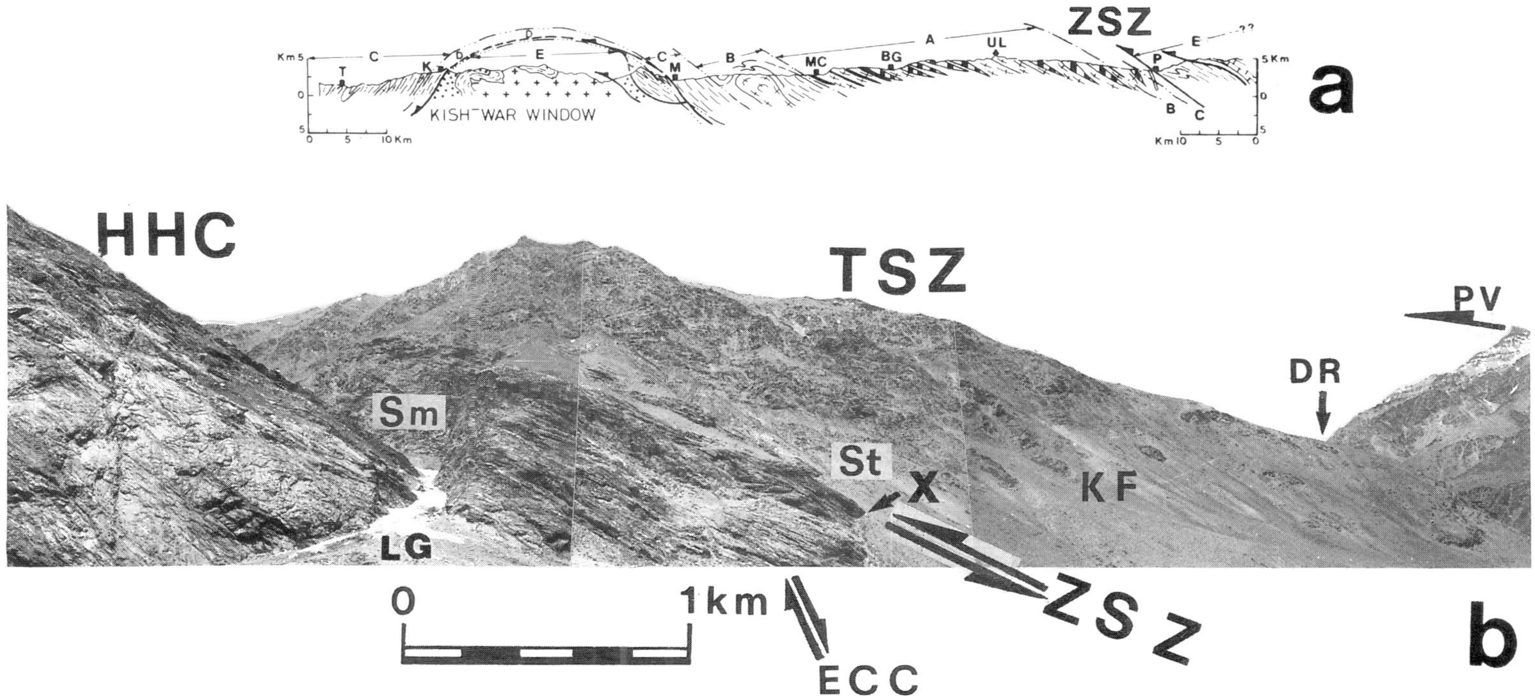

Fig. 2. (**a**) Simplified geological cross-section along Thatri (T)–Kishtwar (K)-Masu (M)–Machel (MC)–Bhuzas Got (BG)–Umasi La (UL)–Padam (P) showing the main structure. Metamorphic isograd boundaries shown as A to E in decreasing order from sillimanite–K-feldspar to biotite grade. (**b**) Panoramic view across the Zanskar Shear Zone (ZSZ) along the left bank of Mulung Tokpo, a tributary of the Doda River (DR). Deformed leucogranite lenses (LG) in sillimanite–muscovite schist/gneiss (Sm) of the HHC are traversed by extensional crenulation foliation (ECC). Note about 200 m thick staurolite-kyanite schist (St) beneath the Tethyan Sedimentary Zone (TSZ) of the Karsha Formation (KF) and the overthrust Phe Volcanics (PV). View looking NW between Ating and Junkul Gompa.

other fault zones, which are documented at almost the same tectonic levels in the Garhwal–Kumaon (Valdiya 1989) and the Annapurna Range, Nepal (Burg *et al.* 1984; Burchfiel & Royden 1985; Royden & Burchfiel 1987).

Deformation pattern

The metamorphosed and deformed HHC pile has been considered as a major 15–20 km wide, NE-dipping ductile shear zone resulting from extreme intracontinental crustal shortening within the Indian plate (Burg *et al.* 1984; Brunel 1986; Coward *et al.* 1986; Jain & Anand 1988). Four major deformational phases have been identified in the HHC of NW Zanskar. The earliest D1 phase has produced rarely noticeable, isoclinal F1 folds with an axial plane foliation S1, which parallels the lithological layering. The subsequent deformation (D2) transposes S1 into the most prominent foliation S2. Reclined to recumbent F2 folds persistently trend NNE/NE with fold axes parallel to the stretching/mineral lineation L2 on the axial plane foliation S2 (Fig. 3a,b).

The S2 foliation is a pervasive ductile S–C shear fabric with a consistent NE- or SW-plunging mineral/stretching lineation (Fig. 4; also Brunel 1986). This lineation plunges at 15–40° with a very high pitch on the S2 foliation surface irrespective of its orientation. It plunges orthogonally to the Himalayan trend and is

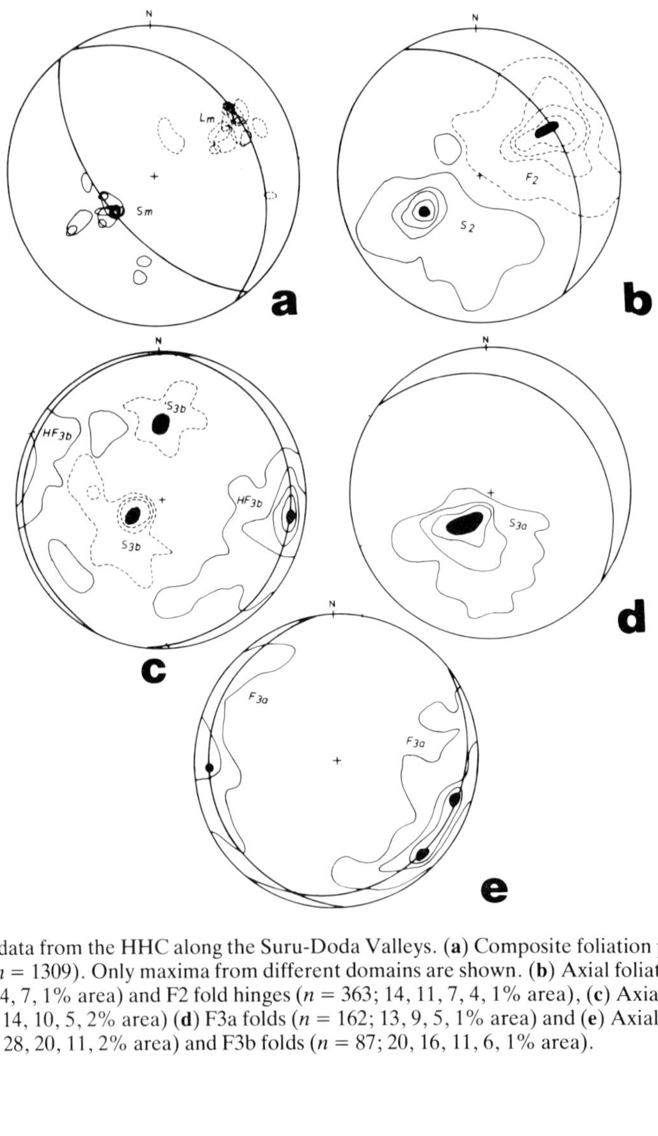

Fig. 3. Structural data from the HHC along the Suru-Doda Valleys. (**a**) Composite foliation poles ($n = 1748$) and lineation L2 ($n = 1309$). Only maxima from different domains are shown. (**b**) Axial foliation S2 poles ($n = 206$; 26, 20, 14, 7, 1% area) and F2 fold hinges ($n = 363$; 14, 11, 7, 4, 1% area), (**c**) Axial plane foliation S3a poles ($n = 98$; 14, 10, 5, 2% area) (**d**) F3a folds ($n = 162$; 13, 9, 5, 1% area) and (**e**) Axial plane foliation S3b poles ($n = 59$; 28, 20, 11, 2% area) and F3b folds ($n = 87$; 20, 16, 11, 6, 1% area).

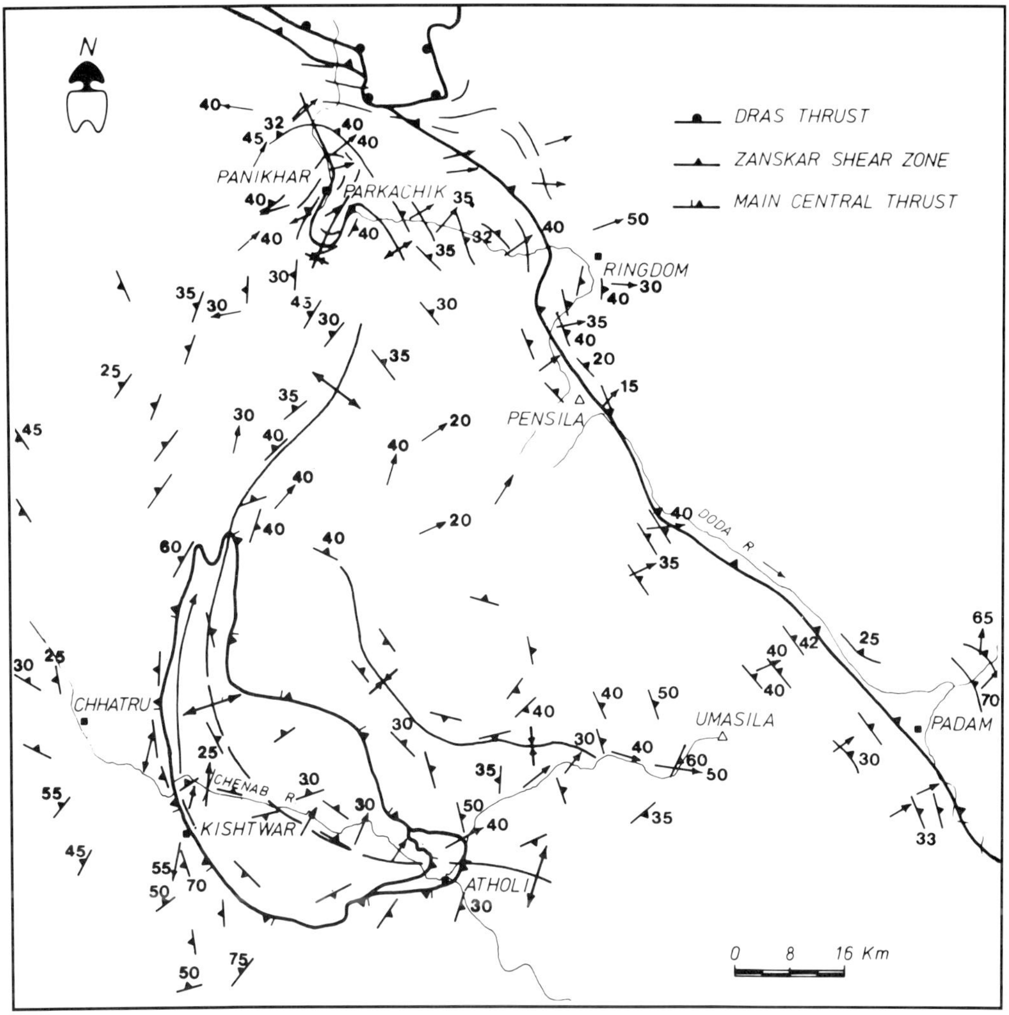

Fig. 4. Regional trends of the main foliations and lineations in SE Jammu and Kashmir across the main tectonic units (compiled mainly from our observations and those of Vohra *et al.* 1982; Herren 1987 and Kundig 1989).

characterized by a preferred orientation of sillimanite, kyanite, staurolite, amphibole, tourmaline and mica and by stretched, elongated augen megacrysts of quartz and feldspar. Many high grade minerals grow along both S- and C-planes. Geothermobarometric data from the HHC reveal peak temperatures of *c.* 780° C and pressures of 1050–1150 MPa from syntectonic garnet cores grown during the main metamorphic and deformation event. The main stretching lineation represents the *x*-direction of the strain ellipsoid during top-to-SW thrusting during peak metamorphic conditions. This interpretation differs from earlier observations in Zanskar and Garhwal, where all the S–C fabrics and the main transverse lineations have been interpreted as being due to NE-orientated ductile normal faulting (Herren 1987) and associated dextral movements along the northern margin of the HHC (Pêcher 1991; Pêcher *et al.* 1992).

During the later D3 deformation, the earlier foliations were folded by isoclinal to close F3a folds with low to moderately NW- or SE-plunging hinges (Fig. 3c,d). Axial surfaces, S3a, of these folds cut across the main foliation in their hinge zones and dip moderately towards the NE (fig. 3c). Continued deformation produced open inclined F3b folds developed coaxially with the F3a folds (Fig. 3e). These have a characteristic crenulation foliation, S3b, that produces a type-3 interference pattern. Kink bands, brittle-ductile discrete shear zones and tension gashes are developed during the last deformation phase (D4).

Shear criteria and sense of displacement

Numerous shear criteria have recently been used to decipher the sense of displacement in orogenic belts (Berthe *et al.* 1979) and reveal complex movement histories during various deformational phases. Such structures have been analysed here, not only from the ZSZ but also from the Kishtwar Window of the Lesser Himalaya, the HHC and the Tethyan Sedimentary Zone for an overall better assessment.

Ductile non-coaxial deformational structures

Ductile non-coaxial deformation structures are present within all tectonic units from the Lesser Himalaya at the base, through the HHC, to the ZSZ at the top. These structures show a consistent top-to-southwest overthrust sense of displacement developed during the pervasive D2 deformation. S–C composite planar fabrics

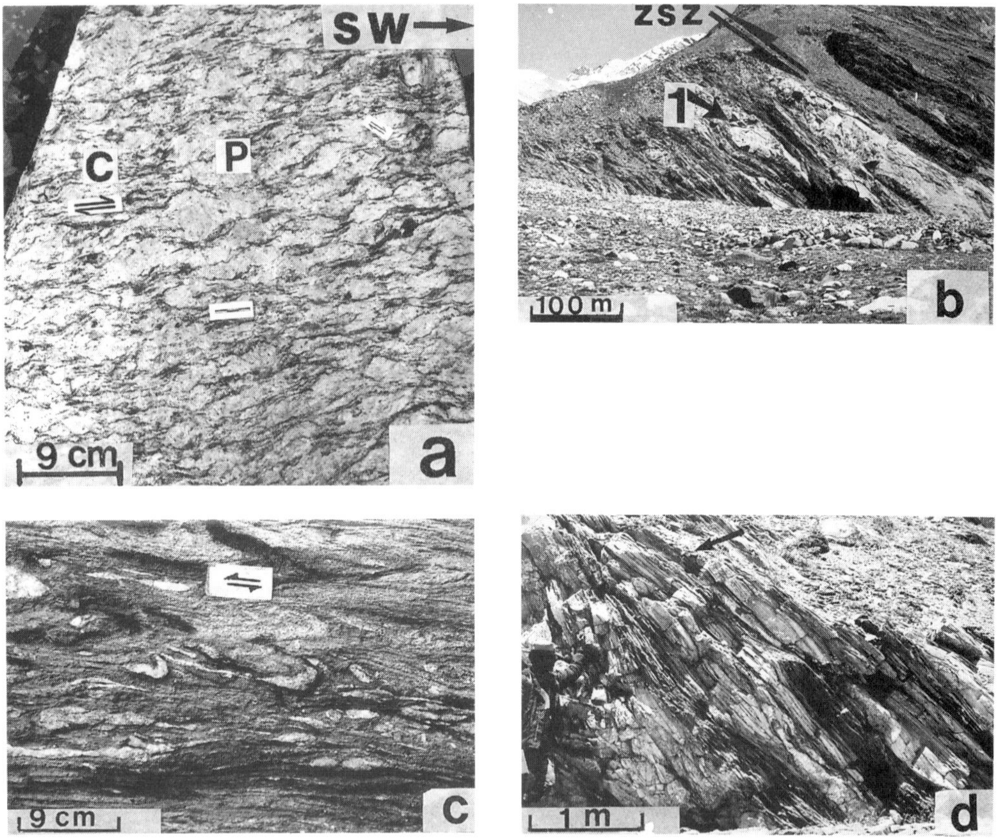

Fig. 5. Ductile non-coaxial deformational structures in SE Jammu and Kashmir revealing top-to-SW overthrust sense of ductile shearing. (**a**) S–C composite planar fabric in porphyroblastic augen gneiss in the HHC. A prominent asymmetric tail/pressure shadow (P) is developed. Loc. Bhuzas along Bhot Nala. (**b**) Asymmetrically sheared leucogranite lenses (1) in gneiss/schist immediately beneath the ZSZ near Shamrakarpo. Here the gently dipping ZSZ separates the Tethyan Sedimentary Zone from the HHC. (**c**) Intrafolial tight to isoclinal folds sheared in the direction of ductile tectonic transport in the MCT zone near Atholi. (**d**) Duplex structure in phyllite and limestone of the Karsha Formation near the ZSZ around Zolidok. Roof thrust is located near arrow. Top left corners of figures except (a) point towards the SW.

(Berthe *et al.* 1979; Lister & Snoke 1984) are extensively distributed within the Lesser Himalaya, HHC, ZSZ and the granitoid body of the Tethyan Sedimentary Zone. The C-surfaces are zones of intense progressive ductile shearing which sigmoidally bend the S-surfaces (Fig. 5a). The average angle between C- and S-surfaces varies from 15–40°. With progressive shearing, both become subparallel to each other. In rocks having an S–C shear fabric, asymmetrical pressure shadows (Simpson & Schmidt 1983) around rigid megacrysts of quartz, feldspar and rotated garnet show sigmoidal tails in the direction of C-shears (Fig. 5a). These shadows are mostly infilled with quartz and mica. Inclusions within garnet porphyroblasts reveal continuous rotational fabrics related to ductile shearing. σ- and δ-types of quartz and feldspar asymmetrical augen/megacryst (Passchier & Simpson 1986) and sheared lenses are common in granite gneiss and its mylonitized derivatives (Fig. 5a). Sheared lenses of deformed granite, pegmatite and of quartzite-limestone are common with the same top-to-southwest sense of displacement in all the tectonic units including the ZSZ (Fig. 5b). Elsewhere, quartz veins and other rigid objects are sheared in the direction of ductile tectonic transport towards the SW within either the Lesser Himalayan rocks or the MCT Zone (Fig. 5c). Asymmetric F3 folds ubiquitous in all tectonic units, verge southwest. In zones of very high strain like the MCT and ZSZ, these folds are even intrafolial with ductile C-shear planes truncating their limbs (Fig. 5c).

Rigid feldspar megacrysts in mylonite along the ZSZ and elsewhere in porphyritic granitoids of the Higher and Lesser Himalayas are fractured and displaced by high-angle mesoscopic fractures. Small-scale displacement along these fractures is antithetic to the overall sense of movement towards SW. These fractures dip mostly at 50–60° in the opposite direction to the main ductile shearing. In the vicinity of subsidiary thrust faults, duplex structures coalesce downwards to a floor thrust and upwards to a roof thrust (Boyer & Elliott 1982). Within the ZSZ and basal parts of the Tethyan Sedimentary zone, limestone and calc-phyllite units are tectonically intercalated by small thrusts which merge into a roof thrust (Fig. 5d). Other small thrusts in phyllite and limestone are sub-parallel to the main thrust and coalesce upwards.

Superposed layer-parallel extensional structures

A consistent NE–SW extension across the HHC and ZSZ is superposed on the earlier ductile non-coaxial deformation and represents a late deformation phase that overprints the D2 structures and orogen-parallel NW–SE-trending D3 phase. Symmetrical and asymmetrical foliation boudins (cf. Platt & Vissers 1980) are developed in migmatite and gneiss on the main foliation (Fig. 6a). In symmetrical foliation boudinage, foliation pinches towards the cuspate region, which is sometimes traversed by irregular fractures, infilled with quartzo-feldspathic material. The asymmetrical foliation boudinage is characterised by ductile shears, which are oblique to the main foliation in the normal-fault sense (Fig. 6a).

Single and conjugate sets of extension crenulation cleavage/foliation (Ecc of Platt & Vissers 1980) in both the HHC and the ZSZ provide undisputed evidence of extension. A single set is prominently developed in schist, gneiss and migmatite, where ecc surfaces are spaced 0.5–5.0 cm apart (Fig. 6b,c). Elsewhere, the foliation changes its orientation indicating gradual rotation during the overall non-coaxial deformation (Platt & Vissers 1980). In the conjugate sets of extensional crenulation foliation, one set dips NE at steeper angles to the main foliation (Fig. 6d) and is more prominent than the westerly, gently-dipping planes. As both of these planes behave like normal-faults (Fig. 6e), the overall movement is one of shortening normal to pre-existing foliations and extension parallel to them. This results in minor warping of the earlier foliations in the shortening domain and development of boudinage in the extensional field (Fig. 6d). In addition, single discrete sets of shear bands (White 1979) are seen in high grade gneisses of the HHC and calc-phyllites of the Tethyan Sedimentary Zone. These bands trend NW–SE and dip NE at about 40–60° indicating a small displacement of the rock mass towards the NE along the ZSZ and within the HHC. Further, brittle–ductile and brittle faults/shear zones along the ZSZ affect migmatite, quartzo-feldspathic gneiss, granite, aplite and amphibolite layers in the HHC (Fig. 6f), where the amount of measurable displacement varies from 0.3–5 cm along the normal-type shear planes (Fig. 6f). In the Mune–Padam area, fault surfaces are, at places, marked with a thin layer of pseudotachylite. These fault zones are parallel to the extension crenulation cleavages and have the same sense of displacement.

Extensional crenulation foliations also contain deformed leucogranite veinlets (Fig. 6c,e and f), indicative that partial melting associated with migmatite formation in the sillimanite-muscovite and sillimanite–K-feldspar zones was synchronous with extension. Here the main

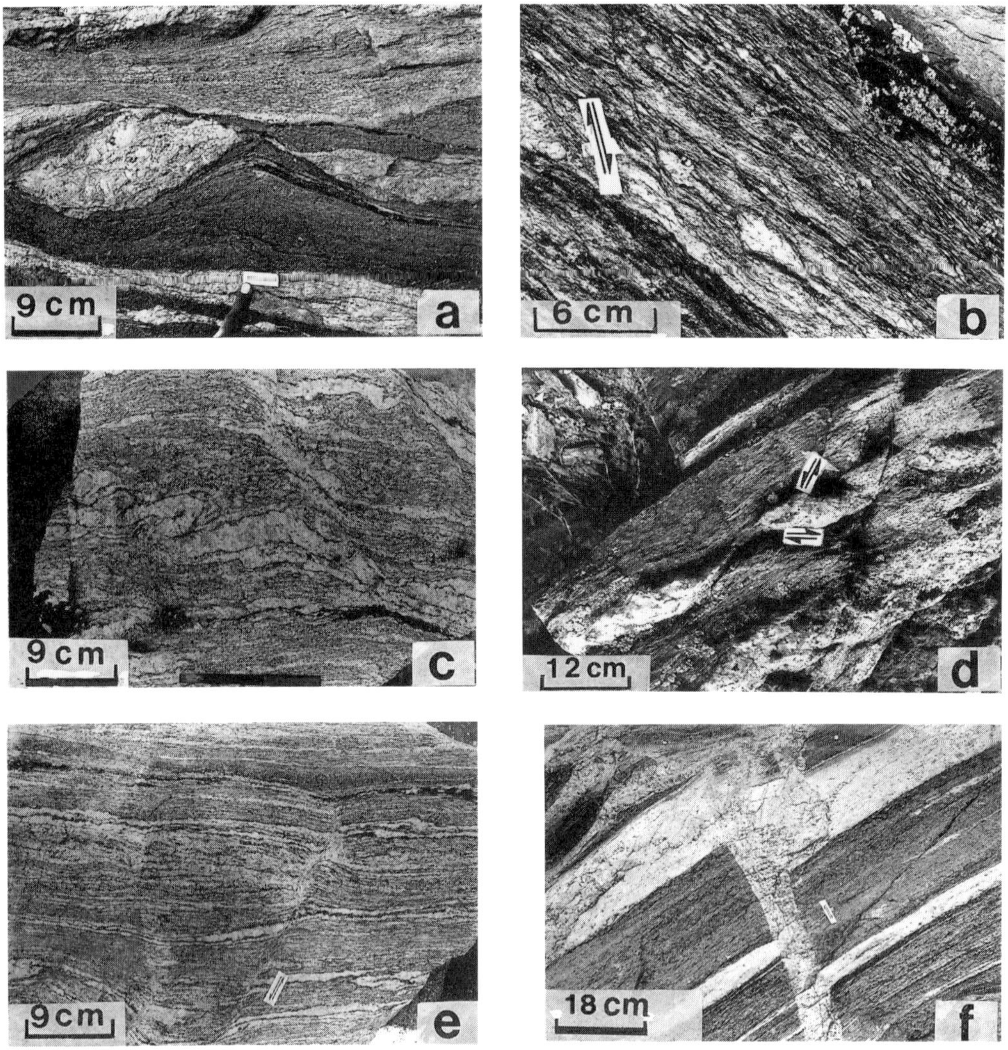

Fig. 6. (**a**) Asymmetrical foliation boudinage due to presence of gently-dipping shears at Sumcham. (**b**) Prominently developed steeply NE-dipping extensional crenulation foliation (arrows) in mylonitised gneiss near Ringdom along Ichi Nala. (**c**) Moderately NE-dipping extensional foliation in migmatite superposed upon F3a folds at Bhuzas. (**d**) Conjugate set of extensional foliations dipping in opposite directions in sillimanite–biotite–muscovite schist and gneiss. Note the presence of boudins due to these planes at Junkul Gompa along Mulung Tokpo. (**e**) Conjugate set of steeply-dipping shear bands having a normal displacement sense in migmatite at Bhuzas. (**f**) Brittle normal fault in migmatite. Interbanded migmatite bands are displaced 12 cm along the fault marked by aplite at Sampuk along Mulung Tokpo. Top right corner indicates NE except (d) and (e).

foliation is either bent sigmoidally or abruptly truncated by the extensional fabric. Mid crustal level extension, synchronous with in situ partial melting, is envisaged for the core of the HHC during its uplift phase.

Orientation of extensional crenulation cleavages

Herren (1987) observed only a single set of extensional planar structures (C-surfaces of S–C fabric) dipping steeply, NE along the ZSZ. Such structures, as well as petrofabric data from deformed leucogranites in southern Tibet, were observed by Burg *et al.* (1984) in support of northerly-directed normal faulting at the top of the HHC, also known as the Tibetan slab in Nepal.

Along the ZSZ, an extensional crenulation foliation is prominently developed in sillimanite–muscovite–biotite gneiss, granite gneiss, tourmaline-bearing leucogranite, mylonite and basal parts of the Tethyan Sedimentary Zone between Temsa Tokpo and Ringdom (Fig. 7). Dip-slip movements have taken place along two sets of these planes and consequently, foliation-parallel extension and foliation-normal compression can be deciphered. One of these planes consistently trends NNW-SSE/NW-SE and dips at between 30 and 60° towards the NNE/ENE. The other plane shows a variable trend between N–S and WNW–ESE and dips gently between 10 and 20° towards the WSW. Two sets of extensional crenulation foliations have been systematically mapped and analysed at Mune–Bardan Gompa, Sani, Mulung Tokpo, Sumche Tokpo, Pensi La (Fig. 7) and their data are summarized in Table 1.

The orientation of principal stresses from the extensional crenulation cleavages indicates that the principal maximum stress $\sigma 1$ plunges steeply towards the SW, whereas the minimum stress $\sigma 3$ consistently plunges towards the NE at low angles. The intermediate stress axis $\sigma 2$ plunges very gently either towards NW or SE.

Himalayan extension tectonic models

Extensional tectonic models of normal-fault zones in the Himalaya have postulated gravity collapse of the high topographic front (Burchfiel & Royden 1985; Herren 1987). Similar models

Table 1. *Extensional crenulation cleavage/foliation ECC1 and ECC2 data from Doda and Suru Valleys, Zanskar, NW Himalaya*

Sector	Sub-area	Fabric	Data points	Orientation (strike & dip)	Stress orientation
(a) Mune–Bardan Gompa	Temsa Tokpo	ECC1	32	N339°, 48°NE	σ1 N254°/70°
		ECC2	22	N297°, 10°SW	σ2 N154°/05°
					σ3 N062°/20°
	Trakkar	ECC1	35	N340°, 50°NE	σ1 N250°/70°
		ECC2	19	N328°, 10°SW	σ2 N168°/03°
					σ3 N070°/20°
(b) Sani	Haptal Tokpo	ECC1	61	N330°, 50°NE	σ1 N246°/70°
		ECC2	25	N310°, 10°SW	σ2 N147°/03°
					σ3 N057°/20°
	Tungri	ECC1	20	N311°, 44°NE	σ1 N211°/73°
		ECC2	13	N336°, 10°SW	σ2 N314°/03°
					σ3 N046°/18°
(c) Mulung Tokpo	Junkul Gompa	ECC1	138	N303°,50°NE	σ1 N205°/62°
		ECC2	52	N03°, 10°W	σ2 N308°/08°
					σ3 N040°/27°
	Ating	ECC1	24	N342°, 40°NE	σ1 N264°/71°
		ECC2	15	N082°, 10°SE	σ2 N152°/07°
					σ3 N060°/16°
(d) Sumche Tokpo	–	ECC1	21	N329°, 60°NE	–
(e) PensiLa	–	ECC1	43	N325°, 58°NE	σ N242°/65°
		ECC2	25	N299°,10°SW	σ2 N148°/04°
					σ3 N050°/24°
(f) Zolidok	–	ECC1	41	N329°, 78°NE	σ1 N240°/52°
		ECC2	24	N039°, 09°NW	σ2 N340°/18°
					σ3 N075°/35°

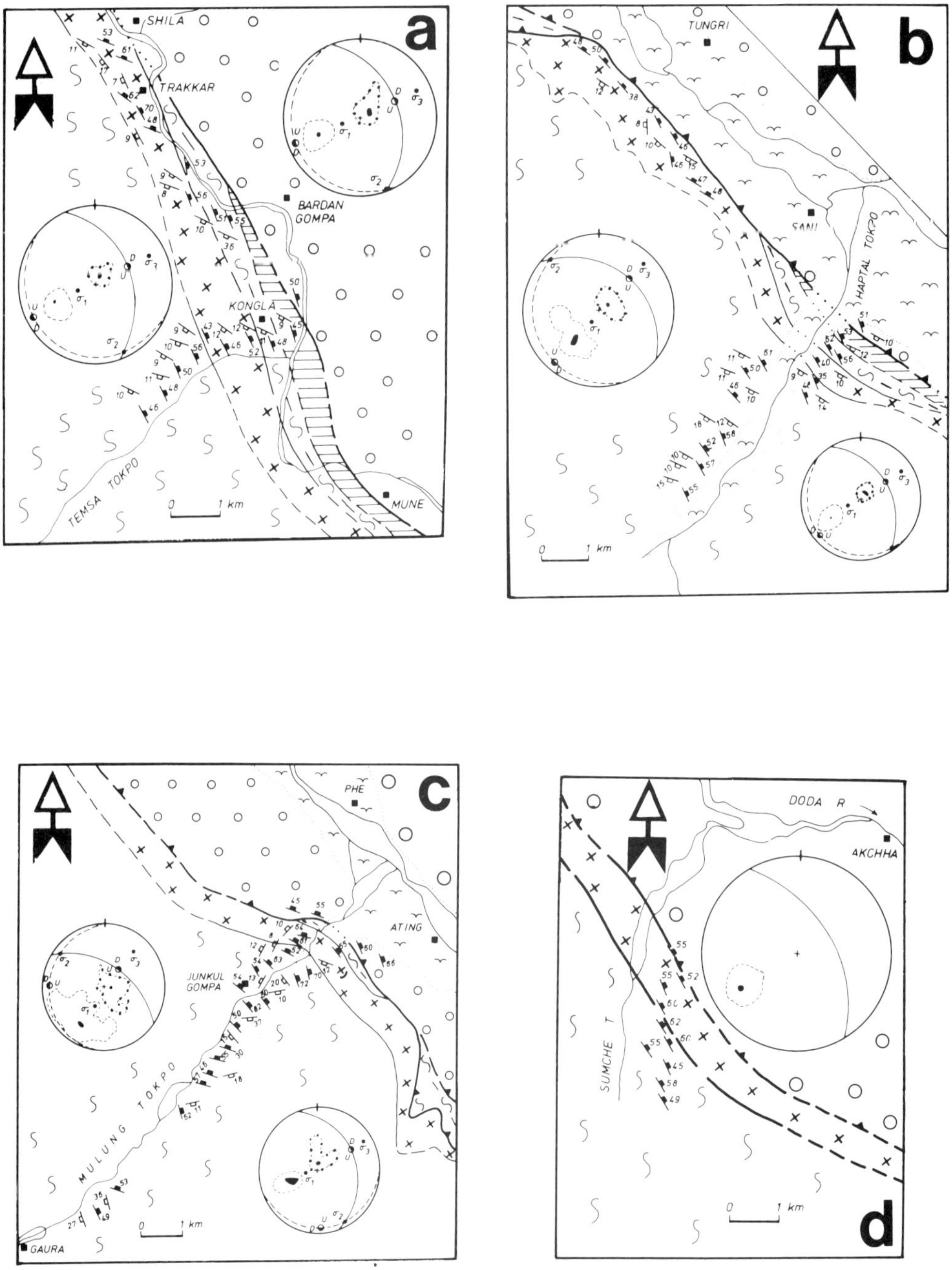
a
SHILA
TRAKKAR
BARDAN GOMPA
KONGLA
MUNE
TEMSA TOKPO
0 1 km
b
TUNGRI
SANI
HAPTAL TOKPO
0 1 km
c
PHE
ATING
JUNKUL GOMPA
MULUNG TOKPO
GAURA
0 1 km
d
DODA R
AKCHHA
SUMCHE T
0 1 km

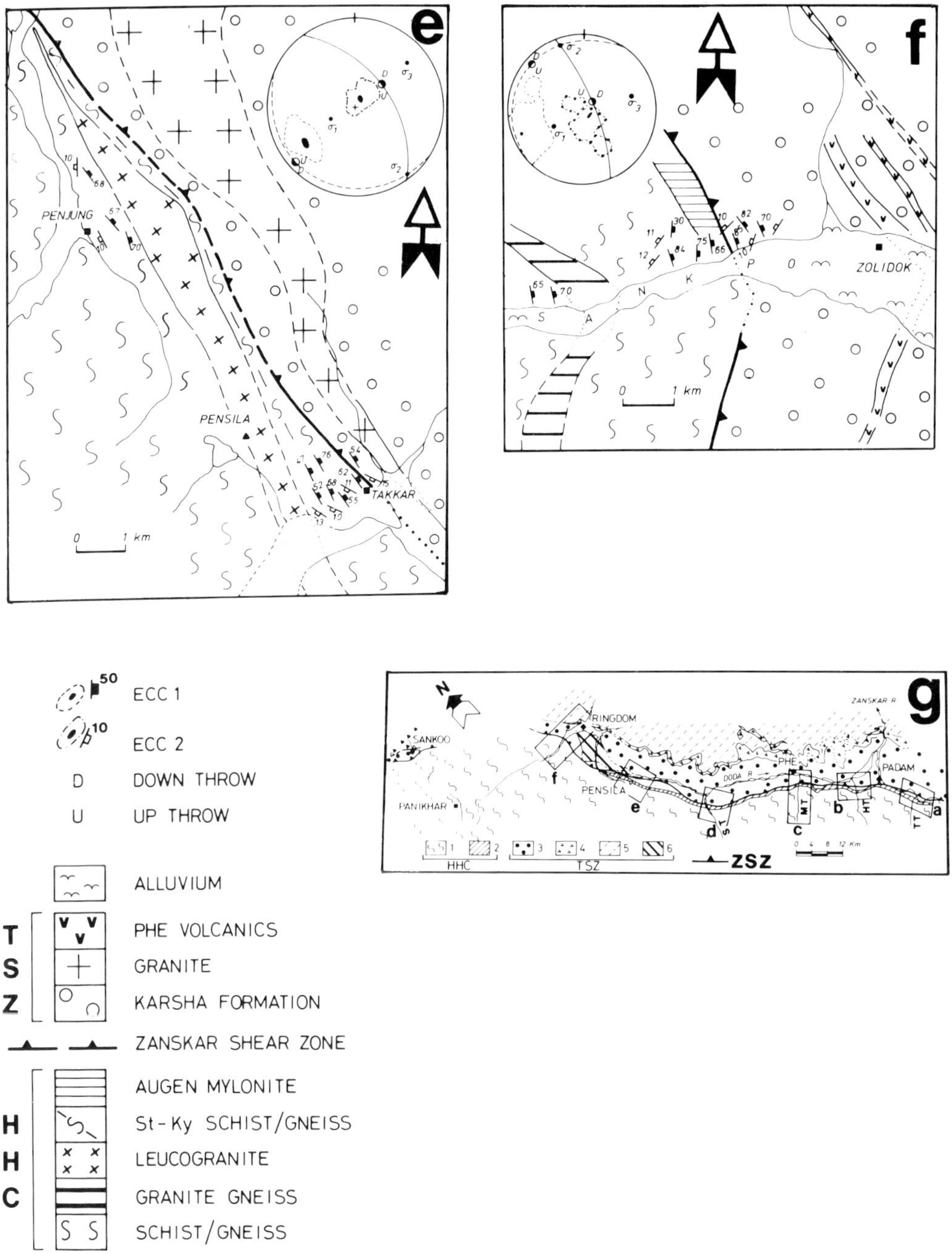

Fig. 7. Structural maps of the extensional crenulation foliations/cleavages in different sectors of the Zanskar Shear Zone, Suru-Doda Valleys (see insert for locations). (**a**) Temsa Tokpo, (**b**) Haptal Tokpo, (**c**) Mulung Tokpo, (**d**) Sumche Tokpo, (**e**) Pensi La and (**f**) Ringdom–Zolidok area. Location of sectors are shown in (**g**) where HHC contains schist, gneiss and migmatite (1) and deformed leucogranite body (2) and the Tethyan Sedimentary Zone contains Karsha Formation (3), Phe Volcanics (4), Lilang Group (5) and granite intrusives (6). Only maxima and minima of ECC1 (dashed) and ECC2 (dash-dot) are shown.

have been applied to many other older belts such as the Scandinavian Caledonides (Seguret *et al.* 1989) and French-Italian western Alps (Philippot 1990).

Evaluation of numerous shear criteria across the HHC and within the ZSZ indicate that the whole Higher Himalayan metamorphic pile has been deformed within a broad non-coaxial ductile shear zone of overthrust-type with a consistent top-to-SW sense of movement with the HHC thrust over the Lesser Himalayan foreland along the MCT (Figs 8 & 9). Like the MCT and numerous other thrusts within the Tethyan Sedimentary Zone (Searle 1986), the ZSZ also reveals an early deformation history involving SW-vergent ductile shearing which involved underthrusting of the HHC beneath its Tethyan cover (Fig. 9). The initial phase of thrust-type high strain ductile shearing along the basement–cover contact in the ZSZ is characterized by the presence of S–C augen mylonites, duplex structures within the cover rocks, truncation of metamorphic isograd boundaries, deformed Tertiary leucogranite bodies and sheared lenses of porphyritic granitoid bodies. The presence of augen mylonite xenoliths within the deformed tourmaline leucogranite tabular sheet are indicative of an early ductile shearing along the ZSZ rather than the emplacement of Late Tertiary leucogranite along this shear zone.

Evidence for superposed layer-parallel NE-orientated extension is ubiquitous across the HHC in rocks of all metamorphic grade (Fig. 9b). Unlike Herren's (1987) observations of C-shears and stretching lineation defining the NE-orientated normal-displacement in the ZSZ, discrete extensional crenulation foliations are superposed upon an earlier S–C fabric. A telescoping of biotite–garnet, staurolite and sillimanite isograds in a 250 m wide zone was indicated by Herren (1987), who calculated 19 km of vertical and 16 km of horizontal displacements respectively on the basis of geometrical relations of dips of shear zone boundary

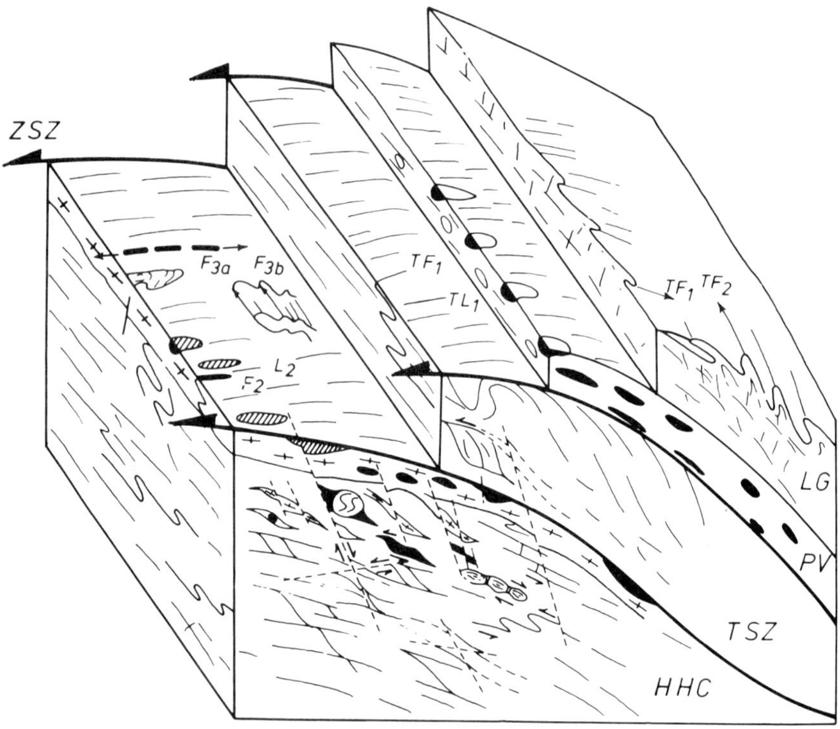

Fig. 8. Isometric diagram for the basement–cover relationship during overthrusting of the Tethyan Sedimentary Zone (TSZ) on the HHC along the Zanskar Shear Zone (ZSZ) and subsequent superposed extensional tectonics. NE-plunging folds (TF1) and mineral/stretching lineation (TL1) in the TSZ (PV, Phe Volcanics; LG, Lilang Group) persist across the tectonic boundary into basement rocks. Numerous shear criteria reveal an initial top-to-SW sense of displacement during ductile deformation and a later phase of layer-parallel extension with σ_3-orientated almost orthogonal to main orogenic trend.

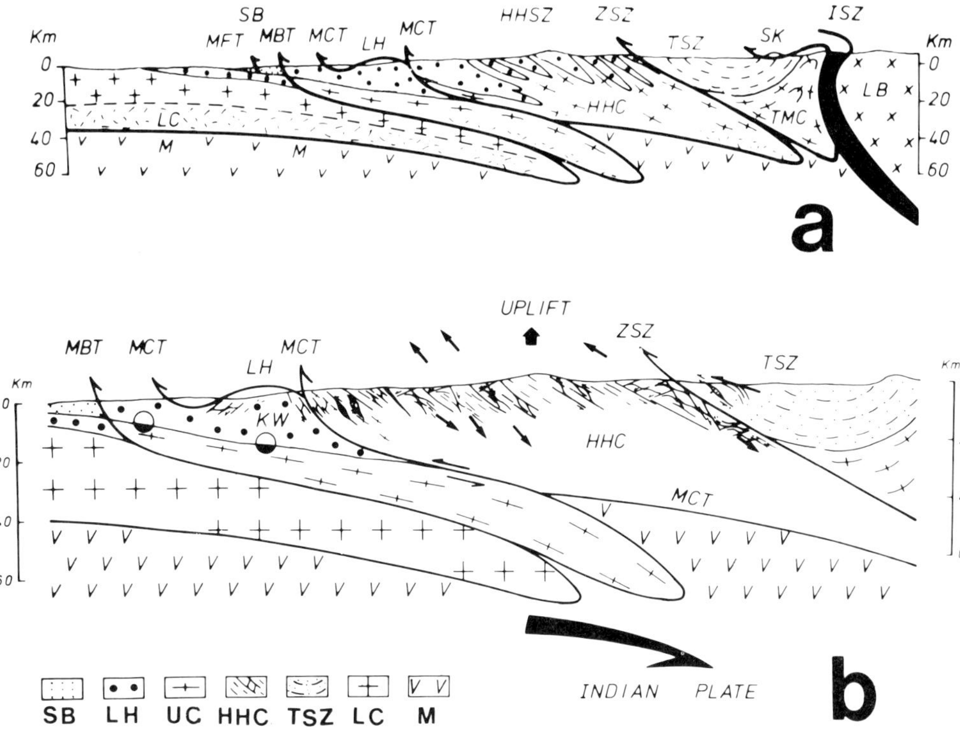

Fig. 9. (**a**) Collisional tectonic model of the NW Himalaya and (**b**) Extensional tectonic model of the NW Himalaya during underthrusting of the Indian plate and ramping of the HHC. SB, Tertiary foreland basins of the Indo-Gangetic Plains and Siwalik; LH, Proterozoic Sedimentary Basin of the Indian Shield and the Lesser Himalaya; UC, underthrust upper crust of the Indian Shield; HHC, Higher Himalayan Crystallines; TSZ, Tethyan Sedimentary basin; LC, Indian continental crust and lower crust; M, upper mantle beneath the Indian Shield and Himalaya. From left to right: MFT, Main Frontal Thrust; MBT, Main Boundary Thrust; HHSZ, Higher Himalayan Shear Zone; TMC, Tso–Morari Crystallines; SK, Spongtang ophiolite klippe; ISZ, Indus Suture Zone; LB, Ladakh Batholith. Basal detachment beneath the Himalaya is approximately constrained from recent earthquakes (shaded circles).

and telescoped isograds during late Himalayan metamorphism. Detailed sampling of the top 200 m of the HHC (X in Fig. 2b) indicates the presence of staurolite–kyanite-grade rocks immediately beneath biotite-grade meta-quartzite and meta-pelite of the Tethyan cover and does not provide evidence for any telescoping of isograd boundaries along the ZSZ.

P–T estimates across the HHC, derived from post-tectonic garnet rims, indicate a sudden isothermal pressure decrease from about 10 kbar to 4 kbar at a temperature of about 625° C (Jain & Manickavasagam 1993). Higher-grade rocks invariably show the development of fibrolite pseudomorphing kyanite and/or replacing muscovite and biotite. The development of quartz–feldspar myrmekite at grain boundaries is common in high grade rocks, possibly suggesting their growth during anatexis. These reaction textures are consistent with late stages of metamorphism, including anatexis, taking place during pressure decrease, a path consistent with garnet zonation profiles that have increases in Mn content and Fe/(Fe + Mg) values towards the rim.

In staurolite–kyanite and garnet grade rocks from between Sankoo and Pensi La, within both the ZSZ and along the northern margin of the HHC, K-Ar hornblende ages from amphibolites (5 samples) fall at between 45.5 ± 2.3 and 30 ± 5 Ma (Searle *et al.* 1992; Sorkhabi *et al.* 1992). Searle *et al.* (1992) published two $^{40}Ar/^{39}Ar$ hornblende plateau ages of 33.9 ± 2.2 and 24.4 ± 1 Ma. K-Ar ages from porphyritic granite gneisses, tourmaline leucogranite and schists from near Padam, Bardan and Raxu in the vicinity of the ZSZ, are at 30 ± 2 Ma to 22 ± 1 Ma for muscovite and 22 ± 1 Ma for biotite

(Searle 1986; Searle *et al.* 1992). These ages, which date cooling through *c.* 350° and 300° C respectively, decrease northwestward towards Ringdom and along the Suru Valley to 24 (muscovite) and 22 to 19 (biotite) Ma respectively (Searle *et al.* 1992; Sorkhabi *et al.* 1992). Zircon and apatite fission track ages represent a cooling history for the ZSZ and the HHC along the Suru Valley through 250° C at *c.* 17 Ma to 120° C at *c.* 9 Ma (Sorkhabi *et al.* 1992).

It is, therefore, apparent that relatively rapid cooling associated with exhumation that was accommodated in part by crustal extension, occurred throughout the HHC. The rates and timing of cooling were variable but earlier along the upper margin and faster and later within the core of the HHC. Cooling from between 500 to 300° C occurred at between 30 and 19 Ma around Sankoo in the ZSZ, whereas within the Suru Valley near Shafat in the HHC it was at between 22 and 15 Ma (Searle *et al.* 1992). As the extensional structures include a range of features and fabrics from ductile melt cored shears to brittle faults, it is apparent that extensional processes were operative throughout the period between 17 and 9 Ma during which cooling took place from conditions suitable for anatexis down to temperatures below 200° C.

In contrast to the late orogenic extensional tectonics in the Higher Himalaya, another type of extensional regime within the HHC and the Tethyan Sedimentary Zone is indicated by the presence of the post-orogenic N–S-trending Holocene Thakhola–Mustang and Upper Sublej graben systems both of which have thick accumulations of young lacustrine deposits. Post-tectonic Holocene extension appears to be associated with E–W extension of Tibet along major strike-slip faults (Mercier *et al.* 1987).

Conclusions

Extensional tectonics in the Cenozoic Himalayan orogenic belt encompasses late orogenic phenomena, which were superposed upon an earlier phase of non-coaxial ductile shearing, associated with continental collision and compressional tectonics (Fig. 9b). Extension was confined not only to narrow shear zones beneath the Phanerozoic Tethyan sedimentary basin, but was extensively distributed across the whole orogenic belt during its exhumation and uplift from deep crustal levels through the brittle-ductile and brittle deformational regimes (Fig. 9b). This extension did not produce any spectacular narrow orogen-parallel zones, as is evident from the complete absence of half-grabens and associated Neogene sedimentary fill along the ZSZ.

Late orogenic NE-directed extension was probably caused by ramping of the overthickened continental crust during underthrusting of the Indian plate along many thrusts/detachments including the MCT. This has resulted in the uplift of the hot HHC and its exhumation from 10 to 4 kbar at an almost constant temperature of about 625° C.

The present investigations form part of the research project 'Evolution of the metamorphic belts in Collision Zones', sponsored by the Department of Science and Technology, New Delhi. The authors (A.K.J. and S.S.) are grateful to D.G. Gee, Lund University for discussions. Logistic back-up support for field investigations by the department, local administration and people in Zanskar are thankfully acknowledged. Critical comments by P.J. Treloar, M.P. Searle and another referee have been extremely useful in modifying the earlier version of the paper. We are grateful to the conference organizers for arranging financial support from the Geological Society of London that enabled A.K. Jain to attend the Oxford meeting.

References

ANDERSEN, T.B., JAMTVEIT, B., DEWEY, J.F. & SWENSSON, E. 1991. Subduction and eduction of continental crust: major mechanism during continent-continent collision and orogenic extensional collapse, a model based on the south Norwegian Caledonides. *Terra Nova,* **3**, 303–310.

BERTHE, D., CHOUKROUNE, P. & JEGOUZO, P. 1979. Orthogneiss, mylonite and non-coaxial deformation of granite: the example of the South Armorican shear zone. *Journal of Structural Geology,* **1**, 31–42.

BOYER, S.F. & ELLIOT, D. 1982. Thrust systems. *Bulletin of American Association of Petroleum Geology,* **66**, 1196–1230.

BRUNEL, M. 1986. Ductile thrusting in the Himalayas: shear sense criteria and stretching lineation. *Tectonics,* **5**, 247–265.

BURCHFIEL, B.C. & ROYDEN, L.H. 1985. North-south extension within the convergent Himalayan region. *Geology,* **13**, 679–682.

BURG, J.P., BRUNEL, M., GAPAIS, D., CHEN, G.M. & LI, G.C. 1984. Deformation of leucogranites of the crystalline Main Central Thrust sheet in southern Tibet (China). *Journal of Structural Geology,* **6**, 535–542.

COWARD, M.P., WINDLEY, B.F., BROUGHTON, R., LUFF, I., PATTERSEN, M.G., PUDSEY, C., REX, D. & KHAN, M.A. 1986. Collision tectonics in the NW-Himalayas. *In*: COWARD, M.P. & RIES, A.C. (eds) *Collision Tectonics.* Geological Society, London, Special Publications, **19**, 203–219.

HERREN, E. 1987. The Zanskar Shear Zone: Northeast-southwest extension within the Higher Himalaya (Ladakh, India). *Geology,* **15**, 409–413.

HONEGGER, K., DIETRICH, V., FRANK, W., GANSSER, A., THONI, M. & TROMMSDORFF, V. 1982. Magmatism and metamorphism in the Ladakh Himalayas (the Indus-Tsangpo Suture Zone). *Earth and Planetary Science Letters*, **60**, 253–292.

JAIN, A.K. & ANAND, A. 1988. Deformational and strain patterns of an intracontinental collision ductile shear zone: an example from the Higher Garhwal Himalaya. *Journal of Structural Geology*, **10**, 717–734.

—— & MANICKAVASAGAM, R.M. 1993. Inverted metamorphism in the intracontinental ductile shear zone during Himalayan collision tectonics. *Geology*, **21**, 407–410.

KUNDIG, R. 1989. Domal structures and high-grade metamorphism in the Higher Himalayan Crystalline, Zanskar region, north-west Himalaya, India. *Journal of Metamorphic Geology*, **7**, 43–55.

LISTER, L.S. & SNOKE, A.W. 1984. S–C mylonites. *Journal of Structural. Geology*, **6**, 617–638.

MALAVIEILLE, J. 1987. Kinematics of compressional and extensional ductile shearing deformation in a metamorphic core complex of the north-eastern Basin and Range. *Journal of Structural Geology*, **9**, 541–554.

MERCIER, J.L., ARMIJO, R., TAPPONNIER, P., GAILHARDIS, E.C. & LIN, H.T. 1987. Change from late Tertiary compression to Quaternary extension in Southern Tibet during the India-Asia collision. *Tectonics*, **6**, 275–304.

PASSCHIER, C.W. & SIMPSON, C. 1986. Porphyroclast systems as kinematic indicators. *Journal of Structural Geology*, **8**, 831–843.

PÊCHER, A. 1991. The contact between the Higher Himalayan Crystallines and the Tibetan Sedimentary Series: Miocene large-scale dextral shearing. *Tectonics*, **10**, 587–598.

——, BOUCHEZ, J.-L. & LE FORT, P. 1992. Miocene dextral shearing between Himalaya and Tibet. *Geology*, **19**, 683–685.

PHILIPPOT, P. 1990. Opposite vergence of nappes and crustal extension in the French-Italian western Alps. *Tectonics*, **9**, 1143–1164.

PLATT, J.P. & VISSERS, R.L.M. 1980. Extensional structures in anisotropic rocks. *Journal of Structural Geology*, **2**, 397–410.

ROYDEN, L.H. & BURCHFIEL, B.C. 1987. Thin-skinned N–S extension within the convergent Himalayan region: gravitational collapse of a Miocene topographic front. *In*: COWARD, M.P., DEWEY, J.F. & HANCOCK (eds) *Continental Extensional Tectonics*. Geological Society, London, Special Publications, **28**, 611–619.

SEARLE, M.P. 1986. Structural evolution and sequence of thrusting in the High Himalayan, Tibetan-Tethys and Indus Suture Zones of Zanskar and Ladakh, Western Himalaya. *Journal of Structural Geology*, **8**, 923–936.

——, COOPER, D.J.W. & REX, A.J. 1988. Collision tectonics of the Ladakh–Zanskar Himalaya. *Philosophical Transactions of Royal Society, London*, **A326**, 117–150.

—— & REX, A.J. 1989. Thermal model for the Zanskar Himalaya. *Journal of Metamorphic Geology*, **7**, 127–134.

——, WATERS, D.J., REX, D.C. & WILSON, R.N. 1992. Pressure, temperature and time constraints on Himalayan metamorphism from eastern Kashmir and western Zanskar. *Journal of Geological Society, London*, **149**, 753–773.

SEGURET, M., SERANNE, M., CHAUVET, A. & BRUNEL, M. 1989. Collapse basin: a new type of extensional sedimentary basin from the Devonian of Norway. *Geology*, **7**, 127–130.

SIMPSON, C. & SCHMIDT, S.M. 1983. An evaluation of criteria to deduce the sense of movement in sheared rocks. *Geological Society of America Bulletin*, **94**, 1281–1288.

SORKHABI, R.B., JAIN, A.K., ITAYA, T., KUMAR, A., LAL, N., MANICKAVASAGAM, R.M. & NISHIMURA, S. 1992. Thermotectonic history of the Higher Himalayan Crystalline (HHC), Zanskar, NW-India: Evidence from Fission-Track (FT) and K-Ar ages (abst.). *29th International Geological Congress*, **3**, 636.

SRIKANTIA, S.V., GANESAN, T.M., RAO, P.M., SINHA, P.K. & TIRKEY, B. 1978. Geology of the Zanskar area, Ladakh Himalaya. *Himalayan Geology*, **8**, 1009–1033.

VALDIYA, K.S. 1989. Trans-Himadri intracrustal fault and basement upwarps south of Indus–Tsangpo Suture Zone. *Geological Society of America, Special Papers*, **232**, 153–168.

VOHRA, C.P., JANGPANGI, B.S., MEHROTRA, P.C., SINGH, D.P., PURI, V.N.K., KAUL, M.K. & MEHTA, P. 1982. Geology of the Warwan–Nunkun area, J. & K. State. *Miscellaneous Publications, Geological Survey of India*, **41**, (II), 56–63.

WHITE, S.H. 1979. Large strain deformation. Report on a tectonic studies group discussion meeting held at Imperial College, London. *Journal of Structural Geology*, **1**, 333–339.

Annapurna detachment fault in the Greater Himalaya of central Nepal

RICHARD L. BROWN & JEFFREY H. NAZARCHUK

Department of Earth Sciences, Carleton University, and Ottawa-Carleton Geoscience Centre, Ottawa, ON, Canada K1S 5B6

Abstract: Displacement on the Annapurna detachment fault (ADF) has exhumed the Greater Himalayan metamorphic sequence in central Nepal. The fault is most likely a continuation of the South Tibetan detachment system recognized elsewhere in the Himalayan orogen.

The characteristics of detachment faulting and the structural and metamorphic histories of the hanging wall and footwall in the Thakkhola region are described. Two stages of normal-sense shearing on the ADF are recognized: the first stage occurred at mid-crustal depth; the most recent detachment faulting on the system was brittle and may have been accompanied by orogen-parallel extension.

First-phase isoclinal folds in the Tibetan sedimentary sequence above the ADF reflect a southwesterly sense of shear and are refolded by spectacular northeasterly verging second-phase folds that have affected a crustal thickness of over 6 km. Third-phase southwesterly directed thrust faults and folds have disrupted the earlier structures but predate the most recent period of normal-sense shearing on the ADF.

The kinematic significance of the northeasterly verging second-phase folds is not clear; these structures may have formed during early motion on the ADF but, alternatively, may reflect older events associated with compression and crustal thickening of the orogen.

Crystalline rocks of the Greater Himalayan metamorphic sequence (known also as the Main Central Thrust Sheet or the Tibetan Slab) have been thrust southwestward along the Main Central Thrust over the lower grade rocks of the Lesser Himalayan sedimentary sequence. Within the Greater Himalayan metamorphic sequence, metamorphic grade increases structurally up-section giving an inversion of isograds (sillimanite above kyanite) in the middle to uppermost structural levels (Swapp & Hollister 1991 and references therein). This metamorphic inversion may be only an apparent one that results from a moderate-pressure, high-temperature Buchan event superimposed on an earlier high-pressure, high-temperature Barrovian event, and not a continuous increase in metamorphic grade (Brunel & Kienast 1986; Hodges & Silverberg 1988; Hodges *et al.* 1988 and references therein; Swapp & Hollister 1991).

The top of the Greater Himalayan metamorphic sequence is bounded in southern Tibet by a northeasterly dipping normal-sense shear zone (Burg *et al.* 1984). It has recently been demonstrated that this shear zone is part of the regional South Tibetan detachment system (Burchfiel *et al.* 1992).

Northeasterly displacement on the South Tibetan detachment system and exhumation of the Greater Himalayan metamorphic sequence have been linked to emplacement of Miocene leucogranites in the upper part of the sequence (Hodges *et al.* 1988; Hubbard & Harrison 1989). Although the age of displacement on the Main Central Thrust is not well constrained along much of its length, in eastern Nepal thrusting appears to have been coeval with leucogranite emplacement (Hubbard & Harrison 1989); thus, it appears that the Main Central Thrust Sheet was extruded southwestward relative to the overlying and underlying strata and was exhumed shortly after generation of the leucogranites. This hypothesis is interpreted by Burchfiel & Royden (1985) in terms of gravitational collapse of the overthickened orogen (see also Dalmayrac & Molnar 1981; England & McKenzie 1982, 1983; Houseman *et al.* 1981). Northeasterly directed thrust faults and crustal-scale folds in the hanging wall of the South Tibetan detachment system have also been attributed by some authors to gravitational sliding or collapse associated with northeastward displacement on the detachment (Caby *et al.* 1983; Pêcher 1991; Burchfiel *et al.* 1992).

In this paper we document the presence of a detachment fault, the Annapurna detachment fault, in the Thakkhola region of central Nepal (Figs 1 and 2), which is presumed to be correlative with the South Tibetan detachment system in southern Tibet. Structures in the hanging wall

From Treloar, P.J. & Searle, M.P. (eds), *Himalayan Tectonics*
Geological Society Special Publication, No. 74, pp. 461–473.

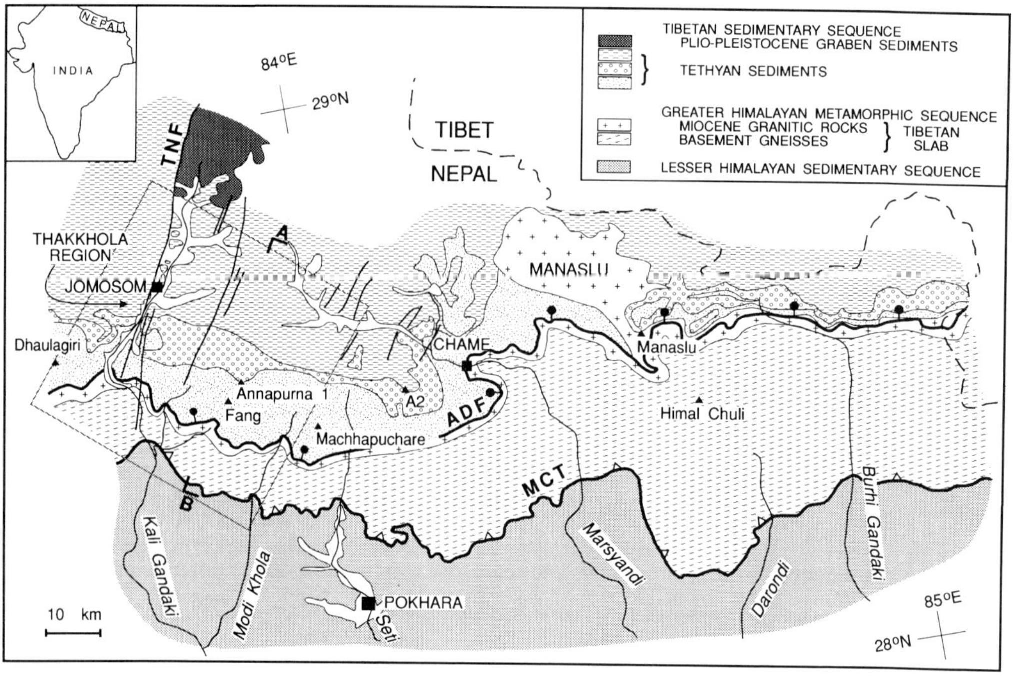

Fig. 1. Geological map of central Nepal simplified and modified after Colchen *et al.* (1986): MCT, Main Central Thrust; ADF, Annapurna detachment fault; TNF, Thakkhola normal fault; box locates Fig. 3; A–B locates section line of Fig. 2.

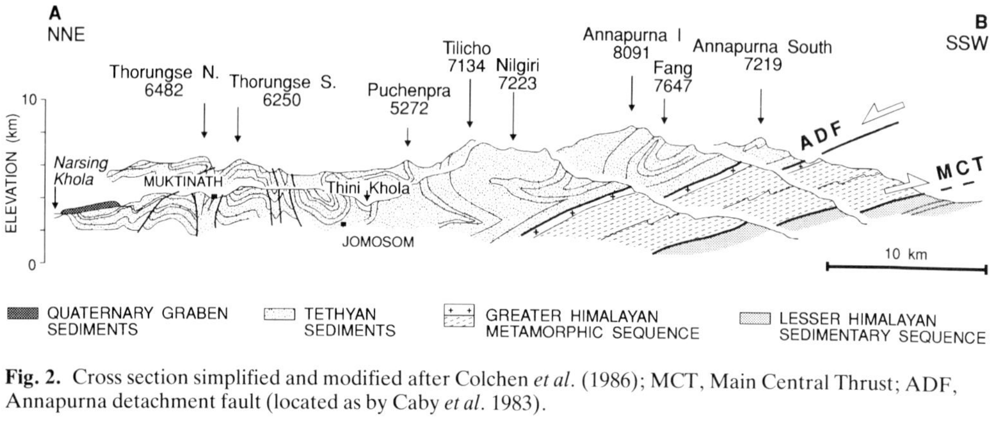

Fig. 2. Cross section simplified and modified after Colchen *et al.* (1986); MCT, Main Central Thrust; ADF, Annapurna detachment fault (located as by Caby *et al.* 1983).

of the detachment appear to be of both extensional and compressional origins. These structures are described and their kinematic significance is discussed. The timing of orogen-parallel extension in the Thakkhola region and its possible correlation with the most recent period of displacement on the Annapurna detachment fault are considered.

Field relations

The Greater Himalayan metamorphic sequence (Main Central Thrust Sheet)

Where the Greater Himalayan metamorphic sequence crosses the Kali Gandaki River in central Nepal the thrust sheet is only about 4 km

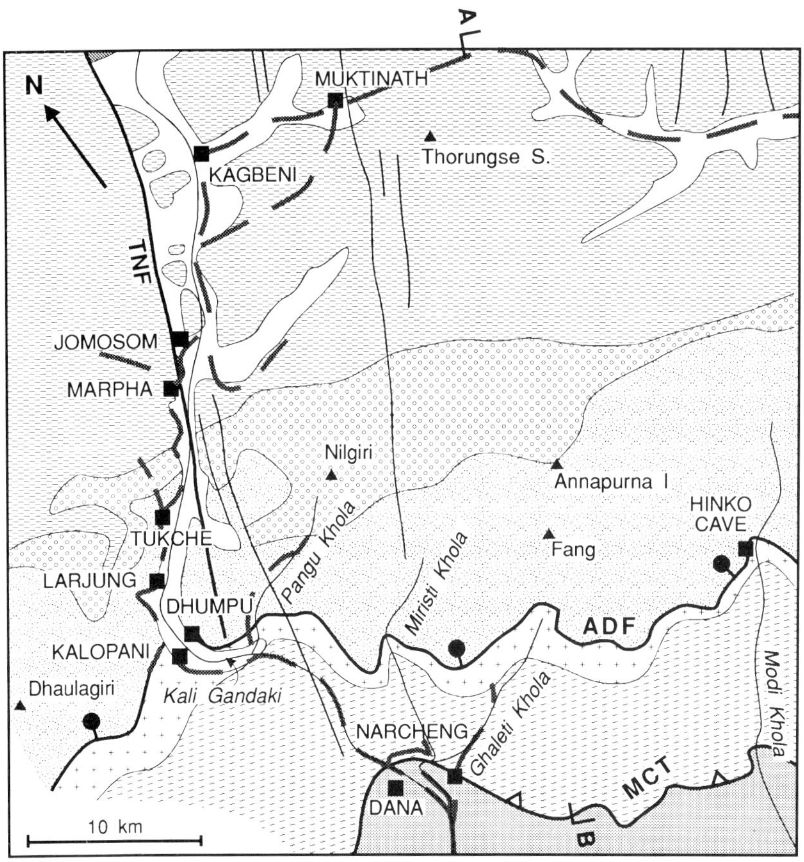

Fig. 3. Location of current study: shaded dashed line follows route of traverses; see Fig. 1 for legend; A–B locates section line of Fig. 2; MCT, Main Central Thrust; ADF, Annapurna detachment fault; TNF, Thakkhola normal fault.

thick. Its base is exposed to the east of Dana on the trail to Narcheng and along the Miristi Khola (Fig. 3), where the boundary is poorly defined by the last appearance of kyanite-bearing pelitic and quartz–feldspar gneisses, which are underlain by phyllite, minor quartzite, and calcareous schist at sub-amphibolite grade. There is clearly post-metamorphic strain in the low-grade rocks of the footwall, but the kyanite-bearing gneisses are annealed and show no sign of retrogression. At this locality it appears that the Main Central Thrust can be located at the boundary between the kyanite-bearing gneisses and the underlying lower grade schists; the most recent motion along this boundary is younger than the metamorphic culmination observed in the hanging-wall gneisses.

A detailed discussion of the fabrics in the vicinity of the Main Central Thrust is beyond the scope of this paper, but we should comment here that kinematic indicators are evident despite considerable post-tectonic recrystallization of the gneisses; not surprisingly, they record a southwesterly direction of shearing of the Greater Himalayan metamorphic sequence relative to the underlying strata of the Lesser Himalayan sedimentary sequence.

The central part of the Greater Himalayan metamorphic sequence is not well exposed in the Kali Gandaki Valley but may be studied in detail by following the mountain trail east of Narcheng along the Ghaleti Khola (Fig. 3). Colchen *et al.* (1986) describe the Greater Himalayan metamorphic sequence (referred to by them as the 'Tibetan Slab') in this region as a sequence with quartz–feldspar gneisses in the lower part and calc-silicate rocks predominating in the upper part. They did not recognize any major folding or faulting within the sequence. We have found quartz–feldspar and pelitic gneisses both below

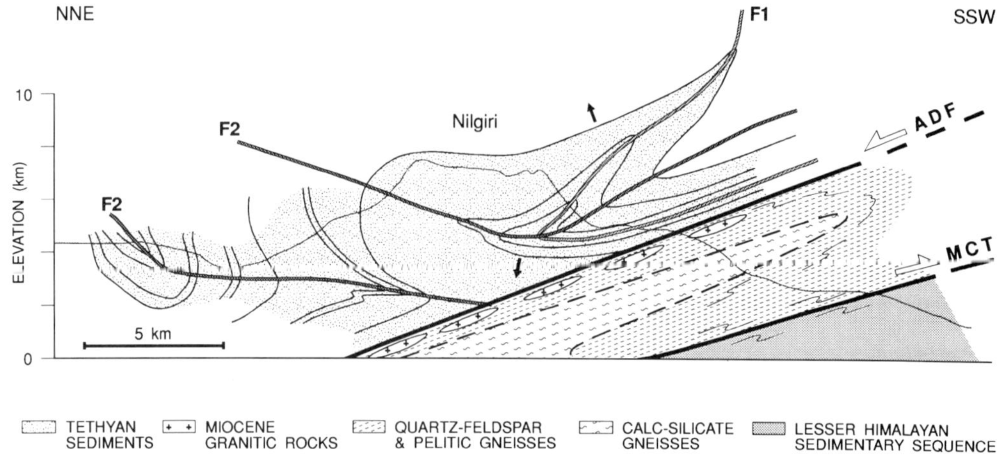

Fig. 4. Interpretation of part of the cross section shown in Fig. 2: MCT, Main Central Thrust; ADF, Annapurna detachment fault; F1, axial surface trace of first-phase southwesterly verging anticline; F2, axial surface trace of second-phase northeasterly verging antiform; solid arrows point in direction of stratigraphic tops.

and above the calc-silicate rocks; sparse minor structural features suggest that the calc-silicate rocks are cored by a second-phase antiform (Fig. 4). Fabric relationships in the hinges of minor folds suggest that kyanite-bearing assemblages predate development of the antiform.

Leucogranite sheets are present toward the top of the Greater Himalayan metamorphic sequence in the upper reaches of the Ghaleti Khola (Fig. 3) and are also exposed on the eastern slopes of the Kali Gandaki Valley in the vicinity of the village of Kalopani (Fig. 3). Migmatitic gneisses adjacent to the leucogranite sheets are kyanite-bearing, and there is no indication of replacement by sillimanite.

In the central portion of the Greater Himalayan metamorphic sequence, hinge lines of minor folds are generally shallow-plunging, and there is no obvious stretching lineation. Leucogranite sheets within the sequence are foliated but only become clearly lineated near the top of the sequence. The leucogranite sheets cut across earlier fabrics in the adjacent migmatitic gneisses.

Annapurna detachment fault

The detachment fault is well exposed in a relatively accessible stream gully west of the village of Dhumpu in the Thakkhola region of the Kali Gandaki River (Fig. 3). We have also crossed the detachment north of Hinko Cave along the Modi Khola (Fig. 3) and at Chame in the Marsyandi valley (Fig. 1), but time constraints and difficulty of access precluded detailed observations in these areas.

The following relationships (see Fig. 5) are clearly displayed in the stream gully on the west side of the Kali Gandaki River west of Dhumpu.

The fault trace is exposed within the stream bed with footwall rocks of the Greater Himalayan metamorphic sequence occurring in the cliffs of the south wall of the gully and hanging-wall rocks of the Tibetan sedimentary sequence (Tethyan assemblage, see Gradstein *et al.* 1992) lying to the north of the stream (Fig. 6).

The footwall is composed of migmatitic quartz–feldspar gneiss and calc-silicate intruded by sheets of two-mica leucogranite. Foliation within the gneisses dips moderately (25–45°) northeast generally parallel to foliation within the leucogranites. Stretching lineations are not well developed, but where observed, the lineation is down the dip of the dominant foliation.

Kinematic indicators are evident in the leucogranite sheets but are not as well displayed within the gneisses. These indicators include S–C fabrics, shear bands, rotated and non-rotated porphyroclasts with tails, asymmetric folds with sub-horizontal hinge lines, and pegmatite dyke arrays that have been extended or shortened according to their orientation with respect to the shear foliation. All these features consistently indicate that the upper member of the shear couple has been displaced to the northeast relative to the lower member (Fig. 5). These shear fabrics have been observed in the footwall from approximately 100 m below to within 5 m of the contact.

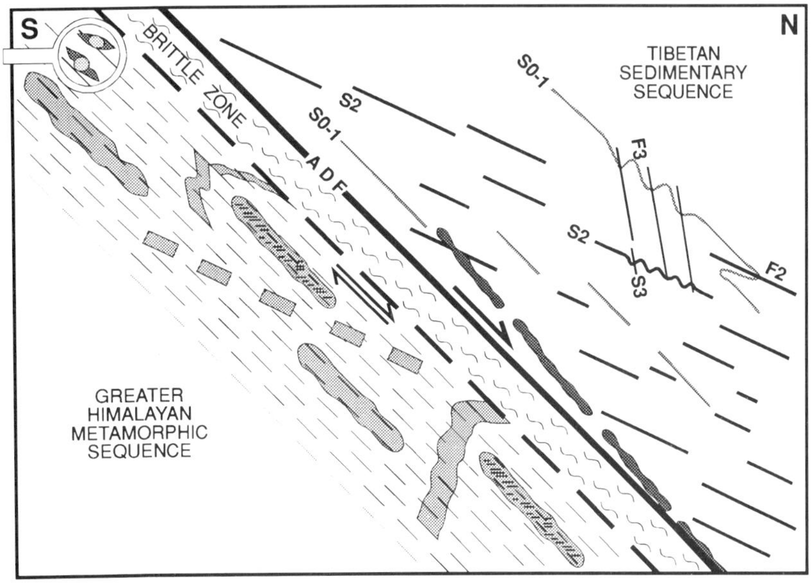

Fig. 5. Schematic diagram (not to scale) of view looking west in the Dhumpu region west of the Kali Gandaki River: ADF, Annapurna detachment fault; arrows indicate sense of shear within footwall rocks; lightly shaded bands in footwall illustrate sheared leucogranite; two orientations of lines within leucogranites illustrate S–C fabrics; magnifying glass illustrates pressure shadows on rotated and unrotated porphyroclasts; S0, orientation of bedding; S1, S2, and S3, first-, second-, and third-phase cleavages; F2 and F3, second- and third-phase folds; darkly shaded lenses above ADF illustrate boudinaged quartz-feldspar dykes. See text for explanation, Fig. 6 for view of ADF, and Figs 7–13 for illustration of structural features in hanging wall.

Fig. 6. View to the west of the Annapurna detachment fault south of Larjung, near Dhumpu along the Kali Gandaki River. Dotted line shows the contact between kyanite-grade migmatite and leucogranite of the Greater Himalayan metamorphic sequence in the footwall and biotite-grade psammitic rocks and micaceous schist of the Tibetan sedimentary sequence in the hanging wall. See Fig. 5 for details of contact relationships.

Across this width of the shear zone there is no sign of retrogression of the peak mineral assemblages. Within pelitic layers of the migmatite, the assemblages include kyanite, garnet, biotite, muscovite, plagioclase and quartz. The 5–10 m thick contact zone is sheared and brecciated; chloritic shear zones occur in pelitic lenses and crushed pegmatite.

Above the brittle zone is an abrupt contact with flaggy fine-grained psammitic rocks, micaceous schists, and phyllites, which grade upwards into fine sandstones and calcareous schists, which have been correlated with the Annapurna Formation by Colchen *et al.* (1986). The hanging-wall rocks have been metamorphosed to biotite grade. Fabrics within these hanging-wall rocks do not penetrate the footwall, and they clearly predate the most recent normal-sense motion on the detachment.

Structures in the hanging wall of the Annapurna detachment fault

Three phases of deformation are readily identified in the strata of the hanging wall.

Cleavage parallel to bedding (S1) is generally present and is defined primarily by preferred orientation or flattening of detrital grains; locally, minor isoclinal folds (F1) have been observed with this cleavage cutting their hinges (Fig. 7). This phase has been recognized by earlier workers (see Le Fort 1975 and references therein) and is illustrated on a crustal scale in cross sections through the Annapurna massif by Colchen *et al.* (1986; see Fig. 2). A major isoclinal nappe is shown closing in the summit of Fang (Varashikhar) (Fig. 2), which according to stratigraphical superposition originated as a south-southwesterly verging anticline (Fig. 4) This is clear since the sequence is right way up on the upper northeast flank of the F1 closure. (Note that shearing on the Annapurna detachment fault has cut out the complementary F1 syncline.)

A second cleavage (S2) oblique to bedding is generally the most obvious planar fabric in the hanging-wall rocks adjacent to the Annapurna detachment fault (Fig. 5). This cleavage is defined by the preferred orientation of micas in pelitic strata and is a spaced cleavage in the more sandy layers (Fig. 8). Folds with this cleavage parallel to their axial surfaces occur as megascopic tight to open folds in the metasediments directly overlying the detachment and continue up through the Tibetan sedimentary sequence as parasitic structures to major northeasterly verging folds (Figs 9 and 10). A spectacular example of these F2 folds may be viewed in the west face of the Nilgiri peaks (Colchen *et al.* 1986; Figs 2

Fig. 7. Photograph shows an F1 isocline outlined by original layering (SO). At this location (south of Larjung), S0, S1, and S2 are all sub-parallel.

Fig. 8. Photo shows decimetre-scale sandstone and pelite with a well developed spaced axial-plane cleavage in the core of a northeasterly verging F2 fold (in the vicinity of Marpha).

Fig. 9. Looking east from Marpha at a northeasterly verging F2 anticline.

and 4). At the Dhumpu locality it is clear that S2 is truncated by the brittle displacement of the Annapurna detachment fault (Fig. 5); its continuation eastward into the slopes of the Nilgiri and Annapurna peaks implies that the major northeasterly verging nappe would also be truncated as shown in Figs 2 and 4.

A third phase of deformation has folded and sheared the pre-existing structures (Figs 11 and 12). The third-phase microscopic to mesoscopic folds consistently verge to the southwest and appear to be coeval with southwesterly directed thrust faults that locally disrupt the Tibetan sedimentary sequence in the Thakkhola region.

Fig. 10. View to the east of a megascopic F2 syncline (Jomosom and the Kali Gandaki River in the foreground). Photograph shows a lower right-way-up long limb and an upper overturned short limb on the northeasterly verging Nilgiri structure.

Fig. 11. F3 folds with a shallow northeasterly dipping axial plane (south of Larjung).

Fig. 12. Looking east at an overturned limb on a northeasterly verging F2 fold along the east bank of the Kali Gandaki River across from Marpha. Photograph shows F1 isoclinal hooks (SO + S1) sub-parallel to S2 and subsequent refolding by southwesterly verging F3 folds. This photograph is a close-up of the overturned limb on the fold in Fig. 9.

Fig. 13. View to the west of a boudinaged, tourmaline-rich, quartz–plagioclase pegmatite dyke in biotite-grade psammitic rocks above the Annapurna detachment fault. The extended dyke dips more steeply than the prominent S2 cleavage and is sub-parallel to the axial planes of F3 folds; this indicates southwestward thrusting.

Southwesterly verging third-phase folds observed at the Dhumpu locality are illustrated in Fig. 5.

A number of northeasterly dipping quartz–feldspar pegmatite dykes occur within a few metres of the Annapurna detachment fault and higher in the Tibetan sedimentary sequence. These pegmatite dykes are sub-parallel to the axial surfaces of F3 minor folds, dip more steeply than S2, and are boudinaged or otherwise extended (Figs 5 and 13). Above the Annapurna detachment fault, high angles are observed between layering (S0 + S1), S2, and weakly boudinaged dykes. Towards the crush zone of the Annapurna detachment fault these fabric elements and dykes are transposed into nearly parallel orientations, and the dykes are progressively more boudinaged and extended (Fig. 5). The importance of these observations lies in the recognition that both the F3 folds and the dykes are younger than the second phase of deformation and older than the brittle extensional shear fabric in the hanging wall of the Annapurna detachment fault.

The youngest sense of shear observed in the footwall rocks below the brittle zone indicates a northeast normal sense of shearing. The most recent fabric in the hanging wall above the brittle zone indicates that a period of southwesterly directed shearing affected these rocks prior to the most recent period of normal faulting on the Annapurna detachment fault.

Thakkhola graben

The youngest structures recognized in the hanging-wall rocks are sub-vertical northeast-trending fractures and normal faults. These features are undoubtedly contemporaneous with formation of the Thakkhola graben, which is topographically expressed by the broad valley of the upper reaches of the Kali Gandaki River (Bordet *et al.* 1971; Colchen *et al.* 1986; Figs 1 and 3). The Thakkhola graben is bounded on its western side by the Thakkhola normal fault (TNF on Figs 1, 3), which in the Mustang area north of Jomosom (Fig. 1) is estimated to have a down-to-the-east vertical component of displacement of approximately 3000 m (Bordet *et al.* 1971). The eastern margin is less well defined and consists of a series of en echelon faults with small displacement, suggesting a half-graben geometry. Displacement on the fault system decreases southward, and northeast-trending faults that have been mapped south of the Annapurna detachment fault within the Greater Himalayan metamorphic sequence display only minor offsets (Colchen *et al.* 1986).

In the upper reaches of the Kali Gandaki River (Fig. 1), Pliocene–Quaternary graben-fill clastic rocks rest unconformably on the down-faulted Mesozoic strata; the trace of the Thakkhola normal fault lies along the western edge of the graben where upturned Quaternary sediments lie in its footwall and Silurian–Devonian strata form the hanging wall. Post-glacial sediments do not appear to have been disrupted by the normal faulting (Colchen *et al.* 1986).

Interpretation

The Annapurna detachment fault

Structural and metamorphic relationships described above demonstrate the existence of a major normal fault at the top of the Greater Himalayan metamorphic sequence in the Thakkhola region. The presence of kyanite-bearing migmatites in the footwall within tens of metres of sub-amphibolite-grade metasediments suggests a vertical component of displacement of at least 10 km. Since observed stretching lineations are down dip, there appears to be no indication of a strike-slip component (although see Pêcher (1991) for discussion of possibly regional strike-slip motions).

Generally, in single-stage normal-sense shearing and exhumation, hot middle or lower crustal rocks of the footwall are progressively brought to higher structural levels and are juxtaposed with cooler upper crustal rocks of the hanging wall. Thus the resulting shear zone may exhibit a ductile mylonitic footwall and a brittle hanging wall. In the contact zone the mylonitic rocks are generally retrograded and brecciated (see Coney 1980).

The absence of a prominent mylonite zone on the footwall of the Annapurna detachment fault requires explanation. The retrograded and brecciated zone is underlain by foliated migmatite and leucogranite with no evidence of extensive mylonitization. However, in outcrop and thin section there is ample evidence of ductile extensional shearing followed by annealing recrystallization. This annealing occurred while the rocks were in the kyanite field and after generation of the leucogranites. These observations are readily explained if a two-stage model of extensional shearing is envisaged. In the initial stage the footwall rocks were in the ductile field at middle crustal depth; extension occurred during generation of leucogranitic melts; this was followed by a period of static annealing. A second stage of extension after cooling of the footwall gave rise to the brittle fabrics.

With the exception of high-angle faults associated with the Thakkhola graben, fabrics of the hanging wall are older than the most recent normal displacement on the Annapurna detachment fault. The third-phase folds were formed during southwesterly directed shearing and generation of a suite of granitic dykes. Attempts at U-Pb dating of these dykes have been unsuccessful, but it is clear that a period of compressional strain occurred after the generation of northeasterly verging phase-two structures and before the most recent extensional shearing on the Annapurna detachment fault.

The northeasterly verging folds

The northeasterly verging folds (F2) in the hanging wall of the Annapurna detachment fault predate southwesterly verging folds (F3), emplacement of a suite of granitic dykes, and the most recent normal displacement on the detachment. The second-phase structures, which include the northeasterly verging Nilgiri nappe, have a well developed axial-plane cleavage that is in part defined by the preferred orientation of metamorphic minerals such as biotite, indicating that the folds developed during metamorphism of the hanging-wall strata. This metamorphic event is undated but must be older than the compressional F3 events, which are post-metamorphic.

It has recently been proposed that structures similar to the Nilgiri nappe elsewhere in the Himalayan orogen are a result of crustal-scale normal-sense shearing related to the South Tibetan detachment system (Burchfiel *et al.* 1992). If this interpretation were accepted for the Thakkhola region, a two-stage model of extension with an intervening period of compression (F3) would be required. It is indeed tempting to suggest that the Nilgiri nappe and associated structures formed during the early ductile phase of extension on the Annapurna detachment fault when both hanging wall and footwall were at middle crustal depth. However, there is no direct evidence in support of an extensional origin for these northeasterly directed structures, and alternatively, they could have been generated during an early period of compression and associated crustal thickening. The resolution of this ambiguity must await additional data; particularly required is a better understanding of the regional significance of vergence reversals elsewhere in the orogen.

Significance of F3 deformation in the Tibetan sedimentary sequence

The southwesterly verging F3 structures and pegmatite dykes in the Tibetan sedimentary sequence clearly overprint and are younger than the F2 Nilgiri structure and associated S2 cleavage. All have been cut by the most recent brittle motion on the Annapurna detachment fault.

If ultimately it is shown that the Nilgiri structure is not genetically related to top-to-the-northeast normal sense of shear between the Greater Himalayan metamorphic sequence and the Tibetan sedimentary sequence (Burchfiel *et al.* 1992), then possibly all three phases of folding predate formation of the Annapurna detachment fault.

If the Nilgiri nappe is considered the result of early crustal thickening and as a structurally higher metamorphic equivalent to the Barrovian event in the Greater Himalayan metamorphic sequence, then the F3 compressional structures and suite of granitic dykes in the Tibetan sedimentary sequence may be documenting a post-Barrovian, southwestward thrusting event that occurred before ductile extensional movement on the Annapurna detachment fault.

Elsewhere in the Himalaya, field relations as well as pressure-temperature paths from garnet-bearing pelitic rocks in the Greater Himalayan metamorphic sequence suggest that the Buchan metamorphism overprinting the earlier Barrovian event was associated with the generation of leucogranites and a tectonic reburial of 5–7 km (Hodges & Silverberg 1988; Hodges *et al.* 1988). This reburial was found to be widespread, but the lack of appropriate structures from detailed mapping in the Tibetan sedimentary sequence hindered explanation. It is suggested here that the F3 compressional structures and suite of granitic dykes in the hanging wall of the Annapurna detachment fault may be documenting the southwestward thrusting event that caused the post-Barrovian tectonic reburial observed in pressure–temperature paths and are possibly related to the Buchan metamorphism and leucogranite generation.

Thakkhola graben

The northeast-trending extensional structure is well developed in the hanging wall of the Annapurna detachment fault but cannot be traced into the footwall. This geometry suggests a kinematic link between the graben and detachment faulting. Could it be that orogen-parallel extension in the Tibetan sedimentary sequence occurred at the same time as northeastward displacement of the assemblage on the Annapurna detachment fault? This suggestion lacks proof at the moment, but if correct, such three-dimensional extension might account for the weak development of stretching lineations in the shear zone of the Annapurna detachment fault and in the Tibetan sedimentary sequence of the hanging wall. Since evidence of orogen-parallel extension has not been recorded in the Greater Himalayan metamorphic sequence, it appears that in this part of the Himalaya the extension has been restricted to upper crustal levels.

Conclusions

The Annapurna detachment fault in the Thakkhola region of Nepal is a discrete north-easterly dipping normal-sense shear zone; it is correlated with the South Tibetan detachment system.

Displacement across the Annapurna detachment fault has juxtaposed mid-crustal rocks of the Greater Himalayan metamorphic sequence in the footwall with upper crustal rocks of the Tibetan sedimentary sequence in the hanging wall.

Early motion on the Annapurna detachment fault was ductile and occurred at mid-crustal levels. Later motion developed a zone of brittle strain that is superimposed on the earlier ductile fabrics.

In the hanging wall of the Annapurna detachment fault, three phases of deformation are recognized: a first phase of mesoscopic to megascopic southwesterly verging isoclines with an associated layer-parallel cleavage, a second phase of northeasterly verging tight to open megascopic folds with an axial-plane cleavage, and a third phase of post-metamorphic southwesterly verging microscopic to mesoscopic folds. The third-phase compressional structures are associated with a suite of granitic dykes, and both are younger than and superimposed on the first- and second-phase structures.

The kinematic significance of the northeasterly verging phase-two Nilgiri nappe and associated structures is not clear; these structures may have formed during the early stage of extension on the Annapurna detachment fault, but alternatively, they may reflect a period of Himalayan crustal thickening.

The most recent period of motion on the Annapurna detachment fault may have been contemporaneous with formation of the Thakkhola graben. The graben is an orogen-parallel structure in the Tibetan sedimentary sequence that was active as recently as the Quaternary.

This research was initially carried out as part of the 1991 international Lost Ocean Expedition, under the leadership of F. Gradstein. R.L.B. returned for a second field season in 1992. We would like to thank the Sherpa Society for their able logistical support. The final draft of this paper has benefitted from constructive reviews by J.P. Burg, L.H. Royden, and P.J. Treloar. We thank L. Hardy for assistance in manuscript preparation and production of figures. This research has been funded by Research Grant 2693 from the National Sciences and Engineering Research Council of Canada to R.L.B.

References

BORDET, P., COLCHEN, M., KRUMMENACHER, D., LE FORT, P., MOUTERDE, R. & REMY, M. 1971. *Recherches géologiques dans L'Himalaya du Népal, région de la Thakkhola*. Editions du Centre National de la Recherche Scientifique, Paris, France.

BRUNEL, M. & KIENAST, J.-R. 1986. Etude petro-structurale des chevauchements ductiles himalayens sur la transversale de l'Everest–Makulu (Nepal oriental). *Canadian Journal of Earth Sciences*, **23**, 1117–1137.

BURCHFIEL, B.C. & ROYDEN, L.H. 1985. North–south extension within the convergent Himalayan region. *Geology*, **13**, 679–682.

——, CHEN, Z., HODGES, K.V., LIU, Y., ROYDEN, L.H., DENG, C. & XU, J. 1992. *The south Tibetan detachment system, Himalayan orogen: extension comtemporaneous with and parallel to shortening in a collisional mountain belt*. Geological Society of America, Special Papers, **269**.

BURG, J.P., BRUNEL, M., GAPAIS, D., CHEN, G.M. & LIU, G.H. 1984. Deformation of leucogranites of the crystalline Main Central Sheet in southern Tibet (China). *Journal of Structural Geology*, **6**, 535–542.

CABY, R., PÊCHER, A. & LE FORT, P. 1983. Le grand chevauchement central himalayen: nouvelles données sur le métamorphisme inverse à la base de la Dalle du Tibet. *Revue de géologie dynamique et de géographie physique*, **24**, 89–100.

COLCHEN, M., LE FORT, P. & PÊCHER, A. 1986. *Geological researches in the Nepal's Himalaya: Annapurna–Manaslu–Ganesh Himal*. Editions du Centre National de la Recherche Scientifique, Paris, France.

CONEY, P. 1980. Cordilleran metamorphic core complexes: an overview. *In*: CRITTENDEN, M.D. (ed.) *Cordilleran Metamorphic Core Complexes*. Geological Society of America, Memoirs, **153**, 7–31.

DALMAYRAC, B. & MOLNAR, P. 1981. Parallel thrust and normal faulting in Peru and constraints on the state of stress. *Earth and Planetary Science Letters*, **55**, 473–481.

ENGLAND, P.C. & MCKENZIE, D. 1982. A thin viscous sheet model for continuous deformation. *Royal Astronomical Society Geophysical Journal*, **70**, 295–321.

—— & —— 1983. Correction to 'A thin viscous sheet model for continuous deformation'. *Royal Astro-*

nomical Society Geophysical Journal, **73**, 523–532.

GRADSTEIN, F.M., VON RAD, U., GIBLING, M.R., JANSA, L.F., KAMINSKI, M.A., KRISTIANSEN, I-L., OGG, J.G., ROHL, U., SARTI, M., THUROW, J.W., WESTERMANN, G.E.G. & WIEDMANN, J. 1992. The Mesozoic continental margin of central Nepal. *Geologisches Jahrbuch*, Reihe B, Heft **77**.

HODGES, K.V. & SILVERBERG, D.S. 1988. Thermal evolution of the Greater Himalaya, Garhwal, India. *Tectonics,* **7**, 583–600.

——, HUBBARD, M.S. & SILVERBERG, D.S. 1988. Metamorphic constraints on the thermal evolution of the central Himalayan Orogen. *Philosophical Transactions of the Royal Society of London,* **A326**, 257–280.

HOUSEMAN, G.A., MCKENZIE, D.P. & MOLNAR, P. 1981. Convective instability of a thickened boundary layer and its relevance for the thermal evolution of continental convergent belts. *Journal of Geophysical Research,* **86**, 6115–6132.

HUBBARD, M.S. & HARRISON, M.T. 1989. $^{40}Ar/^{39}Ar$ age constraints on deformation and metamorphism in the Main Central Thrust zone and Tibetan Slab, eastern Nepal Himalaya. *Tectonics,* **8**, 865–880.

LE FORT, P. 1975. Himalayas: the collided range. Present knowledge of the continental arc. *American Journal of Science,* **275-A**, 1–44.

PÊCHER, A. 1991. The contact between the Higher Himalaya crystallines and the Tibetan sedimentary series: Miocene large-scale dextral shearing. *Tectonics,* **10**, 587–598.

SWAPP, S.M. & HOLLISTER, L.S. 1991. Inverted metamorphism within the Tibetan Slab of Bhutan: evidence for a tectonically transported heat source. *Canadian Mineralogist,* **29**, 1019–1041.

Numerical modelling of the thermal history of the NW Himalayas, Kullu Valley, India

BERNHARD GRASEMANN

Labor für Geochronologie und Isotopengeologie, Franz Grillstr. 9, A-1030 Vienna, Austria

Present address: Institut für Geologie, Universität Wien, Universitätstr 7, A-1010 Vienna, Austria

Abstract: This paper demonstrates two-dimensional time-dependent finite-difference modelling of Himalayan metamorphism illustrated by the Kullu–S Lahul area (Himachal Pradesh) after the collision between India and Asia and the over-thrusting of the Higher Himalayas over the Lesser Himalayas.

It is shown that a one-dimensional model with crustal thickening by a single thrust cannot satisfy the observations present in the section along the Kullu valley. A two-dimensional solution demonstrates that during thrust sheet emplacement at geologically realistic strain rates of several centimetres a year, a temperature inversion does not occur, in contrast to the inversion assumed by the one dimensional instantaneous thrust models. Therefore the variability of two other model parameters were explicitly considered. (i) There is a significant conductivity contrast between the different types of rocks. As conductivity layers function as a barrier to the transport of heat, the conductivity limit should be variable for different regions in the grid. (ii) Rocks in the Higher Himalayas record a phase of metamorphism that preceded the movements on the Main Central Thrust. This observation, together with the knowledge that rocks with different radiogenic heat production and conductivity are not uniformly distributed in the crust, emphasizes that there were considerable horizontal temperature variations across the whole region. For that reason the initial temperature profile, too, has to be adapted for a realistic model.

The area around the Kullu vally in S Lahul (Figs 1 & 2) can be subdivided into three major units (Frank *et al.* 1973, 1977; Thöni 1977):

(i) The Higher Himalayan Crystalline Nappe together with the Tibetan Zone forms a large unit with a relatively simple tectonic structure, which was over-thrusted onto the Lesser Himalaya. The rocks range from unmetamorphosed to sillimanite grade. The southern part of the Higher Himalaya suffered only low to medium greenschist facies conditions. For quite a long distance along the southern front, the Main Central Thrust (MCT) is very close to the Main Boundary Thrust (MBT). Higher metamorphic grades up to sillimanite are only known in the northern part of the Higher Himalayan Crystalline Nappe (HHC).

(ii) Outcropping below the Crystalline Nappe, the Lesser Himalaya in the Larji–Kullu–Rampur Window show metamorphic conditions from the anchizone to greenschist facies. The metamorphic grade increases slightly from SSW to NNE.

(iii) At some locations the Lower Crystalline Nappe (Bajaura Nappe) can be mapped between the Higher Himalayan Crystalline and the Lesser Himalaya. Due to its thickness of only a few tens to one hundred metres and the tectonic characteristics, which are very similar to the overlying HHC, the presence of the Bajaura Nappe will be neglected in the following thermal models.

Across the whole Himalaya the occurrence of inverted metamorphic zones near the MCT can be observed (e.g. Gansser 1964; Frank *et al.* 1973; LeFort 1975). In the investigated area this feature is especially related to the 'root zone' of the MCT. These observations led to the common model involving firstly emplacement of a 'hot' hanging wall over a 'cold' footwall; secondly conductive heating of the uppermost footwall and cooling of the lowermost hanging wall and thirdly thermal equilibration at the MCT.

The main purpose of this paper is to investigate some arguments concerning heat conduction, heat advection and heat production responsible for the cooling history and the inverted metamorphism in this section of the Himalaya.

From Treloar, P.J. & Searle, M.P. (eds), *Himalayan Tectonics*
Geological Society Special Publication, No. 74, pp. 475–484.

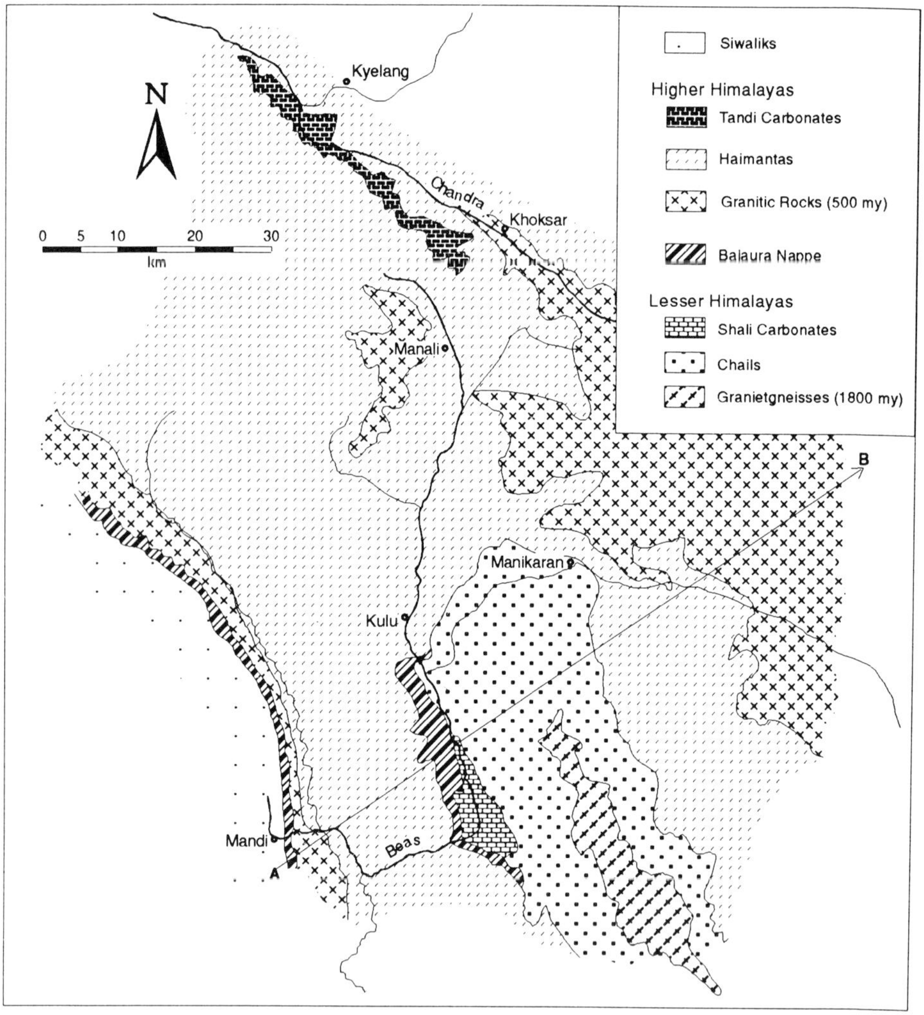

Fig. 1. Simplified map of the S Lahul–Kullu area (modified according to Frank *et al.* 1973). AB indicates the cross section across the Larji–Kullu–Rampur Window shown in Fig. 2.

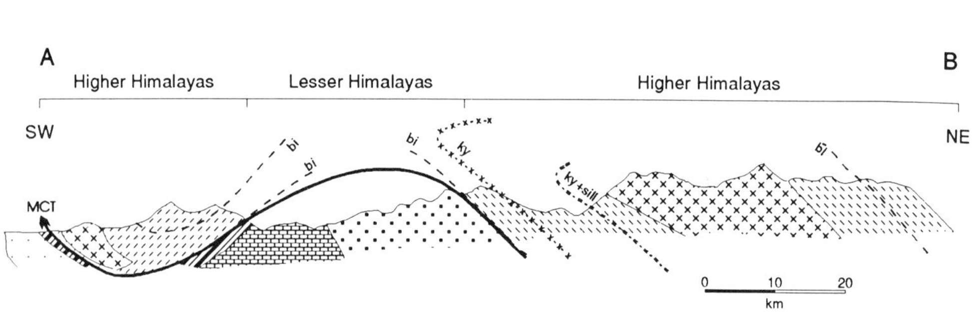

Fig. 2. Cross section of the S Lahul–Kullu area with some metamorphic isograds (modified according to Frank *et al.* 1973). For signatures see map in Fig. 1.

One-dimensional model

Assumptions and solutions of a one dimensional model

Continent–continent collisions usually lead to pronounced regional metamorphism. Today these metamorphic rocks are often underlain by crust of normal thickness (35 km). This has led to the general opinion that during metamorphism the crust was abnormally thick. In order to define a mathematical model for abnormally thick continental crust, two different geometries have been considered (e.g. Oxburgh & Turcotte 1974; England & Thompson 1984, 1986). The first model, the homogeneous thickening model, doubles the initial 35 km thick crust to 70 km in a ductile fashion. For the second model, the sawtooth model, the crust is thickened instantaneously along a single thrust plane. The stacking of the Higher Himalayas over the Lesser Himalayas along the MCT can be seen as an excellent example of the sawtooth model.

The time-dependent heat transfer equation (1) allows us to define a mathematical model for the thermal evolution of thickening geometries:

$$\frac{\partial T}{\partial t} = -V\nabla T + \kappa\nabla^2 T + \frac{A}{\rho C} \tag{1}$$

where T = temperature [K]
t = time [s]
κ = thermal diffusivity [$m^2 s^{-1}$]
V = velocity [$m\ s^{-1}$]
A = volumetric heat production [$W\ m^{-3}$]
C = specific heat [$J\ kg^{-1}\ K^{-1}$]
ρ = density [$kg\ m^{-3}$]

The first term on the right side of equation (1) is the heat advection term, which accounts for heat being transported by a moving medium (e.g. burial, uplift or erosion). The second term describes the heat diffusion. The third term includes all kinds of heat sources such as radiogenetic heat production or shear heating.

The boundary conditions are the heat flux into the base of the crust and the surface temperature at the top of the crust. For a single thrust geometry the initial geotherm is calculated with equation (2) and instantaneously stacked to a sawtooth geotherm (Peacock 1989, equation 2):

$$T = T_s + \frac{q_m z}{k} + \frac{Al}{k}(1 - e^{-z/l}) \tag{2}$$

where T_s = surface temperature [K]
q_m = mantle heat flux [$W\ m^{-2}$]
z = depth [m]
k = thermal conductivity [$W\ m^{-1}\ K^{-1}$]
A = volumetric heat production [$W\ m^{-3}$]
l = scaling depth for radiogenic heat production [m]

The chosen method to solve the one-dimensional time-dependent diffusion problem by finite differences is the implicit scheme by Crank Nicolson which has the main advantage of being unconditionally stable for any time step (Crank & Nicolson 1947).

Application to the situation in the Kullu area

Figure 3 shows the result of a general cooling history of a sawtooth model: the instantaneous stacking of the initial 35 km thick crust doubles the thickness of the whole crust to 70 km. The thrusting event results not only in an initial sawtooth geotherm but also redistributes the heat-producing elements in two layers. In both layers the concentration of heat-producing elements decreases exponentially with depth over a characteristic length scale of 10 km. The thermal evolution of the thickened crust shows abrupt changes in the thermal gradient near the thrust at a depth of 35 km. During the first several million years the heat was rapidly conducted downwards from the hot upper plate into the cool lower plate. Then the temperatures increased slowly as a result of the decay of radioactive elements and the conduction of heat into the base of the crust. The onset of the erosion with 1 mm a^{-1} is assumed after a timelag of 10 Ma. Subsequently, the geotherm evolves towards a steady state geotherm over tens of millions of years, while the thickness of the crust reaches the initial depth of 35 km.

To demonstrate the different metamorphic paths followed by rocks at different depths, Fig. 3 also shows a pressure–temperature–time diagram (*P–T–t*). (i) At a depth of 30 km in the upper plate, (curve 2, Fig. 3b) the rocks cooled rapidly under isobaric conditions, due to the conduction of heat downwards into the cooler lower plate. With the onset of erosion the temperature increased slightly reaching the maximum at 30 Ma after thickening. It is important to note that this maximum temperature was less than the initial temperature. No prograde metamorphism will occur. (ii) Rocks beneath the thrust at 40 km depth (curve 3, Fig. 3b) underwent isobaric heating and continued to heat up with the onset of erosion. The temperature maximum was reached 35 Ma after thickening. However the maximum temperature of

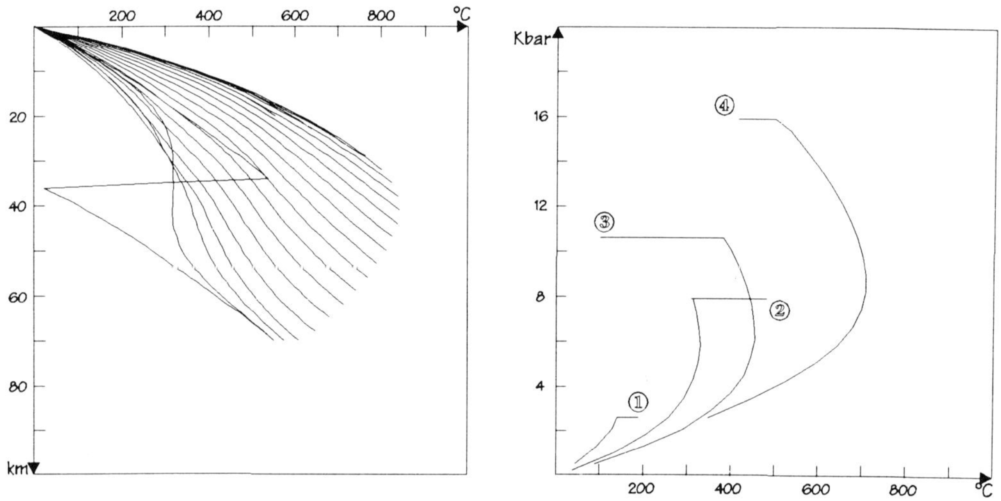

(a) (b)

Fig. 3. (**a**) Temperature–depth and (**b**) pressure–temperature–time (*P–T–t*) diagrams of a single thrust model over 60 Ma using one-dimensional modelling. For parameters used in calculations see Table 1. The heat diffusion equation (1) was solved using the Crank-Nicolson method.

Table 1. *Thermal and geometric parameters used in one-dimensional model*

Parameter	Description	Constant [unit]
T	Temperature	[K]
T_s	Surface temperature	273 [K]
Δ_t	Finite time step	10000 [a]
κ	Thermal diffusivity	1×10^{-6} [m^2 s^{-1}]
v	Erosion rate	-0.001 [m a^{-1}]
A	Volumetric heat production	3×10^{-6} [W m^{-3}]
l	Scaling depth for radiogenic heat production	10000 [m]
C	Specific heat	1000 [J kg^{-1} K^{-1}]
ρ	Density	2800 [kg m^{-3}]
q_m	Mantle heat flux	0.03 [W m^{-2}]
c_t	Thickness of thrust	35000 [m]
t_e	Timelag until erosion	1×10^{-7} [a]
Δ_z	Grid spacing	100 [m]

rocks at different depths was reached at different times.

The *P–T–t* path of the rock samples in the investigated area cannot be satisfactorily explained by this theoretical metamorphic pattern: By conductive heating and radioactive heat production a considerable rise in metamorphic conditions should be obtained in the uppermost footwall beneath the MCT that is only observed at the immediate root zone in the Larji–Kullu–Rampur Window. In the hanging wall, which was rapidly cooled, any second-phase metamorphism recorded by the rocks should be retrograde. However, this feature is only observed in limited parts of the HHC above the MCT. The one dimensional sawtooth model never predicts any temperature and pressure increases within the over-thrusting unit. These observations strongly emphasize that the generally observed inverted metamorphic zonation in this area is mainly produced by a syn to postmetamorphic juxtaposition of the Higher Himalayan Crystalline over the Lesser Himalayas.

Two-dimensional model

Geometry of a two-dimensional model

Modelling regional metamorphism in only one-dimension requires two important but substantial simplifications. Firstly, the isotherms are always assumed to be horizontal which cannot be expected in the pre-collisional, undisturbed continental crust: several authors have demonstrated the influences of inhomogeneously distributed layers with different radiogenetic heat-production or different conductivity parameters, which lead to marked differences of the thermal evolution (Jaupart & Provost 1984; Pinet & Jaupart 1987). Furthermore any pre-collisional thermal structure of the continental margins is neglected. Secondly, only vertical heat transfer is permitted. According to Shi & Wang (1987) a substantial amount of lateral heat conduction occurs during the thrusting event. Molnar & England (1990), however, suggest that for sufficiently large rates of thrusting, greater than 1 mm a^{-1}, lateral conduction of heat is not important compared with advection along the fault or diffusion in the vertical direction. Thirdly, the geometry, especially the dip angle of the thrust, has an important influence on the final temperature distribution. As will be shown later a true 'sawtooth' geotherm, which is presumed for the one-dimensional model is never reached in a two-dimensional situation.

Figure 4 shows the geometry of a two-dimensional situation. The horizontal x distance is 200 km, the vertical z distance is 80 km. The finite difference grid has 1000 nodes in the x direction and 400 nodes in the z direction, which gives an equally spaced grid with a 200 m resolution (this accuracy is more than enough for general geometries). Each node has a velocity field with two components. The velocity field is set to zero except for the overthrusting unit, where the vectors are always parallel to the thrust plane. Deformation details at the corners are neglected and will not affect the thermal Equation (1) is solved by a special ADI method in two dimensions. Initial temperature distributions are calculated with equation (2). To avoid numerical diffusion the advection term is treated separately using a particular upwind differencing scheme (Douglas 1955; Peaceman & Rachford 1955; Anderson *et al.* 1984).

Discussion of the solutions of a two dimensional model

The simulation of the thrusting event over 1 Ma with a thrusting velocity of 5 cm a^{-1} (for further thermal and geometric parameters used in the calculation see Table 2) leads to a very similar result to that shown by Shi & Wang (1987). Figure 5 shows the temperature distribution after modelling and compares the geothermal profile of a location (1) near the root of the thrust and of a location (2) near the leading edge of the thrust. Both curves are very different from the 'sawtoothed' type. The leading edge (2) cools faster than the geotherm (1) near the root zone,

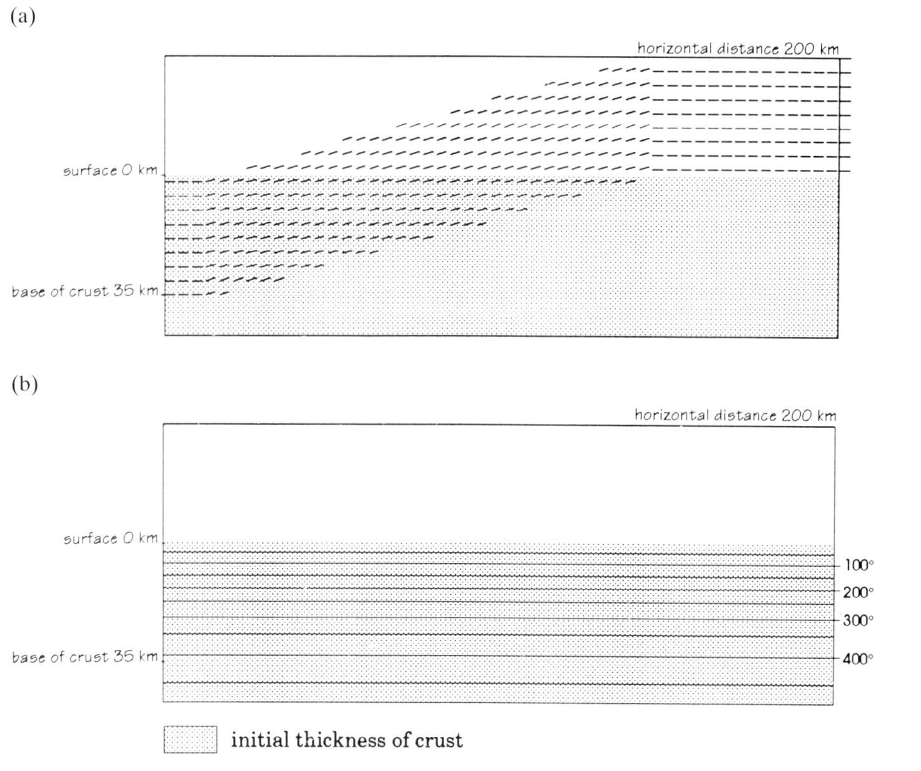

Fig. 4. (**a**) Geometry of a two-dimensional model for single thrust geometry with 15° thrust angle. (**b**) Initial conditions with 0.04 W m^{-2} heat flux to base of lithosphere and 2.5 W m^{-1} K^{-1} thermal conductivity. Detailed explanation is given in the text.

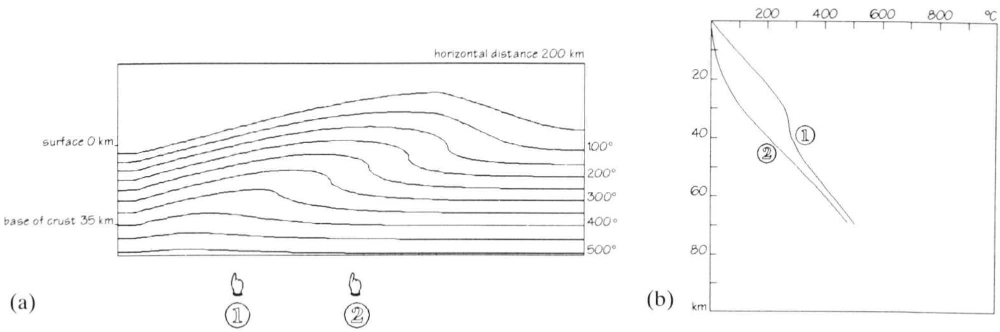

Fig. 5. (**a**) Plot of isotherms after 1 Ma thrusting with 5 cm a^{-1} using the initial conditions and the geometry shown in Fig. 4. The heat diffusion equation (1) was solved using an ADI method. For detailed parameters see Table 2. (**b**) Plot of two geothermal profiles at 50 km and 100 km from the left side of the grid.

Table 2. *Thermal and geometric parameters used in two-dimensional model*

Parameter	Description	Constant [unit]
T	Temperature	[K]
T_s	Surface temperature	273 [K]
Δ_t	Finite time step	10000 [a]
κ	Thermal diffusivity	[$m^2 s^{-1}$]
v, u	Velocity in x, z direction	[m a^{-1}]
A	Volumetric heat production	2×10^{-6} [W m^{-3}]
C	Specific heat	1000 [J kg^{-1} K^{-1}]
ρ_s	Density of sediments	2500 [kg m^{-3}]
ρ_c	Density of crystalline	2800 [kg m^{-3}]
k_s	Conductivity of sediments	1.8 [W m^{-1} K^{-1}]
k_c	Conductivity of crystalline	3.2 [W m^{-1} K^{-1}]
q_m	Mantle heat flux	0.04–0.06 [W m^{-2}]
l	Characteristic length scale	10000 [m]
c_t	Thickness of thrust	35000 [m]
α	Thrust angle	15°
n_{xz}	Number of nodes	400000
x	Horizontal x direction	200 [km]
z	Vertical z direction (depth)	80 [km]
Δ_z	Grid spacing in z direction	200 [m]
Δ_x	Grid spacing in x direction	200 [m]

which results in a temperature difference of more than 100° C between a depth of 10 km and 40 km. There is no location where the temperature decreases with depth. At thrusting velocities close to several centimetres a year, which are upper boundaries for movements on the MCT as shown by palaeomagnetic evidence (Molnar 1987; Dewey *et al.* 1989), the conductive heat transfer removes nearly any temperature disturbance. Only a 'doming' of the isotherms can be observed. A high-velocity explanation, which is needed to result in a 'sawtooth' temperature distribution, is unattractive, because it requires a relative motion several times faster than the plate motions documented.

The most important result of the two-dimensional model is the pronounced difference in temperature distribution between the leading edge and the root zone of the thrust. Such a feature is observed in the temperature increase towards the root zone of the HHC north of the Larji–Kullu–Rampur Window, where the geochronological cooling ages of biotite diminish from 21 Ma at the front of the Crystalline Nappe to a value of 16 Ma close to the root zone north of the Lesser Himalayas (Frank *et al.* 1977).

In addition this parameter configuration again suggests that during thrusting only cooling processes take place in the hanging wall and that the inverse zonation is the result of syn- to post-metamorphic movements. The model is not able to predict a second prograde metamorphism during or after thrusting.

In the Kullu Valley above the MCT Frank *et al.* (1973) and Thöni (1977) describe the formation of a synmetamorphic recumbent fold. Such structures need elevated temperatures during the continuous overturning of material near the base of the overthrusting unit. In the two-dimensional model in Fig. 5, which cannot explain such temperatures, a steady state temperature profile should be reached during and shortly after thrusting, as the conduction of heat is fast enough to remove any temperature inversion. In order to facilitate the transportation of a certain amount of heat during thrusting, which is needed to explain such field observations, a more complex heat transfer mechanism is required, as will be shown later.

Refined model

It is very difficult to explain the transportation of heat in the thrusting unit because the thrusting of cold rock beneath hotter rock should cool the upper block. Several authors try to avoid this

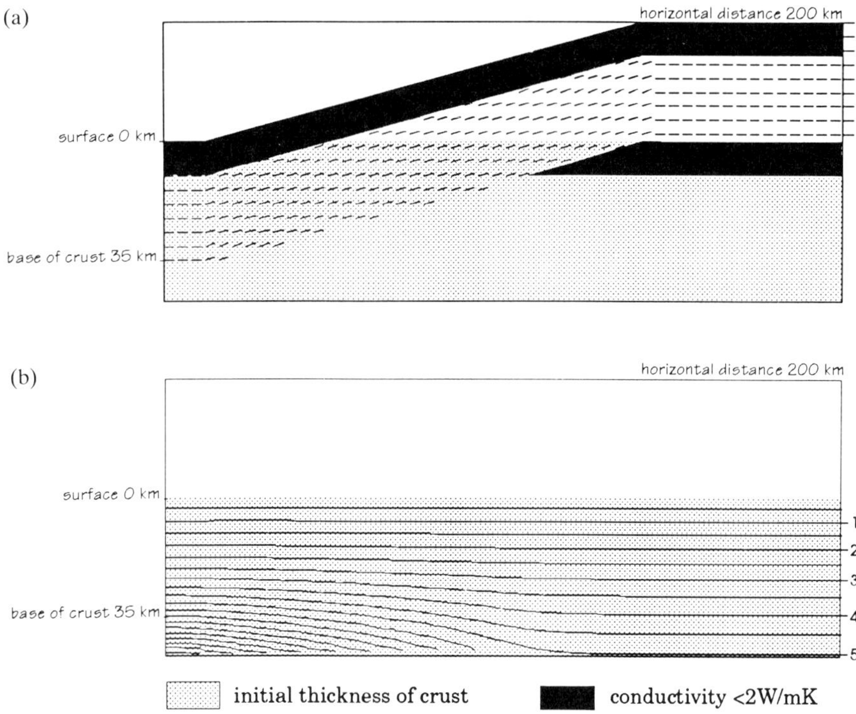

Fig. 6. (**a**) Geometry and (**b**) initial conditions of a two-dimensional model for single thrust geometry. In contrast to the model of Figs 2 and 3 there is a significant change in the thermal conductivity (1.8 W m^{-1} K^{-1} for sediments; 3.2 W m^{-1} K^{-1} for crystalline rocks). The heat flux to the base of lithosphere varies from 0.06 Wm^{-2} on the left side to 0.04 W m^{-2} on the right side of the grid.

problem by invoking an additional heat source, such as shear heating, or elevated radiogenic heat-production. However, if shear heating is the dominant mechanism, one would expect to see the highest temperatures along or close to the MCT, which has never been observed in the Kullu area. The one-dimensional models show that a reasonable temperature increase caused by radioactive heating would need several tens of millions of years (e.g. Bird *et al.* 1975; Toksöz & Bird 1977), which is clearly too long a time scale for thickening processes in the Himalayas. That is why redistribution of heat-producing elements also fails to explain the temperature distribution in the Crystalline Nappe.

One possibility to approximate the result of the model to the observed situation is by using a particular heat transfer mechanism combined with slightly modified initial temperature conditions:

It is generally accepted that a well-developed phase of metamorphism in the Higher Himalayan Crystalline, but not in the Lesser Himalayas, preceded the movements on the MCT (e.g. Pêcher 1989), therefore it is reasonable to raise the temperatures in the overthrust unit before the onset of any movements. Such initial conditions are shown in Fig. 6.

Sedimentary rocks have low thermal conductivities between 1.3 to 2.2 W m^{-1} K^{-1}. Crystalline materials usually exceed these values and range from 2.5 to 3.5 W m^{-1} K^{-1} (Jaupart & Provost 1985). In this model it is now assumed that there are two low conductivity layers (Fig. 6): (i) the Tethyan Zone lying as an 'insulating zone' on the top of the Crystalline Nappe and (ii) the sedimentary rocks of the Lesser Himalayas acting as a barrier between the heat flow from below and the 'hotter' thrusting unit above. The result of this model with the same parameters as used in the previous model (Figs 4 & 5) is shown in Fig. 7. Although no real 'sawtoothed' geotherm is achieved, the temperature values above the shear zone are slightly higher than in the previous simulation. The upper plate is less cooled, due to an 'insulating' sheet of less conductive material of the Lesser Himalaya just beneath the MCT. The less conductive

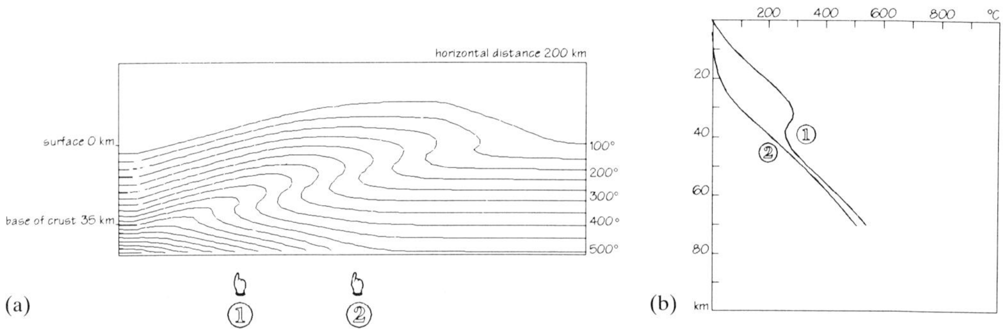

Fig. 7. Plot of isotherms after 1 Ma thrusting with 5 cm a^{-1}. The initial conditions are changed as described in Fig. 4. The heat diffusion equation (1) was solved using an ADI method. For detailed parameters see Table 2. (**b**) Plot of two geothermal profiles at 50 km and 100 km from the left side of the grid.

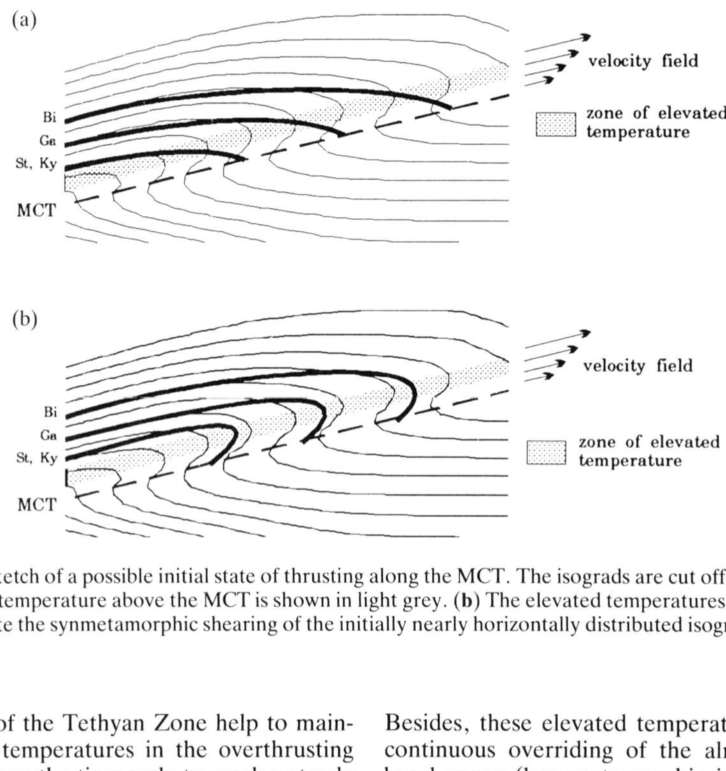

Fig. 8. (**a**) Sketch of a possible initial state of thrusting along the MCT. The isograds are cut off by the thrust. The elevated temperature above the MCT is shown in light grey. (**b**) The elevated temperatures above the MCT facillitate the synmetamorphic shearing of the initially nearly horizontally distributed isograds.

carbonates of the Tethyan Zone help to maintain higher temperatures in the overthrusting unit. Therefore the time scale to reach a steady state condition after thrusting is delayed for several millions of years, which is a much better approximation to the observed cooling pattern.

A very important consequence of such a model is that the temperature maximum in the overriding plate above the MCT is reached after some stage of initial thrusting. This would considerably facilitate synmetamorphic shearing within the HHC with all the consequences of deformation of metamorphic zones (Fig. 8). Besides, these elevated temperatures ease the continuous overriding of the already cooling basal zones (less metamorphic in a more advanced state of thrusting) by still hotter zones with successively higher metamorphic grade. Figure 8 shows a possible initial state of the thrusting process with hypothetical isograds. The elevated temperatures above the MCT are shown in light grey.

An important prerequisite is an appropriate geometry of the isotherm distribution at the initial stage of thrusting as is assumed in this model (Figs 6 & 7).

Conclusion

The two-dimensional modelling of overthrusting using uniformly distributed thermal parameters suggests that it is impossible to transport a significant amount of heat during thrusting into the overthrust unit. A pronounced temperature inversion of several 100° C is never achieved. For that reason a sawtooth geotherm should not be used as initial condition in a one-dimensional model.

Field observations suggest that there was sufficient amount of heat during thrusting, to facilitate synmetamorphic shear folding of metamorphic isograds (Fig. 2). In this paper a special heat transfer mechanism based on different conductivity layers is introduced to the thrusting process. Layers with a low conductivity can work as isolating sheets. They may support the preservation and therefore the transportation of heat in layers with higher conductivities and delay their cooling history.

The first phase of metamorphism, that predated the movements on the MCT, might have caused a considerable horizontal variation in the heat distribution before thrusting. Only with two-dimensional models can such lateral changes in the temperature field be calculated and used as initial conditions.

Because of the complexity of the geological and tectonic situation compared to the oversimplified model, only those parameters that influence the distribution of heat can be evaluated. Therefore no single explanation can account for all the observed metamorphic zones along the Himalayas. Although this model evaluates radioactive heat-production, conductivity, density and two velocity vectors for each node, more parameter variables for each element in the grid will be important for further simulations: Rates of metamorphism, rates of fluid production and fluid flux as a function of space and time can strongly influence the heat distribution pattern. The grid of this model is only dimensioned to the base of the crust. It is widely recognized that the response of the mantle lithosphere to deformation exerts a strong influence on the thermal evolution of orogenetic belts (e.g. McKenzie 1978; Houseman *et al.* 1981). For that reason the grid should be dimensioned in the *z* direction for the whole lithosphere to allow different thickening geometries.

In the present model each node in the thrusting unit has the same absolute velocity: the thrusting velocity. At all other nodes the velocity field is zero. As a result of this simplification the MCT is a distinct plane. In further models the velocity field should be variable in order to simulate a shear zone in the range of 1 km along the MCT. These changes in the velocity field should also help to obtain a better approximation to natural conditions, especially in respect to synmetamorphic folding of the isograds that already existed during the thrusting in the overriding unit. To improve the results from numerical models the adoption of a more detailed geometry of the MCT is needed.

This work is supported by the 'Fonds zur Förderung der wissenschaftlichen Forschung in Österreich', which I gratefully acknowledge. I would especially like to thank W. Frank and Ch. Miller for supporting my field work and for many constructive discussions. I would also like to thank K. Stüwe for many discussions about heat transfer processes. I am highly indebted to T. Reininghaus and F. Kasza for the mathematical support and their help in computer programming.

References

Anderson, D.A., Tannehill, J.C. & Pletcher, R.H. 1984. *Computational fluid mechanics and heat transfer.* Hemisphere Publishing Corporation, New York.

Bird, P., Toksöz, M.N. & Sleep, N.H. 1975. Thermal and mechanical models of continent-continent convergence zones. *Journal of Geophysical Research,* **80**, 4405–4416.

Crank, J. & Nicolson, P. 1947. A practical method for numerical evaluation of solutions of partial differential equations of heat-conduction type. *Proceedings of the Cambridge Philosophical Society,* **43**, 50–67.

Dewey, J.F., Cande, S. & Pitmann, III, W.C. 1989. Tectonic evolution of the India/Eurasia Collision Zone. *Eclogae geologicae Helvetiae.,* **82/3**, 717–734.

Douglas Jr., J. 1955. On the numerical integration of $\partial^2u/\partial x^2 + \partial^2u/\partial y^2 = \partial u/\partial t$ by implicit methods. *Journal of the Indian Mathematical Society,* **3**, 42–65.

England, P.C. & Thompson, A. 1984. Pressure-temperature–time paths of regional metamorphism I. Heat transfer during the evolution of regions of thickened continental crust. *Journal of Petrology,* **25**, 894–928.

—— & —— 1986. Some thermal and tectonic models for crustal melting in continental collision zones. *In*: Coward, M.P. & Ries, A.C. (eds) *Collision Tectonics*. Geological Society, London, Special Publications, **19**, 83–94.

Frank, W., Hoinkes, G., Miller, Ch., Purtscheller, F., Richter, W. & Thöni, M. 1973. Relations between metamorphism and orogeny in a typical section of the Indian Himalayas. *Tschermaks mineralogische und petrographische Mitteilungen,* **20**, 303–332.

Frank, W., Thöni, M. & Purtscheller, F. 1977. Geology and petrography of Kullu–South Lahul area. *Colloques internationaux du CNRS*, **286**, 147–172.

GANSSER, A. 1964. *Geology of the Himalayas*. Interscience Publishers, London.

HOUSEMAN, G.A., MCKENZIE, D.P. & MOLNAR, P. 1981. Convective instability of a thickened boundary layer and its relevance for the thermal evolution of continental convergent belts. *Journal of Geophysical Research,* **86**, 6115–6132.

JAUPART, C. & PROVOST, A. 1985. Heat focussing, granite genesis and inverted metamorphic gradients in continental collision zones. *Earth and Planetary Science Letters,* **73**, 385–397.

LE FORT, P. 1975. Himalayas: The collided range. Present knowledge of the continental arc. *American Journal of Science,* **275-A**, 1–44.

—— 1986. Metamorphism and magmatism during the Himalayan collision. *In*: COWARD, M.P. & RIES, A.C. (eds) *Collision Tectonics*. Geological Society, London, Special Publications, **19**, 159–172.

MCKENZIE, D.P. 1978. Active tectonics of the Alpine-Himalayan belt: the Aegean Sea and surrounding regions. *Geophysical Journal of the Royal Astronomical Society,* **55**, 217–254.

MOLNAR, P. 1987. Inversion of profiles of uplift rates for the geometry of dip-slip faults at depth, with examples from the Alps and the Himalaya. *Annales géophysique,* **5**, 663–670.

—— & ENGLAND, P.C. 1990. Temperatures, heat flux, and frictional stress near major thrust faults. *Journal of Geophysical Research,* **95**, 4833–4856.

OXBURGH, E.R. & TURCOTTE, D.L. 1974. Thermal gradients and regional metamorphism in overthrust terrains with special reference to the eastern Alps. *Schweizerische mineralogische und petrographische Mitteilungen,* **54**, 641–662.

PEACEMAN, D.W. & RACHFORD, H.H. 1955. The numerical solution of parabolic and elliptic differential equations. *Journal of the Indian Mathematical Society,* **3**, 28–41.

PEACOCKS, S.M. 1989. Numerical constraints on rates of metamorphism fluid production and fluid flux during regional metamorphism. *Geological Society of America Bulletin,* **101**, 476–485.

PÊCHER, A. 1989. The metamorphism in the central Himalaya. *Journal of Metamorphic Geology*, **7**, 31–41.

PINET CH. & JAUPART, C. 1987. A thermal model for the distribution in space and time of the Himalayan granites. *Earth and Planetary Science Letters,* **84**, 87–99.

THÖNI, M. 1977. Geology, structural evolution and metamorphic zoning in the Kullu Valley (Himachal Himalayas, India) with special reference to the reversed metamorphism. *Mitteilung der Gesellschaft der Geologie- und Bergbaustudenten in Österreichs,* **24**, 125–187.

TOKSÖZ, M.N. & BIRD, P. 1977. Modelling of temperatures in continental convergence zones. *Tectonophysics,* **41**, 181–193.

SHI, Y. & WANG, C.H. 1987. Two-dimensional modelling of the P–T–t path of regional metamorphism in simple overthrust terrains. *Geology,* **15**, 1048–1051.

Pressure, temperature and time constraints on metamorphism across the Main Central Thrust zone and High Himalayan Slab in the Garhwal Himalaya

R.P. METCALFE

(deceased)

Department of Geology, Leicester University, Leicester LE1 7RH, UK

Abstract: Following the early Eocene collision of India and Asia, continental subduction occurred on the northward-dipping Main Central Thrust (MCT). In western Garhwal, N. India, upper amphibolite–facies gneisses on the High Himalayan Slab are thrust southwards over unmetamorphosed to greenschist facies quartzites, carbonates and metabasics of the Lesser Himalaya. In the Bhagirathi valley, the MCT forms a *c.* 10 km thick shear zone composed of mylonitic augen gneiss, amphibolite and metasediments. Metamorphic grade increases both northwards and with structural height. The MCT zone is bounded to the north by the Vaikrita (roof) Thrust and to the south by the Munsiari (floor) Thrust. The Vaikrita Thrust is a diffuse high-temperature shear zone, whereas the Munsiari Thrust is a relatively discrete fault formed under brittle–ductile conditions. North of the MCT zone, at the top of the High Himalayan Slab a northward-dipping extensional shear zone, the Jhala normal fault, was responsible for the downthrow of the Tethyan sediments to the north with respect to the uplifting High Himalayan Slab gneisses to the south.

Thermobarometic transects reveal a sudden increase in both pressure and temperature across the Vaikrita Thrust from south to north but with subsequent decreases accompanying structural height in the High Himalayan Slab. Temperatures increase going up-structural section from about 500° C to 770° C across the MCT zone, but then decrease again to the north varying between about 550 and 640° C. Similarly, pressures increase sharply up-structural section across the MCT zone from 6 to 12 kbar, then decrease towards the top of the slab to between 7 and 8.9 kbar. The inverted *P–T* gradient across the MCT zone changes to approximately isothermal and isobaric conditions in the top 9 km (horizontal distance) of the High Himalayan slab.

Cooling rates for the upper MCT zone determined from $^{40}Ar/^{39}Ar$ (hornblende) and K-Ar (muscovite and biotite) cooling ages suggest a return to erosion-controlled denudation following extension at the top of the High Himalayan Slab. Additional K-Ar (muscovite) cooling ages from a transect through the MCT zone and High Himalayan Slab are progressively younger towards the south, reflecting the southward propagation of the deformation sequence with time. Hornblende $^{40}Ar/^{39}Ar$ cooling ages from the MCT zone suggest that structurally lower rocks have not been heated above *c.* 500° C since the Precambrian, whilst a 19.8 ± 2.6 Ma hornblende age from the MCT zone dates the latest high-temperature shearing at higher structural levels in the MCT zone and places a minimum age constraint on Himalayan metamorphism in the Garhwal sector of the Himalaya.

The Main Central Thrust (MCT) has long been recognized as a major intra-continental shear zone, with associated inverted metamorphic isograds, that places high-grade gneisses of the High Himalaya to the north over the low-grade Lesser Himalayan thrust sheets to the south (Heim & Gansser 1939; Jain 1971; Valdiya 1978, 1980; Jain & Anand 1988; Pêcher & Scaillet 1989). There has recently been controversy over the exact location of the MCT, mainly due to terminological differences from one area of the Himalaya to the next. Many recent workers believe that the MCT is a wide zone of high strain associated with the inverted metamorphism rather than a single thrust plane. In the Bhagirathi valley, Western Garhwal, the MCT forms a 12 km thick NNE-dipping shear zone that separates Lesser Himalayan Formations from the Vaikrita Group of the High Himalayan Slab (Fig. 1). The MCT zone of deformation is

Due to the death of Paul Metcalfe on 17 March 1992, this paper was compiled by M.P. Searle from his PhD thesis 'A Thermotectonic Evolution of the Main Central Thrust and Higher Himalayas, Western Garhwal, India' submitted to the University of Leicester (Metcalfe 1990).

From TRELOAR, P.J. & SEARLE, M.P. (eds), *Himalayan Tectonics*
Geological Society Special Publication, No. 74, pp. 485–509.

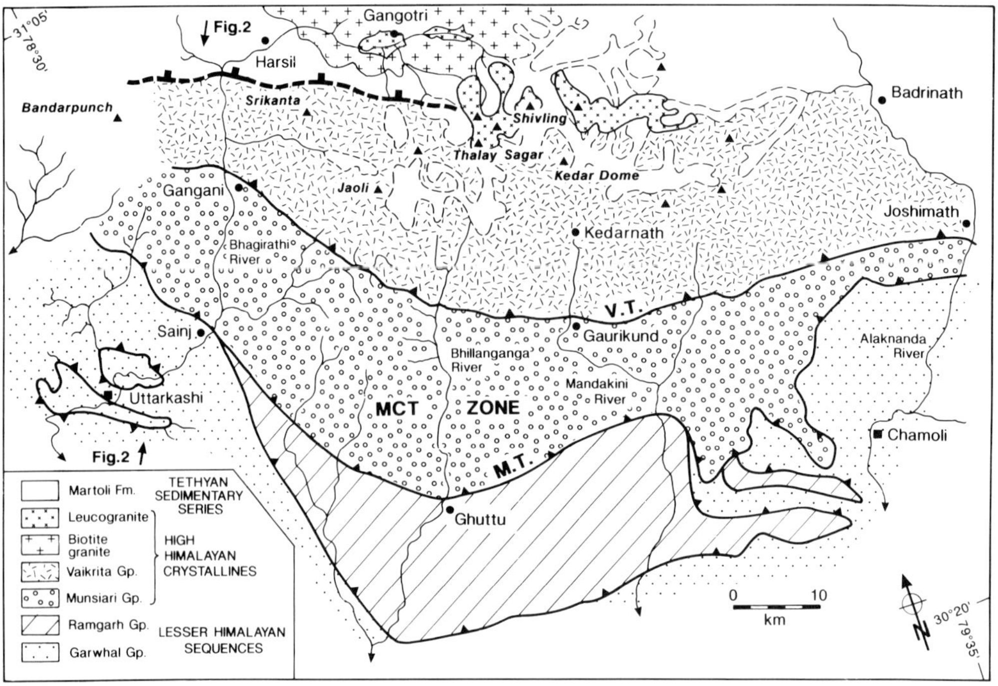

Fig. 1. Lithological and tectonic map of the Garhwal Himalaya compiled from Valdiya (1981), Metcalfe (1990) and Pêcher & Scaillet (1991). MT, Munsiari Thrust; VT, Vaikrita Thrust.

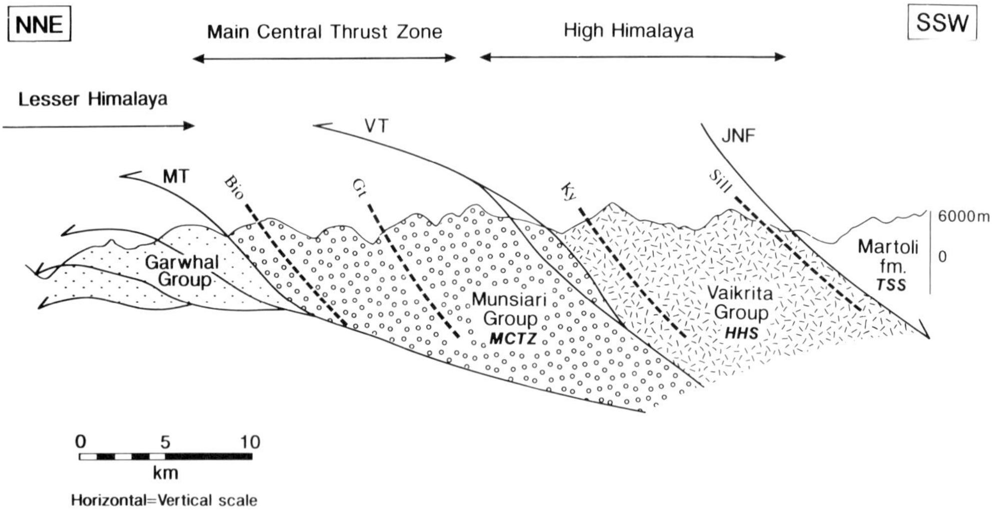

Fig. 2. Schematic structural cross-section of the Bhagirathi River section through the Garhwal Himalaya. Line of section shown on Fig. 1. MT, Munsiari Thrust; VT, Vaikrita Thrust; JNF, Jhala Normal Fault. Isograds are shown as heavy dotted lines: bio, biotite; gt, garnet; ky, kyanite; sill, sillimanite.

bounded by the Munsiari Thrust (MT) below and the Vaikrita Thrust (VT) above (Fig. 2). The latter forms the roof thrust of the MCT zone and marks the boundary between the Vaikrita and Munsiari Groups. At the structural top of the High Himalayan Slab (HHS) a N-dipping fault, the Jhala Normal Fault (JNF), forms the boundary between the HHS and Tibetan–Tethyan (Martoli Fm) sediments. Barrovian-style metamorphism increases in grade upwards towards the top of the High Himalayan Slab (HHS) where sillimanite + K-feldspar gneisses are intercalated with foliated, anatectic granites. Relatively undeformed, later leucogranite dykes become increasingly common towards the top of the HHS, but are intrusive rather than in situ anatectic melts. Large leucogranite bodies, such as the Bhagirathi leucogranite, are located *c.* 10 km along strike, either within upper HHS rocks, or at the interface between Tethyan sediments and the upper Vaikrita complex rocks.

This paper reports the deformation chronology, metamorphic petrography and some thermobarometric data from a traverse across the MCT zone and Higher Himalayan Slab along the Bhagirathi river transect, as well as a cooling history based on $^{40}Ar/^{39}Ar$ and K-Ar geochronology from hornblende, muscovite and biotite samples.

Deformation chronology

Table 1 shows a summary of the relative deformation chronology of the western Garhwal Himalaya, although it must be realised that many of these deformation events probably formed a progressive deformation sequence. The earliest D1 deformation is represented by coaxial strain with no evidence of any rotational component in the HHS. In general terms deformation migrated southwards with time so that the early D1 phase is not present in rocks of the Lesser Himalaya. D2 deformation is represented by non-coaxial strain initiated during movement along the Main Central Thrust zone. The main penetrative shear fabric found throughout the MCT zone is assigned to D2 and numerous kinematic indicators testify to the non-coaxial nature of this event. Within the MCT zone mineral stretching lineations are dominantly north-dipping. Deformation towards the northern part of the MCT zone is entirely ductile, but nearer the base blastomylonitic textures show deformation to have been under brittle–ductile conditions. In the HHS above the Vaikrita Thrust D2 has been subdivided into two. D2a is synchronous with the main south-vergent, MCT-related ductile shear, whilst D2b, also of a non-coaxial nature, shows down-to-the north shear sense along the top of the slab and the base of the overlying Martoli Formation. D3 shows dominantly brittle deformation as thrusting propagated south to the Lesser Himalaya, accompanied by extensional faulting in the north.

Petrography, textures and mineral phases

The metamorphic sequence recorded in pelites of the Bhagirathi valley is inverted with rocks from the highest structural positions in the upper

Table 1. *Summary of relative deformation chronologies for the western Garhwal Himalaya*

Deformation phase	Tectonic unit		
	Lesser Himalaya	MCT zone	High Himalayan Slab
D1		F1 crenulation of S0 only preserved in garnet interiors D1 pressure shadows retained as garnet inclusions	F1 isoclinal (intrafolial) folding of S0 S1 penetrative transposed foliation
D2	F2 asymmetric/recumbent shear folds S2 S–C mylonitic fabric (bedding-parallel) L2 mullion/mineral lineation	S2 penetrative S–C fabric L2 mineral lineation (syn-MCT main phase motion) non-cylindrical L1 parallel folds	Syn MCT F2a asymmetric shear folds (top-to-S shear sense) S2a non-penetrative S–C fabric (deformation of M1 minerals) Syn JNF F2b asymmetric shear folds (down-to-N shear sense) S2b S–C fabric
D3	Brittle thrusts/duplexes	Localized shear and discrete brittle thrusts Chlorite retrogression	Normal-sense brittle offsets Down-to-N duplexes
D4	F4 broad open fault bend folds associated MBT ramps? Hinge culmination/depression	Minor brittle faulting	Sub-vertical fault gouge zones Normal (down-to-N) displacement

amphibolite facies (sillimanite ± K-feldspar), whereas those from the base of the structural pile are unmetamorphosed or have lower greenschist facies mineralogies. Due to the high variance of many of the pelitic assemblages, the relative timing of mineral growth is often ambiguous due to absence of diagnostic specific metamorphic or tectonic events. Metapelitic lithologies contain the assemblage: biotite + muscovite + chlorite ± garnet ± staurolite + Al_2SiO_5 in addition to quartz and feldspars. Garnet occasionally contains spiral inclusion trails of quartz, ilmenite and/or rutile, and also occurs more rarely in the psammitic lithologies. Towards the top of the High Himalayan slab, migmatites appear in which leucosome segregations contain the assemblage: quartz + K-feldspar + plagioclase ± kyanite. Rocks from the MCT zone and lower part of the slab contain garnets that are growth zoned with a strong fractionation of Mn and Ca into core regions. Garnets from above the kyanite isograd in the central and upper part of the slab are typically homogeneous and show features indicative of high temperature (>650° C) diffusional homogenisation of any prior zonation of the major cations, Fe, Mg, Mn and Ca.

Rocks of the MCT zone dip *c.* 50° NNE and possess a mylonitic fabric with a strong element of rotational shear, evident from asymmetric veins and feldspar porphyroclasts. This S–C fabric formed as a result of high-temperature (main D2 phase) motion within the MCT zone. The garnet isograd occurs *c.* 6 km below the Vaikrita Thrust. Typically garnets of the MCT zone are euhedral to subhedral except in quartz-rich lithologies where they are anhedral and poikiloblastic. Many garnet porphyroblasts have undergone rotation during growth (Schoneveld 1977), demonstrated by inclusion of spiral trails of quartz and ilmenite. Some garnet porphyroblasts have overgrown early (F1) crenulations and are surrounded by a discordant external foliation (F2). Garnet growth is therefore pre- or syn-tectonic with respect to D2. Primary chlorite does occur as a matrix component at lower structural levels, but chlorite is most abundant in the MCT zone as a late, undeformed phase associated with garnet retrogression.

A N-dipping planar fabric is also characteristic of the HHS. Early isoclinal fold-hinges (F1) are transposed and therefore rootless, but are occasionally preserved in less pervasively deformed domains within the planar S1 foliation. An S–C fabric (S2a) locally overprints the S1 foliation indicating a later simple shear event (D2a) in the HHS. Kyanite porphyroblasts are parallel to and aligned in the S1 foliation, but were, however, deformed by the D2a shear event. The existence of early kyanite is most commonly demonstrated by its having been deflected asymptotically into the C-plane of the S–C fabric. Extension and breakdown of kyanite to muscovite subsequently took place in this shear regime. Several samples also show muscovite coronas around kyanite despite a lack of S2a foliation. Although several samples display undeformed kyanite, no unambiguous evidence could be found to support a second generation of kyanite growth as reported by Hodges & Silverberg (1988) from east Garhwal.

In the HHS, garnet shows a no ,-rotational growth history and normally displays only a random, isotropic inclusion pattern. The crystals are usually euhedral to subhedral though occasionally flattened varieties occur. Quartz inclusions of <5 μm are common, but biotite and plagioclase are infrequent and kyanite rare. Coarse, quartz-rich lithologies often contain coarsely poikiloblastic or skeletal garnet. Many garnets from near the base of the HHS contain a nearly inclusion-free core, succeeded by an inner zone rich in <5 μm quartz inclusions, which again changes abruptly to an outer inclusion-free rim. This texture may reflect either changes in the surface growth rate of the garnet or a two-stage growth of garnet. Only the larger crystals in these samples possess the inclusion-rich zone: either the smaller garnets belong to a second generation, or this may just be a function of off-centre sectioning of a single garnet generation. As with kyanite, garnet crystals without the possible second generation overgrowth can show brittle fracture and extension in samples affected by the D2a S–C fabrics.

Sillimanite, though only found at the top of the HHS close to the Jhala Normal Fault (JNF), is mostly undeformed and lies unaligned in the S1 foliation plane. However, in outcrop, bands of sillimanite are asymmetrically folded (F2) in a sense consistent with overshear to the N (indicative of normal fault kinematics). One sample shows a vein containing early, folded and boudinaged sillimanite needles oriented at a high angle to a later S2 parallel generation of sillimanite in the host schist.

The textural evidence described above suggests that metamorphism to upper greenschist facies conditions in the MCT zone was accomplished in a single syn-kinematic phase. In the HHS, however, the possible two-stage garnet growth history combined with the existence of both pre- and post-kinematic sillimanite suggests a polymetamorphic history involving at least two prograde events. The D2a S–C fabric in the HHS correlates with the pervasive S–C fabric in the MCT zone, so the M1 Barrovian metamorphism affecting the HHS clearly predates MCT movements (S2a). The post-kinematic, sillimanite2 event (designated M3) in the HHS is likely to be later than that responsible for metamorphism in the MCT zone (M2), because first generation sillimanite is deformed by ductile D2b (syn-MCT [D2]) movements on the Jhala Normal Fault. M3 culmination occurred subsequent to JNF activation. The last event M4, is a retrogressive phase of restricted chlorite and biotite development in both the MCT zone and HHS. It is also manifest in the alteration of kyanite to muscovite.

Thermobarometry

Standard thermobarometric techniques were applied to 13 samples from the MCT zone and HHS, in order to place quantitative constraints on the conditions attained during metamorphism. Table 1 gives microprobe analyses of the relevant mineral phases and sample locations are indicated on Fig. 3.

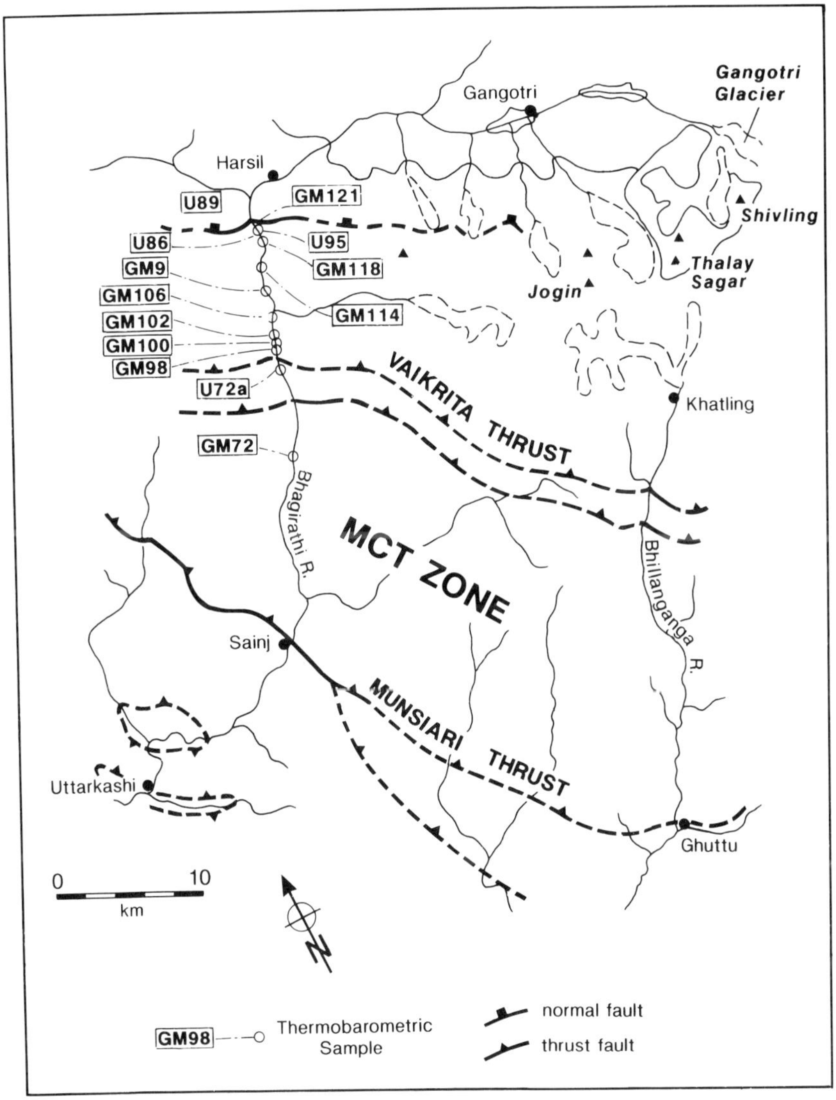

Fig. 3. Tectonic map of the Bhagirathi–Bhillanganga area showing locations of samples used in thermobarometric analysis.

Calibrations

Temperatures were estimated by garnet–biotite geothermometry (Ferry & Spear 1978) which calculates binary Fe-Mg partitioning between the two phases as a function of temperature. The Ferry & Spear (1978) calibration, however, regards the behaviour of garnet in solid solution as ideal. In order to account for non-ideal mixing, Hodges & Spear (1982) modified the existing Ferry & Spear equilibrium constant using the Ganguly & Kennedy (1974) quaternary solution model. An empirically derived Margules parameter of zero was used for the binary Mg-Mn solution ($W_{MgMn} = 0$) within the quaternary system.

Hodges & Spear (1982) suggested caution in the application of this calibration to samples containing high-Ti biotites or garnets with $X_{spess} > 0.2$, due to Ti substitution for Al in biotite and uncertainties in the activity-composition relationships of garnet with plagioclase. Indares & Martignole (1985) provided a recalibration of the Ferry & Spear model but used a non-ideal garnet solution model and corrected for Al and Ti interaction in biotite. However, due to inconsistencies in the published equations, this calibration was not adopted in the present study.

Pressure was calculated through the application of one or more well-calibrated pelite geobarometers solved simultaneously with the garnet-biotite thermometer (Hodges & Spear 1982). The mineral assemblage of the sample determined the use of the available barometers: garnet–biotite–muscovite–plagioclase (GBMP) (Hodges & Crowley 1985), garnet–plagioclase–aluminium silicate–quartz (GPAQ) (Newton & Haselton 1981) or garnet–rutile–aluminium silicate–ilmenite (GRAIL) (Bohlen *et al.* 1983).

Cation exchange in the GBMP equilibrium is between Fe in garnet and biotite, and Ca in garnet and plagioclase (Ghent 1976). However, Ca is usually a minor component ($X_{gros} < 1$) in almandine-rich pelites (Bohlen *et al.* 1983), therefore relatively large errors are involved in analysing for a very small % Ca. The GPAQ solution model also relies on Ca exchange and is prone to the same analytical errors, in addition to uncertainties in the activity of grossular at extreme dilutions (Bohlen *et al.* 1983). Despite these problems the GBMP and GPAQ calibrations (Hodges & Crowley 1985; Newton & Haselton 1981 respectively) are regarded as reliable and produce results that agree well with Al_2SiO_5 constraints (Essene 1982; Hodges & Crowley 1985).

The tight experimental reversal of the GRAIL equilibrium (Table 1) under metamorphically viable conditions makes this the most accurate pelite barometer available (Bohlen *et al.* 1983). In addition, solid solution is effectively between only two phases, garnet and ilmenite, whilst analysis is for the major garnet component (almandine), rather than a minor one, thereby reducing analytical errors. Whilst GPAQ and GBMP are, to some extent, affected by assumed temperatures, the GRAIL barometer behaves almost independently of temperature.

Unfortunately, both rutile and ilmenite are infrequently found in equilibrium together in pelites (Yardley 1989). This is certainly true of samples from the Bhagirathi valley; only two samples contain ilmenite and rutile in textural equilibrium with kyanite or sillimanite and garnet (GM106 and U89) to which the GRAIL barometer could be applied. The accuracy of GBMP may be less than for GPAQ and GRAIL (Hodges & Crowley 1985), but nevertheless internal precision should not be affected and it is the most widely applicable barometer suitable for these assemblages.

Analytical technique

Mineral analyses were performed on the JEOL JXA8600s Superprobe at Leicester University. A beam diameter of 5 μm was used with a 30 nA probe current at an accelerating voltage of 15 KV. Each analysis involved a 20 second peak count and 10 seconds on background. Standardization was conducted against a range of natural and synthetic materials. For rim thermobarometry, phases in mutual contact or close proximity were chosen. Garnet from the MCT and lower HHS prove to be strongly zoned but biotites are usually compositionally homogeneous. Spot analyses were obtained from close to garnet margins in an attempt to estimate the conditions of latest equilibration, which were not necessarily peak metamorphic conditions.

Attempts at placing constraints on the form of the *P–T–t* path, through application of Gibbs Method thermodynamic modelling and inclusion thermobarometry, were prevented by the lack of a suitable garnet inclusion mineralogy. The latter would ideally have indicated the changing paragenesis during garnet growth. Many garnets from the central and upper HHS are homogeneous or show only a narrow margin of inverse zoning attributed to re-equilibration during retrogression. In these instances (GM9, U86, U89, U95) analysis of garnet was performed both in the core and in the retrogression mantle. The biotite/garnet ratio is relatively high in these samples, such that matrix biotite may have diluted the effects of retrograde cation exchange from the relatively minor volume of retrogressed garnet (Tracey *et al.* 1976). Thus matrix biotite analysed in conjunction with garnet cores may allow estimation of conditions closer to the peak of metamorphism. If these are compared with conditions calculated from rim thermobarometry, an indication of at least a portion of the *P–T* path may be obtained.

Results

For each sample the mineral analyses (average of 2–6 analyses) are presented in Table 2, whilst the pressures and temperatures obtained for the relevant calibration are shown in Table 3. All results are plotted with a 2-σ error box on a *P–T* grid (Fig. 4), with GB-GBMP favoured over other barometers in order to maintain internal consistency. Most samples fall in the kyanite field (Al_2SiO_5 triple point and stability fields after Holdaway (1971) using prismatic sillimanite), though two straddle the kyanite/

Table 2. *Mean microprobe analyses of garnet (GT), biotite (BI), muscovite (MU), plagioclase (PL), rutile (RUT) and ilmenite (ILM) from the Bhagirathi valley transect, western Garhwal Himalaya*

	GM72 GT	U72A GT	GM98 GT	GM100 GT	GM102 GT	GM106 GT	GM9 GT	GM114 GT	GM118 GT	U86 GT	U89 GT	GM121 GT	U95 GT
SiO_2	37.35	37.07	37.87	37.74	38.32	38.35	38.88	38.37	37.96	38.09	38.52	38.33	38.16
TiO_2	0.053	0.02	0.055	0.022	1.68	0.095	0.007	0.013	0.015	0.017	0.0175	0.014	0.01
Al_2O_3	21.42	21.4	21.62	21.65	21.83	21.98	22.12	21.95	21.84	21.94	22.055	21.96	21.89
Cr_2O_3	0.03	0.01	0.02	0.02	0.02	0.02	0.04	0.03	0.02	0.01	0.03	0.02	0.03
FeO	36.70	35.45	29.29	34.98	33.18	32.81	32.51	31.73	33.84	31.50	31.92	31.30	33.65
MnO	0.25	0.78	0.82	0.54	0.767	0.82	1.23	2.50	1.94	1.13	0.69	1.13	2.32
MgO	1.93	1.77	1.40	1.82	4.63	4.84	6.77	4.43	3.41	6.98	6.80	7.13	3.46
CaO	2.85	3.73	9.68	4.11	2.61	2.23	0.46	2.11	1.95	0.75	0.99	0.74	2.30
Na_2O	0.017	0.01	0.027	0.022	0.02	0.017	0.02	0.017	0.025	0.023	0.022	0.028	0.01
K_2O	0.007	0.01	0.007	0.01	0.007	0.007	0.01	0.008	0.005	0.003	0.007	0.007	0.006
Total	100.6	100.26	100.8	100.92	101.42	101.18	102.05	101.18	101.01	100.46	101.06	100.66	101.85
Elements normalized to 12 oxygens													
Si	2.997	2.985	2.994	3.006	2.997	2.998	2.995	3.006	3.002	2.976	2.989	2.984	2.996
Ti	0.003	0.001	0.003	0.001	0.001	0.006	0.001	0.001	0.001	0.001	0.001	0.001	0.001
Al	2.026	2.031	2.015	2.032	2.013	2.025	2.008	2.027	2.036	2.021	2.017	2.015	2.025
Cr	0.002	0.001	0.0017	0.001	0.001	0.001	0.002	0.002	0.001	0.001	0.002	0.001	0.001
Fe^{2}	2.462	2.388	1.937	2.33	2.170	2.145	2.094	2.079	2.23	2.058	2.072	2.038	2.209
Mn	0.017	0.053	0.0545	0.0362	0.051	0.0542	0.080	0.166	0.13	0.0743	0.0455	0.0741	0.155
Mg	0.230	0.212	0.164	0.216	0.540	0.564	0.777	0.517	0.401	0.8137	0.787	0.828	0.405
Ca	0.246	0.322	0.819	0.351	0.219	0.187	0.038	0.177	0.165	0.063	0.082	0.61	0.194
Na	0.003	0.001	0.004	0.004	0.003	0.003	0.003	0.003	0.004	0.004	0.004	0.004	0.002
K	0.001	0.001	0.001	0.001	0.001	0.001	0.001	0.001	0.001	0.000	0.001	0.001	0.001
Sum	7.987	7.998	7.996	7.978	7.996	7.985	8.000	7.979	7.98	8.014	8.001	8.009	7.990
X_{Fe}	0.833	0.802	0.650	0.794	0.728	0.727	0.700	0.707	0.762	0.684	0.694	0.679	0.746
X_{Mg}	0.078	0.071	0.055	0.074	0.181	0.191	0.260	0.176	0.137	0.270	0.263	0.276	0.137
X_{Ca}	0.083	0.108	0.275	0.120	0.073	0.063	0.013	0.060	0.056	0.021	0.028	0.020	0.065
X_{Mn}	0.006	0.018	0.018	0.012	0.017	0.018	0.027	0.056	0.044	0.025	0.015	0.025	0.052
Mg#	8.5	8.1	7.8	8.5	20.6	20.8	27.1	19.9	15.2	28.3	27.5	28.9	15.5

Table 2 *Continued.*

	GM72 BI	U72A BI	GM98 BI	GM100 BI	GM102 BI	GM106 BI	GM9 BI	GM114 BI	GM118 BI	U86 BI	U89 BI	GM121 BI	U95 BI
SiO_2	35.26	34.38	35.33	34.64	36.29	36.63	37.9	36.54	35.77	37.69	37.94	37.73	36.29
TiO_2	1.16	1.19	1.72	2.88	1.48	2.52	1.33	2.34	2.24	1.44	1.394	1.39	1.99
Al_2O_3	18.9	18.46	19.48	18.24	18.59	18.54	8.17	18.92	18.85	17.78	17.97	17.82	19.03
Cr_2O_3	0.02	0.02	0.02	0.033	0.031	0.062	0.043	0.04	0.017	0.025	0.052	0.03	0.043
FeO	20.84	20.99	24.0	24.18	18.73	16.94	13.01	16.53	18.64	12.16	12.63	12.42	18.07
MnO	0.045	0.02	0.24	0.052	0.058	0.027	0.02	0.058	0.037	0.04	0.022	0.027	0.043
MgO	8.61	8.15	4.38	5.26	6.37	11.57	15.68	11.31	10.04	15.95	15.97	16.27	10.83
CaO	0.01	0.01	0.02	0.138	0.0114	0.005	0.063	0.008	0.0125	0.015	0.082	0.0214	0.008
Na_2O	0.265	0.22	0.1	0.192	0.181	0.242	0.593	0.262	0.355	0.66	0.634	0.714	0.298
K_2O	9.34	7.94	8.427	8.4	8.54	8.945	8.01	9.455	8.85	8.14	7.36	7.81	8.95
Total	94.44	91.38	93.75	94.025	94.563	95.497	94.81	95.46	94.817	93.9	94.056	94.241	95.6
Elements normalized to 12 oxygens													
Si	5.455	5.472	5.55	5.454	5.508	5.471	5.559	5.464	5.433	5.571	5.579	5.554	5.446
Ti	0.135	0.142	0.204	0.342	0.169	0.283	0.147	0.263	0.255	0.159	0.154	0.154	0.225
Al	3.442	3.463	3.606	3.386	3.32	3.264	3.140	3.334	3.374	3.097	3.115	3.09	3.366
Cr	0.002	0.003	0.002	0.004	0.004	0.007	0.005	0.005	0.002	0.003	0.006	0.003	0.005
Fe^2	2.697	2.794	3.159	3.184	2.378	2.118	1.596	2.0681	2.36825	1.504	1.5534	1.530	2.2685
Mn	0.006	0.003	0.317	0.007	0.007	0.003	0.003	0.007	0.005	0.005	0.003	0.003	0.005
Mg	1.985	1.934	1.027	1.235	2.410	2.574	3.428	2.520	2.274	3.515	3.502	3.569	2.423
Ca	0.002	0.002	0.003	0.023	0.002	0.001	0.01	0.002	0.002	0.002	0.013	0.004	0.001
Na	0.079	0.068	0.030	0.058	0.053	0.07	0.167	0.076	0.104	0.189	0.1806	0.204	0.0868
K	1.843	1.612	1.689	1.687	1.653	1.705	1.50	1.804	1.714	1.535	1.381	1.467	1.722
Sum	15.649	15.493	15.303	15.381	15.512	15.498	15.555	15.555	15.533	15.581	15.487	15.580	15.548
Mg#	42.4	40.9	24.5	27.9	51.00	54.8	54.2	54.9	49.00	70.05	69.28	70.00	51.67

Table 2 *Continued.*

	GM72 MU	U72A MU	GM98 MU	GM100 MU	GM102 MU	GM114 MU	GM118 MU	U95 MU	GM106 RUT	U89 RUT	GM106 ILM	U89 ILM
SiO_2	43.94	45.64	46.53	47.08	47.17	47.27	46.37	46.95	0.11	0.01	1.7	0
TiO_2	0.32	0.27	0.2	1.11	1.2	0.72	0.82	0.72	98.498	99.4	50.475	53.72
Al_2O_3	33.54	34.99	31.48	32.19	31.87	34.14	34.95	35.71	0.8283	0.06	0.14	0.02
Cr_2O_3	0.025	0.06	0.02	0.025	0.03	0.042	0.02	0.015	0.06	0.14	0.04	0.03
FeO	2.24	0.85	3.55	2.41	1.42	1.46	1.11	1.05	0.83	0.46	45.805	45.48
MnO	0.02	0	0.01	0.02	0.02	0.02	0.02	0.02	0.015	0	0.45	0.29
MgO	0.91	0.44	1.09	1.08	1.36	1.01	0.75	0.65	0.03	0	0.21	0.35
CaO	0.01	0.01	0.01	0.01	0.01	0.007	0.01	0.01	0.025	0.02	0.11	0.03
Na_2O	1.74	1.68	0.27	0.49	0.97	0.88	1.22	1.32	0.0283	0.01	0.1	0
K_2O	8.75	8.87	10.73	10.31	9.73	10.06	9.58	9.58	0.01	0	0.005	0.01
Total	91.50	92.81	93.89	94.75	93.79	95.61	94.86	96.04	100.455	100.1	99.035	99.93
Elements normalized to 22 oxygens									*Elements normalized to 2 and 3 oxygens*			
Si	6.109	6.187	6.368	6.335	6.369	6.262	6.177	6.169	0.0013	0	0.0435	0
Ti	0.034	0.028	0.021	0.113	0.1218	0.072	0.082	0.072	0.983	0.995	0.9575	1.012
Al	5.495	5.591	5.078	5.104	5.073	5.332	5.486	5.531	0.0128	0.001	0.0045	0.001
Cr	0.002	0.006	0.002	0.002	0.003	0.004	0.0027	0.0015	0.001	0.001	0.001	0.001
Fe^2	0.262	0.096	0.406	0.272	0.161	0.161	0.124	0.1155	0.0095	0.005	0.967	0.953
Mn	0.002	0	0.001	0.003	0.002	0.002	0.002	0.002	0	0	0.009	0.006
Mg	0.191	0.089	0.222	0.216	0.274	0.198	0.149	0.127	0.001	0	0.008	0.013
Ca	0.001	0.001	0.001	0.001	0.001	0.001	0.001	0.001	0.0003	0	0.003	0.001
Na	0.467	0.442	0.072	0.129	0.254	0.225	0.314	0.336	0.0008	0	0.005	0
K	1.553	1.534	1.873	1.771	1.68	1.700	1.63	1.606	0.001	0	0	0
Sum	4.118	13.974	14.044	13.948	13.937	13.96	13.968	13.96	1.0095	1.004	1.999	1.987
Mg#	41.7	48	35.4	44.00	62.9	55.1	54.6	52.3	6.5		0.8	1.4

Table 2 *Continued.*

	GM72 PL	U72A PL	GM98 PL	GM100 PL	GM102 PL	GM9 PL	GM114 PL	GM118 PL	U86 PL	U89 PL	GM121 PL	U95 PL
SiO_2	62.24	62.08	62.73	65.52	64.62	67.86	63.01	62.94	64.84	65.56	65.08	63.92
TiO_2	0.01	0.02	0.01	0.0125	0.01	0	0.01	0.003	0	0.003	0.008	0.016
Al_2O_3	22.14	23.04	22.26	21.44	21.43	17.57	23.59	23.28	19.36	20.24	20.05	23.34
Cr_2O_3	0.015	0.02	0.015	0.03	0.03	0.04	0.02	0.016	0.04	0.03	0.008	0.013
FeO	0	0	0.135	0.02	0.02	0.07	0.04	0.08	0.055	0.05	0.042	0.023
MnO	0.03	0.03	0.02	0.022	0.02	0.04	0.016	0.023	0.02	0.023	0.018	0.02
MgO	0.005	0	0	0	0	0	0.006	0	0.01	0.003	0.002	0
CaO	4.38	5.03	3.47	2.57	3.03	0.96	5.32	4.65	1.7	1.95	1.80	4.96
Na_2O	9.42	9.15	9.55	10.37	10.21	10.09	8.90	9.25	11.15	10.96	11.02	9.23
K_2O	0.085	0.07	0.22	0.145	0.06	0.05	0.083	0.106	0.06	0.05	0.038	0.08
Total	98.32	99.44	98.41	100.14	99.44	96.68	101.01	100.36	97.24	98.88	98.06	101.61
Elements normalized to 8 oxygens												
Si	2.804	2.77	2.817	2.881	2.866	3.054	2.766	2.779	2.935	2.917	2.9198	2.78
Ti	0	0.001	0	0.001	0.001	0	0.000	0	0	0	0	0.000
Al	1.175	1.212	1.179	1.111	1.120	0.932	1.221	1.212	1.033	1.062	1.06	1.199
Cr	0.001	0.001	0.001	0.001	0.001	0.001	0.000	0.000	0.001	0.001	0.000	0.000
Fe^2	0	0	0.005	0.001	0.001	0.003	0.001	0.003	0.002	0.002	0.001	0.000
Mn	0.001	0.001	0.001	0.001	0.001	0.002	0.000	0.000	0.001	0.000	0.001	0.000
Mg	0.001	0	0	0	0	0	0.000	0	0.001	0.000	0.001	0.000
Ca	0.211	0.241	0.167	0.121	0.144	0.046	0.250	0.22	0.082	0.93	0.086	0
Na	0.823	0.792	0.831	0.884	0.878	0.88	0.758	0.792	0.979	0.946	0.959	0.231
K	0.005	0.004	0.012	0.008	0.003	0.003	0.004	0.006	0.003	0.002	0.002	0.780
Sum	5.021	5.021	5.014	5.009	5.015	4.921	5.004	5.01	5.038	5.025	5.030	5.005
An	20.35	23.2	16.5	12	14.0	5	24.7	21.6	7.75	8.93	8.26	22.8

Table 3. *Results of thermobarometry from three different thermobarometers*

Sample No.	GB-GBMP		GB-GPAQ		GB-GRAIL	
	T(° C)	P (kbar)	T (° C)	P (kbar)	T (° C)	P (kbar)
GM72	500 ± 40	6.45 ± 2.1	–	–	–	–
U72a	628 ± 15	6.3 ± 0.8	–	–	–	–
GM98*	880 ± 30	17.7 ± 2.4	–	–	–	–
GM100*	777 ± 106	13.5 ± 2.9	–	–	–	–
GM102	768 ± 44	12.0 ± 2.4	–	–	–	–
GM106	–	–	–	–	735 ± 10	8.1 ± 0.6
GM9	–	–	584 ± 36	7.7 ± 1.4	–	–
GM114	660 ± 40	7.4 ± 1.2	660 ± 40	7.5 ± 1.5	–	–
GM118	624 ± 37	6.8 ± 1.6	–	–	–	–
U86	–	–	563 ± 19	7.4 ± 0.7	–	–
U89	–	–	600 ± 40	8.9 ± 1.3	588 ± 33	7.6 ± 0.3
GM121	–	–	600 ± 40	8.5 ± 1.2	–	–
U95	623 ± 64	7.8 ± 1.8	628 ± 64	9.1 ± 1.6	–	–

* Unreliable *P* and *T*
Error calculated to 2σ

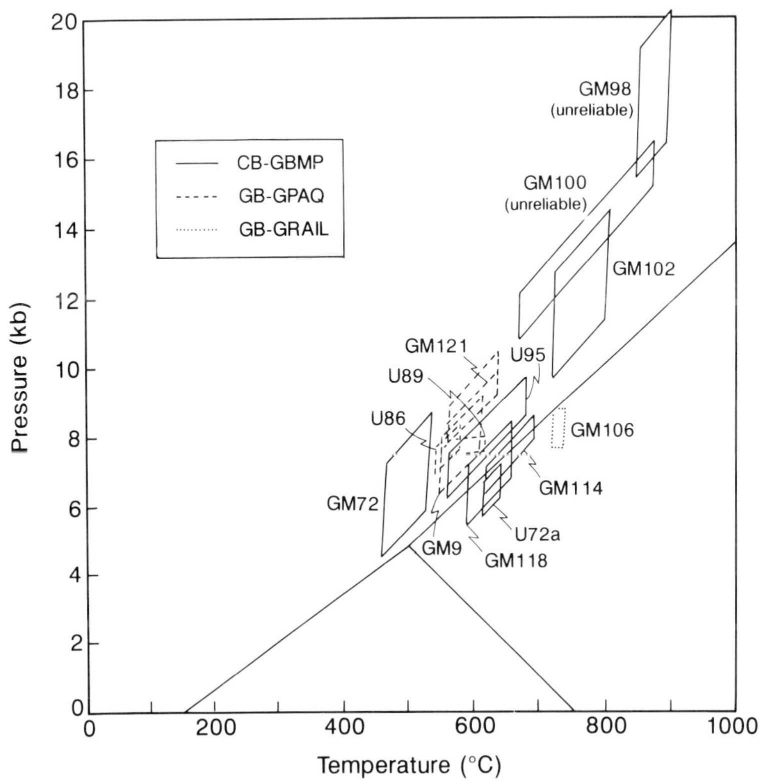

Fig. 4. Pressure–temperature grid showing latest equilibration conditions of metamorphism plotted with 2σ error boxes. Al_2SiO_5 triple point after Holdaway (1971). See Table 2 for the data and Fig. 3 for the locations of samples.

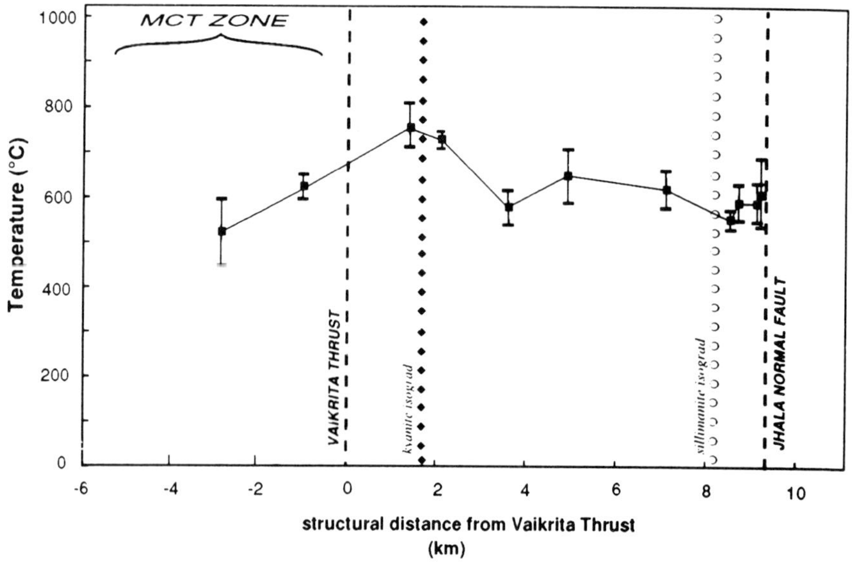

Fig. 5. Calculated temperatures versus structural distance from the Vaikrita Thrust. Error bars calculated to 2σ.

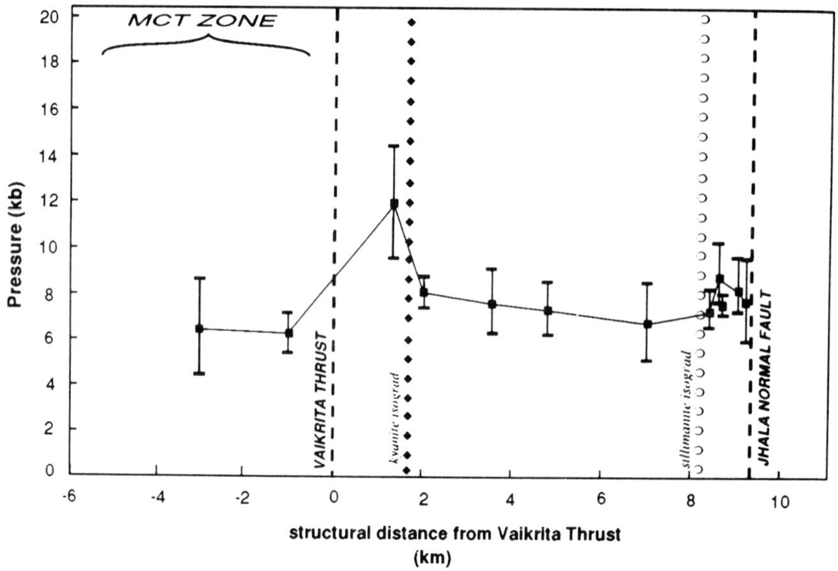

Fig. 6. Calculated pressures versus structural distance from the Vaikrita Thrust. Error bars calculated to 2σ.

sillimanite univariant reaction. Three samples plot at $P > 10$ kbar; GM98 and GM100 give pressure estimates of 17.7 ± 2.4 kbar and 13.5 ± 2.9 kbar respectively (both GB-GBMP), which puts them well above the granite solidus at the calculated temperature. There is, however, no evidence of anatexis having occurred in either of these rocks.

Figures 5 and 6 are plots of calculated temperatures and pressures versus structural distance from the Vaikrita Thrust. The two MCT zone samples (GM72, U72a) produced the lowest calculated pressures whilst GM72 also records the lowest temperatures. Two of the samples showing high-pressures, GM98 and GM100 are discarded as unreliable and are not included. These samples contain very Fe-rich biotite (substantial Fe^{3+}) leading to considerable uncertainty in the temperature determination and one has unusually high Ca in the garnet. These objections cannot be made for sample GM102 which is why this sample is retained in the profiles. It is possible that sample GM102 might represent quenched M1 conditions in the immediate hangingwall of the MCT. Figure 5 displays a *c.* 200° C temperature increase across the Vaikrita Thrust (VT) with a gradual decline to *c.* 620° C in the middle to upper High Himalayan Slab. The trend shows a small increase at the top of the section. However, the temperatures in the upper 5.2 km, despite showing a trend with regard to mean values, are all isothermal within a 2-σ error bracket. The pressure section (Fig. 6) shows one sample (GM102) from the base of the HHS that possess notably higher pressure. A pressure jump of *c.* 4 kbar is therefore recorded across the VT, and thereafter pressures decrease gradually until the top of the HHS where a slight rise is recorded.

Textural evidence shows that the HHS underwent a polymetamorphic history, whereas the MCT zone underwent a single prograde metamorphism. The last prograde event, M3 was

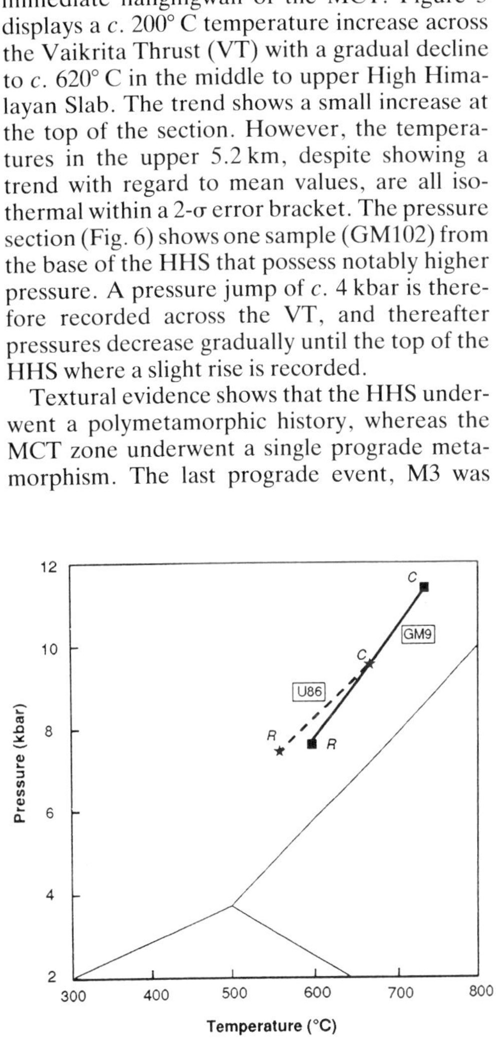

Fig. 7. *P–T* results from two samples from the upper (U86) and central (GM9) part of the High Himalayan slab. Calculations were made using both core (C), and rim (R) garnet analyses. It is suggested that core *P–T* conditions approach those of peak metamorphic grade.

restricted to the upper part of the HHS. M3 conditions are therefore likely to be recorded by thermobarometric results from samples from this section. Pressures and temperatures from the middle and lower part of the HHS probably record the prograde M1 (pre-MCT) or M2 (syn-MCT) conditions. The *P–T* profiles (Figs 5 and 6) therefore probably represent a diachronous thermobarometric profile and do not record a metamorphic geotherm at any one specific time.

Core and rim garnet analyses used for simultaneous solution of garnet–biotite and garnet–plagioclase–Al_2SiO_5–quartz thermobarometers, for both samples GM9 and U86 reveal *P–T* changes with similar gradients (Fig. 7). GM9 retains core conditions of 735° C and 11.4 kbar which evolved to rim conditions of 585° C at 7.6 kbar. This change is equivalent to uplift of the rock of some 10 km. Sample U86 core to rim conditions (660° C at 9.4 kbar to 560° C at 7.4 kbar) suggest uplift of the rock of about 6 km. The difference in peak metamorphic conditions of the two samples may approach the difference in peak *P–T* conditions between M1 (GM9) and M2 (U86).

Geochronology

Cooling age determinations from the MCT zone are available from only two sources, both involving studies of the same section in E Nepal. Krummenacher *et al.* (1978) used the K-Ar method on four samples from the redefined MCT zone (Hubbard 1988, 1989). Ages between 9.0 ± 0.5 Ma and 3.4 ± 0.1 Ma were calculated for biotite from semi-pelitic and augen gneisses. Krummenacher *et al.* (1978) concluded that these young cooling ages were the result of a late thermal event, probably involving movement on the MCT. Four MCT zone samples were analysed by Hubbard & Harrison (1989) using the $^{40}Ar/^{39}Ar$ technique. The results were variable but one muscovite isochron age of 12.0 ± 0.2 Ma from augen gneiss is thought to be reliable, along with a consistent K-feldspar age from the same sample at 8.0 ± 0.2 Ma. The three biotite separates gave excess argon spectra from which no reasonable age could be interpreted. Two adjacent amphibolites gave hornblende ages of 20.9 ± 0.2 Ma (isochron) and 204 ± 5.2 Ma (minimum release age). The older age was, however, interpreted as representing incomplete degassing of a Palaeozoic hornblende rather than excess argon uptake (Hubbard & Harrison 1989). The authors concluded that hornblende ages, with a closure temperature (T_c) of 500 ± 50° C, date the age of peak metamorphism which reached maximum temperatures of *c*. 500° C (Hubbard 1989), with the muscovite (T_c = 350 ± 50° C) recording the time since the MCT rocks cooled through *c*. 350° C.

Analytical techniques

The general procedure followed for K-Ar analysis is outlined in Briden *et al.* (1979) and Faure (1986, chapter 6, pp. 66–92). Total K was measured using a flame photometer calibrated against known standard solutions and the proportion of ^{40}K calculated from the total. ^{40}Ar* (* = radiogenic) measurement was accomplished using an MS10 mass spectrometer by total fusion of a known weight of sample in a molybdenum crucible with an RF furnace. The Leeds method of ^{40}Ar/^{39}Ar analysis used here has been described in Rex *et al.* (1993) and Wartho (1991).

Results

A map of the Bhagirathi valley section showing locations of the samples used for geochronology

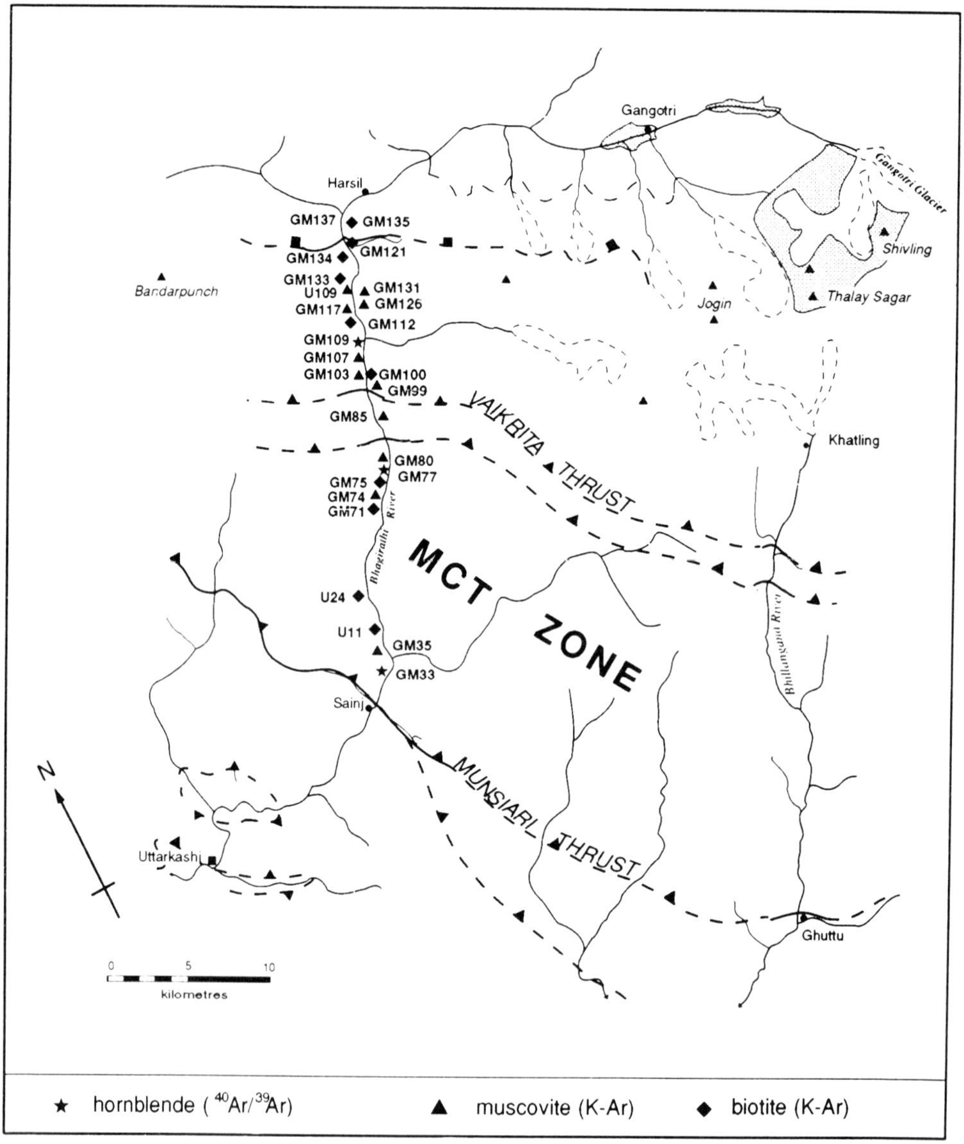

Fig. 8. Structural map of the Bhagirathi–Bhillanganga area showing locations of geochronological samples. See Tables 4 and 5 for the data.

is shown in Fig. 8. Age determinations by both K-Ar and $^{40}Ar/^{39}Ar$ methods on mineral separates from the Bhagirathi section are presented in tabulated form, as well as on structural cross-section plots of the Bhagirathi valley (Tables 4 and 5, Fig. 9). K-Ar and $^{40}Ar/^{39}Ar$ ages reflect not the crystallization age of a rock or mineral, but the time elapsed since argon loss from that mineral lattice became negligible in proportion to the rate of accumulation of radiogenic argon (Harrison 1981; McDougal & Harrison 1988). Thus, the ratio of $^{40}Ar^{*}/^{40}K$ retained in the crystal is a measure of the time since that mineral cooled through a specific temperature at which the lattice became closed to argon loss (Dodson 1973).

The closure temperature for argon loss from biotite is generally agreed to be 300 ± 50° C (Harrison & McDougall 1980*a*; Hubbard & Harrison 1989), although fluctuations of *c.* 100° C are reported (Cliff 1985). For muscovite, however, the closure temperature is 350 ± 50° C (Purdey & Jager 1975). Therefore, ages produced for biotite should be similar to, or slightly younger than, those for muscovite.

Figure 9 (a & b) displays schematic cross-sections of the Bhagirathi valley along with the ages obtained on biotite and muscovite separates by K-Ar analysis. The sample numbers and relevant data are summarised in Table 4. Biotite separates yield ages between 74.3 ± 2.2 Ma and 9.5 ± 0.3 Ma, whilst muscovites fall in the range 22.2 ± 0.7 Ma to 5.7 ± 0.2 Ma. The general trend in the muscovite ages is for younger ages towards the SSW with only a few exceptions. Results from biotites are more complex. Although no mineral pairs from the same sample are dated, several samples were obtained from closely spaced sites and thus might be expected to show roughly similar ages. However, the biotite ages are often much older than on muscovite from an adjacent site. In some cases, for example U11 and GM100, the ages yielded for biotite (70 Ma and 74 Ma respectively) predate the accepted timing for the closure of Tethys (*c.* 50 Ma) and subsequent formation of the Himalaya (Searle *et al.* 1987). Clearly these middle greenschist to upper amphibolite facies rocks must have reached temperatures of >300 ± 50° C during the Himalayan orogeny, unless the pattern of metamorphism accompanying the MCT and HHS zones is considered to predate orogenesis. The latter concept is, however, untenable as muscovite, with its higher closure temperature, records ages subsequent to collision. Thus many of the ages obtained on biotite are geologically meaningless probably owing to the presence of excess argon.

Of the ten biotite separates analysed the ages obtained on six are geologically unreasonable or significantly older than the muscovite age from the equivalent structural position (Fig. 9a & b). Therefore, only four biotite cooling ages from specimens GM75, GM112, GM134 and GM121

Table 4. *K-Ar analytical data from the Garhwal Himalaya*

Sample	Rock type	Phase	%K	Vol ^{40}Ar $\times 10^{-5}$ $cm^3 g^{-1}$	% ^{40}Ar	Age (Ma)	Error (3%)(Ma)
U11	Schist adjacent to aplite	BIO	7.508	2.0908	87.7	70.254	2.11
U24	Semi-pelitic gneiss	BIO	6.927	0.5178	44.9	19.128	0.57
GM71	Mylonitic bio gneiss	BIO	7.624	0.6377	83.1	21.392	0.64
GM75	Mylon. fine-gr porph gneiss	BIO	7.895	0.2934	66.7	9.535	0.29
GM100	Coarse gt-musc gneiss	BIO	6.791	2.0024	84.5	74.304	2.23
GM112	Bio-ky gneiss	BIO	7.369	0.366	68	12.731	0.38
GM133	Ky-mica gneiss	BIO	6.919	0.8624	80.6	31.782	0.95
GM134	Granodioritic augen gneiss	BIO	7.71	0.6353	69.9	21.075	0.63
GM121	Gt-ky-sill gneiss	BIO	5.93	0.4624	67.1	19.947	0.6
GM135	Fine gr semi-pelitic gneiss	BIO	7.777	0.9044	83.3	29.67	0.89
GM35	Mylonitic qz-fsp schist	MU	6.9	0.2161	74.9	8.039	0.24
GM74	Mu schist (tourm)	MU	8.594	0.1982	63.6	5.924	0.18
GM80	Mu schist	MU	7.64	0.4338	75.7	14.898	0.45
GM85	Gt mica schist	MU	6.213	0.1384	51	5.720	0.17
GM99	Bio psammite	MU	8.667	0.474	69.2	14.016	0.42
GM103	Bio-mu gneiss	MU	7.473	0.5476	31.8	18.751	0.75
GM107	Bio-mu-(tourm) gneiss	MU	8.66	0.482	10.5	14.261	0.43
GM117	Tourm pegmatite	MU	8.486	0.7081	13.9	21.338	0.64
GM126	Coarse micaceous gneiss	MU	7.752	0.5339	69.9	17.632	0.53
GM131	Ky-mu-bio migmatite	MU	8.053	0.6631	19.1	21.060	0.63
U109	Tourm pegmatite	MU	7.997	0.6958	18.3	22.245	0.67
GM137	Tourm pegmatite	MU	8.589	0.6977	26.3	20.776	0.62

BIO, biotite; MU, muscovite

Table 5. *$^{40}Ar/^{39}Ar$ analytical data for three hornblende samples from the Bhagirathi valley transect*

Temp ° C	^{39}K ($\times 10^{-9}$ cm^3)	^{37}Ca	^{38}Cl	K/Ca	$^{*40}Ar/^{39}K$	% ^{40}Ar atm	% ^{39}Ar	Age (Ma)	Error
GM 33 Hornblende									
weight = 0.08663 g J value = 0.005945 ± 2.0%									
Atmos $^{40}Ar/^{36}Ar$ = 298.0 ± 2.0% Sensitivity = 1.850 × 10^{-7} cm^3/volt									
770	1.226	1.633	0.028	0.3772161	569.55	8.1	11.8	2667.2	236.39
880	0.568	1.595	0.004	0.1786352	117.159	15.9	5.5	953.6	23.23
940	0.407	4.002	0.011	0.0511091	122.903	13.3	3.9	989.56	21.95
970	1.628	27.938	0.069	0.0292894	254.051	3.1	15.7	1660.52	50.59
1000	1.476	24.331	0.062	0.0304887	219.603	2.7	14.2	1506.99	48.8
1030	1.001	15.627	0.039	0.0321968	206.023	3.5	9.7	1442.69	44.44
1065	0.42	6.367	0.019	0.0331631	200.537	8.3	4.1	1416.04	52.84
1200	2.146	36.847	0.091	0.0292637	213.099	4.5	20.7	1476.48	46.77
1345	1.496	25.246	0.066	0.029769	229.898	9.7	14.4	1554.26	62.21

wt% K = 0.287 $^{*40}Ar$ = 305.21 × 10^{-7} cm^3/g

	Intergrated values	analytic & J error (2−σ)	
Total age	1664.72	64.54	77.82
Plateau age	1491.824		59.72
$^{*40}Ar/^{39}K$	255.036	5.94%	

Temp ° C	^{39}K ($\times 10^{-9}$ cm^3)	^{37}Ca	^{38}Cl	K/Ca	$^{*40}Ar/^{39}K$	% ^{40}Ar atm	% ^{39}Ar	Age (Ma)	Error
GM 77 Hornblende									
weight = 0.09028 g J value = 0.006035 ± 2.0%									
Atmos $^{40}Ar/^{36}Ar$ = 298.0 ± 2.0% Sensitivity 10850 × 10^{-7} cm^3/volt									
750	1.766	0.658	0.008	1.349528	10.276	70.1	5.8	113.19	5.08
900	2.68	7.077	0.031	0.190295	1.967	50.4	8.8	21.29	3.58
930	2.543	12.149	0.058	0.1051967	1.426	54.3	8.4	15.46	4.13
965	5.68	31.717	0.157	0.0899847	1.679	32.6	18.7	18.18	2.16
1000	7.756	43.715	0.22	0.0891504	2.031	13.4	25.6	21.98	2.12
1030	2.216	11.579	0.058	0.0981065	1.344	54.1	7.5	14.58	2.41
1060	1.541	8.505	0.041	0.0910415	1.712	50.5	5.1	18.54	2.59
1150	4.408	29.392	0.137	0.075358	2.539	21.3	14.54	27.54	1.4
1250	0.846	5.702	0.019	0.0745823	5.776	43.7	2.8	61.82	11.78
1340	0.846	5.783	0.018	0.0735024	13.274	41.1	2.8	139.02	13.53

wt% K = 0.794 $^{*40}Ar$ – 9.54 × 10^{-7} cm^3/g

	Intergrated values		analytic & J error (2−σ)
Total age	30.66	1.06	1.62
Plateau age	19.835		2.6
$^{*40}Ar/^{39}K$	2.84	3.5%	

Temp ° C	^{39}K ($\times 10^{-9}$ cm^3)	^{37}Ca	^{38}Cl	K/Ca	$^{*40}Ar/^{39}K$	% ^{40}Ar atm	% ^{39}Ar	Age (Ma)	Error
GM 109 Hornblende									
weight = 0.09188 g J value = 0.007319 ± 2.0%									
Atmos $^{40}Ar/^{36}Ar$ = 298.0 ± 2.0% Sensitivity 1.850 × 10^{-7} cm^3/volt									
740	0.738	6.299	0.005	0.0588685	258.887	23.9	2.4	1917.6	118.02
890	0.744	2.685	0.003	0.1393146	84.487	15.7	2.4	868.52	9.81
946	0.665	2.613	0.003	0.1279918	56.519	12.3	2.1	624.52	11.04
983	1.54	9.163	0.005	0.0844666	28.918	11	4.9	346.36	4.28
1050	10.996	49.367	0.008	0.1119194	8.518	4.9	35.2	109.11	0.73
1075	4.401	19.853	0.004	0.1113834	8.047	9.5	14.1	103.24	2.25
1164	8.294	39.885	0.002	0.1044932	8.386	8.5	26.6	107.46	1.1
1220	1.976	9.758	0.004	0.1017397	11.319	18.8	6.3	143.59	6.22
1336	1.852	10.246	0.005	0.0908348	20.103	28	5.9	247.64	4

wt% K = 0.662 $^{*40}Ar$ = 64.68 × 10^{-7} cm^3/g

	Integrated values	analytic & J error (2−σ)	
Total age	235.41	8.8	12.46
Plateau age	107.638	2.47	
$^{*40}Ar/^{39}K$	19.044	3.99%	

See Fig. 10 for the step heating spectra.
Gas volumes are corrected to standard temperature and pressure (STP)
$^{*40}Ar$, radiogenic Argon 40.

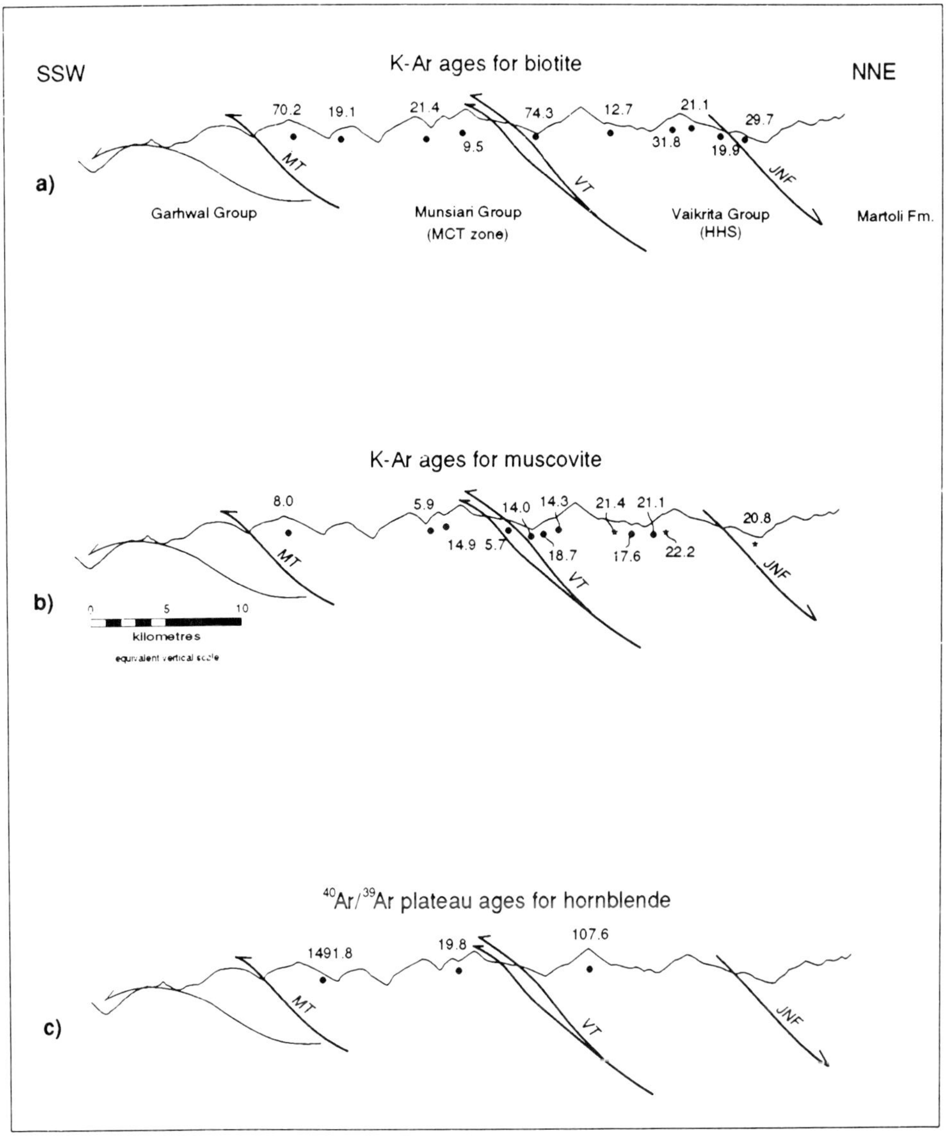

Fig. 9. Schematic structural sections along the Bhagirathi valley transect showing the apparent cooling ages in Ma obtained in this study. (**a**) K-Ar biotite ages, (**b**) K-Ar muscovite ages (**c**) $^{40}Ar/^{39}Ar$ hornblende plateaux ages. MT, Munsiari Thrust; VT, Vaikrita Thrust; JNF, Jhala Normal Fault. Stars indicate samples from leucocratic pegmatites.

are regarded here as being possibly meaningful. Excess argon is obviously a major problem with K-Ar analysis of biotite in a thrust/shear zone where fluid movement has occurred, quartz veins and tourmaline are present and both K and Ar are likely to have been made available at crystal interfaces. Thus only the biotite cooling ages of 9.5 ± 0.3 Ma, 12.7 ± 0.4 Ma, 21.1 ± 0.6 Ma and 19.9 ± 0.6 Ma are taken to be significant.

The muscovite results can, however, be regarded as giving geologically sensible ages. Muscovite is far less susceptible to the inclusion of excess argon in the lattice compared with coexisting biotite (Roddick *et al.* 1980) and, unless thermally disturbed, the ages are likely to reflect an accurate measure of time since cooling. Only by $^{40}Ar/^{39}Ar$ analysis could argon diffusion loss from such a thermal event be identified (Harrison & McDougall 1981; Wijbrans & McDougall 1986).

Interpretation of hornblende $^{40}Ar/^{39}Ar$ spectra

The $^{40}Ar/^{39}Ar$ method described above was used for age determinations on three hornblende samples separated from amphibolites of the Bhagirathi valley. Results from the analysis of GM33, GM77 and GM109 are presented in Table 4 whilst Fig. 10 displays the corresponding release spectra plots. The horizontal length of each box on the spectrum plot is proportional to the amount of ^{39}Ar released during a particular heating step, whilst the vertical height is proportional to the error of that age.

The general shape of the spectrum for GM33 passes from an initial high into a trough before rising again to form a slightly saddle-shaped plateau (Fig. 10a). The very old apparent age of 2667 Ma for this first low temperature release and the rapid drop-off to apparent ages of

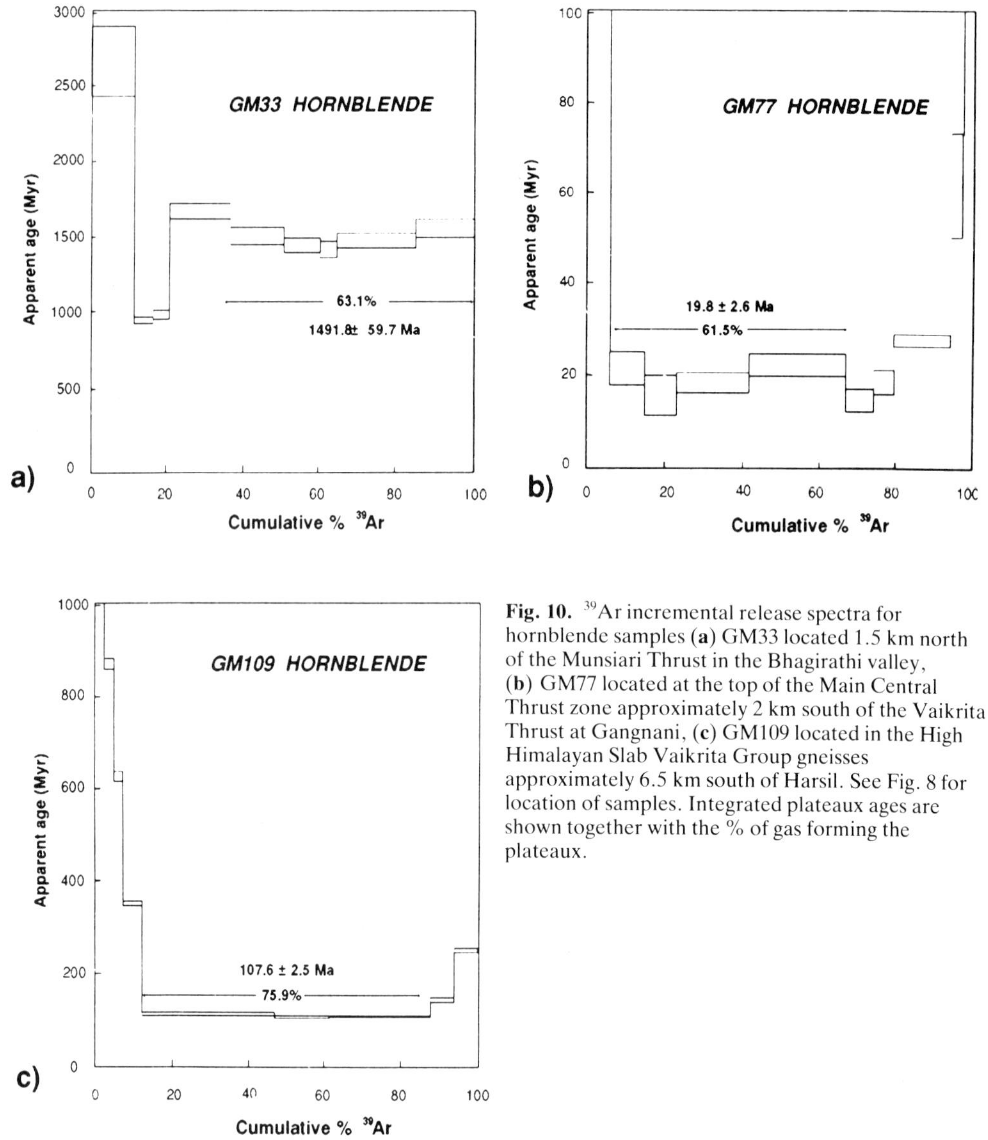

Fig. 10. ^{39}Ar incremental release spectra for hornblende samples (**a**) GM33 located 1.5 km north of the Munsiari Thrust in the Bhagirathi valley, (**b**) GM77 located at the top of the Main Central Thrust zone approximately 2 km south of the Vaikrita Thrust at Gangnani, (**c**) GM109 located in the High Himalayan Slab Vaikrita Group gneisses approximately 6.5 km south of Harsil. See Fig. 8 for location of samples. Integrated plateaux ages are shown together with the % of gas forming the plateaux.

954 Ma and 990 Ma are indicative of excess argon release from low-retention sites close to the margins of the hornblende crystal (Harrison & McDougall 1980*b*). The pattern of the GM33 spectrum as described above and from Figure 10a is similar to that obtained by Harrison & McDougall (1980*b*) for a Palaeozoic hornblende that suffered reheating during the Cretaceous. Therefore, the trough in the spectrum of GM33 may be due to diffusive loss of argon from the crystals during a reheating event of relatively low temperature and short duration, i.e. the temperatures reached were not hot enough to reset the crystal or significantly lower the plateau age. Alternatively it may result from the presence of trace inclusions (of ?biotite) within the hornblende (Rex *et al.* 1993). Higher K/Ca ratios (Table 5) for these anomalous young steps support this contention.

The rough plateau of GM33 accounts for 63.1% of the ^{39}Ar release thereby conforming to the requirement of at least 50% identical (within error) consecutive ages in order to confer a plateau age (Dalrymple & Lanphere 1974). An integrated age for this plateau is calculated at 1491.8 ± 59.7 Ma which is deemed reasonable if the sample conforms to the single-site diffusion model (Turner 1972; Albarède 1978). However, Ar release in hornblendes seems to be controlled by specific hornblende decomposition events rather than by simple volume diffusion (Wartho *et al.* 1991). In this model, gas release is thermally controlled. Experimental work suggests that gas released from vacuum-heating steps at 1030–1060° C may yield a meaningful age (Wartho *et al.* 1991). Hence, the lowest step in the plateau for GM33 at 1416.0 ± 52 Ma may be significant; however, unless acid leaching techniques are employed it is difficult to evaluate the degree of excess ^{40}Ar incorporation.

Hornblende sample GM77 (Table 5) describes a classic saddle-shaped spectrum (Fig. 10b), with both high and low temperature excess argon components (Harrison & McDougall 1981). However, a rough plateau is formed in the trough by 61.5% of ^{39}Ar released which gives an integrated age of 19.8 ± 2.6 Ma. The spectrum reveals no evidence to refute the validity of this age and it is interpreted as a significant cooling age for the foliated second generation hornblende. K/Ca ratios for each heating increment are consistently low for steps 4–10 but are higher for the first three stages. The initial variation might be indicative of excess argon release from inclusion and surface material other than hornblende.

The ^{39}Ar release spectrum from GM109 is L-shaped (Fig. 10c) but with a small rise in apparent age in the last two high temperature steps. A well-defined plateau for 76% ^{39}Ar released yields an integrated plateau age of 107.6 ± 2.5 Ma. A plot of K/Ca against % ^{39}Ar released shows an essentially uniform curve for the plateau forming steps, precluding the presence of phases of markedly different composition. Therefore, the age given by the plateau should not be regarded as a product of phase mixing. Nevertheless, the age of 107.6 Ma is geologically unlikely as this portion of the Indian plate is known from sedimentological and palaeomagnetic records to have formed a passive margin during the Mid-Cretaceous (Searle *et al.* 1987). Textural studies, however, reveal that early hornblende crystals possess overgrowths of finer crystal habit which may mean that the gases from the two generations are mixing during release to give an intermediate age. If one generation of hornblende crystallized or recrystallized at the margin of the other, it is difficult to see how this was accomplished at temperatures that did not permit resetting of the pre-existing crystal. Other alternatives are that (i) the hornblende may have incorporated excess argon that has diffused into the lattice from the exterior causing an anomalously high apparent age plateau with even higher boundary ages or that (ii) the hornblende has undergone only partial resetting of a previous, possibly Palaeozoic age, as described for a hornblende sample from the MCT zone (Hubbard & Harrison 1989).

Significance of cooling ages

Although each age represents cooling of a particular mineral through its respective blocking temperature, it is necessary to establish whether closure of these minerals can be delayed, or reopening implemented on a local scale, without invoking a regional heating event. As temperatures of 250–350° C permit ductile shear and the formation of mylonites (Sibson 1977) these conditions reached in shear zones would allow for the resetting of mica, but not necessarily hornblende. The function of shear partitioning could result in the resetting of micas undergoing crystallographic reorientation in rocks propagating the deformation, whilst little or no argon loss is experienced by minerals from adjacent areas experiencing less deformation. Ages from mica fabrics, therefore, date the last ductile event before uplift through the blocking temperature, but probably not subsequent lower temperature, brittle movements on a fault zone.

If the unreliable cooling ages are disregarded, the overall pattern is one of progressively younger ages towards the Indian foreland in the south. This chronological evidence suggests, therefore, that exhumation and cooling through the respective blocking temperatures proceeded sequentially towards the south (Fig. 9). This is consistent with a foreland-propagating piggy-back style thrust sequence in general terms. Disruptions to the pattern of southward younging ages may represent continuing localized shear or out-of-sequence motion in the upper regions of the MCT zone. Hodges *et al.* (1988) postulated the existence of out-of-sequence thrusts to explain reheating of the HHS in the Garhwal. Hubbard & Harrison (1989) also proposed out-of-sequence thrusts to explain young ages in the Tibetan Slab of eastern Nepal.

The hornblende age of 19.8 ± 2.6 Ma from the upper MCT zone in the Bhagirathi valley dates closure of the second generation hornblende that defines the tectonic fabric and consequently, this must also date the last deformation before these rocks cooled through the 500° C isotherm. What it does not date, however, is the last motions of the MCT, but simply the passage of a rock mass out of the environment which has the capacity for recrystallising hornblende. Subsequent ductile movements may still reorientate biotite and muscovite resetting their ages in the process, but once into the brittle field, movement is confined to discrete planes and would not cause widespread disturbance to crystal lattices. Thus it is likely mineral ages would not be reset once the brittle-ductile transition has been crossed, unless there is significant fluid flux along faults when problems with excess Ar may arise.

The 19.8 ± 2.6 Ma hornblende age along with younger muscovite (*c.* 15 Ma) and biotite (9.5 Ma) ages from adjacent localities are plotted against the relevant closure temperature for each mineral in Fig. 11. The resulting cooling curve obtained for the upper MCT is compared with that found by Hubbard & Harrison (1989) for the MCT zone in eastern Nepal. The upper MCT zone cooling rate of 30° C/Ma for the *c.* 500–350 step, inferred from Fig. 11, was twice as fast as that calculated for the first stage cooling of

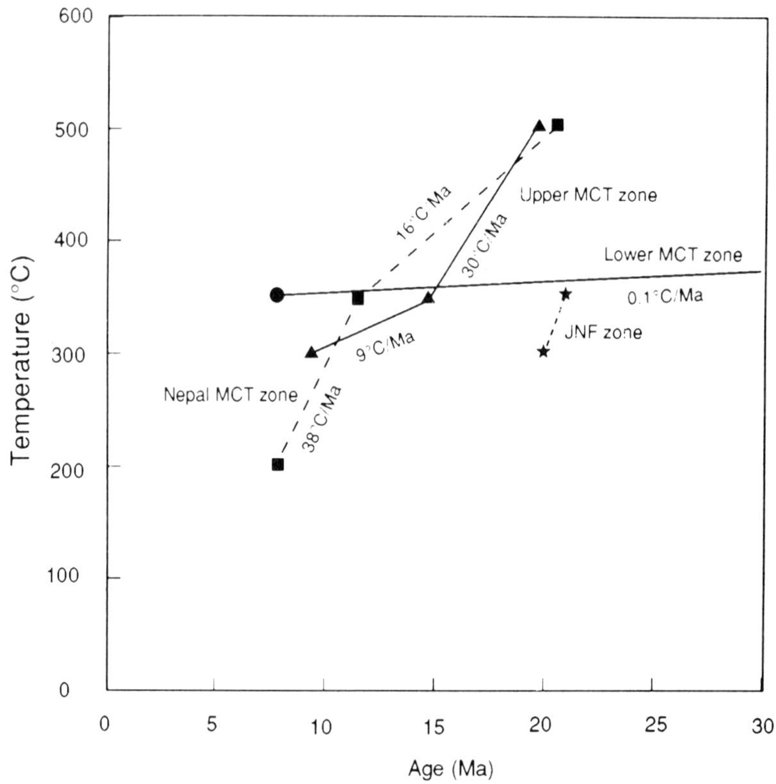

Fig. 11. Approximate cooling ages from samples from different structural locations within the Bhagirathi valley section. Also shown for comparison is the cooling curve obtained from the data of Hubbard & Harrison (1989) for the Nepal MCT in the Everest area.

the Hubbard & Harrison curve. For comparison, the cumulative cooling curve for the lower MCT produces an average cooling rate only 0.1° C Ma^{-1}. The Nepal MCT however, cooled more rapidly during the second stage at 38° C Ma^{-1}, whereas the Garhwal cooling rate decreased to just 9° C Ma^{-1}. The change from a relatively rapid to a slower cooling path is most likely related to tectonic denudation of the hangingwall to the MCT, probably by activation of the Jhala Normal Fault. The slower cooling path would be resumed once erosion-controlled denudation was again dominant. Mica ages of 19.9 Ma (biotite) and 20.8 Ma (muscovite) for the JNF indicate rapid cooling of the zone consistent with tectonic denudation, whereas the relatively slower rates recorded for the MCT are more representative of erosion controlled denudation.

The youngest cooling age of 5.7 Ma (K-Ar, muscovite) comes, not from the lowest portion of the MCT zone, but from garnet-mica schist from immediately beneath the Vaikrita Thrust at Gangnani (upper MCT zone). This locality is significantly close to the emergence point of a thermal spring, suggesting that late circulation of hot fluids may contribute to the young ages and excess Ar artefacts. Alternatively the young ages of recrystallisation of muscovite could be due to localised shear partitioning within the MCT zone.

Ages of *c*. 20–21 Ma were obtained from rocks affected by the Jhala Normal Fault (JNF). GM134 yields a reliable biotite age of 21.1 ± 0.6 Ma which effectively dates the recrystallisation of biotite into an S–C fabric recording a northward, normal sense of shear. Strongly foliated sillimanite + kyanite + garnet gneiss from immediately south of the junction with Martoli metasediments yields a biotite age of 19.9 ± 0.6 Ma. Hence, fabric reorientation in these JNF rocks and cooling through *c*. 350° C was completed by *c*. 20 Ma, whereas the MCT only cooled through this temperature at *c*. 6 Ma.

At the south end of the MCT zone, 2–3 km above the Munsiari Thrust, a hornblende age of 1491.8 ± 59.7 Ma records an early Riphean metamorphic event. A lack of subsequent reheating above *c*. 500° C is demonstrated by this very old age, although temperatures of *c*. 350° C were maintained as late as *c*. 8 Ma (muscovite). This low temperature event may have encouraged diffusion loss of argon from the less retentive hornblende sites in order to produce the trough in the spectrum for GM33 (Fig. 9a). This hornblende age corroborates thermobarometric profiles that reveal that temperatures attained in the lower MCT zone were substantially less than in the structurally higher regions of the MCT (Hubbard 1988).

Figure 12 is a plot of cooling age versus horizontal distance which illustrates the older muscovite cooling ages yielded by samples from the HHS relative to the MCT zone. Also highlighted by this diagram is the stepped pattern of the data with very young ages possibly indicating

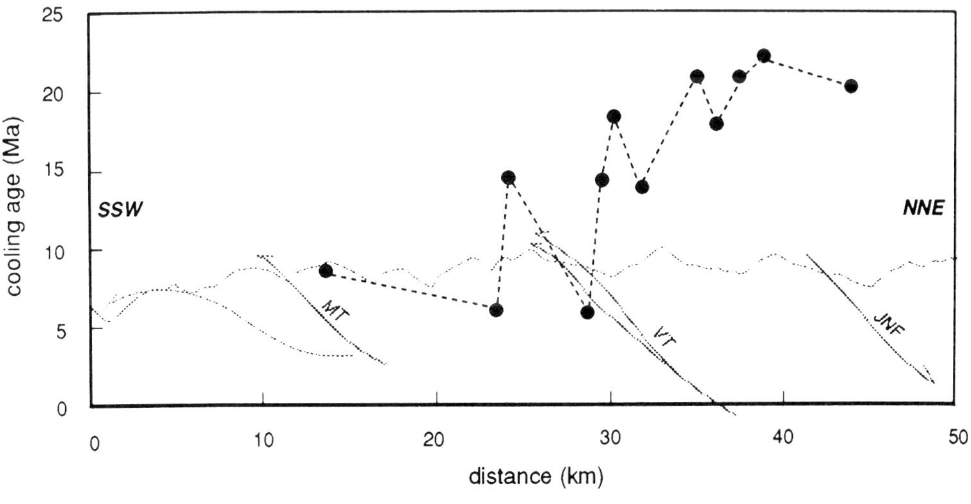

Fig. 12. Schematic structural section across the Bhagirathi valley section of the Garhwal Himalaya plotted with locations of the muscovite thermochronological samples against time. Note the general southward decreasing cooling ages except for the young ages along major thrust fault zones eg. Vaikrita Thrust probably caused by late fluid interactions and excess argon. MT, Munsiari Thrust; VT, Vaikrita Thrust; JNF, Jhala Normal Fault.

late thrust movements. When compared with the ages contained by Krummenacher *et al.* (1978), Hubbard & Harrison (1989) and Maluski *et al.* (1988) close general agreement is found. However, no Precambrian ages are reported from the MCT zone and ages in the upper part of the HHS are 17–18 Ma (muscovite) in Nepal–south Tibet as opposed to 21–22 Ma (muscovite) for Garhwal. One aspect is true of all sections: whilst an overall younging to the south is apparent, the youngest ages are reported from either the lower HHS or the upper MCT zone. This effect is the likely result of lattice disturbances causing age resetting during the latest ductile movements in these zones.

Tectonic implications

Whilst a single phase of prograde metamorphism, synchronous with MCT movements, affected the footwall to the VT, at least two phases of prograde metamorphism are recorded in the hangingwall. The temporal relationships between the various events are not clear. M1 was demonstrably pre-shear fabric development (S2a syn-MCT fabric), although M1 minerals are not ubiquitously deformed and are absent near the base of the hanging wall, thus precluding crucial deformation/textural relationships. M2 is only represented in the basal HHS by an inclusion-free overgrowth to garnet and M3 by restricted post-kinematic sillimanite development at the top of the HHS. A period of cooling occurred simultaneously with M2 footwall heating, evidenced by the muscovite and amphibole coronas to kyanite and pyroxene respectively. Garnet core to rim *P–T* evolution also supports post-M1 cooling. Zoning in garnets from the MCT zone is the result of prograde growth, whilst those from the central and upper part of the HHS are effectively unzoned suggesting a period of elevated temperatures (>600° C; Yardley 1977) at the top of the slab in order to allow diffusional homogenisation. The lower portion of the slab, however, retains growth-zoned garnets.

In order to attempt an explanation of these apparently paradoxical lines of evidence, a modified Le Fort-type model is envisaged (Fig. 13). M1 isograds were dragged down into an S-shaped profile as a result of a cold slab being subducted along the MCT. Peak M1 was, therefore, post-collisional but immediately pre-MCT development. As the isotherms began to relax in the manner described by England & Thompson (1984), rocks at the base of the High Himalayan Slab in the overturned thermal limb, previously at upper greenschist facies, were subjected to prograde metamorphism due to conductive transfer of heat combined with footwall internal heat production. During this M2 episode second generation, inclusion-free garnets were formed simultaneously with syn-kinematic metamorphism to upper greenschist facies in the MCT

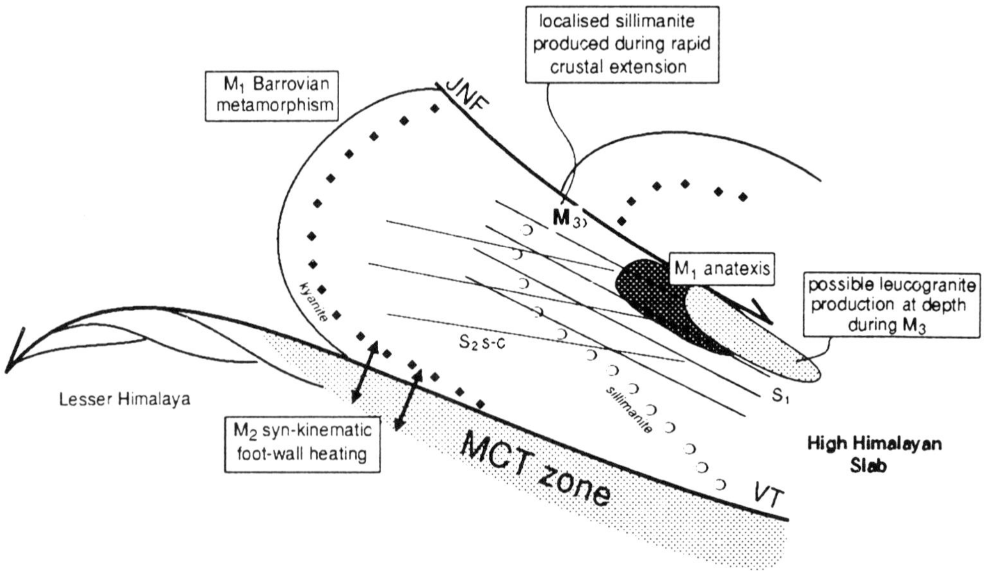

Fig. 13. Model for the tectonic evolution of the western Garhwal Himalaya. See text for discussion. JNF, Jhala Normal Fault.

zone. Subsequent adiabatic cooling was interrupted by M3. This event was the result of rapid pressure release possibly through activation of extensional structures, such as the Jhala Normal Fault, in response to topographic culmination (Burg *et al.* 1984; Searle & Rex 1989) possibly combined with MCT motion to the south over a ramp structure (Pêcher 1989). The effect of rapid pressure release in the upper HHS was to take the *P–T* path into the sillimanite field thus preserving kyanite field temperatures and pressures and recording rapid cooling in the garnet margins. Rapid erosion during M4 followed in order to preserve these mineral relationships. Activation of the Jhala Normal Fault also resulted in the removal of the upper limb of the recumbent M1 palaeothermal fold (Fig. 13).

Conclusions

Textural evidence suggests that the inverted metamorphic sequence in the Bhagirathi area is the cumulative result of polyphase metamorphism. M1 is identified as a tectonically inverted Barrovian sequence from garnet to sillimanite grade restricted to the High Himalayan Slab. M2 was a prograde phase that affected the MCT zone and its syn-tectonic (D2) nature is most clearly demonstrated by rotated garnets in the footwall to the Vaikrita Thrust. The thermal impact of this event decreases towards the south (base of the MCT zone). Unexpectedly the garnets from the base of the High Himalayan Slab show growth zoning, implying that they may also have grown during M2, although syn-D2 shear bands deform M1 kyanite and cause cooling-related retrogression to muscovite. The only unambiguous evidence for M3 is localized, post-kinematic, second generation sillimanite at the top of the slab. M4 was a purely retrogressive phase associated with cooling and uplift.

Conventional thermobarometers were used to calculate pressure and temperature profiles through the MCT zone and High Himalayan Slab. A distinct upward jump in both *P* and *T* is recorded across the Vaikrita Thrust with subsequent decreases accompanying structural height in the High Himalaya. A small local increase at the top of the slab is perhaps related to a later resetting of mineral equilibria during M3. As a consequence of the polyphase metamorphism thermobarometric sections cannot possibly provide snapshots of the *P–T* structure at any one time.

Modifications to the Le Fort model include the production of the inverted metamorphic sequence in two distinct phases. The M1 event was pre-MCT, but presumably post-collisional, whilst M2 was the direct consequence of MCT movements. M1 was tectonically inverted but M2 evolved due to the downward conduction of heat from the still hot slab above. M3 followed as a result of upper slab decompression associated with extension on the Jhala Normal Fault. This metamorphic and deformation chronology is similar to that described from the Zanskar Himalaya further west (Searle & Rex 1989; Searle *et al.* 1992).

K-Ar and $^{40}Ar/^{39}Ar$ mineral age data suggest sequential uplift and cooling towards the south through the High Himalayan Slab and MCT zone which is consistent with a piggy-back style of thrust propagation. Disruptions to the sequence of southward-younging cooling ages are probably due to late out-of-sequence thrust motion (Searle 1986). Cooling below the muscovite closure temperature was completed by *c.* 20 Ma on the Jhala Normal Fault, but as late as *c.* 6 Ma in the MCT zone. The upper MCT zone cooled rapidly at 30° C Ma^{-1} from *c.* 500° C to *c.* 350° C then cooled far more slowly at 9° C Ma^{-1} to *c.* 300° C. Rapid initial cooling of the upper MCT was coeval with cooling and uplift of the upper part of the slab through the *c.* 300° C isotherm. Thus, this particular part of the cooling path is indicative of tectonic denudation along the Jhala Normal Fault, whereas the slower cooling rate reflects a return to erosion-controlled denudation. As the Jhala Normal Fault was active at or before *c.* 20 Ma, thrusts must have been operative further to the south in order to have built up a crustal wedge for the orogenic collapse to have occurred.

This work was funded by NERC post-graduate student grant and supervised by M.P. Searle, A. Barnicoat and B. Windley. This paper was compiled by M.P. Searle following the untimely death of the author. Thanks are due to D. Waters, P. Treloar and A. Barnicoat for detailed and helpful reviews. R. Wilson helped with the microprobe data, and D. Rex, P. Guise and J. Wartho helped with the K-Ar and $^{40}Ar/^{39}Ar$ data. Buddhi Singh Rana, Bali and Karan of Mount Support Trekking, Uttarkashi, are thanked for local help, and C. Jones is thanked for her support.

References

ALBARÈDE, F. 1978. The recovery of spatial isotope distributions from stepwise degassing data. *Earth and Planetary Science Letters,* **39**, 387–397.

BOHLEN, S.R., WALL, V.J. & BOETTCHER, A.L. 1983. Experimental investigations and geological applications of equilibria in the system FeO–TiO_2–Al_2O_3–SiO_2–H_2O. *American Mineralogist,* **68**, 1049–1058.

BRIDEN, J.C., REX, D.C., FALLER, A.M. & TOMBLIN, J.F. 1979. K-Ar geochronology and palaeomagmetics of volcanic rocks in the Lesser Antilles island arc. *Philosophical Transactions Royal Society, London,* **A291**, 485–528.

BURG, J-P., BRUNEL, M., GAPAIS, D., CHEN, G.M. & LI, G.C. 1984. Deformation of leucogranites of the crystalline Main Central Thrust sheet in southern Tibet (China). *Journal of Structural Geology,* **6**, 535–542.

CLIFF, R.A. 1985. Isotopic dating in metamorphic belts. *Journal of the Geological Society, London,* **142**, 97–110.

DALRYMPLE, G.B. & LANPHERE, M.A. 1974. $^{40}Ar/^{39}Ar$ age spectra of some undisturbed terrestrial samples. *Geochimica et Cosmochimica Acta,* **38**, 715–738.

DODSON, M.H. 1973. Closure temperature in cooling geochronological and petrological systems. *Contributions to Mineralogy and Petrology,* **40**, 259–274.

ENGLAND, P.C. & THOMPSON, A.B. 1984. Pressure-temperature-time paths of regional metamorphism. Part I: Heat transfer during the evolution of regions of thickened continental crust. *Journal of Petrology,* **25**, 894–928.

ESSENE, E.J. 1982. Geological thermometry and barometry. *In*: FERRY, J.M. (ed.) *Characterization of Metamorphism Through Mineral Equilibria*, Reviews in Mineralogy. Mineral Society of America, Washington D.C., 153–196.

FAURE, G. 1986. *Principles of isotope geology*. Second edition. John Wiley & Sons.

FERRY, J.M. & SPEAR, F.S. 1978. Experimental calibration of the partitioning of Fe and Mg between garnet and biotite. *Contributions to Mineralogy and Petrology,* **66**, 113–117.

GANGULY, J. & KENNEDY, G.C. 1974. The energetics of natural garnet solid solution. I: mixing of the aluminosilicate end-members. *Contributions to Mineralogy and Petrology,* **48**, 137–148.

GHENT, E.D. 1976. Plagioclase–garnet–Al_2SiO_5–quartz: a potential geothermometer–geobarometer. *American Mineralogist,* **61**, 710–714.

HARRISON, T.M. 1981. Diffusion of ^{40}Ar in hornblende. *Contributions to Mineralogy and Petrology,* **78**, 324–311.

—— & MCDOUGAL, I. 1980*a*. Investigations of an intrusive contact, NW Nelson, New Zealand. I: Thermal, chronological and isotopic constraints. *Geochimica et Cosmochimica Acta,* **44**, 1985–2003.

—— & —— 1980*b*. Investigations of an intrusive contact, NW Nelson, New Zealand. II: Diffusion of radiogenic and excess ^{40}Ar in hornblende revealed by $^{40}Ar/^{39}Ar$ age spectrum analysis. *Geochimica et Cosmochimica Acta,* **44**, 2005–2020.

—— & —— 1981. Excess ^{40}Ar in metamorphic rocks from Broken Hill, New South Wales: implications for $^{40}Ar/^{39}Ar$ age spectra and thermal history of the region. *Earth and Planetary Science Letters,* **55**, 123–149.

HEIM, A. & GANSSER, A. 1939. *Central Himalaya – Geological Observations of the Swiss expedition 1936*. Memoires de la Societé Helvetique des Science Naturelles, **73**.

HODGES, K.V. & CROWLEY, P.D. 1985. Error estimation and empirical geothermobarometry for pelitic systems. *American Mineralogist,* **70**, 702–709.

—— & SILVERBERG, D.S. 1988 Thermal evolution of the Greater Himalaya, Garhwal, India. *Tectonics,* **7**, 583–600.

—— & SPEAR, F.S. 1982. Geothermometry, geobarometry, and the Al_2SiO_5 triple point, at Mt. Moosilauke, New Hampshire. *American Mineralogist,* **67**, 1118–1134.

——, HUBBARD, M.S. & SILVERBERG, D.S. 1988. Metamorphic constraints on the evolution of the central Himalayan orogen. *Philosophical Transactions of the Royal Society of London,* **A326**, 257–280.

HOLDAWAY, M.J. 1971. Stability of andalusite in the aluminosilicate phase diagram. *American Journal of Science,* **271**, 97–131.

HUBBARD, M.S. 1988. *Thermobarometry, $^{40}Ar/^{39}Ar$ geochronology and structure of the Main Central Thrust zone and Tibetan Slab, Eastern Nepal Himalaya.* PhD thesis, Massachusetts Institute of Technology, USA.

—— 1989. Thermobarometric constraints on the thermal history of the Main Central Thrust zone and Tibetan Slab, Eastern Nepal. *Journal of Metamorphic Geology,* **7**, 19–30.

—— & HARRISON, T.M. 1989. $^{40}Ar/^{39}Ar$ age constraints on deformation and metamorphism in the Main Central Thrust Zone and Tibetan Slab, Eastern Nepal Himalaya. *Tectonics,* **8**, 865–880.

INDARES, A. & MARTIGNOLE, J. 1985. Biotite–garnet geothermometry in granulite facies rocks. The influence of Ti and Al in biotite. *American Mineralogist,* **70**, 271–278.

JAIN, A.K. 1971. Stratigraphy and tectonics of the Lesser Himalayan Region of Uttarkashi, Garhwal Himalaya. *Himalayan Geology,* **1**, 25–58.

—— & ANAND, A. 1988. Deformational and strain patterns of an intracontinental collision ductile shear zone – an example from the Higher Garhwal Himalaya. *Journal of Structural Geology,* **10**, 717–134.

KRUMMENACHER, D., BASETT, A.M., KINGERY, F.A. & LAYNE, M.F. 1978. Petrology, metamorphism and K-Ar age determinations in E Nepal. *In*: SAKLANI, P.S. (ed.) *Tectonic geology of the Himalayas*. Today and Tomorrow's Publ. New Delhi, India, 151–166.

MCDOUGAL, I. & HARRISON, T.M. 1988. *Geochronology and thermochronology by the $^{40}Ar/^{39}Ar$ method*. Oxford Monographs on Geology and Geophysics, **9**.

MALUSKI, H., MATTE, P., BRUNEL, M. & XIAO, X. 1988. $^{40}Ar/^{39}Ar$ dating of metamorphic and plutonic events in the North and High Himalayan Belts (S. Tibet–China). *Tectonics,* **7**, 299–326.

METCALFE, R.P. 1990. *A Thermotectonic evolution for the Main Central Thrust and Higher Himalaya,*

Western Garhwal, India. PhD thesis, University of Leicester, U.K.

NEWTON, R.C. & HASELTON, H.T. 1981. Thermodynamics of the garnet-plagioclase-Al_2SiO_5-geobarometer. *In*: NEWTON, R.C., NARVROSKI, A. & WOOD, B.J. (eds) *Thermodynamics of minerals and melts*. Springer-Verlag, Berlin, 125–145.

PÊCHER, A. 1989. The metamorphism in the Central Himalaya. *Journal of Metamorphic Geology*, **7**, 31–41.

—— & SCAILLET, B. 1989. La structure de Haut-Himalaya au Garhwal (Indes). *Eclogae Geologicae Helvetiae*, **82**, 665–668.

PURDEY, J.W. & JAGER, E. 1976. *K-Ar ages on rock forming minerals from the central Alps*. Memorie degli Instituti di Geologia e Minerologia dell'Universita di Padova. **30**.

REX, D.C., GUISE, P.G. & WARTHO, J-A. 1993. Disturbed $^{40}Ar/^{39}Ar$ spectra from hornblendes: thermal loss or contamination. *Chemical Geology, Isotope Geosciences section*, **103**, 271–281.

RODDICK, J.C., CLIFF, R.A. & REX, D.C. 1980. The evolution of excess argon in alpine biotites – an $^{40}Ar/^{39}Ar$ analysis. *Earth and Planetary Science Letters*, **48**, 185–208.

SCHONEVELD, C. 1977. A study of some typical inclusion patterns in strongly paracrystalline rotated garnets. *Tectonophysics*, **39**, 453–471.

SEARLE, M.P. 1986. Structural evolution and sequence of thrusting in the High Himalayan, Tibetan-Tethys and Indus suture zones of Zanskar and Ladakh, Western Himalaya. *Journal of Structural Geology*, **8**, 923–936.

—— & REX, A.J. 1989. Thermal model for the Zanskar Himalaya. *Journal of Metamorphic Geology*, **7**, 127–134.

——, WINDLEY, B.F., COWARD, M.P., COOPER, D.J.W., REX, A.J., REX, D.C., TINGDONG, L., XUCHANG, X., JAN, M.Q., THAKUR, V.C. & KUMAR, S. 1987. The closing of Tethys and the tectonics of the Himalaya. *Geological Society of America Bulletin*, **98**, 679–701.

——, WATERS, D.J., REX, D.C. & WILSON, R.N. 1992. Pressure, temperature and time constraints on Himalayan metamorphism from eastern Kashmir and western Zanskar. *Journal of the Geological Society, London*, **149**, 753–773.

SIBSON, R.H. 1977. Fault rocks and fault mechanisms. *Journal of the Geological Society London*, **133**, 191–213.

SPEAR, F.S., FERRY, J.M. & RUMBLE, III, D. 1982. Analytical formulation of phase equilibria: the Gibbs' Method. *In*: FERRY, J.M. (ed.) *Characterization of Metamorphism Through Mineral Equilibria*. Reviews in Mineralogy. Mineralogical Society of America, Washington D.C., 105–152.

TRACY, R.J., ROBINSON, P. & THOMPSON, A.B. 1976. Garnet composition and zoning in the determination of pressure and temperature of metamorphism, Central Massachusetts. *American Mineralogist*, **61**, 762–775.

TURNER, G., 1972. $^{40}Ar/^{39}Ar$ age and cosmic ray irradiation history of the Apollo 15 anorthosite 15415. *Earth and Planetary Science Letters*, **14**, 169–175.

VALDIYA, K.S. 1978. Outline of the structure of the Kumaun Himalaya. *In*: SAKLANI, P.S. (ed.) *Tectonic Geology of the Himalaya*. Today & Tomorrow's Publ. N. Delhi, 1–14.

—— 1980. *Geology of the Kumaun Lesser Himalaya*. Wadia Institute of Himalayan Geology, Dehra Dun.

WARTHO, J-A. 1991. *Argon isotope systematics and mineralogy of metamorphic hornblendes from the Karakoram*. PhD thesis, University of Leeds.

——, DODSON, M.H., REX, D.C. & GUISE, P.G. 1991. Mechanics of Ar release from Himalayan metamorphic hornblende. *American Mineralogist*, **76**, 1446–8.

WIJBRANS, J.R. & MCDOUGAL, I. 1986. $^{40}Ar/^{39}Ar$ dating of white micas from an Alpine high-pressure metamorphic belt on Naxos (Greece): the resetting of the argon isotope system. *Contributions to Mineralogy and Petrology*, **93**, 187–194.

YARDLEY, B.W.D. 1977. An empirical study of diffusion in garnet. *American Mineralogist*, **62**, 793–800.

—— 1989. *An Introduction to Metamorphic Petrology*. Longmans, Earth Science Series.

Deformation and metamorphism within the Main Central Thrust zone, Arun Tectonic Window, eastern Nepal

K. MEIER[1,2] & E. HILTNER[1]

[1] *Institut für Geologie und Mineralogie der Friedrich-Alexander-Universität Erlangen-Nürnberg, Schlossgarten 5, D-8520 Erlangen, Germany*

[2] *Present address: c.o. Aquaterra consulting, Hydrogeologists Inc., Kastanienweg 2, CH-6353 Weggis, Switzerland*

Abstract: The Lesser Himalayan Arun Tectonic Window comprises the Upper Precambrian, polymetamorphosed metasediments of the Nawakot Complex. These are tectonically overlain by Precambrian gneisses of the Higher Himalayan Kathmandu Complex. The main phase metamorphism within the Higher Himalayan rocks predated thrusting along the Main Central Thrust. By contrast, the main metamorphism (M2) within the Nawakot Complex commenced synchronously with it being underthrust during D2 along the Main Central Thrust Zone beneath the amphibolite facies rocks of the Kathmandu Complex. During D2 rocks on both sides of the Main Central Thrust Zone underwent intensive mylonitization within a several hundred meter wide zone of non-coaxial shearing. Although the stratigraphic order within the Nawakot Complex is the right way up, rocks within the Arun Tectonic Window show a concentric inverted metamorphic zoning, ranging from lower greenschist facies in the core of the window, up to lower amphibolite facies at the outer edge of the window. This 'inverted metamorphism' (M2) of the Nawakot Complex originated as a result of syn- to post-D2 downward heat conduction from the 'hot' hanging wall of the Main Central Thrust onto the 'cold' footwall. This heat flow also caused the strong retrograde overprinting at the base of the Kathmandu Complex. After a static period, a late- to post-M2 NNE-shortening (D3) commenced along the Main Boundary Thrust. The Arun Tectonic Window is the result of doming of the Main Central Thrust during D4 followed by erosional exhumation.

The Arun Tectonic Window (Figs 1 & 2), with a zone of clearly inverted metamorphism, was first described by Bordet (1961) and Hagen (1969). Since then, the origin of the 'inverted metamorphism' across the Main Central Thrust (MCT) as well as the exact location of the thrust which separates the Lesser from the Higher Himalaya has been discussed by many workers. Within the Arun Tectonic Window, Bordet (1961) concluded that the inverted metamorphic zonation resulted from an inversion of the lithostratigraphic succession within large scale recumbent folds. Hagen (1969) explained the inverted metamorphism by the emplacement of the high grade metamorphic rocks of the 'Kathmandu Nappes', onto low grade ones of the 'Nawakot Nappes'.

Akiba & Amma (1973) considered the 'Tumlingtar Window' (their name for the Arun Tectonic Window) and the augen-gneiss klippe around Dhankuta to be of an autochthonous nature (Table 1), whereas Maruo & Kizaki (1981) favoured a nappe-geometry and suggested that two thrusts, an MCT1 and an MCT2 (Table 1), transported the Higher Himalayan Crystalline Unit over the low grade rocks of the meta-sedimentary 'Midland Zone' of the Lesser Himalaya. Andrews (1985), who worked on the eastern side of the Arun valley, classified the low- to medium-grade metasediments of the 'Chainpur Group' and the high-grade rocks of the 'Phakuwa Group' as belonging to the Lesser Himalayan sequence and placed the MCT below the 'Milke Gneiss' (the equivalent of the 'Gneiss éctinites du Barun' of Bordet 1961, or the 'Himalayan Gneiss' of Maruo & Kizaki 1981). According to Brunel (1986), the basement of the Lesser Himalayas consists of a very low metamorphic grade 'Autochthonous–Para-Autochthonous Unit', which is overlain by a complex imbrication of more or less metamorphic 'Lower Nappes'. An enormous monotonous gneiss sheet of the 'Tibetan Slab', representing the high grade rocks of the Higher Himalaya structurally overlies the MCT. Sharma (1990) has published an up-to-date summary of most of the studies that relate to the geology of eastern Nepal.

The aim of this study was to investigate the tectonometamorphic history of the Arun valley by way of small-scale and micro-scale textural analysis. Field mapping was carried out at an exploration-scale from north of the river Tamur

From Treloar, P.J. & Searle, M.P. (eds), *Himalayan Tectonics*
Geological Society Special Publication, No. 74, pp. 511–523.

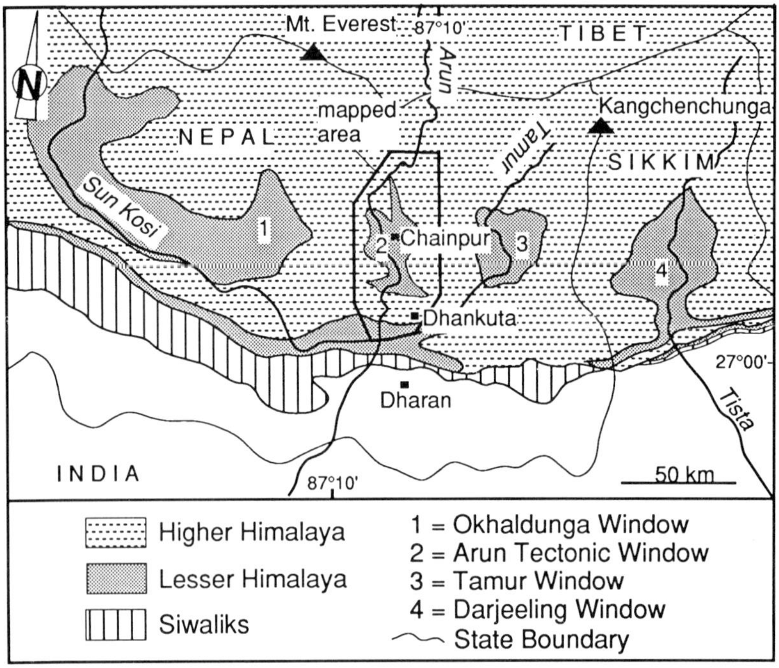

Fig. 1. Regional geological sketch map of eastern Nepal and Sikkim (modified after Fuchs 1981) showing the location of the mapped area.

Table 1. *Lithostratigraphy of the Arun Valley: a comparison*

BORDET 1961 (modified)	AKIBA & AMMA 1973 (modified)	MARUO et al. 1979 (modified)	ANDREWS 1985 (modified)	BRUNEL 1986 (modified)	PRESENT AUTHORS
	-----MCT-----				
ENSEMBLE DU HAUT HIMALAYA	**MIDLAND SCHUPPEN ZONE**	**CRYSTALLINE THRUST NAPPE**	**HIGHER HIMALAYA**	**HIGHER HIMALAYA**	**HIGHER HIMALAYA Kathmandu Complex**
Gneiss éctinites du Barun	Barun Gneiss Schuppe	Himalayan Gneiss	Milke Gneiss	**Tibetan Slab**	banded Gneisses
----- MCT -----	Irkhuwa Crystalline Nappe	---- MCT I ----	---- MCT ----	----- MCT -----	
ENSEMBLE DU BAS HIMALAYA		Calc-Silicate Schist	**LESSER HIMALAYA Phakuwa Group**	**LESSER HIMALAYA**	
Migmatites:			Khitya Khola Formation	Lower Nappes	Granite-Gneiss
·Roche à Phénoblastes"		Irkhuwa Crystalline Nappe	Sikdim Gneiss		
"Roche à Phénoblastes"	Midland Augen-Gneiss Schuppen Zone	Augen-Gneiss	Barabise Gneiss		Augen-Gneiss
"Roche très laminée et schisteuse"		Mylonitic Granite	Khandbari Gneiss		Ultramylonitic Gneiss
		---- MCT II ----			----- MCT -----
Serie de Couverture	**MIDLAND SEDIMENTARY GROUP**	**MIDLAND ZONE: Arun Tectonic Window**	**Chainpur Group**	**AUTOCHTONOUS--PARA-AUTOCH-TONOUS UNIT**	**LESSER HIMALAYA Nawakot-Complex** Arun Tectonic Window
Micaschistes (Quartzites Inférieurs) (Calcschistes)	Gudel Phyllite Schuppe	Kali Khola Schist	Sawanetar Formation Samla Limestone Swachi Formation Basthala Schist Simbuwa Schist	Dolomites Qartzites Limestones	**Metamorphic Unit C:** Staurolite-Kyanite-Micaschistes, mylonitic Quartzites, Marbles
Phyllades Inférieurs Calcschistes Quartzites Inférieurs	Tumlingtar Autochthonous Window	**Tumlingal Unit** Upper Carbonaceous Subunit	Kharang Phyllite Upper Hokse Quartzite Maksuwa graphitic Phyllite	various low grade Metasediments	**Metamorphic Unit B:** garnetiferous Phyllites, graphitic Phyllites, Dolomites, Quartzites
Phyllades Supérieurs Quartzites Supérieurs	Dhankuta Autochthonous Zone (Augen-Gneiss)	Lower Quartzose Subunit	Khekuwa Phyllite Lower Hokse Quarzite Hinuwan Phyllite	various low grade Metasediments	**Metamorphic Unit A:** chloritic Phyllites, Quartzites, Meta-Sandstones

up to the southern border of the 'restricted area' (Figs 1 and 2). The western half of the valley was mapped by Hiltner (1991) and the eastern side by Meier (1989).

Stratigraphy

The lithostratigraphic nomenclature of the Arun Tectonic Window and overlying rocks is confused due to a lack of consistency between

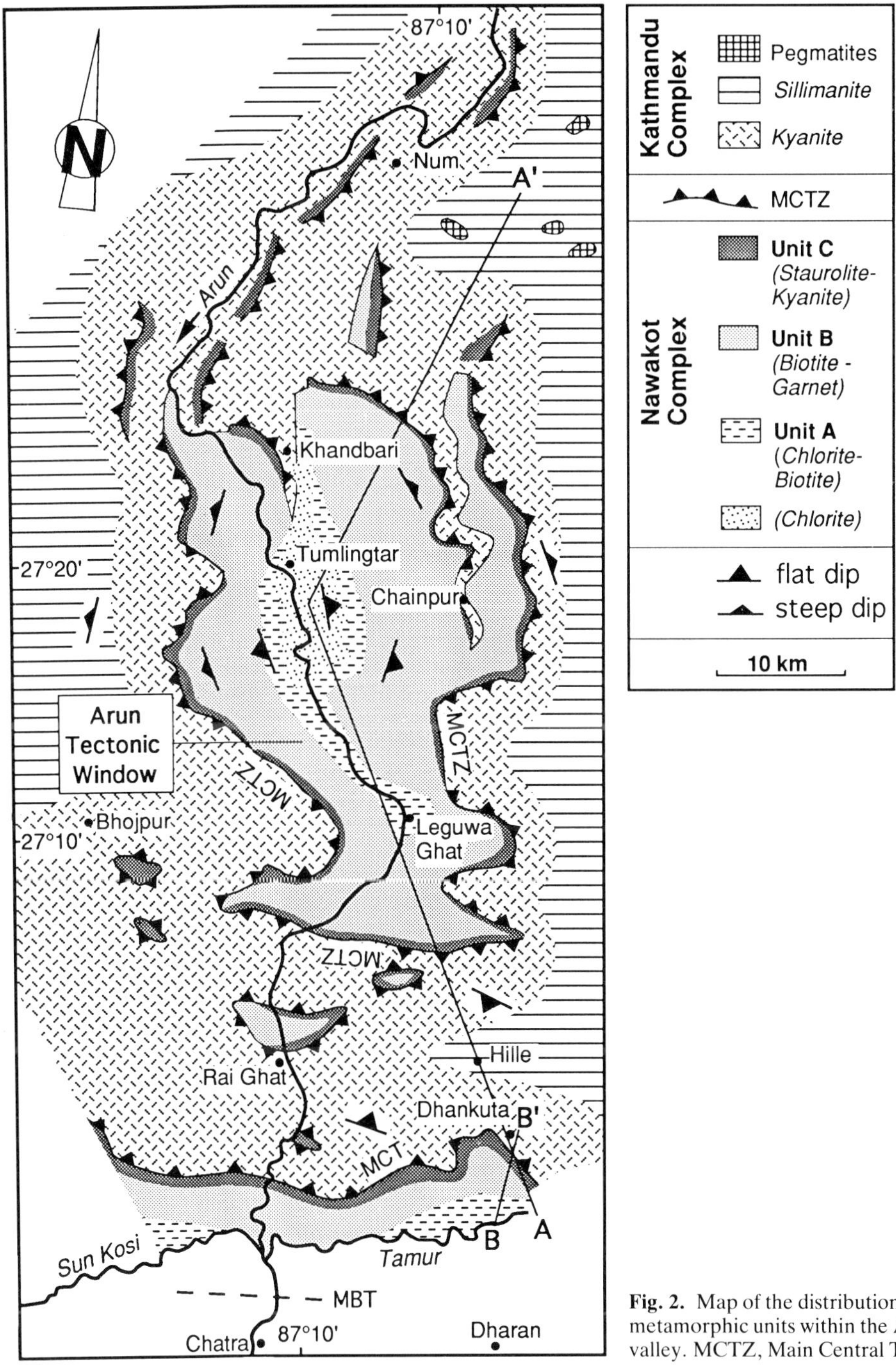

Fig. 2. Map of the distribution of the metamorphic units within the Arun valley. MCTZ, Main Central Thrust.

different authors. A comparison of the various lithostratigraphic frameworks used is shown in Table 1.

Lesser Himalaya

The up to 5 km thick, non-fossiliferous, pelitic and psammitic metasediments of the Arun Tectonic Window are part of the Upper Precambrian to Lower Palaeozoic Nawakot Complex as defined by Stöcklin (1980). This incorporates, variously, the 'Series de Couverture' of Bordet (1961); the metasedimentary 'Nawakot Nappes' of Hagen (1969); the 'Midland Sedimentary Group' of Akiba & Amma (1973); the 'Midland Zone' of Maruo & Kizaki (1981); the metasedimentary formations of the 'Chainpur Group' of Andrews (1985); the 'Autochthonous–Para-Autochthonous Unit' as well as the metasedimentary formations of the 'Lower Nappes' of Brunel (1986). Based on the distribution of Barrovian-type index minerals, three metamorphic units (A, B, C) have been distinguished within the Nawakot Complex, with metamorphic grade increasing upward (Figs 2, 3 and 4b; Table 1). Bordet (1961) suggested that the inverted metamorphism resulted from a post-metamorphic recumbent folding (Fig. 4a). However, way up criteria suggest the stratigraphy not to be inverted (Fig. 4b) and therefore an alternative mechanism has to be invoked to explain the metamorphic inversion.

Unit A is about 1.5–2 km thick and corresponds to the 'Phyllades Superieurs', the 'Calcschistes' and the 'Quartzites Superieurs' of Bordet (1961) and to the 'Hinuwan Phyllite', 'Lower Hokse Quartzite' and 'Khekuwa Phyllite' of Andrews (1985). It contains slightly deformed, light-green epizonal phyllites with subordinate intercalations of thin basic sills, quartzitic and graphitic phyllites as well as medium-grained quartzites and meta-sandstones. Way-up criteria, including graded bedding, cross bedding and ripplemarks, testify to a shallow-water environment of sedimentation for the quartzites and to an upward younging direction for the sedimentary sequence.

Unit B comprises dark-green garnetiferous phyllites and occasional chloritoid-bearing phyllites together with layers of siliceous and graphitic phyllite, which represent the 'Phyllades Inferieurs' and the 'Quartzites Inferieurs' of Bordet (1961) or the 'Maksuwa graphitic Phyllite', 'Upper Hokse Quartzite' and 'Kharang Phyllite' of Andrews (1985). Generally these rocks show a stronger deformation than those of unit A. In the middle part of the unit, lens-shaped quartz-segregations, the frequency of which increases upward, occur within the phyllites. As in unit A, thick bands of quartzite and minor basic sills are common. In many parts of the window, the upper part of unit B consists of an alternation of talcschists, carbonatic and graphitic phyllites. Compared to the northern parts of the window, the thickness of

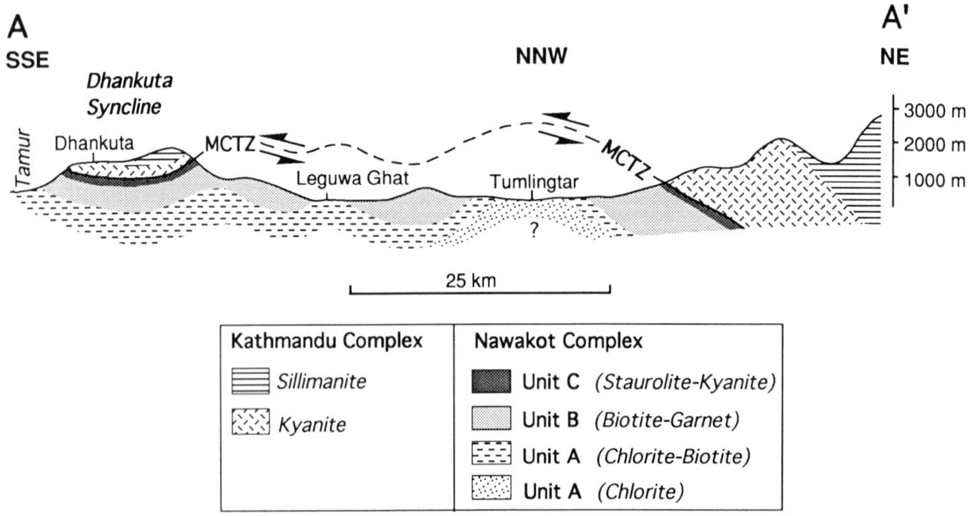

Fig. 3. Cross section A–A′ through the metamorphic units of the Arun Tectonic Window (for line of section see Fig. 2). MCTZ, Main Central Thrust.

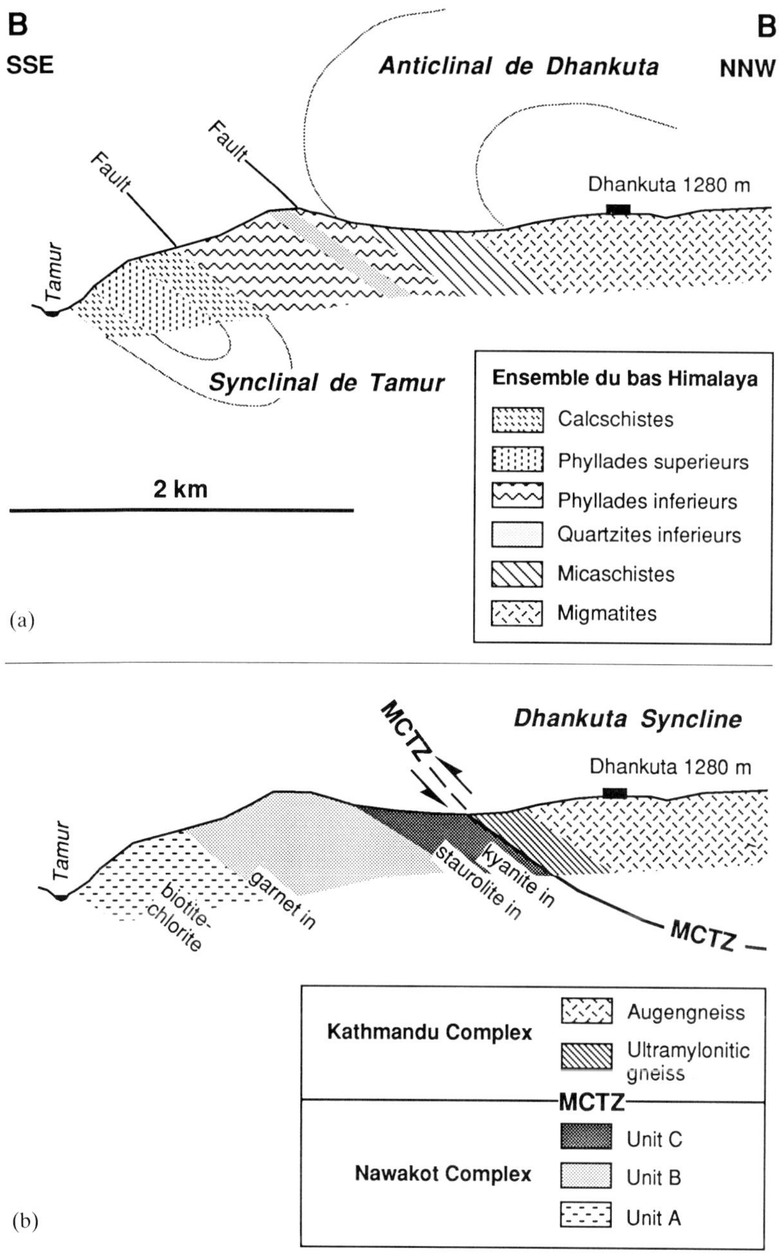

Fig. 4. Cross section B–B′ through the area south of Dhankuta (for line of section see Fig. 2). (**a**) shows Bordet's (1961) explanation of an overturned stratigraphic order in the 'Anticlinal de Dhankuta' and the 'Synclinal de Tamur'. (**b**) shows the authors' interpretation of a normal stratigraphic order and an inverted metamorphic zoning within the southern limb of the 'Dhankuta Synkline' below the MCT.

unit B seems to be reduced south of the Dhankuta Syncline (Figs 3 & 4b).

Unit C corresponds to Bordet's (1961) 'Micaschistes'; to Maruo & Kizaki's (1981) 'Kali Khola Schist'; and to Andrew's (1985) 'Simbuwa Schist', 'Basthala Schist', 'Swachi Formation', 'Samla Limestone' and 'Sawanetar Formation'. It is up to 500 m thick and occurs as a highly mylonitic belt at the structural top of the Arun Tectonic Window. At its base, it contains pelitic staurolite–garnet–mica schists. These are overlain by biotite–kyanite–garnet–plagioclase-bearing micaschists with abundant quartzose segregations. As observed by Bordet (1961), the thickness of the micaschists is variable. In some localities the schists are missing completely, and here strongly sheared kyanite quartzites or mylonitic dolomite-marbles occur. Schistose metasediments of unit C also outcrop in the form of small windows in deeply eroded parts of the Arun valley to the north of the tectonic window, and in some of its tributary valleys (Fig. 2) and appear to be intercalated within the Higher Himalayan gneisses.

Higher Himalaya

The thick thrust sheet of the Precambrian Kathmandu Complex (Stöcklin 1980) structurally overlies the Nawakot Complex (Figs 2, 3 & 4b). This corresponds to the 'Migmatites' of Bordet (1961); the high-grade formations of the 'Kathmandu-Nappes' of Hagen (1969); the 'Midland Schuppen Zone' of Akiba & Amma (1973); the various formations located between the 'MCT1' and 'MCT2' of Maruo & Kizaki (1981); the Phakuwa Group', the 'Barabise Gneiss' and the 'Khandbari Gneiss' of Andrews (1985); and the 'Tibetan Slab' and mylonitic augen-gneisses of the 'Lower Nappes' of Brunel (1986).

The authors were not permitted to enter the 'restricted area' along the Chinese border, where the uppermost part of the Higher Himalayan sequence can be found. According to Brunel (1986) and Brunel & Kienast (1986), however, the topmost part of the Higher Himalayan unit consists of sillimanite–cordierite–garnet gneisses. In the mapped area, the higher portions of the Kathmandu Complex are built up of an alternation of fine-grained, banded paragneisses with recrystallized quartzites and marbles that show a downward transition into migmatitic sillimanite-gneisses with subordinate intercalations of quartzites and amphibolites. In the northeastern corner of the mapped area, swarms of tourmaline–beryl-bearing pegmatites were found.

Below the migmatitic gneisses, light-grey, grano- to porphyroclastic sillimanite–kyanite-bearing granite-gneisses outcrop (= 'Roche à Phénoblastes' of Bordet 1961). These contain idiomorphic K-feldspars within a coarse-grained quartz–plagioclase–muscovite–biotite matrix. Towards the lower levels of the granite-gneisses, the long axes of the K-feldspars are orientated parallel to a distinct micaceous foliation (Fig. 5a).

Augen-gneisses, which make up much of the lower structural levels, appear to have been derived from the granite-gneisses by intense shearing. Brunel (1986) assigned the augen-gneisses to the 'Lower Nappes' (Table 1), as he observed Lesser Himalayan metasediments and micaschists overlying these rocks in the area east of Chainpur (Fig. 2). The textural change from undeformed granite-gneisses (Fig. 5a) into the augen-gneisses takes place gradually downward towards the MCT plane, and marks an increase in shear strain downward through the lower Higher Himalayan crystallines. Initially, asymmetric pressure shadows form at the opposite ends of subidiomorphic K-feldspar crystals (Fig. 5b). The typical augen-gneisses have lensoid, asymmetric K-feldspar augen (Fig. 5c) aligned within a well defined S–C fabric (Lister & Snoke 1984). Many of the augen are cross-cut by the C-surfaces. At the very base of the Higher Himalayan sequence, ultramylonitic kyanite-orthogneisses (the equivalent of the 'Roche très laminée et schisteuse' of Bordet 1961, 'mylonitic granite' of Maruo & Kizaki 1981, or 'mylonitic augen-gneiss' of Brunel 1986) outcrop as a continuous belt around the outer margin of the Arun Tectonic Window. Macroscopically, this ultramylonitic stage is recognised by ductile flattening and shearing of the feldspars within the prominent micaceous fabric (Fig. 5d).

Deformation history

D1

At least one pre-Himalayan ESE–WNW shortening phase has been reported from other regions of the Himalaya (Schwan 1980, 1988), but in the mapped area, no mesoscopic structures were found that would substantiate such pre-Himalayan tectonic activity. However, there is clear micro-tectonic evidence for an 'old' S1 schistosity within both the Lesser and Higher Himalayan rocks that documents the existence of a pre- or early-Himalayan deformation. According to Brunel (1986), this was contemporaneous with the Barrovian amphibolite- to granulite-grade metamorphism of the Higher Himalaya that occurred during an early phase of the Himalayan orogeny. The authors have found no indications to subdivide D1 into two phases, as demonstrated by Brunel (1986).

D2

D2 (D3 of Schwan 1980, 1988 and D3 of Brunel 1986) is documented by the following mesoscopic tectonic features. Phyllites of the upper part of the Nawakot Complex were folded into

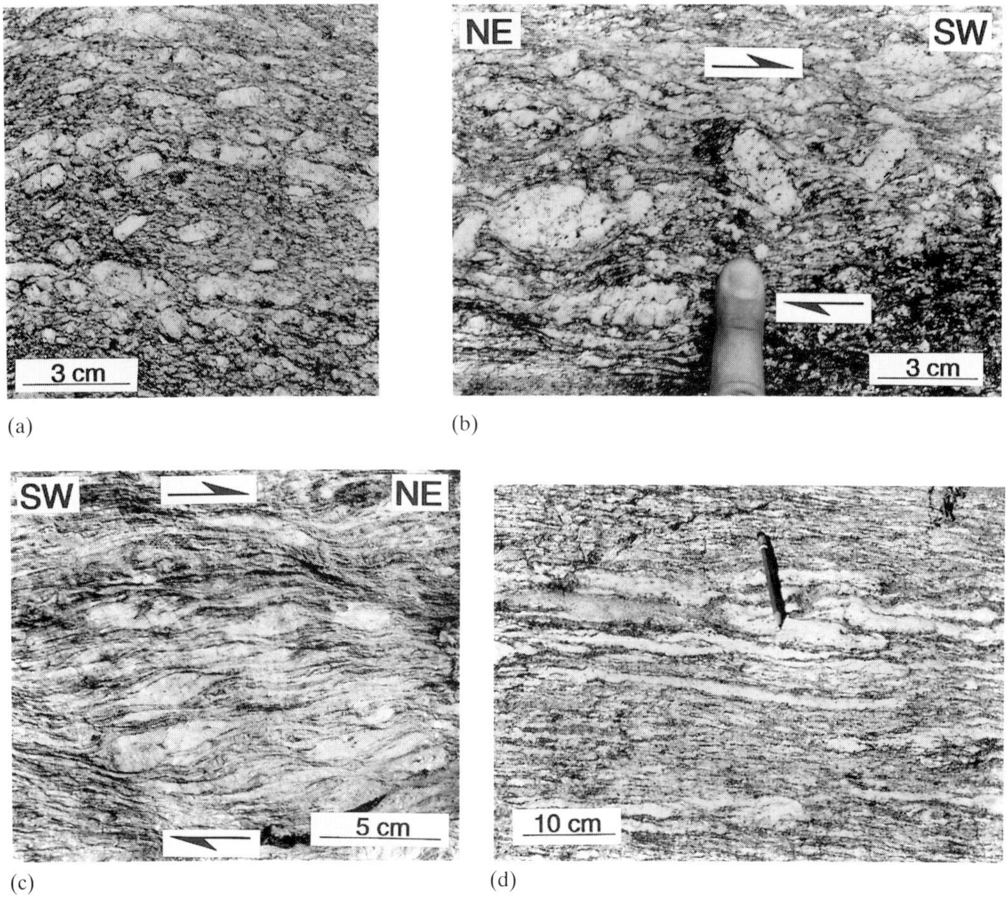

Fig. 5. Progressive deformation of porphyroblastic gneiss due to increasing non-coaxial shear stress downward through the Kathmandu Complex towards the MCT. (**a**) Undeformed porphyroclastic granite-gneiss in the upper portions of the Kathmandu Complex. (**b**) Example of the early stage of the textural transition from granite-gneiss into augen-gneiss. Some of the Subidiomorphic K-feldspars have 'tails' of asymmetric pressure shadows, while other porphyroclasts, which are arranged within the C-foliations exhibit a lens-like shape. The pre-D2 nature of the K-feldspars is indicated by the intersection of the D2-kinetic C-foliations. (**c**) Overall orientation of the lensoidal K-feldspars within the prominent C-foliation within an augen-gneiss. (**d**) Extreme flattening of the K-feldspars within ultramylonitic augen-gneisses occurring at the very base of the Kathmandu Complex.

isoclinal, SW-vergent recumbent folds. By contrast, at the base of the sequence, most F2 folds are upright or gently inclined open folds. The influence of non-coaxial ductile shearing thus clearly increases upwards through the Nawakot Complex towards the MCT plane. At the top of the Nawakot Complex where C-foliations are abundant, the intersections between S-surfaces and axial-plane parallel C-foliations become acute-angled. In places, where intensive non-coaxial shearing has taken place, the upper limbs of recumbent folds have been sheared off along surfaces parallel to their subhorizontal axial planes, giving rise to the formation of mesoscopic thrusts with NE- to NNE-striking slickensides that parallel the mineral stretching lineation. Generally, the thrust planes dip toward the northeast in the northern part and towards the southwest in the southern part of the Arun Tectonic Window. Within the southern limb of the Dhankuta Syncline (Figs 3 & 4b), they again dip to the northeast or north-northeast. The strike-direction of the mesoscopic thrusts parallels both the C-foliations and the Main Central Thrust.

Within the Kathmandu Complex, fold styles show a transition downward towards the MCT plane from SE-trending open F2 folds with vertical or inclined axial planes at the top of the sequence to isoclinal recumbent F2 folds at the base. This rotation is accompanied by a downward increase in shear strain marked by the deformation of K-feldspar crystals described above (Fig. 5).

D3

D3 (D4 of Schwan 1980, 1988) is marked by a NNE–SSW-directed shortening that generated kink bands and chevron folds within phyllitic rocks. The subvertical S3 crenulation cleavage planes invariably have a WNW–ESE strike trend throughout the Arun valley. Most of the F3 structures within the southern part of the mapped area are kink band folds, whereas chevron folds predominate in the northern part. The number of F3 folds clearly decreases northward.

D4

The major domal folds which define the Arun Tectonic Window represent the D4 deformation. Of the two antiformal axes that define the dome one is NNE–SSW-trending and parallels the course of the Arun river. Generally earlier planar structures (sedimentary and tectonic) are more steeply dipping near the margins of the window than in the core (Fig. 2).

Subvertical NNE-striking faults cross-cut the 'Arun Dome'. Bordet (1961) named the most prominent of these as the 'Faille de Khandbari' and considered it to be a thrust fault, which brought the garnetiferous 'Phyllites Inferieurs' (of unit B) into juxtaposition with the chloritic 'Phyllites Superieurs' (of unit A). However, this view conflicts with our mapping which shows the contact between units A and B to be a gradual metamorphic transition. Numerous normal faults, which cut across the metamorphic units, date from a period of WNW–ESE extension which preceded the formation of the Arun Dome. In the western half of the dome, these faults dip ESE, whereas they generally dip WNW in the eastern half. In many sections, the Arun river runs parallel to the strike-direction of these faults. Further north of the mapped area, similar faults have been described by Tapponnier *et al.* (1981).

Metamorphism

Nawakot Complex

M1. The chlorite–biotite-bearing S1 fabric dates from the probably syn-D1 M1 metamorphic event. This bedding plane-parallel schistosity has been overprinted by subsequent Himalayan tectonometamorphic events. These oldest foliations are solely preserved within unit A and the lower parts of unit B, whereas they have been obliterated within the sheared rocks of the upper levels of unit B and within unit C.

M2. Rocks of unit A are demonstrably of low grade greenschist facies. At the base they contain clinozoisite and towards the top pale olive-green biotite, epidote and muscovite. Minor metabasic sills contain blue–green, probably hornblendic, amphiboles. Garnet appears at the base of unit B. The parageneses of the micaschists in unit C are consistent with lower amphibolite facies of metamorphism. Staurolite appears within micaschists at the base of unit C, with abundant kyanite occurring within micaschists and quartzites at the top of the unit. Estimated M2 temperatures, when compared to the variable thicknesses of the three metamorphic units, suggest that there has been a telescoping of metamorphic isograds towards the top of the Nawakot Complex. Many minerals were altered during the late- to post-M2 retrogressive cooling phase. Retrograde reactions included the growth of biotite and titanite after garnet; of Fe-chlorite after biotite; of Fe-chlorite after Mg-chlorite or of chlorite after garnet. Partial pseudomorphic replacement is observed within unit A and at the base of unit B, while complete pseudomorphic replacement predominantly took place at the top of unit B and within unit C. This may imply that the upper part of the Nawakot Complex cooled at a slower rate than the lower portions.

Kathmandu Complex

Brunel & Kienast (1986) demonstrated that metamorphism within the migmatitic gneisses at cordierite–sillimanite grade took place at temperatures of up to 720° C at moderate pressures of 3.5–5 kbar. Sillimanite occurs both as fibrolite bunches and as idiomorphic crystals. The kyanite and sillimanite grade of the granite-gneisses and the upper levels of the augen-gneisses indicates a medium pressure–high temperature metamorphism (Winkler 1979). In these rocks, sillimanite is marginally replaced by retrogressive muscovite and sericite. The basal portions of the augen-gneiss sequence, and the complete succession of the ultramylonitic gneisses, are located within kyanite grade rocks. According to Brunel & Kienast (1986), metamorphic pressures for this tectonic level of between 6–9 kbar are the highest of those re-

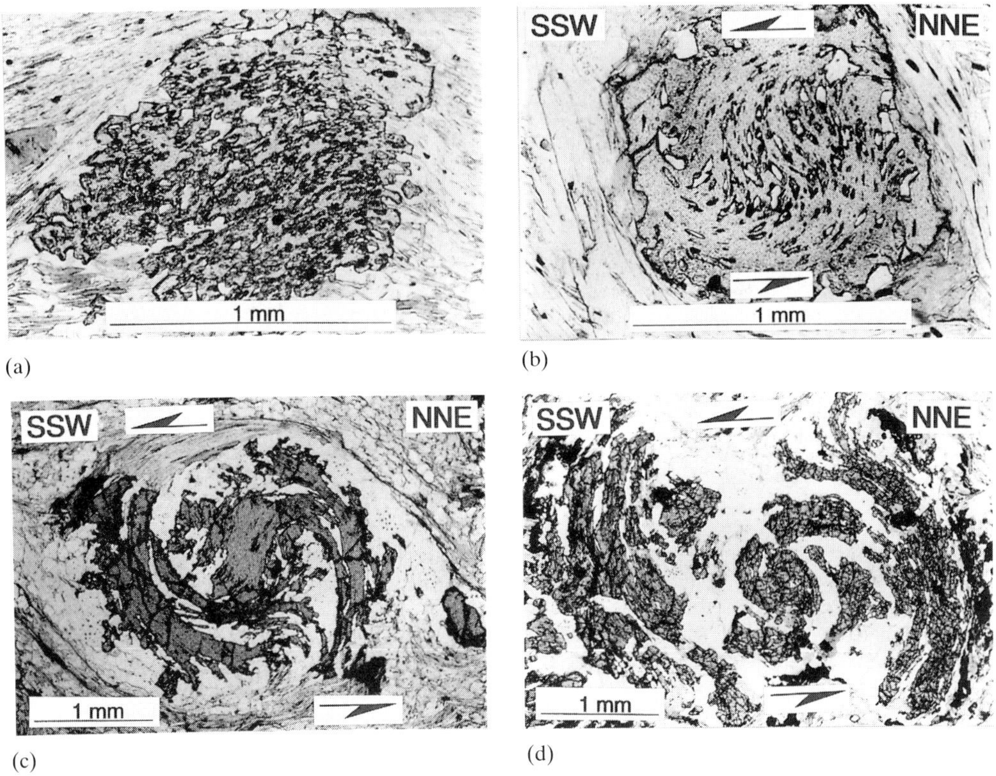

Fig. 6. Increasing rotation of syntectonic garnets upwards through the Nawakot Complex due to non-coaxial shear stress increasing upwards towards the MCT. (**a**) Unrotated syn-D2 garnet at the base of unit B. (**b**) Garnet from the central portions of unit B. This garnet has an angle of rotation of ± 180°, but the speed of rotation was 'slow', because there are no signs of asymmetric pressure shadows (Schonefeld 1977). (**c**) Syn-D2 garnet from the top of unit B, with an angle of rotation of ± 120°. The speed of rotation was 'fast', as the lateral asymmetric pressure shadows are embedded in the internal structure of the garnet (Schonefeld 1977). (**d**) Syn-D2 garnet from the top of unit C, with an angle of rotation of ± 600° and a 'very fast' speed of rotation (Schonefeld 1977) occurring at the top of unit C.

corded within the Higher Himalayan sequence within which pressures decrease upward. Here, sillimanite is completely replaced by muscovite or sericite. Kyanite occurs as deformed and as undeformed varieties. With the exeption of idiomorphic garnets, the garnet fabrics differ completely from those of the Lesser Himalayas, as there occur hypidiomorphic 'ring garnets', poikiloblastic 'atoll garnets', fractured and displaced garnets as well as subidiomorphic 'zoned garnets'.

An increase in the amount of retrograde alteration can be observed within the Kathmandu Complex from the top downward. In the gneisses sericite and clinozoisite replace plagioclase and sericite replaces sillimanite. In the meta-basic rocks green hornblende and calcite replace clinopyroxene; actinolitic hornblende replaces green hornblende; biotite replaces hornblende; chlorite replaces biotite; and epidote replaces clinopyroxene.

Synthesis of the relationship between deformation and metamorphism

Nawakot Complex

M1/D2. Since S1 fabrics are cut by D2 C-surfaces, it is clear that S1 is an older schistosity, presumably of early-Himalayan age (Brunel 1986).

M2/D2. Porphyroblast–fabric relationships clearly show that M2 mineral growth was synchronous with ductile shearing. Garnets are present throughout units A and B. In the weakly sheared, lower parts of unit B they are unrotated or have undergone only little rotation (Fig. 6a).

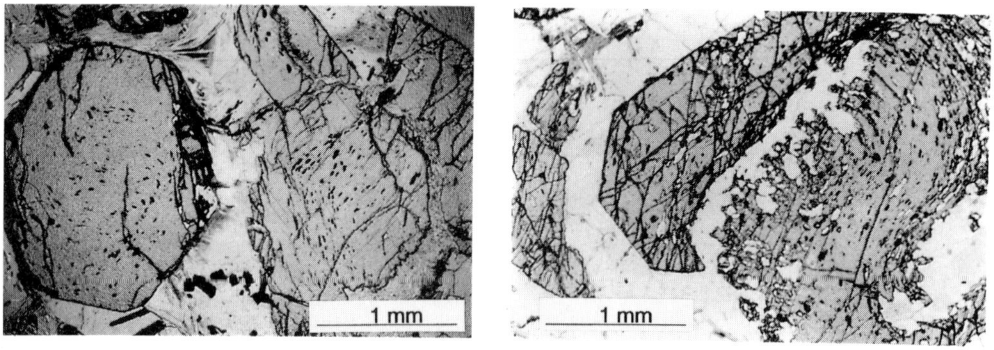

(a) (b)

Fig. 7. (**a**) Post-D2 garnet (left) and staurolite (right) blastesis is documented by the helicitic traces of D2-foliations. (**b**) Polyphase garnet within unit B. The syn-D2 garnet core is surrounded by an idiomorphic post-D2 garnet rim.

'Snow ball garnets' (Spry 1969) within units B and C document an upward increase in shear strain. In the upper levels of unit B the amount of rotation is up to 180° (Fig. 6b, c), whereas at the very top of unit C many garnets have been rotated by up to 600° (Fig. 6d). Based on the findings of Schonefeld (1977), the rotation of the garnets at the top of unit B and within unit C was relatively fast, since their lateral asymmetric pressure shadows were completely embedded during growth (Fig. 6c,d). D2 shearing ended during M2, the peak of which was reached during the static interval between D2 and D3. Helicitic traces of D2 C-fabrics within idiomorphic staurolite, kyanite, plagioclase, garnet and biotite porphyroblasts indicates that growth of these crystals post-dated D2 (Fig. 7a). Often, post-D2 garnet rims overgrew both the syn-D2 garnet cores and their pressure shadows as well as the external fabric (Fig. 7b). Both, the number and the size of the post-D2-kinetic garnets increase toward the top of the Nawakot Complex. Their first occurrence is marked at the base of unit B by sporadic tiny grains, whereas they are abundant and reach diameters of up to 6 cm within unit C.

Kathmandu Complex

M1/D2. High-grade index-minerals, such as K-feldspar and garnets are often cut and displaced along D2 C-surfaces, indicating that they grew during a metamorphic period that pre-dated D2. These crystals often have asymmetric lateral pressure shadows of recrystallized quartzo-feldspathic material. Pre-D2 kyanite-, muscovite- and biotite-flakes were sheared during D2 and are now found as 'mica' (Lister & Snoke 1984) and 'kyanite' fishes.

M2/D2. M2 was a late- to post-D2 metamorphism that was weaker than M1 and which predominantly affected the basal portions of the Kathmandu Complex. During M2 a second generation of kyanite and garnet generation overgrew the D2 C-surfaces. These crystals have no asymmetric lateral pressure shadows or any other signs of ductile deformation. Some D2 garnets overgrew older kyanites. The occurrence of zoned garnets possibly shows that growth took place under two different metamorphic phases.

D2 shear sense criteria within the Nawakot and Kathmandu Complexes

A textural analysis of the megascopic and microscopic strain markers demonstrates the shear sense and the relative intensity of the shear stress (Simpson & Schmid 1983). All of the observed non-coaxial shear sense indicators, such as 'snow ball garnets' (Spry 1969; Fig. 6c,d), 'fish-structures' (Lister & Snoke 1984), 'asymmetric K-feldspar augen' (Berthé *et al.* 1979), 'asymmetric pressure shadows' (Takagi & Ito 1988; Fig. 5b), 'S–C textures' (Lister & Snoke 1984), 'grain boundary sliding' (Drury & Humphries 1988), 'cataclasis' or 'brittle deformation' (Drury & Humphries 1988; Fig. 8), indicate a top to south-southwest shear sense. The shear direction parallels the prominent mineral stretching lineation (Brunel 1986). Shear intensity increases upwards towards the top of the Nawakot Complex as well as downwards to-

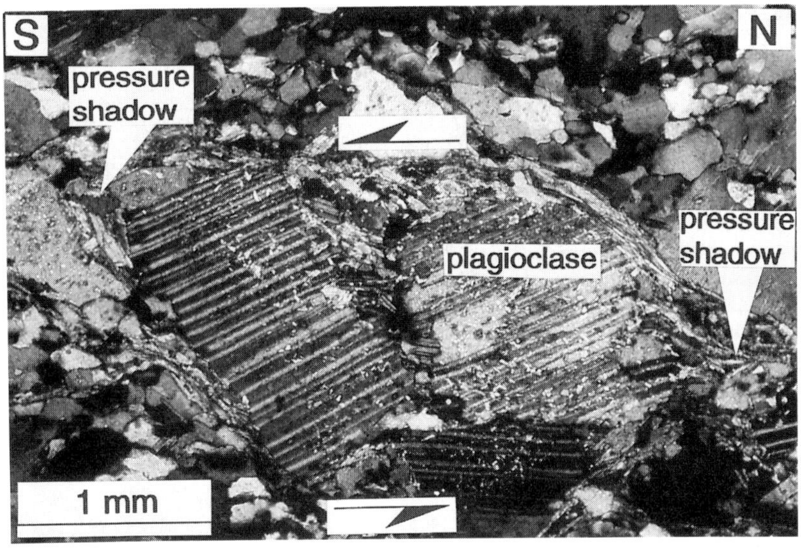

Fig. 8. Brittle deformation ('cataclasis') of a plagioclase in the granite-gneisses of the Kathmandu Complex. The lateral asymmetric pressure shadows originated during syn-D2 shearing.

wards the base of the overlying Kathmandu Complex. The maximum shear strains are found along the sheared contact zone between the two complexes.

D3/M2

The D2 fabrics have been kinked about S3 crenulation cleavages. A study of thin sections from the northern part of the Arun Tectonic Window showed that S3 fabrics have been overgrown by late-M2 retrograde chlorite. Since comparable observations were not made within samples from the southern part, it seems likely that the northern part was still 'warm' when D3 ceased, whereas the southern area was already 'cool', probably as a result of downward heat transfer from the overlying Kathmandu Complex.

D4

D4 was active at a very late stage of the orogeny. The large-scale refolding and faulting of the previous structural fabrics was not accompanied by any metamorphism.

The Main Central Thrust Zone (MCTZ)

The MCTZ, which in the Arun valley is located in the physiogeographic Lesser Himalayas, is the major tectonic feature of the mapped area. It marks the transition from the underthrust metasediments of the Lesser Himalayan Nawakot Complex rocks into the gneisses of the Higher Himalayan Kathmandu Complex. In the authors opinion, the MCTZ is located between the kyanite-micaschists of the Nawakot Complex and the kyanite-gneisses of the Kathmandu Complex. Although there is no significant metamorphic jump across the thrust, the rocks on both sides of the MCTZ have an entirely different tectonometamorphic history. The highest metamorphic grades of the Precambrian Higher Himalayan sequence pre date the underthrusting process (Brunel 1986, Brunel & Kienast 1986), whereas the main metamorphism of the Upper Precambrian to Lower Palaeozoic Lesser Himalayan metasediment pile was syn- to post-kinematic with respect to the underthrusting.

Micaschists of the Nawakot Complex outcrop in small erosional windows up to 100 km north of the southern border of the Kathmandu Complex. Their existence shows that the thrust displacement must have been at least 100 km. The rocks within the MCTZ have been intensly mylonitised within a several hundred meter wide shear zone. There is a gradual increase in both mylonitization and shear intensity towards the top of the Nawakot Complex as well as towards the base of the Kathmandu Complex. The effects of shearing and convergent metamorphic grade blur the lithological distinctions between the Lesser and the Higher Himalayan rocks. It was this transition which caused Bordet (1961) to describe the contact between the 'Mica-schistes' (= unit C of the Nawakot Complex) and the 'Migmatites' (= Kathmandu Complex) as a 'diffuse metamorphic front'. Auden (1935) made comparable observations in the area of the Darjeeling Window (Fig. 1) between the Daling Schist (= unit C of the Nawakot Complex), Darjeeling Thrust (= MCTZ) and Darjeeling Gneiss (= Kathmandu Complex). The deduceable shear sense is consistent with the north-northeasterly directed underthrusting of the Lesser Himalayan metasediments or with the south-southwesterly directed thrusting of the Higher Himalayan gneisses along the MCTZ respectively. The prominent mineral stretching lineation has a remarkably consistent SSW–NNE strike-direction throughout the mapped area that roughly parallels the direction of movement of the Indian Plate (Shackleton & Ries 1984).

In the eastern part of the mapped area, unit C is sandwiched between the mylonitic augen-gneisses of the Kathmandu Complex (Fig. 2). Along cross sections east of Chainpur it can be seen that this lithological repetition is the result of a syn-D2 tectonic imbrication along sub-horizontal shear zones, and not by subvertical post-tectonic faults, as proposed by Bordet (1961).

The origin of inverted metamorphism

In spite of the stratigraphic sequence being the right way up, an increase in metamorphic grade can be noticed from the base towards the top of the Nawakot Complex (Figs 3 & 4b). The dominant influence imposed by the MCTZ on the metamorphic zonation is clearly indicated by the fact that all of the metamorphic isograds are sub-parallel to the thrust zone (Fig. 2). Because the Kathmandu Complex was already at high amphibolite-grade temperatures when D2 shearing began, and because the main phase M2 metamorphism of the Nawakot Complex commenced during D2 shearing, the generation of the inverted metamorphism of the Nawakot Complex is interpreted as a syn- to post-tectonic, thermo-kinematic consequence of the Cenozoic intracontinental underthrusting of the 'cold' Nawakot Complex beneath the 'hot' Kathmandu Complex along the MCTZ (Bird 1978). The reverse metamorphism at the base of the Kathmandu Complex can be explained by the loss of heat during the heat flow into the Lesser Himalaya.

Both sheets suffered strong ductile shearing and mylonitization along the MCTZ and an inverted temperature gradient was created within this zone of high strain (Le Fort 1975). An additional thermal input may have been added by frictional shear heating within the thrust zone. At the juxtaposition of the two slabs a post-D2 thermal equilibrium was reached under kyanite grade conditions. However, as the base of the Nawakot Complex was not affected by heat conduction and metamorphism, uplift and cooling of the orogen must have commenced soon after thrusting and peak metamorphism. This rapid exhumation would also have ensured the freezing in of the inverted metamorphic sequence.

Conclusions

Within the Higher Himalayan Kathmandu Complex, higher amphibolite- to granulite-grades (M1) were reached before D2 thrusting along the Main Central Thrust commenced. By contrast, the Lesser Himalayan Nawakot Complex suffered only a weak M1 metamorphism. During the D2 shearing event, 'cold' metasediments of the Upper Precambrian to Lower Palaeozoic Nawakot Complex were intracontinentally underthrust along the MCTZ beneath the 'hot' Precambrian gneisses of the Kathmandu Complex. The M2 metamorphism of the Nawakot Complex commenced during D2 reaching its peak post-D2. Both shear strain and metamorphic grade converge towards the MCTZ, the transition-zone between the Lesser and Higher Himalayan rocks lying within mylonitic kyanite-grade rocks. All of the mesoscopic and microscopic shear sense indicators parallel the prominent mineral stretching lineation and testify to the north-northeasterly movement of the underthrust sheet.

The Lesser Himalayan Arun Tectonic Window shows a concentric zonation of inverted metamorphic isogrades with lower greenschist facies metasediments at the core of the window, and lower amphibolite facies, meta-pelitic micaschists around its outer edge. The inverted metamorphism (M2) of the Nawakot Complex can be explained by heat flow, during and after D2, across the MCT from the hot hanging wall down into the colder footwall. The loss of heat from the hanging wall gave rise to a strong retrograde overprinting of the base of the Kathmandu Complex. The cooling of the orogen started before the initiation of the shortening along the MBT (= D3) and was faster in the southern part of the mapped area than in the north. Formation of the Arun Dome took place during D4. Doming was associated with extension along NNE-striking subvertical normal faults and was followed by the erosional unroofing of the window.

This paper is dedicated to Professor W. Schwan on the occasion of his 75th birthday. The authors gratefully acknowledge the following people and institutions for their kind support during field work and in the making of this paper: W. Schwan (University of Erlangen-Nürnberg), C.K. Sharma (Water and Energy Commision, Kathmandu), P.S. Saklani (Geol. Dept. Univ. of Dehli), U.B. Upreti (Geol. Dept., University of Kathmandu), K.P. Kaphle (Dept. of Mines, Kathmandu), Konrad-Adenauer-Stiftung e.V. (Bonn), German Academic Exchange Service (Bonn), G. Godizarth and B. Schulz (University of Erlangen-Nürnberg), P.J. Treloar (Kingston University), M.R.W. Johnson and an unnamed reviewer, Temba and Pempasona Sherpa (Hururu), Rinshen Sherpa (Shiruwari), P. Michel and M. Kailides (CSD AG, Berne).

References

AKIBA, C. & AMMA, S. 1973. Arun River Region. *In*: HASHIMOTO, S. (ed.) *Nepal Himalayas*, Hokkaido University, Sapporo, 13–32.

ANDREWS, E. 1985. Stratigraphy of the Sabhaya Khola Region, Sankhuwa Sabha District, eastern Nepal. *Journal of Nepal Geological Society*, **2**, 12–35.

AUDEN, J.B. 1935. Traverses in the Himalaya. *Records of the Geological Survey of India*, **67**, 123–167.

BERTHÉ, D., CHOUKROUNE, P. & JEGOUZO, P. 1979. Ortho-gneiss, mylonite and non-coaxial deformations of granites, the example of the South Armorican Shear Zone. *Journal of Structural Geology*, **1**, 31–42.

BIRD, P. 1978. Initiation of intracontinental Subduction in the Himalaya. *Journal of Geophysical Research*, **80**, 4405–4416.

BORDET, P. 1961. *Récherches Géologiques dans l'Himalaya du Nepal, Région du Makalu*. Edition Centre National Rechèrches Scientifiques.

BRUNEL, M. 1986. Ductile Thrusting in the Himalayas: Shear sense criteria and stretching lineations. *Tectonics*, **5**, 247–265.

—— & KIENAST, J.R. 1986. Étude pétro-structurale des chauvements ductiles himalayens sur la transversale de l'Everest-Makalu (Népal oriental). *Canadian Journal of Earth Sciences*, **23**, 1117–1137.

DRURY, M.R. & HUMPHREYS, F.J. 1988. Microstructural shear criteria associated with grain boundary sliding during ductile deformation. *Journal of Structural Geology*, **10**, 83–89.

FUCHS, G. 1981. Outline of the Geology of the Himalaya. *Mitteilungen der Österreichischen Geologischen Gesellschaft*, **74**, 101–128.

HAGEN, T. 1969. Report on the Geological Survey of Nepal, Vol. 1, Preliminary Reconnaissance. *Denkschriften der Schweizerischen Naturfreunde Gesellschaft*, **86 (1)**, 1–85.

HILTNER, E. 1991. *Untersuchungen zur Geologie und Tektonik des westlichen Aruntales, Koshi Zone, Ostnepal*. PhD Thesis, Universität Erlangen-Nürnberg.

LE FORT, P. 1975. Himalayas: The collided Range. Present knowledge of the continental arc. *American Journal of Science*, **275**, 1–44.

LISTER, G.S. & SNOKE, A.W. 1984. S–C mylonites. *Journal of Structural Geology*, **6**, 617–638.

MARUO, Y. & KIZAKI, K. 1981. Structure and metamorphism in Eastern Himalaya. *In*: SAKLANI, P.S. (ed.) *Metamorphic Tectonites of the Himalaya*, Todays and Tomorrows Printers & Publishers, Delhi, 175–230.

MEIER, K. 1989. *Deformation und inverse Metamorphose im östlichen Aruntal, Ostnepal-Himalaya*. PhD thesis, Univerität Erlangen-Nürnberg.

SCHONEFELD, C. 1977. A study of some typical inclusion patterns in strongly paracrystalline-rotated garnets. *Tectonophysics*, **39**, 453–471.

SCHWAN, W. 1980. *Shortening structures in eastern and northwestern Himalayan rocks*. Todays and Tomorrows Printers & Publishers, Delhi.

—— 1988. Zur Gestaltung und Geschichte der Kompressionsstrukturen im Indischen Himalaya–Fakten und Fragen. *Erlanger Geologische Abhandlungen*, **116**, 123–136.

SHACKLETON, R.M. & RIES, A.C. 1984. The relation between regionally consistent stretching lineations and plate motions. *Journal of Structural Geology*, **2**, 111–117.

SHARMA, C.K. 1990. Geology of Nepal and adjacent areas. Kathmandu, 479 pp.

SIMPSON, C.L. & SCHMID, S.M. 1983. The evaluation of criteria to deduce the sense of movement in sheared rocks. *Geological Society of America Bulletin*, **94**, 1281–1288.

SPRY, A. 1969. The origin and significance of snow ball structures in garnet. *Journal of Petrology*, **4**, 211–222.

STÖCKLIN, J. 1980. Geology of Nepal and its regional frame. *Journal of the Geological Society, London*, **137**, 1–34.

TAKAGI, H. & ITO, M. 1988. The use of asymmetric pressure shadows in mylonites to determine the sense of shear. *Journal of Structural Geology*, **19**, 347–360.

TAPPONNIER, P. *et al.* 1981. The Tibetan side of the India-Eurasia collision. *Nature*, **294**, 405–410.

WINKLER, H.G.F. 1979. *Petrogenesis of Metamorphic Rocks*. Springer Verlag New York.

A study of illite crystallinity and fluid inclusions in the Kathmandu Klippe and the Main Central Thrust zone, Nepal

C.W.K. MORRISON[1,2] & G.J.H. OLIVER[1]

[1] *Geology Division, St Andrews University, St Andrews, Fife, KY16 9ST, UK*

[2] *Halliburton Geodata, 14 Wellheads Terrace, Wellhead Industrial Estate, Dyce, Aberdeen, AB2 0GS, UK*

Abstract: The Kathmandu Klippe is an outlier of the Tibetan crustal slab resting on the Main Central Thrust (MCT); it contains a 20 km thickness of Precambrian and Lower Palaeozoic metasediments and Ordovician granites folded into a large syncline. The regional metamorphism predates granite emplacement and, on the basis of illite crystallinity and index mineralogy, is lowest grade in the core of the syncline (epizone–chlorite zone) and highest on the north and south limbs (garnet zone). The distribution of metamorphic zones is 'normal' for a burial sequence. The klippe rests on the Benighat Slates of the Midland Formations of the Indian crustal slab. Fluid inclusion studies of the syn-MCT quartz veins in the Benighat Slates show H_2O, CO_2, CH_4, Na^+, K^+, Cl′ fluids which were hotter than the country rocks. The MCT contact at Dolalghat does not show the effects of 'reverse' metamorphism expected from the Le Fort model of heat transfer from the hot klippe several kilometres down into the Midland Formations. A study of index minerals, illite crystallinity and illite b_0 indicates that the grade of regional metamorphism within the anticlinal Midland Formations varies between epizone-chlorite and garnet zone with the highest grades occurring in the core of the structure. Further to the N around Barahbise, the contact along the MCT between the Midland Formations and the Tibetan Slab is a 2 km thick zone of highly sheared Benighat Slates. The grade of metamorphism in the Benighat Slates is chlorite-epizone grade i.e. lower than either of the sequences above (kyanite zone) or below (garnet zone). The inference drawn from these relationships is that although the inverse metamorphic zonation in the base of the Tibetan Slab might be explained by cooling against the Indian Slab during the Tertiary Himalayan collision, metamorphism in the Indian plate was largely pre-collision.

This paper reports a study of index minerals, illite crystallinity, illite b_0 and fluid inclusions in metamorphic rocks associated with the Main Central Thrust (MCT) zone in the Kathmandu region of Nepal. Particular attention was paid to the low grade metamorphism of the Lesser Himalayan footwall rocks to the MCT which, it is claimed, were metamorphosed when the 'hot' Tibetan plate was thrust over the 'cold' Indian plate during the Himalayan collision (Le Fort 1975). This gives rise to the famous inverted metamorphic sequence first reported by Oldham in the Kumaun in 1883 (Gansser 1964). We hope to show that the regional metamorphism found in the Lesser Himalayan footwall largely predates the Himalayan collision.

The MCT, extending 2400 km along the length of the Himalayas, represents a major tectonic boundary associated with the collision of India with Asia during the Tertiary. After the initial collision a minimum of 100 km of crustal shortening was taken up by intraplate subduction along the MCT (Hodges *et al.* 1988*a,b*). During the collision the leading edge of the Indian plate (or 'Slab') underthrust the Tibetan plate (or 'Slab'). Though the MCT is still active, intraplate subduction has migrated to the south, and has been centred on the Main Boundary Thrust (MBT) and the Himalayan Frontal Thrust (HFT) since the later Tertiary.

Stratigraphy of Kathmandu region

The geology of the Kathmandu region has been reviewed and revised by Stöcklin (1980) and Stöcklin *et al.* (1980). A simplified geological map of the Kathmandu region is presented in Fig. 1. Cross-sections appropriate to two sampling traverses A (in the E) and B (in the W) are presented in Figs 2 and 3.

Below the MCT

It is confusing that various rocks of the Nawakot Complex (Stöcklin 1980) are also known as the Midland Formations (see Fig. 1 and Pêcher & Le Fort 1986). We will use the latter name. On the basis of stromatolites, conodonts, acritachs and shelly microfauna (Le Fort *et al.* 1986*a*) the age of the Midland formations ranges from Late

From TRELOAR, P.J. & SEARLE, M.P. (eds), *Himalayan Tectonics*
Geological Society Special Publication, No. 74, pp. 525–540.

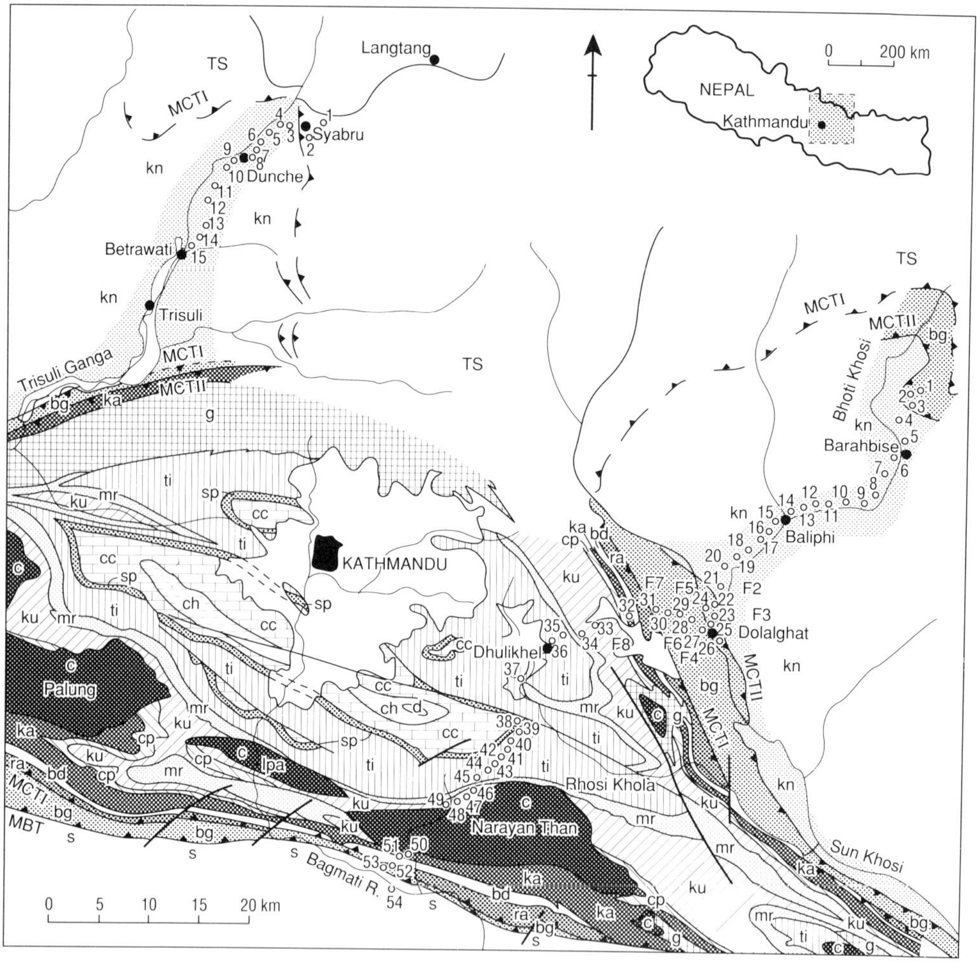

Fig. 1. Geology of the Kathmandu Basin (adapted from Stöcklin 1980; Stöcklin *et al.* 1980). Dark stippled areas (labelled c) are Cambro-Ordovician leucogranites. Lithological units are as follows: Nawakot Complex (Midland Formations): kn, Kuncha Fm.; bg, Benighat Slates; TS, Tibetan Slab (undifferentiated). Kathmandu Complex: ra, Raduwa Fm.; bd, Bhainsedobhan Fm.; ka, Kalitar Fm.; cp, Chisapani Quartzite; ku, Kulikhani Fm.; mr, Markhu Fm.; ti, Tistung Fm.; sp, Sopyang Fm.; cc, Chandragiri Limestone; ch, Chitlang Slates; d, Devonian Limestone; g, undifferentiated garnet, staurolite, kyanite or sillimanite grade schists, gneisses or migmatite equivalents of the Kathmandu Complex. MCTI and MCTII are upper and lower limits of the MCT zone separating the Indian from the Tibetan Slab. MBT, Main Boundary Thrust. Sample traverse A is Barahbise to the Bagmati River (HS sample numbers). Sample traverse B is Syabru to Betrawati (LH sample numbers). FI numbers, fluid inclusion samples.

Precambrian to early Palaeozoic. Correlations have been made with the Krol and Tal Formations of the Lesser Himalaya of Kumaun in India (Brasier & Singh 1987; Bhat 1987). The lower Midlands is dominated by the Kuncha Formation composed of chlorite to garnet grade psammitic and pelitic schists. The Benighat Slates of the upper Midlands were sampled in traverse A. These are dark grey slates and phyllonites, black carbonaceous slates, and calc-phyllites. Benighat Slates were not recognised in traverse B at Sybaru; they are either so tectonically thinned that they were missed or are absent and Tibetan rocks lie directly on Kuncha Formation schists of the Midlands Formations.

Above the MCT

To the north kyanite and sillimanite schists and gneisses of the Bhimphedi Group occur in the

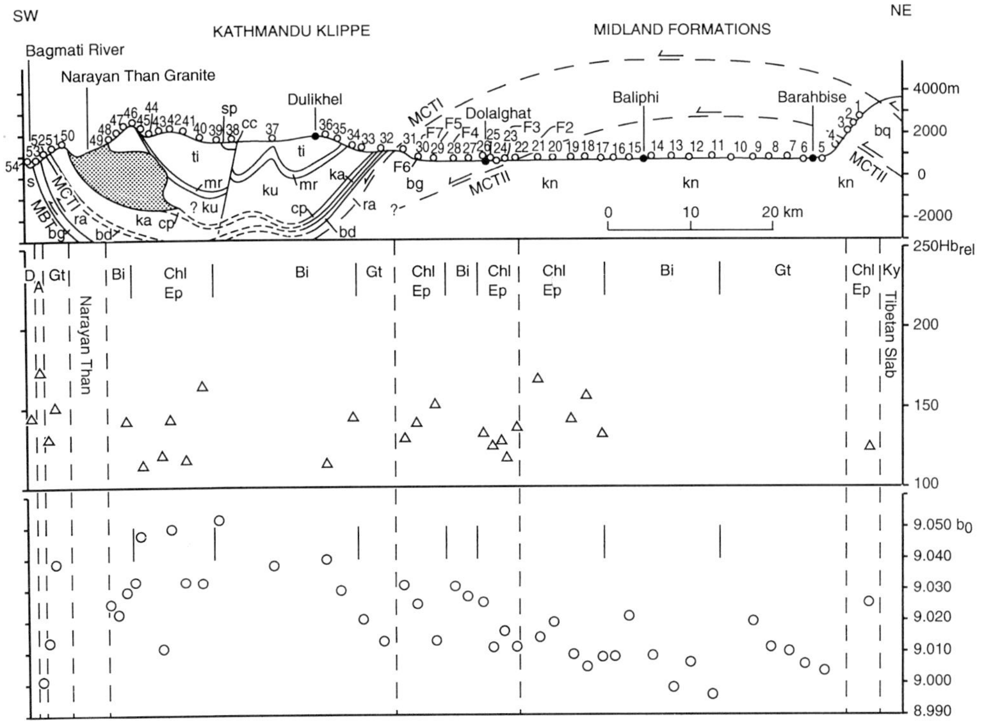

Fig. 2. Traverse A, see Fig. 1, showing geological structure (after Stöcklin 1980; Stöcklin *et al.* 1980) with index minerals, Hb_{rel} and b_o results. For abbreviations see Fig. 1. In addition D, diagenetic zone; A, anchizone; Chl Ep, chlorite-epizone; Bi, biotite zone; Gt, garnet zone; Ky, kyanite zone. Dashed lines indicate lithological boundaries on the sampling traverse. Higher values of Hb_{rel} indicate lower temperatures of metamorphism; higher values of b_o indicate higher pressures of metamorphism (see text for explanation).

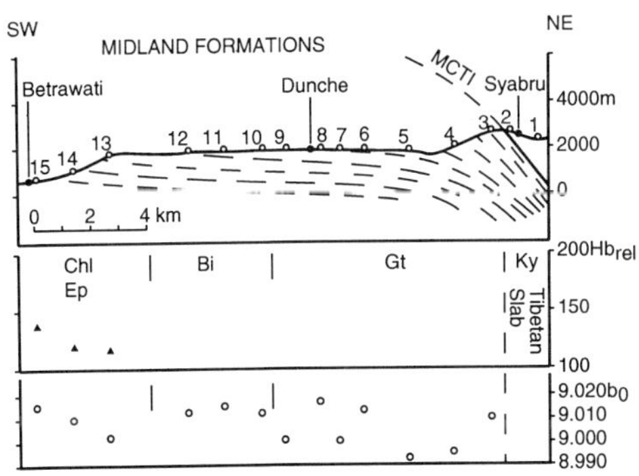

Fig. 3. Traverse B, see Fig. 1, showing geological structure (after Stöcklin 1980) with index minerals, hb_{rel} and b_o results. For abbreviations see Figs 1 & 2.

Tibetan Slab. To the south Tibetan Slab rocks also form the Kathmandu Klippe of the Kathmandu Complex. This complex shows a normal metamorphic succession from the anchizone to garnet zone from the core to the limbs of the Maharabhat Syncline. The Lower Bhimphedi Group is of late Precambrian age. Samples include the garnet schists of the Raduwa Formation, biotite schists of the Kalitar Formation, and garnet schists of the Kulikhani Formation. The Cambrian to Devonian Phulchauki Group, which overlies the Bhimphedi Group conformably, was sampled for biotite schists from the Tistung Formation and mica-bearing marbles of the Chandragiri Limestone. Figures 1 and 2 show that the Kathmandu Klippe is folded into a *c.* 30 km wide synclinorium which complements the *c.* 40 km wide anticlinorium in the Midland Formations just north of the Kathmandu Klippe.

Previous work on regional metamorphism

The thermobarometry of the Nepalese High Himalayas has been well studied: Pecher (1979), Bouchez & Pecher (1981), Arita (1983), Maruo & Kiazaki (1983), Pecher & Le Fort (1986), Pavlis (1986), Le Fort *et al.* (1986*a,b*), Brunel & Kienast (1986), and Hodges *et al.* (1988*a,b*).

Barrovian metamorphism at the base of the Tibetan Slab occurred at 7–8 kbar and 600–700° C and although pressure decreases to *c.* 4 kbar towards the top of the *c.* 12 km thick section the temperature remains constant, perhaps being buffered by the in situ melting of leucogranite, (Hodges *et al.* 1988*a*). In contrast, in the same area, Le Fort *et al.* (1986*a,b*) found no evidence for temperature buffering.

Arita (1983) studied the Annapurna Himal area of central Nepal and analysed pyrope cores in almandine garnets which he associated with high grade Barrovian metamorphism within the Tibetan Slab. Cambrian Rb/Sr whole rock ages for S-type orthogneisses might date the metamorphism. These include 550 ± 16 Ma (Ferrara *et al.* 1983) and 517 ± 62 Ma (Debon *et al.* 1986) for Dudh Kosi gneisses, and 513 ± 30 Ma (Le Fort *et al.* 1986*a*) for Manaslu gneisses. In contrast to the Tibetan Slab, Le Fort (1986) recognises only sub-greenschist grades in the Midland Formations prior to the Tertiary. A second metamorphic event associated with the collision of the Indian and Asian plates took place after 50 Ma. This is believed to have been synchronous with high grade metamorphism of the Tibetan Slab (Le Fort 1986). Arita (1983) associated this second metamorphism with the growth of retrograde almandine margins on pyrope cores in the Tibetan Slab. In the Midland Formations this event may have produced chlorite, biotite and garnet grade metamorphism. Garnets in these rocks have a normal prograde zonation with spessertine-rich cores (Arita 1983; Pecher & Le Fort 1986).

Movement on the MCT produced a third phase of metamorphism, interpreted as involving prograde metamorphism, increasing in grade up the structural succession from chlorite to garnet grade in the Midland Formations (Pecher & Le Fort 1986), and retrogression associated with chloritisation of garnets at the base of the Tibetan Slab (Arita 1983). Latest movements along the MCT zone are associated with brittle shearing.

Inverse metamorphic models

A number of models have been proposed to explain the inverse metamorphic gradient. They can be summarized as follows.

(a) Contact metamorphism from the High Himalayan granites (Auden 1935). Brunel & Kienast (1986) have suggested that the heat source for the kyanite to sillimanite transformation in the High Himalaya was in these granites.

(b) Recumbent folding of Tertiary isograds (Heim & Gansser 1939; Searle *et al.* 1988). Suitable south-verging structures to verify this model have not been mapped in Nepal.

(c) Post-metamorphic thrusting of Tertiary metamorphosed slices (Lal *et al.* 1981; Andrieux *et al.* 1980, 1981; Brunel & Kienast 1986). Johnson & Oliver (1990) proposed that much of the low grade regional metamorphism in the Lower Palaeozoic Kumuan Lesser Himalaya (the footwall of the MCT) was pre-Eocene and pre-Himalayan collision; the collision merely rearranged the metamorphic zones in a thrust stack.

(d) Shear heating during thrusting (Maruo 1979; Boucher & Pecher 1981; Arita 1983; Pavlis 1986). Arita (1983) did not consider that this was an effective heat source, although England *et al.* (1992) do.

(e) Thrusting of the hot Tibetan Slab over the cool Indian Slab (Le Fort 1975). This is perhaps the most popular model; supported by S-shaped isograds formed during rapid syn-metamorphic ductile underthrusting of the Indian Slab under the Tibetan Slab. The speed of underthrusting and uplift prevented re-orientation of the isograds. This model was favoured by Pecher (1979), Stöcklin (1980) and Sinha-Roy (1982). Johnson & Oliver (1990) accepted that reverse metamorphism in the Tibetan Slab could be a Tertiary collision-related event in Kumaun, but doubted that it affected much of the Indian Slab, certainly an order of magnitude less than the 20 km suggested by Searle *et al.* (1988) or the 15 km suggested by Le Fort (1975). Furthermore, Johnson & Oliver (1990) showed that Eocene strata in the Indian Slab were not regionally metamorphosed during the Tertiary collision and therefore the Precambrian and Palaeozoic strata associated with them in thrust slices must have been regionally metamorphosed prior to the Tertiary.

Le Fort's model has been adapated and refined to suit a growing body of data. Pecher & Le Fort (1986) recognized variable tectonometamorphic conditions at different depths on the MCT. Prograde metamorphism was associated with early movement on the MCT. In deep areas initial temperatures were high on both sides of the MCT (600–630° C) and thermal equilibration across a broad zone of inverted isograds was easily achieved. Closer to the thrust front, equilibration to *c.* 500–550° C was achieved by retrogression of the base of the Tibetan Slab and the formation of a narrower zone of inverted isograds under the MCT. Still closer to the thrust front (e.g. the Kathmandu Klippe) hot shearing was replaced by thrusting of cooled gneisses over anchizone to epizone rocks, with a sharper jump in grade. After ductile shearing had finished, shortening in the Indian plate was taken up by brittle fault planes within the MCT zone and below it in the Midland Formations, with the MBT becoming active at this stage.

Position of the MCT

Despite its tectonic importance there has been considerable variation in the choice of the MCT position, see summary of discussions by Searle (1991) and Johnson & Oliver (1991). For the purpose of this study the MCT is taken as lying at the base of the Tibetan Slab (equivalent to the Vaikrita Thrust or MCTI in the Kumaun, see Valdiya 1980). In addition an MCTII is recognised, after Arita (1983), as the base of the zone of thrusting and shearing below the MCTI, and the term MCT zone is used when referring to the general zone of thrusting between the MCTI and MCTII. Note that Arita's (1983) MCT I and II notations have been reversed here so that the MCTI in this paper marks the contact with the Tibetan Slab as in the Kumaun.

Petrography

Thin sections were prepared from pelitic and semipelitic lithologies along traverses A and B (Figs 1, 2 and 3). The occurrence of index minerals such as chlorite, biotite and garnet has been plotted on Figs 2 and 3.

Below the MCTI

On the north side of the Kathmandu Klippe, Kuncha Formation schists and phyllites show an increase in grade across the anticlinal structure from south to north (see Fig. 2). In the north garnet-grade pelites have quartz + albite + garnet + biotite + muscovite ± chlorite ± tourmaline, e.g. samples HS4 to HS10, LH3–9, LH12. One sample, HS18, has garnet porphyroblasts completely pseudomorphed by chlorite. This locality is bounded by the biotite zone to the north and the chlorite zone to the south. This might suggest that the metamorphic zonation has been disrupted by imbricate thrusting. Biotite zone assemblages include quartz + albite + white mica + biotite + chlorite (e.g. HS11 to HS21, LH10–15). In the south, chlorite zone assemblages include quartz + albite + white mica ± chlorite. Retrogressive chlorite is absent from traverse B. Textures indicate a prograde evolution of index minerals. This sequence of zones has been frequently interpreted as part of the zone of inverse metamorphism under the MCT zone (e.g. Arita 1983).

The Benighat Slates (HS1–3, H22–31 and HS53) show a clear jump in grade near Barahbise (Figs 1 & 2) where chlorite schists, carbonate and black carbonaceous slates overlie grey pelitic garnet schists of the Kuncha Formation. Near Dolalghat, green chlorite schists of the Kuncha Formation have a sharp contact with black Benighat Slates (the latter are phyllonites with white mica, ribbon quartz, graphite, carbonate and pyrite). H27 and H28 have biotite as well. Relicts of higher grade mineralogies have not been seen. Pelites of the Benighat Slates appear to have the same index mineralogy as the Kuncha Formation at Dolalghat. However, they have a phyllonite foliation, have more abundant microfolds and are riddled with quartz veins compared to the more schistose Kuncha rocks. Clearly, the Benighat Slates have suffered more intense shearing and Si-infiltration than the Kuncha Formation, a consequence of being situated within the MCT zone. L–S and C–S stretching fabrics indicate south directed thrusting.

Above the MCTI

LH1 and LH2 are samples from near Syabru (Figs 1 and 3) immediately above the MCTI in the hanging wall. LH1 is a garnet–kyanite gneiss with abundant biotite, quartz and feldspar, lesser muscovite, and detrital zircon. There is an absence of retrogressive chlorite. LH2 is an unusual crenulated chlorite–epidote schist containing abundant quartz, lesser muscovite and opaques, and minor biotite and tourmaline.

In the Kathmandu Klippe, distal to the root zone of the MCT, a number of formations were sampled.

Raduwa Formation (HS52). This sample, lying south of the Narayan Than granite and close to the southern limit of the MCTI, is a strongly mylonitic biotite schist. Total recrystallization has occurred with strongly orientated biotite and muscovite crystals, quartz trails and occasional muscovite 'fish'.

Kalitar Formation (HS32, HS50 and HS51). Samples are biotite muscovite schist. HS50 is an augen gneiss, local to the southern margin of the Narayan Than granite. It contains feldspar porphyroblasts in a groundmass of quartz, feldspar, biotite, muscovite, tourmaline and epidote. It may represent part of the tectonised margin of the Narayan Than granite.

Kulikhani Formation (HS33, HS34 and HS46–49). Samples are biotite schist which show little evidence of shearing. Biotite is often very dark brown, and is associated with lesser muscovite and epidote. At higher stratigraphic levels muscovite and chlorite dominate over biotite.

Tistung Formation (HS35–37 and HS39–45). The lower grade biotite and muscovite schists of the Tistung Formation lie higher in the succession. Biotite makes a variable appearance, and minor epidote, tourmaline, calcite and pyrite also occur.

Chadragiri Limestone (HS38). One sample of a massive medium-grained crystalline limestone was sampled. Impurities of detrital quartz and feldspar, and metamorphic opaques and muscovite were present.

Phyllosilicate mineralogy

The <2 μm fraction was examined by XRD for mineralogy, illite crystallinity and illite b_o values. The standard methods and criteria used

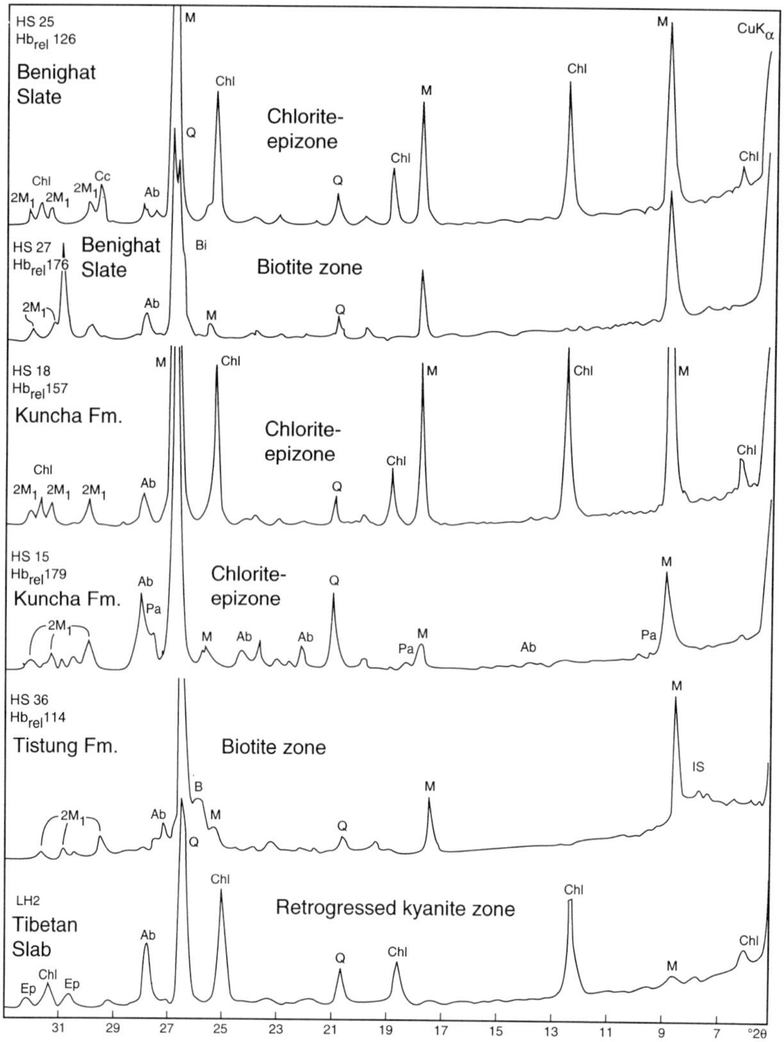

Fig. 4. Representative XRD traces of phyllosilicate mineralogy. Chl, chlorite; M, illite–white mica; Pa, paragonite; Q, quartz; Ab, albite; Ep, epidote; $2M_1$, $2M_1$ muscovite; IS, illite–smectite (due to tropical weathering?); Bi, biotite.

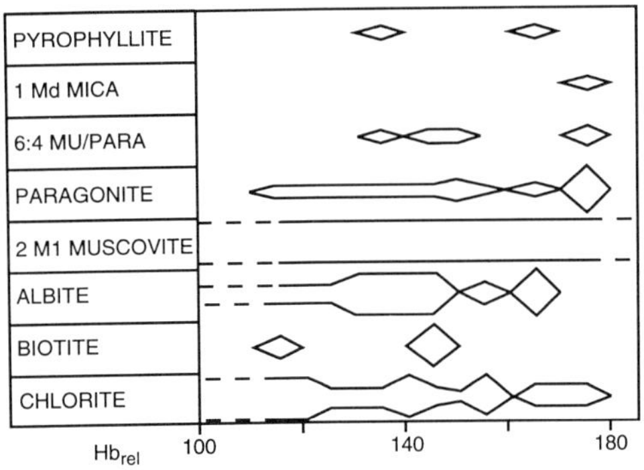

Fig. 5. Mineralogy of the <2 μm size fraction compared to Hb_{rel}. Relative proportions of minerals (shown by width of bars) were visually quantified from the XRD traces as low, medium or high concentration. 6:4 Mu/Para, 6:4 muscovite–paragonite mixed layers (see Frey 1970). These assemblages are typical of the chlorite-epizone (Merriman & Roberts, 1985).

to identify the various minerals are described by Frey (1970), Merriman and Roberts (1985), Roberts *et al.* (1990) and Johnson and Oliver (1990). Representative XRD traces are given in Fig. 4. The mineralogy of the <2 μm size fraction compared to Hb_{rel} is shown on Fig. 5. These assemblages are typical of the chlorite-epizone (Merriman & Roberts 1985).

Below the MCTI

Kuncha Formation. The mineralogy of the Kuncha Formation can be considered in terms of the garnet, biotite and chlorite zones. In the garnet zone the major constituents are $2M_1$ muscovite, albite and quartz. Biotite, chlorite and paragonite are regularly present in small amounts (see Fig. 4, sample HS15). The biotite zone is similar, but has marginally more chlorite and less paragonite. The chlorite zone differs by having greater amounts of chlorite, less albite, and no biotite or garnet (see Fig. 4, sample HS18).

Benighat Slates. The <2 μm mineralogy of the Benighat Slates is dominated by $2M_1$ muscovite and quartz (see Figs 4 & 9). Chlorite and paragonite are usually present in minor amounts. Albite, 6:4 paragonite/muscovite and pyrophyllite occur sporadically. Dolomite and calcite are prominent in some of the samples. Biotite peaks are present in HS27 and HS28 (see Fig. 4). The abundance of $2M_1$ mica in these samples indicates that they are epizonal. HS53, from the most southerly outcrop of the Benighat Slates, contains small amounts of 1Md mica in addition to common $2M_1$ mica, paragonite, chlorite and minor illite/smectite. This sample is assigned as upper anchizonal.

Above the MCTI

The garnet and biotite schists of the Kathmandu Klippe are dominated by $2M_1$ muscovite with more variable amounts of albite, quartz and paragonite. Chlorite is a frequent minor constituent, (see HS36, Fig. 4). Other less common minerals are calcite and dolomite (HS34). The chlorite grade schists contained more abundant albite, less chlorite, paragonite and mixed-layer illite/smectite, and no pyrophyllite or biotite.

Illite crystallinity results

This method of determining metamorphic grade was first described by Kübler (1967). Illite (001) peak widths at half height decrease as stacks of illite 10Å unit cells increase in thickness with increase in metamorphic grade; (see discussion by Merriman & Roberts 1985, and compare the peak widths of the lowest angle white mica peaks of samples HS18 and HS15 in Fig. 4). Kubler (1967) defined the diagenetic zone, anchizone and epizone by setting the anchizone limits at 0.42°2θ (lower) and 0.25°2θ (upper). We follow Weber's (1972) method of standardizing the peak widths relative to a standard quartz (100) peak width:

$$\text{Where } Hb_{rel} = \frac{\text{Width of illite (001) at } \tfrac{1}{2} \text{ height}}{\text{Width of quartz (100) at } \tfrac{1}{2} \text{ height}} \times 100$$

where Hb = Halbertsbreite. This method has the advantage that is allows inter-laboratory comparison because illite peaks are always being compared to a standard quartz peak; instrumental drift is also eliminated. These results were obtained in the same manner as those reported from the Kumaun Lesser Himalayas by Johnson & Oliver (1990). The analyses were made prior to the availability of international standards and before the publication of the guidelines suggested by Kisch (1991). A full description of the St Andrews technique is given in Roberts *et al.* (1990).

Hb_{rel} values decrease with increasing grade. We follow Blenkinsop (1988) and correlated the anchizone boundaries with Hb_{rel} 278 and 147. Hb_{rel} values from samples in which biotite was identified, which has interfered with the 10Å muscovite peak, have been omitted. Results in Hb_{rel}, for the <2 μm size fraction are presented in Figs 2 and 3. From the results the following observations can be made.

(1) Below the MCTII in the Kuncha Formation, metamorphic grade increases steadily northwards towards the MCTII. This is reflected in the limited illite data shown in Fig. 2 on traverses A and B. These show a northwards decrease in Hb_{rel} values in the chlorite zone; values indicate epizonal conditions. North of the chlorite zone the occurrence of biotite and then garnet indicates that grade continues to increase northwards.

(2) The Benighat Slates, between the MCTI and MCTII, outcrop at three localities in traverse A. At Barahbise, proximal to the MCTI root zone, one Hb_{rel} value of 124 was recorded in a muscovite–chlorite schist. This, and the presence of black carbonaceous slates rather than phyllites, indicate that epizonal conditions have not been exceeded. In the Dolalghat area, NE of the Kathmandu Klippe, Hb_{rel} values ranged from 117 to 151, indicating epizone conditions. Hb_{rel} values are relatively uniform along the traverse. The presence of biotite in two samples may reflect variations in grade, or structural imbrication. The most distal occurrence of the Benighat Slates is to the south of the Kathmandu Klippe at the Bagmati River. Slates are bound by the MCTI and MBT. The one sample analysed had a Hb_{rel} of 173 which indicates anchizonal conditions.

Across the three outcrops of the Benighat Slates, which are spaced over 100 km from proximal to distal positions below the MCTI, there is very little change in metamorphic grade. The Benighat Slates reach a maximum grade of chlorite-epizone. Near Barahbise these slates are sandwiched between the MCTI and the garnet grade Kuncha Formation. This implies that the MCTI could not have imposed the inverse zonation on the Kuncha Formation because the grade in the overlying Benighat Slates is lower (see Fig. 2).

(3) Above the MCTI, in the Kathmandu Klippe, the grade of metamorphism decreases upwards through the succession. The Kathmandu Klippe forms a synclinal structure up to 45 km across. In traverse A the grade decreases from garnet grade, immediately above the MCTI, to epizonal conditions in the Tistung Formation, near the core of the syncline (see Fig. 2).

Metamorphism in the Kathmandu Klippe is believed by Stöcklin (1980) to pre-date the emplacement of the Cambro-Ordovician granites of the Lesser Himalaya. It follows that it also pre-dates movement on the MCT zone and folding of the Kathmandu Klippe into a syncline. This evidence is compatible with a decrease in metamorphic grade towards the core of the synclinal Kathmandu Klippe, as described above.

(4) Sample LH1, from immediately above the MCTI in the High Himalaya, lies in the kyanite grade which corresponds to the observation of previous authors (e.g. Le Fort 1975).

b_o spacing results

Sassi & Scolari (1974) showed that the b_o spacing of white mica measured on (060) depends on the rock composition, the celadonite content and the pressure. Kemp *et al.* (1985) showed that as graptolite reflectivity increased and Hb_{rel} decreased with increasing temperature in the Southern Uplands, so the b_o value increased. However, Kemp *et al.* (1985) reported that for the range of the epizone ($Hb_{rel} < 147$ as defined by Blenkinsop (1988), b_o varies only from 9.035 to 9.045. The range in b_o found in epizonal rocks in this study is much larger, i.e. 8.995–9.050. This suggests that the pressure effect on celadonite content is more important than the temperature effect (Guidotti & Sassi 1976). Therefore, provided samples are restricted in composition and are not too different in grade, or temperature, b_o spacing can be a relative measure of pressure. Comparisons of b_o in different metamorphic regimes show that, despite temperature variations, low b_o values correlate with low pressure conditions and high b_o values correlate with high pressures (Fettes *et al.* 1976; Padan *et al.* 1982; Rice *et al.* 1989; and see Fig. 6). b_o spacings were determined on the (060) reflection using the same machine conditions and samples (orientated by sedimentation) used for illite crystallinity (see above). Although peak interference caused by samples being orientated in this way might introduce spurious values (see discussion in Rice *et al.* 1989), the very consistent results of Kemp *et al.* (1985), also analysed at St Andrews, suggest that this is not a practical problem. The quartz (211) peak was used as the internal standard. Two b_o traverses are presented in Figs 2 and 3, and results are included on the cumulative frequency plot (Fig. 6). Biotite-bearing samples are included because the (060) white mica peak is not obscured by a biotite peak. The following observations are noted.

(1) Figures 2 and 3 show consistent trends of b_o results within formations which are compositionally uniform. These trends are continued across boundaries which suggests to us that the compositional differences between

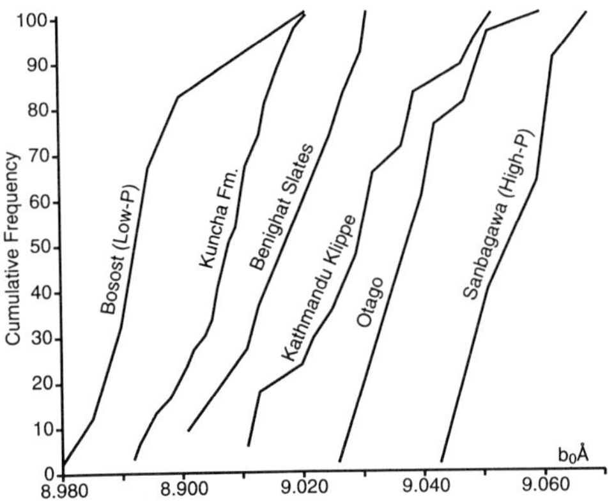

Fig. 6. Cumulative frequency curves of b_o values from Nepal compared with other areas illustrated by Sassi & Scolari (1974), Padan *et al.* (1982). Note the increase in b_o values from the Kuncha Formation, through the Benighat Slates, to the Kathmandu Klippe, i.e. pressure apparently increases up the structural sequence suggesting that the sequence has been stacked by thrusting after metamorphism.

these formations are not big enough to affect the b_o parameter significantly.

(2) A b_o mean value 9.009Å (n = 16, SD = 0.008) in the Kuncha Formation is indicative of low to medium pressure baric conditions intermediate between those found in the Pyrenees (Bosost) and in Otago (see Fig. 6).

(3) The Benighat Slates have a mean b_o of 9.020Å (n = 10, SD = 0.009) suggesting a more medium pressure metamorphism, relatively higher than the Kuncha Formation.

(4) The rocks of the Kathmandu Klippe have a mean b_o of 9.032Å (n = 17, SD = 0.015) similar to Otago Schists (see Fig. 6) and show a trend of increasing b_o towards the core of the syncline contrary to a burial metamorphic sequence (see discussion below). Overall these rocks lie in a relatively medium to high pressure range but higher pressure than either the Kuncha or Benighat Formations.

(5) As a generalization, the rocks of the Klippe are of similar metamorphic grade, or temperature, to those of the Midlands, but the former have illite b_o characteristics that indicate relatively higher pressures than for the latter (see Fig. 6). This does not support one single Himalayan metamorphic event for the Klippe and the Midlands as the pressure relations are apparently inverted.

Previous pressure studies have been limited to the higher grade assemblages of the Tibetan Slab. Hodges *et al.* (1988*a*) show that pressure decreases upwards from 7.2 to 3.8 kbar across 12 km of the succession, without an appreciable change in temperature (*c.* 600° C). Brunel & Kienast (1986) calculated pressures varying from 8 to 5 kbar with temperatures in the range 720–510° C across the lower part of the Tibetan Slab. Brunel & Kienast (1986) also found that pressure–temperature conditions were high in the high grade schists on the north side of the Kathmandu Klippe (6–9 kbar, 550–650° C).

The b_o results from the Midland Formations indicate an increase of pressure of metamorphism up the structural succession in the Dolalghat area (Fig. 3). This cannot be explained by the 'hot slab' model which ought to produce a decrease in pressure up the sequence. In the Barahbise area of Traverse A (Figs 1 & 2) and the Dunche area of Traverse B (Figs 1 & 3) the Benighat Slates do not show significant b_o variation up or down the sequence compared with the neighbouring Kuncha Formation.

Again, it seems that the pressure effects recorded in the rocks are at odds with the 'Hot Slab' model. However, Pecher & Le Fort (1986) have argued that in relatively distal positions, the Kathmandu Klippe would be relatively cool and recrystallisation in the footwall would be restricted to a thin zone. However, this does not explain the jumps in grade across the Benighat Slates. Perhaps more importantly, if Tertiary Himalayan metamorphism below the Kathmandu Klippe is relatively restricted (hundreds of metres thick?) then the kilometres of regionally metamorphosed strata (up to garnet grade) below this must be pre-Himalayan.

Fluid inclusion studies

Fluid inclusions were studied using petrographic, thermometric and extraction techniques on quartz vein material. Thermometric measurements were carried out using a Linkam T_H600 heating and freezing stage following the methods of Touret (1977), Crawford (1981), Roedder (1984) and Shepherd *et al.* (1985). Primary inclusions were used when they had a random distribution and were isolated by at least five times their diameter from the next inclusion. Secondary inclusions were identified as those groups lying along planes. Notes on the types of measurements taken are given in the caption to Table 1. More details concerning the fluid inclusion work can be found in Morrison (1989).

The bulk composition of inclusions was measured by decrepitating quartz separates under vacuum and freezing the volatiles in a liquid nitrogen cold trap. The residual pressure represents non-condensible phases, assumed to be N_2, CH_4, CO and O_2 in that order of

Table 1. *Fluid inclusion results*

Sample	1	2	3	4	5	6	7	7a	7b	8	9	10	11	12
LH1, Tibetan Slab, syn-metamorphic, syn-MCT (?), secondary, necked.														
	A	12.5	–	–	–	–	–	–	–	–	–	–	–	21
F12, Kuncha Fm., post-met, deformed, pop. A = strain orientated, B = undeformed, both secondary. Up to 1 mm.														
A	A	2.0	0.81	9.1	−4.8-L	0.96	253-L	368	485	75.2	1.5	23.3	0.80	18
B	A	5.7	0.92	–	–	–	193-L	303	406	92.4	5.6	2.0	0.98	5
F13, Benighat Slate, post-MCT, undeformed, secondary, up to 250 μm.														
	A	1.5	0.75	0.0	33.3-L	0.50	285-L	408	530	72.1	1.1	26.8	0.65	50
F14, Benighat Slate, post-MCT, undeformed, secondary, up to 150 μm.														
	A	4.0	0.74	–	0.35-L	0.92	287-L	406	531	73.2	3.1	23.7	0.80	25
F15, Benighat Slate, syn-MCT, pop. A = primary, B = secondary, C = CO_2 rich homog to V, *c*. 50 μm.														
A	A	2.0	0.58	14.3	24.3-L	0.71	367-V	523	665	29.6	0.6	69.8	0.35	11
B	A	2.0	0.83	12.0	21.6-L	0.67	240-L	353	467	54.1	1.1	44.8	0.50	11
C	–		–	5.0	26.6-L	–	–	–	–	–	–	–	0.30	1
F16, Benighat Slate, syn-MCT, pop. A = unstrained, B = strained, C = low temperature, D = CO_2 rich, secondary.														
A	–	0.6	0.85	–	26.7-L	0.67	227-L	337	449	91.5	0.5	8.0	0.90	3
B	A	2.8	0.58	4.6	13.1-L	0.85	303-V	433	558	39.8	1.1	59.1	0.50	14
C	–	5.0	0.97	–	21.3-L	0.77	108-L	221	320	91.5	4.8	3.7	0.95	2
D	–	3.9	–	3.0	25.1-L	0.72	–	–	–	9.6	0.4	90.0	0.10	1
F17, Benighat Slate, syn-MCT, pop. A, B = secondary, C = primary, D,E = secondary, <50 μm.														
A	B	2.0	0.80	–	–	–	258-L	376	492	c.90	c.2	c.8	0.90	15
B	B	4.0	0.78	–	−1.1-L	0.94	278-L	395	516	84.7	3.5	11.8	0.90	8
C	B	3.8	0.64	–	–	–	353-L	492	648	c.81	c.3	c.16	0.85	3
D	B	3.8	0.77	–	–	–	284-V	402	524	c.70	c.3	c.27	0.70	3
F18, Kulikhari Fm., post-MCT, pop A, B = secondary, <5 μm.														
A	A	4.5	0.80	2.8	24.1-L	0.73	279-L	396	517	91.1	4.3	4.6	0.95	15

Column 1: Salt system, based on Te, the first melt temperature of H_2O-ice, see Shepherd *et al.* (1985), p. 74. A = H_2O–NaCl–KCl, B = H_2O–KCl.
Column 2: Wt% NaCl equivalent in aqueous solution, from last melt temperature of H_2O-ice and/or clathrate, see Shepherd *et al.* (1985), p. 123.
Column 3: Density of H_2O–NaCl liquid, see Shepherd *et al.* (1985), p. 126.
Column 4: Mol% CH_4 in CO_2 based on temperature of last melt of CO_2-ice, see Shepherd *et al.* (1985), p. 75.
Column 5: Homogenization temperature (C°) for CO_2 inclusion to liquid (L) or vapour (V) Th CO_2.
Column 6: Density of CO_2 phase from homogenization temperature of CO_2 to liquid (L) or vapour (V), see Shepherd *et al.* (1985), p. 75.
Column 7: Average total homogenisation temperature (Th) of the inclusion population to liquid (L) or vapour (V) based on histograms of individual Th, data in Morrison (1989) and summarized on Fig. 7.
Column 7a: Trapping temperature corrected for lithostatic pressure for a depth of 5 km.
Column 7b: same as column 7a for 10 km depth, both assuming rock densities of 2.74 gm cm^{-3} and using wt% NaCl equivalent from column 2, see Potter (1977).

Column 8: $$\text{Wt\% } H_2O = \frac{\text{density } H_2O(\text{col } 3) \times \text{vol } H_2O(\text{col } 11)}{\text{overall density}}$$

overall density = [density CO_2(col 6) × vol CO_2] + [density H_2O(col 3) × vol H_2O].
Column 9: Wt% NaCl equiv. = wt% H_2O(col 8) = wt% NaCl equiv. in aqueous phase (col 2).

Column 10: $$\text{Phase proportion of } CO_2X = \frac{\text{density of } CO_2(\text{col } 6) \times \text{vol } CO_2(\text{col } 11)}{\text{overall density}}$$

Column 11: Degree of fill [$F = vol_{liq}/vol_{liq} + vol_{vap}$] of liquid phase in two phase inclusions, or aqueous phase in three phase inclusions (1.00 = 100%). Qualitative estimate at room temperature using Shepherd *et al.* (1985) pp. 56, 225–226.
Column 12: number of inclusions analysed.

importance (Sheppard *et al.* 1985). The non-condensable phases were analysed on a Carlo-Erba gas chromatograph for CH_4. On removal of the cold trap, first CO_2 and then H_2O were released and their pressures were measured. Finally, the decrepitated grains were leached in H_2O and analysed for Na and K by atomic absorption spectrometry. Details of these methods are given in Batchelor *et al.* (1992); the results are listed in Table 2.

Sampling

Sample descriptions are given in Table 1. Sampling was concentrated on veins in the Benighat Slates in the vicinity of Dolalghat (F13-7) together with one sample from the Kulikhari Formation from the Kathmandu Klippe (F18), one from the Kuncha Formation of the Midland Formations (FI2) and one from the Tibetan Slab from near Dyabru (LH1). Samples in Table 1 are

Table 2. *Fluid extraction results for bulk inclusion populations*

Sample	N–C	CO_2	H_2O	H_2O/CO_2	Atomic Na/K
LH1	1.2	2.1	96.7	46	3.20
F12	0.3	3.0	96.7	32	3.77
F13	0.2	9.0	90.8	10	2.94
F14	3.2*	7.9	88.9	11	3.32
F15	3.2	33.1	63.7	1.9	4.16
F16	3.5	39.8	56.7	1.4	1.70
F17	0.8	4.8	94.4	20	7.34
F18	1.0	2.2	96.8	43	1.39

* CH_4 positively identified by gas chromatography. Values are given as molecular percentages. Non-condensables (N–C) include N_2, CH_4, CO and O_2 in order of likely importance. Atomic Na/K ratio measured by AA analysis.

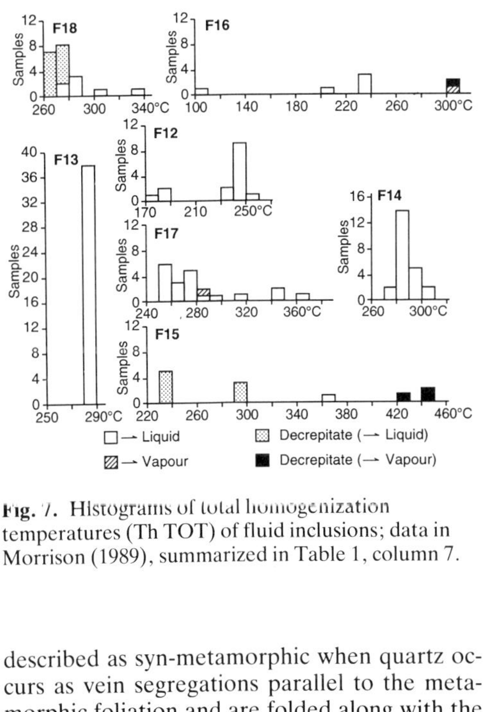

Fig. 7. Histograms of total homogenization temperatures (Th TOT) of fluid inclusions; data in Morrison (1989), summarized in Table 1, column 7.

described as syn-metamorphic when quartz occurs as vein segregations parallel to the metamorphic foliation and are folded along with the foliation. They are described as syn-MCT when quartz segregations lie parallel to (and are folded and/or boudinaged along with) the mylonite foliation within the MCT zone. It is possible that some of the vein material in the MCT zone was early metamorphic and pre-MCT but thoroughly reworked during thrusting. Post-MCT quartz lies in segregations and veins that cross-cut the mylonite foliation and are unfolded.

Fluid inclusion results

In the Benighat Slates, inclusion populations are often very similar in appearance and composition, see Table 1, i.e. F13, F14, F15 and F17B. The highest temperature inclusions homogenize between 300 and 370° C to vapour or liquid. They are low salinity (0.6–0.3 wt% NaCl equiv.) and are rich in CO_2 (*c.* 10–69.8 wt% CO_2). These inclusions are found in syn-ductile MCT zone quartz veins in epizone rocks. Trapping temperatures, when corrected for 5 km of lithostatic pressure, give values of up to 523° C for primary syn-MCT inclusions, (see F14, Table 1), high relative to the peak metamorphic conditions that could be expected in the enclosing chlorite–epizonal grade Benighat Slates. The onset of chlorite-epizone conditions has been estimated at between 270–370° C, with biotite appearing at 425–450° C (Weaver 1984; Kisch 1987). The high trapping temperatures of primary syn-MCT zone quartz veins could be due to the upward migration of hot fluids from deeper in the MCT zone (see Dorobeck 1989). The secondary inclusions in planes homogenised in the temperature range 340–410° C (average = 375° C, after 5 km lithostatic pressure correction, see Table 1). This range is compatible with the chlorite–epizone limits (see above) suggesting that the secondary inclusions are syn-MCT generated. These tend to be low salinity inclusions (1.1–3.1 wt% NaCl equiv.) with significant amounts of CO_2 present (*c.* 8–44.8 wt% CO_2). They are the most common inclusions analysed. Subsequent generations of secondary inclusions mostly homogenize between 200 and 230° C to liquid (two analyses in F16 homogenize at 108° C to liquid), with a mean salinity of 0.5 wt% NaCl equiv., and less CO_2 (2.0–8.0 wt% CO_2).

F12, in the Kuncha Formation has similar inclusion types to the Benighat Slates inclusions (Th TOT of 253° C to liquid, 1.5 wt% NaCl equiv. and 23.3 wt% CO_2, and Th TOT of 193° C to liquid, 5.6 wt% NaCl equiv. and 2.0 wt% CO_2). Above the MCTI, LH1 only provided salinity information (12.5 wt% NaCl equiv.). In F18 secondary inclusions have similar temperatures to the Benighat Slate inclusions (279–316° C, 4.3 wt% NaCl and 4.6 wt% CO_2). From microthermometry there are few differences between inclusions from above and below the MCTI. This is perhaps due to the limited sample population above the MCTI. Salinities are generally low in the Benighat Slates (see Fig. 8).

From extraction data, a plot of mole % H_2O/CO_2 versus atomic Na/K (see Fig. 9) effectively separates the different sample populations on the basis of H_2O/CO_2. Samples from the

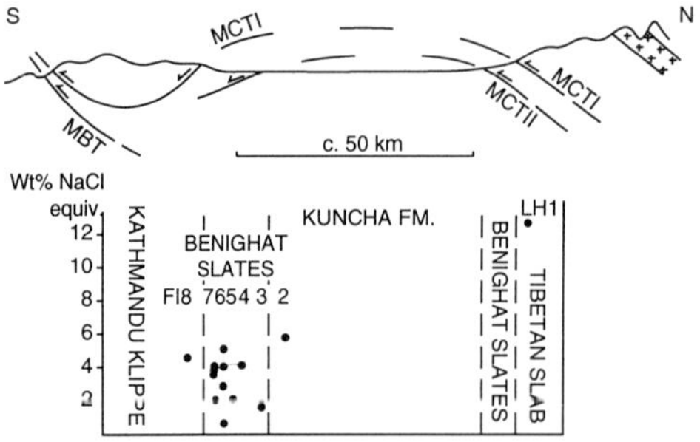

Fig. 8. Sketch cross-section (for explanation see Fig. 2) illustrating variation in salinities (wt% NaCl equivalent from Table 1) of fluid inclusions from Nepal. Note the relatively high value in the Tibetan Slab compared to the Benighat Slates.

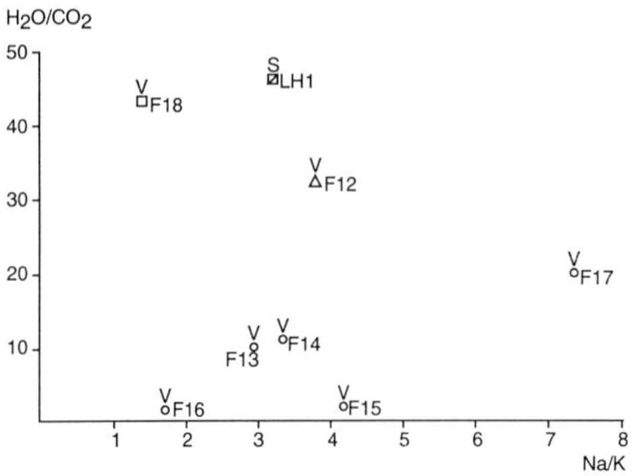

Fig. 9. Chemistry of fluids extracted from bulk quartz samples. V, vein quartz; S, quartz segregation. Circles, Benighat Slates; triangle, Kuncha Fm; square, Kathmandu Klippe; half-filled square, Tibetan Slab. Note that H_2O/CO_2 ratios increase from the Benighat Slates up into the Tibetan Slab.

Benighat Slates are richer in CO_2 than F12 (the Kuncha Formation) which is in turn richer in CO_2 than samples from above the MCTI. Much movement related to the MCT zone has been taken up by the Benighat Slates. It seems possible that shear movement has channelled CO_2 into the MCT zone. Pecher (1979), in studies in the Annapurna and Manaslu areas, found a close correlation between high CO_2 concentrations in inclusions and proximity to the MCTI (i.e. intensity of shearing). The highest CO_2 concentrations occur in the lower part of the Tibetan Slab and the upper Midlands Formations which contains the Benighat Slates. Significantly, the latter contains graphitic phyllonites and carbonate horizons. Sample F14 contains significant CH_4 presumably derived from organic sources.

There is an overall trend of greater wt% CO_2 in inclusions with higher homogenisation temperatures (compare Tables 1 and 2). This may reflect a greater abundance of CO_2 in the earlier (higher temperature) fluids circulating and precipitating vein quartz. Salinities are irregular and low. The Na/K ratio shows limited variation except in F17, close under the MCTI, which is richer in Na.

It is concluded that carbonate- and graphite-rich Benighat Slates are a potential source for CO_2 rich fluids. Movement in the MCT zone, within the Benighat Slates, permitted the upwards migration and channelling of hot CO_2-rich fluids from below. These fluids were around 523° C at times during syn- to post-peak metamorphic shearing, significantly higher temperature than the country rocks as indicated by index minerals and phyllosilicates (<450° C, see below).

Discussion

The key to explaining the inverse metamorphism is the constraining of the relative timing of events in the Himalaya. Unfortunately there is still much uncertainty over this, particularly concerning the timing and nature of metamorphic events below the MCT. The Tibetan Slab was subjected to Cambrian or earlier medium to high grade metamorphism (e.g. Debon *et al.* 1986). This has been recognised from dated Cambrian orthogneisses in the High Himalaya, (Ferrara *et al.* 1983; Brunel & Kienast 1986; Debon *et al.* 1986; Le Fort *et al.* 1986*a*) and from chlorite- to garnet-grade rocks in the Kathmandu Klippe, intruded by dated Cambro-Ordovician leucogranites (Le Fort *et al.* 1983; Debon *et al.* 1986). Further metamorphic events have been related to the collision of India with Asia and are discussed below.

The evidence for pre-MCT zone metamorphism of the Midland Formations has been described above; the distribution of b_o is difficult to reconcile with the hot slab model. Below the MCTI the white mica b_o pressure indicator suggests that pressure is increasing up the structural succession: the hot slab model predicts the opposite. The distribution of index minerals (see Fig. 2) shows a general decrease in temperature up the structural succession towards the MCTI: the hot slab model predicts the opposite. Further evidence can be seen in the Kumaun Himalaya where Johnson & Oliver (1990), using illite crystallinity and b_o spacing, recognized differences in metamorphic grade, and showed that pre-Eocene epizone metamorphism occurred in the equivalent of the Midland Formations. A similar situation is reported from the Tashigang section of East Bhutan (Gansser 1991) where there is a marked jump from lower greenschist facies below, to amphibolite facies above the MCT. Here, rocks along the MCT trace are composed of 'practically unmetamorphosed, banded, siliceous, carbonates' containing Middle to Upper Jurassic palynomorphs (Pantic *et al.* 1981). With this data, Gansser (1991) asks: 'how can one reconcile the hot-iron theory?'

The grade of metamorphism observed in the Benighat Slates directly conflicts with the evidence for the inverse metamorphism in the Midland Formations resulting from overthrusting by the hot Tibetan Slab. The Benighat Slates are confined to the epizone in both proximal and distal positions with respect to the root zone of the MCTI. Since the Benighat Slates are sandwiched between the MCTI and the Kuncha Formation, this precludes heat, either derived by conduction from above or by shearing from within the MCT zone, inversely metamorphosing the Kuncha Formation to garnet grade or higher. This distribution of metamorphic isograds might most easily be explained by the post-metamorphic stacking of Himalayan thrust sheets. Johnson & Oliver (1990) found that the Kumaun inverse metamorphism from biotite to kyanite grade, was restricted to a 1300 m thick zone between the MCTII and MCTI and it should be noted that authors working in other parts of Nepal, adjacent to the High Himalaya, have found a similar zone of inverse metamorphism between the MCTI and MCTII. For example, Hubbard (1989) found this zone to be 3 to 5 km thick and separating kyanite- or sillimanite-grade hanging-wall rocks from garnet- or biotite-grade footwall rocks. It could be that the inverse metamorphism in the MCT hanging wall was caused by cooling from the footwall but that temperatures in the footwall did not exceed and overprint the pre-Himalayan parageneses. However, none of the estimates for the thickness of the zone of Tertiary metamorphism in the footwall to the MCT given in either this paper, Johnson & Oliver (1990) or Gansser (1991) are anywhere near the *c.* 20 km figured by Searle *et al.* (1988) in their folded isograd model or the *c.* 15 km figured by Le Fort's hot slab model (1975, 1988). Metamorphism and tectonism associated with the MCT zone in central Nepal is apparantly intermediate in character to that found in East Bhutan, where it is very thin (*c.* 100 m), and West Nepal and Kumaun, where it is kilometres thick.

Arita (1983) pointed out that the Miocene and Pliocene K/Ar ages for Midland Formation muscovite and biotite close to the MCT

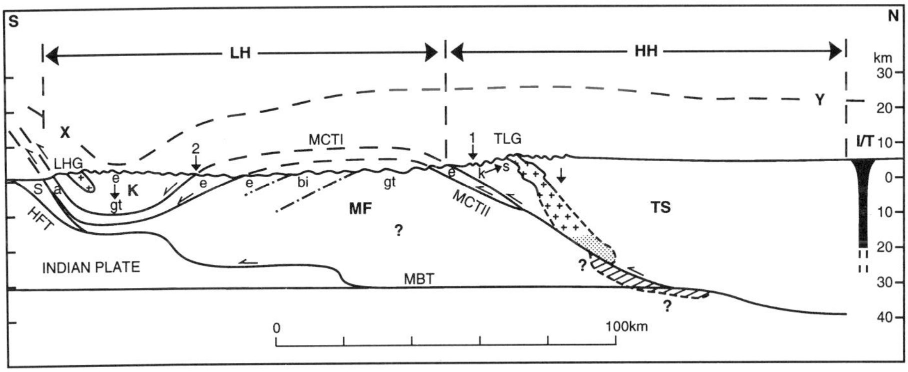

Fig. 10. Metamorphic model and cross-section for the Himalayas of the Kathmandu region, Nepal (partly after Brunel & Kienast 1986). Abbreviations: LH, Lesser Himalaya; HH, Higher Himalaya; S, Siwaliks; K, Kathmandu Klippe; MF, Midland Formations; TS, Tibetan Slab; I/T, Indus/Tsangpo Suture; MCTI & II, Main Central Thrusts I & II; MBT, Main Boundary Thrust; HFT, Himalayan Frontal Thrust; TLG, Tourmaline Leucogranite of the HH; LHG, Lesser Himalayan Granites; a, anchizone; e, chlorite epizone; bi, biotite zone; gt, garnet zone; k, kyanite zone, s, sillimanite zone. Dashed line XY represents depth of cover removed over the last 25 Ma. Arrow 1 points to the proximal MCT zone with inverse metamorphism in the TS; arrow 2 points to more distal MCT zone; neither shows a continuous increase in metamorphic grade up the structure. Note that in this model, metamorphic zones in the MF are Precambrian or Palaeozoic and dip southwards and are unrelated to the northerly dipping Tertiary isograds in the TS. Alternatively, these zones could be thrust bounded and related to Himalayan thrust stacking within the Midlands Formations. TLG's could have originated either by subduction and fluid release in the footwall of the MCT (Le Fort 1975, diagonal shading) or by fluid release and melting in the hanging wall of the MCT (Johnson & Oliver 1990, stippled shading).

measured by Krummenacher *et al.* (1978), indicate that closing temperatures of *c*. 350° C (see Purdy & Jager 1976) were exceeded during MCT thrusting. Without detailed field and petrographic study, this metamorphism, up to garnet grade, might be hard to distinguish from an earlier similar grade regional metamorphism dating from the Palaeozoic or Precambrian.

Figure 10 is a cartoon model for the present-day distribution of metamorphism in the Kathmandu region of the Himalya. Note that in this model, metamorphic zones in the Midland Formations are Precambrian or Palaeozoic and dip southwards, being folded just like the MCT structure, but are unrelated to the northerly dipping Tertiary isograds in the Tibetan Slab. Alternatively, these pre-Himalayan metamorphic zones could be thrust bounded and related to Tertiary Himalayan thrust stacking within the Midland Formations. Clearly, there is a need for a geochronological study of the MCT footwall. Tourmaline leucogranites could have originated either by Tertiary subduction of continental crust and fluid transfer during progressive metamorphism from the footwall of the MCT into the hanging wall, causing melting in metasediments (Le Fort 1975) or by fluid release and melting within the hanging wall of the MCT alone, since much of the fluid in the footwall would have already been lost during the pre-Tertiary metamorphism (Johnson & Oliver 1990).

Conclusions

(1) Normal-way-up Barrovian regional metamorphism in the Kathmandu Klippe of the Tibetan Slab pre-dates Ordovician granites.

(2) The contact between the Tibetan and Indian Slabs around Barahbise is a *c*. 2.5 km thick zone of phyllonitic Benighat Slates of the proximal MCT zone. On the basis of index minerals, phyllosilicates, illite crystallinity and b_o, the grade of metamorphism is chlorite–epizone, lower than either of the sequences above (kyanite zone) or below (garnet zone). This metamorphic sandwich was presumably formed by the stacking and extreme stretching of thrust slices during the Tertiary Himalayan collision.

(3) Fluid-inclusion studies suggest that syn-ductile MCT quartz veining in the Benighat Slates might have formed from relatively hot fluid, (i.e. hotter than the country rocks) that migrated up from deeper in the MCT zone. This fluid could be an 'extra' heat source and should be taken into account in calculations concerning the heat budget in the MCT zone (see Dorobeck 1989).

(4) Between Dolalghat and Dulikhel, a more distal MCT contact between the Kathmandu Klippe and the Midland Formations, there are jumps in metamorphic grade up the structure from chlorite-epizone to biotite zone, and from chlorite-epizone to garnet zone, from N to S across the MCT zone. In between these isograds, b_o results indicate an apparent increase in pressure up the structural sequence. Therefore, there is not a simple pattern of inverse metamorphism below the MCT zone in the Indian Slab. Perhaps Tertiary imbricate stacking can explain the metamorphic pattern.

(5) Although the inverse metamorphic zonation in the Tibetan Slab in the root zone of the MCT might be explained by cooling against the Indian Slab during the Tertiary Himalayan collision, this may not have affected the formation of isograds in the bulk of the Indian Slab which were largely a Precambrian–Palaeozoic development, redistributed during Tertiary thrusting. A similar senario has been proposed for the Kumaun Lesser Himalaya, (Johnson & Oliver 1990).

(6) The normal Barrovian metamorphic sequence found in the Kathmandu Klippe, a distal far travelled, lower metamorphic grade (when compared with the root zone) piece of Tibetan Slab, with a lack of overprinted inverted isograds, suggests that the Klippe was not significantly heated during more than 100 km of Tertiary thrusting.

C.W.K.M. thanks NERC for a PhD Research Studentship and M. Gillespie for field assistance. R.A. Batchelor carried out the fluid extraction experiments. J. Mykura, J. Allen and R. McCormack are thanked for draughting, photography and typing respectively. The paper has benefited greatly through editorial guidance from P.J. Treloar and perceptive reviews by D. Robinson and an anonymous referee.

References

ANDRIEUX, J., ARTHAUD, F., BRUNEL, M. & SAUNIAC, S. 1980. Le caricature postmétamorphique des grandes chevauchements himalayen dans l'Himalaya du Nord-Ouest. Aperçu sur les relations possible avec le metamorphism inverse du Nepal. *Compes Rendus Acadamie Science, Paris,* **291**, 525–528.

——, ——, —— & —— 1981. Geometrie et cinematique des chevauchments, end Himalaya du Nord-Ouest, *Bulleten Societe Geologique de France*, **23**, 651–661.

APRAHAMIAN J. & PARIS, J.L. 1981. Very low grade metamorphism with a reverse gradient induced by an overthrust in Haute-Savoie (France). *In*: McClay, K.R. & Price, N.J. (eds) *Thrust and Nappe Tectonics*. Geological Society, London, Special Publications, **9**, 159–165.

Arita, K. 1983. Origin of the inverted metamorphism of the Lower Himalayas, Central Nepal. *Tectonophysics*, **95**, 43–60.

Auden, J.B. 1935. Traverse in the Himalaya. *Records Geological Survey, India*, **69**, 123–167.

Batchelor, R.A., Armstrong, D.C. & McDonald, M. 1992. Composition of fluids in quartz: discrimination of magma pulses in Caledonian granitoid. *Mineralogical Magazine*, **56**, 335–342.

Bhat, M.I. 1987. Spasmodic rift reaction and its role in the pre-orogenic evolution of the Himalayan region. *Tectonophysics*, **134**, 103–127.

Blenkinsop, T.G. 1988. Definition of low-grade metamorphic zones using illite crystallinity. *Journal Metamorphic Geology*, **6**, 623–636.

Bouchez, U.L. & Pecher, A. 1981. The Himalayan main thrust pile and its quartz-rich tectonites in Central Nepal. *Tectonophysics*, **78**, 23–50.

Brasier, M.D. & Singh, H. 1987. Microfossils and Precambrian-Cambrian boundary stratigraphy at Maldeota, Lesser Himalaya. *Geological Magazine*, **124**, 323–345.

Brunel, M. & Kienast, J. 1986. Etude petrostructurale des chevauchements ductile himalayens sur la transversale de l'Everest-Makalu (Nepal oriental). *Canadian Journal Earth Science*, **23**, 1117–1137.

Crawford, M.L. 1981. Fluid inclusions in metamorphic rocks – Low to medium grade. *In*: Hollister, L.S. & Crawford, M.L. (eds) *Short Course in Fluid Inclusions: Applications to Petrology*. Mineralogical Association Canada.

Debon, F., Le Fort, P., Sheppard, S.M.F. & Sonet, J. 1986. The four plutonic belts of the Transhimalaya-Himalaya: a chemical mineralogical, isotopic and chronological synthesis along a Tibet-Nepal section. *Journal of Petrology*, **27**, part 1, 219–250.

Dorobeck, S. 1989. Migration of orogenic fluids through the Siluro-Devonian Helderberg Group during late Palaeozoic deformation: constraints on fluid sources and implications for the thermal histories of sedimentary basins. *Tectonophysics*, **159**, 25–45.

England, P.C., Le Fort, P., Molnar, P. & Pecher, A. 1992. Heat sources for Tertiary metamorphism and anatexis in the Annapurna-Manaslu region, Central Nepal. *Journal of Geophysical Research*, **97**, 2107–2128.

Ferrara, G., Lombardo, B. & Tonarini, S. 1983. Rb/Sr geochronology of granites and gneisses from the Mount Everest Region, Nepal Himalaya. *Geologish Rundschau*, **72**, 119–136.

Fettes, D.J., Graham, C.M., Sassi, F.P. & Scolari, A. 1976. The basal spacing series variation across the Caledonides. *Scottish Journal of Geology*, **12**, 227–236.

Frey, M. 1970. A mixed-layer paragonite/phengite of low-grade metamorphic origin. *Contributions to Mineralogy & Petrology*, **24**, 63–65.

Gansser, A. 1964. *Geology of the Himalayas*, Interscience, London.

—— 1991. Facts and theories on the Himalayas. *Eclogae geologica Helvetica*, **84**, 33–59.

Guidotti, C.V. & Sassi, F.P. 1976. Muscovite as a petrogenetic indicator mineral in metamorphosed pelites and semi-pelites *Neues Jahrbuch fur Mineralogie Ab.*, **27**, 97–142.

Heim, A. & Gansser, A. 1935 (1975). *Central Himalaya. Geological observations of the Swiss expedition*. Societe Helvetique Science Naturelle Memoire, Zurich **73(1)**. Reprinted: Hindustan Publ. Corp., Delhi.

Hodges, K.V., Le Fort, P. & Pecher, A. 1988*a*. Possible thermal buffering by crustal anatexis in collision orogens: Thermobarometric evidence from the Nepalese Himalaya. *Geology*, **16**, 707–710.

——, Hubbard, M.S. & Silverberg, D.S. 1988*b*. Metamorphic constraints on the thermal evolution of the central Himalayan Orogen. *Philosophical Transactions Royal Society London*, **A326**, 257–280.

Hubbard, M.S. 1989. Thermobarometric constraints on the thermal history of the Main Central Thrust Zone and the Tibetan Slab, eastern Nepal Himalaya. *Tectonics*, **10**, 287–298.

Johnson, M.R.W. & Oliver, G.J.H. 1990. Precollision and postcollision thermal events in the Himalaya. *Geology*, **18**, 753–756.

—— & —— 1991. Comment and Reply on "Precollision and postcollision thermal events in the Himalaya". *Geology*, **19**, 666–667.

Kemp, A.E.S., Oliver, G.J.H. & Baldwin, J.R. 1985. Low grade metamorphism and accretionary tectonics. Southern Uplands terrain, Scotland. *Mineralogical Magazine* **49**, 335–344.

Kisch, H.J. 1987. Correlation between indicators of very low-grade metamorphism. *In*: Frey, M. (ed.) *Low Temperature Metamorphism*. Blackie, Glasgow, 227–303.

—— 1991. Illite crystallinity-recommendations on sample preparation, X-ray diffraction settings, and interlaboratory samples. *Journal of Metamorphic Geology*, **9**, 665–670.

Krummenacher, D., Bassett, A.M., Kingery, F.A. & Layne, H.F. 1978. Petrology, metamorphism and K/Ar age determinations in eastern Nepal. *In*: Saklani, P.S. (ed.) *Tectonic Geology of the Himalaya*. Today and Tomorrow's Printers and Publishers, New Delhi, 151–166.

Kubler, B. 1967. La crystallinite de l'illite et les zones tout a fait superieurs du metamorphisme. *In*: *Etages Tectoniquues*. Ala Baconniere, Neuchatel (Suisse), 105–121.

Lal, R.K., Mukerji, S. & Ackermand, C. 1981. Deformation and Barrovian metamorphism at Takdah, Darjeeling (eastern Himalaya). *In*: Saklani, P.S. (ed.) *Metamorphic tectonites of the Himalaya*. Today and Tomorrows Printers and Publishers, New Delhi, 231–278.

Le Fort, P. 1975. Himalayas: the collided range. Present knowledge of the continental arc. *American Journal of Science*, **275A**, 1–44.

—— 1986. Metamorphism and magmatism during the Himalayan collision. *In*: COWARD, M.P. & RIES, A.C. (eds) *Collision Tectonics*. Geological Society, London, Special Publications, **19**, 159–172.
—— 1988. Granites in the tectonic evolution of the Himalaya, Karakoram and southern Tibet. *Philosophical Transactions of the Royal Society of London*, **A326**, 281–299.
——, DEBON, F. & SONET, J. 1983. The Lower Palaeozoic "Lesser Himalayan" Granitic Belt: emphasis on the Simchar pluton of Central Nepal. *In*: SHAMS, F.A. (ed.) *Granites of Himalaya, Karakorum and Hindu Kush*. Punjab University, Lahore, Pakistan, 235–255.
——, ——, PECHER, A., SONET, J. & VIDAL, P. 1986a. The 500 Ma magmatic event in the Alpine Southern Asia: a thermal episode at Gondwana scale. *Science de la Terre, Memoire*, **47**, 191–209.
——, PECHER, A. & UPRETI, B.N. 1986b. A section through the Tibetan Slab in Central Nepal (Kali Gandaki Valley): Mineral chemistry and thermobarometry of the Main Central Thrust zone. *Science de la Terre, Memoire*, **47**, 211–228.
MARUO, Y. 1979. Geology and metamorphism of the Nanda Devi region, Kumaun Higher Himalaya, India. *Himalayan Geology*, **9(1)**, 3–17.
—— & KIAZAKI, K. 1983. Thermal structure in the nappes of Eastern Nepal Himalayas. *In*: SHAMS, F.A. (ed.) *Granites of Himalaya, Karakoram and Hindu Kush*. Punjab University, Lahore, Pakistan, 271–285.
MERRIMAN, R.J. & ROBERTS, B. 1985. A survey of white mica crystallinity and polytypes in pelitic rocks of Snowdonia and Llyn, N. Wales. *Mineralogical Magazine*, **49**, 305–319.
MORRISON, C.W.K. 1989. *A study of the anchizone-epizone transition*. PhD Thesis, St Andrews University, Scotland.
PADAN, A., KISCH, H.J. & SHARGAM, R. 1982. Use of lattice parameter b_o of dioctahedral illite/muscovite for the characterisation of P/T gradients of incipient metamorphism. *Contributions to Mineralogy & Petrology*, **79**, 85–95.
PANTIC, N., HUCHULI, P.A. & GANSSER, A. 1981. Jurassic palynomorphs below the main central thrust of East Bhutan (Himalayas). *Eclogae geologica Helvetica*, **74**, 883–892.
PAVLIS, T.L. 1986. The role of strain heating in the evolution of megathrusts. *Journal of Geophysical Research*, **91**, B12, 12407–12422.
PECHER, A. 1979. Les inclusions fluides des quartz d'exsudation de la zone du MCT himalayen au Népal central: données sur la phase fluides dans une grande zone de ciasaillement crustal. *Bulleten de Minéralogie*, **102**, 537–554.
—— & LE FORT, P. 1986. The metamorphism in the central Himalaya, its relation with tectonics. *Science de la Terre, Memoire*, **47**, 285–309.
POTTER II, R.W. 1977. Pressure corrections for fluid-inclusion homogenisation temperatures based on the volumetric properties of the system NaCl-H_2O. *United States Geological Survey Resource Journal*, **5**, 603–607.
PURDY, J.W. & JAGER, E. 1976. *K-Ar ages on rock forming minerals from the Central Alps*. Mineralogical Memoirs of the Institute of Geology, University of Padua, 30.
RICE, A.N.H., BEVINS, R.E., ROBINSON, D. & ROBERTS, D. 1989. Evolution of low-grade metamorphic zones in the Caledonides of Finnmark, North Norway. *In*: GAYER, R.A. (ed.) *The Caledonide Geology of Scandinavia*. Graham & Trottman, London, 172–192.
ROBERTS, B., MORRISON, C.W.K. & HIRONS, S. 1990. Low grade metamorphism of the Manx Group, Isle of Man: a comparative study of white mica 'crystallinity' techniques. *Journal of the Geological Society, London*, **147**, 271–278.
ROEDDER, E. 1984. *Fluid inclusions*. Reviews in Mineralogy 12. Mineralogical Society of America.
SASSI, F.P. & SCOLARI, A. 1974. The b_o value of potassic white mica as a barometric indicator in low-grade metamorphism of pelitic schists. *Contributions to Mineralogy & Petrology*, **45**, 143–152.
SEARLE, M.P. 1991. Comment and Reply on "Pre-collision and postcollision thermal events in the Himalaya". *Geology*, **19**, 666–667.
——, COOPER, D.J.W & REX, A.J. 1988. Collision tectonics of the Ladakh–Zanskar Himalayas. *Philosophical Transactions of the Royal Society of London*, **A326**, 117–150.
SHEPHERD, T.J., RANKIN, A.H. & ALDERTON, D.H.M. 1985. *A Practical Guide to Fluid Inclusion Studies*. Blackie, Glasgow.
SINHA-ROY, S. 1982. Himalayan Main Central thrust and its implications for Himalayan inverted metamorphism. *Tectonophysics*, **84**, 197–224.
STÖCKLIN, J. 1980. Geology of Nepal and its regional framework. *Journal Geological Society, London*, **137**, 1–34.
——, BHATTARAI, K.D., SINGH-CHETRI, V. & BHANDARI, A.N. 1980. *Geological map of Kathmandu area and Central Maharabharat Range*. Sajedian Cartographic Co. Ltd. Tehran.
TOURET, J. 1977. The significance of fluid inclusions in metamorphic rocks. *In*: FRASER, D.G. (ed.) *Thermodynamics in Geology*. Riedel Publication Company, Dordrecht, Netherlands, 203–227.
VALDIYA, K.S. 1980. The two intracrustal boundary thrusts of the Himalaya. *Tectonophysics*, **66**, 323–348.
WEAVER, C.E. 1984. *Shale-slate metamorphism in the Southern Appalachians*. Elsevier, Netherlands.
WEBER, K. 1972. Notes on determination of illite crystallinities. *Neues Jahrbuch fur Mineralogie, Monatshefte*, 267–272.

Early stages of foreland basin evolution in the Lesser Himalaya, N India

YANINA NAJMAN, PETER CLIFT, MICHAEL R.W. JOHNSON & ALASTAIR H.F. ROBERTSON

Department of Geology & Geophysics, Grant Institute, University of Edinburgh, West Mains Rd., Edinburgh EH9 3JW, UK

Abstract: In this paper we report preliminary results of an integrated sedimentological and structural study of the Late Cretaceous–Early Tertiary facies exposed in the Lesser Himalaya. The Precambrian–Palaeozoic basement of the Indian plate was initially transgressed in Late Cretaceous times (after *c*. 100 Ma), with the deposition of shallow shelf carbonates, (50 m thick Singtali Formation). These are depositionally overlain by green ferruginous mudstones and thin limestones of the 150 m thick Palaeocene–Eocene (*c*. 65–40 Ma) Subathu Formation. During Late Eocene–Oligocene times (*c*. 40–30 Ma), clastic sediment input increased and red lenticular sandstones, thinner planar-bedded sandstones, mudstones and caliche of the 350 m thick Dagshai Formation were deposited. A semi-arid, meandering fluvial/floodplain setting is envisaged, with thick channelised and overbank sands. Palaeocurrent evidence indicates a general SE progradation. Our detailed field mapping shows that where the entire Subathu and Dagshai Formations are intercalated, this is the result of post-depositional tectonics, rather than primary intertonguing as described in some previous reports. The relative abundance of sandstones greatly increased in upper Dagshai times, continuing into the Early–Mid-Miocene (*c*. 30–10 Ma) when the 250 m thick Kasauli Formation was laid down. This formation is characterized by lenticular and planar-bedded grey sandstones, rich in plant material including occasional logs (but without caliche) interbedded with minor grey mudstones. A rapidly prograding, braided fluvial environment is proposed, with the petrography (e.g. presence of garnets) suggesting erosion and derivation from deeper levels of the Himalayan mountain belt to the north. The climate had by then changed from semi-arid to humid, possibly in response to the onset of the monsoon, initiated when the mountain belt had reached sufficient height to interfere with the jet-stream.

During Late Miocene–Quaternary times (younger than *c*. 15 Ma), the Tertiary foreland basin was deformed and incorporated into three structural levels of a southward migrating thrust stack. At the lowest level, thick successions of the Tertiary Subathu, Dagshai and Kasauli Formations are structurally underlain by fluvial sediments (Siwalik Gp.) and overlain by the Krol Precambrian–Palaeozoic sedimentary nappe. At the intermediate structural level, Early Tertiary sediments unconformably overlie the stratigraphy within this nappe and comprise, often thin (<50 m), successions of mainly Subathu and Dagshai Formation sediments. The highest structural level sediments form small imbricates, highly dismembered units and melange (including evaporites) beneath the major metamorphic Jutogh Nappe.

We interpret the Late Cretaceous–Palaeocene Singtali Formation as pre-collisional transgressive sediments related either to flexural downwarping of the Indian plate due to the obduction of the Spontang ophiolite onto the north Indian plate margin, or to extensional tectonics related to the detachment and drift of India from Gondwana and Neo-Tethyan subduction beneath Asia. This would have been coupled with the Late Cretaceous global eustatic sea-level high stand. The marine Subathu Formation is interpreted as having been deposited between initial and terminal India–Eurasia continental collision, with the overlying fluvial/floodplain Late Eocene–Oligocene Dagshai and Miocene Kasauli Formations as southward prograding foreland basin successions related to progressive stages of India–Eurasia continental collision.

The Himalayan mountain chain, which stretches from Pakistan in the west to Bhutan in the east, is a complicated fold and thrust belt, which can be divided into three units bounded by thrusts: from south to north; the Sub-Himalaya, Lesser Himalaya and Higher Himalaya, (Fig. 1). In India the Mio-Pliocene Siwalik sediments of the Sub-Himalaya are thrust over the Indian craton along the Main Frontal Thrust, and overthrust by Early Tertiary sediments lying beneath the Main Boundary Thrust which marks the Sub-Himalaya/Lesser Himalaya boundary. These

From TRELOAR, P.J. & SEARLE, M.P. (eds), *Himalayan Tectonics*
Geological Society Special Publication, No. 74, pp. 541–558.

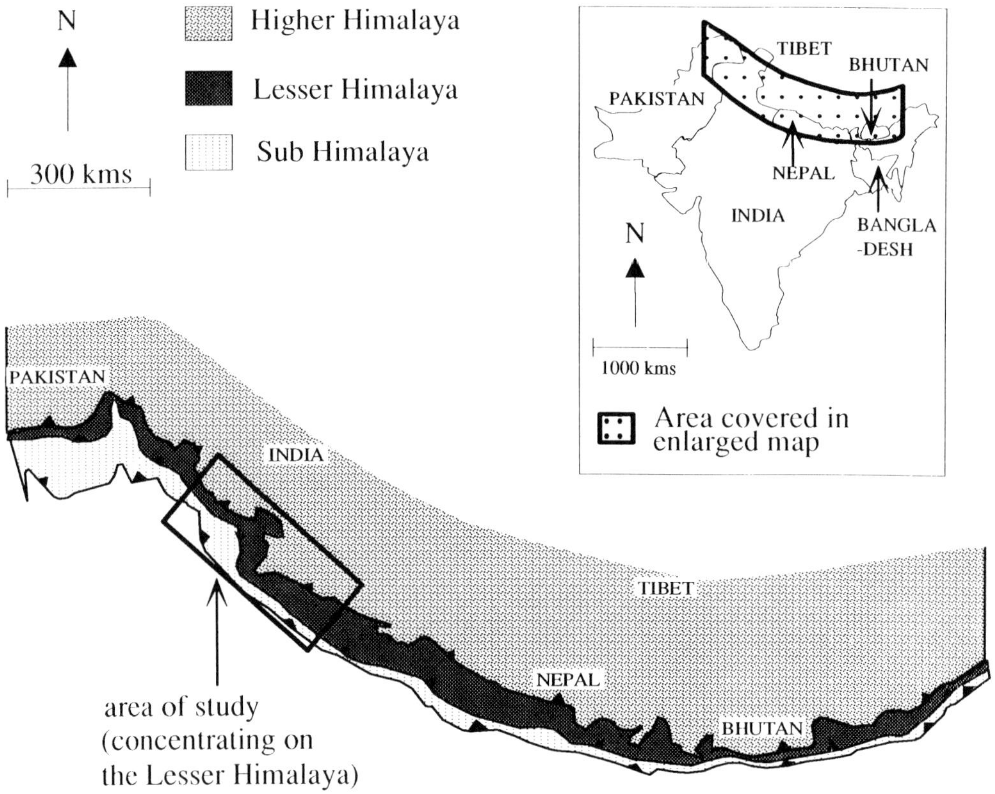

Fig. 1. The Himalayan mountain chain divided into three structural units. Inset: location of the Himalayan mountain belt.

Early Tertiary sediments are overthrust by a sedimentary nappe termed the Krol Nappe (Gansser 1964), which is overthrust by a low grade metamorphic nappe, the Jutogh Nappe (Gansser 1964; Figs 2 & 3d). Above the Jutogh Nappe, high-grade metamorphic rocks of the High Himalaya crop out (e.g. schists, gneisses and granites), separated from the units below by the Main Central Thrust.

This nappe stack resulted from the collision of India with Eurasia sometime between 55 Ma and 40 Ma (Dewey *et al.* 1988; Klootwijk & Peirce 1979) which telescoped the Indian passive continental margin. As the orogen rose, detritus was shed into the basin to the south, a process which continues to the present day as the plates continue to converge, and the mountain range, thrusts and basin migrate southwards. As a consequence, earlier basin sediments have been incorporated into the thrust stack.

This study focuses on the Late Cretaceous and Early Tertiary sediments of the Lesser Himalaya: the Singtali Formation (Valdiya 1980), Subathu, Dagshai and Kasauli Formations (Bhatia 1982), which document the early stages of collision and orogenic uplift, erosion and deposition of the detritus which make up these formations in the basin towards the south.

Previous work has concentrated primarily on the biostratigraphy of the Subathu Formation (e.g. Mathur 1978), although some detailed lithological studies were also carried out on individual formations (e.g. Chaudhri 1971). A general overview of the Early Tertiary sediments and environments was given by Srivastava & Casshyap (1983).

We present new sedimentological and structural data which help provide an integrated model for the evolution of the Himalayan foreland basin and sediment distribution pathways within it.

Tectonostratigraphy

The Lesser Himalaya are composed of Precambrian–Palaeozoic Indian Plate basement

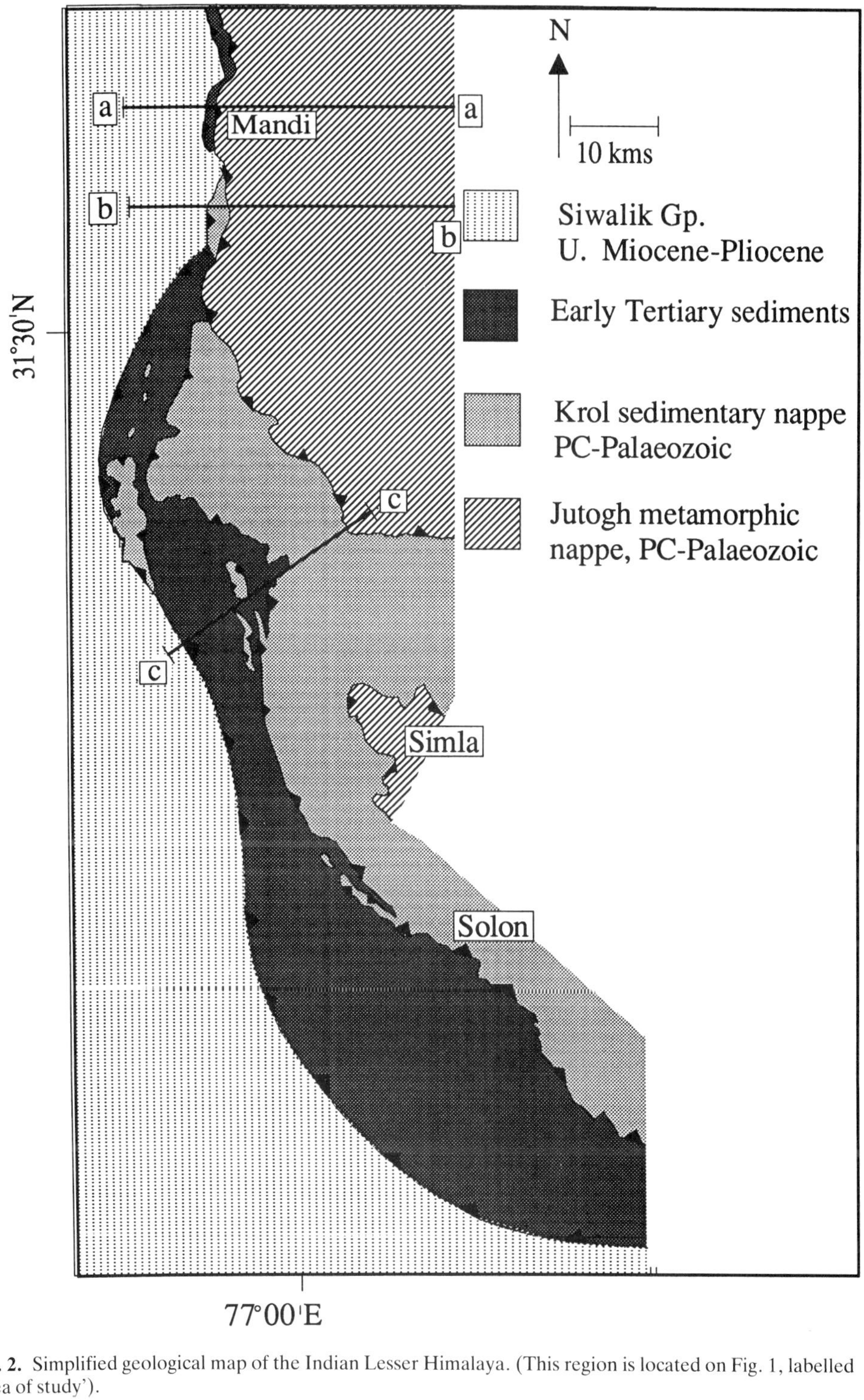

Fig. 2. Simplified geological map of the Indian Lesser Himalaya. (This region is located on Fig. 1, labelled 'area of study').

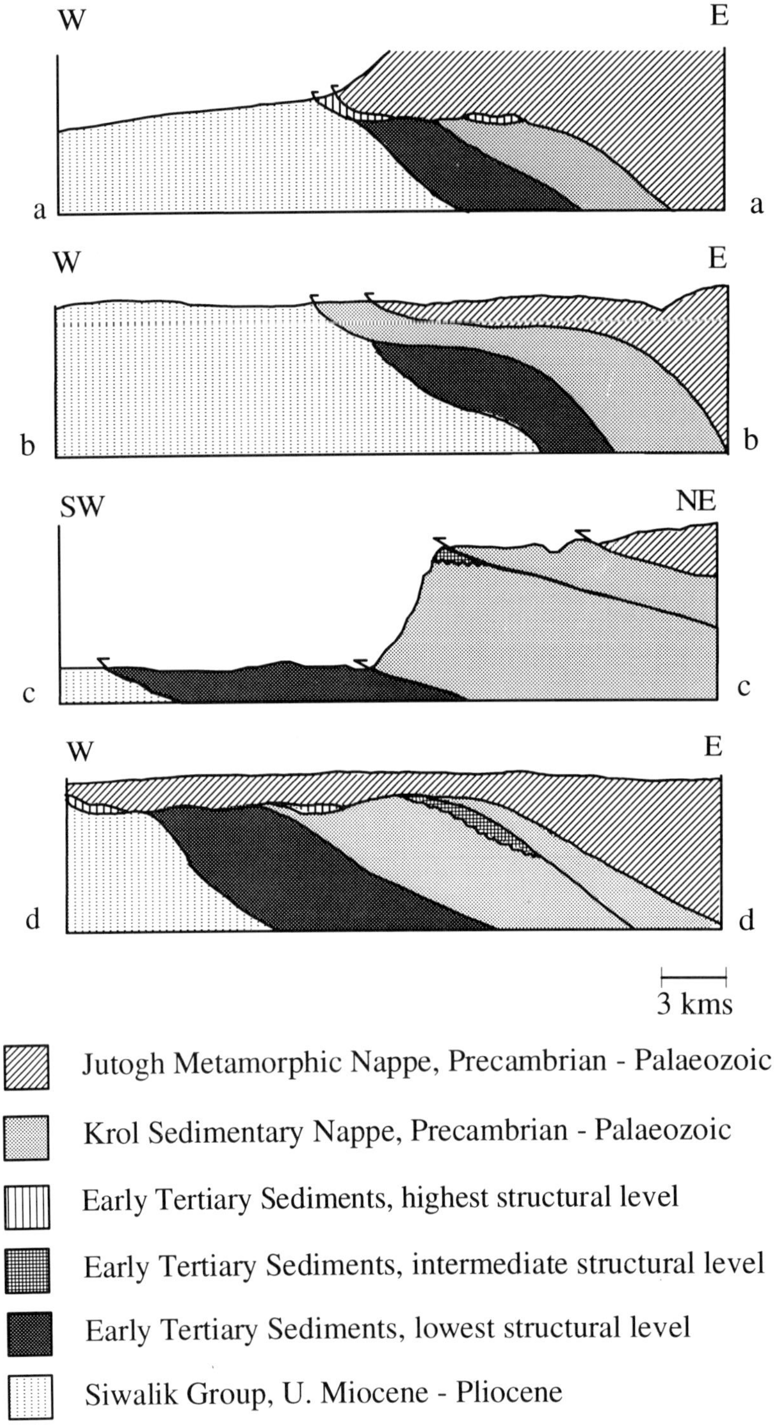

Fig. 3. (**a–c**) Simplified geological cross-sections along lines marked on Fig. 2, showing Early Tertiary sediments at different structural levels. For further explanation, see text. Note: internal deformation within formations is not shown. Line 3c has an exaggerated vertical scale with intermediate level Early Tertiary sediments enlarged for clarity. Although in these sections, thrusts appear to be emplaced towards the west, the overall large-scale view is that of a south-west emplacement direction as described in the text. (**d**) Composite geological cross-section showing structural relationships of the Early Tertiary sediments at different structural levels. This is a combination and summary of (a–c) and cannot be found as a true section in the field.

rocks and Early Tertiary Subathu, Dagshai and Kasauli Formation basin sediments thrusted within a nappe stack emplaced towards the SW. The basin sediments dealt with here are found at three structural levels (Fig. 3a–d).

First, the lowest structural level, shown in all Fig. 3 sections, restores furthest to the south. It is a thick (*c.* 800 m) sequence of Subathu, Dagshai and Kasauli Formations, structurally underlain by the Siwalik Group and structurally overlain by the Precambrian–Palaeozoic Krol sedimentary nappe. These Precambrian–Palaeozoic rocks, originally part of the Indian passive continental margin, are a sequence of limestones, sandstones and slates. The second, intermediate, structural level, which is shown in Fig. 3c & d, comprises, in the west, thin sequences (*c.* 30 m total) of Subathu and Dagshai Formation rocks, whilst in the east, at Dogadda, the Singtali Formation and a thicker succession of Subathu and Dagshai sediments, compared to the western localities, are found. The successions at both the eastern and western localities depositionally overlie the Krol sedimentary nappe and could be correlated. The highest level Early Tertiary sediments, restoring furthest to the north and shown in Fig. 3a & d, consist of small imbricates of highly dismembered units of melange and evaporites lying beneath the metamorphic Jutogh Nappe. Our reasons for distinguishing between the intermediate and highest structural level Tertiary sediments and for correlating the highest structural level with the overlying Jutogh nappe are that, firstly, the Jutogh nappe and Tertiary imbricates below are found at more than one structural position, thrust over more than one unit, often unrelated to the Krol Nappe and intermediate structural level Tertiary sediments. Secondly, the small-scale structures within the highest structural level Early Tertiary sediments give nappe emplacement directions which correlate well with the equivalent structures in the overlying Jutogh Nappe (they both show SSW, S and SE emplacement directions). These vergence directions are distinct from those shown by structures within the Krol sedimentary nappe which have a SW emplacement direction. Thirdly, the highest structural level Early Tertiary sediments show extreme internal deformation unlike the Early Tertiary sediments of the intermediate structural level.

The metamorphic Jutogh Nappe, which is sometimes out of sequence, contains quartzite and schist. It represents deeper levels of the Indian craton and marks the beginning of the structural transition to the High Himalaya to the north.

Lithostratigraphy

The oldest basin sediments, seen at only a few localities in the east, are those belonging to the Singtali Formation, also called the Nilkanth Formation (Azmi & Joshi 1981) and 'Upper Tal' as defined by Mathur (1977) and Bhatia (1980). This Formation is distinct from the basement Tal Formation upon which it rests unconformably (Singh 1979*a,b*; Acharrya 1979; Azmi & Joshi 1981). The underlying basement Tal Formation is, in this area, the topmost unit of the sedimentary Krol Nappe, dated as Pre-Mesozoic by most workers (Valdiya 1980; Azmi & Joshi 1981). The Singtali Formation, which consists primarily of oomicrites, that are sometimes fossiliferous, and minor sandstones, has been assigned a Late Cretaceous–Palaeocene age based on faunal evidence (Mathur 1977; Saklani *et al.* 1977; Bhatia 1980; Singh 1980). The upper contact of the Singtali Formation with the overlying Subathu Formation is controversial. Most workers e.g. Mathur (1977) and Azmi & Joshi (1981) propose an unconformable contact, but others e.g. Bhatia (1980) believe that the sequence is conformable.

The Subathu Formation is a sequence of limestones, green mudrocks and subordinate sandstones, which has been palaeontologically dated as Late Palaeocene–Eocene (Mathur 1978). Where the Singtali Formation is absent, these are the oldest basin sediments and, at the intermediate structural level, an unconformity with the basement sedimentary Krol nappe is visible. The upper contact of the Subathu Formation with the overlying red coloured Dagshai Formation is relatively sharp, but conformable, with a *c.* 15 m thick transitional zone between the two formations.

The red coloured Dagshai Formation is a sequence of mudstones, siltstones, sandstones and caliche. Dating is problematical due to a lack of fossils. A Late Eocene palaeomagnetic age is given by Gautam (1989) for the middle member of the Nepalese Dumri Formation which is equivalent to the Indian Dagshai Formation. These data are supported by palynological evidence which assigns a possible pre-Miocene (Palaeogene) age to the red continental facies of Nepal, equivalent to the Dagshai Formation of India (unpublished confidential oil company report). Using stratigraphic evidence, if the Subathu–Dagshai Formations contact is conformable, the base of the Dagshai Formation will be of Late Eocene age. The top of the Formation is marked by the base of the overlying Kasauli Formation which is of Early–Mid Miocene age (Fiestmantel 1882; Sahni 1953).

The Dagshai Formation shows a gradual transition (*c.* 50 m) to the younger grey coloured Kasauli Formation which is composed primarily of sandstones with minor siltstones and mudstones. A Miocene age has been assigned to this formation based on floral evidence (Fiestmantel 1882; Sahni 1953).

Biostratigraphy

The biostratigraphy of the Subathu Formation and, to a lesser extent the Singtali Formation, is well constrained due to the abundance of fossils. Biostratigraphical dating of the Dagshai and Kasauli Formations is more problematical however as fossils are extremely sparse.

The Singtali Formation has been assigned an age not older than Late Cretaceous on the basis of body fossils described by Mathur (1977) and Saklani *et al.* (1977). Bhatia (1980), assigned a precise age of Maastrichtian to Danian on the basis of his findings which include Calcareous Algae (*Archaeolithothamnium* sp., *Halimeda* sp. & *Neomeris* sp.), Hydrozoa (*Millepora* sp.), Bryozoa (*Diplocava* sp. & *Corymbopora* sp.), Foraminifera (*Edomia* sp & *Dictyconus* sp.), Echinoids, Ostracods and Molluscs. However Singh (1980) identified *Globotruncana, Hetrohelix* and *Hedbergella* from which he deduced a Coniacian age.

Mathur (1978) has defined the base of the Subathu Formation as Late Palaeocene based on the reported occurrences of *Nummulites wadiai, N. planulatus, Ranikothalia sp., Lockhartia haimei, Dictyoconoides flemingi, Dictyokathiana dadhuensis, Operculina patalaensis* and *Nummulites mamilla*.

The presence of *Nummulites beaumonti, N. stamineus, Assilina orientalis, A. spira* and *Dictyioconoides cooki* has enabled a Mid–Late Eocene age to be deduced for the upper limit of the Subathu Formation (Mathur 1978).

Few stratigraphically significant fossils have been found in the Indian Dagshai and Kasauli Formations apart from polospores found in the Dagshai Formation highest structural level which have been dated as Early Tertiary by the Indian Oil and Natural Gas Corporation (Chakraborty *et al.* 1963) and plant remains of *Sabal major* in the Kasauli Formation which have been dated as Early–Mid-Miocene (Fiestmantel 1882; Sahni 1953). However, biostratigraphical dating has been achieved on the red continental facies sequence of Nepal which can be correlated with the Dagshai Formation, and on the Murree Formation of Pakistan which many workers correlate with the Indian Dagshai and Kasauli sequences (Krishnan 1943; Sahni 1953; Gansser 1964). As mentioned above, the red continental facies sequence of Nepal has been assigned a possible pre-Miocene i.e. Palaeogene age based on palynological evidence (unpublished confidential oil company report). Pilgrim (1910) found mammal remains of *Antracotherium bugtiense, Brachyodus africanus* and *Teloceras fatejangense* from which he assigned an Early Oligocene–Mid-Miocene age. Subsequent workers, e.g. Meissner (1974) and Tahirkheli (1982) have accepted an Oligocene–Mid-Miocene age for the Murree rocks, as have Bossart & Ottiger (1989) but only for the southern fluvial Murree lithologies.

Structure

The structure of the Early Tertiary formations is largely related to nappe emplacement. Small scale tectonic structures (centimetre to metre scale) present within the formations, (e.g. folds and duplexes), indicate a SW emplacement direction for the thrust stack. On a larger scale (metre to kilometre), our work has shown that nearly isoclinal folding of the basin sediments has occurred in some areas. The Subathu and Dagshai Formations are sometimes found intercalated and controversy exists between Batra (1989) and Raiverman & Raman (1971) as to the nature of this intercalation. Batra believes that the intercalation is due to tectonics (i.e. folding and faulting) and therefore the Dagshai Formation is younger than the Subathu Formation. Raiverman and Raman are of the opinion that the intercalations are of primary sedimentary origin. In this case no tectonics would be involved and the Subathu and Dagshai Formations would be laterally equivalent intertonguing facies. Our detailed field mapping of Raiverman and Raman's field area, making extensive use of way-up structures and marker beds, shows that the interleaving can be attributed entirely to tectonics (Fig. 4) and therefore that the Dagshai Formation is clearly younger than the Subathu Formation. There is some intercalation of Subathu and Dagshai Formation facies in a few areas but this is confined to a *c.* 15 m thick transitional zone between the two formations rather than the entire sequence.

Lithologies

Singtali Formation

The approximately 50 m thick Singtali Formation sequence is dominated by limestones with subordinate quartz arenites. Medium-bedded massive strata are predominant and no

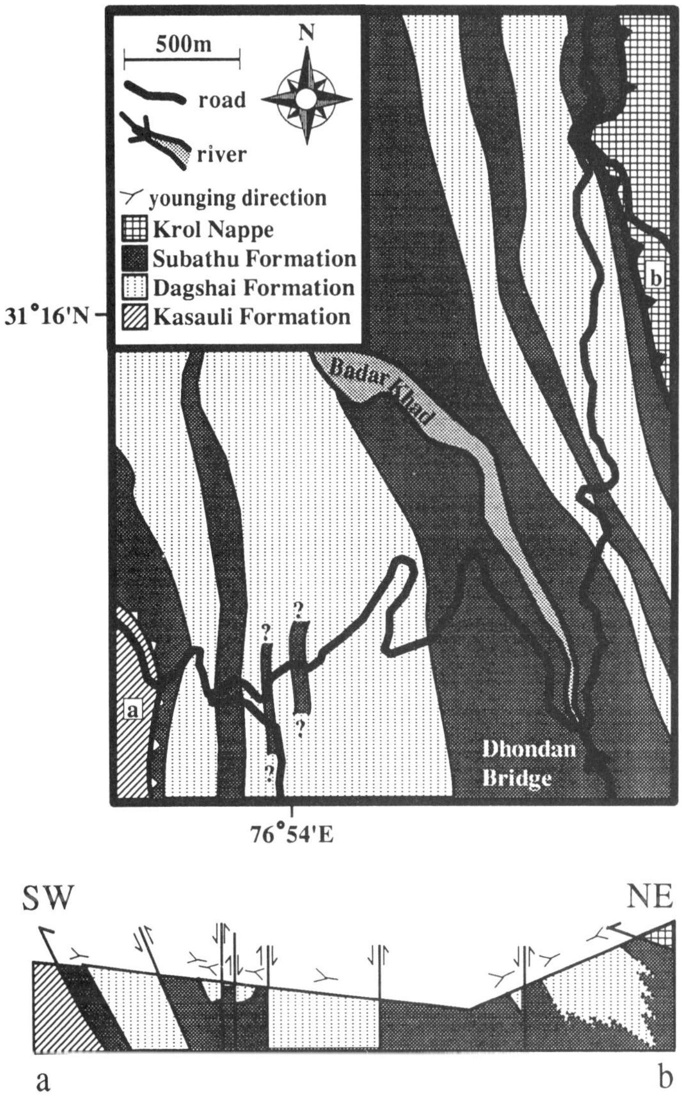

Fig. 4. (**a**) The area where intercalation of the Subathu and Dagshai Formations is interpreted by Raiverman and Raman as being of a sedimentary nature and by Batra as being due to tectonics. Map modified from Raiverman & Raman (1971). (**b**) Geological cross-section of map area (based on this study) from 'a' (SW) to 'b' (NE) approximately following the road. All major intercalations can be accounted for by folding and faulting.

sedimentary structures are visible. In thin section the limestones are seen to be oomicrites which can be rich in quartz grains, opaque minerals, reworked carbonate lithic fragments (one of which contained zircon) and ooids with quartz and an occasional tourmaline nuclei. Rarely, bivalve fragments surrounded by micrite envelopes occur. The sandstones are quartz arenites which contain small amounts of tourmaline and opaque minerals.

Subathu Formation

The Subathu Formation sequence is thickest in the lowest structural level where it comprises limestones, thick green mudstones and subordi-

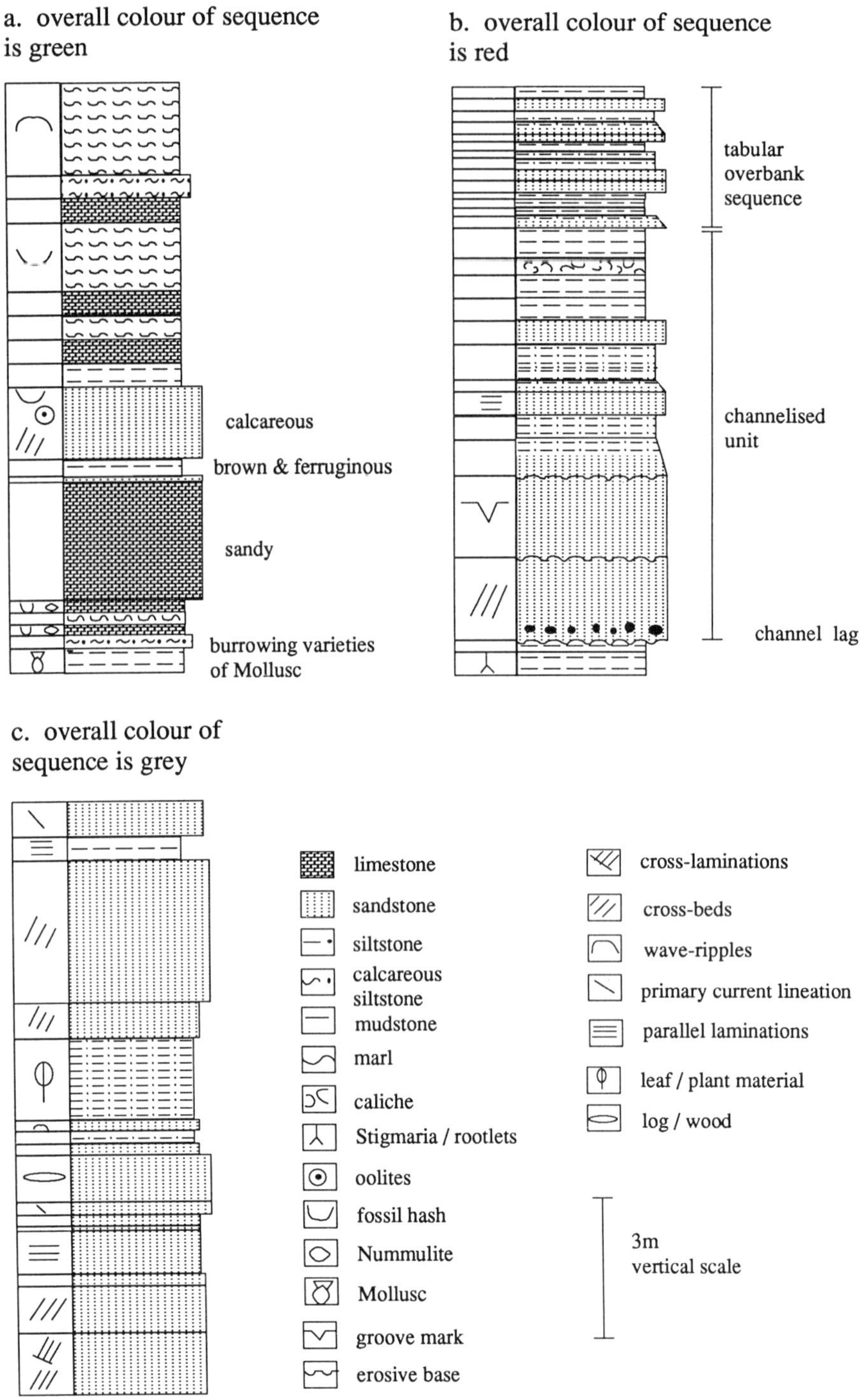

Fig. 5. Composite logs of the Early Tertiary Formations, amalgamating information from more than one locality. (**a**) Subathu Formation. (**b**) Dagshai Formation. (**c**) Kasauli Formation.

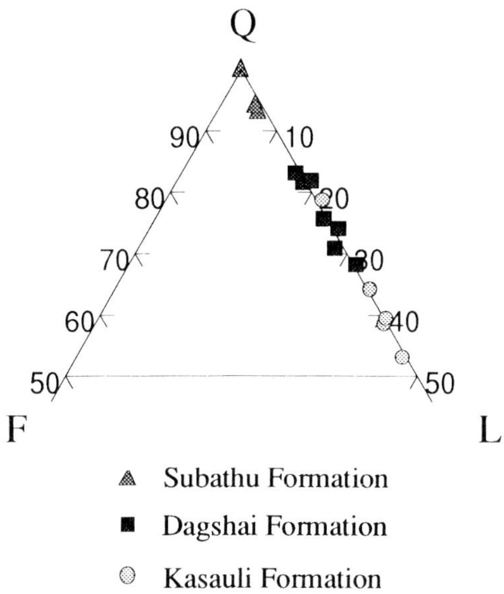

Fig. 6. Components which occur in the sandstones of the Subathu, Dagshai and Kasauli Formations plotted on a quartz–feldspar–lithics triangle. Three discrete fields are shown.

nate sandstones up to 150 m thick (Fig. 5a). The clastic sediments are mainly oolitic calcareous quartz arenites that contain oyster fossils and non-fossiliferous calcareous arenites which contain a high proportion of quartz grains and lesser amounts of feldspar and lithic fragments (Fig. 6). Minor amounts of opaques, chlorite, biotite, glauconite, epidote and tourmaline are present in the last mentioned lithology. The non-arenaceous sediments are of a variable nature ranging from mudstones and siltstones to nummulitic biomicrites. The mudstones and siltstones are green to grey, commonly ferruginous and/or calcareous and can contain quartz, calcite, anatase and chlorite. Well-preserved infaunal bivalve and gastropod faunas are found within the mudstones and the nummulitic biomicrites can contain quartz fragments, glauconite, oysters, echinoid spines and coral. Sedimentary structures are rare apart from parallel laminations. Rare cross-beds indicate a SE to E directed palaeoflow. At the intermediate structural level localities in the west, the Subathu Formation is much thinner (*c.* 15 m) than in the level below, there is a much higher proportion of limestone compared to mudstone and the unconformable contact with the basement is sometimes visible (e.g. at Dalaghat). The unconformity is commonly marked by iron staining, bauxites, laterites and conglomerates which contain clasts of carbonate, chert and laterite, plus bauxite, muscovite and anatase. The limestones are micritic and sometimes contain crinoid ossicles, foraminifera, echinoid plates and spines, micrite envelopes, bivalve fragments and rare quartz grains. The mudstones, which commonly show a gradational contact with siltstones, often contain quartz, calcite, chlorite, illite, kaolinite, lepidolite, anatase and rare muscovite. Organic horizons are common, including coal, seat earth and rootlet beds, pisoliths, limestone rich in bituminous material and iron-rich black mudstones containing ill-defined aluminium oxide hydroxide chloride. At Dogadda, in the east of the study area, the intermediate structural level Subathu Formation sequence is of comparable thickness to that found at the lower structural level and the limestone: mudstone proportions are also more similar. In addition, pyritized wood, pyrite nodules and small carbon particles occur in black mudstone near the contact with the Dagshai Formation and oolitic ironstone containing fluorapatite, siderite, chamosite and hematite crops out close by.

Dagshai Formation

The red-coloured sandstones, siltstones, mudstones and caliche making up the Dagshai Formation are best exposed in the lowest structural level where they are up to 350 m thick (Fig. 5b). Thick lenticular sandstones up to 20 m wide, sometimes with a channel lag at the base, often display a sequence from thick sandstones to thinner bedded sandstones and siltstones to mudstones and finally caliche. In contrast to these lenticular units, thin tabular sheet sand and mudstone deposits also occur. Pipe and nodular caliche is common, with mud rip-up horizons, Stigmaria and rootlet horizons in the mudstones also found.

Sedimentary structures, although indistinct, are often present and include flute and groove marks, cross-laminations and planar cross-beds. The structures show a relatively large amount of variation in palaeocurrent orientations in a general direction towards the SE and SW away from the rising orogen (Fig. 7).

Petrographic analysis of the mudstones and siltstones shows that chlorite, kaolinite, muscovite, lepidolite, quartz, hematite, albite, calcite and illite are present. The sandstones are mainly sublitharenites with a few litharenites, the quartz percentage never dropping below 65% (Fig. 6). The lithic fragments are dominantly schistose, presumably derived from the same source as the tourmaline, epidote, garnet and mica minerals found in small quantities within

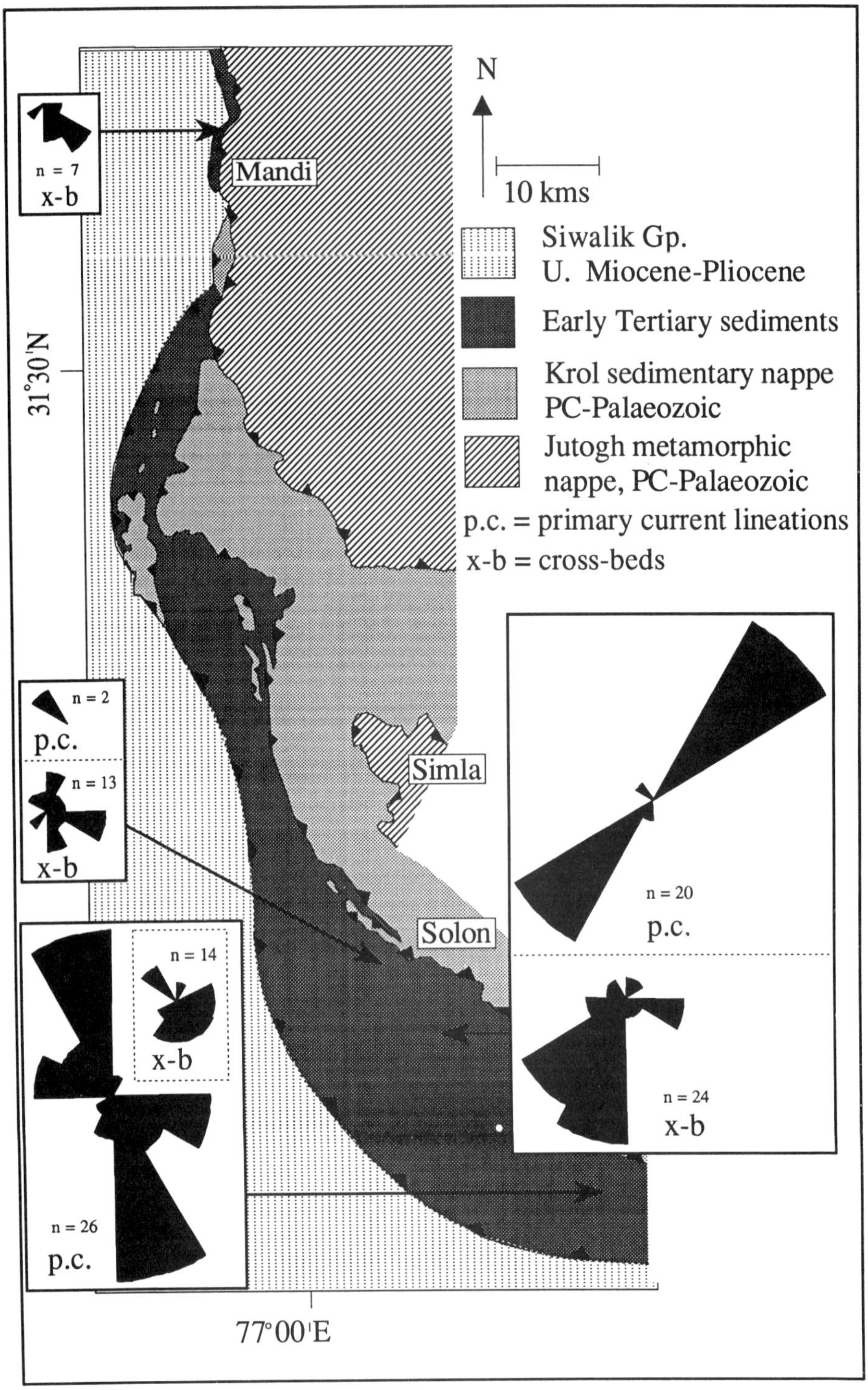

Fig. 7. Rose diagrams displaying palaeocurrent data for the Dagshai Formation. Where primary current lineation diagrams are not mirror images, this indicates that both direction and orientation are known.

the sandstones, i.e. a medium–high grade metamorphic terrain.

Throughout the Dagshai Formation (and continuing into the Kasauli Formation) there is a gradual decrease up-section in the proportion of mudstone within the sediments and an increase in sandstone. At the base of the Dagshai Formation, mudstone is dominant, whereas by the top, sandstone dominates. In addition the uppermost Dagshai Formation transitional sandstones have a higher percentage of lithics, closer in proportion to that of the Kasauli Formation than to that of most of the Dagshai Formation.

In the west of the study area, the intermediate structural level rocks have a much thinner (*c*. 15 m) Dagshai Formation sequence than in the lower structural level. The sequences consist mainly of red mudstones with small amounts of caliche and sandstone. The sandstones are of similar composition to the Dagshai Formation sandstones of the lowest structural level and the mudstones are also similar to their lower structural level counterparts. In the east, at Dogadda, the sequence thickness is greater, approaching the thicknesses found at the lowest structural level, and the sandstones, although still subordinate to the mudstone, make up a greater proportion of the outcrop compared to the west.

At a few localities (e.g. Maigal) the highest structural level of the Dagshai Formation is seen. These sediments are found as imbricates at the base of the Jutogh metamorphic nappe. Lithologically they are unlike the conventional Subathu, Dagshai or Kasauli Formation strata described previously. Instead they consist of a diverse assemblage of lithologies, described below, which, in terms of colour and possible facies, most closely resembles the Dagshai Formation.

At the base of the sequence, microconglomerates, sandstones and a red and white, interbedded sandstone–mudstone unit are found. The conglomerate has a micritic, calcareous matrix with clasts of quartz, calcite and lithic fragments of schist, micrite and quartzite, whilst the sandstone is broadly similar to the lowest level Dagshai Formation sandstones with the addition of amphibole.

Thin (0.5 mm–25 cm), white, parallel- and cross-bedded quartz-rich bands of sandstone and red mudstones of a similar thickness comprise the sandstone–mudstone unit. The quartz-rich bands commonly contain opaque minerals and epidote concentrated at the base of each layer, as well as occasional biotite, feldspar and tourmaline scattered throughout. This unit shows pervasive soft sediment deformation and folding, possibly related to dissolution collapse prior to thrusting, but more likely to be due to collapse caused by fluids associated with and channelled along the thrust plane.

Higher in the sequence, thinly bedded micrites and calc-arenites pass up into dissolution brecciated gypsiferous marl, laminated on a millimetre scale, red, brown and green in colour and often salt-encrusted. Within this lithology are some coarser, parallel- and cross-laminated arenaceous layers.

The overlying sedimentary melange, lying directly beneath the base of the Jutogh metamorphic nappe, contains many angular clasts of the gypsiferous marl along with halite pseudomorphs within a quartz arenaceous matrix. Mineralization by infiltrating fluids and soft sedimentary deformation are extremely common and presumably related to thrusting.

Kasauli Formation

Grey lenticular and planar bedded sandstones with minor amounts of siltstone and mudstone typify the 250 m thick Kasauli Formation (Fig. 5c). The lithology is broadly similar to that of the Dagshai Formation but there are a few important differences.

The rock colour has changed from red to grey and there is a much higher sandstone to mudstone ratio. In addition, caliche is lacking in this formation and wood and plant material is common, including logs up to 9 m long. Palaeocurrent evidence, on the basis of planar cross-bed and cross-lamination data, shows a SSE–SE-directed flow and displays a much more limited scatter of orientations compared to the Dagshai Formation (Fig. 8).

A much higher proportion of metamorphic lithic fragments occur in the Kasauli sandstones (litharenites) when compared to the Dagshai Formation sandstones (Fig. 6). The mineralogy of the mudstones is broadly similar to that of the Dagshai Formation.

Discussion and interpretation

Interpretation of facies

Singtali Formation. The dominance of limestone in the Singtali Formation and sparse fauna would imply shallow marine conditions at this time. A high-energy, agitated environment can be inferred from the presence of ooids.

Subathu Formation. Limestones with normal marine fauna and thick mudstones with well preserved, burrowing-type molluscs indicate a

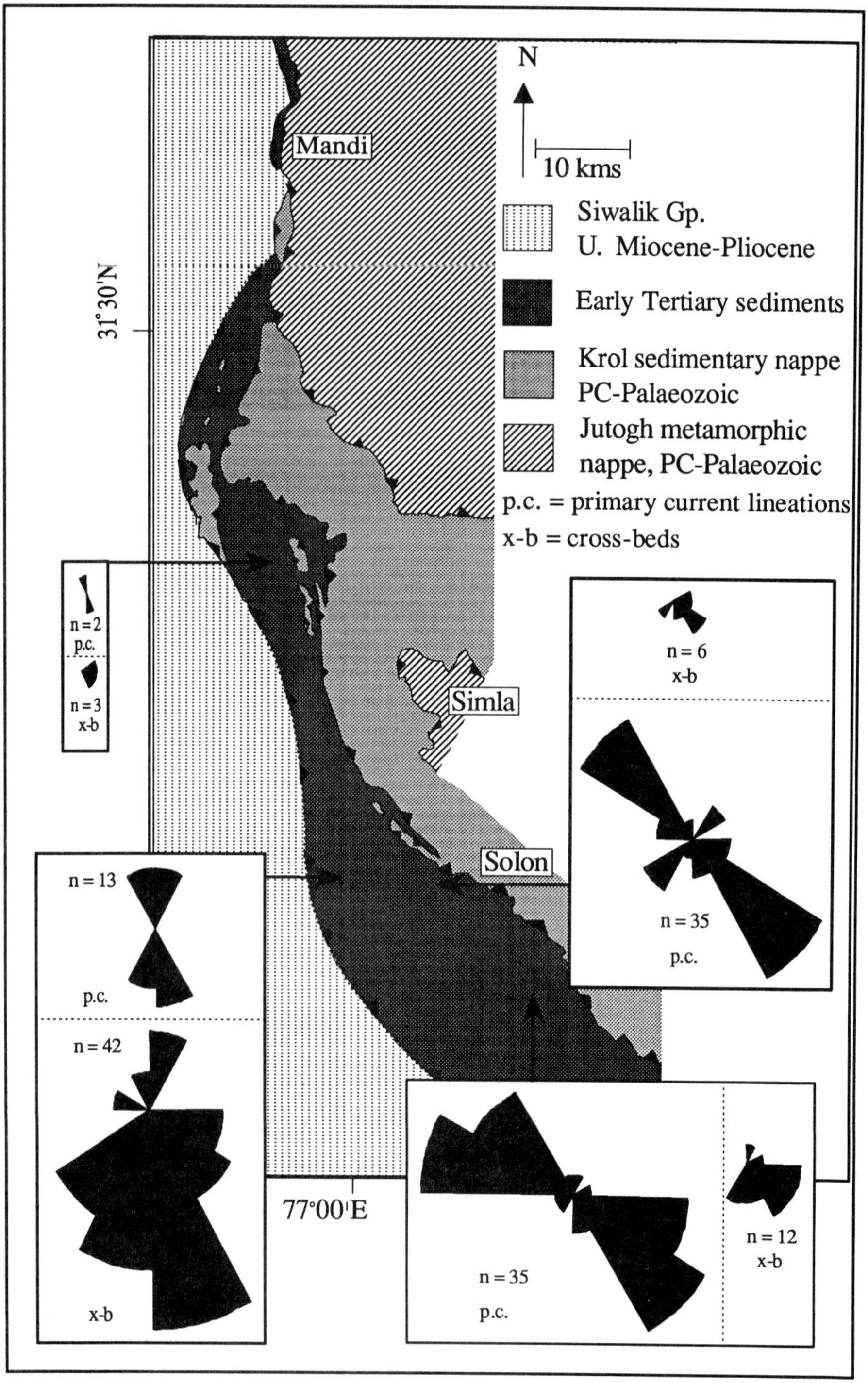

Fig. 8. Rose diagrams displaying palaeocurrent data for the Kasauli Formation. Where primary current lineation diagrams are not mirror images, this indicates that both direction and orientation are known.

quiet, relatively shallow shelf environment during their period of deposition. Oolitic ironstones, found at Dogadda, imply an agitated environment, with weathering of a, probably low lying, continental area (Tucker 1987). In contrast, the pyritized organic material and black mudstones found at the same locality indicate a reducing environment, possibly a swamp or delta. The significantly thinner sequences found at the intermediate structural level eastern localities suggest shallower water depths than for the lower structural level.

Dagshai Formation. The red-coloured channelized and calichified Dagshai Formation sequence was deposited in a continental fluvial, semi-arid setting. The variability of palaeocurrent directions and high proportion of mudstone to sandstone, especially in the lower part of the formation, suggest channels meandering over a wide unconstrained floodplain. Within the channels, the fining upwards pattern reflects the slowing down and silting up of the river with time, whilst the tabular, overbank, crevasse-splay deposits could be indicative of flooding episodes when the river burst its banks (Bluck 1971; Callander 1978; Leeder 1985).

The successions at the highest structural level document a semi-arid fluvial and playa lake environment (Allen & Collinson *in* Reading 1986). The micro-conglomerates, sandstones and interbedded sandstones and mudstones at the base of the sequence are fluvial, whilst the overlying micrites and calc-arenites probably represent an ephemeral lake environment. Evaporitic conditions are implied by the presence of salt encrustations, brecciation due to gypsum dissolution and halite pseudomorphs.

Kasauli Formation The facies of the Kasauli Formation are broadly similar to that of the Dagshai Formation, i.e. fluvial, but the differences in the lithologies and palaeocurrents between these two formations suggest a change in the depositional environment. The colour change to grey and the appearance of abundant wood and plant material indicate a climatic shift from semi-arid to humid conditions. The high channel sandstone to floodplain mudstone ratio and palaeocurrent data, which shows limited variation (Fig. 8) is consistent with a more braided fluvial regime (Miall 1977; Schumm 1968; Schumm *et al*. 1972).

Tectonic interpretations: a foreland basin model

The classic foreland basin model, for example Beaumont (1981) (Fig. 9), involves continental

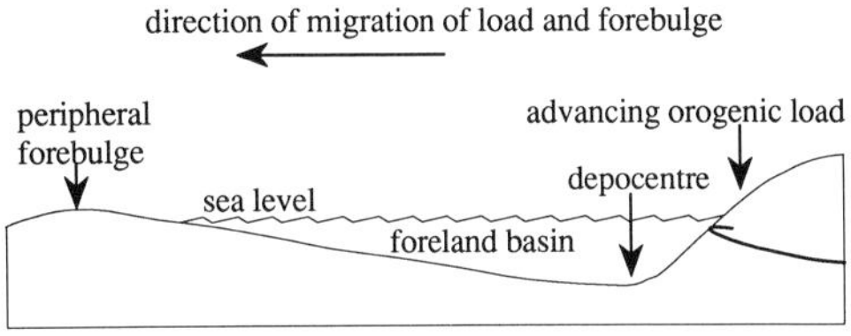

Fig. 9. Theoretical model of foreland basin evolution, (for explanation, see text).

collision where the over-riding plate loads the subducted plate causing downwarping (the foreland basin) and associated updoming cratonwards (the peripheral forebulge). As thrusting occurs, the thrust-belt rises and contributes an increasing amount of eroded material into the foreland basin. This results in a shallowing upward sequence as erosion and deposition exceed subsidence. With continued plate convergence, the thrust belt and associated peripheral forebulge advance and migrate cratonwards, cannibalizing the early foreland basin sediments which become caught up in the thrusting. In this way, the foreland basin will migrate progressively cratonwards, with the depocentre of the basin remaining close to the advancing thrust stack.

The Early Tertiary geology of the Indian Lesser Himalaya conforms well with this model. In Late Cretaceous times, this area of the northern region of the Indian plate finally became submerged after a long period when sub-aerial conditions had dominated. This resulted in deposition of the marine Singtali Formation, of Coniacian or Maastrichtian–Danian age. A possible explanation for this event is that ophiolites, e.g. the Spontang ophiolite, were obducted onto the Indian plate Zanskar continental shelf in the Campanian (Brookfield & Reynolds 1981) or Maastrichtian (Searle 1986), resulting in flexural downwarping of the Indian plate hundreds of kilometres to the south. An alternative explanation for the marine transgression relates to extensional tectonics associated with the Late Albian detachment and start of drifting of India from Gondwana and Neotethyan subduction beneath Asia (Honegger *et al*. 1989; Gaetani & Garzanti 1991). This is coupled with the Late Cretaceous global eustatic sea-level high stand (Vail *et al*. 1977).

Klootwijk *et al*. (1985) consider initial contact between India and Eurasia to have taken place at 62–60 Ma in the NW Himalaya, with terminal collision culminating by 55 Ma in the east. The Subathu Formation rocks were deposited during this period of contact, suturing and initial collision, and, being marine, are consistent with deposition during the initial downwarping stage of the foreland basin model. However, the western intermediate structural level localities show a much thinner Subathu marine sequence compared to the eastern intermediate structural level localities and the lowest structural level. West to east thickness variations could be explained by the progressive suturing of India–Eurasia from northwest to east (Klootwijk *et al.* 1985), with later suturing in the east allowing a longer period where marine conditions could predominate. This progressive suturing would not, however, explain the thickness difference between the lowest and intermediate structural levels. As the intermediate structural level restores further to the north than the lower structural level, northward shallowing of the basin i.e. towards the load is implied. This is at variance with the theoretical model, where the depocentre is close to the load and shallows towards the craton. Jacobi (1981) and Robertson (1987) have noted that in the Appalachians and Oman respectively, old basement faults are reactivated during crustal loading and these exert as major a control over basin morphology as load-induced crustal downflexure. In the Lesser Himalayan early foreland basin, palaeohighs resulting from basement fault reactivation may have been located in the west between the load to the north and the marine Subathu basin to the south. This would result in shallowing towards, and reduced sedimentation on, the palaeohigh, which, coupled with the probable distal nature of the basin, would explain the thin sequences of the western intermediate structural level localities.

In Dagshai Formation times, suturing terminated, collision culminated and indentor tectonics began. Land began to emerge to the north as crustal thickening occurred due to either subduction of continental crust beneath Eurasia or thrust stacking (Searle *et al.* 1988). The marine Subathu Formation to continental Dagshai Formation transitional phase is only seen at Dogadda where reducing facies are present, probably recording the location of an outbuilding delta. This period marks the start of the foreland basin model shallowing upward phase.

With the subaerial emergence of the basin came continental fluvial conditions as recorded by the lower and intermediate structural level Dagshai Formation sequences and a northerly evaporitic playa lake environment as inferred from the Dagshai Formation lithologies of the highest structural level.

With the initiation of the Main Central Thrust after *c.* 30 Ma and as the embryonic Himalaya began to rise and be eroded, metamorphic lithic fragments were transported by rivers in directions varying from SW to SE away from the Early Tertiary orogen and are now found in channel sandstones. The land was still of low relief however and the mountain range in its infancy during deposition of the mud-dominated lower Dagshai sequence, as can be demonstrated by the inferred meandering nature of the rivers, the high percentage of quartz compared to orogen-derived lithics in the sandstones (Fig. 6) and the small proportion of channel sandstones compared to the extent of the floodplain.

There is a gradual increase in the proportion of Dagshai Formation channel sand to floodplain mudstone and a higher lithic fragment percentage within these sandstones up sequence, so that by upper Dagshai Formation times, the sequence has a percentage of channel sand and a lithology which bears a good resemblance to the overlying Kasauli Formation. These changes reflect the increasingly important role that the Himalaya played as a source area through time as the Main Central Thrust influence became stronger and the rate of uplift and erosion increased. 10 km of uplift had occurred by the end of the late Oligocene (Hodges *et al.* 1988).

During Kasauli Formation times (Early–Mid-Miocene), the Main Central Thrust reached its peak of activity at *c.* 20 Ma (Hodges *et al.* 1988) and the thrust stack advanced further south and was of considerable size, sufficient to contribute a substantial amount of medium to high grade metamorphic material to the basin to the south. The Himalaya had also attained sufficient height to influence the weather pattern by acting as a barrier to the jet stream. It was this that probably caused the shift towards a more humid environment and the onset of the monsoon (Kutzbach *et al.* 1993). By this stage the rivers were consistently flowing towards the southeast implying that a major dispersal site, where there was maximum uplift and erosion, lay to the northwest. In addition, the fluvial architecture had taken on a more braided appearance in response to the higher amount of precipitation and the advance and more rapid rise of the orogen with its associated increase in erosion and sedimentation rate. This is reflected in the higher channel sand: floodplain mudstone ratio, and the increased percentage of orogen-derived lithic fragments (Fig. 6), as the Himalaya was eroded to deeper levels.

Summary and conclusions (Table 1)

(1) In Singtali Formation times (Late Cretaceous–Palaeocene), the Indian craton submerged and stable shallow marine conditions

Table 1. *Summary table of lithologies and depositional environments for the Late Cretaceous and Early Tertiary Himalayan foreland basin sediments*

Formation	Age	Approx. age (Ma)	Lithologies	Facies interpretation	Himalayan/Indian events
Kasauli	Lower–Middle Miocene	*c.* 23–10 Ma	Grey sandstones, siltstones & mudstones. Much woody material.	Humid climate, braided fluvial regime.	Peak of MCT activity at *c.* 20 Ma.
Dagshai	Upper Eocene–Upper Oligocene	*c.* 40–23 Ma	Red sandstones, siltstones, mudstones & caliche.	Semi-arid climate, meandering fluvial and floodplain regime.	Initiation of MCT after 30 Ma. Start of thrust-stacking & crustal thickening following on from collision.
Subathu	Palaeocene–Middle Eocene	*c.* 65–40 Ma	Limestones & green mudstones.	Shallow marine environment.	India–Eurasia continental collision.
Singtali	Upper Cretaceous–Palaeocene	*c.* (?88) or 75–65 Ma	Limestones.	Shallow marine environment.	Obduction of Spontang ophiolite onto the Northern Indian plate margin? Neotethyan subduction beneath Asia. Continued northward drift of India after detachment from Gondwana.

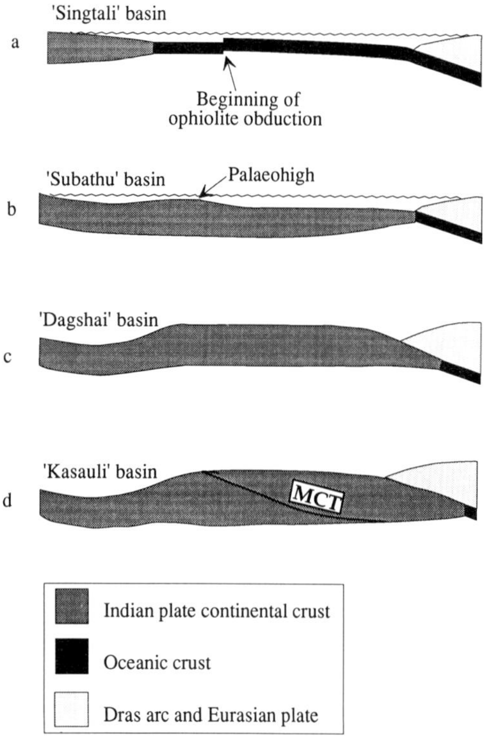

Fig. 10. Model showing the evolution of the Early Tertiary Himalayan foreland basin. For explanation, see text.

ensued (Fig. 10a). This event was possibly flexurally related, i.e. obduction of the Spontang ophiolite, or due to extensional tectonics related to Neotethyan subduction.

(2) The Subathu Formation was deposited during the suturing of India and Eurasia, between initial and terminal collision (Fig. 10b). The inferred pattern of northward shallowing and reduced sedimentation conflicts with classic foreland basin models. However these depositional patterns may reflect basement fault reactivation, giving rise to palaeohighs, rather than simply crustal loading following on from collision.

(3) The mud-dominated lower Dagshai Formation units were deposited as a response to early stages of crustal thickening following continental collision and the beginning of uplift in Late Eocene times (Fig. 10c). Rivers transported detritus southwards into the foreland basin, away from the embryonic but rising orogen. A semi-arid climate prevailed at this time.

(4) During progressive collision, uplift increased and the mountains rose as the thrust front migrated southwards (Fig. 10d). This resulted in increased clastic input to the basin, a climatic change from arid to humid and a switch from inferred meandering to braided fluvial depositional systems. The latest Dagshai Formation and Kasauli Formation units, deposited at this time, represent a period of increasingly rapid uplift and erosion, related to the peak of activity along the Main Central Thrust.

(5) Initiation of the Main Boundary Thrust with continued southward propagation of the fold and thrust belt has resulted in these early foreland basin sediments being incorporated into the thrust stack of the modern Lesser Himalaya.

Y.N. gratefully acknowledges financial support of a Shell studentship from Shell International Petroleum Company Limited, who also assisted the fieldwork of the other authors. A.H.F.R. was partly supported by a Royal Society expedition grant. The authors would also like to thank A. Dubey and N. Mathur of the Wadia Institute of Himalayan Geology, Dehra Dun, India and R. Arya of Chandigarh University, India for their help.

References

ACHARYYA, S.K. 1979. Environment and age of the Tal Formation of Mussoorie and Nilkanth areas of Garhwal Himalaya – comment. *Journal of the Geological Society of India*, **20**, 619–620.

AZMI, R.J. & JOSHI, M.N. 1981. Conodont and other biostratigraphic evidences on the age and evolution of the Krol Belt. *Himalayan Geology*, **11**, 198–223.

BATRA, R.S. 1989. A reinterpretation of the geology and biostratigraphy of the Lower Tertiary Formations exposed along the Bilaspur–Shimla Highway, Himachal Pradesh, India. *Journal of the Geological Society of India*, **33**, 503–523.

BEAUMONT, C. 1981. Foreland basins. *Geophysical Journal of the Royal Astronomical Society*, **65**, 291–329.

BHATIA, S.B. The Tal Tangle in Bhatia & Valdiya (eds) 1980. *Stratigraphy and Correlations of Lesser Himalayan Formations*. Hindustan Publishing Corporation, Delhi, 79–98.

BHATIA, S.B. 1982. Facies, Fauna and Flora in the Lower Tertiary Formations of Northwestern Himalayas: a synthesis. *The Palaeontological Society of India Special Publications*, **1**, 8–20.

BLUCK, B.J. 1971. Sedimentation in the meandering River Endrick. *Scottish Journal of Geology*, **7**, 93–138.

BOSSART, P. & OTTIGER, R. 1989. Rocks of the Murree formation in Northern Pakistan: indicators of a descending foreland basin of late Palaeocene to middle Eocene age. *Eclogae Geologicae Helvetiae*, **82**, 133–165.

BROOKFIELD, M.E. & REYNOLDS, P.H. 1981. Late

Cretaceous emplacement of the Indus suture zone ophiolitic melange and an Eocene-Oligocene magmatic arc on the northern edge of the Indian plate. *Earth & Planetary Science Letters,* **55**, 157–162.

CALLANDER, R.A. 1978. River meandering. *Annual Review of Fluid Mechanics,* **10**, 129–158.

CHAKRABORTY, A., MITRA, R.N. & SASTRY, C.V.S. 1963. Some Tertiary conglomerates and grits of North-West Himalayan foothills. *Geological Survey of India, Miscellaneous Publications,* **15**, 111–116.

CHAUDHRI, R.S. 1971. Petrology of the Lower Tertiary Formations of N.W. Himalayas. *Bulletin of the Indian Geological Association,* **4**, 45–53.

DEWEY, J.F., SHACKLETON, R.M., CHANG CHENGFA & SUN YIYIN 1988. The tectonic evolution of the Tibetan Plateau. *Philosophical Transactions of the Royal Society of London,* **A327**, 379–413.

FIESTMANTEL, O. 1882. Note on the remains of Palm leaves from the Tertiary Murree and Kasauli beds in India. *Records of the Geological Survey of India,* **15**, 51–53.

GAETANI, M. & GARZANTI, E. 1991. Multicyclic History of the Northern India Continental Margin (Northwestern Himalaya). *The American Association of Petroleum Geologists Bulletin,* **75**, 1427–1446.

GANSSER, A. 1964. *Geology of the Himalayas.* Interscience Publishers, London.

GAUTAM, P. 1989. Magnetic properties of some Late Paleozoic to Tertiary sedimentary rocks of Tansen area, Lesser Himalaya, Nepal. *Journal of the Faculty of Science, Hokkaido University, Series IV,* **22**, 467–487.

HODGES, K.V., HUBBARD, M.S. & SILVERBERG, D.S. 1988. Metamorphic constraints on the thermal evolution of the central Himalayan Orogen. *Philosophical Transactions of the Royal Society of London,* **A326**, 257–280.

HONEGGER, K., LE FORT, P., MASCLE, G. & ZIMMERMANN, J.L. 1989. The blueschists along the Indus Suture Zone in Ladakh, NW Himalaya. *Journal of Metamorphic Geology,* **7**, 57–72.

JACOBI, R.D. 1981. Peripheral bulge- a causal mechanism for the Lower/Middle Ordovician unconformity along the western margin of the Northern Appalachians. *Earth & Planetary Science Letters,* **56**, 245–251.

KLOOTWIJK, C.T. & PEIRCE, J.W. 1979. India's and Australia's pole path since the late Mesozoic and the India-Asia collision. *Nature,* **282**, 605–607.

——, CONAGHAN, P.J. & POWELL, C.MCA. 1985. The Himalayan Arc: large-scale continental subduction, oroclinal bending and back-arc spreading. *Earth & Planetary Science Letters,* **75**, 167–183.

KRISHNAN, M.S. 1982. *Geology of India and Burma.* 6th ed. CBS Publishers and Distributers, Delhi, India.

KUTZBACH, J.E., PRELL, W.E. & RUDDIMAN, W.F. 1993. Sensitivity of Eurasian climate to surface uplift of the Tibetan Plateau. *Journal of Geology,* **101**, 177–190.

LEEDER, M.R. 1985. *Sedimentology, Process and Product.* George Allen & Unwin (Publishers) Ltd., London.

MATHUR, N.S. 1977. Age of the Tal and Subathu Formations in the Garhwal Region, Uttar Pradesh, India. *Bulletin of the Indian Geological Association,* **10**, 21–27.

—— 1978. Biostratigraphical aspects of the Subathu Formation, Kumaun Himalaya. *Recent Researches in Geology,* **5**, 96–112.

MIALL, A.D. 1977. A review of the braided river depositional environment. *Earth Science Reviews,* **13**, 1–62.

MEISSNER, C.R. 1974. *Stratigraphy of the Kohat Quadrangle, Pakistan.* US Geological Survey Professional Papers, **716–D**.

PILGRIM, G.E. 1910. Preliminary note on a revised classification of Tertiary Freshwater deposits of India. *Records of the Geological Society of India,* **40/3**, 187–196.

RAIVERMAN, V. & RAMAN, K.S. 1971. Facies relations in the Subathu sediments, Simla Hills, N.W. Himalaya, India. *Geological Magazine,* **108**, 329–341.

READING, H.G. (ed.) 1986. *Sedimentary Environments and Facies.* Blackwell Scientific Publications, London.

ROBERTSON, A.H.F. 1987. Upper Cretaceous Muti Formation: transition of a Mesozoic carbonate platform to a foreland basin in the Oman Mountains. *Sedimentology,* **34**, 1123–1142.

SAHNI, B. 1953. Angiosperm leaf impressions from the Kasauli beds, N.W. Himalaya. *The Palaeobotanist,* **2**, 83–87.

SAKLANI, P.S., KHANDURI, H.C. & SINGH, S. 1977. A reappraisal of geology of a part of Mussoorie Syncline around Santengal, Garhwal Himalaya. *Chayanica Geologica,* **3**, 95–103.

SCHUMM, S.A. 1968. *River adjustment to altered hydrologic regimen – Murrumbidgee River and paleochannels, Australia.* United States Geological Survey Professional Papers, **598**.

——, KHAN, H.R., WINKLEY, B.R. & ROBBINS, L.G. 1972. Variability of river patterns. *Nature Physical Science,* **237**, 75–76.

SEARLE, M.P. 1986. Structural evolution and sequence of thrusting in the High Himalayan Tibetan-Tethys and Indus Suture Zones of Zanskar and Ladakh, western Himalayas. *Journal of Structural Geology,* **8**, 923–938.

——, COOPER, D.J.W. & REX, A.J. 1988. Collision tectonics of the Ladakh-Zanskar Himalaya. *Philosophical Transactions of the Royal Society of London,* **A326**, 117–150.

SINGH, I.B. 1979*a*. Some thoughts on the evolution of the Himalaya and the northern limit of the Indian Shield. *Geological Research,* **68**, 342–350.

SINGH, I.B. 1979*b*. Recognition of a sedimentological break between Quartzite and Limestone Members of the Tal Formation, Lesser Himalaya, India. *Current Science,* **48**, 206–208.

SINGH, P. 1980. Microfauna, age, palaeoenvironment and palaeobiogeography of the Tal Formation, Lesser Himalaya, Garhwal, U.P., India. *Current Science,* **49**, 255–261.

SRIVASTAVA, V.K. & CASSHYAP, S.M. 1983. Evolution of the Pre-Siwalik Tertiary basin of Himachal Himalaya. *Journal of the Geological Society of India*, **24**, 134–147.

TAHIRKHELI KHAN, R.A. 1982. *Geology of the Himalaya, Karakoram, and Hindukush in Pakistan*. Geological Bulletin of the University of Peshawar, Special Issue **15**.

TUCKER, M.E. 1987. *Sedimentary Petrology*. Blackwell Scientific Publications, London.

VAIL, P.R., MITCHUM, R.M. & THOMPSON, S. 1977. Global cycles of relative changes in sea level. *In*: PAYTON, C.E. (ed.) *Seismic Stratigraphy – applications to hydrocarbon exploration*. American Association of Petroleum Geologists, Memoirs, **26**, 83–98.

VALDIYA, K.S. 1980. *Geology of Kumaun Lesser Himalaya*. Wadia Institute of Himalayan Geology, Dehra Dun, India.

Compression- and transpression-related deformation in the Kohat Plateau, NW Pakistan

DAVID A. PIVNIK[1,2] & WILLIAM J. SERCOMBE[2]

[1] *Department of Earth Sciences, Dartmouth College, Hanover, NH, 03755, USA. Now at Amoco.*

[2] *Amoco Production Company, 501 Westlake Park Blvd, PO Box 3092, Houston, TX 77253, USA*

Abstract: Detailed surface mapping and recently acquired subsurface data from the Kohat Plateau of northwestern Pakistan show complex structural styles created by the overprinting by transpression-related structures of pre-existing compression-related structures. The Kohat Plateau is roughly divided into two structurally distinct regions; a northwest region (NW) and a southeast region (SE). Surface structures in the Plateau include doubly-overturned folds (both limbs overturned) which incorporate a lower Eocene to Pleistocene stratigraphic succession. Intensely folded with the stratigraphy in the NW region is a relict thrust belt, representing an earlier compressional episode of deformation. In the SE region many folds (related to wrench faulting at depth) have had either their southern or northern limbs faulted out by high-angle reverse faults which flatten near the surface along a Lower Eocene shale and evaporite horizon. In the NW region, lower Eocene evaporites and limestones are replaced by thick shales which served as a fault-flat forming horizon. This is believed to have hindered the surfaceward propagation of faults, leaving detachment folds exposed at the surface. Steeply dipping strata at depths of over 4 km, steeply dipping faults that become shallow near the surface forming positive flower structures, large vertical displacements across faults, and left-stepping folds suggest transpressional rather than purely compressional deformation. Regions adjacent to the Kohat Plateau also contain transpression-related structures, rather than thin-skinned imbricate thrusts. In view of the position of the Kohat Plateau near a major syntaxial bend in the Himalayan orogen, both north–south and east–west convergence played a dominant role in creating structural styles, and the combination of these two convergence directions influenced the development during the Plio-Pleistocene of transpression-related structures.

The deformational zone in the northernmost portion of the Pakistan foreland basin, which includes the Kohat Plateau and the Northern Potwar Deformation Zone (NDPZ), has been described as a foreland fold and thrust belt whose geometry is controlled partly by an Eo-cambrian evaporite sequence that served as a basal decollment horizon (Jaumé & Lillie 1988). Previous workers have presented structural analyses of the Potwar Plateau and Salt Range which included the construction of balanced structural cross-sections (Baker *et al.* 1988; Lillie *et al.* 1987; Pennock *et al.* 1989). Two United States Geological Survey publications (Meissner *et al.* 1974, 1975) and three recent papers (Abassi & McElroy 1991; McDougall & Hussain 1990; McDougall *et al.* 1993) address the surface and subsurface structure of the eastern parts of the Kohat Plateau. Two PhD theses, by Izatt (1990) and Wells (1984), also contain references to structural styles exhibited in the Kohat Plateau. This paper presents previously un-addressed details of styles and timing of the several phases of deformation which have affected the Kohat Plateau, as well as new interpretations of the structural style of the region.

Recent and current petroleum exploration by Amoco Pakistan Exploration Company in the Kohat Plateau (1990 to present) has provided an extensive collection of new surface and subsurface data. This paper, which focuses on the details of the surficial structural geology of the Plateau, is based largely on new, highly detailed surface data collected during two field seasons (9 months) connected with Amoco's exploration program: August 1990 to February 1991 and July to November 1991. Geological maps (at a scale of 1:50 000) and structural cross-sections (at a scale of 1:20 000) spanning much of the Kohat Plateau were constructed, and portions of these will be presented here to illustrate various structural styles and relationships (a 1:50 000 scale geological map of the Kohat Plateau can be found in Pivnik 1992). The data set used to construct these maps and cross-sections includes over 7000 dip measurements, formation-contact

From TRELOAR, P.J. & SEARLE, M.P. (eds), *Himalayan Tectonics*
Geological Society Special Publication, No. 74, pp. 559–580.

locations, fault orientations and motion-fabric orientations collected during traverses of over 600 km of seismic lines throughout an approximately 7000 km^2 area within the Kohat Plateau. Newly acquired seismic and well data are proprietary, although some preliminary interpretations will be presented in this paper. A well that was drilled recently by Amoco in 1991 in the Kohat Plateau represents the deepest penetration of the stratigraphic sequence of the region. The new seismic data are of better quality and are more regionally extensive than those previously collected by Amoco and other petroleum companies. Interpretations made from this new data set suggest that deformational styles within the Kohat Plateau are the result of both compressional and transpressional tectonics. Many of the features previously interpreted as being related to imbricated thrust faults and associated folds are in fact attributable to oblique convergence resulting in a hybrid terrain which has experienced polyphase thrust and wrench faulting. Reexamination of older, as well as newly-acquired seismic data from the NPDZ suggests that the NPDZ shares a similar structural history as the Kohat Plateau. These new interpretations, emphasizing the role of transpression, seem to complement observations and tectonic interpretations made from surrounding regions in Pakistan, and will hopefully provide impetus for new research and rethinking of the effects of Himalayan convergence in this portion of the foreland.

Regional geology and stratigraphy

Continuing convergence of the Indian subcontinent with the Asian plate and other minor plates during the Tertiary has produced several distinct tectonomorphic terrains in northern Pakistan (Fig. 1). In the northernmost areas of Pakistan, the Main Karakoram thrust (MKT) or Shyok suture separates rocks of the Asian continental shelf/slope and the accreted island-arc terrain of Kohistan, and is believed to have been active during the Cretaceous (Tahirkheli 1979). The Kohistan Arc terrain is interpreted to have been thrust southward over igneous, sedimentary and metasedimentary rocks of the northern Indian continental margin along the Main Mantle thrust (MMT) before 50 Ma (Treloar *et al.* 1990). Precambrian to Tertiary aged rocks in the footwall of the MMT are interpreted to have been thrust southward over continental-shelf and foreland-basin strata of the Kohat and Potwar Plateaux along the Main Boundary thrust (MBT), probably during the late Miocene or early Pliocene (*c.* 8.0 Ma; Izatt 1990). The southernmost thrust, the Salt Range thrust (SRT), juxtaposes Eocambrian to Pleistocene continental shelf and foreland-basin sequences of the Salt Range and Trans-Indus Salt Ranges southward over the Indo-Gangetic plain. Activity on the SRT most likely began at approximately 1.0 Ma in the Trans-Indus region (Pivnik 1992; Pivnik & Khan 1992), and perhaps at approximately 5.0 Ma in the Potwar portion of the Salt Range (Burbank & Beck 1989). Evidence for deformation as young as 0.4 Ma has been described by Yeats *et al.* (1984) for the southeastern Salt Range, near Jhelum. Yeats & Lillie (1991) have identified a 'blind' thrust south of the SRT that could represent the most recently developed structure of this portion of the northern Pakistani foreland. This paper focuses on the structural geology of the Kohat Plateau, a terrain of folded and faulted Tertiary aged rocks located in the footwall of the MBT and situated on the western margin of the northwestern syntaxis (Figs 1 & 2), at the western extremity of the northern Pakistan foreland.

Stratigraphically, the Kohat Plateau contains predominantly Eocene and younger rocks that record first the formation of a tectonically (?) restricted basin, followed by an Oligocene hiatus and the initiation of terrestrial foreland-basin deposition in the Miocene (Fig. 3; Meissner *et al.* 1974, 1975; Pivnik unpublished data; Wells 1984). Palaeocene and Lower Eocene black, grey and green shales and ribbon limestones of the Patala Formation and Panoba Shale were most likely deposited in a small, restricted marine basin that had limited outlet to the closing Tethys Sea. Later in the Lower Eocene, a carbonate platform developed in the northeast and an evaporite plateau, or sabkha, developed in the south and southeast (Shekhan Limestone and Jatta Gypsum, respectively). In the north and western portions of the basin, grey and green shales containing turbidites were being deposited (Panoba Shale). The late Lower Eocene Mami Khel Formation records the southward progradation of a fluvial/deltaic system that introduced red shales, sandstones and local conglomerates into the basin. In the Middle Eocene, shallow-water carbonate deposition occurred across the entire region (Kohat Formation). An unconformity between the carbonate shelf and Oligo (?)-Miocene synorogenic deposits mark the end of the last purely marine phase of deposition in the basin. The Murree Formation (Oligocene–Miocene) contains deposits of a mixed marine and fluvial depositional system which most likely formed in an estuarine environment. Miocene to Pliocene

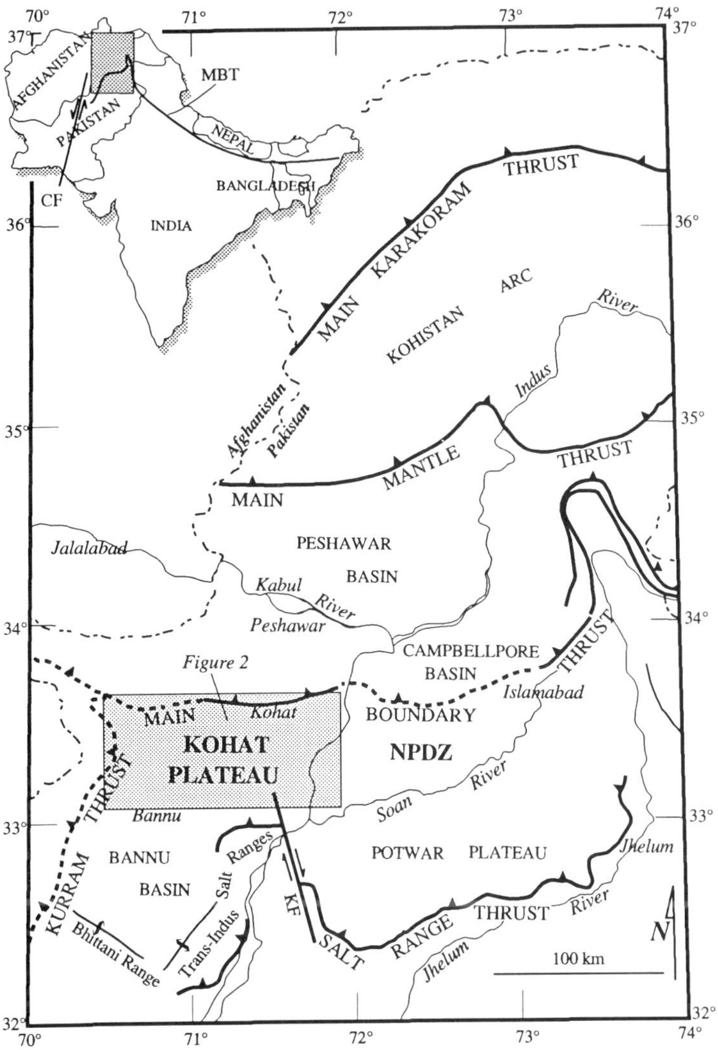

Fig. 1. Regional map of Pakistan showing major tectonomorphic terrains. Shaded box shows location of Figure 2. KF, Kalabagh Fault; CF, Chaman Fault; NPDZ, Northern Potwar Deformation Zone. Modified from Lillie *et al.* (1987).

aged foreland deposits of the Kamlial, Chinji, Nagri and Dhok Pathan Formations (Siwalik Group) contain the deposits of a south-flowing palaeo-Indus River that has served as a major foreland-basin trunk stream from the Miocene until the present (Bonis 1985; Pivnik, unpublished data). Plio-Pleistocene synorogenic deposits (Upper Siwalik Group) were deposited as the deformational front of the Himalayan convergence zone encroached upon the foreland basin (Pivnik 1992). Pleistocene and younger (?) gravelly terrace deposits are located throughout the Plateau, reaching thicknesses of greater than 50 m locally. Travertine mounds are exposed adjacent to many faults or within basal synorogenic units which are tightly folded. Jurassic through Palaeocene aged limestones, quartzites and minor shales outcrop on the hanging wall of the MBT, however, the structure and stratigraphy of these rocks remain poorly understood. To the south, the Trans-Indus Salt Ranges contain a Cambrian through Pleistocene stratigraphic section.

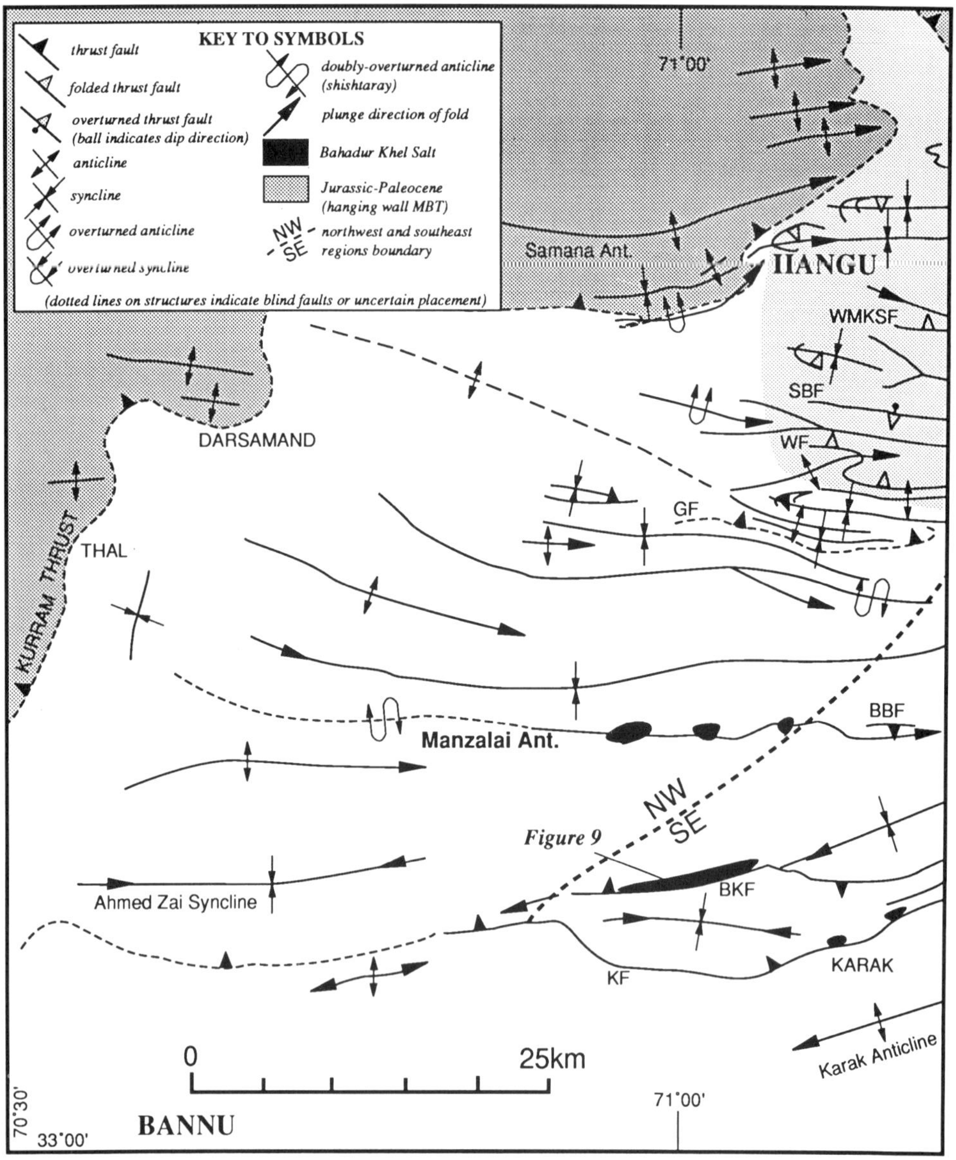

Fig. 2. Tectonic map of the Kohat Plateau showing major structural features and towns. Positions of structures are based on recent mapping, satellite images and Meissner *et al.* (1974, 1975). Lightly shaded area denotes the MKSTB. Locations of other figures in the paper are shown.

Structural geology

The Kohat Plateau contains structures that are related to thrust faulting and high-angle, oblique, reverse faulting. The predominant style of deformation at the surface of the Kohat Plateau is tight, east-west trending, gently to steeply and doubly plunging, doubly overturned (both limbs overturned) folds that can be traced for tens of kilometres along strike. The Plateau can be divided into two regions based on the structural styles exhibited in each region: a northwestern (NW) and southeastern (SE) region (Fig. 2). In the SE region, folds and both south- and north-dipping reverse faults are present, but in the NW region these types of

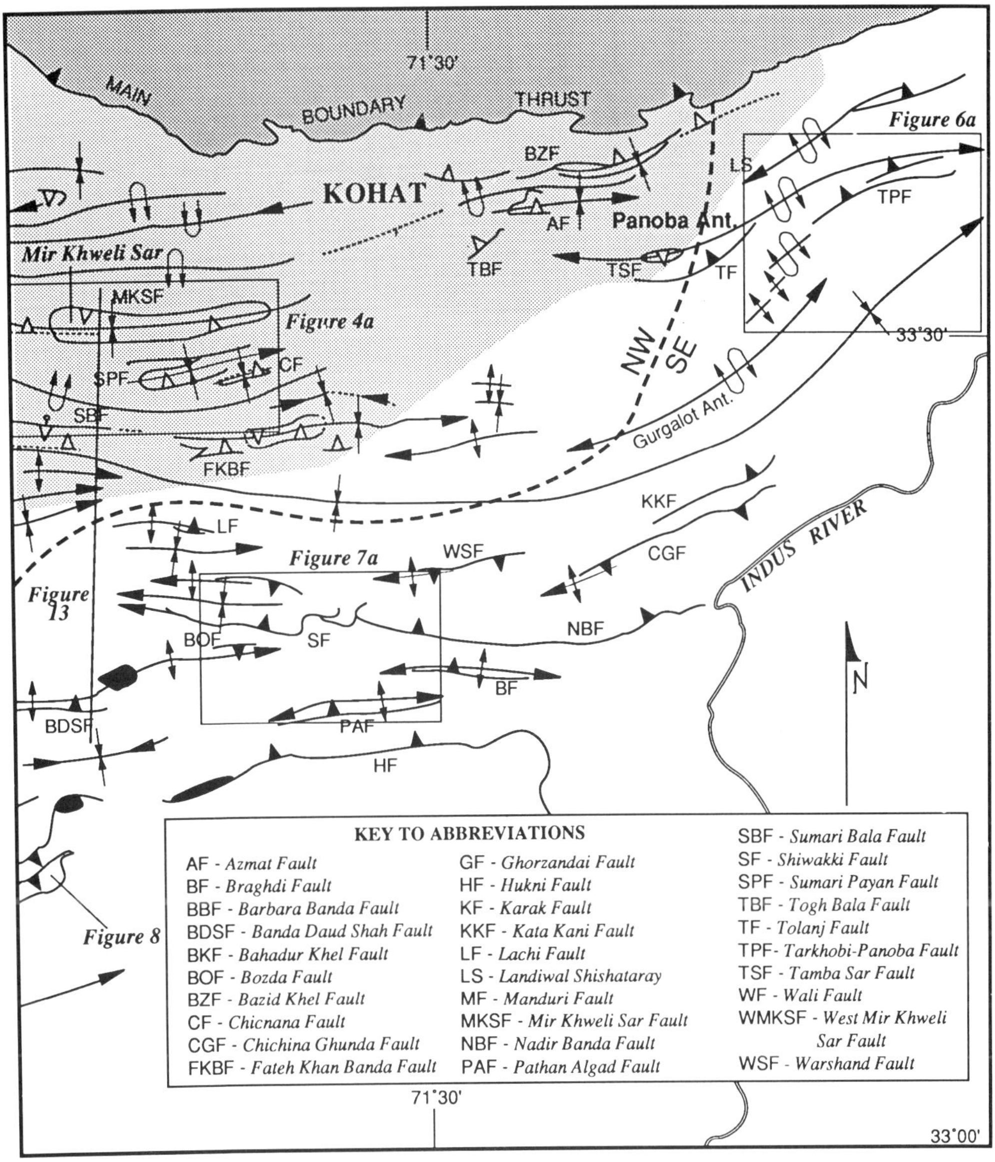

faults are not present, tight folding being the main deformational feature. However, folded with the rest of the Eocene and younger stratigraphy in the NW region is a relict thrust belt herein named the Mir Khweli Sar thrust belt (MKSTB). The MKSTB includes what were originally low-angle thrust faults and local, associated folds.

Mir Khweli Sar thrust belt: relict thrust belt

The MKSTB is characterized by folded, commonly overturned, E–W-trending faults that sole into a decollement in the Mami Khel Formation or lower Kohat Formation and that tip out in Miocene synorogenic deposits. Table 1 lists the faults and klippen that are part of the

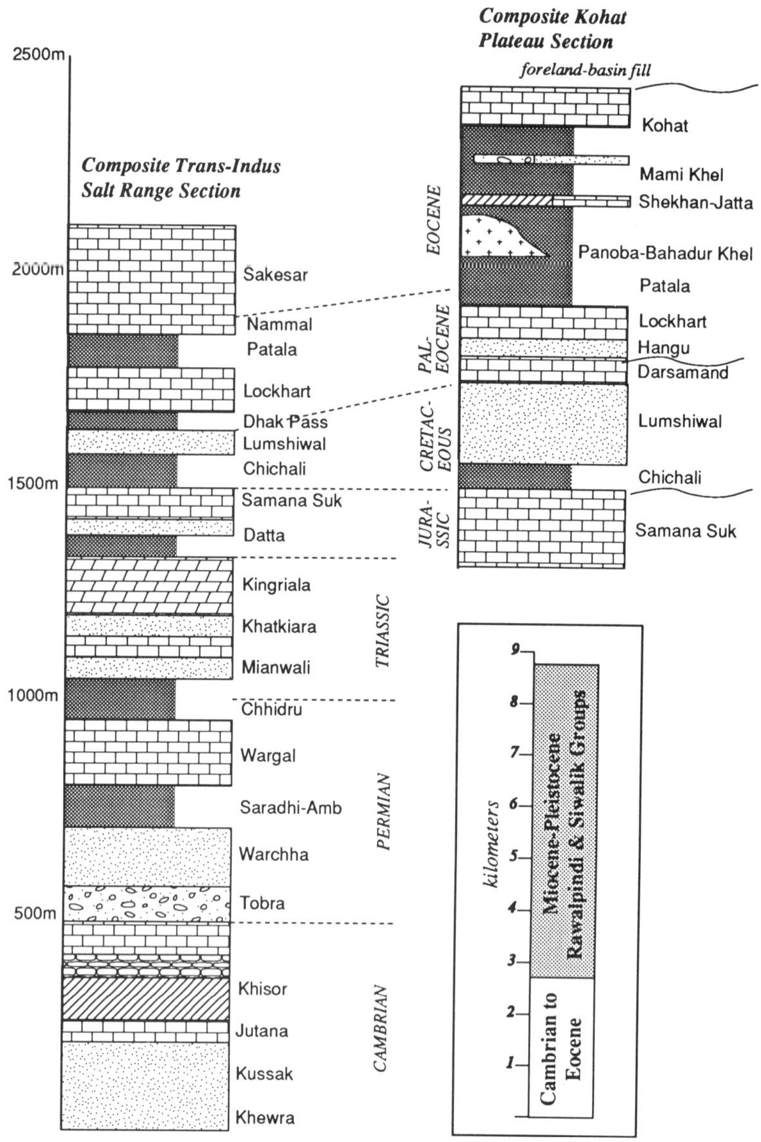

Fig. 3. (**a**) Stratigraphic column. Inset is a generalized Cambrian through Pleistocene section which shows the relationship between molasse and pre-molasse stratigraphic thicknesses. Data from recent mapping and sections previously measured by Amoco.
(**b**) Generalized basin cross-sections for the Palaeocene through Miocene of the Kohat Plateau. The true thicknesses of the Panoba Shale (Tp) in the western portion of the Plateau, and Bahadur Khel Salt (Tb) are unknown. The Jatta Gypsum (Tj) is present only in the southeastern portion of the Plateau. The Panoba Shale develops sandstone and conglomerate facies in the western portion of the field area, while the Mami Khel Formation (Tmk) develops a conglomeratic facies in the north-central portion, southeast of Hangu. Other abbreviations include: Tl, Lockhart Limestone; Tpa, Patala Formation; Ts, Shekhan Limestone; Tko, Kohat Formation; Tm, Murree Formation; and Tk, Kamlial Formation.

(b)

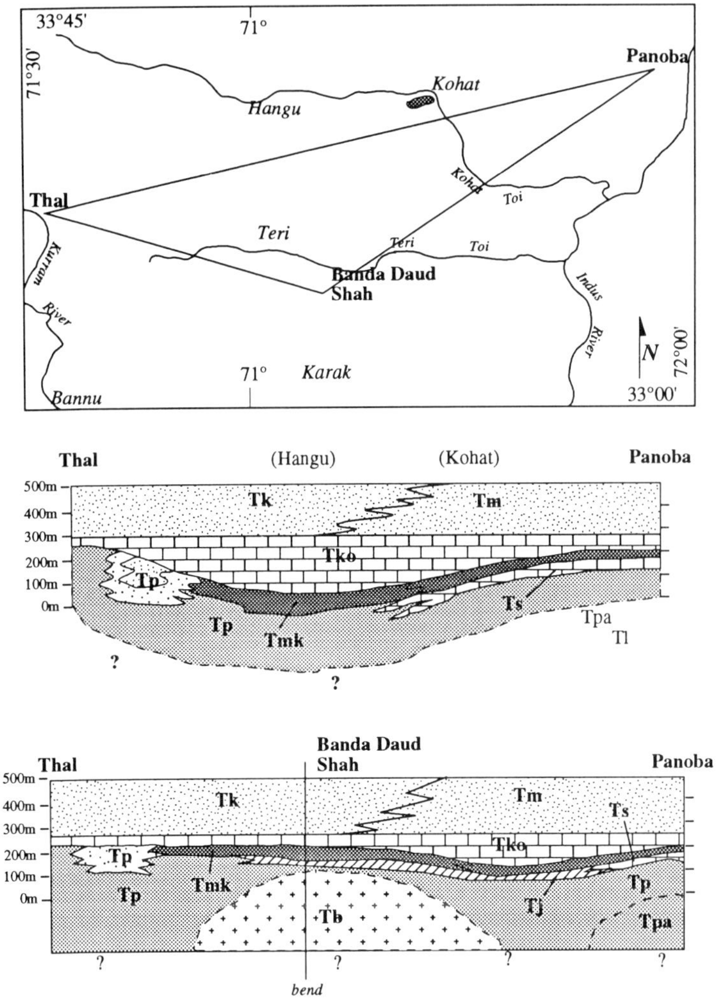

Table 1. *Thrust faults and klippen of the Mir Khweli Sar thrust belt**

Thrust faults	Klippen
Mir Khweli Sar fault	Mir Khweli Sar klippe
Sumari Payan fault	Sumari Payan klippe
Chichana fault	Chichana klippe
Sumari Bala fault	
Walai fault	
West Mir Khweli fault	
Fateh Khan Banda fault	Fateh Khan Banda klippe
Manduri fault	Manduri klippe
Bazid Khel fault	Bazid Khel klippe
Azmat fault	
Tamba Sar fault	Tamba Sar klippe
Togh Bala fault	

* See Fig. 2 for abbreviations and locations of faults and klippen.

MKSTB. The fault system occurs in the NW region of the Plateau and is named after excellent exposures of a repeated section of Mami Khel to Murree Formations atop Mir Khweli Sar (1417 m elevation). At this locality, the Mir Khweli Sar fault is folded into a syncline, forming a large klippe with Murree Formation in the hanging wall (Fig. 4). Because the MKSTB is folded, many of the faults are easily mapped as paired ridges of Kohat limestone (Fig. 4). Traverses across all of these paired ridges have identified repeated stratigraphies and facing directions that confirm the position of the faults. After the faults are retrodeformed and unfolded, they appear shallow dipping, commonly dipping less than 30° to the north. Lateral thrust tips are well-exposed, expressed as sudden, along strike terminations of one in a pair of

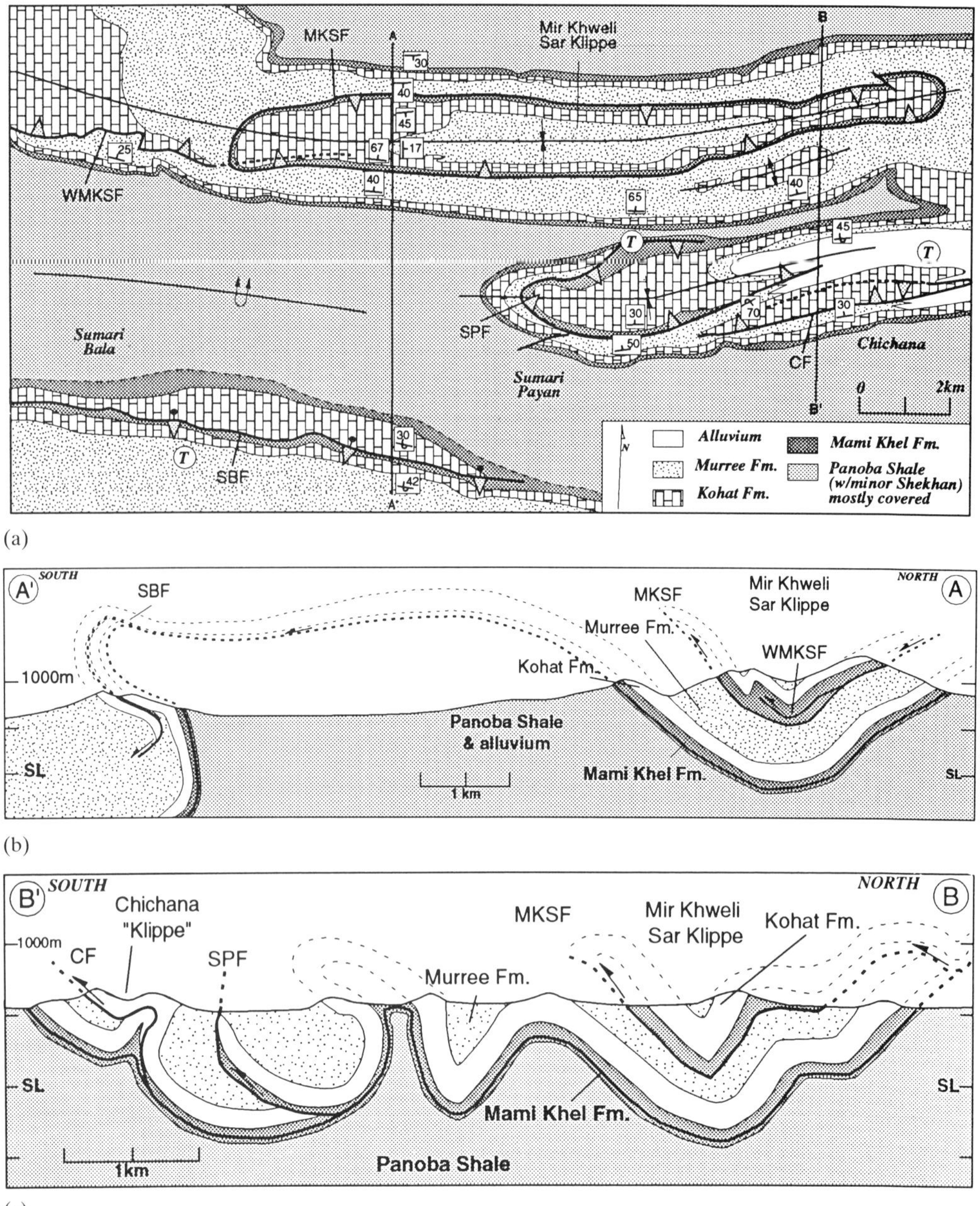

Fig. 4. (**a**) Geological map of the Mir Khweli Sar area showing folded thrust faults and folds. Structural symbols and abbreviations for faults are the same as those used in Fig. 2. (**b**) and (**c**) Cross-sections A–A′ and B–B′ as shown in (4a). Fault tips and cut off angles are estimated. SBF, Sumari Bala Fault; MKSF, Mir Khweli Sar Fault; WMKSF, West Mir Khweli Sar Fault; CF, Chichana Fault; SPF, Sumari Payan Fault. Topography is controlled by detailed surveying as part of Amoco's recent exploration program. No vertical exaggeration.

vertical ridges of Kohat limestone (Fig. 4). Many thrust ramps and tips have been inferred in the cross sections drawn across the region, however, there are several locations where ramps have been observed in the field (Fig. 4). Because of overprinting effects of later, similarly trending folds that characterize the entire Kohat Plateau, only a few examples of folding associated with

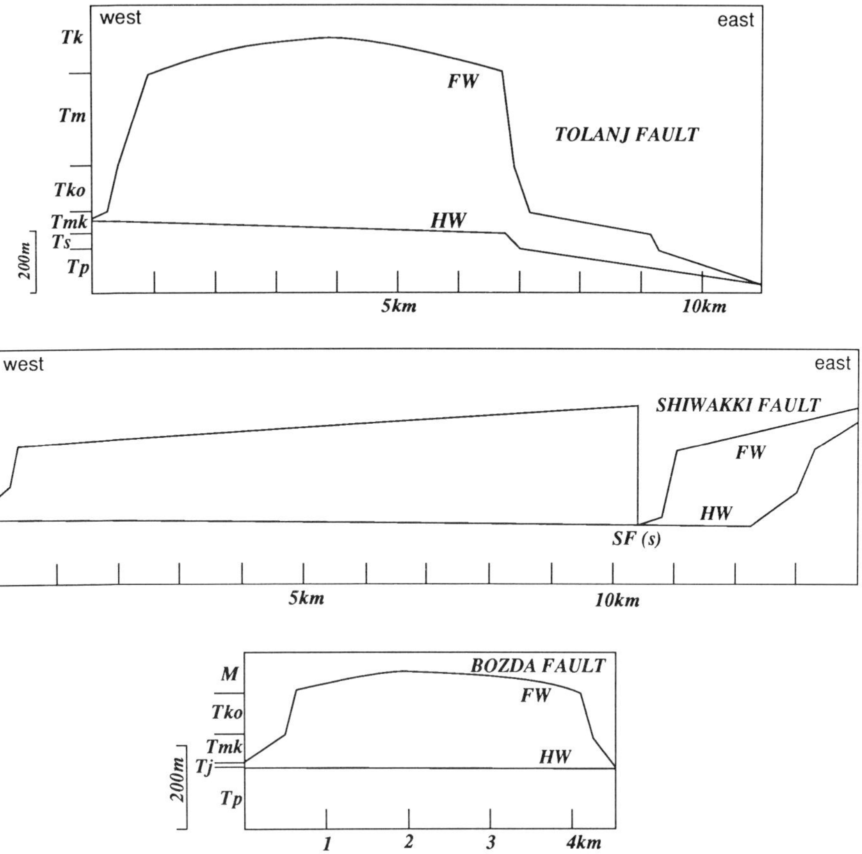

Fig. 5. Stratigraphic separation diagrams for faults in the SE region of the Kohat Plateau. At the surface, the Tolanj fault cuts upsection from rocks at least as deep as the Panoba Shale, and the Mami Khel Formation that is exposed along the majority of the length of the trace of the hanging wall (HW) is located on the southern limb of a fault-propagation fold (the cross section of the Panoba fault, shown in Fig. 6b, shows the same structure). Ramp-forming horizons are the Shekhan Limestone (Ts) and the Kohat Formation (Tko), while flats have formed in the Panoba Shale (Tp), the Mami Khel Formation (Tmk) and the upper Murree and lower Kamlial Formations (Tm & Tk). The Shiwakki and Bozda faults both have formed flats near the top of the Panoba Shale, just a few meters below the Jatta Gypsum (Tj). The sudden jump in the trace of the footwall (FW) of the Shiwakki fault is where the fault intersects a splay. The eastern termination of the Shiwakki fault is a lateral ramp (also see Fig. 7a & b).

the MKSTB have been recognized. The few examples of these types of folds are: a fault-propagation fold in the Mir Khweli Sar klippe and fault-bend folds associated with ramps of the Chichana and Bazid Khel faults (Figs 2 & 4).

The West Mir Khweli Sar fault is relatively undeformed, and could be the closest remaining representation of the pre-folded MKSTB (Fig. 4). The geometry of the faults of the MKSTB indicate that this was a relict, thin-skinned thrust belt, formed during an episode of compression-related tectonics. Travertine mounds are associated with many of the faults of the MKSTB.

Kohat Plateau faulting and folding

Tight folding and faulting have affected both the NW and SE regions of the Kohat Plateau. However, faults are predominant in the SE region, and only folds are exposed in the NW region (with the exception of the MKSTB).

All faults in the SE region place Lower Eocene rocks over Miocene synorogenic deposits. There are two basic varieties of faults. The first variety is typified by the Tolanj and Panoba faults (Figs 2, 5 & 6). They both juxtapose the Mami Khel Formation on the

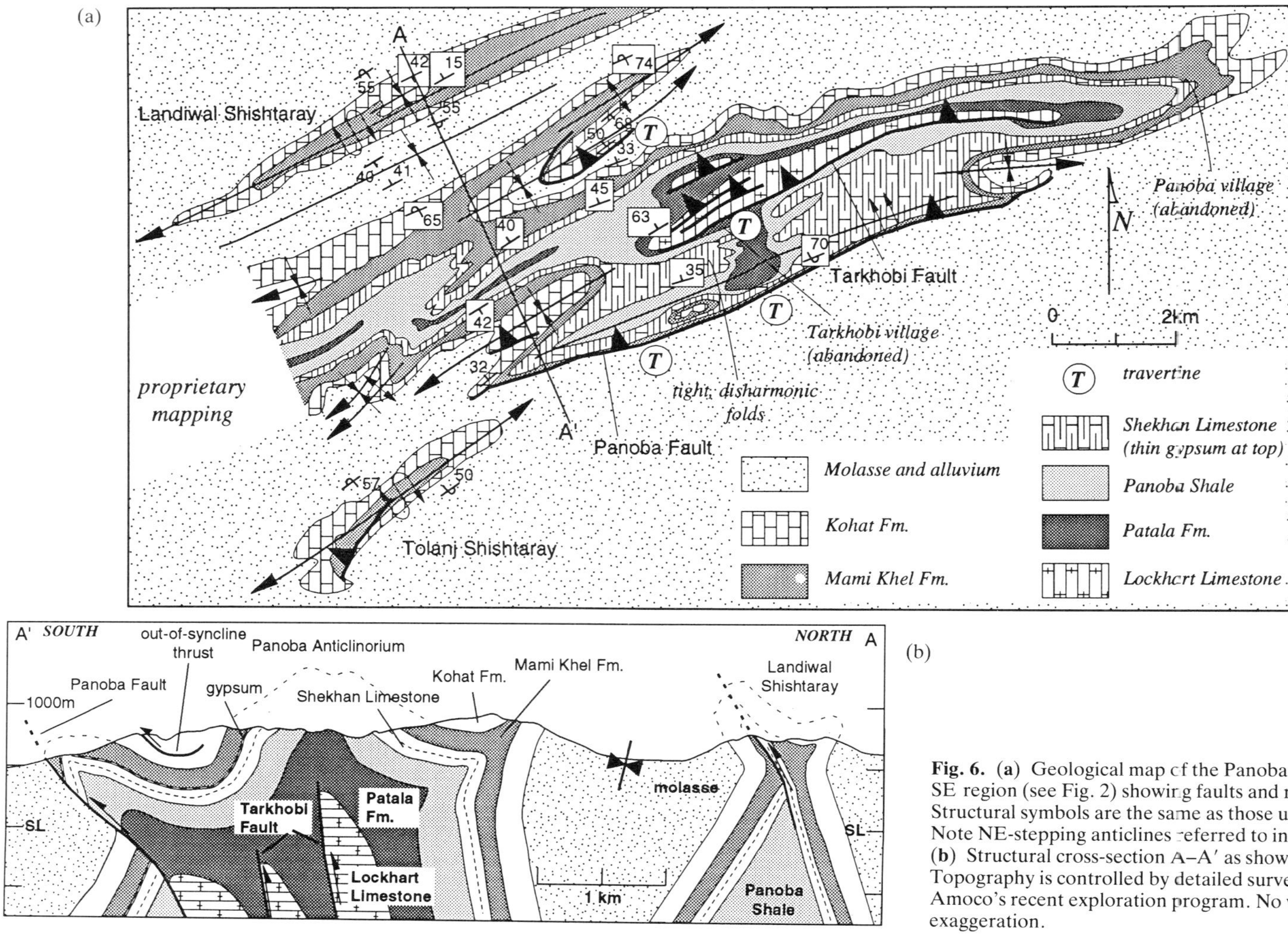

Fig. 6. (**a**) Geological map of the Panoba fault area, SE region (see Fig. 2) showing faults and related folds. Structural symbols are the same as those used in Figure 2. Note NE-stepping anticlines referred to in text. (**b**) Structural cross-section A–A′ as shown in (a). Topography is controlled by detailed surveying as part of Amoco's recent exploration program. No vertical exaggeration.

(a)

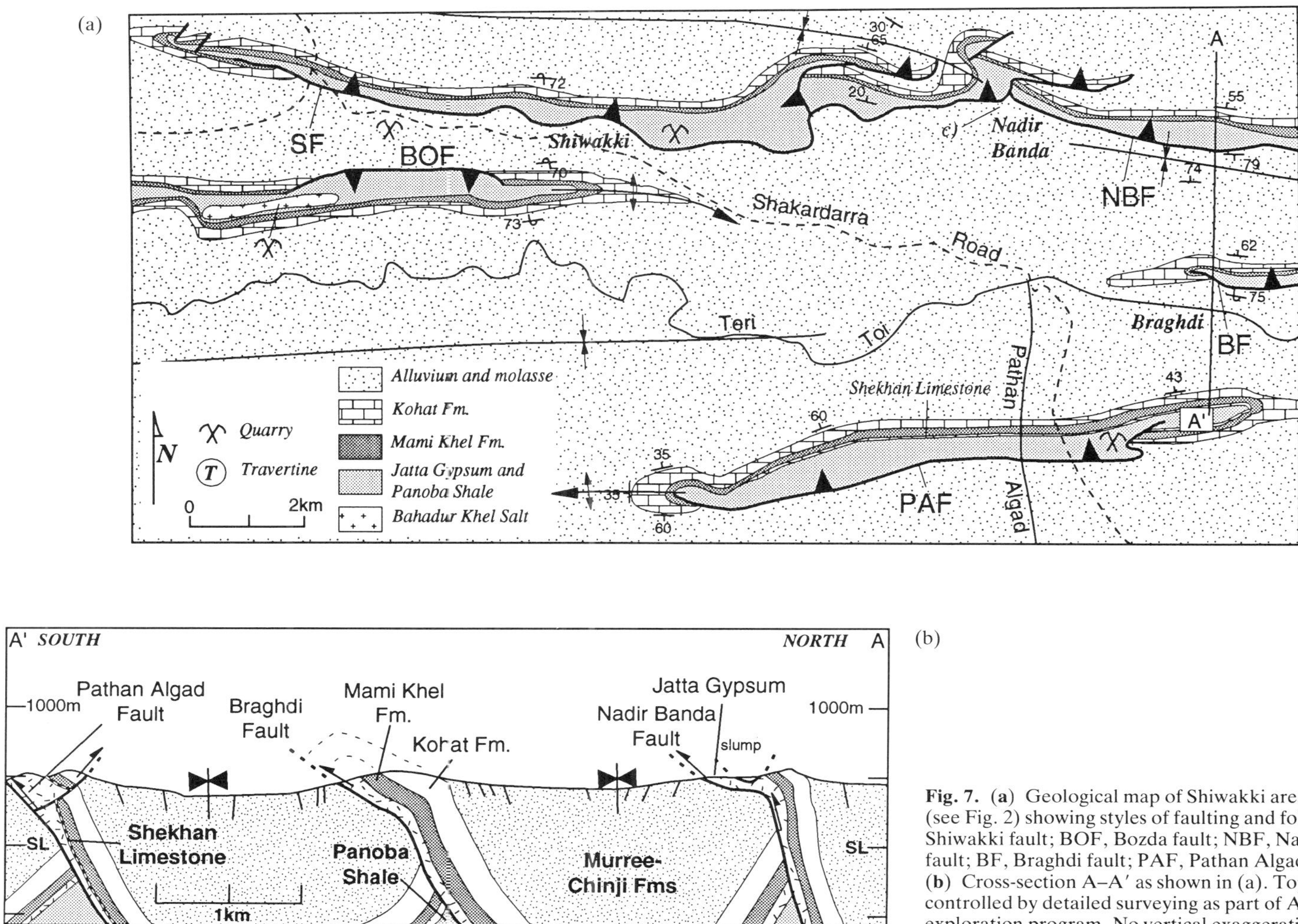

(b)

Fig. 7. (**a**) Geological map of Shiwakki area, SE region (see Fig. 2) showing styles of faulting and folding. SF, Shiwakki fault; BOF, Bozda fault; NBF, Nadir Banda fault; BF, Braghdi fault; PAF, Pathan Algad fault. (**b**) Cross-section A–A′ as shown in (a). Topography is controlled by detailed surveying as part of Amoco's recent exploration program. No vertical exaggeration.

southern limb of an overturned anticline on their hanging walls southward over the upper Murree and lower Kamlial Formations on their footwalls (Figs 5 & 6). In the cross-section shown in Fig. 6b, the Lockhart Limestone is complexly faulted in the core of the Panoba Anticlinorium. The frontal fold above the thrust is interpreted as a drape fold (Zolnai 1988) which has formed above a NE–SW-trending, high-angle reverse fault. The Tolanj and Panoba faults occur in the northeastern section of the SE region of the Plateau where the Eocene section contains two competent formations; the Kohat Formation and Shekhan Limestone. The Panoba and Tolanj faults have cut upsection through these competent units; the faults are at high angles to the Shekhan Limestone in their hanging walls (Fig. 6b). Because the Shekhan pinches out into the Panoba Shale to the west, faults did not cut up-section west of the Panoba Anticlinorium, but rather flattened out within the Panoba Shale (as described below). This first type of fault is restricted to the Panoba Anticlinorium, where the Palaeocene/Eocene section includes competent, ramp-forming lithologies.

The second variety of faults are both north verging and south verging and are exposed in the southern part of the SE region. Figure 7 shows several of these faults. Their positions coincide with the lateral distribution of the Jatta Gypsum and Bahadur Khel Salt, which outcrop only in the SE region (Figs 2 & 3). The uppermost Panoba Shale, directly underneath the Jatta Gypsum, served as a decollment horizon for these faults, as is graphically displayed on the stratigraphic-separation diagrams shown in Fig. 5. These faults cut upsection through the Jatta Gypsum, which commonly has a hard calcrete or dolomite unit at its base. The Jatta Gypsum, which crops out on the hanging walls of these faults, is commonly imbricated and folded with slivers of Panoba Shale underneath the gypsum, suggesting that flats formed high in the shale, underneath the *relatively* more competent Jatta (Figs 7 & 8). The imbricated thrusts and fault-propagation folds exposed locally within the gypsum are geometrically similar to structures shown in several geometric and kinematic models of fault-propagation folding (Fig. 8; Chester *et al.* 1991; Jamison 1987; Mitra 1990; Suppe & Medwedeff 1990), and represent the uppermost portions of high-angle faults that shallow near the surface. The Jatta Gypsum is folded disharmonically, and poorly exposed sections have left previous workers with the impression that it is several times thicker than its

Fig. 8. Outcrop photograph of Jatta Gypsum (Tj) and Panoba Shale (Tp) faulted northward over the Chinji Formation (Tc) along the Karak fault northeast of Karak. View is to the east. Location of the outcrop is shown on Fig. 2. View of Fig. 12a is from left side of picture.

Fig. 9. Photograph of salt diapir at Bahadur Khel Salt Quarry (see Fig. 2). Panoba Shale (Tp) outcrops above the salt (Tb), and a complete section of the salt through Miocene synorogenic deposits are exposed to the left of the photograph. Dome is approximately 50 m high. View is to the east.

actual thickness of 5–20 m. Nearly all of the faults in the SE region tip out laterally into anticlines (Fig. 7a), suggesting that fault-propagation folds developed on their hanging walls as the faults cut upsection and that continued fault propagation truncated these folds.

The Bahadur Khel Salt has behaved diapirically and forms the cores of both faulted and unfaulted anticlines (Figs 2 & 9). Commonly, a thin unit (*c.* 5 km) of Panoba Shale and the Jatta Gypsum, which overlies the shale, are folded above the salt diapirs (Fig. 9). As an example, the Bahadur Khel Salt forms diapirs within the core of the Manzalai structure (Fig. 2). The anticlinorium has remained largely unfaulted at the surface. The true thickness of the Bahadur Khel salt cannot be determined because its lower contact is not exposed, and no wells have completely penetrated its true stratigraphic thickness.

Out-of-syncline thrusts also occur in the plateau. These faults occur within synclines and have only local extent. They most commonly have a lower detachment in the Mami Khel or lower Kohat Formations. It appears that these faults formed during folding to accommodate diminishing space in the cores of tight folds.

The varieties of faults described above (excluding out-of-syncline faults) are not present in the NW region of the Kohat Plateau (Fig. 2). Instead, the region is characterised by tight folds and associated out-of-syncline faults. In the NW region, the Shekhan Limestone, Jatta Gypsum and Bahadur Khel Salt are not present or are greatly reduced in thickness, as compared to the SE region (Fig. 3). Instead, the Panoba Shale crops out in the cores of most anticlines. Based on seismic data and the fact that the base of the Panoba Shale is never exposed in the NW region, the Panoba Shale is as much as 2 km thick in the NW region, as opposed to approximately 150 m thick in the SE region. The tight folds that are present in the NW region are interpreted as drape folds above wrench faults or detachment folds (Jamison 1987) that have formed above flats in these faults. The stratigraphy differs significantly enough between the SE and NW regions to control the absence or presence of faults at the surface (Fig. 10). The Panoba Shale, as much as 2 km thick in the NW region, served as the detachment horizon in which faults flattened near the surface in the NW region, preventing them from being exposed at the surface (Fig. 10). Diapirism of the Panoba Shale (and Bahadur Khel Salt) most likely occurred along a detachment located above the more competent Palaeocene and older section.

Travertine mounds are exposed adjacent to many faults in the SE region and to many out-of-syncline faults in the NW region.

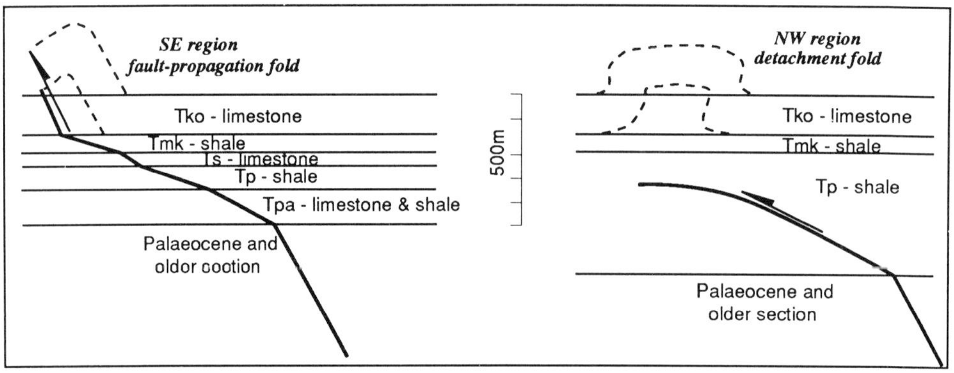

Fig. 10. Diagrammatic sketch showing differing stratigraphies in the NW and SE regions and its effects on the propagation of faults. In the NW region, faults flatten in the relatively thick Panoba Shale, and detachment folds form above them. In the SE region, fault ramps are formed in several horizons, resulting in surfaceward propagation of faults and the development of fault-propagation folds. Abbreviations for formations are the same as in all other figures.

Evidence for transpressional tectonics in the Kohat Plateau

Previous models of deformation in the NPDZ and eastern Kohat Plateau have used imbricate thrust and passive-roof duplex geometries in the construction of balanced, structural cross-sections (Abbasi & McElroy 1991; Baker *et al.* 1988; McDougall & Hussain 1991). These models assume, among other things, that out-of-section transport of material was not significant and that the predominant style of faulting was low-angle, imbricate thrusting with a passive roof thrust capping a blind, northward dipping duplex. However, the interpretations of the subsurface geology of the eastern Kohat Plateau have been based on seismic data of 'poor' quality, or only very limited amounts of seismic data which provided 'no constraint on the internal geometry of the stack', or blind duplex (Abbasi & McElroy 1991, p. 322); used stratigraphic data from regions external to the Kohat Plateau and contained no seismic-velocity data from within the Kohat area (McDougall & Hussain 1991). Our detailed mapping and basin analyses of the Palaeocene and Eocene rocks exposed in the Kohat Plateau have shown that facies are quite variable across the Potwar–Kohat region, as well as within the Kohat Plateau itself (Pivnik unpublished data; Wells 1984), thus emphasizing the importance of using stratigraphic data collected within the area. Initial interpretations of the structure of the Kohat Plateau also utilized the assumptions listed above. The original interpretations of the seismic data resulted in cross sections which

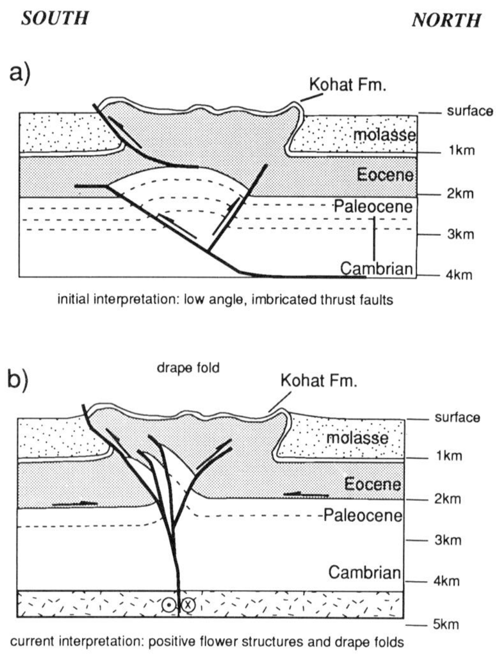

Fig. 11. (**a**) Cartoon sketch of original interpretations of subsurface structures in the Kohat Plateau. This model assumed that deformation was dominated by thin-skinned thrusting and the development of low-amplitude folds that formed above bends in thrust faults.
(**b**) Cartoon sketch of interpretations of subsurface structures after well was drilled, and based on seismic data. This model shows steep dips on bedding and faults at depths of over 3 km, as well as offsets in seismic basement and the development of positive flower structures.

included north-dipping, imbricate thrusts and broad, low-relief structures at depth which presumably involved the Palaeozoic and Mesozoic section. This was accompanied by disharmonic folding of the shale-rich Tertiary section above (Fig. 11a). These initial interpretations were not consistent with subsequently acquired well data. There are several lines of evidence that show that in the Kohat Plateau, fault and bed dips are steep at great depths, throw on faults is extremely high, and that an important degree of wrench faulting and lateral transport has taken place. These pieces of evidence are presented below.

A wildcat well recently drilled approximately 25 km east of Kohat at the village of Tolanj encountered dips of over 60° from spud to depths of over 3 km. Seismic data from this area are still proprietary and cannot be graphically presented here. Migrated seismic and well data suggest high-angle fault trajectories that flatten out near the surface, rather than low angle imbricated thrusts. Faults in the SE region are interpreted as being different surficial expressions of a generation of deep-seated, high-angle reverse faults. The Lockhart Limestone and older section is involved in this faulting (e.g., Tarkhobi and Panoba faults; Fig. 6), and the Eocene shale- and evaporite-rich section has served as an upper incompetent horizon in which these faults have shallowed near the surface. Near-surface shallowing, high-angle, reverse faults have been termed 'positive flower structures' (Wilcox *et al.* 1973) or 'palm-tree structures' (Sylvester 1988). The geometries of the faults as drawn on cross sections based on the well data and seismic data are similar to positive flower structures, common in terrains subjected to oblique compression, or transpression (Harding 1985). Offsets in seismic basement (crystalline rocks of the Indian craton) on the order of 100 milliseconds are also present, suggesting that basement is involved in faulting with throws of up to 500 m. Proprietary aeromagnetic and gravity data from the Kohat Plateau show linear, conjugate sets of faults and basement uplifts, which are structures uncommon in thrust belts (Sercombe unpublished data).

Vertical separations along faults that are greater than 4 km have been observed in previously- and recently-acquired seismic and well data. For example, a well near Karak, south of the Karak–Hukni fault, which is the southernmost fault in the Kohat Plateau (Fig. 2; Abassi & McElroy, 1991 used the term Hukni fault), did not reach the bottom of the Miocene Chinji Formation before it was abandoned at over 4 km depth (the Chinji Formation has a stratigraphic thickness of approximately 1 km). Just a few kilometres north of the well, the lower Eocene Jatta Gypsum crops out on the hanging wall of the Karak–Hukni fault. This suggests that there is a vertical displacement across the fault of well over 4 km, because the Jatta lies stratigraphically approximately 1 km below the base of the Chinji Formation, and there are approximately 2 km of Siwalik Group strata that overlie the Chinji. The Karak-Hukni fault is vertical in the subsurface and flattens at the surface where it flips in vergence from north to south (Fig. 2). Newly-acquired seismic data collected near Bannu show that the western equivalent of the Karak-Hukni fault has uplifted the Kohat Formation more than 4 km relative to its position at depth in the footwall of the fault. These kinds of vertical separations and seismic images are characteristic not of shallowly-dipping, imbricate thrusts, but of steeply-dipping reverse faults.

Small displacement, strike-slip deformation is expressed on the surface as northeast-stepping folds. In the hanging wall of the Bahadur Khel fault, several northeast-stepping synclines deform rocks as young as the Plio-Pleistocene Upper Siwalik Group (Fig. 2); east of Kohat, three northeast-stepping anticlines separate the Gurgalot and Panoba anticlinoria (Figs 2 & 6); and, the hanging wall of the MBT contains several left-stepping anticlines between Thal and Hangu, as well as NE-stepping trends throughout the Kohat Range (Fig. 2; unpublished satellite imagery; Meissner *et al.* 1974, 1975).

Orientations of motion-fabrics such as slickensides and fault-plane scratches exposed throughout the Kohat Plateau indicate predominantly north over south transport of rocks, related to bedding-plane slip normal to E–W trending fold axes. However, there are several localities which contain horizontal fabrics that indicate lateral motion, including horizontal slickensides on nearly vertical bedding planes of the Shekhan Limestone near the village of Tolanj and folds with north–south-trending fold axes along the Karak fault northeast of Karak (Fig. 12a). The high-angle faults and positive flower structures used in the construction of cross-sections (Fig. 11b), also suggest that a component of lateral motion, as a result of transpression rather than pure compression, has controlled the geometries of the structures in the Kohat Plateau. The use of these types of structures in cross section makes it difficult to assume plane strain, one of the primary constraints for area-balanced cross-sections (Harding 1985).

High-angle, oblique-slip faults that flatten out

Fig. 12. Photographs of mesoscopic motion fabrics. (**a**) Outcrop photograph of north–south-trending fold on north verging Karak fault, indicating lateral, east–west motion as well as north–south motion on the fault. See Fig. 8 for a N–S cross section of fault. (**b**) Fault-plane scratches in early Palaeozoic slates on Karakoram Highway indicating east–west, strike slip motion. View is to north.

to form thrust faults near the surface (i.e. positive flower structures) have been presented for a variety of strike-slip and oblique-slip terrains (for the San Andreas fault see Allen 1957; Burke 1979; Sylvester & Smith 1976; Willis 1938; for the Alpine fault see Wellman 1955; for Svalbard see Kellog 1975; for the Spitsbergen fracture zone see Lowell 1972; for a comprehensive overview see Sylvester 1988). Wrench faulting simulated in laboratory models using clay (Lowell 1972), sand (Emmons 1969; Naylor *et al.* 1986) and limestone (Bartlett *et al.* 1981) also resulted in the formation of near-vertical faults that flattened near the surface forming thrust faults. According to Allen (1965), it is common to interpret these types of structures as being the result of purely compressional tectonics because the faults that outcrop at the surface can resemble a series of imbricated thrust faults, rather than faults related to lateral displacements (e.g., Fig. 8). Compounding the difficulty of interpretation, the relationship between the deep-seated, high-angle portions of these faults and their near-surface, shallow extensions can often not be determined from outcrop studies. However, detailed outcrop studies in the Mecca Hills by Sylvester & Smith (1976) documented the relationship between competent basement rocks uplifted by high-angle, oblique-slip faults and the sedimentary rocks that overlie them. In particular, the Painted Canyon fault is shown to cut steeply upsection through competent basement rocks, and then to flatten out in the sedimentary cover sequence, which is complexly folded (Sylvester & Smith 1976, Figs 15 & 16). This geometry is analogous to the relatively competent Precambrian, Palaeozoic and Mesozoic section and overlying, shale- and evaporite-rich Eocene section of the Kohat Plateau (Fig. 11b). The geometries of the folds exposed at the surface above the oblique-slip reverse faults in the Mecca Hills grossly resemble those of folds in the Kohat Plateau (Sylvester & Smith 1976, Fig. 10).

The rationale for interpreting the folds and faults of the Kohat Plateau (excluding the MKSTB) as being transpression-related rather than purely compression-related is also rooted in regional tectonics and structural trends. The notion that major strike-slip deformation is associated with Himalayan convergence is not a new one. Tapponnier *et al.* (1982, 1986) used plasticene models and field studies to examine the role of lateral transport of material around the edges of the Indian plate indentor during collision of the Indian and Asian continents. The Kohat Plateau is located in the northwestern corner of a major syntaxial bend of the Himalayan suture zone and near the NW corner of the Indian plate indentor (Fig. 1). The western boundary of the Plateau is the Kurram thrust, a series of low- to high-angle, west dipping faults that carried Cretaceous and older strata eastward from depths of greater than 6 km to the surface in a horizontal distance of 5–10 km. West of the Kurram thrust, in Afghanistan, is the Chaman fault, a north–south-trending sinistral strike-slip fault that represents the western suture between the Indian and Afghan plates (Lawrence *et al.* 1981).

Verma *et al.* (1980) presented focal-mechanism solutions for 4 earthquakes (depths of 17–43 km) in the Kohat Plateau and in the mountains on the hanging wall of the MBT. Each of these solutions suggest left-lateral strike-slip faulting, and no thrust faulting. Focal-mechanism solutions from earthquakes in the Sulaiman Ranges (previously interpreted as an east-verging, imbricate thrust system) also consistently yielded strike-slip motion (Verma *et al.* 1980).

Thus, the Kohat Plateau is seen to be located in an area where north-south compression, east-west compression (Kurram thrust) and sinistral strike-slip motion (Chaman fault) are all likely to have played a role in controlling the evolution of structural deformation. Models utilizing purely compressional, south-vergent imbricate thrusting do not apply to this portion of the foreland. Structures within the Kohat Plateau reflect polyphase, multidirectional deformation, and constitute a hybrid terrain of compressional and wrench features.

It should be noted however, that the difficulty in absolutely identifying transpressional surface features in the Kohat Plateau is compounded by the presence of compression-related structures (MKSTB) and by the presence of tight folds that were created by diapirism of salt or shale associated with wrench faulting (Harding 1985, 1990). At present, no large-displacement strike-slip faults have been identified at the surface in the Kohat Plateau, but rather short-displacement foreland wrench-adjustment features such as left-stepping structures which are most likely related to a strike-slip component of motion. A single strike-slip fault does not exist, but instead, a series of steeply-plunging wrench 'welts' (Sylvester 1988) or pressure ridges (Biddle & Christie-Blick 1985) that formed over positive flower structures are expressed at the surface as reverse faults and tight folds. The Karak–Hukni fault, which forms the southern margin of the plateau is perhaps the best example of an exposed wrench fault. In addition to the large vertical displacements along the fault and motion fabrics that indicate E–W transport (discussed above; Fig. 12a), the fault is traceable for over 100 km, and flips from being south to north vergent (Fig. 2) and can be identified on seismic data as a high-angle reverse fault (Sercombe unpublished data).

Compressional and transpressional deformation

The distribution of structures in the Kohat Plateau shows that compressional structures represented by the MKSTB have been overprinted by transpression-related structures that have affected the entire Kohat Plateau. As

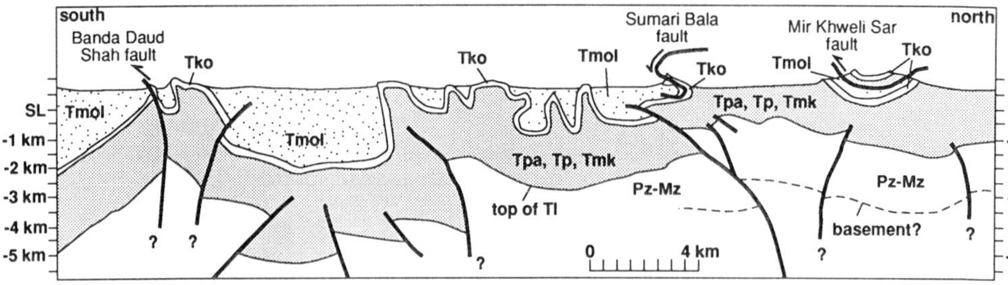

Fig. 13. Regional cross-section across the Kohat Plateau, based on proprietary seismic data and surface mapping. Northern portion of section corresponds with Fig. 4b. Line of section is marked on Fig. 2. Tmol, Rawalpindi & Siwalik Groups; Tko, Kohat Fm; Tmk, Mami Khel Fm; Tp, Panoba shale; Tpa, Patala Fm; Tl, Lockhart Fm; Pz–Mz, Palaeozoic to Mesozoic.

shown in the regional structural cross-section of Fig. 13, the faults in the MKSTB have been incorporated into drape or detachment folds related to deep-seated, high-angle reverse faulting. Formerly low-angle thrust faults of the MKSTB suggest that these relict structures were the result of generally north-south compression. Steep dips of bedding both at the surface and at depth (>4 km), positive flower structures, large vertical displacements on faults, and left-stepping folds all indicate that the Kohat Plateau has experienced deformation in response to a combination of north–south compression and regional north–south oriented, sinistral strike-slip motion. In addition, we were not able to map duplexes, passive roof thrusts, antiformal stacks or any other structures that would indicate that the plateau has undergone purely north to south, progressive thrusting. Syndeformational deposits associated with the tranpressive deformation are Plio-Pleistocene in age (Pivnik, in prep.).

What may be gleaned from the structures exposed in the Kohat Plateau is the relative importance of compression and transpression during what seems to be a continuum of deformational episodes. At this point a complex history of compression and transpression seems to characterize not only the Kohat Plateau, but also regions as far north as the Chinese border, and south in the Trans-Indus Salt Ranges and also southern Waziristan. On a recent reconnaissance trip through the Karakoram, we found that orientations of slickensides and fault scratches from Peshawar to the Khunjerab Pass indicated predominantly horizontal motion along steeply-dipping, east–west-trending faults and bedding planes (Fig. 12b); and recent mapping and motion-fabric orientation data collected south of the Kohat Plateau in the Trans-Indus Salt Ranges also reveal a component of strike-slip deformation as well as thrusting (P. Blisniuk unpublished data).

Transpressive structures elsewhere in the northern Pakistan foreland

The types of structures described from within the Kohat Plateau are also distributed throughout the northern part of the foreland region. Of particular importance to this study are: NE-stepping pressure ridges in the southern Peshawar Basin; NE-stepping pressure ridges in the southern Campbellpore Basin; tight, doubly-overturned folds and NE-trending structures in the Kohat Range, Attock–Cherat Range and Kala Chitta Range; tight, doubly overturned folds and high-angle faults in the NPDZ (Figs 1 & 14).

Peshawar Basin, Kohat Range and Attock–Cherat Range. Yeats & Hussain (1989) describe a series of NE-stepping (en-echelon) pressure ridges bounded by high-angle reverse faults occurring in the southern Peshawar Basin (Fig. 14). Low-angle slickensides are present on one of the faults, suggesting a component of strike-slip motion. Mapping of the Kohat Range and western Attock–Cherat Range using SPOT satellite images, as well as mapping by Yeats & Hussain (1989) has revealed NE-trending lineations that parallel the pressure ridges and that are discordant to the east–west-trending (thrust?) faults in the ranges (which are locally displaced by NE-trending, right-lateral strike-slip faults). One of these lineations is marked by a north-dipping reverse fault (Yeats & Hussain 1989). The entire Attock–Cherat Range may also be a large-scale version of the pressure ridges, and the high-angle faults can be compared to strike-slip faults in Afghanistan and China and to right-lateral strike-slip reactivation

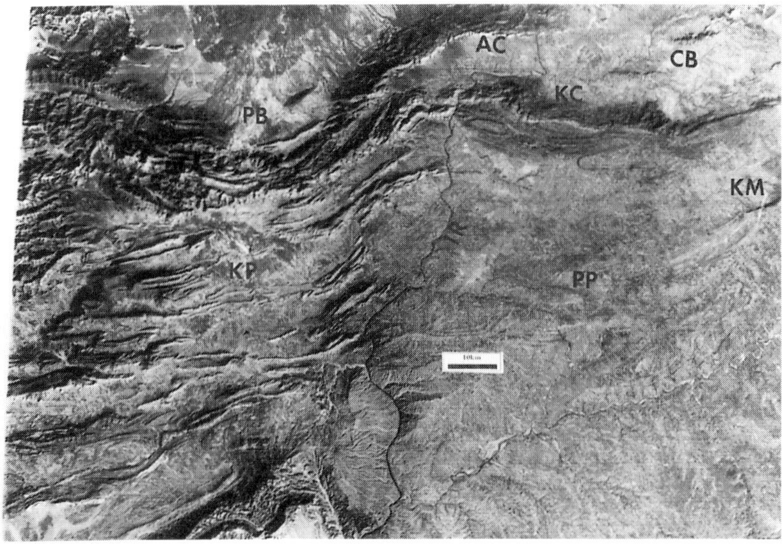

Fig. 14. Satellite image of Kohat & Potwar Plateaux (KP and PP respectively) showing vastly different degrees of exposure. PB, Peshawar Basin; CB, Campbellpore Basin; KM, Khair-i-Murat structure; AC, Attock-Cherat Range; KC, Kala Chitta Range; IR, Indus River. Scale bar is 10 km.

of the Main Central Thrust in the lesser Himalayas (Yeats & Hussain 1989). Reconnaissance transects through the Kohat Range in the Dara Adam Khel tribal area have revealed tight folds and high-angle faults (this study and Asif Khan pers. comm. 1992). That the faults are vertical and cut down into basement rocks is supported by their linear (map) traces and instrumental seismicity (Yeats & Hussain 1989).

Campbellpore Basin and Kala Chitta Range. Structural cross-sections of the Kala Chitta Range and southern Campbellpore Basin drawn by geologists over an almost 80 year period show structural styles similar to those of the Attock–Cherat Range and Peshawar Basin (Cotter 1933; Izatt 1990; Pascoe 1912). Structures in the Kala Chitta Range resemble the tight folds observed in the Kohat Range. The Kawa Ghar and Kheramar Hills are another example of NE-stepping ridges that are probably bounded by high-angle reverse faults (Fig. 14).

The NPDZ. Previous structural studies within the Potwar Plateau, adjacent to the Kohat Plateau, have presented balanced structural cross-sections that involve blind duplexes containing imbricated, north-dipping thrust faults and horses (Abassi & McElroy 1991; Baker *et al.* 1988; Izatt 1990; Lillie *et al.* 1987; McDougall & Hussain 1990; Pennock *et al.* 1989). However, seismic data from the Potwar Plateau show structures similar to those in the Kohat Plateau (Sercombe unpublished data). According to these data, the Potwar Plateau is a hybrid terrain consisting of thrust faults and a series of short lateral-displacement, high vertical-displacement pressure ridges associated with offsets in the basement. Perhaps the best exposed structure in the NPDZ is the Khair-i-Murat uplift, south of Fatehjong (Fig. 14). This structure is a doubly-overturned anticline that is (ambiguously) bounded on its southern flank by a high-angle reverse fault (Cotter 1933; Izatt 1990; Pascoe 1912; Wynne 1877). Pascoe (1912) described the uplift as part of a system of NE-stepping faults that can be traced from the Kohat Plateau (e.g., Shiwakki fault) across the northern Potwar Plateau. He also noted that the structure contains en-echelon faults and folds within it.

The Riwat fault, south of Rawalpindi near the town of Riwat, has been described by Wadia (1928) as being a 'strike fault of the normal variety with a very low hade and a throw of 6000 feet', meaning that it is a high-angle reverse fault. Many of the other structures of the Potwar Plateau have similar geometries (e.g., steep dips and along-strike changes in vergence) as the faults and folds described for the Kohat Plateau and other areas, but discussion of these is beyond the scope of this paper and should become a subject of further investigation.

Differing interpretations of structural styles in the Kohat Plateau and NPDZ

Why is there a major difference in the interpretation of structural styles of the Kohat Plateau and NPDZ? Differences in the surface-geological data sets are related to the degree and depth of exposure in the Kohat versus the Potwar Plateau. Much of the Potwar Plateau is covered by Miocene and younger synorogenic deposits and Quaternary alluvium, but in the Kohat Plateau, many of the structures are mappable because the depth of exposure is much greater; rocks as old as Palaeocene are exposed (Fig. 14). Thus, the rocks exposed in the Kohat Plateau can serve as an up-plunge projection of the subsurface structures in the NPDZ. We have found no evidence for imbricate thrusting (aside from the MKSTB) or duplex formation in the rocks that outcrop in the Kohat Plateau, rocks which are believed to be involved in such deformation in the subsurface of the NPDZ. It is possible that the surface-data set used here might have inspired very different conclusions from those derived from surface mapping in the Potwar.

If it is the case that structural styles do indeed change rapidly from one side of the Indus River to the other, there may be several reasons for this, and the present location of the Indus may be a reflection of one or more of these reasons. The proximity of the Kohat Plateau to the eastward verging Kurram thrust and the western, sinistral suture (Chaman fault) could have been a main control on the development of differing structures in these two regions: transpression-related structures near the suture, in Kohat, and compressional structures further from the strike-slip suture, in the NPDZ where north–south compression may have predominated. Facies in the basal portion of the stratigraphic sections of the Kohat and Potwar Plateaux also could have controlled the development of low-angle thrusts versus high-angle, reverse faults and flower structures even if the influence of oblique convergence was equal in both areas. Richard & Cobbold (1989) showed that weak material at depth in association with oblique stresses would reduce basal drag and lead to the development of detached faults such as thrust faults. The absence of a weak horizon at depth would mean more basal drag and might have encouraged the formation of flower structures. Eocambrian salt deposits are thought to have served as a decollement in the Potwar Plateau, however, Eocambrian evaporites have not yet been identified in the Kohat region. The present position of the Indus River, which separates the Kohat and Potwar Plateaux (Fig. 14), could reflect the subsurface position of the margin of the Eocambrian evaporite basin. We suggest that this question of differing structural styles in these two adjacent regions should become a topic of concentrated future research, and that the role of transpression throughout the Himalayan orogen in northwest Pakistan be examined more closely.

Conclusions

The structural styles observed in the Kohat Plateau are the surficial manifestations of a hybrid terrain consisting of high-angle, oblique-reverse and thrust faults. A continuum of episodes of deformation has been recognized in the Kohat Plateau. The first episode included south-verging, compression-related thrust faulting involving Middle Eocene and Miocene aged rocks. During the second episode of deformation, these thrust faults were overprinted by transpression-related wrench faulting and were incorporated into tight, often doubly-overturned folds which are pervasive throughout the Kohat Plateau. The age of molasse deposits deformed by these folds and faults dates this deformation as being Plio-Pleistocene in age. In the SE region of the Plateau, Lower Eocene evaporites and limestones served as ramp-forming horizons, allowing high-angle, oblique-reverse faults to cut upsection, truncating folds which were forming during fault propagation. These faults flatten near the surface to form positive flower structures. In the northwest region of the Plateau, where thick sequences of green and grey shales replace the evaporites and limestones of the Lower Eocene portion of the section in the SE region, these faults most likely tip out in the shales and are manifested at the surface as tight folds, and are interpreted as drape folds that formed above positive flower structures.

Evidence for deformation related to transpression rather than pure compression includes steeply-dipping bedding and faults at depths of over 3 km, large vertical displacements across faults, diapirism of shales and evaporites above wrench faults, northward and southward vergence of faults at the surface, tight, doubly overturned folds, left-stepping folds and motion fabrics that indicate a component of strike-slip displacement in addition to dip-slip. There are also numerous examples of high-angle, reverse faults in regions adjacent to the Kohat Plateau. Styles of deformation within the Kohat Plateau have been influenced by the Plateau's position in the western corner of a major syntaxial bend in the Himalayan orogen.

We would like to thank SPOT Image for the permission to present satellite images. Field work in the Kohat Plateau was coordinated by Amoco Pakistan Exploration Company, and J.K. Barndt is thanked for these matters. Additional field work was supported by grants to DAP from the Stoiber Field Fund of Dartmouth College, the Albert Cass Traveling Fellowship, the Geological Society of America, Sigma Xi, and the American Association of Petroleum Geologists.

References

ABBASI, I.A. & MCELROY, R. 1991. Thrust kinematics of the Kohat Plateau, Trans Indus Salt Range, Pakistan. *Journal of Structural Geology,* **13**, 319–327.

ALLEN, C.R. 1957. San Andreas fault zone in San Gorgonio Pass, southern California. *Geological Society of America Bulletin,* **68**, 315–350.

—— 1965. Transcurrent faults in continental areas. *Philosophical Transactions of the Royal Society of London,* **258**, 82–89.

BAKER, D., LILLIE, R., YEATS, R., JOHNSON, G., YOUSUF, M. & ZAMIN, A. 1988. Development of the Himalayan frontal thrust zone: Salt Range, Pakistan. *Geology,* **16**, 3–7.

BARTLETT, W.L., FRIEDMAN, M. & LOGAN, J.M. 1981. Experimental folding and faulting of rocks under confining pressure, Part IX. Wrench Faults in limestone layers. *Tectonophysics,* **79**, 255–277.

BIDDLE, K.T. & CHRISTIE-BLICK, N. 1985. Glossary – Strike-slip deformation, basin formation and sedimentation. *In*: BIDDLE, K.T. & CHRISTIE-BLICK, N. (eds) *Strike-slip Deformation, Basin Formation and Sedimentation*. Society of Economic Paleontologists and Mineralogists, Tulsa, 375–386.

BONIS, N.R. 1985. *Paleocurrent analysis of Siwalik strata in the Trans-Indus Region, North western Pakistan*. Thesis, Dartmouth College, Hanover.

BURBANK, D.W. & BECK, R.A. 1989. Early Pliocene uplift of the Salt Range; Temporal constraints on thrust wedge development, northwest Himalaya, Pakistan. *In*: MALINCONICO, L.L. & LILLIE, R.J. (eds) *Tectonics of the Western Himalayas*. Geological Society of America, Special Papers, **232**, 113–128.

BURKE, D.B. 1979. *Log of a trench in the Garlock fault zone, Fremont Valley, California*. United States Geological Survey Miscellaneous Field Studies Map MF-1028.

CHESTER, J.S., LOGAN, J.M. & SPANG, J.H. 1991. Influence of layering and boundary conditions on fault-bend and fault propagation folding. *Geological Society America Bulletin,* **103**, 1059–1072.

COTTER, G.DEP. 1933. The geology of the part of the Attock District west of longitude 72°45′. *Memoirs of the Geological Survey of India,* **55**, 63–161.

EMMONS, R.C. 1969. Strike-slip rupture patterns in sand models. *Tectonophysics,* **7**, 71–87.

HARDING, T.P. 1985. Seismic characteristics and identification of negative flower structures, positive flower structures, and positive structural inversion. *American Association of Petroleum Geologists Bulletin,* **69**, 582–600.

—— 1990. Identification of wrench faults using subsurface structural data: criteria and pitfalls. *American Association of Petroleum Geologists Bulletin,* **74**, 1590–1609.

IZATT, C.N. 1990. *Variation in thrust front geometry across the Potwar Plateau and Hazara Kala Chitta Hill Ranges, northern Pakistan*. PhD thesis, Imperial College.

JAMISON, W.R. 1987. Geometric analysis of fold development in overthrust terranes. *Journal of Structural Geology,* **9**, 207–219.

JAUMÉ, S. & LILLIE, R. 1988. Mechanics of the Salt Range–Potwar Plateau, Pakistan: A fold and thrust belt underlain by evaporites. *Tectonics,* **5**, 57–71.

KELLOG, H.E. 1975. Tertiary stratigraphy and tectonism in Svalbard and continental drift: *American Association of Petroleum Geologists Bulletin,* **59**, 465–485.

LAWRENCE, R.D., KHAN, S.H., DEJONG, K.A. & YEATS, R.S. 1981. Thrust and strike slip fault interaction along the Chaman transform zone, Pakistan. *In*: MCCLAY, K.R. & PRICE, N.J. (eds) *Thrust and Nappe Tectonics*. Geological Society, London, Special Publications, **9**, 363–370.

LILLIE, R., JOHNSON, G., YOUSUF, M., ZAMIN, A. & YEATS, R. 1987. Structural development within the Himalayan foreland fold-and-thrust belt of Pakistan. *In*: BEAUMONT, C. & TANKARD, A. (eds) *Sedimentary Basins and Basin-Forming Mechanisms*. Canadian Society of Petroleum Geologists, Halifax, Memoirs, **12**, 379–392.

LOWELL, J.D. 1972. Spitsbergen Tertiary orogenic belt and the Spitsbergen fracture zone. *Geological Society of America Bulletin,* **83**, 3091–3102.

MCDOUGALL, J.W. & HUSSAIN, A. 1991. Fold and thrust propagation in the western Himalaya based on a balanced cross section of the Surghar Range and Kohat Plateau, Pakistan. *American Association of Petroleum Geologists Bulletin,* **75**, 463–478.

——, HUSSAIN, A. & YEATS, R.S. 1993. The Main Boundary Thrust and propagation of deformation into the foreland fold-thrust belt in northern Pakistan near the Indus River. *This volume*.

MEISSNER, C.R., MASTER, J.M., RASHID, M.A. & HUSSAIN, M. 1974. *Stratigraphy of the Kohat Quadrangle*, Pakistan. United States Geological Survey Professional Paper **716–D**.

——, HUSSAIN, M., RASHID, M.A. & SETHI, U.B. 1975. *Geology of the Parachinar Quadrangle, Pakistan*. United States Geological Survey Professional Paper **716–F**.

MITRA, S. 1990. Fault-propagation folds: Geometry, kinematic evolution, and hydrocarbon traps. *American Association of Petroleum Geologists Bulletin,* **74**, 921–945.

NAYLOR, M.A., MANDL, G. & SIJPESTIJN, C.H.K. 1986. Fault geometries in basement-induced wrench faulting under different initial stress states. *Journal of Structural Geology,* **8**, 737–752.

PASCOE, M.A. 1912. Petroleum in the Punjab and

Northwest Frontier Province. *Memoirs of the Geological Survey of India,* **40**, 331–493.

PENNOCK, E., LILLIE, R., ZAMAN, A. & YOUSAF, M. 1989. Structural interpretation of seismic reflection data from the eastern Salt Range and Potwar Plateau, Pakistan. *American Association of Petroleum Geologists Bulletin,* **73**, 841–857.

PIVNIK, D.A. 1992. *Depositional response to encroachment of Himalayan compressional and transpressional deformation on the northern Pakistan foreland.* PhD thesis, Dartmouth College.

—— & KHAN, M.J. 1992. Sedimentological record of foreland- to piggyback-basin transition, Trans-Indus Salt Ranges, NW Pakistan. *Abstracts with Programs, Geological Society of America Annual Meeting,* **A53**.

RICHARD, P. & COBBOLD, P. 1989. Structures en fleur positives et décrochments crustaux: modélisation analogique et interprétation mécanique. *Comptes-Rendus L'Academie des Sciences,* Paris, Serie 2, **308**, 553–560.

SUPPE, J. & MEDWEDEFF, D.A. 1990. Geometry and kinematics of fault-propagation folding. *Eclogae Geologicae Helvetiae,* **83**, 409–454.

SYLVESTER, A.G. 1988. Strike-slip faults. *Geological Society of America Bulletin,* **100**, 1666–1703.

—— & SMITH, R.R. 1976. Tectonic transpression and basement-controlled deformation in the San Andreas fault zone, Salton Trough, California. *American Association of Petroleum Geologists Bulletin,* **60**, 2081–2102.

TAHIRKHELI, R.A.K. 1979. Geology of Kohistan and adjoining Eurasian and Indo-Pakistan continents. *Geological Bulletin of the University of Peshawar,* **11**, 1–30.

TAPPONNIER, P., PELTZER, G., LEDAIN, A.Y., ARMIJO, R. & COBBOLD, P. 1982. Propagating extrusion tectonics in Asia: New insights from simple experiments in plasticene. *Geology,* **10**, 611–616.

——, PELTZER, G. & ARMIJO, R. 1986. On the mechanics of the collision between India and Asia. *In*: COWARD, M.P. & RIES, A.C. (eds) *Collision Tectonics.* Geological Society, London, Special Publications, **19**, 115–157.

TRELOAR, P.J., COWARD, M.P. & WILLIAMS, M.P. 1990. Metamorphism and crustal stacking in the north Indian Plate, north Pakistan. *Tectonophysics,* **165**, 167–184.

VERMA, R.K., MUKHOPADHYAY, M. & BHANJA, A.K. 1980. Seismotectonics of the Hindukush and Baluchistan Arc. *Tectonophysics,* **66**, 301–322.

WADIA, D.N. 1928. The geology of the Poonch state (Kashmir) and adjacent portions of the Punjab. *Memoirs of the Geological Survey of India,* **51**, 185–370.

WELLMAN, H.W. 1955. *The geology between Bruce Bay and Haast River, South Westland.* New Zealand Geological Survey Bulletin, **48** (n.s.) (2nd edition).

WELLS, N.A. 1984. *Marine and continental sedimentation in the early Cenozoic Kohat Basin and northwestern Indo-Pakistan.* PhD thesis, University of Michigan.

WILCOX, R.E., HARDING, T.P. & SEELY, D.R. 1973. Basic wrench tectonics. *American Association of Petroleum Geologists Bulletin,* **57**, 74–96.

WILLIS, B. 1938. San Andreas rift, California. *Journal of Geology,* **46**, 793–827.

WYNNE, A.B. 1877. Note on the Tertiary zone and underlying rocks of the north-west Panjab. *Records of the Geological Survey of India,* **10**, 107–132.

YEATS, R.S. & HUSSAIN, A. 1989. Zone of Late Quaternary deformation in the southern Peshawar Basin. *In*: MALINCONICO, L.L. & LILLIE, R.J. (eds) *Tectonics of the Western Himalayas.* Geological Society of America, Special Paper, **232**, 265–274.

YEATS, R.S. & LILLIE, R.J. 1991. Contemporary tectonics of the Himalayan frontal fault system: folds, blind thrusts and the 1905 Kangra earthquake. *Journal of Structural Geology,* **13**, 215–225.

——, KHAN, S.H. & AKHTAR, M. 1984. Late Quaternary deformation of the Salt Range of Pakistan. *Geological Society of America Bulletin,* **95**, 958–966.

ZOLNAI, G. 1988. *Continental Wrench-Tectonics and Hydrocarbon Habitat.* American Association of Petroleum Geologists Continuing Education Course Note Series No. **30**.

The Main Boundary Thrust and propagation of deformation into the foreland fold-and-thrust belt in northern Pakistan near the Indus River

JAMES W. McDOUGALL[1], AHMAD HUSSAIN[2] & ROBERT S. YEATS[3]

[1] *Research School of Earth Sciences, Victoria University, Wellington, New Zealand*
Present address: Department of Geography and Earth Sciences, University of North Carolina, Charlotte, NC 28223, USA

[2] *Geological Survey of Pakistan, Peshawar, Pakistan*

[3] *Department of Geosciences, Oregon State University, Corvallis 97331–5506, USA*

Abstract: The Main Boundary Thrust in northwestern Pakistan is a floor thrust along which a thrust system incorporating Precambrian and Phanerozoic rocks of the Kala Chitta and Attock–Cherat Ranges was emplaced over Cenozoic strata of the northern Kohat and Potwar Plateaus. The MBT and successive thrusts toward the foreland are interpreted as low angle to flat decollement thrusts at 8–10 km depths that bound thrust sheets with large lateral dimensions. Admissible cross sections indicate over 60 km of shortening along the largely pre-Palaeocene Khairabad, Cherat, and Hissartang thrusts within the MBT allochthon. In the toe of the MBT allochthon, contraction is manifest at the surface as tight folds in largely Mesozoic and lower Tertiary strata, decoupled in the upper third of the allochthon. Displacement of the MBT is estimated at over 40 km and south of the MBT, shortening is interpreted along blind thrusts accompanied by shallow (0–3 km deep) backthrusting. The MBT allochthon was deformed prior to its low angle displacement over the northern Kohat and Potwar Plateaus, and folded and eroded during or after its emplacement. The MBT thrust front may participate in its own later deformation, but not as an emergent low angle thrust through the Quaternary.

The Main Boundary Thrust (MBT) extends westward from the front of the main Himalayan Range around the Hazara–Kashmir syntaxial bend, and thrusts the hill ranges over the northern foreland fold–thrust belt in northern Pakistan. The MBT is mapped to the south and southwest from Hazara as the Murree fault and westward as the MBT along the southern edge of the Kala Chitta Range (Fig. 1). Deformation in the thrust system that includes the MBT juxtaposes stratigraphic assemblages ranging in age from Precambrian in the north to successively younger folded strata southward, dominated by folded early to mid-Cenozoic strata in the northern Kohat and Potwar plateaus. Farther to the south the Salt Range thrust and Surghar Range thrust bring Eocambrian and younger strata over the Quaternary floodplain deposits of the Indus and Jhelum rivers. The MBT is thrust over a 1–2° N-dipping basement and crustal studies of seismicity indicate a nearly flat decollement thrust beneath the Peshawar Basin, over a 10° N-dipping slab, broken at the Hazara seismic zone beneath the northern hill ranges or Margala Hills (Seeber *et al.* 1981). Gravity data indicate a similar gently N-dipping slab of Indian continental basement below the MBT (Duroy *et al.* 1989).

Propagation of thrusting of the MBT into the northern Kohat and Potwar plateaus occurred from 2.1 to 1.9 Ma and ponded sediments behind the MBT in the Campbellpore intermontane basin, between Rawalpindi and the Indus River (Burbank & Raynolds 1988). Deformation within the Kala Chitta fold and thrust belt pre-dates thrusting of the Kala Chitta Range over the Kohat and Potwar plateaus (Yeats & Hussain 1987). Quaternary thrusting south of the MBT occurred more recently than 0.6 Ma in the Salt Range (Yeats *et al.* 1984) and Trans-Indus Salt Range. Out-of-sequence thrusting in the northern Kalabagh fault zone, north of the Surghar Range followed initial ramping of the Surghar Range thrust (McDougall & Khan 1990; McDougall & Hussain 1991). Active faulting occurs to the north of the MBT along ENE-trending high angle faults and associated pressure ridges in the southern Peshawar Basin (Yeats & Hussain 1989). The MBT, while active as recently as 1.9 Ma, is folded (McDougall & Hussain 1991) and was thus deformed after final emplacement.

From TRELOAR, P.J. & SEARLE, M.P. (eds), *Himalayan Tectonics*
Geological Society Special Publication, No. 74, pp. 581–588.

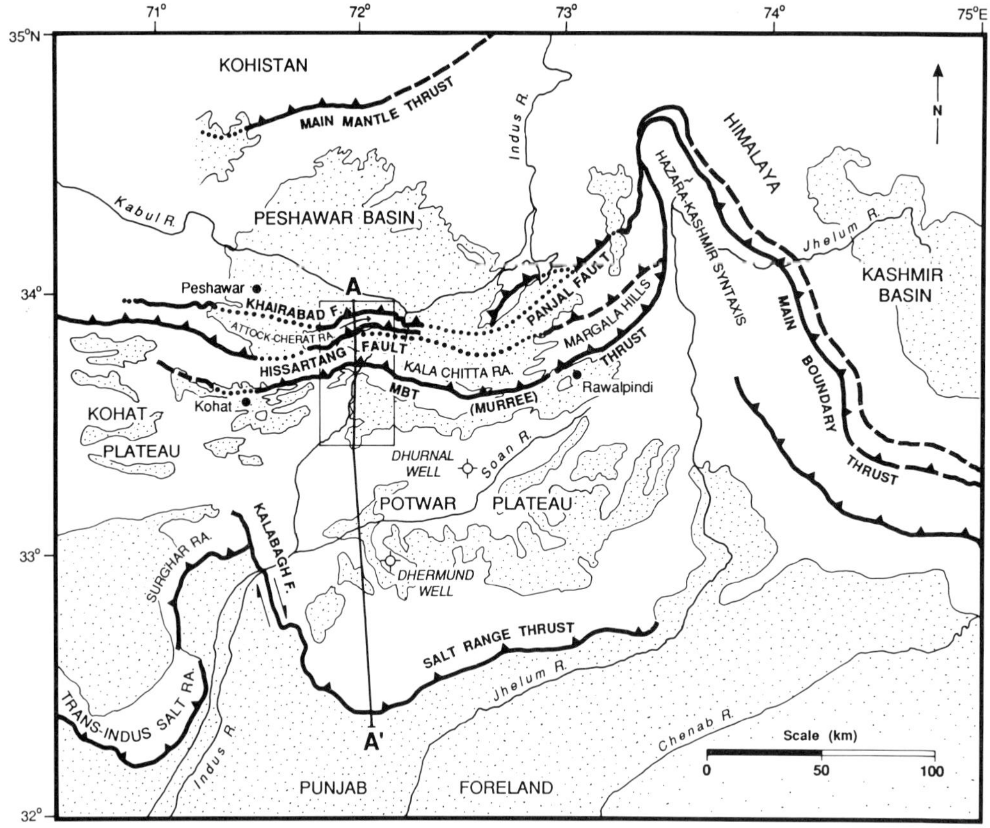

Fig. 1. Index map locating major faults in the Himalayan foreland of northwest Pakistan and surrounding regions. Lowlands and Quaternary deposits are in a stippled pattern. The rectangle (left center) is the area of the geological map of Fig. 3 and A–A′ is the line of the geological cross section of Fig. 4.

A passive roof or backthrust component of deformation accompanying southward-directed emplacement of major thrust sheets was described in retrodeformable cross sections south of the western MBT (McDougall & Hussain 1991). The general kinematics, overthrusting of synorogenic deposits, lateral variations in stratal thickness, and interaction of the MBT with the foreland fold thrust belt are addressed in this paper.

Stratigraphy and implications

The stratigraphic sections of Fig. 2 include 571 m of Palaeozoic strata in the north central Potwar Plateau (McDougall & Hussain 1991), south of the MBT, that thickens northwestward toward the Attock–Cherat Range to almost 2 km north of the Hissartang fault (Yeats & Hussain 1987). A Mesozoic section about 1 km thick in the Surghar Range (Meissner *et al.* 1974), 60 km south of the MBT, thins northward to about 650 m in the Kala Chitta Range immediately north of the MBT, and is cut out in an unconformity in the Attock–Cherat Range north of the Hissartang fault (Yeats & Hussain 1987). The lower Tertiary section, underlying synorogenic deposits of the Murree and Siwalik groups, is about 1 km thick in the Surghar Range (southern Kohat Plateau) and thins both to the north and northeast toward the Kala Chitta Range and to the southeast toward the southern Potwar Plateau and Salt Range. Palaeocene strata below the Murree Fm. are about 100 m thick in the Attock–Cherat Range to the north of the MBT (Yeats & Hussain 1987). The synorogenic Murree–Kamlial–Siwalik section is nearly 4 km thick in the northern Potwar Plateau and north of the Surghar Range, and varies greatly in preserved thickness.

A geological map (Fig. 3) and subsurface data are used to constrain stratigraphic thickness

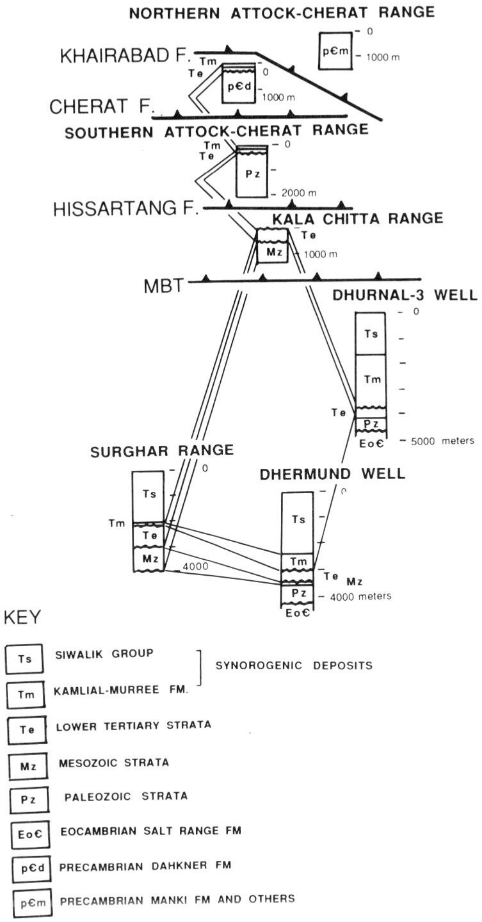

Fig. 2. Scaled representative stratigraphic sections, from areas located in figures 1 and 3. The Surghar Range section is from Meissner *et al.* (1974), stratigraphic sections north of the MBT are discussed in Yeats & Hussain (1987). Thicknesses of strata from the Dhurnal and Dhermund wells are from McDougall & Hussain (1991).

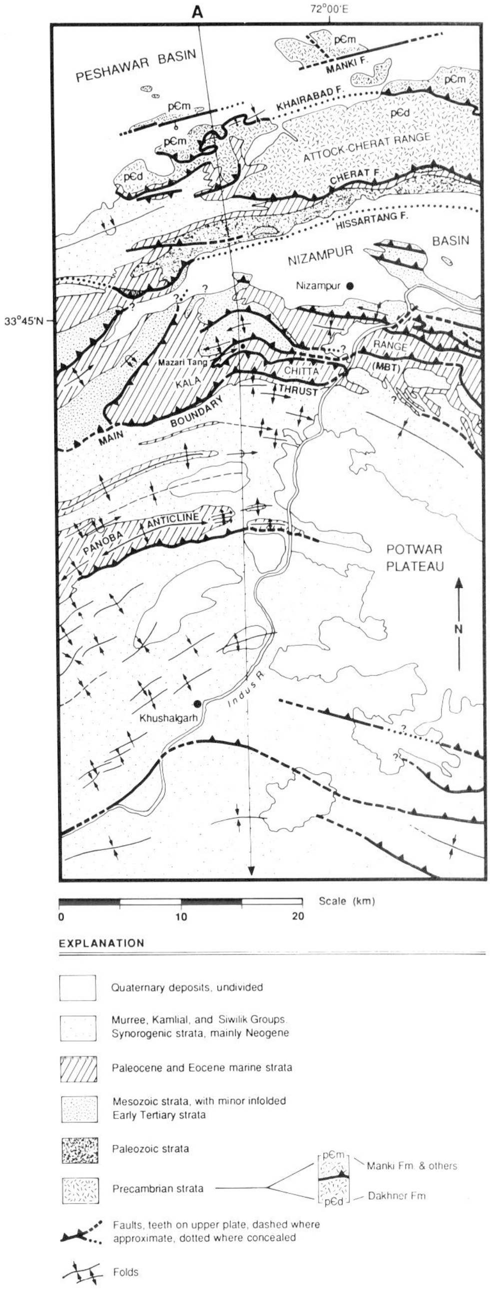

Fig. 3. Geological map of the NE Kohat and NW Potwar plateaus, the western Kala Chitta Range, and western Attock–Cherat Range. The MBT is folded and the lower plate is exposed at Mazari Tang, near the slight bend in the line of cross section.

changes in cross sections (Fig. 4) across the northern foreland fold–thrust belt and MBT allochthon of northwestern Pakistan. Palaeozoic strata increase in thickness to the north and the Mesozoic section decreases in thickness northward, disappearing completely north of the Hissartang fault. These changes in the stratigraphic section imply pronounced telescoping by thrusting and associated deformation north of the MBT toward the Attock–Cherat Range. The combination of stratigraphic thickness changes from north to south in the Palaeozoic, Mesozoic and Lower Tertiary section is less between separate thrust sheets in the foreland fold–thrust belt, to the south of the MBT where deformation is largely post 2 Ma. The overall stratigraphic

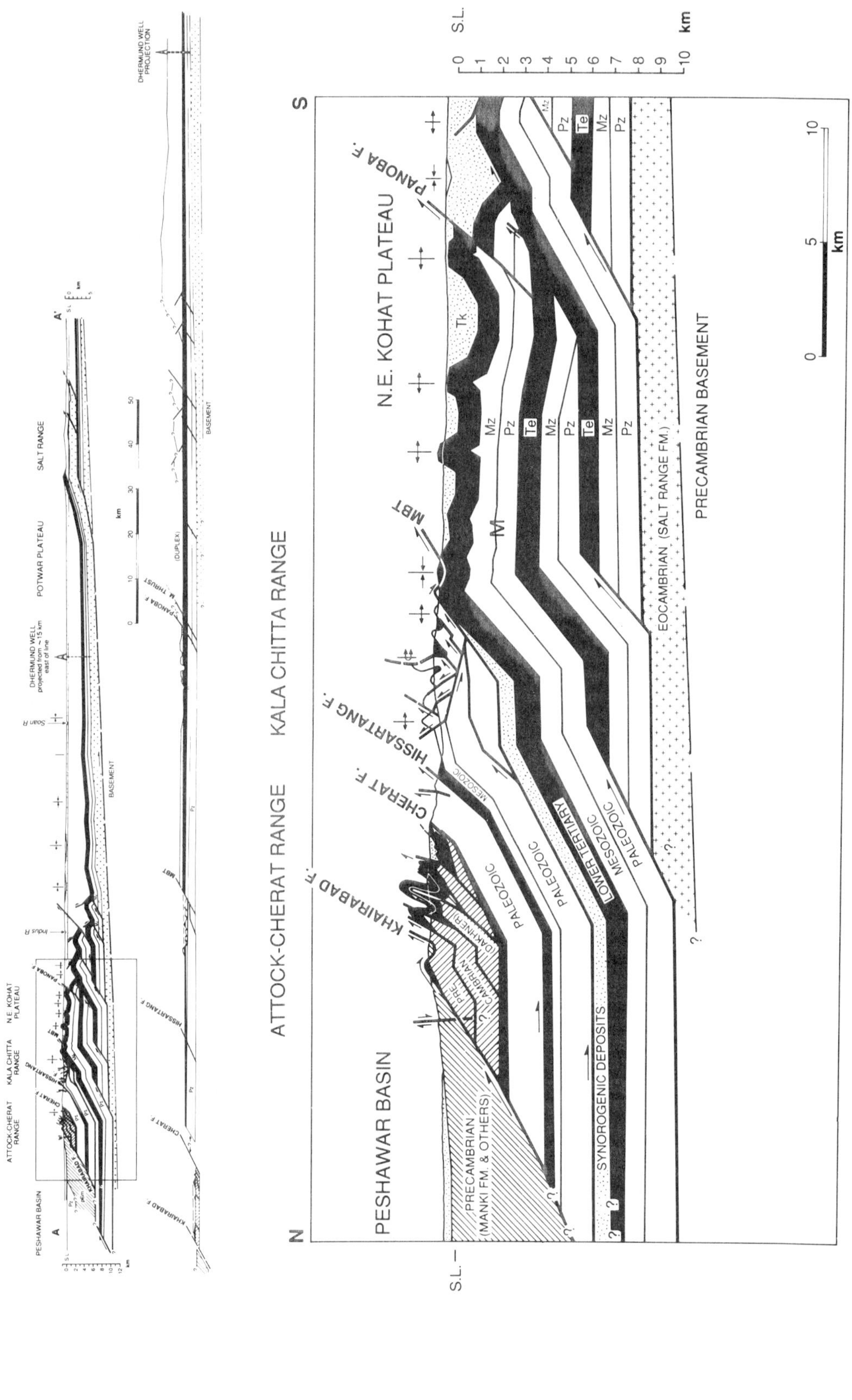
PESHAWAR BASIN
ATTOCK-CHERAT RANGE
KALA CHITTA RANGE
N.E. KOHAT PLATEAU
POTWAR PLATEAU
SALT RANGE
DHERMUND WELL projected from ~15 km east of line
DHERMUND WELL PROJECTION
BASEMENT
(DUPLEX)
KHAIRABAD F.
CHERAT F.
HISSARTANG F.
MBT
PANOBA F.
N
S
S.L.
PRECAMBRIAN (MANKI FM. & OTHERS)
PALEOZOIC
MESOZOIC
LOWER TERTIARY
SYNOROGENIC DEPOSITS
EOCAMBRIAN (SALT RANGE FM.)
PRECAMBRIAN BASEMENT
Mz
Pz
Te
Tk
km

variation is a <1 km increase in the combined thicknesses of Paleozoic and Mesozoic strata to the north of the surface trace of the MBT thrust.

Uncertainties in stratal thicknesses affect the depth of the MBT beneath the Attock–Cherat Range. The greatest uncertainty in stratigraphic thickness is in the synorogenic deposits of the Murree Formation which underlie the MBT, and which are interpreted as having an average thickness of about 1 km on the immediate footwall of the MBT in the line of cross section (Fig. 4). The maximum thickness of the Murree Formation below the MBT could, however, be more than 2 km or similar to that of the Murree–Kamlial sequence in the northern Potwar Plateau at the Dhurnal well (for example, McDougall & Hussain 1991). Preservation of the clastic wedge of the Murree, Kamlial, and Siwalik formations below the MBT is relatively unknown because of its distribution and variable thickness at the surface, both north and south of the MBT.

Kinematics of the MBT thrust front and foreland fold–thrust belt

The kinematics of thrusting vary from north to south in the MBT allochthon and foreland fold-thrust belt to the south, with closely spaced fault-cored folds in the Cherat block below the Khairabad roof thrust; steep and overturned N-vergent tight folds in Mesozoic strata in the toe of the MBT allochthon; and more symmetrical lift-off or box folds that are exposed south of the MBT in the northern Kohat Plateau (Figs 3 & 4).

Thrusting along the Cherat, Khairabad, and Hissartang faults in early Tertiary time emplaced blocks of Precambrian strata over Palaeozoic and Mesozoic rocks, subsequently uplifted in the MBT allochthon. The earlier thrusting produced fault-cored folds that may have been rotated to higher dips by regional backthrusting during emplacement of subjacent thrust sheets southward. The Khairabad fault is a folded roof thrust for the duplex of thrust slices in the basement underlain by the Cherat fault. The duplex was emplaced between the N-dipping Cherat and Khairabad thrusts during Palaeozoic time (Yeats & Hussain 1987), and these thrusts, that plucked Precambrian rocks from the top of the basement, project downdip toward the subsurface projection of the MBT floor thrust. Both the Cherat and Hissartang faults project eastward to the southern Kherimar Hills and farther east merge along the Nathia Gali fault in the northern Margala Hills (Hylland 1990).

Tight folds in Mesozoic and early Tertiary strata between the Hissartang and Main Boundary thrusts include N-verging overturned folds and associated S-dipping thrusts. The wavelength of these folds, many less than 1 km, and stratigraphic similarity in the deformed section indicate decoupling at a depth of 1–2 km in Mesozoic rocks. Higher shear strains probably occurred in Jurassic strata which occupy the cores of many of these folds. This shallow horizontal shortening in Mesozoic and lower Tertiary rocks is compensated between 1 and 4 km depths by thrusting along a horse located in Palaeozoic rocks near the toe of the MBT thrust (Fig. 4). The Mesozoic section may also be repeated beneath the shallow decollement at 1–2 km depths in the Kala Chitta Range. In this case delamination would then be required between the Palaeozoic and Mesozoic sections downdip below the Hissartang fault (Fig. 4), bending the fault at its intersection with the Khairabad fault, or else delamination would be necessary in the MBT allochthon to the south. In the southern Kala Chitta Range, the MBT is folded over the early Tertiary section of the northern Kohat Plateau at Mazari Tang (McDougall & Hussain 1991) above a hinge in the underlying plate (Fig. 4). Folding of the MBT near the Indus River probably occurred during emplacement of the underlying thrust sheet interpreted in the subsurface by McDougall & Hussain (1991).

Immediately south of the MBT, the northern Kohat Plateau is characterized by nearly symmetrical lift-off or box folds and broader synclines trending E to ENE and having a fold wavelength of several kilometres. The Panoba anticline is typical of these folds (Fig. 3), with a minor syncline along its crest (McDougall & Hussain 1991). Box-shaped anticlines or fan folds have been distorted by ductile deformation in the cores of folds that include beds of the

Fig. 4. (**a**) Cross section A–A′, extending from the Peshawar Basin to the western Salt Range, with detail in rectangle, located in Figs 1 and 3. The M thrust sheet is directly below the MBT. The lower portion of the figure is the retrodeformed configuration of A–A′. (**b**) Detailed enlargement of the MBT and the duplex on its footwall that deforms the MBT Area of enlargement is the box in (a).

Eocene Panoba and Patala shales, along which the deformed Tertiary sequence may be decoupled. Final emplacement of the MBT frontal thrust was probably synchronous with near-surface deformation of the northern Kohat Plateau.

Discussion

Large overlaps measuring tens of kilometres on horizontally-directed thrust sheets, including the MBT allochthon and related thrusts, are shown in partially retrodeformable cross sections (Fig. 4). The northward extent of the subthrust Murree Formation 35 km north of Muzaffarabad in the Hazara–Kashmir syntaxis (for example, Bossart & Ottiger 1990), about 40 km north of the strike of the Murree fault (MBT) in the Margala Hills agrees with a low N-dipping to flat MBT floor thrust system below the Attock–Cherat Range with the Murree Formation in its footwall. This interpreted overlap of the Murree Formation is also supported by extensive outcrops of the Murree Formation within the MBT allochthon in the southern Peshawar Basin and Attock–Cherat Range (Fig. 3). Seismic reflection data across the northern Kohat Plateau indicate broad horizontal overthrusts (McDougall & Hussain 1991) and gravity data support a very shallow 1–2° northward dip on the top of the basement below the MBT (Duroy *et al.* 1989). Decoupling in the upper 2 km of the MBT allochthon is required to produce tight folding and near vertical faults between the Hissartang fault and the MBT, where Palaeocene deformation was overprinted by deformation in the MBT allochthon during its emplacement by thrusting.

The MBT is the floor thrust for major thrusts in the MBT allochthon, including the Hissartang fault and Cherat fault (Yeats & Hussain 1987) and the MBT has the greatest projected dip displacement of >40 km. Overall shortening, described by area balancing of the MBT allochthon and the foreland fold and thrust wedge, is similar to that described previously by Coward & Butler (1985) and others in the eastern hill ranges, near the Hazara syntaxis. Coward & Butler (1985) showed steep thrusts in the toe of the MBT allochthon to propagate directly from the top of the basement in the eastern hill ranges. In contrast, our cross sections show the shallow-dipping MBT to be thrust over synorogenic deposits of the Murree Fm beneath the Peshawar Basin, involving a large horizontal displacement over the basement. The true thickness and northward limit of synorogenic deposits preserved in the footwall of the MBT below the Peshawar Basin is not known.

The Khairabad roof thrust in the MBT allochthon is displaced 25 km or more and folded over a block of imbricate thrusts in Precambrian rocks that are unconformably overlain by Tertiary strata. The initial ramp from the basement of the Khairabad fault must project to the north of the MBT, unless the Khairabad fault dips to the south in the subsurface. The Cherat fault forms a horse below the Khairabad roof thrust, with a minimum dip displacement of about 12 km. However, the Cherat fault juxtaposes Precambrian strata with Palaeozoic strata to the south, and its dip displacement may be substantially greater. The location of the initial stepup or ramp thrust of the Cherat fault from the basement is unknown. The Hissartang fault, with a dip displacement of 25 km or more, is a thrust along which Palaeozoic strata are emplaced over deformed Mesozoic strata, with a faulted back limb at the surface. Between the Hissartang fault and the MBT, tight folds and steeply dipping faults in Mesozoic and early Tertiary strata characterize the Kala Chitta Range to the east of the Indus River, and this belt continues north and northeast of Rawalpindi and Islamabad in the Margala and Murree hills.

Dip displacement of 60 km or more is interpreted as a minimum shortening along major thrusts mapped over a geographic distance of about 20 km (N–S) in the MBT allochthon, exclusive of the MBT. A calculation of horizontal strain must consider the potential error in locating initial stepups or ramps from the top of the basement for the Hissartang, Cherat, and Khairabad thrust faults. These displacements along the above three major emergent thrusts and associated compressional deformation are considered to be a minimum, with greater shortening indicated during thrusting along the MBT. Incorporating MBT displacement, the undeformed line length of the southern MBT allochthon may be 100 km or more, now shortened to 25% or less of its original N–S horizontal dimension. However, cumulative thrust stacking may be at a maximum near the toe of the MBT allochthon and regional contractile strain should be calculated over a greater line length. In the foreland fold-thrust belt south of the MBT, to the Surghar Range, about 50% shortening characterizes an interpreted system of blind thrusts overlying a nearly flat 8–10 km deep basement over a distance of 70 km (N–S), (McDougall & Hussain 1991).

An alternative interpretation of cross section line A–A′ in the northeastern Kohat Plateau

(Fig. 4) would replace the duplex with a single thicker Phanerozoic section below the M thrust, with thick synorogenic strata of the Murree, Kamlial, and Siwalik groups. While this alternate interpretation is possible on the restored stratigraphic template of Fig. 4, the absence of the duplex would lengthen the subjacent thrust sheet, and much more synorogenic clastic strata would be preserved on the top of these thrust sheets that make up the northern Potwar deformed zone. This interpretation does not allow either the relatively small displacement on the emergent Panoba fault, which dies out along strike, or the lack of evidence of a major emergent thrust fault 0–5 km south of the Panoba anticline that would be necessary to justify 20 km or more of dip displacement. The northern Kohat Plateau is relatively flat and characterized by rocks exposed at a similar stratigraphic level, suggesting a flat footwall block at depth. A duplex of Palaeozoic to early Tertiary strata below the M thrust sheet is therefore favored, with shallow backthrusting in the synorogenic strata during emplacement. The Murree, Kamlial and possibly Siwalik formations may also underlie the M thrust near the toe of the duplex, decreasing overall horizontal contraction slightly as a line-length measurement across the foreland fold-thrust belt, although this configuration also requires increased displacement along the Panoba fault.

A cumulative dip displacement of more than 40 km along the MBT is great enough that internal folding of the major allochthon probably occurred, while stratigraphic thicknesses of major thrust sheets in the allochthon may be maintained, as shown in Fig. 4. Indirect geophysical observations such as seismicity data from small earthquakes centred 50–150 km east of cross section A–A′ indicate decoupling at about 10 km below the MBT allochthon (Seeber *et al.* 1981). Gravity data (for example, Duroy *et al.* 1989) indicate a 10–12 km depth of the basement below the frontal MBT allochthon. Present day displacement as expressed at the surface has propagated toward the foreland from the MBT, otherwise, backthrusting would be the most essential component if there were modern thrusting at the front of the MBT.

Major ramps from the top of the basement underlie the emergent northern Salt Range (Baker *et al.* 1988) and Surghar Range (McDougall & Hussain 1991) and these are associated with N-facing fault scarps in the basement. A major ramp from the basement is interpreted in the footwall of the folded MBT, directly below the Hissartang fault (Fig. 4). The step at the base of the Palaeozoic section at this ramp, and thinning of the Palaeozoic to Lower Tertiary strata to the south suggest that this ramp system in the footwall of the MBT may override southward-thickening evaporites of the Eocambrian Salt Range Formation, as shown in cross section. The ramp from the basement below the Panoba fault is supported by interpretations of seismic reflection data (for example, McDougall & Hussain 1991) and is a step-up on the underlying thrust sheet. The Precambrian basement may also be stepped up to the south by faulting below the ramp system in the footwall of the MBT.

Conclusions

Thrust sheets were emplaced and deformed over the Cherat, Khairabad, and Hissartang faults, respectively, and later displaced 40 km or more toward the Himalayan foreland in the MBT allochthon. The MBT was folded in association with late emplacement of its allochthon and deformation in the northern Kohat Plateau, just prior to 1.9 Ma. Styles of deformation and the age of the deformed section vary between the southern MBT allochthon and the foreland fold-thrust belt to the south. Fault-cored steeply-dipping and overturned folds occur in Tertiary strata over the basement in the duplex between the Khairabad and the Cherat faults and tight steeply N- to S-dipping faults in dominantly Mesozoic strata crop out in the toe of the MBT allochthon, south of the Hissartang fault. South of the MBT, folds are nearly symmetrical box or lift-off folds over decoupled horizons at 1–2 km depths.

Kinematics of major thrusts, including those within the MBT allochthon and the MBT are based on constant stratigraphic thickness changes laterally (N–S), as derived by retrodeformation of structural cross sections. Near-surface faults branch from sole thrusts at shallow depths in the toe of the MBT allochthon rather than directly from the top of the basement, and thrust fault trajectories are shown as ramps and flats. The approximate depth of the basement below the MBT is derived from gravity data and inferred from seismicity data, where decoupling is a product of active Himalayan convergence.

The MBT is not active in the present day as a thrust, however, regional contraction is reflected by thrust emplacement to the south and contraction may affect the entire foreland fold–thrust belt. The wedge is building critical taper in the foreland fold-thrust belt north of the Surghar Range (McDougall & Hussain 1991). The large overlap of synorogenic strata by thrusting along the MBT is not interpreted to have occurred beneath the M thrust sheet to the south. The inferred distribution of the Precambrian Salt Range Formation below the M thrust suggests that thrust propagation may have occurred under a thicker sedimentary section that underlies the northern Kohat Plateau.

The authors are grateful for discussions with our many colleagues and support through the Geological Survey of Pakistan, the University of Peshawar, and the Oil and Gas Development Corporation of Pakistan during the decade in which research was undertaken providing ideas put forth in this paper. J.W. McD. especially appreciates financial support through the US Geological Survey during 1989–90 for geological mapping including field investigations with A. Bhatti and Feroze-ud-din of the Geological Survey of Pakistan. Victoria University of Wellington generously provided assistance for J.W. McD. during a postdoctoral fellowship in the Research School of Earth Sciences.

References

BAKER, D.M., LILLIE, R.J., YEATS, R.S., JOHNSON, G.D., YOUSUF, M. & ZAMIN, A.S.H. 1988. Development of the Himalayan frontal thrust zone: Salt Range, Pakistan. *Geology,* **16**, 3–7.

BOSSART, P. & OTTIGER, R. 1990. Rock magnetic properties and structural development in the core of the Hazara-Kashmir syntaxis, NE Pakistan. *Tectonics,* **9**, 103–121.

BURBANK, D.W. & RAYNOLDS, R.G.H. 1988. Stratigraphic keys to the timing of thrusting in terrestrial foreland basins: applications to the northwest Himalaya. *In*: KLEINSPEHN, K. & PAOLA, C. (eds) *New perspectives in basin analysis.* Springer-Verlag, New York, 331–351.

COWARD, M.P. & BUTLER, R.W.H. 1985. Thrust tectonics and the deep structure of the Pakistan Himalaya, 1985. *Geology,* **13**, 417–420.

DUROY, Y.A., FARAH, A. & LILLIE, R.J. 1989. Subsurface densities and lithospheric flexure of the Himalayan foreland in Pakistan. *In*: MALINCONICO, L.L. & LILLIE, R.J. (eds) *Tectonics of the western Himalayas*. Geological Society of America Special Papers, **232**, 217–236.

HYLLAND, M.D. 1990. *Geology of the Southern Gandghar Range and Kherimar Hills, Northern Pakistan.* MS thesis, Oregon State University, Corvallis.

MCDOUGALL & HUSSAIN, A. 1991. Fold and thrust propagation in the western Himalaya based on a balanced cross section of the Surghar Range and Kohat Plateau, Pakistan. *American Association of Petroleum Geologists Bulletin,* **75**, 463–478.

—— & KHAN, S.H. 1990. Strike-slip faulting in a foreland fold-thrust belt: the Kalabagh fault and western Salt Range, Pakistan. *Tectonics,* **9**, 1061–1075.

MEISSNER, C.R., MASTER, J.M., RASHID, M.A. & HUSSAIN, M. 1974. *Stratigraphy of the Kohat Quadrangle, Pakistan.* US Geological Survey Professional Papers, **716–D**.

SEEBER, L., ARMBRUSTER, J.G. & QUITTMEYER, R.C. 1981. Seismicity and continental subduction in the Himalayan arc. *In*: GUPTA, H.K. & DELANEY, F.M. (eds) *Zagros–Hindu Kush–Himalaya geodynamic evolution*. American Geophysical Union Geodynamics Series, **3**, 215–242.

YEATS, R.S. & HUSSAIN, A. 1987. Timing of structural events in the Himalayan foothills of northwestern Pakistan. *Geological Society of America Bulletin,* **99**, 161–176.

—— & —— 1989. Zone of Late Quaternary deformation in the southern Peshawar Basin, Pakistan. *In*: MALINCONICO, L.L. & LILLIE, R.J. (eds) *Tectonics of the western Himalayas*. Geological Society of America Special Papers, **232**, 265–274.

——, KHAN, S.H. & AKHTAR, M. 1984. Late Quaternary deformation of the Salt Range of Pakistan. *Geological Society of America Bulletin,* **95**, 958–966.

Evolution of foreland structures: an example from the Sulaiman thrust lobe of Pakistan, southwest of the Himalayas

I.A.K. JADOON[1], R.D. LAWRENCE[2] & R.J. LILLIE[2]

[1] *Department of Earth Sciences, Quaid-i-Azam University, Islamabad, Pakistan*

[2] *Department of Geosciences, Oregon State University, Corvallis, OR 97331, USA*

Abstract: Three cross-sections based on integrated surface geology, borehole and seismic reflection data have been used to interpret the style and evolution of the evolving Sulaiman foreland. A 10 km thick stratigraphic section outcrops at the deformation front. Available evidence suggests detachment of nearly all the stratigraphic sections from the crystalline basement. The decollement is believed to be located along the brittle–ductile boundary at the base of the wedge in the Sulaiman lobe of gentler topography. Coherent stratigraphy and structural uplift of 4–8 km is interpreted to be a result of a passive-roof duplex style of deformation. The duplex sequence consists of competent Jurassic and older strata. It has a floor thrust (decollement) in Palaeozoic rocks and a roof thrust (passive-roof backthrust) in thick Cretaceous shales. The roof sequence preserved over several horses along length of the section shows hindward translation.

The structural variation in cross-sections has led to the modification of an earlier interpretation that folding precedes duplexing. The present chronology of structures is the: (1) development of an incipient (*c.* 3.5 km half-wavelength) Uch fault-propagation fold at the tip of the decollement; (2) formation of a low amplitude, broad (*c.* 20 km half-wavelength) concentric detachment fold (Sul); (3) increase in amplitude of a detachment fold (Loti/Zin) to a critical level for development of ramp and duplex structures.

In recent years exploration for hydrocarbons has provided a large amount of seismic reflection and borehole data along mountain fronts. Integrated surface and subsurface data has allowed the production of well-constrained geological cross-sections. These sections display a variety of thin-skinned structures from concentric folding (Dahlstrom 1970; Dixon & Liu 1992) to piggyback thrusting (Dahlstrom 1970; Boyer & Elliot 1982; Butler 1982), and duplex style structures (Jones 1982; Banks & Warburton 1986; Hobson 1986; Mitra 1986; Vann *et al.* 1986) in the foreland of mountain systems. More detailed studies of the geometry of individual structures show that folds may be related to faults at depth. Such folds are recognized as fault-bend folds with stair-case trajectories (Suppe 1983) to fault-propagation folds developed at the decollement tip (Suppe & Medwedeff 1984; Mitra 1990). In other cases a ductile bead or concentric folding is recognized in front of the thrusts (Elliot 1976; Hossack 1983; Cooper & Trayner 1986).

In this paper, three structural cross-sections based on surface geology, seismic reflection profiles, and well data across the active Sulaiman Mountain front have been used to: (1) determine the geometry, style of deformation, and structural variation; and (2) to understand the evolution of the foreland structures.

Tectonic setting

The Sulaiman lobe to the SW of the Himalayas is a tectonically active, broad (>300 km) fold-and-thrust belt. Its surface geology is dominated by platform rocks bordered by ophiolites and flysch in the rear and molasse strata in the foredeep (Fig. 1). The main structural elements in the Sulaiman fold belt are E–W-trending arcuate, convex to the south folds in the foreland (Figs 1 & 2). Imbricate faults visible at the surface in the hinterland gradually disappear towards the deformation front.

Three different structural models have been proposed for the evolution of the Sulaiman/Kirthar Ranges in Pakistan (Banks & Warburton 1986; Bannert *et al.* 1989; Ahmed & Ali 1991). Banks & Warburton (1986) suggested a duplex style of deformation for the western Sulaiman/Kirthar Ranges (E–E in Fig. 1). The tectonic map of Pakistan (Kazmi & Rana 1982) shows faults exposed as imbricate structures in the central and northern part of the fold belt. Bannert *et al.* (1989) interpreted these faults as nappes. Ahmed & Ali (1991) drew a speculative cross-section across the northern Kirthar Ranges that proposed basement involvement in the foreland of the Kirthar Ranges. Recent balanced cross sections (D–D′ & F–F′ in Fig. 1) based on seismic reflection data favour a duplex

From TRELOAR, P.J. & SEARLE, M.P. (eds), *Himalayan Tectonics*
Geological Society Special Publication, No. 74, pp. 589–602.

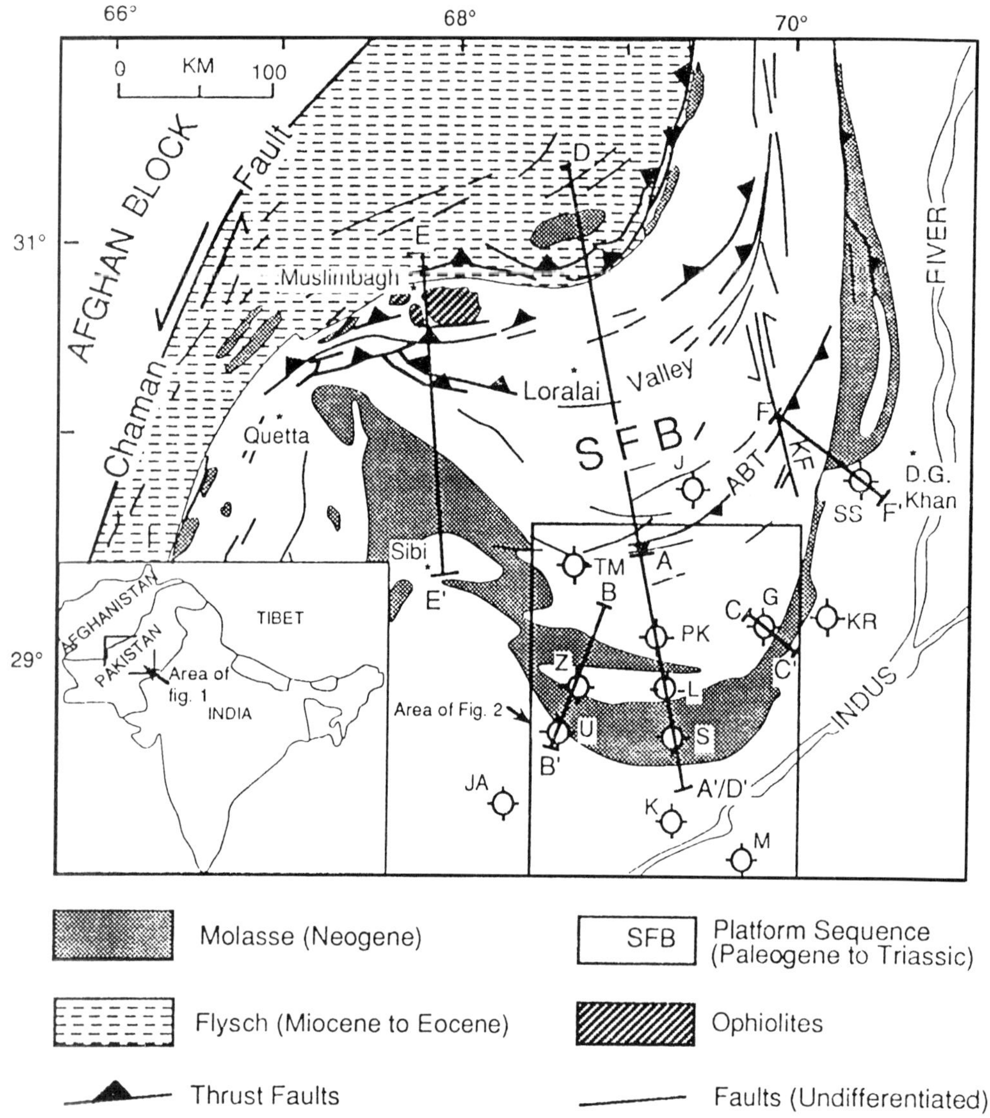

Fig. 1. Location of the Sulaiman fold-and-thrust belt in Pakistan (box). Structural cross sections A–A′, B–B′, and C–C′ are shown in figures 5, 7, & 8 respectively. Cross sections D–D′, E–E′, and F–F′ shows the location of balanced cross-sections constructed by Jadoon (1991), Banks & Warburton (1986) and Humayon *et al.* (1991) respectively. Abbreviations: ABT, Andari Backthrust; KF, Kingri fault. Well abbreviations: G, Giandari; J, Jandran; JA, Jacobabad; K, Kandkot; KR, Kot-Rum; L, Loti; M, Mari; PK, Pirkoh; S, Sui; SS, Sakhi Sarwar; TM, Tadri Main; U, Uch; Z, Zin.

style of deformation (Humayon *et al.* 1991; Jadoon 1991; Jadoon *et al.* 1992).

Deformation in the Sulaiman lobe probably became significant during the Miocene (20 ± 5 Ma) with the initiation of the Chaman fault system and the deposition of the molasse deposits (Khan *et al.* 1991). Ongoing, prograde deformation continuously reworked the molasse strata so that the centre of deposition migrated to the south and east. This is similar to the foreland translation of the Pine Mountain thrust sheet of the central Appalachians (Harris & Milici 1977), Jura Mountains of Europe (Laubscher 1981) and the Salt Range/Potwar Plateau of Pakistan (Raynolds & Johnson 1985). That active deformation continues to occur is

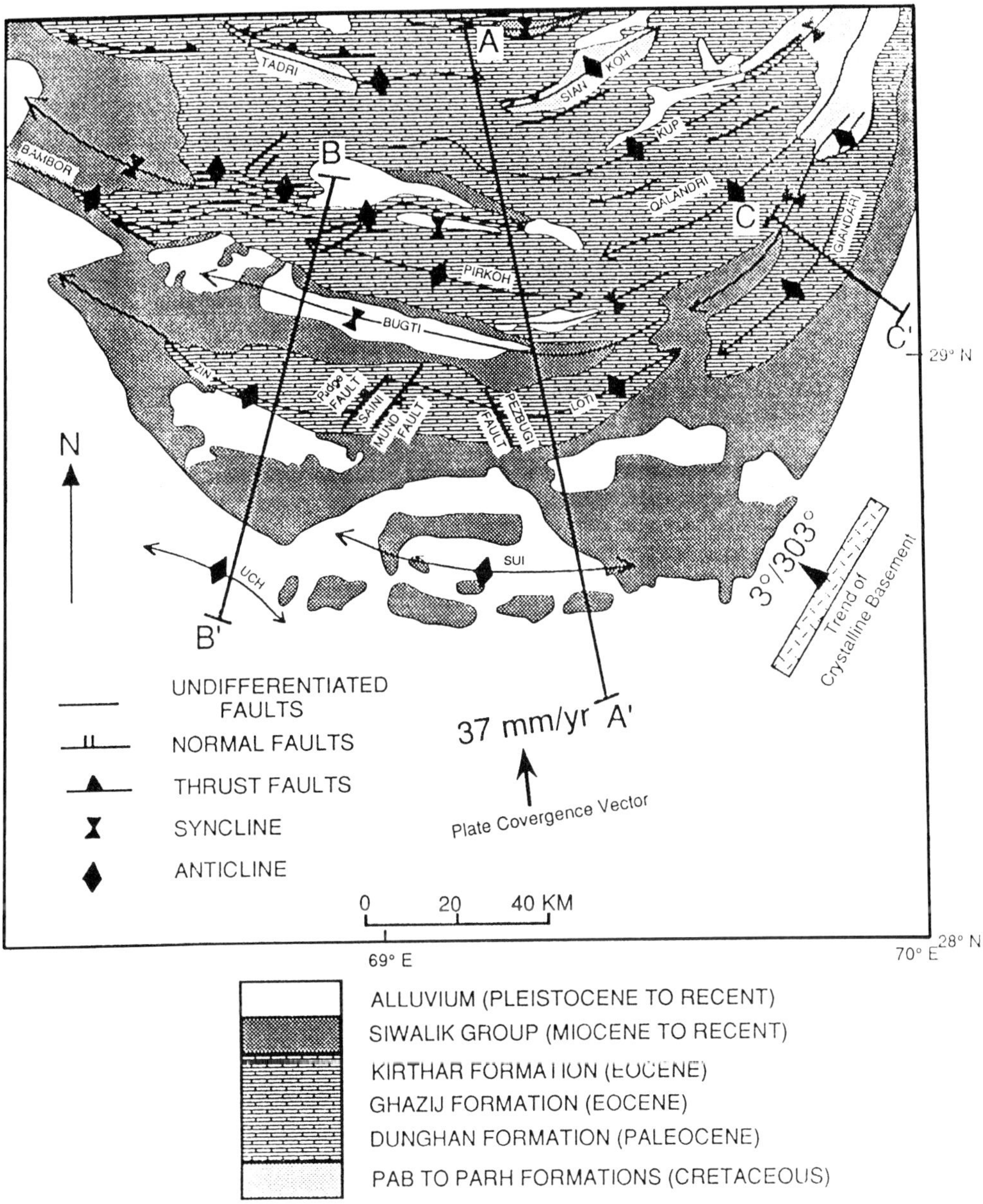

Fig. 2. Generalized geological map of the southern Sulaiman lobe (modified from Jadoon *et al.* 1992). Notice the folds as dominant foreland structures. Plate convergence vector is adopted from Minster *et al.* (1974). See text for the attitude of the basement. A–A′, B–B′, and C–C′ are the locations of the structural cross sections.

evidenced by recent unconformities from the southern Sulaiman Range (Tainsh *et al.* 1959) and local seismicity (Quittmeyer *et al.* 1979). Age dating by magnetostratigraphy from the eastern Sulaiman mountain front (Ahmad & Khan 1990) shows that the continental Siwaliks, deposited between 0.7 Ma to 50 000 years, are overlain by alluvial fan deposits. The latter are tilted along the eastern Sulaiman front. The discussion to follow deals mainly with the evolution of the foreland structures at the evolving mountain front.

Discussion of surface and subsurface data

Surface expression of the foreland

The frontal Sulaiman ranges are composed largely of exposed Neogene molasse, Palaeogene shales, and carbonates. Dominant structures of the foreland are mostly broad folds (Fig. 2). Progressively older rocks are exposed in the cores of the folds toward the hinterland. However, these exposed rocks in the foreland show a coherent stratigraphy that is not disrupted by the emergent thrusts. The Sui and Loti folds are two gentle (dips of limbs range from 5° to 15°) and broad (*c*. 20 km half-wavelength) doubly plunging folds at the mountain front. The Pirkoh fold north of Loti has a fairly steep (35°–75°) southern limb and a relatively flat hinge zone of about 15 km. The overall expression of the Pirkoh-box fold is of a foreland dipping monocline. The Uch fold is another relatively small doubly plunging anticline located at the mountain front. Axes of the Sui and Uch anticlines form an en echelon pattern relative to each other (Fig. 2). A similar en echelon pattern is depicted by the Loti and Giandari folds. This en echelon pattern, along with deflection of fold axes (Kup fold), is widely observed in the Sulaiman foreland (Fig. 2). This lateral discontinuity may be related to buried ramps.

In addition to folds, oblique normal faults (Ridge, Sani Mund, and Pezbugi) are dominant structures of the foreland. They show a dip-slip offset of about 20 m with a significant component of strike-slip displacement. Due to these faults, the axis of the Loti anticline is displaced for several hundred metres.

Seismic reflection profiles and boreholes

Extensive seismic reflection and borehole data have been used to constrain the structural cross-sections A–A′, B–B′, and C–C′ (Figs 2 & 3). The industry seismic profiles provided good coverage for the foreland of the Sulaiman lobe.

The subsurface data has been used to resolve: (1) trend and depth to the top of the basement in order to determine stratigraphic and tectonic thicknesses; (2) the major decollement; and (3) the geometry of structures along structural cross-sections. The first two factors are vital to constrain the geometry of structures and style of deformation.

(i) Crystalline basement and sedimentary package. It is important to locate the top of the basement in a fold-and-thrust belt in order to evaluate: (1) total thickness (stratigraphic and tectonic) of the sedimentary wedge above the basement; (2) location and nature of the decollement at the base of the wedge or in some younger horizon; (3) basement slope, which is important in understanding the mechanics of thrusting (Davis *et al.* 1983; Davis & Engelder 1985); and (4) the role of basement structures in controlling the deformation (Jackson 1980; Lillie 1984).

Most of the seismic reflection lines shown in Fig. 3 include data of 5 seconds two way travel time (TWTT). Due to the extreme thickness of the sediments, basement can only be seen on lines W-15-BP from the eastern, and 834-SAJ-22 from the southern Sulaiman foredeep. Seismic reflection data (834-SAJ-22) shows that the Precambrian to Quaternary rocks are about 6 km thick in the foredeep near the Mari gas fields and that they thicken stratigraphically to about 8 km along the axis of the Indus river (Jadoon *et al.* 1992). At the deformation front, the basement reaches depths beyond 5 seconds two-way travel time. However, the basement configuration has been interpreted by extrapolating the layercake stratigraphy into the fold-and-thrust belt from the foredeep region (using bold lines in Fig. 3). Seismic reflection interpretation (Jadoon *et al.* 1992) shows a stratigraphic thickness of about 10 km at the deformation front. Extrapolating the top basement dip (2°–3°) to the north suggests a tectonic thickness of about 20 km in the hinterland. Tabular layercake stratigraphy and broad structures as far north as the Bugti syncline suggest a planar basement surface.

Seismic reflection data constrains the attitude of the top of the crystalline basement. Seismic reflection lines show that northwestward dipping basement in the eastern (Raza *et al.* 1989; Humayon *et al.* 1991) and southern Sulaiman foredeep is at a depth of about 8 km along the axis of the Indus River. It is inferred at a depth of about 13 km underneath the Tadri anticline along cross section A–A′ in Fig. 2. This data implies an attitude on top of the crystalline basement of 213°/3°. This interpretation is supported by additional 3-point constructions using boreholes (Karampur, Bahawalpur East, and Marot-1) in which basement has been encountered. These observations show that the E–W trend of the folds at the southern Sulaiman lobe is oblique to the orientation of the underlying NW-dipping basement.

(ii) Location of major decollement. A stratigraphic column (Jadoon *et al.* 1992) based on subsurface data in the foreland and the surface

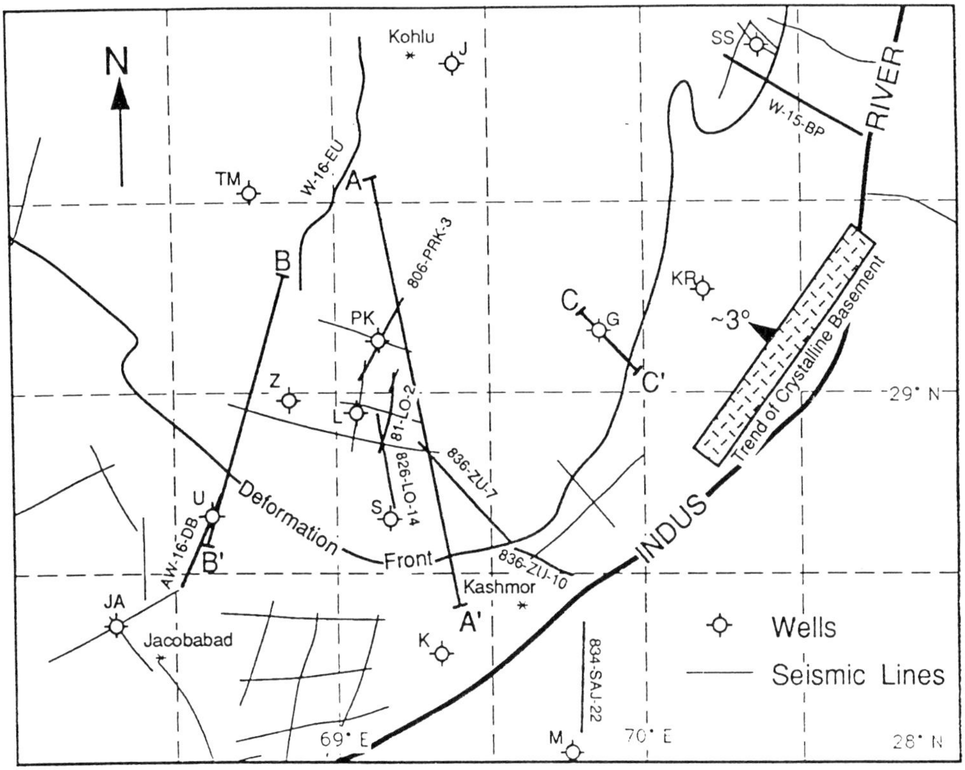

Fig. 3. Seismic layout and borehole map to show the location of the seismic reflection lines available for this study. Bold lines with text shows the location of the composite seismic line described by Jadoon *et al.* (1992). A–A′, B–B′, and C–C′ are the location of Figs 5, 7 & 8 respectively. Well abbreviations are the same as given in Fig. 1.

geology in the hinterland indicates that potential decollement horizons can occur in the Eocene, Cretaceous and Palaeozoic pelitic strata. However, seismic reflection profiles show that the complete stratigraphic section is detached from the basement in the southernmost Sui and Loti anticlines. Thus, the major decollement remains in Palaeozoic rocks at the interface between crystalline basement and the sedimentary package at the deformation front. A similar level for the decollement horizon has been inferred for both the western (Banks & Warburton 1986) and eastern (Humayon *et al.* 1991) Sulaiman Range.

(iii) Nature of major decollement. The gross geometry of the overthrust wedge, including gentle topography (<1°) and broad width (>300 km), is compatible with that proposed by Davis & Engelder (1985) for thrust belts developed over a weak decollement. A thin-skinned style of deformation is supported by the seismic reflection data from the Sulaiman fold belt (Humayon *et al.* 1991; Jadoon *et al.* 1992). However, there is evidence that the Eocambrian evaporite sequence that provides an effective zone of decoupling at the base of the section in the Salt Range and Potwar Plateau (Lillie *et al.* 1987; Jaume & Lillie 1988) may not be present underneath the Sulaiman fold belt. This evidence includes: (a) absence of salt related diapiric structures (e.g., tight anticlines, broad synclines and disharmonic folding); (b) the nearest occurrence of Eocambrian evaporites in wells and seismic lines is about 200 km east of the deformation front; and (c) the seismic reflection signature of the ductile zone (salt pillows) associated with the evaporites has not been observed on the seismic lines from the southern Sulaiman foredeep. Unlike the 2 to 4 km depths in the frontal Salt Range/Potwar Plateau, seismic reflection data from the Sulaiman fold belt show that the decollement is about 10 km deep at the base of the wedge at the deformation

front. Khan & Raza (1986) and Raza *et al.* (1989) calculated an average goethermal gradient of about 30° C/km in boreholes from the Sulaiman foreland and adjacent foredeep. Davis & Engelder (1985) show that with such a geothermal gradient, limestones are as weak as evaporites at depths of about 10–12 km. That limestones may have marked the decoupling zone in the Sulaiman fold belt is indicated by ductile deformation of fine-grained Triassic limestones and shales exposed in the core zones of small folds in the hinterland of the Sulaiman fold belt. The signature of this ductile deformation is also seen in seismic lines across the core zones of the frontal anticlines (Sui and Loti). This suggests that the effective zone of weak decoupling in the Sulaiman may be in fine-grained carbonate rocks at their brittle/ductile transition at depths of 10–15 km. The stratigraphic section suggests that abundant pelitic rocks at this level might provide a weak zone through dewatering and/or recrystallization. Thus, at this depth, fine-grained sedimentary rocks may provide a weak detachment similar to the evaporites at depths of 1 to 3 km beneath the Salt Range/Potwar Plateau region.

Foreland structures and style of deformation

Cross-section A–A′

Structural cross-section A–A′ is modified from Jadoon *et al.* (1992) and is shown to elaborate the foreland structures. One of the most important observations resulting from the study of the composite seismic line from the Mari well (line 834-SAJ-22) in the Sulaiman foredeep to Kohlu (line W-16-EU) in the central Sulaiman is the interpreted depth to the top of the crystalline basement. The basement located at a depth of 6 km in the foredeep descends northwards with a gentle inclination to attain a depth of about 13 km below the Tadri anticline.

In the foreland, progressively older rocks exposed in the core zones of doubly plunging anticlines (e.g., the Sui anticline exposes molasse at the surface; the Loti, Pirkoh and Danda folds have Eocene strata exposed in their cores, while farther north, the Kurdan and Tadri anticlines are cored by Palaeocene and Cretaceous strata) are structurally uplifted. These exposed rocks show a coherent stratigraphy as far north as the Tadri syncline, and are not disrupted by significant thrust faults (Fig. 2). This is consistent with borehole data that penetrates a normal stratigraphic sequence as deep as Jurassic in the Sulaiman foreland.

Seismic reflection interpretation of the previously described Sui and Loi anticlines is of low amplitude, concentric detachment folds located at the deformation front (Fig. 4). These concentric anticlines maintain their layer-parallel thickness above the Triassic. Excess space in the cores of these anticlines below the Jurassic horizon is considered to be filled by ductile flow of fine carbonates and pelitic strata at the decollement level. The exposed Triassic section in the hinter-

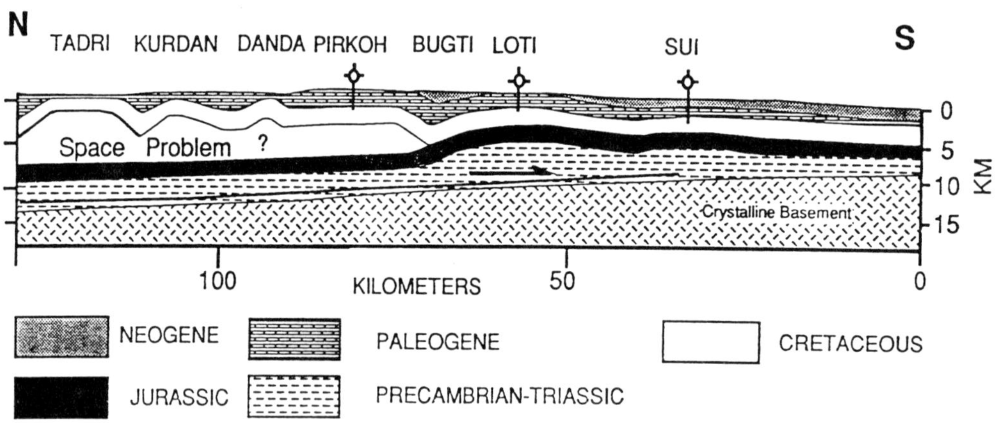

Fig. 4. Partial geological cross-section (A–A′) of the southern Sulaiman foreland. See Fig. 2 for location and Jadoon *et al.* (1992) for seismic coverage. The section shows the extrapolated stratigraphy and top to the crystalline basement, decollement zone, and a space problem due to tectonic uplift of the Cretaceous and younger strata as a coherent slab, over the older strata. Note that low amplitude broad concentric folding is the structural style of the foreland. The space in the cores of these folds (Sui and Loti) is considered to be filled by ductile flow of material at the deep decollement level. Names for individual folds mapped at surface are from local mountains.

land shows feature related to ductile deformation. The excess ductile material in the cores of the Sui and Loti anticlines is believed to be similar to a ductile bead in front of propagating thrusts (Elliot 1976; Hossack 1983; Cooper & Trayner 1986).

The surface expression of the Pirkoh anticline (third fold from the deformation front) is as a foreland-dipping monocline with dips of between 35° and 75°. It has almost flat strata (nodular Eocene limestone) along its hinge area for about 15 km at outcrop. The northern limb of the Pirkoh fold is concealed below an anticline, and gives the entire structure a box-like form. A depth section based on seismic reflection data shows a zone of Cretaceous and younger strata uplifted by 4 km under the hinge zone of the Pirkoh fold (Fig. 4). The structural relief of about 4 km below the Pirkoh structure increases northward to about 8 km below the Tadri anticline. As the stratigraphy is not disrupted by major faults above the Jurassic, this structural uplift is interpreted as due to a thin-skinned, passive-roof duplex style of deformation (Fig. 5). The duplex sequence consists of Jurassic and older rocks, bounded between a floor thrust at the base of the wedge and a roof thrust in the Cretaceous shales. The roof sequence is interpreted as remaining passive during the foreland propagation of the underlying duplex horses and is regarded as a passive-roof sequence (Banks & Warburton 1986) with a decollement in thick Cretaceous (Sembar) shales. In cross-section A–A′ (Fig. 5) the duplex sequence is concealed by a continuous roof sequence. The broad folds at the surface reflect the shape of deep structures associated with faults in the duplex sequence which never break the surface in this zone of blind faults. However, the duplex structures are exposed in the western Sulaiman fold belt. In this framework, the Pirkoh structure is a duplex related, fault-bend fold (Jadoon *et al.* 1992). The near horizontal dips along the hinge zone of the Pirkoh anticline are due to the juxtaposition of the hanging-wall flat of the Pirkoh duplex horse over the footwall flat. The steep foreland dipping limb of a monocline, like the Pirkoh fault-bend fold, may represent a culmination wall over the foreland propagating duplexes. In contrast, the Tadri anticline cored by two duplex horses is an anticlinal stack.

The interpreted chronology of structures along the transect A–A′ is: (1) growth of low amplitude broad concentric detachment folds in the foreland: (2) development of a duplex horse bounded between the basal decollement and a passive-roof backthrust; (3) propagation of the duplex as critical taper is achieved; and (4) development of tear and extensional normal faults (flexural faults) within the overthickened wedge. The chronological order in this section suggests folding prior to faulting and is similar to that proposed for the Canadian Rockies (Dahlstrom 1970) and demonstrated by experimental modelling (Dixon & Liu 1992).

Cross-section B–B′

Cross-section B–B′ (Fig. 2) extends 86 km into the foreland. The frontal doubly plunging Uch anticline observed in this section is located farther to the south in the foredeep. It has an en echelon relationship with the Sui fold. Unlike the broad (half-wavelength is *c.* 20 km) Sui and Loti folds, the Uch is a relatively small (half-wavelength is *c.* 3.5 km) fold. Industry seismic reflection line AW-16-DB (Fig. 3) is located over the Uch fold. Along this line the crystalline basement is deeper than 5 s TWTT. Seismic reflection data between SP-316 to 193 show that

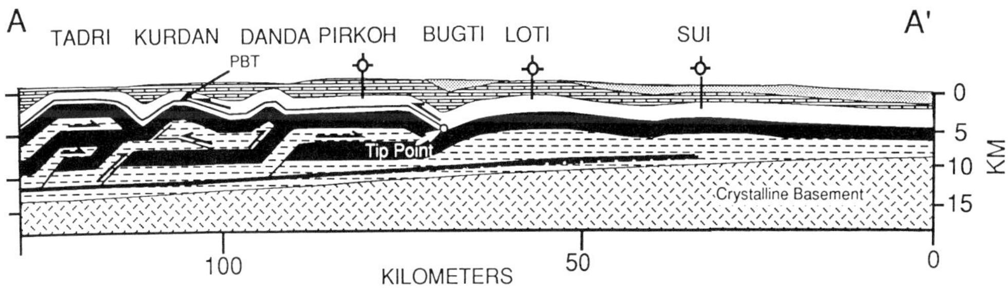

Fig. 5. Complete geological cross-section (A–A′) of the southern Sulaiman foreland (modified from Jadoon *et al.* 1992) suggesting a duplex style of deformation to solve the space problem shown in Fig. 4. The duplex sequence of Jurassic and older rocks is bounded between a floor thrust (decollement) above the crystalline basement and a roof thrust (passive-rock backthrust) in Cretaceous Sembar shales. Folds at the surface reflect the shape of the duplex horses below. PBT, Passive-roof backthrust. Ornament as in Fig. 4.

the decollement beneath the Uch anticline is in Palaeozoic strata at a depth of more than 4 s TWTT. This depth to decollement is consistent with the structural cross-section A–A′ in Fig. 5. Our interpretation of the seismic reflection data (Fig. 6) suggests that the Uch fold is an incipient, active fault-propagation fold developed as a result of simple shear at the tip of the decollement. The thrust has not propagated up section through the incompetent unit to offset the competent Jurassic limestone. The active deformation is evidenced by growth sequences in molasse strata (Fig. 6), an unconformity between Holocene and Recent deposits (Tainsh *et al.* 1959), and seismicity (Quittmeyer *et al.* 1979). Similar folds from the eastern Sulaiman front are less than 50 000 years old (Ahmad & Khan 1990).

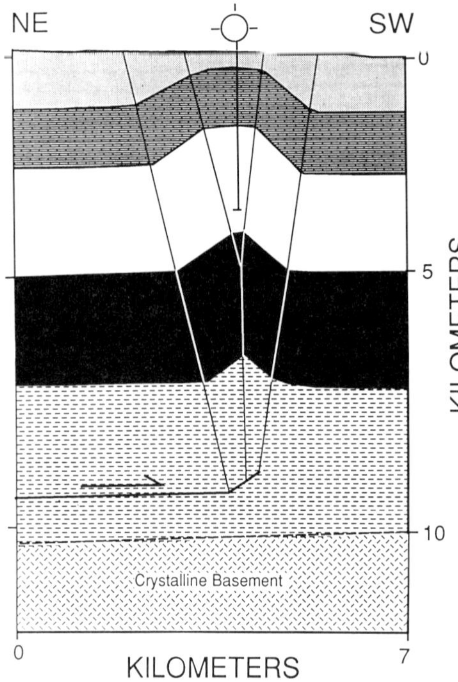

Fig. 6. Depth section of seismic reflection line AW-16-DB between SP193–316 over the Uch frontal anticline that interprets the Uch fold as a fault-propagation fold. Notice the decollement in the Paleozoic section consistent with A–A′. Compare the narrow wavelength of the Uch anticline with the broad folds (Sui and Loti) in Fig. 5. Ornament as in Fig. 4.

Extrapolation of limited seismic reflection profiles along B–B′ indicates that north of the frontal Uch fault-propagation fold, the broad (*c.* 20 km half-wavelength), low-amplitude Zin anticline is a detachment fold cored by excess ductile strata at the decollement horizon (Figs 2 & 7). The Zin anticline is the westward extension of the Loti anticline (Figs 5 & 7). North of these fault-tip-line folds are the duplex structures similar to those shown in section A–A′ while the relatively tight folds at the surface may be interpreted as detachment folds in the passive-roof sequence.

The interpreted chronology of structures along B–B′ is: (1) a small wavelength (*c.* 3.5 km) fault-propagation Uch fold; (2) development of a low amplitude, broad concentric detachment fold (Zin/Loti); (3) development and propagation of the duplex with some tight detachment folds in the roof-sequence.

Cross-section C–C′: Giandari fold

Cross section C–C′ (Figs 2 & 8) is drawn across the Giandari anticline at the deformation front. This anticline with a half-wavelength of about 15 km has the surface expression of a monocline with dips of *c.* 40° towards the foredeep. This is in contrast to the low amplitude detachment folding developed at the tip of the decollement

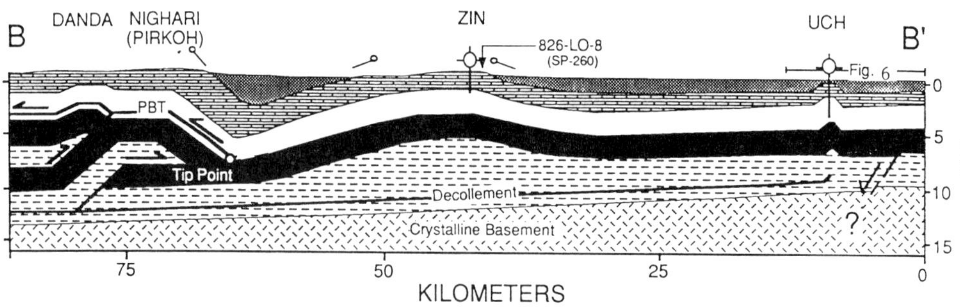

Fig. 7. Geological cross section (B–B′) in the Sulaiman foreland. Notice a fault-propagation fold at the tip of the decollement preceding the broad concentric fold and duplex structures. Ornament as in Fig. 4.

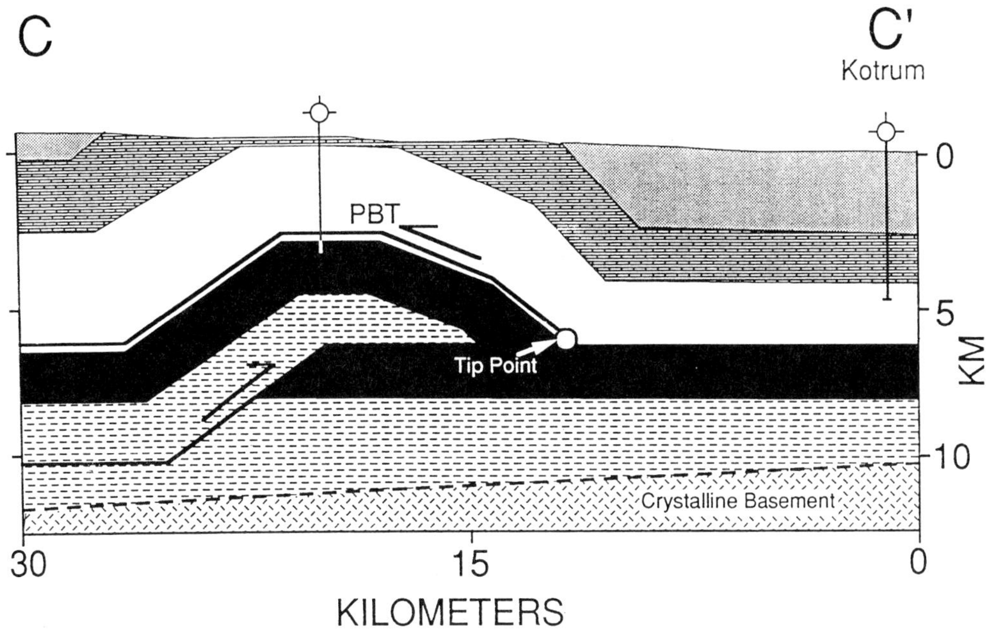

Fig. 8. Geological cross section (C–C′) in the Sulaiman foreland. Interpretation suggests that the Giandari anticline is a duplex related fault-bend fold at the deformation front. Ornament as in Fig. 4.

in A–A′ (Fig. 5). It should be noted that C–C′ is oblique to the plate convergence vector (Fig. 2). However, the strike of the surface structures (NE–SW) is similar to that of the crystalline basement.

Seismic data are not available across this fold. The crystalline basement has been extrapolated northwestwards from the axis of the Indus river where it is at a depth of about 8 km. Two wells, Kot-Rum in the foredeep and the Giandari in the fold belt provide additional subsurface information. The Kot-Rum well in the foredeep is drilled to a depth of 4798 m. It penetrated 2780 m of Neogene molasse strata to the Cretaceous. In the Giandari anticline, the well penetrated a normal section to a depth of 3659 m. The top of the Pab sandstone (Cretaceous) in the Giandari well was hit at a depth of 30 m. The same sandstone in the Kot-Rum well in the adjacent foredeep was penetrated at a depth of 4425 m. This observation shows an uplift of more than 4 km of Cretaceous and younger rocks in the Giandari anticline. In the absence of a fault in the Cretaceous and younger rocks, this structural relief is interpreted as due to duplication of Jurassic and older strata, along a blind fault (Fig. 8). Thus, the Giandari anticline is interpreted as a duplex related fault-bend fold at the deformation front. The roof thrust is proposed to be in the Sembar shale (Cretaceous). Note that about 1700 m of this shale has been drilled in the Giandari well. The structural duplication is consistent with that observed in the other sections.

Discussion

Duplex geometry

The term ‘duplex’ was first used by Dahlstrom (1970) for structures in the Canadian Rocky Mountains. The duplex geometry requires duplication of strata along blind thrusts bounded between a floor and a roof thrust (Boyer & Elliot 1982). Theoretical examples and case studies of duplex geometry considered in this study are shown in Fig. 9. The duplex geometry in Fig. 9A is a kinematically inadmissible section due to inequivalent shortening between the duplex and the roof sequence. On the other hand Figs 9B and C are kinematically admissible sections for duplex geometry, suggesting, respectively, forward and hindward translation of the roof sequence. The duplex geometry shown in Fig. 9B (Boyer & Elliot 1982) and termed an ‘emergent allochtonous roof duplex’ by Geiser (1988) shows the emergent roof sequence along foreland vergent faults in the foredeep (Fig. 9Bi).

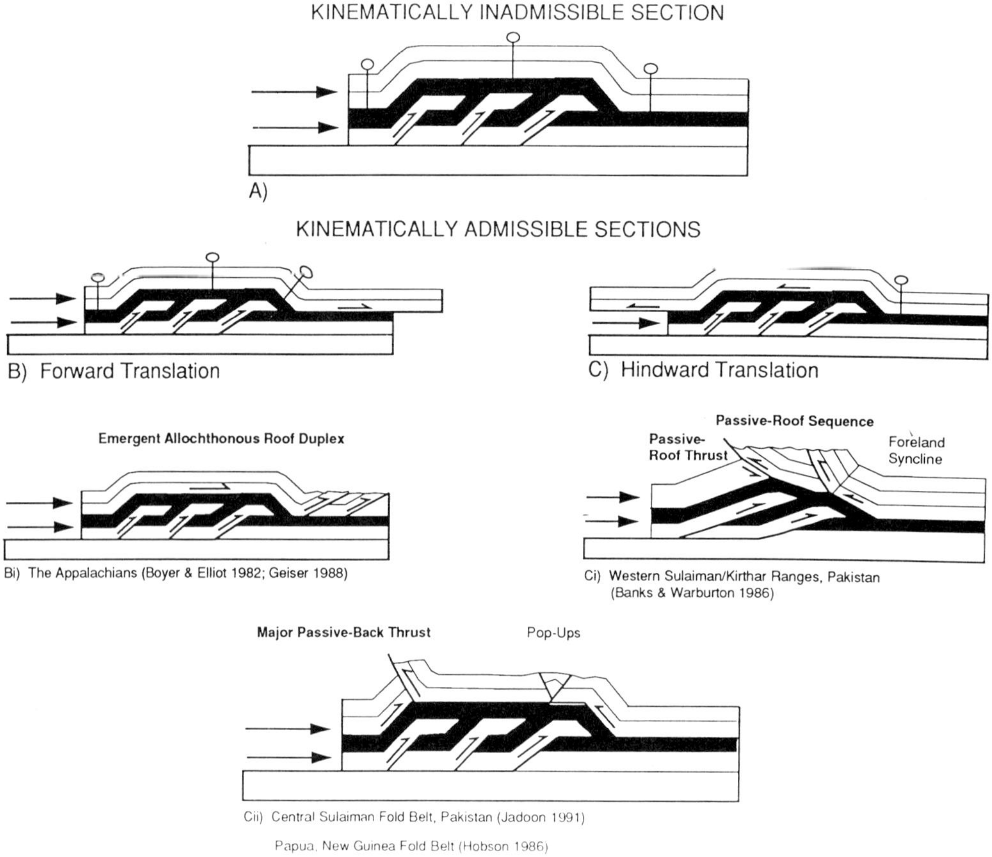

Fig. 9. Schematic diagram showing kinematically inadmissible (**A**) and kinematically admissible (**B** & **C**) sections of duplex geometry. Notice the forward (B) and hindward (C) translation of the roof layer with respect to the duplex layer.

of this type besides the Appalachians are the Jura and Swiss Plain, and the Mackenzie Mountains area of Canada (Boyer & Elliot 1982; Vann *et al.* 1986). The other geometry shown (Fig. 9c) and termed as 'passive-roof duplex geometry' (Banks & Warburton 1986) suggests a hindward vergence for the passive-roof sequence (Fig. 9Ci). Some examples of this type are the western Sulaiman fold belt (Banks & Warburton 1986), Mackenzie Mountains, Alberta foothills of Rocky Mountains, and Athabasca Valley foothills (Jones 1982; Vann *et al.* 1986). Banks & Warburton (1986) proposed that rather than inferring a continuous roof sequence which extends over several duplex horses, it is logical to consider an imbricated roof sequence with multiple backthrusts (Fig. 9Ci). Each backthrust can accommodate several kilometres of shortening by passive uplift and erosion of the roof sequence over a foreland propagating duplex.

The structural cross-sections in this study show a passive-roof duplex geometry consistent with the western (Banks & Warburton 1986) and eastern (Humayon *et al.* 1991) Sulaiman fold belt. However, contrary to the western Sulaiman, cross-sections from the southern (this study) and eastern (Humayon *et al.* 1991) Sulaiman Range preserve a continuous roof sequence of about 150 km (Jadoon 1991) that lacks prominent overstep-backthrusts. Other examples of preserved roof sequences have been shown by Jones (1982) and Hobson (1986). Our structural model for the Sulaiman Range based on surface geology and seismic reflection data is shown in Fig. 9Cii. This model shows a hindward emergent roof sequence with pop-ups and associated out-of-sequence thrusting (Jadoon & Khan 1990). Clear examples of a preserved roof sequence along D–D′ in Fig. 1 and in the eastern (Humayon *et al.* 1991) Sulaiman demonstrate the presence of a continuous roof sequence over several duplex horses at an early stage of evolution of such structures. The resultant structural geometry (Fig. 9Cii) is similar to Fig. 9Bi in that both have a long roof sequence. The floor (decollement) and the roof thrust (passive-roof backthrust) in the present model may be equivalent to a set of conjugate faults.

Evolution of foreland structures

The data discussed in the previous sections allow a detailed analysis of the southward propagation of an active mountain front. The low amplitude, broad concentric Sui and Loti folds are interpreted as detachment folds cored by ductile material at the decollement horizon. A ductile bead of deformation formed at the tip of the decollement during the propagation of thrust sheets as discussed by Elliot (1976), Hossack (1983), and Cooper & Trayner (1986), while Dixon & Liu (1992) have experimentally demonstrated the formation of similar folds preceding faulting. Such a similar process is believed to have occurred in the Sulaiman fold-and-thrust belt (see Fig. 5). In contrast to the Sui and Loti folds, the Uch is a fault-propagation fold in front of a low amplitude broad concentric fold (Zin) and duplex (see Fig. 7). This shows that faulting preceded folding. These observations suggest that the Sui, Loti, and Zin folds may have been initiated by a structural perturbation similar to the Uch fold before subsequently attaining their present structural configuration of low amplitude, broad concentric folds. Further, following Dixon & Liu (1992), it can be envisaged that during the development of flat and ramp structure, ramp propagation will be through locally strained competent strata in the forelimb of the detachment fold.

Fault-propagation and fault-bend folding in the fold-and-thrust systems

Fault-propagation and fault-bend folding (Suppe 1983; Suppe & Medwedeff 1984; Mitra 1990) are common folding mechanisms in fold-and-thrust belts. In fault-propagation folds a thrust loses slip up section by transferring shortening to a fold developing at the fault tip. In contrast fault-bend folding requires a stair-case trajectory to transfer slip from a lower to an upper decollement horizon. Bending of the hanging wall block along such a non-planar surface results in fault-bend folding (Suppe 1983). Examples of these structures are widely described in the literature.

The structural pattern and sequential development of fault-propagation and fault-bend folds is shown in Fig. 10. The structural cross-sections from the foreland of the Sulaiman lobe (Figs 5, 7, & 8) show dominance of fault-bend folding related to bending of duplex horses over non-planar surfaces. Fault-propagation folds in the Sulaiman as well as other areas (e.g. the Wyoming thrust belt, Lamerson 1982) are observed to form at two locations as shown in Fig. 10. Firstly at the leading edge of the decollement and secondly in the interior of the fold-and-thrust belts. Folds related to these two types are recognized as primary and secondary fault-propagation folds respectively (Mitra 1990). The Uch anticline in the Sulaiman foredeep and Lodgepole anticline at the leading edge of the Absaroka thrust sheet (Lamerson 1982) are primary fault-propagation folds, whereas those formed in the interior of the fold-and-thrust belts such as the Jandran anticline in the central Sulaiman Range (Jadoon & Khan 1990; Jadoon 1991) and the Cave Creek anticline in the Wyoming thrust sheet (Mitra 1990) are recognized as secondary fault-propagation folds. The secondary fault-propagation folds in the internal part of the fold-and-thrust belts may be associated with secondary deformation at a time when prograde deformation is occurring at the leading edge of the decollement. Such secondary structures are termed out-of-sequence structures (Morley 1988) and the Jandran anticline in the central Sulaiman has been described as such a structure (Jadoon & Khan 1990). Sequential evolution of out-of-sequence structures increases the critical taper coupled with internal deformation of a wedge (see Fig. 10) in order to enable it to grow outwards following the bulldozer model (Davis *et al.* 1983; Dahlen *et al.* 1984; Davis & Engelder 1985). A wedge at critical taper can slide along its basal decollement without internal deformation. It is suggested that out-of-sequence deformation may occur to produce internal deformation in order to create the appropriate critical taper for thrusting.

Translation of a fault-propagation fold to a fault-bend fold is described by Mitra (1990) with examples from the Appalachians and the Cordilleran, Absaroka thrust sheet. We notice such translation of fault-propagation folds at the leading edge into fault-bend folds. Such in-sequence translation of fault-propagation folds into fault-bend folds at the leading edge may produce a fold-and-thrust belt dominated with stair-case trajectories such as the Sulaiman and

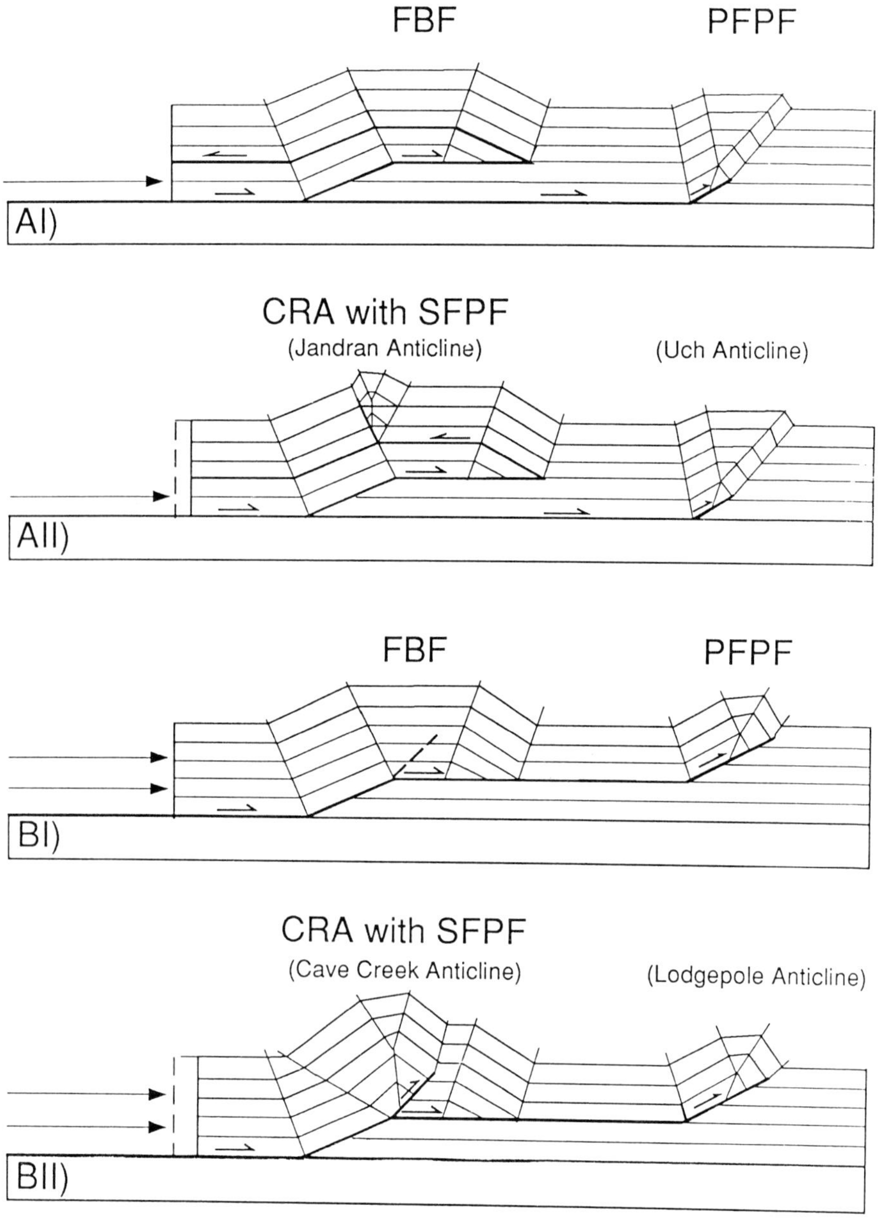

Fig. 10. Structural position and sequential evolution of primary and secondary fault-propagation and fault-bend folds. Model A is based on structures in the Sulaiman, SW Himalaya (Jadoon 1991). Model B is based on structure in the Rocky Mountains (modified from Lamerson 1982: Mitra 1990). Abbreviations: CRA, compound ramp anticline (from Lamerson 1982); FBF, fault-bend fold; PFPF, primary fault-propagation fold; SFPF, secondary fault-propagation fold.

other fold-and-thrust belts. Alternately, a propagating thrust may cut up section through all the layers retaining its listric or planar geometry to produce an emergent thrust sheet.

Conclusions

Surface and subsurface data have been investigated from the southern Sulaiman to analyse the evolution of active foreland structures along the western margin of the Indian plate. The important conclusions are summarized as follows.

(1) Compressive deformation along the frontal part of the Sulaiman fold belt is accommodated by development of a passive-roof duplex, the floor thrust of which is situated at about 10 km depth above the crystalline basement in Palaeozoic strata and the roof thrust in thick Cretaceous shales.

(2) The passive-roof sequence has hindward translation relative to the foreland propagating duplex. The roof sequence described in the present study extending over several duplex horses is not breached in the foreland. Thus, such a continuous roof sequence may be a feature of thrust belts during their early stages of evolution.

(3) Jadoon *et al.* (1992) proposed progressive structural development with long wavelength, low amplitude concentric detachment folding preceding duplexing. Presence of an incipient fault-propagation fold (Uch) in front of a low amplitude broad concentric fold (Loti/Zin) suggests that detachment folds (Sui, Loti/Zin) were initiated by fault-propagation folding as a result of simple shear at the tip of the decollement. Future evolution of these structures involves the propagation of a ramp through the localized strained competent unit in the forelimb of the detachment fold and development of the duplex structures. These data suggest evolution of the Sulaiman foreland structures from an incipient fault-propagation fold at the tip of the decollement followed by low amplitude, broad concentric detachment folding, with lastly the development of the duplex complex.

(4) In the Sulaiman fold-and-thrust belt, both primary and secondary (in-sequence and out-of-sequence respectively) fault-propagation folds are recognizable. The primary fault-propagation folds occur at the tip of the decollement, whereas secondary fault-propagation folds may form in the interior of the thrust belt. Out-of-sequence deformation may occur to create critical taper for an outward growth of the foreland fold-and-thrust belts. The geometry of a primary fault-propagation fold may be modified by imbrication or by translation from a lower to an upper decollement. Imbrication along an up section propagating blind thrust forms an emergent thrust sheet. In the case of translation along an upper decollement, a fault-bend fold is produced.

We gratefully acknowledge the collaboration of Oil and Gas Development Corporation of Pakistan Hydrocarbon Development Institute of Pakistan, Amoco Exploration Company, Texaco Overseas, and Shahid Hassan Khan of the Geological Survey of Pakistan (GSP). I.A.K.J. appreciates the grant of study leave by Quaid-i-Azam University under the United States Agency for International Development fellowship. National Science Foundation grants INT-86-09914 and EAR-8816962 to Oregon State University projects in northern Pakistan provided partial support. We are grateful to G. Huftile and R.S. Yeats for reading an early draft of the manuscript. Comments and suggestions by P.J. Treloar and G. Akhtar helped improve the revised manuscript. A.A. Khawaja, C.J. Banks, and G.D. Williams are gratefully acknowledged for a critical review of the manuscript. I.A.K.J. is indebted to V. Langer, M.S. Baig and M. Qayyum for many interesting discussions.

References

AHMAD, W. & KHAN, M.J. 1990. Sedimentologic and magnetostratigraphic studies of the Upper Siwalik Group, Sulaiman Range, Pakistan (abs). *Second Pakistan Geologic Congress (2–4 September 1990), Department of Geology, University of Peshawar, Pakistan,* 33–34.

AHMED, R. & ALI, S.M. 1991. Tectonic and structural development of the eastern part of Kirthar fold belt and its hydrocarbon potential. *Pakistan Journal of Hydrocarbon Research,* **3**, 19–31.

BANKS, C.J. & WARBURTON, J. 1986. 'Passive-roof' duplex geometry in the frontal structures of the Kirthar and Sulaiman mountain belt, Pakistan. *Journal of Structural Geology,* **8**, 229–237.

BANNERT, D., CHEEMA, A. & AHMAD, A. 1989. *Interpretation of Landsat-MSS Imagery of the Sulaiman and Kirthar Mountain Ranges in western Pakistan.* A technical report of a cooperative project between Bundesanstalt für Geowissenschaften und Rohstoffe, Hannover, West Germany and Hydrocarbon Development Institute of Pakistan.

BOYER, S.E. & Elliot, D. 1982. Thrust Systems. *American Association of Petroleum Geologist Bulletin,* **66**, 196–1230.

BUTLER, R.W.H. 1982. The terminology of structures in thrust belts. *Journal of Structural Geology,* **4**, 239–245.

COOPER, M.A. & TRAYNER, P.M. 1986. Thrust-surface geometry: implications for thrust-belt evolution and section balancing techniques. *Journal of Structural Geology,* **8**, 305–312.

DAHLEN, F.A., SUPPE, J. & DAVIS, D.M. 1984. Mechanics of fold-and-thrust belts and accretionary wedges: cohesive Coulomb theory. *Journal of Geophysical Research,* **89**, 10087–10101.

DAHLSTROM, C.D.A. 1970. Structural geology in the eastern margin of the Canadian Rocky Mountains. *Bulletin of Canadian Petroleum Geology,* **18**, 332–406.

DAVIS, D., SUPPE, J. & DAHLEN, F.A. 1983. Mechanics of fold-and-thrust belts and accretionary wedges. *Journal of Geophysical Research,* **88**, 1153–1172.

—— & ENGELDER, T. 1985. The role of salt in fold-and-thrust belts. *Tectonophysics,* **119**, 67–88.

DIXON, J.M. & LIU, S. 1992. Centrifuge modelling of

the propagation of thrust belts. *In*: McClay, K.R. (ed.) *Thrust Tectonics*. Chapman & Hall, London, 53–70.

Elliot, D. 1976. The energy balance and deformation mechanisms of thrust sheets. *Philosophical Transactions Royal Society, London,* **A283**, 289–312.

Geiser, P.A. 1988. The role of kinematics in the construction and analysis of geological cross sections in deformed terranes. *In*: Mitra, G. & Wojtal, S. (eds) *Geometries and mechanisms of thrusting with special reference to the Appalachians.* Geological Society of America, Special Papers, **222**, 47–76.

Harris, L.D. & Milici, R.C. 1977. *Characteristics of thin-skinned style of deformation in the southern Appalachian and potential hydrocarbon traps.* United States Geological Survey, Professional Papers, **1018**.

Hobson, D.M. 1986. A thin skinned model for the Papuan thrust belt and some implications for hydrocarbon exploration. *Australian Petroleum Exploration Association Journal,* **26**, 214–224.

Hossack, J.R. 1983. A cross-section through the Scandinavian Caledonides constructed with the aid of branch-line maps. *Journal of Structural Geology,* **5**, 103–111.

Humayon, M., Lillie, R.J. & Lawrence, R.D. 1991. Structural interpretation of eastern Sulaiman fold belt and foredeep, Pakistan. *Tectonics,* **10**, 299–324.

Jackson, J.A. 1980. Reactivation of basement faults and crustal shortening in orogenic belts. *Nature,* **283**, 343–346.

Jadoon, I.A.K. 1991. *Thin-skinned tectonics on continent/ocean transitional crust, Sulaiman Range, Pakistan*. PhD thesis, Oregon State University, Corvallis, Oregon.

—— & Khan, S.H. 1990. Pop-up structures in the central Sulaiman fold belt of Pakistan. *EOS, Transactions, American Geophysical Union,* **71**, 1592.

——, Lawrence, R.D. & Lillie, R.J. 1992. Balanced and retrodeformed geological cross section from the frontal Sulaiman Lobe, Pakistan: Duplex development in thick strata along the western margin of the Indian plate. *In*: McClay, K.R. (ed.) *Thrust Tectonics*. Chapman & Hall, London, 343–356.

Jaume, S.C. & Lillie, R.J. 1988. Mechanics of the Salt Range-Potwar Plateau, Pakistan. A fold and thrust belt underlain by evaporites. *Tectonics,* **7**, 57–71.

Jones, P.B. 1982. Oil and gas beneath east-dipping thrust faults in the Alberta Foothills. *In*: *Rocky Mountain Association of Geologists Guidebook, Denver, Colorado,* 61–74.

Kazmi, A.H. & Rana, R.A. 1982. *Tectonic map of Pakistan, 1:2,000,000*. Geological Survey of Pakistan, Quetta.

Khan, M.A. & Raza, H.A. 1986. The role of geothermal gradients in hydrocarbon exploration in Pakistan. *Journal of Petroleum Geology,* **9**, 245–258.

Khan, S.H., Lawrence, R.D. & Nakata, T. 1991. *Chaman fault, Pakistan, Afghanistan*. Report of the Geological Survey of Pakistan.

Lamerson, P.R. 1982. The fossil basin area and its relationship to the Absaroka thrust fault system. *In:* Powers, R.B. (eds) *Geologic Studies of the Cordilleran Thrust Belt.* Rocky Mountains Association of Geologists, 279–340.

Laubscher, H.P. 1981. The 3D propagation of decollement in the Jura. *In*: McClay, K.R. & Price, N.J. (eds) *Thrust and Nappe Tectonics*. Geological Society, London, Special Publications, **9**, 311–318.

Lillie, R.J. 1984. Tectonic implications of subthrust structures revealed by seismic profiling of the Appalachian-Quachitas Orogenic Belt. *Tectonics,* **3**, 619–646.

——, Johnson, G.D., Yousaf, M., Zamin, A.S.H. & Yeats, R.S. 1987. Structural development within the Himalayan foreland fold-and-thrust belt of Pakistan. *In:* Beaumont, C. & Tankard, A.J. (eds) *Sedimentary Basins and Basin-forming Mechanisms.* Canadian Society of Petroleum Geologist, Memoirs, **12**, 379–392.

Minister, J.B., Jordan, T.H., Molnar, P. & Haines, E. 1974. Numerical modelling of instantaneous plate tectonics. *Royal Astronomical Society Geophysics Journal,* **36**, 541–576.

Mitra, S. 1986. Duplex structures and imbricate thrust systems: Geometry, structural position and hydrocarbon potential. *American Association of Petroleum Geologist Bulletin,* **70**, 1087–1112.

—— 1990. Fault-propagation folds: Geometry, Kinematic Evolution, and Hydrocarbon Traps. *American Association of Petroleum Geologist Bulletin,* **74**, 921–945.

Morley, C. 1988. Out of sequence thrust. *Tectonics,* **7**, 539–561.

Quittmeyer, R.C., Farah, A. & Jacob, K.H. 1979. The seismicity of Pakistan and its relation to surface faults. *In*: Farah, A. & DeJong, K.A. (eds) *Geodynamics of Pakistan.* Geological Survey of Pakistan, Quetta, 351–358.

Raynolds, R.G.H. & Johnson, G.D. 1985. Rates of Neogene depositional and deformational processes, northwest Himalayan foredeep margin, Pakistan. *In*: Snelling, N.J. (ed.) *The Chronology of the Geological Record*, Geological Society of London. Memoirs, **10**, 297–311.

Raza, H.A., Ahmed, R., Alam, S. & Ali, S.M. 1989. Petroleum Prospects: Sulaiman Sub-Basin. Pakistan. *Pakistan Journal of Hydrocarbon Research,* **1**, 21–56.

Suppe, J. 1983. Geometry and kinematics of fault bend folding. *American Journal of Science,* **283**, 684–721.

—— & Medwedeff, D.A. 1984. Fault-propagation folding. *Geological Society of America (Abs),* **16**, 670.

Tainsh, H.R., Stringer, K.V. & Azad, J. 1959. Major gas fields of west Pakistan. *American Association of Petroleum Geologist Bulletin,* **43**, 2675–2700.

Vann, I.R., Graham, R.H. & Hayward, A.B. 1986. The structure of mountain fronts. *Journal of Structural Geology,* **8**, 215–228.

Evolution of the Himalaya since Miocene time: isotopic and sedimentological evidence from the Bengal Fan

CHRISTIAN FRANCE-LANORD[1], LOUIS DERRY[1] & ANNIE MICHARD[2]

[1] *CRPG-CNRS, BP 20, 54501 Vandoeuvre les Nancy, France*

[2] *LGE-CNRS, Université St Jérôme, CO 431, 13397 Marseille, France*

Abstract: We report Sr, Nd, O, and H isotopic data and clay mineral abundances for turbidite sediments recovered in ODP Leg 116 cores from the Bengal Fan at 1° S. The samples studied cover the period between *c.* 17 Ma and the present. We also present new and compiled data on the isotopic compositions of potential source regions for the Bengal Fan sediments. $\varepsilon_{Nd}(0)$ values in the Bengal Fan sediments (all samples) define a narrow range about −16.0. $^{87}Sr/^{86}Sr$ values (all samples) are also in a narrow range near 0.741. $\delta^{18}O$ values in quartz separates define a narrow range at +12.8 ± 0.5‰. Coarse biotite–chlorite separates give $\delta^{18}O$ = 3.6–5.6‰. Combined $\delta^{18}O$ values of quartz and biotites indicate a metamorphic source. Clay mineral abundances define two clay facies: an illite–chlorite-rich assemblage (IC) and a smectite-kaolinite-rich assemblage (SK). $\delta^{18}O$ in the IC clay fractions is 11.5–15‰, while SK clays are 18.2–22.6‰. The narrow range of isotopic values throughout the deposition history implies that the source of the Bengal Fan sediments has not changed since the early Miocene, despite changes in sedimentation rate, sedimentary facies, tectonic history and climactic regime. The difference between $\delta^{18}O$ in the IC and SK clay fractions represents different alteration histories of the same source material. The SK clays appear to have been altered at low T in the Indo-Gangetic Plain, while the IC clays and coarse fractions preserve metamorphic signatures. The narrow range of the Sr values, despite wide variation in Rb/Sr ratio, also argues for a source that underwent isotopic homogenization shortly before erosion and deposition of the sediment. The source that meets these criteria is the High Himalayan Crystalline series (HHC) or a close analogue, although subordinate contributions (probably <20%) from the Lesser Himalaya (LH) and Tibetan Sedimentary Series (TSS) are possible.

A model in which the HHC are exposed to erosion since the early Miocene on the south flank of the orogen by thrusting along the MCT, while the TSS is simultaneously removed by northward-directed normal faulting satisfies the constraints above. The results of this study require that the Himalaya have been a significant topographic feature since at least the early Miocene. Independent evidence supports this contention. Variations in the sedimentation style in the Bengal Fan since that time appear to represent a combination of factors, including tectonic activity and the coupled effects of climate and sea-level changes.

The Bengal Fan is the world's largest submarine fan, extending some 3000 km south into the Indian Ocean from Bangla Desh. The Bengal Fan has developed largely in response to the collision of India with Asia and the uplift of the Himalaya in Tertiary time (Gansser 1966; Curray & Moore 1971). Curray & Munasinghe (1989) have divided the Bengal Fan into precollisional and postcollisional sediments, based on the presence of an Early Eocene unconformity. The Fan reaches a maximum thickness of over 22 km and the mass of the postcollision sediments is estimated at 2.9×10^{16} tons (Curray 1991). The Bengal Fan is fed by the Ganges–Brahmaputra river system, which currently drains both the north and south slopes of the Himalaya. The great relief, monsoon climate and extensive glaciation of the Himalayan region combine to produce some of the world's highest erosion rates. The Ganges–Brahmaputra system provides 12% of the river particulate flux to the world oceans, while providing less than 3% of the water flux (Milliman & Meade 1983). Chemical erosion is also important; the chemical denudation rates are two to three times higher than the world average in the Ganges–Brahmaputra watershed (Sarin *et al.* 1989).

The Bengal Fan provides an unparalleled record of Himalayan erosion, which in turn is related to the collision history of India and Asia, and the uplift history of the Himalaya. Erosional processes are also related to climate, thus the Bengal Fan may record information about climate change in the regions that supplied it with sediment. The Bengal Fan has been sampled by underwater drilling, permitting a reconstruction of the erosion history of the Himalaya in Neogene time. We wish to address several questions: What were the sources of sediment to the Bengal Fan, and have these sources varied in time? How

From Treloar, P.J. & Searle, M.P. (eds), *Himalayan Tectonics*
Geological Society Special Publication, No. 74, pp. 603–621.

can we use the sedimentological record to constrain the erosion history of the source regions? What is the relative importance of physical versus chemical erosion in the denudation of the Himalaya? What role does climate change play in the sedimentation history of the Bengal Fan, and by inference, erosion in the Himalaya?

In this paper, we present sedimentological and Nd, Sr, O and H isotopic data on samples recovered from the Bengal Fan by the Ocean Drilling Program during Leg 116. These samples yield a record of sedimentation in the Bengal Fan beginning in Early Miocene time. These data, along with sedimentological data from DSDP Leg 22 and new and compiled data on some potential source regions are used to address the questions listed above. The sedimentological record of the Bengal Fan is an integrated record, sampling the Himalayas across the wide area of the Ganges–Brahmaputra watershed (1.48 million km^2, Milliman & Meade 1983). Thus the Bengal Fan record can yield important constraints on processes in the Himalayan source region unavailable for terrestrial studies. The stratigraphic resolution of the sedimentary record also permits a detailed comparison in time of sedimentary events and possible forcing factors, such as tectonic events or climate change.

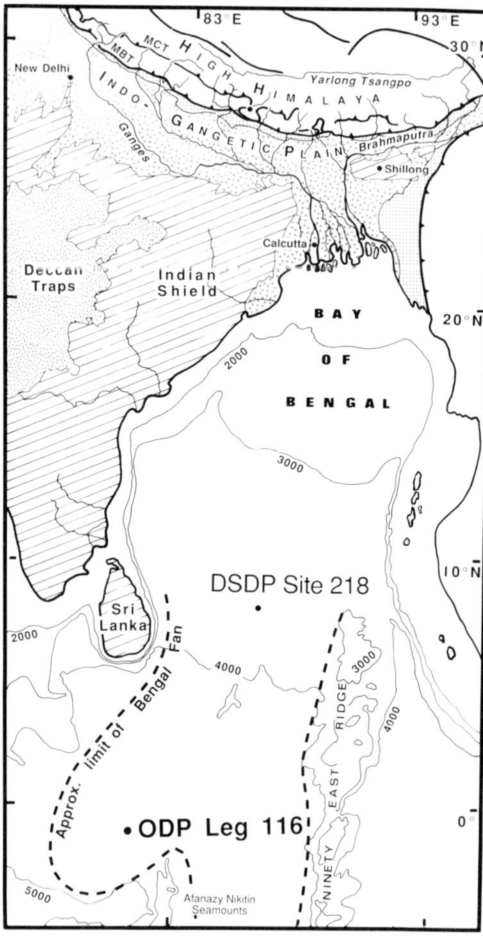

Fig. 1. Location of ODP Leg 116 Sites and DSDP Site 218 in the Bengal Basin.

Geology of Leg 116 Sites

Ocean Drilling Project (ODP) sites 717, 718 and 719, drilled during Leg 116 are located in the distal portion of the Bengal Fan (Fig. 1). The holes are approximately 8 km apart. Hole 717C penetrated 828 mbsf (metres below sea floor) bottoming in late Miocene sediments, while Hole 718C penetrated 935 mbsf, bottoming in early Miocene sediments. Hole 719B penetrated 466 mbsf, bottoming in late Miocene sediments. The principal lithologies encountered in all holes were silt and mud turbidites, with lesser but variable amounts of biogenic mud turbidites (Shipboard Scientific Party 1989). Correlations between the holes are based primarily on lithostratigraphy and nannofossil biostratigraphy (Stow *et al.* 1990; Gartner 1990). A number of potential problems exist with the assignment of biostratigraphic ages to these sediments. Most of the sequence appears to have been deposited near or below the CCD, thus the state of preservation of calcareous fossil material is often poor. Second, it is probable that most of the fossils in the turbidites of the Bengal Fan have been redeposited, and thus fossil-bearing horizons would actually be younger than indicated. However, the succession of highest and lowest occurrences of fauna is both self-consistent and closely comparable with that observed in contemporaneous pelagic sediments, suggesting that redeposition has not seriously disturbed the biostratigraphic record (Gartner 1990).

Stow *et al.* (1990) have presented a detailed summary of the sedimentology of Sites 717, 718 and 719. Lower to upper Miocene sediments are characterized by silt and mud turbidites (Fig. 2). The latest Miocene, Pliocene, and early Pleistocene are characterized by mud turbidites, with interbedded silt turbidites, biogenic turbidites and pelagic clays. In the middle-upper Pleistocene section, sandy and silt- to mud turbidites dominate, except for the last few metres of Holocene pelagic oozes. Variations in sedimentation rates derived from biostratigraphy are consistent with the lithologic variations. Over-

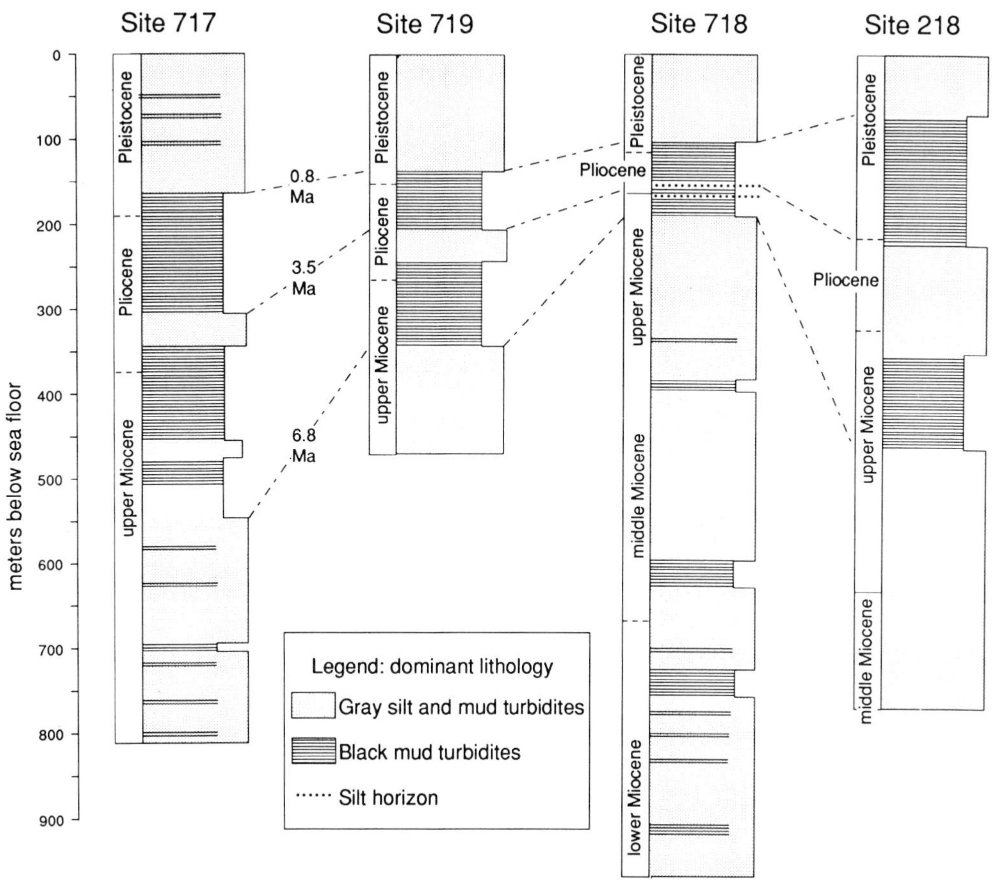

Fig. 2. Stratigraphic correlation of the Leg 116 Holes 717C, 718C and 719A and Leg 22 Hole 218. Lithological logs are after Shipboard Scientific Party (1974, 1989).

all, sedimentation rates were high (70–200 m/Ma) during the early and mid-Miocene, low (20–70 m Ma^{-1}) during the late Miocene–early Pleistocene, and very high (>200 m Ma^{-1}) from the early Pleistocene to the Holocene (Gartner 1990). A wide variety of detrital minerals have been identified in the Bengal Fan silts, the most common of which are quartz, feldspars, carbonates, micas, amphiboles and epidotes (Yokoyama *et al.* 1990). Two clay facies have been identified in the <2 μm Leg 116 sediments: an illite–chlorite-rich assemblage (IC) and a smectite–kaolinite-rich assemblage (SK) (Bouquillon *et al.* 1990; Brass & Raman 1990). The IC assemblage is dominant from the lower Miocene to upper Miocene, the SK assemblage from the upper Miocene to the lower Pleistocene, and the IC assemblage above the lower Pleistocene.

Figure 3 presents a summary of the above data for Hole 717C. Also shown are maximum quartz grain size (MQGS) data obtained from smear slide analysis (Shipboard Scientific Party 1989). Changes in lithology, clay mineralogy, MQGS and sedimentation rate are correlated. Prior to *c.* 11 Ma the sedimentation record in Hole 718C is marked by silt turbidites and coarse MQGS, although the overall sedimentation rate near the bottom of Hole 718C is apparently lower than during other intervals of similar lithology. Cochran (1990) argued that an increase in abundance of pelagic facies near the bottom of 718C indicates an early stage of submarine fan development. Thus the lower sedimentation rate below about 650 mbsf may simply reflect less sediment supply to this distal portion of the Fan in its early stages of development. The increase in sedimentation rate after 11 Ma correlates with a sea level drop at about 10.5 Ma, and with the establishment of the 'modern'

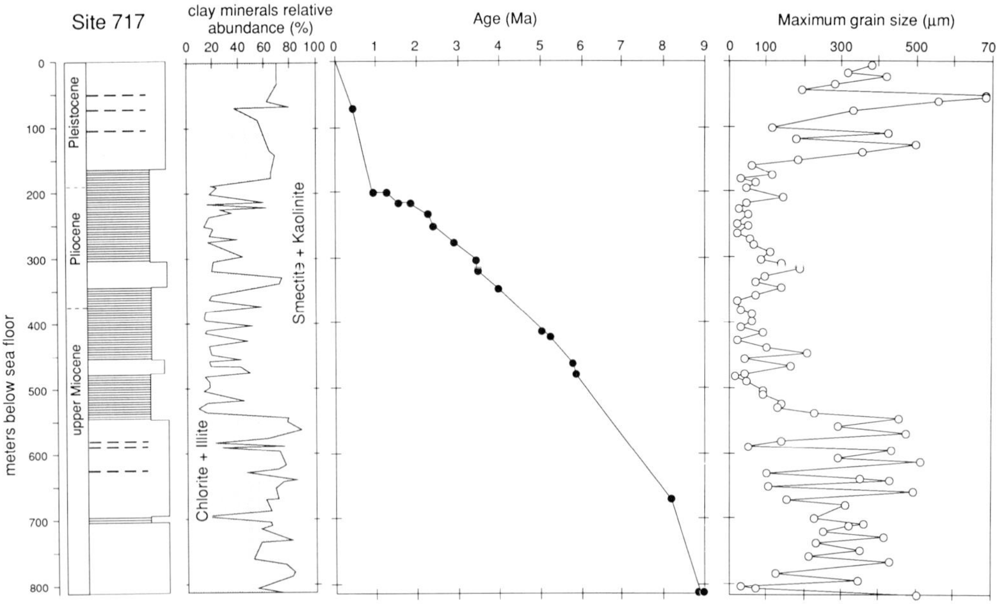

Fig. 3. Depth versus lithology, chronology (Gartner 1990), clay mineralogy (Bouquillon *et al.* 1990) and maximum grain size (Shipboard Scientific Party 1989) at Site 717. The period dominated by smectite and kaolinite sedimentation is characterized by lower rates of deposition and smaller grain size.

Ganges delta (Haq *et al.* 1987; Cochran 1990; Lindsay *et al.* 1991). The Pliocene section is better developed at Sites 717 and 719 than at Site 718. The poorly developed Pliocene section at Site 718 results from non-deposition or erosion. All holes show a drop in sedimentation rate at *c.* 6.8 Ma (Fig. 4). This change is evident in the lithology and the MQGS data. It also corresponds to a marked change in clay mineralogy at all three sites, from the IC assemblage to the SK assemblage (Fig. 2). Coarser clastic deposition recurred for a brief period prior to 3.5 Ma. The sedimentation rate increased markedly at all three Sites at 0.8 Ma, accompanied by a return to coarse clastic sedimentation and the IC clay assemblage. Lithostratigraphic and biostratigraphic correlations between the three Sites are good at this scale of study, providing further confidence in the biostratigraphic ages assignments (e.g. Shipboard Scientific Party 1989).

DSDP Site 218, Leg 22, is located approximately 1100 km north of the Leg 116 Sites (Fig. 1). Drilling encountered similar lithologies, bottoming at 773 mbsf in Middle Miocene turbidite sediments (Figs 1, 2 and 4) (Shipboard Scientific Party 1974). We correlate Site 218 with Sites 717, 718, and 719 based on lithostratigraphy and biostratigraphy (Shipboard Scientific Party 1974; Gartner *et al.* 1974). At Site 218, changes in lithology, grain size, and clay mineralogy similar to those at the Leg 116 Sites support this correlation (Pimm 1974; Bouquillon *et al.* 1989).

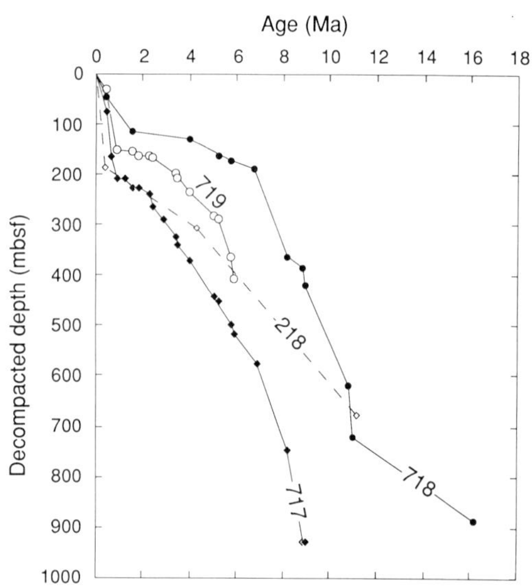

Fig. 4. Age–depth relations for Holes 218, 717C, 718C, and 719A. Depth (metres below sea floor) is recalculated to correct for the effect of sediment compaction. Biostratigraphic ages are from Gartner *et al.* (1974) and Gartner (1990).

The occurrence of correlative lithologies, grain size distributions, sedimentation rate and clay mineral assemblages at three different Leg 116 Sites, located near 1° S, and at Site 218, located at 8° N, demonstrates that the important variations in the sedimentation history of the DSDP and ODP sites reflect regional phenomenon. Pimm (1974) also noted that clastic sediment pulses were correlative over long distances both within the Bengal Fan and into the Nicobar Fan (Andaman sea). Such a wide geographic dispersion precludes channel switching or other changes in the location of depocentres as an explanation for the variations in sedimentation rate and lithology. It is also improbable that simply changing the channel network in the Fan, without a change in sediment source, could result in significant changes in the mineralogical composition of the <2 μm clay fraction. Finally, studies of other submarine fans suggest that levee construction take place on a 10^5 year time scale, but the major variations in sediment composition observed at the Leg 116 sites and Site 218 are of 10^6 years duration, implying that simple construction of underwater structures is not the process responsible for the sedimentological variations (e.g. Damuth *et al.* 1988; Manley & Flood 1988). The implications of the basin-wide nature of stratigraphic events in the Bengal Fan are that the sedimentological variations are caused by (1) changes in provenance, and/or (2) changes in transport phenomena in the Ganges watershed or near the head of the Bengal Fan complex.

Analytical methods

Samples of sediment from ODP collections were selected to provide a range in age and sedimentary facies. Samples were first disaggregated in distilled water. The coarse fraction (>63 μm) was separated by screening, and the remainder either analysed as the <63 μm fraction or subject to further size separation. Mono-mineralic separates were prepared from the >63 μm fraction by hand picking. The preparation of <2 μm and 2–50 μm fractions generally followed the methods described in Bartoli *et al.* (1991) and Feller *et al.* (1991). Briefly, the samples (including both those previously screened at 63 μm, or those previously untreated) were disaggregated by agitation with Amberlite cation exchange resin (Na form) for 16 hours. This treatment disperses the fine-grained particles because of the high exchange capacity of the resin. The samples were screened to remove >50 μm particles. The remaining fraction was allowed to settle for 24 hours, effectively separating the <2 μm fraction. The suspended 2 μm clays were decanted, and 1 ml of a clean $MgCl_2$ solution (Prolabo Normapur®) was added to flocculate the clays. This sequence was repeated as necessary. The 2–50 μm and <2 μm fractions were then separated from solution by centrifugation. The quality of the grain size separations was periodically checked using a microscope-mounted image processing system. In all cases the size separation technique was found to be highly effective. A portion of the <2 μm fraction was analysed by standard X-ray techniques for clay mineral identification.

Clay fractions were decarbonated by leaching with 10% acetic acid in an ultrasonic bath. Samples for Sr and Nd isotopic analysis were roasted at 700° C in platinum crucibles to remove organic matter. Samples were weighed and spiked, and dissolved in a HF–HNO_3–$HClO_4$ mixture. Separation procedures for Rb, Sr, Sm and Nd are described in Alibert *et al.* (1983). Sr and Nd isotopic compositions were measured on a Finnigan-MAT 262 mass spectrometer equipped with an SEM and 6 faraday detectors. Blanks are estimated at ca. 0.5 ng for Nd and 1 ng for Sr (blank $^{87}Sr/^{86}Sr \approx 0.710$). These high blanks result from the complex sample preparation procedure. However, sample sizes were sufficiently large that blank corrections are not necessary.

Oxygen from silicates were extracted by fluorination following the technique of Clayton & Mayeda (1963). CO_2 gas was analysed on a VG 602D mass spectrometer. All extractions were duplicated and average reproducibility on clay samples is ±0.3‰.

Results

Mineralogical, isotopic and major element data are listed in Tables 1 and 2 for whole rock samples, mineral separates and granulometric fractions. Figures 5 and 6 show composite logs of the isotopic compositions of Nd, Sr, O and H as a function of depth at Sites 717 and 718 for the different mineral separates or size fractions analysed. Isotopic analyses of coarse mineral separates and most XRD analyses are from Bouquillon *et al.* (1990). Isotopic analyses of Sr for the <2 μm fraction samples given in Bouquillon *et al.* (1990) were not retained here because these samples were not leached with acetic acid prior to analysis, and are thus not directly comparable.

Neodymium and strontium

The isotopic compositions of Nd and Sr are remarkably uniform over the whole stratigraphic column. $\varepsilon_{Nd}(0)$ values are between -17 and -14; the entire range is from -13.8 to -19.3 (mean $= -16.0$, std. dev. $= 1.2$). The two lowest values are from SK samples; however, no clear difference is evident overall between SK and IC samples (Fig. 7a). $^{87}Sr/^{86}Sr$ ratios for all samples vary between 0.720 and 0.762 (mean = 0.741, std. dev. = 0.009). For both the whole rock and granulometric fractions $^{87}Sr/^{86}Sr$ is independent of Sr concentration, which supports the efficiency of the leaching technique in removing

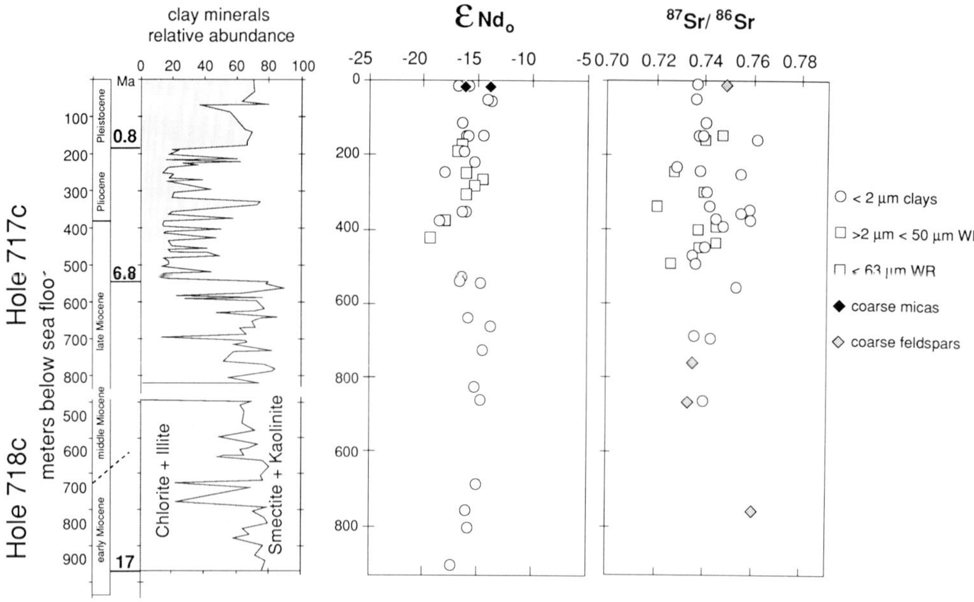

Fig. 5. Depth profiles of isotopic composition of Nd and Sr for Holes 717C and bottom of 718C. Data are listed in Table 1 or from Bouquillon *et al.* (1990). Clay mineralogical profiles are simplified after Bouquillon *et al.* (1990).

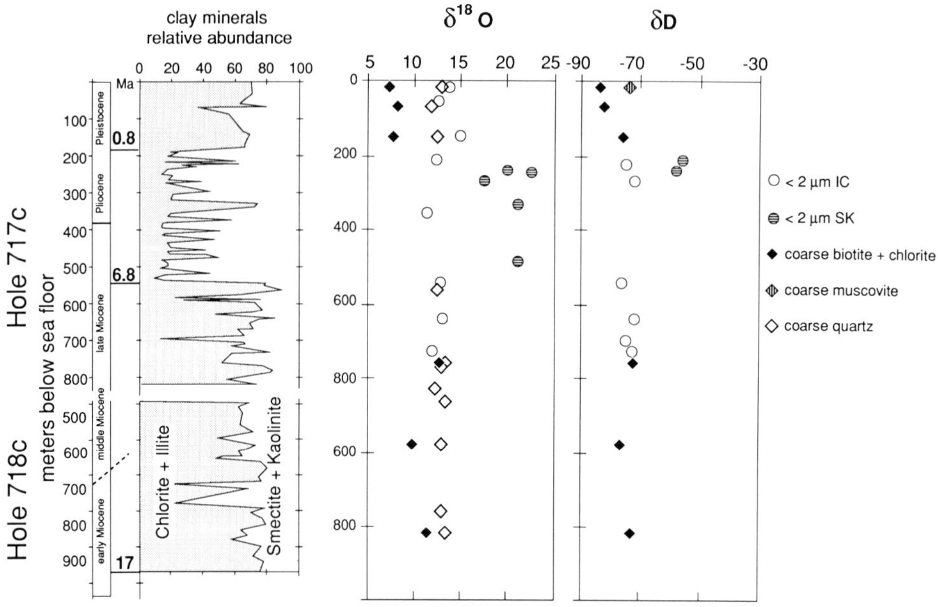

Fig. 6. Depth profiles of the isotopic composition of O and H for Holes 717C and bottom of 718C. Data are listed in Table 1 or from Bouquillon *et al.* (1990). Clay mineralogical profiles are simplified after Bouquillon *et al.* (1990).

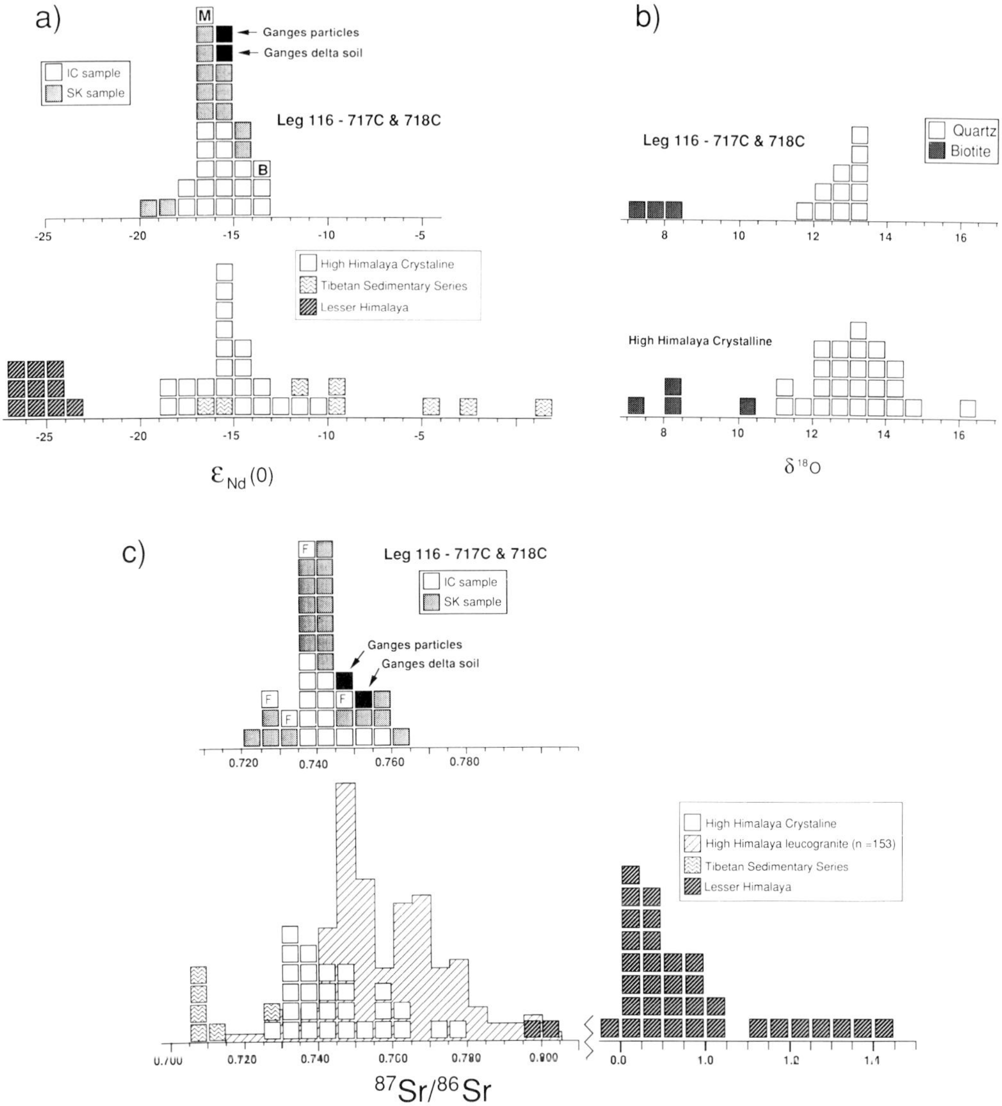

Fig. 7. Histograms of Nd-, O- and Sr-isotope data of the Leg 116 sediments compared to the Himalayan formations. Average values are similar for the HHC and the Leg 116 samples. (**a**) Nd isotope data; **B**, **M** represent respectively the values of the coarse biotite + chlorite, and the coarse muscovite separated from sample (116-717C-2H-01, 02). (**b**) Oxygen isotope data. (**c**) Sr isotope data, **F** represents values of coarse feldspar separates. Data are from Table 1 and 2 and: Blattner *et al.* (1983), Dietrich & Gansser (1981), Deniel *et al.* (1986, 1987), France-Lanord *et al.* (1988), France-Lanord & Le Fort, (1988), Goldstein (1988), Goldstein *et al.* (1984), Parrish *et al.* (1992), Scaillet *et al.* (1990), Trivedi *et al.* (1984), Vidal *et al.* (1982, 1984).

carbonates containing seawater-derived Sr. SK and IC samples have similar $^{87}Sr/^{86}Sr$ ratios. Not only are $^{87}Sr/^{86}Sr$ ratios relatively homogeneous with depth, but also across a variety of mineralogies, despite large variations in Rb/Sr ratio.

Stable isotopes

$\delta^{18}O$ values of coarse quartz are remarkably constant with depth at 12.8‰ ± 0.5. Mixtures of coarse biotite and chlorite vary from 7.2 to 9.8‰

Table 1. *H, O, Sr and Nd isotope data and relative abundances of clay minerals in Leg 116 sediments*

	mbsf	Ø µm	δD ‰	H_2O^+ %	$\delta^{18}O$ ‰	Rb ppm	Sr ppm	$^{87}Sr/^{86}Sr$	$^{87}Rb/^{86}Sr$	$^{143}Nd/^{144}Nd$	±	$\varepsilon_{Nd}(0)$	Sm ppm	Nd ppm	% abundance of clay in <2 µm: Ch	Ill	MxL	Sm	Ka
Hole 717C																			
2X-1 (80–81)	15.8	<2			13.5	232	91	0.73723	7.39	0.511824	30	−15.9			35	40	–	10	10
		<2								*0.511774*	*26*	*−16.8*	*6.4*	*32.5*					
6X-2 (27–28)	49.3	<2				267	79	0.73810	9.88	0.511920	16	−14.0	5.16	26.61					
7X-1 (85–86)	56.4	<2			*12.7*					*0.511929*	*17*	*−13.8*	*6.5*	*32.9*	*15*	*41*	*9*	*25*	*10*
15X-2 (49–2–50)	114.5	<2				209	82	0.74048	7.32	*0.511791*	*25*	*−16.5*	*4.7*	*23.5*	35	45	–	12	8
19X-CC (5–6)	150.5	<2			15.0	235	87	0.74006	7.81	0.511893	23	−14.5			30	50	–	10	10
		<2								*0.511817*	*27*	*−16.0*	*7.9*	*40.7*					
20X-1 (76–77)	151.3	<2						0.73791		*0.511832*	*28*	*−15.8*	*6.5*	*33.2*	*16*	*41*	*18*	*20*	*5*
		2–50				159	251	0.74731	1.83										
21X-3 (34–55)	163.3	2–50				126	208	0.74064	1.75										
		<2						0.76224							*3*	*12*	*13*	*62*	*10*
22X-4 (100–101)	175	<63								0.511800	25	−16.4			*2*	*10*	*13*	*62*	*13*
24X-4 (66–67)	193.7	<63								0.511776	12	−16.8			*2*	*8*	*10*	*70*	*10*
		<2								*0.511806*	*18*	*−16.2*	*5.7*	*27.9*					
26X-4 (85–86)	214.4	<2	*−55.6*	*3.4*	*12.5*	13													
27X-5 (70–71)	223.7	<2	*−74.3*	*5.6*	12.5					*0.511855*	*22*	*−15.3*	*3.5*	*18.4*					
28X-6 (19–20)	235.7	<2				209.5	40	0.72798	15.06										
29X-3 (110–111)	240.1	<2				112	24	0.75484	13.68	0.511722	14	−17.9		28.1					
29X-5 (70–71)	242.7	WR	*−57.4*	*5.4*	*20.1*														
30X-1 (116–117)	246.7	<2			22.6			0.73851							*3*	*5*	*16*	*60*	*16*
		2–50				78	131	0.72791	1.72										
30X-3 (20–21)	248.7	<63								0.511815	13	−16.0			*6*	*7*	*14*	*58*	*15*
32X-3 (118–119)	268.7	WR	*−72*	*6.7*	17.7					*0.511898*	*26*	*−14.5*	*5.0*	*24.0*					
34X-2 (76–77)	285.8	<63								0.511858	13	−15.2			*4*	*8*	*13*	*55*	*20*
35X-6 (28–29)	301.4	<2						0.74067							*2*	*11*	*9*	*61*	*17*
		2–50				105	114	0.74008	2.66										
36X-3 (136–137)	306.9	<63								0.511815	9	−16.0			*1*	*9*	*15*	*57*	*18*
40X-2 (90–91)	342.9	<2			21.3	72	24	0.74220	8.58						*3*	*7*	*15*	*57*	*18*
		2–50				77	138	0.72086	1.62										
40X-6 (30–31)	348.3	<2				64	25	0.75877	7.34						*3*	*7*	*12*	*65*	*13*
41X-CC (33–34)	359.2	<2			11.5		73	0.75512		0.511796	117	−16.4			40	45	–	5	10
		<2								*0.511818*	*34*	*−16.0*	*7.2*	*37.2*					
43X-5 (100–101)	376	<2				62	13	0.74476	14.10						*2*	*6*	*9*	*68*	*15*
44X-2 (80–81)	380.8	<2						0.75891		0.511689	22	−18.5			*2*	*10*	*8*	*60*	*20*
		2–50								0.511719	19	−17.9							

Table 1 *Continued*

	mbsf	Ø µm	δD ‰	H_2O^+ %	$\delta^{18}O$ ‰	Rb ppm	Sr ppm	$^{87}Sr/^{86}Sr$	$^{87}Rb/^{86}Sr$	$^{143}Nd/^{144}Nd$	±	$\varepsilon_{Nd}(0)$	Sm ppm	Nd ppm	% abundance of clay in <2 µm Ch	Ill	MxL	Sm	Ka
Hole 717C																			
45X-7 (4–5)	397	<2				56	15	0.74836	10.66						*3*	*6*	*11*	*65*	*15*
		2–50						0.74500											
46X-3 (39–40)	400.9	<63				63	43	0.73729	4.26						*2*	*5*	*15*	*60*	*18*
49X-2 (40–41)	427.9	<63								0.511650	13	−19.2			*3*	*5*	*20*	*59*	*13*
50X-1 (141–142)	436.9	2–50				139	127	0.74479	3.16						*2*	*10*	*18*	*55*	*15*
51X-4 (120–121)	450.7	<2						0.74058							*2*	*8*	*15*	*53*	*22*
		2–50				116	120	0.73780	2.80										
54X-1 (59–60)	474.1	<2				56	29	0.73540	5.67						*3*	*3*	*14*	*63*	*17*
55X-CC (45–46)	492.1	<2			21.1	75	26	0.73665	8.45						*3*	*7*	*15*	*55*	*20*
		2–50				98	129	0.72685	2.19										
59X-6 (57–58)	529.1	<2								*0.511798*	*34*	*−16.4*	*4.8*	*22.9*	*2*	*4*	*11*	*73*	*10*
61X-1 (115–116)	541.2	<2	*−76.2*	*2.2*	12.9					*0.511788*	*30*	*−16.6*	*5.1*	*25.8*	*22*	*48*	*22*	*1*	*7*
63X-1 (104–105)	569.5	<2				301	17	0.75284	50.62	0.511900	11	−14.4		31.5					
71X-4 (125–126)	640.8	<2	*−71.3*	*2.2*	13.0					*0.511828*	*23*	*−15.8*	*5.0*	*27.2*	*25*	*45*	*25*	*3*	*2*
75X-4 (25–26)	679.3	<2				234	56	0.73628	12.03	0.511924	27	−13.9	5.1	26.3					
77X-5 (132–133)	699.3	<2						0.74333							*2*	*3*	*10*	*75*	*10*
78X-1 (14–15)	701.6	<2	*−74.2*	*2.2*															
81X-CC (29–30)	729.7	<2	*−72.5*	*2.9*	12.0					*0.511895*	*29*	*−14.5*	*4.7*	*23.9*	*25*	*50*	*20*	*2*	*3*
91X-4 (110–111)	830.6	<2								*0.511855*	*32*	*−15.3*	*7.9*	*41.1*	*27*	*41*	*10*	*19*	*3*
Hole 718C																			
49X-1 (66–67)	466	<2								*0.511891*	*17*	*−14.6*	*9.1*	*47.4*	*22*	*39*	*14*	*18*	*7*
57X-5 (70–71)	473.5	<2				234	64	0.74003	10.62										
72X-6 (93–94)	692.2	<2								*0.511857*	*26*	*−15.0*	*4.9*	*25.2*	*2*	*8*	*10*	*55*	*25*
78X-4 (14–15)	745.4	<2													*6*	*10*	*9*	*50*	*25*
79X-CC (27–28)	759.8	<2								*0.511805*	*24*	*−16.0*	*5.6*	*28.8*	*17*	*56*	*12*	*10*	*5*
84X-CC (29–30)	807.3	<2								*0.511821*	*23*	*−15.9*	*5.4*	*28.0*	*27*	*47*	*13*	*8*	*5*
96X-1 (31–32)	907.6	<2								*0.511744*	*44*	*−17.4*	*5.0*	*25.7*	*25*	*47*	*12*	*10*	*6*

Data in italics are from Bouquillon *et al.* (1990). Ch, chlorite; Ill, illite; MxL, mixed layers; Sm, smectite; Kao, kaolinite.
Ø column indicates the granulometric fraction in µm.
$^{87}Sr/^{86}Sr$ normalized to $^{87}Sr/^{86}Sr = 0.1194$. Average value of NBS 987 during the period of analysis was 0.71020. Analysis errors ± 0.00001 $^{143}Nd/^{144}Nd$ normalized to $^{146}Nd/^{144}Nd = 0.7219$.
Average value of LaJolla Nd during the period of analysis was 0.511847. $\delta^{18}O$ of NBS 28 = 9.60‰. Data reported relative V-SMOW.

and are depleted in ^{18}O with regard to quartz. The δ difference between quartz and biotite is 3.6 to 5.6‰, which corresponds to an isotopic equilibrium temperature between 500 and 650° C (e.g. Botinga & Javoy 1975). Despite the fact that these analyses represent average values of grains of different origin this average 'fractionation' between quartz and biotite is likely to reflect high-temperature fractionation in the source rock.

The <2 μm fraction samples display two groups of $\delta^{18}O$ values. IC samples are between 11.5 to 15‰, whereas SK samples are more variable and significantly enriched in ^{18}O at 18.2 to 22.6‰. The values for IC are extremely low for clay minerals, especially in marine sediment. 'Normal' values of marine clays are in the range of 16 to 25‰, with most between 20 and 25‰ (Savin & Epstein 1970; Yeh & Savin 1977; Perry *et al.* 1976; Laurence *et al.* 1979; Taieb 1990). High values are due to large isotopic fractionation between clay and water at low temperature (e.g. Yeh & Savin 1977). The IC assemblage clays have not been enriched in ^{18}O by low temperature exchange. Like the coarse micas, the IC clays reflect the high temperature of their source rock, whereas smectite and kaolinite were formed at low temperature.

Discussion

Sources of sediment in the Bengal Fan

The Nd, Sr, O and H isotopic data impose several constraints on the sources of the sediments in the Bengal fan.

(1) The remarkably constant Nd, Sr and O isotopic values with time are a strong indication that the material has been derived from the same major source or melange of sources since 17 Ma. The isotopic data are consistent with the overall monotony of the mineralogy. The stratigraphic levels rich in SK clays have higher $\delta^{18}O$ values than elsewhere in the column, but have the same Nd and Sr values as the coarse fraction and samples of the IC clay assemblage.

(2) The source(s) of this material is old, with a long crustal residence time. Nd model ages are *c.* 2.0 Ga in the sediments at all levels, and $^{87}Sr/^{86}Sr$ cluster about 0.74. The $\delta^{18}O$ values of quartz and micas are also typical of silicic crystalline crustal sources, such as paragneisses and granites derived from sedimentary sources.

(3) In the IC levels $\delta^{18}O$ values ($\approx$ 13‰) are very low for the <2 μm clay fraction when compared to normal deep sea clays. The $\delta^{18}O$ data reinforce the observation that low temperature alteration minerals are present only in minor abundance in IC levels, which implies that erosion was dominated by physical rather than chemical processes.

(4) The SK levels are enriched in D and ^{18}O relative to IC levels, but have the same Nd and Sr signature. This implies that smectite and kaolinite are derived from the same material as the illite and chlorite. The SK clays appear to be altered equivalents of the IC material, which preserves the radiogenic isotopic compositions but results in higher $\delta^{18}O$ and δD values.

(5) The present-day isotopic composition of Sr in unaltered samples (IC clays and coarse feldspars) are quite homogeneous, despite variable Rb/Sr ratios (Fig. 7c, Table 1; Bouquillon *et al.* 1990). This requires a recent metamorphic/magmatic event in the source to isotopically homogenize the Rb-Sr system shortly before erosion and deposition of the sediments.

We now examine potential sources for the Bengal Fan sediments around the Bengal Basin. Early studies of the Bengal Fan concluded that the Himalaya were the primary source for detrital material (Curray & Moore 1971; Shipboard Scientific Party 1974; Ingersoll & Suczek 1979). Several authors have used mineralogical criteria to identify the presence of several additional components in the Bengal Fan (Bouquillon *et al.* 1989; Brass & Raman 1990; Stow *et al.* 1990; Yokoyama *et al.* 1990; Amano & Taira 1992). The most important of these include the Precambrian terranes of south India and the Deccan Traps. Both of these terranes are a potential source of the smectite and kaolinite clays which are common in some Bengal Fan strata. Stow *et al.* (1990) proposed that dark-grey and organic rich turbidites, which correspond to some SK levels drilled on Leg 116, are derived either from the Deccan Traps or the granulite terrains of India and Sri Lanka. This is not supported by the isotopic composition of Sr and Nd in these dark-grey turbidites (e.g. 717C-24X, 66–67 cm and 717C-35X, 28–29 cm) which are indistinguishable from other facies. A detrital component derived from these granulites should be easily resolved isotopically, with $\varepsilon_{Nd}(0) < -30$ and $^{87}Sr/^{86}Sr \approx 0.716$ (e.g. Peucat *et al.* 1989). The Deccan Traps are equally distinctive with $\varepsilon_{Nd}(0) > 0$ and $^{87}Sr/^{86}Sr < 0.708$, except for volumetrically minor early flows (e.g. Mahoney *et al.* 1982; Lightfoot *et al.* 1990).

Three arguments suggest that non-Himalayan sources are minor. First, the similarity of Nd and Sr isotope ratios between IC and SK levels imply that the smectite and kaolinite have the same source as the illite, chlorite and coarser material. Furthermore, soils from the Ganges delta (Table 2) and the suspended matter of the Ganges and

Table 2. *Sr and Nd isotope data for soil from Bangla Desh and rocks from the Lesser Himalaya and High Himalaya Sedimentary Series*

	Type <ø μm	Rb ppm	Sr ppm	$^{87}Sr/^{86}Sr$	$^{87}Rb/^{86}Sr$	$^{143}Nd/^{144}Nd$	±	$\varepsilon_{Nd}(0)$	Sm ppm	Nd ppm	T_{DM} Ga
Soil from the Ganges delta											
Batra series	<2 μm		32	0.75236	22.18	0.511855	15	−15.3	4.7	23.6	
High Himalaya Sedimentary Series											
NA 178 Jurassic schist	W.R.					*0.511795*	*28*	*−16.4*	*4.0*	*20.3*	*2.00*
NA 131 Cretaceous (volcaniclastic)	W.R.					*0.512405*	*28*	*−4.5*	*12.6*	*59.0*	*1.26*
87 28 Cretaceous	W.R.	42	61	0.71266	1.90	0.512093	13	−10.6	9.1	45.5	1.53
87 32 Trias	W.R.	19	45	0.70899	1.19	0.511824	17	−15.9	3.6	18.6	1.93
LA 194 Indus margin	W.R.	11	299	0.70981	0.10	0.511883	9	−14.7	5.9	33.2	1.64
LA 158 Indus margin	W.R.	96	50	0.72903	5.29	0.512046	37	−11.6	5.0	23.2	1.77
25 1 Suture Ladak	W.R.	7	662	0.70762	0.03	0.512313	34	−6.3	1.6	6.4	1.68
89 1 Suture Ladak	W.R.	44	258	0.70606	0.47	0.512580	24	−1.1	3.0	13.9	0.88
Lesser Himalaya (Central Nepal)											
AP 346 Kunchha pelites	W.R.	*142.6*	*11.1*	*1.68090*	*38.07*	*0.511397*	*28*	*−24.0*	*8.1*	*44.0*	*2.52*
AP 385 Kunchha pelites	W.R.	*286*	*63*	*1.03010*	*13.48*	*0.511331*	*25*	*−25.0*	*6.6*	*36.6*	*2.60*

Data in italics from France-Lanord & Le Fort (1988) and Bouquillon *et al.* (1990).
See Table 1 for analytical parameters.

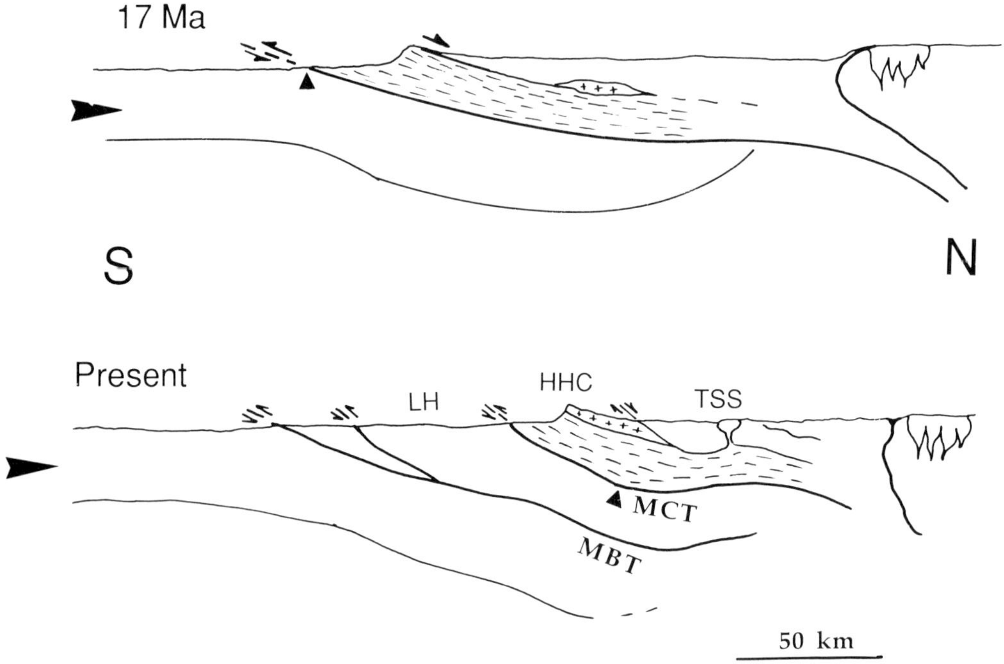

Fig. 8. Schematic cross-section of the Himalaya at 17 Ma and the present day. Thrusting has maintained high elevation in the HHC, while normal faulting has continuously removed the TSS to the north. This scheme satisfies the requirement that the HHC has been exposed to rapid erosion since the early Miocene, while the TSS has contributed little to the sediment flux. The black triangle represents a reference point in the Lesser Himalaya.

Brahmaputra (Goldstein 1988; Goldstein *et al.* 1984) have Nd and Sr isotopic compositions identical with those of the Bengal sediments (Fig. 7). Second, the relief and the rivers from Sri Lanka and eastern India are not likely to supply an amount of sediment sufficient to account for the large volume of SK sediments in the Fan. Finally, the same type of smectite and kaolinite parageneses associated with variable proportions of illite and chlorite are observed (1) in the Tertiary sediments of the Siwaliks (e.g. Mukherjee 1964), (2) in the Ganges delta soils (Rahman 1987), (3) in the Ganges river suspended matter (Sarin *et al.* 1989) and (4) in the proximal Bengal Fan (Bouquillon *et al.* 1989). Thus it is not necessary to call on the presence of components derived from the southern Indian shield or Sri Lanka to explain the clay mineral abundances found in the Bengal Fan sediments, as proposed by previous workers (Brass & Raman 1990; Stow *et al.* 1990). While material derived from south India, Sri Lanka and the Deccan Traps is certainly present in the Bengal Fan, the isotopic data demonstrate conclusively that such components are of minor volumetric importance.

We can compare the isotopic compositions of the Bengal Fan sediments with potential Himalayan sources in some detail (Fig. 7). Despite the large area of the Ganges–Brahmaputra watershed, the gross structure of the Himalaya is relatively simple and regular throughout the range (e.g. Gansser 1964; Le Fort 1989). The three main Himalayan structural/stratigraphic units have very distinct isotopic features. The Lesser Himalaya (LH), dominated by old Himalayan metasediments (>1.8 Ga; Deniel 1985), have very high Sr isotope ratios and $\varepsilon_{Nd}(0)$ around −25, reflecting their long crustal evolution (France-Lanord & Le Fort 1988; Parrish *et al.* 1992; Table 2). The High Himalaya Crystalline metasedimentary series (HHC), located above the Main Central Thrust (MCT), have been intensely investigated to examine the genetic relations between the paragneisses and the overlying leucogranites (e.g. Scaillet *et al.* 1990). Averages of $\varepsilon_{Nd}(0)$ and $^{87}Sr/^{86}Sr$ are −15 and 0.740 respectively (Fig. 7). The Himalayan leucogranites, which are derived from the HHC, have similar ranges. The High Himalayan Sedimentary Series (TSS) are Cambrian to Eocene in age, and include limestones, sandstone, shales and some volcanic material. Nd and Sr isotopic data for TSS sediments are given in Table 2. These data do not provide a comprehensive view of the TSS but indicate that this unit is isotopically variable, as would be expected of a sedimentary sequence with a wide age range in an evolving tectonic setting. They have higher $\varepsilon_{Nd}(0)$ and lower $^{87}Sr/^{86}Sr$ than the HHC (Fig. 7), indicative of their younger source ages and, for some, the presence of volcaniclastic components. On histograms of isotopic compositions of Nd, Sr, O and H, the HHC paragneisses appear extraordinarily concordant with the Bengal sediments (Fig. 7). In contrast, the LH and TSS lie outside the narrow range of values shown by the Bengal Fan. Thus, the major source for the detrital material has been an equivalent to the modern HHC, and the LH and the TSS must be subordinate to the HHC. This is consistent with the present structure of the Himalaya (Fig. 8). The HHC is exposed on the southern flank of the high ranges of the Himalaya, which is the region of highest relief and precipitation, hence of most rapid erosion. Conversely, the TSS is exposed on the north flank of the range, and is undergoing less erosion. The LH contributes relatively little to the Bengal sediments, because it is at a lower elevation, is not glaciated, and is under extensive plant cover.

A more quantitative estimate of the relative importance of the Lesser Himalaya, the High Himalaya Crystalline series, and the Tibetan Sedimentary Series in supplying Bengal fan sediment may be obtained from isotopic mass balance. Taking average values of each of the three groups, Nd-Sr mass balance suggests that the contribution of LH and TSS sediments to the Bengal Fan could have varied between 0 and 40%, but was most often closer to the lower of these two estimates. In principle, the Nd and Sr signature of the Bengal Fan could be produced by a simple mixture of LH and TSS. Two points mitigate against this interpretation. First, the ranges of Nd and Sr values in the Bengal sediments are very narrow at all times. The LH and TSS lie in structurally and topographically distinct parts of the Himalayan Range. It is highly improbable that a mixture of LH and TSS would always fall in the same narrow range, despite important changes in sedimentation rate, mineralogy, and tectonic and climatic history. Second, the sediments have highly variable Rb/Sr ratios. This variability would give rise to a wide range in $^{87}Sr/^{86}Sr$ values in the sediments over the time scale of the age of the TSS and the LH. In fact, $^{87}Sr/^{86}Sr$ varies only slightly in the Bengal Fan, implying that the sediment source was isotopically homogeneous on the order of 10–100 Ma before present. This constraint is not satisfied by the TSS or the LH because the Himalayan metamorphism *sensu lato* is weak or only locally developed in these units (e.g. Le Fort 1986). However, the total amount of erosion of the TSS may be underestimated by the record of clastic sediments of the Bengal Fan. At least 50% of the TSS is composed of platform carbonates that are mostly dissolved during the erosional process. The abundance of detrital grains of dolomite before ≈10 Ma in Hole 718C (Yokoyama *et al.* 1990) reflects a higher proportion of carbonate rocks in the source during the early stage of Fan development which could be a by-product of early Miocene erosion of the TSS. Amano & Taira (1992) argue for a change in sediment source at 15.2 Ma on the basis of heavy mineral abundances. They propose that a mix of LH and TSS was the source prior to 15.2 Ma, and that the HHC became important only after. However Nd isotopic values in Hole 718C show no evidence of change near this time. Prior to 15 Ma the TSS may have been a more important component of Fan sedimentation, but it is unlikely that it was dominant, and less likely that the LH was a significant source during any time interval recorded in the Bengal Fan.

Implications for Himalayan tectonics since 17 Ma

The primary conclusion from the isotopic data is that an equivalent of the High Himalayan Crystalline series has been the dominant source of sediment supply to the Bengal fan since before 17 Ma. This situation has apparently not varied significantly since that time, despite changes in sedimentation rate in the Fan, alteration history of the sediments, and tectonic and/or climatic changes that may have affected the erosional history of the Himalayan source regions. In a few samples, there is a hint of low $\varepsilon_{Nd}(0)$ material, presumably from either an Indian or LH source. However, this component is always minor, and in most cases not resolvable. The presence of large volumes of sediment derived from the HHC since at least as early as 17 Ma has significant implications for the tectonic history of the Himalaya. Clearly, the sedimentary record requires that the HHC have been extensively exposed at the surface since that time. Further more, as argued above, the mineral assemblage (e.g. Yokoyama *et al.* 1990), $^{87}Sr/^{86}Sr$ and stable isotopic values in Bengal sediments are indicative of a metamorphic source which experienced isotopic homogenization of Sr shortly before erosion and deposition. The majority of single grain $^{40}Ar/^{39}Ar$ dates of feldspar and muscovite of Leg 116 sediments are in the range of 10 to 50 Ma (Copeland & Harrison 1990). This evidence is also consistent with a HHC source, and implies that the high temperature metamorphic rocks from the HHC were being eroded at 17 Ma. The large volume of sediment supplied to the Bengal Fan requires that the Himalaya were a significant topographic feature throughout this time. Furthermore, the observation that the sources of Bengal sediment have not varied implies that the main lithologies present in the high range were closely analogous to the present HHC. Therefore the present general structure of the high range has remained grossly similar since the early Miocene. The HHC are exposed to rapid erosion, while the TSS and LH have remained in less susceptible positions. A plausible hypothesis is that thrusting has continuously brought the HHC to high elevations, while the overlying TSS have been simultaneously removed by north-directed normal faulting (Fig. 8). In this way the TSS has been protected from erosion, while the HHC has formed the region of high elevations and relief.

The source of the SK assemblage

The Nd and Sr data show that the IC and SK clay assemblages have the same source, but their O and H isotope compositions are different (Figs 5, 6 and 7). The IC and SK assemblages therefore represent differences in alteration history of the same material. The $\delta^{18}O$ values of +20‰ and δD values of −65‰ in the SK samples are consistent with alteration by meteoric water in the Indo-Gangetic Plain at *c.* 20° C. It is essentially variation in the smectite/illite ratio that is responsible for the mineralogical difference between the two clay assemblages. Kaolinite is found in both IC and SK samples (Table 1), and in lakes and river sediments of the High Himalayan chain (CRPG unpublished data; Sarin *et al.* 1989). Conversely, smectite only appears downstream of the Siwaliks in the Indo-Gangetic plain. The ratio of smectite to illite in the Ganges river sediments increases downstream, from zero in the highlands to 6–8 in the Indo-Gangetic Plain (Sarin *et al.* 1989). Sarin *et al.* (1989) attributed the increase to the addition of smectite by tributaries draining the Deccan Traps to the south. However, Nd isotopic data from the <2 μm fraction of the SK sediments in the Bengal Fan show no sign of a contribution from this region. Thus, while some smectite is certainly derived from the Deccan, it cannot represent an important component in the SK sediments. Estimates of $\delta^{18}O$ and δD in Bengal smectites correspond with those of clays generated during continental alteration, but are too low to be generated by reaction with cold seawater in the deep ocean (Bouquillon *et al.* 1990). While carbonate precipitation has occurred in the section, there is no evidence that diagenetic processes have significantly affected the clastic material in Bengal Fan sediments (e.g. Bouquillon *et al.* 1989). We conclude that the alteration which has produced the SK assemblage occurred during transport in the Ganges–Brahmaputra river system. The intervals rich in SK clays represent periods when the eroded material was altered at low temperature prior to its sedimentation. This process could take place in several ways. Clays could be altered during river transport, they could be deposited in the foreland basin and re-eroded under a warm climate regime, or they could be altered during residence in the Indo-Gangetic plain.

Variation in sedimentation history of the Bengal Fan

Between 6.5 and 0.8 Ma, the change in sediment facies, high abundance of smectite, fine grain size, and lower accumulation rates (Fig. 3) indicate a major change in the supply of sediment to the Bengal Fan. Such an event may be related to tectonic evolution of the Himalaya, climatic variations, sea level changes or some combination of these factors.

Major eustatic sea-level changes have occurred several times since the Miocene (Haq *et al.* 1987), and Cochran (1990) proposed that this has been the main control on sedimentation in the distal Fan. During periods of low sea level stand, the delta should be exposed to erosion, and sediments will be redeposited in the deep sea. During periods of high sea level, sediments should be impounded in the delta, and coarse clastic deposition cut off from the distal Fan. A major sea level lowstand at 10.5 Ma resulted in major erosion of the Ganges–Brahmaputra delta (Lindsay *et al.* 1991). The drop in sea level observed at 10.5 Ma and attributed to Antarctic ice sheet growth (Shackelton & Kennett 1975; Miller *et al.* 1987) corresponds to an increase in the accumulation rate in the Hole 718C. However, despite an important rise in sea level after this time, which corresponds to the beginning of growth of the 'modern' delta (Lindsay *et al.* 1991), deposition rates and sediment facies in the distal Bengal Fan remain essentially unchanged (Fig. 4). The sedimentation rate at the Leg 116 sites does begin to drop around 6.8 Ma with the appearance of the SK assemblage, which corresponds to a new rise in sea level (e.g. Cochran 1990). However, other equally important variations in sea level are not followed by changes in sedimentation, especially between 3 and 1 Ma. The Bengal Fan does show a striking change in sedimentation at about 0.8 Ma. This event apparently correlates with the onset of major glacial cycles in the mid-Pleistocene. It seems unlikely that the prominent stratigraphic break at 0.8 Ma merely reflects a (possibly) somewhat more pronounced drop in sea level. Rather, the data suggest that Himalayan erosion rates increased at 0.8 Ma, perhaps due to increased alpine glaciation, perhaps due to renewed tectonic activity (cf. Burbank & Johnson 1983). Furthermore, large Quaternary glacial-interglacial sea level changes are not evident in the distal Fan. Bouquillon *et al.* (1989) find only minor changes in the smectite–illite ratio of clays from the delta and proximal Fan during Pleistocene glacial–interglacial cycles. If the relation between sea level and erosion in the Ganges–Brahmaputra delta is relatively clear, the relation between sea level and sedimentation in the distal Fan is evidently not so simple. Additional evidence against a simple relation between sea level and Fan sedimentation comes from the early and mid-Miocene. The drop in sea level at 10.5 Ma resulted in major erosion of the Ganges delta (Lindsay *et al.* 1991), but only a brief change in sedimentation in the Fan. Nor is a major change in the marine $\delta^{18}O$ record at 14 Ma, the largest in mid-Tertiary time (Miller *et al.* 1987), reflected in the sedimentary record. We agree with the assessment of Amano & Taira (1992) that sea-level changes are only a secondary control on sedimentation in the Fan. This implies that other processes are important for controlling sedimentation in the distal Fan, particularly changes in the sediment supply from the river system. It is also important to remember that Neogene sea level changes have been driven by ice volume changes in many instances, therefore they correspond to episodes of climate change (e.g. Miller *et al.* 1987; Crowley & North 1991). Climate change in the Himalayan source region of the Bengal Fan sediments can easily result in changes in erosion rate and thus sediment supply. The problem of distinguishing the effects of sea level change from the effects of climate change which may itself have been the cause of sea level variation is not easily resolved. This problem is of a similar nature to the 'chicken and egg' problem outlined by Molnar & England (1990), trying to distinguish between faunal change in response to changing elevation or changing climate.

Studies of both the terrestrial record in Pakistan and the marine record in the Arabian Sea suggest that the monsoonal weather pattern intensified *c.* 7 Ma (Quade *et al.* 1989; Kroon *et al.* 1991). This climactic change appears to be directly reflected in the sedimentation record of the Bengal Fan (Figs 2 & 3). The sedimentation rate decreases, while the clay mineralogy shifts to the SK assemblage. However, the relation between the monsoon climate and sedimentation in the Bengal Fan is complex. While the appearance of SK clays may simply be related to a warmer and more humid climate in the Indo-Gangetic Plain, monsoon conditions would be expected to increase erosion rates in the High Himalaya, and thus sedimentation rates in the deep sea. This apparent paradox remains to be resolved. One possible solution is that sediment was stored in the Siwaliks, rather than transported to the Fan. Such a scenario may be consistent with Late Miocene–Pliocene tectonic subsidence in the Siwalik basin (Burbank 1992). He has proposed that the geometry of basin subsidence in the Indian region changed sometime in Plio-Pleistocene time. If sedimentation in the Bengal Fan was partially controlled by the behavior of the foreland basin, the renewal of rapid Fan sedimentation rates at 0.8 Ma would suggest that the change in basin subsidence style occurred at about this time.

Ruddiman & Raymo (1988) and Ruddiman & Kutzbach (1989) have proposed that late Neogene uplift of the Himalaya and Tibetan plateau has driven regional and global climate change, including causing the onset of the monsoon. The Bengal Fan records no evidence of renewed uplift and increased erosion coincident with the intensification of the monsoon climate at 7–8 Ma. In fact, the Leg 116 results suggest the opposite; a decrease in sedimentation rate is possibly related to a decrease in erosion in the Himalaya. Rather, our results require the presence of the Himalaya as a significant topographic feature since at least 17 Ma. This implies that the late Miocene establishment or enhancement of monsoonal circulation in the Indian region is related to factors other than uplift in the Himalayan chain alone.

Tectonic changes are likely to cause important variation in the supply of detrital sediment to the Indo-Gangetic plain. For example, a decrease in the relief and general elevation of the range would directly result in decreased erosion. Episodic variations of the exhumation rate could well be related to motion on the MCT between

20 and 10 Ma, and on the MBT since at least 5 Ma (e.g. Le Fort 1989). Direct evidence in favour of this hypothesis is, however, lacking. Various studies have suggested pulses of uplift based on cooling ages or faunal studies (e.g. Gansser 1983). However, these data have proven very difficult to interpet in a unique way (Molnar & England 1990; England & Molnar 1990). A simple link between tectonics and sedimentation in the Bengal Fan is therefore not easy to demonstrate because to date, no reliable record of the past elevation history of the Himalaya is available. The Bengal fan itself is probably the best indication that the Himalaya was a highly elevated range since the early Miocene. Independent evidence to support high elevations in the mid–early Miocence comes from two sources. France-Lanord *et al.* (1988) have show that rare chloritization of the Manaslu leucogranite was effected by meteoric water (δD values of chlorite are *c.* −190‰). This episode of chloritization has been dated at 14.7 ± 0.2 Ma ($^{40}Ar/^{39}Ar$, Copeland *et al.* 1990). The δD value of the chloritizing water is estimated at <−140‰, markedly lower than modern meteoric water on the southern flank of the High Himalaya (δD – 80‰, France-Lanord *et al.* 1988), and very similar to meteoric water in Tibet (Yu *et al.* 1984; Grimaud *et al.* 1985). Such extremely low values imply that very strong altitude and temperature effects on the isotopic composition of precipitation were already in place at the tropical latitude of the Himalaya in the mid-Miocene. The second independent indication comes from evidence for large scale normal faulting coaxial with the axis of compression at around 20 Ma. Pêcher (1991) observed stretching lineations in the upper Manaslu granite which record movement on an overlying normal fault while the granite was still hot, *c.* 21 Ma. Hodges *et al.* (1992) find evidence for significant extension along the Qomolangma detachment at the same time. Such coeval and parallel extension and compression suggests gravitational collapse of an already highly elevated region due to limitations in crustal strength (Molnar & Lyon-Caen 1988). We believe that the Himalayan Range had already achieved an elevation of the same order as the present day by 17 Ma.

Conclusion

Turbidite sedimentation in the distal Bengal Fan began in the early Miocene, before 17 Ma. Sedimentological data from ODP Leg 116 and DSDP Leg 22 define three main intervals of depositional history. From 17 to 6.8 Ma, deposition was characterized by coarse clastic debris, high sedimentation rates, and clay mineralogy dominated by an illite–chlorite assemblage. Between 6.8 and 0.8 Ma, deposition rates decreased, along with grain size, and clays were composed primarily of a smectite–kaolinite assemblage. After 0.8 Ma there was an abrupt return to coarse clastic sedimentation, very high accumulation rates, and the illite–chlorite clay assemblage.

Nd, Sr and O isotopic measurements of Bengal Fan sediments from all stratigraphic levels are very uniform, implying a single dominant source for the entire period since 17 Ma. The Rb-Sr isotopic system in this source was apparently homogenized not long before erosion and deposition of the sediment. Oxygen isotopic systematics on coarse quartz and micas also imply a high temperature metamorphic source, consistent with the observed mineralogy of the detrital sediments (e.g. Yokoyama *et al.* 1990). A comparison of possible sources for the Bengal Fan sediments shows that the High Himalayan Crystalline series closely satisfies the above requirements, and that alternative sources do not. Isotopic mass balance suggests that only 20% or less of the Bengal Fan sediment was derived from other Himalayan sources such as the Lesser Himalaya or Tibetan Sedimentary Series, although occasionally this fraction may have been as large as 40%. The isotope data show no resolvable contribution from either the Precambrian rocks of the Indian subcontinent or the Deccan Traps.

Bengal Fan sediments from IC periods are unaffected by low temperature alteration, implying that physical erosion has been dominant over chemical erosion. The smectite–kaolinite clay mineral assemblage is apparently produced from the same source material by low temperature meteoric alteration in the Indo-Gangetic Plain. The appearance of this assemblage in the Bengal fan appears to correlate with an intensification of the monsoon climate around 7 Ma. Paradoxically, the sedimentation rate and grain size in the Fan both decrease at the same time, implying that other factors are important as well.

The relative lack of LH and TSS-derived material in the Bengal Fan has implications for the structure of the Himalayan Range since the early Miocene. Southward-directed thrusting on the MCT–MBT system has been accompanied by simultaneous removal of the TSS by northward-directed normal faulting. The large quantity of Bengal Fan sediment which began arriving in the Indian Ocean before 17 Ma requires that the Himalaya have been a significant topographic feature at least since that time.

Independent evidence supports this hypothesis, and we believe that the Himalaya have been at or near their present elevation since early Miocene time. Evidence for significant uplift in the Himalaya during late Miocene and Pliocene time is absent in the Bengal Fan sedimentary record.

This study was carried out under the auspices of the French program 'Dynamique et Bilans de la Terre – Fleuve et érosion'. We thank P. Copeland, P. Le Fort, A. Galy and A. Moore for helpful discussion and review. We are grateful to J.J. Tiercelin, G. Mascle and T. Heulin who provided samples for this study. G. Burtin is thanked for his efficient help in clay separation. R. Taieb, D. Dautel and F. Lothe are thanked for their analytical support. CRPG contribution N° 951. CNRS-INSU-DBT contribution N° 490.

References

ALIBERT, C., MICHARD, A. & ALBAREDE, F. 1983. The transition from alkali basalts to kimberlies: isotope and trace element evidence from melilites. *Contribution to Mineralogy and Petrology,* **82**, 176–186.

AMANO, K. & TAIRA, A. 1992. Two-phase uplift of Higher Himalayas since 17 Ma. *Geology,* **20**, 391–394.

BARTOLI, F., BURTIN, G. & HERBILLON, A.J. 1991. Disagregation and clay dispersion of oxisols: Na resin, a recommended methodology. *Geoderma,* **49**, 301–317.

BLATTNER, P., DIETRICH, V. & GANSSER, A. 1983. Contrasting 18-O enrichment and origins of High Himalayan and Transhimalayan intrusives. *Earth and Planetary Science Letters,* **65**, 276–86.

BOTTINGA, Y. & JAVOY, M. 1973. Oxygen isotope partitioning among the minerals in igneous and metamorphic rocks. *Review of Geophysics and Space Physics,* **13**, 401–418.

BOUQUILLON, A., CHAMLEY, H. & FRÖHLICH, F. 1989. Sédimentation argileuse au Cénozoïque supérieur dans l'océan Indien Nord-oriental. *Oceanologica Acta,* **12**, 133–147.

——, FRANCE-LANORD, C., MICHARD, A. & TIERCELIN, J.-J. 1990. Sedimentology and isotopic chemistry of the Bengal fan sediments: the denudation of the Himalaya. *In*: COCHRAN, J.R., STOW, D.A.V. *et al. Proceedings of the Ocean Drilling Program, Scientific Results,* **116**, College Station, TX (Ocean Drilling Program) 43–58.

BRASS, G.W. & RAMAN, C.V. 1990. Clay mineralogy of sediments from the Bengal fan. *In*: COCHRAN, J.R., STOW, D.A.V *et al. Proceedings of the Ocean Drilling Program, Scientific Results,* **116**, College Station, TX (Ocean Drilling Program) 35–41.

BURBANK, D.W. 1992. Causes of recent Himalayan uplift deduced from deposited patterns in the Ganges basin. *Nature,* **357**, 680–683.

BURBANK, D.W. & JOHNSON, G.D. 1983. The late Cenozoic chronologic and stratigraphic development of the Kashmir intermontane basin, Northwestern Himalaya. *Paleogeography, Palaeoclimatology, Paleoecology,* **43**, 205–235.

CLAYTON, R.N. & MAYEDA, T.D. 1963. The use of bromine pentafluoride in the extraction of oxygen from oxides and silicates for isotopic analysis. *Geochimica et Cosmochimica Acta,* **27**, 43–52.

COCHRAN, J.R. 1990. Himalayan uplift, sea level, and the record of Bengal fan sedimentation at the ODP Leg 116 Sites. *In*: COCHRAN, J.R., STOW, D.A.V. *et al. Proceedings of the Ocean Drilling Program, Scientific Results,* **116**, 397–414.

COPELAND, P. & HARRISON, T.M. 1990. Episodic rapid uplift in the Himalaya revealed by $^{40}Ar/^{39}Ar$ analysis of detrital K-feldspar and muscovite, Bengal fan. *Geology,* **18**, 354–357.

——, HARRISON, T.M. & LE FORT, P. 1990. Age and cooling history of the Manaslu Granite: Implications for Himalayan tectonics. *Journal of Volcanology and Geothermal Research,* **44**, 33–50.

CROWLEY, T.J. & NORTH, G.R. 1991. *Paleoclimatology.* Oxford Monographs on Geology and Geophysics. Oxford University Press, Oxford.

CURRAY, J.R. 1991. Postcollision sediments in the Bay of Bengal. *Eos,* **72**, 250.

—— & MOORE, D.G. 1971. Growth of the Bengal Deep-Sea Fan and denudation in the Himalayas. *Geological Society of America Bulletin,* **82**, 563–572.

—— & MUNASINGHE, T. 1989. Timing of intraplate deformation, northeast Indian Ocean. *Earth and Planetary Science Letters,* **94**, 71–77.

DAMUTH, J.E., FLOOD, R.D., KOWSMANN, R.O., BELDERSON, R.H. & GORINI, M.A. 1988. Anatomy and growth pattern of Amazon deep-sea fan as revealed by long-range side-scan sonar (GLORIA) and high-resolution seismic studies. *American Association of Petroleum Geologists Bulletin,* **75**, 885–911.

DENIEL, C. 1985. *Apport des isotopes du Sr, du Nd et du Pb à la connaissance de l'âge et de l'origine des leucogranites himalayens. Exemple du Manaslu (Himalaya, Népal).* Thesis, Université de Clermont-Ferrand, France.

——, VIDAL, P. & LE FORT, P. 1986. Les leucogranites himalayens et leur région source probable: les gneiss de la "Dalle du Tibet". *Compte Rendu de l'Académie des Sciences, Paris,* **303**, II, n° 1, 57–60.

——, ——, FERNANDEZ, A. & LE FORT, P. 1987. Isotopic study of the Manaslu granite (Himalaya, Nepal): Inferences on the age and source of Himalayan leucogranites. *Contributions to Mineralogy and Petrology,* **96**, 78–92.

DIETRICH, V. & GANSSER, A. 1981. The leucogranites of the Bhutan Himalaya (crustal anatexis versus mantle melting). *Bulletin Suisse de Minéralogie et Pétrographie,* **61**, 177–202.

ENGLAND, P. & MOLNAR, P. 1990. Surface uplift, uplift of rocks, and exhumation of rocks. *Geology,* **18**, 1173–1177.

FELLER, C., BURTIN, G., GÉRARD, B. & BALESDENT, J. 1991. Utilisation des résines sodiques et des ultrasons dans le fractionnement granulométrique de la matière organique des sols. Intérêt et limites. *Sciences du Sol*, **29**, 77–93.

FRANCE-LANORD, C. & LE FORT, P. 1988. Crustal melting and granite genesis during the Himalayan collision orogenesis. *Transactions of the Royal Society of Edinburgh Earth Sciences*, **79**, 183–195.

——, SHEPPARD, S.M.F. & LE FORT, P. 1988. Hydrogen and oxygen isotope variations in the High Himalaya peraluminous Manaslu leucogranite: evidence for heterogeneous sedimentary sources. *Geochimica et Cosmochimica Acta*, **52**, 513–526.

GANSSER, A. 1964. *Geology of the Himalayas*. Interscience Publishers, L.U. de Sitter Ed.

—— 1966. The Indian Ocean and the Himalayas A geological interpretation. *Eclogae geologicae Helvetiae*, **59**, 831–848.

—— 1983. The morphogenic phase of mountain building. *In*: HSÜ, K.J. (ed.) *Mountain building processes*. Academic Press, London, 221–228.

GARTNER, S. 1990. Neogene calcareous nannofossil biostratigraphy, Leg 116 (Central Indian Ocean). *In:* COCHRAN, J.R., STOW, D.A.V *et al. Proceedings of the Ocean Drilling Program, Scientific Results*, **116**, College Station, TX (Ocean Drilling Program), 165–187.

——, JOHNSON, D.A. & MCGROWRAN, B. 1974. Paleontology synthesis of deep sea drilling results from Leg 22 in the Northeastern Indian Ocean. *In*: VON DER BORCH, C.C., SCLATER, J.C. *et al. Initial Reports DSDP*, **22**, Washington, U.S. Government Printing Office, 805–813.

GOLDSTEIN, S.L. 1988. Decoupled evolution of Nd and Sr isotopes in the continental crust and the mantle, *Nature*, **336**, 733–738.

——, O'NIONS, R.K. & HAMILTON, P.J. 1984. A Sm-Nd isotopic study of atmospheric dusts and particulates from major river systems. *Earth and Planetary Science Letters*, **70**, 221–236.

GRIMAUD, D., HUANG, S., MICHARD, G. & ZHENG, K. 1985. Chemical study of geothermal waters of Central Tibet (China). *Geothermics*, **14**, 35–48.

HAQ, B.U., HARDENBOL, J. & VAIL, P.R. 1987. Chronology of fluctuating sea levels since the Triassic. *Science*, **235**, 1156–1167.

HODGES, K.V., BURCHFIELD, B.C., CHEN, Z., HOUST, T., LUX, D., PARRISH, R.R. & ROYDEN, L.H. 1992. Rapid Early Miocene tectonic unroofing of the metamorphic core of the Himalaya: Evidence from the Qomolangma (Everest) region, Tibet. *Abstracts of the 7th Himalaya-Karakoram-Tibet workshop, Oxford University, 6–8 April 1992*, 39.

INGERSOLL, R.V. & SUCZEK, C.A. 1979. Petrology and provenance of Neogene sand from Nicobar and Bengal Fans, DSDP Sites 211 and 218. *Journal of Sedimentary Petrology*, **49**, 1217–1228.

KROON, D., STEENS, T. & TROELSTRA, S.R. 1991. Onset of monsoonal related upwelling in the western Arabian Sea as revealed by planktonic foraminifers. *In*: PRELL, W.L., NIITSUMA, N., *et al. Proceedings of the Ocean Drilling Program, Scientific Results*, **117**, College Station, TX (Ocean Drilling Program), 257–263.

LAWRENCE, J.R., DREVER, J.I., ANDERSON, T.F. & BRUECKNER, H.K. 1979. Importance of alteration of volcanic material in the sediments of Deep Sea Drilling Site 323: chemistry, $^{18}O/^{16}O$ and $^{87}Sr/^{86}Sr$. *Geochimica et Cosmochimica Acta*, **36**, 573–588.

LE FORT, P. 1986. Metamorphism and magmatism during the Himalayan collision. *In*: COWARD, M.P. & RIES, A.C. (eds) *Collision Tectonics. Geological Society, London, Special Publication*, **19**, 159–172.

—— 1989. The Himalayan orogenic segment. *In*: SENGÖR, A.M.C. (ed.) *Tectonic evolution of the Tethyan regions. Proceedings of the NATO ASI meeting, Istanbul, October 1985*. Reidel Publ. Co.

LINDSAY, J.F., HOLLIDAY, D.W. & HULBERT, A.G. 1991. Sequence stratigraphy and the evolution of the Ganges-Brahmaputra Delta Complex. *American Association of Petroleum Geologist Bulletin*, **75**, 1233–1254.

LIGHTFOOT, P.C., HAWKESWORTH, C.J., DEVEY, C.W., ROGERS, N.W. & VAN CALSTEREN, P.W.C. 1990. Source and differentiation of Deccan Traps Lavas: implications of geochemical and mineral chemical variations. *Journal of Petrology*, **31**, 1165–1200.

MAHONEY, J.J., MACDOUGALL, J.D., LUGMAIR, G.W., MURALI, A.V., SANKAR, DAS, M. & GOPALAN, K. 1982. Origin of the Deccan Trap flows at Mahabaleshwar inferred from the Nd and Sr isotopic and chemical evidence. *Earth and Planetary Sciences Letters*, **60**, 47–60.

MANLEY, P.L. & FLOOD, R.D. 1988. Cyclic sediment deposition within Amazon deep-sea fan. *American Association of Petroleum Geologists Bulletin*, **75**, 912–925.

MILLER, K.G., FAIRBANKS, R.G. & MOUNTAIN, G.S. 1987. Tertiary oxygen isotope synthesis, sea level history, and continental margin erosion. *Paleoceanography*, **2**, 1–19.

MILLIMAN, J.D. & MEADE, R.H. 1983. World Delivery of River Sediment to the Oceans, *Journal of Geology*, **1**, 1–21.

MOLNAR, P. & ENGLAND, P. 1990. Late Cenozoic uplift of mountain ranges and global climate change: chicken or egg? *Nature*, **346**, 29–34.

MOLNAR, P. & LYON-CAEN, H. 1988. Some simple physical aspects of the support, structure, and evolution of mountain belts. *Geological Society of America Special Papers*, **218**, 179–207.

MUKHERJEE, B. 1964. Clay minerals in argillaceous sediments of the Himalayan zone. *Clay Minerals Bulletin*, **5**, 363–372.

PARRISH, R.R., HODGES, K.V. & MACFARLANE, A. 1992. U-Pb geochronology of igneous and metamorphic rocks near the Main Central Thrust in the Langtang area, Central Nepal Himalaya. *Abstracts of the 7th Himalaya-Karakoram-Tibet workshop, Oxford University, 6–8 April 1992*, 67–68.

PÊCHER, A. 1991. The contact between the Higher Himalayan Crystallines and the Tibetan Sedimentary Series: Miocene large-scale dextral shearing. *Tectonics*, **10**, 587–598.

PERRY, E.A., GIESKES, J.M. & LAWRENCE, J.R. 1976.

Mg, Ca and O^{18}/O^{16} exchange in the sediment pore water system, Hole 149, DSDP. *Geochimica et Cosmochimica Acta,* **40**, 413–423.

PEUCAT, J.J., VIDAL, P., BERNARD-GRIFFITHS, J. & CONDIE, K.C. 1989. Sr, Nd and Pb isotopic systematics in the Archean low-to-high-grade transition zone of southern India: Syn-accretion granulites. *Journal of Geology,* **97**, 537–550.

PIMM, A.C. 1974. Mineralization and trace element variation in deep-sea pelagic sediments of the Wharton basin, Indian Ocean. *In*: VON DER BORCH, C.C., SCLATER, J.C., *et al. Initial Reports DSDP*, **22**, Washington, U.S. Government Printing Office, 469–476.

QUADE, J., CERLING, T.E. & BOWMAN, J.R. 1989. Development of Asian monsoon revealed by marked ecological shift during the latest Miocene in northern Pakistan. *Nature,* **342**, 163–166.

RAHMAN, M.A.H.M. 1987. *Amélioration de la fixation d'azote dans la rhizosphére du riz cultivé sur différents sols du Bangladesh.* Thesis Université de Nancy 1.

RUDDIMAN, W.F. & KUTZBACH, J.E. 1989. Forcing of the late Cenozoic northern hemisphere climate by plateau uplift in southeast Asia and the American southwest. *Journal of Geophysical Research,* **94**, 18409–18427.

—— & RAYMO, M.E. 1988. Northern hemisphere climate regimes during the past 3 Ma: possible tectonic connections. *Philosophical Transactions of the Royal Society (London)*, **B 318**, 411–430.

SARIN, M.M., KRISHNASWAMI, S., DILLI, K., OMAYAJULU, B.L.K. & MOORE, W.S. 1989. Major ion chemistry of the Ganga-Brahmaputra river system: Weathering processes and fluxes to the Bay of Bengal. *Geochimica et Cosmochimica Acta,* **53**, 997–1009.

SAVIN, S.M. & EPSTEIN, S. 1970. The oxygen and hydrogen isotope geochemistry of ocean sediments and shales. *Geochimica et Cosmochimica Acta,* **34**, 43–63.

SCAILLET, B., FRANCE-LANORD, C. & LE FORT, P. 1990. Badrinath-Gangotri plutons (Garhwal, India) Petrological and geochemical evidence for fractionation processes in a High Himalayan Leucogranite. *Journal of Volcanology and Geothermal Research,* **44**, 163–188.

SHACKELTON, N.J. & KENNETT, J.P. 1975. Paleotemperature history of the Cenozoic and the initiation of Antarctic glaciation: oxygen and carbon isotope analyses in DSDP Sites 277, 279 and 281. *In*: KENNETT, J.P., HOUTZ, R.E., *et al. Initial Reports DSDP,* **29**, Washington (U.S. Government Printing Office), 743–756.

SHIPBOARD SCIENTIFIC PARTY 1974. Sites 2182 *In*: VON DER BORCH, C.C., SCLATER, J.C., *et al. Initial Reports DSDP,* **22**, Washington, U.S. Government Printing Office, 325–367.

—— 1989. Initial Reports Sites 717-718-719 Distal Bengal Fan, *Proceedings of the Ocean Drilling Program, Initial Reports,* **116**, College Station, TX (Ocean Drilling Program).

STOW, D.A.V., AMANO, K., BALSON, P.S., BRASS, G.W., CORRIGAN, J., RAMAN, C.V., TIERCELIN, J.-J., TOWNSEND, M. & WIJAYANANDA, N.P. 1990. Sediment facies and processes on the distal Bengal Fan, Leg 116. *In*: COCHRAN, J.R., STOW, D.A.V *et al. Proceedings of the Ocean Drilling Program, Scientific Results,* **116**, College Station, TX (Ocean Drilling Program), 377–396.

TAIEB, R. 1990. *Les isotopes de l'hydrogène, du carbone et de l'oxygène dans les minéraux argileux et les eaux de formation.* Thèse, Institut National Polytechnique de Lorraine, Nancy, France.

TRIVEDI, J.R., GOPALAN, K. & VALDIYA, K.S. 1984. Rb-Sr ages of granitic rocks within the Lesser Himalayan nappes, Kumaun, India. *Journal of the Geological Society of India,* **25**, 641–654.

VIDAL, PH., BERNARD-GRIFFITHS, J., COCHERIE, A., LE FORT, P., PEUCAT, J.J. & SHEPPARD, S.M.F 1984. Geochemical comparison between Himalayan and Hercynian leucogranites. *Physics of the Earth and Planetary Interiors,* **35**, 179–90.

——, COCHERIE, A. & LE FORT, P. 1982. Geochemical investigations of the origin of the Manaslu leucogranite (Himalaya, Nepal). *Geochimica et Cosmochimica Acta,* **46**, 2279–92.

YEH, H. & SAVIN, S.M. 1976. The extent of oxygen isotope exchange between clay minerals and sea water. *Geochimica et Cosmochimica Acta,* **40**, 743–748.

YOKOYAMA, K., AMANO, K., TAIRA, A. & SAITO, Y. 1990. Mineralogy of silts from the Bengal fan. *In*: COCHRAN, J.R., STOW, D.A.V. *et al. Proceedings of the Ocean Drilling Program, Scientific Results,* **116**, College Station, TX (Ocean Drilling Program), 59–73.

YU, J., ZHANG, H., YU, F. & LIU, D. 1984. Oxygen and hydrogen isotopic compositions of meteoric waters in the Eastern part of Xizang. *Geochemistry,* **3**, 93–101.

Index

Abbottabad Group 223–4
Afghanistan 69–87
  orogenic evidence 49, 50
  suturing evidence 70–1
Afghanistan block
  collision and deformation 79–84
  history 72–4
Aghil Formation 24, 40
Aghil Group
  biostratigraphy 42–3
  lithostratigraphy 39–40
Alpurai Schist Group 208, 213, 216–17
amphiboles 361–2, 363
Andarab fault 81
Annapurna Detachment Fault
  setting 464–6
  significance 470–1
  structures 466–70
anorthosite 116
Ar isotope studies 497
Arun tectonic window
  deformation 516–18
  deformation/metamorphism interrelationship 519–21
  inverse metamorphism 522
  metamorphism 518–19
  setting 511–12
  stratigraphy 513–16
  tectonic junctions 521–3
Ashtigar Formation
  biostratigraphy 44
  lithostratigraphy 40
  petrography 44–5
Attock Cherat Range 575–6, 582, 584
Azmat fault 563, 565

Babusar amphibolite 113, 125
Bahadur Khel fault 563
Bahadur Khel Salt 570–1
Bajaura nappe 475
Baltoro plutonic unit 54, 55
  petrogenesis 61–2
Banda Daud Shah fault 563
Barabise Gneiss 345, 512, 516
Baralacha La dyke swarm
  geochemistry 242, 243, 244
  setting 239–40
Baraul Banda Slate Formation
  depositional model 147–50
  lithology 142–4
Barbara Banda fault 563
Barun gneiss 348
Basthala Schist Formation 345, 512, 515
Bazid Khel fault 563, 565
Bazid Khel klippe 565
Bela ophiolite 101
  leucocratic suites
    ages 96–7
    formation 98–9
    geochemistry 93–6
    mineralogy 92–3
    origin 99
    setting 89–92
Belar keratophyre 92
Bengal fan
  evolution 616–18
  sediment analysis
    methods 607
    results 607–12
    results discussed 612–15
    significance 615–16
  sedimentary history 604–7
  setting 603–4
Benighat Slates 526, 531, 535
Berinag thrust sheet 432
Bhagirathi granite
  petrogenetic model 439–41
  petrography 437
  setting 433–7, 487
  source 437–8
Bimphedi Group 342, 526–7
biostratigraphy
  Aghil Formation 42–3
  Ashtigar Formation 44
  Dagshai Formation 546
  Kasauli Formation 546
  Lamayuru complex 302–3
  Northern Karakoram Terrain 42–6
  Singtali Formation 546
  Subathu Formation 546
  Yashkuk Formation 44
Black gneiss 346, 349
Bouguer anomalies 9
  methods of data collection 10–11
  results 11–13, 18–20
  results discussed 13–16
Bozda fault 563
Braghdi fault 563
Burawai Formation 225

calcite–dolomite geothermometry 364–8
Cambro–Ordovician magmatism 251
Campbellpore Basin 576, 581
Carboniferous studies
  Kaghan Valley 224
  Lower Swat 216
Chainpur Group 345, 512
Chalt Volcanic Group 123, 140
Chaman fault 69, 73, 80–1, 590
Chandragiri Limestone 527, 530
Chapursan Valley 41
  deformation studies 22–8
  faulting 28–30
  thrusting 28
chemistry *see* geochemical studies
Chichana fault 563, 565
Chichana klippe 565
Chichina Ghunda fault 563
Chikkim Formation 282, 291
Chilas complex 113, 123, 125–6, 140
  geochemistry 131–3
Chinji Formation 561

Choga gneiss 210
Chomolungma-Makalu transect 345–52
chromite
  Sapat complex 116
  Shangla study
    mineral chemistry 104–6
    setting 103–4
    significance 110–11
clay mineralogy
  Bengal fan 605, 606
    methods of analysis 607
    results 607–12
    results discussed 612–15
    significance 616
Cretaceous studies
  magmatism
    Afghanistan 72
    Zanskar Range 291–2
  orogeny 50–1
  sediments
    Giumal Group 279–83
    Karakoram 22, 46–7
    Lamayuru complex 303
crustal structure and thickness 9, 64–6

Dagshai Formation 542
  biostratigraphy 546
  depositional environment 556
  facies analysis 553
  lithology 549–51
  lithostratigraphy 545
  tectonic setting 542–5, 554
Danda fold 595
Darband Formation 47
Dargai ophiolite 113
Dark Band Formation 361
deformation studies
  Chapursan Valley 22–8
  Garhwal Himalaya 487
  High Himalaya Crystallines 448–50
  High Himalaya Tibetan Zone 265–73
  Kathmandu complex 517–18
  Nawakot complex 517
  Nilgiri Limestone 370–1
  Shimshal Valley 30–5
Dhok Pathan Formation 561
Dir thrust 150
Dir Volcanic Group 123
  lithology 142–54
  setting 139, 140–2
  significance 155–8
Dopialo Sar gneiss 210
Dras-Kohistan-Nuristan arc 1
Dunghai Formation 591
dunites 116
duplex geometry 597–9

Eo-Cimmerian orogeny 49
Eocene studies
  sediments 432, 545, 560
  tectonics 81–3
extensional structures 273

Farad block 71, 73
Fateh Khan Banda fault 563, 565
Fateh Khan Banda klippe 565
Fatu La Formation 282, 291
faulting studies 28–30
  normal
    causes 405–7
    relation to thrust faults 401–3, 405
    stress fields 403–4
  thrust 401–3
  wrench 28–30
Ferghana 9
Ferruginous Oolite Formation 291
flake tectonics 274
flexural modelling 9
fluid inclusion studies 533–6

Ganga Basin 9
Ganmachidam Formation 238
Garhwal Group 431, 432, 486
Garhwal Himalaya 432
  deformation 487
  magmatism
    petrogenesis 439–41
    petrography 437
    setting 433–7, 487
    source 437–8
  metamorphism
    geochronology 497–506
    petrography 487–9
    thermobarometry 489–97
  regional tectonics 506–7
geochemical studies
  Baralacha La dyke swarm 242
  Chilas complex 131–3
  High Himalaya Crystallines 165, 331–6
  Hunza granites 58–61
  Indus Confluence granites 177–8
  Kamila amphibolite 127–31
  Khuzdar leucocratics 93–6
  Kohistan granites 177–9
  Nanga Parbat–Haramosh massif 165
  Panjal Traps 241–6
  Parri granites 179
  Sapat complex 118–19
  Shangla chromite 104–10
  Sumayar pluton 60–1
  Yunam granite 257–9
  Yunam microgabbro 244, 245
geochronology 55, 96–7, 187, 393–6, 497–506
geothermobarometry
  Garhwal Himalaya 489–97
  High Himalaya 331, 333–6
  Nanga Parbat–Haramosh Syntaxis 165–9
geothermometry 364–8
Ghazij Formation 591
Ghorzandi fault 563
Gilgit paragneiss 123
Giandari fold 592, 596–7
Gircha Formation 27, 28
Giumal Group
  eustatic setting 290
  sedimentology 287
  sequence stratigraphy 291
  setting 279, 282

glaucophosphorite 279–82, 289, 290, 291
Gophu La granite 393, 394
granite studies
  Bhagirathi 433–41
  Hunza Valley Complex 53, 55–61
  Indus Confluence 175–80
  Khuzdar 89–99
  Kohistan arc 175–89
  Langtang Valley 383–4
  Manaslu 393, 394, 415–25
  Nanga Parbat–Haramosh massif 184–9, 392
gravity anomalies
  methods of data collection 10–11
  results 11–13, 18–20
  results discussed 13–16
Guhjal unit 21, 24, 41
Gumburajun pluton 251

Hazara Formation 222–3
Hazara Kashmir syntaxis
  metamorphism 299–30
  stratigraphy 221–6
  structure 226–8
  tectonic model 230–3
Helmand block 71, 73
Herat fault 71, 80–1
Herat–Central Badakhstan fault system 69
High Himalaya (Tibetan Slab) 1, 4–5, 228, 229, 251
  crenulation cleavage 453
  deformation phases 448–50
  exhumation model 338
  geochemistry 165, 331–6
  geothermobarometry 165–9
  leucogranite studies 337, 347, 392–8
    Bhagirathi 433–41
    Manaslu 415–25
  lithotectonic units 346
  metamorphic assemblage 169–70, 327–31, 348–9
  migmatization 337
  partial melting evidence 350–2
  setting 163–5, 323–5, 343
  shear patterns 450–3
  tectonic modelling 453–8
  zonation 336–7
Himalayan Frontal Thrust 525, 541
Hinuwan phyllite 512, 514
Hissartang fault 582, 584
Hokse quartzite 512, 514
Hukni fault 563
Hunza dykes 56
  geochemistry 58–60
  mineralogy 57
  petrogenesis 61–2
Hunza Valley
  crustal structure 64–6
  plutonic complex 55–7
    geochemistry 58–61
    mineralogy 57–8
    petrogenesis 61–4
  setting 55

illite crystallinity 531–3
Indian Plate motions 81–4, 158
Indus Confluence granites
  age 179
  geochemistry 177–8
  isotope analysis 179–80, 182
  setting 175–7
Indus flysch 299, 302
Indus–Tsangpo Suture Zone (Main Mantle Thrust) 1, 53, 113, 174–5, 251, 560
isotope analyses
  radio 55, 96–7, 179–82, 187, 393–6, 497–506, 607–9
  stable 385, 609–12

Jafar Kandao Formation 216
Jatta gypsum 560, 570
Jhala fault 432, 487
Jijal ophiolite 101, 113, 123, 125, 140
Jobra Formation 209, 213, 216
Jurassic studies
  orogeny 48–50
  sediments
    Karakoram 22–46
    Lamayuru complex 303
Jutogh nappe 542, 545

K/Ar geochronology 497–506
Kabul block 76, 81
Kabul ophiolite 76, 81
Kaghan Valley
  metamorphism 229–30
  stratigraphy 221–6
  structure 226–8
  tectonic model 230–3
Kala Chitta Range 576, 582, 584
Kalam Group 142
Kali Khola schist 512, 515
Kalitar Formation 527, 529
Kamila amphibolite 123, 140
  geochemistry 127–31
Kamlial Formation 561
Kangi La Formation 282, 291
Kangmar–Lhagoi Kangri domes 53
Karak fault 563
Karakoram 2
  Bouguer gravity measurements
    data collection 10–11
    results 11–13, 18–20
    results discussed 13–16
  crustal thickness 9
  see also Northern Karakoram Terrain
Karakoram Axial Batholith (KAB) 53–4
Karapa greenschist 216
Karku Jhal granite 92
Karsha Formation 238
Kasauli Formation 542
  biostratigraphy 546
  depoositional environment 556
  facies analysis 553
  lithology 551
  lithostratigraphy 546
  tectonic setting 542–5, 554
Kashala Formation 209, 210, 215, 216
Kata fault 563
Katawaz basin 76–9

Kathmandu complex 342
  deformation 517–18
  metamorphism 518–19
  metamorphism/deformation interrelationship 520
  stratigraphy 512, 516
Kathmandu klippe 527
  fluid inclusion studies 533–6
  metamorphism 527–9
  petrography 529–33
Khandbari gneiss 512, 516
Kharang phyllite 512, 514
Khekuwa phyllite 512, 514
Khost ophiolite 76
Khuzdar plagiogranites
  geochemistry 93–6
  geochronology 97
  mineralogy 92–3
  setting 89–91
  significance 97–9
Kioto Limestone Group 291
Kirthar Formation 591
Kohat Formation 560
Kohat plateau 560
  stratigraphy 560–1
  structure 562–3, 575–6
    compression evidence 574–5
    folds 567–71
    thrusts 563–7
    transpression evidence 571–4
Kohat Range 575–6
Kohistan batholith 123, 140
Kohistan Island Arc Terrane 53, 113
  accretionary history 157–8
  Chilas complex 125–6
    geochemistry 131–3
  Dir Group 123, 139
    lithology 142–54
    significance 155–8
  evolution 133–6
  granite magmatism 175–83, 187–9
  Kamila amphibolite 125
    geochemistry 127–31
  Sapat complex 114–15
    lithology 115–17
    mineral chemistry 118–19
    petrography 117–18
    significance 119–20
  setting 123, 125, 140–2, 560
Konar fault 69
Krol nappe 542, 545
Krol thrust sheet 432
Kulikhani Formation 527, 529–30
Kuling Formation 238
Kullu Valley 475
  thermal history modelling
    one-dimensional 477–8
    two-dimensional 478–80
    model refinement 480–2
Kuncha Formation 526, 531, 535
Kurgiakh Formation 238

Labar La arenite 279, 289, 290, 291
Lachi fault 563
Ladakh batholith 265
Lamayuru complex
  biostratigraphy 302–3
  depositional environment 313–17
  facies analysis 303–13
  lithostratigraphy 302
  setting 300–1
Lamayuru flysch 320
Landiwal fault 563
LANDSAT imaging
  methods 197–201
  results 202–4
Langtang Valley
  granites 383–4, 393, 394
  lithological units 377–83
  regional synthesis 384–7
Laptal Formation 291
Laya Garr intrusives 92
Lesser Himalaya
  biostratigraphy 546
  facies analysis 551–3
  foreland basin modelling 553–4
  lithologies 546–51
  lithostratigraphy 545–6
  metamorphism 230, 331
    inversion 336
    mineral chemistry 336
  setting 2, 6–7, 327, 342, 348
  tectonostratigraphy 542–5
    Arun tectonic window 514–16
    Kullu Valley 475
    lithotectonic units 345
    tectonic elements 227–8, 251
    see also Kohat Plateau
leucogranites 337, 347
  High Himalaya
    isotope analysis 393–6
    melt modelling 396–8
    melting behaviour 392–3
    trace element analysis 393
  Karakoram 53, 55, 62
lineament analysis
  definition 196–7
  methods 197–201
  results 202–4
Lipak Formation 238
lithostratigraphic studies
  Aghil Group 39–40
  Ashtigar Formation 40
  Dagshai Formation 545
  Kasauli Formation 546
  Lamayuru complex 302
  Northern Karakoram Terrain 39–42
  Singtali Formation 545
  Subathu Formation 545
Loe Sar gneiss 210
Lohi Jhal trondhjemite 92
Loti fold 592, 594–5
Lower Swat
  history of research 208–9
  regional setting 207
  stratigraphy 213–16, 216–18
  structure 210–13
Lower Swat-Buner Schistose Group 208

magmatism
  Afghanistan 72, 73
  Kohistan 175–83
  Nanga Parbat–Haramosh massif 184–7
  Zanskar Range 291–2
magnetic susceptibility
  Manaslu granite 420–3
  measurement 413–15
Main Boundary Thrust
  activity 2, 6–7, 525, 560
  initiation 556
  kinematics 585–7
  setting 227, 230, 251, 541, 581–2
Main Central Thrust 1, 2, 251, 325, 347, 475, 485, 529
  activity 405, 525
  initiation 554
  metamorphic setting 5–6, 228, 229–30, 432, 463
Main Central Thrust Zone 345–6
  see also Arun window; Garhwal Himalaya; Kathmandu klippe; Kullu Valley
Main Karakoram Thrust (Shyok Suture) 17, 560
Main Mantle Thrust (Indus Suture) 1, 53, 113, 174–5, 251, 560
Makalu transect 345–52
Makalu granite 405
Maksuwa phyllite 512, 514
Mami Khet Formation 560
Mamir–Khushhal Garr plagiogranite 92
Manaslu granite 393, 394
  contact features 415–17
  fabric 417–19
  magnetic susceptibility 420–3
  microstructure 419–20
  regional significance 423–5
  setting 415
Manduri fault 563, 565
Manduri klippe 565
Manglaur Schist Formation 208, 210, 213, 216
Mansehra granite 208, 224
Marghazar Formation 209, 213–15, 216
Marpo Formation 48, 282, 291
Martoli Formation 431, 433, 486, 487
metamorphic studies
  Arun window 518–19
  Higher Himalaya 169–70, 327–31, 331–6, 348–9
  Kaghan Valley 229–30
  Kathmandu klippe 527–9
  Langtang Valley 377–83
  Lesser Himalaya 331, 336, 348
  Nanga Parbat–Haramosh massif 169–70
  Nawakot complex 518
  Nilgiri Limestone 360–4
  Tibetan Series 331, 336, 349
Midland Formations 525–6
migmatization 337
Milke gneiss 512
mineralogy
  Hunza granites 57–8
  Khuzdar leucocratics 92–3
  Shangla chromites 106–10
Miocene studies
  sediments 541, 546, 560
  tectonics 81–3
  see also Bengal fan
Mir Khweli Sar thrust belt 563–7
Mir Khwlei Sar klippe 565
Misgar fault 28
Misgar unit 22, 41
Mt Jolmo Lungma (= Chomolungma) Formation 346
Munsiari Group 431, 486
Munsiari thrust 432, 487
Murree Brewery Limestone 158
Murree fault 227
Murree Formation 226, 560
Muth Formation 238

Nadir Banda fault 563
Nagar leucogranite 59
Nagri Formation 561
Nal Limestone Formation 92
Namche migmatite orthogneiss 346, 348
Namika-La flysch 320
Nanga Parbat–Haramosh massif
  granites
    age 185, 392
    geochemistry 184
    isotope systematics 186–7
  petrography 184
    regional significance 189
  lineament analysis
    definition 196–7
    methods 197–201
    results 202–4
  setting 161–3, 193
Nanga Parbat–Haramosh syntaxis
  field relations 163–5
  geochemistry 165
  geothermobarometry 165–9
  metamorphism 169–70
Naran Formation 225
Nawakot complex
  deformation 517
  metamorphism 518
  metamorphism/deformation inter-relationship 519–20
  stratigraphy 512, 514
Nawakot nappes 342
Nd isotope studies 96–7, 186–7, 609–11
Nerak glaucophosphorite 279–82, 289, 290, 291
Nikanai Ghar Marble Formation 209, 215–16
Nilgiri Limestone (Formation de Larjung)
  CO_2 levels 369
  deformation history 370–1
  metamorphism 360–4
  setting 358
Nilkanth Formation 545
North Col Formation 331, 336, 345, 346
Northern Karakoram Terrane 21–2
  biostratigraphy 42–6
  lithostratigraphy 39–42, 46–7
  tectonics 48–51
    Chapursan Valley 22–30
    Shimshal Valley 30–5
Northern Potwar Deformation Zone 559, 560, 571, 576
Northern Suture Zone *see* Shyok Suture Zone
Num orthogneiss 345
Nuristan block 73, 76

Nyimaling Shear Zone 272
Nyimaling–Tsarap nappe 267–72

$\delta^{18}O$ measurements 385, 611–14
Oji pluton 251
Oligocene studies
  sediments 560
  tectonics 81–3
Oma Chu glaucophosphorite 279–82, 289, 290, 291
ophiolites
  Bela 101
  Jijal 101, 113, 123, 125, 140
  Kabul 76, 81
  Khost 76
  Shangla 101, 104–10, 113
  Spontang 553, 556
  Waziristan 101
  Zhob Valley 101

Palaeocene studies
  sediments 226, 432, 545, 560
  tectonics 81
Palaeozoic magmatism 72, 251
Pamirs 9
Panjal Formation 224–5
Panjal Traps 1, 225
  geochemistry 241–6
  regional significance 246–8
  setting 237
  subgroups 237–41
Panjao suture 71
Panoba fault 567, 570, 584
Panoba Shale 560, 570–1
Panshah Formation 28
Parri granites
  ages 179
  geochemistry 179
  isotopic analysis 180–2
  setting 175–7
Patala Formation 560
Pathan Algad fault 563
peridotite 116
Permian sediments 217
Peshawar basin 217–18, 575–6, 584
petrogenetic studies
  Baltoro plutonic unit 54, 55
  Hunza dykes 61–2
  Sumayar pluton 62–4
petrographic analysis
  Ashtigar Formation 44–5
  Garhwal Himalaya 487–9
  Kathmandu klippe 529–33
  Sapat Complex 117–18
  Yashkuk Formation 45
  Yunam granite 255
Pezbugi fault 592
Phakuwa Group 512, 516
phlogopite 360–1
Phulchauki Group 342, 527
Pingdon La Formation
  age 289
  sedimentation rate 289
  sedimentology 284, 285, 286
  sequence stratigraphy 291
  setting 279
Pirkoh fold 592, 595
plagiogranites
  ages 96–7
  formation 98–9
  geochemistry 93–6
  mineralogy 92–3
  setting 92
plate tectonic settings 81–4, 158
Pliocene studies
  sediments 541
  tectonics 83
Po Formation 238
Potwar plateau 582
Precambrian studies 222–3
Purwait Bhut granite 92
pyroxenes 362–3
pyroxenites 116

Raduwra Formation 527, 529
Ramgarth Group 431, 432, 486
Rara Formation 225–6
Rb/Sr geochronology 96–7, 187, 393–6
Reshit Formation 42, 44
reverse metamorphism 522
Ridge fault 592
Rongbuk Formation 346, 349
Rushan-Pshart Suture 21

Saidu Schist Formation 208, 215
Salkhala Formation 221–2
Salt Range thrust 560
Samandh Jhal quartz diorite 92
Samla Limestone 512, 515
Sani Muid fault 592
Sapat complex 114–15
  lithology 115–17
  mineral chemistry 118–19
  petrography 117–18
  regional significance 119–20
Sapat gali complex 125
Sarchu Shear Zone 272
Sassi granites 183–4
Sawanetar Formation 512, 515
sequence stratigraphy 291
Shaigalu Formation 77
Shamran volcanics 154–5
Shandur thrust 150
Shangla ophiolite 101, 113
  mineral chemistry 104–10
  regional significance 110–11
  setting 101–3
Shekhan Limestone 560
Shengus gneiss 163
Shikar Beh nappe 265–7
Shimshal Valley deformation studies
  intermediate zone 34–5
  northern zone 35
  southern zone 30–4
Shiwakki fault 563
Shyok (Northern) Suture Zone 1, 21, 53
Shyok Suture (Main Karakoram Thrust) 17, 560
Sikdim gneiss 512
Sikdim orthogneiss 345

Simbuwa schist 345, 512, 515
Singtali Formation 545
  biostratigraphy 546
  depositional environment 554–6
  facies analysis 551
  lithology 546–7
  lithostratigraphy 545
  tectonic setting 542–5, 553
Siwalik Group 541
  stratigraphy 561
  tectonic setting 545, 591
Siwalik Hills 342
Skhakot Qila ophiolite 101
Sm/Nd ratios 96–7
Sost unit 21–2, 24, 41
South Tibetan Detachment 347, 348, 377
Spiti Shale
  age 289
  eustatic environment 289–90
  sediment accumulation rate 289
  sedimentology 284, 286, 287
  setting 279
Spontang ophiolite 553, 556
Sr isotope studies 96–7, 186–7, 385, 609–11
Stak Valley
  field relations 163–5
  geochemistry 165
  geothermobarometry 165–9
  metamorphism 169–70
  setting 161–3
Stumpata quartzarenite 282, 291
Subathu Formation 432, 542
  biostratigraphy 546
  depositional environment 556
  facies analysis 551–3
  lithology 547–9
  lithostratigraphy 545
  tectonic setting 542–5, 553–4
Subhimalayan tectonic element 227, 230, 251
Sui fold 592, 594–5
Sulaiman fold belt
  deformation style 594–7
  seismic data 592–4
  setting 589–91
  structural interpretation 597–601
  structures 592
Sumari Bala fault 563, 565
Sumari klippe 565
Sumari Payan fault 563, 565
Sumayar pluton 56
  geochemistry 60–1
  mineralogy 57
  petrogenesis 62–4
Surghar Range 582
Swachi Formation 512, 515
Swat granites and gneisses 208

Tadri fold 595
Takh Formation
  age 289
  eustatic environment 290
  sediment accumulation rate 289
  sedimentology 284, 285, 286
  sequence stratigraphy 291
  setting 279
Tal Formation 545
Talemazar fault 81
Tamba Sar fault 563, 565
Tamba Sar klippe 565
Tanawal Formation 223
Tandi syncline 265–7
Tanol (= Tanawal) Formation 223
Tarkhobi Panoba fault 563
tectonic studies
  continental collisions 273
  extension measures 273
  model studies 453–8
  nappes 265–72
  regional synthesis 274
  shear zones 272–3
Tertiary studies 72, 251
  see also Palaeocene; Eocene; Oligocene; Miocene; Pliocene
Tethys Himalaya
  age 289
  deformation studies
    continent collision evidence 273
    extension 273
    nappes 265–72
    regional synthesis 274
    shear zones 272–3
  eustatic/tectonic signal 289–90
  Lamayuru complex 300–1
    biostratigraphy 302–3
    depositional environment 313–17
    facies analysis 303–13
    lithostratigraphy 302
  magmatism 291–2, 392
    Panjal Traps 241–8
    Yunam granite 255–62
  sediment accumulation rate 289
  sedimentology 284–8
  sequence stratigraphy 291
  setting 277–9
  stratigraphy 279–83
Thakkhola Graben 470, 471–2
Thaple Formation 238
thermal modelling
  one-dimensional 477–8
  two-dimensional 478–80
  model refinement 480–2
thermobarometry
  Garhwal Himalaya 489–97
  High Himalaya 331, 333–6
  Nanga Parbat–Haramosh Syntaxis 165–9
thrusting studies
  Chapusan Valley 28
  Kohat plateau 563–7
Tibetan Series
  deformation phases 370–1
  geothermometry 364–8
  lithology 357–60
  lithotectonic units 346
  metamorphic assemblage 331, 349
    CO_2 estimates 369–70
    contact 337–8
    pressure estimates 368–9
    prograde 360–3

retrograde 363–4
mineral chemistry 336
setting 325–7
summary 371–2
Tibetan Slab *see* High Himalaya
Tilicho Lake Formation 360, 371
Tilicho Pass Formation 360
Tisting Formation 527, 530
Togh Bala fault 563, 565
Tolanj fault 563, 567, 570
Tora Tigga complex 113, 125
trace element analysis 243–6, 393
Trans Himalayan Volcanic Arc 155
Triassic studies
magmatism 72
sediments
Kaghan Valley 226
Lamayuru complex 302–3
Lower Swat 216
troctolite 117
Tupop Formation 22, 46–7
Turmik Valley 162, 163

U/Pb dating 55
Uch fold 592
UMA association 126, 132–3
underthrusting 522
Utror Volcanic Formation
facies model 153–4
lithology 150–2

Vaikrita Group 431, 486
Vaikrita thrust 432, 487

Wali fault 563, 565
Warshand fault 563
Waziristan ophiolite 101
wehrlite 116
West Mir Khweli fault 563, 565

Yashkuk Formation
biostratigraphy 44
lithology 40, 42
petrography 45
Yasin Group 123, 140
Yunam granite
geochemistry 255–7
geochronology
K/Ar 259–60
U/Pb 257–9
petrography 255
regional significance 260–2
setting 240, 254
Yunam microgabbro
geochemistry 244, 245
setting 240–1

Zanskar region studies *see* Tethys Himalaya studies
Zanskar Range
age 289
magmatism 291–2
sediment accumulation rate 289
sedimentology 284–8
sequence stratigraphy 291
setting 277–9
stratigraphy 279–83
tectonic/eustatic signal 289–90
Zanskar Shear Zone 445
Zhob Valley ophiolite 101
Zin fold 596, 599
zircon studies 257–9